Mathematical Principles of the Internet

Volume 1
Engineering Fundamentals

Mathematical Principles of the Internet

Volume 1: Engineering Fundamentals

Nirdosh Bhatnagar

CRC Press is an imprint of the
Taylor & Francis Group, an **informa** business

CRC Press
Taylor & Francis Group
6000 Broken Sound Parkway NW, Suite 300
Boca Raton, FL 33487-2742

Printed on acid-free paper

International Standard Book Number Volume I: **978-1-1385-05483** (Hardback)
International Standard Book Number Volume II: **978-1-1385-05513** (Hardback)

Visit the Taylor & Francis Web site at
http://www.taylorandfrancis.com

and the CRC Press Web site at
http://www.crcpress.com

...to the memory of my parents:

Smt. Shakuntla Bhatnagar & Shri Rai Chandulal Bhatnagar

Contents

Preface

The author wrote this two-volume set of books to teach himself about the foundation and building blocks of the Internet. To both his delight and dismay, the author found that this is indeed a continual process. It is delightful, because there is so much to learn, and dismal because there is so little time. Nevertheless, this work is a partial report of the author's efforts to learn the *elements* (in a Euclidean sense) of Internet engineering.

As with any grand scientific endeavor, the Internet is a product of several persons. The true beauty of it is that this experiment, called the Internet is still evolving. It is the thesis of this work that mathematics provides a strong basis upon which the Internet can be studied, analyzed, and improved upon.

Rationale for the work

The purpose of this work is to provide a mathematical basis for some aspects of Internet engineering. The Internet is a network of autonomous networks. It is also a young, unprecedented, unparalleled, and *continuously evolving* man-made commercial and social phenomenon. Thus, the Internet is in a state of *constant flux*. This italicized phrase appears to be an oxymoron, but it is indeed a recurring and prominent theme in the description of the Internet. However, the Internet has matured enough so that its structure deserves a careful study. In order to study the form and function of the Internet only an elementary knowledge of mathematics is required. Even though the form and functionality of any physical phenomenon are basically intertwined, the emphasis in this work is upon the form and its structure. In order to strive towards this goal, the universal language: *mathematics* is used. Moreover, as mathematician Richard Hamming opined, mathematics acts as a conscience of large projects (such as the Internet).

The goal of modeling any natural or man-made phenomenon is to advance our understanding of it. More specifically, the aim is to search for *invariants* that describe a phenomenon by using the *principle of parsimony* (or simplicity). The principle of parsimony states that the description of a phenomenon should be as simple as possible, be evocative, and use the least number of assumptions. There is certainly no other tool, language, or discipline besides mathematics, which is capable of meeting these requirements. The goal of this work is to offer the reader a glimpse of hidden laws buried deep inside the complex network called the Internet. As usual, this phenomenon of the Internet did not occur overnight, and so its success is indeed a testament to the human spirit.

Information theory, algebraic coding theory, cryptography, Internet traffic, dynamics and control of Internet congestion, and queueing theory are discussed in this work. In addition, stochastic networks, graph-theoretic algorithms, application of game theory to the Internet, Internet economics, data mining and knowledge discovery; and quantum computation, communication, and cryptography are also discussed. Some of the topics like information theory, algebraic coding theory, cryptography, queueing theory, and graph-theoretic algorithms are mature and relevant in both understanding and designing the tapestry of the Internet. On other hand, some subjects like, Internet traffic, dynamics and control of Internet congestion, stochastic structure of the Internet and the World Wide Web, application of game theory to the Internet, Internet economics, and data mining techniques are still evolving. However, the application of quantum computation, communication, and cryptography is still in its infancy.

Note that this work by no means attempts, or is ambitious enough, to provide a complete description of the mathematical foundations upon which the Internet is based. It simply offers a partial panorama and perspective of the principles upon which the Internet was envisioned. There are several outstanding treatises, books, and monographs on different topics covered in this work. However these are all targeted towards experts in the field. Furthermore, for those who are interested in simply the basic principles, these are much too detailed. In summary, no attempt was made in this work to be encyclopedic.

Contents of the books

As of today, the Internet is still in its infancy. Therefore, if it has to be regarded as a scientific discipline, then its edifice has to be built upon sound mathematical principles. This is essentially the *premise* and *raison d' être* of this work. This assumption is not new. In his exultation, Eugene P. Wigner a physicist by profession (physics Nobel Prize 1963) wrote an article: "The Unreasonable Effectiveness of Mathematics in the Natural Sciences," in the year 1959. He was sufficiently impressed and awed by the utility of mathematics in modeling physical phenomena.

The selection of topics in describing certain fundamental aspects of Internet engineering is entirely based upon the author's wisdom or lack thereof. Others could have selected entirely different themes to study. However, the author chose to study the subjects outlined below in more detail, simply because of their utility value, importance, and elegance; perhaps not in that order.

Different analytical tools, which are used in describing the Internet are also listed, discussed, and elaborated upon. These tools enable a more precise and deeper understanding of the concepts upon which the Internet is built. An effort has been made to make this work succinct yet *self-contained,* complete and readable. This work assumes that the reader is familiar with basic concepts in computer networking. Although a certain level of mathematical maturity and aptitude is certainly beneficial, all the relevant details have been summarized in this work.

In order to study the structure and function of the Internet only a basic knowledge of number theory, abstract algebra, matrices and determinants, graph theory, geometry, applied analysis, optimization, stability theory, and chaos theory; probability theory, and stochastic processes is required. These mathematical disciplines are defined and developed in this work to the extent that is needed to develop and justify their application to Internet engineering. The effort to do so is often substantial and not elementary. Occasionally, applications of the intermix of these different mathematical disciplines should also be evident in this work. A striking example is Wigner's semicircle law of random matrices, which uses elements of probability theory and matrices.

This work has been divided into two main parts. These are: certain fundamental aspects of Internet engineering, and the corresponding mathematical prerequisites. In general, the chapters on mathematical concepts complement those on the engineering fundamentals. An effort has been made to provide, justification and informal *proofs* of different assertions and propositions, as much as possible. The word "proof" has been italicized, because there appears to be no universally accepted definition of it. It is hoped that the reader does not accept the mathematical results on faith. He or she is encouraged to step through the proofs of the results. In order to lighten up the presentation, each chapter begins with a brief description of the life and works of a prominent mathematician or scientist or engineer.

The Internet engineering fundamentals are described in Volume 1. Each chapter in this volume essentially describes a single Internet engineering body of knowledge. The supporting mathematics is outlined in Volume 2. It is hoped that these later chapters provide *basic tools* to analyze different aspects of Internet engineering. These chapters on mathematics also go a long way towards making the chapters on Internet engineering self-contained. Occasionally, additional material has been included to improve the comprehensiveness of various definitions and theorems. Each volume has a list of commonly used symbols. A list of Greek symbols is also provided for ready reference.

The two-volume set is titled *Mathematical Principles of the Internet*. The Volume 1 is titled *Engineering Fundamentals* and Volume 2 *Mathematical Concepts.* The contents of the respective volumes are:

Volume 1. Engineering Fundamentals

- List of symbols
- Greek symbols
- Prologue
- Chapter 1. Information theory
- Chapter 2. Algebraic coding theory
- Chapter 3. Cryptography
- Chapter 4. Internet traffic
- Chapter 5. Dynamics, control, and management of Internet congestion
- Chapter 6. Queueing theory
- Chapter 7. Stochastic structure of the Internet and the World Wide Web
- Chapter 8. Graph-theoretic algorithms
- Chapter 9. Game theory and the Internet
- Chapter 10. Internet economics
- Chapter 11. Data mining and knowledge discovery
- Chapter 12. Quantum computation, communication, and cryptography
- Index

Volume 2. Mathematical Concepts

- List of symbols
- Greek symbols
- Chapter 1. Number theory
- Chapter 2. Abstract algebra
- Chapter 3. Matrices and determinants
- Chapter 4. Graph theory
- Chapter 5. Geometry
- Chapter 6. Applied analysis
- Chapter 7. Optimization, stability, and chaos theory
- Chapter 8. Probability theory

The chapter details, along with their dependencies on other chapters are listed below.

Volume 1 - Engineering Fundamentals

Prologue: The prologue is an introduction to modeling the Internet. It provides a brief history of communication, and an introduction to telecommunication networks, and the World Wide Web. Some empirical laws about the value of a network are also outlined. This is followed by a rationale for mathematical modeling of the Internet.

Dependency: This is independent of all other chapters.

Chapter 1. Information theory: The Internet is a vehicle for transportation of information. Therefore concepts developed in information theory are immediately applicable. Information theory as developed by Claude E. Shannon determines the ultimate transmission rate of a communication channel, which is its channel capacity. It also aids in determining the ultimate data compression rate. Concepts developed in information theory are also applicable to cryptography, and data mining.

Dependency: Applied analysis; optimization, stability, and chaos theory; probability theory, and stochastic processes.

Chapter 2. Algebraic coding theory: When information is transmitted on a noisy communication channel, packets become error prone. Consequently error correcting codes merit a careful study. This theory is based upon algebraic concepts developed by the likes of N. H. Abel and E. Galois.

Dependency: Number theory, abstract algebra, matrices and determinants, applied analysis, and basics of information theory.

Chapter 3. Cryptography: The study of cryptographic techniques is essential in order to have secure communication. The science and art of cryptographic techniques is as old as the history of mankind itself. Since time immemorial cryptographic techniques have been practiced by both good and bad persons (and organizations).

Dependency: Number theory, abstract algebra, matrices and determinants, applied analysis, probability theory, and basics of information theory.

Chapter 4. Internet traffic: The predecessor of the Internet in the United States of America was the telephone network. In the pre-Internet days it was designed to essentially carry voice traffic. Voice traffic was generally described by Poisson processes and exponential distributions. In contrast Internet traffic is self-similar and characterized by long-range dependence. The mathematical details required to model this type of traffic are significantly more complicated than those of the voice traffic. We also describe a model for Internet packet generation via nonlinear chaotic maps.

Dependency: Applied analysis; optimization, stability, and chaos theory; probability theory, and stochastic processes.

Chapter 5. Dynamics, control, and management of Internet congestion: Congestion control in the network addresses the problem of fair allocation of resources, so that the network can operate at an acceptable performance level when the traffic demand is very near the capacity of the network. This is accomplished via a feedback mechanism. If Internet is modeled as a dynamical system, then the stability of the feedback schemes should be studied carefully. In the midst of chaos and congestion of the Internet, it is still possible to successfully manage traffic flow with different characteristics and requirements. This is done via intelligent scheduling of packets. An idealized scheme of packet scheduling called the generalized processor sharing (GPS) is discussed. Elements of network calculus are also outlined. Wireless networks are studied by using stochastic geometry.

Dependency: Applied analysis; optimization, stability, and chaos theory; probability theory, and stochastic processes.

Chapter 6. Queueing theory: Internet can be viewed as an extremely large queueing network. Given a specific level of traffic, queueing theory evaluates the utilization of physical resources. It also helps in modeling the response time characteristics of different types of Internet traffic. This discipline requires a sound knowledge of stochastic processes, including Markov processes, some knowledge of classical transformation techniques, and large deviation theory.

Dependency: Matrices and determinants, applied analysis; optimization, stability, and chaos theory; probability theory, stochastic processes, and basics of Internet traffic.

Chapter 7. Stochastic structure of the Internet and the World Wide Web: In order to study the structure of the Internet and the Web graph, it is assumed that these networks are stochastic. Use of stochastic and nondeterministic models might perhaps be an expression of our ignorance. Nevertheless, these models provide some useful insight into the structure of the Internet and the Web. The spread of computer virus, and its vulnerability to external attacks are also studied.

Dependency: Matrices and determinants, graph theory, geometry, applied analysis, probability theory, stochastic processes; and optimization, stability, and chaos theory.

Chapter 8. Graph-theoretic algorithms: A network can be modeled abstractly as a graph, where the vertices of the graph are the hosts and/or routers and its arcs are the transmission links. Using this model, routing algorithms for transporting packets can be developed. Furthermore, the reliability of the network layout can also be evaluated. The network coding technique is also described.

Dependency: Matrices and determinants, graph theory, geometry, applied analysis, and probability theory.

Chapter 9. Game theory and the Internet: Game theory is the study of nonaltruistic interaction between different entities (agents or users). This field of knowledge has found application in political science, sociology, moral philosophy, economics, evolution, and many other disciplines of human endeavor. Due to the distributed nature of the Internet, game theory appears to be eminently suitable to explore the selfish behavior of the Internet entities.

Dependency: Matrices and determinants, applied analysis; optimization, stability, and chaos theory; and probability theory.

Chapter 10. Internet economics: Economics has to be one of the important drivers of the Internet evolution. Interaction between network resources, congestion, and pricing are studied in this chapter. In addition, service differentiation techniques; applications of auction theory, coalitional game theory, and bargaining games are also elaborated upon.

Dependency: Matrices and determinants, applied analysis; optimization, stability, and chaos theory; and probability theory.

Chapter 11. Data mining and knowledge discovery: Internet makes the accessibility of information so much more convenient. It also generates and stores enormous amounts of information. Data mining and knowledge discovery processes help in extracting useful nuggets of information from this immense amount of data.

Dependency: Matrices and determinants, graph theory, applied analysis; optimization, stability, and chaos theory; probability theory, stochastic processes, and basics of information theory.

Chapter 12. Quantum computation, communication, and cryptography: The author is by no means suggesting that quantum computation, communication, and cryptography are the wave of the future. However, the study of this discipline and its application to the Internet is interesting in its own right. This paradigm offers immense parallelism. Furthermore, the algorithms developed under this discipline are fascinating. Peter Shor's factorization algorithm with possible application to cryptography and Lov K. Grover's search algorithm offer enough food for thought.

Dependency: Number theory, matrices and determinants, applied analysis, probability theory, and basics of information theory.

Volume 2 - Mathematical Concepts

Chapter 1. Number theory: Number theory is presumably the foundation of all branches of mathematics. Some mathematicians might also opine that it is the purest of all mathematical disciplines. However this subject has found several important applications, most notably in cryptography, signal processing, and error correcting codes. It has also made foray into quantum computation, communication, and cryptography. Therefore to the dismay of the purists and to the delight of the pragmatists the subject of number theory has indeed found practical applications.

Dependency: Abstract algebra, and matrices and determinants.

Chapter 2. Abstract algebra: This subject is used in developing error detecting and correcting codes, cryptography, and quantum computation and communication.

Dependency: Number theory, and applied analysis.

Chapter 3. Matrices and determinants: Elementary as it might appear, applications of matrices and determinants have produced several elegant results. The applications of matrices and determinants can be found in error detection and correction codes, cryptography, graph-theoretic algorithms, study of stochastic networks, queueing theory, game theory and the Internet, data mining and knowledge discovery, and quantum computation and communication.

Dependency: Applied analysis, and probability theory.

Chapter 4. Graph theory: The subject of graph theory lays down the foundation for graph-theoretic algorithms. It also includes study of routing algorithms, random graphs, power-law graphs, and network flow algorithms.

Dependency: Matrices and determinants.

Chapter 5. Geometry: Geometry is one of the most ancient branch of mathematics. This subject, more specifically hyperbolic geometry has found application in routing algorithms, and the study of structure of the Internet.

Dependency: Matrices and determinants, applied analysis; and optimization, stability, and chaos theory.

Chapter 6. Applied analysis: Analysis is a deep mathematical discipline. Its applications pervade all the topics discussed in this work including information theory, Internet traffic, queueing theory, and dynamics and control of Internet congestion. Other prominent applications of this subject are the application of game theory to the Internet, Internet economics, and data mining and knowledge discovery.

Dependency: Number theory, and matrices and determinants.

Chapter 7. Optimization, stability, and chaos theory: Concepts related to optimization, stability, and chaos theory can be found in information theory, application of game theory to the routing of packets, modeling packet dynamics, control of Internet congestion, Internet economics, and data mining and knowledge discovery.

Dependency: Matrices and determinants, and applied analysis.

Chapter 8. Probability theory: Probability theory, like any other theory has three parts. These are formal axiomatics and definitions, applications, and intuitive background. This subject provides the foundation for stochastic processes.

Dependency: Applied analysis; and optimization, stability, and chaos theory.

Chapter 9. Stochastic processes: This subject is used in describing the laws and results of information theory, Internet traffic, structure of the Internet and the Web graphs, queueing theory, and data mining and knowledge discovery.

Dependency: Applied analysis, and probability theory.

Why read this work?

The main features of this work are listed below.

(a) *Uniqueness of the work*: What is so special about this work? It is the author's humble belief, that the material presented in this work is the first of its type to describe the Internet. To the author's knowledge, several topics are discussed for the very first time in a student-friendly textbook form.

(b) *Precise presentation*: All the axioms, definitions, lemmas, theorems, corollaries, observations, and examples are clearly delineated in the chapters on mathematics. Proofs of important lemmas, theorems, and corollaries are also provided. These results are used to precisely state, prove, and develop significant engineering results. Important facts are presented as observations.

(c) *Self-contained*: The author has strived to make this work self-contained as much as possible. That is, sufficient mathematical background is provided to make the discussion of different topics complete. The author also does not assume that the reader has any prior knowledge of the subject. Sufficient mathematical details are also provided, so that the results of Internet related mathematics can be stated with plausible proofs.

(d) *Independence of chapters*: All chapters on Internet engineering are largely independent of each other. Each chapter is organized into sections and subsections. Concepts are made precise via the use of definitions. Notation is generally introduced in the definitions. Relatively easy consequences of the definitions are listed as observations, and important results are stated as theorems.

(e) *Historical introduction*: Prologue and each chapter begin with a brief biography of a mathematician or a scientist or an engineer. The purpose of providing biographies is two-fold. The first purpose is to provide inspiration to the student, and the second is to inform him or her about the lives of persons who developed important mathematical and engineering concepts. It is left to the individual student to evaluate the impact of these ideas upon the description and evolution of the Internet.

(f) *List of observations*: Several significant results are listed precisely as observations. Proofs of some of these observations are outlined in the problem section provided at the end of each chapter.

(g) *Examples and figures*: Each chapter is interspersed with several examples. These examples serve to clarify, enhance, and sometimes motivate different results. It is the author's belief that examples play a crucial role in getting a firm grasp of the subject. If and where necessary, figures are provided to improve the clarity of the exposition.

(h) *Algorithms*: Algorithms are provided in a pseudocode format for ready implementation in any high-level language.

(i) *Reference notes*: Each chapter is provided with a reference-notes section. This section lists the primary sources of the technical literature used in writing the chapter. These sources generally constitute the main citations. Sources for further study are also provided.

(j) *Problems*: Each chapter is provided with a problem section. Besides, enhancing the material presented in the main chapter, each problem states a significant result. A majority of the problems are provided with sufficient hints. In order to keep the continuity and not clutter with too many details, proofs of important observations made in the chapter are relegated to the problem section. It is strongly suggested that the reader peruse the problem section.
(k) *List of references*: A list of references is provided at the end of each chapter.
(l) *List of symbols*: A list of commonly used symbols and Greek letters is provided.
(m) *User-friendly index*: A comprehensive and user-friendly index of topics is provided at the end of each volume.

Target audience of the work

This work evolved as the author taught courses on network modeling and related topics. These books can be used either in an upper-level undergraduate or first-year graduate class in electrical engineering or computer science. The choice and depth of topics actually depends upon the preparation of the students.

These books should also serve as a useful reference for networking researchers, professionals and practitioners of the art of Internet engineering, that will provide a better understanding and appreciation for the deeper engineering principles which are central to Internet engineering. This, in turn would provide the necessary guidelines for designing the next generation of the Internet.

Commissions and omissions

The pseudocode for the different algorithms has been written with especial care, but is not guaranteed for any specific purpose. It is also quite possible, that the author has not provided complete credit to the different contributors of the subject. To them, the author offers a sincere apology for any such inadvertent omission. Receiving information about errors, and suggestions for improvement of the books will be greatly appreciated.

Acknowledgments

Invention of the Internet was perhaps, one of the greatest technical achievements of the last century. Therefore, first and foremost I offer my sincere thanks to the inventors and builders of the Internet for their heroic work and dedication. I am also grateful to all the authors whose names appear in the list of references provided at the end of each chapter.

Further, these books were written in LaTeX, which is an extension of TeX. TeXis a trademark of American Mathematical Society. So a very special thanks to their respective creators: Leslie Lamport and Donald E. Knuth.

The impetus for the publication of this work was provided by Sartaj Sahni. I am forever grateful to him for his initial interest and consideration. His own work has certainly been a source of inspiration. My first contact at Taylor and Francis Publications was with Randi Cohen. She had been most courteous and extremely patient. The other staff members at Taylor and Francis Publications, including Robin Lloyd-Starkes, Alice Mulhern, and Veronica Rodriguez have been most cooperative. Expert typesetting was provided by Shashi Kumar. The constructive comments and suggestions provided by anonymous reviewers are also appreciated. Encouragement provided by Shyam P. Parekh is certainly treasured.

This work would not have been possible without the inspiration of my students and esteemed teachers. A very special thanks is extended to family, friends, and colleagues for their unwavering support. Occasional help in diction and artwork was provided by Rishi who makes all the work worthwhile.

NB
San Jose, California
Email address: nbhatnagar@alumni.stanford.edu

List of Symbols

Different types of commonly used symbols are categorized as:

1. Logical operators
2. Set operators
3. Sets of numbers
4. Basic arithmetic operators
5. More arithmetic operators
6. Arithmetical relationships
7. Analysis
8. Complex numbers
9. Vectors
10. Matrices
11. Mappings
12. Combinatorial functions
13. Probability theory
14. Mathematical constants
15. Notation used in describing the algorithms

Logical Operators

$\wedge$	logical and operator
$\vee$	logical or operator
$\leftarrow$	assignment operator
$\rightarrow$, $\Rightarrow$	logical implication
$\Leftrightarrow$, iff	if and only if

Set Operators

$\in$	belongs to
$\notin$	does not belong to, negation of $\in$
$\ni$	such that
$\forall$	universal quantifier, for all
$\exists$	existential quantifier, there exists
$\nexists$	there does not exist
$\cap$	set intersection operator
$\cup$	set union operator
$\backslash$	set difference operator
$\subset$	proper subset containment operator
$\subseteq$	subset operator
$\varnothing$	empty set
$\oplus$	set addition operator
$\square$	end of: proof, definition, example, or observation
$\{\cdot, \cdots, \cdot\}$	set list
$\sim$	equivalence between sets
$A^c, \overline{A}$	complement of the set A
$\vert A\vert$	cardinality of the set A
$A^\perp$	set orthogonal to the set A
$A \times B$	Cartesian product of sets A and B
$A^{(n)}, A^n$	Cartesian product of the set A with itself, n times over
$\{x \mid R(x)\}$, $\{x : R(x)\}$	set of all x for which the relationship $R(x)$ is true

Sets of Numbers

$\mathbb{B}$	binary field, also $\mathbb{F}_2, GF_2$
$\mathbb{C}$	set of complex numbers
$\widehat{\mathbb{C}}$	set of nonzero complex numbers $\mathbb{C}\backslash \{0\}$
$\mathbb{C}^*$	set of extended complex numbers $\mathbb{C} \cup \{+\infty\}$
$\mathbb{F}_q$	finite field of size q, also GF_q, $q = 2, 3, \ldots$
GF_q	Galois field of size q, also $\mathbb{F}_q$, $q = 2, 3, \ldots$
$\mathbb{P}$	set of positive numbers $\{1, 2, 3, \ldots\}$
$\mathbb{N}$	set of natural numbers $\{0, 1, 2, 3, \ldots\}$
$\mathbb{Q}$	set of rational numbers
$\mathbb{R}$	set of real numbers
$\widehat{\mathbb{R}}$	set of nonzero real numbers $\mathbb{R}\backslash \{0\}$
$\mathbb{R}_0^+$	set of nonnegative real numbers $\mathbb{R}^+ \cup \{0\}$
$\mathbb{R}^+$	set of positive real numbers
$\mathbb{R}^*$	set of extended real numbers $\mathbb{R} \cup \{+\infty\}$
$\mathbb{R}^n$	n-dimensional real vector space, where $n \in \mathbb{P}$
$\mathbb{Z}$	set of integers $\{\ldots, -2, -1, 0, 1, 2, \ldots\}$
$\mathbb{Z}_n$	set of integers modulo n, the set $\{0, 1, 2, \ldots, n-1\}$
$\mathbb{Z}_n^*$	set of nonzero integers in the set $\mathbb{Z}_n$, which are relatively prime to n

Basic Arithmetic Operators

$+$	addition operator
$-$	subtraction operator
$\times, \cdot$	multiplication operator
$\div, /$	division operator
$\pm$	plus or minus operator
$\sqrt{\cdot}$	square root operator
$\lceil\cdot\rceil$	ceiling operator; for $x \in \mathbb{R}$, $\lceil x\rceil =$ least integer greater than or equal to x
$\lfloor\cdot\rfloor$	floor operator; for $x \in \mathbb{R}$, $\lfloor x\rfloor =$ greatest integer less than or equal to x
$[\cdot]$	round-off operator; for $x \in \mathbb{R}$, $[x] =$ integer closest to x
$\cdot \mid \cdot$	divisibility operator; $a \mid m$ means nonzero integer a can divide integer m
$\cdot \nmid \cdot$	nondivisibility operator; $a \nmid m$ means nonzero integer a cannot divide integer m

More Arithmetic Operators

$\|a\|$	absolute value (magnitude) of $a \in \mathbb{R}$
$\langle n \rangle_p$	modulus operator $n \pmod p$, $p \in \mathbb{P}$
$\sum$	discrete summation operator
$\prod$	product operator
$*$	convolution operator
$\gcd(a, b)$	greatest common divisor of a and b; $a, b \in \mathbb{P}$
$\max\{\ldots\}, \max(\ldots)$	maximum operator
$\min\{\ldots\}, \min(\ldots)$	minimum operator
$\max(a, b)$	maximum of a and b; $a, b \in \mathbb{R}$
$\min(a, b)$	minimum of a and b; $a, b \in \mathbb{R}$
$\bmod$	modulo operator
a^+	$\max(0, a)$, $a \in \mathbb{R}$
a^-	$\max(0, -a)$, $a \in \mathbb{R}$
$\exp(\cdot)$	exponential function with base e
$\ln(\cdot)$	natural logarithm
$\log_a(\cdot)$	logarithm to the base a, where $a \in \mathbb{R}^+$
$sgn(\cdot)$	signum function

Arithmetical Relationships

$=$	equality operator
$\neq$	not equal to
$\sim$	asymptotically equal
$\simeq$	approximate relationship between functions
$\approx$	approximate relationship between numbers
$\asymp$	approximate relationship between functions within a constant
$\geq$	greater than or equal to
$\leq$	less than or equal to
$\gg$	much greater than
$\ll$	much less than
$\gtrless$	greater or less than
$\rightarrow$	approaches, tends towards
$\propto$	proportional to
$\equiv$	congruent to
$\not\equiv$	not congruent to
$\cong$	isomorphism

Analysis

∞	infinity
$\lim$	limit
$\frac{d}{dt}$	differentiation operator
$f'(t), \dot{f}(t)$	$\frac{d}{dt}f(t), t \in \mathbb{R}$
$\frac{\partial}{\partial t}$	partial differentiation operator
$\int$	integration operator
$\lVert\cdot\rVert$	norm of a vector
l^2	square summable sequences
$L^2(\mathbb{R})$	set of square-integrable functions
$\leftrightarrow$	Fourier transform pair
$\arg\max_x f(x)$	$\{x \mid f(y) \leq f(x) \ \forall\, y\}$
$\arg\min_x f(x)$	$\{x \mid f(y) \geq f(x) \ \forall\, y\}$
$\delta_{ij};\ i,j \in \mathbb{Z}$	Kronecker's delta function.
$f \circ g(\cdot)$	$f(g(\cdot))$ function composition
$\circ,\ \langle\cdot,\cdot\rangle,\ \langle\cdot \mid \cdot\rangle$	inner (dot) product operators
$\times$	cross product operator

Complex Numbers

i	$\sqrt{-1}$
$\overline{z}$	complex conjugate of $z \in \mathbb{C}$
$\lvert z\rvert$	magnitude of $z \in \mathbb{C}$
$\mathrm{Re}(z)$	real part of $z \in \mathbb{C}$
$\mathrm{Im}(z)$	imaginary part of $z \in \mathbb{C}$
$\arg(z)$	argument of $z \in \mathbb{C}$

Vectors

$\boxplus$	vector addition
$\otimes$	vector multiplication
$u^{\perp}$	a vector orthogonal to vector u
$x \perp y$	vectors x and y are orthogonal

Matrices

A^T	transpose of matrix A
$A^\dagger$	Hermitian transpose of matrix A
A^{-1}	inverse of square matrix A
I	identity matrix
$[a_{ij}]$	matrix with entries a_{ij}
$tr\,(A)$	trace of the square matrix A
$\det A, \lvert A\rvert$	determinant of the square matrix A

Mappings

$f : A \to B$	f is a mapping from the set A to the set B
$f\,(x)$	image of $x \in A$ under the mapping f
$f\,(X)$	$\{f\,(x) \mid x \in X\}$ for $f : A \to B$ and $X \subset A$
$\triangleq$	definition, or alternate notation

Combinatorial Functions

$n!$	$n \in \mathbb{N}$, factorial of n
$\binom{n}{k}$	$k, n \in \mathbb{N}, 0 \leq k \leq n,$ binomial coefficient
$(n)_k$	$n \in \mathbb{N}, 0 \leq k \leq n,$ falling factorial

Probability Theory

$\stackrel{d}{=}$	equality in distribution
$\stackrel{d}{\to}$	convergence in distribution
$P\,(\cdot)$	probability function
$\sim$	distribution of a random variable
$\mathcal{E}\,(X), \mathcal{E}\,[X]$	expectation of random variable X
$Var\,(X)$	variance of random variable X
$Cov\,(X, Y)$	covariance between random variables X and Y

Mathematical Constants

π	$3.141592653\ldots$
e	$2.718281828\ldots,$ Euler's number
ϕ_g	$\left(\sqrt{5}+1\right)/2,$ golden ratio
γ	$0.577215664\ldots,$ Euler's constant

Notation Used in Describing the Algorithms

The algorithms are given in a pseudo-C language. Comments inside the algorithm are italicized. The language elements are: **all, begin, do, end, find, for, for all, go to label, if** . . . **then, else, label, procedure, stop, such that, while, with, Case, Subcase, Step.**

Greek Symbols

Greek letters are often used as symbols in mathematics and engineering. For example the English equivalent of the symbol Σ is S. Therefore the symbol Σ is used for summation. Similarly, the English equivalent of the symbol Π is P. Therefore the symbol Π is used for product. A list of lower- and upper-case Greek letters and their spelling in English language follows.

Lower Case	Upper Case	Name	Lower Case	Upper Case	Name
α	A	alpha	ν	N	nu
β	B	beta	ξ	Ξ	xi
γ	Γ	gamma	o	O	omicron
δ	Δ	delta	π	Π	pi
ϵ, ε	E	epsilon	ρ	P	rho
ζ	Z	zeta	σ, ς	Σ	sigma
η	H	eta	τ	T	tau
θ, ϑ	Θ	theta	υ	Υ	upsilon
ι	I	iota	ϕ, φ	Φ	phi
κ	K	kappa	χ	X	chi
λ	Λ	lambda	ψ	Ψ	psi
μ	M	mu	ω	Ω	omega

Prologue

$$F_{21} = -G \frac{m_1 m_2}{|r_{12}|^2} \widehat{r}_{12}$$

Universal law of gravitation

Isaac Newton. Isaac Newton was born on 4 January, 1643 in Woolsthorpe, Lincolnshire, England. He was regarded as the greatest scientific mind of his era. Newton's father died before he was born, and his mother remarried three years later. Newton's early upbringing was therefore left to his maternal grandmother.

Newton entered Trinity College, Cambridge, in June 1661. The University closed in the summer of 1665 due to plague, and Newton was forced to leave Cambridge. The next eighteen months in the life of Newton proved to be most revolutionary. During this period he made fundamental contributions to optics, mechanics, and calculus. Newton became the Lucasian Professor of Mathematics at Cambridge University at the age of 27.

Newton is generally regarded as the codiscoverer of calculus along with Gottfried Wilhelm von Leibniz (1646-1716). Newton termed his discovery *method of fluxions.* He also propounded the theory of universal gravitation, and stated the laws of motion. Using the universal law of gravitation, the laws of motion, and the method of fluxions, Newton was able to explain Kepler's laws of planetary motion. In the year 1687, Newton published *Principia Mathematica.* This book is generally regarded as the single most significant and influential work in all of physics. He also published *Optics* in 1704.

In 1696 Isaac Newton became Warden of the Royal Mint, and was promoted to the lucrative position of the Master of the Mint in 1699.

The following statement is generally attributed to Newton. *I do not know what I may appear to the world, but to myself I seem to have been only like a boy playing on the sea-shore, and diverting myself in now and then finding a smoother pebble or a prettier shell than ordinary, whilst the great ocean of truth lay all undiscovered before me.*

Newton is also well-known for the quote: *If I have seen further, it is by standing on the shoulders of giants.*

Newton was knighted in 1708, and died on 31 March, 1727 in London, England.

1 Introduction

The following topics are discussed briefly in this prologue: a brief history of communication, telecommunication networks, internetworking, principle of layering, and the World Wide Web and its function. Empirical models of the value of networking are also explored. The search for form and structure, and the role of mathematics in this exploration are also examined. Not surprisingly, it can be inferred that mathematics has been and is reasonably effective in this inquiry.

What is the Internet? There are several answers to this question. Internet, short for Internetwork, is a collection of connected individual autonomous networking systems, which provide an infrastructure for information storage and retrieval. The connection is facilitated by a common set of standards and compatible interfaces between network devices. Internet is also a physical medium for interaction and cooperative intellectual endeavors, between groups or individuals and their computers irrespective of their geographical location. In addition, Internet also serves as a vehicle for commerce and social networking.

Using technical jargon, Internet is defined as a universal set of interconnected networks that rely heavily upon the TCP/IP protocol. A protocol is a uniform set of rules. TCP stands for the Transmission Control Protocol, and IP stands for the Internet Protocol. These two protocols are at the heart of the Internet. Cumulatively, they are responsible for efficient routing of packets in the Internet, where a packet is defined as a block of data for efficient transmission of information.

A Web is a shortened name of the World Wide Web (WWW). It is a actually a client-server system running on the Internet which can deliver voice, data, image, and video. Recall that servers are powerful computers, whereas the clients are workstations, desktop personal computers, or intelligent wireless devices. In an extended definition of a client-server system, a client is a program which runs on one host computer, and requests and receives service from a server program running on another host computer.

The power of the Internet is largely due to its flexibility. Only the network protocol has to be uniform, but each node in the network can be designed to have its own hardware platform, own operating system and programming language, and its own intranet structure and organization.

2 A Brief History of Communication

In order to make progress in any field of human endeavor, a knowledge of its history is essential. Furthermore, history is said to be both interesting and instructive. Therefore a brief account of history of communication among human beings is given. The art of communication is unique to human beings among all creatures on the planet earth. It has been stated that the homo-sapiens of the upper Paleolithic period (35,000 BC to 10,000 BC) were users of language. The cave paintings of Lascaux (France) are conjectured to have been created between 28,000 BC and 22,000 BC. These paintings suggest some use of language among homo-sapiens of that age. First evidence of writing dates back to the Sumerians of approximately 4,000 BC. Perhaps the best indication of a language in written form is that of the Egyptian hieroglyphics dated about 3000 BC. A more simplified expression than those of the hieroglyphics evolved around the year 1500 BC.

Another prime example of communication and human spirit was the first marathon run. A ferocious battle was being waged on the coast of the Aegean Sea, near the town of Marathon, between a small army of Greeks and a numerically larger army of the Persians in the September of 490 BC. The Greek army eventually defeated the mighty Persian army. In his exuberance, the commander of the Greek army dispatched a courier to the city of Athens with the message of victory. The unfortunate and now romanticized messenger, jogged a distance of approximately 26 miles and 385 yards, reached the city streets of Athens, uttered the words: "Rejoice! We have won the war!" and collapsed. Perhaps a more efficient form of communication would have been to place multiple posts with men on each of them, who would communicate with each other with some predetermined arm signals. In modern networking terminology, the predetermined arm signals would be called a *protocol.*

Denizens of ancient China used a Tamtam as a telecommunication instrument. Tamtam is a huge free hanging circular plate of metal. Tappings on this plate would produce an audible tone which can travel large distances. Also the natives of Africa, and tropical America used drum telegraphy in heavily forested areas. The arrival of jungle explorers was preceded by roar of such signal drums. American Indians are credited with the use of smoke signals to exchange information. Smoke signals were also said to have been used in the year 200 BC by Hannibal, a Carthaginian general.

A significant advance in communication was the invention of a moveable-type printing press around the year 1450 by Johannes Gutenberg. The importance of this invention is comparable to that of telegraph, telephone, and of course the Internet.

In the year 1791, the Chappe brothers when they were in high school in France, developed a signaling system so that they could send messages to each other. The teenaged brothers' semaphore system was made up of moveable arms on a pole. The position of these arms would indicate letters of the French alphabet. It was said that, this scheme impressed Napoleon Bonaparte (1769-1821). This semaphore signaling system was used in Germany, Italy, Russia, across the pond in the United States of America, Canada, and several other countries. This scheme worked only in the absence of fog. Wigwag, a form of communication was invented in the United States in the nineteenth century. This is a form of semaphore system of communication used during the Civil War in the United States of America. Finally, in mediaeval ages pigeons were used in fairy tales and real life to exchange romantic letters between near and dear ones!

The modern era of telecommunications can be said to have begun in the 1830's. Carl Friedrich Gauss (1777-1855) and Wilhelm Eduard Weber (1804-1891) developed a small-scale electrical telegraph (tele = distant, and graph = writing) in Göttingen, Germany. Samuel F. B. Morse (1791-1872) patented a practical telegraph in 1840. Morse of course invented Morse-code, which uses dots and dashes to transmit signal. This was a precursor to the use of binary symbols 0 and 1 in modern computers. The telegraph permitted high-speed communication over long distances. However, an interpreter was required at both the transmitting and receiving ends. Transatlantic cable was laid between the years 1858 and 1866. It permitted communication directly across the Atlantic ocean. Modern day cables connecting different continents are still the main carriers of information.

Clarendon Press published *A Treatise on Electricity and Magnetism* written by James Clerk Maxwell (1831-1879) in the year 1871. It is regarded as perhaps the most well-known and important book in physics after Isaac Newton's *Principia.* In the year 1876, Alexander Graham Bell (1847-1922) invented the telephone. Elisha Gray (1835-1901) filed a patent application three hours later than Bell. This is a prime example in the history of science, where the importance of speed and timeliness cannot be stressed enough. The telephone allowed the sender and receiver of messages to communicate directly and seemingly simultaneously. Modern telephone exchanges provide the bulk of Internet connections today.

Heinrich Hertz (1857-1894) generated the first man-made radio waves in 1887. The dial telephone was introduced in 1891. Introduction of this phone made the placing of a phone call much more efficient, as an operator was not required. Jagadish Chandra Bose (1858-1937) built a prototype millimeter-wave radio. He invented the mercury coherer (a form of radio signal detector), together with the telephone receiver in 1895. Using the receiver developed by Bose, Gugliemo Marconi (1874-1937) was able to receive the radio signal in his first radio communication from Poldhu, United Kingdom to Newfoundland, St. Johns in December 1901. The distance between these two sites is about 2000 miles. Lee de Forest (1873-1961) invented the vacuum tube in 1906. This invention marked the beginning of the modern electronic age. Vladimir K. Zworykin (1889-1982) filed several patents on electronic scanning television using the celebrated iconoscope in the year 1923.

John V. Atanasoff (1903-1995) and Clifford E. Berry (1918-1963) built the world's first electronic digital computer at Iowa State University between 1932 and 1942. Their work spawned several innovations in computing. For example, they used binary system for arithmetic, parallel processing, separated memory and computing functions, and regenerative memory.

The year 1939 also saw the invention of pulse code modulation (PCM). This was an essential foundation for digitized voice transmission. The transistor was invented in 1947 by John Bardeen (1908-1991) and Walter H. Brattain (1902-1987). William B. Shockley (1910-1989) realized its importance. It can safely be said that this invention heralded the development of modern computers, calculators and other electronic devices. It also played a significant role in today's telecommunication network.

In the year 1948 Claude. E. Shannon published landmark papers on information theory, and laid the foundations for data compression and transmission, and error detection and correction. During the same year, use of microwaves permitted the phone signals to be transmitted across inhospitable and remote terrain that ordinary telephone wires couldn't cover.

The very first transatlantic telephone cable was laid in the year 1958. Jack St. Clair Kilby (1923-2005) of Texas Instruments and Robert Noyce (1927-1990) of Fairchild Semiconductors developed the first integrated circuit in 1956.

Electronic switching was invented in 1960. In contrast to mechanical switches, electronic switches were faster and allowed services such as voice messaging, speed dialing, and caller identification.

Soviet Union launched the first artificial earth satellite in the year 1957. Not to be taken down by any challenge, United States of America formed the Advanced Research Project Agency (ARPA) within the Department of Defense (DoD) to explore the military application of science and technology. Thus the first communication satellite, Telstar was launched into earth's orbit in 1962. Launching of this satellite truly signalled the beginning of global communication. Today, satellites play a truly prominent role in transmitting data.

Perhaps there are several more important inventions which have not been given space here. However, against this stellar array of ideas and inventions, Internet came to be in the second half of the twentieth century.

L. Kleinrock formulated a mathematical theory of packet networks in 1961-62. J. C. R. Licklider first proposed the idea of a global network of computers in the year 1962. Internet depends on packets to transfer data. That is, data is fragmented into small packets for transmission. Furthermore, these packets which are generated from the same source are permitted to take different routes between the source and the destination. The idea of packet switching has a militaristic slant. It was presumed that it will be hard to eavesdrop on packetized messages. In the early 1960's, during the days of the Cold War, Paul Baran conceived of a distributed network which would survive an attack, even if some of its nodes and links would be destroyed. He also developed the concept of dividing data into "message blocks" before sending them out across the network.

Doug Englebart invented the computer mouse in 1963, while working at Stanford Research Institute in California. Thomas Merrill and Larry Roberts in 1965, connected a computer in Massachusetts to another in California with a low speed dial-up telephone line. This implementation can be regarded as the first wide-area computer network ever assembled. Marcian Edward "Ted" Hoff, an Intel Corporation employee created the first computer-on-a-chip, the microprocessor, in 1968. This invention was instrumental in the development of personal computers.

Kenneth Thompson and Dennis Ritchie, of the Bell Laboratories teamed together to invent the Unix operating system and the C programming language in 1969. Their inventions paved the way to multiple advances in networking, computer hardware, and software.

The year 1969 also witnessed the birth of the Internet. ARPANET (ARPA network) was created by DoD for doing research in the nascent field of computer networking. However, the phrase "Internet" was not used until 1974. A demonstration of a 40 machine ARPANET took place in the year 1972. Also the File Transfer Protocol (FTP) was specified in the year 1973. It set up rules for sending and receiving data on computers.

A public packet data service, Telnet was inaugurated in the year 1974. It was a commercial version of the ARPANET. The same year also witnessed the specification of the Transmission Control Protocol. However, the TCP and the Internet Protocol, now commonly known as the TCP/IP was introduced in the year 1982. Vinton G. Cerf and Robert E. Kahn were pioneers in the development of this protocol. David Clark and his colleagues at Massachusetts Institute of Technology implemented a compact and simple version of TCP for personal workstations.

The initial goal of ARPANET was resource sharing, but the first application on this network was the electronic-mail. This was first implemented by Raymond S. Tomlinson in 1971. The Ethernet LAN was invented by Robert Metcalfe in the year 1973. It permitted the connection of several computers and printers in a geographically localized area. This invention provided the impetus for the growth of the personal computer industry, which in turn spurred the growth of the Internet. The personal computer industry also got a significant boost from the entrepreneurial acumen and skills of William "Bill" Gates and Paul Allen.

Dennis P. Hayes invented the modem (modulator-demodulator) for the personal computer (PC) in 1977. This invention allowed both individuals and businesses to access the Internet.

By the year 1989, there were over 100,000 hosts (computers) on the Internet. The World Wide Web as we know it today was inaugurated in the year 1991 by the truly noble efforts of Tim Berners-Lee. This invention made the Internet widely available to consumers and businesses.

Data storage technology is a significant driving force in the growth of the Internet. Some examples of these are: magnetic, optical, magneto-optical, and solid-state storage. Further, data transmission technology like optical fiber is also an important factor in the evolution of the Internet. Optical fiber is a cylindrical waveguide which is used to transmit light by using the principle of total internal reflection. Optical fiber is flexible and can be bundled as cables. These can be used for both long and short distance communication.

Wireless technologies are also revolutionizing access to the Internet. This technology transfers information over air by using electromagnetic waves. It does not require the use of cables or wiring. Hence the term "wireless." Wireless communication has been in use since the year 1876. Only in recent days has the technology found a dramatic increase in its use. Some of the applications of wireless technology are in cellular telephones, personal digital assistants, wireless networking, computer mice, keyboards, satellite television, GPS (Global Positioning System) units, and deep-space radio communications.

Such technologies have brought Internet closer to persons from different walks of life, where physical access to the Internet is difficult, and economically prohibitive.

3 Telecommunication Networks

Telecommunication networks can be classified based upon their geographical disbursement. There are three major kinds of such networks. These are the Local Area Networks (LANs), the Metropolitan Area Networks (MANs), and the Wide Area Networks (WANs).

A LAN is a network of computers and associated peripherals that are connected by communication links. As the name suggests, this network is geographically localized. A LAN can be present either in a single residence, or it can connect several thousand users, their computers, peripherals, and servers in a business environment. The three most well-known LAN technologies are the Ethernet, token-ring, and the Fiber Distributed Data Interface (FDDI). Of these three technologies, Ethernet is the most widely used LAN. LANs typically span less than one kilometer.

The MAN is next in the hierarchy. These networks are usually spread across a city. The physical span of MANs is usually of the order of tens of kilometers. A WAN interconnects geographically dispersed computers or networks spread over several distant cities.

A well-known example of a WAN is the Internet. The Internet is a conglomeration of computer systems interconnected via telecommunication links. It is a cooperative and public entity. The Internet is accessed by hundreds of millions of people world wide. Its success depends upon a host of protocols, the most prominent of which is the TCP/IP.

From a hardware perspective, the Internet consists of hosts, routers, and links. The hosts are the end systems, routers are external/internal switching points, and links are transmission channels which connect hosts and/or routers.

The most famous application of the Internet is the World Wide Web, sometimes simply called the Web. Millions and millions of pages of information can be accessed by using the Web. The basis of the Web is the HyperText Transfer Protocol (HTTP). The use of the Web is facilitated by the presence of the browsers. A browser is a computer program for the Web which provides several configuration and navigational features.

Another type of important network is the Storage Area Network (SAN). It is specially designed to provide storage of information. With the advent of search technologies and social networks, cost-effective performance of SANs is a must. There are also other types of networks like: campus area networks, and personal area networks.

In the next section, we explore the purpose and requirements for internetworking. Internetworking simply means connecting. Technical requirements in the design of the Internet are specified. Further, the overall design principles of the Internet are also outlined.

4 Internetworking

The initial purpose of the Internet was *internetworking*. This means effective and coherent networking of often heterogeneous and autonomous networks. The Internet was expected to provide: robustness, heterogeneity of services, heterogeneity of networks, distributed management, reasonable cost, and accountability. In order to achieve these seemingly disparate goals, certain overall design philosophy was required. Key requirements in the design of the Internet are initially listed. This is followed by a brief discussion of the design principles of the Internet.

Requirements in the Design of the Internet

Some important requirements in the design of the Internet are: connectivity, reliability, performance, security, flexibility, network management, and cost. These are next elaborated.

(a) *Connectivity*: Perhaps the primary reason for the existence of the Internet is the requirement of connectivity. That is, the network should be able to provide connection between users on different sites, using diverse technologies.
(b) *Reliability*: Users should be able to communicate and share with each other all types of resources in a reliable manner. The Internet must be able to provide reliable communication even in the presence of failed networks, routers, and transmission links. Perhaps an equivalent term for reliability is availability.
(c) *Performance*: The Internet should provide tolerable level of performance. Some of the possible metrics are response times and throughputs. In addition, the resources used in the network must be accountable.
(d) *Security*: The complete network, and the user connections should be secure.
(e) *Flexibility*: The Internet should have sufficient flexibility to accommodate new users, applications, services, and foster interoperability. It should be able to support both heterogeneous services and networks.
(f) *Network management*: Network management implies network configuration, troubleshooting, and support. It might also imply all of the above requirements. The network management should be distributed.
(g) *Cost*: All of the above requirements should be balanced by cost.

Design Principles of the Internet

The guiding principles in the successful operation and design of the Internet are next stated. As per Vinton G. Cerf, an Internet pioneer, the design philosophy of the Internet is based upon three simple principles: end-to-end design, layered architecture, and open standards.

(a) *End-to-end design*: In this approach, the intelligence and control functions generally reside with the users at the "edges" of the network and not at the core of the network.
(b) *Layering*: The purpose of a layered architecture is to have a simple yet flexible architecture. It is based upon the principle of modularity.
(c) *Open standards*: The purpose of open standards was to allow any user to create not only innovative applications but also independent networks that could successfully exist on the Internet.

These principles, in turn give the consumer or the end-user, choice and control over their use of the network. The above three guidelines are again based upon the belief that there should be no central gatekeeper who would have a control over the Internet. That is, neither the Internet service providers nor government should impose any type of restriction upon access to the Internet. It is claimed that this type of *neutral network* fosters innovation and discourages fragmentation of the network.

Therefore these three principles provide a metaphoric resonance to the mantra of network neutrality. As per the principle of network neutrality, priority should not be offered to the traffic flow of an entity on the Internet, if it contributes monetarily to the Internet service providers. On the other hand, opponents of network neutrality argue that the principle of network neutrality stifles innovation and inhibits competition.

5 Principle of Layering

The principle of layering is used to organize computer programs, which successfully run the Internet. Each layer of a computer program performs its special function. As per this principle, computer programs in a particular layer communicate with computer programs, which are either in a layer immediately below or above it.

The Internet uses the Open Systems Interconnection (OSI) model for multilayered communication. All communication programs have seven layers in this model. Each layer performs its distinct function. It is convenient to think of these layers as stacked one layer on top of the other. If there are two communicating programs on two different computers, a computer program on a particular computer system communicates with its peer program on a different computer at the same level of layer. An outcome of the layering approach is the creation of modular "building blocks," where the building blocks of a particular layer can be modified with minimal impact on other layers. In summary, layering is an organized and flexible way of transmitting messages between two computers.

The actual hardware and software which supports these seven layers, is a combination of applications such as Web browser, computer operating system, network protocols like TCP/IP, and the physical components of the network. In this seven-layered hierarchy, the bottom most layer is the layer number one. On top of it is the layer number two, then on top of the layer number two is layer number three and so on. The top most layer, is layer number seven. These seven layers can be further grouped into two categories. The bottom three layers are used when a message passes through the host computer. The upper four layers are used for message transfer between users. The flow of information occurs as follows. User messages are created at layer seven of a program in a computer system. These messages percolate through its seven layers of the communication program and then pass via the telecommunication links to a receiving computer. If these messages are meant for this computer, then the message is processed at layer one level, then at layer two level, and finally at the layer seven level. However, if the message is addressed to another computer, then these messages are processed at only the lower three layers, and then redirected to another computer. The Figure 1 shows the OSI seven-layered architecture. More specifically, the seven layers in the OSI model are:

Layer 1. *The physical layer*: This layer deals with transmission of messages at the *bit* (hardware or physical) level.

Layer 2. *The data-link layer*: This layer sequences bits to form a *frame* in hardware. Network adapters and device drivers residing in the computer operating system perform this function.

Layer 3. *The network layer*: This layer provides the routing intelligence to the messages. The unit of information at this level is called a *packet*.

Layer 4. *The transport layer*: This layer furnishes the end-to-end control of data transfer. It is responsible for error checking, and determines whether all the packets have arrived at the right destination. The unit of information at this level is called a *message*. This protocol layer, and the remaining higher layers run only on the end hosts and not on the intermediate nodes (switches).

Layer 5. *The session layer*: As the name suggests, this layer is responsible for initiating, sustaining, and terminating sessions between two computers.

Layer 6. *The presentation layer*: This layer is concerned with the format and syntax of data exchange between two computers.

Layer 7. *The application layer*: Communication entities are identified at this layer. Furthermore, security and quality of service considerations are also addressed.

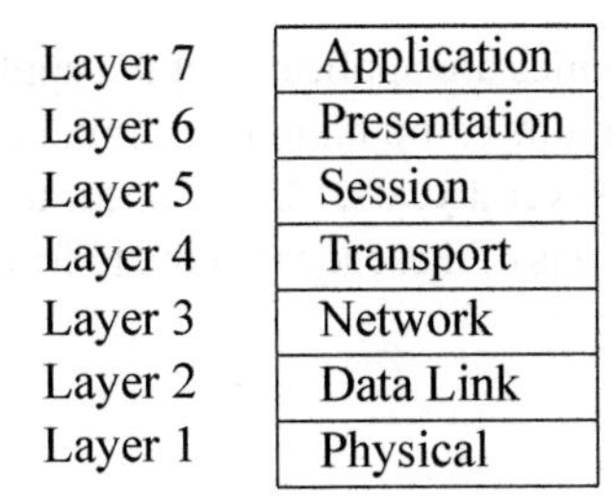

Figure 1. OSI's seven-layered architecture.

In addition to the seven-layered OSI architecture, there other equivalent multilayered architectures. In summary, layering provides modularity and organization to the Internet communication system.

Mapping of Textbook Chapters to Different Layers

Some examples of topics covered in the book are next mapped to these different layers. This mapping is not necessarily concrete, as distinction between successive layers is often blurred in practice.

1. Information theory is directly applicable at the physical layer. For example, it can be used to determine the raw bit rate of a physical communication channel. Data compression techniques, which use information-theoretic concepts occur at the presentation layer. Information theory is also applicable at the application layer. It is also useful in data mining algorithms, and in describing World Wide Web graphs.
2. Coding theory is applicable at the transport layer, network layer, and data-link layer to code data packets/frames, and also at the presentation layer.
3. Cryptographic applications occur at the application, presentation, session, transport, network, and data link layers. That is, encryption and decryption occur at these layers. Actually management of security-concerns occur at all the seven layers.
4. Internet traffic is studied at both the data-link and network layers.
5. Network congestion management occurs at the transport layer. Stochastic geometry of wireless networks is studied at the lower three layers of the OSI architecture.
6. Queueing theory is the study resource management. Buffer management occurs at the network layer. Furthermore, this theory helps in quantifying user response times. Therefore, study of this subject is applicable at all the seven layers.
7. Stochastic structure of the Internet is examined at both the router and the interdomain level. These are the physical networks. Such networks are studied at the network layer. Study of the network of World Wide Web pages occurs at the application layer.
8. An example, of a graph-theoretic algorithm is the routing of packets. Interdomain routing of packets occurs at the transport layer. Routing of packets at the intradomain level occurs at the network layer. Some applications of these techniques also occur at the application layer as well.
9. Game theory can be used to design routing algorithms, and manage resources. Thus, use of game-theoretic techniques is useful at the network and data link layers of the OSI architecture. These techniques can also be used in advertisement, and electronic commerce. These occur at the application layer.
10. Use of data mining and knowledge discovery techniques occur at the application layer.

11. Principles of Internet economics are applicable at the application layer.
12. Theoretical aspects of quantum computation, communication, and cryptography have made rapid progress over the last several years. These appear to show immense promise in principle. If the practical implementations of these ideas come to fruition, then this technology will indeed be "disruptive."

---

6 The World Wide Web and its Function

The Internet is physically made up of computers (hosts), routers, and cables connecting them. A router is a complex internetworking device. It is responsible for routing a frame from one network to another network. It works at the network layer of the OSI model.

The Web is an application which resides on the Internet. It cannot exist without the Internet. However the World Wide Web made the Internet a really useful entity. One can access documents, music, image, and video via the Web. These items are primarily information, and occasionally trinkets of knowledge and wisdom!

Summarizing, a Web is an interface for delivering voice, data, image, and video to the end user (client) on the Internet. In this section, physical components necessary to use the Web: Web server and service, and HyperText Markup Language (HTML) are discussed.

Physical Components Necessary to Use the Web

The Web is made up of the following components. These are required for the existence of the Web.

(a) The end users (clients). They are located either in homes or business centers.
(b) The access network. This is sometimes called the last mile. The end user is connected to the outside world via a telephone line, Digital Subscriber Line (DSL), or a cable.
(c) The Internet Service Providers (ISPs). The ISPs provide the infrastructure to provide caching, traffic load balancing, and bandwidth management.
(d) Backbone. The Internet backbone is a conglomeration of high-speed interconnected networks. These are generally made up of satellites, high capacity lines, and caching proxies.
(e) Web servers. These are the repositories of information that is sought.
(f) Web browser. The Internet or the Web is accessible to most people via a graphical client program called the Web browser. The Web browser lets the user open a Uniform Resource Locator (URL). Its purpose is to provide information about the location of the objects on the Web.

Web Server and Service

Web server is a combination of hardware (server), operating system, server software, and its contents. Its contents are primarily: documents, audio, video, graphics, and data. The server, the storage devices, and the networks are not transparent to the Web users. All these hardware and software items provide services. A Web service is basically a transaction, a computational resource, a business process. Generally Web service gets impaired due to: overloaded servers, uneven distribution of server loads, insufficient capacity, and shortage of connections between application and database servers.

HTML

HTML stands for HyperText Markup Language. It is the software language in which a Web site performs its functions. This language permits user document headings, and hyperlinks to other similar documents. Furthermore, it allows the integration of software applications, and audio and video objects on the Web. Documents written in HTML are text and multimedia objects. HTML uses predefined tags and attributes. Extensible Markup Language (XML) has the capability to allow user defined tags.

7 Value of Networking

We describe several empirical models of the utility of networking. The value or utility of a network is allowed to be an abstract or a vague entity in this section. The value of a network is defined to be a function of the total number of members of the communication network. Let this number be n. Ramifications of utility laws which are linear, quadratic, or exponential functions of n are explored. A network utility law which also depend upon Zipf's law is also discussed. This law is named after the twentieth century Harvard-linguist George Kingsley Zipf (1902-1950). A network utility law based upon generalized Zipf's distribution is also considered.

(a) *Linear law*: In the linear law, the value of a network is simply proportional to n. This law is attributed to the legendary David Sarnoff (1891-1971), a RCA (Radio Corporation of America) television executive and entrepreneur. According to this law, the value of a broadcast network grows linearly with n.

(b) *Quadratic law*: This empirical law is generally attributed to Robert M. Metcalfe, the inventor of the Ethernet LAN. Consider a communication network in which each communication or contact is *equally valuable.* Each member of the network can communicate with $(n-1)$ other members of the network. Therefore total value of the network is proportional to

$$n\left(n-1\right)$$

which is approximately equal to n^2. Thus this law says that the value of a network is a quadratic function of the number of its members.

(c) *Exponential law*: David P. Reed, a computer networking and software pioneer, opined that the value of a network is proportional to 2^n. As per this exponential law, the value of a network is doubled by simply increasing n by unity. As in the quadratic law, it is again assumed that each member of a communicating network is equally valuable. This law might possibly be applicable to Internet chat rooms and mailing lists.

(d) *A utility law based upon Zipf distribution*: Bob Briscoe, Andrew Odlyzko, and Benjamin Tilly proposed that the value of a network is proportional to $n \ln n$. In this formulation, all members of the networking group *do not have equal value*. Justification for this law is based upon Zipf's law. Suppose a member of the communication network ranks the value of the remaining $(n-1)$

persons in order of decreasing value. Of the $(n-1)$ persons, assume that the first person has a value proportional to $1/1$, the second person has a value proportional to $1/2$, the kth person has a value proportional to $1/k$, and finally the $(n-1)$th person has a value proportional to $1/(n-1)$. Therefore the cumulative value of the $(n-1)$ persons is proportional to

$$\frac{1}{1}+\frac{1}{2}+\ldots+\frac{1}{(n-1)}$$

which can be approximated by $\ln(n-1) \simeq \ln n$. Thus, the value of the network to a single person is proportional to $\ln n$. However, there are a total of n persons in the network. Therefore the value of the complete network is proportional to $n \ln n$. This estimate is more conservative than the quadratic law.

(e) *A utility law based upon generalized Zipf distribution*: As we shall study in the rest of the work that the Zipf's law is ubiquitous in the Internet. Actually a generalized form of Zipf's law is more prevalent in the Internet. In the generalized form of the Zipf's law, the kth person has a value proportional to $1/k^{\gamma}$, where γ is a positive constant greater than or equal to unity. Therefore, the cumulative value of the $(n-1)$ persons is proportional to

$$\frac{1}{1^{\gamma}}+\frac{1}{2^{\gamma}}+\ldots+\frac{1}{(n-1)^{\gamma}}$$

If γ is equal to unity, then we have the Briscoe, Odlyzko, and Tilly law. However, if γ is greater than unity, then this summation can be approximated to

$$\frac{\left\{1-(n-1)^{-\gamma+1}\right\}}{(\gamma-1)} \simeq \frac{\left(1-n^{-\gamma+1}\right)}{(\gamma-1)}$$

As there are n members in the network, the value of the communication network is proportional to

$$n\frac{\left(1-n^{-\gamma+1}\right)}{(\gamma-1)}$$

where $\gamma > 1$. The astute reader would notice that $n^{-\gamma+1}$ tends to zero for appropriate values of n and γ. In this case, the value of the communication network turns out to be linear in n. Thus this result appears to be in consonance with the Sarnoff's law, and more conservative than the Briscoe, Odlyzko, and Tilly's law.

All of the above models are growth laws. That is, these laws simply cannot predict the value of a network based solely upon its size. However, if the value of a network is known at a specified value of n, then its value can be estimated at say, another value m provided all other parameters remain unchanged. The reader should be forewarned that all the proponents of the above different laws were aware of the strengths and limitations of their respective models. So in the final analysis, which law is correct? It depends!

It has been implicitly assumed in the above analyses that the principle of network neutrality holds. In brief, this principle implies that all users of the Internet have the freedom to access any Web-site or service they want equally. Thus the principle of network neutrality is a strong social metric of the Internet.

The reader is forewarned that the approach used in this section in defining the value of a network is one of several possible metrics used in defining the value of a network.

8 The Search for Structure

The Internet has reached a critical mass, where it is both important and necessary to search for its structure, and mathematically model its different facets. A mathematical model is supposed to be an icon of knowledge in any scientific discipline. It is generally based upon a set of measurements, intuition, and creative thinking of minds both *past* and *present*.

The purpose of modeling in any engineering discipline is to search and provide a deeper understanding of the structure of the phenomenon. In other words, the goal is to discover order in an apparently amorphous body of knowledge. This is often accomplished by developing simple, elementary, and parsimonious (compact) models to provide insight into the behavior of the phenomenon. It is the purpose of parsimonious models to condense maximum of information with a minimum of axiomatics.

The principle of parsimony is also called Ockham's (also Occam's) razor. It is a principle attributed to William of Ockham (1285-1349), who hypothesized that explanation of a phenomenon should make as few assumptions as possible. Alternately stated, simplest explanation of a physical phenomenon are usually the best.

In the development of such models, details of least importance are often left out. The advantage is that such abstraction may lead to tractable mathematical analysis, yield closed-form results, and ultimately provide useful finger-prints of the structure. It is also hoped that these models have predictive capabilities. The disadvantage is that the models may possibly be not completely applicable.

On the other hand, more complicated models may neither provide closed-form results nor insight into the phenomenon. However, the model may yield to numerical calculations and may directly be applicable. It is the task of the analyst or the modeler to obtain a reasonable balance between these different aspects of a modeling effort. Ultimately, the quintessential goal of a model is to understand the past and current phenomenon, and predict and control accurately the future behavior. This, indeed is a daunting task!

Due to heroic efforts of several persons, the quest for structure of the Internet has opened up many frontiers of research. These capture to some extent the essence of the form and function of the Internet. Some of these topics are information theory, theory of error detecting and correcting codes, cryptography, graph-theoretic algorithms, queueing theory, and data mining techniques. Some of these topics predate the birth of the Internet, however their applications are still current and reverberant. Concrete progress has also been made in description of the Internet traffic, Internet congestion, and the stochastic structure of the Internet and the World Wide Web. Game theory has also found recent application in the Internet. Certain basic principles of Internet economics have also been discussed. In looking at the future, some theoretical advances have also been made in the area of quantum computation, communication, and cryptography.

Therefore if there is a quest for improvement in the engineering of the Internet, then the search for structure in it has to be a continual process.

9 Effectiveness of Mathematics

We next examine the effectiveness of mathematics in search of structure in the Internet. The phrase "effectiveness of mathematics" was used by the physicist Eugene P. Wigner (1902-1995) in describing the role of mathematics in explaining natural phenomena. Mathematics is generally described as the science of study of patterns. Furthermore, it is the general perception that mathematics is the language of natural sciences. The Internet does not turn out to be an exception. Mathematics has been surprisingly effective in describing a multitude of physical phenomena. Mathematics is used in science and engineering to discover invariances. Besides, it reduces ideas to common sense.

The Internet is a network of computers, and computers are built upon mathematical principles. Therefore, not surprisingly several situations in the Internet are amenable to mathematical analysis. Indeed, a colorful tapestry of mathematics is woven in mathematical analysis with multivariegated strands of several different mathematical disciplines.

To date, the mathematical mosaic provided by elements of number theory, abstract algebra, applied analysis, matrices, determinants, geometry, graph theory, probability theory, stochastic processes, optimization principles, stability theory, and chaos theory are required to understand the many facets of Internet engineering. These mathematical topics are discussed in different chapters of Volume 2, and may be consulted by the reader as the need arises.

In the final analysis, the different forms and functions of the Internet may or may not yield to the principle of Ockham's razor. That is, explanations should not be multiplied beyond the basic requirements. It is also quite possible that a paradigm shift is necessary in the spirit of the scientist-philosopher Thomas S. Kuhn (1922-1996), in qualitatively and quantitatively examining the panorama provided by the Internet.

In the exercise of modeling the Internet, there are usually a myriad questions, many more speculations, and very few answers. Some of these answers are partially addressed in this work. It should however be remembered that modeling of any phenomenon evolves by initially examining its own errors and then possibly seeking unanticipated connections. Perhaps an apt tool for this purpose is mathematics.

Reference Notes

An entertaining account of the history of telecommunication can be found in the textbook by Oslin (1992). Discussion about the contributions of Bose can be found in Bondyopadhyay (1998), Emerson (1997), and Sarkar, and Sengupta (1997).

The initial design principles of the Internet have been eloquently laid out in Clark (1988). A description of the early days of the Internet can be found in Kleinrock (1975). An entertaining and instructive account of the creation of the World Wide Web was written by its inventor Berners-Lee (1999). A first-hand account of the evolution of the Internet can be found in Day (2008). The section on power of networking is based upon a thought provoking paper by Briscoe, Odlyzko, and Tilly (2006).

An excellent article about the usefulness of mathematics in natural sciences can be found in Wigner (1979). In this article, he writes about the surprising effectiveness of mathematics in explaining natural phenomenon.

A general introduction to computer networks can be found in the classics by Tanenbaum (2002), Kurose, and Ross (2009), Peterson, and Davie (2007), Walrand, and Varaiya (2000), Forouzan (2007), and Comer (2009). The book by Kumar, Manjunath, and Kuri (2004), provides an analytical approach to communication networking.

Problems

1. There are two parts in the problem.
   (a) If you were to design a newer version of the Internet, what design principles from the current Internet would you retain? Which principles would you discard? What additional principles would you incorporate in the new design? Explain and justify all your statements.
   (b) Describe how your design can evolve from the current Internet. Specify all the constraints and limitations.
2. Compare and contrast the Internet revolution with other significant revolutions in the history of mankind. For example, how does it compare with the invention of the printing press, or the harness of electricity.
3. Technological revolutions impact human lives in several different ways. What are some of the ethical and social requirements of a technological revolution?
   This is an open-ended question. Answers to such question are often hotly debated, both inside and outside classrooms.

References

1. Berners-Lee, T. with Fischetti, M., 1999. *Weaving the Web*, Harper Collins Publishers, New York, New York.
2. Bondyopadhyay, P. K., 1998. "Sir J. C. Bose's Diode Detector Received Marconi's First Transatlantic Wireless Signal of December 1901 (The "Italian Navy Coherer" Scandal Revisited)," Proc. IEEE, Vol. 86, No. 1, pp. 259-285.
3. Briscoe, B., Odlyzko, A., and Tilly, B., 2006. "Metcalfe's Law is Wrong," IEEE Spectrum, Vol. 43, pp. 34-39.
4. Bronowski, J., 1978. *The Origins of Knowledge*, Yale University Press, New Haven and London.
5. Bush, V., 1945. "As We May Think," Atlantic Monthly, July.
6. Cerf, V. G., 2006. "Network Neutrality," Prepared Statement for U.S. Senate Committee on Commerce, Science, and Transportation.
7. Clark, D. D., 1988. "The Design Philosophy of the DARPA Internet Protocols," Proc. of the ACM SIGCOMM'88, in: ACM Computer Communication Reviews, Vol. 18, No. 4, pp. 106-114.
8. Comer, D. E., 2009. *Computer Networks and Internets*, Pearson Prentice-Hall, Upper Saddle River, New Jersey.
9. Day, J., 2008. *Patterns in Network Architecture: A Return to Fundamentals*, Pearson Prentice-Hall, Upper Saddle River, New Jersey.
10. Emerson, D. T., 1997. "The Work of Jagadish Chandra Bose: 100 Years of Millimeter-Wave Research," IEEE Transactions on Microwave Theory and Techniques, Vol. 45, No. 12, pp. 2267-2273.
11. Forouzan, B. A., 2007. *Data Communication and Networking*, Fourth Edition, McGraw-Hill Book Company, New York, New York.

12. Gebali, F., 2008. *Analysis of Computer and Communication Networks*, Springer-Verlag, Berlin, Germany.
13. Hadamard, J., 1945. *The Psychology of Invention in the Mathematical Field*, Dover Publications, Inc., New York, New York.
14. Kleinrock, L., 1975. *Queueing Systems, Volumes I and II,* John Wiley & Sons, Inc., New York, New York.
15. Kuhn, T. S., 1970. *The Structure of Scientific Revolutions*, Second Edition, Enlarged, The University of Chicago Press, Chicago, Illinois.
16. Kumar, A., Manjunath, D., and Kuri, J., 2004. *Communication Networking, An Analytical Approach,* Morgan Kaufmann Publishers, San Francisco, California.
17. Kurose, J. F., and Ross, K. W., 2009. *Computer Networking*, Fifth Edition, Addison-Wesley Publishing Company, New York, New York.
18. Oslin, G. P., 1992. *The Story of Telecommunications*, Mercer University Press, Macon, Georgia.
19. Peterson, L. L., and Davie, B. S., 2007. *Computer Networks*, Fourth Edition, Morgan Kaufmann Publishers, San Francisco, California.
20. Poincaré, J. H., 1952. *Science and Hypothesis*, Dover Publications, Inc., New York, New York.
21. Sarkar, T. K., and Sengupta, D. L., 1997. "An Appreciation of J. C. Bose's Pioneering Work in Millimeter Wave," IEEE Antenna and Propagation Magazine, Vol. 39, No. 5, pp. 55-63.
22. Tanenbaum, A. S., 2002. *Computer Networks*, Fourth Edition, Prentice-Hall, Englewood Cliffs, New Jersey.
23. Walrand, J., and Varaiya, P., 2000. *High-Performance Communication Networks*, Morgan Kaufmann Publishers, San Francisco, California.
24. Wigner, E. P., 1979. *Symmetries and Reflections*, Ox Bow Press, Woodbridge, Connecticut.

CHAPTER 1

Information Theory

$$H(X) = -\sum_{j=1}^{n} p(x_j) \log_2 p(x_j)$$

Entropy of a discrete random variable

Claude Elwood Shannon. Claude Elwood Shannon was born on April 30, 1916 in Gaylord, Michigan, U.S.A. Shannon is regarded as the father of information theory. He graduated from Massachusetts Institute of Technology in 1940 with a master's degree and a doctorate in mathematics. Shannon applied Boolean algebra to switching circuits in his master's thesis.

In the year 1948 Shannon published a landmark paper: *A Mathematical Theory of Communication* in Bell System Technical Journal. In this pioneering paper he introduced the concept of information entropy. This paper had a profound impact on the development of information processing machines and computer communication systems. It deals with the methods and techniques of transmitting data efficiently on a communication channel. One can unequivocally assert that all packet transfers, routers and switches, and communication channels on the Internet have the stamp of Shannon's seminal contributions. Shannon passed away on February 27, 2001 in Medford, Massachusetts.

1.1 Introduction

Information theory is the study of techniques for efficient, reliable, and secure transmission of data over communication channels. A basic artifice in the study of information theory is the concept of entropy. In this chapter, entropy is defined mathematically and applied to different but related areas of noiseless coding for memoryless source, data compression, and transmission of information over noisy communication channels. Some well-known information capacity theorems are also developed. The following topics are specifically discussed in this chapter: entropy, mutual information and relative entropy, noiseless coding for memoryless source, data compression, communication channels, differential entropy, capacity of communication channels, rate distortion theory, rudiments of network information theory, and maximum entropy principle.

Information theory is essentially mathematical in nature. Furthermore, it describes the mathematical laws that control the transmission and processing of information. These laws are stated in the language of probability theory. It also strengthens the belief of several philosophers, scientists, and engineers that the modern world we live in is primarily probabilistic. The ground work for information theory was laid by Claude E. Shannon (1916-2001). His paper published in *Bell System Technical Journal* in July and October of 1948, is generally regarded as the Magna Carta of the modern information age. Information theory provides answers to some extremely important questions in communication theory. For example, it allows us to quantify the maximum possible data compression. This quantification is specified in terms of entropy. Information theory also provides a framework to determine the ultimate transmission rate of a communication channel. This is specified in terms of the channel capacity.

Consider a communication system described by Shannon. In his vision, a communication system consists of six major components. These are a source, transmitter, communication channel, noise source, receiver and destination. The *source* generates information. More specifically, the source produces *messages*. The *transmitter* transforms the messages suitable for transmission over the *communication channel*. The transformed message is called a *signal*. The channel next transports the signal from the transmitter to the *receiver*. The *noise source* emits unfriendly signals. Its purpose is to obliterate (or corrupt) the original signal. The receiver, if possible converts the received signal into a form suitable for interpretation by the *destination*. The destination can be either a person or receiving equipment. See Figure 1.1.

The block diagram in Figure 1.1 is generally regarded as one of the important contributions of Shannon. In a single stroke of genius, he abstracted all the important entities necessary to represent communication between a source and destination pair.

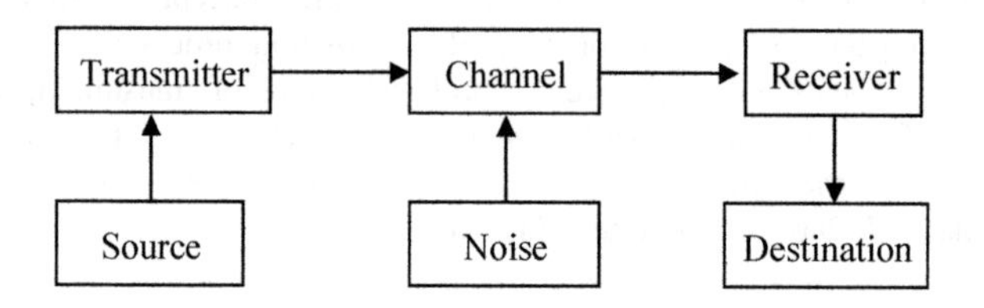

Figure 1.1. A canonical communication system.

A convenient vehicle to formulate information theory is the concept of *entropy*. This term has been borrowed from thermodynamics. The term "entropy" was apparently coined by the physicist Rudolph Julius Emmanuel Clausius (1822-1888) in the year 1865. The concept of entropy appears to be all embracing. It has often been stated that along with the concepts of distance, mass, mechanical and electrical forces, entropy can be explained in the language of arithmetic. However, it also finds itself in the company of abstract concepts such as beauty and melody.

The building blocks of information theory such as entropy, mutual information, relative entropy, and redundancy were introduced by Shannon in his work on cryptographic systems, and not on communication channels. Application of information-theoretic concepts to cryptographic systems is discussed in the chapter on cryptography.

The reader is assumed to have a basic knowledge of applied analysis, matrices, probability theory, stochastic processes, and optimization techniques. It is assumed in this chapter that $0 \log_2 0 \to 0$, $x \log_2 (x/0) \to \infty$ for $x > 0$, and $0 \log_2 (0/0) \to 0$. It can be shown that these results follow from a continuity argument. The expectation and variance operators of a random variable are denoted by $\mathcal{E}(\cdot)$ and $Var(\cdot)$ respectively.

1.2 Entropy

Entropy of a discrete random variable measures the amount of information obtained by observing it. Uncertainty, surprise, information, and entropy are identical concepts. The notion of self-information is first developed. Entropy is defined in terms of logarithm with respect to base 2. Therefore unit of entropy (information) is in bits. A *bit* is a binary digit. It takes a value of either 0 or 1.

Definition 1.1. *Let X be a discrete random variable, which takes real values $x_j, j = 1, 2, \ldots, n$. Also $P(X = x_j) = p(x_j), j = 1, 2, \ldots, n$ are the corresponding probabilities, such that* $0 <$

$p(x_j) \leq 1$, and $\sum_{j=1}^{n} p(x_j) = 1$. The information gained after observing that the discrete random variable X takes the value x_j is $\widetilde{\vartheta}(x_j)$, where

$$\widetilde{\vartheta}(x_j) = \log_2 \left(\frac{1}{p(x_j)} \right), \qquad 1 \leq j \leq n \tag{1.1}$$

The quantity $\widetilde{\vartheta}(x_j) = -\log_2 p(x_j)$ is called the self-information of the event $X = x_j$. Its units are in bits. □

Following are some observations about the self-information function $\widetilde{\vartheta}(\cdot)$.

Observations 1.1. Self-information related observations.

1. $\widetilde{\vartheta}(x_j) \geq 0$ for $1 \leq j \leq n$. That is, self-information of an event is greater than or equal to zero.
2. $\widetilde{\vartheta}(x_j) = 0$ if $p(x_j) = 1$. The information content of a certain event is equal to zero.
3. $\widetilde{\vartheta}(x_j) > \widetilde{\vartheta}(x_k)$ if $p(x_j) < p(x_k)$, where $j \neq k$. That is, a less probable event has more information. □

Entropy is next defined. Average value of self-information of a discrete random variable is called its entropy.

Definitions 1.2. *Some entropy related definitions.*

1. *The entropy or uncertainty of the discrete random variable X is defined to be $H(X)$, where*

$$H(X) = -\sum_{j=1}^{n} p(x_j) \log_2 p(x_j) \tag{1.2}$$

where, $P(X = x_j) = p(x_j)$, $1 \leq j \leq n$; $\sum_{j=1}^{n} p(x_j) = 1$, and $p(x_j)\log_2 p(x_j)$ is defined to be equal to zero, if $p(x_j) = 0$.
Sometimes $H(X)$ is also denoted by $H(p)$.

2. *Let X and Y be two jointly distributed discrete random variables which take real values. Their probability distribution function $P(\cdot,\cdot)$ is specified by $P(X = x, Y = y)$, where $\sum_{x,y} P(X = x, Y = y) = 1$.*
   (a) *The joint entropy of random variables X and Y is $H(X,Y)$ where*

$$H(X,Y) = -\sum_{x,y} P(X = x, Y = y) \log_2 P(X = x, Y = y) \tag{1.3}$$

   where the summation extends over all possible values of X and Y.
   (b) *The conditional entropy of X given $Y = y$, is $H(X \mid Y = y)$.*

$$H(X \mid Y = y) = -\sum_{x} P(X = x \mid Y = y) \log_2 P(X = x \mid Y = y) \tag{1.4}$$

   (c) *The conditional entropy of X given Y, is $H(X \mid Y)$.*

$$H(X \mid Y) = \sum_{y} P(Y = y) H(X \mid Y = y) \tag{1.5}$$

   The entropy $H(X \mid Y)$, measures the uncertainty remaining in the discrete random variable X after the discrete random variable Y has been observed. □

Example 1.1. Let X be a Bernoulli distributed random variable with parameter p. Then $H(p) = -p\log_2 p - (1-p)\log_2(1-p)$. See Figure 1.2 for a plot of this function.

Observe that $H(0) = H(1) = 0$, and $H(0.5) = 1$. Its maximum occurs at $p = 0.5$. In addition, $H(\cdot)$ is a concave function of p, and symmetric about $p = 0.5$.

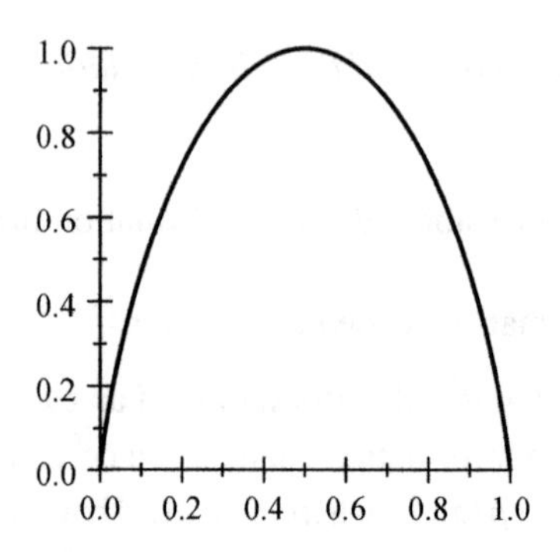

Figure 1.2. $H(p)$ versus p.

□

Example 1.2. Let X be a discrete random variable, which takes values $1, 2, \ldots 5$. The corresponding probabilities are $p(1) = 1/2, p(2) = 1/4, p(3) = 1/8, p(4) = 1/16$, and $p(5) = 1/16$. The entropy $H(X)$ of random variable X is given by

$$\begin{aligned} H(X) &= -\sum_{j=1}^{5} p(j)\log_2 p(j) \\ &= -\left(2^{-1}\log_2 2^{-1} + 2^{-2}\log_2 2^{-2} + 2^{-3}\log_2 2^{-3} + 2^{-4}\log_2 2^{-4}\right. \\ &\quad \left. +2^{-4}\log_2 2^{-4}\right) \\ &= 1.875 \end{aligned}$$

Therefore $H(X) = 1.875$ bits. □

Observations 1.2. In the following observations, the jointly distributed discrete random variables X, Y, and Z take only a finite set of values.

1. $0 \le H(X) \le \log_2 n$, where the random variable X takes n different values.
   (a) $H(X) = 0$, if $p(x_j) = 1$ for some x_j, and $p(x_i) = 0$ for all $i \ne j, 1 \le i, j \le n$. That is, the random variable X is a constant.
   (b) $H(X) = \log_2 n$, if $p(x_j) = 1/n$ for all $j, 1 \le j \le n$. If $n = 2^m$, then $H(X) = m$.
2. $H(X,Y) = H(Y,X)$.
3. $H(X,Y) \le H(X) + H(Y)$, where the equality is observed if and only if the discrete random variables X and Y are independent of each other. Similarly, let $X_1, X_2, \ldots, X_n$ be discrete random variables, then

$$H(X_1, X_2, \ldots, X_n) \le \sum_{i=1}^{n} H(X_i)$$

with equality if and only if the n random variables are independent of each other. This is called the law of *subadditivity*.

4. $H(X,Y) = H(Y) + H(X \mid Y) = H(X) + H(Y \mid X) = H(Y,X)$. Thus $H(X) \leq H(X,Y)$, with equality if and only if Y is a function of X.
5. $H(X \mid X) = 0$.
6. $H(X \mid Y) \geq 0$.
7. $H(X \mid Y) \leq H(X)$, where the equality is observed if and only if the discrete random variables X and Y are independent of each other. Similarly $H(X \mid Y, Z) \leq H(X \mid Y)$. These inequalities imply that *conditioning might reduce entropy*.
8. *Chain rule for entropy*: $X_1, X_2, \ldots, X_n$ are jointly distributed discrete random variables. Then
   (a) $H(X_1, X_2) = H(X_1) + H(X_2 \mid X_1)$
   (b) $H(X_1, X_2, X_3) = H(X_1) + H(X_2, X_3 \mid X_1)$
   $= H(X_1) + H(X_2 \mid X_1) + H(X_3 \mid X_2, X_1)$
   (c) $H(X_1, X_2, \ldots, X_n)$
   $= H(X_1) + H(X_2 \mid X_1) + \ldots + H(X_n \mid X_{n-1}, \ldots, X_1)$
9. Entropy is a *concave function* of its arguments $p(x_j), 1 \leq j \leq n$. This implies that for any two probability distributions $\{p'(x_j) \mid 1 \leq j \leq n\}$ and $\{p''(x_j) \mid 1 \leq j \leq n\}$, and for any $\lambda \in [0,1]$
$$H[\lambda p' + (1-\lambda) p''] \geq \lambda H(p') + (1-\lambda) H(p'')$$
   This inequality has a physical interpretation. It implies that mixing probability distributions results in an increase in entropy.
10. *Change of base of logarithm*: Let $a, b \in \mathbb{R}^+$,
$$H_a(X) = -\sum_{j=1}^{n} p(x_j) \log_a p(x_j), \text{ and } H_b(X) = -\sum_{j=1}^{n} p(x_j) \log_b p(x_j)$$
   Then
$$H_b(X) = (\log_b a) H_a(X)$$

□

1.3 Mutual Information and Relative Entropy

Mutual information and relative entropy are studied in this section. Recall that the entropy of a random variable is sometimes referred to as its self-information. Analogously, the amount of information that a given random variable contains about another random variable can be quantified. This measure is specified by the *mutual information* of the two random variables X and Y. The *relative entropy* quantifies the distance between two probability distribution functions.

1.3.1 Mutual Information

The entropy of a discrete random variable measures its uncertainty. Mutual information between two discrete random variables measures the uncertainty between them. Equivalently, it measures the information the two discrete random variables have in common. The concept of mutual information between two discrete random variables is useful in describing noisy communication channels.

Definitions 1.3. *Mutual information between jointly distributed random variables.*

1. *Let X and Y be two jointly distributed discrete random variables. The mutual information between these two random variables is*

$$I(X;Y) = H(X) - H(X \mid Y) \tag{1.6}$$

2. *Let X, Y, and Z be jointly distributed discrete random variables. The mutual information between the random variable X and the random variables Y and Z is*

$$I(X;Y,Z) = H(X) - H(X \mid Y,Z) \tag{1.7}$$

□

Therefore, mutual information $I(X;Y)$ is the decrease in uncertainty of X due to the knowledge of Y. Alternately, the mutual information $I(X;Y)$ may be considered to be the information that the discrete random variable Y reveals about the discrete random variable X. The following observations can be noted.

Observations 1.3. Let X and Y be two jointly distributed discrete random variables.

1. An alternate definition of mutual information:

$$\begin{aligned} I(X;Y) &= \sum_{x,y} P(X=x, Y=y) \log_2 \left\{ \frac{P(X=x \mid Y=y)}{P(X=x)} \right\} \\ &= \sum_{x,y} P(X=x, Y=y) \log_2 \left\{ \frac{P(X=x, Y=y)}{P(X=x)P(Y=y)} \right\} \end{aligned}$$

2. $I(X;Y) \geq 0$, with equality if and only if X and Y are independent of each other.
3. $I(X;Y) = I(Y;X)$, where $I(Y;X) = H(Y) - H(Y \mid X)$.
4. $I(X;Y) \leq \min(H(X), H(Y))$.
5. $I(X;X) = H(X)$. That is, the entropy of a random variable is equal to the mutual information of the random variable with itself. Therefore, entropy of a random variable is occasionally referred to as its self-information.
6. $I(X;Y) = H(X) + H(Y) - H(X,Y)$.
7. The joint probability distribution of the random variable pair (X,Y) is specified as

$$P(X=x, Y=y) = P(X=x) P(Y=y \mid X=x)$$

For brevity in notation, let

$$P(X=x, Y=y) \triangleq p_{X,Y}(x,y), \quad P(X=x) \triangleq p_X(x),$$
$$P(Y=y \mid X=x) \triangleq p_{Y|X}(y \mid x)$$

Therefore, $p_{X,Y}(x,y) = p_X(x) p_{Y|X}(y \mid x)$.

(a) For fixed $p_{Y|X}(y \mid x)$, the mutual information $I(X;Y)$ is a concave function of $p_X(x)$.

(b) For fixed $p_X(x)$, the mutual information $I(X;Y)$ is a convex function of $p_{Y|X}(y \mid x)$. □

A Venn-diagrammatic representation of the relationship between entropy and mutual information is shown in Figure 1.3. A Venn diagram shows relationship between sets. In this diagram, X and Y are two random variables. Observe that their mutual information $I(X;Y)$ is at the intersection of the entropy of these two random variables. Furthermore, the joint entropy $H(X,Y)$ of these two random variables is the union of their individual entropies.

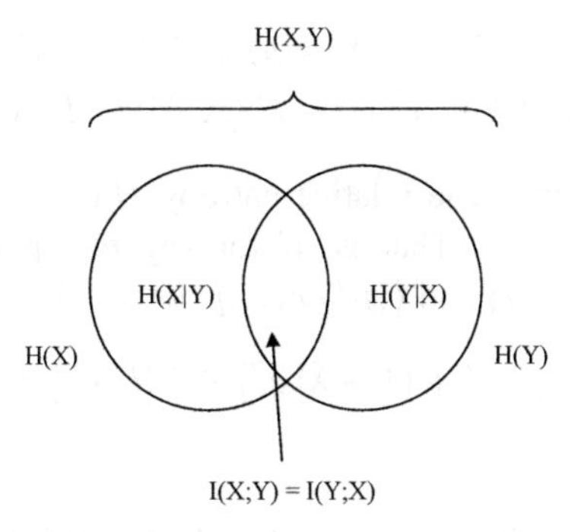

Figure 1.3. A Venn diagram representation of the relationship between entropy, conditional entropy, and mutual information.

1.3.2 Relative Entropy

Relative entropy measures the distance between a pair of discrete probability distribution functions defined on the same index set. That is, the two random variables take values in a single set. Relative entropy is also related to mutual information.

Definition 1.4. *Let*

$$P(X = x_j) = p(x_j), \ 1 \le j \le n, \ \textit{and} \ P(X = x_j) = q(x_j), \ 1 \le j \le n$$

be two distinct distribution functions. The relative entropy of $p(x)$ *to* $q(x)$ *is given by*

$$\begin{aligned} H(p(x) \parallel q(x)) &= \sum_{j=1}^{n} p(x_j) \log_2 \frac{p(x_j)}{q(x_j)} \\ &= -H(X) - \sum_{j=1}^{n} p(x_j) \log_2 q(x_j) \end{aligned} \tag{1.8}$$

□

Relative entropy is sometimes called the *Kullback-Leibler distance* or *divergence*. Note that the relative entropy can be infinite if zero value of probability is permitted.

Observations 1.4. Some basic observations about relative entropy are listed below.

1. Relative entropy is nonnegative. Therefore it can possibly be called a distance, however it is not a true measure of distance between two probability distributions. This is true because the Kullback-Leibler distance is not symmetric, and does not satisfy the triangle inequality.
2. Relative entropy $H(p(x) \parallel q(x))$ is equal to zero, if the two distributions are identical, that is $p(x_j) = q(x_j)$, for $1 \le j \le n$.
3. In general $H(p(x) \parallel q(x)) \ne H(q(x) \parallel p(x))$.
4. Let X and Y be two jointly distributed discrete random variables, with probability distribution function specified by $P(X = x, Y = y)$'s. Then

$$
\begin{aligned}
& H\left(P(X=x,Y=y) \parallel P(X=x)P(Y=y)\right) \\
& = H(X) + H(Y) - H(X,Y) = I(X;Y)
\end{aligned}
$$

5. Convexity of relative entropy. The relative entropy $H(p(x) \parallel q(x))$, is convex in the pair $(p(x_j), q(x_j)), 1 \le j \le n$. That is, if for any two pairs of probability distributions $\{(p'(x_j), q'(x_j)) \mid 1 \le j \le n\}$ and $\{(p''(x_j), q''(x_j)) \mid 1 \le j \le n\}$; and for each $\lambda \in [0,1]$

$$
H\left[\lambda p' + (1-\lambda) p'' \parallel \lambda q' + (1-\lambda) q''\right] \le \lambda H(p' \parallel q') + (1-\lambda) H(p'' \parallel q'')
$$

□

Using the above observations, it can be proved again that $H(X,Y) \le H(X) + H(Y)$, where the equality is observed if and only if X and Y are independent of each other. An important result in statistics which uses the concept of relative entropy is the celebrated Sanov's theorem. This theorem is next examined.

Sanov's Theorem

Sanov's theorem can be derived as an application of the Kullback-Leibler distance. This theorem is named after I. N. Sanov. Sanov's theorem specifies the closeness of two distributions. It also has applications in large deviation theory. Let $X_1, X_2, \ldots$ be a sequence of independent, and identically distributed random variables. Each of these random variables takes values in the discrete finite set $\mathfrak{X} = \{x_1, x_2, \ldots, x_d\}$, where

$$
P(X_i = x_j) = p(x_j) > 0, \ \forall\, x_j \in \mathfrak{X}; \text{ and } \sum_{j=1}^{d} p(x_j) = 1
$$

Let $\{e_1, e_2, \ldots, e_d\}$ be the set of standard basis of the space $\mathbb{R}^d$, where

$$
e_j = (0, 0, \ldots, 0, 1, 0, \ldots, 0, 0)^T \in \mathbb{R}^d
$$

and the digit 1 is in the jth position for each $j \in \{1, 2, \ldots, d\}$. The random variables can be represented by the random vectors in $\mathbb{R}^d$ as follows. Define

$$
z_i = \sum_{j=1}^{d} e_j I_{\{X_i = x_j\}}
$$

where $I_{\{X_i = x_j\}}$ is the indicator function of the event $\{X_i = x_j\}$. That is, if $X_i = x_j$, then all components of the vector z_i are 0's except the jth component which is equal to 1. Thus

$$
\widehat{z}_n = \frac{1}{n} \sum_{i=1}^{n} z_i
$$

is an empirical measure induced on $\mathfrak{X}$ by the sequence $X_1, X_2, X_3, \ldots, X_n$. Also observe that $z_1, z_2, z_3, \ldots, z_n$ is a sequence of independent, and identically distributed random vectors. Let

$$
\mathcal{E}(z_1) = (p(x_1), p(x_2), \ldots, p(x_d))^T \triangleq \widetilde{p}
$$

The strong law of large numbers implies that $P\left(\lim_{n\to\infty} \widehat{z}_n = \widetilde{p}\right) = 1$. Another probability distribution on the set $\mathfrak{X}$ is defined as

$$\widetilde{q} = (q(x_1), q(x_2), \ldots, q(x_d))^T$$

$$q(x_i) > 0, \ 1 \leq i \leq d, \ \sum_{i=1}^{d} q(x_i) = 1$$

We evaluate the probability $P(\widehat{z}_n = \widetilde{q})$. Observe that each component of the vector $\widehat{z}_n$ is a multiple of $1/n$. Therefore use of multinomial distribution yields

$$P(\widehat{z}_n = \widetilde{q}) = \frac{n!}{\prod_{i=1}^{d} \{nq(x_i)\}!} \prod_{i=1}^{d} \{p(x_i)\}^{nq(x_i)}$$

Use of Stirling's approximation for the factorial of an integer yields

$$P(\widehat{z}_n = \widetilde{q}) \simeq \left\{ \prod_{i=1}^{d} \left(\frac{p(x_i)}{q(x_i)} \right)^{q(x_i)} \right\}^n = \exp\left\{ -n \sum_{i=1}^{d} q(x_i) \ln \frac{q(x_i)}{p(x_i)} \right\}$$

In the above approximation, the nonexponential terms were ignored. Thus

$$P(\widehat{z}_n = \widetilde{q}) \simeq \exp\{-nH(q(x) \parallel p(x))\}$$

The above discussion is summarized in the following discrete version of Sanov's theorem.

Theorem 1.1. *Let $X_1, X_2, \ldots$ be a sequence of independent, and identically distributed random variables which take values in the discrete finite set $\mathfrak{X} = \{x_1, x_2, \ldots, x_d\}$. The probability mass function of a random variable X_i is $p(x), x \in \mathfrak{X}$. Also $\widehat{z}_n$ is an empirical measure induced on $\mathfrak{X}$ by the sequence $X_1, X_2, X_3, \ldots, X_n$. Another probability distribution on the set $\mathfrak{X}$ is defined as $q(x), x \in \mathfrak{X}$. If $\widetilde{q} = (q(x_1), q(x_2), \ldots, q(x_d))^T$, where $q(x_i) > 0$, for $1 \leq i \leq d$, then $P(\widehat{z}_n = \widetilde{q}) \simeq \exp\{-nH(q(x) \parallel p(x))\}$.* □

More refined versions of this theorem can be found in the literature.

1.4 Noiseless Coding for Memoryless Source

Coding techniques for signals from a memoryless source are examined in this section. A source is said to be zero-memory or memoryless, if it does not remember the symbol (word) it transmitted the previous instant (or last time). These signals are assumed to be noise-free. That is, there is no noise signal on the transmission channel. Note once again that the noise is a signal that tries to corrupt the transmitted signal. Consider an information source S, which emits a stream of *words* from the set $W = \{w_1, w_2, \ldots, w_K\}$, where $|W| = K$. Define W^* to be the set of all finite strings of elements of W. Members of the set W^* are called messages. A message $m \in W^*$ is of the form

$$m = w_{i_1} w_{i_2} \ldots w_{i_j}$$

This message m has j words. Let U_i be the ith word produced by the source. As per the hypothesis, the probability $P(U_i = w_k) = p(w_k)$ is independent of i. That is, under the assumption of a *memoryless source*, the probability $P(U_i = w_k)$ is independent of words emitted earlier or to be emitted in future. This implies that the sequence of discrete random variables $U_1, U_2, \ldots$ are independent, and identically distributed. For simplicity, denote these random variables by U.

Assume that $p(w_k) > 0$. However, if $p(w_k) = 0$, then the word w_k can be discarded from the set W. The entropy of this memoryless source is defined as

$$H(U) = -\sum_{k=1}^{K} p(w_k) \log_2 p(w_k)$$

The words emitted by the source need to be encoded for efficient transmission over a channel. The encoder uses characters (symbols or letters or elements) from the *alphabet*

$$\mathcal{X} = \{x_1, x_2, \ldots, x_D\}$$

Each of these characters takes equal time for transmission over a channel. For example this set can be $\{0, 1\}$. Define $\mathcal{X}^*$ to be the set of all finite strings of elements of $\mathcal{X}$. The encoder performs a mapping operation $f(\cdot)$ from the set W to the set $\mathcal{X}^*$. For each source word $w_k \in W$, the string $f(w_k) \in \mathcal{X}^*$ is a *codeword*. Mapping of different words are assumed to be different. Let the number of characters in $f(w_k)$ be denoted by $|f(w_k)| \triangleq n_k$, for $1 \leq k \leq K$. The average length of this code is given by $\overline{n}$.

$$\overline{n} = \sum_{k=1}^{K} p(w_k) n_k$$

The goal of this section is to provide insight and develop coding schemes to minimize $\overline{n}$. Note that the message $m = w_{i_1} w_{i_2} \ldots w_{i_j} \in W^*$ is coded as a concatenation $f(w_{i_1}) f(w_{i_2}) \ldots f(w_{i_j})$.

For a coding scheme to be useful, it should be uniquely decipherable. A code is said to be uniquely decodable, if every element of $\mathcal{X}^*$ is the image of at most one message in W^*.

A subset of uniquely decodable codes are called *prefix-free codes*. A code is said to be a prefix-free code, if $f(w_i)$ is not a prefix of $f(w_j)$ for all $i \neq j$, where $1 \leq i, j \leq K$. For example, if $a, b, c \in \mathcal{X}^*$, then a is said to be a prefix of c if $ab = c$. Note that the prefix-free codes are *instantaneous*. That is, the codes can be decoded "on the fly" without having a knowledge of the future words.

Example 1.3. Let $W = \{w_1, w_2, w_3, w_4, w_5\}$, and $\mathcal{X} = \{0, 1\}$. Define the coding function $f(\cdot)$ as

$$f(w_1) = 1, \; f(w_2) = 01, \; f(w_3) = 001, \; f(w_4) = 0001, \text{ and } f(w_5) = 00001$$

This coding scheme is a prefix-free code, because the message

0000100010110011

can be decoded as $w_5 w_4 w_2 w_1 w_3 w_1$. □

1.4.1 Pair of Inequalities

A pair of inequalities useful in proving Shannon's results on noiseless coding, is next studied. These inequalities are called Kraft's and McMillan's inequalities. The following theorem is due to L. G. Kraft. The corresponding inequality is called Kraft's inequality in his honor.

Theorem 1.2. *Let W be the set of words which belong to a memoryless source $\mathcal{S}$. Also $\mathcal{X}$ is the set of the encoding alphabet, and $\mathcal{X}^*$ is the set of all finite strings of elements of $\mathcal{X}$. Let $|W| = K$ and $|\mathcal{X}| = D$, where $D > 1$. Consider a code $f : W \to \mathcal{X}^*$ with codeword-length $|f(w_k)| = n_k$, for $1 \le k \le K$. Then the code $f(\cdot)$ is a prefix-free code if and only if*

$$\sum_{k=1}^{K} D^{-n_k} \le 1 \tag{1.9}$$

The above inequality is called Kraft's inequality.

Proof. Arrange the codeword-lengths in increasing order as $n_1 \le n_2 \le \ldots \le n_K$.

The proof of this theorem is facilitated by considering a D-ary tree. This tree has a root node. Root node is said to be at level-0 of the tree. Beginning at this node, the tree expands in D different branches to create D different nodes at level-1 of the tree. Thus, there are D nodes at level-1 of the tree. Each of these D nodes in turn create D new branches and nodes at level-2. Therefore, there are D^2 nodes at level-2 . This process continues till level-n_K. At level-n_K, there are D^{n_K} nodes.

In this tree generation process, each of the D different branches from a single node represents a single encoding character of the set $\mathcal{X}$.

Note that, the nodes at level-k are candidates to represent codewords of length k. Assign a node at level n_1 to the first codeword. The prefix condition on the codewords imposes the condition that no codeword is an ancestor node of any other codeword. Therefore, the choice of the first codeword disables all its descendents nodes in the D-ary tree. Therefore $D^{n_K - n_1}$ nodes at level-n_K are disabled. As there are D^{n_K} nodes at level-n_K, D^{-n_1} fraction of the nodes at level-n_K are disabled. When the second codeword is assigned to a node at level n_2, it in turn disables D^{-n_2} fraction of the nodes at level-n_K. Therefore the two codewords with smallest lengths have disabled $(D^{-n_1} + D^{-n_2})$ fraction of the nodes at level-n_K. Continuing this process for all the K codewords results in the required inequality.

Conversely, assume that the codeword-lengths satisfy the condition $n_1 \le n_2 \le \ldots \le n_K$, the stated inequality, and that a D-ary tree can be constructed. Select the first node (lexicographically) at level-n_1 of the tree and assign it to the first codeword, and then remove all of its descendents. Next assign the first remaining node at level n_2 to the second codeword. Continuation of this process creates a prefix-free code with lengths $n_1, n_2, \ldots, n_K$. □

The above theorem shows that the Kraft's inequality is necessary and sufficient for the existence of a prefix-free code. However, a stronger result can be proved. It states that the existence of Kraft's inequality is necessary and sufficient for the existence of a uniquely decipherable code. This result is due to B. McMillan. An elegant proof of which has been provided by J. Karush.

Theorem 1.3. *Let W be the set of words which belong to a memoryless source $\mathcal{S}$. Also $\mathcal{X}$ is the set of encoding alphabet, and $\mathcal{X}^*$ is the set of all finite strings of elements of $\mathcal{X}$. Let $|W| = K$ and $|\mathcal{X}| = D$, where $D > 1$. Consider a code $f : W \to \mathcal{X}^*$ with codeword-length $|f(w_k)| = n_k$, for $1 \le k \le K$. Then the code is uniquely decipherable if and only if the codeword-lengths satisfy the Kraft's inequality.*

Proof. Arrange the codeword-lengths of a uniquely decipherable code in increasing order as $n_1 \le n_2 \le \ldots \le n_K$. Let N be any arbitrary positive integer. Then

$$\left(\sum_{k=1}^{K} D^{-n_k}\right)^N = \sum_{k_1=1}^{K} \cdots \sum_{k_N=1}^{K} D^{-\left(n_{k_1}+\ldots+n_{k_N}\right)}$$

Observe that $(n_{k_1} + \ldots + n_{k_N})$ is the length of a sequence of N codewords. The length of this sequence of codewords is anywhere between Nn_1 and Nn_K. Let B_j be the number of sequences of N codewords of total length j. Thus

$$\left(\sum_{k=1}^{K} D^{-n_k}\right)^N = \sum_{j=Nn_1}^{Nn_K} B_j D^{-j}$$

Unique decodablity of the codewords implies that the sequences of N codewords of total length j are distinct. Therefore B_j can be upper-bounded by D^j, that is $B_j \leq D^j$. This yields

$$\left(\sum_{k=1}^{K} D^{-n_k}\right)^N \leq \sum_{j=Nn_1}^{Nn_K} D^j D^{-j} = N\left(n_K - n_1\right) + 1$$

If $\sum_{k=1}^{K} D^{-n_k} > 1$ then the left side of the above inequality grows exponentially with N, however the right side grows linearly with N. Since N can be arbitrarily large this is not possible. Thus Kraft's inequality $\sum_{k=1}^{K} D^{-n_k} \leq 1$ should hold. Therefore for a uniquely decipherable code with codeword-lengths $n_1, n_2, \ldots, n_K$ Kraft's inequality must be satisfied.

Conversely, given any set of $n_1, n_2, \ldots, n_K$ values which satisfy the Kraft's inequality, a procedure to construct a prefix-free code has been demonstrated in the proof of the previous theorem. Since a prefix-free code is uniquely decodable, a uniquely decodable code has indeed been constructed. □

Thus a code which is uniquely decodable, can be replaced by a prefix-free code with identical word-lengths, and vice-versa.

1.4.2 Shannon-Fano Coding Theorem

The following noiseless coding theorem is due to C. E. Shannon and R. M. Fano. This theorem provides lower and upper bounds for the average codeword-length of a uniquely decipherable code. These bounds are stated in terms of the entropy of the source word, and the size of the encoding alphabet.

Theorem 1.4. *Let $W = \{w_1, w_2, \ldots, w_K\}$ be the set of words which belong to a memoryless source $\mathcal{S}$, where $|W| = K$. Let the corresponding probabilities of occurrence of these words be $p(w_1), p(w_2), \ldots, p(w_K)$; where $p(w_k) > 0$, for $1 \leq k \leq K$. Also let U be a random variable which takes values $w_k, 1 \leq k \leq K$; and the entropy of this source be $H(U)$.*

Also $\mathcal{X}$ is the set of encoding alphabet, and $\mathcal{X}^*$ is the set of all finite strings of elements of $\mathcal{X}$. Let $|\mathcal{X}| = D$, where $D > 1$. Consider a code $f : W \to \mathcal{X}^*$ with codeword-length $|f(w_k)| = n_k$, for $1 \leq k \leq K$. Let the average length of the codeword be $\overline{n}$. Assume that the code is uniquely decipherable. Then

$$\frac{H(U)}{\log_2 D} \leq \overline{n} < \frac{H(U)}{\log_2 D} + 1 \tag{1.10}$$

Proof. Observe that

$$H(U) - \overline{n}\log_2 D = -\sum_{k=1}^{K} p(w_k)\log_2 p(w_k) - \{\log_2 D\}\sum_{k=1}^{K} p(w_k)\, n_k$$
$$= \sum_{k=1}^{K} p(w_k)\log_2\left(\frac{D^{-n_k}}{p(w_k)}\right)$$

Using the relationships $\log_2 x = \log_2 e \ln x$, and $\ln x \leq (x-1)$ for $x > 0$ yields

$$H(U) - \overline{n}\log_2 D \leq (\log_2 e)\sum_{k=1}^{K} p(w_k)\left(\frac{D^{-n_k}}{p(w_k)} - 1\right)$$
$$= (\log_2 e)\left\{\left(\sum_{k=1}^{K} D^{-n_k}\right) - 1\right\}$$

Since the code is prefix-free and uniquely decipherable, the Kraft's inequality should be satisfied. Therefore $H(U) - \overline{n}\log_2 D \leq 0$, which yields the lower bound on $\overline{n}$. The upper bound is proved as follows.

As per the hypothesis, the probabilities $p(w_k)$'s are greater than zero. Select the codeword for w_k, such that $n_k = \lceil -\log_D p(w_k) \rceil$ for $1 \leq k \leq K$. Consequently $n_k \geq -\log_D p(w_k)$. Therefore $D^{-n_k} \leq p(w_k)$, and $\sum_{k=1}^{K} D^{-n_k} \leq \sum_{k=1}^{K} p(w_k) = 1$. Therefore the Kraft's inequality is satisfied when this choice of n_k's is made.

Since $n_k = \lceil -\log_D p(w_k) \rceil \Rightarrow n_k < (-\log_D p(w_k) + 1)$ we have

$$\overline{n} = \sum_{k=1}^{K} p(w_k)\, n_k < \sum_{k=1}^{K} p(w_k)\left(-\log_D p(w_k) + 1\right) = \frac{H(U)}{\log_2 D} + 1$$

This completes the proof. □

The above results are used in data compression. This topic is next discussed briefly.

1.5 Data Compression

The mathematical basis for formulating the principles of data compression precisely is information theory. Data compression is possible when some sequences of data symbols emitted by an information source are more likely than others. Therefore data compression seeks to eliminate or reduce redundancy in data. That is, data compression techniques try to reduce the number of bits used to either store or transmit information.

The data compression that can be achieved depends upon two factors. The first factor is the redundancy in the data source. And the second factor is the efficiency of the data compression algorithm.

There are two types of data compression. These are *lossless* and *lossy* compression. After the data is compressed, the data has to be decompressed for eventual use. Lossless compression techniques recover the original data exactly. Lossy compression is said to occur, if the decompressed data is not exactly same as the original data. Note however, that generally more data compression is possible with lossy than lossless compression.

Three source coding algorithms are discussed in this section. These are: Huffman's coding, arithmetic coding, and Lempel-Ziv coding algorithms. These codes are lossless data compression schemes. Furthermore, these techniques output codeword sequences, which asymptotically approach source entropy.

The Huffman coding algorithm generates a new codeword for each source word. The length of these codewords (in number of bits) can be different for different source words. These codes are developed by building a tree. In addition, this algorithm is developed such that the coded words satisfy the prefix condition.

In contrast, in arithmetic coding a stream of source words are processed as a single *data unit.* A data unit is defined to be a finite sequence of source words. A single codeword is assigned to each data set. Furthermore, each codeword is a half-open interval, and a subset of the interval $[0,1)$. Two different subintervals are distinguished from one another by assigning enough precision to the subintervals. Thus arithmetic coding maps a stream of source words into a single floating-point output number. If the length of the source word stream is longer, then more bits are required to code it as a floating-point number. A big drawback of arithmetic coding is the presence of multiplication and division operations in the algorithm.

Lempel-Ziv coding is a dictionary-based compression scheme. This scheme is named after its inventors Abraham Lempel and Jacob Ziv. Huffman's and arithmetic coding schemes are based upon a statistical model of the alphabet used by the information source. In contrast, the Lempel-Ziv scheme depends upon the use of a dictionary. The Lempel-Ziv scheme creates a dictionary, which stores strings of characters of an alphabet. Compression is achieved by using indices of strings in the dictionary to encode repetitive character-string patterns. Thus compression is achieved by eliminating redundancies.

If the probability of occurrence of a symbol is p, then as per Shannon, maximum compression is obtained by coding the symbol in an average of $-\log_2 p$ bits.

1.5.1 Huffman's Coding

Let S be a memoryless source which emits a stream of *words* from the set W, where $W = \{w_1, w_2, \ldots, w_K\}$. The emitted word is a discrete random variable U. Furthermore, the probability distribution of emission of these words is $p(w_k), 1 \leq k \leq K$. These words are encoded by an encoder which uses the symbols from the binary set $\mathfrak{B} = \{0,1\}$. Recall that a binary digit is called a bit. Huffman's coding technique creates a uniquely decodable prefix-free code for each word $w_k, 1 \leq k \leq K$. Let the codeword corresponding to the word w_k be $f(w_k)$, and the corresponding codeword-length $|f(w_k)| = n_k$. These codes are binary strings of finite length. Furthermore, this technique ensures that the average word-length of the code $\overline{n}$, is minimized. Then for binary encoding, the coding theorem of Shannon and Fano yields $H(U) \leq \overline{n} < (H(U)+1)$. Thus, the goal of coding is to determine a uniquely decipherable code with average word-length as small as possible. Coding schemes of this type are said to be optimum and compact.

Without loss of generality assume that $p(w_1) \geq p(w_2) \geq \ldots \geq p(w_K) > 0$. A coding scheme is also considered to be optimal if $\overline{n}$, is minimized. Huffman's coding scheme is based upon the following lemmas.

Lemma 1.1. *For any probability distribution of words from a memoryless source, there exists an optimal assignment of prefix-free code with the following characteristics*:

(a) $p(w_i) > p(w_j)$ *implies* $n_i \leq n_j$.

(b) *The two longest words* w_{K-1} *and* w_K *have the same length. That is,* $n_{K-1} = n_K$.
(c) *The codewords* $f(w_{K-1})$ *and* $f(w_K)$ *differ in only the last bit.*

Proof. The above statements are proved by contradiction.

(a) Let $p(w_i) > p(w_j)$ such that $n_i > n_j$. Swap the codewords for the words w_i and w_j. Denote the change in the average codeword-length by $\Delta\overline{n}$. Then

$$\begin{aligned}\Delta\overline{n} &= (p(w_i)\,n_j + p(w_j)\,n_i) - (p(w_i)\,n_i + p(w_j)\,n_j) \\ &= (n_i - n_j)\,(p(w_j) - p(w_i))\end{aligned}$$

Note that $(n_i - n_j)$ is positive and $(p(w_j) - p(w_i))$ is negative. Therefore $\Delta\overline{n}$ is negative. This implies that a new coding scheme is possible by swapping the codewords w_i and w_j. This contradicts the initial hypothesis of an optimum code. Therefore $p(w_i) > p(w_j)$ implies $n_i \leq n_j$. Consequently, for an optimum prefix-free code $n_1 \leq n_2 \leq \ldots \leq n_K$.
(b) Assume that $n_{K-1} < n_K$ is true. Then the last $(n_K - n_{K-1})$ bits of $f(w_K)$ can be trimmed without affecting the requirement of a prefix-free code. This results in a new coding scheme with a smaller $\overline{n}$. This contradicts the assumption of optimum code. Therefore the two longest codewords have the same length.
(c) Assume the opposite of the statement of property (c). Then there might be another codeword of length n_K, for example $f(w_{K-2}), f(w_{K-3}), \ldots$, which might differ from $f(w_K)$ in the very last bit. If this codeword is swapped with $f(w_{K-1})$ then the statement of the property (c) is satisfied. However, if such a codeword is nonexistent, then the last bit of $f(w_K)$ can be trimmed without affecting the requirement of a prefix-free code. However, this trimming decreases the value of $\overline{n}$. This condition results in a contradiction of the statement in the property (c). Consequently, with proper rearrangement, codewords $f(w_{K-1})$ and $f(w_K)$ differ in only the last bit. □

Following lemma is developed for the recursive construction of the Huffman code.

Lemma 1.2. *Let* S *be a memoryless source which uses a word set* $W = \{w_1, w_2, \ldots, w_K\}$. *Let the probability distribution of these words be* $p(w_k), 1 \leq k \leq K$. *Also, a memoryless source* S' *has a word set* $W' = \{w'_1, w'_2, \ldots, w'_{K-1}\}$. *Let the probability distribution of these words be* $p(w'_k), 1 \leq k \leq (K-1)$. *Assume that* $p(w'_k) = p(w_k)$, *for* $1 \leq k \leq (K-2)$ *and* $p(w'_{K-1}) = (p(w_{K-1}) + p(w_K))$. *Let* $\{f(w'_1), f(w'_2), \ldots, f(w'_{K-1})\}$ *be optimum prefix-free code for* S', *then the following prefix-free code is optimum for the source* S.

$$f(w_k) = f(w'_k), \quad 1 \leq k \leq (K-2) \tag{1.11a}$$

$$f(w_{K-1}) = f(w'_{K-1})\,0 \tag{1.11b}$$

$$f(w_K) = f(w'_{K-1})\,1 \tag{1.11c}$$

where $f(w'_{K-1})\,0$ *is the concatenation of* $f(w'_{K-1})$ *with* 0, *and* $f(w'_{K-1})\,1$ *is the concatenation of* $f(w'_{K-1})$ *with* 1.

Proof. Define $n'_k = |f(w'_k)|$ for $1 \leq k \leq (K-1)$; and $\overline{n}' = \sum_{k=1}^{K-1} p(w'_k)\,n'_k$. Observe that $n_{K-1} = n_K = (n'_{K-1} + 1)$. This implies

$$(\overline{n} - \overline{n}') = (p(w_{K-1}) + p(w_K))$$

That is, the difference between $\overline{n}$ and $\overline{n}'$ is a constant. Therefore, if the value of $\overline{n}$ can be improved, then the value of $\overline{n}'$ can also be improved. This is not possible, since the code for the source words from the set $\mathcal{S}'$ is indeed optimum (cannot be improved). Consequently the construction of the code for the words in $\mathcal{S}$ is optimum. □

The above lemmas suggest a scheme to develop optimum prefix-code for a given source. Initially assume that the source $\mathcal{S}$ has K words. This problem can be reduced to finding an optimum prefix-code for a source $\mathcal{S}'$ with $(K-1)$ words. This process can be continued recursively till the new source has only two words. The code for words in a source with two words is 0 and 1.

The following notation is used to indicate the concatenation of 0's and 1's. Let $f(w)$ be a finite string of 0's and 1's; and $f(w) = z_1 z_2 \ldots z_m$ where $z_i = 0$ or 1 for $1 \leq i \leq m$. Then the concatenation $0f(w)$ is equal to $0z_1 z_2 \ldots z_m$ and $1f(w)$ equal to $1z_1 z_2 \ldots z_m$. If $f(w)$ does not contain any string of 0's and 1's then it is denoted by $\varnothing$. Therefore trivially, $0 = 0\varnothing$ and $1 = 1\varnothing$.

An algorithm for Huffman's coding scheme is next outlined. Only the main concepts are depicted. Redundancy in notation has been introduced in the description of the algorithm for clarity.

Algorithm 1.1. *Huffman's Binary Coding.*

Input: The source $\mathcal{S}$ is memoryless.
Let $W = \{w_1, w_2, \ldots, w_K\}$ be the word set of the source $\mathcal{S}$.
Let the corresponding positive probabilities of occurrence of these words be $p(w_1), p(w_2), \ldots, p(w_K)$.
Output: Let $\mathfrak{B} = \{0, 1\}$ be the encoding alphabet.
$\mathfrak{B}^*$ is the set of all finite binary strings.
Output is the encoding function $f : W \to \mathfrak{B}^*$,
where $|f(w_k)| = n_k, 1 \leq k \leq K$.
Also $\overline{n} = \sum_{k=1}^{K} n_k p(w_k)$ is a minimum.
This code is uniquely decipherable.
begin
  (*initialization*)
  $J \leftarrow K$
  for $j = 1$ **to** J **do**
  begin
    $p_{i_j} \leftarrow p(w_j)$
    $\omega_{i_j} \leftarrow \{w_j\}$
    $f(w_j) \leftarrow \varnothing$
  end (*end of j for-loop*)
  label A
  begin
    (*begin iteration*)
    Step 1: Sort the probabilities in the set $\{p_{i_j} \mid 1 \leq j \leq J\}$ in decreasing order. That is, $p_{m_1} \geq p_{m_2} \geq \ldots \geq p_{m_J}$. Determine the corresponding ω_{m_j} for $1 \leq j \leq J$.
    Step 2: (*concatenate* 0 *with* $f(w_k)$, $\forall\, w_k \in \omega_{m_J}$)
      Let $f(w_k) \leftarrow 0f(w_k), \forall\, w_k \in \omega_{m_J}$
      (*concatenate* 1 *with* $f(w_k)$, $\forall\, w_k \in \omega_{m_{J-1}}$)
      Let $f(w_k) \leftarrow 1f(w_k), \forall\, w_k \in \omega_{m_{J-1}}$

if $J = 2$ **stop**
Step 3: Let $p_{i_j} \leftarrow p_{m_j}$ and $\omega_{i_j} \leftarrow \omega_{m_j}$ for $1 \leq j \leq (J-2)$
Let $\omega \leftarrow \omega_{m_{J-1}} \cup \omega_{m_J}$
$p_{i_{J-1}} \leftarrow \sum_{w_k \in \omega} p(w_k)$
$\omega_{i_{J-1}} \leftarrow \omega$
Step 4: $J \leftarrow (J-1)$
go to label A
end (*end of iteration*)
end (*end of Huffman's binary coding algorithm*)

This algorithm is elucidated via the following example.

Example 1.4. A memoryless source of 5 words is given. Its word set is $W = \{w_1, w_2, \ldots, w_5\}$, where

$$p(w_1) = 0.1,\ p(w_2) = 0.05,\ p(w_3) = 0.05,\ p(w_4) = 0.5,\ \text{and}\ p(w_5) = 0.3$$

An optimum uniquely decipherable code $f(\cdot)$ of binary strings has to be determined. Huffman's coding scheme is used to find $f(w_k)$, for $1 \leq k \leq 5$.
Initialization: $J \leftarrow 5$

$$p_{i_1} = 0.1,\ p_{i_2} = 0.05,\ p_{i_3} = 0.05,\ p_{i_4} = 0.5,\ \text{and}\ p_{i_5} = 0.3$$
$$\omega_{i_1} = \{w_1\},\ \omega_{i_2} = \{w_2\},\ \omega_{i_3} = \{w_3\},\ \omega_{i_4} = \{w_4\},\ \text{and}\ \omega_{i_5} = \{w_5\}$$

and $f(w_j) = \varnothing$, for $1 \leq j \leq 5$.
Iteration 1:
Step 1: Sort the probabilities $p_{i_1}, p_{i_2}, p_{i_3}, p_{i_4}$, and p_{i_5} in decreasing order.

$$p_{m_1} = 0.5,\ p_{m_2} = 0.3,\ p_{m_3} = 0.1,\ p_{m_4} = 0.05,\ \text{and}\ p_{m_5} = 0.05$$
$$\omega_{m_1} = \{w_4\},\ \omega_{m_2} = \{w_5\},\ \omega_{m_3} = \{w_1\},\ \omega_{m_4} = \{w_2\},\ \text{and}\ \omega_{m_5} = \{w_3\}$$

Step 2: $f(w_3) \leftarrow 0$, and $f(w_2) \leftarrow 1$.
Step 3:

$$\begin{aligned} p_{i_1} &= 0.5,\ p_{i_2} = 0.3,\ \text{and}\ p_{i_3} = 0.1 \\ \omega_{i_1} &= \{w_4\},\ \omega_{i_2} = \{w_5\},\ \text{and}\ \omega_{i_3} = \{w_1\} \\ \omega &\leftarrow \{w_2, w_3\},\ p_{i_4} \leftarrow 0.1,\ \text{and}\ \omega_{i_4} \leftarrow \{w_2, w_3\} \end{aligned}$$

Step 4: $J \leftarrow 4$
Iteration 2:
Step 1: Sort the probabilities $p_{i_1}, p_{i_2}, p_{i_3}$, and p_{i_4} in decreasing order.

$$p_{m_1} = 0.5,\ p_{m_2} = 0.3,\ p_{m_3} = 0.1,\ \text{and}\ p_{m_4} = 0.1$$
$$\omega_{m_1} = \{w_4\},\ \omega_{m_2} = \{w_5\},\ \omega_{m_3} = \{w_2, w_3\},\ \text{and}\ \omega_{m_4} = \{w_1\}$$

Step 2: $f(w_1) \leftarrow 0$, $f(w_2) \leftarrow 11$, and $f(w_3) \leftarrow 10$.

Step 3:

$$\begin{aligned} p_{i_1} &= 0.5, \text{ and } p_{i_2} = 0.3 \\ \omega_{i_1} &= \{w_4\}, \text{ and } \omega_{i_2} = \{w_5\} \\ \omega &\leftarrow \{w_1, w_2, w_3\}, \ p_{i_3} \leftarrow 0.2, \text{ and } \omega_{i_3} \leftarrow \{w_1, w_2, w_3\} \end{aligned}$$

Step 4: $J \leftarrow 3$
Iteration 3:
Step 1: Sort the probabilities p_{i_1}, p_{i_2}, and p_{i_3} in decreasing order.

$$\begin{aligned} p_{m_1} &= 0.5, \ p_{m_2} = 0.3, \text{ and } p_{m_3} = 0.2 \\ \omega_{m_1} &= \{w_4\}, \ \omega_{m_2} = \{w_5\}, \text{ and } \omega_{m_3} = \{w_1, w_2, w_3\} \end{aligned}$$

Step 2: $f(w_1) \leftarrow 00$, $f(w_2) \leftarrow 011$, $f(w_3) \leftarrow 010$, and $f(w_5) \leftarrow 1$.
Step 3:

$$\begin{aligned} p_{i_1} &= 0.5 \\ \omega_{i_1} &= \{w_4\} \\ \omega &\leftarrow \{w_1, w_2, w_3, w_5\}, \ p_{i_2} \leftarrow 0.5, \text{ and } \omega_{i_2} \leftarrow \{w_1, w_2, w_3, w_5\} \end{aligned}$$

Step 4: $J \leftarrow 2$
Iteration 4:
Step 1: Sort the probabilities p_{i_1} and p_{i_2} in decreasing order.

$$\begin{aligned} p_{m_1} &= 0.5, \text{ and } p_{m_2} = 0.5 \\ \omega_{m_1} &= \{w_1, w_2, w_3, w_5\}, \ \omega_{m_2} = \{w_4\} \end{aligned}$$

Step 2: $f(w_4) = 0$, $f(w_1) \leftarrow 100$, $f(w_2) \leftarrow 1011$, $f(w_3) \leftarrow 1010$, and $f(w_5) \leftarrow 11$.
Since $J = 2$, stop.
Output: $f(w_1) = 100$, $f(w_2) = 1011$, $f(w_3) = 1010$, $f(w_4) = 0$, and $f(w_5) = 11$.
For this mapping $\overline{n} = \sum_{k=1}^{5} p(w_k) n_k = 1.8$. The entropy of the source words is

$$H(U) = -\sum_{k=1}^{5} p(w_k) \log_2 p(w_k) = 1.7855$$

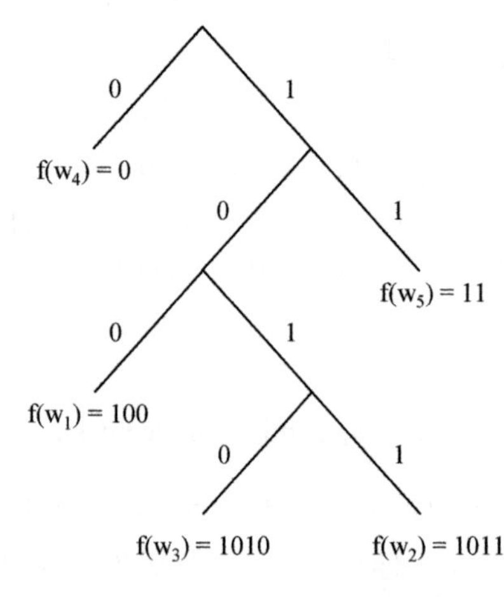

Figure 1.4. Graphical depiction of Huffman's scheme.

It can be observed that the average codeword-length and the entropy are approximately equal. See Figure 1.4 for a graphical depiction of Huffman's coding scheme. The astute reader must have noticed that even though the coding is uniquely decipherable, the encodings are not necessarily unique. □

1.5.2 Arithmetic Coding

A message is coded as a floating-point number in arithmetic coding. Recall that a message is a sequence of words of finite length. As in the last subsection, assume that S is a memoryless source which emits a stream of words from the set $W = \{w_1, w_2, \ldots, w_K\}$. Recall that W^* is the set of all finite strings of elements of W. Members of the set W^* are called messages. A message $m \in W^*$ of length L, is of the form $m = w_{i_1} w_{i_2} \ldots w_{i_L}$, where $w_{i_j} \in W$, for $1 \le j \le L$. This message m is also said to have L words.

The source words are independent, and identically distributed random variables. Therefore the emitted word can be represented as a discrete random variable U. Assume that $P(U = w_k) = p(w_k)$ for $1 \le k \le K$. The encoder uses members of the set $\{0, 1\}$. Note that in order to code the word w_k, an average of $-\log_2 p(w_k)$ bits are required.

The purpose of the arithmetic coding scheme is to represent a message m, by a floating-point number $\widehat{\theta}$ in the interval $[0, 1)$.

Assume that $p(w_k) > 0$ for $1 \le k \le K$. If $p(w_k) = 0$, then the word w_k can be removed from the set W. Introduce two variables $a(w_k)$ and $b(w_k)$, $1 \le k \le K$. These are defined as:

$$
\begin{aligned}
a(w_1) &= 0, \quad b(w_1) = p(w_1), \\
a(w_k) &= \sum_{j=1}^{k-1} p(w_j), \quad b(w_k) = \sum_{j=1}^{k} p(w_j), \qquad 2 \le k \le K
\end{aligned}
$$

Note that $b(w_k)$ for $1 \le k \le K$, is equal to the cumulative distribution function of the random variable U. Also $b(w_{k-1}) = a(w_k)$ for $2 \le k \le K$, and $b(w_K) = 1$.

The arithmetic coding process codes the message $m = w_{i_1} w_{i_2} \ldots w_{i_L}$ as a sequence of nested intervals of the form $\Psi_j = [\alpha_j, \beta_j) \subseteq [0, 1)$ for $1 \le j \le L$. Define the length of the interval $[\alpha_j, \beta_j)$ as $l_j = (\beta_j - \alpha_j)$ for $1 \le j \le L$. The upper and lower limits of the intervals can be computed recursively as follows. Let

$$
\Psi_0 = [\alpha_0, \beta_0), \quad \alpha_0 = 0, \ \beta_0 = 1, \ l_0 = 1
$$

Then for $1 \le j \le L$

$$
\begin{aligned}
\alpha_j &= \alpha_{j-1} + l_{j-1} a(w_{i_j}) \\
\beta_j &= \alpha_{j-1} + l_{j-1} b(w_{i_j}) \\
l_j &= (\beta_j - \alpha_j)
\end{aligned}
$$

Following observations can be made about the above computations:

(a) $\alpha_j = \sum_{k=1}^{j} l_{k-1} a(w_{i_k})$ for $1 \le j \le L$.
(b) $\beta_j = (\alpha_j + l_j)$ for $1 \le j \le L$.
(c) $0 < \alpha_j \le \alpha_{j+1}$ for $1 \le j \le (L-1)$.
(d) $\beta_{j+1} \le \beta_j < 1$ for $1 \le j \le (L-1)$.
(e) $l_j = \prod_{n=1}^{j} p(w_{i_n})$ for $1 \le j \le L$.

(f) $\Psi_j \subseteq \Psi_{j-1}$ for $1 \leq j \leq L$.

The final step in the arithmetic coding process is to determine the message m representation $\widehat{\theta} \in \Psi_L$, such that $\widehat{\theta}$ has the shortest binary representation. The binary representation of α_L and β_L is next determined. Let these be

$$\alpha_L = 0.d_1 d_2 \ldots d_{\tau-1} 0 \ldots$$
$$\beta_L = 0.d_1 d_2 \ldots d_{\tau-1} 1 \ldots$$

where d_i's are equal to either 0 or 1. The first $(\tau - 1)$ bits of α_L and β_L are identical. The binary representation of $\widehat{\theta}$ has τ bits, and it is set to $0.d_1 d_2 \ldots d_{\tau-1} 1$. Therefore $\widehat{\theta}$ is coded in τ number of bits. Also $\widehat{\theta} \in \Psi_j$ for $0 \leq j \leq L$. Thus, the message m has been transformed to a single floating point number $\widehat{\theta}$. The different steps are shown in the arithmetic coding algorithm.

The reverse process of decoding is also an iterative process. Let the decoded message be $\widehat{m} = \widehat{w}_{i_1} \widehat{w}_{i_2} \ldots \widehat{w}_{i_L}$, where $\widehat{w}_{i_j} \in W$ for $1 \leq j \leq L$. The decoding process recovers the original message words in the same order in which they were coded. This is illustrated by the decoding algorithm.

Algorithm 1.2a. *Arithmetic Coding.*

Input: The source $\mathcal{S}$ is memoryless.
Let $W = \{w_1, w_2, \ldots, w_K\}$ be the word set of the source $\mathcal{S}$.
Let the probability of occurrence of the word w_i be $p(w_i)$,
where $1 \leq i \leq K$.
The message m to be coded is of length L, where $m = w_{i_1} w_{i_2} \ldots w_{i_L}$.
Output: The binary representation of the coded value of m is $\widehat{\theta}$.
It is equal to $0.d_1 d_2 \ldots d_{\tau-1} 1$.
begin
  (*initialization*)
  (*compute* $a(w_k)$ *and* $b(w_k)$ *for* $1 \leq k \leq K$)
  $a(w_1) \leftarrow 0$
  $b(w_1) \leftarrow p(w_1)$
  $a(w_k) \leftarrow \sum_{j=1}^{k-1} p(w_j), 2 \leq k \leq K$
  $b(w_k) \leftarrow \sum_{j=1}^{k} p(w_j), 2 \leq k \leq K$
  $\alpha_0 \leftarrow 0, \beta_0 \leftarrow 1, l_0 \leftarrow 1$
  for $j = 1$ **to** L **do**
  begin
    $\alpha_j \leftarrow \alpha_{j-1} + l_{j-1} a(w_{i_j})$
    $\beta_j \leftarrow \alpha_{j-1} + l_{j-1} b(w_{i_j})$
    $l_j \leftarrow (\beta_j - \alpha_j)$
  end (*end of j for-loop*)
  (*find the binary representation of* α_L *and* β_L)
  (*the first* $(\tau - 1)$ *bits of* α_L *and* β_L *are identical*)
  $\alpha_L \leftarrow 0.d_1 d_2 \ldots d_{\tau-1} 0 \ldots$
  $\beta_L \leftarrow 0.d_1 d_2 \ldots d_{\tau-1} 1 \ldots$
  $\widehat{\theta} \leftarrow 0.d_1 d_2 \ldots d_{\tau-1} 1$

end (*end of arithmetic coding algorithm*)

Algorithm 1.2b. *Arithmetic Decoding.*

Input: The source $\mathcal{S}$ is memoryless.
Let $W = \{w_1, w_2, \ldots, w_K\}$ be the word set of the source $\mathcal{S}$.
Let the probability of occurrence of the word w_i be $p(w_i)$,
where $1 \leq i \leq K$.
The number of words in the original message m, is equal to L.
The binary representation of the coded value of message m, is
$\widehat{\theta} = 0.d_1 d_2 \ldots d_{\tau-1} 1$.
Output: The decoded message, $\widehat{m} = \widehat{w}_{i_1} \widehat{w}_{i_2} \ldots \widehat{w}_{i_L}$,
where $\widehat{w}_{i_j} \in W$ for $1 \leq j \leq L$.
begin
(*initialization*)
(*compute* $a(w_k)$ *and* $b(w_k)$ *for* $1 \leq k \leq K$)
$a(w_1) \leftarrow 0$
$b(w_1) \leftarrow p(w_1)$
$a(w_k) \leftarrow \sum_{j=1}^{k-1} p(w_j), \; 2 \leq k \leq K$
$b(w_k) \leftarrow \sum_{j=1}^{k} p(w_j), \; 2 \leq k \leq K$
$\alpha_0 \leftarrow 0, \; \beta_0 \leftarrow 1, \; l_0 \leftarrow 1$
$j \leftarrow 0$
while $j < L$ **do**
begin
find n **such that**
begin

$$a(w_n) \leq \frac{\left(\widehat{\theta} - \alpha_j\right)}{\left(\beta_j - \alpha_j\right)} \leq b(w_n), \;\; 1 \leq n \leq K$$

end
$j \leftarrow j + 1$
$\widehat{w}_{i_j} \leftarrow w_n$
$\alpha_j \leftarrow \alpha_{j-1} + l_{j-1} a(w_n)$
$\beta_j \leftarrow \alpha_{j-1} + l_{j-1} b(w_n)$
$l_j \leftarrow \left(\beta_j - \alpha_j\right)$
end (*end of while-loop*)
$\widehat{m} \leftarrow \widehat{w}_{i_1} \widehat{w}_{i_2} \ldots \widehat{w}_{i_L}$
end (*end of arithmetic decoding algorithm*)

Optimality of Arithmetic Coding

The optimality of arithmetic coding is next examined. The source $\mathcal{S}$ is memoryless. Therefore, the average number of bits required to code the source words $w_k \in W$, cannot be smaller than the entropy $H(U)$, of the random variable U. The possible values of the random variable U are $w_k, 1 \leq k \leq K$. Also

$$H(U) = -\sum_{k=1}^{K} p(w_k) \log_2 p(w_k) \quad \text{bits per word}$$

Observe in the coding algorithm that the minimum number of bits required to code the message is $\lceil -\log_2 l_L \rceil$ bits. During the data compression process, there is some overhead associated with arithmetic coding. For example, the probabilities $p(w_k), 1 \leq k \leq K$ and the message length L, also have to be transmitted. Assume that the total overhead for coding and transmitting the message is ρ bits. Let $\mathcal{B}$ be the number of bits used per word for coding the words from source $\mathcal{S}$. Therefore

$$\mathcal{B} \leq \frac{(\rho - \log_2 l_L)}{L}$$

As $l_L = \prod_{n=1}^{L} p(U = w_{i_n})$

$$\mathcal{B} \leq \frac{\left(\rho - \sum_{n=1}^{L} \log_2 p(U = w_{i_n})\right)}{L}$$

Taking expectations on both sides results in

$$\begin{aligned}\mathcal{E}(\mathcal{B}) &\leq \frac{\left(\rho - \sum_{n=1}^{L} \mathcal{E}\{\log_2 p(U = w_{i_n})\}\right)}{L} \\ &= \frac{\left(\rho - \sum_{n=1}^{L} \sum_{k=1}^{K} p(w_k) \log_2 p(w_k)\right)}{L} \\ &= H(U) + \frac{\rho}{L}\end{aligned}$$

Also since the average number of bits per word cannot be smaller than its entropy, $H(U) \leq \mathcal{E}(\mathcal{B})$. Consequently

$$H(U) \leq \mathcal{E}(\mathcal{B}) \leq H(U) + \frac{\rho}{L}$$

Therefore if $L \to \infty$, $\lim_{L\to\infty} \mathcal{E}(\mathcal{B}) = H(U)$. This limit implies that the arithmetic coding, does indeed provide optimal compression. This discussion is next illustrated with the following two examples.

Example 1.5. The source $\mathcal{S}$ is memoryless. Let $W = \{w_1, w_2, w_3, w_4\}$, and $K = 4$. The probability of occurrence of these words are $p(w_1) = 0.2, p(w_2) = 0.4, p(w_3) = 0.3$, and $p(w_4) = 0.1$. The message $m = w_2 w_3 w_2 w_1 w_4$. The entropy of these words is

$$H(U) = -\sum_{k=1}^{4} p(w_k) \log_2 p(w_k) = 1.84644 \quad \text{bits per word}$$

Note that $L = 5$. The message $m \in W^*$ is coded as follows.

Initialization:
$a(w_1) = 0.0$, $a(w_2) = 0.2$, $a(w_3) = 0.6$, and $a(w_4) = 0.9$
$b(w_1) = 0.2$, $b(w_2) = 0.6$, $b(w_3) = 0.9$, and $b(w_4) = 1.0$
$\alpha_0 = 0$, $\beta_0 = 1$, $l_0 = 1$
Iteration 1: $j = 1, w_{i_1} = w_2$
$\alpha_1 = 0.2, \beta_1 = 0.6$, and $l_1 = 0.4$
Iteration 2: $j = 2, w_{i_2} = w_3$
$\alpha_2 = 0.44, \beta_2 = 0.56$, and $l_2 = 0.12$
Iteration 3: $j = 3, w_{i_3} = w_2$
$\alpha_3 = 0.464, \beta_3 = 0.512$, and $l_3 = 0.048$
Iteration 4: $j = 4, w_{i_4} = w_1$
$\alpha_4 = 0.464, \beta_4 = 0.4736$, and $l_4 = 0.0096$
Iteration 5: $j = 5, w_{i_5} = w_4$
$\alpha_5 = 0.47264, \beta_5 = 0.4736$, and $l_5 = 0.00096$
End of iterations.
Binary representation of $\alpha_5 = 0.01111000\ldots$
Binary representation of $\beta_5 = 0.01111001\ldots$
Binary representation of $\widehat{\theta} = 0.01111001$.
Decimal representation of $\widehat{\theta} = 0.47265625 \in \Psi_5$.
The number of bits required to represent $\widehat{\theta}$ is equal to $\tau = 8$.
Output: $\widehat{\theta} = 0.01111001$.

Note that the number of bits needed to transmit the message, without any special coding is equal to

$$\begin{aligned}
-\log_2 P(m) &= -\log_2 P(w_2 w_3 w_2 w_1 w_4) \\
&= -\log_2 \{p(w_2)\, p(w_3)\, p(w_2)\, p(w_1)\, p(w_4)\} \\
&= -\log_2 (0.00096) = 10.0247 \text{ bits}
\end{aligned}$$

However the use of arithmetic coding required only 8 bits. Note that for messages of very large length, the average number of bits required to code a single word from this source tends towards $H(U) = 1.84644$ bits. □

Example 1.6. The source $\mathcal{S}$ is memoryless. Let $W = \{w_1, w_2, w_3, w_4\}$, and $K = 4$. The probability of occurrence of these words are $p(w_1) = 0.2, p(w_2) = 0.4, p(w_3) = 0.3$, and $p(w_4) = 0.1$. The possibility of recovering the coded message of the previous example is tested in this example. A message m was coded to be $\widehat{\theta} = 0.47265625$. The message has 5 words $(L = 5)$. The message m is decoded as follows.

Initialization:
$a(w_1) = 0.0$, $a(w_2) = 0.2$, $a(w_3) = 0.6$, and $a(w_4) = 0.9$
$b(w_1) = 0.2$, $b(w_2) = 0.6$, $b(w_3) = 0.9$, and $b(w_4) = 1.0$
$\alpha_0 = 0$, $\beta_0 = 1$, $l_0 = 1$
$j = 0$
Iteration 1:
$\left(\widehat{\theta} - \alpha_0\right) / (\beta_0 - \alpha_0) = 0.47265625$
$a(w_2) \leq 0.47265625 \leq b(w_2), n = 2$
$j = 1$

$\widehat{w}_{i_1} = w_2$
$\alpha_1 = 0.2, \beta_1 = 0.6$, and $l_1 = 0.4$
Iteration 2:
$\left(\widehat{\theta} - \alpha_1\right) / (\beta_1 - \alpha_1) = 0.681640625$
$a(w_3) \leq 0.681640625 \leq b(w_3), n = 3$
$j = 2$
$\widehat{w}_{i_2} = w_3$
$\alpha_2 = 0.44, \beta_2 = 0.56$, and $l_2 = 0.12$
Iteration 3:
$\left(\widehat{\theta} - \alpha_2\right) / (\beta_2 - \alpha_2) = 0.272135417$
$a(w_2) \leq 0.272135417 \leq b(w_2), n = 2$
$j = 3$
$\widehat{w}_{i_3} = w_2$
$\alpha_3 = 0.464, \beta_3 = 0.512$, and $l_3 = 0.048$
Iteration 4:
$\left(\widehat{\theta} - \alpha_3\right) / (\beta_3 - \alpha_3) = 0.180338542$
$a(w_1) \leq 0.180338542 \leq b(w_1), n = 1$
$j = 4$
$\widehat{w}_{i_4} = w_1$
$\alpha_4 = 0.464, \beta_4 = 0.4736$, and $l_4 = 0.0096$
Iteration 5:
$\left(\widehat{\theta} - \alpha_4\right) / (\beta_4 - \alpha_4) = 0.901692708$
$a(w_4) \leq 0.901692708 \leq b(w_4), n = 4$
$j = 5$
$\widehat{w}_{i_5} = w_4$
$\alpha_5 = 0.47264, \beta_5 = 0.4736$, and $l_5 = 0.00096$
End of iterations.
Output: $\widehat{m} \leftarrow w_2 w_3 w_2 w_1 w_4$
This is indeed the message m, of the previous example. □

1.5.3 Lempel-Ziv Coding

The Lempel-Ziv coding is a dictionary-based data compression scheme. This compression technique is also called a universal coding scheme, because it works without a knowledge of the source statistics. Lempel and Ziv published their landmark papers in the years 1977 and 1978. The algorithms for data compression described in these papers are called LZ77 and LZ78 respectively. We describe the LZ78 algorithm in this subsection.

The input to the LZ78 coding scheme consists of a stream of characters, which are members of an alphabet. The characters of the alphabet are also called letters or symbols. A finite contiguous sequence of characters is called a phrase, or a string of characters, or simply a string. The number of characters in a string is called its length.

This scheme uses a dictionary, which is an array or table of strings. Each element of the array is such string. The position of a given string in the dictionary is specified by an index. The empty or null character $\star$ always has an index 0 in the dictionary. After index 0, the dictionary has index 1, and then index 2, and so on. The LZ78 encoding scheme achieves compression by storing repetitive strings of characters in the dictionary. It then uses the indices of these strings for outputting codewords. The encoding algorithm reads the input string, determines a repetitive phrase of largest possible length, and then outputs its index in the dictionary. It is also possible for a new phrase in the input character-stream to be emitted by the coder (or encoder), and added to the dictionary. The output stream of the coder is a sequence of codewords, which are ordered two-tuples. Each such ordered two-tuple consists of an index in the dictionary, and a single character. It is not necessary to transmit the dictionary in this scheme. The dictionary is created during both the coding and decoding processes. Compression is achieved if the ratio of number of bits required to represent the compressed stream of characters, and the input stream of characters is less than unity. We next formalize these notions.

Definitions 1.5. *Let $\mathcal{X}$ be the source alphabet, where $|\mathcal{X}| \in \mathbb{P}\backslash\{1\}$. The members of the set $\mathcal{X}$ are called characters, or letters, or symbols. Let $\star \notin \mathcal{X}$ denote the empty or null character.*

1. *A sequence or string of characters is $z = z_1 z_2 \ldots z_{m-1} z_m$, where $z_i \in \mathcal{X}$ for $1 \leq i \leq m$. The length of this string is denoted by $|z|$. It is equal to $m \in \mathbb{P}$.*
2. *A contiguous sequence of characters w is a subsequence of the string z, if it is of the form $z = z_1 z_2 \ldots w \ldots z_{m-1} z_m$.*
3. *A phrase in a string x is a subsequence in it.*
4. *For the given sequence of characters z; the sequence $z_1 z_2 \ldots z_{m-1}$ is the prefix of the sequence z, where $m \geq 2$. The prefix of the sequence z is denoted by prefix(z).*
5. *The concatenation of two strings of characters w_1 and w_2 is denoted by either $w_1 + w_2$ or simply $w_1 w_2$. The concatenation of the empty character $\star$ and another character $w \in \mathcal{X}$ is $(\star + w) = w$. Also $(w + \star) = w$.*
6. *The dictionary $\mathcal{D}$ is a table of sequences (or strings) of characters. The index set of the sequences in the dictionary is $\mathbb{N}$. The index of the sequence $z \in \mathcal{D}$, is denoted by index$(z) = y \in \mathbb{N}$. The index of the empty character $\star$ is 0.*
7. *A codeword in the output stream of the compression algorithm is a two-tuple (y, z). In this codeword $y \in \mathbb{N}$ is the index of the sequence w in the dictionary $\mathcal{D}$, $z \in \mathcal{X}$ is a single character, and wz is a phrase in the input sequence x which is to be compressed.* □

During the compression process, the following scheme is adapted for creating a codeword.

- Let z be a single character which is not in the dictionary $\mathcal{D}$. The corresponding codeword is $(0, z)$.
- Let z be a string of characters of length greater than 1, and it is not in the dictionary $\mathcal{D}$. Let $z = ab$, where $a = \text{prefix}(z)$ is a sequence of characters which is in the dictionary, and b is a single character. The corresponding codeword is $(\text{index}(a), b)$.
- Let x be the string of characters which has to be compressed. Let the *last phrase* z in x be in the dictionary. That is, $x = wz$, where w is a subsequence of characters of x. The corresponding codeword is $(\text{index}(z), \star)$

Description of the LZ78 coding scheme requires a careful definition of parsing of a given sequence of characters.

Definition 1.6. *The source alphabet is $\mathcal{X}$, and $x = x_1 x_2 \ldots x_n$ is a string of characters of length n, where $x_i \in \mathcal{X}$, for $1 \leq i \leq n$. A parsing of the string x is a division of x into phrases which are separated by commas. A distinct parsing of x is a parsing of x into phrases which are all different.* □

Let the source alphabet be $\mathfrak{B} = \{0, 1\}$, and consider a binary string 110000. One of its distinct parsing is $1, 100, 00$. However the parsing, $11, 00, 00$ is not distinct because the sequence 00 occurs twice. Some basic concepts about the LZ78 compression scheme are demonstrated via following examples.

Examples 1.7. Assume that the source alphabet is $\mathfrak{B} = \{0, 1\}$.

1. Let the input sequence be

$$01001011001111$$

This string is parsed from left to right. It is parsed as

$$0, 1, 00, 10, 11, 001, 111$$

Notice that after placing each comma in the sequence, we scanned the input string, and placed a comma after the shortest substring that has not yet been seen before. As this is the shortest substring, its prefix must have occurred earlier in the scan. Recall that a prefix of a string is a contiguous sequence of characters that precede a single character.
For the given input sequence, the algorithm inserts a comma after the first bit, which is 0, because it has not yet been marked off. Then 1 occurs. A comma is placed after it because it has not occurred earlier. Then a comma is placed after 00, as it is the shortest and newest phrase, and so on.
Let $c(n)$ be the number of distinct phrases in the input string of length n. In this example, $n = 14$, and the number of distinct phrases is equal to $c(14) = 7$. Therefore, we need

$$\lceil \log_2 (c(n) + 1) \rceil = 3 \text{ bits}$$

to index these phrases in the dictionary. In the above expression, the number 1 is added to include index 0 in the dictionary. As we shall see in the description of the LZ78 algorithm, the empty character $\star$ is stored at index 0. For the given input sequence, we list the phrases, and their indices in decimal and binary notation.

parsed source:	0	1	00	10	11	001	111
index in decimal:	1	2	3	4	5	6	7
index in binary:	001	010	011	100	101	110	111

The codeword to transmit each phrase of the input string is specified as follows. The codeword of each phrase is the index of its prefix, which takes 3 bits, followed by 1 bit to specify the last bit in its phrase. For example, the codeword of the phrase 001 is the index of 00, which is 3 (in decimal) and 011 (in binary); followed by the last bit of the phrase, which is 1. Thus the codeword for the phrase 001 is $(011, 1)$. Finally, the input string is coded as

$$(000, 0)\,(000, 1)\,(001, 0)\,(010, 0)\,(010, 1)\,(011, 1)\,(101, 1)$$

The output string coded by the algorithm is actually without parentheses and commas.
It appears that the LZ78 algorithm, might need two passes. The first pass to determine the number of distinct phrases $c(n)$, and the second pass to output the codewords. At the end of the first pass, the number of bits required to specify the index in binary is determined. This is equal to $\lceil \log_2 (c(n) + 1) \rceil$ bits. The algorithm can however be modified to require only a single

pass in order to generate the coded output string. See the problem section for its demonstration. The decoding of the coded sequences appears to be relatively simple.

It also appears in this example, that the coded output has more bits than in the input string. This is so, because the input string is short, and there does not appears to be sufficient redundancy in it. In practice, if the input strings are long, the distinct phrases might be longer, and might require fewer bits to index them. This will lead to efficient compression.

2. This example is a slight modification of the last example. Let the input sequence be

$$0100101100111$$

This string is parsed from left to right. It is parsed as

$$0, 1, 00, 10, 11, 001, 11$$

Observe that, in this parsing the last phrase 11 occurs twice, and all other phrases occur only once. In this example, $n = 13$, and the number of distinct phrases is equal to $c(13) = (7-1) = 6$ (note that the substring 11 occurred twice). Therefore, we need $\lceil \log_2 (c(n) + 1) \rceil = 3$ bits to index these phrases. For the given input sequence, we list the phrases and their indices in decimal and binary notation.

parsed source:	0	1	00	10	11	001	11
index in decimal:	1	2	3	4	5	6	5
index in binary:	001	010	011	100	101	110	101

The codewords required to transmit the first six phrases is the same as in the last example. The last phrase 11 occurs twice. Its index in the dictionary is 5 (in decimal notation). Therefore the corresponding codeword is $(101, \star)$. The input string is coded as

$$(000, 0)\,(000, 1)\,(001, 0)\,(010, 0)\,(010, 1)\,(011, 1)\,(101, \star)$$

The output string coded by the algorithm is actually without parentheses and commas. □

It is next demonstrated that this algorithm is asymptotically optimal. In addition, the algorithm does not have any prior knowledge of the statistics of its alphabet. The important characteristics of this algorithm are summarized in the following set of observations.

Observations 1.5. Some observations about the LZ78 compression scheme. Let the source alphabet be $\mathfrak{B} = \{0, 1\}$, and $x = x_1 x_2 \ldots x_n$ be a string of characters of length n, where $x_i \in \mathfrak{B}$, for $1 \leq i \leq n$.

1. Let the number of phrases in a distinct parsing of the binary sequence x be $c(n)$. Then

$$c(n) \leq \frac{n}{(1-\epsilon_n)\log_2 n}, \quad \text{where} \quad \epsilon_n \in (0, 1)$$

for all $n \geq n_0$, and n_0 is some positive integer. Also $\epsilon_n \to 0$ as $n \to \infty$.

2. The total length of the compressed sequence (output) is upper-bounded by

$$\ell(n) = c(n)\{\lceil \log_2 (c(n) + 1) \rceil + 1\}$$

bits. In this expression, the first occurrence of 1 takes into account the use of index 0 in the dictionary. The second occurrence of 1 accounts for the bit needed to specify the last symbol in a phrase.

3. Let X be a random variable which takes values in the set $\mathfrak{B}$. Assume that the properties of the random variable do not change over time (stationarity). The entropy of the random variable X, also called the entropy rate of the source, is $H(X)$. Then $\lim_{n\to\infty} \ell(n)/n \leq H(X)$. That is, the length per source symbol is asymptotically upper-bounded by the entropy rate of the stationary source. □

The proofs of the above observations are provided in the problem section. The LZ78 coding and decoding algorithms are outlined below. The coding and decoding algorithms process characters from the alphabet

$$\mathcal{X} = \{x_{a_0}, x_{a_1}, \ldots, x_{a_{D-1}}\}, \quad \text{where } D \geq 2$$

An example of the set $\mathcal{X}$ is the binary set $\mathfrak{B}$. The goal of the coding algorithm is to compress the sequence of characters

$$x = x_1 x_2 \ldots x_n, \quad \text{where } x_i \in \mathcal{X}, \text{ for } 1 \leq i \leq n$$

The coding algorithm creates a dictionary $\mathcal{D}$ of distinct parsings of characters from the input character-stream x. The indices of distinct parsings in the dictionary take values in the set $\mathbb{N}$. The empty character $\star$ is stored at index 0 in the dictionary. The output of the algorithm is a compressed representation of strings of characters. More precisely, the output is a sequence of codewords of ordered pairs (index(w), z). In this ordered pair, w is a phrase in the dictionary $\mathcal{D}$, and index(w) is its index in $\mathcal{D}$, and $z \in \mathcal{X}$ is a character which follows the phrase w in the input sequence x.

Algorithm 1.3a. *LZ78 Coding.*

Input: Stream x of source characters from the alphabet $\mathcal{X}$.
An empty dictionary $\mathcal{D}$ with empty character $\star$ at index 0.
Output: Compressed sequence of ordered pairs (index (w), z),
where w is a phrase in dictionary $\mathcal{D}$, and index(w) is its index in $\mathcal{D}$,
$z \in \mathcal{X}$ is a character in the input stream x,
which follows the phrase w in it.
begin
    (*initialization*)
    $w \leftarrow \star$
    ($\star$ *is the empty character*)
    while (there are yet unscanned input characters in x) **do**
    begin
        $z \leftarrow \eta_{next\_char}()$
        ($\eta_{next\_char}()$ *is the next character in the input stream* x)
        if $((w + z) \in \mathcal{D})$ **then** $w \leftarrow (w + z)$
        (*the addition sign* $+$ *means concatenation*)
        else
        begin
            output the ordered pair (index (w), z)
            add $(w + z)$ to the dictionary at the next available location
            $w \leftarrow \star$
        end (*end of if statement*)

end (*end of while-loop*)
if ($w \neq \star$) **then**
begin
output the ordered pair (index (w), $\star$)
end (*end of if statement*)
end (*end of LZ78 coding*)

The input to the LZ78 decoding algorithm is a sequence of ordered pairs (y, z). In this ordered pair, $y \in \mathbb{N}$ is the index of the phrase w in the dictionary. That is, y = index(w), and w is a phrase in x. Also, $z \in \mathcal{X}$ is a character which follows the phrase w in the original sequence x. The decoding algorithm reads one ordered pair (y, z) at a time. It also maintains a dictionary $\mathcal{D}$, just as in the coding algorithm. The output of the decoding algorithm is the sequence x. It is the original uncompressed sequence of characters.

Algorithm 1.3b. *LZ78 Decoding.*

Input: Sequence of ordered pairs (y, z); where
$y \in \mathbb{N}$ is the index of the phrase w in the dictionary, that is, $y = \text{index}(w)$;
and, $z \in \mathcal{X}$ is the character immediately after the phrase w in the original sequence x. An empty dictionary $\mathcal{D}$ with empty character $\star$ at index 0.
Output: Sequence of decoded characters, which is x.
Updated dictionary $\mathcal{D}$ of distinct phrases.
while (there are yet unscanned codewords) **do**
begin
$y \leftarrow \eta_{next\_index}()$
$z \leftarrow \eta_{next\_char}()$
($\eta_{next\_index}()$ *is an index in the dictionary*)
($\eta_{next\_char}()$ *is the next character*)
output $\mathcal{D}(y) + z$
if $z \neq \star$ **then**
add $\mathcal{D}(y) + z$ to the dictionary at the next available location
(*if* $z = \star$ *then* $\mathcal{D}(y) + z$ *is already present in the dictionary*)
end (*end of while-loop*)
end (*end of LZ78 decoding*)

The LZ78 coding scheme is popular because of its simplicity, speed, and efficiency. Speed is achieved because the input to the algorithm can be a sequence of characters, rather than simply a bit-stream. The size of the dictionary can become very large in this coding scheme. As the size of a dictionary is finite, overflow might occur, and some phrases might have to be reinserted.

1.6 Communication Channels

The purpose of a communication channel is to transmit information. Therefore, a communication channel can be described in the language of entropy and related concepts. The coding theorems discussed in this section provide a bound for the transmission of information over a communication channel.

In this section: discrete memoryless channel, a symmetric channel, and an extension of a discrete memoryless channel are described. Basics of coding and decoding are also discussed. Noiseless and noisy communication channel coding theorems are also stated and proved.

1.6.1 Discrete Memoryless Channel

Recall that an information or communication channel is a black box which accepts strings of symbols from the set of input alphabet $A = \{x_1, x_2, \ldots, x_K\}$ and emits symbols from the set of output alphabet $B = \{y_1, y_2, \ldots, y_J\}$. For mathematical tractability, the channel is assumed to be a *discrete memoryless channel* (DMC). The transmission channel is discrete, because it is assumed to operate only at discrete instants of time. The transmission channel is said to be memoryless, because it is assumed that the probability distribution of the output symbol depends only on its input at that time, and is conditionally independent of past channel input and output symbols. It is possible for this communication channel to be noisy.

Denote the ith input and output random variables by X_i and Y_i respectively. Because of the memoryless property, the probability

$$P(Y_i = y_j \mid X_i = x_k) = p(y_j \mid x_k), \; 1 \le k \le K, \; 1 \le j \le J$$

is independent of each i. Therefore a DMC is completely represented by a channel matrix P_{DMC}. This matrix has K rows and J columns. Also $P_{DMC} = [p(y_j \mid x_k)]$. The sum of the elements in each row of this matrix is equal to unity. That is, $\sum_{j=1}^{J} p(y_j \mid x_k) = 1$ for $1 \le k \le K$. Therefore P_{DMC} is a stochastic matrix. An information channel is next defined.

Definition 1.7. *A communication channel (or information channel) is a triple* (A, B, P_{DMC}). *In this triple,* A *and* B *are the sets of input and output alphabets respectively, where* $|A| = K$, *and* $|B| = J$. *The transmission behavior of a memoryless channel is specified by a* $K \times J$ *probability matrix* P_{DMC}, *where the sum of each row is equal to unity.* □

This channel description is useful in modeling the transmission of information in presence of noise. Observe that if the sequence of input symbols are $(x_{k_1}, x_{k_2}, \ldots, x_{k_n})$ and the corresponding output symbols are $(y_{j_1}, y_{j_2}, \ldots, y_{j_n})$, then the probability

$$\begin{aligned} &P(Y_1 = y_{j_1}, Y_2 = y_{j_2}, \ldots, Y_n = y_{j_n} \mid X_1 = x_{k_1}, X_2 = x_{k_2}, \ldots, X_n = x_{k_n}) \\ &= p(y_{j_1}, y_{j_2}, \ldots, y_{j_n} \mid x_{k_1}, x_{k_2}, \ldots, x_{k_n}) = \prod_{i=1}^{n} p(y_{j_i} \mid x_{k_i}) \end{aligned}$$

Furthermore, if the probability distribution of the occurrence of the input symbol $P(X = x_k)$ for $1 \le k \le K$ is known, then the probability distribution of the occurrence of the output symbol

$P(Y = y_j)$ for $1 \leq j \leq J$ can be determined. This is possible via the use of the channel matrix P_{DMC}. Let $P(X = x_k) = p(x_k)$ for $1 \leq k \leq K$ and $P(Y = y_j) = p(y_j)$ for $1 \leq j \leq J$. Then

$$p(y_j) = \sum_{k=1}^{K} p(x_k)\, p(y_j \mid x_k), \quad 1 \leq j \leq J$$

A channel is next characterized by introducing new terminology. This is done in entropy-theoretic language.

(a) $H(X)$ is the entropy or information of the source of the input alphabet.
(b) $H(Y)$ is the entropy or information of the received symbols.
(c) $H(X, Y)$ is the entropy or information per pair of transmitted and received symbols
(d) $H(Y \mid X)$ is the entropy of the received symbols, when it is known that X is transmitted. This measure is an indication of the noise on the communication channel.
(e) $H(X \mid Y)$ is the entropy of the symbols at the source, where it is known that Y is received. This measure is an indication of how efficiently one can recover an input symbol from the output.
(f) $I(X; Y)$, the mutual information is a measure of the information transmitted on the communication channel.

The performance of a communication channel can be quantified in terms of *channel capacity*. The channel capacity is a measure of the communication channel to transmit information reliably in presence of noise.

Definition 1.8. *The channel capacity of a discrete communication system is C, where*

$$C = \max I(X; Y) \tag{1.12}$$

and maximization is with respect to the probability distribution of the input alphabet. The unit of C, is in bits per transmission. □

That is, the unit of channel capacity is in bits per symbol. Computation of capacity of a general channel is tedious. In its complete generality, this computation is a constrained optimization problem.

Observations 1.6. Some observations related to channel capacity.

1. $C \geq 0$. This follows from the fact that $I(X; Y) \geq 0$.
2. $C \leq \min\{\log_2 K, \log_2 J\}$. See the problem section for its proof. □

Examples 1.8. Following examples describe different types of channels.

1. *Lossless Channel*: A channel is said to be lossless, if the input is determined uniquely from the output. That is, $H(X \mid Y) = 0$. Consequently, $I(X; Y) = H(X)$, and $C = \log_2 K$, where K is the size of the input alphabet.
2. *Deterministic Channel*: In a deterministic channel, the output is uniquely determined from the input. That is, $H(Y \mid X) = 0$. Thus $I(X; Y) = H(Y)$, and $C = \log_2 J$, where J is the size of the output alphabet.
3. *Useless Channel*: A channel is useless if and only if the discrete random variables X and Y are independent of each other. That is, $H(X \mid Y) = H(X)$. Therefore $I(X; Y) = C = 0$.

□

A symmetric channel is next studied in greater detail.

1.6.2 Symmetric Channel

A channel of special interest is the symmetric channel. A DMC is said to be symmetric, if each row of the channel matrix contains the same set of numbers $\alpha_1, \alpha_2, \ldots, \alpha_J$, such that their sum is equal to unity. Furthermore, each column contains the same set of numbers $\beta_1, \beta_2, \ldots, \beta_K$. The sum of β_j's is not necessarily equal to unity. Let

$$P(X_i = x_k, Y_i = y_j) \triangleq p(x_k, y_j), \quad 1 \le k \le K, \quad 1 \le j \le J$$

Note that this probability is independent of i, because of the memoryless property.

Lemma 1.3. *For a symmetric DMC, $H(Y \mid X)$ is independent of the distribution of the input symbols. It is completely determined by it channel matrix P_{DMC}.*

Proof. Consider

$$\begin{aligned} H(Y \mid X) &= -\sum_{k=1}^{K}\sum_{j=1}^{J} p(x_k, y_j) \log_2 p(y_j \mid x_k) \\ &= -\sum_{k=1}^{K} p(x_k) \sum_{j=1}^{J} p(y_j \mid x_k) \log_2 p(y_j \mid x_k) \\ &= -\sum_{k=1}^{K} p(x_k) \sum_{j=1}^{J} \alpha_j \log_2 \alpha_j \end{aligned}$$

where $\alpha_j = p(y_j \mid x_k)$ for $1 \le j \le J$. Finally, $\sum_{k=1}^{K} p(x_k) = 1$ implies $H(Y \mid X) = -\sum_{j=1}^{J} \alpha_j \log_2 \alpha_j$, which is independent of the input symbol distribution. □

Lemma 1.4. *The channel capacity C of a symmetric DMC is given by*

$$C = \log_2 J + \sum_{j=1}^{J} \alpha_j \log_2 \alpha_j \tag{1.13}$$

where $\alpha_j, 1 \le j \le J$ are the elements of a row of the channel matrix.

Proof. The channel capacity is determined by maximizing

$$I(X;Y) = H(Y) - H(Y \mid X)$$

over the input symbol distribution. In the last lemma, $H(Y \mid X)$ was shown to be independent of the input distribution. Therefore, of immediate concern is the need to maximize $H(Y)$ over the input symbol distribution. Note however that the maximum value of $H(Y)$ is equal to $\log_2 J$. It can be checked that this is true if the output symbols have a uniform distribution. This follows if X has a uniform distribution. That is, $p(x_k) = 1/K$ for $1 \le k \le K$. This in turn is a consequence of the fact that the columns of the channel matrix of a symmetric channel are the permutations of a set of numbers $\beta_1, \beta_2, \ldots, \beta_K$ (as per its definition). Therefore

$$p(y_j) = \sum_{k=1}^{K} p(x_k) p(y_j \mid x_k) = \frac{1}{K} \sum_{k=1}^{K} \beta_k, \quad 1 \le j \le J$$

That is, $p(y_j)$ is a constant for $1 \leq j \leq J$. Consequently, $p(y_j) = 1/J$ for $1 \leq j \leq J$. The result follows immediately. □

Binary symmetric channel

In a communication system with binary symmetric channel, the set of input and output alphabets are $A = B = \{0, 1\}$ respectively. Also the channel matrix is

$$P_{DMC} = \begin{bmatrix} q & p \\ p & q \end{bmatrix}$$
$$p + q = 1$$

Note that the probability of error is p for each transmitted symbol. The channel capacity of this communication system can be obtained from

$$h(p) = -p \log_2 p - q \log_2 q$$
$$C = 1 - h(p) \quad \text{bits per symbol}$$

Also note that if $p = 0$ or 1, then $C = 1$; and if $p = 0.5$, then $C = 0$ as expected.

1.6.3 Extension of a Discrete Memoryless Channel

The concept of discrete memoryless channel can be extended to code, transmit, receive, and decode more than a single character at a time. In this case, the DMC is said to be *extended.* A precise definition of this communication system is given below.

Definition 1.9. *Let $A = \{x_1, x_2, \ldots, x_K\}$ and $B = \{y_1, y_2, \ldots, y_J\}$ be the set of input and output alphabets respectively of a DMC. The corresponding channel matrix is $P_{DMC} = [p(y_j \mid x_k)]$. This matrix has K rows and J columns.*

The n-th extension of this channel has the sets $A^{(n)}$ and $B^{(n)}$ as the input and output alphabets respectively. Its channel matrix is $P_{DMC}^{(n)}$. This matrix has K^n rows and J^n columns. Then

$$x = (x_{i_1}, x_{i_2}, \ldots, x_{i_n}) \in A^{(n)}, \quad x_{i_m} \in A, \; 1 \leq m \leq n \tag{1.14a}$$
$$y = (y_{i_1}, y_{i_2}, \ldots, y_{i_n}) \in B^{(n)}, \quad y_{i_m} \in B, \; 1 \leq m \leq n \tag{1.14b}$$

where x and y can be considered to be vectors. The matrix element of $P_{DMC}^{(n)}$ corresponding to these input and output vectors is

$$p(y \mid x) = p(y_{i_1} \mid x_{i_1}) p(y_{i_2} \mid x_{i_2}) \ldots p(y_{i_n} \mid x_{i_n}) \tag{1.14c}$$

□

Let $\widetilde{X} = (X_1, X_2, \ldots, X_n)$ and $\widetilde{Y} = (Y_1, Y_2, \ldots, Y_n)$ be two random vectors. Instances of the vectors $\widetilde{X}$ and $\widetilde{Y}$ are x and y respectively.

Theorem 1.5. *Let the capacity of a discrete memoryless channel be C, then its n-th extension has a capacity nC.*

Proof. Let $\widetilde{X}$ and $\widetilde{Y}$ be the input and output vectors of a memoryless communication channel. Define the capacity of the n-th extension of the channel by C_n. Then

$$C_n = \max I\left(\widetilde{X}; \widetilde{Y}\right)$$

where maximization is with respect to the probability distribution of the random vector $\widetilde{X}$. Recall that

$$H\left(\widetilde{X}\right) - H\left(\widetilde{X} \mid \widetilde{Y}\right) = H\left(\widetilde{Y}\right) - H\left(\widetilde{Y} \mid \widetilde{X}\right)$$

Also

$$H\left(\widetilde{Y} \mid \widetilde{X}\right) = \sum_{x} P\left(\widetilde{X} = x\right) H\left(\widetilde{Y} \mid \widetilde{X} = x\right)$$

$$H\left(\widetilde{Y} \mid \widetilde{X} = x\right) = \sum_{m=1}^{n} H\left(Y_m \mid \widetilde{X} = x\right) = \sum_{m=1}^{n} H\left(Y_m \mid X_m = x_{i_m}\right)$$

In the last equation, the memorylessness property of the communication channel has been used. Therefore

$$\begin{aligned} H\left(\widetilde{Y} \mid \widetilde{X}\right) &= \sum_{x} P\left(\widetilde{X} = x\right) \sum_{m=1}^{n} H\left(Y_m \mid X_m = x_{i_m}\right) \\ &= \sum_{m=1}^{n} \sum_{x_{i_m}} H\left(Y_m \mid X_m = x_{i_m}\right) P\left(X_m = x_{i_m}\right) \\ &= \sum_{m=1}^{n} H\left(Y_m \mid X_m\right) \end{aligned}$$

Also

$$H\left(\widetilde{Y}\right) \leq \sum_{m=1}^{n} H\left(Y_m\right)$$

Equality holds, if the discrete random variables $Y_1, Y_2, \ldots, Y_n$ are mutually independent of each other. This can be obtained by taking $X_1, X_2, \ldots, X_n$ to be mutually independent of each other. Hence

$$\begin{aligned} I\left(\widetilde{X}; \widetilde{Y}\right) &= H\left(\widetilde{X}\right) - H\left(\widetilde{X} \mid \widetilde{Y}\right) = H\left(\widetilde{Y}\right) - H\left(\widetilde{Y} \mid \widetilde{X}\right) \\ &\leq \sum_{m=1}^{n} H\left(Y_m\right) - \sum_{m=1}^{n} H\left(Y_m \mid X_m\right) \\ &= \sum_{m=1}^{n} I\left(X_m; Y_m\right) \end{aligned}$$

Therefore $C_n = nC$. □

1.6.4 Basics of Coding and Decoding

Consider a communication system in which the input and output symbol sets are A and B respectively. Also assume that this system has a discrete memoryless channel. It is possible that noise on the communication channel might introduce error in the received n-tuple. Let $A^{(n)}$ denote the set of n-tuples with entries in A. Members of the set $A^{(n)}$ are called words. A *code* of *length* n, denoted by $\mathcal{C}$, is any collection of distinct n-sequences. Members of the set $\mathcal{C} \subseteq A^{(n)}$ are called the *codewords*. Let the cardinality of the set $\mathcal{C}$ be equal to M, and $\mathcal{C} = \{c_j \mid 1 \leq j \leq M\}$, where c_j is a codeword. Before a description of the decoding rule is given, *Hamming distance* is defined.

Definition 1.10. *Let $u, v \in A^{(n)}$. The Hamming distance $d(u, v)$ between the n-tuples u and v is the number of entries in which they differ.* □

The Hamming distance satisfies the following properties. Let $u, v, w \in A^{(n)}$. Then $d(u, v) \geq 0$, with equality iff $u = v$, $d(u, v) = d(v, u)$, and the triangle inequality: $d(u, v) + d(v, w) \geq d(u, w)$. Therefore, the Hamming distance is a true distance measure.

Example 1.9. Let $A = \{0, 1\}$, $n = 6$ and

$$u = (110001), v = (010111), \text{ and } w = (111001)$$

Note that $d(u, v) = d(v, u) = 3 > 0$; $d(v, w) = 4$, and $d(u, w) = 1$. The triangle inequality can also be verified. Observe that $d(u, v) + d(v, w) = 7 > 1 = d(u, w)$. □

A decoding rule is obtained by partitioning the set of possible received sequences of length n into disjoint sets $L_1, L_2, \ldots, L_M$, where if the received n-sequence $r \in L_j$, it is decoded as $c_j \in \mathcal{C}$. An ideal rule to decode is the following. Let the transmitted and received words (n-tuples) be c_j and r respectively. Then the received n-tuple is decoded as c_j if

$$P(c_j \mid r) \geq P(c_i \mid r), \quad 1 \leq i \leq M$$

where $P(\cdot)$ is the probability function. This rule is called the *minimum error probability decoding rule*. However, this scheme cannot be used without a knowledge of the probabilities of the use of c_j's. Therefore, *maximum-likelihood decoding rule* is used in practice. In this scheme, if an n-tuple r is received, and there is a unique $c_j \in \mathcal{C}$ such that $d(r, c_j)$ is a minimum, then correct r to c_j. However, if no such $c_j \in \mathcal{C}$ exists, then either an error is reported, or arbitrarily select one of the codewords $c \in \mathcal{C}$ nearest to r, and correct r to c. The following lemma can be stated about minimum-distance decoding rule.

Lemma 1.5. *In a binary symmetric channel, let the probability of error in a bit be $p \leq 0.5$. Then the minimum-distance decoding rule and maximum-likelihood decoding rules are equivalent.*

Proof. Let $u, v \in A^{(n)}$, where u is the transmitted vector, and v is the corresponding received vector. Also, let the Hamming distance $d(u, v) = m$. Then

$$P(v \mid u) = p^m q^{n-m}$$

where $q = (1 - p)$. For values of $p < 0.5$, this probability is a maximum when m is a minimum. □

The next lemma due to Fano is useful in proving Shannon's theorem for noisy channel.

Lemma 1.6. *Consider a communication system in which the input and output symbol set is A. Furthermore, this system has a DMC. Let $A^{(n)}$ denote the set of n-tuples with entries in the set A.*

The codewords of this communication system are members of the set $\mathcal{C} \subseteq A^{(n)}$, *where* $|\mathcal{C}| = M$. *Let* X *be the transmitted random codeword, and the corresponding received* n*-tuple be* Y. *The probability that* $X \neq Y$ *is defined as* p_e. *Then*

$$H(X \mid Y) \le h(p_e) + p_e \log_2 (M-1) \tag{1.15a}$$

where

$$h(p_e) = -p_e \log_2 p_e - (1-p_e)\log_2 (1-p_e) \tag{1.15b}$$

Proof. Let E be a discrete random variable defined as

$$E = \begin{cases} 1, & X \neq Y \\ 0, & X = Y \end{cases}$$

That is, $P(E=1) = p_e$, and $P(E=0) = (1-p_e)$. Using the chain rule of entropies, $H(E, X \mid Y)$ is expanded as

$$H(E, X \mid Y) = H(X \mid Y) + H(E \mid X, Y)$$

As $H(E \mid X, Y) = 0$

$$H(E, X \mid Y) = H(X \mid Y)$$

Again expanding $H(E, X \mid Y)$ via the chain rule of entropies yields

$$H(E, X \mid Y) = H(E \mid Y) + H(X \mid E, Y)$$

As conditioning might reduce entropy, $H(E \mid Y) \le H(E) = h(p_e)$. Also

$$\begin{aligned} H(X \mid E, Y) &= P(E=0)\, H(X \mid E=0, Y) + P(E=1)\, H(X \mid E=1, Y) \\ &\le (1-p_e)\, 0 + p_e \log_2 (M-1) \end{aligned}$$

In the above inequality $H(X \mid E=1, Y)$ is upper bounded by the logarithm of the remaining outcomes, which is $(M-1)$. Finally

$$\begin{aligned} H(X \mid Y) &= H(E, X \mid Y) \\ &\le h(p_e) + p_e \log_2 (M-1) \end{aligned}$$

The inequality derived in this lemma is called Fano's inequality. Also note that $p_e = 0 \Rightarrow H(X \mid Y) = 0$ as expected. □

1.6.5 Noiseless Communication Channel Coding Theorem

The noiseless communication channel coding theorem is due to Shannon. This theorem quantifies the minimum requirements for a noiseless communication channel to transmit reliably a source signal with a specified entropy. Shannon's theorem is elucidated in terms of *typical* and *atypical sequences*.

Consider an information source $\mathcal{S}$. This source emits a sequence of symbols from the set $\mathcal{X}$, each with a specified probability. The emission of symbols from the source can be described by a sequence of random variables $X_1, X_2, \ldots$. These random variables take values in the set $\mathcal{X}$.

Assume that these random variables are independent, and identically distributed. Consider a sequence of random variables $(X_1, X_2, \ldots, X_n)$. These take values $(x_1, x_2, \ldots, x_n)$, where $x_i \in \mathcal{X}$ for $1 \leq i \leq n$. Denote the sequence of random variables by $\widetilde{X} = (X_1, X_2, \ldots, X_n)$ and their instance by $x = (x_1, x_2, \ldots, x_n)$. Let $H(X_i)$ be the entropy of the random variable X_i for $1 \leq i \leq n$. Also, let the functions $P(\cdot)$ and $p(\cdot)$ denote the probability and the probability mass function respectively of the sequence $\widetilde{X}$. Further assume that, $H(X) = H(X_1) = H(X_2) = \ldots = H(X_n)$, and X is a representative random variable of the source random variables X_i's.

A motivation for the definition of a typical sequence is next provided. Let X_i's be Bernoulli distributed random variables, each with parameter p, then for typical sequences

$$p(x) = p(x_1) p(x_2) \ldots p(x_n) \simeq p^{np} (1-p)^{n(1-p)}$$

Taking logarithm on both sides yields

$$\log_2 p(x) = n \{p \log_2 p + (1-p) \log_2 (1-p)\} = -nH(X)$$

where $H(X) = -\{p \log_2 p + (1-p) \log_2 (1-p)\}$. Thus

$$P\left(\widetilde{X} = x\right) = p(x) = p(x_1) p(x_2) \ldots p(x_n) \simeq 2^{-nH(X)}$$

The value $H(X)$ is called the *entropy rate* of the source. Observe that the number of typical sequences is approximately $2^{nH(X)}$. Sequences which are not typical are called atypical sequences. That is, atypical sequences are less likely to occur.

Definition 1.11. *Let $\widetilde{X} = (X_1, X_2, \ldots, X_n)$ be a sequence of independent, and identically distributed random variables. Also let $H(X)$ be the entropy rate of the source. Let $\epsilon > 0$.*

(a) *A sequence $x = (x_1, x_2, \ldots, x_n) \in \widetilde{X}$ is ϵ-typical if*

$$2^{-n(H(X)+\epsilon)} \leq p(x) \leq 2^{-n(H(X)-\epsilon)} \tag{1.16}$$

A sequence which is not ϵ-typical is an atypical sequence.

(b) *The set of all ϵ-typical sequences of length n is denoted by $V(n, \epsilon)$.* □

Observe that equivalent conditions for an ϵ-typical sequence are

$$H(X) - \epsilon \leq -\frac{1}{n} \log_2 p(x) \leq H(X) + \epsilon$$

and

$$\left| H(X) + \frac{1}{n} \log_2 p(x) \right| \leq \epsilon$$

These statements quantify the so-called *asymptotic equipartition property* of a sequence of random numbers. This property is similar to the weak law of large numbers. According to this property $\{-\log_2 p(x)\} / n \to H(X)$ for large n. A theorem about typical sequences due to Shannon is next stated.

Theorem 1.6. *Typical sequence.*

(a) *Let $\epsilon > 0$ be fixed. For any $\delta > 0$, and sufficiently large n, the probability that a sequence is ϵ-typical is at least $(1-\delta)$.*

(b) *Let $\epsilon > 0$ and $\delta > 0$ be fixed. For sufficiently large n, the cardinality of the set $V(n, \epsilon)$ satisfies*

$$(1 - \delta) 2^{n(H(X)-\epsilon)} \leq |V(n, \epsilon)| \leq 2^{n(H(X)+\epsilon)} \tag{1.17}$$

where $V(n, \epsilon)$ is the set of all ϵ-typical sequences of length n.

(c) *Let $R < H(X)$ be fixed, and $S(n)$ be a set of sequences of length n such that $|S(n)| \leq 2^{nR}$. For any $\delta > 0$ and sufficiently large n, $\sum_{x \in S(n)} p(x) \leq \delta$.*

Proof. See the problem section. □

A theorem about average codeword-length is given below. It is a direct consequence of the above theorem.

Theorem 1.7. *Average codeword-length. Suppose that $\widetilde{X}$ is a sequence of independent, and identically distributed random variables $X_i, 1 \leq i \leq n$. Also let $H(X)$ be the entropy rate of the source $\mathcal{S}$. Then for any $\epsilon' > 0$, and sufficiently large values of n, there exists a code which maps one-to-one, a sequence x into a binary string. If the length of this binary string is denoted by $\ell(x)$, then*

$$\mathcal{E}\left\{\frac{1}{n}\ell\left(\widetilde{X}\right)\right\} \leq H(X) + \epsilon' \tag{1.18}$$

Proof. See the problem section. □

The above theorem asserts that, on the average a sequence $\widetilde{X}$ of length n, can be represented using $nH(X)$ bits.

Before Shannon's noiseless communication channel coding theorem is stated in the next subsection, the concepts of *rate* of data compression, and *reliability* of data compression and decompression scheme are explained. Assume again that $\widetilde{X} = (X_1, X_2, \ldots, X_n)$ is a sequence of independent, and identically distributed random variables. A data compression technique of rate R is a map $M^n_{comp}(\cdot)$ which transforms a source symbol-sequence $x = (x_1, x_2, \ldots, x_n)$ to a bit string of length $\lceil nR \rceil = m$. The output of this compression scheme is the bit string $M^n_{comp}(x)$. The corresponding decompression process is a map $M^n_{decomp}(\cdot)$ which takes the compressed bit sequence of length m and attempts to transforms it to the original sequence x. Thus the output of the decompression scheme is $M^n_{decomp}(M^n_{comp}(x))$. A compression-decompression scheme

$$\left(M^n_{comp}, M^n_{decomp}\right)$$

is said to be reliable if the probability of the event $M^n_{decomp}(M^n_{comp}(x)) = x$ tends to unity for all sequences $x \in \widetilde{X}$, and very large values of n. That is,

$$P\left(M^n_{decomp}\left(M^n_{comp}(x)\right) = x\right) \to 1, \ \forall\, x \in \widetilde{X} \ \text{ as } n \to \infty$$

Shannon's noiseless communication channel coding theorem specifies the values of rate R for which a compression-decompression scheme is reliable.

Theorem 1.8. *Let $X_1, X_2, \ldots$ be a sequence of independent, and identically distributed information source random variables (symbols), each with entropy $H(X)$.*

(a) *If $R > H(X)$ then there exists a reliable compression-decompression scheme of rate R.*

(b) *Conversely, if $R < H(X)$ then any type of compression-decompression scheme is not reliable.*

Proof. See the problem section. □

Coding results about noiseless communication channel were studied in this subsection. The next subsection addresses transmission of information over noisy communication channel.

1.6.6 Noisy Communication Channel Coding Theorem

Some observations can be made about transmission of information over noisy channels. These were first stated by Shannon. Shannon's noisy communication channel coding theorem asserts that arbitrarily high reliable transmission of information is possible, provided the transmission rate is below the channel capacity. However, this theorem does not indicate a procedure to develop a coding scheme to achieve this reliability.

Let $\mathcal{C} = \{c_j \mid 1 \leq j \leq M\}$ be a code, where each codeword c_j is of length n. Assume that each codeword is equally likely to have been transmitted over the channel. Then for a specified decoding scheme, the average probability that a codeword has been received in error is equal to e_{avg}. Similarly, the maximum error probability is defined as $e_{\max}$.

$$\begin{aligned} e_{avg} &= \frac{1}{M} \sum_{j=1}^{M} P\left(\text{error} \mid \text{codeword } c_j \text{ is transmitted}\right) \\ e_{\max} &= \max_{1 \leq j \leq M} \left\{P\left(\text{error} \mid \text{codeword } c_j \text{ is transmitted}\right)\right\} \end{aligned}$$

Consider a binary symmetric DMC communication system with the input and output symbol set $A = \{0, 1\}$. Let $A^{(n)}$ denote the set of n-tuples with entries in A. The codewords of this communication system are members of the set $\mathcal{C} \subseteq A^{(n)}$. In proving Shannon's theorem, it is assumed that $|\mathcal{C}| = M$ codewords are chosen at random from the set of words in $A^{(n)}$. This type of coding is called *random coding.*

Note that, for a binary symmetric DMC, the decoding rule is the maximum-likelihood decoding, or equivalently minimum-distance rule. This decoding rule works as follows. Let $c_i \in \mathcal{C}$, be a transmitted codeword, and the corresponding received word be $y \in A^{(n)}$. For a fixed integer $k > 0$, define a set $S_k(y)$, called a *k-sphere.*

$$S_k(y) = \left\{w \mid w \in A^{(n)}, d(y, w) \leq k\right\}$$

For a received word y, the decoded codeword is $c_j \in \mathcal{C}$ if c_j is the only element of the set $S_k(y)$; otherwise y is decoded as any arbitrary codeword $c_l \in S_k(y)$. Following is the well-known Shannon's noisy coding theorem.

Theorem 1.9. *Consider a binary symmetric DMC of capacity C, where*

$$C = 1 - h(p) \tag{1.19a}$$

$$h(p) = -p \log_2 p - q \log_2 q \tag{1.19b}$$

$$q = 1 - p, \quad p < \frac{1}{2} \tag{1.19c}$$

and the probability of error in a bit is p. For any R with $0 < R < C$, if $1 \leq M \leq 2^{Rn}$, for $n \in \mathbb{P}$, and any $\epsilon > 0$; there exists a code $\mathcal{C}$, with codewords of length n, where $|\mathcal{C}| = M$, and a positive integer $\eta(\epsilon)$, such that $e_{\max} \leq \epsilon$ for all $n \geq \eta(\epsilon)$.

Proof. Assume a binary symmetric DMC communication system with the input and output symbol set $A = \{0, 1\}$. Recall that these symbols are also called bits. The set of n-tuples with entries in A is denoted by $A^{(n)}$. Furthermore, the codewords of this communication system are members of the set $\mathcal{C} \subseteq A^{(n)}$, where $|\mathcal{C}| = M$. In addition, members of the code $\mathcal{C}$ are selected at random from the set $A^{(n)}$.

Let $c \in \mathcal{C}$, be a transmitted codeword, and the corresponding received word be a discrete random variable Y, which takes a value in the set $A^{(n)}$. The event, that an error has been made in decoding is denoted by E. As per the minimum-distance decoding rule, for a fixed integer $k > 0$, the received word is in error because of the following two possibilities.

(a) If $d(c, Y) > k$. Denote this event by E_1.
(b) If $d(c, Y) \leq k$ and $d(c', Y) \leq k$ for $c \neq c'$. Denote this event by E_2.

Then $E = E_1 \cup E_2$. This implies $P(E) \leq P(E_1) + P(E_2)$.
The event E_2 occurs if the following two events occur.

(a) At most k errors occur in transmission. Denote this event by E_{21}.
(b) There is a codeword $c' \neq c$ such that $d(c', Y) \leq k$. Denote this event by E_{22}.

Observe that $E_2 = E_{21} \cap E_{22}$. Therefore $P(E_2) \leq P(E_{22})$. Consequently $P(E) \leq P(E_1) + P(E_{22})$. An upper estimate of $P(E_1)$ is first obtained. Let Q be the random number of symbols in error when the transmitted codeword is c. This discrete random variable Q has a binomial distribution with parameters n and p. Let $k = \lfloor np + n\epsilon \rfloor \leq n$, where $\epsilon > 0$. Then use of Bienaymé-Chebyshev inequality results in

$$P(E_1) = P(Q > np + n\epsilon) \leq P(|Q - np| > n\epsilon) \leq Var(Q) / (n\epsilon)^2$$

Variance of the random variable Q is equal to npq, where $q = (1 - p)$. Therefore $P(E_1) \leq pq/n\epsilon^2$.

The probability $P(E_{22})$, is next upper-bounded. Since the M codewords are chosen at random from 2^n possible codewords of length n, the probability that a specific codeword c' is within distance k is given by $\sum_{0 \leq j \leq k} \binom{n}{j} 2^{-n}$. Furthermore, there are $(M - 1)$ possibilities for the choice of the codeword c' which are not equal to c. Thus

$$P(E_{22}) \leq \frac{(M-1)}{2^n} \sum_{0 \leq j \leq k} \binom{n}{j}$$

If $\epsilon > 0$, and $k = \lfloor np + n\epsilon \rfloor \leq n$, the summation $\sum_{0 \leq j \leq k} \binom{n}{j}$ can be upper bounded by $2^{nh(p+\epsilon)}$. See the problem section for a proof of this result. Consequently

$$P(E_2) \leq P(E_{22}) \leq M 2^{-n\{1-h(p+\epsilon)\}}$$

Combining the upper bounds for $P(E_1)$ and $P(E_2)$ results in

$$P(E) \leq P(E_1) + P(E_2) \leq \frac{pq}{n\epsilon^2} + M 2^{-n\{1-h(p+\epsilon)\}}$$

Therefore, as $\epsilon > 0$, the probability $P(E)$ can be made arbitrarily small for sufficiently large values of n, if M does not grow faster than $2^{n\{1-h(p)\}}$. Note that, in the above discussion the *average* error probability $P(E)$ has been bounded.

However it has to be shown that the *maximum* error probability of the received word is bounded by ϵ, and demonstrate the existence of a code $\mathcal{C}$ with codewords of length n such that $|\mathcal{C}| = M \leq 2^{nR}$, where $0 < R < C$.

In the next step, let $\epsilon' = \epsilon/2$ and $M' = 2M$. If $M \leq 2^{nR}$, then there exists R', with $R < R' < C$, and $\eta'(\epsilon)$ such that; for $n \geq \eta'(\epsilon)$ there exists a code $\mathcal{C}'$ of codeword-length n, $|\mathcal{C}'| = M' \leq 2^{nR'}$, and average error probability less than ϵ'. Let $\mathcal{C}' = \{c'_j \mid 1 \leq j \leq M'\}$, then

$$\sum_{j=1}^{M'} P\left(E \mid \text{codeword } c'_j \text{ is transmitted}\right) \leq \epsilon' M'$$

Therefore at least half of these codewords c'_j's must observe the relationship

$$P\left(E \mid \text{codeword } c'_j \text{ is transmitted}\right) \leq 2\epsilon' = \epsilon$$

Select the codewords which belong to the set $\mathcal{C}$, from the codewords that satisfy the above relationship. This is indeed the required code. It satisfies the condition that for sufficiently large values of n, the maximum error probability of a codeword in $\mathcal{C}$ is upper-bounded by ϵ. □

The above theorem was established for a binary symmetric DMC. A similar result holds true for a general channel. The converse of the above theorem, using Fano's inequality is next proved. This converse states that, it is impossible to transmit information reliably on a noisy channel at a rate greater than its capacity.

Theorem 1.10. *Consider a discrete memoryless channel of capacity C. Then there cannot exist a code $\mathcal{C}$, of codeword-length n, with the property that $|\mathcal{C}| = 2^{nR}$, where $R > C$, and error probability p_e that tends to 0 as $n \to \infty$.*

Proof. Let $R = C + \epsilon$, where $\epsilon > 0$. Let X be a uniformly distributed random variable, which takes values in the set of codewords $\mathcal{C}$. Therefore, $H(X) = \log_2 |\mathcal{C}| = nR = n(C + \epsilon)$. Let Y be the output random variable corresponding to X. Since the codewords are of length n, using the result of the nth extension of a memoryless channel yields

$$\begin{aligned} nC &\geq H(X) - H(X \mid Y) \\ &= n(C + \epsilon) - H(X \mid Y) \end{aligned}$$

Therefore

$$H(X \mid Y) \geq n\epsilon$$

The probability that $X \neq Y$ is defined as p_e. Then application of Fano's inequality yields

$$H(X \mid Y) \leq h(p_e) + p_e \log_2 (M - 1)$$

where

$$h(p_e) = -p_e \log_2 p_e - (1 - p_e) \log_2 (1 - p_e), \quad \text{and} \quad M = |\mathcal{C}| = 2^{nR}$$

Consequently,

$$\begin{aligned} n\epsilon &\leq H(X \mid Y) \leq h(p_e) + p_e \log_2 (M - 1) \\ &\leq h(p_e) + p_e \log_2 M \\ &= h(p_e) + p_e nR = h(p_e) + p_e n(C + \epsilon) \end{aligned}$$

That is,

$$p_e \geq \frac{n\epsilon - h(p_e)}{n(C+\epsilon)}$$

Thus, as $n \to \infty$, p_e does not tend to zero. Therefore the code $\mathcal{C}$ with the stated properties does not exist as $n \to \infty$. □

1.7 Differential Entropy and Capacity of Communication Channels

Differential entropy is an extension of the concept of entropy of a discrete random variable to continuously distributed random variable. This notion of differential entropy is used in proving the well-known information capacity theorem. The celebrated *sampling theorem* is also stated and proved. It is subsequently used in proving the information capacity theorem. Parallel Gaussian-noise communication channels are also examined.

1.7.1 Differential Entropy

Differential entropy is initially defined. This is followed by a discussion of its application.

Definitions 1.12. *Some definitions related to differential entropy.*

1. *Let X be a continuously distributed random variable, with probability density function $f_X(x)$, where $x \in \mathbb{R}$. The differential entropy of this random variable X is given by*

$$H_c(X) = -\int_{-\infty}^{\infty} f_X(x) \log_2 f_X(x)\, dx \tag{1.20}$$

The integral in the above equation is assumed to exist.

2. *Let X and Y be two jointly distributed continuous random variables, with joint probability density function $f_{X,Y}(x,y)$, where $x, y \in \mathbb{R}$. Their joint differential entropy $H_c(X,Y)$ is given by*

$$H_c(X,Y) = -\int_{-\infty}^{\infty}\int_{-\infty}^{\infty} f_{X,Y}(x,y) \log_2 f_{X,Y}(x,y)\, dxdy \tag{1.21}$$

The double integral in the above equation is assumed to exist. □

In contrast to the entropy of discrete random variable, differential entropy can be negative. Therefore, the interpretation of entropy as the amount of uncertainty associated with the continuously distributed random variable X is not correct. Consequently care should be exercised in its interpretation and use.

Example 1.10. Let X be a uniformly distributed random variable. Its probability density function is given by

$$f_X(x) = \begin{cases} \dfrac{1}{(b-a)}, & x \in [a,b] \\ 0, & \text{otherwise} \end{cases}$$

where $a, b \in \mathbb{R}$, and $a < b$. Its differential entropy is $H_c(X) = \log_2(b-a)$.

It can be observed that the differential entropy, $H_c(X)$ can be negative, zero, or positive; if $(b-a)$ is less than 1, equal to 1, or greater than 1 respectively. □

Example 1.11. Let X be a normally distributed random variable. Its probability density function is given by

$$f_X(x) = \frac{1}{\sqrt{2\pi}\sigma} \exp\left\{-\frac{1}{2}\left(\frac{x-\mu}{\sigma}\right)^2\right\}, \qquad x \in \mathbb{R}$$

where $\mu \in \mathbb{R}$, and $\sigma \in \mathbb{R}^+$. Also $\mathcal{E}(X) = \mu$, and $Var(X) = \sigma^2$. The differential entropy of this random variable is found to be

$$H_c(X) = \frac{1}{2}\log_2\left(2\pi e\sigma^2\right)$$

Observe that the differential entropy of a normally distributed random variable is independent of its mean. Furthermore, depending on the value of σ, the differential entropy can be negative, zero, or positive. □

Relationship between discrete entropy and differential entropy is next explored.

Relationship between Discrete Entropy and Differential Entropy

The difference between the entropy of a discrete random variable, and the differential entropy of a continuous random variable X, is next explored. Let the probability density function of the random variable X, be $f_X(x)$ where $x \in \mathbb{R}$. The difference arises because the differential entropy is not the limiting case of entropy of a discrete random variable which approximates the continuous random variable. Define a discrete random variable X_Δ which takes values $x_j = j\Delta x$, where $P(X = x_j) = f(x_j)\Delta x$ for $j \in \mathbb{Z}$, and $\Delta x > 0$.

$$\begin{aligned} H(X_\Delta) &= -\sum_{j\in\mathbb{Z}} f(x_j)\Delta x \log_2\left(f(x_j)\Delta x\right) \\ &= -\sum_{j\in\mathbb{Z}} \left[f(x_j)\log_2 f(x_j)\right]\Delta x - \log_2 \Delta x \sum_{j\in\mathbb{Z}} f(x_j)\Delta x \\ &= -\sum_{j\in\mathbb{Z}} \left[f(x_j)\log_2 f(x_j)\right]\Delta x - \log_2 \Delta x \end{aligned}$$

As $\Delta x \to 0$, $-\sum_{j\in\mathbb{Z}}\left[f(x_j)\log_2(f(x_j))\right]\Delta x \to H_c(X)$, but $-\log_2 \Delta x \to \infty$. Therefore, it can be inferred that the presence of this infinitely large term in the above equation produces the strange behavior of differential entropy. Alternately, we can state that

$$H(X_\Delta) + \log_2 \Delta x \to H_c(X) \text{ as } \Delta x \to 0$$

A useful observation about the differential entropy of a continuous random variable is next established. We prove that the normal random variable has the largest differential entropy attainable by any continuous random variable for a specified value of variance.

Lemma 1.7. *Let X and Y be continuous random variables, with probability density functions, $f_X(t)$ and $f_Y(t)$ respectively, where $t \in \mathbb{R}$. Then*

$$-\int_{-\infty}^{\infty} f_X(t) \log_2 f_X(t)\, dt \leq -\int_{-\infty}^{\infty} f_X(t) \log_2 f_Y(t)\, dt \tag{1.22}$$

Equality holds iff $f_X(t) = f_Y(t)$ for all values of $t \in \mathbb{R}$.

Proof. See the problem section. □

Lemma 1.8. *Let X be a continuous random variable with variance σ^2, then*

$$H_c(X) \leq \frac{1}{2} \log_2 \left(2\pi e \sigma^2\right) \tag{1.23}$$

Proof. Let $\mathcal{E}(X) = \mu$. Use the previous lemma, and let Y be a normally distributed random variable with mean μ and variance σ^2. □

This result implies that the normal distribution has the largest differential entropy for fixed variance. It is because of this result, that Gaussian channels are used to model communication systems. The concept of conditional entropy and mutual information, defined for discrete random variables can be extended to continuous random variables. In these definitions, $f_{X|Y}(x \mid y)$ is the conditional probability density function of X, given $Y = y$.

Definitions 1.13. *Conditional differential entropy, and mutual information.*

1. *Let X and Y be two jointly distributed continuous random variables. The conditional differential entropy of X given Y, is denoted by $H_c(X \mid Y)$.*

$$H_c(X \mid Y) = -\int_{-\infty}^{\infty}\int_{-\infty}^{\infty} f_{X,Y}(x,y) \log_2 f_{X|Y}(x \mid y)\, dxdy \tag{1.24a}$$

$$= -\int_{-\infty}^{\infty}\int_{-\infty}^{\infty} f_{X,Y}(x,y) \log_2 \left[\frac{f_{X,Y}(x,y)}{f_Y(y)}\right] dxdy \tag{1.24b}$$

where $f_Y(y)$ is positive for all values of y.

2. *The mutual information between two jointly distributed continuous random variables X and Y is denoted by $I_c(X;Y)$.*

$$I_c(X;Y) = \int_{-\infty}^{\infty}\int_{-\infty}^{\infty} f_{X,Y}(x,y) \log_2 \left[\frac{f_{X,Y}(x,y)}{f_X(x)\, f_Y(y)}\right] dxdy \tag{1.25}$$

where $f_X(x)$ and $f_Y(y)$ are positive for all values of x and y respectively. □

The following observations are similar to those of discrete random variables.

Observations 1.7. Some observations related to differential entropy.

1. $I_c(X;Y) \geq 0$, with equality iff X and Y are independent of each other.
2. $I_c(X;Y) = I_c(Y;X)$.
3. $I_c(X;Y) = H_c(X) - H_c(X \mid Y) = H_c(Y) - H_c(Y \mid X)$.
4. $I_c(X;Y) = H_c(X) + H_c(Y) - H_c(X,Y)$.
5. $H_c(X) \geq H_c(X \mid Y)$, with equality iff X and Y are independent of each other.

6. $H_c(X,Y) \leq H_c(X) + H_c(Y)$.
7. Let $a \in \mathbb{R}$.
   (a) Let $Y = (X + a)$, then $H_c(Y) = H_c(X)$. That is, translation does not alter the differential entropy.
   (b) Let $Y = aX$, then $H_c(Y) = H_c(X) + \log_2 |a|$. □

Sampling Theorem

Sampling theorem is one of the most fundamental results in communication engineering. It is the basis for conversion of a continuous signal to a discrete signal, and vice-versa.

Consider a signal $f(t), t \in \mathbb{R}$, where t is the time variable. This signal is limited to a frequency content of at most ω_c. Define $\omega_c = 2\pi f_c$, and $T_c = 1/f_c$.

Assume that the signal $f(t)$ is sampled at every T_s units of time. Define $\omega_s = 2\pi f_s$, where $T_s = 1/f_s$. Let the sampled signal be $f_n = f(nT_s), n \in \mathbb{Z}$.

As per the sampling theorem, if a sequence of sampled points $\{f_n \mid n \in \mathbb{Z}\}$ is given, then it is possible to recover the original signal $f(t), t \in \mathbb{R}$ provided, $2\omega_c \leq \omega_s$, that is $T_s \leq T_c/2$. The sampling theorem is next stated.

Theorem 1.11. *Let $f(t)$ and $F(\omega)$ be a Fourier transform pair, that is $f(t) \leftrightarrow F(\omega)$, where $t, \omega \in \mathbb{R}$. Also $F(\omega) = 0$ for $|\omega| > \omega_c$. Then $f(t)$ can be uniquely determined from*

$$f(t) = \sum_{n \in \mathbb{Z}} f_n \, sinc(\omega_c t - n\pi), \quad f_n = f(nT_s), \quad \forall n \in \mathbb{Z} \tag{1.26}$$

where $\omega_c = 2\pi/T_c, T_s = T_c/2$, and

$$sinc(t) \triangleq \frac{\sin t}{t}, \quad t \in \mathbb{R}$$

Proof.

$$f(t) = \frac{1}{2\pi}\int_{-\infty}^{\infty} F(\omega) e^{i\omega t} d\omega = \frac{1}{2\pi}\int_{-\omega_c}^{\omega_c} F(\omega) e^{i\omega t} d\omega$$

Then

$$f_n = f\left(\frac{n\pi}{\omega_c}\right) = \frac{1}{2\pi}\int_{-\omega_c}^{\omega_c} F(\omega) e^{in\pi\omega/\omega_c} d\omega, \quad \forall n \in \mathbb{Z}$$

Expand $F(\omega)$ into a Fourier series in the interval $(-\omega_c, \omega_c)$. Note that the length of this period is equal to $2\omega_c$. Thus

$$F(\omega) = \sum_{n \in \mathbb{Z}} \alpha_n e^{in\pi\omega/\omega_c}, \quad \alpha_n = \frac{1}{2\omega_c}\int_{-\omega_c}^{\omega_c} F(\omega) e^{-in\pi\omega/\omega_c} d\omega, \quad \forall n \in \mathbb{Z}$$

From the above equations it follows that

$$\alpha_n = \frac{\pi f_{-n}}{\omega_c}, \quad \forall n \in \mathbb{Z}$$

Thus

$$F(\omega) = \frac{\pi}{\omega_c}\sum_{n \in \mathbb{Z}} f_{-n} e^{in\pi\omega/\omega_c} = \frac{\pi}{\omega_c}\sum_{n \in \mathbb{Z}} f_n e^{-in\pi\omega/\omega_c}$$

Define a gate function $g_{\omega_c}(\cdot)$, for $\omega_c > 0$, where

$$g_{\omega_c}(\omega) = \begin{cases} 1, & |\omega| < \omega_c \\ 0, & |\omega| > \omega_c \end{cases}$$

Note that $F(\omega)$ is band-limited, and $g_{\omega_c}(\cdot)$ is the gate function. Thus use of the above equations results in

$$F(\omega)\, g_{\omega_c}(\omega) = \frac{\pi}{\omega_c} \sum_{n \in \mathbb{Z}} f_n g_{\omega_c}(\omega)\, e^{-in\pi\omega/\omega_c}$$

The stated result follows by applying the inverse Fourier transform operation on both sides of the above equation, and noting that

$$\frac{\omega_c}{\pi} sinc(\omega_c t - n\pi) \leftrightarrow e^{-in\pi\omega/\omega_c} g_{\omega_c}(\omega), \quad n \in \mathbb{Z}$$

□

It is proved in the problem section, that the sequence of functions

$$\{\beta_n(t) = sinc(\omega_c t - n\pi) \mid n \in \mathbb{Z}\}$$

are orthogonal over $t \in \mathbb{R}$. The above sampling theorem has been proved in time domain. A similar theorem can be formulated in the frequency domain, for time limited function. The sampling theorem is used in the next subsection in establishing the information capacity theorem.

1.7.2 Information Capacity Theorem

The well-known information capacity theorem due to Shannon is next stated and examined. The information capacity theorem states a fundamental limit on the rate of error-free transmission on a communication channel. This theorem assumes, that the transmission channel is band-limited, and the perturbing signal is described by Gaussian white noise process. A band-limited channel implies that this channel cannot transmit signals beyond a certain fixed frequency (band) value. In order to prove this theorem, assume that the channel is a black-box. Information is transmitted through this channel in the form of discrete symbols, at equi-spaced instants, but the distribution of the input symbols is assumed to be a continuous random variable. Consequently, this channel is said to be *discrete-time continuous-amplitude*.

Assume that the channel is band-limited to B_{bw} samples per unit time. This implies, that the channel cannot transmit signals which have frequencies higher than B_{bw} samples per unit time. For simplicity, assume that the continuous time input signal has a maximum frequency of $2\pi B_{bw}$. Therefore, as per the sampling theorem, the input signal should be sampled at a rate of at least $2B_{bw}$ samples per unit time.

The channel receives a sampled input X_i at time i, which is a continuous random variable. This channel produces a corresponding output Y_i. This output is a perturbed version of the input signal. The perturbing signal, generally called noise, is denoted by N_i. Therefore

$$Y_i = X_i + N_i, \quad i = 1, 2, 3, \ldots$$

In this model, noise is said to be additive because it adds directly to the input signal. As the noise is white, the N_i's are independent, and identically distributed random variables. Assume that the input samples, that is the X_i's are independent, and identically distributed random variables. Also assume that the input and noise sequences are independent of each other. Denote the generic input, noise, and output sequences by X, N, and Y respectively. The information capacity theorem makes the following assumptions about these sequences.

Characteristics of the input sequence

(a) The input samples are independent of each other. Furthermore, these samples have identical probability distribution. Thus, $\mathcal{E}(X) = 0$, and $Var(X) = \sigma_X^2 < \infty$. Sometimes σ_X^2 is also denoted by P_{pow}. The quantity P_{pow}, is also called the *average transmitted power.*
(b) Sampling rate of the input sequence is $2B_{bw}$ samples per unit time.

Characteristics of the noise sequence

(a) The noise process is white. Further, the N_i's are independent of each other, and have identical probability distribution.
(b) The noise sequence is additive.
(c) The noise process is independent of the input sequence.
(d) The random variable N, has a Gaussian distribution, where $\mathcal{E}(N) = 0$, and $Var(N) = \sigma_N^2$. The variance of the random variable N is also called the *noise power.*
(e) The differential entropy of the noise sample is $H_c(N) = \frac{1}{2}\log_2\left(2\pi e\sigma_N^2\right)$.
(f) Define $\sigma_N^2 = N_0 B_{bw}$, where $N_0/2$ is said to be *power spectral density of the noise sequence.* That is

$$\int_{-B_{bw}}^{B_{bw}} \frac{N_0}{2} df = N_0 B_{bw} = \sigma_N^2$$

Characteristics of the output sequence

(a) $\mathcal{E}(Y) = 0, Var(Y) = \sigma_Y^2 = \left(\sigma_X^2 + \sigma_N^2\right)$.
(b) $H_c(Y \mid X) = H_c(N)$, as the input and noise sequences are independent of each other.
(c) $H_c(Y) \le \frac{1}{2}\log_2\left(2\pi e\sigma_Y^2\right)$.

Information capacity of the channel is next defined.

Definition 1.14. *The channel capacity of a discrete communication system is C, and*

$$C = \max I_c(X;Y) \tag{1.27}$$

where X and Y are the input and output random variables of the channel respectively. Maximization is with respect to the probability distribution of the input alphabet. The units of C are in bits per transmission (sample). □

Lemma 1.9. *The capacity (bits per transmission) of a discrete-time continuous-amplitude channel, perturbed by Gaussian noise is given by*

$$C = \frac{1}{2}\log_2\left(1 + \frac{P_{pow}}{\sigma_N^2}\right) \text{ bits per transmission sample} \tag{1.28}$$

where $\sigma_N^2 > 0$, and P_{pow}/σ_N^2 is called the signal-to-noise ratio.

Proof. Use of the definition of channel capacity yields

$$\begin{aligned} C &= \max I_c(X;Y) \\ &= \max\{H_c(Y) - H_c(Y \mid X)\} = \max\{H_c(Y) - H_c(N)\} \\ &= \max\left\{H_c(Y) - \frac{1}{2}\log_2\left(2\pi e\sigma_N^2\right)\right\} \\ &= \frac{1}{2}\log_2\left(2\pi e\sigma_Y^2\right) - \frac{1}{2}\log_2\left(2\pi e\sigma_N^2\right) \end{aligned}$$

Simplification yields the stated result. The maximum occurs when $X \sim \mathcal{N}(0, P_{pow})$. □

Observe that, if the noise variance σ_N^2 is equal to zero, the receiver receives the transmitted signal without any error, as expected. Similar observation follows if the input signal is unconstrained, that is if $P_{pow} \to \infty$. Shannon's celebrated information capacity theorem follows immediately from the above lemma.

Theorem 1.12. *The capacity C_t (bits per unit time) of a discrete-time and continuous-amplitude channel, perturbed by Gaussian noise is given by*

$$C_t = B_{bw}\log_2\left(1 + \frac{P_{pow}}{N_0 B_{bw}}\right) \text{ bits per unit time} \tag{1.29}$$

Proof. The proof follows directly from the above lemma and the definitions. □

Corollary 1.1.

$$C_t \to \left(\frac{P_{pow}}{N_0}\right)\log_2 e \text{ bits per unit time, as } B_{bw} \to \infty \tag{1.30}$$

Therefore for very large values of B_{bw}, the capacity C_t increases linearly with P_{pow}. □

Decoding of the received signal

Let the alphabet of a code $\mathcal{C}$ be $A = \{0,1\}$, and the codeword be a symbol string of length n. The code $\mathcal{C}$ is a subset of $A^{(n)}$, and $|\mathcal{C}| \triangleq M$, where $M \leq 2^n$. This code is also called an $[n, M]$-code. The *information rate*, or the *code rate*, or the efficiency of this code is given by ϑ, where

$$\vartheta = \frac{\log_2 M}{n} \tag{1.31}$$

Therefore if $\vartheta \leq C$, the received codeword c is decoded into $c_i \in \mathcal{C}$ which is closest to c. That is, minimum distance decoding scheme is used. In addition, if n is sufficiently large, the probability of decoding error can be made arbitrarily small. A general definition of information or code rate is provided.

Definition 1.15. *Information or code rate. Let the alphabet of a code $\mathcal{C}$ be $A = \{0,1\}$, and the codeword (message) be a symbol string of length n. The code $\mathcal{C}$ is a subset of $A^{(n)}$, and $|\mathcal{C}| \triangleq M$, where $M \leq 2^n$. This code is also called an $[n, M]$-code. Let $H(M)$ be the entropy of the source message. It is the average number of bits transmitted for each source message. The information rate, or the code rate, or the efficiency of this code is given by ϑ, where*

$$\vartheta = \frac{H(M)}{n} \tag{1.32}$$

If each message is equally likely, then $H(M) = \log_2 M$. □

1.7.3 Parallel Gaussian-Noise Communication Channels

Consider n number of discrete-time, memoryless, additive Gaussian-noise channels working in parallel. These communication channels operate under a common power constraint. The goal is to apportion the total power P_{pow} among all the n channels so that the total capacity of the channels is maximized. This analysis is summarized in the following observation.

Observation 1.8. Determination of the capacity of $n \in \mathbb{P}$ parallel, discrete-time, memoryless Gaussian-noise communication channels. The input to channel j is described by the random variable X_j, where $\mathcal{E}(X_j) = 0$, and $Var(X_j) = \sigma_{X_j}^2 \triangleq P_{pow_j} < \infty$. The quantity $P_{pow_j} = \mathcal{E}(X_j^2)$, is also called the *average transmitted power over channel* j. Note that $j = 1, 2, \ldots, n$.

The noise on channel j is described by the Gaussian random variable $N_j \sim \mathcal{N}\left(0, \sigma_{N_j}^2\right)$, where $\sigma_{N_j}^2 \in \mathbb{R}^+$ and $1 \leq j \leq n$. The random variable N_j is independent of the random variable N_k, where $j \neq k$, and $1 \leq j, k \leq n$.

The output of the channel j is Y_j, where $Y_j = (X_j + N_j)$, and $1 \leq j \leq n$.

Also the sequence of random variables $X_1, X_2, \ldots, X_n$ is independent of the sequence of random variables $N_1, N_2, \ldots, N_n$.

The channel powers are constrained to have a net power of P_{pow}. That is

$$\sum_{j=1}^{n} P_{pow_j} \leq P_{pow}$$

Let $\widetilde{X} = (X_1, X_2, \ldots, X_n)$, and $\widetilde{Y} = (Y_1, Y_2, \ldots, Y_n)$. An instance of the random vector $\widetilde{X}$ is x. Let the joint probability density function of $\widetilde{X}$ be $f_{\widetilde{X}}(\cdot)$. The information capacity C of all the n parallel channels is

$$C = \max_{f_{\widetilde{X}}(x)} I_c\left(\widetilde{X}, \widetilde{Y}\right), \quad \text{where} \sum_{j=1}^{n} P_{pow_j} \leq P_{pow}$$

Let $a^+ = \max(0, a)$, $a \in \mathbb{R}$. Then the capacity of the set of parallel channels is

$$C \leq \sum_{j=1}^{n} \frac{1}{2} \log_2 \left(1 + \frac{P_{pow_j}}{\sigma_{N_j}^2}\right)$$

where $\sum_{j=1}^{n} P_{pow_j} = P_{pow}$.

Equality in the above result occurs if $X_j \sim \mathcal{N}\left(0, \sigma_{X_j}^2\right)$, where $\sigma_{X_j}^2 \in \mathbb{R}^+$ for $1 \leq j \leq n$; and the random variable X_j is independent of random variable X_k, for $j \neq k$, and $1 \leq j, k \leq n$. Further,

$$P_{pow_j} = \left(\theta - \sigma_{N_j}^2\right)^+, \quad \text{for } 1 \leq j \leq n$$

The value θ is chosen so that $\sum_{j=1}^{n} P_{pow_j} = P_{pow}$. □

See the problem section for details of the above observation.

1.8 Rate Distortion Theory

Rate distortion theory is the study of approximation of functions in the language of information theory. It addresses the problem and consequences of distortion that a signal undergoes, when it is transmitted over a communication channel. This theory also provides an information-theoretic foundation for lossy data compression. Rate distortion theory is applicable to both discrete and continuous random variables.

At an abstract level, the communication channel, and the data compression algorithm, are both considered as black boxes in this section. Typically, input-symbol probability distribution, and a distortion measure are specified. Based upon these specifications, the rate distortion theory determines the information or code rate that a channel must possess so that the average distortion does not exceed a threshold $D \in \mathbb{R}_0^+$. Alternately, based upon these specifications, and a specific rate, the rate distortion theory determines the achievable minimum expected distortion.

The basics of this topic are initially studied by examining the distortion introduced by quantization of a continuous signal. This is followed by a description of rate distortion function and its properties. Representation of independent Gaussian random variables is also discussed. Subsequently, a scheme is outlined to compute the rate distortion function.

1.8.1 Quantization

The basics of distortion theory are initially demonstrated via an example on quantization. Quantization is the process of representing a continuous and real-valued sample with finite number of bits. In principle, such continuous random sample would require infinite number of bits for its faithful representation. Therefore an inexact representation of a random sample with finite number of bits induces a distortion. The degree of approximation of the sample is specified by defining a distortion measure.

Assume that a single sample from a source has to be represented. This sample occurs with a certain probability distribution. Let the corresponding random variable be X. Denote its representation with finite number of bits by $\widehat{X}(X)$. Suppose that m number of bits are required to specify $\widehat{X}$. Therefore, $\widehat{X}$ takes 2^m number of possible values. The goal is to determine the optimum values of $\widehat{X}$ along with their domain values. The optimum values are based upon certain prespecified error metric.

For example, let X have a normal distribution with zero mean, and standard deviation σ. Also assume that only a single bit ($m = 1$) is available for its representation, and the expression

$$\mathcal{E}\left[\left(X - \widehat{X}(X)\right)^2\right]$$

is minimized. It is shown in the problem section, that

$$\widehat{X}(x) = \begin{cases} \sqrt{\dfrac{2}{\pi}}\sigma, & x \geq 0 \\ -\sqrt{\dfrac{2}{\pi}}\sigma, & x < 0 \end{cases}$$

The corresponding expected distortion is $((\pi - 2)/\pi)\,\sigma^2$.

1.8.2 Rate Distortion Function

Let the input or source alphabet for a communication channel be $A = \{x_1, x_2, \ldots, x_K\}$. Assume that the source is discrete, and memoryless. The corresponding symbol emission probabilities are $p(x_i) > 0$, $1 \leq i \leq K$. Note that, if $p(x_i)$ is equal to zero, then the letter x_i can be removed from the alphabet A. The input-symbol random variable is denoted by X. This random variable takes values from the input (source) alphabet A.

The letters (or symbols) from the source will in general appear as symbols from an output alphabet $B = \{\widehat{x}_1, \widehat{x}_2, \ldots, \widehat{x}_J\}$ at the destination. The output alphabet is sometimes also called a *representation* or *code word alphabet*. The output-symbol random variable is denoted by $\widehat{X}$. This random variable takes values from the output alphabet B. It is possible for the two alphabets A and B to be identical.

In the absence of distortion (in the case of data transmission), or in the absence of lossiness (in the case of data compression), there will be a perfect representation of an input symbol x_i by an output symbol $\widehat{x}_j$.

However, it is quite possible that a source symbol may not always be mapped to its proper output symbol. This might occur due to the rate of information being more than the capacity of the communication channel, or because channel is noisy. It might also occur in a lossy data compression process.

These situations might lead to errors and distortions in the representation of an input symbol x_i by an output symbol $\widehat{x}_j$. The corresponding distortion is denoted by $d(x_i, \widehat{x}_j) \in \mathbb{R}_0^+$. A communication model is initially described, to further study the distortion phenomenon.

Definition 1.16. *Consider a communication channel system. Let the input (source) alphabet be* $A = \{x_1, x_2, \ldots, x_K\}$, *and the output (representation or code word) alphabet be* $B = \{\widehat{x}_1, \widehat{x}_2, \ldots, \widehat{x}_J\}$. *The random variables corresponding to the input and output symbols are* X *and* $\widehat{X}$ *respectively. The random variable* X *takes values from the set* A, *and the random variable* $\widehat{X}$ *takes values from the set* B.

(a) *The input-symbol probability distribution is* $p(x_i)$, $1 \leq i \leq K$, *where the* $p(x_i)$*'s are positive.*
(b) *The output-symbol probability distribution is* $p(\widehat{x}_j)$, $1 \leq j \leq J$.
(c) *Transition probabilities. The probability of receiving a symbol* $\widehat{x}_j \in B$, *when a symbol* $x_i \in A$ *is transmitted is* $p(\widehat{x}_j \mid x_i)$, *where* $1 \leq i \leq K$, *and* $1 \leq j \leq J$.
(d) *Probability transition matrix. Let* $G = [g_{ij}]$ *be the* $K \times J$ *probability transition matrix, where* $g_{ij} = p(\widehat{x}_j \mid x_i)$; *for* $1 \leq i \leq K$, *and* $1 \leq j \leq J$.
(e) *A single symbol distortion measure or function* $d(\cdot, \cdot)$ *is*

$$d : A \times B \to \mathbb{R}_0^+ \tag{1.33}$$

That is, $d(x_i, \widehat{x}_j) \in \mathbb{R}_0^+$ *is a measure of the cost incurred in representing the source symbol* $x_i \in A$ *by the representation symbol* $\widehat{x}_j \in B$. *Generally,* $d(x_i, \widehat{x}_j) = 0$ *implies a faithful representation of* $x_i \in A$ *by* $\widehat{x}_j \in B$. □

Based upon the above definition, some elementary observations readily follow.

Observations 1.9. Some observations concerning probabilities.

1. The representation word probability is $p(\widehat{x}_j) = \sum_{i=1}^{K} p(x_i)\, p(\widehat{x}_j \mid x_i)$, where $1 \leq j \leq J$.

2. $\sum_{j=1}^{J} p(\widehat{x}_j \mid x_i) = 1$ for each i, where $1 \leq i \leq K$. Equivalently, the sum of probabilities in each row of the matrix G is equal to unity. □

A proper definition of average distortion measure is in order.

Definition 1.17. *Average distortion measure. Consider a communication channel system. The input alphabet is* $A = \{x_1, x_2, \ldots, x_K\}$, *and the output alphabet is* $B = \{\widehat{x}_1, \widehat{x}_2, \ldots, \widehat{x}_J\}$. *The input-symbol emission probabilities are* $p(x_i)$, $1 \leq i \leq K$. *The transition probabilities are* $p(\widehat{x}_j \mid x_i)$, *where* $1 \leq i \leq K$, *and* $1 \leq j \leq J$. *The transition probability matrix is* G.

Also, $d(x_i, \widehat{x}_j) \in \mathbb{R}_0^+$ *is the distortion induced by the output symbol* $\widehat{x}_j$ *when input symbol* x_i *is transmitted, where* $1 \leq i \leq K$, *and* $1 \leq j \leq J$. *The average distortion* $\overline{d}(G)$ *is*

$$\overline{d}(G) = \sum_{i=1}^{K} \sum_{j=1}^{J} p(x_i)\, p(\widehat{x}_j \mid x_i)\, d(x_i, \widehat{x}_j) \tag{1.34}$$

□

Examples of distortion measures are: *Hamming distortion*, and *squared error distortion*. Let $x \in A$, and $\widehat{x} \in B$. The Hamming distortion is

$$d(x, \widehat{x}) = \begin{cases} 0, & x = \widehat{x} \\ 1, & x \neq \widehat{x} \end{cases}$$

The squared error distortion is

$$d(x, \widehat{x}) = (x - \widehat{x})^2$$

Assume that the input-symbol emission probability distribution $\{p(x_i) \mid 1 \leq i \leq K\}$ is fixed. Also assume that $D \in \mathbb{R}_0^+$ is the largest possible average distortion that is permitted. All the possible conditional probability assignments (the $p(\widehat{x}_j \mid x_i)$'s) for a given set of input and output alphabets, are of interest. An assignment is considered to be D-admissible, if the average distortion $\overline{d}(G)$ is at most D. Let the set of all D-admissible transition matrices be

$$G_D = \{G \mid \overline{d}(G) \leq D\}$$

For each input-symbol probability distribution and D-admissible probability transition matrix $G \in G_D$, it is possible to have a mutual information, or information rate. The mutual information is

$$I\left(X; \widehat{X}\right) = \sum_{i=1}^{K} \sum_{j=1}^{J} p(x_i)\, p(\widehat{x}_j \mid x_i) \log_2 \frac{p(\widehat{x}_j \mid x_i)}{p(\widehat{x}_j)}$$

The rate distortion function is next defined.

Definition 1.18. *Rate distortion function. Consider a communication channel system. Let the input and output alphabets be* A *and* B *respectively. A single symbol distortion measure is* $d(\cdot, \cdot)$. *Let* $p(x_i)$*'s be the input-symbol emission probabilities,* $p(\widehat{x}_j)$*'s are the representation word probabilities, and* $p(\widehat{x}_j \mid x_i)$*'s are the channel or transition probabilities. The probability transition matrix is* G. *The average distortion is* $\overline{d}(G)$.

For a fixed $D \in \mathbb{R}_0^+$, the set of all D-admissible probability transition matrices is

$$G_D = \{G \mid \overline{d}(G) \leq D\} \tag{1.35a}$$

For each D-admissible probability transition matrix, there is a mutual information, or information rate. It is

$$I\left(X;\widehat{X}\right) = \sum_{i=1}^{K}\sum_{j=1}^{J} p(x_i)\, p(\widehat{x}_j \mid x_i) \log_2 \frac{p(\widehat{x}_j \mid x_i)}{p(\widehat{x}_j)} \tag{1.35b}$$

For a given value of D, and fixed source emission probability distribution $\{p(x_i) \mid 1 \leq i \leq K\}$, the rate distortion function $R(D)$ is the smallest mutual information with D-admissible probability transition matrix. That is,

$$R(D) = \min_{G \in G_D} I\left(X;\widehat{X}\right) \text{ bits per symbol} \tag{1.35c}$$

subject to $\sum_{j=1}^{J} p(\widehat{x}_j \mid x_i) = 1$ for each $i \in [1, K]$. □

The above definition of rate distortion function implies that, a larger value of D could possibly permit a lower information rate. For example, it is possible to have transmission, at a lower bit rate via lossy data compression. On the other hand, if the information rate of the channel is increased, lower distortion might be possible.

1.8.3 Properties of Rate Distortion Function

Properties of the rate distortion function $R(D)$ are examined in this subsection. A discrete memoryless information source is assumed for simplicity.

Recall that the rate distortion function is obtained by the minimization of mutual entropies, subject to an upper bound on the average distortion.

- Observe that the smallest possible value of mutual information is zero. Therefore the smallest possible value of $R(D)$ is zero.
- In addition, the maximum possible value of $R(D)$ should be equal to $H(X)$. This should occur at the smallest possible value of D. Let it be equal to D_{min}. It is established subsequently that $R(D_{min}) = H(X)$.

As the average distortion $\overline{d}(G) \leq D$, finding D_{min} is equivalent to finding the elements of the transition-probability matrix G which minimize $\overline{d}(G)$, where

$$\overline{d}(G) = \sum_{i=1}^{K}\sum_{j=1}^{J} p(x_i)\, p(\widehat{x}_j \mid x_i)\, d(x_i, \widehat{x}_j)$$

The average distortion $\overline{d}(G)$ is minimized, if for each source symbol x_i, an output symbol $\widehat{x}_j$ is chosen that minimizes the distortion $d(x_i, \widehat{x}_j)$ by assigning that transition probability $p(\widehat{x}_j \mid x_i) = 1$, and assigning all other transition probabilities the value of zero.

More concretely, let $j(x_i) = \arg\min_{1 \leq j \leq J} d(x_i, \widehat{x}_j)$. Therefore for the input symbol x_i, the output symbol $\widehat{x}_{j(x_i)}$ yields minimum distortion. Assign $p\left(\widehat{x}_{j(x_i)} \mid x_i\right) = 1$, and all other conditional probabilities $p(\widehat{x}_j \mid x_i) = 0$, for $j \neq j(x_i)$, where $1 \leq j \leq J$. Thus, for each x_i let

$$d(x_i) = \min_{1 \le j \le J} d(x_i, \widehat{x}_j), \quad 1 \le i \le K$$

Therefore

$$D_{min} = \sum_{i=1}^{K} p(x_i)\, d(x_i)$$

Without loss of generality, assume that $D_{min} = 0$. Observe that for each source symbol, the assignment of the output symbol is unique. That is, there is a perfect reconstruction of the output symbols (no uncertainty). Therefore $H\left(\widehat{X} \mid X\right) = 0$, which in turn implies $H\left(X \mid \widehat{X}\right) = 0$. Consequently, the mutual information in this case

$$I\left(X; \widehat{X}\right) = H(X), \quad \text{and} \quad R(D_{min}) = R(0) = H(X)$$

An expression for D_{max}, is next determined. It is the smallest possible value of D at which the rate distortion function takes a zero value. That is, D_{max} is the value of D such that $R(D) = 0$ for $D \ge D_{max}$. Thus $R(D_{max}) = 0$. In this case

$$I\left(X; \widehat{X}\right) = H(X) - H\left(X \mid \widehat{X}\right) = 0$$

This implies

$$H(X) = H\left(X \mid \widehat{X}\right)$$

This occurs, if the random variables X and $\widehat{X}$ are independent of each other. Therefore $p(\widehat{x}_j \mid x_i) = p(\widehat{x}_j)$. The corresponding average distortion is

$$\overline{d}(G) = \sum_{j=1}^{J} p(\widehat{x}_j) \left\{ \sum_{i=1}^{K} p(x_i)\, d(x_i, \widehat{x}_j) \right\}$$

As mentioned earlier, D_{max} is the smallest possible value of D at which the rate distortion function is equal to zero. Therefore, D_{max} is achieved by minimizing the right hand side of the above expression by varying the probability assignments $\{p(\widehat{x}_j) \mid 1 \le j \le J\}$. Let

$$j^* = \arg \min_{1 \le j \le J} \sum_{i=1}^{K} p(x_i)\, d(x_i, \widehat{x}_j)$$

The minimum value of $\overline{d}(G)$ occurs at $p(\widehat{x}_{j^*}) = 1$; and $p(\widehat{x}_j) = 0$, for $j \ne j^*$ and $1 \le j \le J$. Therefore

$$D_{max} = \min_{1 \le j \le J} \sum_{i=1}^{K} p(x_i)\, d(x_i, \widehat{x}_j)$$

The above discussion is summarized in the following observation.

Observation 1.10. The rate distortion function $R(D)$ varies between zero and $H(X)$ (the entropy of input symbols). That is

$$0 \le R(D) \le H(X) \tag{1.36}$$

The upper bound of $R(D)$ occurs at $D = D_{min}$, and $R(D_{min}) = H(X)$. Typically $D_{min} = 0$. The smallest value of D at which $R(D) = 0$ is D_{max}. That is, $R(D) = 0$ for $D \geq D_{max}$, where $D_{max} = \min_{1 \leq j \leq J} \sum_{i=1}^{K} p(x_i)\, d(x_i, \widehat{x}_j)$. □

It is next established that the rate distortion function $R(D)$ is a nonincreasing convex function of D. See Figure 1.5 for the plot of a typical rate distortion function. Note that, $R(D)$ is obtained by minimizing mutual information. Therefore, if the domain of the mutual information function is increased, then it is possible for its smallest value, the value of $R(D)$; to either remain same, or decrease. Consequently, $R(D)$ is a nonincreasing function of D.

Convexity of the rate distortion function is next established. Consider the rate distortion pair $(R(D'), D')$ and $(R(D''), D'')$ on the rate-distortion curve. Let the corresponding joint probability distributions be $p'(x, \widehat{x}) = p(x)\, p'(\widehat{x} \mid x)$ and $p''(x, \widehat{x}) = p(x)\, p''(\widehat{x} \mid x)$ respectively. Also consider a distribution $p^{\lambda}(x, \widehat{x}) = \lambda p'(x, \widehat{x}) + (1 - \lambda)\, p''(x, \widehat{x})$, where $\lambda \in [0, 1]$. As the distortion is a linear function of the distribution, $D^{\lambda}(p^{\lambda}) = \lambda D'(p') + (1 - \lambda)\, D''(p'')$. It is also known that the mutual information is a convex function of the conditional distribution $p(\widehat{x} \mid x)$, for fixed $p(x)$. Therefore

$$I^{\lambda}\left(X; \widehat{X}\right) \leq \lambda I'\left(X; \widehat{X}\right) + (1 - \lambda)\, I''\left(X; \widehat{X}\right)$$

Use of the definition of rate distortion function leads to

$$\begin{aligned} R(D^{\lambda}) &\leq I^{\lambda}\left(X; \widehat{X}\right) \\ &\leq \lambda I'\left(X; \widehat{X}\right) + (1 - \lambda)\, I''\left(X; \widehat{X}\right) \\ &= \lambda R(D') + (1 - \lambda)\, R(D'') \end{aligned}$$

Thus $R(D)$ is a convex function of D. The above discussion is summarized in the following observation.

Observation 1.11. The rate distortion function $R(D)$ is a nonincreasing convex function of D. □

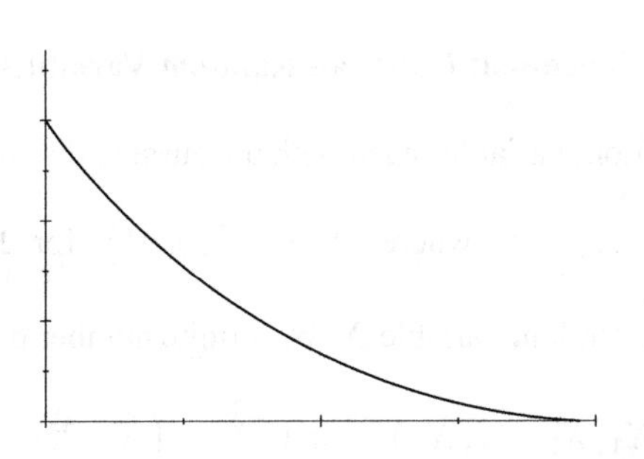

Figure 1.5. A typical rate distortion function $R(D)$ versus D.

It is also possible to determine upper and lower bounds for rate distortion function for a continuous random variable with zero mean and variance σ^2, where squared error distortion is assumed.

Observation 1.12. Consider a continuous random variable X, where $\mathcal{E}(X) = 0$, $Var(X) = \sigma^2$, and entropy equal to $H_c(X)$. For squared error distortion, and distortion threshold $D \in \mathbb{R}^{+}$, the rate distortion $R(D)$ is bounded as

$$H_c(X) - \frac{1}{2}\log_2(2\pi e D) \leq R(D) \leq \frac{1}{2}\log_2\frac{\sigma^2}{D} \tag{1.37}$$

□

The above observation is established in the problem section.

Observations 1.13. Explicit results on computation of rate distortion function.

1. Gaussian source. Consider a source which emits symbols at discrete instants of time. Furthermore, the source is memoryless. However the symbol values take continuous values, and have a Gaussian distribution with zero mean and variance σ^2. As the value of the input symbol $x \in \mathbb{R}$ is continuous, it needs to be discretized for either transmission or storage. Let $\widehat{x}$ be its discretized representative value, and it belongs to the representation alphabet B. Thus, $\widehat{x}$ is a quantized version of x.
   For, a Gaussian source $\mathcal{N}(0,\sigma^2)$ with a squared error distortion $d(x,\widehat{x}) = (x-\widehat{x})^2$, the rate distortion function is
   $$R(D) = \begin{cases} \frac{1}{2}\log_2\frac{\sigma^2}{D}, & 0 < D \leq \sigma^2 \\ 0, & D > \sigma^2 \end{cases}$$
2. Binary source. Consider a source which emits symbols at discrete instants of time. Furthermore, the source is memoryless, and the input symbols are binary. That is, the input alphabet $A = \{0,1\}$. The output alphabet B, is also equal to $\{0,1\}$. The input symbols have a Bernoulli distribution with parameter $\xi \in (0,1)$. For Hamming distortion, the rate distortion function is
   $$R(D) = \begin{cases} H(\xi) - H(D), & 0 \leq D \leq \min\{\xi, 1-\xi\} \\ 0, & D > \min\{\xi, 1-\xi\} \end{cases}$$
   where $H(X) = -\xi\log_2\xi - (1-\xi)\log_2(1-\xi) \triangleq H(\xi)$. □

1.8.4 Representation of Independent Gaussian Random Variables

Consider n independent random variables each with a Gaussian distribution. Let these be

$$X_1, X_2, \ldots, X_n; \quad \text{where} \quad X_i \sim \mathcal{N}(0,\sigma_i^2) \quad \text{for} \quad 1 \leq i \leq n$$

Let the representation of the random variable X_i by a finite number of bits be $\widehat{X}_i$. Also let

$$\widetilde{X} = (X_1, X_2, \ldots, X_n), \quad \text{and} \quad \widetilde{\widehat{X}} = \left(\widehat{X}_1, \widehat{X}_2, \ldots, \widehat{X}_n\right)$$

Let instances of the random vectors $\widetilde{X}$, and $\widetilde{\widehat{X}}$ be x, and $\widehat{x}$ respectively. The cumulative distortion measure is specified as

$$d(x,\widehat{x}) = \sum_{i=1}^{n}(x_i - \widehat{x}_i)^2$$

Further, a finite number of number of bits are prespecified to represent the random vector $\widetilde{X}$. How should these bits be allocated to different elements of the vector $\widetilde{X}$ so that the total distortion $D \in \mathbb{R}_0^+$ is minimized? For a fixed D, the rate distortion function for this vector case is

$$R(D) = \min_{f(\widehat{x}|x)} I_c\left(\widetilde{X}; \widetilde{\widehat{X}}\right), \text{ such that } \int_x \int_{\widehat{x}} f(\widehat{x} \mid x) f(x) d(x, \widehat{x}) \, d\widehat{x} \, dx \leq D$$

Observe that

$$\begin{aligned}
& I_c\left(\widetilde{X}; \widetilde{\widehat{X}}\right) \\
& = H_c\left(\widetilde{X}\right) - H_c\left(\widetilde{X} \mid \widetilde{\widehat{X}}\right) \\
& = \sum_{i=1}^{n} H_c(X_i) - \left\{ H_c\left(X_1 \mid \widetilde{\widehat{X}}\right) + H_c\left(X_2 \mid X_1, \widetilde{\widehat{X}}\right) + H_c\left(X_3 \mid X_1, X_2, \widetilde{\widehat{X}}\right) + \right. \\
& \quad \left. \ldots + H_c\left(X_n \mid X_1, X_2, \ldots, X_{n-1}, \widetilde{\widehat{X}}\right) \right\} \\
& \geq \sum_{i=1}^{n} H_c(X_i) - \sum_{i=1}^{n} H_c\left(X_i \mid \widehat{X}_i\right)
\end{aligned}$$

The last inequality follows because, conditioning might reduce entropy. Equality in the above result can be achieved by selecting $f(x \mid \widehat{x}) = \prod_{i=1}^{n} f(x_i \mid \widehat{x}_i)$. Continuing on

$$\begin{aligned}
I_c\left(\widetilde{X}; \widetilde{\widehat{X}}\right) & \geq \sum_{i=1}^{n} H_c(X_i) - \sum_{i=1}^{n} H_c\left(X_i \mid \widehat{X}_i\right) \\
& = \sum_{i=1}^{n} I_c\left(X_i; \widehat{X}_i\right) \geq \sum_{i=1}^{n} R(D_i)
\end{aligned}$$

In the last inequality, equality can be achieved by letting $\widehat{X}_i \sim \mathcal{N}\left(0, \sigma_i^2 - D_i\right)$, for $1 \leq i \leq n$. Use of the notation $a^+ = \max(0, a)$, for $a \in \mathbb{R}$ leads to

$$\begin{aligned}
I_c\left(\widetilde{X}; \widetilde{\widehat{X}}\right) & \geq \sum_{i=1}^{n} R(D_i) \\
& = \sum_{i=1}^{n} \left(\frac{1}{2} \log_2 \frac{\sigma_i^2}{D_i}\right)^+
\end{aligned}$$

Thus

$$R(D) = \min_{\sum_{i=1}^{n} D_i = D} \sum_{i=1}^{n} \left(\frac{1}{2} \log_2 \frac{\sigma_i^2}{D_i}\right)^+$$

This is a constrained minimization problem, which can be solved via the method of Lagrange multipliers. Let $\widetilde{D} = (D_1, D_2, \ldots, D_n)$. Consider the following nonlinear minimization problem.

$$\mathop{Minimize}_{\widetilde{D}} \sum_{i=1}^{n} \left(\frac{1}{2} \log_2 \frac{\sigma_i^2}{D_i}\right)^+$$

Subject to:

$$\sum_{i=1}^{n} D_i = D$$

where $D_i \geq 0$, for $1 \leq i \leq n$.

The Lagrangian functional of this problem is

$$\mathcal{L}\left(\widetilde{D}, \lambda\right) = \sum_{i=1}^{n} \left(\frac{1}{2} \log_2 \frac{\sigma_i^2}{D_i}\right)^+ + \lambda \left(\sum_{i=1}^{n} D_i - D\right)$$

where λ is a Lagrangian multiplier. This Lagrangian is minimized by varying the D_i's where $D_i \geq 0,\ 1 \leq i \leq n$. Consider two cases.

Case 1: $D_i < \sigma_i^2$. The corresponding Karush-Kuhn-Tucker conditions are:

$$\begin{aligned} \frac{\partial \mathcal{L}}{\partial D_i} &\geq 0, \quad 1 \leq i \leq n \\ D_i \frac{\partial \mathcal{L}}{\partial D_i} &= 0, \quad 1 \leq i \leq n \\ \frac{\partial \mathcal{L}}{\partial \lambda} &= 0 \\ D_i &\geq 0, \quad 1 \leq i \leq n \end{aligned}$$

Note that $\partial \mathcal{L}/\partial \lambda = 0$ implies $\sum_{i=1}^{n} D_i = D$. Let $\log_2 e \triangleq \varrho$, and assume $D_i > 0$.

$$\frac{\partial \mathcal{L}\left(\widetilde{D}, \lambda\right)}{\partial D_i} = -\frac{\varrho}{2D_i} + \lambda = 0$$

This implies

$$D_i = \beta$$

where $\beta = \varrho / (2\lambda)$ and $\lambda > 0$. Thus $D_i = \beta$, for $1 \leq i \leq n$. That is, $D_i = \beta$ if $\beta < \sigma_i^2$, for $1 \leq i \leq n$.

Case 2: $D_i \geq \sigma_i^2$. In this case

$$\frac{\partial \mathcal{L}}{\partial D_i} \leq 0, \quad 1 \leq i \leq n$$

The solution occurs on the boundary of the feasible domain of distortions. Select $D_i = \sigma_i^2$, if $\beta \geq \sigma_i^2$. The above discussion is encapsulated in the following observation.

Observation 1.14. Rate distortion function for parallel, discrete-time, memoryless, and independent Gaussian sources. Consider n independent random variables each with a Gaussian distribution. Let these be $X_1, X_2, \ldots, X_n$; where $X_i \sim \mathcal{N}\left(0, \sigma_i^2\right)$ for $1 \leq i \leq n$. Let the representation of the random variable X_i by a finite number of bits be $\widehat{X}_i$. The distortion measure is

$$d\left(x, \widehat{x}\right) = \sum_{i=1}^{n} \left(x_i - \widehat{x}_i\right)^2 \tag{1.38a}$$

Then the rate distortion function for a given $D \in \mathbb{R}_0^+$ is

$$R(D) = \sum_{i=1}^{n} \frac{1}{2} \log_2 \frac{\sigma_i^2}{D_i} \tag{1.38b}$$

where $D_i = \min\left(\beta, \sigma_i^2\right)$, for $1 \leq i \leq n$; and $\sum_{i=1}^{n} D_i = D$. □

1.8.5 Computation of Rate Distortion Function

The problem of computation of the rate distortion function $R(D)$ is addressed in this subsection. This problem is first formulated. Recall that the mutual information between the input-symbol and output-symbol random variables X and $\widehat{X}$ is

$$I\left(X;\widehat{X}\right)=\sum_{i=1}^{K}\sum_{j=1}^{J}p(x_i)\,p(\widehat{x}_j\mid x_i)\log_2\frac{p(\widehat{x}_j\mid x_i)}{p(\widehat{x}_j)}$$

where

$$p(\widehat{x}_j)=\sum_{i=1}^{K}p(x_i)\,p(\widehat{x}_j\mid x_i),\quad 1\le j\le J$$

Therefore

$$I\left(X;\widehat{X}\right)=\sum_{i=1}^{K}\sum_{j=1}^{J}p(x_i)\,p(\widehat{x}_j\mid x_i)\log_2\frac{p(\widehat{x}_j\mid x_i)}{\sum_{k=1}^{K}p(x_k)\,p(\widehat{x}_j\mid x_k)}\triangleq f(G)$$

The rate distortion function is obtained by minimizing the above expression for mutual information under the constraint that the average distortion is less than or equal to the distortion threshold D. Before this problem is addressed, the notation is modified for brevity. Let

$d(x_i,\widehat{x}_j)\triangleq d_{ij}\in\mathbb{R}_0^+$ for $1\le i\le K, 1\le j\le J$.
$p(x_i)\triangleq p_i$ for $1\le i\le K$, where $\sum_{i=1}^{K}p_i=1$. Let the row matrix representation of these probabilities be $\begin{bmatrix}p_1 \; p_2 \; \cdots \; p_K\end{bmatrix}\triangleq p$.
$p(\widehat{x}_j)\triangleq q_j$ for $1\le j\le J$, where $\sum_{j=1}^{J}q_j=1$. Let the row matrix representation of these probabilities be $\begin{bmatrix}q_1 \; q_2 \; \cdots \; q_J\end{bmatrix}\triangleq q$.
$p(\widehat{x}_j\mid x_i)\triangleq q_{j|i}$ for $1\le i\le K, 1\le j\le J$; where $\sum_{j=1}^{J}q_{j|i}=1, 1\le i\le K$.
$\sum_{k=1}^{K}p_k q_{j|k}=q_j$, for $1\le j\le J$.
The probability transition matrix is $G=[g_{ij}]$, where $g_{ij}=q_{j|i}$ for $1\le i\le K, 1\le j\le J$. Thus $pG=q$.
The mean distortion constraint is: $\overline{d}(G)=\sum_{i=1}^{K}\sum_{j=1}^{J}p_i q_{j|i}d_{ij}\le D$.
The mutual information between the input-symbol and output-symbol random variables X and $\widehat{X}$ in this notation is $f(G)$, where

$$f(G)\triangleq\sum_{i=1}^{K}\sum_{j=1}^{J}p_i q_{j|i}\log_2\frac{q_{j|i}}{\sum_{k=1}^{K}p_k q_{j|k}}$$

$$=\sum_{i=1}^{K}\sum_{j=1}^{J}p_i q_{j|i}\log_2\frac{q_{j|i}}{q_j}$$

Problem: Goal is to compute the rate distortion function $R(D)$. This is obtained by minimizing $f(G)$ over all matrices $G\in G_D$. Also $D\in\mathbb{R}_0^+$, and $d_{ij}\in\mathbb{R}_0^+$ for $1\le i\le K, 1\le j\le J$. The input symbol distribution, $\{p_i\mid p_i>0, 1\le i\le K\}$ is fixed, where $\sum_{i=1}^{K}p_i=1$.

$$\underset{G\in G_D}{\mathit{Minimize}}\; f(G)=\sum_{i=1}^{K}\sum_{j=1}^{J}p_i q_{j|i}\log_2\frac{q_{j|i}}{q_j}$$

Subject to:

$$\sum_{j=1}^{J} q_{j|i} = 1, \quad 1 \leq i \leq K$$

$$\overline{d}(G) = \sum_{i=1}^{K} \sum_{j=1}^{J} p_i q_{j|i} d_{ij} \leq D$$

where $q_{j|i} \geq 0$, $1 \leq i \leq K, 1 \leq j \leq J$. The minimum value of the objective function $f(G)$, is obtained by varying the $q_{j|i}$'s. The optimum conditional distribution, the $q_{j|i}$'s are first determined. The optimum $q_{j|i}$'s are subsequently substituted in the expression for $f(G)$ to yield $R(D)$. □

This is a nonlinear optimization problem with inequality and equality constraints, with nonnegative variables. The method of Lagrange multipliers is used to solve this problem. Let

$$(\nu_1, \nu_2, \ldots, \nu_K) \triangleq \nu$$

The Lagrangian functional for this problem is

$$\mathcal{L}(G, \nu, \lambda) = f(G) + \lambda \{\overline{d}(G) - D\} + \sum_{i=1}^{K} \nu_i \left\{ \sum_{j=1}^{J} q_{j|i} - 1 \right\}$$

In the above Lagrangian functional, λ and the ν_i's are called the Lagrange multipliers. Observe that for fixed p_i's the mutual information $f(G)$ is a continuously differentiable, and convex function of the $q_{j|i}$'s. Also $\{\overline{d}(G) - D\}$ and the $\left\{\sum_{j=1}^{J} q_{j|i} - 1\right\}$'s are linear functions of the $q_{j|i}$'s. Therefore, the Lagrangian is minimized by varying the $q_{j|i}$'s, where $q_{j|i} \geq 0, 1 \leq i \leq K, 1 \leq j \leq J$. At the optimum values of the $q_{j|i}$'s, the Karush-Kuhn-Tucker (KKT) conditions are:

$$\frac{\partial \mathcal{L}}{\partial q_{j|i}} \geq 0, \quad 1 \leq i \leq K, \ 1 \leq j \leq J$$

$$q_{j|i} \frac{\partial \mathcal{L}}{\partial q_{j|i}} = 0, \quad 1 \leq i \leq K, \ 1 \leq j \leq J$$

$$\overline{d}(G) \leq D, \quad \lambda \geq 0, \quad \lambda \{\overline{d}(G) - D\} = 0$$

$$\sum_{j=1}^{J} q_{j|i} = 1, \quad 1 \leq i \leq K$$

$$q_{j|i} \geq 0, \quad 1 \leq i \leq K, \ 1 \leq j \leq J$$

Next compute $\partial \mathcal{L} / \partial q_{j|i}$, where $1 \leq i \leq K, 1 \leq j \leq J$. Let $\log_2 e \triangleq \varrho$. Observe that

$$\frac{\partial f(G)}{\partial q_{j|i}} = p_i \log_2 q_{j|i} + \varrho p_i - p_i \log_2 q_j - \left\{ \sum_{i=1}^{K} p_i q_{j|i} \right\} \frac{\varrho p_i}{q_j} = p_i \log_2 \frac{q_{j|i}}{q_j}$$

$$\frac{\partial \overline{d}(G)}{\partial q_{j|i}} = p_i d_{ij}$$

$$\frac{\partial}{\partial q_{j|i}}\left\{\sum_{j=1}^{J} q_{j|i} - 1\right\} = 1$$

Therefore

$$\frac{\partial \mathcal{L}}{\partial q_{j|i}} = p_i\left\{\log_2 \frac{q_{j|i}}{q_j} + \lambda d_{ij} + \frac{\nu_i}{p_i}\right\}$$

Let $\nu_i/p_i \triangleq \log_2 \mu_i$, for $1 \le i \le K$. This leads to

$$\frac{\partial \mathcal{L}}{\partial q_{j|i}} = p_i\left\{\log_2 \frac{q_{j|i}}{q_j} + \lambda d_{ij} + \log_2 \mu_i\right\}, \quad 1 \le i \le K,\ 1 \le j \le J$$

Also

$$\frac{\partial \mathcal{L}}{\partial \lambda} = \overline{d}\,(G) - D$$

$$\frac{\partial \mathcal{L}}{\partial \nu_i} = \left\{\sum_{j=1}^{J} q_{j|i} - 1\right\}, \quad 1 \le i \le K$$

Next assume that $\partial \mathcal{L}/\partial q_{j|i} = 0$ and $q_{j|i} \ge 0$. This leads to

$$q_{j|i} = \frac{q_j}{\mu_i}\exp\left(-\lambda d_{ij}/\varrho\right)$$

Also let $\partial \mathcal{L}/\partial \nu_i = 0$. This implies $\sum_{j=1}^{J} q_{j|i} = 1$. Combining this condition, and the above expression for $q_{j|i}$ yield

$$\mu_i = \sum_{l=1}^{J} q_l \exp\left(-\lambda d_{il}/\varrho\right)$$

$$q_{j|i} = \frac{q_j \exp\left(-\lambda d_{ij}/\varrho\right)}{\sum_{l=1}^{J} q_l \exp\left(-\lambda d_{il}/\varrho\right)}$$

Substitution of the above expression for $q_{j|i}$ in the result $q_j = \sum_{i=1}^{K} p_i q_{j|i}$ leads to

$$q_j = q_j \sum_{i=1}^{K} p_i \frac{\exp\left(-\lambda d_{ij}/\varrho\right)}{\sum_{l=1}^{J} q_l \exp\left(-\lambda d_{il}/\varrho\right)}, \quad 1 \le j \le J$$

If $q_j > 0$, then

$$\sum_{i=1}^{K} p_i \frac{\exp\left(-\lambda d_{ij}/\varrho\right)}{\sum_{l=1}^{J} q_l \exp\left(-\lambda d_{il}/\varrho\right)} = 1, \quad 1 \le j \le J$$

Thus there are J number of equations. However there are $(J+1)$ number of unknowns, which are q_j, $1 \le j \le J$ and λ. Another relationship can be obtained by letting $\partial \mathcal{L}/\partial \lambda = 0$. That is, $\left\{D - \overline{d}\,(G)\right\} = 0$. Therefore

$$D = \overline{d}\,(G) = \sum_{i=1}^{K}\sum_{j=1}^{J} p_i q_{j|i} d_{ij}$$

$$= \sum_{i=1}^{K}\sum_{j=1}^{J} p_i d_{ij} \frac{q_j \exp\left(-\lambda d_{ij}/\varrho\right)}{\sum_{l=1}^{J} q_l \exp\left(-\lambda d_{il}/\varrho\right)}$$

Using the above analysis, steps needed to determine the elements of the transition matrix G, and the rate distortion function $R(D)$ are summarized.

Summary of results for determining the optimal transition probabilities and the rate distortion function

Input:
The input-symbol emission probabilities: p_i, $1 \le i \le K$; where p_i's are positive.
Also $\sum_{i=1}^{K} p_i = 1$.
The distortion values: $d_{ij} \in \mathbb{R}_0^+$, where $1 \le i \le K, 1 \le j \le J$.
The distortion threshold $D \in \mathbb{R}_0^+$. Let $\log_2 e \triangleq \varrho$.

Assumptions:
The transition probabilities $q_{j|i} > 0$, for $1 \le i \le K, 1 \le j \le J$.
Also $\sum_{j=1}^{J} q_{j|i} = 1$ for $1 \le i \le K$.

Algorithm:

Step 1: We have

$$\sum_{i=1}^{K} p_i \frac{\exp\left(-\lambda d_{ij}/\varrho\right)}{\sum_{l=1}^{J} q_l \exp\left(-\lambda d_{il}/\varrho\right)} = 1, \quad 1 \le j \le J \tag{1.39a}$$

$$D = \overline{d}(G) = \sum_{i=1}^{K}\sum_{j=1}^{J} p_i d_{ij} \frac{q_j \exp\left(-\lambda d_{ij}/\varrho\right)}{\sum_{l=1}^{J} q_l \exp\left(-\lambda d_{il}/\varrho\right)} \tag{1.39b}$$

From the above $(J+1)$ number of equations, determine the $(J+1)$ unknowns; q_j, $1 \le j \le J$; and λ. This is usually done via an iterative procedure.

Step 2: Having determined, q_j, $1 \le j \le J$; and λ in the last step, the transition probabilities are computed as

$$q_{j|i} = \frac{q_j \exp\left(-\lambda d_{ij}/\varrho\right)}{\sum_{l=1}^{J} q_l \exp\left(-\lambda d_{il}/\varrho\right)}, \quad 1 \le i \le K,\ 1 \le j \le J \tag{1.39c}$$

Step 3: Finally, the rate distortion function is determined as

$$R(D) = \sum_{i=1}^{K}\sum_{j=1}^{J} p_i q_{j|i} \log_2 \frac{q_{j|i}}{q_j} \tag{1.39d}$$

□

In the above algorithm, the transition probabilities, $q_{j|i}$'s were assumed to be positive. It is possible for these probabilities to be equal to zero. As per the KKT conditions

$$\begin{aligned} \frac{\partial \mathcal{L}}{\partial q_{j|i}} &= 0, \text{ if } q_{j|i} > 0 \\ &\ge 0, \text{ if } q_{j|i} = 0 \end{aligned}$$

Subsequent analysis leads to

$$\sum_{i=1}^{K} p_i \frac{\exp\left(-\lambda d_{ij}/\varrho\right)}{\sum_{l=1}^{J} q_l \exp\left(-\lambda d_{il}/\varrho\right)} = 1, \text{ if } q_j > 0$$
$$\leq 1, \text{ if } q_j = 0$$

Computation of a parametrized rate distortion function

A parametrized rate distortion function is determined. The parameter is the Lagrangian multiplier λ, from the earlier analysis. Assume that a specific value of $\lambda \geq 0$ is given. Then determine q_l, $1 \leq l \leq J$ from the following J equations.

$$\sum_{i=1}^{K} p_i \frac{\exp\left(-\lambda d_{ij}/\varrho\right)}{\sum_{l=1}^{J} q_l \exp\left(-\lambda d_{il}/\varrho\right)} = 1, \quad 1 \leq j \leq J$$

Once q_j, for $1 \leq j \leq J$ are determined, $q_{j|i}$'s can be evaluated from

$$q_{j|i} = \frac{q_j \exp\left(-\lambda d_{ij}/\varrho\right)}{\sum_{l=1}^{J} q_l \exp\left(-\lambda d_{il}/\varrho\right)}, \quad 1 \leq i \leq K, \ 1 \leq j \leq J$$

The $q_{j|i}$'s can also be expressed as

$$q_{j|i} = \frac{q_j}{\mu_i} \exp\left(-\lambda d_{ij}/\varrho\right), \text{ where } \mu_i = \sum_{l=1}^{J} q_l \exp\left(-\lambda d_{il}/\varrho\right)$$

Also

$$\overline{d}\left(G\right) = \sum_{i=1}^{K} \sum_{j=1}^{J} p_i q_{j|i} d_{ij}$$

The minimized value of the mutual information is the rate distortion function $R\left(D\right)$. The parameterized rate distortion function $R\left(D\right)$ is

$$\begin{aligned}
R\left(D\right) &= \sum_{i=1}^{K} \sum_{j=1}^{J} p_i q_{j|i} \log_2 \frac{q_{j|i}}{q_j} \\
&= \sum_{i=1}^{K} \sum_{j=1}^{J} p_i q_{j|i} \log_2 \frac{\exp\left(-\lambda d_{ij}/\varrho\right)}{\mu_i} \\
&= \sum_{i=1}^{K} \sum_{j=1}^{J} p_i q_{j|i} \left\{-\lambda d_{ij} - \log_2 \mu_i\right\} \\
&= -\lambda \overline{d}\left(G\right) - \sum_{i=1}^{K} p_i \log_2 \mu_i \sum_{j=1}^{J} q_{j|i} \\
&= -\lambda \overline{d}\left(G\right) - \sum_{i=1}^{K} p_i \log_2 \mu_i
\end{aligned}$$

The constrained minimization of the mutual information requires that

$$D = \overline{d}(G) = \sum_{i=1}^{K} \sum_{j=1}^{J} p_i q_{j|i} d_{ij}$$

Therefore

$$R(D) = -\lambda D - \sum_{i=1}^{K} p_i \log_2 \mu_i$$

Once, the parameter λ is determined, the rate distortion function $R(D)$ can be determined from the above expression. See the problem section for the determination of the rate distortion function of a binary source. It uses the above computational approach.

1.9 Network Information Theory

Network information theory studies systems in which there are multiple transmitters and receivers. These give rise to issues such as feedback, and constructive and destructive interference among different channels. The communication between different entities (transmitters and receivers) in a network can be described by a probability transition matrix. The aim of network information theory is to determine the possibility of transmission between different transmitters and receivers. This problem in its complete generality has not yet been addressed successfully by the research community. In this chapter, we only discuss two elementary examples of network information theory. These are the multi-access Gaussian communication channel and the Gaussian broadcast channels.

1.9.1 Multi-access Gaussian Communication Channel

The multi-access Gaussian communication channel is a generalization of the single Gaussian communication channel discussed in the last section. This model has several transmitters and a single receiver. Consider m channels, where a single transmitter is associated with each channel. See Figure 1.6. It is assumed that each transmitter has its own code $\mathcal{C}_j, 1 \le j \le m$ and acts independently of the other $(m-1)$ transmitters. Let ϑ_j be the code rate of the jth code, where $1 \le j \le m$. Furthermore, the codeword-length of each code is equal to n. Let $X(j)$ be the Gaussian input from transmitter j, where $\mathcal{E}(X(j)) = 0$, and $Var(X(j)) = P_{pow} < \infty$, for $1 \le j \le m$. Let

$$Y = \sum_{j=1}^{m} X(j) + N$$

where the random variable N represents the effect of additive noise and Y is the signal observed at the receiver. Assume that this random variable N has a Gaussian distribution with zero mean, and variance equal to σ_N^2. For simplicity in notation, denote the signal-to-noise ratio P_{pow}/σ_N^2 by x. Therefore, in this notation the capacity of a single user Gaussian channel with a signal-to-noise ratio of x is

$$C(x) = \frac{1}{2} \log_2 (1 + x) \tag{1.40}$$

The feasible region of operation of the multi-access Gaussian communication channel is

$$\vartheta_i < C(x), \quad 1 \le i \le m \tag{1.41a}$$

$$\vartheta_i + \vartheta_j < C(2x), \quad i \ne j, \quad 1 \le i, j \le m \tag{1.41b}$$

$$\vartheta_i + \vartheta_j + \vartheta_k < C(3x), \quad i \neq j \neq k, \; 1 \leq i, j, k \leq m \tag{1.41c}$$

$$\vdots$$

$$\sum_{j=1}^{m} \vartheta_j < C(mx) \tag{1.41d}$$

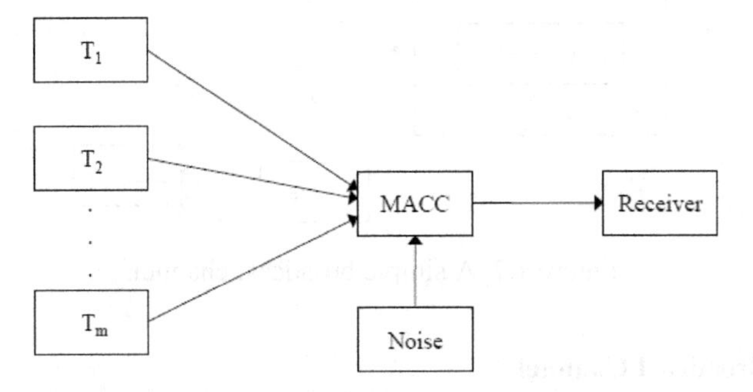

Figure 1.6. Multi-access communication channel (MACC).

Observe that if all the code rates are equal, the last inequality is more dominant than others. Let the receiver observe a vector $\Upsilon = (Y_1, Y_2, \ldots, Y_n)$ of length n. Optimal decoding is performed by searching for the m codewords, one from each code (or channel), and using the minimum distance decoding rule on the vector sum of these m codewords, and the received vector Υ. Therefore, if the code rates $\vartheta_j, 1 \leq j \leq m$ satisfy the above set of constraints, the probability of decoding error goes to zero, as $n \to \infty$.

1.9.2 Gaussian Broadcast Channels

A broadcast channel has a single transmitter, and two or more receivers. Two types of broadcast channels are studied in this subsection: a simple broadcast channel, and a degraded broadcast channel. In each of these configurations, there is a single transmitter and two receivers.

A Simple Broadcast Channel

See Figure 1.7. Let X be the random variable which represents the broadcast signal, where $\mathcal{E}(X) = 0$, and $Var(X) = P_{pow} < \infty$. Let the Gaussian random variables which represent noise on the two channels be N_1 and N_2 respectively, where $\mathcal{E}(N_i) = 0, Var(N_i) = \sigma^2_{N_i}$, and $i \in \{1, 2\}$. Furthermore, the random variables N_1, and N_2 are arbitrarily correlated. The output of the corresponding broadcast channels is Y_1 and Y_2 respectively. Thus

$$Y_1 = (X + N_1)$$
$$Y_2 = (X + N_2)$$

The corresponding individual capacities of the two communication channels are

$$C_1 = \frac{1}{2}\log_2\left(1 + \frac{P_{pow}}{\sigma^2_{N_1}}\right) \tag{1.42a}$$

$$C_2 = \frac{1}{2}\log_2\left(1 + \frac{P_{pow}}{\sigma^2_{N_2}}\right) \tag{1.42b}$$

Therefore, if the code rate is below both these capacities, the probability of decoding error goes to zero, as the length of the codeword goes to infinity.

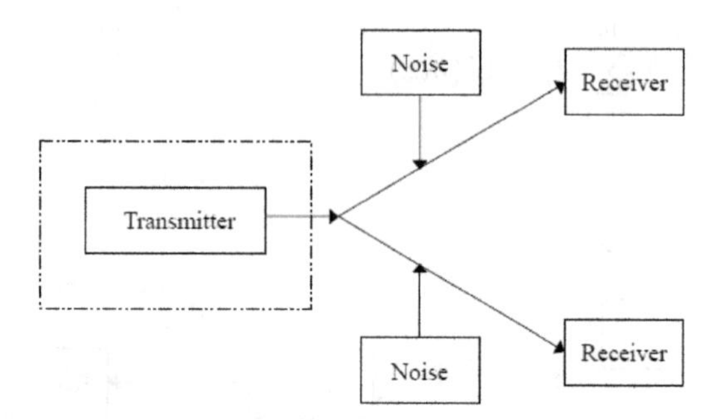

Figure 1.7. A simple broadcast channel.

A Degraded Broadcast Channel

Next consider a broadcast scenario, in which the single transmitter works with two distinct codes $\mathcal{C}_1$ and $\mathcal{C}_2$, with the code rates ϑ_1 and ϑ_2 respectively. Therefore a given transmitted message X consists of a superposition of two messages X_1 and X_2, such that $X = (X_1 + X_2)$. The message X_i uses codewords from the code $\mathcal{C}_i$, and is destined for receiver-i, where $i \in \{1, 2\}$. The average transmitted power associated with the messages X_1 and X_2, are αP_{pow} and $(1 - \alpha) P_{pow}$ respectively, where $\alpha \in [0, 1]$. Let the Gaussian noise on the two channels be N_1 and N_2 respectively, such that $\mathcal{E}(N_i) = 0, Var(N_i) = \sigma_{N_i}^2$, and $i \in \{1, 2\}$. The random variables N_1, and N_2 are arbitrarily correlated. Without loss of generality assume that $\sigma_{N_1}^2 < \sigma_{N_2}^2$. Let the output of the two broadcast channels be Y_1 and Y_2. The broadcast channels are configured such that

$$Y_1 = (X + N_1)$$
$$Y_2 = (X + N_2)$$

Let $C_1(\alpha)$ and $C_2(\alpha)$ be the channel capacities of the two channels. The code rate ϑ_i is selected such that $\vartheta_i < C_i(\alpha)$, for $i \in \{1, 2\}$. These code rates ϑ_1 and ϑ_2 are varied by selecting appropriate value of α in the interval $[0, 1]$.

The decoding is done as follows. The receiver-1 decodes the codeword destined for receiver-2. This is possible, because $\sigma_{N_1}^2 < \sigma_{N_2}^2$. Let this decoded codeword be Ξ_2. Also let the received codeword at this receiver be Υ_1. The receiver-1 then subtracts the decoded codeword Ξ_2 from Υ_1. The receiver-1 then looks for a codeword in the first code $\mathcal{C}_1$ which is closest to $(\Upsilon_1 - \Xi_2)$. The probability of this decoding error can be made as small as possible. For receiver-2 the effective noise variance is $\left(\alpha P_{pow} + \sigma_{N_2}^2\right)$. Thus

$$C_1(\alpha) = \frac{1}{2}\log_2\left\{1 + \frac{\alpha P_{pow}}{\sigma_{N_1}^2}\right\} \tag{1.43a}$$

$$C_2(\alpha) = \frac{1}{2}\log_2\left\{1 + \frac{(1-\alpha) P_{pow}}{\left(\alpha P_{pow} + \sigma_{N_2}^2\right)}\right\} \tag{1.43b}$$

A pleasant outcome of this scheme is that, in addition to the message destined for itself, the receiver-1 with lower noise variance always knows about the message destined for its more noisy counterpart.

1.10 Maximum Entropy Principle

The principle of maximum entropy (ME) can be used in estimating the parameters of a probability distribution. Given a set of statistics derived from the collected data, this principle makes minimal number of assumptions in estimating the parameters of a probability distribution.

In this section, we also explore the relationship between the methods of maximum entropy and maximum-likelihood (ML) to estimate the parameters of a probability distribution.

1.10.1 Maximum-Likelihood Estimation

Maximum-likelihood estimation is a method for deriving the parameters of a probability distribution from a set of observations (data points) drawn from that distribution. This data is assumed to be most likely (probable).

Let $\mathcal{Y} = \{y_i \mid y_i \in \mathbb{R}, 1 \leq i \leq m\}$ be the set of observations. Assume that the m observations are independent of each other. We consider both discrete and continuous random variables. Let $\theta = (\theta_1, \theta_2, \ldots, \theta_r) \in \mathbb{R}^r$ be the vector of parameters of the probability distribution which are to be estimated in either case.

Continuous Random Variable

Let $y \in \mathbb{R}$ be a data point observed from a probability density function $g(y, \theta)$. The function

$$\ell(\theta, \mathcal{Y}) = \prod_{i=1}^{m} g(y_i, \theta) \tag{1.44}$$

is called the *likelihood function*. The likelihood function is considered a function of θ, with the set of m observations $\mathcal{Y}$ fixed. Therefore, $\ell(\theta, \mathcal{Y})$ is often simply expressed as $\ell_{\mathcal{Y}}(\theta)$. The maximum-likelihood estimate of the parameters is the value of θ which maximizes the likelihood function. Thus

$$\theta_{ML} = \arg\max_{\theta} \{\ell_{\mathcal{Y}}(\theta)\} \tag{1.45a}$$

It is often convenient to determine θ_{ML} from maximizing the value of $\ln \ell_{\mathcal{Y}}(\theta) \triangleq L_{\mathcal{Y}}(\theta)$. The function $L_{\mathcal{Y}}(\theta)$ is called the log-likelihood function. As logarithm is a monotonically increasing function, maximizing $\ell_{\mathcal{Y}}(\cdot)$ is equivalent to maximizing $L_{\mathcal{Y}}(\cdot)$. Thus

$$\theta_{ML} = \arg\max_{\theta} \{L_{\mathcal{Y}}(\theta)\} \tag{1.45b}$$

If $L_{\mathcal{Y}}(\theta)$ is a continuously differentiable function of the parameters θ, and θ_{ML} lies in the interior range, then a necessary, though not sufficient condition for maximization is

$$\frac{\partial L_{\mathcal{Y}}(\theta)}{\partial \theta_j} = 0, \quad 1 \leq j \leq r$$

In its complete generality, the log-likelihood function can possess more than a single maxima, or it may not have any solution at all. Further, the log-likelihood function may not also be differentiable, or it might be numerically challenging to determine θ_{ML}.

Discrete Random Variable

Let X be a discrete random variable, which takes real values $x_j, j = 1, 2, \ldots, n$. Also $P(X = x_j) \triangleq p(x_j), j = 1, 2, \ldots, n$ are the corresponding probabilities, such that $0 \leq p(x_j) \leq 1$, and $\sum_{j=1}^{n} p(x_j) = 1$. The vector of parameters of this discrete random variable is $\theta \in \mathbb{R}^r$. This is to be estimated from the set of data points $\mathcal{Y}$, where $|\mathcal{Y}| = m$.

The empirical estimates of the probabilities, $\widetilde{p}(x_j)$, for $1 \leq j \leq n$, can be estimated from the set of data points $\mathcal{Y}$ as follows. Define δ_{ab} equal to unity if $a = b$, and equal to zero if $a \neq b$. Then

$$\widetilde{p}(x_j) = \frac{1}{m} \sum_{y \in \mathcal{Y}} \delta_{yx_j}, \quad 1 \leq j \leq n \tag{1.46a}$$

Note that in the above equation, $\sum_{y \in \mathcal{Y}} \delta_{yx_j}$ is equal to the number of times x_j occurs in the set of data points $\mathcal{Y}$. For a given set of data points $\mathcal{Y}$, the likelihood function is $\ell_{\mathcal{Y}}(\theta)$. It is

$$\ell_{\mathcal{Y}}(\theta) = \prod_{j=1}^{n} p(x_j, \theta)^{m\widetilde{p}(x_j)} \tag{1.46b}$$

The corresponding log-likelihood function is $\ln \ell_{\mathcal{Y}}(\theta) \triangleq L_{\mathcal{Y}}(\theta)$. As in the case of continuous random variable, the maximum-likelihood estimate of the parameters is the value of θ which maximizes the log-likelihood function. Thus $\theta_{ML} = \arg\max_{\theta} \{L_{\mathcal{Y}}(\theta)\}$. It is obtained via the optimization problem

$$\max_{\theta} \sum_{j=1}^{n} \widetilde{p}(x_j) \ln p(x_j, \theta) \tag{1.46c}$$

1.10.2 Maximum Entropy Estimation

Method of maximum entropy (ME) estimation is discussed in this subsection. We restrict our attention to the estimation of parameters of a discrete random variable. Extension of the technique to continuous random variables is straightforward.

Let X be a discrete random variable, which takes real values $x_j, j = 1, 2, \ldots, n$. Also $P(X = x_j) \triangleq p(x_j), j = 1, 2, \ldots, n$ are the corresponding probabilities, such that $0 \leq p(x_j) \leq 1$, and $\sum_{j=1}^{n} p(x_j) = 1$. Assume that the random variable X is specified by $r \in \mathbb{P}$ number of parameters. Let $\theta \in \mathbb{R}^r$ be the vector of parameters of the probability distribution which are to be estimated. The entropy of the random variable X with logarithm in base e (Euler's number) is denoted by

$$H_e(X) = -\sum_{j=1}^{n} p(x_j) \ln p(x_j)$$

The properties of the random variable X are evaluated by a set of features (observables, measures). Let this set of features be $\{f_i \mid 1 \leq i \leq r\}$. Also let $f_i(x_j)$ be the value of $f_i(\cdot)$, at x_j, where $1 \leq j \leq n$ and $1 \leq i \leq r$. The expected value of the random variable $f_i(X)$ is

$$\mathcal{E}(f_i(X)) = \sum_{j=1}^{n} f_i(x_j) p(x_j), \quad 1 \leq i \leq r$$

The measured value of $\mathcal{E}\left(f_i\left(X\right)\right)$ is $\overline{f}_i \in \mathbb{R}$, for $1 \leq i \leq r$. That is,

$$\sum_{j=1}^{n} f_i\left(x_j\right) \widetilde{p}\left(x_j\right) = \overline{f}_i, \quad 1 \leq i \leq r$$

The probabilities $p\left(x_j\right), 1 \leq j \leq n$ are determined by maximizing the entropy $H_e\left(X\right)$, subject to certain constraints. The corresponding optimization problem is

$$\underset{p(x_j), 1 \leq j \leq n}{\textit{Maximize}} \; H_e\left(X\right) = -\sum_{j=1}^{n} p\left(x_j\right) \ln p\left(x_j\right)$$

Subject to:

$$\sum_{j=1}^{n} p\left(x_j\right) = 1$$

$$\sum_{j=1}^{n} f_i\left(x_j\right) p\left(x_j\right) = \overline{f}_i, \quad 1 \leq i \leq r$$

and $p\left(x_j\right) \geq 0, 1 \leq j \leq n$. This nonlinear optimization problem is called $\mathcal{P}_{ME}$ for future reference. We address this problem by using the method of Lagrange multipliers. The nonnegativity constraints of the probabilities, $p\left(x_j\right) \geq 0$, for $1 \leq j \leq n$ are initially ignored. Let $\alpha \in \mathbb{R}$, and $\beta = \left(\beta_1, \beta_2, \ldots, \beta_r\right) \in \mathbb{R}^r$ be the Lagrange multipliers. The Lagrangian is

$$\mathcal{L}\left(p, \alpha, \beta\right) = H_e\left(X\right) + \alpha \left\{ \sum_{j=1}^{n} p\left(x_j\right) - 1 \right\} + \sum_{i=1}^{r} \beta_i \left\{ \sum_{j=1}^{n} f_i\left(x_j\right) p\left(x_j\right) - \overline{f}_i \right\}$$

Taking partial derivatives of $\mathcal{L}\left(p, \alpha, \beta\right)$ with respect to $p\left(x_j\right), 1 \leq j \leq n$, α, and $\beta_i, 1 \leq i \leq r$ yields

$$\frac{\partial \mathcal{L}}{\partial p\left(x_j\right)} = -\ln p\left(x_j\right) - 1 + \alpha + \sum_{i=1}^{r} \beta_i f_i\left(x_j\right), \quad 1 \leq j \leq n$$

$$\frac{\partial \mathcal{L}}{\partial \alpha} = \sum_{j=1}^{n} p\left(x_j\right) - 1$$

$$\frac{\partial \mathcal{L}}{\partial \beta_i} = \sum_{j=1}^{n} f_i\left(x_j\right) p\left(x_j\right) - \overline{f}_i, \quad 1 \leq i \leq r$$

Equate the above partial derivatives to zero, and define

$$\mathcal{H}\left(x_j\right) = \sum_{i=1}^{r} \beta_i f_i\left(x_j\right), \quad 1 \leq j \leq n$$

This yields $p\left(x_j\right) = e^{\alpha - 1 + \mathcal{H}(x_j)}$, $1 \leq j \leq n$. Further, it fortunately turns out that $p\left(x_j\right) \geq 0$, for $1 \leq j \leq n$. As $\sum_{j=1}^{n} p\left(x_j\right) = 1$, we obtain

$$p(x_j) = \frac{e^{\mathcal{H}(x_j)}}{Z(\beta)}, \quad \text{where} \quad Z(\beta) = \sum_{j=1}^{n} e^{\mathcal{H}(x_j)}, \quad 1 \le j \le n \tag{1.47}$$

and

$$\ln Z(\beta) = (1 - \alpha)$$

The vector β is selected so that the constraints are satisfied. Because of the form of the $p(x_j)$'s, such distributions belong to the family of exponential distributions. Let w be another quantity of interest in the model. The expected value of this quantity is given by

$$\mathcal{E}(w(X)) = \sum_{j=1}^{n} w(x_j) p(x_j)$$

Computation of any $\mathcal{E}(w(X))$ is generally hard in this model. However, it can be shown that

$$\mathcal{E}(f_i(X)) = \sum_{j=1}^{n} f_i(x_j) p(x_j) = \frac{\partial \ln Z(\beta)}{\partial \beta_i}, \quad 1 \le i \le r$$

Relationship between ML and ME Estimation Methods

We next invoke the duality theory of constrained nonlinear optimization to relate the ML and ME methods of parameter estimation of a probability distribution. If the constrained optimization problem $\mathcal{P}_{ME}$ stated earlier in the subsection is primal, then its dual $\mathcal{D}_{ME}$ is obtained as follows. In the expression for the Lagrangian $\mathcal{L}(p, \alpha, \beta)$, substitute $p(x_j) = e^{\mathcal{H}(x_j)}/Z(\beta)$, for $1 \le j \le n$. This yields the expression $\Psi(\beta)$, where

$$\Psi(\beta) = \ln Z(\beta) - \sum_{i=1}^{r} \beta_i \overline{f}_i$$

The dual optimization problem $\mathcal{D}_{ME}$ is thus

$$\min_{\beta} \Psi(\beta)$$

where $\beta \in \mathbb{R}^r$. It turns out that this is an unconstrained optimization minimization problem. Equivalently, this optimization problem can be stated as a maximization problem $\mathcal{D}'_{ME}$.

$$\max_{\beta} \left\{ -\ln Z(\beta) + \sum_{i=1}^{r} \beta_i \overline{f}_i \right\}$$

The objective function of this optimization problem $\{-\ln Z(\beta) + \sum_{i=1}^{r} \beta_i \overline{f}_i\}$ can be rewritten by using the expressions $\sum_{j=1}^{n} f_i(x_j) \widetilde{p}(x_j) = \overline{f}_i$, $1 \le i \le r$. This yields

$$-\ln Z(\beta) + \sum_{i=1}^{r} \beta_i \overline{f}_i = \sum_{j=1}^{n} \widetilde{p}(x_j) \ln p(x_j, \beta)$$

As the maximization of the objective function is with respect to the vector β, we have shown explicitly the dependence of the probabilities upon it. That is

$$\max_{\beta}\left\{-\ln Z(\beta)+\sum_{i=1}^{r}\beta_i\overline{f}_i\right\}=\max_{\beta}\sum_{j=1}^{n}\widetilde{p}(x_j)\ln p(x_j,\beta)$$

Observe, that the vector β in the above optimization solves for the vector θ in the maximum log-likelihood estimation technique. Thus, the estimates of the parameter θ via ML and ME procedures for exponential family of probability distributions yield identical results. Some examples of exponential family of distributions are: normal and exponential distributions.

Examples 1.12. We show some examples of probability distributions achieved via entropy maximization.

1. Tossing of a dice has a sample space $\{1,2,3,4,5,6\}$. Uniform distribution maximizes the entropy. That is, the probability of getting any of the six numbers on a toss is equal to $1/6$.
2. Let X be a continuously distributed random variable defined over the interval $[a,b]$, where $a,b\in\mathbb{R}$ and $a<b$. A uniform distribution defined over this range maximizes the entropy.
3. Let X be a discrete random variable, which takes values in the set $\mathbb{N}$. The expected value of the random variable X is $\mathcal{E}(X)=\mu\in\mathbb{R}^{+}$, and its entropy is $H(X)$. Then

$$H(X)\leq(\mu+1)\log_2(\mu+1)-\mu\log_2\mu$$

   The right-hand expression in the above inequality is the entropy of a geometrically distributed random variable with parameter $(\mu+1)^{-1}$.
4. Let X be a continuously distributed random variable defined over $\mathbb{R}_0^{+}$. It is also given that $\mathcal{E}(X)=\mu$, where $\mu\in\mathbb{R}^{+}$. The entropy maximizing distribution is exponentially distributed with parameter $1/\mu$.
5. Let X be a continuously distributed random variable defined over $\mathbb{R}$. It is also given that $\mathcal{E}(X)=\mu$, and $Var(X)=\sigma^2$; where $\mu\in\mathbb{R}$ and $\sigma^2\in\mathbb{R}^{+}$. The entropy maximizing distribution is normally distributed with the given mean and variance as parameters.
6. It is not always possible to get an entropy maximizing probability distribution. Consider a random variable defined over $\mathbb{R}$. It is also given that $\mathcal{E}(X)=\mu$, and no other features. This constraint results in infinite maximum entropy. This can be visualized by letting the variance of a normally distributed random variable tend to infinity. □

Reference Notes

The seminal work on information theory can be found in the book by Shannon, and Weaver (1972). A nontechnical introduction to this subject has also been provided in a special chapter in this book. This exclusive chapter was written by Weaver. The books by Gallager (1968), and Cover, and Thomas (1991) are considered to be standard textbooks on information theory. A very readable account of this subject is provided in the book by Reza (1994). Other useful books on information theory are by Abramson (1963), Ash (1965), Blahut (1987), Bruen, and Forcinito (2005), Hamming (1986), Han, and Kobayashi (2002), Jones, and Jones (2000), Khinchin (1957), Mansuripur (1987), MacKay (2003), Sayood (2003), Togneri, and de Silva (2003), van der Lubbe (1997), Wells (1999), Welsh (1988), and Yeung (2002).

The notes on Sanov's theorem are based upon Shwartz, and Weiss (1995). More detailed results on this theorem can be found in the elegant textbook by Cover, and Thomas (1991). The subsection on Lempel-Ziv coding is based upon the papers of Ziv, and Lempel (1977, and 1978), and Cover and Thomas (1991). The LZW algorithm is a popular extension of the LZ78 algorithm. It is due to Welch (1984).

Problems

1. Let X be a random variable which takes k values, where $P(X = x_i) = p(x_i), i = 1, 2, \ldots, k$; and $\sum_{i=1}^{k} p(x_i) = 1$. Define

$$H(X) = -\sum_{i=1}^{k} p(x_i) \log_2 p(x_i)$$

Let n be an integer, which is greater than or equal to k. Prove that the multinomial coefficient

$$\binom{n}{np(x_1), np(x_2), \ldots, np(x_k)} \triangleq \frac{n!}{\prod_{i=1}^{k} (np(x_i))!} \simeq 2^{nH(X)}$$

where $(np(x_i))! \triangleq \Gamma(np(x_i) + 1), 1 \leq i \leq k$, and $\Gamma(\cdot)$ is the gamma function. Hint: Use Stirling's approximation $n! \simeq (n/e)^n$.

2. For any real number α, with $0 \leq \alpha \leq 0.5$, and

$$h(\alpha) = -\{\alpha \log_2 \alpha + (1 - \alpha) \log_2 (1 - \alpha)\}$$

prove that

$$\sum_{k \leq \alpha n} \binom{n}{k} \leq 2^{nh(\alpha)}$$

Hint: Observe that

$$\begin{aligned} 1 &= \{\alpha + (1 - \alpha)\}^n \\ &\geq \sum_{k \leq \alpha n} \binom{n}{k} \alpha^k (1 - \alpha)^{n-k} = (1 - \alpha)^n \sum_{k \leq \alpha n} \binom{n}{k} \left(\frac{\alpha}{1 - \alpha}\right)^k \end{aligned}$$

Note that the range of values of $\alpha / (1 - \alpha)$ for $0 \leq \alpha \leq 0.5$ is from 0 to 1. Furthermore, it has a positive slope in this range of values of α. Therefore

$$\begin{aligned} 1 &\geq (1 - \alpha)^n \sum_{k \leq \alpha n} \binom{n}{k} \left(\frac{\alpha}{1 - \alpha}\right)^{\alpha n} \\ &\geq \alpha^{\alpha n} (1 - \alpha)^{n(1-\alpha)} \sum_{k \leq \alpha n} \binom{n}{k} \end{aligned}$$

This yields

$$\sum_{k\leq \alpha n}\binom{n}{k} \leq \alpha^{-\alpha n}(1-\alpha)^{-n(1-\alpha)} = 2^{nh(\alpha)}$$

Compare the above result with the identity $\sum_{k=0}^{n}\binom{n}{k} = 2^n$.

3. Prove that, if $H(X) = -\sum_{i=1}^{n} p(x_i)\log_2 p(x_i)$, then $0 \leq H(X) \leq \log_2 n$.
Hint: Arbitrarily select $p(x_n)$ as a variable depending upon

$$p(x_1), p(x_2), \ldots, p(x_{n-1})$$

Then for any k where, $1 \leq k \leq (n-1)$

$$\begin{aligned}
&\frac{dH(X)}{dp(x_k)} \\
&= \sum_{i=1}^{n} \frac{\partial H(X)}{\partial p(x_i)} \frac{\partial p(x_i)}{\partial p(x_k)} \\
&= -\frac{d}{dp(x_k)}(p(x_k)\log_2 p(x_k)) - \frac{d}{dp(x_n)}(p(x_n)\log_2 p(x_n))\frac{\partial p(x_n)}{\partial p(x_k)}
\end{aligned}$$

Use of the relationship,

$$p(x_n) = 1 - \sum_{i=1}^{n-1} p(x_i)$$

yields

$$\frac{\partial p(x_k)}{\partial p(x_n)} = -1$$

Consequently,

$$\frac{dH(X)}{dp(x_k)} = -(\log_2 p(x_k) + \log_2 e) + (\log_2 p(x_n) + \log_2 e) = 0$$

That is, $p(x_k) = p(x_n)$, for $1 \leq k \leq (n-1)$. This in turn yields $p(x_k) = 1/n$ for $1 \leq k \leq n$. For this set of values of $p(x_k)$'s, $H(X) = \log_2 n$.
Observe that:
(a) $H(X) = 0$, if $p(x_j) = 1$ for some x_j, and $p(x_i) = 0$ for all $i \neq j, 1 \leq i, j \leq n$.
(b) $H(X) = \log_2 n$, if $p(x_j) = 1/n$ for all $j, 1 \leq j \leq n$.
Alternative proofs can be provided via Jensen's inequality or the method of Lagrange multipliers, or simply proceed as follows. In this alternate proof, we use the inequality: $\ln x \leq (x-1)$, $\forall\, x \in \mathbb{R}^+$.

$$\begin{aligned}
H(X) - \log_2 n &= -\sum_{i=1}^{n} p(x_i)\log_2 p(x_i) - \sum_{i=1}^{n} p(x_i)\log_2 n \\
&= \sum_{i=1}^{n} p(x_i)\log_2\left\{\frac{1}{np(x_i)}\right\} \\
&\leq \log_2 e \sum_{i=1}^{n} p(x_i)\left\{\frac{1}{np(x_i)} - 1\right\} \\
&= \log_2 e \left\{\sum_{i=1}^{n}\frac{1}{n} - \sum_{i=1}^{n} p(x_i)\right\} = 0
\end{aligned}$$

Therefore $H(X) \leq \log_2 n$, with equality if and only if the outcomes are equiprobable.

4. Let X and Y be two discrete random variables which take only a finite set of values. Prove that $H(X,Y) \leq H(X) + H(Y)$, where the equality is observed if and only if the random variables X and Y are independent of each other.
Hint: Assume that X takes values x_i, for $1 \leq i \leq m$, and Y takes values y_j, for $1 \leq j \leq n$. Let $P(X = x_i) \triangleq p_X(x_i)$, for $1 \leq i \leq m$ and $P(Y = y_j) \triangleq p_Y(y_j)$, for $1 \leq j \leq n$. Also let $P(X = x_i, Y = y_j) \triangleq p_{X,Y}(x_i, y_j)$, for $1 \leq i \leq m$ and $1 \leq j \leq n$. It follows that $p_X(x_i) = \sum_{j=1}^{n} p_{X,Y}(x_i, y_j)$ for $1 \leq i \leq m$, and $p_Y(y_j) = \sum_{i=1}^{m} p_{X,Y}(x_i, y_j)$ for $1 \leq j \leq n$. Note that

$$H(X) = -\sum_{i=1}^{m} p_X(x_i) \log_2 p_X(x_i) = -\sum_{i=1}^{m}\sum_{j=1}^{n} p_{X,Y}(x_i, y_j) \log_2 p_X(x_i)$$

$$H(Y) = -\sum_{j=1}^{n} p_Y(y_j) \log_2 p_Y(y_j) = -\sum_{i=1}^{m}\sum_{j=1}^{n} p_{X,Y}(x_i, y_j) \log_2 p_Y(y_j)$$

$$H(X,Y) = -\sum_{i=1}^{m}\sum_{j=1}^{n} p_{X,Y}(x_i, y_j) \log_2 p_{X,Y}(x_i, y_j)$$

Then

$$H(X,Y) - H(X) - H(Y) = \sum_{i=1}^{m}\sum_{j=1}^{n} p_{X,Y}(x_i, y_j) \log_2 \left\{ \frac{p_X(x_i)\, p_Y(y_j)}{p_{X,Y}(x_i, y_j)} \right\}$$

$$\leq \log_2 \left\{ \sum_{i=1}^{m}\sum_{j=1}^{n} p_X(x_i)\, p_Y(y_j) \right\}$$

$$= \log_2 1 = 0$$

In the last step, Jensen's inequality was used by noting that $\log_2(\cdot)$ is a concave function over $\mathbb{R}^+$ and

$$\sum_{i=1}^{m}\sum_{j=1}^{n} p_{X,Y}(x_i, y_j) = 1$$

Consequently, the inequality part of the result follows.
Equality occurs if $p_X(x_i)\, p_Y(y_j) / p_{X,Y}(x_i, y_j) = k$, where k is a positive constant. As

$$1 = \sum_{i=1}^{m}\sum_{j=1}^{n} p_{X,Y}(x_i, y_j) = \frac{1}{k}\sum_{i=1}^{m}\sum_{j=1}^{n} p_X(x_i)\, p_Y(y_j) = \frac{1}{k}$$

it follows that $k = 1$. Therefore equality occurs if and only if $p_{X,Y}(x_i, y_j) = p_X(x_i)\, p_Y(y_j)$ for $1 \leq i \leq m$ and $1 \leq j \leq n$. This implies that the random variables X and Y are independent of each other.

5. Let X and Y be two discrete random variables which take only a finite set of values. Prove that

$$H(X,Y) = H(Y) + H(X \mid Y)$$

Hint: Use of the notation in the last problem results in

$$
\begin{aligned}
H(X,Y) &= -\sum_{i=1}^{m}\sum_{j=1}^{n} p_{X,Y}(x_i,y_j)\log_2 p_{X,Y}(x_i,y_j) \\
&= -\sum_{i=1}^{m}\sum_{j=1}^{n} p_{X,Y}(x_i,y_j)\log_2 \{p_{X|Y}(x_i \mid y_j)\,p_Y(y_j)\} \\
&= -\sum_{i=1}^{m}\sum_{j=1}^{n} p_{X,Y}(x_i,y_j)\{\log_2 p_{X|Y}(x_i \mid y_j) + \log_2 p_Y(y_j)\} \\
&= -\sum_{i=1}^{m}\sum_{j=1}^{n} p_{X,Y}(x_i,y_j)\log_2 p_{X|Y}(x_i \mid y_j) + H(Y) \\
&= -\sum_{i=1}^{m}\sum_{j=1}^{n} p_Y(y_j)\,p_{X|Y}(x_i \mid y_j)\log_2 p_{X|Y}(x_i \mid y_j) + H(Y) \\
&= -\sum_{j=1}^{n} p_Y(y_j)\sum_{i=1}^{m} p_{X|Y}(x_i \mid y_j)\log_2 p_{X|Y}(x_i \mid y_j) + H(Y) \\
&= \sum_{j=1}^{n} p_Y(y_j)\,H(X \mid Y = y_j) + H(Y) \\
&= H(X \mid Y) + H(Y)
\end{aligned}
$$

6. Let X and Y be two discrete random variables which take only a finite set of values. Prove that $H(X \mid Y) \leq H(X)$, where the equality is observed if and only if the discrete random variables X and Y are independent of each other. .
   Hint: $H(X \mid Y) = H(X,Y) - H(Y)$, from the last problem. Further, $H(X,Y) \leq H(X) + H(Y)$, with equality iff the random variables are independent. The result follows.
7. Entropy is a concave function of its arguments. Prove this assertion.
   Hint: See Peres (1993). Observe that for $\lambda \in [0,1]$

$$
\frac{d^2 H[\lambda p' + (1-\lambda)p'']}{d\lambda^2} = -\{\log_2 e\}\sum_{j=1}^{n} \frac{(p'(x_j) - p''(x_j))^2}{\{\lambda p'(x_j) + (1-\lambda)p''(x_j)\}} \leq 0
$$

   The above relationship is a sufficient condition for a function to be concave. Furthermore, this second derivative vanishes only if $p'(x_j) = p''(x_j)$ for $1 \leq j \leq n$.
8. Let X be a discrete random variable, and $f(\cdot)$ be a function of X. Show that $H(f(X)) \leq H(X)$, where equality occurs iff the function $f(\cdot)$ is one-to-one.
   Hint: We establish the result in two steps.
   Step 1: It is shown in this step that $H(f(X) \mid X) = 0$. Note that for any fixed value of $X = x$, $P(f(X) = f(x) \mid X = x) = 1$. Therefore

$$
\begin{aligned}
&H(f(X) \mid X = x) \\
&= -P(f(X) = f(x) \mid X = x)\log_2 P(f(X) = f(x) \mid X = x) = 0
\end{aligned}
$$

   Thus

$$
H(f(X) \mid X) = -\sum_{x} P(X = x) H(f(X) \mid X = x) = 0
$$

   Step 2: Observe that $H(X \mid f(X)) \geq 0$, where equality occurs iff the function $f(\cdot)$ is one-to-one. Thus, use of the chain rule of entropies results in

$$H(X, f(X)) = H(f(X)) + H(X \mid f(X)) \geq H(f(X))$$

Again use of chain rule of entropies and Step 1 yields

$$H(X, f(X)) = H(X) + H(f(X) \mid X) = H(X)$$

Combination of the above results gives $H(f(X)) \leq H(X)$.

9. Let X and Y be two discrete random variables. The mutual information of these two random variables is $I(X;Y)$. Prove that

$$I(X;Y) = \sum_{x,y} P(X=x, Y=y) \log_2 \left(\frac{P(X=x, Y=y)}{P(X=x)P(Y=y)} \right)$$

Hint: See Cover, and Thomas (1991). Using the relationships

$$H(X) = -\sum_{x,y} P(X=x, Y=y) \log_2 P(X=x)$$

$$H(X \mid Y) = -\sum_{x,y} P(X=x, Y=y) \log_2 P(X=x \mid Y=y)$$

we obtain

$$\begin{aligned} I(X;Y) &= H(X) - H(X \mid Y) \\ &= \sum_{x,y} P(X=x, Y=y) \log_2 \left\{ \frac{P(X=x \mid Y=y)}{P(X=x)} \right\} \\ &= \sum_{x,y} P(X=x, Y=y) \log_2 \left\{ \frac{P(X=x, Y=y)}{P(X=x)P(Y=y)} \right\} \end{aligned}$$

This is the required result.

10. Prove that $I(X;Y) \geq 0$ with equality iff X and Y are independent.
Hint: $I(X;Y) = H(P(X=x, Y=y) \parallel P(X=x)P(Y=y)) \geq 0$.

11. Establish the nonnegativity of the relative entropy.
Hint: See Nielsen, and Chuang (2000). We establish that $H(p(x) \parallel q(x)) \geq 0$, with equality if and only if $p(x) = q(x)$ for all values of x. It is known that $\ln x \leq (x-1)$ for all $x > 0$. Since $\ln x = \ln 2 \log_2 x$,

$$\begin{aligned} H(p(x) \parallel q(x)) &= -\sum_{j=1}^{n} p(x_j) \log_2 \frac{q(x_j)}{p(x_j)} \geq \frac{1}{\ln 2} \sum_{j=1}^{n} p(x_j) \left\{ 1 - \frac{q(x_j)}{p(x_j)} \right\} \\ &= \frac{1}{\ln 2} \sum_{j=1}^{n} (p(x_j) - q(x_j)) = 0 \end{aligned}$$

12. Prove that the relative entropy is convex in its arguments.
Hint: See Cover and Thomas (1991). Apply the log sum inequality to a single term of $H[\lambda p' + (1-\lambda) p'' \parallel \lambda q' + (1-\lambda) q'']$. That is

$$\begin{aligned} &\{\lambda p'(x) + (1-\lambda) p''(x)\} \log_2 \frac{\{\lambda p'(x) + (1-\lambda) p''(x)\}}{\{\lambda q'(x) + (1-\lambda) q''(x)\}} \\ &\leq \lambda p'(x) \log_2 \frac{\lambda p'(x)}{\lambda q'(x)} + (1-\lambda) p''(x) \log_2 \frac{(1-\lambda) p''(x)}{(1-\lambda) q''(x)} \end{aligned}$$

Summing the above inequality over all the stated values of x leads to the stated result.

13. The joint probability distribution of the random variable pair (X,Y) is specified as $P(X = x, Y = y) = P(X = x)P(Y = y \mid X = x)$. For brevity in notation, let $P(X = x, Y = y) \triangleq p_{X,Y}(x,y)$, $P(X = x) \triangleq p_X(x)$, and $P(Y = y \mid X = x) \triangleq p_{Y|X}(y \mid x)$. Therefore, $p_{X,Y}(x,y) = p_X(x)\, p_{Y|X}(y \mid x)$. Establish the following results.
    (a) For fixed $p_{Y|X}(y \mid x)$, the mutual information $I(X;Y)$ is a concave function of $p_X(x)$.
    (b) For fixed $p_X(x)$, the mutual information $I(X;Y)$ is a convex function of $p_{Y|X}(y \mid x)$.

    Hint: See Cover and Thomas (1991).
    (a) It is known that

$$I(X;Y) = H(Y) - H(Y \mid X) = H(Y) - \sum_x p_X(x)\, H(Y \mid X = x)$$

Also observe that

$$p_Y(y) = \sum_x p_{X,Y}(x,y) = \sum_x p_{Y|X}(y \mid x)\, p_X(x)$$

Therefore, as $p_{Y|X}(y \mid x)$ is fixed; $p_Y(y)$ is a linear function of $p_X(x)$. Consequently, as $H(Y)$ is a concave function of $p_Y(y)$, is also a concave function of $p_X(x)$. The second term in the expression for mutual information $I(X;Y)$, is also a linear function of $p_X(x)$. Consequently as $I(X;Y) \geq 0$, the difference of the two concave functions is also a concave function of $p_X(x)$, for fixed $p_{Y|X}(y \mid x)$.

(b) In this part $p_X(x)$ is fixed. This result is proved via the convexity property of relative entropy. For fixed $p_X(x)$ consider two different joint distributions, $p'_{X,Y}(x,y)$ and $p''_{X,Y}(x,y)$. Associated with these distributions, the marginal distributions are $p'_Y(y)$ and $p''_Y(y)$ respectively; and conditional distributions are $p'_{Y|X}(y \mid x)$ and $p''_{Y|X}(y \mid x)$ respectively. Let

$$p'_{X,Y}(x,y) \triangleq p_X(x)\, p'_{Y|X}(y \mid x), \quad \text{and} \quad p''_{X,Y}(x,y) \triangleq p_X(x)\, p''_{Y|X}(y \mid x)$$

Consider another distribution with parameter $\lambda \in [0,1]$, where we define

$$p^{\lambda}_{Y|X}(y \mid x) \triangleq \lambda p'_{Y|X}(y \mid x) + (1-\lambda)\, p''_{Y|X}(y \mid x)$$
$$p^{\lambda}_{X,Y}(x,y) \triangleq p_X(x)\, p^{\lambda}_{Y|X}(y \mid x)$$

This leads to

$$p^{\lambda}_{X,Y}(x,y) = \lambda p'_{X,Y}(x,y) + (1-\lambda)\, p''_{X,Y}(x,y)$$
$$p^{\lambda}_Y(y) = \lambda p'_Y(y) + (1-\lambda)\, p''_Y(y)$$

Also define

$$q'_{X,Y}(x,y) \triangleq p_X(x)\, p'_Y(y)$$
$$q''_{X,Y}(x,y) \triangleq p_X(x)\, p''_Y(y)$$
$$q^{\lambda}_{X,Y}(x,y) \triangleq p_X(x)\, p^{\lambda}_Y(y)$$

This leads to

$$q^{\lambda}_{X,Y}(x,y) = \lambda q'_{X,Y}(x,y) + (1-\lambda)\, q''_{X,Y}(x,y)$$

Let the mutual information corresponding to the distributions $p'_{X,Y}(x,y)$ and $q'_{X,Y}(x,y)$ be $I'(X;Y)$. Similarly, let the mutual information corresponding to the distributions $p''_{X,Y}(x,y)$ and $q''_{X,Y}(x,y)$ be $I''(X;Y)$. Also, let the mutual information corresponding to the distributions $p^{\lambda}_{X,Y}(x,y)$ and $q^{\lambda}_{X,Y}(x,y)$ be $I^{\lambda}(X;Y)$. In order to establish the stated result, it is required to prove that

$$I^{\lambda}(X;Y) \le \lambda I'(X;Y) + (1-\lambda) I''(X;Y)$$

We have

$$I^{\lambda}(X;Y) = H\left(p^{\lambda}_{X,Y}(x,y) \parallel q^{\lambda}_{X,Y}(x,y)\right)$$

Use of the expressions for $p^{\lambda}_{X,Y}(x,y)$, and $q^{\lambda}_{X,Y}(x,y)$ results in

$$\begin{aligned}
&I^{\lambda}(X;Y) \\
&= H\left(\lambda p'_{X,Y}(x,y) + (1-\lambda) p''_{X,Y}(x,y) \parallel \lambda q'_{X,Y}(x,y) + (1-\lambda) q''_{X,Y}(x,y)\right) \\
&\le \lambda H\left(p'_{X,Y}(x,y) \parallel q'_{X,Y}(x,y)\right) + (1-\lambda) H\left(p''_{X,Y}(x,y) \parallel q''_{X,Y}(x,y)\right) \\
&= \lambda I'(X;Y) + (1-\lambda) I''(X;Y)
\end{aligned}$$

The above inequality follows because of the convexity of the relative entropy.

14. Establish the validity of observations about the LZ78 compression scheme.

Observation 1: Hint: See Cover, and Thomas (1991), and Moser (2014). Let c_k be equal to the total number of phrases of length at most $k \in \mathbb{P}$. Also let the corresponding total length of all these phrases be n_k. Then

$$c_k = 2^1 + 2^2 + \cdots + 2^k = \left(2^{k+1} - 2\right)$$

$$n_k = 1 \cdot 2^1 + 2 \cdot 2^2 + \cdots + k \cdot 2^k = \sum_{j=1}^{k} j2^j = (k-1)\, 2^{k+1} + 2$$

This yields $(n_{k+1} - n_k) = (k+1)\, 2^{k+1}$ and

$$2^{k+1} = \frac{n_k - 2}{(k-1)} \le \frac{n_k}{(k-1)}, \quad \text{for } \ k \ge 2$$

In order to obtain worst-case bound, we maximize $c(n)$ the number of phrases in a distinct parsing of a binary sequence of length n. This is done by making the length of phrases as small as possible. Consider the following two cases.

(a) If $n = n_k$ for some k, then

$$c(n) = c_k = \left(2^{k+1} - 2\right) < 2^{k+1} \le \frac{n_k}{(k-1)}, \quad \text{for } \ k \ge 2$$

(b) If $n_k < n < n_{k+1}$ for some k, then let $n = (n_k + \Delta)$. Observe that

$$n < n_{k+1} \Rightarrow (n_k + \Delta) < n_{k+1} \Rightarrow \Delta < (n_{k+1} - n_k) = (k+1)\, 2^{k+1}$$

That is, $\Delta < (k+1)\, 2^{k+1}$. In the worst case scenario, where distinct parsing occurs into shortest phrases, the initial n_k bits of the sequence are split into phrases of length at most k, and the remaining Δ bits into $\Delta / (k+1)$ phrases of length $(k+1)$. Thus

$$c(n) \leq \frac{n_k}{(k-1)} + \frac{\Delta}{(k+1)} \leq \frac{n_k + \Delta}{(k-1)} = \frac{n}{(k-1)}$$

We next obtain a bound on k for a specified value of n. Let $n_k \leq n < n_{k+1}$. Then

$$n \geq n_k = (k-1)\, 2^{k+1} + 2 \geq 2^k \Rightarrow \log_2 n \geq k$$

Also

$$n < n_{k+1} = \left(k 2^{k+2} + 2\right) \leq (k+1)\, 2^{k+2} \leq (\log_2 n + 1)\, 2^{k+2}$$

Therefore

$$(k+2) \geq \log_2 n - \log_2 ((\log_2 n + 1))$$

This implies

$$\begin{aligned}(k-1) &\geq \log_2 n - \log_2 ((\log_2 n + 1)) - 3 \\ &= \left\{1 - \frac{\log_2 ((\log_2 n + 1)) + 3}{\log_2 n}\right\} \log_2 n \\ &\triangleq (1 - \epsilon_n) \log_2 n\end{aligned}$$

where $\epsilon_n = \{\log_2 ((\log_2 n + 1)) + 3\} / \log_2 n$. Observe that $\epsilon_n < 1$ if $n \geq 55$. Also, if $n \geq 128$, then $\epsilon_n \leq 6/7$, and $(1 - \epsilon_n) \log_2 n \geq 1$. Furthermore, $\epsilon_n \to 0$ as $n \to \infty$. Finally combining this result with the inequality $c(n) \leq n/(k-1)$ yields the stated result.

Observation 2: Note that $\lceil \log_2 (c(n) + 1) \rceil$ bits are required to specify an index in the dictionary. The number of bits required to represent a codeword is equal to $\{\lceil \log_2 (c(n) + 1) \rceil + 1\}$.

Observation 3: Hint: See Cover and Thomas (1991), and the class notes of Shor (2005).

The source emits a sequence of symbols from the set $\mathfrak{B}$, each with a specified probability. The emission of symbols from the source can be described by a sequence of random variables $X_1, X_2, \ldots$. These random variables take values in the set $\mathfrak{B}$. Assume that these random variables are independent, and identically distributed. Consider a sequence of random variables $(X_1, X_2, \ldots, X_n)$. These take values $(x_1, x_2, \ldots, x_n)$, where $x_i \in \mathfrak{B}$ for $1 \leq i \leq n$. Let X be a representative random variable of the source random variables X_i's, and its entropy be $H(X)$. Probability of occurrence of the sequence x is

$$p(x) = \prod_{i=1}^{n} p(x_i) \simeq 2^{-nH(X)}$$

Let $c(n) \in \mathbb{P}$ be the number of distinct phrases in a string of length n. The sequence (or string) x is broken into distinct phrases as

$$x = y_1 y_2 \cdots y_{c(n)}$$

Therefore

$$p(x) = \prod_{i=1}^{c(n)} p(y_i)$$

Let $c_l \in \mathbb{P}$ be the number of distinct phrases y_i of length $l \in \mathbb{P}$. Therefore $\sum_l c_l = c(n) \triangleq c$, and $\sum_l l c_l = n$. Also

$$p(x) = \prod_l \prod_{|y_i| = l} p(y_i)$$

The number of characters in the string y_i is denoted by $|y_i|$. As the y_i's with length l are all distinct, these are mutually independent events. Therefore we have

$$\sum_{|y_i|=l} p(y_i) \leq 1$$

With the above constraint, the product $\prod_{|y_i|=l} p(y_i)$ is maximized, if all its terms are equal, and there are c_l such terms. Therefore

$$\prod_{|y_i|=l} p(y_i) \leq \left(\frac{1}{c_l}\right)^{c_l}$$

This implies

$$-\log_2 p(x) \geq \sum_l c_l \log_2 c_l$$

We next write $\sum_l c_l \log_2 c_l$ as

$$\sum_l c_l \log_2 c_l = c \log_2 c + c \sum_l \frac{c_l}{c} \log_2 \frac{c_l}{c}$$

The above equation can be written as

$$\sum_l c_l \log_2 c_l = c \log_2 c - cH(Z)$$

where Z is a discrete random variable with probability mass function $p_l = c_l/c$, where the l's are positive integers. The average value of the random variable Z is $\mathcal{E}(Z) = \sum_l l p_l = n/c$, and its entropy is equal to $H(Z)$. It can be shown that the entropy of the random variable Z defined on positive integers, with expected value n/c is upper-bounded by $O(\log_2(n/c))$. Therefore

$$-\log_2 p(x) \geq \sum_l c_l \log_2 c_l \geq c \log_2 c - cO\left(\log_2\left(\frac{n}{c}\right)\right)$$

As $\log_2 p(x) \simeq -nH(X)$, we have

$$H(X) \geq \frac{c \log_2 c}{n} - \frac{c}{n} O\left(\log_2\left(\frac{n}{c}\right)\right)$$

Observe that $(c/n)\log_2(n/c)$ is bounded by $e^{-1}\log_2 e$. Also from Observation 1, we have $c/n \leq (1 + o(1)) / \log_2 n$. Therefore

$$\frac{c}{n} \log_2\left(\frac{n}{c}\right) \leq O\left(\frac{\log \log n}{\log n}\right)$$

The stated result follows as $n \to \infty$.

15. Show how the LZ78 algorithm can be modified to produce codewords of a given sequence of characters in a single pass.
Hint: Let the source alphabet be $\mathcal{X} = \{x_{a_0}, x_{a_1}, \ldots, x_{a_{D-1}}\}$, where $|\mathcal{X}| = D$, and $D \geq 2$. For convenience in notation, let the empty character $\star \triangleq x_{a_D}$. Let the decimal representation of the character x_{a_j} be $j \in \mathbb{N}$ for $0 \leq j \leq D$. Also let $\mathcal{X}^{\star} = \mathcal{X} \cup \{\star\}$.

The sequence of characters $x = x_1 x_2 \ldots x_n$, where $x_i \in \mathcal{X}$ for $1 \le i \le n$, is to be compressed. The sequence (or string) x is broken into $c(n) \triangleq c$ distinct phrases as $x = y_1 y_2 \ldots y_c$. Let the codeword that corresponds to the phrase y_j be (m, x_{a_k}), where $m \in \mathbb{N}$ and $x_{a_k} \in \mathcal{X}$, for $1 \le j \le (c-1)$. The codeword that corresponds to the phrase y_c is (m, x_{a_k}), where $m \in \mathbb{N}$ and $x_{a_k} \in \mathcal{X}^{\star}$. That is, it is possible for the second coordinate of the codeword of the last phrase to be the empty character.

The codeword is converted into a single binary number as follows. We map the codeword two-tuple (m, x_{a_k}) to the integer $\{m(D+1)+k\}$. This integer is converted into binary number with sufficient number of zeros padded on the left so that a predetermined number of bits are used in its representation. The total number of bits used in the representation of $\{m(D+1)+k\}$ is determined as follows.

Let the codeword that corresponds to the phrase y_j be (m, x_{a_k}). The largest possible values of m and k are $(j-1)$ and D respectively. Therefore the largest possible value of $\{m(D+1)+k\}$ is

$$\{(j-1)(D+1)+D\} = \{j(D+1)-1\}$$

The number of bits in the binary expansion of $\{j(D+1)-1\}$ is

$$\lceil \log_2 \{j(D+1)\} \rceil \triangleq B_j, \quad \text{where} \quad 1 \le j \le c$$

The binary representation of $\{m(D+1)+k\}$ is made to have B_j number of bits, by padding with zeros on the left side if necessary. The decoder is aware that the jth codeword is B_j number of bits long for $1 \le j \le c$. Therefore the decoding is straightforward. If $\{m(D+1)+k\}$ is divided by $(D+1)$, we obtain the remainder k, and the quotient m. Consequently, the phrase y_j is the string with index(m) concatenated with the character x_{a_k}.

16. In a communication system, let the number of input and output symbols be K and J respectively. Prove that $C \le \min\{\log_2 K, \log_2 J\}$.
Hint: $C = \max I(X;Y)$, and $\max I(X;Y) \le \max H(X) = \log_2 K$, when the expression $I(X;Y) = H(X) - H(X \mid Y)$ is considered. Similarly $\max I(X;Y) \le \max H(Y) = \log_2 J$, when the expression $I(X;Y) = H(Y) - H(Y \mid X)$ is considered.
17. Prove the theorem about typical sequence.
Hint: See Nielsen, and Chuang (2000).
    (a) Observe that $-\log_2 P(X_i)$'s are independent, and identically distributed random variables. Define $\mu = \mathcal{E}(-\log_2 P(X))$, that is $\mu = H(X)$. Using the weak law of large numbers for any $\epsilon > 0$ and $\delta > 0$ and sufficiently large value of n yields

$$P\left(\left| \frac{1}{n} \sum_{i=1}^{n} \{-\log_2 P(X_i)\} - \mu \right| \le \epsilon \right) \ge 1 - \delta$$

Since $\sum_{i=1}^{n} \log_2 P(X_i) = \log_2 P(X_1, X_2, \ldots, X_n)$

$$P\left(\left| -\frac{1}{n} \log_2 P(X_1, X_2, \ldots, X_n) - H(X) \right| \le \epsilon \right) \ge 1 - \delta$$

which is the probability that a sequence is ϵ-typical is at least $(1-\delta)$.
    (b) Observe that the definition of typical sequences, the last result, and $\delta > 0$ imply $(1-\delta) \le \sum_{x \in V(n,\epsilon)} p(x) \le 1$. Thus

$$(1-\delta) \le \sum_{x\in V(n,\epsilon)} p(x) \le \sum_{x\in V(n,\epsilon)} 2^{-n(H(X)-\epsilon)}$$
$$= |V(n,\epsilon)|\, 2^{-n(H(X)-\epsilon)}$$

This in turn implies $(1-\delta)\, 2^{n(H(X)-\epsilon)} \le |V(n,\epsilon)|$. Similarly

$$1 \ge \sum_{x\in V(n,\epsilon)} p(x) \ge \sum_{x\in V(n,\epsilon)} 2^{-n(H(X)+\epsilon)} = |V(n,\epsilon)|\, 2^{-n(H(X)+\epsilon)}$$

implies $2^{n(H(X)+\epsilon)} \ge |V(n,\epsilon)|$. The result follows.

(c) Let $S(n) = S_a(n) \cup S_t(n)$, where $S_a(n)$ and $S_t(n)$ are the sets of atypical and typical sequences respectively in $S(n)$. As n becomes large the probability of existence of sequences in $S_a(n)$ becomes small. Furthermore, $|S_t(n)| \le |S(n)| \le 2^{nR}$ and the probability of each typical sequence is $\simeq 2^{-nH(X)}$. Thus the net probability of typical sequences in $S(n)$ is $\simeq 2^{n(R-H(X))}$. If $R < H(X)$ and n is very large this probability tends to zero.

18. Prove the theorem about the average codeword-length for a noiseless communication channel.
Hint: See Cover, and Thomas (1991). Sequences can be divided into atypical and typical sequences.
Note that for large values of n, the ϵ-typical sequences occur with probability at least $(1-\epsilon)$, where $\epsilon > 0$. Therefore atypical sequences occur with probability at most ϵ.
Observe that the cardinality of the set of typical sequences is $|V(n,\epsilon)| \le 2^{n(H(X)+\epsilon)}$. Assume that typical sequences are coded in binary strings. Indexing of these binary strings may not need more than $\{n(H(X)+\epsilon)+1\}$ bits. The extra 1 bit may be necessary, because $n(H(X)+\epsilon)$ may not be an integer. These sequences are prefixed by 0. Thus up to

$$\{n(H(X)+\epsilon)+2\}$$

bits are used to represent typical sequences. Observe that an atypical sequence from the source is encoded by using no more than $(n\log_2|\mathcal{X}|+1)$ bits, where $\mathcal{X}$ is the set of the source alphabet. These codewords have a prefix of 1. As $V(n,\epsilon)$ is the set of all ϵ-typical sequences of length n, define $W(n,\epsilon)$ as the set of all atypical sequences of length n. Thus

$$\mathcal{E}\left\{\ell\left(\widetilde{X}\right)\right\} = \sum_{x\in V(n,\epsilon)} p(x)\,\ell(x) + \sum_{x\in W(n,\epsilon)} p(x)\,\ell(x)$$

$$\mathcal{E}\left\{\ell\left(\widetilde{X}\right)\right\} \le \sum_{x\in V(n,\epsilon)} p(x)\,\{n(H(X)+\epsilon)+2\}$$
$$+ \sum_{x\in W(n,\epsilon)} p(x)\,(n\log_2|\mathcal{X}|+2)$$
$$\le n(H(X)+\epsilon) + \epsilon n\log_2|\mathcal{X}| + 2 = n(H(X)+\epsilon')$$

where $\epsilon' = \epsilon + \epsilon\log_2|\mathcal{X}| + 2/n$ can be made arbitrarily small by appropriate choices of ϵ and n.

19. Prove Shannon's noiseless channel coding theorem.
Hint: See Nielsen, and Chuang (2000).

(a) Let $R > H(X)$, and select $R > H(X) + \epsilon$ where $\epsilon > 0$. Use of the theorem about typical sequences for any $\delta > 0$ and for sufficiently large n results in $2^{n(H(X)+\epsilon)} < 2^{nR}$ ϵ-typical sequences. That is, $|V(n,\epsilon)| < 2^{nR}$, and the probability of this event is at least $(1-\delta)$. The compression scheme consists in checking the output of the source to see if it is a ϵ-typical sequence or not.
If the output of the sources is ϵ-typical then, using $\lceil nR \rceil$ bits the output is stored as an index for the particular sequence. This permits unambiguous recovery of the compressed sequence later on. However, if the output is not ϵ-typical then, the string is represented by some fixed string of length $\lceil nR \rceil$ bits. In this case, decompression is performed by producing some random sequence $(x_1', x_2', \ldots, x_n')$, where $x_i' \in \mathcal{X}$ for $1 \le i \le n$. Thus data compression is a failure in the case of atypical sequences.

(b) Assume that $R < H(X)$. Under this condition, the compression and decompression process has at most 2^{nR} unambiguous outputs. Thus only 2^{nR} sequences can be compressed and decompressed uniquely. However the number of typical sequences is $\simeq 2^{nH(X)}$. Since $R < H(X)$, the compression-decompression process is not reliable.

20. Let X be a normally distributed random variable, with mean μ and variance σ^2. Prove that its differential entropy is given by

$$H_c(X) = \frac{1}{2}\log_2\left(2\pi e\sigma^2\right)$$

21. Let X and Y be continuous random variables, with probability density functions, $f_X(t)$ and $f_Y(t)$ respectively, for $t \in \mathbb{R}$. Prove that

$$-\int_{-\infty}^{\infty} f_X(t)\log_2 f_X(t)\,dt \le -\int_{-\infty}^{\infty} f_X(t)\log_2 f_Y(t)\,dt$$

Equality holds iff $f_X(t) = f_Y(t)$ for all values of $t \in \mathbb{R}$.
Hint: Use the inequality $\log_2 x \le (x-1)\log_2 e$, for $x > 0$.

$$\begin{aligned} &-\int_{-\infty}^{\infty} f_X(t)\log_2 f_X(t)\,dt + \int_{-\infty}^{\infty} f_X(t)\log_2 f_Y(t)\,dt \\ &= \int_{-\infty}^{\infty} f_X(t)\log_2 \frac{f_Y(t)}{f_X(t)}dt \le \log_2 e \int_{-\infty}^{\infty} f_X(t)\left(\frac{f_Y(t)}{f_X(t)} - 1\right)dt = 0 \end{aligned}$$

It can be observed that equality holds iff $f_X(t) = f_Y(t)$ for all values of $t \in \mathbb{R}$.

22. Prove the following results.

$$\begin{aligned} &I_c(X;Y) \ge 0 \\ &I_c(X;Y) = I_c(Y;X) \\ &I_c(X;Y) = H_c(X) - H_c(X \mid Y) = H_c(Y) - H_c(Y \mid X) \end{aligned}$$

Hint: See Cover, and Thomas (1991). As $\log_2(\cdot)$ is a concave function over $\mathbb{R}^+$, use of Jensen's inequality results in

$$\begin{aligned} -I_c(X;Y) &= \int_{-\infty}^{\infty}\int_{-\infty}^{\infty} f_{X,Y}(x,y)\log_2\left[\frac{f_X(x)\,f_Y(y)}{f_{X,Y}(x,y)}\right]dxdy \\ &\le \log_2 \int_{-\infty}^{\infty}\int_{-\infty}^{\infty} f_X(x)\,f_Y(y)\,dxdy = \log_2 1 = 0 \end{aligned}$$

Therefore $I_c(X;Y) \ge 0$. The other results follow from the definition of $I_c(X;Y)$.

23. Let $H_c(X)$ be the differential entropy of a continuously distributed random variable X. Prove that the differential entropy $H_c(Y)$ of the random variable $Y = aX$, where $a \in \mathbb{R}$ is given by $H_c(Y) = H_c(X) + \log_2 |a|$.
Hint: See Cover, and Thomas (1991). Let the probability density function of the random variables X and Y be $f_X(x)$ and $f_Y(y)$ respectively. Then

$$f_Y(y) = \frac{1}{|a|} f_X\left(\frac{y}{a}\right)$$

The result follows by using the definition of differential entropy.
24. Let $\beta_n(t) = sinc(\omega_c t - n\pi)$, for each $n \in \mathbb{Z}$, $t \in \mathbb{R}$. Prove that the sequence of functions $\{\beta_n(t) \mid n \in \mathbb{Z}\}$ are orthogonal over $t \in \mathbb{R}$.
Hint: We prove that

$$\int_{-\infty}^{\infty} \beta_m(t)\beta_n(t)\,dt = \frac{\pi}{\omega_c}\delta_{mn}, \quad m, n \in \mathbb{Z}$$

In the above equation, δ_{mn} is the Kronecker's delta function, where $\delta_{mn} = 1$ if $m = n$, and equal to 0 otherwise. Note the Fourier transform pair

$$sinc(\omega_c t - n\pi) \leftrightarrow \frac{\pi}{\omega_c} e^{-in\pi\omega/\omega_c} g_{\omega_c}(\omega),$$

where $g_{\omega_c}(\cdot)$ is the gate function. In the next step use Parseval's relationship

$$\int_{-\infty}^{\infty} f_1(t) f_2(t)\,dt = \frac{1}{2\pi}\int_{-\infty}^{\infty} F_1(-\omega) F_2(\omega)\,d\omega$$

where $f_1(t) \leftrightarrow F_1(\omega)$, and $f_2(t) \leftrightarrow F_2(\omega)$ are Fourier transform pairs. Substitute $f_1(t) = \beta_m(t)$, and $f_2(t) = \beta_n(t)$ in the above equation. The result follows immediately.
25. Establish the observation about the capacity of $n \in \mathbb{P}$ parallel, discrete-time, memoryless Gaussian-noise communication channels.
Hint: See Cover and Thomas (1991), and Blahut (1987). Let $\widetilde{N} = (N_1, N_2, \ldots, N_n)$. Use of the hypotheses, and properties of entropy and mutual information lead to

$$\begin{aligned}
I_c\left(\widetilde{X}, \widetilde{Y}\right) &= H_c\left(\widetilde{Y}\right) - H_c\left(\widetilde{Y} \mid \widetilde{X}\right) = H_c\left(\widetilde{Y}\right) - H_c\left(\widetilde{N} \mid \widetilde{X}\right) \\
&= H_c\left(\widetilde{Y}\right) - H_c\left(\widetilde{N}\right) = H_c\left(\widetilde{Y}\right) - \sum_{j=1}^{n} H_c(N_j) \\
&\le \sum_{j=1}^{n} H_c(Y_j) - \sum_{j=1}^{n} H_c(N_j) = \sum_{j=1}^{n} \{H_c(Y_j) - H_c(N_j)\} \\
&\le \sum_{j=1}^{n} \frac{1}{2}\log_2\left(1 + \frac{P_{pow_j}}{\sigma_{N_j}^2}\right)
\end{aligned}$$

Equality in the above result occurs if the random variable X_j is independent of the random variable X_k, for $j \neq k$, and $1 \le j, k \le n$; and if $X_j \sim \mathcal{N}\left(0, \sigma_{X_j}^2\right)$, where $\sigma_{X_j}^2 \in \mathbb{R}^+$ for $1 \le j \le n$.
In the next step, the power allotment to each channel is determined, so that the capacity is maximized and $\sum_{j=1}^{n} P_{pow_j} \le P_{pow}$. This is a constrained optimization problem, which is

addressed via the method of Lagrange multipliers. Let $(P_{pow_1}, P_{pow_2}, \ldots, P_{pow_n}) \triangleq \Pi_{pow}$. The goal is to

$$\underset{\Pi_{pow}}{Maximize} \sum_{j=1}^{n} \frac{1}{2} \log_2 \left(1 + \frac{P_{pow_j}}{\sigma_{N_j}^2}\right)$$

Subject to:

$$\sum_{j=1}^{n} P_{pow_j} \leq P_{pow}$$

where $P_{pow_j} \geq 0$, for $1 \leq j \leq n$.
Observe that the objective function is concave. The Lagrangian functional of this problem is

$$\mathcal{L}(\Pi_{pow}, \lambda) = \sum_{j=1}^{n} \frac{1}{2} \log_2 \left(1 + \frac{P_{pow_j}}{\sigma_{N_j}^2}\right) + \lambda \left(P_{pow} - \sum_{j=1}^{n} P_{pow_j}\right)$$

where λ is a Lagrangian multiplier. This Lagrangian is maximized by varying the P_{pow_j}'s where $P_{pow_j} \geq 0$, $1 \leq j \leq n$. The corresponding Karush-Kuhn-Tucker conditions are:

$$\frac{\partial \mathcal{L}}{\partial P_{pow_j}} \leq 0, \quad 1 \leq j \leq n$$

$$P_{pow_j} \frac{\partial \mathcal{L}}{\partial P_{pow_j}} = 0, \quad 1 \leq j \leq n$$

$$\lambda \frac{\partial \mathcal{L}}{\partial \lambda} = 0, \quad \lambda \geq 0$$

$$P_{pow_j} \geq 0, \quad 1 \leq j \leq n$$

If $\lambda > 0$, then $\partial \mathcal{L}/\partial \lambda = 0$ implies $\sum_{j=1}^{n} P_{pow_j} = P_{pow}$. Let $\log_2 e \triangleq \varrho$, and

$$\frac{\partial \mathcal{L}(\Pi_{pow}, \lambda)}{\partial P_{pow_j}} = \frac{\varrho}{2\left(\sigma_{N_j}^2 + P_{pow_j}\right)} - \lambda = 0$$

This implies

$$P_{pow_j} = \theta - \sigma_{N_j}^2$$

where $\theta = \varrho/(2\lambda)$. As $P_{pow_j} \geq 0$, for $1 \leq j \leq n$; let $P_{pow_j} = \left(\theta - \sigma_{N_j}^2\right)^+$. Recall that $x^+ = \max(0, x)$, for $x \in \mathbb{R}$. The value θ is selected so that

$$P_{pow} = \sum_{j=1}^{n} P_{pow_j} = \sum_{j=1}^{n} \left(\theta - \sigma_{N_j}^2\right)^+$$

26. Let a random variable X have a normal distribution with zero mean, and standard deviation σ. Also assume that only a single bit is available for its representation. Determine $\widehat{X}(x)$, $x \in \mathbb{R}$, so that the expected squared error is minimized. That is,

$$\mathcal{E}\left[\left(X - \widehat{X}(X)\right)^2\right]$$

is minimized. Also determine the corresponding expected distortion.

Hint: See Cover, and Thomas (1991). The probability density function of X is given by

$$f_X(x) = \frac{1}{\sqrt{2\pi}\sigma} \exp\left\{ -\frac{1}{2}\left(\frac{x}{\sigma}\right)^2 \right\}, \quad x \in \mathbb{R}$$

where $\sigma \in \mathbb{R}^+$. As only a single bit is available, the reconstruction regions are the positive and negative parts of the real-line. Let $\widehat{X}(X)$ take value β for X positive, and value $-\beta$ for X negative. Observe that

$$\begin{aligned}
&\mathcal{E}\left[\left(X - \widehat{X}(X)\right)^2\right] \\
&= \int_{-\infty}^{0} (x+\beta)^2 f_X(x)\,dx + \int_{0}^{\infty} (x-\beta)^2 f_X(x)\,dx \\
&= 2\int_{0}^{\infty} (x-\beta)^2 f_X(x)\,dx \\
&= 2\int_{0}^{\infty} x^2 f_X(x)\,dx - 4\beta\int_{0}^{\infty} x f_X(x)\,dx + 2\beta^2\int_{0}^{\infty} f_X(x)\,dx \\
&= \left(\sigma^2+\beta^2\right) - 4\beta\int_{0}^{\infty} x f_X(x)\,dx \\
&= \left(\sigma^2+\beta^2\right) - 4\beta\frac{\sigma}{\sqrt{2\pi}}
\end{aligned}$$

Therefore

$$\left(\sigma^2+\beta^2\right) - 4\beta\frac{\sigma}{\sqrt{2\pi}}$$

is minimized by varying β. It occurs at $\beta = \sqrt{2/\pi}\sigma$. Thus

$$\widehat{X}(x) = \begin{cases} \sqrt{\frac{2}{\pi}}\sigma, & x \geq 0 \\ -\sqrt{\frac{2}{\pi}}\sigma, & x < 0 \end{cases}$$

The corresponding expected distortion is

$$\begin{aligned}
\mathcal{E}\left[\left(X - \widehat{X}(X)\right)^2\right] &= \left(\sigma^2+\beta^2\right) - 4\beta\frac{\sigma}{\sqrt{2\pi}}\bigg|_{\beta=\sqrt{2/\pi}\sigma} \\
&= \frac{\pi-2}{\pi}\sigma^2
\end{aligned}$$

27. Consider a continuous random variable X, where $\mathcal{E}(X) = 0$, $Var(X) = \sigma^2$, and entropy equal to $H_c(X)$. For squared error distortion, and distortion threshold $D \in \mathbb{R}^+$, the rate distortion $R(D)$ is bounded as

$$H_c(X) - \frac{1}{2}\log_2(2\pi e D) \leq R(D) \leq \frac{1}{2}\log_2\frac{\sigma^2}{D}$$

Establish this assertion.
Hint: See Cover, and Thomas (1991). The lower bound is first established. Assume

$$\mathcal{E}\left(\left(X-\widehat{X}\right)^2\right) \le D$$

where $\widehat{X}$ is the representation random variable which corresponds to the random variable X. Following facts are used in establishing a lower bound.
Fact 1: Conditioning might reduce entropy.
Fact 2: Let $Y = \left(X - \widehat{X}\right)$, and $\mathcal{E}\left(Y^2\right) \triangleq \sigma_Y^2$. Then $H_c(Y) \le \frac{1}{2}\log_2\left\{2\pi e\sigma_Y^2\right\}$.
Fact 3: $\mathcal{E}\left[\left(X-\widehat{X}\right)^2\right] \le D$.
Thus

$$\begin{aligned}
I_c\left(X;\widehat{X}\right) &= H_c(X) - H_c\left(X \mid \widehat{X}\right) \\
&= H_c(X) - H_c\left(X - \widehat{X} \mid \widehat{X}\right) \\
&\ge H_c(X) - H_c\left(X - \widehat{X}\right) && \text{(via Fact 1)} \\
&\ge H_c(X) - \frac{1}{2}\log_2\left(2\pi e\sigma_Y^2\right) && \text{(via Fact 2)} \\
&= H_c(X) - \frac{1}{2}\log_2\left\{2\pi e\mathcal{E}\left[\left(X-\widehat{X}\right)^2\right]\right\} && \text{(via Fact 2)} \\
&\ge H_c(X) - \frac{1}{2}\log_2\left\{2\pi eD\right\} && \text{(via Fact 3)}
\end{aligned}$$

Therefore, a lower bound for $I_c\left(X;\widehat{X}\right)$ is

$$H_c(X) - \frac{1}{2}\log_2\left\{2\pi eD\right\} \le I_c\left(X;\widehat{X}\right)$$

In order to establish an upper bound for $I_c\left(X;\widehat{X}\right)$, let

$$\widehat{X} = c(X+Z),\ c \in \mathbb{R};\ \text{and}\ \ Z \sim \mathcal{N}\left(0, b\sigma^2\right),\ \ b \in \mathbb{R}^+$$

where X is independent of Z. The relationship between the unknowns b and c; and σ^2 and D is determined subsequently. Observe that

$$\mathcal{E}\left(\widehat{X}\right) = 0,\ \text{and}\ \sigma_{\widehat{X}}^2 = \mathcal{E}\left(\widehat{X}^2\right) = (1+b)\,c^2\sigma^2$$

Let $Y = \left(X - \widehat{X}\right)$, then

$$\mathcal{E}(Y) = 0,\ \ \text{and}\ \ \mathcal{E}\left(Y^2\right) = \left\{bc^2 + (1-c)^2\right\}\sigma^2$$

Following facts are used in establishing an upper bound for $I_c\left(X;\widehat{X}\right)$.
Fact 4: As random variables X and Z are independent,

$$H_c\left(\widehat{X} \mid X\right) = H_c(c(X+Z) \mid X) = H_c(cZ)$$

Fact 5: $H_c\left(\widehat{X}\right) \le \frac{1}{2}\log_2\left\{2\pi e\sigma_{\widehat{X}}^2\right\}$, and $H_c(cZ) = \frac{1}{2}\log_2\left\{2\pi ebc^2\sigma^2\right\}$.

Thus

$$\begin{aligned}
I_c\left(X;\widehat{X}\right) &= I_c\left(\widehat{X};X\right) = H_c\left(\widehat{X}\right) - H_c\left(\widehat{X} \mid X\right) \\
&= H_c\left(\widehat{X}\right) - H_c\left(cZ\right) \qquad \text{(via Fact 4)} \\
&\le \frac{1}{2}\log_2\left\{2\pi e\sigma_{\widehat{X}}^2\right\} - \frac{1}{2}\log_2\left\{2\pi e b c^2\sigma^2\right\} \qquad \text{(via Fact 5)} \\
&= \frac{1}{2}\log_2\left\{\frac{\sigma_{\widehat{X}}^2}{bc^2\sigma^2}\right\} = \frac{1}{2}\log_2\left\{\frac{b+1}{b}\right\}
\end{aligned}$$

Therefore, an upper bound for $I_c\left(X;\widehat{X}\right)$ is

$$I_c\left(X;\widehat{X}\right) \le \frac{1}{2}\log_2\left\{\frac{b+1}{b}\right\}$$

In the next step, the relationship between b, and c; and σ^2 and D is determined. Let

$$\sigma^2 = \sigma_{\widehat{X}}^2 + D, \text{ and } 1 + \frac{1}{b} = \frac{\sigma^2}{D}$$

Use of the above results lead to

$$c = 1 - \frac{D}{\sigma^2}, \text{ and } b = \frac{D}{\sigma^2 - D}, \quad \text{for } D < \sigma^2$$

Finally, an upper bound for $I_c\left(X;\widehat{X}\right)$ in terms of the known parameters is

$$I_c\left(X;\widehat{X}\right) \le \frac{1}{2}\log_2\frac{\sigma^2}{D}, \quad \text{for } D < \sigma^2$$

28. Prove that for a Gaussian source $\mathcal{N}\left(0,\sigma^2\right)$, with a squared error distortion, the rate distortion function is

$$R(D) = \begin{cases} \frac{1}{2}\log_2\frac{\sigma^2}{D}, & 0 < D \le \sigma^2 \\ 0, & D > \sigma^2 \end{cases}$$

Hint: See Cover, and Thomas (1991). The input symbol has a Gaussian distribution. Denote this random variable by X, and its representation by $\widehat{X}(X)$. The rate distortion function for the Gaussian source is

$$R(D) = \min_{f(\widehat{x}\mid x)} I_c\left(X;\widehat{X}\right), \text{ such that } \int_x\int_{\widehat{x}} f\left(\widehat{x}\mid x\right) f(x)\, d\left(x,\widehat{x}\right) d\widehat{x}\, dx \le D$$

where the distortion $d\left(x,\widehat{x}\right) = \left(x-\widehat{x}\right)^2$. The rate distortion $R(D)$ is evaluated in two steps. In the first step, a lower bound for $R(D)$ is obtained. In the next step, it is shown that the lower bound is achievable by some conditional density $f\left(\widehat{x}\mid x\right)$.

Step 1: The following facts are used in the derivation.

Fact 1: The differential entropy of a Gaussian random variable X, with variance σ^2 is $H_c(X) = \frac{1}{2}\log_2\left(2\pi e\sigma^2\right)$.

Fact 2: Conditioning might reduce entropy.

Fact 3: Let $Y = \left(X - \widehat{X}\right)$, and $\sigma_Y^2 = \mathcal{E}\left[\left(X - \widehat{X}\right)^2\right]$, then $H_c(Y) \le \frac{1}{2}\log_2\left(2\pi e\sigma_Y^2\right)$.
Fact 4: $\mathcal{E}\left[\left(X - \widehat{X}\right)^2\right] \le D$.
Fact 5: Mutual information is always nonnegative.
Thus

$$
\begin{aligned}
I_c\left(X;\widehat{X}\right) &= H_c(X) - H_c\left(X \mid \widehat{X}\right) \\
&= \frac{1}{2}\log_2\left(2\pi e\sigma^2\right) - H_c\left(X - \widehat{X} \mid \widehat{X}\right) \quad \text{(via Fact 1)} \\
&\ge \frac{1}{2}\log_2\left(2\pi e\sigma^2\right) - H_c\left(X - \widehat{X}\right) \quad \text{(via Fact 2)} \\
&\ge \frac{1}{2}\log_2\left(2\pi e\sigma^2\right) - \frac{1}{2}\log_2\left\{2\pi e\mathcal{E}\left[\left(X - \widehat{X}\right)^2\right]\right\} \quad \text{(via Fact 3)} \\
&\ge \frac{1}{2}\log_2\left(2\pi e\sigma^2\right) - \frac{1}{2}\log_2\{2\pi eD\} \quad \text{(via Fact 4)} \\
&= \frac{1}{2}\log_2\frac{\sigma^2}{D}
\end{aligned}
$$

Therefore

$$R(D) \ge \frac{1}{2}\log_2\frac{\sigma^2}{D},\quad D \le \sigma^2 \quad \text{(via Fact 5)}$$

As the smallest possible value of rate distortion function is zero, $R(D) = 0$ for $D > \sigma^2$,
Step 2: In order to find the conditional density $f(\widehat{x} \mid x)$ with the stated lower bound, it is convenient to examine the conditional density $f(x \mid \widehat{x})$. If $D \le \sigma^2$, then let

$$Z \sim \mathcal{N}(0, D),\quad \widehat{X} \sim \mathcal{N}\left(0, \sigma^2 - D\right),\quad \text{and}\quad X = \widehat{X} + Z$$

Further $\widehat{X}$ and Z are independent of each other. In this case $\mathcal{E}\left(\left(X - \widehat{X}\right)^2\right) = D$, and

$$I_c\left(X;\widehat{X}\right) = \frac{1}{2}\log_2\frac{\sigma^2}{D}$$

This achieves the lower bound derived in Step 1, if $D \le \sigma^2$. If $D > \sigma^2$, select $\widehat{X} = 0$ with probability 1; which in turn implies $R(D) = 0$.

29. Consider a source which emits symbols at discrete instants of time. Furthermore, the source is memoryless, and the input symbols are binary. That is, the input alphabet $A = \{0, 1\}$. The output alphabet B, is also equal to $\{0, 1\}$. The input symbols have a Bernoulli distribution with parameter $\xi \in (0, 1)$. For Hamming distortion, prove that the rate distortion function is

$$
R(D) = \begin{cases} H(\xi) - H(D), & 0 \le D \le \min\{\xi, 1-\xi\} \\ 0, & D > \min\{\xi, 1-\xi\} \end{cases}
$$

where $H(X) = -\xi\log_2\xi - (1-\xi)\log_2(1-\xi) \triangleq H(\xi)$.
Hint: See Cover, and Thomas (1991). The input symbol has a Bernoulli distribution. Denote this random variable by X, and its representation by $\widehat{X}(X)$. Without loss of generality, assume that $\xi < 1/2$. For brevity in notation, let $\oplus$ denote modulo 2 addition. Therefore $X \oplus \widehat{X} = 1$

implies $X \neq \widehat{X}$. The rate distortion $R(D)$ is evaluated in two steps. In the first step, a lower bound for $R(D)$ is obtained. In the second step it is demonstrated that this lower bound is feasible.

Step 1: The following facts are used in the derivation.

Fact 1: The entropy of the Bernoulli random variable X is

$$H(X) = -\xi \log_2 \xi - (1-\xi)\log_2(1-\xi) \triangleq H(\xi)$$

Fact 2: Conditioning might reduce entropy.

Fact 3: Let $X \oplus \widehat{X} \triangleq Z$. It is required that the average distortion is less than or equal to D. The output alphabet B, is also equal to $\{0,1\}$. The average value of Hamming distortion is

$$\overline{d}(G) = \sum_{x,\widehat{x}\in\{0,1\}} p(x,\widehat{x})\, d(x,\widehat{x}) = p(0,1) + p(1,0) = P(Z=1)$$

As $\overline{d}(G) \leq D$, we have $P(Z=1) \leq D$. This implies $P\left(X \neq \widehat{X}\right) \leq D$. If $0 \leq D \leq 1/2$, then $H(D)$ is an increasing function of D. Thus

$$\begin{aligned}
I\left(X;\widehat{X}\right) &= H(X) - H\left(X \mid \widehat{X}\right) \\
&= H(\xi) - H\left(X - \widehat{X} \mid \widehat{X}\right) \quad \text{(via Fact 1)} \\
&= H(\xi) - H\left(X \oplus \widehat{X} \mid \widehat{X}\right) \\
&\geq H(\xi) - H\left(X \oplus \widehat{X}\right) \quad \text{(via Fact 2)} \\
&\geq H(\xi) - H(D), \quad \text{for } D \leq 1/2 \quad \text{(via Fact 3)}
\end{aligned}$$

Thus $I\left(X;\widehat{X}\right) \geq H(\xi) - H(D)$, for $D \leq 1/2$.

Step 2: It is shown in this step that, the lower bound for $I\left(X;\widehat{X}\right)$ obtained in Step 1 is actually feasible. It is actually shown that it is possible to have a rate distortion function with $R(D) = H(\xi) - H(D)$, for $0 \leq D \leq \xi$. Note that the right hand side expression for $R(D)$ resembles the expression for the mutual information between the input and output symbols of a binary symmetric channel. In this imitation of the binary symmetric channel, the input-symbol and output-symbol random variables are $\widehat{X}$ and X respectively, where

$$\begin{aligned}
P\left(X=1 \mid \widehat{X}=0\right) &= P\left(X=0 \mid \widehat{X}=1\right) = D \\
P\left(X=0 \mid \widehat{X}=0\right) &= P\left(X=1 \mid \widehat{X}=1\right) = 1-D
\end{aligned}$$

Let $P\left(\widehat{X}=1\right) = r$, and $P\left(\widehat{X}=0\right) = (1-r)$. It is known that, $P(X=0) = (1-\xi)$, and $P(X=1) = \xi$. Therefore

$$\begin{aligned}
\xi &= P(X=1) \\
&= P\left(X=1 \mid \widehat{X}=0\right) P\left(\widehat{X}=0\right) + P\left(X=1 \mid \widehat{X}=1\right) P\left(\widehat{X}=1\right) \\
&= D(1-r) + (1-D)\, r
\end{aligned}$$

That is

$$r = \frac{\xi - D}{1 - 2D}$$

Therefore

$$(1 - r) = \frac{1 - \xi - D}{1 - 2D}$$

Assume $D < 1/2$. Then $r \geq 0$ implies $D \leq \xi$. Similarly, $(1 - r) \geq 0$ implies $D \leq (1 - \xi)$. Therefore $D \leq \xi$ and $D \leq (1 - \xi)$ together imply $D \leq \min\{\xi, 1 - \xi\}$. Thus

$$I\left(X; \widehat{X}\right) = H(X) - H\left(X \mid \widehat{X}\right) = H(\xi) - H(D)$$

and the expected distortion is $P\left(X \neq \widehat{X}\right) = D$.

If $D \geq \xi$, then it is possible to have $R(D) = 0$ by letting $\widehat{X} = 0$ with probability 1. Actually, in this case, $D = \xi$ and $I\left(X; \widehat{X}\right) = 0$. Analogously, for $D \geq (1 - \xi)$ it is possible to have $R(D) = 0$ by letting $\widehat{X} = 1$ with probability 1. The result follows.

30. The results of the last problem are duplicated by using Lagrange's optimization technique. The problem is restated for completeness.

Consider a source which emits symbols at discrete instants of time. Furthermore, the source is memoryless, and the input symbols are binary. That is, the input alphabet $A = \{0, 1\}$. The output alphabet B, is also equal to $\{0, 1\}$. The input symbols have a Bernoulli distribution with parameter $\xi \in (0, 1)$. For Hamming distortion, determine the rate distortion function.

Hint: See Cover, and Thomas (1991); and van der Lubbe (1997). Denote the input-symbol random variable by X, and its representation by $\widehat{X}(X)$. We have

$$P(X = 1) \triangleq p(x_1) \triangleq p_1 = \xi, \text{ and } P(X = 0) \triangleq p(x_2) \triangleq p_2 = (1 - \xi)$$

Also

$$P\left(\widehat{X} = 1\right) \triangleq p(\widehat{x}_1) \triangleq q_1, \text{ and } P\left(\widehat{X} = 0\right) = p(\widehat{x}_2) \triangleq q_2$$

The problem is addressed in several steps. Without loss of generality, assume that $0 < \xi \leq 1/2$. Define the distortion matrix $\mathcal{D}$ as

$$\mathcal{D} = \begin{bmatrix} d_{11} & d_{12} \\ d_{21} & d_{22} \end{bmatrix} = \begin{bmatrix} 0 & 1 \\ 1 & 0 \end{bmatrix}$$

The probability transition matrix G is

$$G = \begin{bmatrix} q_{1|1} & q_{2|1} \\ q_{1|2} & d_{2|2} \end{bmatrix}$$

Note that

$$p = \begin{bmatrix} p_1 & p_2 \end{bmatrix}, \ q = \begin{bmatrix} q_1 & q_2 \end{bmatrix}, \text{ and } pG = q$$

Therefore

$$p = \begin{bmatrix} \xi & (1 - \xi) \end{bmatrix}$$

Step 1: In this step, at the smallest value of $D = D_{min}$; the probability transition matrix G, the average distortion $\overline{d}(G)$, and the corresponding rate distortion function $R(D_{min})$ are determined. Thus

$$j(x_1) = \arg\min_{1\le j\le 2} d_{1j} = 1, \text{ and } j(x_2) = \arg\min_{1\le j\le 2} d_{2j} = 2$$

Therefore, the corresponding probability transition matrix is

$$G = \begin{bmatrix} 1 & 0 \\ 0 & 1 \end{bmatrix}$$

Also

$$d(x_1) = \min_{1\le j\le 2} d_{1j} = 0, \text{ and } d(x_2) = \min_{1\le j\le 2} d_{2j} = 0$$

Therefore

$$D_{min} = \sum_{i=1}^{2} p_i d(x_i) = 0$$

The corresponding rate distortion function $R(D_{min})$ is next determined. Use of the relationship $pG = q$, leads to

$$q = \begin{bmatrix} q_1 & q_2 \end{bmatrix} = \begin{bmatrix} \xi & (1-\xi) \end{bmatrix}$$

and

$$R(D) = \sum_{i=1}^{2}\sum_{j=1}^{2} p_i q_{j|i} \log_2 \frac{q_{j|i}}{q_j} = -\xi\log_2\xi - (1-\xi)\log_2(1-\xi)$$
$$= H(\xi) = H(X)$$

Step 2: In this step, at D_{max}; the probability transition matrix G, and the average distortion $\overline{d}(G)$ are determined. As per the definition of D_{max}, $R(D_{max}) = 0$. Let

$$j^* = \arg\min_{1\le j\le 2} \sum_{i=1}^{2} p_i d_{ij} = 2$$

The minimum value of $\overline{d}(G)$ occurs at $q_{j^*} = q_2 = 1$ and $q_1 = 0$. Thus

$$D_{max} = \min_{1\le j\le 2} \sum_{i=1}^{2} p_i d_{ij} = \xi$$

Therefore, $q = \begin{bmatrix} q_1 & q_2 \end{bmatrix} = \begin{bmatrix} 0 & 1 \end{bmatrix}$. As $p = \begin{bmatrix} \xi\ (1-\xi) \end{bmatrix}$, and $pG = q$, the corresponding probability transition matrix is

$$G = \begin{bmatrix} 0 & 1 \\ 0 & 1 \end{bmatrix}$$

It can indeed be verified that

$$R(D_{max}) = \sum_{i=1}^{2}\sum_{j=1}^{2} p_i q_{j|i} \log_2 \frac{q_{j|i}}{q_j} = 0$$

Step 3: In this step, the transition probability matrix G, and the corresponding rate distortion function $R(D)$ for $D_{min} < D < D_{max}$ is determined. That is, $0 < D < \xi$. This is done in several substeps. The results of Lagrangian optimization are used. Let

$$e^{-\lambda/\varrho} \triangleq a; \text{ and } \mu_i \triangleq \frac{1}{\beta_i}, \text{ for } i = 1, 2$$

Step 3.1: The algorithm specifies the relationships

$$\sum_{i=1}^{2} \beta_i p_i a^{d_{ij}} = 1, \quad \text{for } j = 1, 2$$

The above set of equations yield

$$\beta_1 = \frac{1}{(a+1)\,\xi}, \quad \beta_2 = \frac{1}{(a+1)\,(1-\xi)}$$

This implies

$$\mu_1 = (a+1)\,\xi, \quad \mu_2 = (a+1)\,(1-\xi)$$

Step 3.2: Determine q_1 and q_2 from the relationships

$$\mu_i = \sum_{l=1}^{2} q_l a^{d_{il}}, \quad \text{for } i = 1, 2$$

The above set of equations yield

$$q_1 = \frac{\xi - a\,(1-\xi)}{(1-a)}, \quad q_2 = \frac{(1 - \xi - a\xi)}{(1-a)}$$

Observe that $(q_1 + q_2) = 1$.
Step 3.3: An explicit expression for the average distortion $\overline{d}\,(G)$ is determined.

$$\begin{aligned} D = \overline{d}\,(G) &= \sum_{i=1}^{2} \sum_{j=1}^{2} p_i q_{j|i} d_{ij} \\ &= \sum_{i=1}^{2} \sum_{j=1}^{2} p_i \beta_i q_j a^{d_{ij}} d_{ij} \\ &= \frac{a}{1+a} \end{aligned}$$

This also implies

$$a = \frac{D}{1-D}$$

Step 3.4: Expressions for q_1 and q_2 can also be obtained in terms of ξ and D by using results from Steps 3.2 and 3.3. This yields

$$q_1 = \frac{\xi - D}{(1-2D)}, \quad q_2 = \frac{(1-\xi) - D}{(1-2D)}$$

These probabilities were also obtained in the last problem.
Step 3.5: An expression for $R\,(D)$ is determined. As $D = \overline{d}\,(G)$,

$$R\,(D) = -\lambda D - \sum_{i=1}^{2} p_i \log_2 \mu_i$$

The relationship $a = e^{-\lambda/\varrho}$ implies $\lambda = -\log_2 a$. Therefore $-\lambda D = D \log_2 a$. Thus

$$\begin{aligned} R(D) &= D \log_2 a - \xi \log_2 \{(a+1)\xi\} - (1-\xi)\log_2 \{(a+1)(1-\xi)\} \\ &= -\xi \log_2 \xi - (1-\xi)\log_2 (1-\xi) + D\log_2 a - \log_2 (a+1) \\ &= H(\xi) + D\log_2 a - \log_2 (a+1) \end{aligned}$$

Substitute $a = D/(1-D)$ in the above result. This leads to

$$\begin{aligned} R(D) &= H(\xi) + D\log_2 D + (1-D)\log_2 (1-D) \\ &= H(\xi) - H(D) \end{aligned}$$

This is the expression for $R(D)$ which was obtained in the last problem.
Step 3.6: The elements of the probability transition matrix G are determined. Note that

$$q_{j|i} = q_j \beta_i a^{d_{ij}}, \quad \text{for } 1 \le i \le 2, 1 \le j \le 2$$

Use of results from Steps 3.1 and 3.2 leads to

$$\begin{aligned} q_{1|1} &= \frac{\xi - (1-\xi)a}{\xi(1-a^2)}, \quad & q_{2|1} &= \frac{-a^2\xi + (1-\xi)a}{\xi(1-a^2)} \\ q_{1|2} &= \frac{\xi a - (1-\xi)a^2}{(1-\xi)(1-a^2)}, \quad & q_{2|2} &= \frac{-\xi a + (1-\xi)}{(1-\xi)(1-a^2)} \end{aligned}$$

It can be checked that $(q_{1|1} + q_{2|1}) = 1$, and $(q_{1|2} + q_{2|2}) = 1$. The elements of the probability transition matrix G can also be expressed in terms of D. Substitute $a = D/(1-D)$ in the above result. This leads to

$$\begin{aligned} q_{1|1} &= \frac{(1-D)(\xi - D)}{\xi(1-2D)}, \quad & q_{2|1} &= \frac{D(1-\xi-D)}{\xi(1-2D)} \\ q_{1|2} &= \frac{D(\xi - D)}{(1-\xi)(1-2D)}, \quad & q_{2|2} &= \frac{(1-D)(1-\xi-D)}{(1-\xi)(1-2D)} \end{aligned}$$

31. Consider two channels C_{ch1} and C_{ch2} in cascade (series). The output of channel C_{ch1} is connected to the input of channel C_{ch2}. The channel C_{ch1} is connected to a memoryless source. Let the input and output of channel C_{ch1} be X and Y respectively. As the two channels are in cascade, the input to channel C_{ch2} is Y. Let the output of channel C_{ch2} be Z. Prove that

$$I(X;Z) \le I(X;Y), \quad \text{and} \quad I(X;Z) \le I(Y;Z)$$

This result is called the data processing theorem. It establishes the relationship between the overall transmission rate and the transmission rates per sub-channel.
Hint: See van der Lubbe (1997). Let $P(X = x, Y = y, Z = z) \triangleq p(x, y, z)$. In order to prove $I(X;Z) \le I(X;Y)$, we establish that $H(X \mid Y) \le H(X \mid Z)$. As the input to channel C_{ch1} is memoryless, $p(z \mid y, x) = p(z \mid y)$, and $p(x \mid y, z) = p(x \mid z)$. It is next proved that $p(x \mid y, z) = p(x \mid y)$.

$$\begin{aligned} p(x \mid y, z) &= \frac{p(x, y, z)}{p(y, z)} \\ &= \frac{p(z \mid x, y)\, p(x, y)}{p(z \mid y)\, p(y)} = p(x \mid y) \end{aligned}$$

The following inequality is also used. It is known that if $\alpha \in \mathbb{R}^+$, $\ln \alpha \leq (\alpha - 1)$. Consequently $\log_2 \alpha \leq (\alpha - 1) \log_2 e$. Thus

$$\begin{aligned}
&H(X \mid Z) - H(X \mid Y) \\
&= -\sum_x \sum_z p(x,z) \log_2 p(x \mid z) + \sum_x \sum_y p(x,y) \log_2 p(x \mid y) \\
&= -\sum_x \sum_y \sum_z p(x,y,z) \log_2 p(x \mid z) + \sum_x \sum_y \sum_z p(x,y,z) \log_2 p(x \mid y) \\
&= -\sum_x \sum_y \sum_z p(x,y,z) \log_2 \frac{p(x \mid z)}{p(x \mid y)} \\
&= -\sum_x \sum_y \sum_z p(x,y,z) \log_2 \frac{p(x \mid z)}{p(x \mid y,z)} \\
&= -\sum_y \sum_z p(y,z) \sum_x p(x \mid y,z) \log_2 \frac{p(x \mid z)}{p(x \mid y,z)} \\
&\geq -(\log_2 e) \sum_y \sum_z p(y,z) \sum_x p(x \mid y,z) \left\{ \frac{p(x \mid z)}{p(x \mid y,z)} - 1 \right\} \\
&= -(\log_2 e) \sum_y \sum_z p(y,z) \sum_x \{p(x \mid z) - p(x \mid y,z)\} \\
&= -(\log_2 e) \sum_y \sum_z p(y,z) \{1 - 1\} = 0
\end{aligned}$$

Therefore

$$H(X \mid Z) - H(X \mid Y) \geq 0 \Rightarrow I(X;Z) \leq I(X;Y)$$

The second inequality is similarly proved.

32. Suppose that m independent observations $\mathcal{Y} = \{y_1, y_2, \ldots, y_m\}$ are drawn from a normal probability density function with parameters (μ, σ^2). Use the principle of maximum-likelihood estimation, to show that $\theta_{ML} = (\mu_{ML}, \sigma^2_{ML})$, where

$$\mu_{ML} = \frac{1}{m} \sum_{i=1}^{m} y_i,$$

$$\sigma^2_{ML} = \frac{1}{m} \sum_{i=1}^{m} (y_i - \mu_{ML})^2$$

Hint: See Spiegel, Schiller, and Srinivasan (2000), and Mood, Graybill, and Boes (1974).

References

1. Abramson, N., 1963. *Information Theory and Coding*, McGraw-Hill Book Company, New York, New York.

2. Ash, R. B., 1965. *Information Theory*, Dover Publications, Inc., New York, New York.
3. Blahut, R. E., 1987. *Principles and Practice of Information Theory*, Addison-Wesley Publishing Company, New York, New York.
4. Bruen, A. A., and Forcinito, M. A., 2005. *Cryptography, Information Theory, and Error-Correction*, John Wiley & Sons, Inc., New York, New York.
5. Cover, T. M., and Thomas, J. A., 1991. *Elements of Information Theory*, John Wiley & Sons, Inc., New York, New York.
6. Csiszár, I., and Körner, J., 2011. *Information Theory: Coding Theorems for Discrete Memoryless Systems*, Cambridge University Press, Cambridge, Great Britain.
7. El Gamal, A., and Kim, Y., 2011. *Network Information Theory*, Cambridge University Press, Cambridge, Great Britain.
8. Gallager, R. G., 1968. *Information Theory and Reliable Communication*, John Wiley & Sons, Inc., New York, New York.
9. Gallager, R. G., 2008. *Principles of Digital Communication*, Cambridge University Press, Cambridge, Great Britain.
10. Greven, A., Keller, G., and Warnecke, G., 2003. *Entropy*, Princeton University Press, Princeton, New Jersey.
11. Hamming, R. W., 1986. *Coding and Information Theory*, Second Edition, Prentice-Hall, Englewood Cliffs, New Jersey.
12. Han, T. S. and Kobayashi, K., 2002. *Mathematics of Information and Coding*, American Mathematical Society, Providence, Rhode Island.
13. Hankerson, D., Harris, G. A., and Johnson, P. D., 2003. *Introduction to Information Theory and Data Compression*, Second Edition, Chapman and Hall/CRC Press, New York, New York.
14. Haykin, S., 2001. *Communication Systems*, Fourth Edition, John Wiley & Sons, Inc., New York, New York.
15. Huffman, D. A., 1952. "A Method for the Construction of Minimum Redundancy Codes," Proc. IRE, Vol. 40, No. 9, pp. 1098-1101.
16. Jones, G. A., and Jones, J. M., 2000. *Information and Coding Theory*, Springer-Verlag, Berlin, Germany.
17. Khinchin, A. I., 1957. *Mathematical Foundations of Information Theory*, Dover Publications, Inc., New York, New York.
18. MacKay, D. J. C., 2003. *Information Theory, Inference and Learning Algorithms*, Cambridge University Press, Cambridge, Great Britain.
19. Mansuripur, M. 1987. *Introduction to Information Theory*, Prentice-Hall, Englewood Cliffs, New Jersey.
20. Mood, A. M., Graybill, F. A., and Boes, D. C., 1974. *Introduction to the Theory of Statistics,* Third Edition, McGraw-Hill Book Company, New York, New York.
21. Moser, S., 2014. "Information Theory," available at http://moser-isi.ethz.ch/docs/it_script_v42.pdf. Retrieved February 16, 2015.
22. Nielsen, M. A., and Chuang, I. L., 2000. *Quantum Computation and Quantum Information*, Cambridge University Press, Cambridge, Great Britain.
23. Peres, A., 1993. *Quantum Theory: Concepts and Methods*, Kluwer Academic Publishers, Norwell, Massachusetts.
24. Pu, I. M., 2006. *Fundamental Data Compression*, Butterworth-Heinemann, Great Britain.
25. Reza, F. M., 1994. *An Introduction to Information Theory*, Dover Publications, Inc., New York, New York.
26. Sanov, I., 1957. "On the Probability of Large Deviations of Random Variables," (in Russian), Mat. Sb., Vol. 42 (84), pp. 11-44. (English Translation in: *Selected Translations in Mathematical Statistics*: I, pp. 213-244, 1961).
27. Sayood, K., Editor, 2003. *Lossless Compression Handbook*, Academic Press, New York, New York.

28. Shannon, C. E., and Weaver, W., 1972. *The Mathematical Theory of Communication*, University of Illinois Press, Urbana, Illinois.
29. Shor, P. W., 2005. "Lempel-Ziv," available at http://www-math.mit.edu/~shor/PAM/lempel_ziv_notes.pdf. Retrieved February 16, 2015.
30. Shwartz, A., and Weiss, A., 1995. *Large Deviations for Performance Analysis*, Chapman and Hall/CRC Press, New York, New York.
31. Spiegel, M. R., Schiller, J., and Srinivasan, R. A., 2000. *Probability and Statistics*, Second Edition, Schaum's Outline Series, McGraw-Hill Book Company, New York, New York.
32. Togneri, R., and deSilva C. J. S., 2002. *Fundamentals of Information Theory and Coding Design*, Chapman & Hall/CRC Press: New York, New York.
33. van der Lubbe, J. C. A., 1997. *Information Theory*, Cambridge University Press, Cambridge, Great Britain.
34. Verdú, S., 1998. "Fifty Years of Shannon Theory," IEEE Transactions on Information Theory, Vol. 44, No. 6, pp. 2057-2078.
35. Welch, T. A., 1984. "A Technique for High Performance Data Compression." IEEE Computer, Vol. 17, No. 6, pp. 8-19.
36. Wells, R. B., 1999. *Applied Coding and Information Theory for Engineers*, Prentice-Hall, Englewood Cliffs, New Jersey.
37. Welsh, D., 1988. *Codes and Cryptography*, Oxford University Press, Oxford, Great Britain.
38. Yeung, R. W., 2002. *A First Course in Information Theory*, Kluwer Academic Publishers, Norwell, Massachusetts.
39. Ziv, J., and Lempel, A., 1977. "A Universal Algorithm for Sequential Data Compression." IEEE Transactions on Information Theory, Vol. 23, No. 3, pp. 337-343.
40. Ziv, J., and Lempel, A., 1978. "Compression of Individual Sequences via Variable Rate Coding." IEEE Transactions on Information Theory, Vol. 24, No. 5, pp. 530-536.

CHAPTER 2

Algebraic Coding Theory

$$GF_{p^n} = \mathbb{Z}_p[x] / (f(x))$$

Galois field

Évariste Galois. Galois was born on 25 October, 1811 in Bourg La Reine (near Paris), France. Galois' life was indeed very tragic. His father committed suicide, when he was very young. He floundered in his school work and twice failed to gain admission to the famed École Polytechnique. Galois developed a novel technique to study the solubility of polynomial equations which is currently called group theory. He demonstrated that polynomial equations of degree higher than four are in general not soluble in terms of finite number of rational operations and root extractions. The phrase, "Galois group" was coined in his honor.

Galois was killed in a duel, under mysterious circumstances. The duel was supposedly over a woman. He died on 31 May, 1832 in Paris, France. If ever there is a prime example of a genius, it would be Galois, whose life was unfortunately abbreviated by such an event.

2.1 Introduction

Algebraic coding theory is the science of detecting and correcting errors on noisy information channels via algebraic techniques. The following topics are discussed in this chapter: basics of coding theory, vector or linear space over a binary field, linear codes, cyclic codes, Hamming codes, and cyclic redundancy check codes. In addition, Reed-Muller codes, Bose-Chaudhuri-Hocquenghem codes, Reed-Solomon codes, convolutional codes, and turbo codes are also discussed.

Typically, the source data is encoded and transmitted over a communication channel. Often times, the originally transmitted information gets perturbed by external and unwanted sources on the communication channel. This external disturbance on the channel is called *noise*. In order to possibly recover the original signal at the receiver, the data encoding process at the transmitter adds calculated redundancy to the transmitted code. When the transmitted information is perturbed by noise, this redundancy enables the possible recovery of the original message at the receiver. This redundancy occurs in the form of error-detecting and error-correcting codes. Therefore, it can be said that error-detecting and error-correcting codes are implemented to possibly detect and correct errors when data is transmitted over noisy communication channels.

A code is a string of symbols or words which represent source words or symbols. For example, modern computers work with binary numbers. Consequently decimal numbers are represented in binary form inside the computer. Samuel F. B. Morse, the inventor of telegraph used a variable length ternary code to represent characters of the English language. His alphabet (set of symbols) was made up of three characters: dot, dash, and space. Louis Braille developed the Braille system for blind people when he was only fifteen years old. He used dots in a 3×2 matrix to represent letters, numbers, and punctuation symbols. This code led to $2^6 = 64$ different representations.

There are several types of communication channels: radio channels, satellite channels, fiber channels, magnetic storage devices, compact discs, cellular telephones, and the list goes on. All these channels are susceptible to errors. There is a well-understood discipline to detect and correct errors. This field was born in 1948 with the publication of Claude E. Shannon's landmark paper

"A Mathematical Theory of Communication." One of the celebrated results of Shannon, states that arbitrarily reliable communication is possible over a channel, if the information is transmitted at a rate below the channel capacity. This result also implies that if the raw data is encoded with a specified redundancy and transmitted over a noisy channel, the received data can be decoded to a specified degree of accuracy. Simply stated: By adding redundancy to data, it is possible to detect errors and sometimes possibly correct them. This error detection and correction is done via the use of error detecting and correcting codes. It might be added that, Shannon's proof of the noisy channel coding theorem is nonconstructive. That is, it does not specify a technique to construct such codes.

Among several applications, the science and art of error-detection and error-correction is applied to voice, data, image, and video transmission, encryption of electronic messages, compact disc recording, other storage technologies, satellite transmission, and packet transmission on the Internet. We initially consider the simple examples of triple-repetition code, parity-check code, and checksum code. In these examples, and elsewhere in the textbook, a *bit* is a binary digit. It takes a value of either 0 or 1. The phrase "bit" was coined by John W. Tukey (1915-2000) in the year 1947. It was subsequently used by Claude E. Shannon in his landmark paper: *Mathematical Theory of Communication* in the year 1948.

Triple-Repetition Code

Assume that the source words are represented by strings of 0's and 1's. In this coding scheme, each occurrence of 0 is replaced by a string of three 0's, and an occurrence of 1 is replaced by a string of three 1's. For example the word (100110) is coded as (111 000 000 111 111 000) and transmitted. Note that blanks were inserted in the coded word for readability. This encoded word gets contaminated by noise during its channel transmission. That is, a 1 might be received as a 0, or a 0 might be received as a 1. Let the received codeword be $(101\ 000\ 100\ 110\ 111\ 001)$. It is decoded by using a simple majority rule. Notice that error occurred in four positions, yet we were lucky enough to decode the received block correctly. For example, the received triple 101 corresponds to a 1 of the original word. Similarly, a 100 corresponds to a 0 of the original word. Some of the disadvantages in this coding scheme can be observed immediately.

(a) The code is very inefficient. The encoded message is three times as long as the original message. Therefore the effective information transfer rate reduces to a third of what it would have been possible on a noise-free channel.
(b) It can correct only single errors in a block of three. For example if there were two errors in a block, then the interpreted bit of the received word is incorrect.

In summary, the triple-repetition code has poor information transfer rate and only a single error-correcting capability.

Parity-Check Code

Recall that a byte has 8 bits. The first seven bits of a byte can be used to represent source words. The eighth bit of each byte can be used for error protection. It is called the parity bit. Let the source word be $a = (a_1, a_2, \ldots, a_7)$ where $a_i = 0$ or 1 for $1 \leq i \leq 7$. Then the 8-bit encoded word is $b = (b_1, b_2, \ldots, b_8)$, where $b_i = a_i$ for $1 \leq i \leq 7$, and $b_8 \equiv \sum_{i=1}^{7} a_i \pmod 2$. This coding scheme implies that the encoded word can be transmitted as a byte, and the number of 1's in a byte is either zero or an even number. Note that:

(a) The code is very efficient. The encoded message is only a seventh longer than the original word.
(b) The code cannot correct errors, but it can detect errors. That is, the location of errors in the received byte cannot be determined. Therefore, this code can only be used if the receiver requests for a retransmission, provided it detects an error in the received byte.
(c) This code cannot tolerate more than one error per byte. Therefore probability of error in the transmission channel should be low.

Checksum Code

Checksum code is a generalization of the parity-check codes. It is also an error detecting code only. Assume that a message is made up of several message codewords, and that the message codewords are of fixed length (in number of bits). An example of a message codeword would be a byte. A checksum of a message is the modular sum of all the message codewords in the message.

Let $m \in \mathbb{P}$; and $a_1, a_2, \ldots, a_n \in [0, m)$ be a sequence of integers. Also let

$$w = a_0 + \sum_{i=1}^{n} a_i$$

where the integer $a_0 \in [0, m)$ is selected so that $w \equiv 0 \pmod{m}$. Then w is called the checksum. The transmitter transmits the sequence $a_0, a_1, \ldots, a_n$ to the receiver. The receiver receives the corresponding sequence $a'_0, a'_1, \ldots, a'_n$. The receiver then computes

$$w' \equiv \sum_{i=0}^{n} a'_i \pmod{m}$$

If $w' \equiv 0 \pmod{m}$, then the receiver assumes that the received sequence is indeed correct. A similar checksum scheme is used in the IP header using binary field arithmetic.

Codes which are more sophisticated and efficient than the triple-repetition, parity-check codes, and checksum code are discussed in the rest of this chapter.

It is assumed in this chapter that the reader is familiar with basic concepts from elementary number theory, abstract algebra, applied analysis, and matrices and determinants. It is advisable, but not necessary for the reader to get familiar with the basics of information theory. Information theory has been covered in a different chapter.

2.2 Basics of Coding Theory

In this section: information (or code) rate, Hamming distance, decoding rules, some interesting bounds on codes, perfect codes, and error probabilities are discussed. This list of topics is essentially the basics of coding theory. It also provides a framework for describing different characteristics of codes.

2.2.1 Information Rate

It is assumed in this chapter that the data (message) is representable by a string of characters (symbols or letters or elements) which belong to a finite set A, called *alphabet*. Denote the cardinality of this set by $|A| = q$.

The string of elements which represent a message (raw data) are assumed to be of fixed length k. The basic idea behind error-correcting codes is to add redundancy to this data word. This data word of length k symbols is encoded as a word of length n symbols, where $n > k$. The extra $(n - k)$ symbols help in detecting and correcting possible errors in the transmitted data. The $(n - k)$ symbols are also called *parity symbols*. The string of k message-word symbols is called an *information or message word*, and the string of n codeword symbols is called a *codeword*. Therefore, a codeword is made up of a string of information and parity symbols.

The disadvantage of letting $n > k$ is that the effective information transfer rate decreases due to the inclusion of this redundancy. Therefore the communication channel must be able transmit symbols at a faster rate than the source can produce. Also let M be the number of n-tuples which represent the encoded data. These M codewords are a subset of the set $A^{(n)}$, where

$$A^{(n)} = \underbrace{A \times A \times \cdots \times A}_{n \text{ times}} \tag{2.1}$$

This type of code is called an $[n, M]$-code over A. Such codes are also called *block-codes* of length n. The value n is called the block-length. The largest value of M is equal to q^k.

A *channel-encoder* at the transmitter is said to map the source-message of length k-symbols into a codeword of length n-symbols. Similarly, the *channel-decoder* at the receiver is required to detect and correct errors (if any). The channel decoder can process the received codeword in two different ways. These are the *forward error-correction* (FEC), and the *automatic-repeat request* (ARQ) schemes. In the FEC scheme, the channel-decoder is capable of both detecting and correcting errors. In the ARQ strategy, channel decoder can detect errors, and use a feedback channel (from the receiver, back to the transmitter) to request a repeat of data transmission, till the codeword has been received without any detectable errors.

The information rate of a block-code is next defined. The *information* or *code rate* ϑ, is a measure of the relative information transmitted in each codeword. It is also a measure of the nonredundancy in the code.

Definitions 2.1. *Basic definitions.*

1. *Let A be the alphabet, where $|A| = q$. The message word and the codeword are strings of k and n symbols respectively. The code C is a subset of $A^{(n)}$, and $|C| = M$. Therefore this code is also called an $[n, M]$-code.*
   The parameter k is also called the message-word length, and the parameter n the codeword-length or block-length.
2. *The information rate, or the code rate, or the efficiency of this code is given by ϑ, where*

$$\vartheta = \frac{\log_q M}{n} \tag{2.2}$$

□

The information rate ϑ of a block-code is a positive fraction. Note that

$$0 < \vartheta \leq \frac{k}{n} \leq 1 \tag{2.3}$$

In the digital world, the alphabet $A = \{0, 1\}$. The corresponding block-codes are called *binary block-codes*.

2.2.2 Hamming Distance

An important concept in algebraic coding theory is the *Hamming distance.* The Hamming distance is a metric, named after Richard W. Hamming (1915-1998). Hamming is generally regarded as a cofounder of algebraic coding theory. The concept of Hamming distance has also been described in the chapter on information theory. The *Hamming distance of a block-code* $\mathcal{C}$ is defined below. This metric is used in quantifying the distance between two codewords.

Definitions 2.2. *Let* $\mathcal{C}$ *be an* $[n, M]$*-block-code over the alphabet* A.

1. *Hamming distance* $d(a, b) \in \mathbb{N}$ *between two* n*-tuples* $a, b \in A^{(n)}$ *is the number of positions in which they differ.*
2. *The Hamming distance* d_H*, of the code* $\mathcal{C}$ *is the smallest Hamming distance between all possible pairs of different codewords. That is*

$$d_H = \min_{a,b \in \mathcal{C},\ a \neq b} \{d(a, b)\} \tag{2.4}$$

The block-code $\mathcal{C}$ *is generally specified by* $[n, M, d_H]$. □

The next theorem proves that the Hamming distance satisfies the distance axioms.

Theorem 2.1. *For any* $u, v, w \in \mathcal{C}$*, we have:*

(a) $d(u, v) \geq 0$*, with equality if and only if* $u = v$.
(b) $d(u, v) = d(v, u)$.
(c) $d(u, w) \leq d(u, v) + d(v, w)$*. This is the triangle inequality.*

Proof. See the problem section. □

Before making some useful observations about these codes, few more concepts and definitions are introduced. Note that the notation $[n, M, d_H]$ is applicable to all types of codes.

Definitions 2.3. *Basic definitions.*

1. *Let* $\tau, \rho \in A^{(n)}$ *be the transmitted and the received codewords respectively. The number of errors in the received codeword is* $d(\tau, \rho)$.
2. *A code* $\mathcal{C}$ *is said to detect* u *errors per codeword, if the decoder at the receiver-end is able to detect at most* u *errors in the received codeword.*
3. *A code* $\mathcal{C}$ *is said to correct* u *errors per codeword, if the decoder at the receiver-end is able to correct at most* u *errors in the received codeword.* □

Lemma 2.1. *Let* $\mathcal{C}$ *be an* $[n, M, d_H]$*-code. This code can detect at most* $(d_H - 1)$ *errors per codeword, and it can correct at most* $\lfloor (d_H - 1)/2 \rfloor$ *errors per codeword.*
Proof. The proof is left to the reader. □

One of the main goals of coding theory is to discover codes with large code rate and large Hamming distance. Large value of the code rate ϑ is required for efficiency, and large value of the Hamming distance d_H is necessary for correcting errors. However these two requirements are not harmonious.

Example 2.1. Consider, the code:

$$\mathcal{C} = \{(11100101), (00110010), (10101010), (11001100)\}$$

This is a $[8, 4, 3]$-code. The information rate of this code is $\vartheta = (\log_2 4)/8 = 0.25$, and $d_H = 3$. This code can detect up to 2 errors, and correct 1 error. □

2.2.3 Decoding Rules

Assume that the transmitted word is τ, and the received codeword is ρ. The goal of the decoder is to determine if error has occurred or not. Therefore

(a) If no error has occurred, it accepts $\rho = \tau$ as the transmitted codeword.
(b) If errors have occurred:
   (i) If possible, it corrects codeword ρ to codeword τ.
   (ii) Correction is impossible.

Different decoding rules have been discussed in the chapter on information theory. These were minimum-error decoding rule, and maximum-likelihood decoding rule. The later incorporated a Hamming-distance based decoding rule for a binary symmetric channel.

2.2.4 Bounds on Codes

A code $\mathcal{C}$ with symbols from the set A is specified by $[n, M, d_H]$, where the actual message length is k. The following question can be asked about this code. Given A, n, and d_H, what is the maximum possible value of M? Before this question is addressed, the following definitions are necessary.

Definitions 2.4. *A code $\mathcal{C}$ is specified by $[n, M, d_H]$. The alphabet of this code is A, and $|A| = q$.*

1. *$A_q(n, d_H)$ is the maximum value of M.*
2. *$V_q(n, \delta)$ is the number of codewords which are at most a distance δ from a specific codeword.* □

Observations 2.1. Some observations about $A_q(n, d_H)$ are listed below.

1. If $n \geq 1$, then $A_q(n, 1) = q^n$.
2. If $n \geq 1$, then $A_q(n, n) = q$.
3. If $n, d_H \geq 2$, then $A_q(n, d_H) \leq A_q(n-1, d_H - 1)$.
4. If d_H is even, then $A_2(n, d_H) = A_2(n-1, d_H - 1)$.
5. If $n \geq 2$, then $A_q(n, d_H) \leq q A_q(n-1, d_H)$. □

Some of the above observations are proved in the problem section. Using the above observations, it can be concluded that $A_2(n, 2) = 2^{n-1}$.

Observations 2.2. The following observations are about $V_q(n, \delta)$.

1. If the alphabet $A = \{0, 1\}$, then $q = 2$, and

$$V_2(n, \delta) = \binom{n}{0} + \binom{n}{1} + \cdots + \binom{n}{\delta} = \sum_{j=0}^{\delta} \binom{n}{j}$$

2. Let A be any alphabet. Then

$$V_q(n, \delta) = \binom{n}{0} + \binom{n}{1}(q-1) + \cdots + \binom{n}{\delta}(q-1)^{\delta} = \sum_{j=0}^{\delta} \binom{n}{j}(q-1)^j$$

□

The first observation follows from the second observation by substituting $q = 2$ in it. See the problem section for the proof of the second observation.

Bounds for $A_q(n, d_H)$, which is the maximum value of M, are next obtained. These bounds are called: Hamming bound, Singleton bound, and Gilbert-Varshamov bound. These bounds are named after the mathematicians R. W. Hamming, R. C. Singleton, E. N. Gilbert, and R. R. Varshamov. Having different upper bounds, allows us to select the smallest possible upper bound.

Theorem 2.2. (*Hamming or sphere-packing bound*) *The code $\mathcal{C}$ is specified by $[n, M, d_H]$. Its alphabet is A, $|A| = q$, and it can correct at most $t = \lfloor (d_H - 1)/2 \rfloor$ errors. Then*

$$A_q(n, d_H) \le \frac{q^n}{V_q(n, t)} \tag{2.5}$$

Proof. Consider a ball of radius δ around $u \in \mathcal{C}$. If $\delta \le t$, then all such balls should be disjoint. Consequently

$$MV_q(n, t) \le q^n$$

The last inequality is obtained by considering that there are M spheres in the space of codewords, each with a volume $V_q(n, t)$. The result follows. □

Theorem 2.3. (*Singleton bound*) *$\mathcal{C}$ is a code defined over the alphabet A. It is specified by $[n, M, d_H]$, and $|A| = q$. Then*

$$A_q(n, d_H) \le q^{n-d_H+1} \tag{2.6}$$

Proof. Observe that $A_q(n, n) = q$, which satisfies the above inequality. Assume that $n > d_H$. It has been proved that $A_q(n, d_H) \le qA_q(n-1, d_H)$. A repeated application of this inequality (which can be proved inductively) yields $A_q(n, d_H) \le q^{n-d_H} A_q(d_H, d_H) = q^{n-d_H+1}$. □

A code for which equality holds in the Singleton bound is called a *maximum-distance separable* (MDS) code. Another bound for M was obtained by Gilbert and Varshamov. It provides a lower bound on the number of codewords. This bound is called the *Gilbert-Varshamov bound.* Before this bound is stated, the concept of *maximal code* is explained.

Definition 2.5. *A code $\mathcal{C}$ is maximal, if no codeword can be added to it without reducing the minimum distance.* □

Maximal codes are more efficient, because for the same value of n and d_H they provide maximum-possible value of M, and consequently the code rate ϑ. Equivalently, the MDS code for a specified value of n and M provides the maximum value of d_H. The Gilbert-Varshamov bound is specified in the following theorem.

Theorem 2.4. (*Gilbert-Varshamov bound*) *Assume that the code C is maximal. It is specified by $[n, M, d_H]$, and its alphabet is A, where $|A| = q$. Then*

$$\frac{q^n}{V_q(n, d_H - 1)} \leq A_q(n, d_H) \tag{2.7}$$

Proof. Since the code C is maximal, there can be no codeword that is distance $(d_H - 1)$ or less from at least a single codeword of C. Consider spheres of radius $(d_H - 1)$ about each codeword cover C. Therefore

$$MV_q(n, d_H - 1) \geq q^n$$

The result follows. □

Using the above bounds, the following result about code rate ϑ is obtained.

Corollary 2.1. *Consider a code C specified by $[n, M, d_H]$, alphabet A, $|A| = q$, an error-correcting capability of $t = \lfloor (d_H - 1)/2 \rfloor$ errors, and a code rate ϑ. The*:

(a) *Hamming's sphere-packing bound yields*:

$$\vartheta \leq \frac{(n - \log_q V_q(n, t))}{n} \tag{2.8a}$$

(b) *Singleton bound gives*:

$$\vartheta \leq \frac{(n - d_H + 1)}{n} \tag{2.8b}$$

(c) *If the code C is maximal, Gilbert-Varshamov bound gives*:

$$\vartheta \geq \frac{(n - \log_q V_q(n, d_H - 1))}{n} \tag{2.8c}$$

□

2.2.5 Perfect Codes

A code for which equality holds in the Hamming bound is called a *perfect code.*

Definition 2.6. *A code C is defined over the alphabet A. It is specified by $[n, M, d_H]$, and has an error-correcting capability of $t = \lfloor (d_H - 1)/2 \rfloor$ errors. This code is perfect if the t-spheres around the codewords are disjoint, but their union contains all the codewords in C.* □

Alternately, a code with a Hamming distance d_H is perfect if its every word is within distance $t = \lfloor (d_H - 1)/2 \rfloor$ of some codeword. Some observations can be made about perfect codes.

Observations 2.3. A perfect code C specified by $[n, M, d_H]$ has the following properties.

1. The minimum Hamming distance $d_H = (2t+1)$ is an odd number.
2. The code $\mathcal{C}$ is maximal and it meets the Hamming bound. That is

$$M = \frac{q^n}{V_q(n,t)}, \qquad t = \left\lfloor \frac{(d_H - 1)}{2} \right\rfloor$$

Thus

$$M \sum_{j=0}^{t} \binom{n}{j} (q-1)^j = q^n$$

3. The code $\mathcal{C}$ has maximum code rate, but minimum redundancy for correcting t errors. □

Examples 2.2. Some trivial and not very interesting examples of perfect codes are:

1. A code with precisely one codeword.
2. $\mathcal{C}$ is a code defined over a finite field $\mathbb{F}_q$, where the order of the field is equal to q. Its codeword-length is n. The code $\mathcal{C}$ is perfect if $|\mathcal{C}| = q^n$.
3. A code with q codewords, where each word is a string of identical symbols. For example, if $q = 2$, then the codewords are $(00\ldots0)$, and $(11\ldots1)$. It is assumed that these codewords have an odd length. □

Hamming codes are examples of perfect codes. These codes are discussed later in the chapter. The *binary Golay code* $(23,12,7)$-code defined over binary field is an example of perfect code. The *ternary Golay code* $(11,6,5)$-code defined over $\mathbb{F}_3 = \{0,1,2\}$ is also a perfect code. Golay codes are not discussed in this chapter.

2.2.6 Error Probabilities

For obvious reasons, communication channels which have small channel error-probabilities are desirable. If the code is only required for error-detection, then it is important to have a knowledge of the probability of undetected error. However, if the code is error-correcting, then a knowledge of the block error-probability is preferable. Assume that the alphabet of a code $\mathcal{C}$ is $A = \{0,1\}$, where the code is specified by $[n, M, d_H]$. Let the probability of a bit error between the message and decoded message be p.

An expression for the probability of undetected block-error is initially obtained. Let this probability be p_u. Also let the set of codewords be

$$\{u_i \mid u_i \in \mathcal{C}, 1 \le i \le M\}$$

Assume that $P(u_i)$ is the nonzero probability that the codeword u_i is transmitted, and $P(u_j \mid u_i)$ is the probability that the transmitted codeword u_i is detected as u_j, for $1 \le i,j \le M$. For ease in notation denote the Hamming distance $d(u_i, u_j) \triangleq \delta_{ij}$ for $1 \le i,j \le M$. Then

$$\begin{aligned} p_u &= \sum_{\substack{1 \le i,j \le M \\ i \ne j}} P(u_i) P(u_j \mid u_i) \\ &= \sum_{\substack{1 \le i,j \le M \\ i \ne j}} P(u_i) p^{\delta_{ij}} (1-p)^{n-\delta_{ij}} \end{aligned}$$

If the probability of occurrence of the codeword is uniform, then $P(u_i) = 1/M$ for $1 \le i \le M$. Then the expression for p_u simplifies to

$$p_u = \frac{1}{M} \sum_{\substack{1 \le i,j \le M \\ i \ne j}} p^{\delta_{ij}} (1-p)^{n-\delta_{ij}}$$

Block error-probability of a t-bit error-correcting code $\mathcal{C}$, is next determined. Let the probability of the block error be p_c, and $t = \lfloor (d_H - 1)/2 \rfloor$. Then

$$p_c = 1 - \sum_{j=0}^{t} \binom{n}{j} p^j (1-p)^{n-j}$$

The proof of this result is left to the reader.

Example 2.3. Let the original message be of length $k = 2$ bits, and $M = 4$. Consider a binary code of block-length $n = 5$, $d_H = 3$, and $p = 10^{-4}$.

Therefore $q = 2$, and $t = 1$. The block error-probability p_c is given by

$$\begin{aligned} p_c &= 1 - \binom{5}{0} (1-p)^5 - \binom{5}{1} p (1-p)^4 \\ &\simeq 10p^2 = 10^{-7} \end{aligned}$$

Observe that if there were no error-correcting capability, the probability of error in a message of length 2, is equal to

$$\left\{1 - (1-p)^2\right\} \simeq 2p = 2 \times 10^{-4}$$

However, due to the use of error-correction, the probability of an error in the received message has been decreased to 10^{-7}. Note that this has been achieved at the expense of decreased code rate. The code rate has decreased from 1 to 0.4. □

2.3 Vector Spaces over Binary Field

In this section, vector spaces defined over a binary field are introduced. In this vector space, the field elements are $\{0, 1\}$ and the arithmetic is modulo 2, consequently this field is $\mathbb{Z}_2$. For convenience in notation, this field is denoted by $\mathbb{B}$, and the vector space of n-tuples over the binary field is denoted by $\mathbb{B}^{(n)}$. The total number of n-tuples in this vector space is equal to 2^n. The addition of two vectors in this space is done component-wise. Multiplication is done as follows: let $b \in \mathbb{B}^{(n)}$, then $0b = 0$, and $1b = b$.

Notation. A vector in $\mathbb{B}^{(n)}$ is written as an n-tuple, or as an n-tuple without any commas between any adjacent elements, or as a row or a column vector. The choice of representation of a vector in this space should be evident from its context. □

The concept of Hamming weight is introduced in the following definition.

Definition 2.7. *The Hamming weight, or simply the weight of a vector in $\mathbb{B}^{(n)}$ is the number of ones in it.* □

Examples 2.4. Some elementary examples.

1. The arithmetic in the field $\mathbb{B}$ is modulo 2. Therefore

$$\begin{aligned}0+0 &\equiv 0 \pmod 2, \quad 0+1 \equiv 1 \pmod 2, \\ 1+0 &\equiv 1 \pmod 2, \quad 1+1 \equiv 0 \pmod 2 \\ 0\times 0 &\equiv 0 \pmod 2, \quad 0\times 1 \equiv 0 \pmod 2, \\ 1\times 0 &\equiv 0 \pmod 2, \quad 1\times 1 \equiv 1 \pmod 2\end{aligned}$$

2. The elements of the vector space $\mathbb{B}^{(3)}$ are:

$$\{(000), (001), (010), (011), (100), (101), (110), (111)\}$$

Also

$$\begin{aligned}(010)+(011) &= (001) \\ 0\,(110) &= (000) \\ 1\,(110) &= (110)\end{aligned}$$

3. The vectors

$$(000), (001), (010), (011), (100), (101), (110), \text{ and } (111)$$

belong to the vector space $\mathbb{B}^{(3)}$. The weights of these vectors are

$$0, 1, 1, 2, 1, 2, 2, \text{ and } 3$$

respectively.

4. The Hamming distance between the vectors $(1101), (0110) \in \mathbb{B}^{(4)}$ is equal to 3. This is true because the vector elements (numbered from right to left) differ in the first, the second, and the fourth coordinates.
   Similarly, the Hamming distance between the vectors $(1001), (0110) \in \mathbb{B}^{(4)}$ is equal to 4.
5. Some examples of linear subspaces of $\mathbb{B}^{(3)}$ are

$$\{(000)\}, \{(000), (111)\}, \text{ and } \{(000), (001), (100), (101)\}$$

□

2.4 Linear Codes

A linear code is essentially a subspace of a finite-dimensional vector space defined over a finite field. Linear codes are easier to encode and decode than nonlinear codes. Furthermore, linear codes are particularly attractive because they are based upon matrix computations. This in turn implies easy implementability. Moreover nonlinear codes are harder to discover and characterize.

2.4.1 Basics of Linear Codes

Some basic facts about linear codes are outlined in this subsection.

Definitions 2.8. *Basic definitions.*

1. $\mathbb{F}_q$ *or equivalently the Galois field* GF_q *is a finite field with* q *elements. The set of* n*-tuples of elements from* $\mathbb{F}_q$ *is a vector space* $\mathbb{F}_q^{(n)}$. *These* n*-tuples are called either vectors or words.*
2. *A linear code* C *is a vector subspace of* $\mathbb{F}_q^{(n)}$. *If the dimension of this subspace is* k, *then* C *is also called an* (n, k)*-linear code over* $\mathbb{F}_q$. *The value* k *is also sometimes referred to as the dimension of the code* C. *The block-length of this code is* n. *Sometimes, this code is also specified as* (n, k, d_H), *where* d_H *is the Hamming distance of the code* C. *Note that this notation is only applicable to linear codes.*
3. *The Hamming weight* $w(c)$ *of a codeword* $c \in \mathbb{F}_q^{(n)}$ *is equal to the number of nonzero components in* c. □

From these definitions, it can be immediately concluded that the (n, k, d_H) linear code defined over the field $\mathbb{F}_q$ is the $[n, q^k, d_H]$ block-code. Moreover, the information rate of this code is equal to $\vartheta = k/n$.

Theorem 2.5. *Let* C *be a linear code, with a Hamming distance* d_H. *Then*

$$d_H = \min_{c \in C, c \neq 0} w(c) \tag{2.9}$$

Proof.

$$d_H = \min_{\substack{c_i, c_j \in C \\ i \neq j}} d(c_i, c_j) = \min_{\substack{c_i, c_j \in C \\ i \neq j}} d(0, c_i - c_j) = \min_{c \in C, c \neq 0} w(c)$$

Thus $d_H = \min_{c \in C, c \neq 0} w(c)$. □

Corollary 2.2. *If* C *is a linear code, with a Hamming distance* d_H, *then* $d_H \leq w(c)$ *for all* $c \in C$. □

The above corollary implies that, in order to find the d_H of a linear code, it is convenient to first find its minimum weight. This requires a check of at most M codewords. Whereas, finding the minimum distance between all pairs of codewords requires $O(M^2)$ checks.

It should be emphasized that a linear code cannot be obtained by simply selecting any M codewords from the q^n possible n-tuples. The procedure to generate an (n, k)-linear code over $\mathbb{F}_q$ is as follows. Select k codewords $b_1, b_2, \ldots, b_k$ in $\mathbb{F}_q^{(n)}$ which form a basis of the code C. Then all the linear combinations of these basis codewords form the code C. This is conveniently done via a *generator matrix*. This technique is a compact way to determine a linear code.

Definitions 2.9. *Linear code and generator matrix.*

1. C *is an* (n, k)*-linear code over* $\mathbb{F}_q$. *The generator matrix of this code is a* $k \times n$ *matrix* G, *whose entries are from the finite field* $\mathbb{F}_q$. *Furthermore, the rows of this matrix form a basis of the code* C. *Also* $C = \left\{ fG \mid f \in \mathbb{F}_q^{(k)} \right\}$.

2. *An* (n, k)*-linear code over* $\mathbb{F}_q$*, with a Hamming distance* d_H *is given. A different code with the same parameters can be obtained by selecting two components and transposing the symbols in these two components of every codeword. This new code is said to be equivalent to the original code.*
   In summary, two codes are equivalent, if there is a specified permutation of the coordinate positions which map one code to the other code.
3. *The simplest and canonical form of the generator matrix is* $G = [I_k \mid B]$*, where* I_k *is a* $k \times k$ *identity matrix, and* B *is a* $k \times (n-k)$ *matrix with entries from the finite field* $\mathbb{F}_q$*. This matrix is called the standard form of the generator matrix. The code* $\mathcal{C}$ *generated by this generator matrix is called a systematic code.*
   Therefore, in a systematic code each codeword begins with k *information symbols. These information symbols are followed by* $(n-k)$ *parity symbols.* □

Equivalent codes have same Hamming distance, and therefore identical error-correcting properties. It is possible for a linear code to have more than a single generator matrix. A matrix G' obtained from the generator matrix G via elementary row operations produces the same linear code. Elementary row operations do not change the set of codewords.

The generator matrices of equivalent codes are related. Each generator matrix can be obtained from the other by interchanging columns. Therefore the generator matrix G of a code can be transformed into a generator matrix G_c of equivalent code in canonical form. This is done by a combination of elementary row operations and column interchanges.

The *parity-check matrix* is next studied. This matrix is used to check if a word belongs to the code.

Definitions 2.10. *Parity-check matrix, and canonical parity-check matrix.*

1. Let $\mathcal{C}$ *be an* (n, k)*-linear code over* $\mathbb{F}_q$*. This code has a* $k \times n$ *generator matrix* G*. The parity-check matrix of this linear code* $\mathcal{C}$ *is a* $(n-k) \times n$ *matrix* H*, such that* $GH^T = 0$*, where* H^T *is the transpose of the matrix* H *and* 0 *is the* $k \times (n-k)$ *zero matrix.*
2. *Let the generator matrix* G *be in canonical form. This generator matrix is called canonical generator matrix, and the corresponding parity-check matrix is called the canonical parity-check matrix.* □

Observations 2.4. Some basic observations.

1. If the matrix $G = [I_k \mid B]$, then $H = \left[-B^T \mid I_{n-k}\right]$, or $H = \left[B^T \mid -I_{n-k}\right]$.
2. If the matrix G is not in canonical form, the corresponding parity-check matrix H is obtained via the following steps.
   (a) First reduce G to canonical form G_c via elementary row operations and/or column interchanges. Elementary row operations do not change the set of codewords. Column interchanges simply create a generator matrix of an equivalent code. Then find the corresponding parity-check matrix H_c.
   (b) In the next step, apply the reversed column operations used in step (a), to the matrix H_c to obtain the parity-check matrix H.
3. If $c \in \mathcal{C}$, then $cH^T = 0$, where 0 is a row vector of size $(n-k)$. This observation suggests that the parity-check matrix can be used to determine if a word $c \in \mathbb{F}_q^{(n)}$ belongs to the code $\mathcal{C}$.

□

The next logical step is to study the code generated by the parity-check matrix H, and its relationship with the code $\mathcal{C}$. The code generated by the matrix H is labeled $\mathcal{C}^{\perp}$ (read "C perp"). It is also called the *dual code* of $\mathcal{C}$. The reason for this notation becomes clear from the following definitions. Recall that the inner product of two vectors $u, v \in \mathbb{F}_q^{(n)}$ is defined as $u \circ v = \sum_{j=1}^{n} u_j v_j$, where $u = (u_1, u_2, \ldots, u_n)$, $v = (v_1, v_2, \ldots, v_n)$, and each element of the vectors u and v belongs to $\mathbb{F}_q$.

Definitions 2.11. *Orthogonal complement of a code, and self-orthogonal code.*

1. *$\mathcal{C}$ is an (n, k)-linear code over $\mathbb{F}_q$. The orthogonal complement of the code $\mathcal{C}$ is denoted by $\mathcal{C}^{\perp}$. It is the set of vectors orthogonal to all vectors in the code $\mathcal{C}$.*

$$\mathcal{C}^{\perp} = \left\{ u \in \mathbb{F}_q^{(n)} \mid u \circ v = 0, \ \forall \, v \in \mathcal{C} \right\} \tag{2.10}$$

The code $\mathcal{C}^{\perp}$ is called the dual code of $\mathcal{C}$.

2. *A linear code $\mathcal{C}$ is self-orthogonal if and only if $\mathcal{C} \subseteq \mathcal{C}^{\perp}$. Furthermore, it is self-dual if $\mathcal{C} = \mathcal{C}^{\perp}$.* □

Observations 2.5. Let $\mathcal{C}$ be an (n, k)-linear code over $\mathbb{F}_q$. Its generator matrix is G, its parity-check matrix is H, and its dual code is $\mathcal{C}^{\perp}$.

1. $\mathcal{C}^{\perp}$ is an $(n, n-k)$-linear code over $\mathbb{F}_q$. It is generated by the matrix H.
2. The dual code of $\mathcal{C}^{\perp}$ is the code $\mathcal{C}$.
3. The code $\mathcal{C}$ is self-orthogonal iff $GG^T = 0$, where 0 is a $k \times k$ matrix.
4. The code $\mathcal{C}$ is self-dual iff it is self-orthogonal and $n = 2k$. □

Theorem 2.6. *The Hamming distance d_H of any (n, k) linear code satisfies $d_H \leq (n - k + 1)$. This relationship is also sometimes referred to as the Singleton bound.*

Proof. It has been already established that $A_q(n, d_H) \leq q^{n-d_H+1}$. This yields the required inequality. □

Therefore for maximum-distance separable linear codes $d_H = (n - k + 1)$.

Examples 2.5. Assume that the field $\mathbb{F}_q = \mathbb{B}$.

1. A linear subspace of $\mathbb{B}^{(3)}$ is $\{(000), (001), (100), (101)\}$. The minimum weight of the nonzero codewords is equal to 1, which in turn is equal to d_H as expected.
2. A $(4, 2)$-linear code $\mathcal{C}$ defined over the field $\mathbb{B}$ is generated. This code is also required to be systematic. In this code, the length of the codewords is $n = 4$, and the dimension of a subspace of $\mathbb{B}^{(4)}$ is $k = 2$. Furthermore, the number of codewords in this code is equal to $M = 2^k = 4$. The generator matrix G of this code has $k = 2$ rows. The rows of this matrix form a basis for the code $\mathcal{C}$. Let

$$G = \begin{bmatrix} 1 & 0 & 1 & 1 \\ 0 & 1 & 0 & 1 \end{bmatrix}$$

The rows of this matrix form a basis, because the rows are independent of each other. The codewords are generated via the matrix multiplication operation cG, where $c \in \mathbb{B}^{(2)}$. This yields

$$\begin{bmatrix} 0 & 0 \end{bmatrix} \begin{bmatrix} 1 & 0 & 1 & 1 \\ 0 & 1 & 0 & 1 \end{bmatrix} = \begin{bmatrix} 0 & 0 & 0 & 0 \end{bmatrix}$$

$$\begin{bmatrix} 0 & 1 \end{bmatrix} \begin{bmatrix} 1 & 0 & 1 & 1 \\ 0 & 1 & 0 & 1 \end{bmatrix} = \begin{bmatrix} 0 & 1 & 0 & 1 \end{bmatrix}$$

$$\begin{bmatrix} 1 & 0 \end{bmatrix} \begin{bmatrix} 1 & 0 & 1 & 1 \\ 0 & 1 & 0 & 1 \end{bmatrix} = \begin{bmatrix} 1 & 0 & 1 & 1 \end{bmatrix}$$

$$\begin{bmatrix} 1 & 1 \end{bmatrix} \begin{bmatrix} 1 & 0 & 1 & 1 \\ 0 & 1 & 0 & 1 \end{bmatrix} = \begin{bmatrix} 1 & 1 & 1 & 0 \end{bmatrix}$$

The code $\mathcal{C} = \{(0000), (0101), (1011), (1110)\}$. In the generator matrix G, the submatrix B is

$$B = \begin{bmatrix} 1 & 1 \\ 0 & 1 \end{bmatrix}$$

The parity-check matrix $H = \begin{bmatrix} -B^T \mid I_2 \end{bmatrix}$, where I_2 is the 2×2 identity matrix. Then

$$H = \begin{bmatrix} 1 & 0 & 1 & 0 \\ 1 & 1 & 0 & 1 \end{bmatrix}$$

It can be checked that $GH^T = 0$, where 0 is a 2×2 matrix with all zero entries. It can also be checked, that $cH^T = 0$ for all $c \in \mathcal{C}$, where 0 is a row vector of size 2.
Let $c = (1010) \notin \mathcal{C}$, then $cH^T = \begin{bmatrix} 0 & 1 \end{bmatrix} \neq \begin{bmatrix} 0 & 0 \end{bmatrix}$, similarly if $c = (1111) \notin \mathcal{C}$, then $cH^T = \begin{bmatrix} 0 & 1 \end{bmatrix} \neq \begin{bmatrix} 0 & 0 \end{bmatrix}$. It can be shown in general, that for any $c \notin \mathcal{C}, cH^T \neq \begin{bmatrix} 0 & 0 \end{bmatrix}$. The code generated by the matrix H yields

$$\mathcal{C}^{\perp} = \{(0000), (1101), (1010), (0111)\}$$

It can verified that the codewords in the codes $\mathcal{C}$ and $\mathcal{C}^{\perp}$ are orthogonal to each other.

3. A $(4, 2)$-linear code $\mathcal{C}$ defined over the field $\mathbb{B}$ is generated. In this code, $n = 4, k = 2$, and $M = 2^k = 4$. The generator matrix G of this code has $k = 2$ rows. Let

$$G = \begin{bmatrix} 0 & 1 & 1 & 0 \\ 1 & 0 & 1 & 1 \end{bmatrix}$$

The codewords are generated via the matrix multiplication operation cG, where $c \in \mathbb{B}^{(2)}$. This yields, $\mathcal{C} = \{(0000), (1011), (0110), (1101)\}$. A canonical generator matrix G_c is obtained by interchanging the two rows in the matrix G. Thus

$$G_c = \begin{bmatrix} 1 & 0 & 1 & 1 \\ 0 & 1 & 1 & 0 \end{bmatrix}$$

The corresponding parity-check matrix H_c is

$$H_c = \begin{bmatrix} 1 & 1 & 1 & 0 \\ 1 & 0 & 0 & 1 \end{bmatrix}$$

In this example, the parity-check matrix H for the code $\mathcal{C}$ can simply be taken to be the matrix H_c, that is $H = H_c$. It can be verified that $GH^T = 0$, where 0 is a 2×2 matrix with all zero entries. Use of this matrix H, results in $\mathcal{C}^{\perp} = \{(0000), (1110), (1001), (0111)\}$. It can be verified that the codewords in the codes $\mathcal{C}$ and $\mathcal{C}^{\perp}$ are orthogonal to each other. □

2.4.2 Minimum-Distance Decoding

Let $\mathcal{C}$ be an (n, k, d_H)-linear code over $\mathbb{F}_q$. This linear code is a k-dimensional subspace of the linear space $\mathbb{F}_q^{(n)}$. Note that the code $\mathcal{C}$ is an Abelian subgroup of the additive group $\mathbb{F}_q^{(n)}$. The minimum-distance decoding of the linear codes at the receiver can be interpreted in terms of the cosets $x + \mathcal{C}$. Therefore some properties of cosets are next summarized. This technique of decoding linear codes is also called *syndrome decoding*. The decoding scheme is called syndrome decoding, because it offers the symptoms of the errors.

Definitions 2.12. *Let $\mathcal{C}$ be an (n, k, d_H)-linear code over field $\mathbb{F}_q$.*

1. *For any element $u \in \mathbb{F}_q^{(n)}$, the coset of the code $\mathcal{C}$ determined by u is the set $\mathcal{C} + u = \{c + u \mid c \in \mathcal{C}\}$.*
2. *Coset leader of a coset of the linear code $\mathcal{C}$ is a coset member of smallest weight. If there are two or more members with the smallest weight, then arbitrarily select anyone of them.* □

Observations 2.6. Some basic observations.

1. For any $u \in \mathbb{F}_q^{(n)}$, $u \in \mathcal{C} + u$.
2. If u is equal to the 0 codeword, then the coset is simply the code $\mathcal{C}$.
3. For any $u, v \in \mathbb{F}_q^{(n)}$, u and v belong to the same coset iff $(v - u) \in \mathcal{C}$. This implies that each word in a coset determines that particular coset.
4. For any $u, v \in \mathbb{F}_q^{(n)}$, if $v \in \mathcal{C} + u$, then $\mathcal{C} + u = \mathcal{C} + v$.
5. Each coset has q^k codewords, and there are q^{n-k} disjoint cosets. □

The decoding of the received codeword is next discussed.

Definition 2.13. *Let the transmitted and received words be τ and ρ respectively, where $\tau \in \mathcal{C}$ and $\rho \in \mathbb{F}_q^{(n)}$. Denote the possible error vector of ρ as ε. That is, $\varepsilon = (\rho - \tau)$.* □

Lemma 2.2. *Let ρ be the received vector, then the set of possible error vectors is the coset $\rho + \mathcal{C}$.*

Proof. It is given that the received vector is $\rho \in \mathbb{F}_q^{(n)}$. Then ε is an error vector iff there exists a codeword $\tau \in \mathcal{C}$ such that $\varepsilon = (\rho - \tau)$. Define $-\tau = \tau'$, then $\tau' \in \mathcal{C}$ because $\mathcal{C}$ is a subspace of $\mathbb{F}_q^{(n)}$. Therefore $\varepsilon = (\rho + \tau')$, this in turn implies that $\varepsilon \in \rho + \mathcal{C}$. □

Given the vector ρ, the decoder's strategy is to select a minimum weight vector ε in the coset containing ρ, and then decode ρ as $\widehat{\tau} = (\rho - \varepsilon)$. The minimum weight vector is indeed the coset leader of the coset. The task of finding the coset leader might be time consuming. However, the process of decoding can be speeded up by using the concept of syndrome decoding.

Definition 2.14. *Let $\mathcal{C}$ be an (n, k, d_H)-linear code over field $\mathbb{F}_q$, and H be its parity-check matrix. Also let $\tau \in \mathbb{F}_q^{(n)}$, then the syndrome of τ is the vector τH^T.* □

Observations 2.7. Properties of syndrome vector.

1. The length of the syndrome vector is $(n - k)$.
2. A vector $\tau \in \mathbb{F}_q^{(n)}$, is a codeword iff its syndrome is 0, that is $\tau H^T = 0$.

3. The vectors $u, v \in \mathbb{F}_q^{(n)}$ belong to the same coset of $\mathcal{C}$ iff $uH^T = vH^T$. That is, the two vectors u and v which belong to the same coset of $\mathcal{C}$ have the same syndrome. The proof of this observation is in the problem section. □

An algorithm for the minimum-distance (syndrome) decoding of a received vector is next outlined. This is followed by a proof of the correctness of the algorithm.

Algorithm 2.1. *Minimum-Distance (Syndrome) Decoding of a Received Vector.*

Input: Let $\mathcal{C}$ be an (n, k, d_H)-linear code over field $\mathbb{F}_q$.
Also let H be its parity-check matrix. The received vector is $\rho \in \mathbb{F}_q^{(n)}$.
Output: The presumably transmitted vector $\widehat{\tau}$.
begin
  (*initialization*)
  For each coset of the code $\mathcal{C}$:
  find its coset leader and the corresponding syndrome.
  (*main part of the algorithm*)
  begin
    Step 1: Determine the syndrome of the received vector ρ.
      It is $s = \rho H^T$.
    Step 2: Determine the coset leader ε associated with s.
    Step 3: Compute $\widehat{\tau} = (\rho - \varepsilon)$.
  end (*end of the main part of the algorithm*)
end (*end of syndrome decoding algorithm*)

Proof of the correctness of the algorithm. It is proved that this algorithm is indeed the minimum-distance decoding scheme for the linear code $\mathcal{C}$. Observe that ρ and ε are in the same coset, therefore $(\rho - \varepsilon) = \widehat{\tau}$ is in $\mathcal{C}$.

The final result is proved by contradiction. Suppose there exists a codeword $c_0 \in \mathcal{C}$ such that $d(\rho, \rho - \varepsilon) > d(\rho, c_0)$. This implies that $d(\varepsilon, 0) > d(\rho - c_0, 0)$, that is the weight of ε is greater than the weight of $(\rho - c_0)$. Also

$$
\begin{aligned}
&(\rho - c_0) H^T \\
&= (\rho H^T - c_0 H^T) \\
&= \rho H^T
\end{aligned}
$$

Therefore $(\rho - c_0)$ and ρ have the same syndrome, and therefore belong to the same coset. Moreover, the weight of $(\rho - c_0)$ is smaller than the weight of the coset leader ε. This is a contradiction. □

Examples 2.6. Some illustrative examples.

1. $\mathcal{C} = \{(000), (001), (100), (101)\}$ is a subspace of $\mathbb{B}^{(3)}$. The cosets of $\mathcal{C}$ are:

$$\begin{aligned}(000) + \mathcal{C} &= (001) + \mathcal{C} = (100) + \mathcal{C} = (101) + \mathcal{C} = \mathcal{C} \\ (010) + \mathcal{C} &= (011) + \mathcal{C} = (110) + \mathcal{C} = (111) + \mathcal{C} \\ &= \{(010), (011), (110), (111)\}\end{aligned}$$

The coset leader of the coset $\mathcal{C}$ is (000). Similarly, the coset leader of the coset

$$\{(010), (011), (110), (111)\}$$

is (010).

2. The generator matrix G of a $(5, 3)$ code defined over the binary field $\mathbb{B}$, is

$$G = \begin{bmatrix} 1 & 0 & 0 & 0 & 1 \\ 0 & 1 & 0 & 1 & 0 \\ 0 & 0 & 1 & 1 & 1 \end{bmatrix}$$

The corresponding parity-check matrix H is

$$H = \begin{bmatrix} 0 & 1 & 1 & 1 & 0 \\ 1 & 0 & 1 & 0 & 1 \end{bmatrix}$$

It can be checked that $GH^T = 0$, where 0 is a 3×2 matrix. The linear code is given by

$$\begin{aligned}\mathcal{C} = \{&(00000), (01010), (10001), (00111), \\ &(01101), (10110), (11100), (11011)\}\end{aligned}$$

The four cosets are

$$\begin{aligned}&\{(00000), (01010), (10001), (00111), (01101), (10110), (11100), (11011)\} \\ &\{(00001), (01011), (10000), (00110), (01100), (10111), (11101), (11010)\} \\ &\{(00010), (01000), (10011), (00101), (01111), (10100), (11110), (11001)\} \\ &\{(00100), (01110), (10101), (00011), (01001), (10010), (11000), (11111)\}\end{aligned}$$

The list of their coset leaders and syndromes are:

coset leader	(00000)	(00001)	(00010)	(00100)
syndrome	(00)	(01)	(10)	(11)

(a) Assume that the received word is $\rho = (11011)$. Its syndrome is $s = \rho H^T = (00)$. The corresponding coset leader is (00000). Therefore $\varepsilon = (00000)$, that is the received word did not have any error in it. Thus the decoded word is

$$\widehat{\tau} = (\rho - \varepsilon) = (11011) \in \mathcal{C}$$

(b) Assume that the received word is $\rho = (00110)$. Its syndrome is $s = \rho H^T = (01)$. The corresponding coset leader is (00001). Therefore $\varepsilon = (00001)$, and the decoded word is

$$\widehat{\tau} = (\rho - \varepsilon) = (00111) \in \mathcal{C}$$

□

2.5 Cyclic Codes

A cyclic code is a linear code, in which a codeword is obtained by the cyclic shift of another codeword. Cyclic codes are popular because they are easy to encode. We initially discuss the basics of cyclic codes. This is followed by a description of check polynomial and the parity-check matrix of a cyclic code. Finally, the encoding and decoding technique of these codes is specified. Bose-Chaudhuri-Hocquenghem and Reed-Solomon codes are also examples of cyclic codes. Hamming codes can also be studied via the theory of cyclic codes. These codes are discussed later in the chapter.

2.5.1 Basics of Cyclic Codes

The framework of cyclic codes can best be described via polynomials defined over a finite field $\mathbb{F}_q$, where q is equal to the number of elements in the finite field. Before a description of cyclic codes is given, the reader should make a note of the following elementary, yet useful observations.

Observations 2.8. Some basic observations.

1. The vector $(a_0, a_1, \ldots, a_{n-1}) \in \mathbb{F}_q^{(n)}$ can be represented by the polynomial in the indeterminate x as $\sum_{j=0}^{n-1} a_j x^j$.
2. $\mathbb{F}_q[x] / (x^n - 1)$ is the ring of polynomials modulo $(x^n - 1)$. This ring of polynomials is isomorphic to the vector space $\mathbb{F}_q^{(n)}$.
3. If $g(x) \in \mathbb{F}_q[x] / (x^n - 1)$ is used to represent an n-tuple, then $xg(x) \in \mathbb{F}_q[x] / (x^n - 1)$ represents a cyclic shift of this n-tuple. □

In this section and the rest of the chapter, we permit ourselves a little misuse of notation. If $a = (a_0, a_1, \ldots, a_{n-1}) \in \mathcal{C}$, then its equivalent polynomial representation in the indeterminate x is denoted by $a(x) = \sum_{j=0}^{n-1} a_j x^j$.

The last observation provides a convenient vehicle to study cyclic codes in terms of *principal ideal polynomial* of the ring polynomial $\mathbb{F}_q[x] / (x^n - 1)$. A constructive definition of this ideal is next given.

Definition 2.15. *Denote the ring polynomial* $\mathbb{F}_q[x] / (x^n - 1)$ *by* $\mathcal{R}_n$. *Let* $g(x) \in \mathcal{R}_n$, *then the ideal generated by the polynomial* $g(x)$ *is*

$$\{a(x) g(x) \mid a(x) \in \mathbb{F}_q[x] / (x^n - 1)\} \tag{2.11}$$

This ideal is denoted by $\langle g(x) \rangle$. □

This definition yields the following facts.

Observations 2.9. Some basic observations.

1. The ideal of the ring polynomial $\mathcal{R}_n$ is isomorphic to a vector subspace of $\mathbb{F}_q^{(n)}$.
2. If $g(x)$ is a member of an ideal, then so is $xg(x)$.

3. An ideal may have more than one generator. That is, an ideal may have several elements which can generate it. □

Example 2.7. Consider the ring $\mathbb{B}[x] / (x^3 - 1)$. Then a subset

$$\{0, (1+x), (1+x^2), (x+x^2)\}$$

is an ideal of the ring. It can be checked that the members of this set are closed under addition and multiplication. □

A cyclic code and its generator are next defined.

Definitions 2.16. *We define a cyclic code, and its generator polynomial.*

1. *Let $\mathbb{F}_q$ be a finite field, where the order of this field is equal to q. A cyclic code $\mathcal{C}$ is a linear code of length $n \in \mathbb{P}$, if $(a_0, a_1, \ldots, a_{n-1})$ is a codeword, where $a_j \in \mathbb{F}_q$ for $0 \le j \le (n-1)$, so is a cyclic shift of this codeword $(a_{n-1}, a_0, a_1, \ldots, a_{n-2}) \in \mathcal{C}$.*
2. *Let $\mathcal{C}$ be a nonzero cyclic code in $\mathcal{R}_n$. A monic polynomial $g(x)$ of minimal degree which generates the codewords in $\mathcal{C}$ is called the generator polynomial of the code $\mathcal{C}$.* □

In the above definition, a polynomial of minimal degree means a polynomial of least degree. Using this definition, it can be concluded that the polynomial representation of the cyclic code $\mathcal{C}$ is identical to an ideal of the polynomial ring $\mathcal{R}_n$. If the ideal is $\langle g(x) \rangle$, then the code $\mathcal{C}$ is said to be *generated* by $g(x)$. The theory of cyclic codes is essentially summarized in the following theorem.

Theorem 2.7. *$\mathcal{C}$ is a nonzero (n,k)-cyclic code in $\mathcal{R}_n$. This is a cyclic code of length n. For any $c \in \mathcal{C}$, let $c = \begin{bmatrix} c_0 & c_1 & \cdots & \cdots & c_{n-1} \end{bmatrix}$, and $c(x) = \sum_{j=0}^{n-1} c_j x^j$. Then*

(a) *There exists a unique monic polynomial $g(x) \in \langle g(x) \rangle$ of minimal degree.*
(b) *The code $\mathcal{C}$ is generated by the ideal $\langle g(x) \rangle$. That is, the polynomial $g(x)$ is the generator of the cyclic code $\mathcal{C}$.*
(c) *$g(x) \mid (x^n - 1)$ is in $\mathbb{F}_q[x]$.*
(d) *Let $\deg g(x) = \theta = (n-k)$. Also let $f(x)$ be a polynomial representation of the message word, where $f(x) = 0$ or $\deg f(x) \le (k-1)$. Then any codeword $c \in \mathcal{C}$ can be uniquely expressed as $c(x) = f(x)g(x)$ in $\mathbb{F}_q[x]$. This implies that $c(x) \equiv 0 \pmod{g(x)}$.*
(e) *If $g(x) = \sum_{j=0}^{\theta} g_j x^j$, where $g_\theta = 1$, and $\theta = (n-k)$, then a basis for $\langle g(x) \rangle$ is the set of polynomials*

$$\{g(x)x^j \pmod{(x^n - 1)} \mid 0 \le j \le (n - \theta - 1)\} \tag{2.12a}$$

and a $k \times n$ generator matrix G for the code $\mathcal{C}$ is

$$G = \begin{bmatrix} g_0 & g_1 & g_2 & \cdots & \cdots & g_\theta & 0 & 0 & \cdots & 0 \\ 0 & g_0 & g_1 & \cdots & \cdots & g_{\theta-1} & g_\theta & 0 & \cdots & 0 \\ 0 & 0 & g_0 & \cdots & \cdots & g_{\theta-2} & g_{\theta-1} & g_\theta & \cdots & 0 \\ \vdots & \vdots & \vdots & \vdots & \vdots & \vdots & \vdots & \vdots & \ddots & \vdots \\ 0 & 0 & \cdots & 0 & g_0 & \cdots & \cdots & \cdots & \cdots & g_\theta \end{bmatrix} \tag{2.12b}$$

Proof. See the problem section. □

A convenient and alternate representation of the generator matrix G is:

$$G \triangleq \begin{bmatrix} g(x) & & & & \\ & xg(x) & & & \\ & & \cdots & & \\ & & & \cdots & \\ & & & & x^{n-\theta-1}g(x) \end{bmatrix}$$

The polynomial $f(x)$ is sometimes referred to as either the *message* or *information polynomial.* It also follows from the above theorem that the generator polynomial $g(x)$ is the polynomial representation of the codeword $g \in \mathcal{C}$. This observation leads us to a quick observation about the distance d_H of the code. That is, the weight of the codeword g is $\geq d_H$ for cyclic codes.

2.5.2 Check Polynomial and Parity-Check Matrix

A technique to develop the $(n-k) \times n$ parity-check matrix H, via the *check polynomial* is outlined in this subsection. From the last theorem, the polynomial $(x^n - 1)$ is divisible by the generator polynomial $g(x)$, that is $g(x) \mid (x^n - 1)$ in $\mathbb{F}_q[x]$. This leads naturally to the definition of the check polynomial.

Definition 2.17. *The check polynomial of an (n,k)-code $\mathcal{C}$, generated by a polynomial $g(x)$ is denoted by $h(x)$. It is*

$$h(x) = \frac{(x^n - 1)}{g(x)}, \quad \deg h(x) = k \tag{2.13}$$

and $h(x) \in \mathbb{F}_q[x]$. □

The check polynomial is so named because it is used in determining the parity-check matrix H of the cyclic code. Let $h(x) = \sum_{j=0}^{k} h_j x^j$. Since the generator polynomial is monic, the check polynomial is also monic. This implies that $h_k = 1$. Also from the last theorem, any $c \in \mathcal{C}$ can be uniquely expressed as

$$c(x) = f(x)\,g(x) \in \mathbb{F}_q[x]$$

where, if $f(x) \neq 0$, $\deg f(x) \leq (k-1)$ and $\deg c(x) \leq (n-1)$. Multiplying both sides of the above equation by $h(x)$ yields

$$\begin{aligned} c(x)\,h(x) &= f(x)\,g(x)\,h(x) \\ &= f(x)\,(x^n - 1) \end{aligned}$$

Substituting

$$c(x) = \sum_{j=0}^{n-1} c_j x^j, \; h(x) = \sum_{j=0}^{k} h_j x^j, \text{ and } f(x) = \sum_{j=0}^{k-1} f_j x^j$$

in the above equation yields

$$\sum_{j=0}^{n-1} c_j x^j \sum_{j=0}^{k} h_j x^j = (x^n - 1) \sum_{j=0}^{k-1} f_j x^j$$

Observe in the right-hand side expansion that the terms $x^k, x^{k+1}, \ldots, x^{n-1}$ are absent. Comparing the coefficients of these terms on both sides results in

$$
\begin{aligned}
c_0 h_k + c_1 h_{k-1} + \cdots + c_k h_0 &= 0 \\
c_1 h_k + c_2 h_{k-1} + \cdots + c_{k+1} h_0 &= 0 \\
\vdots \quad & \quad \vdots \\
c_{n-k-1} h_k + c_{n-k} h_{k-1} + \cdots + c_{n-1} h_0 &= 0
\end{aligned}
$$

These equations can be written compactly as

$$\sum_{j=0}^{k} h_j c_{n-i-j} = 0, \quad 1 \leq i \leq n-k$$

It is also illuminating to write the above equations in matrix form. It is $cH^T = 0$, where 0 is a row vector of size $(n-k)$, $c = \begin{bmatrix} c_0 \; c_1 \; \cdots \; \cdots \; c_{n-1} \end{bmatrix}$ and

$$H = \begin{bmatrix} h_k & h_{k-1} & h_{k-2} & \cdots & \cdots & h_0 & 0 & 0 & \cdots & 0 \\ 0 & h_k & h_{k-1} & \cdots & \cdots & h_1 & h_0 & 0 & \cdots & 0 \\ 0 & 0 & h_k & \cdots & \cdots & h_2 & h_1 & h_0 & \cdots & 0 \\ \vdots & \vdots & \vdots & \vdots & \vdots & \vdots & \vdots & \vdots & \ddots & \vdots \\ 0 & 0 & \cdots & 0 & h_k & \cdots & \cdots & \cdots & \cdots & h_0 \end{bmatrix} \tag{2.14}$$

The matrix H is of size $(n-k) \times n$. The equation $cH^T = 0$ implies that the codeword c is orthogonal to each and every row of the matrix H. Also the relationship $cH^T = 0$ is true for any codeword $c \in \mathcal{C}$. This implies that $GH^T = 0$. Consequently, the matrix H is the parity-check matrix. This matrix generates the dual code $\mathcal{C}^{\perp}$, and its generator polynomial is $h^*(x) = x^k h(x^{-1})/h_0$, where $h_0 \neq 0$. It can also be observed from the matrix H that the code $\mathcal{C}^{\perp}$ is cyclic. This discussion is summarized in the following theorem.

Theorem 2.8. *Let $g(x)$ and G be the generator polynomial and the generator matrix respectively of the cyclic code $\mathcal{C}$. Let $h(x) = (x^n - 1)/g(x)$, $h^*(x) = x^k h(x^{-1})/h_0$, and $h_0 \neq 0$. Then $h^*(x)$ and H are the generator polynomial and the generator matrix respectively of the cyclic code $\mathcal{C}^{\perp}$.* □

2.5.3 Encoding and Decoding Process

The codeword polynomial $\tau(x)$ is derived form the information (message) polynomial $f(x)$ by multiplying it by the generator polynomial $g(x)$. Assume that the received word is represented by the polynomial $\rho(x)$. Due to the presence of noise on the transmission channel, $\rho(x)$ is a corrupted version of the polynomial $\tau(x)$. Let

$$\rho(x) = (\tau(x) + \varepsilon(x)), \quad \deg \rho(x) \leq (n-1)$$

where $\varepsilon(x)$ is called the *error polynomial*. As $\tau(x) = f(x)g(x)$, we have

$$\rho(x) = (f(x)g(x) + \varepsilon(x))$$

Therefore

$$\rho(x) \equiv \varepsilon(x) \pmod{g(x)}$$

It can also be shown that

$$\rho(x) h(x) \equiv \varepsilon(x) h(x) \pmod{(x^n - 1)}$$

The syndrome of a received word is $s = \rho H^T$. The corresponding *syndrome polynomial* is $s(x)$. It is

$$s(x) \equiv \rho(x) \pmod{g(x)} \equiv \varepsilon(x) \pmod{g(x)}$$

Therefore, if the codeword polynomial is uncorrupted by noise on the channel, the error and the corresponding syndrome polynomials are each equal to zero.

Example 2.8. Let $\mathcal{C}$ be a code defined over the binary field $\mathbb{B}$. Also let the codewords be of length 9. The factorization of $(x^9 - 1)$ over the binary field is

$$(x^9 - 1) = (x + 1)(x^2 + x + 1)(x^6 + x^3 + 1)$$

The monic divisors of this polynomial are

$$\begin{aligned}
g_1(x) &= 1 \\
g_2(x) &= (x + 1) \\
g_3(x) &= (x^2 + x + 1) \\
g_4(x) &= (x^6 + x^3 + 1) \\
g_5(x) &= (x + 1)(x^2 + x + 1) = (x^3 + 1) \\
g_6(x) &= (x + 1)(x^6 + x^3 + 1) = (x^7 + x^6 + x^4 + x^3 + x + 1) \\
g_7(x) &= (x^2 + x + 1)(x^6 + x^3 + 1) = \sum_{j=0}^{8} x^j \\
g_8(x) &= (x^9 + 1)
\end{aligned}$$

From the above set of equations, the candidates for the generator polynomials of a cyclic code are $g_j(x)$ for $2 \le j \le 7$. Let the generator polynomial be $g(x) = (x^6 + x^3 + 1)$. This produces a $(9, 3)$-cyclic code. Its 3×9 generator matrix G is

$$G = \begin{bmatrix} 1 & 0 & 0 & 1 & 0 & 0 & 1 & 0 & 0 \\ 0 & 1 & 0 & 0 & 1 & 0 & 0 & 1 & 0 \\ 0 & 0 & 1 & 0 & 0 & 1 & 0 & 0 & 1 \end{bmatrix}$$

The matrix G generates $2^3 = 8$ codewords, each of length 9. These codewords are

$$\begin{aligned}
\mathcal{C} = &\{(000000000), (001001001), (010010010), (011011011)\} \cup \\
&\{(100100100), (101101101), (110110110), (111111111)\}
\end{aligned}$$

The polynomial

$$h(x) = (x + 1)(x^2 + x + 1) = (x^3 + 1), \text{ and } h^*(x) = (x^3 + 1)$$

The 6×9 parity-check matrix H is

$$H = \begin{bmatrix} 1 & 0 & 0 & 1 & 0 & 0 & 0 & 0 & 0 \\ 0 & 1 & 0 & 0 & 1 & 0 & 0 & 0 & 0 \\ 0 & 0 & 1 & 0 & 0 & 1 & 0 & 0 & 0 \\ 0 & 0 & 0 & 1 & 0 & 0 & 1 & 0 & 0 \\ 0 & 0 & 0 & 0 & 1 & 0 & 0 & 1 & 0 \\ 0 & 0 & 0 & 0 & 0 & 1 & 0 & 0 & 1 \end{bmatrix}$$

It can be verified that $GH^T = 0$. The parity-check matrix H can be used to generate the dual code $\mathcal{C}^{\perp}$. This matrix generates $2^6 = 64$ codewords, each of length 9. A complete generation of the dual code $\mathcal{C}^{\perp}$ is left to the reader. It can also be verified that any codeword in $\mathcal{C}$ is orthogonal to any codeword in $\mathcal{C}^{\perp}$. □

2.6 Hamming Codes

Hamming codes are linear single error-correcting perfect codes. These are named after their discoverer R. W. Hamming, one of the foremost pioneers of coding theory. Basics of Hamming codes, and Hamming codes via the cyclic-code paradigm are discussed in this section.

2.6.1 Basics of Hamming Codes

Elements of Hamming code are outlined in this subsection. Its definition, some important properties, and a decoding algorithm for the code is given. This code is defined in terms of the parity-check matrix.

Definitions 2.18. *Hamming code.*

1. *Hamming code over a finite field $\mathbb{F}_q$, is an (n, k, d_H)-code, where the order of the field is equal to q, $n = (q^m - 1) / (q - 1), k = (n - m), m \geq 2$, and d_H is the distance of the code. The parity-check matrix has nonzero columns. Furthermore, any two columns are not scalar multiples of each other.*
2. *The parity-check matrix H of an (n, k, d_H)-binary Hamming code is an $m \times n$ matrix, where $m = (n - k), n = (2^m - 1), m \geq 2$, and d_H is the distance of the code. The columns of this matrix consists of all n nonzero binary vectors of length m, each occurring once.* □

It can be observed that the Hamming code defined over the binary field $\mathbb{B}$, is a specific case of the Hamming code defined over a finite field $\mathbb{F}_q$. Following observations can be made about the Hamming codes.

Observations 2.10. Let the (n, k, d_H) Hamming code be defined over a finite field $\mathbb{F}_q$.

1. The number of codewords in the code is equal to $M = q^k$.
2. The minimum distance $d_H = 3$. Therefore it can correct $t = 1$ error, if necessary. That is, the Hamming code is a single error-correcting code.
3. The Hamming code is perfect. □

Some of the above observations for the case of binary field are proved in the problem section. If the Hamming code is defined over a binary field $\mathbb{B}$, the property of perfect code yields $n = (2^m - 1)$, where $m = (n - k)$ is the number of parity bits. Therefore the binary Hamming code $\mathcal{C}$ is a $(2^m - 1, 2^m - m - 1, 3)$-code. Since $d_H = 3$, we have $m \geq 2$, and the code is a single error-correcting code.

We next outline a simple scheme for correcting an error in the received word, if a binary Hamming code is used. Observe that the binary matrix H^T is $n \times m$. In this matrix, the elements in a specific row are the binary representation of that row number.

Therefore if the received word is ρ, then its syndrome is $s = \rho H^T$. Thus if the received word ρ is error-free then $s = 0$. However, if the received word ρ has an error in position r (the position numbers are counted from left to right), then the elements of the syndrome s are the binary representation of the number r, where $1 \leq r \leq n$. This observation translates into the following simple scheme for correcting (if necessary) the received word.

If $s \neq 0$, then obtain a decimal representation of the elements of the syndrome. It gives the position number r of the bit in error in the received word. The rth binary element of the received word is changed from 0 to 1, or from 1 to 0. This technique is best illustrated via an example.

Example 2.9. Consider a binary Hamming code where the number of parity bits is equal to 3, that is $m = 3$. This is a $(7, 4, 3)$-code. Its parity-check matrix H is defined as

$$H = \begin{bmatrix} 0 & 0 & 0 & 1 & 1 & 1 & 1 \\ 0 & 1 & 1 & 0 & 0 & 1 & 1 \\ 1 & 0 & 1 & 0 & 1 & 0 & 1 \end{bmatrix}$$

The canonical parity-check matrix H_c is obtained by interchanging columns 1 and 7, columns 2 and 6, and columns 4 and 5.

$$H_c = \begin{bmatrix} 1 & 1 & 0 & 1 & 1 & 0 & 0 \\ 1 & 1 & 1 & 0 & 0 & 1 & 0 \\ 1 & 0 & 1 & 1 & 0 & 0 & 1 \end{bmatrix}$$

Note that the three right-most columns of this matrix form an identity matrix of size 3. Thus H_c is in the form $\left[-B^T \mid I_3\right]$. The corresponding canonical generator matrix G_c is equal to $[I_4 \mid B]$.

$$G_c = \begin{bmatrix} 1 & 0 & 0 & 0 & 1 & 1 & 1 \\ 0 & 1 & 0 & 0 & 1 & 1 & 0 \\ 0 & 0 & 1 & 0 & 0 & 1 & 1 \\ 0 & 0 & 0 & 1 & 1 & 0 & 1 \end{bmatrix}$$

The generator matrix G corresponding to the parity-check matrix H is obtained by interchanging columns 1 and 7, columns 2 and 6, and columns 4 and 5 of the matrix G_c.

$$G = \begin{bmatrix} 1 & 1 & 0 & 1 & 0 & 0 & 1 \\ 0 & 1 & 0 & 1 & 0 & 1 & 0 \\ 1 & 1 & 1 & 0 & 0 & 0 & 0 \\ 1 & 0 & 0 & 1 & 1 & 0 & 0 \end{bmatrix}$$

It can indeed be checked that $GH^T = G_c H_c^T = 0$. The set $\mathcal{C}$ of $2^4 = 16$ codewords is given below. Note that the codeword-length is equal to 7.

$$\begin{aligned}\mathcal{C} = \{&(0000000), (1001100), (1110000), (0111100), (0101010), (1100110),\\ &(1011010), (0010110), (1101001), (0100101), (0011001), (1010101),\\ &(1000011), (0001111), (0110011), (1111111)\}\end{aligned}$$

Let the received word be $\rho = (0110011) \in \mathcal{C}$, then it can be checked that the syndrome $s = \rho H^T = 0$. However, if the received word is $\rho = (0110010)$, then $s = \rho H^T = \begin{bmatrix} 1\,1\,1 \end{bmatrix}$. This implies that the 7th received bit (counting from left) is corrupted. Therefore the corrected received word is (0110011). □

The definition of the Hamming code via its parity-check matrix yields to an elegant decoding algorithm. The decoding algorithm for the binary Hamming code is next outlined. It assumes single error, and maximum-likelihood (minimum distance) decoding scheme.

Algorithm 2.2. *Decoding Binary Hamming Code.*

Input: Codeword ρ belongs to the binary Hamming code (n, k, d_H).
Output: Correct codeword ρ'.
begin
  Step 1: If the word ρ is received, calculate its syndrome $s = \rho H^T$.
    Compute its equivalent decimal representation.
    Let it be the decimal integer j.
  Step 2: Make the following decision:
    if $j = 0$ **then** $\rho' \leftarrow \rho$
    (*no error occurred, and* ρ *is the transmitted codeword*)
    else
    begin
    ($j \neq 0$, *error occurred*)
      Correct the error in the jth position of the received
      word ρ and set it to ρ'. The position numbers are
      counted from left to right.
    end (*end of if statement*)
end (*end of decoding algorithm for binary Hamming code*)

2.6.2 Hamming Codes via Cyclic Codes

Recall that cyclic codes have the advantage of easy encoding. Furthermore, cyclic Hamming codes are perfect single error-correcting codes. In addition, these are easy to both encode and decode. Cyclic Hamming codes have the property that these can be determined via the use of primitive polynomials. Only cyclic Hamming codes defined over a binary field are described in this chapter.

Consider a binary Hamming code $\mathcal{C}$ with parameters (n, k, d_H), where $k = (n - m)$, $d_H = 3$, and $n = (2^m - 1)$. Let its parity-check matrix be H. It is of size $m \times n$.

Also let α be a primitive element of the finite field GF_{2^m}, that is this element generates all nonzero elements of this finite field. The nonzero field elements $1, \alpha, \alpha^2, \ldots, \alpha^{n-1}$ are all distinct. Each field element can be represented as a binary m-tuple. Let the minimal polynomial associated with the element α be $\eta_\alpha(x)$. It is also a primitive polynomial, because α is a primitive element. Therefore $\deg \eta_\alpha(x) = m$.

Let the nonzero field elements of GF_{2^m} be the columns of the parity-check matrix H. That is

$$H = \begin{bmatrix} 1 \; \alpha \; \alpha^2 \; \cdots \; \alpha^{n-1} \end{bmatrix}$$

Note that this matrix can be written as a binary matrix, as each nonzero element of GF_{2^m} has a binary polynomial representation $v(x) = \sum_{i=0}^{(m-1)} v_i x^i$ of degree smaller than m, where $v_i \in \mathbb{B}$ for $0 \leq i \leq (m-1)$. Equivalently, $v(x)$ can also be represented as a column vector

$$v(x) \triangleq \begin{bmatrix} v_{m-1} \\ v_{m-2} \\ \cdots \\ v_0 \end{bmatrix}$$

Each α^j for $0 \leq j \leq (n-1)$ has a binary column representation. Therefore, this construction yields an $m \times n$ parity-check matrix H. The columns of this matrix H are nonzero and pair-wise distinct.

Let the received codeword ρ be $\begin{bmatrix} \rho_0 \; \rho_1 \; \rho_2 \; \cdots \; \rho_{n-1} \end{bmatrix}$ where $\rho_j \in \mathbb{B}$ for $0 \leq j \leq (n-1)$. If the received codeword $\rho \in \mathcal{C}$ then $H\rho^T = 0$. This implies $\rho_0 + \rho_1\alpha + \cdots + \rho_{n-1}\alpha^{n-1} = 0$. Define

$$\rho(\alpha) = \sum_{j=0}^{(n-1)} \rho_j \alpha^j$$

That is, $\rho(\alpha)$ is a polynomial in α over $\mathbb{B}$. Therefore ρ is a codeword iff α is a root of the polynomial $\rho(x)$. Using a property of minimal polynomials it can be inferred that if α is a root of the polynomial $\rho(\alpha)$ then $\eta_\alpha(x) \mid \rho(x)$. That is, $\eta_\alpha(x)$ divides $\rho(x)$, which is a polynomial representation of the codeword ρ. Thus $\eta_\alpha(x)$ divides every $\rho(x)$. Moreover the degree of the minimal polynomial $\eta_\alpha(x)$ is equal to m. Thus each polynomial representation $\rho(x)$ of a codeword is a multiple of $\eta_\alpha(x)$. So by an important result (theorem) of cyclic codes, this code is indeed cyclic, and the primitive polynomial $\eta_\alpha(x)$ is its generator polynomial. Therefore

$$G \triangleq \begin{bmatrix} \eta_\alpha(x) & & & & \\ & x\eta_\alpha(x) & & & \\ & & \cdots & & \\ & & & \cdots & \\ & & & & x^{n-m-1}\eta_\alpha(x) \end{bmatrix}$$

This discussion is condensed in the following theorem.

Theorem 2.9. *A primitive polynomial of degree m of the finite field GF_{2^m}, is the generator polynomial of a cyclic binary Hamming code $\mathcal{C}$, of codeword-length $(2^m - 1)$.* □

Examples 2.10. Some illustrative examples.

1. Recall that a primitive polynomial of degree n, over $\mathbb{B}$ always divides $(x^j - 1)$, $j = (2^n - 1)$, but the primitive polynomial does not divide $(x^m - 1)$ for any $m < j$. Primitive polynomials over $\mathbb{B}$ which can be used as the generator polynomials of cyclic binary Hamming codes are listed below.

Degree	Primitive Polynomials
2	$(x^2 + x + 1)$
3	$(x^3 + x + 1)$
4	$(x^4 + x + 1)$
5	$(x^5 + x^2 + 1)$
6	$(x^6 + x + 1)$
7	$(x^7 + x^3 + 1)$

2. A Hamming code is generated by using the primitive polynomial of the finite field GF_8. In this case, $m = 3$ and $n = (2^m - 1) = 7$. From the previous example, its primitive polynomial is $\eta_\alpha(x) = (x^3 + x + 1)$. Let the primitive element be α. Therefore $\eta_\alpha(\alpha) = 0$. Then the parity-check matrix is:

$$H = \begin{bmatrix} 1 \ \alpha \ \alpha^2 \ \alpha^3 \ \alpha^4 \ \alpha^5 \ \alpha^6 \end{bmatrix}$$

The binary representations of the column vectors in the above matrix are

$$1 = (001), \ \alpha = (010), \ \alpha^2 = (100), \ \alpha^3 = (011), \\ \alpha^4 = (110), \ \alpha^5 = (111), \ \alpha^6 = (101)$$

This yields the parity-check matrix H. It is

$$H = \begin{bmatrix} 0 & 0 & 1 & 0 & 1 & 1 & 1 \\ 0 & 1 & 0 & 1 & 1 & 1 & 0 \\ 1 & 0 & 0 & 1 & 0 & 1 & 1 \end{bmatrix}$$

The corresponding generator matrix G is

$$G \triangleq \begin{bmatrix} \eta_\alpha(x) & & & \\ & x\eta_\alpha(x) & & \\ & & x^2\eta_\alpha(x) & \\ & & & x^3\eta_\alpha(x) \end{bmatrix}$$
$$= \begin{bmatrix} 1 & 1 & 0 & 1 & 0 & 0 & 0 \\ 0 & 1 & 1 & 0 & 1 & 0 & 0 \\ 0 & 0 & 1 & 1 & 0 & 1 & 0 \\ 0 & 0 & 0 & 1 & 1 & 0 & 1 \end{bmatrix}$$

It can be verified that $GH^T = 0$. □

2.7 Cyclic Redundancy Check Codes

Cyclic redundancy check (CRC) codes are error-detecting codes. That is, these codes do not have any error-correcting capabilities. Consequently they are used with automatic repeat request systems, or along with an error-correcting code to detect the limits of its error-correcting scheme. Therefore when the CRC coding scheme is used, the syndrome of the received word is computed. If the syndrome is equal to zero, then no error has been detected, but if it is nonzero then the syndrome cannot be used for the purpose of error-correction.

Also note that the CRC codes can detect bursts of errors, but cannot correct them. An example of use of the CRC code is in the packet of an Ethernet LAN (local area network). Some properties of the CRC codes are listed below.

Properties of CRC Codes

(a) Let α be a primitive element of the finite field GF_q, and the corresponding primitive polynomial be $\eta_\alpha(x)$. Let the generator polynomial of the CRC code be $g(x)$, where $\deg g(x) = \gamma$. Then

$$g(x) = (x+1)\,\eta_\alpha(x)$$

(b) The codeword-length of this code is $n \le (2^{\gamma-1} - 1)$, and the message-word length is $k = (n - \gamma)$. □

In practice, large values of γ are selected, and the codeword-length is slightly less than the maximum. This process is termed *shortening* the code. Recall that $\tau(x), \rho(x)$, and $\varepsilon(x)$ are the transmitted codeword, received word, and error word polynomials respectively. These polynomials are all of degree at most $(n-1)$. Furthermore, $\rho(x) = (\tau(x) + \varepsilon(x))$. It has been established that since $\tau(x) \equiv 0 \pmod{g(x)}$, we have $\rho(x) \equiv \varepsilon(x) \pmod{g(x)}$. The relationship between the error polynomial $\varepsilon(x)$ and burst errors is next studied. Consider the following cases.

(a) Let $\varepsilon(x) = 0$, then no error has been detected.
(b) Let $\varepsilon(x) = x^j$ for any j, where $0 \le j \le (n-1)$. Assume that $g(x)$ has two or more terms. In this case a single bit error is detected, because $g(x)$ does not divide x^j.
(c) Let $\varepsilon(x) = x^j + x^l$ for any j, where $0 \le j < l \le (n-1)$. Assume that $g(x)$ has two or more terms, and does not divide $(1 + x^r)$ for $r = 1, 2, \ldots, n$. In this case $g(x)$ does not divide the error polynomial. Consequently two bits in error can be detected.
(d) Let $\varepsilon(x) = (x^j + x^{j+1} + \cdots + x^{j+\zeta-1})$ for some j, where $0 \le j \le (n-\zeta)$, then ζ successive bits are in error. That is, there is a burst of ζ errors. Write the error polynomial $\varepsilon(x)$ as $x^j \sum_{i=0}^{\zeta-1} x^i$. Consequently, if the $\deg g(x) = \zeta$, then it does not divide $\varepsilon(x)$. Therefore, it can be concluded that if $\deg g(x) = \gamma$ then all error bursts of length less than or equal to γ are detected.

Example 2.11. Consider a primitive polynomial $(x^4 + x + 1)$ over a binary field. Its degree is 4. Then

$$g(x) = (x+1)(x^4 + x + 1) = x^5 + x^4 + x^2 + 1$$

Therefore $\deg g(x) = \gamma = 5$, and $n \le (2^{\gamma-1} - 1) = 15$. Select the codeword-length as $n = 9$. This implies that the message-word length is $k = (n - \gamma) = 4$. The 4×9 generator matrix G is

$$G = \begin{bmatrix} 1 & 0 & 1 & 0 & 1 & 1 & 0 & 0 & 0 \\ 0 & 1 & 0 & 1 & 0 & 1 & 1 & 0 & 0 \\ 0 & 0 & 1 & 0 & 1 & 0 & 1 & 1 & 0 \\ 0 & 0 & 0 & 1 & 0 & 1 & 0 & 1 & 1 \end{bmatrix}$$

This matrix is converted to the canonical generator matrix G_c by replacing row one by the sum of rows one and three. Similarly replace row two by the sum of rows two and four. This results in

$$G_c = \begin{bmatrix} 1 & 0 & 0 & 0 & 0 & 1 & 1 & 1 & 0 \\ 0 & 1 & 0 & 0 & 0 & 0 & 1 & 1 & 1 \\ 0 & 0 & 1 & 0 & 1 & 0 & 1 & 1 & 0 \\ 0 & 0 & 0 & 1 & 0 & 1 & 0 & 1 & 1 \end{bmatrix}$$

The corresponding parity-check matrix H_c is a 5×9 matrix.

$$H_c = \begin{bmatrix} 0 & 0 & 1 & 0 & 1 & 0 & 0 & 0 & 0 \\ 1 & 0 & 0 & 1 & 0 & 1 & 0 & 0 & 0 \\ 1 & 1 & 1 & 0 & 0 & 0 & 1 & 0 & 0 \\ 1 & 1 & 1 & 1 & 0 & 0 & 0 & 1 & 0 \\ 0 & 1 & 0 & 1 & 0 & 0 & 0 & 0 & 1 \end{bmatrix}$$

Since the matrix G_c was obtained from the matrix G by only row operations, the parity-check matrix H corresponding to the original generator matrix G is equal to H_c. That is, $H = H_c$ It can be checked that indeed $GH^T = 0$.

Let the message word be $f = (1011)$ then the corresponding codeword $\tau = fG = (100100101)$. Recall that the syndrome s of a received codeword ρ is specified by ρH^T. It can directly be checked that if $\rho = \tau$, then $s = \rho H^T = 0$. The different cases of corruption of the codeword τ are examined below. In the following cases, the bit numbers are counted from the left to right.

(a) Assume that the fourth bit (counting from left) of the codeword τ is corrupted, that is $\rho = (100000101)$, then the syndrome $s = \rho H^T = (01011)$, which is nonzero.
(b) The fourth and sixth bits of the codeword τ are corrupted, that is $\rho = (100001101)$. Then the syndrome $s = \rho H^T = (00011)$, which is nonzero.
(c) Five successive bits, starting from bit number two are corrupted due to noise on the channel. The received word $\rho = (111011101)$. The corresponding syndrome word is (00010). Therefore the code was able to detect an error burst of length 5.
(d) Six successive bits, starting from bit number two are corrupted due to noise on the channel. The received word $\rho = (111011001)$. The corresponding syndrome word is (00110). Fortunately the code was able to detect an error burst of length 6.
(e) Six bits are corrupted due to noise on the channel. These are bit numbers $2, 3, 4, 5, 6$, and 8. The received word $\rho = (111011111)$. The corresponding syndrome word is (00000)! Therefore the code was unable to detect this sequence of error. □

2.8 Reed-Muller Codes

Reed-Muller codes were discovered by D. E. Muller in 1954, and their special decoding algorithm was discovered by I. S. Reed in the same year. These are linear codes defined over the binary field, and are easily decodable by a simple voting technique. However, their minimum distance values d_H are generally not as good as other codes.

These codes are defined over the binary field $\mathbb{B}$, and are parametrized by two integers, r and m; where $m, r \in \mathbb{N}$, and $0 \leq r \leq m$. Furthermore, the length of the codeword in these codes is equal to $n = 2^m$. Denote these codes by $\mathcal{R}(r, m)$. Reed-Muller codes can be defined by constructing a generator matrix.

Let $a, b \in \mathbb{B}^{(n)}$, where $a = (a_1, a_2, \ldots, a_n)$ and $b = (b_1, b_2, \ldots, b_n)$. The *product* $(\cdot)$ of these two vectors is defined by a component-wise multiplication. That is

$$a \cdot b \triangleq (a_1 b_1, a_2 b_2, \ldots, a_n b_n) \in \mathbb{B}^{(n)}$$

A matrix $A(r, m)$ which is associated with the code $\mathcal{R}(r, m)$ for $0 \le r \le m$ is next defined. This matrix $A(r, m)$ is an array of blocks.

$$A(r, m) = \begin{bmatrix} G_0 \\ G_1 \\ \vdots \\ G_r \end{bmatrix}$$

The matrices G_j for $0 \le j \le r$ are computed as follows.

(a) The matrix G_0 is $1 \times n$, and all its elements are 1's.
(b) The matrix G_1 is $m \times n$. It has binary m-tuple appearing once per column. The first column is all 0's. The second column has a 1 in its first element, and the remaining elements are all 0's; and so on. The last column has all 1's as its elements.
(c) The computation of the matrix G_j for $2 \le j \le r$ uses the above mentioned vector product. This matrix is of size $\binom{m}{j} \times n$. The rows of this matrix are determined by selecting j rows of matrix G_1 at a time and computing their component-wise product.

The number of rows in the matrix $A(r, m)$ is given by $k = \sum_{j=0}^{r} \binom{m}{j}$, where k is equal to the number of bits in the message. Therefore the size of the matrix $A(r, m)$ is $k \times n$. It is shown subsequently that $\mathcal{R}(r, m)$ is a linear code in $\mathbb{B}^{(n)}$ whose basis vectors are the rows of the matrix $A(r, m)$.

Example 2.12. Let $m = 3$, and $n = 2^m = 8$. Then

$$G_0 = \begin{bmatrix} 1 & 1 & 1 & 1 & 1 & 1 & 1 & 1 \end{bmatrix} = [u_0]$$

$$G_1 = \begin{bmatrix} 0 & 1 & 0 & 1 & 0 & 1 & 0 & 1 \\ 0 & 0 & 1 & 1 & 0 & 0 & 1 & 1 \\ 0 & 0 & 0 & 0 & 1 & 1 & 1 & 1 \end{bmatrix} = \begin{bmatrix} u_1 \\ u_2 \\ u_3 \end{bmatrix}$$

$$G_2 = \begin{bmatrix} 0 & 0 & 0 & 1 & 0 & 0 & 0 & 1 \\ 0 & 0 & 0 & 0 & 0 & 1 & 0 & 1 \\ 0 & 0 & 0 & 0 & 0 & 0 & 1 & 1 \end{bmatrix} = \begin{bmatrix} u_1 \cdot u_2 \\ u_1 \cdot u_3 \\ u_2 \cdot u_3 \end{bmatrix}$$

$$G_3 = \begin{bmatrix} 0 & 0 & 0 & 0 & 0 & 0 & 0 & 1 \end{bmatrix} = [u_1 \cdot u_2 \cdot u_3]$$

The matrices G_1, G_2, and G_3 have $\binom{3}{1} = 3$, $\binom{3}{2} = 3$, and $\binom{3}{3} = 1$ rows respectively. Observe that the row vector u_0 is the basis of the code $\mathcal{R}(0, 3)$. Similarly, the row vectors u_0, u_1, u_2, and u_3 are the basis of the code $\mathcal{R}(1, 3)$. □

We next describe an alternate, yet equivalent method to define these codes. It is done via their set of codewords.

Definition 2.19. *Let $m, r \in \mathbb{N}$ such that $0 \leq r \leq m$. The codes are defined in the binary field $\mathbb{B}$, and are of length $n = 2^m$. For each length n, there are $(m+1)$ linear codes denoted by $\mathcal{R}(r, m)$. The code $\mathcal{R}(r, m)$ is called the rth order Reed-Muller code. These codes are recursively defined:*

(a) *$\mathcal{R}(0, m)$ is a binary repetition code. $\mathcal{R}(0, m) = \{(00\ldots0), (11\ldots1)\}$, and $\mathcal{R}(m, m)$ is the entire space $\mathbb{B}^{(2^m)}$.*

(b) $\mathcal{R}(r, m) = \{(u, u+v) \mid u \in \mathcal{R}(r, m-1), v \in \mathcal{R}(r-1, m-1)\}, 0 < r < m.$ □

These codes can also be described recursively via their generator matrices $G(r, m)$. Note that the matrix $G(r, m)$ has 2^m columns for all values of r. Using the definition of the Reed-Muller codes, the generator matrices of these codes are:

$$G(0, m) = \begin{bmatrix} 1 & 1 & \cdots & 1 \end{bmatrix}, \quad m \in \mathbb{N} \tag{2.15a}$$

$$G(m, m) = \begin{bmatrix} G(m-1, m) \\ 0 \ \cdots \ 0 \ \ 1 \end{bmatrix}, \quad m \in \mathbb{P} \tag{2.15b}$$

$$G(r, m) = \begin{bmatrix} G(r, m-1) & G(r, m-1) \\ 0 & G(r-1, m-1) \end{bmatrix}, \quad 0 < r < m, \ m \in \mathbb{P} \backslash \{1\} \tag{2.15c}$$

In the $G(r, m)$ matrix, the 0 entry is actually a matrix with all elements equal to zero. Furthermore, the size of this all-zero matrix is identical to that of the matrix $G(r-1, m-1)$. Actually the rows of the matrix $G(r, m)$ can be obtained by a permutation of rows of the matrix $A(r, m)$. This can most easily be observed via elementary examples.

Examples 2.13. Some illustrative examples.

1. Let $m = 0, n = 1$.

$$G_0 = [1]$$
$$A(0,0) = [1]$$
$$G(0,0) = \begin{bmatrix} 1 \end{bmatrix} = A(0,0)$$
$$\mathcal{R}(0,0) = \{(0), (1)\}$$

.

2. Let $m = 1, n = 2$.

$$G_0 = \begin{bmatrix} 1 & 1 \end{bmatrix}$$
$$G_1 = \begin{bmatrix} 0 & 1 \end{bmatrix}$$

(a) For $r = 0$

$$A(0,1) = G_0 = \begin{bmatrix} 1 & 1 \end{bmatrix}$$
$$G(0,1) = \begin{bmatrix} 1 & 1 \end{bmatrix} = A(0,1)$$
$$\mathcal{R}(0,1) = \{(00), (11)\}$$

(b) For $r = 1$

$$A(1,1) = \begin{bmatrix} G_0 \\ G_1 \end{bmatrix} = \begin{bmatrix} 1 & 1 \\ 0 & 1 \end{bmatrix}$$

$$G(1,1) = \begin{bmatrix} G(0,1) \\ 0 \ 1 \end{bmatrix} = \begin{bmatrix} 1 & 1 \\ 0 & 1 \end{bmatrix} = A(1,1)$$

$$\mathcal{R}(1,1) = \mathbb{B}^{(2)} = \{(00),(01),(10),(11)\}$$

3. Let $m = 2, n = 2^2 = 4$.

$$G_0 = \begin{bmatrix} 1 & 1 & 1 & 1 \end{bmatrix}$$

$$G_1 = \begin{bmatrix} 0 & 1 & 0 & 1 \\ 0 & 0 & 1 & 1 \end{bmatrix}$$

$$G_2 = \begin{bmatrix} 0 & 0 & 0 & 1 \end{bmatrix}$$

(a) For $r = 0$

$$A(0,2) = G_0 = \begin{bmatrix} 1 & 1 & 1 & 1 \end{bmatrix}$$

$$G(0,2) = \begin{bmatrix} 1 & 1 & 1 & 1 \end{bmatrix} = A(0,2)$$

$$\mathcal{R}(0,2) = \{(0000),(1111)\}$$

(b) For $r = 1$

$$A(1,2) = \begin{bmatrix} G_0 \\ G_1 \end{bmatrix} = \begin{bmatrix} 1 & 1 & 1 & 1 \\ 0 & 1 & 0 & 1 \\ 0 & 0 & 1 & 1 \end{bmatrix}$$

$$G(1,2) = \begin{bmatrix} G(1,1) & G(1,1) \\ 0 & G(0,1) \end{bmatrix} = A(1,2)$$

$$\mathcal{R}(1,2) = \{(0000),(0011),(0101),(0110),$$
$$(1001),(1010),(1100),(1111)\}$$

(c) For $r = 2$

$$A(2,2) = \begin{bmatrix} G_0 \\ G_1 \\ G_2 \end{bmatrix} = \begin{bmatrix} 1 & 1 & 1 & 1 \\ 0 & 1 & 0 & 1 \\ 0 & 0 & 1 & 1 \\ 0 & 0 & 0 & 1 \end{bmatrix}$$

$$G(2,2) = \begin{bmatrix} G(1,2) \\ 0 \ 0 \ 0 \ 1 \end{bmatrix} = A(2,2)$$

$$\mathcal{R}(2,2) = \mathbb{B}^{(4)}$$

4. Let $m = 3, n = 2^3 = 8$.

$$G_0 = \begin{bmatrix} 1 & 1 & 1 & 1 & 1 & 1 & 1 & 1 \end{bmatrix}$$

$$G_1 = \begin{bmatrix} 0 & 1 & 0 & 1 & 0 & 1 & 0 & 1 \\ 0 & 0 & 1 & 1 & 0 & 0 & 1 & 1 \\ 0 & 0 & 0 & 0 & 1 & 1 & 1 & 1 \end{bmatrix}$$

$$G_2 = \begin{bmatrix} 0 & 0 & 0 & 1 & 0 & 0 & 0 & 1 \\ 0 & 0 & 0 & 0 & 0 & 1 & 0 & 1 \\ 0 & 0 & 0 & 0 & 0 & 0 & 1 & 1 \end{bmatrix}$$

$$G_3 = \begin{bmatrix} 0 & 0 & 0 & 0 & 0 & 0 & 0 & 1 \end{bmatrix}$$

(a) For $r = 0$

$$A(0,3) = G_0 = \begin{bmatrix} 1 & 1 & 1 & 1 & 1 & 1 & 1 & 1 \end{bmatrix}$$
$$G(0,3) = \begin{bmatrix} 1 & 1 & 1 & 1 & 1 & 1 & 1 & 1 \end{bmatrix} = A(0,3)$$
$$\mathcal{R}(0,3) = \{(00000000), (11111111)\}$$

(b) For $r = 1$

$$A(1,3) = \begin{bmatrix} G_0 \\ G_1 \end{bmatrix} = \begin{bmatrix} 1 & 1 & 1 & 1 & 1 & 1 & 1 & 1 \\ 0 & 1 & 0 & 1 & 0 & 1 & 0 & 1 \\ 0 & 0 & 1 & 1 & 0 & 0 & 1 & 1 \\ 0 & 0 & 0 & 0 & 1 & 1 & 1 & 1 \end{bmatrix}$$

$$G(1,3) = \begin{bmatrix} G(1,2) & G(1,2) \\ 0 & G(0,2) \end{bmatrix} = A(1,3)$$

Listing of $\mathcal{R}(1,3)$ is left to the reader.

(c) For $r = 2$

$$A(2,3) = \begin{bmatrix} G_0 \\ G_1 \\ G_2 \end{bmatrix} = \begin{bmatrix} 1 & 1 & 1 & 1 & 1 & 1 & 1 & 1 \\ 0 & 1 & 0 & 1 & 0 & 1 & 0 & 1 \\ 0 & 0 & 1 & 1 & 0 & 0 & 1 & 1 \\ 0 & 0 & 0 & 0 & 1 & 1 & 1 & 1 \\ 0 & 0 & 0 & 1 & 0 & 0 & 0 & 1 \\ 0 & 0 & 0 & 0 & 0 & 1 & 0 & 1 \\ 0 & 0 & 0 & 0 & 0 & 0 & 1 & 1 \end{bmatrix}$$

$$G(2,3) = \begin{bmatrix} G(2,2) & G(2,2) \\ 0 & G(1,2) \end{bmatrix} = \begin{bmatrix} 1 & 1 & 1 & 1 & 1 & 1 & 1 & 1 \\ 0 & 1 & 0 & 1 & 0 & 1 & 0 & 1 \\ 0 & 0 & 1 & 1 & 0 & 0 & 1 & 1 \\ 0 & 0 & 0 & 1 & 0 & 0 & 0 & 1 \\ 0 & 0 & 0 & 0 & 1 & 1 & 1 & 1 \\ 0 & 0 & 0 & 0 & 0 & 1 & 0 & 1 \\ 0 & 0 & 0 & 0 & 0 & 0 & 1 & 1 \end{bmatrix}$$

Note that $G(2,3) \neq A(2,3)$, however the matrix $G(2,3)$ can be obtained from $A(2,3)$ by an interchange of rows 4 and 5. Listing of $\mathcal{R}(2,3)$ is left to the reader.

(d) For $r = 3$

$$A(3,3) = \begin{bmatrix} G_0 \\ G_1 \\ G_2 \\ G_3 \end{bmatrix}, \quad G(3,3) = \begin{bmatrix} & & & G(2,3) & & & & \\ 0 & 0 & 0 & 0 & 0 & 0 & 0 & 1 \end{bmatrix}$$

The 8×8 matrices $A(3,3)$ and $G(3,3)$ are not equal. However the matrix $G(3,3)$ can be obtained from $A(3,3)$ by an interchange of rows 4 and 5. Also $\mathcal{R}(3,3) = \mathbb{B}^{(8)}$. □

Some observations about these codes are listed below.

Observations 2.11. Let $\mathcal{R}(r,m)$ be a rth order Reed-Muller code, where $0 \leq r \leq m$. The length of the codeword is given by $n = 2^m$. Denote the dual of this code by $\mathcal{R}^{\perp}(r,m)$.

1. $\mathcal{R}(r-1,m) \subseteq \mathcal{R}(r,m)$ for $r > 0$. Equivalently $\mathcal{R}(i,m) \subseteq \mathcal{R}(j,m)$ for $0 \leq i \leq j \leq m$.

2. The dimension of this code is equal to $k = \sum_{j=0}^{r} \binom{m}{j}$.
3. The Hamming distance of this code is $d_H = 2^{m-r}$.
4. Dual codes: $\mathcal{R}^{\perp}(m, m) = \{(00\ldots0)\}$, and if $0 \le r < m$, then

$$\mathcal{R}^{\perp}(r, m) = \mathcal{R}(m - r - 1, m)$$

□

These results can be established by induction. See the problem section for proofs of these observations. A special algorithm for decoding the Reed-Muller codes was developed by Reed. It uses a technique called *majority logic decoding*. This algorithm is not described due to space limitation. However, the general syndrome technique of decoding linear codes is applicable to these codes.

The next section describes the well-known Bose-Chaudhuri-Hocquenghem family of codes. These are cyclic codes with several tunable parameters. In addition, such codes can correct both multiple and random errors. Several efficient decoding algorithms also exist for these codes.

2.9 Bose-Chaudhuri-Hocquenghem Codes

The binary Bose-Chaudhuri-Hocquenghem (BCH) codes were independently discovered by R. C. Bose and D. Ray-Chaudhuri in 1960; and A. Hocquenghem in 1959. D. C. Gorenstein and N. Zierler extended these codes to all finite fields in 1961. The BCH codes are powerful multiple and random error-correcting codes. The Reed-Solomon codes are a special case of the BCH codes, which are described later in the chapter.

The BCH codes are cyclic codes defined over a finite field $\mathbb{F}_q$, where the order of the field is equal to q. These codes are capable of correcting all random patterns of t errors by a simple decoding algorithm. Let the codeword-length of these codes be n. The generator polynomial $g(x)$, of this code is the least common multiple of the minimal polynomials of successive powers of a primitive nth root of unity in an extension field $\mathbb{F}_{q^m}$ of $\mathbb{F}_q$ where $m \in \mathbb{P}$. The order of the field $\mathbb{F}_{q^m}$ is equal to q^m.

Definitions 2.20. *Definition of the BCH code*:

1. *A BCH code is a cyclic code $\mathcal{C}$, defined over the finite field $\mathbb{F}_q$, where the order of this field is equal to q. The codeword-length is n such that* $\gcd(n, q) = 1$. *Let α be a primitive nth root of unity in an extension field of $\mathbb{F}_q$, say $\mathbb{F}_{q^m}$. The order of the field $\mathbb{F}_{q^m}$ is equal to q^m, where $m \in \mathbb{P}$ is the multiplicative order of q modulo n. That is, n is a divisor of $(q^m - 1)$. The generator polynomial $g(x)$ of the code is the least common multiple of the minimal polynomials of $\alpha^l, \alpha^{l+1}, \ldots, \alpha^{l+\delta_{BCH}-2}$ for some integers $l \in \mathbb{N}$ and $\delta_{BCH} \in \mathbb{P}$, where $2 \le \delta_{BCH} \le n$. The parameter δ_{BCH} is called the designed distance of the code.*
2. *If $l = 1$, the BCH code is narrow-sense.*
3. *If $n = (q^m - 1)$, for some $m \in \mathbb{P}$, the BCH code is primitive. In this case, α is a primitive element in the extension field $\mathbb{F}_{q^m}$.* □

It also follows from the definition of the generator polynomial $g(x)$ of the BCH code, that its roots are $\alpha^l, \alpha^{l+1}, \ldots, \alpha^{l+\delta_{BCH}-2}$.

Proceeding as in the case of discussion of Hamming codes via the cyclic code paradigm, a matrix $\widetilde{H}$ can be determined, such that its rows are orthogonal to the codewords. Let $c \in \mathcal{C}$ be a codeword and the corresponding codeword polynomial be $c(x) = \sum_{j=0}^{n-1} c_j x^j$, where $c_j \in \mathbb{F}_q$. A row vector representation of the codeword is

$$\begin{bmatrix} c_0 & c_1 & \cdots & c_{n-1} \end{bmatrix}$$

Then $c\left(\alpha^{l+i}\right) = 0$, for $0 \leq i \leq (\delta_{BCH} - 2)$. This is true because the roots of the generator polynomial $g(x)$ are $\alpha^l, \alpha^{l+1}, \ldots, \alpha^{l+\delta_{BCH}-2}$, and $c(x)$ is a multiple of $g(x)$. This set of equations can be compactly stated in matrix form as $\widetilde{H}c^T = 0$, where

$$\widetilde{H} = \begin{bmatrix} \left(\alpha^l\right)^0 & \alpha^l & \alpha^{2l} & \cdots & \alpha^{(n-1)l} \\ \left(\alpha^{l+1}\right)^0 & \alpha^{l+1} & \alpha^{2(l+1)} & \cdots & \alpha^{(n-1)(l+1)} \\ \vdots & \vdots & \vdots & \ddots & \vdots \\ \left(\alpha^{l+\delta_{BCH}-2}\right)^0 & \alpha^{l+\delta_{BCH}-2} & \alpha^{2(l+\delta_{BCH}-2)} & \cdots & \alpha^{(n-1)(l+\delta_{BCH}-2)} \end{bmatrix}$$

This matrix is actually $(\delta_{BCH} - 1)m \times n$, where each entry in the matrix $\widetilde{H}$ as shown in the above equation, is a column vector of length m, and the vector elements belong to the field $\mathbb{F}_q$. The exponents of elements in the first column of the matrix $\widetilde{H}$ are each equal to 0. These are so expressed to indicate that each such element is a column vector of length m.

The so-called *BCH bound* is next determined. This result is established via a Vandermonde determinant defined over a finite field.

Theorem 2.10. *Let d_H and δ_{BCH} be the Hamming and the designed distances of the BCH code respectively. Then $\delta_{BCH} \leq d_H$.*

Proof. It is proved that $w(c) \geq \delta_{BCH}$ for all nonzero $c \in \mathcal{C}$, where $w(c)$ is the weight of the codeword c.

Assume that there is a codeword $c \in \mathcal{C} \setminus \{0\}$ such that $w(c) \leq (\delta_{BCH} - 1)$. It is proved that $c = 0$, which is a contradiction. For simplicity in notation let $w(c) \triangleq \omega$.

Since $\omega \leq (\delta_{BCH} - 1)$, the indices $j_1, j_2, \ldots, j_\omega$ can be determined such that $c_j \neq 0$ for all values of indices $j \in \{j_1, j_2, \ldots, j_\omega\}$. Using the system of equations, $\widetilde{H}c^T = 0$, obtain a system of equations by deleting all columns of $\widetilde{H}$, except those with the set of indices $\{j_1, j_2, \ldots, j_\omega\}$. This yields

$$\begin{bmatrix} \alpha^{j_1 l} & \alpha^{j_2 l} & \cdots & \alpha^{j_\omega l} \\ \alpha^{j_1(l+1)} & \alpha^{j_2(l+1)} & \cdots & \alpha^{j_\omega(l+1)} \\ \alpha^{j_1(l+2)} & \alpha^{j_2(l+2)} & \cdots & \alpha^{j_\omega(l+2)} \\ \vdots & \vdots & \ddots & \vdots \\ \alpha^{j_1(l+\delta_{BCH}-2)} & \alpha^{j_2(l+\delta_{BCH}-2)} & \cdots & \alpha^{j_\omega(l+\delta_{BCH}-2)} \end{bmatrix} \begin{bmatrix} c_{j_1} \\ c_{j_2} \\ c_{j_3} \\ \vdots \\ c_{j_\omega} \end{bmatrix} = \begin{bmatrix} 0 \\ 0 \\ 0 \\ \vdots \\ 0 \end{bmatrix}$$

From the above $(\delta_{BCH} - 1)$ set of equations, select the first ω equations.

$$\begin{bmatrix} \alpha^{j_1 l} & \alpha^{j_2 l} & \cdots & \alpha^{j_\omega l} \\ \alpha^{j_1(l+1)} & \alpha^{j_2(l+1)} & \cdots & \alpha^{j_\omega(l+1)} \\ \alpha^{j_1(l+2)} & \alpha^{j_2(l+2)} & \cdots & \alpha^{j_\omega(l+2)} \\ \vdots & \vdots & \ddots & \vdots \\ \alpha^{j_1(l+\omega-1)} & \alpha^{j_2(l+\omega-1)} & \cdots & \alpha^{j_\omega(l+\omega-1)} \end{bmatrix} \begin{bmatrix} c_{j_1} \\ c_{j_2} \\ c_{j_3} \\ \vdots \\ c_{j_\omega} \end{bmatrix} = \begin{bmatrix} 0 \\ 0 \\ 0 \\ \vdots \\ 0 \end{bmatrix}$$

It can be shown that the determinant of the $\omega \times \omega$ matrix on the left-hand side of the above equation can be determined by using the expression for a Vandermonde determinant. The determinant of this square matrix is equal to

$$\alpha^{(j_1+\cdots+j_\omega)l} \prod_{\omega \geq u > v \geq 1} \left(\alpha^{j_u} - \alpha^{j_v}\right)$$

As α is a primitive nth root of unity, this expression is not equal to zero because $\alpha^{j_1}, \alpha^{j_2}, \ldots, \alpha^{j_\omega}$ are pair-wise distinct. This implies that $c = 0$, which is a contradiction. □

The meaning of the term designed distance δ_{BCH}, will become clear shortly. If the BCH code can correct t errors, then $d_H \geq (2t+1)$. Also from the above theorem $d_H \geq \delta_{BCH}$. A possible candidate for the value of the designed distance δ_{BCH} is $(2t+1)$.

Also note that the matrix $\widetilde{H}$ has $m(\delta_{BCH} - 1)$ rows. This value is generally higher than $(n-k)$, the number of rows in the parity-check matrix. Therefore $(n-k) \leq m(\delta_{BCH} - 1)$. This implies that $n - m(\delta_{BCH} - 1) \leq k$. This is not a very good bound for k, the dimension of the BCH code. A tighter bound for k is obtained for binary BCH codes in the next theorem.

Theorem 2.11. *If $m \in \mathbb{P}, t \leq \left(2^{m-1} - 1\right)$, and $\delta_{BCH} = (2t+1)$, there exists a binary narrow-sense BCH t error-correcting code of length $n = (2^m - 1)$ such that the dimension of the code $k \geq (n - mt)$.*

Proof. Let α be a primitive element of the Galois field GF_{2^m}, and $\mathcal{C}$ be a cyclic code with roots $\alpha, \alpha^2, \ldots, \alpha^{2t} \in GF_{2^m}$, where the designed distance $\delta_{BCH} = (2t+1)$. Also let the minimal polynomials of these elements be $\eta_\alpha(x), \eta_{\alpha^2}(x), \ldots, \eta_{\alpha^{2t}}(x)$ respectively. The generator polynomial $g(x)$ of this code is the least common multiple of these minimal polynomials. Actually, the generator polynomial $g(x)$ of this code is the least common multiple of minimal polynomials of field elements which are odd powers of α. This is because the minimal polynomial of an element which is an even power of α, is the minimal polynomial of some element which is an odd power of α, when the BCH code is binary. For example, if $\eta_\alpha(x)$ is a minimal polynomial of α, then $\alpha^2, \alpha^4, \alpha^8, \ldots$ are also roots of $\eta_\alpha(x)$. Therefore $g(x) = \operatorname{lcm}\{\eta_\alpha(x), \eta_{\alpha^3}(x), \ldots, \eta_{\alpha^{2t-1}}(x)\}$. The degree of each of these polynomials is less than or equal to m. Thus $(n-k) = \deg g(x) \leq mt$. That is, $k \geq (n - mt)$. □

The above theorem is one of the reasons for the popularity of the binary BCH codes.

2.9.1 Binary BCH Codes

The parameters for the specification of the binary BCH codes are summarized in this subsection. Further, guidelines and examples are provided to elucidate important steps in the design of binary BCH codes.

Specification of Binary Narrow-Sense BCH Codes

(a) Let $\mathcal{C}$ be a t error-correcting narrow-sense BCH code, where the code symbols belong to the field $\mathbb{B}$. The dimension of this code is $k \in \mathbb{P}$.
(b) The codeword-length n is chosen such that $n = (2^m - 1)$, where $m \in \mathbb{P}$.
(c) Assume that $m \geq 2$ and $t < n/2$.
(d) The number of parity bits is equal to $(n-k) \in \mathbb{N}$, and $0 \leq (n-k) \leq mt$.

(e) Assume that the designed distance $\delta_{BCH} = (2t+1) \le d_H$.

Steps to Construct Binary BCH Codes

(a) Find a primitive field element $\alpha \in GF_{2^m}$. Using the primitive polynomial $\eta_\alpha(x)$ of degree m, construct the field GF_{2^m}.
(b) Determine the minimal polynomials $\eta_\alpha(x), \eta_{\alpha^3}(x), \ldots, \eta_{\alpha^{2t-1}}(x)$.
(c) The generator polynomial of this cyclic code is

$$g(x) = \text{lcm}\{\eta_\alpha(x), \eta_{\alpha^3}(x), \ldots, \eta_{\alpha^{2t-1}}(x)\}$$

(d) Find the dimension of the code $k = (n - \deg g(x))$.
(e) The binary representation of the generator polynomial $g(x)$ is a codeword. Therefore distance of the code d_H observes the relationship

$$(2t+1) \le d_H \le w(g)$$

□

These steps are illustrated via an example.

Example 2.14. Let $m = 4$, then the codeword-length $n = 15$. The minimal polynomials of the elements of the finite field GF_{16} are first determined. The polynomial representation of the nonzero elements of this field is also determined. Denote the set of elements $GF_{16} \backslash \{0\} = \{\alpha^j \mid 0 \le j \le 14\}$. A primitive polynomial of the element α is $\eta_\alpha(x) = (x^4 + x + 1)$. Therefore $(\alpha^4 + \alpha + 1) = 0$. The polynomial representation of α^j is obtained by computing $x^j \pmod{(x^4 + x + 1)}$ for $0 \le j \le 14$. The table of the field elements and the corresponding minimal polynomials is shown below.

Field elements	Minimal polynomial
0	x
1	$x+1$
$\alpha, \alpha^2, \alpha^4, \alpha^8$	x^4+x+1
$\alpha^3, \alpha^6, \alpha^9, \alpha^{12}$	$x^4+x^3+x^2+x+1$
α^5, α^{10}	x^2+x+1
$\alpha^7, \alpha^{11}, \alpha^{13}, \alpha^{14}$	x^4+x^3+1

The primitive polynomials are

$$\eta_{\alpha^0}(x) = (x+1)$$
$$\eta_\alpha(x) = (x^4+x+1),\ \eta_{\alpha^3}(x) = (x^4+x^3+x^2+x+1),$$
$$\eta_{\alpha^5}(x) = (x^2+x+1),\ \eta_{\alpha^7}(x) = (x^4+x^3+1)$$

Denote the generator polynomial of a t error-correcting code by $g_t(x)$, where $1 \le t \le 7$. Also $k = (n - \deg g_t(x))$. Furthermore, the relationship $0 \le (n-k) \le mt$, yields $0 \le (15-k) \le 4t$. The code distance d_H follows the relationship $(2t+1) \le d_H \le w(g_t)$.

(a) Let $t = 1$, then $g_1(x) = \eta_\alpha(x) = (x^4+x+1)$. Also $\deg g_1(x) = 4$, therefore $k = (15-4) = 11$. Furthermore, $w(g_1) = 3$, therefore $d_H = 3$.

(b) Let $t = 2$, then

$$\begin{aligned} g_2(x) &= \text{lcm}\{\eta_\alpha(x), \eta_{\alpha^3}(x)\} = \eta_\alpha(x)\eta_{\alpha^3}(x) \\ &= g_1(x)\eta_{\alpha^3}(x) = \left(x^8 + x^7 + x^6 + x^4 + 1\right) \end{aligned}$$

Also $\deg g_2(x) = 8$, then $k = (15 - 8) = 7$. Since $w(g_2) = 5$, the value of $d_H = 5$. The relationship $0 \leq (n - k) \leq mt$ is also satisfied.

(c) Let $t = 3$, then

$$\begin{aligned} g_3(x) &= \text{lcm}\{\eta_\alpha(x), \eta_{\alpha^3}(x), \eta_{\alpha^5}(x)\} = \eta_\alpha(x)\eta_{\alpha^3}(x)\eta_{\alpha^5}(x) \\ &= g_2(x)\eta_{\alpha^5}(x) = \left(x^{10} + x^8 + x^5 + x^4 + x^2 + x + 1\right) \end{aligned}$$

Also $\deg g_3(x) = 10$, then $k = (15 - 10) = 5$. Since $w(g_3) = 7$, the value of $d_H = 7$. The relationship $0 \leq (n - k) \leq mt$ is also satisfied.

(d) Let $t = 4$, then

$$\begin{aligned} g_4(x) &= \text{lcm}\{\eta_\alpha(x), \eta_{\alpha^3}(x), \eta_{\alpha^5}(x), \eta_{\alpha^7}(x)\} \\ &= \eta_\alpha(x)\eta_{\alpha^3}(x)\eta_{\alpha^5}(x)\eta_{\alpha^7}(x) = g_3(x)\eta_{\alpha^7}(x) = \sum_{j=0}^{14} x^j \end{aligned}$$

Also $\deg g_4(x) = 14$, then $k = (15 - 14) = 1$. Since $w(g_4) = 15$ and $k = 1$, the value of $d_H = 15$. The relationship $0 \leq (n - k) \leq mt$ is also satisfied. Since $k = 1$, this is a repetition code.

(e) For $t = 5, 6$, and 7, the generator polynomial is also $\sum_{j=0}^{14} x^j$. □

2.9.2 Decoding of BCH Codes

BCH codes are cyclic codes. Therefore these codes can be decoded by any algorithm for decoding cyclic codes. However, because of the special structure of these codes, there are efficient algorithms for its decoding. An algorithm known after W. W. Peterson, D. C. Gorenstein, and N. Zierler (PGZ) is presented in this subsection.

Let the BCH code $\mathcal{C}$ be a t error-correcting code of codeword-length n, with the designed distance $\delta_{BCH} = (2t + 1)$. Assume that the transmitted and received words are τ and ρ respectively, where $\tau \in \mathcal{C}$. The error word (vector) is ε, that is $\varepsilon = (\rho - \tau)$. The corresponding transmitted, received, and error polynomials are $\tau(x), \rho(x)$, and $\varepsilon(x)$ respectively. Furthermore, the degree of each polynomial is at most $(n - 1)$. These polynomials are related by $\rho(x) = (\tau(x) + \varepsilon(x))$. Let the received and error polynomials be $\rho(x) = \sum_{j=0}^{n-1} \rho_j x^j$ and $\varepsilon(x) = \sum_{j=0}^{n-1} \varepsilon_j x^j$ respectively.

Also $\tau(x) = 0$, at $x = \alpha^l, \alpha^{l+1}, \ldots, \alpha^{l+2t-1}$ for some integer l, where α is a primitive nth root of unity. For simplicity assume that $l = 1$. Consider the matrix $\widetilde{H}$, where

$$\widetilde{H} = \begin{bmatrix} (\alpha)^0 & \alpha & \alpha^2 & \cdots & \alpha^{(n-1)} \\ (\alpha^2)^0 & \alpha^2 & (\alpha^2)^2 & \cdots & (\alpha^2)^{(n-1)} \\ \vdots & \vdots & \vdots & \ddots & \vdots \\ (\alpha^{2t})^0 & \alpha^{2t} & (\alpha^{2t})^2 & \cdots & (\alpha^{2t})^{(n-1)} \end{bmatrix}$$

The syndrome of the received vector is defined by $s = \rho\widetilde{H}^T$. Since $\tau\widetilde{H}^T = 0$, we have $s = \varepsilon\widetilde{H}^T$. The corresponding polynomial is given by $s(x) = \sum_{u=0}^{(2t-1)} s_u x^u$. Also

$$s_u = \sum_{j=0}^{n-1} \varepsilon_j \alpha^{j(u+1)}, \quad 0 \le u \le (2t-1)$$

Thus

$$s_u = \varepsilon\left(\alpha^{u+1}\right) = \rho\left(\alpha^{u+1}\right), \qquad 0 \le u \le (2t-1)$$

Assume that ν errors occurred, where $0 \le \nu \le t$, and these errors occurred at locations $j_1, j_2, \ldots, j_\nu$. Note that ν is the Hamming weight of ε. The nonzero components of ε are $\varepsilon_{j_1}, \varepsilon_{j_2}, \ldots, \varepsilon_{j_\nu}$. That is, ε_{j_i} is the magnitude of the i-th error. For a binary BCH code $\varepsilon_{j_i} = 1$ for $1 \le i \le \nu$. The error polynomial can also be expressed as $\varepsilon(x) = \sum_{i=1}^{\nu} \varepsilon_{j_i} x^{j_i}$. For ease in notation define $S_u = s_{u-1}$ for $1 \le u \le 2t$, and $S = \begin{bmatrix} S_1 & S_2 & \cdots & S_{2t} \end{bmatrix}$.

Then $S = \rho \widetilde{H}^T$ and

$$S_u = \rho\left(\alpha^u\right) = \varepsilon\left(\alpha^u\right) = \sum_{i=1}^{\nu} \varepsilon_{j_i} \alpha^{u j_i}, \qquad 1 \le u \le 2t$$

The computation of syndromes results in a system of equations involving unknown error locations and unknown error magnitudes. The notation in the above set of equations can be further simplified.

Let $X_i = \alpha^{j_i}$ be the *error-location number* for $1 \le i \le \nu$. It is the *field element* corresponding to *the error location* j_i for $1 \le i \le \nu$. Also define $Y_i = \varepsilon_{j_i}$ as the *error magnitude at coordinate* j_i for $1 \le i \le \nu$. Since α is the nth root of unity, knowledge of X_i's lets us determine uniquely the error location j_i. Using this notation the above set of equations become

$$S_u = \sum_{i=1}^{\nu} Y_i X_i^u, \qquad 1 \le u \le 2t$$

In this system of equations, the knowns are the syndromes S_u's, and the unknowns are the Y_i's. Observe that the system of equations is nonlinear in the X_i's. To overcome this problem, intermediate variables are used. Define the *error-locator polynomial* $\sigma(x)$ as

$$\sigma(x) = \prod_{i=1}^{\nu} (1 - xX_i) = 1 + \sum_{i=1}^{\nu} \sigma_i x^i$$

The roots of the polynomial $\sigma(x)$ are the inverse of the error-location numbers X_i's. That is, $\sigma\left(X_i^{-1}\right) = 0$ for $1 \le i \le \nu$. Furthermore,

$$\sigma\left(X_j^{-1}\right) = 1 + \sum_{i=1}^{\nu} \sigma_i X_j^{-i} = 0, \quad 1 \le j \le \nu$$

Multiplying both sides by $Y_j X_j^{\lambda+\nu}$ yields

$$Y_j X_j^{\lambda+\nu} + \sum_{i=1}^{\nu} \sigma_i Y_j X_j^{\lambda+\nu-i} = 0, \quad 1 \le j \le \nu, \quad 1 \le \lambda \le \nu$$

Summing the above equation for values of j from 1 through ν results in

$$\sum_{j=1}^{\nu} Y_j X_j^{\lambda+\nu} + \sum_{i=1}^{\nu} \sigma_i \sum_{j=1}^{\nu} Y_j X_j^{\lambda+\nu-i} = 0, \quad 1 \le \lambda \le \nu$$

In these equations the summations over the index j are the syndromes, as $\nu \le t$. That is

$$S_{\lambda+\nu} + \sum_{i=1}^{\nu} \sigma_i S_{\lambda+\nu-i} = 0, \quad 1 \le \lambda \le \nu$$

or

$$\sum_{i=1}^{\nu} \sigma_i S_{\lambda+\nu-i} = -S_{\lambda+\nu}, \quad 1 \le \lambda \le \nu$$

This is the set of equations relating the coefficients of the error-locator polynomial and the syndromes. Write these equations in matrix form as

$$\begin{bmatrix} S_1 & S_2 & S_3 & \cdots & S_{\nu-1} & S_\nu \\ S_2 & S_3 & S_4 & \cdots & S_\nu & S_{\nu+1} \\ S_3 & S_4 & S_5 & \cdots & S_{\nu+1} & S_{\nu+2} \\ \vdots & \vdots & \vdots & \ddots & \vdots & \vdots \\ S_\nu & S_{\nu+1} & S_{\nu+2} & \cdots & S_{2\nu-2} & S_{2\nu-1} \end{bmatrix} \begin{bmatrix} \sigma_\nu \\ \sigma_{\nu-1} \\ \sigma_{\nu-2} \\ \vdots \\ \sigma_1 \end{bmatrix} = \begin{bmatrix} -S_{\nu+1} \\ -S_{\nu+2} \\ -S_{\nu+3} \\ \vdots \\ -S_{2\nu} \end{bmatrix}$$

This matrix equation can be used to solve for $\sigma_1, \sigma_2, \ldots, \sigma_\nu$, which are the coefficients of the error-locator polynomial. This is possible if the $\nu \times \nu$ matrix in the above equation is nonsingular (invertible). It is next proved that this matrix is nonsingular if there are indeed ν errors.

Theorem 2.12. *Let $\mu, \nu \le t$ and the matrix of syndromes $\mathcal{W}_\mu$ is given by*

$$\mathcal{W}_\mu = \begin{bmatrix} S_1 & S_2 & \cdots & S_\mu \\ S_2 & S_3 & \cdots & S_{\mu+1} \\ \vdots & \vdots & \ddots & \vdots \\ S_\mu & S_{\mu+1} & \cdots & S_{2\mu-1} \end{bmatrix} \tag{2.16}$$

This matrix $\mathcal{W}_\mu$ is nonsingular if $\mu = \nu$ and singular if $\mu > \nu$, and ν is the number of errors in the received word.

Proof. If $\mu > \nu$, set

$$Y_{\nu+1} = Y_{\nu+2} = \cdots = Y_\mu = 0$$

Define A_μ as a Vandermonde matrix, and B_μ as a diagonal matrix.

$$A_\mu = \begin{bmatrix} 1 & 1 & \cdots & 1 \\ X_1 & X_2 & \cdots & X_\mu \\ \vdots & \vdots & \ddots & \vdots \\ X_1^{\mu-1} & X_2^{\mu-1} & \cdots & X_\mu^{\mu-1} \end{bmatrix}, \quad B_\mu = \begin{bmatrix} Y_1X_1 & 0 & \cdots & 0 \\ 0 & Y_2X_2 & \cdots & 0 \\ \vdots & \vdots & \ddots & \vdots \\ 0 & 0 & \cdots & Y_\mu X_\mu \end{bmatrix}$$

It can be verified that $\mathcal{W}_\mu = A_\mu B_\mu A_\mu^T$. Therefore

$$\det \mathcal{W}_\mu = \det A_\mu \det B_\mu \det A_\mu^T$$

Note that if $\mu > \nu$, then $\det B_\mu = 0$, consequently $\det \mathcal{W}_\mu = 0$. However if $\mu = \nu$, then $\det B_\mu \neq 0$ because all the diagonal elements are nonzero. Furthermore, $\det A_\mu \neq 0$ because the elements $X_1, X_2, \ldots, X_\mu$ are all distinct. Therefore $\det \mathcal{W}_\mu \neq 0$, that is the matrix $\mathcal{W}_\mu$ is nonsingular. □

This theorem enables us to describe the PGZ decoding algorithm. For simplicity, assume that the BCH code is narrow-sense. Before the PGZ algorithm for decoding is described, the attributes of a BCH code are summarized for easy reference.

Summary of Attributes of the Narrow-Sense BCH Code

(a) Let C be a t error-correcting narrow-sense BCH code, where the code symbols belong to the field $\mathbb{F}_q$, and the number of elements in this field is equal to q. Also let the dimension of this code be $k \in \mathbb{P}$, that is the message length is k symbols. The codeword-length of this code is n.
(b) Its designed and Hamming distances are δ_{BCH} and d_H respectively. Also $\delta_{BCH} \leq d_H$ and $(2t+1) \leq d_H$.
(c) Let α be a primitive nth root of unity in the extension field $\mathbb{F}_{q^m}$ of $\mathbb{F}_q$ where $m \in \mathbb{P}$. The generator polynomial $g(x)$ of the code is the least common multiple of the minimal polynomials of $\alpha^1, \alpha^2, \ldots, \alpha^{\delta_{BCH}-1}$.
(d) The codeword-length is chosen such that $n = (q^m - 1)$.
(e) The received polynomial $\rho(x) \in \mathbb{F}_q[x] / (x^n - 1)$ has ν errors.
(f) The error-locator polynomial is $\sigma(x), \deg \sigma(x) = \nu$, and σ_i is the coefficient of x^i in this polynomial, where $1 \leq i \leq \nu$. The error-location numbers and the error-magnitude values are X_i and Y_i respectively for $1 \leq i \leq \nu$. □

Algorithm 2.3. *Peterson-Gorenstein-Zierler's Decoding of BCH Code.*

Input: This algorithm *assumes* that $0 \leq \nu \leq t$. That is, at most t errors occur. The received polynomial $\rho(x) \in \mathbb{F}_q[x] / (x^n - 1)$, with possible errors at ν coordinates.
Output: The error-location number X_i, and the error-magnitude value Y_i respectively of the received polynomial $\rho(x)$, for $1 \leq i \leq \nu$.
begin
 Step 1: Compute the syndromes $S_u = \rho(\alpha^u)$ for $1 \leq u \leq 2t$.
  If all syndromes are 0, then the received vector is indeed a codeword, and $\tau(x) = \rho(x)$.
  No further processing is required, otherwise go to Step 2.
 Step 2: This step determines the value of ν.
  $\mu \leftarrow t$
  label A
  (*determine if the matrix* W_μ *is singular.*)
  Compute det $\mathcal{W}_\mu$
  if det $\mathcal{W}_\mu = 0$
  begin
   $\mu \leftarrow (\mu - 1)$
   go to label A
  end
  else $\nu \leftarrow \mu$
 Step 3: The coefficients of the error-locator polynomial $\sigma(x)$ are determined by inverting the matrix $\mathcal{W}_\nu$ and using

the syndrome values.

$$\begin{bmatrix} \sigma_\nu \\ \sigma_{\nu-1} \\ \sigma_{\nu-2} \\ \vdots \\ \sigma_1 \end{bmatrix} = \mathcal{W}_\nu^{-1} \begin{bmatrix} -S_{\nu+1} \\ -S_{\nu+2} \\ -S_{\nu+3} \\ \vdots \\ -S_{2\nu} \end{bmatrix}$$

Step 4: Find the ν roots of the polynomial $\sigma(x)$.
The roots of the polynomial are found by an exhaustive search by computing $\sigma(\alpha^i)$ for $0 \le i \le n$.
Inversion of the roots yields the error-location numbers $X_i, 1 \le i \le \nu$.

Step 5: The error magnitudes $Y_i, 1 \le i \le \nu$ are obtained by solving the following ν linear equations: $S_u = \sum_{i=1}^{\nu} Y_i X_i^u, 1 \le u \le \nu$.

end (*end of PGZ decoding algorithm*)

Note that if the code is binary, then all error magnitudes are 1's. Therefore, Step 5 of the algorithm can be skipped for such BCH codes. In actual practice, the above algorithm should be modified to consider the case when $\nu > t$. This is certainly a possibility, because the number of errors in a received word is not known in advance, and the condition $\nu \le t$ cannot be guaranteed. Besides the above PGZ algorithm, there are other efficient decoding techniques which have not been discussed in this chapter. The PGZ decoding algorithm is next illustrated via examples.

Examples 2.15. Some illustrative examples.

1. Let $\mathcal{C}$ be narrow-sense BCH code defined over a binary field. It is a $(15, 5)$ code and designed distance $\delta_{BCH} = 7$. That is, the code is triple error-correcting, with $t = 3$. The generator polynomial of this code has been derived in an earlier example to be

$$g(x) = \left(x^{10} + x^8 + x^5 + x^4 + x^2 + x + 1\right)$$

Let the received word polynomial be

$$\rho(x) = \left(x^{11} + x^9 + x^6 + x^4 + x^3 + 1\right)$$

Step 1: The syndromes $S_u = \rho(\alpha^u), 1 \le u \le 6$ are computed. The computation of these values is facilitated by the use of binary representation of elements of the field GF_{16}.

$$S_1 = \alpha,\ S_2 = \alpha^2,\ S_3 = \alpha^7,\ S_4 = \alpha^4,\ S_5 = 1, \text{ and } S_6 = \alpha^{14}$$

Step 2: Set $\mu = 3$, then

$$\mathcal{W}_3 = \begin{bmatrix} S_1 & S_2 & S_3 \\ S_2 & S_3 & S_4 \\ S_3 & S_4 & S_5 \end{bmatrix} = \begin{bmatrix} \alpha & \alpha^2 & \alpha^7 \\ \alpha^2 & \alpha^7 & \alpha^4 \\ \alpha^7 & \alpha^4 & 1 \end{bmatrix}$$

It can be checked that $\det \mathcal{W}_3 = 0$. Next set $\mu = 2$, then

$$\mathcal{W}_2 = \begin{bmatrix} S_1 & S_2 \\ S_2 & S_3 \end{bmatrix} = \begin{bmatrix} \alpha & \alpha^2 \\ \alpha^2 & \alpha^7 \end{bmatrix}$$

The matrix $\mathcal{W}_2$ is nonsingular with inverse

$$\mathcal{W}_2^{-1} = \begin{bmatrix} \alpha^2 & \alpha^{12} \\ \alpha^{12} & \alpha^{11} \end{bmatrix}$$

Therefore $\nu = 2$ errors have occurred in the transmitted codeword.
Step 3: The coefficients of the error-locator polynomial $\sigma(x)$ are next determined.

$$\begin{aligned} \begin{bmatrix} \sigma_2 \\ \sigma_1 \end{bmatrix} &= \mathcal{W}_2^{-1} \begin{bmatrix} -S_3 \\ -S_4 \end{bmatrix} \\ &= \begin{bmatrix} \alpha^2 & \alpha^{12} \\ \alpha^{12} & \alpha^{11} \end{bmatrix} \begin{bmatrix} -\alpha^7 \\ -\alpha^4 \end{bmatrix} = \begin{bmatrix} \alpha^3 \\ \alpha \end{bmatrix} \end{aligned}$$

This yields $\sigma_1 = \alpha$ and $\sigma_2 = \alpha^3$. Therefore

$$\sigma(x) = \left(\sigma_2 x^2 + \sigma_1 x + 1\right) = \left(\alpha^3 x^2 + \alpha x + 1\right)$$

Step 4: Then $\sigma(x)$ can be factored as

$$\sigma(x) = \left(\alpha^{10} x + 1\right)\left(\alpha^8 x + 1\right) = \alpha^3 \left(x - \alpha^5\right)\left(x - \alpha^7\right)$$

The roots of the error-locator polynomial $\sigma(x)$ are α^5 and α^7. The error locations are reciprocal of these roots. Therefore errors occurred at the eighth and the tenth components. That is, $\varepsilon(x) = \left(x^{10} + x^8\right)$. Therefore the transmitted codeword over the binary field is

$$\begin{aligned} \tau(x) &= (\rho(x) + \varepsilon(x)) \\ &= \left(x^{11} + x^{10} + x^9 + x^8 + x^6 + x^4 + x^3 + 1\right) \end{aligned}$$

It turns out that $\tau(x) = (x+1)\, g(x)$.

2. Let C be a narrow-sense BCH code defined over a binary field. It is a $(15, 7)$ code and designed distance $\delta_{BCH} = 5$. That is, the code is double error-correcting, with $t = 2$. The generator polynomial of this code has been derived in an earlier example to be $g(x) = \left(x^8 + x^7 + x^6 + x^4 + 1\right)$.
Let the received word polynomial be $\rho(x) = \left(x^4 + x + 1\right)$.
Step 1: The syndromes $S_u = \rho(\alpha^u)$ are computed, where $1 \le u \le 4$.

$$S_1 = 0,\ S_2 = 0,\ S_3 = \alpha^5, \text{ and } S_4 = 0$$

Note that the syndromes are not all zeros.
Step 2: Set $\mu = 2$, then

$$\mathcal{W}_2 = \begin{bmatrix} S_1 & S_2 \\ S_2 & S_3 \end{bmatrix} = \begin{bmatrix} 0 & 0 \\ 0 & \alpha^5 \end{bmatrix}$$

It can be checked that $\det \mathcal{W}_2 = 0$. Next set $\mu = 1$, then $\mathcal{W}_1 = [S_1] = [0]$. The algorithm cannot make progress. Consequently it can be inferred that more than two errors occurred in the transmitted codeword. □

2.10 Reed-Solomon Codes

Reed-Solomon (RS) codes were discovered by I. S. Reed and G. Solomon in 1960. These codes are an important, simple, and practical class of *nonbinary* BCH codes which can deal with bursts of errors. Since RS codes are a subset of BCH codes, these can use the same encoding and decoding algorithms as the BCH codes. The RS codes are also useful in building up other codes. An example of use of the RS code is in mass storage.

The designed distance of the RS code denoted by δ_{RS}, has the same meaning as the designed distance δ_{BCH} of a BCH code.

Definition 2.21. *A Reed-Solomon code defined over a finite field $\mathbb{F}_q$, is a primitive BCH code. The order of this finite field is equal to q, and the codeword-length is $n = (q-1)$.* □

Note that the roots of $(x^n - 1)$ are precisely the nonzero elements of $\mathbb{F}_q$, and a primitive nth root of unity is indeed a primitive element of $\mathbb{F}_q$. Therefore RS codes can be obtained from the BCH codes by substituting $m = 1$. Note that $\mathbb{F}_q$ cannot be a binary field, in which case $q = 2$ and $n = 1$. These values are not very interesting.

Theorem 2.13. *Let α be a primitive element of $\mathbb{F}_q$. An (n,k) RS code with designed distance δ_{RS} and Hamming distance d_H has a generator polynomial*

$$g(x) = \left(x - \alpha^l\right)\left(x - \alpha^{l+1}\right)\ldots\left(x - \alpha^{l+\delta_{RS}-2}\right) \tag{2.17}$$

where $l \in \mathbb{N}$, $\delta_{RS} \in \mathbb{P}$, $(n-k) = (\delta_{RS} - 1)$, and $\delta_{RS} = d_H$.

Proof. Let α be a primitive element of $\mathbb{F}_q$. Therefore the elements

$$\alpha^l, \alpha^{l+1}, \ldots, \alpha^{l+\delta_{RS}-2}$$

for $(\delta_{RS} - 1) < n$ are all distinct. The minimal polynomial of the element α^i over the field $\mathbb{F}_q$ is $\eta_{\alpha^i}(x) = \left(x - \alpha^i\right)$. Thus

$$\begin{aligned} g(x) &= \operatorname{lcm}\left\{\left(x - \alpha^l\right), \left(x - \alpha^{l+1}\right), \ldots, \left(x - \alpha^{l+\delta_{RS}-2}\right)\right\} \\ &= \left(x - \alpha^l\right)\left(x - \alpha^{l+1}\right)\ldots\left(x - \alpha^{l+\delta_{RS}-2}\right) \end{aligned}$$

as $\alpha^l, \alpha^{l+1}, \ldots, \alpha^{l+\delta_{RS}-2}$ are pair-wise distinct. Since $\deg g(x) = (\delta_{RS} - 1)$, we have $(n-k) = (\delta_{RS} - 1)$. Also as per the Singleton's bound, $d_H \le (n - k + 1)$. This implies $d_H \le \delta_{RS}$. However, for any BCH code $\delta_{RS} \le d_H$. Thus it can be concluded that $\delta_{RS} = d_H$. □

Generally the value of l is equal to unity, but not always. The above theorem immediately yields the following important result.

Corollary 2.3. *For a fixed value of n and k, no linear code can have a greater Hamming distance d_H than the RS code.* □

Therefore the RS codes are optimum as per the Singleton-bound criteria. Consequently the RS code is a maximum-distance separable code.

Burst-Error Correction with RS Codes

The RS codes can also be used for burst-error correction. Consider a nonbinary (n, k) RS code $\mathcal{C}$ defined over $\mathbb{F}_q$, where $q = 2^{\varphi}, \varphi \in \mathbb{P}$. Assume the code to be a t error-correction code, where $t = \lfloor (d_H - 1)/2 \rfloor$. Using this code, a new code $\mathcal{C}^*$ is derived. Each nonbinary codeword $c \in \mathcal{C}$ can be represented by a binary φ-tuple $(c_{\varphi-1} \ldots c_1 c_0)$.

Note that the codeword-length in the code $\mathcal{C}$ is $n = (q-1) = (2^{\varphi} - 1)$, but the codeword-length in the code $\mathcal{C}^*$ is $n\varphi$. These derived codes are significant because they can correct error bursts of length up to $\zeta = \{(t-1)\varphi + 1\}$ bits. An error burst of this value in the derived code $\mathcal{C}^*$ can effect at most t adjacent codewords of the original code $\mathcal{C}$.

The attributes and properties of a RS code are summarized below for easy reference.

Summary of Attributes of the RS Code.

(a) The RS codes are linear, cyclic, and a subclass of the nonbinary BCH codes. Therefore, the encoding and decoding algorithms for the RS codes are identical to those of the BCH codes.
(b) It can also be used to correct burst errors.
(c) Let $\mathcal{C}$ be a RS code, where the code symbols belong to the field $\mathbb{F}_q$, and the order of this field is equal to q. Also let the dimension of this code be $k \in \mathbb{P}$, that is the message length is k symbols. The codeword-length of this code is $n = (q-1)$.
(d) Its designed and Hamming distances are δ_{RS} and d_H respectively, such that

$$(n - k + 1) = \delta_{RS} = d_H$$

Therefore the RS code is a maximum-distance separable code.
(e) Let α be a primitive element in the field $\mathbb{F}_q$. The generator polynomial of the code is $g(x) = \prod_{i=0}^{\delta_{RS}-2} (x - \alpha^{l+i})$, where $l \in \mathbb{N}$.
(f) For given values of n and k, the RS code has the largest Hamming distance of any linear code.

□

Examples 2.16. Some illustrative examples.

1. The RS code is defined over GF_8. Therefore, $q = 8$ and $n = 7$. Let $l = 1$, and the primitive element in the field GF_8 be α, that is $\alpha^7 = 1$.
   (a) Let the value of k be equal to 5. Therefore, this is a $(7,5)$ code, where $t = 1$, and $\delta_{RS} = d_H = 3$. Its generator polynomial $g(x)$ is given by

$$\begin{aligned} g(x) &= (x - \alpha)(x - \alpha^2) \\ &= (x^2 + \alpha^4 x + \alpha^3) \end{aligned}$$

   (b) Let the value of k be equal to 3. Therefore, this is a $(7,3)$ code, where $t = 2$, and $\delta_{RS} = d_H = 5$. Its generator polynomial $g(x)$ is given by

$$\begin{aligned} g(x) &= (x - \alpha)(x - \alpha^2)(x - \alpha^3)(x - \alpha^4) \\ &= (x^4 + \alpha^3 x^3 + x^2 + \alpha x + \alpha^3) \end{aligned}$$

2. The RS code is defined over GF_{16}. Therefore, $q = 16$ and $n = 15$. Let $l = 1$, and the primitive element in the field GF_{16} be α, that is $\alpha^{15} = 1$.

(a) Let the value of k be equal to 13. Therefore, this is a $(15, 13)$ code, where $t = 1$, and $\delta_{RS} = d_H = 3$. Its generator polynomial $g(x)$ is given by

$$g(x) = (x - \alpha)(x - \alpha^2) = (x^2 + \alpha^5 x + \alpha^3)$$

(b) Let the value of k be equal to 11. Therefore, this is a $(15, 11)$ code, where $t = 2$, and $\delta_{RS} = d_H = 5$. Its generator polynomial $g(x)$ is given by

$$\begin{aligned} g(x) &= (x - \alpha)(x - \alpha^2)(x - \alpha^3)(x - \alpha^4) \\ &= (x^4 + \alpha^{13} x^3 + \alpha^6 x^2 + \alpha^3 x + \alpha^{10}) \end{aligned}$$

□

2.11 Convolutional Codes

All the linear codes which have been discussed in the earlier sections of this chapter are called block-codes. These codes are specified by the two-tuple (n, k), where a message of k information symbols is mapped into a code of n symbols. These codes do not have any memory. In contrast, the convolutional codes use an encoding scheme which uses not only the current message frame, but also a specific number of preceding message frames. Therefore the convolutional coding schemes are based on *memory*. The concept of convolutional codes was introduced by P. Elias (1954) and developed by J. M. Wozencraft (1957) and others. Convolution codes are used extensively in satellite and radio links. These have also found application in digital video, and mobile communication.

The classical block-codes can correct a certain number of errors within the block. If the number of errors is above this threshold value, the erroneous symbols are not guaranteed to be corrected. Furthermore, these codes can have large values of minimum distance, and are algebraically decoded. Block-codes are also not sensitive to the position of errors in the block. Therefore it is possible to construct block-codes that are very resilient to long bursts of errors. This is in direct contrast to the convolutional codes which are extremely sensitive to the error-pattern within a sequence of symbols. Consequently the convolutional codes can correct randomly-occurring errors, but are intolerant to long bursts of errors. In addition, the stream nature of convolutional codes allows for continuous decoding. Convolutional codes are generally discovered via computer search. In summary, it can be stated that block and convolutional codes have complementary strengths.

Consider a stream of incoming symbols, which are segmented into frames of size $k_0 \in \mathbb{P}$ symbols. These are called *information or message frames*. The convolutional coding scheme transforms this single information frame into a *coded frame* of length n_0 symbols, by using not only the current but also the past $\beta \in \mathbb{P}$ information frames. Evidently n_0 is greater than k_0. Consequently the communication channel transmits n_0 coded symbols for the corresponding k_0 information symbols. That is, the code rate of this coding scheme is $\vartheta = k_0/n_0$. Some terminology is next introduced.

Definitions 2.22. *Let the size of the message and code frames be k_0 and n_0 symbols respectively. The number of past frames used in coding is equal to β. The variables k_0, n_0, and β are positive integers.*

1. *The constraint length of the convolutional code is equal to $\psi = \beta k_0$.*

2. *The length of the information word of the convolutional code is equal to* $k = (\beta + 1) k_0$.
3. *The codeword-length or the block-length of the convolutional code is equal to* $n = (\beta + 1) n_0$.

□

Use of the above definitions yields the length of the information word to be $k = (\psi + k_0)$. The reader should note that there are several different definitions of constraint length in the literature. Convolutional codes are time-invariant and linear. Also, the notion of systematic code is similar to other linear codes.

Definitions 2.23. *Notions of time-invariance and linearity of a code are defined below. Another useful concept is that of a systematic code.*

1. *A code* $\mathcal{C}$ *is time-invariant, if two sequences of information symbols are identical except for a time shift by an integer number of frames, then the coded sequences are also identical except for a time shift by the same number of frames.*
2. *A code* $\mathcal{C}$ *is linear, if the coded stream of a linear combination of two information streams is equal to the linear combination of the individually coded streams.*
3. *In a systematic code, the symbols in the information frame occupy the first* k_0 *positions of the codeword frame which is* n_0 *symbols in length.* □

A formal definition of a convolution code $\mathcal{C}$ is as follows.

Definition 2.24. *An* (n, k) *convolutional code* $\mathcal{C}$ *defined over a finite field* $\mathbb{F}_q$, *is linear and time-invariant. In this code:*

(a) *The number of elements in the field* $\mathbb{F}_q$ *is equal to* $q \in \mathbb{P}$.
(b) *Information (message) frame size is equal to* k_0 *symbols.*
(c) *Coded frame size is equal to* n_0 *symbols.*
(d) *Information (message) word length* $k = (\beta + 1) k_0$ *symbols.*
(e) *Codeword-length* $n = (\beta + 1) n_0$ *symbols.* □

2.11.1 Polynomial Representation

Since a convolutional coding scheme has memory, it is convenient to describe it in terms of polynomials. For example, consider a stream of information symbols $u_0, u_1, u_2, \ldots,$ where u_i belongs to the set of information symbols A. For convenience assume that these information symbols are emitted by the source at equal intervals of time. These instants are indexed by numbers $0, 1, 2, \ldots$.. Therefore, this stream can be represented by a polynomial $u(x) = \sum_{i \in \mathbb{N}} u_i x^i$.

The notion of generator and parity-check matrices for convolutional codes is next developed. In contrast to the block-codes, the elements of these matrices are polynomials. The notions of the constraint length, information-word length, and codeword-length are further refined and extended.

Definitions 2.25. *Let* $((\beta + 1) n_0, (\beta + 1) k_0)$ *be a convolutional code defined over the Galois field* GF_q, *where* $k_0, n_0, \beta \in \mathbb{P}$. *Consider the information frame of* k_0 *symbols to be in parallel. Similarly, consider the coded frame of* n_0 *symbols also to be in parallel.*

1. *The generator-polynomial matrix is a* $k_0 \times n_0$ *matrix of polynomials specified by*

$$\mathcal{G}(x) = [g_{ij}(x)], \quad 0 \le i \le (k_0 - 1), \; 0 \le j \le (n_0 - 1) \tag{2.18}$$

The $g_{ij}(x)$*'s are called the generator polynomials. Also* $\max_{i,j} \deg g_{ij}(x) = \beta$. *If* $k_0 > 1$, *then it is possible for some of the generator polynomials to be the zero polynomial.*

2. *The constraint length of this code is*

$$\psi = \sum_{i=0}^{(k_0-1)} \max_{0\le j\le(n_0-1)} \{\deg g_{ij}(x)\} \tag{2.19a}$$

Observe that $\max_{0\le j\le(n_0-1)} \{\deg g_{ij}(x)\}$ *is the maximum degree of the polynomial in the* ith *row of the matrix* $\mathcal{G}(x)$. *It is equal to the storage required by the encoder for the* ith *input stream. Thus the constraint length* ψ *is equal to the storage required by all the input streams.*

3. *The information-word length is*

$$k = k_0 \max_{\substack{0\le i\le(k_0-1)\\ 0\le j\le(n_0-1)}} \{\deg g_{ij}(x) + 1\} \tag{2.19b}$$

4. *The codeword-length or block-length is*

$$n = n_0 \max_{\substack{0\le i\le(k_0-1)\\ 0\le j\le(n_0-1)}} \{\deg g_{ij}(x) + 1\} \tag{2.19c}$$

5. *Consider information frames as a sequence of* k_0 *symbols in parallel. The successive frames (in time) are represented by* k_0 *information polynomials*

$$f_i(x), \quad \deg f_i(x) \le \beta, \quad 0 \le i \le (k_0 - 1) \tag{2.20a}$$

$$\widetilde{f}(x) = \begin{bmatrix} f_0(x) & f_1(x) & \ldots & \ldots & f_{k_0-1}(x) \end{bmatrix} \tag{2.20b}$$

where $\widetilde{f}(x)$ *is a vector of polynomials.*

6. *Let*

$$\widetilde{f}(x)\,\mathcal{G}(x) = \widetilde{c}(x) \tag{2.21a}$$

where

$$\widetilde{c}(x) = \begin{bmatrix} c_0(x) & c_1(x) & \ldots & \ldots & c_{n_0-1}(x) \end{bmatrix} \tag{2.21b}$$

Note that $\widetilde{c}(x)$ *is a vector of polynomials, and* $\deg c_j(x) \le 2\beta,\ 0 \le j \le (n_0 - 1)$. *The coefficients of the codeword polynomials* $c_j(x), 0 \le j \le (n_0 - 1)$ *are interleaved before transmission on the communication channel. The polynomial representation of the* n*-symbols long information stream can be determined from* $c'(x)$.

$$c'(x) = \sum_{j=0}^{(n_0-1)} x^j c_j(x^{n_0}) \tag{2.21c}$$

Also $\deg c'(x) \le \{n_0(2\beta + 1) - 1\}$.

7. *The parity-check-polynomial matrix* $\mathcal{H}(x)$ *is a* $(n_0 - k_0) \times n_0$ *matrix of polynomials, such that*

$$\mathcal{G}(x)\,\mathcal{H}(x)^T = 0 \tag{2.22}$$

where 0 *on the right-hand side of the above equation is a* $k_0 \times (n_0 - k_0)$ *matrix of all zeros.*

8. *The received vector of polynomials corresponding to the transmitted coded polynomial vector* $\widetilde{c}(x)$ *is denoted by* $\rho(x)$. *The syndrome-polynomial vector is denoted by* $s(x)$. *The polynomial vectors* $\rho(x)$ *and* $s(x)$, *with polynomial elements are of size* n_0 *and* $(n_0 - k_0)$ *respectively. Thus*

$$s(x) = \rho(x)\mathcal{H}(x)^T \tag{2.23}$$

□

The generator-polynomial matrix of a systematic encoder of a convolutional coder has the form

$$\mathcal{G}(x) = [I_{k_0} \mid \mathcal{P}(x)]$$

where I_{k_0} is an identity matrix of size k_0, and $\mathcal{P}(x)$ is a $k_0 \times (n_0 - k_0)$ matrix of polynomials. The parity-check-polynomial matrix is

$$\mathcal{H}(x) = \left[-\mathcal{P}(x)^T \mid I_{n_0-k_0}\right] \quad \text{or} \quad \mathcal{H}(x) = \left[\mathcal{P}(x)^T \mid -I_{n_0-k_0}\right]$$

where $I_{n_0-k_0}$ is an identity matrix of size $(n_0 - k_0)$. It can be verified that $\mathcal{G}(x)\mathcal{H}(x)^T = 0$, where 0 is a $k_0 \times (n_0 - k_0)$ matrix of all zeros. The following notation is introduced to obtain a corresponding representation in terms of matrices. Let

$$\begin{aligned}
g_{ij}(x) &= \sum_{l=0}^{\beta} a_{ij}(l)\,x^l, \quad 0 \le i \le (k_0 - 1),\ 0 \le j \le (n_0 - 1) \\
G_l &= [a_{ij}(l)], \quad 0 \le i \le (k_0 - 1),\ 0 \le j \le (n_0 - 1); \ \text{and}\ 0 \le l \le \beta \\
\mathcal{G}(x) &= \sum_{l=0}^{\beta} G_l x^l
\end{aligned}$$

where the elements of the matrices G_l's are in GF_q.

Special case of $k_0 = 1$: The notation in the above discussion is modified for convenience if $k_0 = 1$. Let

$$\begin{aligned}
\mathcal{G}(x) &= \left[g_0(x)\ \ g_1(x)\ \ \cdots\ \ g_{n_0-1}(x)\right] \\
G(x) &= \sum_{j=0}^{(n_0-1)} x^j g_j(x^{n_0})
\end{aligned}$$

Also let $\widetilde{f}(x) = [f_0(x)]$, and $f_0(x) \triangleq f(x)$. Thus

$$\begin{aligned}
f(x)\mathcal{G}(x) &= \left[c_0(x)\ \ c_1(x)\ \ \cdots\ \ c_{n_0-1}(x)\right] \\
c_j(x) &= f(x)\,g_j(x), \qquad 0 \le j \le (n_0 - 1) \\
c'(x) &= f(x^{n_0})\,G(x)
\end{aligned}$$

If the code $\mathcal{C}$ is systematic, then $g_0(x) = 1$.

Further conditions are imposed on the polynomials $g_j(x)$ where $0 \le j \le (n_0 - 1)$. If these polynomials are selected such that they are relatively prime in pairs, then

$$\gcd(g_0(x), g_1(x), \ldots, g_{n_0-1}(x)) = 1$$

Using the extended Euclidean algorithm for polynomials, it can be inferred that there exist polynomials $a_0(x), a_1(x), \ldots, a_{n_0-1}(x)$, such that

$$a_0(x)\, g_0(x) + a_1(x)\, g_1(x) + \ldots + a_{n_0-1}(x)\, g_{n_0-1}(x) = 1$$

If there are no errors in the communication channel, the information in the message polynomial $f(x)$ can be recovered from the coded information in the polynomials $c_j(x), 0 \le j \le (n_0 - 1)$.

The following operation is performed at the decoder: $\sum_{j=0}^{(n_0-1)} a_j(x)\, c_j(x)$. It can be demonstrated that this summation is indeed equal to $f(x)$. That is

$$\sum_{j=0}^{(n_0-1)} a_j(x)\, c_j(x) = \sum_{j=0}^{(n_0-1)} a_j(x)\, f(x)\, g_j(x) = f(x)$$

This type of code is called a *noncatastrophic convolutional code*. This concept is formalized in the following definition.

Definition 2.26. *A convolutional code (n, k), where $k_0 = 1$, has generator polynomials $g_0(x), g_1(x), \ldots, g_{n_0-1}(x)$, such that*

$$\gcd\left(g_0(x), g_1(x), \ldots, g_{n_0-1}(x)\right) = x^{\varrho} \tag{2.24}$$

for some $\varrho \in \mathbb{N}$ is called a noncatastrophic convolutional code, otherwise, it is a catastrophic convolutional code. □

Note that x^{ϱ} corresponds to a simple delay in the encoder. Therefore, without any loss of significant generality it can be assumed that $x^{\varrho} = 1$. In summary, decoding of noncatastrophic convolutional code in the absence of noise in the channel is possible by using extended Euclidean algorithm. The concept of noncatastrophic and catastrophic convolutional code is also applicable to $k_0 \ge 2$.

2.11.2 Error-Correction and Distance Measures

The codewords encoded by a convolutional coder are very long (infinite) in length. But the decoding of the received stream of symbols is based upon codeword segments of finite length. Therefore, if a current codeword segment is not decoded correctly, it can affect the correct decoding of subsequent codeword segments. Consequently, it is quite possible for a single decoding-error in the current codeword segment to create extremely large number of errors in decoding subsequent codeword segments. This phenomenon is called *error propagation.* This error propagation can occur either due to the design of the decoding algorithm, or it may be due to the improper choice of generator polynomials. In the first case, the errors are said to be due to *ordinary error propagation,* while the errors due to the later case are said to be due to *catastrophic error propagation.* A good convolutional coding scheme should overcome these two types of error propagation.

The number of symbols which a decoder can store is called its *decoding-window width.* Its value should be greater than or equal to the block-length n. Therefore, the minimum distance of these codes depends upon the length of the initial codeword segment over which the minimum distance is measured. Recall that the Hamming distance of a block-code is a measure of its error-correcting capability. An analogous distance measure of the convolutional codes which measures its error-correcting capabilities is called *free distance.*

Definition 2.27. *Let (n, k) be a convolutional code $\mathcal{C}$, where $n = n_0(\beta + 1)$. Denote the Hamming distance of this code by d_n. Then a sequence of codes can be generated by increasing β, which is equal to the number of past frames. The limiting distance*

$$\lim_{\beta \to \infty} d_n = d_{free} \tag{2.25}$$

is called the free distance of the sequence of codes. However it is generally referred to simply as the free distance of the convolutional code $\mathcal{C}$. □

This definition yields the following set of observations.

Observations 2.12. Some basic observations.

1. It should be evident that $d_n \le d_{n+n_0} \le d_{n+2n_0} \le \ldots \le d_{free}$.
2. The free distance of a convolutional code $\mathcal{C}$ over the binary field $\mathbb{B}$ is the minimum weight of any nonzero codeword of $\mathcal{C}$.
3. A convolutional code $\mathcal{C}$ can correct at most t errors iff $(2t + 1) \le d_{free}$. □

The proof of the last observation is left to the reader.

2.11.3 Matrix Representation

There is a convenient representation of the convolutional codes in terms of matrices. The codeword-length of a convolutional code $\mathcal{C}$ is very large (infinite). This is possible if the input stream of symbols to be coded is large (infinite) in length. Furthermore, the number of codewords in $\mathcal{C}$ is large (infinite). The generator matrix G of this code is given below.

$$G = \begin{bmatrix} G_0 & G_1 & G_2 & \cdots & G_\beta & 0 & 0 & 0 & 0 & \cdots \\ 0 & G_0 & G_1 & \cdots & G_{\beta-1} & G_\beta & 0 & 0 & 0 & \cdots \\ 0 & 0 & G_0 & \cdots & G_{\beta-2} & G_{\beta-1} & G_\beta & 0 & 0 & \cdots \\ \vdots & \vdots & \vdots & \ddots & \vdots & \vdots & \vdots & \vdots & \vdots & \ddots \end{bmatrix}$$

This matrix is semi-infinite in length. All its submatrices are zeros except those along the diagonal band. The nonzero matrices are

$$G_l = [a_{ij}(l)], \quad 0 \le i \le (k_0 - 1), \quad 0 \le j \le (n_0 - 1), \quad \text{and} \quad 0 \le l \le \beta$$

Each 0 in the matrix G is a $k_0 \times n_0$ matrix of zeros. For an (n, k) convolutional code $\mathcal{C}$, the generator matrix G of a convolutional code truncated to a block-code of length n is given by

$$G = \begin{bmatrix} G_0 & G_1 & G_2 & \cdots & G_\beta \\ 0 & G_0 & G_1 & \cdots & G_{\beta-1} \\ 0 & 0 & G_0 & \cdots & G_{\beta-2} \\ \vdots & \vdots & \vdots & \ddots & \vdots \\ 0 & 0 & 0 & \cdots & G_0 \end{bmatrix}$$

The code produced is $c = fG$, where f and c are vectors of lengths k and n respectively. This code can also be generated by using the coefficients of $x^i, 0 \le i \le 2\beta$ in the polynomials $c_j(x)$

for $0 \leq j \leq (n_0 - 1)$. In a systematic convolutional encoder, let P_l be a $k_0 \times (n_0 - k_0)$ matrix for $0 \leq l \leq \beta$.

$$\begin{aligned} G_0 &= [I_{k_0} \mid P_0] \\ G_l &= [0 \mid P_l], \quad 1 \leq l \leq \beta \end{aligned}$$

where I_{k_0} is an identity matrix of size k_0, and 0 is a $k_0 \times k_0$ matrix of all zeros. The corresponding parity-check submatrices are defined as

$$\begin{aligned} H_0 &= \left[P_0^T \mid -I_{n_0-k_0}\right] \\ H_l &= \left[P_l^T \mid 0\right], \quad 1 \leq l \leq \beta \end{aligned}$$

where $I_{n_0-k_0}$ is an identity matrix of size $(n_0 - k_0)$, and 0 is a $(n_0 - k_0) \times (n_0 - k_0)$ matrix of all zeros. Thus $G_l H_l^T = 0$, where 0 is a $k_0 \times (n_0 - k_0)$ matrix of all zeros, and $0 \leq l \leq \beta$. Thus the parity-check matrix for the convolutional code truncated to a block-code of length n is given by

$$H = \begin{bmatrix} H_0 & 0 & 0 & \cdots & 0 \\ H_1 & H_0 & 0 & \cdots & 0 \\ H_2 & H_1 & H_0 & \cdots & 0 \\ \vdots & \vdots & \vdots & \ddots & \vdots \\ H_\beta & H_{\beta-1} & H_{\beta-2} & \cdots & H_0 \end{bmatrix}$$

The generator and parity-check matrices for the convolutional code truncated to a block-code of length n can be written explicitly as

$$G = \begin{bmatrix} I_{k_0} & P_0 & 0 & P_1 & 0 & P_2 & \cdots & 0 & P_\beta \\ 0 & 0 & I_{k_0} & P_0 & 0 & P_1 & \cdots & 0 & P_{\beta-1} \\ 0 & 0 & 0 & 0 & I_{k_0} & P_0 & \cdots & 0 & P_{\beta-2} \\ \vdots & \vdots & \vdots & \vdots & \vdots & \vdots & \ddots & \vdots & \vdots \\ 0 & 0 & 0 & 0 & 0 & 0 & \cdots & I_{k_0} & P_0 \end{bmatrix}$$

$$H = \begin{bmatrix} P_0^T & -I_{n_0-k_0} & 0 & 0 & 0 & 0 & \cdots & 0 & 0 \\ P_1^T & 0 & P_0^T & -I_{n_0-k_0} & 0 & 0 & \cdots & 0 & 0 \\ P_2^T & 0 & P_1^T & 0 & P_0^T & -I_{n_0-k_0} & \cdots & 0 & 0 \\ \vdots & \vdots & \vdots & \vdots & \vdots & \vdots & \ddots & \vdots & \vdots \\ P_\beta^T & 0 & P_{\beta-1}^T & 0 & P_{\beta-2}^T & 0 & \cdots & P_0^T & -I_{n_0-k_0} \end{bmatrix}$$

It can indeed be verified that $GH^T = 0$. The above details are illustrated via an example.

Example 2.17. Consider a convolutional code $(9, 3)$ over field $\mathbb{B}$. This code is truncated to a block-length of 9. In this code $n = 9$, $k = 3$, and $\beta = 2$. This yields $k_0 = 1$ and $n_0 = 3$. This code is generated by

$$\mathcal{G}(x) = \left[g_0(x) \;\; g_1(x) \;\; g_2(x)\right] = \sum_{l=0}^{2} G_l x^l$$

where

$$g_0(x) = (x^2+1), \; g_1(x) = (x^2+x+1), \text{ and } g_2(x) = (x^2+x+1)$$
$$G_0 = \begin{bmatrix}1 & 1 & 1\end{bmatrix}, \; G_1 = \begin{bmatrix}0 & 1 & 1\end{bmatrix}, \text{ and } G_2 = \begin{bmatrix}1 & 1 & 1\end{bmatrix}$$

The generator matrix of this convolutional code truncated to a block-length of 9 is

$$G = \begin{bmatrix} G_0 & G_1 & G_2 \\ 0 & G_0 & G_1 \\ 0 & 0 & G_0 \end{bmatrix} = \begin{bmatrix} 1&1&1&0&1&1&1&1&1 \\ 0&0&0&1&1&1&0&1&1 \\ 0&0&0&0&0&0&1&1&1 \end{bmatrix}$$

Let the input data be represented by the polynomial

$$f_0(x) = (b_0 + b_1 x + b_2 x^2)$$

Therefore $\widetilde{f}(x) = [f_0(x)]$, $f_0(x) \triangleq f(x)$, and

$$\widetilde{c}(x) = \widetilde{f}(x)\mathcal{G}(x) = \begin{bmatrix} f(x)g_0(x) & f(x)g_1(x) & f(x)g_2(x) \end{bmatrix}$$

That is

$$c_0(x) = f(x)g_0(x), \; c_1(x) = f(x)g_1(x), \text{ and } c_2(x) = f(x)g_2(x)$$
$$c'(x) = c_0(x^3) + xc_1(x^3) + x^2c_2(x^3)$$

Also

$$c_0(x) = b_0 + b_1x + (b_0+b_2)x^2 + b_1x^3 + b_2x^4$$
$$c_1(x) = b_0 + (b_0+b_1)x + (b_0+b_1+b_2)x^2 + (b_1+b_2)x^3 + b_2x^4$$
$$c_2(x) = b_0 + (b_0+b_1)x + (b_0+b_1+b_2)x^2 + (b_1+b_2)x^3 + b_2x^4$$

The output of the convolutional encoder is obtained by interleaving the coefficients of the above three polynomials by considering the coefficients of $x^i, 0 \le i \le 2$. Therefore the output sequence is

$$b_0, b_0, b_0, b_1, (b_0+b_1), (b_0+b_1), (b_0+b_2), (b_0+b_1+b_2), \text{ and } (b_0+b_1+b_2)$$

It can also be checked that if $f = \begin{bmatrix} b_0 & b_1 & b_2 \end{bmatrix}$, the product $fG = c$ yields the above sequence of symbols. □

2.11.4 Wyner-Ash Convolutional Codes

The best known convolutional codes are generally discovered by computer search. However, a single error-correcting convolutional code due to A. D. Wyner and R. B. Ash is described in this subsection. These codes are similar to a set of Hamming codes. Furthermore, these are defined in terms of the parity-check matrix of a Hamming code.

Definition 2.28. *Let $k_0, n_0, \beta \in \mathbb{P}$, and (n,k) be a convolutional code $\mathcal{C}$ defined over the binary field $\mathbb{B}$, where $n_0 = 2^\beta$, $k_0 = (n_0 - 1)$, $n = (\beta+1)n_0$, and $k = (\beta+1)k_0$. Also, consider a $(2^\beta - 1, 2^\beta - 1 - \beta)$ Hamming code with parity-check matrix H_h. This is a $\beta \times (2^\beta - 1)$ matrix in which all columns are distinct and nonzero.*

The parity-check matrix H of the code $\mathcal{C}$ is defined in terms of the $1 \times (2^\beta - 1)$ matrices P_l, $0 \le l \le \beta$.

(a) P_0^T *is a row vector of size* $\left(2^\beta - 1\right)$ *whose elements are all ones.*
(b) *Use the rows of the matrix* H_h *to define* $P_1^T, P_2^T, \ldots, P_\beta^T$. *The number of elements in each row is equal to* $\left(2^\beta - 1\right)$.

The parity-check matrix of the Wyner-Ash code is

$$H = \begin{bmatrix} P_0^T & 1 & 0 & 0 & 0 & 0 & \cdots & 0 & 0 \\ P_1^T & 0 & P_0^T & 1 & 0 & 0 & \cdots & 0 & 0 \\ P_2^T & 0 & P_1^T & 0 & P_0^T & 1 & \cdots & 0 & 0 \\ \vdots & \vdots & \vdots & \vdots & \vdots & \vdots & \ddots & \vdots & \vdots \\ P_\beta^T & 0 & P_{\beta-1}^T & 0 & P_{\beta-2}^T & 0 & \cdots & P_0^T & 1 \end{bmatrix} \tag{2.26}$$

where all 1*'s are elements of the field* $\mathbb{B}$, *the* 0*'s with a single* 1 *in its column are also elements of the field* $\mathbb{B}$, *and each remaining* 0 *is an all-zero row vector of size* $\left(2^\beta - 1\right)$. □

Theorem 2.14. *The minimum distance of a Wyner-Ash code* $\mathcal{C}$ *is equal to* 3, *and therefore it is a single error-correcting convolutional code.*
Proof. See the problem section. □

Example 2.18. The parity-check and generator matrices of the $(12, 9)$ Wyner-Ash convolutional code over $\mathbb{B}$, where $\beta = 2$ are determined. In this code, $n = 12, n_0 = 4, k = 9$, and $k_0 = 3$. Note that $n_0 = 2^\beta = 4$, and $(n_0 - k_0) = 1$. The parity-check matrix H_h of the Hamming code $(3, 1)$ is

$$H_h = \begin{bmatrix} 0 & 1 & 1 \\ 1 & 0 & 1 \end{bmatrix}$$

Then

$$P_0^T = \begin{bmatrix} 1 & 1 & 1 \end{bmatrix}, \; P_1^T = \begin{bmatrix} 0 & 1 & 1 \end{bmatrix}, \text{ and } P_2^T = \begin{bmatrix} 1 & 0 & 1 \end{bmatrix}$$

The 3×12 truncated parity-check matrix of the Wyner-Ash code is

$$H = \begin{bmatrix} P_0^T & 1 & 0 & 0 & 0 & 0 \\ P_1^T & 0 & P_0^T & 1 & 0 & 0 \\ P_2^T & 0 & P_1^T & 0 & P_0^T & 1 \end{bmatrix}$$

where 1 is an element of the field $\mathbb{B}$, and the 0's with a single 1 in a column also belong to the field $\mathbb{B}$. The remaining 0's are all-zero vectors of size 3. The corresponding 9×12 truncated generator matrix is

$$G = \begin{bmatrix} I_3 & P_0 & 0 & P_1 & 0 & P_2 \\ 0 & 0 & I_3 & P_0 & 0 & P_1 \\ 0 & 0 & 0 & 0 & I_3 & P_0 \end{bmatrix}$$

where I_3 is an identity matrix of size 3, and the 0 entry below it is a 3×3 submatrix of all zeros. The truncated parity-check matrix is explicitly shown below.

$$H = \begin{bmatrix} 1&1&1&1&0&0&0&0&0&0&0&0 \\ 0&1&1&0&1&1&1&1&0&0&0&0 \\ 1&0&1&0&0&1&1&0&1&1&1&1 \end{bmatrix}$$

The corresponding generator matrix is

$$G=\begin{bmatrix}1&0&0&1&0&0&0&0&0&0&0&1\\0&1&0&1&0&0&0&1&0&0&0&0\\0&0&1&1&0&0&0&1&0&0&0&1\\0&0&0&0&1&0&0&1&0&0&0&0\\0&0&0&0&0&1&0&1&0&0&0&1\\0&0&0&0&0&0&1&1&0&0&0&1\\0&0&0&0&0&0&0&0&1&0&0&1\\0&0&0&0&0&0&0&0&0&1&0&1\\0&0&0&0&0&0&0&0&0&0&1&1\end{bmatrix}$$

It can indeed be checked that $GH^T = 0$. □

2.11.5 Shift Registers

Convolutional codes defined over the binary field $\mathbb{B}$ can be implemented in hardware devices called shift registers. A shift register is a device which can conveniently store, manipulate, and perform arithmetic in the field $\mathbb{B}$. The arithmetic operations in the field $\mathbb{B}$ are modulo-2. An s-stage shift register consists of s delay elements. Each delay element has a single input, a single output, and a clock which controls the state of this element. This delay element can be in two states: either in a 0-state or a 1-state. Therefore the delay element, commonly called a flip-flop, can store either a 0 or a 1 element of the field $\mathbb{B}$.

Recall that the elements of the field $\mathbb{B}$ are called bits. During each increment of time, the shift register can perform the following operations.

(a) The new input bit is fed into a delay element.
(b) The contents of the delay elements are shifted into a neighboring element.
(c) The new input bit and the values in some of the delay elements are added modulo 2 to create an output bit.

The output of the s-stage shift register can be specified by a polynomial $a(x) \in \mathbb{B}[x]$, where $\deg a(x) \leq s$. If $a(x) = \sum_{i=0}^{s} a_i x^i$, then the coefficient $a_i = 1$ if the ith delay element is used in the modulo 2 sum operation that occurs in the output, and 0 otherwise.

Shift registers are convenient devices to implement convolution codes over the binary field $\mathbb{B}$, because if the input stream to the shift register is described by a polynomial $f(x)$, the output stream can be described by a polynomial $c(x) = f(x)a(x)$. The coefficients of these polynomials are either 0 or 1, that is $f(x), c(x) \in \mathbb{B}[x]$. It is also known that the multiplication of two polynomials corresponds to the convolution operation in the time domain.

2.11.6 Decoding

A codeword transmitted on a communication channel is susceptible to errors. The codewords created by using convolutional codes are infinite in length, however the decoder detects errors based upon the finite width of its decoding-window. A simple procedure to decode convolutional code is as follows. Assume that the decoding window width is greater than the block-length n. Create a dictionary of all codewords of length n_0, and compare the received word with the words in it. The transmitted word is assumed to be the word in this dictionary which is closest to the received word.

This is typically done via the use of the Hamming metric. Using this technique, the first information frame of n_0 symbols is determined. Then the next n_0 received-symbols are moved in to the decoder and the process is repeated to determine the next information frame, and so on. Albeit simple, this procedure is computationally expensive. Therefore this technique called the *minimum-distance decoding*, is useful for short codes and codes with low rates.

A. Viterbi (1967) discovered a decoding technique for convolutional codes based upon the principles of dynamic programming. This technique can be interpreted as a shortest-path computation algorithm. It can also be cast in terms of a maximum-likelihood decoding technique. Each newly arriving noisy frame is decoded such that the *distance* between the noisy sequence received up to that point in time, and the estimated codeword is minimized. That is, the algorithm sequentially decodes the most likely codeword. The distance metric used in this algorithm is the Hamming distance.

An (n, k) convolutional encoder defined over the finite field $\mathbb{F}_q$ has $\beta k_0 = (k - k_0)$ number of internal storage elements. The number of elements in the field $\mathbb{F}_q$ is equal to q. Therefore there are $q^{\beta k_0} = B$ different states. Number these states as $0, 1, \ldots, (B-1)$. Define the set of states to be Λ. Thus $\Lambda = \{0, 1, \ldots, (B-1)\}$.

The state $\xi_j \in \Lambda$ of the encoder at a given time $j \in \mathbb{N}$ is represented by β number of k_0-tuples. Any new k_0-tuple which is input to the encoder changes the state of the encoder and emits a particular n_0-tuple. These transitions can be represented in a state-table or graphically in a state-diagram. The graphical representation of these transitions is called a *trellis-diagram*. It is so named, because this diagram looks like a trellis. More precisely, a trellis-diagram is a graphical representation of the internal states of the encoder and the corresponding transitions produced by all the different input sequences at all instants of time. Therefore the trellis-diagram is an extension of the state-diagram in time. In the trellis-diagram, there are several paths which begin in state ξ_0 at time 0 and are in state ξ_j at time j. Assume that ξ_0 is a state with all-zero symbols. Denote the n_0-tuple emitted by the encoder, when there is a transition from state $\xi_j = \delta$ to $\xi_{j+1} = \kappa$ by $\partial_{\delta\kappa} \in \mathbb{F}_q^{(n_0)}$, where $\delta, \kappa \in \Lambda$. Note that it is quite possible that $\partial_{\delta\kappa}$ may not exist, because there may not be a valid transition from state δ to state κ. In this instance let $\partial_{\delta\kappa} = \varnothing$. The sequence of these states $(\xi_0, \xi_1, \ldots, \xi_j)$ up to time j also represent a path on the trellis diagram.

It should be noted that the trellis-diagram expands to include all B states after βk_0 time units. The input sequence of symbols in practice is typically finite in length. Therefore, this finite sequence is generally appended with sufficient number of zero-symbols so that the last symbol can pass though the encoder. Consequently, the last βk_0 stages of the trellis represent the termination of the codeword and indicate an all-zero symbol input. This implies that Viterbi's algorithm begins and ends at the same state ξ_0 after, say J units of time.

In the decoding process, assume that the decoder receives an n_0-tuple at each time instant $j \in \mathbb{P}$. Assume that the decoding process runs from time 0 to time $J \in \mathbb{P}$. Let the received and output sequences be denoted by a and b respectively, where

$$a = (a_1, a_2, \ldots, a_J); \;\; a_j \in \mathbb{F}_q^{(n_0)}, \; 1 \le j \le J$$
$$b = (b_1, b_2, \ldots, b_J); \; b_j \in \mathbb{F}_q^{(k_0)}, \; 1 \le j \le J$$

Two more variables are introduced in the decoding process. Denote the path on the trellis-diagram with the smallest deviation from the received sequence of the n_0-tuple at time j by $\Omega(\xi_j)$, and the value of the corresponding deviation by $\mathcal{D}(\xi_j) \in \mathbb{N}$. The sequence $\Omega(\xi_j)$ and deviation $\mathcal{D}(\xi_j)$ are called the *survivor-sequence* and the *survivor-length* of the state ξ_j respectively.

At any time j, only the information about all the B survivors has to be maintained. If this were not the case, the received symbols are not decoded as per the assumption of the shortest path on the trellis diagram. In this algorithm, $d(\cdot,\cdot)$ is the Hamming distance. If the algorithm starts in the state ξ_0, and sufficient number of zeros are appended at the end of the finite sequence of message frames before the encoding operation, the state $\xi_J = \xi_0$. If after J iterations, several states end in ξ_0 with identical value of the deviation $\mathcal{D}(\xi_0)$, then uncorrectable errors have been detected. See description of the Viterbi's decoding algorithm.

Algorithm 2.4. *Viterbi's Decoding of Convolution Code.*

Input: The (n,k) convolutional code $\mathcal{C}$ over finite field $\mathbb{F}_q$.
The code is specified by its generator matrix G.
The input sequence to be decoded is $(a_1, a_2, \ldots, a_J)$,
$a_j \in \mathbb{F}_q^{(n_0)}$, for $1 \le j \le J$. Assume that $2\beta k_0 \le J$.
Output: The output sequence of the decoder is $(b_1, b_2, \ldots, b_J)$,
$b_j \in \mathbb{F}_q^{(k_0)}$, for $1 \le j \le J$.
(*initialization*)
Assume that the starting state is $\xi_0 \in \Lambda$.
begin
  $j \leftarrow 0$
  $\Omega(\xi_0) \leftarrow \xi_0$, and $\Omega(\xi')$ is arbitrary $\forall\, \xi' \in \Lambda \backslash \{\xi_0\}$
  $\mathcal{D}(\xi_0) \leftarrow 0$, and $\mathcal{D}(\xi') \leftarrow \infty$, $\forall\, \xi' \in \Lambda \backslash \{\xi_0\}$
end
(*repetitive step: compute survivor-sequence and survivor-length of each state*)
for $j = 1$ **to** J **do**
begin
  for all $\xi_j \in \Lambda$
  (ξ_j *is a state at time instant* j)
  $\kappa \leftarrow \xi_j$
  begin
    for all $\xi_{j-1} \in \Lambda$
    (ξ_{j-1} *is a state at time instant* $(j-1)$)
    begin
      $\delta \leftarrow \xi_{j-1}$
      if $\partial_{\delta\kappa} \neq \varnothing$ **then** $\phi_\delta \leftarrow \mathcal{D}(\delta) + d(a_j, \partial_{\delta\kappa})$
      else $\phi_\delta \leftarrow \infty$
    end (*end of for-loop*)
    $\mathcal{D}(\kappa) \leftarrow \min_{\xi_{j-1} \in \Lambda} \phi_{\xi_{j-1}}$
    (*if there is a tie choose arbitrarily*)
    (*let the minimizing* ξ_{j-1} *value be* $\widetilde{\xi}_{j-1}$)
    if $\mathcal{D}(\kappa) < \infty$ **then** $\Omega(\kappa) \leftarrow \left(\Omega\left(\widetilde{\xi}_{j-1}\right), \kappa\right)$
  end (*end of for-loop*)
end (*end of* j *for-loop*)
(*output*)
The b_j's for $1 \le j \le J$ can be determined by using the state-table and

the survivor-sequence $\Omega(\xi_J)$. The state $\xi_0 = \xi_J$.

In the above algorithm, the statement $\partial_{\delta\kappa} \neq \varnothing$ implies that there is a transition from state δ to state κ. This algorithm is best illustrated via an example.

Example 2.19. Consider a $(6,3)$ convolutional code defined over the binary field $\mathbb{B}$. For this code $n = 6, k = 3$. Let $k_0 = 1$, then $\beta = 2$, and $n_0 = 2$. The rate of this code is $\vartheta = 1/2$. The generator-polynomial matrix $\mathcal{G}(x)$ is

$$\mathcal{G}(x) = \begin{bmatrix} g_0(x) & g_1(x) \end{bmatrix}; \quad g_0(x) = (x^2+1), g_1(x) = (x^2+x+1)$$

The generator matrix G is

$$G = \begin{bmatrix} G_0 & G_1 & G_2 & 0 & 0 & 0 & 0 & \cdots \\ 0 & G_0 & G_1 & G_2 & 0 & 0 & 0 & \cdots \\ 0 & 0 & G_0 & G_1 & G_2 & 0 & 0 & \cdots \\ \vdots & & & & & & & \end{bmatrix}$$

$$G_0 = \begin{bmatrix} 1 & 1 \end{bmatrix}, \quad G_1 = \begin{bmatrix} 0 & 1 \end{bmatrix}, \quad \text{and } G_2 = \begin{bmatrix} 1 & 1 \end{bmatrix}$$

If the input sequence is described by the polynomial $\widetilde{f}(x) = [f_0(x)]$, $f_0(x) \triangleq f(x)$, then

$$\widetilde{f}(x)\mathcal{G}(x) = \widetilde{c}(x) = \begin{bmatrix} c_0(x) & c_1(x) \end{bmatrix}, \text{ where } c_0(x) = f(x)g_0(x), \quad c_1(x) = f(x)g_1(x)$$

A shift register schematic of the encoder is shown in Figure 2.1. In this coding scheme $q = |\mathbb{B}| = 2$, and the number of states is equal to $B = 4$. Note that the state of each of the delay elements of the shift register is either a 0 or 1. The states of this encoder are $00, 01, 10$, and 11. For simplicity in notation, denote these states by s_0, s_1, s_2, and s_3 respectively. With an input of either 0 or 1, the encoder changes its state and emits a 2-tuple encoded output.

The encoding operation implemented via shift-registers can also be described by a state diagram. See Figure 2.2. In this figure, the four rectangular boxes represent the states of the decoder. For example, the box in the north-corner represents the state 00, which is s_0, and the box on the east-corner represents the state 10, which is s_2. The arrows represent the state transitions. The arrows with solid line represent state transition with a 0 at the input, and the arrows with dotted line represent state transition with a 1 at the input. The two-tuple besides the transition arrows represent the output (c_0, c_1) of the shift register.

A tabular representation of the state-diagram is summarized in Table 2.1. This table can be read as follows. For example, if the shift-register is in state s_0, then with a 0 at the input the coded output is 00 and it transitions to state s_0. However, with a 1 at the input the coded output is 11 and it transitions to state s_2. Assume that the encoder is in state s_0. Let the input sequence be $101101\ldots$, the corresponding output sequence can be obtained by using the state transition diagram. The encoder input and output sequences are tabulated in Table 2.2.

The encoded symbol sequence is 11 01 00 10 10 00.... The input sequence of symbols in practice is of finite length. Therefore, this finite sequence is generally appended with sufficient number of zeros so that the last symbol (bit) can pass through the encoding shift register. For the given sequence of six input symbols, two more zeros are appended. The corresponding encoded output sequence is 11 01 00 10 10 00 01 11. Assume that this is the transmitted sequence. Next apply the Viterbi's decoding algorithm to the received sequence 11 01 01 10 10 00 01 11. Note that the third 2-tuple has an error in its second position. Assume that at each time instant, starting at time $j = 1$, a 2-tuple is received. Let the starting state be s_0. Viterbi's decoding algorithm is next used to recover the transmitted sequence. In this example $J = 8$.

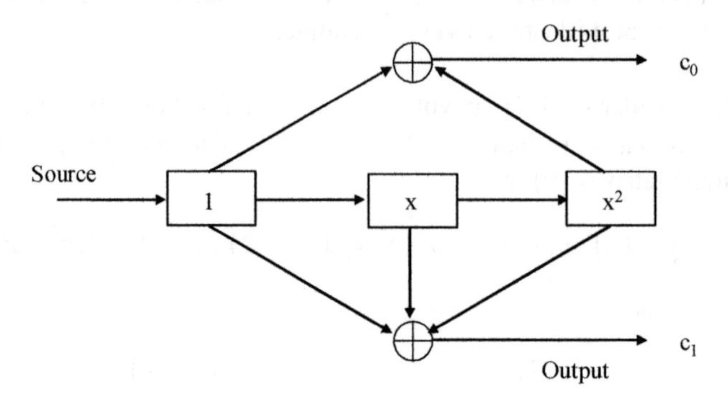

Figure 2.1. Encoding shift register.

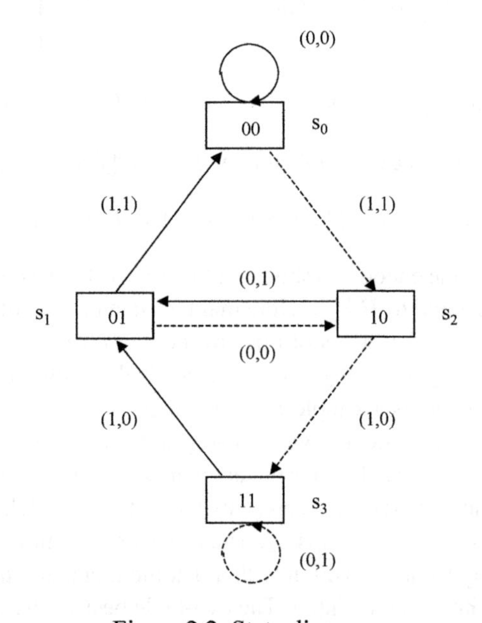

Figure 2.2. State diagram.

States	Input-0	Input-1
s_0	s_0	s_2
s_1	s_0	s_2
s_2	s_1	s_3
s_3	s_1	s_3

Transition

States	Input-0	Input-1
s_0	00	11
s_1	11	00
s_2	01	10
s_3	10	01

Output

Table 2.1. Tabular representation of the state-diagram.
Transition and output.

Time	0	1	2	3	4	5	6	7	8
Input	1	0	1	1	0	1	0	0	
State	s_0	s_2	s_1	s_2	s_3	s_1	s_2	s_1	s_0
Output		11	01	00	10	10	00	01	11

Table 2.2. The encoder input and output.

Received sequence is 11 01 01 10 10 00 01 11 at time instants j equal to 1 through 8 respectively. The Viterbi's decoding algorithm is applied to this sequence.

Time $j = 0$. This is initialization.

$$\Omega(s_0) = (s_0), \ \mathcal{D}(s_0) = 0$$
$$\Omega(s_i) = \text{arbitrary}, \ \mathcal{D}(s_i) \leftarrow \infty, \quad 1 \le i \le 3$$

As per the algorithm $\xi_0 = s_0$.

Time $j = 1$. Input to the decoder is $a_1 = 11$.

$$\Omega(s_0) = (s_0, s_0), \ \mathcal{D}(s_0) = 2$$
$$\Omega(s_2) = (s_0, s_2), \ \mathcal{D}(s_2) = 0$$

The survivor-sequence and survivor-length of states s_1 and s_3 remain unchanged.

Time $j = 2$. Input to the decoder is $a_2 = 01$.

$$\Omega(s_0) = (s_0, s_0, s_0), \ \mathcal{D}(s_0) = 3$$
$$\Omega(s_1) = (s_0, s_2, s_1), \ \mathcal{D}(s_1) = 0$$
$$\Omega(s_2) = (s_0, s_0, s_2), \ \mathcal{D}(s_2) = 3$$
$$\Omega(s_3) = (s_0, s_2, s_3), \ \mathcal{D}(s_3) = 2$$

Time $j = 3$. Input to the decoder is $a_3 = 01$.

$$\Omega(s_0) = (s_0, s_2, s_1, s_0), \ \mathcal{D}(s_0) = 1$$
$$\Omega(s_1) = (s_0, s_0, s_2, s_1), \ \mathcal{D}(s_1) = 3$$
$$\Omega(s_2) = (s_0, s_2, s_1, s_2), \ \mathcal{D}(s_2) = 1$$
$$\Omega(s_3) = (s_0, s_2, s_3, s_3), \ \mathcal{D}(s_3) = 2$$

Time $j = 4$. Input to the decoder is $a_4 = 10$.

$$\Omega(s_0) = (s_0, s_2, s_1, s_0, s_0), \ \mathcal{D}(s_0) = 2$$
$$\Omega(s_1) = (s_0, s_2, s_3, s_3, s_1), \ \mathcal{D}(s_1) = 2$$
$$\Omega(s_2) = (s_0, s_2, s_1, s_0, s_2), \ \mathcal{D}(s_2) = 2$$
$$\Omega(s_3) = (s_0, s_2, s_1, s_2, s_3), \ \mathcal{D}(s_3) = 1$$

Time $j = 5$. Input to the decoder is $a_5 = 10$.

$$\Omega(s_0) = (s_0, s_2, s_1, s_0, s_0, s_0), \ \mathcal{D}(s_0) = 3$$
$$\Omega(s_1) = (s_0, s_2, s_1, s_2, s_3, s_1), \ \mathcal{D}(s_1) = 1$$
$$\Omega(s_2) = (s_0, s_2, s_3, s_3, s_1, s_2), \ \mathcal{D}(s_2) = 3$$
$$\Omega(s_3) = (s_0, s_2, s_1, s_0, s_2, s_3), \ \mathcal{D}(s_3) = 2$$

Time $j = 6$. Input to the decoder is $a_6 = 00$.

$$\begin{aligned}
\Omega(s_0) &= (s_0, s_2, s_1, s_0, s_0, s_0, s_0), \ \mathcal{D}(s_0) = 3 \\
\Omega(s_1) &= (s_0, s_2, s_1, s_0, s_2, s_3, s_1), \ \mathcal{D}(s_1) = 3 \\
\Omega(s_2) &= (s_0, s_2, s_1, s_2, s_3, s_1, s_2), \ \mathcal{D}(s_2) = 1 \\
\Omega(s_3) &= (s_0, s_2, s_1, s_0, s_2, s_3, s_3), \ \mathcal{D}(s_3) = 3
\end{aligned}$$

Time $j = 7$. Input to the decoder is $a_7 = 01$.

$$\begin{aligned}
\Omega(s_0) &= (s_0, s_2, s_1, s_0, s_0, s_0, s_0, s_0), \ \mathcal{D}(s_0) = 4 \\
\Omega(s_1) &= (s_0, s_2, s_1, s_2, s_3, s_1, s_2, s_1), \ \mathcal{D}(s_1) = 1 \\
\Omega(s_2) &= (s_0, s_2, s_1, s_0, s_0, s_0, s_0, s_2), \ \mathcal{D}(s_2) = 4 \\
\Omega(s_3) &= (s_0, s_2, s_1, s_2, s_3, s_1, s_2, s_3), \ \mathcal{D}(s_3) = 3
\end{aligned}$$

Time $j = 8$. Input to the decoder is $a_8 = 11$.

$$\begin{aligned}
\Omega(s_0) &= (s_0, s_2, s_1, s_2, s_3, s_1, s_2, s_1, s_0), \ \mathcal{D}(s_0) = 1 \\
\Omega(s_1) &= (s_0, s_2, s_1, s_2, s_3, s_1, s_2, s_3, s_1), \ \mathcal{D}(s_1) = 4 \\
\Omega(s_2) &= (s_0, s_2, s_1, s_2, s_3, s_1, s_2, s_1, s_2), \ \mathcal{D}(s_2) = 3 \\
\Omega(s_3) &= (s_0, s_2, s_1, s_2, s_3, s_1, s_2, s_3, s_3), \ \mathcal{D}(s_3) = 4
\end{aligned}$$

As the last result corresponds to the state $\xi_0 = s_0 = \xi_J$, the algorithm terminates.

Output. The sequence of states in the survival-state sequence $\Omega(s_0)$ results in the correct decoded sequence: 10110100. Note that the Hamming distance between the sequence to be decoded, and the corrected sequence is equal to $\mathcal{D}(s_0)$. □

2.12 Turbo Codes

The turbo codes are important from both theoretical and practical point of view. Turbo codes were discovered by Claude Berrou, Alain Glavieux, and P. Thitimajshima in 1993. Turbo codes are built using either conventional block or convolutional codes. These codes asymptotically approach the theoretical performance of codes predicted by Shannon's theorem. The performance is quantified in terms of bit-error probability.

A turbo code essentially consists of a serial or parallel *encoder* which uses an *iterative decoding* scheme. Further, *turbo* refers to the principle of iterative decoding. The improved performance of turbo codes is due to the introduction of an *interleaver* (also called *permuter*). It changes the order of the input bits, thus facilitating the operation of two encoders which operate on two different input sequences. However, the two input sequences are related by a permutation operation. This results in improved error-correcting performance.

Some possible applications of these codes are: mobile telephony, terrestrial mobile television systems, satellite communication, wireless metropolitan networks, and deep space exploration.

Reference Notes

A lucid description of error-correcting codes can be found in the classical books by Berlekamp (1968), Hamming (1986), and MacWilliams, and Sloane (1988). A comprehensive treatise on this subject is the book by Huffman, and Pless (2003). Wicker (1995), Togneri, and de Silva (2002), and Moon (2005) provide a readable account of the subject of code design. The textbook by Roth (2006) is a scholarly introduction to this subject. Equally good and readable account is provided in Blahut (1984), Hankerson, Hoffman, Leonard, Lindner, Phelps, Rodger, and Wall (2000), Jones, and Jones (2000), Pretzel (1992), Rosen (2018), van Lint (1992), Vermani (1996), Wells (1999), and Welsh (1988).

Problems

1. Prove that the Hamming distance satisfies the distance axioms.
2. If $n, d_H \geq 2$, then prove that $A_q(n, d_H) \leq A_q(n-1, d_H - 1)$.
   Hint: See Huffman, and Pless (2003). Let $\mathcal{C}$ be a code with M codewords, and minimum distance d_H. Puncture any single coordinate of the codewords to get a new code $\mathcal{C}^*$. This new code also has M codewords. If this were not true, that is if $\mathcal{C}^*$ had fewer distinct codewords, there would exist two codewords $u, v \in \mathcal{C}$ such that $d(u, v) = 1$, but this contradicts the hypothesis that $d_H \geq 2$.
3. If d_H is even, then prove that $A_2(n, d_H) = A_2(n-1, d_H - 1)$.
   Hint: See Huffman, and Pless (2003). Only $A_2(n, d_H) \geq A_2(n-1, d_H - 1)$ has to be established. Let $\mathcal{C}$ be a binary $[n-1, M, d_H - 1]$-code. Extend the code $\mathcal{C}$ by appending an overall parity-check bit to obtain a binary code $\widetilde{\mathcal{C}}$ specified by $[n, M, d_H]$-code, since $(d_H - 1)$ is odd by hypothesis. As the code $\widetilde{\mathcal{C}}$ has M codewords, $A_2(n, d_H) \geq A_2(n-1, d_H - 1)$. The final result follows by using the last problem.
4. Prove that $A_q(n, d_H) \leq qA_q(n-1, d_H)$, for $n \geq 2$.
   Hint: See Huffman, and Pless (2003). Let $\mathcal{C}$ be an $[n, M, d_H]$-code, such that $A_q(n, d_H) = M$. Also let $\mathcal{C}_\alpha$ be a subcode of $\mathcal{C}$ such that every codeword has α in coordinate n. Then for some $\alpha \in A$, the code $\mathcal{C}_\alpha$ has at least M/q codewords. If this code is punctured in coordinate n, then a code of length $(n-1)$ and Hamming distance d_H is obtained. Consequently $M/q \leq A_q(n-1, d_H)$, which yields the desired result.
5. Prove that $A_2(n, 2) = 2^{n-1}$.
   Hint: See Huffman, and Pless (2003). $A_2(n, 2) = A_2(n-1, 1) = 2^{n-1}$.
6. A code $\mathcal{C}$ is specified by $[n, M, d_H]$. Its alphabet is A, and $|A| = q$. Also, $V_q(n, \delta)$ is the number of codewords which have a Hamming distance at most δ from a specific codeword. Prove that

$$V_q(n, \delta) = \binom{n}{0} + \binom{n}{1}(q-1) + \cdots + \binom{n}{\delta}(q-1)^\delta = \sum_{j=0}^{\delta} \binom{n}{j}(q-1)^j$$

Hint: Let u be any word in $A^{(n)}$, then a word with distance j from it can be chosen in $\binom{n}{j}$ different ways, where $0 \le j \le n$. Denote this word by v. Select the j positions in which the words u and v differ. Once such positions have been determined, each differing position can be filled in $(q-1)$ different ways. There are a total of $(q-1)^j$ possibilities once the differing positions are selected. Therefore, there are $\binom{n}{j}(q-1)^j$ codewords at a distance j from u. The result follows.

7. A code $\mathcal{C}$ is specified by $[n, M, d_H]$. Its alphabet is A, and $|A| = q$. The number of codewords which are at most a distance δ from a specific codeword is $V_q(n, \delta)$, where

$$V_q(n,\delta) = \sum_{j=0}^{\delta} \binom{n}{j} (q-1)^j$$

An upper bound for $V_q(n, \delta)$ can be obtained in terms of $H_q(\cdot)$, the q-ary Hilbert entropy function, $H_q : [0,1] \to [0,1]$. It is defined as

$$H_q(x) = x \log_q (q-1) - x \log_q x - (1-x) \log_q (1-x), \quad \text{for} \quad 0 \le x \le 1$$

where it is assumed that $0 \log_q 0 \to 0$. If $q \ge 2$, prove that

$$V_q(n,\delta) \le q^{nH_q(\delta/n)}, \quad \text{for} \quad 0 \le \delta/n \le (1 - 1/q)$$

Hint: See Huffman, and Pless (2003), and Roth (2006). Note that $H_2(\cdot)$ is the binary entropy function. Furthermore, $H_q(0) = 0$, and $H_q(1) = \log_q(q-1)$. It can be verified that $H_q(x)$ is nonnegative, and attains a maximum value of 1 at $x = (1 - 1/q)$. Its derivative with respect to x is positive in the interval $(0, 1 - 1/q)$, and $H_q(\cdot)$ is also continuous, increasing, and strictly concave in this interval. The inequality is true if $\delta = 0$. Consider the case, when $\delta > 0$. Let $\theta = \delta/n$. Thus

$$\begin{aligned}
\frac{V_q(n,\delta)}{q^{nH_q(\theta)}} &= \frac{\sum_{j=0}^{\delta}\binom{n}{j}(q-1)^j}{(q-1)^{\delta}\theta^{-\delta}(1-\theta)^{-n+\delta}} \\
&= \sum_{j=0}^{\delta}\binom{n}{j}(q-1)^j(1-\theta)^n\left\{\frac{\theta}{(q-1)(1-\theta)}\right\}^{\delta}
\end{aligned}$$

The condition $0 < \theta \le (1 - 1/q)$ implies

$$\frac{\theta}{(q-1)} \le \frac{1}{q} \le (1-\theta)$$

This in turn implies

$$\frac{\theta}{(q-1)(1-\theta)} \le 1$$

Thus

$$\begin{aligned}
\frac{V_q(n,\delta)}{q^{nH_q(\theta)}} &\le \sum_{j=0}^{\delta}\binom{n}{j}(q-1)^j(1-\theta)^n\left\{\frac{\theta}{(q-1)(1-\theta)}\right\}^{j} \\
&= \sum_{j=0}^{\delta}\binom{n}{j}\theta^j(1-\theta)^{n-j} \le \sum_{j=0}^{n}\binom{n}{j}\theta^j(1-\theta)^{n-j} = 1
\end{aligned}$$

The result is immediate.

8. Let $\mathcal{C}$ be an (n,k)-linear code over $\mathbb{F}_q$. If $c \in \mathcal{C}$, then prove that $cH^T = 0$, where H is the parity-check matrix.
Hint: Since $c \in \mathcal{C}$, then there exists a $f \in \mathbb{F}_q^{(k)}$ such that $c = fG$. Therefore $cH^T = fGH^T = 0$.
9. Prove that $u, v \in \mathbb{F}_q^{(n)}$ belong to the same coset of $\mathcal{C}$ iff $uH^T = vH^T$.
Hint: The vectors u and v belong to the same coset iff there exists a codeword $c \in \mathcal{C}$ such that $u = (v + c)$. The result follows by noting that $cH^T = 0$.
10. Let $\mathcal{C}$ be an (n, k, d_H)-linear code over $\mathbb{F}_q$. Its parity-check matrix is H. Prove that:
    (a) $d_H \geq d$ if and only if any $(d-1)$ columns of H are linearly independent.
    (b) $d_H \leq d$ if and only if the matrix H has d columns which are linearly dependent.
    (c) Any $(d_H - 1)$ columns of H are linearly independent and has d_H columns that are linearly dependent.

    Hint: See Ling, and Xing (2004).
    (a) Let $c = \begin{bmatrix} c_0 \; c_1 \; \cdots \; \cdots \; c_{n-1} \end{bmatrix} \in \mathcal{C}$ be a codeword of weight $\omega > 0$. Let the nonzero coordinates of c be in positions $i_1, i_2, \ldots, i_\omega$. That is, $c_i \neq 0$ if $i \in \{i_1, i_2, \ldots, i_\omega\}$ and $c_i = 0$ if $i \notin \{i_1, i_2, \ldots, i_\omega\}$. Also let ith column of H be $H_i, 0 \leq i \leq (n-1)$. Since $c \in \mathcal{C}$,

    $$0 = cH^T = \sum_{j=1}^{\omega} c_{i_j} H_{i_j}^T$$

    The 0 on the left-hand side of the above equation is a zero vector of size $(n-k)$. The above equation is true if and only if there exist ω columns, $H_{i_1}, H_{i_2}, \ldots, H_{i_\omega}$ in the matrix H that are linearly dependent.
    Denote the weight of the codeword $c \in \mathcal{C}$ by $w(c)$.
    Also $d_H \geq d \Leftrightarrow \nexists\, c' \in \mathcal{C}\backslash\{0\}$ such that $w(c') \leq (d-1) \Leftrightarrow$ any $j \leq (d-1)$ columns of H are linearly independent.
    (b) As in part (a) of the problem:
    $d_H \leq d \Leftrightarrow \exists\, c \in \mathcal{C}\backslash\{0\}$ such that $w(c) \leq d \Leftrightarrow j \leq d$ columns of H are linearly dependent.
    (c) Follows from parts (a) and (b) of the problem.
11. The parity-check matrix H of an (n, k, d_H)-binary Hamming code is an $m \times n$ matrix, where $m = (n-k), n = (2^m - 1), m \geq 2$, and d_H is the distance of the code. The columns of this matrix consists of all n nonzero binary vectors of length m, each occurring once. Prove that:
    (a) $k = (2^m - 1 - m)$.
    (b) $d_H = 3$.
    (c) The Hamming code (n, k, d_H) is a perfect code.

    Hint: See Ling, and Xing (2004).
    (a) $k = (n-m) = (2^m - 1 - m)$.
    (b) Observe that all columns of the matrix H are different. Therefore any two columns of H are linearly independent. Also note that the three columns

    $$\begin{bmatrix} 1\,0\,0 \cdots 0 \end{bmatrix}^T, \; \begin{bmatrix} 0\,1\,0 \cdots 0 \end{bmatrix}^T, \text{ and } \begin{bmatrix} 1\,1\,0 \cdots 0 \end{bmatrix}^T$$

    form a linearly dependent set. Therefore from part (c) of the last problem $d_H = 3$. Since $d_H = 3$, this code can correct only $t = 1$ error.
    (c) The result follows by using the definition of perfect code.

12. $\mathcal{C}$ is a nonzero cyclic code in $\mathcal{R}_n$. This is a cyclic code of length n. For any $c \in \mathcal{C}$, let $c = \begin{bmatrix} c_0 & c_1 & \cdots & \cdots & c_{n-1} \end{bmatrix}$, and $c(x) = \sum_{j=0}^{n-1} c_j x^j$. Prove the following statements.
    (a) There exists a unique monic polynomial $g(x) \in \langle g(x) \rangle$ of minimal degree.
    (b) The code $\mathcal{C}$ is generated by the ideal $\langle g(x) \rangle$. That is, the polynomial $g(x)$ is the generator of the cyclic code $\mathcal{C}$.
    (c) $g(x) \mid (x^n - 1)$ in $\mathbb{F}_q[x]$.
    (d) Let $\deg g(x) = \theta$. Any $c \in \mathcal{C}$ can be uniquely expressed as

$$c(x) = f(x) g(x) \in \mathbb{F}_q[x],$$
$$\text{where } f(x) = 0 \text{ or } \deg f(x) \leq (n - \theta - 1)$$

    This implies that $c(x) \equiv 0 \pmod{g(x)}$.
    (e) If $g(x) = \sum_{j=0}^{\theta} g_j x^j$, where $g_\theta = 1$, and $\theta = (n-k)$, then a basis for $\langle g(x) \rangle$ is the set of polynomials

$$\left\{ g(x) x^j \pmod{(x^n - 1)} \mid 0 \leq j \leq (n - \theta - 1) \right\}$$

Hint: See MacWilliams, and Sloane (1988), and Huffman, and Pless (2003). Following is the outline of the proofs
    (a) Since $\mathcal{C}$ is nonzero, a monic polynomial of minimal degree exists. Let $g(x), h(x) \in \langle g(x) \rangle$ be both monic and have minimal degree θ, then $(g(x) - h(x)) \in \langle g(x) \rangle$ has a lower degree. This is a contradiction. Therefore the monic polynomial of minimum degree must indeed be unique.
    (b) Let $c \in \mathcal{C}$. Write $c(x) = a(x) g(x) + b(x)$ in $\mathcal{R}_n$, where $\deg b(x) < \theta$ or $b(x) = 0$. But $c(x)$ and $a(x) g(x)$ are codewords, and the code is linear. Therefore, the polynomial $b(x)$ represents a codeword in $\mathcal{C}$. This contradicts the requirement that the polynomial $g(x)$ is of minimal degree, thus $b(x) = 0$. That is, $\mathcal{C}$ is generated by the polynomial $g(x)$.
    (c) Write $(x^n - 1) = h(x) g(x) + b(x)$ in $\mathbb{F}_q[x]$, where $\deg b(x) < \theta$. That is $b(x) \equiv -h(x) g(x) \pmod{(x^n - 1)}$, represents a codeword in $\mathcal{C}$, which is a contradiction unless $b(x) = 0$.
    (d) Let $(x^n - 1) = h(x) g(x)$. From (b) for any $c \in \mathcal{C}$, $c(x) = a(x) g(x)$ in $\mathcal{R}_n$, and $\deg c(x) < n$. Consequently

$$\begin{aligned} c(x) &= a(x) g(x) + d(x)(x^n - 1) \in \mathbb{F}_q[x] \\ &= (a(x) + d(x) h(x)) g(x) \in \mathbb{F}_q[x] \\ &= f(x) g(x) \in \mathbb{F}_q[x], \quad \deg f(x) \leq (n - \theta - 1) \end{aligned}$$

    Therefore $c(x)$ can be expressed as multiples of polynomial $g(x)$ evaluated in $\mathbb{F}_q[x]$. Moreover, the multiplying polynomial $f(x)$ has a degree $\leq (n - \theta - 1)$.
    (e) The multiplication of $g(x)$ by x modulo $(x^n - 1)$ produces a codeword shifted to the right by one place. Similarly, the multiplication of $g(x)$ by x^j modulo $(x^n - 1)$ produces a codeword shifted to the right by j places. That is, $g(x) x^j \pmod{(x^n - 1)}$, for $j \geq 0$ produces a codeword of $\mathcal{C}$. Therefore multiplying $g(x)$ by any other polynomial is equivalent to multiplying $g(x)$ by various powers of x and summing them, where the polynomial arithmetic is done modulo $(x^n - 1)$ over the field $\mathbb{F}_q$.
    Observe that since the degree of the polynomial $g(x)$ is θ, the set of polynomials $\left\{ g(x) x^j \pmod{(x^n - 1)} \mid 0 \leq j \leq (n - \theta - 1) \right\}$ form a basis of the ideal $\langle g(x) \rangle$. That is, a codeword is formed by a linear combination of these basis polynomials. The basis vectors are the rows of the matrix G. Thus the dimension of the code is $(n - \theta)$.

13. Let $\mathcal{R}(r,m)$ be a rth order Reed-Muller code, where $0 \le r \le m$. The length of the codeword is given by $n = 2^m$. The dual of this code is denoted by $\mathcal{R}^{\perp}(r,m)$. Prove the following results.
   (a) $\mathcal{R}(r-1,m) \subseteq \mathcal{R}(r,m)$ for $r > 0$.
   (b) The dimension of the code $\mathcal{R}(r,m)$ is equal to $\sum_{j=0}^{r}\binom{m}{j}$.
   (c) The Hamming distance of this code is $d_H = 2^{m-r}$.
   (d) $\mathcal{R}^{\perp}(m,m) = \{(00\ldots0)\}$, and if $0 \le r < m$, then

$$\mathcal{R}^{\perp}(r,m) = \mathcal{R}(m-r-1,m)$$

   Hint: See Hankerson, Hoffman, Leonard, Lindner, Phelps, Rodger, and Wall (2000). These results are established by induction.
   (a) It is first shown that $\mathcal{R}(0,m) \subseteq \mathcal{R}(1,m)$. Note that

$$G(1,m) = \begin{bmatrix} G(1,m-1) & G(1,m-1) \\ 0 & G(0,m-1) \end{bmatrix}$$

   The first row of $G(1,m-1)$ has all 1's (the number of 1's is equal to 2^{m-1}). Therefore the first row of $\left[G(1,m-1)\ G(1,m-1)\right]$ has all 1's (the number of 1's is equal to 2^m). Consequently $\mathcal{R}(0,m) = \{(00\ldots0),(11\ldots1)\}$ is contained in $\mathcal{R}(1,m)$.
   Furthermore, $G(r-1,m-1)$ is a submatrix of $G(r,m-1)$. Also, $G(r-2,m-1)$ is a submatrix of $G(r-1,m-1)$, and

$$G(r-1,m) = \begin{bmatrix} G(r-1,m-1) & G(r-1,m-1) \\ 0 & G(r-2,m-1) \end{bmatrix}$$

   is a submatrix of $G(r,m)$. Therefore $\mathcal{R}(r-1,m) \subseteq \mathcal{R}(r,m)$ by induction.
   (b) The result is true for $r = m$, since $\mathcal{R}(m,m) = \mathbb{B}^{(2^m)}$. It can also be checked that the statement is true for $m = 1$. Assume that the statement is true for the dimension of the code $\mathcal{R}(r,m-1)$, for $0 \le r \le (m-1)$. The recursive equation of $G(r,m)$ yields: Dimension of $\mathcal{R}(r,m)$ (equal to k) is equal to the sum of dimensions of the codes $\mathcal{R}(r,m-1)$ and $\mathcal{R}(r-1,m-1)$. Therefore

$$k = \sum_{j=0}^{r}\binom{m-1}{j} + \sum_{j=0}^{r-1}\binom{m-1}{j}$$

   Using the binomial identities

$$\binom{m}{j} = \binom{m-1}{j} + \binom{m-1}{j-1}, \quad 1 \le j \le (m-1)$$
$$\binom{m}{0} = \binom{m-1}{0} = 1$$

   yields the stated result.
   (c) It can be checked that the distance value d_H is correct for $m = 1$ and for both $r = 0$ and $r = m$. Recall that

$$\mathcal{R}(r,m) = \{(u,u+v) \mid u \in \mathcal{R}(r,m-1), v \in \mathcal{R}(r-1,m-1)\}$$

   and $\mathcal{R}(r-1,m-1) \subseteq \mathcal{R}(r,m-1)$; therefore $u+v \in \mathcal{R}(r,m-1)$. Let the function $w(\cdot)$ denote the weight of a codeword. Consider the two cases.

(i) If $u \neq v$, then as per the inductive hypothesis, $w(u+v) \geq 2^{m-1-r}$, and $w(u) \geq 2^{m-1-r}$. Therefore $w(u, u+v) = w(u) + w(u+v) \geq 2(2^{m-1-r}) = 2^{m-r}$.

(ii) If $u = v$, then $(u, u+v) = (v, 0)$ but $v \in \mathcal{R}(r-1, m-1)$, therefore $w(v, 0) = w(v) \geq 2^{(m-1)-(r-1)} = 2^{m-r}$.

(d) It is known that
$\mathcal{R}(r,m) = \{(u, u+v) \mid u \in \mathcal{R}(r, m-1), v \in \mathcal{R}(r-1, m-1)\}$
$\mathcal{R}(m-r-1, m) = \{(x, x+y) \mid x \in \mathcal{R}(m-r-1, m-1),$
$y \in \mathcal{R}(m-r-2, m-1)\}$
The dual of $\mathcal{R}(r, m-1)$ is $\mathcal{R}(m-r-2, m-1)$. This follows by using the induction hypothesis. Similarly the dual of $\mathcal{R}(r-1, m-1)$ is $\mathcal{R}(m-r-1, m-1)$. Consequently $u \cdot y = x \cdot v = 0$. Furthermore, $v \cdot y = 0$ because $\mathcal{R}(r-1, m-1) \subseteq \mathcal{R}(r, m-1)$. The component-wise multiplication yields

$$\begin{aligned}(u, u+v) \cdot (x, x+y) &= u \cdot x + (u+v) \cdot (x+y) \\ &= 2(u \cdot x) + u \cdot y + v \cdot x + v \cdot y = 0\end{aligned}$$

Note that $2(u \cdot x) = 0$ in binary arithmetic. Thus any vector in $\mathcal{R}(r,m)$ is orthogonal to any vector in $\mathcal{R}(m-r-1, m)$. Also the sum of the dimensions of the codes $\mathcal{R}(r,m)$ and $\mathcal{R}(m-r-1, m)$ is

$$\begin{aligned}\sum_{j=0}^{r} \binom{m}{j} + \sum_{j=0}^{m-r-1} \binom{m}{j} &= \sum_{j=0}^{r} \binom{m}{m-j} + \sum_{j=0}^{m-r-1} \binom{m}{j} \\ &= \sum_{j=0}^{m} \binom{m}{j} = 2^m\end{aligned}$$

The result follows.

14. Prove that the minimum distance of a Wyner-Ash convolutional code C is equal to 3.
Hint: See Blahut (1984). As the code is linear, the minimum distance is equal to the minimum weight of a codeword. Assume that the codeword $c \in C$ is minimum weight and nonzero in its first frame. If H is the parity-check matrix of the code, then $cH^T = 0$. As the top row of H is nonzero in the first frame and all zeros in all others, there must be at least two ones in the first frame. Furthermore, since any two columns of the matrix H in the first frame are linearly independent, there must be at least another 1 somewhere in the codeword. Thus the minimum distance is at least 3.
In order to show that the minimum distance is precisely equal to three, it has to be shown that there are three columns which add to zero. Consider the column, where in the row vectors

$$P_0^T, P_1^T, \ldots, P_\beta^T$$

a one occurs in the same column. Add this column in the first frame, with the corresponding column in the second frame. This sum is equal to

$$\begin{bmatrix} 1 \, 0 \, 0 \cdots 0 \end{bmatrix}^T$$

and this column indeed occurs in the first frame. Thus there are three columns which add up to zero.

References

1. Abramson, N., 1963. *Information Theory and Coding*, McGraw-Hill Book Company, New York, New York.
2. Berlekamp, E. R., 1968. *Algebraic Coding Theory*, McGraw-Hill Book Company, New York, New York.
3. Berrou, C., Glavieux, and Thitimajshima, P., 1993. "Near Shannon Limit Error-Correcting Coding and Decoding: Turbo Codes," Proc. of the 1993 IEEE International Communications Conference, Geneva, Switzerland, pp. 1064-1070.
4. Bierbrauer, J., 2005. *Introduction to Coding Theory*, Chapman and Hall/CRC Press, New York, New York.
5. Blahut, R. E., 1984. *Theory and Practice of Error Control Codes*, Addison-Wesley Publishing Company, New York, New York.
6. Bruen, A. A., and Forcinito, M. A., 2005. *Cryptography, Information Theory, and Error-Correction*, John Wiley & Sons, Inc., New York, New York.
7. Hamming, R. W., 1986. *Coding and Information Theory*, Second Edition, Prentice-Hall, Englewood Cliffs, New Jersey.
8. Hankerson, D., Hoffman, D. G., Leonard, D. A., Lindner, C. C., Phelps, K. T., Rodger, C. A., and Wall, J. R., 2000. *Coding Theory and Cryptography, The Essentials*, Second Edition, Marcel Dekker, Inc., New York, New York.
9. Hardy, D. W., and Walker, C. L., 2003. *Applied Algebra, Codes, Ciphers, and Discrete Algorithms*, Prentice-Hall, Upper Saddle River, New Jersey.
10. Huffman, W. C., and Pless, V., 2003. *Fundamentals of Error-Correcting Codes*, Cambridge University Press, Cambridge, Great Britain.
11. Jones, G. A., and Jones, J. M., 2000. *Information and Coding Theory*, Springer-Verlag, Berlin, Germany.
12. Justesen, J., and Høholdt, T., 2004. *A Course in Error-Correcting Codes*, European Mathematical Society, Zürich, Switzerland.
13. Lidl, R., and Niederreiter, H., 1986. *Introduction to Finite Fields and Their Applications*, Cambridge University Press, Cambridge, Great Britain.
14. Ling, S., and Xing, C., 2004. *Coding Theory A First Course*, Cambridge University Press, Cambridge, Great Britain.
15. MacWilliams, F. J., and Sloane, N. J. A., 1988. *The Theory of Error-Correcting Codes*, North-Holland, New York, New York.
16. McEliece, R. J., 1987. *Finite Fields for Computer Scientists and Engineers*, Kluwer Academic Publishers, Norwell, Massachusetts.
17. Moon, T. K., 2005. *Error Correction Coding Mathematical Methods and Algorithms*, John Wiley & Sons, Inc., New York, New York.
18. Niederreiter, H., and Xing, C. 2009. *Algebraic Geometry in Coding Theory and Cryptography*, Princeton University Press, Princeton, New Jersey.
19. Pless, V., 1998. *Introduction to the Theory of Error-Correcting Codes*, Third Edition, John Wiley & Sons, Inc., New York, New York.
20. Pretzel, O., 1992. *Error-Correcting Codes and Finite Fields*, Oxford University Press, Oxford, Great Britain.
21. Rosen, K. H., Editor-in-Chief, 2018. *Handbook of Discrete and Combinatorial Mathematics,* Second Edition, CRC Press: Boca Raton, Florida.

22. Roth, R. M., 2006. *Introduction to Coding Theory*, Cambridge University Press, Cambridge, Great Britain.
23. Togneri, R., and deSilva C. J. S., 2002. *Fundamentals of Information Theory and Coding Design*, Chapman & Hall/CRC Press: New York, New York.
24. van Lint, J. H., 1992. *Introduction to Coding Theory*, Second Edition, Springer-Verlag, Berlin, Germany.
25. Vermani, L. R., 1996. *Elements of Algebraic Coding Theory*, Chapman and Hall/CRC Press, New York, New York.
26. Viterbi, A. J., 1967. "Error Bounds for Convolutional Codes and an Asymptotically Optimal Decoding Algorithm," IEEE Trans. on Information Theory, Vol. IT-13, No. 2, pp. 260-269.
27. Wells, R. B., 1999. *Applied Coding and Information Theory for Engineers*, Prentice-Hall, Englewood Cliffs, New Jersey.
28. Welsh, D., 1988. *Codes and Cryptography*, Oxford University Press, Oxford, Great Britain.
29. Wicker, S. B., 1995. *Error Control Systems for Digital Communication and Storage*, Prentice-Hall, Englewood Cliffs, New Jersey.

CHAPTER 3

Cryptography

$$x^n + y^n \neq z^n$$

$$x, y, z \in \mathbb{P}; \quad n = 3, 4, 5, \ldots$$

Fermat's last theorem

Pierre de Fermat. Fermat was born on 17 August, 1601 in Beaumont-de-Lomagne, France. He was a lawyer by profession, but pursued mathematics as a pastime. Fermat anticipated the work of Descartes in analytic geometry, and that of Leibniz and Newton in differential calculus but did not publish his work. However, his lasting fame rests upon his work in number theory. His great renown essentially rests upon a note he scribbled in the margin of a book by Diophantus. In it he wrote that he had discovered a proof of an apparently simple Diophantine equation. It states:

For any positive integer n greater than two, it is impossible to find three positive integers x, y, and z which satisfy: x to the n-th power plus y to the n-th power is equal to z to the n-th power.

The assertion of this statement is often called Fermat's conjecture or more popularly, Fermat's last theorem. Andrew Wiles gave a remarkable proof of this statement in 1994. Fermat died on 12 January, 1665 in Castres, France.

3.1 Introduction

This chapter discusses techniques for secure transmission of information over a communication channel. It is quite possible that there is an unfriendly intruder on this communication channel. The goal of *cryptography* is to keep the transmission of information secure in presence of such unfavorable intrusion. Thus the basic aim of cryptography is to protect data of any form. Some classical ciphering schemes, cryptanalysis, information theory based cryptanalysis, public-key cryptography, probabilistic encryption, homomorphic encryption, and digital signatures are discussed in this chapter. Key distribution problem, elliptic and hyperelliptic curve based cryptosystems, hashing, authentication, the Data Encryption Standard (DES), and the Advanced Encryption Standard (AES) are also discussed. Finally, a probabilistic perspective on cryptographic security is provided.

Since time immemorial there has been a need to protect information from unauthorized access. The message to be protected (from an enemy) is generally called *plaintext,* or *message.* The entire set of messages is called the *message space*. An *encryption algorithm* converts the plaintext into *ciphertext.* It is hoped that this ciphertext is intelligible only to the designated receiving parties. The implementation of the encryption algorithm is called *ciphering* operation or simply *encipherment.* The ciphering process (encrypting algorithm) uses *keys* to scramble the plaintext. The intended recipient *deciphers* the received message via a *decryption algorithm*. The true receiver uses the corresponding decrypting key to recover the plaintext message.

The primary purpose of *cryptanalysis* is the use of mathematical techniques to overcome the embedded protection mechanism provided by cryptography. It tries to decode the ciphertext without any key information. *Cryptology* is the branch of knowledge which deals with both cryptography and cryptanalysis.

The word cryptography is derived from the Greek words *kryptos* and *logos*. The word kryptos means "hidden," and logos means "word." On a historical note, it should be noted that the word *cipher* is derived from the Latin word *cifra,* which in turn was derived from the Arabic *sifr*. The

Arabic word sifr has its origin in the Sanskrit word *sunya*, which means "empty" (or zero). We next enumerate: services provided by cryptography, canonical description of a cryptosystem, requirements of a cryptosystem, and different types of security offered by a cryptosystem.

Services Provided by Cryptography

Consider a canonical communication system, in which information is transmitted from the source to the receiver via a communication channel. Cryptographic techniques provide the following services to this communication ensemble.

(a) *Privacy or secrecy or confidentiality*: This is the nonavailability of the transmitted information to parties other than the intended parties.
(b) *Data integrity*: The recipient of the message should be able to verify that the message has neither been altered nor substituted in transit.
(c) *Authentication*: The receiver of the message should be able to ensure that the sender of the message is as claimed (*entity authentication*), and that the origin of the data is as specified (*data-origin authentication*).
(d) *Nonrepudiation*: The sender of a message should not falsely retract a message he or she has sent.

Canonical Description of a Cryptosystem

A *cryptosystem* is a mechanism which implements a cryptographic scheme. This terminology is elucidated via an example. Alice and Bob are two persons who want to communicate with each other over a transmission channel. This channel is assumed to be insecure. Assume that Alice wants to send Bob a message. This message is encrypted at Alice's transmitter, and decrypted at Bob's receiver. The message which Alice wants to send is called the plaintext. Alice encrypts this plaintext by using a key before transmitting it. The encrypted message which Bob receives is called the ciphertext. Bob decrypts the message because he is aware of the key used by Alice.

There is another person, called Oscar (or Eve) who would like to intercept this message. Oscar is called the adversary (enemy, intruder, opponent, or eavesdropper). Alice, Bob, and Oscar are sometimes referred to as entities in this chapter. The aim of the encryption scheme is to thwart eavesdropping of this message by Oscar. See Figure 3.1.

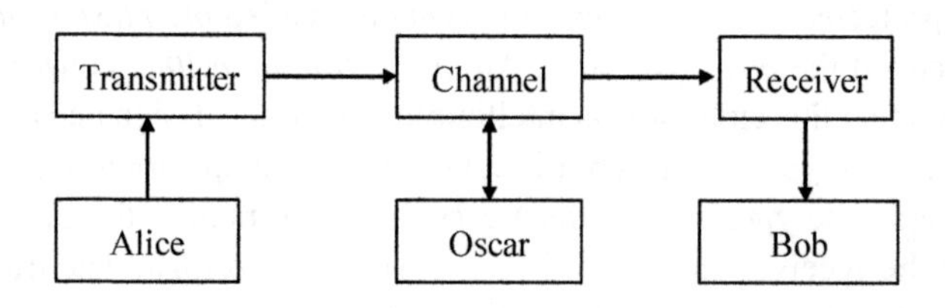

Figure 3.1. A communication system with Alice, Bob, and Oscar.

Definition 3.1. *A cryptosystem is a five-tuple* $(\mathcal{P}, \mathcal{C}, \mathcal{K}, \mathcal{E}, \mathcal{D})$ *where*:

(a) $\mathcal{P}$ *is the finite set of plaintexts.*
(b) $\mathcal{C}$ *is the finite set of ciphertexts.*
(c) $\mathcal{K}$ *is the finite set of keys. This is the key space.*

(d) *$\mathcal{E}$ is a family of invertible mappings indexed by $k \in \mathcal{K}$. These mappings are the encrypting algorithms. Let the mapping corresponding to the key k be denoted by $E_k \in \mathcal{E}$. For any $p \in \mathcal{P}$, let $E_k(p) = c$, where $c \in \mathcal{C}$.*

(e) *$\mathcal{D}$ is a family of invertible mappings indexed by $k \in \mathcal{K}$. These mappings are called decrypting algorithms. For each encrypting key $k \in \mathcal{K}$, there is a mapping $D_k \in \mathcal{D}$, such that $D_k(E_k(p)) = p$, for all $p \in \mathcal{P}$. Note that D_k is a decrypting algorithm corresponding to the key k.* □

Requirements of a Cryptosystem

Based upon the above definition, the following requirements for a cryptosystem can be stated.

(a) Secrecy of the key: The adversary should not be able to determine the value of a key k, by observing plaintext-ciphertext pairs.
(b) Secrecy of the message: An adversary should not be able to determine the plaintext p, by observing any ciphertext c.
(c) Authentication requirement: An adversary should not be able to determine a ciphertext c, for a given plaintext p.
(d) Confusion and diffusion: Confusion means that the effect of small changes in the plaintext p on the ciphertext c should not be easily predictable. Diffusion implies that the effect of small changes in the plaintext p should effect a substantial part of the ciphertext c.
(e) Size of the key space: In general, the larger the size of the key space $|\mathcal{K}|$, the harder is the task of the adversary to decode the message. However, a larger value of $|\mathcal{K}|$ by itself, does not guarantee an increase in security. The encryption algorithm also plays an important role in the security.

Each encrypting function should be an injective function (one-to-one). Furthermore, if $\mathcal{P} = \mathcal{C}$, then each encryption is merely a permutation operation. A basic assumption in cryptography is that the five-tuple $(\mathcal{P}, \mathcal{C}, \mathcal{K}, \mathcal{E}, \mathcal{D})$ is public knowledge. However, when two persons Alice and Bob want to communicate via classical cryptographic techniques, the encrypting key and the corresponding decrypting key are kept secret from Oscar the eavesdropper. Therefore, in order for two entities to secretly communicate with each other, this pair must initially determine the key secretly via a secure communication channel. This is indeed a paradox of classical cryptography.

Types of Security

A cryptosystem can be classified based upon the level of security it can offer. These are:

(a) *Unconditional security*: A cryptosystem is said to be unconditionally secure if it is unbreakable. In this cryptosystem, Oscar is allowed to possess an infinite amount of computational resources and time.
(b) *Complexity-theoretic security*: In this type of security, the adversary is assumed to have polynomial computational power. Polynomial attacks appear to be feasible in principle, but might be computationally infeasible in practice.
This type of security is generally theoretical and uses complexity theory for its assessment of security. However, it helps in understanding the power and limitations of the cryptographic scheme.

(c) *Computational security*: In a computationally secure cryptosystem, it takes a very long time (may be years) to decipher a message with or without the aid of computing machinery. This type of security is sometimes called *practical security*.
(d) *Perfect security*: A cryptographic communication system is perfectly secure if the ciphertext provides no information about the plaintext in a probabilistic sense.
(e) *Provable security*: A cryptosystem is said to be provably secure if its security can be related to some other system. This other system might have embedded in it a mathematically intractable problem. Evidently, such security is relative, and not absolute.
(f) *Ad hoc security*: This is also known as heuristic security. It puts forward several types of convincing arguments that every successful attack requires more resources than the ones available to Oscar the adversary. In this scenario, unforeseen attacks remain a threat to the security.

Basics of number theory, algebra, analysis, matrices and determinants, probability theory, and information theory are required to study this chapter. Reader should note that an identical symbol is used to denote an n-tuple and its row vector representation of length n.

3.2 Some Classical Ciphering Schemes

Some well-known ciphering schemes are discussed in this section. These are: the shift cipher, the affine cryptosystem, the polyalphabetic cipher, the Hill cipher, the permutation cipher, and one-time pad.

3.2.1 The Shift Cipher

The shift cipher is a very old ciphering scheme. This scheme is defined over integers modulo 26. These integers are the set of integers $\mathbb{Z}_{26} = \{0, 1, 2, \ldots, 25\}$. The number 26 is chosen because there are 26 letters in the English language. The letters $A, B, C, \ldots, Z$ are represented by the numbers $0, 1, 2, \ldots, 25$ respectively. There is no numerical representation for blank space. The shift cipher is defined below.

Definition 3.2. *The sets $\mathcal{P}, \mathcal{C}$, and $\mathcal{K}$ are each equal to $\mathbb{Z}_{26}$. For $p \in \mathcal{P}, c \in \mathcal{C}$, and $k \in \mathcal{K}$ define $c = E_k(p)$ where*

$$E_k(p) \equiv (p + k) \pmod{26} \tag{3.1a}$$

$$D_k(c) \equiv (c - k) \pmod{26} \tag{3.1b}$$

and $p, c \in \mathbb{Z}_{26}$. □

It can be observed in the above definition that indeed $D_k(E_k(p)) = p$. This cipher was supposedly used in ancient Rome by Julius Caesar. He chose a value of $k = 3$.

Example 3.1. The plaintext message

PEACE ON EARTH

becomes

15 4 0 2 4 14 13 4 0 17 19 7

The ciphering operation with $k = 3$ yields

18 7 3 5 7 17 16 7 3 20 22 10

After converting these numerics to letters the ciphertext is obtained. The text actually mailed by Alice is

S H D F H R Q H D U W K

When Bob receives this text, he converts it to numerical values and then performs the decrypting operation. Finally, the decrypted numbers are converted to text. For example, the first letter S is converted to a numerical value, which is 18. Its decrypted value is $(18 - 3) \pmod{26}$, which is 15 $\pmod{26}$. Its text equivalent is the letter P, which is indeed equal to the first letter of the plaintext message. The remaining ciphertext is similarly decrypted. □

We next ask ourselves the question: Is this ciphering system secure? The answer is negative for the following reasons.

(a) As there are only 26 possible values (modulo 26) of a key, the transmitted message can be decrypted by an exhaustive search of the key. The lesson learned from this scheme is that the key-space should be very large. This by itself may not be sufficient for a secure message.
(b) It has been known that the most frequently occurring letter in English language is e, followed by t, a, o, i, n, s, h, ... The original message can be recovered, by using this information and performing frequency analysis of the encrypted message.

3.2.2 The Affine Cryptosystem

The affine cryptosystem is a generalization of the shift cipher. Let $a \in \mathbb{Z}_{26}$, then $b \in \mathbb{Z}_{26}$ is said to be the multiplicative inverse of a if $ab \equiv ba \equiv 1 \pmod{26}$. The multiplicative inverse of a is sometimes denoted by a^{-1}. For arithmetic in $\mathbb{Z}_{26}$ it can be checked that $1^{-1} \equiv 1 \pmod{26}$, $3^{-1} \equiv 9 \pmod{26}$, and $5^{-1} \equiv 21 \pmod{26}$. The affine cryptosystem for English language text is defined below.

Definition 3.3. *The sets $\mathcal{P}$ and $\mathcal{C}$ are each equal to $\mathbb{Z}_{26}$. The key set is*

$$\mathcal{K} = \{(a, b) \mid a, b \in \mathbb{Z}_{26},\ \gcd(a, 26) = 1\} \tag{3.2a}$$

For $p \in \mathcal{P}$ and $c \in \mathcal{C}$, $c = E_k(p)$ where

$$E_k(p) \equiv (ap + b) \pmod{26} \tag{3.2b}$$
$$D_k(c) \equiv a^{-1}(c - b) \pmod{26} \tag{3.2c}$$

and a^{-1} is the multiplicative inverse of $a \in \mathbb{Z}_{26}$. □

The condition $\gcd(a, 26) = 1$ is required for $E_k(\cdot)$ to be a one-to-one function. It can be checked in the above definition that $D_k(E_k(p)) = p$.

Example 3.2. Consider the affine transformation with the key $k = (5, 19)$. It has been noted that $5^{-1} \equiv 21 \pmod{26}$. The encrypting and decrypting operations are:

$$E_k(p) \equiv (5p + 19) \pmod{26}$$
$$D_k(c) \equiv 21(c - 19) \pmod{26} \equiv (21c - 9) \pmod{26}$$

Note that $D_k(E_k(p)) \equiv \{21(5p + 19) - 9\} \pmod{26} \equiv p \pmod{26}$. Let the message received by Bob be

B L G T W H F T

Its numerical values are

1 11 6 19 22 7 5 19

Using the decrypting transformation $D_k(c) \equiv (21c - 9) \pmod{26}$ yields

12 14 13 0 11 8 18 0

The transmitted plaintext message is

MONA LISA

□

3.2.3 A Polyalphabetic Cipher

The shift cipher and the affine cipher are both *monoalphabetic cryptosystems*. This means that a key maps each alphabetic character to another unique character. These ciphers are susceptible to frequency based cryptanalytic attacks. Therefore these are not inherently safe. Block or *polyalphabetic* ciphers were invented to overcome this drawback. In block ciphers, mapping is performed on groups of letters. Batista Belaso first described this cipher in the year 1553. This cipher is generally named after Blaise de Vigenère.

This ciphering scheme encrypts m alphabetic characters at a time, where m is a positive integer greater than 1. Define a set of m-tuples in $\mathbb{Z}_{26}$ as

$$\mathbb{Z}_{26}^{(m)} = \{(n_1, n_2, \ldots, n_m) \mid n_i \in \mathbb{Z}_{26}, 1 \leq i \leq m\} \tag{3.3}$$

Notice that the total number of m-tuples is equal to 26^m.

Definition 3.4. *Let the sets $\mathcal{P}$, $\mathcal{C}$, and $\mathcal{K}$ be each equal to $\mathbb{Z}_{26}^{(m)}$, $m \geq 2$. Also let*

$$p = (p_1, p_2, \ldots, p_m) \in \mathcal{P} \tag{3.4a}$$
$$c = (c_1, c_2, \ldots, c_m) \in \mathcal{C} \tag{3.4b}$$
$$k = (k_1, k_2, \ldots, k_m) \in \mathcal{K} \tag{3.4c}$$

then $c = E_k(p)$ where

$$E_k(p) \equiv (p_1 + k_1, p_2 + k_2, \ldots, p_m + k_m) \pmod{26} \tag{3.5a}$$
$$D_k(c) \equiv (c_1 - k_1, c_2 - k_2, \ldots, c_m - k_m) \pmod{26} \tag{3.5b}$$

and the arithmetic is performed in $\mathbb{Z}_{26}$. □

The reader can verify that $D_k(E_k(p)) = p$. In this definition k is called a *keyword*. As there are 26^m keys, a cryptanalytic attack on this cipher system with exhaustive search is relatively more time consuming than either the shift cipher or the affine cryptosystem. Evidently, this cryptosystem degenerates to the shift cipher if m takes a value of unity.

Example 3.3. Let $m = 4$, and the keyword is CODE. The numerical equivalent of the keyword is $(2, 14, 3, 4)$. The plaintext to be encrypted is

SHERLOCK HOLMES

Its numerical equivalent is

18 7 4 17 11 14 2 10 7 14 11 12 4 18

Note that the plaintext has 14 letters, which is not a multiple of $m = 4$. Therefore two integers, each equal to 0 are appended to the numerical equivalent of the plaintext message. Thus the following message is encrypted.

18 7 4 17 11 14 2 10 7 14 11 12 4 18 0 0

The encryption definition yields

20 21 7 21 13 2 5 14 9 2 14 16 6 6 3 4

The beginning 4 numbers are obtained as follows:

$$(18 + 2) \equiv 20 \pmod{26}$$
$$(7 + 14) \equiv 21 \pmod{26}$$
$$(4 + 3) \equiv 7 \pmod{26}$$
$$(17 + 4) \equiv 21 \pmod{26}$$

The next 4 numbers are

$$(11 + 2) \equiv 13 \pmod{26}$$
$$(14 + 14) \equiv 2 \pmod{26}$$
$$(2 + 3) \equiv 5 \pmod{26}$$
$$(10 + 4) \equiv 14 \pmod{26}$$

The next 4 numbers are

$$(7 + 2) \equiv 9 \pmod{26}$$
$$(14 + 14) \equiv 2 \pmod{26}$$
$$(11 + 3) \equiv 14 \pmod{26}$$
$$(12 + 4) \equiv 16 \pmod{26}$$

Finally the last 4 numbers are

$$(4 + 2) \equiv 6 \pmod{26}$$
$$(18 + 14) \equiv 6 \pmod{26}$$
$$(0 + 3) \equiv 3 \pmod{26}$$
$$(0 + 4) \equiv 4 \pmod{26}$$

The corresponding ciphertext is

U V H V N C F O J C O Q G G D E

□

3.2.4 The Hill Cipher

The Hill cipher is also a polyalphabetic cryptosystem. It was invented in 1929 by the mathematician Lester S. Hill. This cipher also uses arithmetic in $\mathbb{Z}_{26}^{(m)}$, for $m \geq 2$. However, the keys in this ciphering scheme are $m \times m$ invertible matrices.

Definition 3.5. *Let the sets $\mathcal{P}$ and $\mathcal{C}$ be each equal to $\mathbb{Z}_{26}^{(m)}$, $m \geq 2$. Also*

$$\mathcal{K} = \{K \mid K \text{ is } m \times m \text{ invertible matrix over } \mathbb{Z}_{26}\} \tag{3.6a}$$

and

$$p = (p_1, p_2, \ldots, p_m) \in \mathbb{Z}_{26}^{(m)} \tag{3.6b}$$

$$c = (c_1, c_2, \ldots, c_m) \in \mathbb{Z}_{26}^{(m)} \tag{3.6c}$$

Then

$$c = E_K(p) = pK \tag{3.7a}$$

$$D_K(c) = cK^{-1} \tag{3.7b}$$

where the arithmetic is performed in $\mathbb{Z}_{26}$. □

It can be observed that $D_K(E_K(p)) = p$.

Example 3.4. Let $m = 2$, and the matrix key K be

$$K = \begin{bmatrix} 3 & 4 \\ 2 & 5 \end{bmatrix}$$

The inverse of this matrix K^{-1} in $\mathbb{Z}_{26}$ is

$$K^{-1} = \begin{bmatrix} 23 & 18 \\ 22 & 19 \end{bmatrix}$$

The plaintext to be encrypted is

SWEET HOME

Its numerical equivalent is

18 22 4 4 19 7 14 12 4

These values are taken two at a time and transformed. The pair $(18, 22)$ is encrypted as $(20, 0)$. This follows from

$$\begin{bmatrix} 18 & 22 \end{bmatrix} \begin{bmatrix} 3 & 4 \\ 2 & 5 \end{bmatrix} = \begin{bmatrix} 20 & 0 \end{bmatrix}$$

Similarly, the pair $(4, 4)$ is encrypted as $(20, 10)$; $(19, 7)$ is encrypted as $(19, 7)$; and $(14, 12)$ is encrypted as $(14, 12)$. Since the original text has odd number of characters, a 2-tuple is formed by appending a 0 to 4 (the last number in the numerical value of the plaintext). Therefore $(4, 0)$ is encrypted to $(12, 16)$. Thus the encrypted sequence of numbers is

20 0 20 10 19 7 14 12 12 16

The corresponding ciphertext is

U A U K T H O M M Q

Decryption of this ciphertext is performed by using the matrix K^{-1}. □

3.2.5 The Permutation Cipher

The permutation cipher is also known as the *transposition cipher*. This ciphering scheme keeps the plaintext characters unchanged, but transposes their positions within a specified block. This ciphering scheme is a special case of the Hill cipher. A formal definition of this scheme is given below.

Definition 3.6. *Let the sets $\mathcal{P}$ and $\mathcal{C}$ be each equal to $\mathbb{Z}_{26}^{(m)}$ for $m \geq 2$. The set of keys $\mathcal{K}$ consists of all the permutations of the members of the set $\{1, 2, \ldots, m\}$. Let*

$$p = (p_1, p_2, \ldots, p_m) \in \mathbb{Z}_{26}^{(m)} \tag{3.8a}$$

$$c = (c_1, c_2, \ldots, c_m) \in \mathbb{Z}_{26}^{(m)} \tag{3.8b}$$

For any key $k \in \mathcal{K}$

$$c_i = k(p_i), \ 1 \leq i \leq m \tag{3.9a}$$

$$p_i = k^{-1}(c_i), \ 1 \leq i \leq m \tag{3.9b}$$

where $k^{-1} \in \mathcal{K}$ is the inverse mapping of the permutation k. Note that, if $\widetilde{M}_k$ and $\widetilde{M}_{k^{-1}}$ are the matrix representations of the permutations k and k^{-1} respectively, then

$$c = p\widetilde{M}_k \tag{3.10a}$$

$$p = c\widetilde{M}_{k^{-1}} \tag{3.10b}$$

and $\widetilde{M}_k^{-1} = \widetilde{M}_{k^{-1}}$. □

Observe that there are no arithmetic operations performed in these encryption and decryption operations. Therefore, it is more convenient to use alphabetic characters directly, instead of numbers from the set $\mathbb{Z}_{26}^{(m)}$.

Example 3.5. Let $m = 5$, and the key k is the permutation

$$k = \begin{pmatrix} 1 & 2 & 3 & 4 & 5 \\ 3 & 5 & 1 & 2 & 4 \end{pmatrix}$$

The inverse permutation of the key k is

$$k^{-1} = \begin{pmatrix} 1 & 2 & 3 & 4 & 5 \\ 3 & 4 & 1 & 5 & 2 \end{pmatrix}$$

Let the plaintext be

FYODOR DOSTOYEVSKY

This text is grouped into groups of five letters. Notice that the text has 17 letters and requires four blocks to encrypt. The text is padded at the end by three letters to fill in the last three empty positions. The padded plaintext message is

FYODO RDOST OYEVS KYAAA

Each group of five letters is rearranged according to the permutation k. This results in

OOFYD OTRDS ESOYV AAKYA

Therefore the ciphertext is

OOFYDOTRDSESOYVAAKYA

Decryption of the received message can be done similarly, where the permutation k^{-1} is used. □

3.2.6 One-Time Pad

The one-time pad cryptosystem was discovered by George Vernam in 1917 for encryption and decryption of telegraph messages. This ciphering scheme encrypts m bits at a time, where $m \in \mathbb{P}$. Define a set of m-tuples in $\mathbb{Z}_2$ as

$$\mathbb{Z}_2^{(m)} = \{(n_1, n_2, \ldots, n_m) \mid n_i \in \mathbb{Z}_2, 1 \leq i \leq m\} \tag{3.11}$$

Notice that the total number of m-tuples is equal to 2^m.

Definition 3.7. *Let the sets $\mathcal{P}$, $\mathcal{C}$, and $\mathcal{K}$ be each equal to $\mathbb{Z}_2^{(m)}$, $m \in \mathbb{P}$. Also let*

$$p = (p_1, p_2, \ldots, p_m) \in \mathcal{P} \tag{3.12a}$$

$$c = (c_1, c_2, \ldots, c_m) \in \mathcal{C} \tag{3.12b}$$

$$k = (k_1, k_2, \ldots, k_m) \in \mathcal{K} \tag{3.12c}$$

then $c = E_k(p)$ where

$$E_k(p) \equiv (p_1 + k_1, p_2 + k_2, \ldots, p_m + k_m) \pmod 2 \tag{3.13a}$$

$$D_k(c) \equiv (c_1 - k_1, c_2 - k_2, \ldots, c_m - k_m) \pmod 2 \tag{3.13b}$$

and the arithmetic is performed in $\mathbb{Z}_2$. □

The reader can readily verify that $D_k(E_k(p)) = p$. This cryptosystem is remarkable for its simplicity. Note that the number of bits in the keyword is equal to the number of bits in the plaintext. Further, the keyword is easily recoverable once the palintext $p \in \mathcal{P}$ and the corresponding $c \in \mathcal{C}$ are known. It is for this very reason that in this cryptosystem a new keyword is used for each new encryption. Hence the name one-time pad. It also implies that a new keyword has to be generated for each new encryption and shared between Alice and Bob via a separate and secure communication channel. This results in inefficient and tedious key management. Nevertheless, this cryptographic scheme has been employed in diplomatic and military communications.

3.3 Cryptanalysis

Cryptanalysis is the art and science of determining the plaintext from ciphertext. Design of any new cryptographic scheme should take into account every possible cryptanalytic technique against which it is vulnerable. The practice of this art is carried by Oscar the enemy. In describing different cryptanalytic techniques, assume that Oscar has a knowledge of the cryptosystem being used, but not the ciphering key. This principle is generally called the *Kerckhoff's assumption* or *desiderata*, after the mathematician Auguste Kerckhoff (1835-1903). However, if Oscar is not aware of the cryptosystem being used, his job becomes doubly difficult. Oscar can attack a cryptosystem, in the following different ways.

(a) *Ciphertext-only attack*: Oscar possesses a string c of the ciphertext, where $c \in \mathcal{C}$. He tries to determine the corresponding plaintext and possibly the decryption key.
(b) *Known-plaintext attack*: Oscar knows for a $p \in \mathcal{P}$, the corresponding $c \in \mathcal{C}$; and he tries to derive the corresponding key.
(c) *Chosen-plaintext attack*: Oscar the enemy has a temporary access to the encrypting machinery. He selects a plaintext $p_i \in \mathcal{P}$ of his choice, and constructs the corresponding ciphertext $c_i \in \mathcal{C}$ for $1 \leq i \leq n$. Oscar's goal is to determine the corresponding encrypting key.
(d) *Chosen-ciphertext attack*: Oscar has a temporary access to the decrypting machinery. He selects a ciphertext $c_i \in \mathcal{C}$ of his choice, and constructs the corresponding plaintext $p_i \in \mathcal{P}$, for $1 \leq i \leq n$. Oscar's goal is to determine the corresponding decrypting key.

The above four types of attacks have been listed in increasing order of severity of the attack. The goal of Oscar is to determine the key in each of the four cases. A well-designed cryptosystem should be able to resist against both plaintext and ciphertext attacks.

Some of the possible cryptanalytic techniques are: statistical analysis of English language, and exhaustive key-search.

3.4 Information-Theoretic Cryptanalysis

In this section, a probabilistic notion of a cryptosystem is developed. It is then used to define perfect secrecy precisely. This is followed by applying entropy-theoretic concepts to this probabilistic cryptosystem. The merits of combining cryptosystems are then explored. Finally, language entropy, redundancy and unicity distance are discussed. Claude E. Shannon, founder of information theory, applied information-theoretic concepts to cryptography. Shannon also propounded that a cryptographic scheme is practically secure if it can be broken after $\widetilde{T}$ years, where $\widetilde{T}$ is large. The actual value of $\widetilde{T}$ depends upon the existing technology and the requirements of security.

3.4.1 Probabilistic Cryptosystem

Consider the cryptosystem $(\mathcal{P}, \mathcal{C}, \mathcal{K}, \mathcal{E}, \mathcal{D})$. Let

$$\begin{aligned}\mathcal{P} &= \{x_1, x_2, \ldots, x_m\}\\ \mathcal{C} &= \{y_1, y_2, \ldots, y_n\}\\ \mathcal{K} &= \{z_1, z_2, \ldots, z_l\}\end{aligned}$$

Thus

$$|\mathcal{P}| = m, \quad |\mathcal{C}| = n, \quad \text{and} \quad |\mathcal{K}| = l$$

Since it is possible to have the same ciphertext produced by different combinations of plaintext messages and keys, we have $n \leq lm$. Let X, Y, and Z be random variables which take values in the sets $\mathcal{P}$, $\mathcal{C}$, and $\mathcal{K}$ respectively. For convenience the following notation is used to denote the probabilities.

$$P(X = x_i) \triangleq p_X(x_i), \quad \forall\, x_i \in \mathcal{P}$$
$$P(Y = y_j) \triangleq p_Y(y_j), \quad \forall\, y_j \in \mathcal{C}$$
$$P(Z = z_k) \triangleq p_Z(z_k), \quad \forall\, z_k \in \mathcal{K}$$

It is also assumed that the above probabilities are nonzero, and that the plaintext and the key random-variables are independent of each other. The probability distributions of the plaintext and key random variables (X and Z respectively) are said to induce a probability distribution on the ciphertext random variable Y. For a key $z \in \mathcal{K}$ define

$$\widetilde{\mathcal{C}}(z) = \{E_z(x) \mid x \in \mathcal{P},\ E_z \in \mathcal{E}\}$$

Note that $\widetilde{\mathcal{C}}(z)$ is the set of candidate ciphertexts if the key z is used. Thus

$$\begin{aligned} & P(Y = y_j) \\ &= p_Y(y_j) \\ &= \sum_{\{z_k \mid z_k \in \mathcal{K},\ y_j \in \widetilde{\mathcal{C}}(z_k)\}} p_Z(z_k)\, p_X(D_{z_k}(y_j)), \quad \forall\, y_j \in \mathcal{C} \\ & P(Y = y_j \mid X = x_i) \\ &= \sum_{\{z_k \mid x_i = D_{z_k}(y_j), z_k \in \mathcal{K}\}} p_Z(z_k), \quad \forall\, x_i \in \mathcal{P},\ \forall\, y_j \in \mathcal{C} \end{aligned}$$

Recall that $D_{z_k}(\cdot) \in \mathcal{D}$ is the decrypting function which corresponds to the key $z_k \in \mathcal{K}$. Also

$$\begin{aligned} & P(X = x_i \mid Y = y_j) \\ &= \frac{P(X = x_i)\, P(Y = y_j \mid X = x_i)}{P(Y = y_j)}, \quad \forall\, x_i \in \mathcal{P},\ \forall\, y_j \in \mathcal{C} \end{aligned}$$

where $P(Y = y_j) \neq 0, \forall\, y_j \in \mathcal{C}$. Consider the following illustrative example.

Example 3.6. In a cryptosystem

$$\mathcal{P} = \{x_1, x_2, x_3\}, \quad \mathcal{C} = \{y_1, y_2, y_3, y_4\}, \quad \mathcal{K} = \{z_1, z_2\}$$
$$p_X(x_1) = \frac{1}{2}, \ p_X(x_2) = \frac{1}{4}, \ p_X(x_3) = \frac{1}{4}$$
$$p_Z(z_1) = \frac{3}{4}, \ p_Z(z_2) = \frac{1}{4}$$

The encryption mapping is

$$E_{z_1}(x_1) = y_1, \ E_{z_1}(x_2) = y_3, \ E_{z_1}(x_3) = y_2$$
$$E_{z_2}(x_1) = y_2, \ E_{z_2}(x_2) = y_4, \ E_{z_2}(x_3) = y_3$$

The computed values of the probabilities $p_Y(y_j)$, for $1 \leq j \leq 4$ are as follows.

$$p_Y(y_1) = \frac{3}{8}, \ p_Y(y_2) = \frac{5}{16}, \ p_Y(y_3) = \frac{1}{4}, \ p_Y(y_4) = \frac{1}{16}$$

The probabilities $P(X = x_i \mid Y = y_j)$ for $1 \leq i \leq 3$, and $1 \leq j \leq 4$ are next computed.

$$
\begin{aligned}
&P(X = x_1 \mid Y = y_1) = 1,\ P(X = x_2 \mid Y = y_1) = 0, \\
&P(X = x_3 \mid Y = y_1) = 0 \\
&P(X = x_1 \mid Y = y_2) = \frac{2}{5},\ P(X = x_2 \mid Y = y_2) = 0, \\
&P(X = x_3 \mid Y = y_2) = \frac{3}{5} \\
&P(X = x_1 \mid Y = y_3) = 0,\ P(X = x_2 \mid Y = y_3) = \frac{3}{4}, \\
&P(X = x_3 \mid Y = y_3) = \frac{1}{4} \\
&P(X = x_1 \mid Y = y_4) = 0,\ P(X = x_2 \mid Y = y_4) = 1, \\
&P(X = x_3 \mid Y = y_4) = 0
\end{aligned}
$$

Notice that the sum of probabilities in each row in the above set of equations is equal to unity. The probabilities $P(Z = z_k \mid Y = y_j)$ for $1 \le k \le 2$, and $1 \le j \le 4$ are also computed below.

$$
\begin{aligned}
&P(Z = z_1 \mid Y = y_1) = 1,\ P(Z = z_2 \mid Y = y_1) = 0 \\
&P(Z = z_1 \mid Y = y_2) = \frac{3}{5},\ P(Z = z_2 \mid Y = y_2) = \frac{2}{5} \\
&P(Z = z_1 \mid Y = y_3) = \frac{3}{4},\ P(Z = z_2 \mid Y = y_3) = \frac{1}{4} \\
&P(Z = z_1 \mid Y = y_4) = 0,\ P(Z = z_2 \mid Y = y_4) = 1
\end{aligned}
$$

The sum of probabilities in each row in the above set of equations is also equal to unity. □

3.4.2 Perfect Secrecy

Perfect secrecy implies that Oscar the opponent cannot obtain any information about the transmitted text by simply observing the ciphertext in this probabilistic framework.

Definition 3.8. *Consider a cryptosystem* $(\mathcal{P}, \mathcal{C}, \mathcal{K}, \mathcal{E}, \mathcal{D})$. *The random variables* X *and* Y *take values in the sets* $\mathcal{P}$ *and* $\mathcal{C}$ *respectively. This cryptosystem has perfect secrecy, if*

$$P(X = x \mid Y = y) = P(X = x),\ P(Y = y) > 0,\ \forall\, x \in \mathcal{P},\ \forall\, y \in \mathcal{C} \tag{3.14}$$

□

Perfectly secret cryptosystems rarely occur in practice.

Observations 3.1. Let a cryptosystem have perfect secrecy.

1. $P(Y = y \mid X = x) = P(Y = y)$, $\forall\, x \in \mathcal{P}$ and $\forall\, y \in \mathcal{C}$.
2. $|\mathcal{P}| \le |\mathcal{C}| \le |\mathcal{K}|$.
3. Perfect secrecy is achieved by the one-time pad cryptosystem if the random variable corresponding to the keyword is uniformly distributed. □

The first observation follows from the definition of perfect secrecy of a cryptosystem. The next two observations are established in the problem section. The following result about a perfectly secret cryptosystem is due to Claude E. Shannon.

Theorem 3.1. *Let* $(\mathcal{P}, \mathcal{C}, \mathcal{K}, \mathcal{E}, \mathcal{D})$ *be a cryptosystem, such that* $|\mathcal{P}| = |\mathcal{C}| = |\mathcal{K}| = m$. *The random variables* X, Y *and* Z *take values in the sets* $\mathcal{P}$, $\mathcal{C}$, *and* $\mathcal{K}$ *respectively. Assume that the random variables* X *and* Z *are independent of each other. This cryptosystem has perfect secrecy iff every key is used with probability* m^{-1}, *and for every value of* $x \in \mathcal{P}$ *and every value of* $y \in \mathcal{C}$, *there is a unique key* $z \in \mathcal{K}$, *such that* $E_z(x) = y$.

Proof. See the problem section. □

3.4.3 Entropy

The concept of entropy developed in the chapter on information theory can be applied to the cryptosystem $(\mathcal{P}, \mathcal{C}, \mathcal{K}, \mathcal{E}, \mathcal{D})$. Let X, Y and Z be the random variables which take values in the sets $\mathcal{P}$, $\mathcal{C}$, and $\mathcal{K}$ respectively. Assume that the random variables X and Z are independent of each other. Let the entropy of the random variables X, Y and Z be $H(X), H(Y)$ and $H(Z)$ respectively. The conditional entropy $H(Z \mid Y)$, is called the *key-equivocation.* It is a measure of the uncertainty remaining in the key, given that the ciphertext is known.

Observations 3.2. The following observations can be made about these entropies.

1. $H(X \mid Z) = H(X)$. This result follows from the observation that the random variables X and Z are independent of each other.
2. For a cryptosystem with perfect secrecy:
   (a) $H(X \mid Y) = H(X)$. This statement is often taken as the definition of perfect secrecy. It implies that the eavesdropper learns absolutely nothing about the plaintext from a knowledge of the ciphertext.
   (b) Similarly $H(Y \mid X) = H(Y)$.
3. $H(Z \mid Y) = H(X) + H(Z) - H(Y)$. This result is called the *key-equivocation principle* (or theorem). □

See the problem section for a proof of the key-equivocation principle.

Example 3.7. For the probabilities given in the previous example, it can be determined that $H(X) = 1.5, H(Z) = 0.81128$, and $H(Y) = 1.8054$. Using the probabilities

$$P(Z = z_k \mid Y = y_j), \quad 1 \leq k \leq 2, \ 1 \leq j \leq 4$$

$H(Z \mid Y)$ can be explicitly evaluated to be equal to 0.50624. It can be numerically checked that these entropies indeed satisfy the key-equivocation principle. □

3.4.4 Combination of Cryptosystems

Assume that two cryptosystems Θ_1 and Θ_2 are given. A new cryptosystem Θ can be obtained by combining these two cryptosystems. This new cryptosystem can have higher security than either Θ_1 or Θ_2. Furthermore, this is a practical technique to design a cryptosystem with higher security. Three types of such cryptosystems are outlined below. These are: weighted sum of cryptosystems, product of cryptosystems, and hybrid of weighted and product type of cryptosystems.

Weighted Sum of Cryptosystem

Let

$$\Theta_1 = (\mathcal{P}, \mathcal{C}_1, \mathcal{K}_1, \mathcal{E}_1, \mathcal{D}_1), \text{ and } \Theta_2 = (\mathcal{P}, \mathcal{C}_2, \mathcal{K}_2, \mathcal{E}_2, \mathcal{D}_2)$$

be two cryptosystems with the same plaintext set $\mathcal{P}$. A weighted cryptosystem is defined as $\Theta = p\Theta_1 + q\Theta_2$, where $0 < p < 1$ and $q = (1 - p)$. The cryptosystem $\Theta = (\mathcal{P}, \mathcal{C}, \mathcal{K}, \mathcal{E}, \mathcal{D})$ is obtained by selecting Θ_1 with probability p, and the Θ_2 with probability q. It can also be observed that $|\mathcal{K}| = (|\mathcal{K}_1| + |\mathcal{K}_2|)$. This technique can clearly be extended to the weighted sum of more than two cryptosystems.

Product Cryptosystem

For simplicity assume that Θ_1 and Θ_2 are two cryptosystems with the same plaintext and the ciphertext set $\mathcal{P}$. That is, $\Theta_1 = (\mathcal{P}, \mathcal{P}, \mathcal{K}_1, \mathcal{E}_1, \mathcal{D}_1)$, and $\Theta_2 = (\mathcal{P}, \mathcal{P}, \mathcal{K}_2, \mathcal{E}_2, \mathcal{D}_2)$. The product cryptosystem $\Theta_1 \times \Theta_2$ is defined as $\Theta = (\mathcal{P}, \mathcal{P}, \mathcal{K}, \mathcal{E}, \mathcal{D})$, with the following properties.

(a) $\mathcal{K} = \mathcal{K}_1 \times \mathcal{K}_2, k = (k_1, k_2) \in \mathcal{K}$, where $k_1 \in \mathcal{K}_1, k_2 \in \mathcal{K}_2$.
(b) Encryption is done as: $E_k(p) = E_{k_2}(E_{k_1}(p)), p \in \mathcal{P}, E_{k_1} \in \mathcal{E}_1, E_{k_2} \in \mathcal{E}_2$, and $E_k \in \mathcal{E}$.
(c) Decryption is done as: $D_k(c) = D_{k_1}(D_{k_2}(c)), c \in \mathcal{P}, D_{k_1} \in \mathcal{D}_1, D_{k_2} \in \mathcal{D}_2$, and $D_k \in \mathcal{D}$.
(d) $|\mathcal{K}| = |\mathcal{K}_1| \, |\mathcal{K}_2|$.
(e) In such cryptosystems, the ciphertext set of Θ_1 should be the same as the plaintext set of Θ_2.

The product of more than two cryptosystem can also be extended suitably.

Hybrid Weighted-Product Cryptosystems

Let Θ_1, Θ_2, and Θ_3 be three cryptosystems. Also let $0 < p < 1$, and $q = (1 - p)$. Then the following cryptosystems are well-defined.

(a) $\Theta_3 \times (p\Theta_1 + q\Theta_2)$
(b) $(p\Theta_1 + q\Theta_2) \times \Theta_3$

These composite hybrid cryptosystems have the following properties:

(a) $\Theta_3 \times (p\Theta_1 + q\Theta_2) = p(\Theta_3 \times \Theta_1) + q(\Theta_3 \times \Theta_2)$
(b) $(p\Theta_1 + q\Theta_2) \times \Theta_3 = p(\Theta_1 \times \Theta_3) + q(\Theta_2 \times \Theta_3)$
(c) $\Theta_1 \times (\Theta_2 \times \Theta_3) = (\Theta_1 \times \Theta_2) \times \Theta_3$
(d) $\Theta_1 \times \Theta_2 \neq \Theta_2 \times \Theta_1$ in general.

3.4.5 Unicity Distance

Given a string of ciphertext symbols, how many keys are required on the average to decrypt it to a meaningful text? An attempt is made to answer this question via entropy-theoretic concepts. In the cryptosystem $(\mathcal{P}, \mathcal{C}, \mathcal{K}, \mathcal{E}, \mathcal{D})$, let X, Y and Z be the random variables which take values in the sets $\mathcal{P}$, $\mathcal{C}$, and $\mathcal{K}$ respectively. Recall that the random variables X and Z are independent of each other. Also, the entropies of the random variables X, Y and Z are $H(X), H(Y)$ and $H(Z)$ respectively.

Recall from an earlier subsection on entropies that, $H(Z \mid Y)$ was defined as key-equivocation. It measures the information disclosed by the ciphertext about the key. If $H(Z \mid Y) = 0$, then Oscar the enemy has broken the cryptosystem. However if $H(Z \mid Y) > 0$, then more than one key is a candidate for the mapping of the encryption scheme. Of the several candidates for the key, only one key is a correct key, and the remaining keys are called *spurious keys*.

An example of the plaintext might be the English language. A lower bound on the expected number spurious keys is next obtained. This is accomplished by first defining the entropy of a natural language.

Definition 3.9. *Denote a natural language by L. Let $X_i, i = 1, 2, 3, \ldots$ be random variables which take values in the set of plaintexts $\mathcal{P}$. The entropy and redundancy of L are denoted by H_L and R_L respectively.*

$$H_L \triangleq \lim_{N\to\infty} \frac{H(X_1, X_2, \ldots, X_N)}{N} \tag{3.15a}$$

$$R_L \triangleq 1 - \frac{H_L}{\log_2 |\mathcal{P}|} \tag{3.15b}$$

□

A random language has an entropy of $H_L = \log_2 |\mathcal{P}|$, and $R_L = 0$. For English language, $|\mathcal{P}| = 26$, $\log_2 |\mathcal{P}| \approx 4.70$, $H(X) \approx 4.19$, $1.0 \le H_L \le 1.5$, and $R_L \approx 0.75$. Note that the values of $H(X)$ and H_L are obtained experimentally. These in turn are derived from the probability distribution of the letters and their strings.

Consider N-tuple random variables

$$X^{(N)} = (X_1, X_2, \ldots, X_N)$$
$$Y^{(N)} = (Y_1, Y_2, \ldots, Y_N)$$

Instances of $X^{(N)}$ and $Y^{(N)}$ are $x^{(N)}$ and $y^{(N)}$ respectively, where

$$x^{(N)} = (x_{i_1}, x_{i_2}, \ldots, x_{i_N}),\ x_{i_\omega} \in \mathcal{P},\ 1 \le \omega \le N$$
$$y^{(N)} = (y_{j_1}, y_{j_2}, \ldots, y_{j_N}),\ y_{j_\omega} \in \mathcal{C},\ 1 \le \omega \le N$$

Define the set

$$\kappa\left(y^{(N)}\right) = \{z \mid z \in \mathcal{K},\ E_z(x_{i_\omega}) = y_{j_\omega},\ x_{i_\omega} \in \mathcal{P},\ y_{j_\omega} \in \mathcal{C},\ 1 \le \omega \le N\}$$

Note that $\kappa\left(y^{(N)}\right)$ is the set of possible keys used in enciphering the plaintext N-tuples. Assume that $P\left(X^{(N)}\right) > 0$. Let S_N be the number of spurious keys observed by N-tuple ciphered symbols. If $y^{(N)}$ is the observed N-tuple of the ciphertext, then the number of spurious keys is $\left\{\left|\kappa\left(y^{(N)}\right)\right| - 1\right\}$, as only one of the "possible" keys is the right key. Denote the expected (average) number of spurious keys by ϱ_N. This yields

$$\varrho_N = \sum_{y^{(N)}} P\left(Y^{(N)} = y^{(N)}\right) \left\{\left|\kappa\left(y^{(N)}\right)\right| - 1\right\}$$

Thus

$$1 + \varrho_N = \sum_{y^{(N)}} P\left(Y^{(N)} = y^{(N)}\right) \left|\kappa\left(y^{(N)}\right)\right|$$

The following theorem states a bound on the expected number of spurious keys obtained by observing N ciphered symbols. It relates the expected number of spurious keys to the size of the plaintext set, the size of the key space, the redundancy of the language, and the number of observed ciphered symbols in a tuple.

Theorem 3.2. *Consider a cryptosystem $(\mathcal{P}, \mathcal{C}, \mathcal{K}, \mathcal{E}, \mathcal{D})$, where the keys occur with equal probability, and $|\mathcal{P}| = |\mathcal{C}|$. Denote the redundancy of the plaintext, and the number of spurious keys observed by N ciphered symbols by R_L and S_N respectively. The average number of spurious keys is ϱ_N. For sufficiently large values of N*

$$\varrho_N \geq \frac{|\mathcal{K}|}{|\mathcal{P}|^{NR_L}} - 1 \tag{3.16}$$

Proof. The key-equivocation principle can be extended to N symbols. Thus

$$H\left(Z \mid Y^{(N)}\right) = H\left(X^{(N)}\right) + H(Z) - H\left(Y^{(N)}\right)$$

Also

$$\begin{aligned}
H\left(Y^{(N)}\right) &\leq N \log_2 |\mathcal{C}| = N \log_2 |\mathcal{P}| \\
H\left(X^{(N)}\right) &\simeq NH_L = N(1 - R_L)\log_2 |\mathcal{P}| \\
H(Z) &\simeq \log_2 |\mathcal{K}|
\end{aligned}$$

The last result follows from the equi-probability of the key use. Also

$$\begin{aligned}
H\left(Z \mid Y^{(N)}\right) &= \sum_{y^{(N)}} P\left(Y^{(N)} = y^{(N)}\right) H\left(Z \mid Y^{(N)} = y^{(N)}\right) \\
&\leq \sum_{y^{(N)}} P\left(Y^{(N)} = y^{(N)}\right) \log_2 \left|\kappa\left(y^{(N)}\right)\right| \\
&\leq \log_2 \left\{ \sum_{y^{(N)}} P\left(Y^{(N)} = y^{(N)}\right) \left|\kappa\left(y^{(N)}\right)\right| \right\} \\
&= \log_2 (1 + \varrho_N)
\end{aligned}$$

The last inequality uses Jensen's result. Combining the above results yields

$$\log_2 (1 + \varrho_N) \geq N(1 - R_L)\log_2 |\mathcal{P}| + \log_2 |\mathcal{K}| - N \log_2 |\mathcal{P}|$$

The stated result follows. □

Note that ϱ_N approaches 0 as N increases. It should be noted that this may not be an accurate bound for small values of N.

Unicity distance is next defined. It is the value of N at which the expected number of spurious keys approaches zero. Thus unicity distance N_0 of a cryptosystem is the length of ciphertext at which there is possibly a unique meaningful plaintext that can be determined by Oscar. Therefore, if Oscar the enemy has infinite computing power, unicity distance is equal to the average number of ciphertext symbols required to uniquely determine the key.

Definition 3.10. *The unicity distance of a cryptosystem is defined to be the value of N, for which the average number of spurious keys ϱ_N is equal to zero. It is denoted by N_0.* □

It follows that

$$N_0 \simeq \frac{\log_2 |\mathcal{K}|}{R_L \log_2 |\mathcal{P}|}$$

The above result should only be used to provide a rough approximate value of N_0.

Example 3.8. The unicity distance of a polyalphabetic cryptosystem is estimated, where the keyword length is 4, and the plaintext symbols are the letters of the English language. Assume that the language redundancy $R_L = 0.75$.

Note that $|\mathcal{P}| = 26$, and $|\mathcal{K}| = 26^4$. Substituting in the expression for N_0 results in

$$N_0 \approx \frac{4 \log_2 26}{0.75 \log_2 26} \approx 5.333 \text{ ciphertext symbols}$$

Therefore, a ciphertext string of approximate length 6 is required on the average to determine the key uniquely. This result is obviously a rough estimate. However, it clearly indicates that only a few ciphertext symbols are necessary to determine a unique decryption in several cases. □

3.5 Public-Key Cryptography

The basic idea behind public-key cryptography was proposed by W. Diffie and M. Hellman in 1976. Their work is regarded as a major milestone in the science of public-key cryptography. It addresses the basic problem of secure communication between members of a group. Each member of the group has two keys: one public key and another private key. The public key is known to all members of the group. In contrast, the private key of a member is known only to the member, and no one else. Assume that Alice and Bob are members of this group. If Alice wants to send a message to Bob, she uses the public key of Bob to encrypt her message. When Bob receives her message, he uses his private key to decrypt this message. The fundamental assumption in this scheme is that, if public key of a person is known, then it is computationally difficult to determine the corresponding private key. Some terminology is next introduced more precisely.

If in a cryptographic scheme, the decryption key d is either equal to or easily derivable from the encryption key k, then the cryptographic scheme is said to be *symmetric-key*. Such schemes are also called *private-key cryptosystems*, as the keys are private to the users. The shift cipher is an example of symmetric-key cryptosystem. One disadvantage of this scheme is that the key between Alice and Bob has to be initially shared via a secure communication channel prior to any form of cryptographic communication between them.

In *asymmetric-key cryptosystems*, the enciphering key for coding a message destined for a particular person is determined from a published public file. However the corresponding deciphering key cannot be determined in a computationally feasible time by simply a knowledge of the enciphering key. Therefore anyone in public can perform the encryption transformation, however the decryption operation is nearly impractical for anyone other than the intended person (or party). Consequently, in this system, each user has two keys: *a public key* and a *private key*.

More precisely, in asymmetric-key cryptosystems, the encryption key k and the decryption key d are different. Furthermore, the computation of d from k is computationally infeasible. Therefore the encryption key k can be made public. If Bob wants to receive encrypted messages, he publishes his encryption key k, and does not disclose the corresponding decrypting key d to anyone. Anyone with access to Bob's public key k can send an encrypted message to Bob. Therefore, the key k is called the public key. The key d is private to Bob, consequently it is called the private or secret key. Such asymmetric-key cryptosystems are also called *public-key cryptosystems.*

The concept of public-key cryptosystem is based upon *trapdoor one-way function.* A basic understanding of *one-way function* and trapdoor one-way function is essential to the understanding of public-key cryptography. The following definitions are intuitive and not meant to be rigorous.

Definition 3.11. *A function $g(\cdot)$ is defined from the set T to the set S. It is said to be a one-way function if $s = g(t)$ is "relatively easy" to compute for all values of $t \in T$, where $s \in S$. However, the computation of t for "almost all" values of s in the image of $g(\cdot)$ is "computationally expensive" for all values of $t \in T$ such that $s = g(t)$.* □

In this definition, the phrases "relatively easy" and "computationally expensive" are not defined rigorously.

Example 3.9. Let

$$T = S = \{1, 2, 3, 4, 5, 6\} \text{ and } s = g(t), \text{ where } g(t) \equiv 3^t \pmod 7$$

The (t, s)-tuples are $(1,3), (2,2), (3,6), (4,4), (5,5)$, and $(6,1)$. Given a number $t \in T$, it is easy to compute $g(t)$. However if a value of s is specified, then it is generally not straightforward to compute t, unless the complete (t, s)-tuple list is known. □

Definition 3.12. *A function $g(\cdot)$ is defined from the set T to the set S. It is said to be a trapdoor one-way function if it is a one-way function with the additional information (sometimes called the trapdoor information) that makes it "computationally feasible" to compute t for "almost all" values of s in the image of $g(\cdot)$ for all values of $t \in T$ such that $s = g(t)$.* □

The significance of one-way and trapdoor one-way functions will become transparent in the following subsections.

3.5.1 The RSA Cryptosystem

The RSA cryptosystem is the most well-known and widely used public-key cryptosystem. It is named after its inventors R. Rivest, A. Shamir, and L. Adleman. This cryptosystem was invented in 1978. The security of this cryptosystem is based upon the difficulty of factoring integers. The RSA cryptosystem performs arithmetic in $\mathbb{Z}_n$, where $n = pq$, and p and q are two distinct odd prime numbers. Therefore $\varphi(n) = (p-1)(q-1)$, where $\varphi(\cdot)$ is the Euler's totient function. The complete description of the RSA algorithm consists of a procedure for key generation, and procedures for encryption and decryption. Each member (entity) of the group which wants to receive a secure message generates his or her public and private keys. Thus each entity A of the group should execute the following algorithm.

Algorithm 3.1a. *Key Generation for the RSA Cryptographic Scheme.*

Output: Key generation for entity A. That is, generate A's public and private keys.
begin

Step 1: Generate two large, random and distinct prime numbers p and q. These two prime numbers should be approximately of the same size.
Step 2: Compute $n = pq$ and $\varphi(n) = (p-1)(q-1)$.
Step 3: Generate a random integer b, such that $1 < b < \varphi(n)$ and $\gcd(b, \varphi(n)) = 1$.
Step 4: Compute a such that $ab \equiv 1 \pmod{\varphi(n)}$.
This is accomplished by using the extended Euclidean algorithm.
Step 5: A's public key is (n, b) and the private key is (p, q, a).

end (*end of RSA key generation algorithm*)

The next algorithm is for encryption and decryption operations of the RSA cryptographic scheme.

Algorithm 3.1b. *Encryption and Decryption in the RSA Cryptographic Scheme.*

Specification: The RSA cryptosystem is $(\mathcal{P}, \mathcal{C}, \mathcal{K}, \mathcal{E}, \mathcal{D})$.
Let $\mathcal{P} = \mathcal{C} = \mathbb{Z}_n$, where n is the product of two odd prime numbers p and q, of comparable size. Also let

$$\mathcal{K} = \{(n, p, q, a, b) \mid n = pq; 1 < b < \varphi(n), \gcd(b, \varphi(n)) = 1; \\ ab \equiv 1 \pmod{\varphi(n)}\},$$

and $K \triangleq (n, p, q, a, b) \in \mathcal{K}$.
A message $x \in \mathcal{P}$ is encrypted into ciphertext $y \in \mathcal{C}$.

Encryption Operation. Entity B encrypts a message $x \in \mathcal{P}$ destined for entity A.
Input: The plaintext message $x \in \mathcal{P}$ of entity B.
Output: Entity B transmits, and entity A receives $y \in \mathcal{C}$.
begin

Step 1: Entity B obtains entity A's public key (n, b). The entity B wants to transmit the message $x \in \mathcal{P}$ to entity A.
Step 2: It computes $E_K(x) \equiv x^b \pmod{n} \equiv y \pmod{n}$.
Step 3: Entity B transmits the ciphertext $y \in \mathcal{C}$ to the entity A.

end (*end of encryption operation*)

Decryption Operation. Entity A decrypts the message $y \in \mathcal{C}$.
Input: The encrypted message $y \in \mathcal{C}$ with entity A.
Output: Entity A generates the plaintext message $x \in \mathcal{P}$.
begin
Step 1: Entity A receives the encrypted message y.
Step 2: Entity A uses its private key to compute
$D_K(y) \equiv y^a \pmod n$. This value is equal to $x \in \mathcal{P}$.
end (*end of decryption operation*)

Note that the arithmetic performed in these operations is in $\mathbb{Z}_n$. The number n is called the *RSA modulus*, b is called the *encryption exponent*, and a is called the *decryption exponent.* Observe that the decryption exponent a can be computed from the encryption exponent b via the extended Euclidean algorithm, provided $\varphi(n)$ or the prime factors p and q are known. However it is computationally easier to compute a from a knowledge of p and q, and not directly from either n or $\varphi(n)$.

This cryptosystem has a one-way function with a trapdoor. $E_K(\cdot)$ is a one-way trapdoor function with the trapdoor a. This follows from the fact that it is computationally infeasible to determine $x \in \mathcal{P}$ from $y \in \mathcal{C}$ without the knowledge of the trapdoor $a \in \mathbb{Z}_n$. The exponentiation process in this algorithm can be implemented via a binary expansion of the exponent and the principle of repeated-squaring operation. This algorithm has been discussed in the chapter on number theory. Furthermore, the multiplicative inverse of $b \in \mathbb{Z}_{\varphi(n)}$, which is equal to a, can be obtained via the extended Euclidean algorithm.

Proof of the correctness of the algorithm. It has to be established that $D_K(y) = x$ for $x \in \mathbb{Z}_n$. If $x = 0$, then $y = 0$. In this case the verification is evident. Assume that $x \in \mathbb{Z}_n \backslash \{0\}$.

Observe that $ab \equiv 1 \pmod{\varphi(n)}$ implies $ab = (k\varphi(n) + 1)$ for an integer k. This implies

$$\left(x^b\right)^a = x^{ab} = x^{k\varphi(n)+1} = x\left(x^{k(p-1)(q-1)}\right)$$

It can be claimed that

$$x\left(x^{k(p-1)(q-1)}\right) \equiv x \pmod p$$

If $\gcd(x,p) = 1$, then the result follows from Fermat's little theorem. However if $\gcd(x,p) = p$, then both sides of the congruence are equal to $0 \pmod p$. Similarly it can be proved that $\left(x^b\right)^a \equiv x \pmod q$. Therefore $\left\{\left(x^b\right)^a - x\right\}$ is an integer multiple of both p and q. Since p and q are distinct prime numbers, it follows that $\left\{\left(x^b\right)^a - x\right\}$ is an integer multiple of $pq = n$. This observation yields

$$\left(x^b\right)^a \equiv x \pmod n$$

The last step is also a especial application of the Chinese remainder theorem. The final result follows. □

The use of this scheme is demonstrated in the following example. The values chosen are small for the purpose of illustration. This cipher is definitely not secure.

Example 3.10. Let $p = 7$ and $q = 11$, then $n = 77$ and $\varphi(n) = 60$. Let $b = 37$, where $1 < b < \varphi(n)$ and $\gcd(b, \varphi(n)) = 1$. The value of $a = 13$. It can be checked that $ab \equiv 1 \pmod{\varphi(n)}$. The public key is $(n, b) = (77, 37)$ and the private key is $(p, q, a) = (7, 11, 13)$.

The message to be coded is $x \in \mathbb{Z}_{77}$. Let $x = 18$, then

$$y = E_K(x) \equiv x^b \pmod{n} \equiv 18^{37} \pmod{77} \equiv 39 \pmod{77}$$

Therefore the ciphered value of 18 is 39. It can be verified that the original value of x can be recovered.

$$D_K(y) \equiv y^a \pmod{n} \equiv 39^{13} \pmod{77} \equiv 18 \pmod{77} = x$$

□

Possible Methods of Attacking the RSA Cryptosystem

Assume that the entity which wishes to break the RSA cryptosystem has the following information.

(a) The knowledge of the public key, which is (n, b).
(b) Several pairs of plaintext and the corresponding ciphertext. These pairs are

$$(x_i, y_i), \quad 0 \le x_i, y_i \le (n-1), \quad i = 1, 2, \ldots$$

The goal of this person is to determine the private key, (p, q, a). Two possible approaches for attacking this cryptosystem are:

(a) *Factorization of n.* The private key is completely determined, once the integer n is factored. As the factors of n are p and q, Euler's totient-function $\varphi(\cdot)$ can be computed easily. Subsequently the integer a can be computed via extended Euclidean algorithm.
(b) *Knowledge of $\varphi(n)$ and n yields p and q.* In this method, assume that Euler's totient-function $\varphi(\cdot)$ is some how determined. Once $\varphi(n)$ is known, p and q can be determined. Observe that

$$\begin{aligned}(p+q) &= (n+1-\varphi(n)) \\ (p-q) &= \left\{(p+q)^2 - 4n\right\}^{1/2} \\ p &= \frac{1}{2}\{(p+q)+(p-q)\} \\ q &= \frac{1}{2}\{(p+q)-(p-q)\}\end{aligned}$$

Therefore let

$$\begin{aligned}\widetilde{\alpha} &= (n+1-\varphi(n)), \\ \widetilde{\beta} &= \left\{\widetilde{\alpha}^2 - 4n\right\}^{1/2}\end{aligned}$$

Then

$$\begin{aligned}p &= \frac{1}{2}\left(\widetilde{\alpha} + \widetilde{\beta}\right) \\ q &= \frac{1}{2}\left(\widetilde{\alpha} - \widetilde{\beta}\right)\end{aligned}$$

3.5.2 The El Gamal Cryptosystem

The El Gamal public-key cryptosystem was invented by T. El Gamal in 1985. Its security is based upon the difficulty of computing discrete logarithm. Recall that a group $\mathcal{G} = (G, *)$ is said to be cyclic, if there exists an element $a \in G$ such that for each $b \in G$ there exists an integer n such that $b = a^n$. The element a is called the generator of the group, and n is called the discrete logarithm of b to the base a. This is indicated formally as $n = \log_a b$. If the element $a \in G$ is not a generator of the group, then discrete logarithm to the base a is not defined for every $b \in G$.

For example, consider a multiplicative group of order $(p-1)$, where p is a prime number, $G = \mathbb{Z}_p^* \triangleq \mathbb{Z}_p \backslash \{0\}$, and the group operation is the usual (arithmetic) multiplication operation modulo p. For example, assume that the prime number p is equal to 7. A generator of this group is 5, then $5^4 \equiv 2 \pmod 7$. Therefore the discrete logarithm of 2 to the base 5 is equal to 4. Alternately $4 = \log_5 2$ in this notation. The so-called the *discrete logarithm problem* (DLP) in the multiplicative group $(\mathbb{Z}_p^*, \times)$, where p is a prime number, is next defined.

Definition 3.13. *Discrete logarithm problem. Consider the multiplicative group* $(\mathbb{Z}_p^*, \times)$ *where* p *is a prime number, and the group operation* $\times$ *is multiplication modulo* p. *Let* $a \in \mathbb{Z}_p^*$ *be a generator of the group. For a given* $b \in \mathbb{Z}_p^*$, *it is required to find an integer* $n, 1 \le n \le (p-2)$, *such that* $a^n \equiv b \pmod p$. □

The El Gamal cryptosystem assumes that computation of the discrete logarithm in $\mathbb{Z}_p^*$ is computationally very expensive. However the inverse operation of exponentiation is assumed to be relatively less expensive. Exponentiation operation can be performed via repeated square-and-multiply operations. The implementation of this operation has been described in the chapter on number theory.

Each member (entity) of the group who wants to receive a secure message, generates his or her public and private keys. Thus each entity A should execute the key generation algorithm. The key generation algorithm, and algorithms for encryption and decryption operations of the El Gamal cryptographic scheme are subsequently outlined.

Algorithm 3.2a. *Key Generation for the El Gamal Cryptographic Scheme.*

Output: Key generation for entity A. That is, generate A's public and private keys.

begin

Step 1: $(\mathbb{Z}_p^*, \times)$ is a group, where p is a large random prime number, a is a generator of this group. It is assumed that the computation of discrete logarithm in this group is computationally very expensive.

Step 2: Pick a random integer $n, 1 \le n \le (p-2)$, and compute $a^n \equiv b \pmod p$.

Step 3: A's public key is (p, a, b) and the corresponding private key is n.

end (*end of El Gamal key generation algorithm*)

Algorithm 3.2b. *Encryption and Decryption in the El Gamal Cryptographic Scheme.*

Specification: The El Gamal cryptosystem is $(\mathcal{P}, \mathcal{C}, \mathcal{K}, \mathcal{E}, \mathcal{D})$.
$\mathcal{P} = \mathbb{Z}_p^*$ and $\mathcal{C} = \mathbb{Z}_p^* \times \mathbb{Z}_p^*$, where p is a prime number. Also let
$\mathcal{K} = \{(p, a, n, b) \mid a^n \equiv b \pmod{p}, 1 \le n \le (p-2)\}$,
and $K \triangleq (p, a, n, b) \in \mathcal{K}$.
A message $x \in \mathcal{P}$ is encrypted into $(y_1, y_2) \in \mathcal{C}$.

Encryption Operation. Entity B encrypts a message $x \in \mathcal{P}$ destined for entity A.
Input: The plaintext message $x \in \mathcal{P}$ with entity B.
Output: Entity B transmits, and entity A receives $(y_1, y_2) \in \mathcal{C}$.
begin
Step 1: Entity B obtains entity A's public key (p, a, b).
The entity B wants to transmit the message $x \in \mathcal{P}$ to the entity A.
Step 2: Generate a secret random integer k, $1 \le k \le (p-2)$.
Step 3: Compute $y_1 \equiv a^k \pmod{p}$, and $y_2 \equiv xb^k \pmod{p}$.
Step 4: The encrypting map is $E_K(x) = (y_1, y_2)$.
Step 5: Entity B transmits the ciphertext $(y_1, y_2) \in \mathcal{C}$ to the entity A.
end (*end of encryption operation*)

Decryption Operation. Entity A decrypts the ciphertext $(y_1, y_2) \in \mathcal{C}$.
Input: The encrypted message $(y_1, y_2) \in \mathcal{C}$ with entity A.
Output: Entity A generates the plaintext message $x \in \mathcal{P}$.
begin
Step 1: Entity A receives the encrypted message (y_1, y_2).
Step 2: Entity A uses its private key to compute
$D_K(y_1, y_2) \equiv y_2 (y_1^n)^{-1} \pmod{p}$.
This value is equal to $x \in \mathcal{P}$.
end (*end of decryption operation*)

Proof of the correctness of the algorithm. Observe that

$$\begin{aligned} D_K(y_1, y_2) &\equiv y_2 (y_1^n)^{-1} \pmod{p} \equiv xb^k \left(\left(a^k\right)^n \right)^{-1} \pmod{p} \\ &\equiv x (a^n)^k \left((a^n)^k \right)^{-1} \pmod{p} \equiv x \pmod{p} \end{aligned}$$

□

Observe that this cryptosystem is nondeterministic, as it depends upon the parameter k, which is a random integer. Consequently, there can actually be $(p-2)$ mappings for the same plaintext. This is also associated with a disadvantage of the El Gamal encryption scheme. The ciphertext is twice the length of the corresponding plaintext message. This cryptographic scheme is next illustrated via an example. Small parameter values have been chosen in this example for pedagogical convenience.

Example 3.11. Encrypting and decrypting in the El Gamal system is demonstrated with small parameters. Let the prime number $p = 13$. Therefore the group $(\mathbb{Z}_{13}^*, \times)$ is cyclic, where the operation $\times$ is multiplication modulo 13. A generator of this group is $a = 2$. Let the message to be encrypted be $x = 6$, where $x \in \{1, 2, \ldots, 12\}$. The message is sent by entity B to entity A.

Key Creation: Let A's private key be $n = 8$. Therefore $b \equiv a^n \pmod{p} \equiv 2^8 \pmod{13} \equiv 9 \pmod{13}$, and A's public key is $(p, a, b) = (13, 2, 9)$.

Encryption: Select a random number k from the list $\{1, 2, \ldots, 11\}$. Let k be equal to 5. Entity B gets entity A's public key, and performs the following operations. Therefore $y_1 \equiv a^k \pmod{p} \equiv 2^5 \pmod{13} \equiv 6 \pmod{13}$, and $y_2 \equiv xb^k \pmod{p} \equiv 6\,(9)^5 \pmod{13} \equiv 5 \pmod{13}$.

The transmitted message is $(y_1, y_2) = (6, 5)$.

Decryption: The message decrypted by entity A is

$$\begin{aligned} D_K(y_1, y_2) &\equiv y_2 \left(y_1^n\right)^{-1} \pmod{p} \equiv 5 \left(6^8\right)^{-1} \pmod{13} \\ &\equiv 5\,(3)^{-1} \pmod{13} \equiv 5\,(9) \pmod{13} \equiv 6 \pmod{13} \end{aligned}$$

It is indeed equal to the original message x. □

The El Gamal encrypting scheme can be extended to any finite cyclic group, instead of the multiplicative group $(\mathbb{Z}_p^*, \times)$. Candidates for this scheme are multiplicative group $\mathbb{F}_{2^m}^*$ of finite fields of characteristic two, and the group of points on an elliptic (or hyperelliptic) curve over a finite field.

3.6 Probabilistic Encryption

Consider a public-key cryptosystem in which the set of plaintexts is $\mathcal{P}$. In this cryptosystem the total number of possible plaintexts is small. That is, $|\mathcal{P}|$ is small. Alice wants to send an encrypted plaintext to Bob. Assume that Eve (or Oscar) the adversary is able to intercept this ciphertext. Eve can subsequently encrypt a plaintext by using Bob's public key, and then compare this ciphertext with the one sent by Alice to Bob till a match is found. This is certainly computationally feasible because of the assumption that $|\mathcal{P}|$ is small. Probabilistic encryption can be used to partially overcome this scenario. It incorporates randomness to increase the level of security. In addition, it provides semantic security. A public-key cryptosystem is said to be semantically secure if the adversary is not able to recover information about the plaintext, either partially or wholly in polynomial-time. A probabilistic cryptosystem is based upon the *indistinguishability* of probability distributions. This is next explained.

A probabilistic public-key cryptosystem works as follows. Alice encodes the plaintext $p \in \mathcal{P}$ along with a random word $r \in \mathcal{R}$. The word r is called a randomizer, and $\mathcal{R}$ is the set of randomizers. Alice uses Bob's public key to encrypt (p, r). It is hoped that, as r varies, the corresponding ciphertext c also varies over the values in $\mathcal{C}$. Furthermore, for $p, p' \in \mathcal{P}$, $p \neq p'$, and key $K \in \mathcal{K}$;

the ciphered values $E_K(p, r)$ and $E_K(p', r)$ are statistically indistinguishable (close) for different values of $r \in \mathcal{R}$. It is not required that Bob recover (p, r). He only needs to recover the plaintext $p \in \mathcal{P}$. We next clarify the notion of indistinguishability of two probability distributions (or random variables) more concretely.

Definition 3.14. *Let U and V be two random variables which take values in the set S. If*

$$|P(U \in W) - P(V \in W)| \leq \varepsilon, \quad \forall\, W \subseteq S \tag{3.17}$$

then the random variables U and V, and their corresponding probability distributions are said to be ε-indistinguishable, where $\varepsilon \in [0, 1]$. □

A probabilistic public-key cryptosystem uses the artifice of ε-indistinguishable probability distributions.

Definition 3.15. *A probabilistic public-key cryptosystem with security parameter ε is a six-tuple $(\mathcal{P}, \mathcal{C}, \mathcal{K}, \mathcal{E}, \mathcal{D}, \mathcal{R})$ where:*

(a) *$\mathcal{P}$ is the finite set of plaintexts.*
(b) *$\mathcal{C}$ is the finite set of ciphertexts.*
(c) *$\mathcal{K}$ is the finite set of keys. This is the key space.*
(d) *$\mathcal{R}$ is the set of randomizers.*
(e) *$\mathcal{E}$ is a family of invertible mappings indexed by $K \in \mathcal{K}$. These mappings are the encrypting algorithms. Let the mapping corresponding to the key K be denoted by $E_K \in \mathcal{E}$. For any $p \in \mathcal{P}$, and $r \in \mathcal{R}$ let $E_K(p, r) = c$, where $c \in \mathcal{C}$. The encryption algorithm is known to the public.*
Note that if $p, p' \in \mathcal{P}$, and $p \neq p'$ then $E_K(p, r) \neq E_K(p', r)$.
(f) *$\mathcal{D}$ is a family of invertible mappings indexed by $K \in \mathcal{K}$. These mappings are called decrypting algorithms. For each encrypting key $K \in \mathcal{K}$, there is a mapping $D_K \in \mathcal{D}$, such that $D_K(E_K(p, r)) = p$, for all $p \in \mathcal{P}$ and for all $r \in \mathcal{R}$. Note that D_K is a decrypting algorithm corresponding to the key K. The decryption algorithm is private.*
(g) *The security parameter is $\varepsilon \in [0, 1]$. For a fixed key $K \in \mathcal{K}$ and any $p \in \mathcal{P}$, define a probability distribution (mass) function $p_{K,p}(\cdot)$ on $\mathcal{C}$, where $p_{K,p}(c)$ is the probability that $c \in \mathcal{C}$ is the ciphertext for plaintext p and key K. This probability distribution occurs due to the random choices $r \in \mathcal{R}$.*
Let $p, p' \in \mathcal{P}$, $p \neq p'$ and $K \in \mathcal{K}$. Then the probability distributions $p_{K,p}(\cdot)$ and $p_{K,p'}(\cdot)$ are ε-indistinguishable in polynomial-time. □

The ε-indistinguishability (or ε-close) implies that an adversary, equipped with computational resources bounded in polynomial-time, cannot distinguish randomized ciphertexts with a probability bounded by ε. That is, the randomized ciphertexts of any two different plaintexts are indistinguishable by an adversary which uses algorithms with polynomial-time complexity to decipher ciphertexts, provided ε is small.

The security parameter ε should be a small number. It is typically equal to $c_{const}/|\mathcal{R}|$ where $c_{const} > 0$ is small. Goldwasser and Micali devised a probabilistic public-key cryptosystem which is based upon the intractability of the quadratic residuosity problem. The quadratic residuosity problem is next discussed.

Quadratic Residuosity Problem

Let $n \geq 3$ be an odd integer, and Q_n be the set of all quadratic residues modulo n. Also

$$J_n = \left\{ \theta \in \mathbb{Z}_n^* \mid \left(\frac{\theta}{n} \right) = 1 \right\}$$

where $\left(\frac{\cdot}{n}\right)$ is the Jacobi symbol, and $\mathbb{Z}_n^*$ is the set of nonzero integers in the set $\mathbb{Z}_n$ which are relatively prime to n. The set of pseudosquares modulo n is $\widetilde{Q}_n = (J_n - Q_n)$. The so-called *quadratic residuosity problem* (QRP) is next defined.

Definition 3.16. *Quadratic residuosity problem. Let $n \geq 3$ be an odd integer, and Q_n be the set of all quadratic residues modulo n. The set J_n is equal to $\left\{\theta \in \mathbb{Z}_n^* \mid \left(\frac{\theta}{n}\right) = 1\right\}$, where $\left(\frac{\cdot}{n}\right)$ is the Jacobi symbol. Also $\widetilde{Q}_n = (J_n - Q_n)$.*
If in addition, n is a composite integer and $\theta \in J_n$, determine if $\theta \in Q_n$ or $\theta \in \widetilde{Q}_n$. □

Let $\theta \in \mathbb{Z}_n^*$. Observe that, if n is a prime number, then $\theta \in Q_n$ if and only if $\left(\frac{\theta}{n}\right) = 1$. Efficient algorithm exists for computing the Legendre symbol $\left(\frac{\theta}{n}\right)$. Next consider the case in which $n = pq$, where p and q are distinct odd prime numbers. Therefore, if $\theta \in J_n$ then $\theta \in Q_n$ if and only if $\left(\frac{\theta}{p}\right) = 1$.

Thus the membership of θ in the set Q_n can be determined easily by computing $\left(\frac{\theta}{p}\right)$, if the factors of n are known. Therefore it is reasonable to assume that the quadratic residuosity problem is at most as hard as the factoring of very large integers.

A complete description of the Goldwasser and Micali probabilistic public-key cryptosystem consists of a procedure for key generation, and procedures for encryption and decryption. Each member (entity) of the group which wants to receive a secure message generates his or her public and private keys. Thus each entity A of the group should execute the following key generation algorithm.

Algorithm 3.3a. *Key Generation for the Goldwasser-Micali Probabilistic Cryptographic Scheme.*

Output: Key generation for entity A. That is, generate A's public and private keys.
begin
 Step 1: Generate two large random and distinct prime numbers p and q. These two prime numbers should be approximately of the same size.
 Step 2: Compute $n = pq$.
 Step 3: Determine a pseudosquare $s \in \mathbb{Z}_n$ as follows.
  Find a quadratic nonresidue a modulo p. That is, $a \in \overline{Q}_p$.
  ($\overline{Q}_p$ is the set of quadratic nonresidues modulo p)
  Find a quadratic nonresidue b modulo q. That is, $b \in \overline{Q}_q$.
  ($\overline{Q}_q$ is the set of quadratic nonresidues modulo q)

Using Chinese remainder theorem, determine $s \in \mathbb{Z}_n$ which satisfies simultaneously $s \equiv a \pmod{p}$, and $s \equiv b \pmod{q}$.
Therefore $\left(\frac{s}{n}\right) = \left(\frac{s}{p}\right)\left(\frac{s}{q}\right) = (-1)(-1) = 1$.

Step 4: A's public key is (n, s) and the private key is (p, q).

end (*end of Goldwasser-Micali key generation algorithm*)

The encryption and decryption operations of the Goldwasser-Micali probabilistic cryptographic scheme are outlined in the next algorithm. In this scheme, the set of plaintexts $\mathcal{P} = \mathbb{Z}_2^{(m)}$, where $m \in \mathbb{P}$. Thus, a plaintext is simply a binary string of length m. This scheme encrypts each bit of the plaintext independently. The corresponding set of ciphertexts is $\mathcal{C} = \mathbb{Z}_n^{*(m)}$, where $m \in \mathbb{P}$ and n is a product of two large distinct primes.

Algorithm 3.3b. *Encryption and Decryption in the Goldwasser-Micali Probabilistic Cryptographic Scheme.*

Specification: The Goldwasser-Micali cryptosystem is specified by $(\mathcal{P}, \mathcal{C}, \mathcal{K}, \mathcal{E}, \mathcal{D}, \mathcal{R})$, where $\mathcal{P} = \mathbb{Z}_2^{(m)}$, $\mathcal{C} = \mathbb{Z}_n^{*(m)}$, and $m \in \mathbb{P}$.
n is the product of two odd distinct prime numbers p and q of comparable size.
$\mathcal{K} = \{(n, p, q, s) \mid n = pq;$ p and q are large random and distinct prime numbers, $s \in \mathbb{Z}_n^*$ is a pseudosquare$\}$,
and $K \triangleq (n, p, q, s) \in \mathcal{K}$. The randomizer set $\mathcal{R} = \mathbb{Z}_n^{*(m)}$.
A plaintext $x \in \mathcal{P}$ is encrypted into the ciphertext $y \in \mathcal{C}$.

Encryption Operation. Entity B encrypts a plaintext $x \in \mathcal{P}$ destined for entity A.

Input: Entity B's plaintext $x \in \mathcal{P}$, where $x = (x_1, x_2, \ldots, x_m)$, $x_i \in \mathbb{Z}_2$ for $1 \le i \le m$.

Output: Entity B transmits, and entity A receives $y \in \mathcal{C}$, where $y = (y_1, y_2, \ldots, y_m)$, $y_i \in \mathbb{Z}_n^*$ for $1 \le i \le m$.

begin

Step 1: Entity B obtains entity A's public key (n, s).

for $i = 1$ **to** m **do**
begin
select $w_i \in \mathbb{Z}_n^*$ at random
if $x_i = 1$ **then** $y_i \leftarrow s w_i^2 \pmod{n}$
else $y_i \leftarrow w_i^2 \pmod{n}$
end (*end of i for-loop*)
$(w \triangleq (w_1, w_2, \ldots, w_m) \in \mathcal{R})$
Thus $E_K(x, w) = y$.

Step 2: Entity B transmits the ciphertext $y \in \mathcal{C}$ to the entity A.

end (*end of encryption operation*)

Decryption Operation. Entity A decrypts the ciphertext $y \in \mathcal{C}$.
Input: The encrypted ciphertext $y \in \mathcal{C}$ with entity A.
Output: Entity A generates the plaintext $x \in \mathcal{P}$.
begin
  Step 1: Entity A receives the ciphertext y. It performs
    for $i = 1$ **to** m **do**
    begin
      $e_i \leftarrow \left(\frac{y_i}{p}\right)$
      (compute the Legendre symbol)
      if $e_i = 1$ **then** $x_i \leftarrow 0$
      else $x_i \leftarrow 1$
    end (*end of i for-loop*)
    Thus $D_K(y) = x$.
  Step 2: The plaintext is equal to $x \in \mathcal{P}$.
end (*end of decryption operation*)

Proof of the correctness of the algorithm. If the plaintext bit $x_i = 0$, then $y_i = w_i^2 \pmod n$ is a quadratic residue modulo n, where $w_i \in \mathbb{Z}_n^*$ is randomly selected. However, if the plaintext bit $x_i = 1$, then $y_i = sw_i^2 \pmod n$. In this case y_i is a pseudosquare modulo n, as s is a pseudosquare modulo n. As $n = pq$, and y_i a pseudosquare modulo n implies that, y_i is a quadratic nonresidue modulo p and also modulo q. Therefore since p is known to entity A, it can compute the Legendre symbol $\left(\frac{y_i}{p}\right)$ and determine the corresponding $x_i \in \mathbb{Z}_2$. □

We next examine the security provided by the Goldwasser-Micali probabilistic cryptographic scheme. In the above algorithm, w_i is selected at random from the set $\mathbb{Z}_n^*$ for each $i \in [1, m]$. As $w_i^2 \pmod n$ is a quadratic residue, and $sw_i^2 \pmod n$ is a random pseudosquare; the adversary encounters multiple quadratic residues and pseudosquares modulo n. If it is assumed that the quadratic residuosity problem is intractable, the adversary would be left with a fair amount of guessing for each plaintext bit. This in turn provides semantic security.

Associated with this advantage comes a drawback. This is in the size of the ciphertext. This cryptographic scheme uses $\lceil \log_2 n \rceil$ bits in the ciphertext for each plaintext bit. This is essentially the price of using a probabilistic cryptographic scheme. However, it has the possibility of several ciphertexts for the same plaintext.

3.7 Homomorphic Encryption

Recall that encryption is the process of obscuring data so that it is indecipherable to anyone except the intended recipient. Homomorphic encryption is a special form encryption, which permits certain computations to be executed on the ciphertext and generate a ciphertext, which upon subsequent decryption would match the results of similar operations performed on the plaintext.

Thus, homomorphic encryption is the conversion of data into ciphertext that can be analyzed and worked with as if it were still in its original plaintext form. That is, homomorphic encryptions allow complex mathematical operations to be performed on encrypted data without compromising the encryption.

So why is homomorphic encryption important? Use of homomorphic encryption allows confidentiality of processed data. Homomorphism decouples the encryption operation of the plaintext from the necessity to view the actual plaintext. Thus manipulation of secret data can be outsourced to an untrusted third party. This is important in scenarios where the ciphertexts are in an insecure location. In summary, homomorphic encryption permits manipulation of the secret ciphertexts in a potentially useful manner.

3.7.1 Applications

Some applications of homomorphic encryption are: electronic-cash, electronic-voting, cloud computing, and collision-resistant hash functions. It can also be used for private information retrieval schemes for financial transactions, and medical records.

Example 3.12. We describe a simple example, where a homomorphic encryption scheme might work in cloud storage. A cloud is supposedly a safe haven for data storage. The encryption mapping in this example is definitely insecure.

Step 1: Business Boondoggle has a very important data set $\mathcal{S}$ that consists of the numbers 6 and 12. To encrypt the data set, Business Boondoggle multiplies each element in the set by 2, thus creating the encrypted set $\mathcal{T}$ whose members are 12 and 24.

Step 2: Business Boondoggle sends the set $\mathcal{T}$ to the cloud for safe storage. A few months later, the government contacts Business Boondoggle and requests the sum of elements in the set $\mathcal{S}$.

Step 3: Business Boondoggle is very busy, so it asks the cloud provider to perform the operation. The cloud provider, who only has access to the encrypted data set $\mathcal{T}$, finds the sum of $12 + 24$ and returns the answer 36 to Business Boondoggle.

Step 4: Business Boondoggle decrypts the cloud provider's reply and provides the government with the decrypted answer, which is 18. □

3.7.2 Algebraic Homomorphism

In mathematics, homomorphism describes the mapping of one data set into another while preserving relationships between elements in both sets. The term *homomorphism* is derived from the Greek words for "similar structure." As the data in a homomorphic encryption scheme preserves similar structure, mathematical operations performed on either encrypted or decrypted data yield equivalent results. We define two types of algebraic homomorphisms. These are group and ring homomorphisms. Groups and rings are algebraic structures.

A group $(G, *)$ is an algebraic structure, where the set G is closed under the binary operation $*$. Furthermore, the operation $*$ is associative, G has an identity element, and each element of G has an inverse. The group G is called Abelian, if $a * b = b * a$ for all elements $a, b \in G$.

A ring $(R, +, \circ)$ is an algebraic structure, where the set R is closed under the two binary operations $+$ (addition operator) and $\circ$ (multiplication operator). In addition, $(R, +)$ is an Abelian group, and the operation $\circ$ is associative over all elements of R. Also, the left and right distributive laws for multiplication over addition for all elements of R are also satisfied.

Definitions 3.17. *Group and ring homomorphisms.*

1. *Let* $(G, *_1)$ *and* $(H, *_2)$ *be two groups, and* $\varphi : G \to H$ *be a mapping such that*

$$\varphi(a *_1 b) = \varphi(a) *_2 \varphi(b), \quad \forall\, a, b \in G \tag{3.18}$$

then φ *is a group homomorphism.*

2. *Let* $(R, +_1, \circ_1)$ *and* $(S, +_2, \circ_2)$ *be two rings, and* $\varphi : R \to S$ *be a mapping such that*

$$\varphi(a +_1 b) = \varphi(a) +_2 \varphi(b), \quad \forall\, a, b \in R \tag{3.19a}$$
$$\varphi(a \circ_1 b) = \varphi(a) \circ_2 \varphi(b), \quad \forall\, a, b \in R \tag{3.19b}$$

then φ *is a ring homomorphism.* □

3.7.3 Homomorphic Cryptosystems

Several cryptosystems are defined over algebraic structures such as groups and rings. A cryptosystem defined over a group supports a single operation. These operations are usually denoted as either multiplication or addition. Similarly, a cryptosystem defined over a ring supports two operations. These operations are usually denoted as addition and multiplication.

Definition 3.18. *Group homomorphic cryptosystem. Let* $(\mathcal{P}, \mathcal{C}, \mathcal{K}, \mathcal{E}, \mathcal{D})$ *be a cryptosystem, where:*

(a) $(\mathcal{P}, *)$ *and* $(\mathcal{C}, \circledast)$ *are groups.*
(b) *Let* $k \in \mathcal{K}$, $E_k \in \mathcal{E}$ *and* $D_k \in \mathcal{D}$.
(c) *For any* $k \in \mathcal{K}$, *and any two ciphertexts*

$$c_1 = E_k(p_1), \quad c_2 = E_k(p_2), \quad \textit{where } p_1, p_2 \in \mathcal{P} \tag{3.20a}$$

the following condition is true:

$$D_k(c_1 \circledast c_2) = p_1 * p_2 \tag{3.20b}$$

□

If the operator $*$ is equal to the addition operator $+$, we have an *additive group homomorphic cryptosystem.* However, if $*$ is equal to the multiplication operator $\times$, we have a *multiplicative group homomorphic cryptosystem*. A ring homomorphic cryptosystem is similarly defined.

Definition 3.19. *Ring homomorphic cryptosystem. Let* $(\mathcal{P}, \mathcal{C}, \mathcal{K}, \mathcal{E}, \mathcal{D})$ *be a cryptosystem, where:*

(a) $(\mathcal{P}, +, \times)$ *and* $(\mathcal{C}, \oplus, \otimes)$ *are rings.*
(b) *Let* $k \in \mathcal{K}$, $E_k \in \mathcal{E}$ *and* $D_k \in \mathcal{D}$.
(c) *For any* $k \in \mathcal{K}$, *and any two ciphertexts*

$$c_1 = E_k(p_1), \quad c_2 = E_k(p_2), \quad \textit{where } p_1, p_2 \in \mathcal{P} \tag{3.21a}$$

the following conditions are true:

$$D_k(c_1 \oplus c_2) = p_1 + p_2 \tag{3.21b}$$
$$D_k(c_1 \otimes c_2) = p_1 \times p_2 \tag{3.21c}$$

□

A cryptosystem is *additively homomorphic*, if there exists a binary operator $\oplus$ such that

$$(E_k(p_1) \oplus E_k(p_2)) = E_k(p_1 + p_2), \quad \forall\ p_1, p_2 \in \mathcal{P}$$

Similarly, a cryptosystem is *multiplicatively homomorphic*, if there exists a binary operator $\otimes$ such that

$$E_k(p_1) \otimes E_k(p_2) = E_k(p_1 \times p_2), \quad \forall\ p_1, p_2 \in \mathcal{P}$$

Types of Homomorphic Cryptosystems

There are three types of homomorphic cryptosystems. These are:

- Partially homomorphic cryptosystems.
- Fully homomorphic cryptosystems.
- Somewhat homomorphic cryptosystems.

A cryptosystem is homomorphic, if it exhibits either additive or multiplicative homomorphism. Some examples of partially homomorphic cryptosystems are: RSA and El Gamal cryptosystems. The probabilistic Goldwasser-Micali cryptographic scheme is also partially homomorphic. The homomorphic operation in the RSA scheme is a multiplication operation. The homomorphic operations in the El Gamal scheme are multiplication and exponentiation operations. The homomorphic operations in the Goldwasser-Micali scheme is modulo-2 arithmetic.

A cryptosystem is fully homomorphic, if it exhibits both additive and multiplicative homomorphism. A first such system was demonstrated by Craig Gentry in his doctoral dissertation in the year 2009. This lattice-based scheme is not yet computationally viable.

In a somewhat homomorphic cryptosystem, it is possible to perform operations on the data with respect to the two operators, only a limited number of times. For example, it is possible to perform additions and multiplications, a limited number of times before the cryptosystem fails. The limitation occurs from an error term which increases with each operation. Decryption is not possible, if the error term exceeds a certain threshold value.

Examples 3.13. Some examples of partially homomorphic cryptosystems.

1. *RSA cryptosystem.* Let $n = pq$, where p and q are large prime numbers. Also let $ab \equiv 1 \,(\text{mod}\, \varphi(n))$, where $1 < b < \varphi(n)$, and $\varphi(\cdot)$ is Euler's totient function. Actually $\varphi(n) = (p-1)(q-1)$. The public key is (b, n), and private key is (a, n).
   The encryption, decryption exponent pair is (b, a). For $K \in \mathcal{K}$, let the encrypting mapping be $E_K(\cdot)$, and $p_1, p_2 \in \mathcal{P}$ be plaintexts. In this scheme
   $$E_K(p_1) = p_1^b, \quad E_K(p_2) = p_2^b \text{ and } E_K(p_1 p_2) = (p_1 p_2)^b$$
   Therefore
   $$E_K(p_1 p_2) = E_K(p_1) E_K(p_2)$$
   which is the multiplicative homomorphic property. Note that the arithmetic in this scheme is actually modulo n.

2. *El Gamal cryptosystem.* $(\mathbb{Z}_p^*, \times)$ is a group, where p is a large random prime number, and a is a generator of this group. Pick a random integer $n, 1 \le n \le (p-2)$, and compute $a^n \equiv b \pmod{p}$. The public key is (p, a, b) and the corresponding private key is n.
The encrypter generates a secret random integer $k, 1 \le k \le (p-2)$. It computes

$$y_1 \equiv a^k \pmod{p}, \quad \text{and} \quad y_2 \equiv xb^k \pmod{p}$$

The encrypting map is $E_K(x) = (y_1, y_2)$, where $x \in \mathcal{P}$.
Using the same public key, the plain texts x_1 and x_2 are encrypted as

$$E_K(x_1) = \left(a^{k_1} \pmod{p}, x_1 b^{k_1} \pmod{p}\right)$$
$$E_K(x_2) = \left(a^{k_2} \pmod{p}, x_2 b^{k_2} \pmod{p}\right)$$

where $1 \le k_1, k_2 \le (p-2)$. Thus

$$\begin{aligned} & E_K(x_1) \circledast E_K(x_2) \\ &= \left(a^{k_1} \pmod{p}, x_1 b^{k_1} \pmod{p}\right) \circledast \left(a^{k_2} \pmod{p}, x_2 b^{k_2} \pmod{p}\right) \\ &\triangleq \left(a^{k_1} a^{k_2} \pmod{p}, x_1 x_2 b^{k_1} b^{k_2} \pmod{p}\right) \\ &= \left(a^{k_1+k_2} \pmod{p}, x_1 x_2 b^{k_1+k_2} \pmod{p}\right) \\ &= E_K(x_1 x_2) \end{aligned}$$

which is the multiplicative homomorphic property. The homomorphic operations are multiplication and exponentiation.

3. *Goldwasser-Micali cryptosystem.* Let $n = pq$, where p and q are odd and distinct prime numbers. Also let $s \in \mathbb{Z}_n$ be a pseudosquare. The public key is (n, s), and private key is (p, q). Let the bit to be encrypted be $x \in \mathbb{Z}_2$. For simplicity assume that the set of plaintext $\mathcal{P} = \mathbb{Z}_2$. Also let $w \in \mathbb{Z}_n^*$ be selected randomly. For $K \in \mathcal{K}$, let the encrypting mapping be $E_K(\cdot)$. Then $E_K(x) = s^x w^2 \pmod{n}$. Let $x_1, x_2 \in \mathcal{P}$, and $w_1, w_2 \in \mathbb{Z}_n^*$. In this scheme

$$E_K(x_1) = s^{x_1} w_1^2 \pmod{n}, \quad E_K(x_2) = s^{x_2} w_2^2 \pmod{n}$$

Let $(x_1 + x_2) \pmod{2} \triangleq y$, then

$$\begin{aligned} E_K(x_1) E_K(x_2) &= s^{x_1+x_2} (w_1 w_2)^2 \pmod{n} = E_K(y) \\ &= E_K((x_1 + x_2) \pmod{2}) \end{aligned}$$

which is a homomorphic property. The homomorphic operation is modulo 2 arithmetic. □

3.8 Digital Signatures

The signature known to all of us is the hand written signature. It is usually appended to a document to signify that the person who pens the signature is responsible for the document. Similarly in the digital world, digital signatures are used to sign documents which are either stored or transmitted in electronic form. It is essentially a data string, which confirms a relationship between a digital message and some entity (for example, a person).

Digital signatures are possible via both symmetric-key techniques, and public-key cryptography. However, use of symmetric-key techniques requires the use of a reliable and trustworthy third-party. Consequently, digital signatures use public-key cryptographic schemes. Further, nonrepudiation is possible with the use of digital signatures.

More precisely, if Bob wants to send a message to Alice, he uses a secret key and computes a signature. This signature depends upon the message and the secret key. When Alice receives the message; she uses the corresponding public key to verify that the digital signature is indeed Bob's signature. Therefore, signatures must be easy for the transmitting entity to sign, and be easily verifiable by the receiving entity. A formal definition of a signature scheme is given below.

Definition 3.20. *A digital signature scheme* $(\mathcal{M}, \mathcal{A}, \mathcal{K}, \mathcal{S}, \mathcal{V})$ *is a five-tuple where:*

(a) $\mathcal{M}$ *is a finite set of messages which can be signed.*

(b) $\mathcal{A}$ *is the finite set of signatures. These can be binary strings of fixed length.*

(c) $\mathcal{K}$ *is the finite set of keys. This is the key space. Each member (entity) of the community is associated with a key.*

(i) *For every* $K \in \mathcal{K}$, *there is a signing algorithm (or transformation) denoted by* $S_K \in \mathcal{S}$. *For each* $K \in \mathcal{K}$, S_K *is actually a mapping* $S_K : \mathcal{M} \rightarrow \mathcal{A}$. *The transformation* S_K *is kept secret by the entity which transmits and signs the message from* $\mathcal{M}$.

(ii) *For every* $K \in \mathcal{K}$, *there is a verifying algorithm (or transformation) denoted by* $V_K \in \mathcal{V}$. *For each* $K \in \mathcal{K}$, V_K *is actually a mapping:* $V_K : \mathcal{M} \times \mathcal{A} \rightarrow \{true, false\}$. *The transformation* V_K *is publicly known, and is used by other entities to verify the signature of the transmitting entity.*

(d) *Signing Procedure: An entity or the person which wants to transmit a message* $m \in \mathcal{M}$ *performs the following steps:*

(i) *Compute the signature* $s = S_K(m)$. *Note that* $s \in \mathcal{A}$.

(ii) *Transmit* (m, s). *This two-tuple is called a signed message.*

(e) *Verification Procedure: The verifier receives* (m', s); *where* m' *may or may not be equal to* m. *Oscar might modify* m, *however he cannot modify* s *because he does not know the key* K. *The verifier also knows that the message has been received from an entity with key* $K \in \mathcal{K}$. *However, he or she needs to confirm by checking its signature.*

(i) *The verifier computes* $c = V_K(m', s)$. *Note that* $c \in \{true, false\}$, *where* $c =$ *true, if* $s = S_K(m')$; *and* $c =$ *false, if* $s \neq S_K(m')$.

(ii) *If* $c =$ *true, the signature is accepted, else the signature is rejected* ($c =$ *false*). □

Requirements for the Transformations S_K and V_K

For all $K \in \mathcal{K}$, the transformations S_K and V_K must be computable in polynomial-time. Furthermore, the verification of a received message by an entity other than the intended-entity should be computationally infeasible.

The RSA and El Gamal public-key based digital signature schemes are next discussed.

3.8.1 The RSA Signature Scheme

The RSA signature key generation is identical to the RSA cryptosystem key generation. Each entity (member) of the group which wants to transmit and receive a signed message performs the following steps to create keys. Let p and q be two large distinct prime numbers. The number b is chosen randomly such that $1 < b < \varphi(n)$ and $\gcd(b, \varphi(n)) = 1$. Find an integer a such that $ab \equiv 1 \pmod{\varphi(n)}$.

The Signature Scheme. Let p and q be large distinct prime numbers, and $n = pq$. Also let $\mathcal{M} = \mathcal{A} = \mathbb{Z}_n$. Entity A signs a message $m \in \mathcal{M}$, and entity B wants to verify A's digital signature $s \in \mathcal{A}$, and retrieve the message m. Also

$$\mathcal{K} = \{(n, p, q, a, b) \mid n = pq; 1 < b < \varphi(n), \gcd(b, \varphi(n)) = 1; ab \equiv 1 \pmod{\varphi(n)}\},$$

and $K \triangleq (n, p, q, a, b) \in \mathcal{K}$. A's public key is (n, b) and the private key is (p, q, a).

Signature Creation. Entity A generates the signature $s \in \mathbb{Z}_n$ for the message $m \in \mathbb{Z}_n$.

1. It computes $S_K(m) \equiv m^a \pmod{n} \equiv s \pmod{n}$.
2. The transmitted information is (m, s).

Signature Verification. Entity B receives (m', s), where m' may or may not be equal to m. Entity B performs the following steps.

1. Obtain entity A's public key (n, b).
2. Compute $\widetilde{m} \equiv s^b \pmod{n}$.
3. If $m' = \widetilde{m}$, then $V_K(m', s) = $ *true*; accept the message. Otherwise, the message is rejected. □

Note that the arithmetic is performed in $\mathbb{Z}_n$. The proof that RSA's signature verification process works is left as an exercise to the reader. Also observe that if the message is large, then it can be split into several blocks, and each block is signed independent of other block signatures. Alternately, a technique called hashing can be used. The principle and ramifications of the use of hashing are discussed later in the chapter.

3.8.2 The El Gamal Signature Scheme

The El Gamal signature scheme is a nondeterministic signature scheme. This implies that there are several valid signatures for any specific message. Accordingly, the verification procedure should be designed such that it is capable of accepting any of these signatures as authentic. The key generation in this scheme is identical to the El Gamal cryptosystem key generation. This was discussed in an earlier section. Each entity of the group which wants to send a signed message generates a public key and a private key. Therefore, each entity of the group should execute the following steps.

Initially, each entity A of the group generates a large random prime number p such that the DLP in $(\mathbb{Z}_p^*, \times)$ is computationally very expensive, where $\mathbb{Z}_p^* \triangleq \mathbb{Z}_p \backslash \{0\}$. Let $a \in \mathbb{Z}_p^*$ be a generator of this multiplicative group, where the group operation $\times$ is multiplication modulo p. Next, pick a random integer n, such that $1 \leq n \leq (p-2)$, and compute $a^n \equiv b \pmod{p}$.

Signature Scheme. Let $\mathcal{M} = \mathbb{Z}_p^*$, and $\mathcal{A} = \mathbb{Z}_p^* \times \mathbb{Z}_{p-1}$. Entity A signs a message $m \in \mathcal{M}$, and entity B wants to verify A's digital signature $s \in \mathcal{A}$, and retrieve the message m. Also

$$\mathcal{K} = \{(p, a, n, b) \mid a^n \equiv b \pmod{p}, 1 \leq n \leq (p-2)\},$$

and $K \triangleq (p, a, n, b) \in \mathcal{K}$.

A's public key is (p, a, b) and the corresponding private key is n.

Signature Creation. Entity A executes the following steps. It generates the signature $s \in \mathcal{A}$ for the message $m \in \mathcal{M}$.

1. Select a random integer $k \in \mathbb{Z}_{p-1}^*$, such that $\gcd(k, p-1) = 1$.
2. Compute $j \in \mathbb{Z}_{p-1}^*$, such that $j \equiv k^{-1} \pmod{(p-1)}$ using the extended Euclidean algorithm.
3. Compute $a^k \equiv \beta \pmod{p}$.
4. Compute $j(m - n\beta) \equiv \gamma \pmod{(p-1)}$.
5. A's signature for the message m is the two-tuple $s = (\beta, \gamma)$.

Signature Verification. Entity B receives (m', s), where m' may or may not be equal to m. Entity B performs the following steps.

1. Obtain entity A's public key (p, a, b).
2. Check if $\beta \in \mathbb{Z}_p^*$ and $\gamma \in \mathbb{Z}_{p-1}$. If this is not true, then reject the signature.
3. Compute $z_1 \equiv b^{\beta}\beta^{\gamma} \pmod{p}$.
4. Compute $z_2 \equiv a^{m'} \pmod{p}$.
5. Accept the signature iff $z_1 \equiv z_2 \pmod{p}$, otherwise reject the signature. □

Proof of the correctness of the signature scheme. We show that $z_1 \equiv z_2 \pmod{p}$ if $m' = m$. Observe that

$$\begin{aligned} z_1 &\equiv b^{\beta}\beta^{\gamma} \pmod{p} \equiv a^{n\beta}a^{kj(m-n\beta)} \pmod{p} \\ &\equiv a^{n\beta}a^{(m-n\beta)} \pmod{p} \equiv a^m \pmod{p} \equiv z_2 \pmod{p} \end{aligned}$$

Thus $z_1 \equiv z_2 \pmod{p}$. □

Example 3.14. The El Gamal digital signature scheme is demonstrated with small parameters. Let the prime number $p = 13$. Also the group $(\mathbb{Z}_{13}^*, \times)$ is cyclic, where the $\times$ operation is multiplication modulo 13. A generator of this group is $a = 2$. Also let the message to be signed be $m = 6$, where $m \in \{1, 2, \ldots, 12\}$.

Key Creation: Let the private key be $n = 8$. Then $b \equiv a^n \pmod{p} \equiv 2^8 \pmod{13} \equiv 9 \pmod{13}$. Therefore the public key is $(p, a, b) = (13, 2, 9)$.

Signature creation: Entity A executes the following steps:

1. Select a random integer $k \in \mathbb{Z}_{12}^* = \{1, 2, \ldots, 11\}$, such that $\gcd(k, 12) = 1$. Let $k = 5$.
2. Compute $j = k^{-1} \in \mathbb{Z}_{12}^*$ using extended Euclidean algorithm. This yields $j = 5$.
3. Compute $\beta \equiv 2^5 \pmod{13} \equiv 6 \pmod{13}$.
4. Compute $\gamma \equiv j(m - n\beta) \pmod{(p-1)} \equiv 5(6 - 8 \times 6) \pmod{12} \equiv 6 \pmod{12}$.
5. A's signature for the message m is the two-tuple $s = (\beta, \gamma) = (6, 6)$.

The signed message is $(m, (\beta, \gamma)) = (6, (6, 6))$.

Signature verification: Entity B receives the signed message. It is $(m, (\beta, \gamma)) = (6, (6, 6))$. Entity B then performs the following steps:

1. Obtain entity A's public key $(p, a, b) = (13, 2, 9)$.
2. Check if $\beta = 6 \in \mathbb{Z}_{13}^*$, and $\gamma = 6 \in \mathbb{Z}_{12}$. This is true.
3. Compute $z_1 \equiv b^{\beta}\beta^{\gamma} \pmod{p} \equiv (9^6)(6^6) \pmod{13} \equiv 12 \pmod{13}$.
4. Compute $z_2 \equiv a^m \pmod{p} \equiv 2^6 \pmod{13} \equiv 12 \pmod{13}$.
5. $z_1 \equiv z_2 \equiv 12 \pmod{13}$. Accept the signature. □

There are several other variations of the El Gamal signature scheme. These can be found in the vast literature on the subject.

3.9 Key Distribution

In order to have secure communication, the keys which are to be used by a specific group of users have to be distributed securely. This is called the key distribution problem. Key distribution can be accomplished via a key distribution center. A key distribution center is a trusted third party, which is designated to deliver short-term secret keys to the community of users for secure communication. Key delivery can also be accomplished via public-key cryptosystem.

In the year 1976, W. Diffie and M. Hellman discovered a practical solution to the key distribution problem. Their pioneering work provided a protocol for exchanging secret keys over insecure channels. Keys that are used in symmetric key cryptosystems, like Data Encryption Standard (DES), and Advanced Encryption Standard (AES), can be established over insecure communication channels using this scheme.

This protocol also provided the ground work for the El Gamal cryptosystem. The Diffie-Hellman protocol has the following properties:

(a) It permits two parties who have never met before to create a shared secret by exchanging message over an open (public) channel.
(b) The security of this scheme depends upon the DLP.

The Diffie-Hellman Key Exchange Algorithm

Two entities A and B establish a shared secret key K, over a public channel. The following steps are executed.

1. *Initial setup*: Select a large prime number p, and a generator a of the group $(\mathbb{Z}_p^*, \times)$, where the operation $\times$ is multiplication modulo p, and $2 \leq a \leq (p-2)$. The numbers p and a are published.
2. Entity A selects a random secret number u, where $1 \leq u \leq (p-2)$, and sends $\alpha \equiv a^u \pmod{p}$ to entity B. The values u and α are A's private and public key respectively.
3. Entity B selects a random secret number v, where $1 \leq v \leq (p-2)$, and sends $\beta \equiv a^v \pmod{p}$ to entity A. The values v and β are B's private and public key respectively.
4. Entity B receives α and computes the shared key as

$$K \equiv \alpha^v \pmod{p} \equiv (a^u)^v \pmod{p}$$

5. Entity A receives β and computes the shared key as

$$K \equiv \beta^u \pmod{p} \equiv (a^v)^u \pmod{p}$$

□

Oscar the eavesdropper, can learn about the integers p, a, α, and β; but not about u and v. Consequently, he cannot determine the key K, unless he solves the DLP in modulo p arithmetic. This protocol can also be implemented by using another group in which DLP is computationally hard and exponentiation is efficient.

3.10 Elliptic Curve Based Cryptosystems

The elliptic curve based cryptosystems were discovered independently by Neal Koblitz and Victor S. Miller in 1985. They suggested that the public-key algorithms can be implemented using elliptic curves. The elliptic curves defined over finite fields have sufficient properties to form a group via an "exotic" group operation $\uplus$. This group operation is discussed at length in the chapter on abstract algebra. Furthermore, these groups also have sufficient properties to resist cryptanalytic attacks. A major advantage of *elliptic curve* (EC) cryptography over RSA scheme is that it offers the same security for smaller key size.

Analogy between operations in EC and RSA cryptosystems is next established. The "addition operation" $\uplus$ of EC cryptosystem is analogous to the RSA cryptosystem's multiplication operation in modulo arithmetic. Multiple addition operations of the elliptic curve group is analogous to the exponentiation operation of modulo arithmetic.

The security of El Gamal cryptosystem was based upon the discrete logarithm problem. There is an analogous DLP for elliptic curves. Define the set of points $\mathbb{E}_p(a,b)$ to be the union of the zero-point O, and the set of points on the elliptic curve $y^2 \equiv (x^3 + ax + b) \pmod p$, where $a, b, x, y \in \mathbb{Z}_p$ and $p > 3$ is a prime number. In order to define a group structure, it is also required that $(x^3 + ax + b)$ has no repeated factors. This implies that $(4a^3 + 27b^2) \not\equiv 0 \pmod p$.

Let $P, Q \in \mathbb{E}_p(a,b)$, and

$$Q = \underbrace{P \uplus P \uplus \ldots \uplus P}_{n \text{ terms}} \triangleq [n]P, \quad n \in \mathbb{P} \backslash \{1\}$$

Observe that it is easy to compute Q, given P and n. However, it is relatively hard to determine n, given P and Q. This is the DLP for the elliptic curves. For the purpose of elucidation, a "toy" example discussed in the chapter on abstract algebra is discussed below.

Example 3.15. Consider the congruence

$$y^2 \equiv (x^3 + 2x + 6) \pmod 7, \quad x, y \in \mathbb{Z}_7$$

The number of points in the set $\mathbb{E}_7(2,6)$, including the zero-point O, is equal to $N = 11$. Therefore the Abelian group $(\mathbb{E}_7(2,6), \uplus)$ has a prime order. Consequently it is a cyclic group isomorphic to $(\mathbb{Z}_{11}, +)$. Assume that a generator of this group is $P = \beta = (1,3)$. Use of arithmetic modulo 7 yields:

$$\begin{gathered}\beta = (1,3), [2]\beta = (2,2), [3]\beta = (5,1), [4]\beta = (3,5), [5]\beta = (4,1), \\ [6]\beta = (4,6), [7]\beta = (3,2), [8]\beta = (5,6), [9]\beta = (2,5), [10]\beta = (1,4)\end{gathered}$$

Note that

$$[11]\beta = \beta \uplus [10]\beta = (1,3) \uplus (1,4) = O$$

Let $Q = [5]\beta = (4,1)$, then the discrete logarithm of the point Q to the base P is equal to 5. Similarly, the logarithm of the point $(2,5)$ to the base $(1,3)$ is equal to 9. □

In the rest of this section, elliptic curve based El Gamal cryptosystem, an algorithm due to Menezes and Vanstone, and finally the Diffie-Hellman elliptic curve based key exchange algorithm are described.

3.10.1 Elliptic Curve Based El Gamal Cryptosystem

The elliptic-curve El Gamal ciphering scheme is an abstract version of the El Gamal cipher. It assumes that the DLP is computationally expensive in the group of points generated by an elliptic curve. However, the group operations in the elliptic curve are assumed to be relatively less expensive to implement. Consider the elliptic curve

$$y^2 \equiv \left(x^3 + ax + b\right) \pmod{p}\,; \quad a, b \in \mathbb{Z}_p,\; x, y \in \mathbb{Z}_p$$

where p is a prime number greater than 3. Let the group generated by this elliptic curve be denoted by $\mathcal{G}_{\mathbb{E}_p} = \left(\mathbb{E}_p\left(a, b\right), \uplus\right)$, where a and b are the parameters of the elliptic curve, and $\uplus$ is the group operation. Denote the set of points in $\mathbb{E}_p\left(a, b\right)$ without the zero-point O, by $\mathbb{E}_p^*\left(a, b\right)$.

Each member (entity) of the group who wants to receive a secure message, generates his or her public and private keys. Thus each entity A should execute the key generation algorithm. The key generation algorithm, and algorithms for encryption and decryption operations of the EC based El Gamal cryptographic scheme are subsequently outlined.

Algorithm 3.4a. *Key Generation for EC Based El Gamal Scheme.*

Output: Key generation for entity A. That is, generate A's public and private keys.

begin

Step 1: Let $\mathcal{J} = (J, \uplus)$ be a cyclic subgroup of $\mathcal{G}_{\mathbb{E}_p}$.
Let $P \in J$ be a generator of this subgroup.
It is assumed that the DLP in this subgroup is computation- -ally very expensive.

Step 2: Pick a random integer $n, 1 \le n \le \left(|J| - 1\right)$, and compute $Q = [n]\,P$.

Step 3: A's public key is $\left(\mathbb{E}_p^*\left(a, b\right), P, Q\right)$, and a description of the $\uplus$ operation on the elements of J. The corresponding private key is n.

end (*end of El Gamal EC based key generation algorithm*)

Algorithm 3.4b. *Encryption and Decryption in the EC Based El Gamal Scheme.*

Specification: The EC based El Gamal cryptosystem is $\left(\mathcal{P}, \mathcal{C}, \mathcal{K}, \mathcal{E}, \mathcal{D}\right)$.
$\mathcal{P} = \mathbb{E}_p^*\left(a, b\right)$ and $\mathcal{C} = \mathbb{E}_p^*\left(a, b\right) \times \mathbb{E}_p^*\left(a, b\right)$, where $p > 3$ is a prime number. Let $\mathcal{J} = (J, \uplus)$ be a cyclic subgroup of $\mathcal{G}_{\mathbb{E}_p}$. Also let $\mathcal{K} = \left\{\left(\mathbb{E}_p^*\left(a, b\right), P, n, Q\right) \mid Q = [n]\,P, 1 \le n \le \left(|J| - 1\right)\right\}$, and $K \triangleq \left(\mathbb{E}_p^*\left(a, b\right), P, n, Q\right) \in \mathcal{K}$. A message $m \in \mathcal{P}$ is encrypted into $\left(\mu_1, \mu_2\right) \in \mathcal{C}$.

Encryption Operation. Entity B encrypts a message $m \in \mathcal{P}$ destined for entity A.
Input: The plaintext message $m \in \mathcal{P}$ with entity B.
Output: Entity B transmits and entity A receives $(\mu_1, \mu_2) \in \mathcal{C}$.
begin

Step 1: Entity B obtains entity A's public key $\left(\mathbb{E}_p^*(a,b), P, Q\right)$.
The entity B wants to transmit the message $m \in \mathcal{P}$ to entity A.
Step 2: Generate a secret random integer k, where $1 \le k \le (|J| - 1)$.
Step 3: Compute $\mu_1 = [k] P$, and $\mu_2 = (m \uplus [k] Q)$.
Step 4: The encrypting map is $E_K(m) = (\mu_1, \mu_2)$.
Step 5: Entity B sends the ciphertext (μ_1, μ_2) to the entity A.

end (*end of encryption operation*)

Decryption Operation. Entity A decrypts the ciphertext $(\mu_1, \mu_2) \in \mathcal{C}$.
Input: The encrypted message $(\mu_1, \mu_2) \in \mathcal{C}$ with entity A.
Output: Entity A generates the plaintext message $m \in \mathcal{P}$.
begin

Step 1: Entity A receives the encrypted message (μ_1, μ_2).
Step 2: Entity A uses its private key to compute
$D_K(\mu_1, \mu_2) = \mu_2 \uplus (-[n]\mu_1)$, where "$-$" is the negative operation in $\uplus$ arithmetic.
This value is equal to $m \in \mathcal{P}$.

end (*end of decryption operation*)

Proof of the correctness of the algorithm. Observe that

$$\begin{aligned} D_K(\mu_1, \mu_2) &= \mu_2 \uplus (-[n]\mu_1) = (m \uplus [k] Q) \uplus (-[nk] P) \\ &= (m \uplus [kn] P) \uplus (-[nk] P) = m \end{aligned}$$

□

Observe that this cryptosystem is nondeterministic, since it depends upon the parameter k, which is a random integer. The next example illustrates the use of this scheme.

Example 3.16. Encrypting and decrypting operations in the elliptic-curve El Gamal cryptosystem is demonstrated with small parameters. Due to the small values of the parameters, this cryptosystem is definitely not secure. The following congruence is used.

$$y^2 \equiv \left(x^3 + 2x + 6\right) \pmod 7, \quad x, y \in \mathbb{Z}_7$$

The number of points in the set $\mathbb{E}_7(2,6)$, which includes the zero-point O, is equal to $N = 11$. Therefore the Abelian group $(\mathbb{E}_7(2,6), \uplus)$ has a prime order. Consequently it is a cyclic group isomorphic to $(\mathbb{Z}_{11}, +)$. Assume that a generator of this group is $P = \beta = (1,3)$. Other members of this group which can be generated from β are listed in an earlier example. Select the cyclic subgroup $\mathcal{J}$ to be equal to $\mathcal{G}_{\mathbb{E}_7}$. Let the message to be encrypted be $m = [3]\beta = (5,1)$. It is to be sent by entity B to entity A.

Key generation: Select a random number n, between 1 and 10. Let $n = 4$, then $Q = [n]P = [4]\beta = (3,5)$. A's public key is $(\mathbb{E}_7^*(2,6), P, Q)$ and the corresponding private key is 4.

Encryption: Entity B wants to send entity A, message m. Therefore entity B retrieves entity A's public key. Entity B then selects a secret random integer $k, 1 \leq k \leq 10$. Let $k = 7$. Compute μ_1 and μ_2 as follows.

$$\mu_1 = [k]P = [7]P = [7]\beta = (3,2)$$
$$\mu_2 = (m \uplus [k]Q) = (5,1) \uplus [7](3,5) = [3]\beta \uplus [28]\beta = [31]\beta = [9]\beta = (2,5)$$

Thus the encrypting map is

$$E_K(m) = (\mu_1, \mu_2)$$
$$E_K((5,1)) = ((3,2),(2,5))$$

Entity B sends the ciphertext $(\mu_1, \mu_2) = ((3,2),(2,5))$ to the entity A.

Decryption: The ciphertext is decrypted by entity A. Its private key is $n = 4$. The decrypted message is

$$D_K(\mu_1, \mu_2) = \mu_2 \uplus (-[n]\mu_1)$$
$$D_K((3,2),(2,5)) = (2,5) \uplus \{-[4](3,2)\} = [9]\beta \uplus \{-[4]([7]\beta)\}$$
$$= -[19]\beta = [3]\beta = (5,1)$$

Therefore $D_K((3,2),(2,5))$ is indeed equal to the original message m. □

From the above discussion of the elliptic-curve El Gamal cryptosystem, it can be observed that El Gamal's scheme can be made to work with any finite cyclic group in which the discrete logarithm problem is computationally infeasible.

3.10.2 Menezes-Vanstone Elliptic Curve Cryptosystem

The El Gamal elliptic curve cryptosystem uses message elements in the set $\mathbb{E}_p^*(a,b)$. However, it is not convenient to generate the elements of the group $\mathcal{G}_{\mathbb{E}_p} = (\mathbb{E}_p(a,b), \uplus)$; where p is a prime number greater than 3, a and b are the parameters of the elliptic curve, and $\uplus$ is the group operation. A. Menezes and S. Vanstone suggested a scheme for an elliptic curve cryptosystem where messages are chosen from the space $\mathbb{Z}_p^* \times \mathbb{Z}_p^*$.

The key generation algorithm uses a cyclic subgroup of $\mathcal{G}_{\mathbb{E}_p}$. Thus the key generation algorithm is identical to the one used for the EC based El Gamal cryptosystem. Let $\mathcal{J} = (J, \uplus)$ be a cyclic subgroup of $\mathcal{G}_{\mathbb{E}_p}$. It is also assumed that the DLP in this subgroup is computationally very expensive. Also let $P \in J$ be a generator of this subgroup.

Algorithm 3.5. *Encryption and Decryption in the Menezes-Vanstone Scheme.*

Specification: The Menezes-Vanstone cryptosystem is $(\mathcal{P}, \mathcal{C}, \mathcal{K}, \mathcal{E}, \mathcal{D})$.

$\mathcal{P} = \mathbb{Z}_p^* \times \mathbb{Z}_p^*$ and $\mathcal{C} = \mathbb{E}_p^*(a,b) \times \mathbb{Z}_p^* \times \mathbb{Z}_p^*$,
where $p > 3$ is a prime number.
Let $\mathcal{J} = (J, \uplus)$ be a cyclic subgroup of $\mathcal{G}_{\mathbb{E}_p}$.
Also let $\mathcal{K} = \{(\mathbb{E}_p^*(a,b), P, n, Q) \mid Q = [n]P, 1 \le n \le (|J|-1)\}$, and
$K \triangleq (\mathbb{E}_p^*(a,b), P, n, Q) \in \mathcal{K}$.
A message $m \in \mathcal{P}, m = (m_1, m_2)$, and $m_1, m_2 \in \mathbb{Z}_p^*$.
It is encrypted into $(\mu_0, \mu_1, \mu_2) \in \mathcal{C}$.

Encryption Operation. The entity B wants to transmit the message $m \in \mathcal{P}$ to entity A. Entity B performs the following steps.
Input: The plaintext message $m \in \mathcal{P}$ with entity B.
Output: Entity B transmits, and entity A receives $(\mu_0, \mu_1, \mu_2) \in \mathcal{C}$.
begin
Step 1: Entity B obtains entity A's public key $(\mathbb{E}_p^*(a,b), P, Q)$.
Step 2: Generate a secret random integer k, where $1 \le k \le (|J|-1)$.
Step 3: Compute $\mu_0 = [k]P$, and $(\eta_1, \eta_2) = [k]Q$.
Step 4: Compute $\mu_1 \equiv \eta_1 m_1 \pmod p$ and $\mu_2 \equiv \eta_2 m_2 \pmod p$
Step 5: The encrypting map is $E_K(m) = (\mu_0, \mu_1, \mu_2)$.
Entity B sends the ciphertext (μ_0, μ_1, μ_2) to the entity A.
end (*end of encryption operation*)

Decryption Operation. Entity A receives the encrypted message (μ_0, μ_1, μ_2). Entity A uses its private key and performs the following operations.
Input: The encrypted message $(\mu_0, \mu_1, \mu_2) \in \mathcal{C}$ with entity A.
Output: Entity A generates the plaintext message $m \in \mathcal{P}$.
begin
Step 1: Compute $[n]\mu_0 = (\delta_1, \delta_2) \in \mathbb{E}_p^*(a,b)$.
Step 2: Compute $\theta_1 \equiv \mu_1 \delta_1^{-1} \pmod p$ and $\theta_2 \equiv \mu_2 \delta_2^{-1} \pmod p$.
Step 3: $D_K(\mu_0, \mu_1, \mu_2) = (\theta_1, \theta_2) = (m_1, m_2) = m$.
end (*end of decryption operation*)

Proof of the correctness of the algorithm. Observe that

$$(\delta_1, \delta_2) = [n]\mu_0 = [nk]P = (\eta_1, \eta_2)$$

Also $\theta_1 \equiv \mu_1 \delta_1^{-1} \pmod p \equiv (\eta_1 m_1)\eta_1^{-1} \pmod p \equiv m_1$. Similarly $\theta_2 = m_2$. □

Example 3.17. Encrypting and decrypting in the Menezes-Vanstone elliptic curve cryptosystem is demonstrated with small parameters. Use the congruence

$$y^2 \equiv (x^3 + 2x + 6) \pmod 7, \quad x, y \in \mathbb{Z}_7$$

The number of points in the set $\mathbb{E}_7(2,6)$, which includes the zero-point $\mathcal{O}$, is equal to $N = 11$. Therefore the Abelian group $(\mathbb{E}_7(2,6), \uplus)$ has a prime order. Consequently it is a cyclic group isomorphic to $(\mathbb{Z}_{11}, +)$. Assume that a generator of this group is $P = \beta = (1,3)$. Other members of this group which can be generated from β are listed in an earlier example. Select the cyclic subgroup $\mathcal{J}$ to be equal to $\mathcal{G}_{\mathbb{E}_7}$. Let the message to be encrypted be $m = (m_1, m_2) = (2, 4) \in \mathbb{Z}_7^* \times \mathbb{Z}_7^*$. It is to be coded and transmitted by entity B to entity A.

Key generation: Select a random number n between 1 and 10. Let $n = 4$, then $Q = [n] P = [4] \beta = (3, 5)$. A's public key is $(\mathbb{E}_7^* (2, 6), P, Q)$ and the corresponding private key is 4.

Encryption: Entity B wants to send entity A, message m. Note that $|J| = 11$. Entity B gets entity A's public key. Subsequently, entity B generates a secret random integer k, lying in the interval $1 \leq k \leq 10$. Let $k = 7$. Compute $\mu_0 = [k] P$, and $(\eta_1, \eta_2) = [k] Q$ as follows.

$$\mu_0 = [k] P = [7] P = [7] \beta = (3, 2)$$
$$(\eta_1, \eta_2) = [k] Q = [7] (3, 5) = [7] ([4] \beta) = [28] \beta = [6] \beta = (4, 6)$$

This yields

$$\mu_1 \equiv \eta_1 m_1 \pmod p \equiv 4 (2) \pmod 7 \equiv 1 \pmod 7$$
$$\mu_2 \equiv \eta_2 m_2 \pmod p \equiv 6 (4) \pmod 7 \equiv 3 \pmod 7$$

Then the encrypting map is $E_K (m) = (\mu_0, \mu_1, \mu_2) = ((3, 2), 1, 3)$. Entity B sends the ciphertext $((3, 2), 1, 3)$ to the entity A.

Decryption: The ciphertext is decrypted by entity A.

Compute

$$(\delta_1, \delta_2) = [n] \mu_0 = [4] (3, 2) = [28] \beta = [6] \beta = (4, 6)$$

Next compute

$$\theta_1 \equiv \mu_1 \delta_1^{-1} \pmod p \equiv 1 (4)^{-1} \pmod 7 \equiv 1 (2) \pmod 7 \equiv 2 \pmod 7$$
$$\theta_2 \equiv \mu_2 \delta_2^{-1} \pmod p \equiv 3 (6)^{-1} \pmod 7 \equiv 3 (6) \pmod 7 \equiv 4 \pmod 7$$

Therefore the decrypted message is equal to $(\theta_1, \theta_2) = (2, 4)$, which is indeed equal to the original message m. □

3.10.3 Elliptic Curve Based Key Exchange Algorithm

An elliptic curve based key exchange algorithm is described. This is essentially the application of groups defined on elliptic curves to the Diffie-Hellman key exchange algorithm.

The Elliptic Curve Based Diffie-Hellman Key Exchange Algorithm

Two entities A and B wish to establish a shared secret key K, over a public channel. The following steps are executed.

1. *Initial setup*: Select a large number q such that either q is a large prime number p, or $q = 2^m$.
   (a) If q is a large prime number $p > 3$, then $\mathbb{E}_q (a, b)$ is the union of the zero-point O, and the set of points on the elliptic curve $y^2 \equiv (x^3 + ax + b) \pmod p$, where $a, b, x, y \in \mathbb{Z}_p$, $(4a^3 + 27b^2) \not\equiv 0 \pmod p$, and $p > 3$.
   (b) If $q = 2^m$, then $\mathbb{E}_q (a, b)$ is the union of the zero-point O, and the set of points on the elliptic curve $y^2 + xy = (x^3 + ax^2 + b)$, where $a, b, x, y \in GF_q$, and $b \neq 0$.

Select a *base point* $P \in \mathbb{E}_q(a, b)$ such that its order n is very large, where $[n]P = O$. The published parameters are $\mathbb{E}_q(a, b)$ and the point P.

2. Entity A selects a random secret number u, $1 \leq u < n$, and sends $\alpha = [u]P$ to entity B. The values u and α are A's private and public key respectively.
3. Entity B selects a random secret number v, $1 \leq v < n$, and sends $\beta = [v]P$ to entity A. The values v and β are B's private and public key respectively.
4. Entity A receives β and then computes the shared key as $K = [u]\beta = [u][v]P$.
5. Entity B receives α and then computes the shared key as $K = [v]\alpha = [v][u]P = [u][v]P$. □

This key exchange algorithm implies that the Diffie-Hellman key exchange algorithm can be used in any group in which the DLP is hard, and the exponentiation operation is efficient.

3.11 Hyperelliptic Curve Based Cryptosystem

The first proponent for the use of hyperelliptic curves in cryptography was Neal Koblitz (1989). He suggested that the public-key algorithms can be implemented using hyperelliptic curves. Actually, hyperelliptic curves are a generalization of elliptic curves. Hyperelliptic curves defined over finite fields are considered for cryptographic applications. This curve has a Jacobian group associated with it. Its group operation is discussed at length in the chapter on abstract algebra. Furthermore, these groups have sufficient properties to resist cryptanalytic attacks. As in the case of elliptic curve cryptography, hyperelliptic curve cryptographic schemes offer the same level of security as the RSA scheme, albeit with smaller key sizes. We cover basic concepts that lead to the definition of the Jacobian group of a hyperelliptic curve, and indicate its use in cryptography.

Consider a field $\mathbb{F}$ and let $\overline{\mathbb{F}}$ be its algebraic closure. A hyperelliptic curve C of genus $g \in \mathbb{P}$ over the field $\mathbb{F}$ is:

$$C : y^2 + h(x)y = f(x) \quad \text{in} \quad \mathbb{F}[x, y]$$

where:

(a) $h(x) \in \mathbb{F}[x]$ is a polynomial of $\deg h(x) \leq g$.
(b) $f(x) \in \mathbb{F}[x]$ is a monic polynomial of $\deg f(x) = (2g + 1)$.
(c) There are no solutions $(x, y) \in \overline{\mathbb{F}} \times \overline{\mathbb{F}}$ that simultaneously satisfy the equation $(y^2 + h(x)y) = f(x)$, and the partial derivative equations

$$2y + h(x) = 0, \quad \text{and} \quad h'(x)y = f'(x)$$

where $h'(x)$ and $f'(x)$ are first derivatives of $h(x)$ and $f(x)$ respectively, with respect to x.

If the genus g is equal to unity, then C is simply an elliptic curve. In contrast to points on elliptic curves, there is no simple technique to provide a group structure for points on the curve C for genus $g \geq 2$. An alternate scheme would be to introduce a different entity which is related to the curve C, and then define the group operation on such entities. This entity, and its associated group operation is called the Jacobian of C. Jacobians of hyperelliptic curves in turn are studied by studying an entity called divisor. A divisor D is a formal sum of points on C, where

$$D = \sum_{P \in C} m_P P, \quad m_P \in \mathbb{Z}$$

and only a finite number of m_P's are nonzero. The degree of divisor D is denoted by $\deg D$. It is equal to the integer sum $\sum_{P \in C} m_P$. The set of all divisors is denoted by $\mathbb{D}(C)$, and the set of all divisors of degree 0 is denoted by $\mathbb{D}^0(C)$. That is,

$$\mathbb{D}^0(C) = \{D \in \mathbb{D}(C) \mid \deg D = 0\}$$

The set of all divisors $\mathbb{D}(C)$ form an additive Abelian group under the divisor sum operation. Also, $\mathbb{D}^0(C)$ is a subgroup of $\mathbb{D}(C)$. The coordinate ring of the hyperelliptic curve C over the field $\overline{\mathbb{F}}$ is denoted by $\overline{\mathbb{F}}[C]$. It is the quotient ring

$$\overline{\mathbb{F}}[C] = \overline{\mathbb{F}}[x, y] \,/\, (r(x, y))$$

where the polynomial $r(x, y) = \left(y^2 + h(x)y - f(x)\right)$ is irreducible over the field $\overline{\mathbb{F}}$. An element of $\overline{\mathbb{F}}[C]$ is called a polynomial function on C.

The function field $\mathbb{F}(C)$ of the hyperelliptic curve C over $\mathbb{F}$ is the field of fractions of $\mathbb{F}[C]$. Similarly, the function field $\overline{\mathbb{F}}(C)$ of C over $\overline{\mathbb{F}}$ is the field of fractions of $\overline{\mathbb{F}}[C]$. The elements of $\overline{\mathbb{F}}(C)$ are called rational functions on C.

Let $R \in \overline{\mathbb{F}}(C)$, and $P \in C$, where P is not the point at infinity, which is P_∞. The rational function R is defined at P, if there exist polynomial functions $G, H \in \overline{\mathbb{F}}[C]$ so that $R = G/H$ and $H(P) \neq 0$. However, if no such $G, H \in \overline{\mathbb{F}}[C]$ exist, then R is not defined at P. If R is defined at the point P, the value of R at P is $R(P) = G(P)/H(P)$. We are now ready to describe an entity which will help us specify the Jacobian group of the hyperelliptic curve C.

A divisor $D \in \mathbb{D}^0(C)$ is a principal divisor, if $D = \operatorname{div}(R)$ for some nonzero rational function $R \in \overline{\mathbb{F}}(C)$. Denote the set of all principal divisors by $\mathbb{P}_{pd}(C)$. Observe that the set of all principal divisors $\mathbb{P}_{pd}(C)$ is a subgroup of $\mathbb{D}^0(C)$. The Jacobian of the curve C is the quotient group $\mathbb{J}(C) = \mathbb{D}^0(C)/\mathbb{P}_{pd}(C)$. Therefore, the identity element of this group is the coset of $[P_\infty]$, where P_∞ is the point at infinity. The binary operation $\oplus$ of this group is best described by an algorithm due to Cantor and Koblitz. Mumford representation of a divisor is a computationally convenient and elegant technique to work with divisors.

Assume that the hyperelliptic curve is defined over a finite field $\mathbb{F}_q$; where $q = p^n$, p is a prime number, and $n \in \mathbb{P}$. Let J be the set of points on which this group operation acts. In this notation, the group $\mathbb{J}(C) = (J, \oplus)$. Denote the cardinality of the set J by $|J| = N$. Thus the order of the Jacobian is N. An estimate of the cardinality of the set J is given by the Hasse-Weil theorem. It is named after the mathematicians H. Hasse and André Weil. It is

$$(\sqrt{q} - 1)^{2g} \leq N \leq (\sqrt{q} + 1)^{2g}$$

For large values of q, $N \simeq q^g$.

Analogous to the security of the elliptic curve based cryptosystems, the security of the hyperelliptic curve based cryptosystems is based upon the discrete logarithm problem. Let $D, D_1 \in J$, and

$$D_1 = \underbrace{D + D + \cdots + D}_{n \text{ terms}} \triangleq [n]D, \quad n \in \mathbb{P}\backslash\{1\}$$

Observe that it is easy to compute D_1, given D and n. However, it is relatively hard to determine n, given D and D_1. This is the DLP for the hyperelliptic curves. This is the basis for considering the hyperelliptic curve arithmetic in designing public key cryptosystems.

3.12 Hash Functions

Generally, digital signature schemes permit small messages to be signed. Therefore, if the message is very long, it can be split into several blocks, and then each block can be signed separately. However, this entails huge overhead, both in transmitting the large message with several signatures, and also in generating the signatures. Recall that secure digital signatures take large computational time. Therefore, in order to decrease the transmission time, and preserve the integrity of the entire message, a hash function can be used.

In the field of cryptography, hash functions have found widespread use in digital signatures. More precisely, such functions are called cryptographic hash functions. A hash function maps bitstrings of arbitrary length to strings of small fixed-length. A hash function is said to output a *digital fingerprint* of the message. Alternate terms for it are *imprint* of the message, and *message digest.*

Assume that σ is an alphabet. For example, the set of symbols $\{0, 1\}$ of a bitstring represent an alphabet. Denote the set of sequences of symbols of length n by $\sigma^{(n)}$, and the set of sequences of symbols of any length by σ^*.

Definition 3.21. *A hash function is a map $h : \sigma^* \to \sigma^{(n)}$, where $n \in \mathbb{P}$. Also:*

(a) *The hash function performs compression. That is, it maps an input $m \in \sigma^*$ of arbitrary finite length, to an output $h(m)$ of fixed length n. Note that m is a message, and $h(m)$ is called its hash-value.*

(b) *The function $h(\cdot)$ is one-way.*

(c) *The function $h(\cdot)$ is easily computable for a given value of m.* □

It is also required that the output of the hashed function be sufficiently pseudorandom. The one-way property of hash function ensures that, for a given hash value, it is hard to find the corresponding input string. Let us assume that Bob wants to send Alice a signed message m. Bob first creates an imprint of the message m, by computing its hash-value $h(m) = \lambda$. He then creates its digital signature $s = S_K(\lambda)$ by using his private key. Bob then transmits the ordered pair (m, s) over the communication channel to Alice. Note that the hash functions are publicly available. Due to the presence of eavesdropper Oscar on the communication channel, Alice receives (m', s), where m' may not necessarily equal to m. When Alice receives the two-tuple (m', s), she reconstructs the message digest $\lambda' = h(m')$. In the next step, she performs the verification operation $V_K(\lambda', s)$ by using Bob's public key. If its value is true, she accepts the message, otherwise she rejects it. One has to be cautious in this scheme, because it is the message digest that is signed and not the message. We next discuss collision-free hash functions.

Creation of hash-functions is a compression operation. Collision is said to occur if a pair of different strings are mapped to the same output by a hash function. Therefore, care should be taken to prevent *forgery.* Oscar the opponent can execute forgery the following way. Oscar has in his possession, a valid signed message (m, s). Assume that this signed message was sent by Bob to Alice. Oscar can find another message m', such that $h(m) = h(m')$. Then (m', s) is a valid signed message, albeit a forged one. Therefore, in order to prevent such attack, the hash-function is required to have certain so-called collision-free properties.

Definitions 3.22. *Weakly and strongly collision-free hash functions.*

1. *A weakly collision-free hash function* $h(\cdot)$ *has the following characteristics: If message* m *is given, then it is computationally infeasible to find another message* $m' \neq m$ *such that* $h(m') = h(m)$. *Note that in this requirement, the message* m *is specified.*
2. *A strongly collision-free hash function is a weakly collision-free hash function and it also satisfies the following condition: It is computationally infeasible to find any two messages* $m_1 \neq m_2$, *such that* $h(m_1) = h(m_2)$. *Note that in this requirement,* m_1 *and* m_2 *are any two distinct messages.* □

A hash-function, which is weakly collision-free is also susceptible to forgery. Most hash functions are one-way functions. A hash-function $h(\cdot)$ is said to be a one-way function, if for say hash-value h_0, it is computationally infeasible to find a message m', such that $h(m') = h_0$, where m' is not necessarily equal to m. It is generally a requirement that a strongly collision-free hash function be a one-way function.

3.13 Authentication

Authentication is the confirmation or verification, that an entity (a person or party), or the source of a message, is as claimed. As mentioned earlier, there are two types of authentication. These are:

(a) Entity authentication
(b) Data-origin authentication.

Entity authentication is also referred to as *identification.* For example, Alice and Bob would like to identify each other in real time. This can be done by Alice issuing Bob a challenge to which only Bob can answer correctly. Bob can reciprocate by sending Alice a similar challenge.

Data-origin authentication is also referred to as *message authentication.* Let entity A send a message to entity B. An authentication of this message is provided to the receiver B, by appending additional information to the message about the entity A which originated the message. Note that data-origin authentication also provide data integrity.

3.13.1 Identification

The purpose of identification is to authenticate an entity's identity. The two entities A and B share a secret key ahead of time. Entity A wishes to identify unknown entity Ω. It wants to confirm whether entity Ω is indeed entity B. The entity A is called the *challenger*, and the entity Ω is called the *responder*. Identification schemes via one-way functions, public-key cryptosystem, and a scheme due to L. Guillou and J.-J. Quisquater are described. The later scheme is referred to as the GQ identification scheme.

Identification Using One-Way Functions

Following steps can be executed for identification via one-way function. Recall that one-way functions are relatively easy to compute, but difficult to invert efficiently.

1. Entity A transmits a time-varying number to the unidentified entity Ω. Entity A wants to know whether the unidentified entity Ω is indeed entity B.

2. Upon receiving this number, entity Ω replies to entity A with a message. This message is a one-way function of the number it received from entity A, and the shared key it possessed (if any).
3. If the message received by entity A from entity Ω was as expected, then the entity Ω is the entity B. However, if the message received by entity A from entity Ω was not as expected, then the entity Ω remains unknown.

A possible candidate for one-way function is a keyed one-way hash function, or a one-way function of the shared secret key.

Identification via Public-Key Cryptosystem

The two entities A and B share a secret key ahead of time. Entity A wishes to confirm whether unknown entity Ω is indeed entity B. This is done via public-key cryptosystem as follows. For a key K; $E_K(\cdot)$ and $D_K(\cdot)$ are the encrypting and decrypting functions respectively.

1. Entity A selects a pseudorandom number x. It generates another number r such that $r = E_K(x)$. Entity A sends r to entity Ω.
2. When entity Ω receives message r, it uses its own private key to decrypt r, and then sends $D_K(r)$ to entity A on the transmission channel.
3. Entity A checks if $E_K(D_K(r)) = r$. If this is true, then entity Ω is indeed entity B, otherwise entity Ω remains unknown.

The GQ identification scheme is specified below.

GQ Identification Scheme

The GQ identification scheme requires the use of a *trusted third party* (TTP). A TTP is a person who is not an entity requiring an identification. Member entities of a group which need to be identified are initially issued identities by the TTP. These identities are issued as per the following rules.

1. *System parameters*:
   (a) A TTP selects two prime numbers p and q, and computes $n = pq$. Assume that it is computationally expensive to factor n.
   (b) TTP selects an exponent $a \geq 3$, so that $\gcd(a, \varphi(n)) = 1$, and $\varphi(\cdot)$ is Euler's totient function. TTP also computes $s \equiv a^{-1} \pmod{\varphi(n)}$.
   (c) The values n and a are public, while the prime numbers p and q are kept secret by TTP. The integer s is also kept secret by TTP.
2. *Parameters for each user of the group*:
   (a) Each entity Υ which belongs to the group has a unique identity ϑ_Υ from which an integer J_Υ is derived such that $1 < J_\Upsilon < n$. As factoring n is difficult, it is implicitly assumed that $\gcd(J_\Upsilon, \varphi(n)) = 1$. In addition, the integer J_Υ is public.
   (b) For each entity Υ, the TTP computes

$$U_\Upsilon \equiv J_\Upsilon^s \pmod{n}, \text{ and } W_\Upsilon \equiv U_\Upsilon^{-1} \pmod{n}$$

Thus

$$J_{\Upsilon} W_{\Upsilon}^{a} \equiv 1 \pmod{n}$$

The TTP gives the entity Υ, the secret W_{Υ}.

3. *Protocol actions*: The identification scheme works as follows. It provides unilateral identification of entity Ω to entity A. Thus the entity A is the verifier (challenger), and the unknown entity Ω is the prover (entity whose identity is to be confirmed). Entity A wants to confirm whether the entity Ω is actually entity B.
   (a) Entity Ω selects a random integer $r, 1 < r < (n-1)$, and computes $\beta \equiv r^{a} \pmod{n}$.
   (b) Entity Ω sends to entity A, the two-tuple $(\vartheta_{\Omega}, \beta)$.
   (c) Entity A selects a random integer d such that, $1 \leq d \leq (a-1)$ and sends it to entity Ω. The integer d is called the *challenge.*
   (d) Entity Ω computes $t \equiv r W_{\Omega}^{d} \pmod{n}$, and sends t to entity A. The integer t is called the *response.*
   (e) Upon receiving t, entity A computes $\alpha \equiv J_{\Omega}^{d} t^{a} \pmod{n}$. If $\alpha = \beta$ and $\alpha \neq 0$, then the entity A concludes that the entity Ω is indeed entity B. The condition $\alpha \neq 0$ eliminates the possibility that the adversary succeeds by selecting $r = 0$.
   (f) For extra precaution, steps (a) though (e) are executed, say $\varkappa$ number of times. The identification is true, if it succeeds in all $\varkappa$ cases. □

We make a note of the following set of observations about this identification scheme.

Observations 3.3. Observations about the GQ identification scheme.

1. Each time a new request for identification is made by entity A, a new random integer r should be generated.
2. The security of this scheme rests on the fact that the factors of integer n are not known. If the factors of integer n are known, then computation of a-th root of integers modulo n is easier.
3. Oscar the opponent has approximately only one in a chance of subverting this identification scheme by correctly guessing a value of d. □

3.13.2 Message Authentication

As mentioned earlier, data-origin authentication is also called message authentication. Any specific digital signature scheme or public-key scheme can be used for message authentication. However, the following simple scheme is described which uses *message authentication code* (MAC). This technique uses a secret key to generate a small fixed-size block of data, which is appended to the message. This block is called MAC or a *cryptographic checksum.*

Consider two entities A and B who want to communicate with each other. Let these entities share a secret key K. Also let the MAC function be $C_K(\cdot)$. Assume that the entity A wants to send a message m to entity B.

1. The entity A computes $C_K(m) = c$.
2. The entity A sends the two-tuple (m, c) to entity B.
3. Entity B possibly receives the two-tuple (m', c), due to tampering by Oscar the opponent. Note that Oscar can alter the message, but not the MAC, because he does not know the secret key.
4. The entity B computes $C_K(m') = d$.
5. If $d = c$, then entity B is assured that the message is from entity A. □

The success of this scheme depends upon the secure sharing of the secret key K, between the two entities A and B.

3.14 The Data Encryption Standard

The Data Encryption Standard or the DES is a symmetric-key encryption algorithm. It is also known as the Data Encryption Algorithm. This algorithm has been adopted by the government of United States of America (USA) and several other countries. This scheme is a joint work of International Business Machine (IBM) Corporation and US National Security Agency (NSA). However, the implementation details of this algorithm are public knowledge.

DES works on the principle of *confusion* and *diffusion.* As per Claude E. Shannon, this is a requirement of a good cipher. Confusion scrambles the relationship between the key and the ciphertext. On the other hand, diffusion disperses the redundancy of the plaintext over the ciphertext. DES achieves confusion and diffusion, via substitution and permutation operations respectively. DES implements these by an initial permutation, followed by sixteen rounds of substitution, and then finally a permutation. Each of these sixteen rounds of substitution is implemented in a *standard building block* (SBB).

DES is a block ciphering scheme. That is, each block of plaintext is coded independently of past and future blocks of data. The size of a block is 64 bits. Therefore 2^{64} different blocks of plaintext are possible. The length of the key in this coding scheme is 56 bits. The remaining 8 bits are used as parity bits (for error detection). Therefore, there are a total of 2^{56} possible keys. From a 56-bit key, 16 subkeys $K_1, K_2, \ldots, K_{16}$ are selected deterministically, where a subkey is a 48 bit subword of the 56 bit key. Each of these subkeys is used in a SBB to transform 64 bits of input into 64 bits of output. Since there are 16 SBB's, a single subkey is designated to work with a single SBB.

The DES uses modulo 2 arithmetic. This facilitates the implementation of the DES algorithm in hardware. The addition operator in this arithmetic is denoted by $\oplus$, and is termed "exclusive-or" operator by hardware engineers. Let $\mathbb{Z}_2 = \{0, 1\}$, then

$$0 \oplus 0 = 0, \quad 0 \oplus 1 = 1,$$
$$1 \oplus 0 = 1, \quad \text{and} \quad 1 \oplus 1 = 0$$

The basic idea behind the implementation of the DES algorithm is simple. It depends upon the fact that the solution of algebraic equations in modulo 2 arithmetic is hard. Note that these algebraic equations are not required to be linear. Consider the set of polynomials

$$p_j(x_1, x_2, \ldots, x_k) \in \mathbb{Z}_2[x_1, x_2, \ldots, x_k], \quad \text{for} \quad 1 \le j \le n,$$

where the coefficients of these polynomials are in $\mathbb{Z}_2$. It is known that the problem of finding a k-tuple $(x_1, x_2, \ldots, x_k) \in \mathbb{Z}_2^{(k)}$, in polynomial-time such that $p_j(x_1, x_2, \ldots, x_k) = 0$, for $1 \le j \le n$ is computationally intractable for large values of k and n.

A brief outline of the DES algorithm is next provided. This algorithm has three stages. An initial permutation stage. This is followed by sixteen iterations in the SBB's. Lastly there is an inverse permutation operation. The algorithm ciphers a 64 bit plaintext $m \in \mathcal{P}$ into a ciphertext $c \in \mathcal{C}$ of length 64 bits.

3.14.1 The DES Algorithm

The DES encryption algorithm is next described. An outline of the encryption algorithm is first given. This is followed by an algorithm for subkey generation, and the decryption algorithm.

Outline of the Encryption Algorithm

A message m of length 64 bits, and a key K of length 56 bits are the input to the algorithm. The output is a ciphertext c of length 64 bits.

1. The message block m is transformed by an initial permutation $IP(\cdot)$ into a bitstring m_0 of length 64 bits. Denote this transformation by $IP(m) = m_0$. Let $m_0 = L_0 R_0$, where L_0 and R_0 are the first and last 32 bits of m_0 respectively.
2. The following operations are next executed sixteen times. That is, for $1 \leq i \leq 16$

$$L_i = R_{i-1}$$
$$R_i = L_{i-1} \oplus f(R_{i-1}, K_i)$$

where the L_i's and R_i's are each 32-bit strings, and the operation $\oplus$ is performed componentwise. The function $f(\cdot,\cdot)$ is defined separately. Recall that the subkeys K_i for $1 \leq i \leq 16$ are 48 bits each. Each of these subkeys is derived from a single key K of length 56 bits. An algorithm to derive these subkeys is given separately.
3. Finally, an inverse permutation $IP^{-1}(\cdot)$ is performed on the bitstring $R_{16}L_{16}$. That is, $IP^{-1}(R_{16}, L_{16}) = c$, the ciphertext. Observe the order of the strings R_{16} and L_{16} in this mapping. □

The function $f(\cdot,\cdot)$, which is the evaluation of $f(R_{i-1}, K_i)$ is next described.

Evaluation of $f(R_{i-1}, K_i)$

1. The bitstring R_{i-1} is expanded from 32 bits to 48 bits, according to an *expansion function* $E(\cdot)$. The mapping $E(R_{i-1})$ results in an output with the bits from R_{i-1} permuted in some order, with 16 of the bits occurring twice.
2. Next evaluate $E(R_{i-1}) \oplus K_i$. This output is a bitstring of length 48 bits. Denote it by $(B_1 B_2 \ldots B_8)$. Each of the B_j's is a bitstring of length 6 bits.
3. The 6-bit B_j-string is transformed by a substitution-box called the S-box S_j for $1 \leq j \leq 8$. The output of the box S_j is a bitstring D_j, of length 4 bits. Therefore $D_j = S_j(B_j)$ for $1 \leq j \leq 8$. The S_j box can be considered to be an array of 4 rows and 16 columns. Let $B_j = b_1 b_2 b_3 b_4 b_5 b_6$, then the bits b_1 and b_6 determine the row, and the bits b_2, b_3, b_4, and b_5 determine the column of the S_j box. Actually the row $\widetilde{r} = 2b_1 + b_6$, and $0 \leq \widetilde{r} \leq 3$. The binary representation of the column $\widetilde{c}$ is $b_2 b_3 b_4 b_5$, and $0 \leq \widetilde{c} \leq 15$. The $(\widetilde{r}, \widetilde{c})$ entry in the S_j box is D_j. This $(\widetilde{r}, \widetilde{c})$ entry is 4 bits.
4. Therefore the 8 S-boxes yield $D_j = S_j(B_j)$ for $1 \leq j \leq 8$. The bitstring $D = D_1 D_2 \ldots D_8$ is of length 32 bits.
5. The bitstring D is permuted via a permutation operation $P(\cdot)$. Consequently $P(D) = f(R_{i-1}, K_i)$.

The subkey generation algorithm is outlined below.

Subkey Generation Algorithm

A key K of length 64 bits is given. The output of this subkey generation algorithm is sixteen 48-bit keys $K_i, 1 \leq i \leq 16$.

1. The key K is actually a 56 bit key, with 8 parity-check bits for error detection. The parity bits are in positions $8, 16, \ldots, 64$. The parity bits are discarded to obtain a 56 bit key, K_e.
2. Perform a permutation map $PC1(K_e)$. Its output is (C_0, D_0), where C_0, and D_0 are strings of length 28 bits each.
3. For j from 1 to 16, compute

$$C_j = LS_j(C_{j-1}), D_j = LS_j(D_{j-1}), \text{ and } K_j = PC2(C_j, D_j)$$

where LS_j is a circular shift (to the left) operator. It shifts one position, if $j = 1, 2, 9$, or 16. It shifts two positions for all other values of j. Also, $PC2(\cdot, \cdot)$ is a fixed permutation map.

A complete specification of the mappings such as $IP(,), IP^{-1}(,), E(\cdot)$, the S_j's, $P(\cdot)$, $PC1(\cdot)$ and $PC2(\cdot, \cdot)$ can be found in the U.S. Federal Information Processing Standards Publication 46 (dated January 15, 1977).

Decryption Algorithm

The decryption operation is performed by using the encryption algorithm with the same key, but the subkeys are used in the reverse order. That is, the subkey sequence is: $K_{16}, K_{15}, \ldots, K_1$. □

Observations 3.4. Some observations about the DES scheme are listed below.

1. The initial permutation and the final permutation do not add anything to the security of the encrypting scheme. The purpose of the IP is to transpose the input block. This transposition facilitates the loading of the plaintext and ciphertext into the DES hardware in units of bytes (one byte is 8 bits).
2. *Complementation property*: Let x' be the bit-wise complement of a number x. This is obtained by replacing all the 0's by 1's and 1's by 0's in a bitstring. Let m, c, and K be the plaintext message, the ciphertext, and the key respectively. Also let the corresponding bit-wise complement numbers be m', c', and K' respectively. Then

$$E_K(m) = c, \quad \text{and } E_{K'}(m') = c'$$

That is, the complement of a key encrypts the complement of the plaintext into the complement of the ciphertext. This property implies that a chosen-plaintext attack against DES has to test only 2^{55} keys, instead of 2^{56} keys.
3. *Algebraic structure*: If the DES mapping were *closed*, then

$$E_{K_2}(E_{K_1}(m)) = E_{K_3}(m); \;\; m \in \mathcal{P}, \text{ and } K_1, K_2, K_3 \in \mathcal{K}$$

This condition implies that the DES encryption would form a group. Moreover if this were true, then the DES would be more vulnerable to attacks. Fortunately, it has been proven that the DES mapping does not form a group.
4. The size of the key, which is 56 bits has been the subject of intense scrutiny and debate among cryptographers. With modern technology, a brute-force attack against DES is possible. This can be a variation of an exhaustive search. Multiple encryption appears to be the answer for this small key-size problem. A suggested scheme is *triple-encryption*.
A triple encryption is defined as $E^{(3)}_{K_3}\left(E^{(2)}_{K_2}\left(E^{(1)}_{K_1}(m)\right)\right)$, where $E^{(j)}_K(\cdot)$ for $1 \le j \le 3$ is defined as either $E_K(\cdot)$ or $D_K(\cdot)$. For example $E_{K_3}(D_{K_2}(E_{K_1}(m)))$ is called the *E-D-E* triple-encryption scheme.

5. *Nonlinearity*: The S-boxes in the DES scheme are nonlinear. This nonlinearity by itself does not ensure strong cryptographic properties. The S-boxes cannot be approximated by a linear transformation. Therefore these are not amenable to the so-called *linear cryptanalysis*. But the design of S-box is susceptible to attacks by *differential cryptanalysis.* In the realm of DES, "difference" is defined using the exclusive-or operation. In this cryptanalytic scheme, exclusive-or of two plaintexts is compared with the exclusive-or of the corresponding ciphertexts. Note that the plaintexts are encrypted with the same key. The differential cryptanalytic scheme studies the evolution of these differences as the plaintext propagates through the different stages of DES.
6. The most successful attack on the DES is an exhaustive search of the key space. This is possible by assembling special purpose hardware. □

3.14.2 Modes of Operation of DES

DES has been designed to work in five modes. These are: *electronic code book* (ECB), *cipher block chaining* (CBC), *cipher feedback* (CFB), *output feedback* (OFB), and *counter* (CTR) modes. In all these modes, let the secret key be K, and the encryption and decryption functions be $E_K(\cdot)$ and $D_K(\cdot)$ respectively.

1. **Electronic Code Book Mode.** The plaintext message m, is split into 64-bit blocks, and then encrypted. For example, let $m = m_1 m_2 \ldots m_n$, where each m_i is a 64-bit block for $1 \leq i \leq n$. The ciphertext corresponding to the message block m_i is c_i for $1 \leq i \leq n$. Thus, in this mode of operation,

$$\begin{aligned} c_i &= E_K(m_i), \quad 1 \leq i \leq n \\ m_i &= D_K(c_i), \quad 1 \leq i \leq n \end{aligned}$$

   An advantage of the ECB mode, is that the transmission errors of a block, do not affect subsequent blocks.
2. **Cipher Block Chaining Mode.** In this scheme, the plaintext message m, is split into 64-bit blocks as in the ECB mode. Also, each ciphertext block is dependent upon earlier plaintext blocks. Let the bitstring c_0 be initialized by a 64-bit initialization vector IV, that is $c_0 = IV$. This IV value is a random number known to both the transmitter and the receiver. Then

$$\begin{aligned} c_i &= E_K(m_i \oplus c_{i-1}), \quad 1 \leq i \leq n \\ m_i &= c_{i-1} \oplus D_K(c_i), \quad 1 \leq i \leq n \end{aligned}$$

   Different values of initial vector results in different values of the ciphertext.
3. **Cipher Feedback Mode.** In the CBC mode, plaintext messages are ciphered 64-bits at a time. However, in some applications t-bit plaintext units are ciphered, and transmitted without delay, where $1 \leq t \leq 64$. Typically $t = 1$ or $t = 8$. The message is split into t-bit plaintext blocks. Let, $m = m_1 m_2 \ldots m_v$, where each m_i is a t-bits block for $1 \leq i \leq v$. The ciphertext corresponding to the message block m_i is c_i for $1 \leq i \leq v$. Encryption is done as follows. Let $W_1 = IV$, where IV is a 64-bit initialization vector. This IV value is a random 64-bit number known to both the transmitter and the receiver. It is also not necessarily a secret. Let W_i be the value of a shift register where $1 \leq i \leq v$. The length of the shift register is 64 bits. Perform the following operations for $1 \leq i \leq v$:

(a) $\psi_i = E_K(W_i)$
(b) $u_i =$ the t leftmost bits of ψ_i.
(c) $c_i = m_i \oplus u_i$. This is the t-bit ciphertext, which is transmitted.
(d) $W_{i+1} \equiv (2^t W_i + c_i) \pmod{2^{64}}$. That is, shift left the register contents of W_i by t positions and then transfer c_i into the t rightmost positions of the register.

Decryption is performed as follows. Let $W_1 = IV$, then for $1 \le i \le v$, do: $m_i = c_i \oplus u_i$; where u_i, ψ_i, and W_i are computed as in the encryption stage. Note that the block cipher is operated in the encryption mode at both the transmitter and the receiver. Furthermore, in this scheme an error in transmission may affect several message blocks.

4. **Output Feedback Mode.** This scheme is used when no error propagation is specified. The CBC and CFB modes of operation are susceptible to error propagation during the time required to transmit a block of the ciphertext. As in the case of CFB mode, this block cipher is operated in the encryption mode at both the transmitter and the receiver. Also as in the CFB mode, the message is split into t-bit plaintext blocks. If $m = m_1 m_2 \ldots m_v$, where each m_i is a t-bits block for $1 \le i \le v$, then the ciphertext corresponding to the message block m_i is c_i for $1 \le i \le v$. Evidently, higher values of t provide more security. Let $W_1 = IV$, where IV is a 64-bit random initialization vector. It is known to both the transmitter and the receiver, and not necessarily a secret. However, a new IV should be used, if the key K is reused. Let W_i be the value of a 64-bit shift register for $1 \le i \le v$. Perform the following operations for $1 \le i \le \nu$:
(a) $\psi_i = E_K(W_i)$
(b) $u_i =$ the t leftmost bits of ψ_i.
(c) $c_i = m_i \oplus u_i$. This is the t-bit ciphertext, which is transmitted.
(d) $W_{i+1} \equiv (2^t W_i + u_i) \pmod{2^{64}}$. That is, shift left the register contents of W_i by t positions and then transfer u_i into the t rightmost positions of the register.

Observe that the W_i's can be calculated ahead of time. Decryption is similar to the CFB mode. Let $W_1 = IV$, then for $1 \le i \le v$, do: $m_i = c_i \oplus u_i$; where u_i, ψ_i, and W_i are computed as in the encryption stage. There is no error propagation in this scheme. For example, if errors occur in some bit-positions in c_i when it is delivered to the receiver, then only the corresponding bits in the plaintext are in error upon decryption.
The above algorithm used t-bit feedback. Another version of the OFB algorithm uses, 64-bit feedback. This complete 64-bit feedback algorithm is more secure than the t-bit feedback algorithm.

5. **Counter Mode.** In this mode of operation, the message is split into t-bit plaintext blocks. If $m = m_1 m_2 \ldots m_v$, where each m_i is a t-bits block for $1 \le i \le v$, then the ciphertext corresponding to the message block m_i is c_i for $1 \le i \le v$. Let $W_1 = IV$, where IV is a 64-bit random initialization vector. Also, let W_i be the value of a 64-bit shift register for $1 \le i \le v$. Perform the following operations for $1 \le i \le \nu$:
(a) $\psi_i = E_K(W_i)$
(b) $u_i =$ the t leftmost bits of ψ_i.
(c) $c_i = m_i \oplus u_i$. This is the t-bit ciphertext, which is transmitted.
(d) $W_{i+1} \equiv (W_i + 1) \pmod{2^{64}}$.

Similar to OFB mode, the W_i's can be calculated ahead of time. The advantage of CTR mode over OFB mode results from the fact that the c_i's may be calculated in parallel. Consequently, the CTR mode is an ideal candidate for parallelizing.

The commercial successor of the DES is the Advanced Encryption Standard. This is the subject of next section.

3.15 The Advanced Encryption Standard

Special purpose hardware to crack DES has made it almost antiquated. Only triple-DES is still in current use. The Advanced Encryption Standard (AES) can be considered to be the successor of DES. This symmetric iterative cipher is the work of the Belgian mathematicians Joan Daemen and Vincent Rijmen. This new encryption scheme was approved by *National Institute of Standards and Technology* (NIST) in the year 2000. Actually, it was selected in an open worldwide competition. It is also called the Rijndael block cipher (named after its inventors). One of the several preferred pronunciations of Rijndael is "Rain Doll."

The Rijndael block cipher was selected because of its security, performance, computational efficiency, implementability, and flexibility. Observe that all of these characteristics are interrelated.

This block cipher supports keys of size 128 bits, 192 bits, or 256 bits. Similar to the DES algorithm, this cryptographic scheme is iterative. Denote the number of iterations or rounds to code a plaintext block by N_r. The number of iterations $N_r = 10$ if the key size is 128 bits, $N_r = 12$ if the key size is 192 bits, and $N_r = 14$ if the key size is 256 bits. For simplicity in presentation, the algorithm discussed in this chapter is restricted to a key size of 128 bits. The corresponding block size is 128 bits.

3.15.1 Arithmetic Used in AES

The AES algorithm which uses 128-bit key, uses arithmetic in the finite field GF_{256}. This algorithm operates upon bytes, and also on 32-bitstrings.

Recall that a byte is an eight-bitstring. Therefore, the set of all bytes has a cardinality of 256 elements. A byte can be specified as $(b_7 b_6 \ldots b_0)$, where each b_i is either equal to 0 or 1. This is actually a polynomial representation of a byte in base 2. A byte can also be represented as a polynomial of degree less than 8, with coefficients in $\mathbb{Z}_2$. The polynomial representation of $(b_7 b_6 \ldots b_0)$, is

$$\sum_{j=0}^{7} b_j x^j, \quad b_j \in \mathbb{Z}_2, \quad 0 \le j \le 7, \quad x = 2$$

For example, the polynomial representation of the byte (01101011) is

$$\left(x^6 + x^5 + x^3 + x + 1\right)$$

Note that a byte can also be represented in hexadecimal notation, which is a polynomial representation of a number in base 16. The finite field GF_{256} can be represented by $\mathbb{F}_{2^8}$ where

$$\mathbb{F}_{2^8} = \mathbb{Z}_2 [x] / (\xi (x))$$
$$\xi (x) = \left(x^8 + x^4 + x^3 + x + 1\right)$$

The polynomial $\xi (x)$ is irreducible over GF_2. Recall that a polynomial is irreducible if its only divisors are 1 and itself. The irreducible polynomial $\xi (x)$, was chosen by the inventors of the Rijndael algorithm. Perhaps other choices of an irreducible polynomial of degree 8 could have given equally good results. Its is established in the problem section that the polynomial $\xi (x)$ is irreducible over GF_2.

Arithmetical Operations in $\mathbb{F}_{2^8}$

Let the polynomials $a(x), b(x) \in \mathbb{Z}_2[x] / (\xi(x))$ where $\xi(x)$ is an irreducible polynomial of degree 8. Therefore GF_{256} is isomorphic to $\mathbb{Z}_2[x] / (\xi(x))$. Thus:

1. The sum of the polynomials $a(x)$ and $b(x)$ is $c(x) \equiv (a(x) + b(x)) \pmod{\xi(x)}$.
2. The negative of a polynomial $a(x)$, is $a(x)$ itself, because the coefficients of the polynomial $a(x)$ are in $\mathbb{Z}_2$.
3. The product of the polynomials $a(x)$ and $b(x)$ is $d(x) \equiv a(x) b(x) \pmod{\xi(x)}$.
4. The inverse of a polynomial $a(x)$ can be obtained by applying the extended Euclidean algorithm to polynomials. The extended Euclidean algorithm for polynomials is similar to the extended Euclidean algorithm for integers. For a given polynomial $a(x)$, using the extended Euclidean algorithm for polynomials obtain the polynomials $e(x), f(x) \in \mathbb{F}_{2^8}$ such that
$$e(x) a(x) + f(x) \xi(x) = 1$$
Hence $e(x) a(x) \equiv 1 \pmod{\xi(x)}$, and $\{a(x)\}^{-1} \equiv e(x) \pmod{\xi(x)}$.

Note that the arithmetic operations in $\mathbb{Z}_2$ can also be modeled by the exclusive-or operator $\oplus$.

Polynomials over GF_{256}

Assume that α is a root of the irreducible polynomial
$$\xi(x) = \left(x^8 + x^4 + x^3 + x + 1\right)$$
Elements of the finite field GF_{256} can be represented as polynomials in α of degree less than 8. If $a(x) \in GF_{256}[x]$, and $n \in \mathbb{P}$, then
$$a(x) = \sum_{j=0}^{n} a_j x^j; \quad a_j \in GF_{256}, \;\; 0 \le j \le n$$
where a_j's are appropriate polynomials in α.

Example 3.18. Let
$$\begin{aligned} a(x) &= \left(\alpha^6 + \alpha^4 + \alpha + 1\right) x^3 + \left(\alpha^5 + \alpha^3\right) x + \alpha^2 \\ b(x) &= \left(\alpha^5 + \alpha^3 + \alpha\right) x^3 + \left(\alpha^4 + \alpha\right) x^2 + \alpha \end{aligned}$$
Then
$$\begin{aligned} a(x) + b(x) = &\left(\alpha^6 + \alpha^5 + \alpha^4 + \alpha^3 + 1\right) x^3 \\ &+ \left(\alpha^4 + \alpha\right) x^2 + \left(\alpha^5 + \alpha^3\right) x + \alpha^2 + \alpha \end{aligned}$$
and
$$\begin{aligned} a(x) b(x) = &\left(\alpha^7 + \alpha^2 + \alpha\right) x^6 + \left(\alpha^7 + \alpha^6 + \alpha^5 + 1\right) x^5 \\ &+ \left(\alpha^6 + \alpha^5 + \alpha^4 + \alpha^3 + \alpha^2\right) x^4 \\ &+ \left(\alpha^7 + \alpha^6 + \alpha^5 + \alpha^3\right) x^3 + \left(\alpha^6 + \alpha^3\right) x^2 \\ &+ \left(\alpha^6 + \alpha^4\right) x + \alpha^3 \end{aligned}$$

□

3.15.2 The AES Algorithm

The AES encryption algorithm which uses 128-bit key has 10 rounds of iterations. The plaintext and ciphertext size is also 128 bits. This algorithm uses four basic operations, called *layers*, that form a round (iteration). These are:

1. The ByteSub transformation (BS).
2. The ShiftRow transformation (SR).
3. The MixColumn transformation (MC).
4. The AddRoundKey operation (ARK).

The purpose of the BS transformation is to introduce resistance to linear and differential cryptanalytic attacks. This layer is also nonlinear. The SR and MC transformations cause diffusion of the bits over several rounds. Finally, the ARK operation provides the keying.

A single iteration in the AES encrypting algorithm is made up of the BS transformation, followed by the SR transformation, then the MC transformation, and finally the ARK operation. These transformations and operation are discussed in detail in the next subsection. Following is an outline of the AES encryption algorithm.

Algorithm 3.6a. *AES Encryption Algorithm.*

Input: The plaintext $m \in \mathcal{P}$, and the key $k \in \mathcal{K}$.
Output: The output is the ciphertext $c \in \mathcal{C}$.
begin
Step 1: Implement the ARK operation, using the 0-th round key.
Step 2: In each of the rounds from 1 through 9 implement:
BS, SR, MC, and ARK transformations in order. Use round keys 1 through 9 respectively.
Step 3: The round number 10, is slightly different than the previous rounds. Implement BS, SR, and ARK transformations.
Use the key which corresponds to round 10.
end (*end of AES encryption algorithm*)

The decryption is done by implementing the inverse of the above four transformations (operations). These are:

1. The InvByteSub transformation (IBS).
2. The InvShiftRow transformation (ISR).
3. The InvMixColumn transformation (IMC).
4. The AddRoundKey operation is self-inverse. That is, it is its own inverse.

The decryption steps are implemented by using the inverses of the BS, SR, MC, and ARK operations in the reverse order of the encryption algorithm. The input to this algorithm is the ciphertext $c \in \mathcal{C}$, and the output is the plaintext $m \in \mathcal{P}$. Following is an outline of the algorithm. Note that the ARK operations of a round in the decrypting process should be the same as the ARK operation of the corresponding encrypting round.

Algorithm 3.6b. *AES Decryption Algorithm.*

Input: The ciphertext $c \in \mathcal{C}$, and the key $k \in \mathcal{K}$.
Output: The output is the plaintext $m \in \mathcal{P}$.
begin
Step 1: Implement the ARK operation.
Use the 10-th round key, followed by ISR, and IBS.
Step 2: In each of the rounds from 9 through 1 implement:
ARK, IMC, ISR, and IBS transformations.
Step 3: Use the ARK operation which belongs to the 0-th round.
end (*end of AES decryption algorithm*)

This decryption algorithm can also be modified to make it look like the encryption algorithm.

3.15.3 Different Layers

All the four layer operations: the ByteSub transformation, the ShiftRow transformation, the MixColumn transformation, and the AddRoundKey operation are discussed in this subsection. This is followed by an outline of the key scheduling algorithm.

The processing of information in AES is byte oriented. Also observe that the elements of the finite field GF_{256} can be represented by bytes. The plaintext m consists of 16 bytes, which are $m_0, m_1, \ldots, m_{15}$. Note that each m_i is a byte where $0 \leq i \leq 15$. These 16 bytes are arranged in a 4×4 matrix $\widetilde{S}$, as follows:

$$\widetilde{S} = \begin{bmatrix} s_{00} & s_{01} & s_{02} & s_{03} \\ s_{10} & s_{11} & s_{12} & s_{13} \\ s_{20} & s_{21} & s_{22} & s_{23} \\ s_{30} & s_{31} & s_{32} & s_{33} \end{bmatrix} \leftarrow \begin{bmatrix} m_0 & m_4 & m_8 & m_{12} \\ m_1 & m_5 & m_9 & m_{13} \\ m_2 & m_6 & m_{10} & m_{14} \\ m_3 & m_7 & m_{11} & m_{15} \end{bmatrix}$$

In the rest of the discussion, assume that α is a root of the polynomial

$$\xi(x) = \left(x^8 + x^4 + x^3 + x + 1\right)$$

The ByteSub Transformation

The BS transformation, transforms the $\widetilde{S}$ matrix into another matrix $\widetilde{B}$. The matrix $\widetilde{B}$ is also a 4×4 matrix. Its elements are bytes. Denote this transformation by $T_{bs}(\cdot)$, that is $T_{bs}\left(\widetilde{S}\right) = \widetilde{B}$.

$$\widetilde{B} = T_{bs}\left(\widetilde{S}\right) = \begin{bmatrix} b_{00} & b_{01} & b_{02} & b_{03} \\ b_{10} & b_{11} & b_{12} & b_{13} \\ b_{20} & b_{21} & b_{22} & b_{23} \\ b_{30} & b_{31} & b_{32} & b_{33} \end{bmatrix}$$

The b_{ij} element of the matrix $\widetilde{B}$ is obtained from the s_{ij} element of the matrix $\widetilde{S}$, where $0 \le i, j \le 3$.

For all values of i and j do:

1. Let $s \leftarrow s_{ij}$, and

$$q = \begin{cases} s, & s = 0 \\ s^{-1} \in GF_{256}, & s \neq 0 \end{cases}$$

This inversion of nonzero element s introduces nonlinearity.
2. Let $q = \sum_{k=0}^{7} q_k \alpha^k = (q_7 q_6 \dots q_0)$, where $q_k \in \mathbb{Z}_2$ for $0 \le k \le 7$.
3. $(g_7 g_6 g_5 g_4 g_3 g_2 g_1 g_0) \leftarrow (01100011)$.
4. All subscripts in the following equation are computed modulo 8.

$$d_k \equiv (q_k + q_{k+4} + q_{k+5} + q_{k+6} + q_{k+7} + g_k) \pmod 2, \qquad 0 \le k \le 7$$

5. Let $d \leftarrow (d_7 d_6 d_5 d_4 d_3 d_2 d_1 d_0)$
6. Finally, $b_{ij} \leftarrow d$. End of computation of b_{ij}.

A matrix formulation of this transformation is as follows.

$$\begin{bmatrix} d_0 \\ d_1 \\ d_2 \\ d_3 \\ d_4 \\ d_5 \\ d_6 \\ d_7 \end{bmatrix} = \begin{bmatrix} 1&0&0&0&1&1&1&1 \\ 1&1&0&0&0&1&1&1 \\ 1&1&1&0&0&0&1&1 \\ 1&1&1&1&0&0&0&1 \\ 1&1&1&1&1&0&0&0 \\ 0&1&1&1&1&1&0&0 \\ 0&0&1&1&1&1&1&0 \\ 0&0&0&1&1&1&1&1 \end{bmatrix} \begin{bmatrix} q_0 \\ q_1 \\ q_2 \\ q_3 \\ q_4 \\ q_5 \\ q_6 \\ q_7 \end{bmatrix} + \begin{bmatrix} 1 \\ 1 \\ 0 \\ 0 \\ 0 \\ 1 \\ 1 \\ 0 \end{bmatrix}$$

where the arithmetic is in $\mathbb{Z}_2$. This transformation is invertible, because

$$\begin{bmatrix} 1&0&0&0&1&1&1&1 \\ 1&1&0&0&0&1&1&1 \\ 1&1&1&0&0&0&1&1 \\ 1&1&1&1&0&0&0&1 \\ 1&1&1&1&1&0&0&0 \\ 0&1&1&1&1&1&0&0 \\ 0&0&1&1&1&1&1&0 \\ 0&0&0&1&1&1&1&1 \end{bmatrix}^{-1} = \begin{bmatrix} 0&0&1&0&0&1&0&1 \\ 1&0&0&1&0&0&1&0 \\ 0&1&0&0&1&0&0&1 \\ 1&0&1&0&0&1&0&0 \\ 0&1&0&1&0&0&1&0 \\ 0&0&1&0&1&0&0&1 \\ 1&0&0&1&0&1&0&0 \\ 0&1&0&0&1&0&1&0 \end{bmatrix}$$

Example 3.19. Let $s = (01100010)$. Since it is nonzero, its inverse in GF_{256} has to be computed. The representation of s in the finite field GF_{256} is $(x^6 + x^5 + x)$. Its inverse in this field modulo the irreducible polynomial $\xi(x)$, is $(x^7 + x^5 + x^3 + x^2 + x + 1)$. Its binary representation is $q = (q_7 q_6 \dots q_0) = (10101111)$. This yields $d = (d_7 d_6 d_5 d_4 d_3 d_2 d_1 d_0)$, which is equal to (10101010). □

The ShiftRow Transformation

The ShiftRow transformation transforms a 4×4 matrix $\widetilde{B}$ as follows. Denote this SR transformation by $\widetilde{U} = T_{sr}\left(\widetilde{B}\right)$ then

$$\widetilde{U} = \begin{bmatrix} u_{00} & u_{01} & u_{02} & u_{03} \\ u_{10} & u_{11} & u_{12} & u_{13} \\ u_{20} & u_{21} & u_{22} & u_{23} \\ u_{30} & u_{31} & u_{32} & u_{33} \end{bmatrix} = \begin{bmatrix} b_{00} & b_{01} & b_{02} & b_{03} \\ b_{11} & b_{12} & b_{13} & b_{10} \\ b_{22} & b_{23} & b_{20} & b_{21} \\ b_{33} & b_{30} & b_{31} & b_{32} \end{bmatrix}$$

In this transformation, the first row remains unchanged, the second row is rotated by one cell to the left along with a wrap around. The third row is rotated by two cells to the left along with a wrap around, and the fourth row is rotated by three cells to the left along with a wrap around. The inverse transformation is obtained by rotating the rows in the opposite direction.

The MixColumn Transformation

The MixColumn transformation is a linear transformation. It transforms the 4×4 matrix $\widetilde{U}$. Denote the MC transformation by

$$\widetilde{V} = T_{mc}\left(\widetilde{U}\right)$$

where

$$\widetilde{V} = T_{mc}\left(\widetilde{U}\right) = \begin{bmatrix} v_{00} & v_{01} & v_{02} & v_{03} \\ v_{10} & v_{11} & v_{12} & v_{13} \\ v_{20} & v_{21} & v_{22} & v_{23} \\ v_{30} & v_{31} & v_{32} & v_{33} \end{bmatrix}$$

$$= \begin{bmatrix} (00000010) & (00000011) & (00000001) & (00000001) \\ (00000001) & (00000010) & (00000011) & (00000001) \\ (00000001) & (00000001) & (00000010) & (00000011) \\ (00000011) & (00000001) & (00000001) & (00000010) \end{bmatrix} \widetilde{U}$$

Note that the elements of the transformation matrix are the binary representations of the elements which belong to the finite field GF_{256}. In this matrix transformation, the multiplication and addition operations are those of the field GF_{256}. This transformation matrix can also be represented by

$$\begin{bmatrix} \alpha & (\alpha+1) & 1 & 1 \\ 1 & \alpha & (\alpha+1) & 1 \\ 1 & 1 & \alpha & (\alpha+1) \\ (\alpha+1) & 1 & 1 & \alpha \end{bmatrix}$$

The inverse of the above matrix is

$$\begin{bmatrix} (\alpha^3+\alpha^2+\alpha) & (\alpha^3+\alpha+1) & (\alpha^3+\alpha^2+1) & (\alpha^3+1) \\ (\alpha^3+1) & (\alpha^3+\alpha^2+\alpha) & (\alpha^3+\alpha+1) & (\alpha^3+\alpha^2+1) \\ (\alpha^3+\alpha^2+1) & (\alpha^3+1) & (\alpha^3+\alpha^2+\alpha) & (\alpha^3+\alpha+1) \\ (\alpha^3+\alpha+1) & (\alpha^3+\alpha^2+1) & (\alpha^3+1) & (\alpha^3+\alpha^2+\alpha) \end{bmatrix}$$

In the above transformation $\alpha = (00000010)$. Thus the inverse transformation is

$$\begin{bmatrix} (00001110) & (00001011) & (00001101) & (00001001) \\ (00001001) & (00001110) & (00001011) & (00001101) \\ (00001101) & (00001001) & (00001110) & (00001011) \\ (00001011) & (00001101) & (00001001) & (00001110) \end{bmatrix}$$

AddRoundKey Operation

The AddRoundKey operation is a transformation of a 4×4 matrix $\widetilde{V}$. Denote this ARK transformation by $\widetilde{W} = T_{ark}\left(\widetilde{V}\right)$. Define $\widetilde{K}$ as a 4×4 key matrix

$$\widetilde{K} = \begin{bmatrix} k_{00} & k_{01} & k_{02} & k_{03} \\ k_{10} & k_{11} & k_{12} & k_{13} \\ k_{20} & k_{21} & k_{22} & k_{23} \\ k_{30} & k_{31} & k_{32} & k_{33} \end{bmatrix}$$

Then

$$\widetilde{W} \equiv \left(\widetilde{V} + \widetilde{K}\right) \pmod 2$$

That is, if $\widetilde{W} = [w_{ij}]$, $\widetilde{V} = [v_{ij}]$, and $\widetilde{K} = [k_{ij}]$, then $w_{ij} = (v_{ij} + k_{ij}) \pmod 2$ for $0 \le i, j \le 3$. Note that the inverse transformation is also addition by matrix $\widetilde{K}$. Furthermore, the matrix $\widetilde{K}$ is different for each of the 10 rounds.

Key Schedule

The key schedule for the AES algorithm is as follows. Let the 128-bit cipher key be $k \in \mathcal{K}$. Its byte representation is $k_0, k_1, \ldots, k_{15}$. Note that each $k_i, 0 \le i \le 15$ is a byte. These 16 bytes are arranged in a 4×4 matrix as follows:

$$\begin{bmatrix} k_{00} & k_{01} & k_{02} & k_{03} \\ k_{10} & k_{11} & k_{12} & k_{13} \\ k_{20} & k_{21} & k_{22} & k_{23} \\ k_{30} & k_{31} & k_{32} & k_{33} \end{bmatrix} \leftarrow \begin{bmatrix} k_0 & k_4 & k_8 & k_{12} \\ k_1 & k_5 & k_9 & k_{13} \\ k_2 & k_6 & k_{10} & k_{14} \\ k_3 & k_7 & k_{11} & k_{15} \end{bmatrix}$$

This newly generated 4×4 matrix is expanded by adding 40 more columns. Denote all these columns by $\widetilde{Y}(i)$ for $0 \le i \le 43$. Let the first 4 columns be $\widetilde{Y}(0), \widetilde{Y}(1), \widetilde{Y}(2)$, and $\widetilde{Y}(3)$. These first four columns are the four columns of the above matrix.

For $i = 4$ to 43, perform the following steps:

1. If i is not a multiple of 4, then $\widetilde{Y}(i) \equiv \left\{\widetilde{Y}(i-4) + \widetilde{Y}(i-1)\right\} \pmod 2$.
2. If i is a multiple of 4, then do:
   (a) Select column $\widetilde{Y}(i-1)$. Let its elements be $\beta_1, \beta_2, \beta_3$, and β_4. Rotate these elements to obtain $\beta_2, \beta_3, \beta_4$, and β_1.
   (b) Apply the BS transformation to each of these bytes, to get the bytes $\gamma_1, \gamma_2, \gamma_3$, and γ_4 respectively.
   (c) Compute $\delta = (00000010)^{(i-4)/4}$ in GF_{256}. Compute the column vector $\widetilde{X}(i-1)$ as:

   $$\widetilde{X}(i-1) = [(\delta + \gamma_1) \pmod 2, \gamma_2, \gamma_3, \gamma_4]^T$$

   The purpose of the constant δ is to get rid of symmetries in the encryption process, so that each round is different.
   (d) Compute $\widetilde{Y}(i) \equiv \left\{\widetilde{Y}(i-4) + \widetilde{X}(i-1)\right\} \pmod 2$.
3. This completes the steps to generate column number i.

The *round key* for the round number i is made up of the following columns.

$$\widetilde{Y}(4i), \widetilde{Y}(4i+1), \widetilde{Y}(4i+2), \text{ and } \widetilde{Y}(4i+3)$$

where $0 \leq i \leq 10$. □

Observations 3.5. Some important observations about the AES scheme are listed below.

1. In the DES algorithm, half the bits are shifted, but unchanged. However, in the AES algorithm all the bits are transformed uniformly. This diffuses the message bits faster.
2. As mentioned earlier, the BS transformation of this scheme are highly nonlinear. Therefore, it provides resistance against linear and differential cryptanalytic attacks.
3. The SR and MC transformations are a source of diffusion among the input bytes.
4. The Key Schedule also introduces nonlinear mixing of the key bits, because it uses the BS transformation. □

The hallmark of the AES algorithm appears to be its simplicity and versatility.

3.16 Probabilistic Perspective on Cryptographic Security

We develop a probabilistic perspective on cryptographic security in this section. Recall that the cryptographic key is generally selected at random from the key space. Further, the encryption process is also sometimes randomized. Oscar or Eve the adversary might also use probabilistic techniques for deciphering the intercepted ciphertexts. Therefore the computational cryptographic security has to be assessed from a probabilistic point of view.

The basics to discuss probabilistic cryptographic security are initially described in this section. This is followed by an examination of probabilistic security of symmetric-key and asymmetric-key cryptosystems.

Notation. Let A be a set, and $x \in A$. The notation $x \xleftarrow{R} A$ implies that the element $x \in A$ is selected uniformly at random from the set A. □

3.16.1 Basics of Probabilistic Security

A framework to specify computational complexity of deterministic and probabilistic algorithms is initially specified. This is then applied to study security of cryptographic schemes.

Framework to Specify Computational Complexity

In order to evaluate the computational complexity of ciphering and deciphering process, the notion of an algorithm and its running (execution) time in a deterministic and probabilistic setting has to be studied.

An algorithm is a finite sequence of bit operations. The algorithm is deterministic if it does not use any randomization of its parameters. The "input size" of an algorithm is the number of bits required to represent its input. Let the input size of the algorithm be equal to $n \in \mathbb{P}$. Also, let the finite number of steps required to execute a deterministic algorithm A_D be $T_{A_D}(n) \in \mathbb{N}$. If for each input size $n \in \mathbb{P}$, $T_{A_D}(n) = O(n^c)$ for some constant $c \in \mathbb{P}$, the algorithm A_D is said to have polynomial-time computational complexity. We next specify these concepts more carefully.

Let $\mathfrak{B} = \{0, 1\}$ be an alphabet with binary elements (symbols, letters). A notation to describe a sequence of symbols from this set is next introduced. Let $\mathfrak{B}^k$ be the set of all binary strings of length k over $\mathfrak{B}$, where $k \in \mathbb{P}$. These strings are also called *words*. Let $\mathfrak{B}^0 = \{\epsilon\}$, where ϵ is called the *empty word*. The empty word has no letters. Also define

$$\mathfrak{B}^* = \mathfrak{B}^0 \cup \mathfrak{B}^1 \cup \mathfrak{B}^2 \cup \ldots$$

The set $\mathfrak{B}^*$ is the dictionary of words. The length of a word $x \in \mathfrak{B}^*$, denoted by $|x|$, is the number of symbols that make up the word.

Definitions 3.23. *Let $\mathfrak{B} = \{0, 1\}$ be an alphabet, and $\mathfrak{B}^*$ be the corresponding dictionary of words. The length of a word $x \in \mathfrak{B}^*$, is the number of symbols in it. It is denoted by $|x|$.*

1. *Algorithm. Let $f_{A_D} : \mathfrak{B}^* \to \mathfrak{B}^*$. The deterministic algorithm A_D computes $y = f_{A_D}(x)$ in a finite number of steps, where $x, y \in \mathfrak{B}^*$.*
2. *Input size of an instance of the algorithm: The input size of an instance of the algorithm A_D is equal to the number of bits required to represent its input $x \in \mathfrak{B}^*$. It is equal to $|x| \triangleq n \in \mathbb{P}$.*
3. *Run time of the algorithmA_D. Let $T_{A_D} : \mathbb{N} \to \mathbb{N}$. For input $x \in \mathfrak{B}^*$, the algorithm A_D halts after $T_{A_D}(|x|)$ time units or steps. This is the run time of the algorithm A_D for input x.*
4. *Polynomial-time complexity of the algorithm: Algorithm A_D with input size $n \in \mathbb{P}$ runs in polynomial-time, if it runs in time $T_{A_D}(n) = O(n^c)$ for some constant $c \in \mathbb{P}$.*
5. *If the deterministic algorithm A_D computes $f_{A_D}(x)$ in polynomial-time, $\forall\, x \in \mathfrak{B}^*$, then A_D is a polynomial-time algorithm.* □

Algorithms which run in polynomial-time are generally regarded as efficient or feasible. The notion of computational complexity within a probabilistic framework can also be developed. The concept of a randomized algorithm and probabilistic polynomial-time algorithm are next stated.

Definitions 3.24. *Randomizer, randomized algorithm, and probabilistic polynomial-time (PPT) complexity algorithm. Let $\mathfrak{B} = \{0, 1\}$ be an alphabet, $\mathfrak{B}^*$ be the corresponding dictionary of words.*

1. *Let $\mathcal{R} \subseteq \mathfrak{B}^*$. Each element $r \in \mathcal{R}$ occurs probabilistically. An element r is called a randomizer, and $\mathcal{R}$ is the set of randomizers.*
2. *Let $f_{A_R} : \mathfrak{B}^* \times \mathcal{R} \to \mathfrak{B}^*$. The randomized algorithm A_R computes $y = f_{A_R}(x, r)$ in a finite number of steps, where $r \in \mathcal{R}$ and $x, y \in \mathfrak{B}^*$. Computation of $f_{A_R}(x, r)$ is deterministic.*
3. *If the randomized algorithm A_R computes $f_{A_R}(x, r)$ in polynomial-time $\forall\, x \in \mathfrak{B}^*$ and $\forall\, r \in \mathcal{R}$, then A_R is a PPT-complexity algorithm.* □

We can specify this randomized computation by a random variable $Y_R \triangleq f_{A_R}(x, \cdot)$. It corresponds to the probabilistic output of the algorithm for the deterministic input x and random $r \in \mathcal{R}$. The random variable Y_R takes the value $y = f_{A_R}(x, r)$.

Security of a cryptographic scheme can be specified by using the above terminology. We study:

- Machine-specific (computational) security
- Complexity-theoretic based security.

Machine-Specific Security

It is generally desired that a cryptographic scheme be indecipherable by the adversary from a practical point of view. Assuming that the adversary has finite computational power and finite time, we would like to know if he or she is able to decipher the cryptographic scheme in a prespecified time with a very small probability.

Definition 3.25. *A cryptographic scheme is (t, ε)-secure, if every adversary can decipher the scheme in at most $t \in \mathbb{R}_0^+$ time units with probability at most ε, where $\varepsilon \in [0, 1]$.* □

The above definition of security is machine-specific. Nevertheless, it can be useful in practice.

Complexity-Theoretic Based Security

One of the alternatives to machine-specific security is complexity-theoretic (or asymptotic) based security. Asymptotic security is specified in terms of certain important parameters of the cryptographic scheme. In this approach, the time (number of steps) required by the adversary to decipher a ciphertext with a certain prespecified probability are evaluated in terms of parameters of the cryptographic scheme. Let the value of a significant parameter of the cryptographic scheme be $n \in \mathbb{P}$. It is called its *security parameter.* The computational complexity of the ciphering and deciphering process is also generally evaluated in terms of this value.

Definition 3.26. *The security parameter of the cryptographic scheme represents its significant parameter. Its value is $n \in \mathbb{P}$. It is typically represented as 1^n, which is a string of n number of 1's. It is assumed that the adversary is aware of this value.* □

We are now ready to discuss asymptotic security of a cryptographic scheme. Let $n \in \mathbb{P}$ be the security parameter of a cryptographic scheme. It is assumed that the ciphering scheme is run in PPT, and the adversary also uses PPT-complexity algorithm for deciphering the ciphertext. That is, these algorithms run in time an^c, where a and c are some constants. For large enough values of n, it is assumed that the adversary is successful with probability smaller than n^{-c}. Therefore for sufficiently large values of n, this value can be negligible. That is, a cryptographic scheme is asymptotically secure if every adversary which uses PPT-complexity deciphering algorithms, succeeds with negligible probability. The notion of negligible probability can be made more precise by defining negligible function.

Definition 3.27. *Negligible function. The function $\epsilon : \mathbb{N} \to \mathbb{R}$ is negligible, if for every positive polynomial $p(\cdot)$, there exists an $n_0 \in \mathbb{N}$ such that for all integers $n > n_0$, we have $|\epsilon(n)| < (p(n))^{-1}$.* □

Observations 3.6. Let $\epsilon_1(\cdot)$, and $\epsilon_2(\cdot)$ be negligible functions.

1. Let $\lambda_1, \lambda_2 \in \mathbb{R}$. The function $u(\cdot)$ defined by $u(n) = (\lambda_1 \epsilon_1(n) + \lambda_2 \epsilon_2(n))$ is negligible.
2. The function $u(\cdot)$ defined by $u(n) = \epsilon_1(n)\, \epsilon_2(n)$ is negligible.
3. Let $q(\cdot)$ be any polynomial. The function $u(\cdot)$ defined by $u(n) = q(n)\, \epsilon_1(n)$ is negligible. □

Asymptotic security of a cryptographic scheme can be defined precisely.

Definition 3.28. *Asymptotic security. Consider a cryptographic scheme in which each adversary uses PPT-complexity algorithm to decipher it. The cryptographic scheme is asymptotically secure, if for every positive polynomial* $p(\cdot)$ *there exists an integer* $n_0 \in \mathbb{N}$ *such that the probability that each such adversary succeeds in this deciphering process is less than* $(p(n))^{-1}$ *for every* $n > n_0$. □

Observe that the cryptographic scheme may or may not be secure for $n \leq n_0$. In order to demonstrate that cryptographic scheme is secure, it has to be established that each adversary which uses PPT-complexity algorithm can decipher it with negligible probability.

3.16.2 Security of a Symmetric-Key Cryptosystem

Several examples of both classical and modern symmetric-key cryptosystems were discussed in this chapter. In each of these cryptosystems, the encrypting key was randomly selected. Furthermore, the encrypting algorithm can also be probabilistic. We therefore examine the security of these cryptosystems within a probabilistic framework.

Assume that the set of plaintexts $\mathcal{P}$ is a subset of the dictionary of binary strings $\mathfrak{B}^*$, where $\mathfrak{B} = \{0, 1\}$. As the key generation algorithm is randomized, it is listed as a distinct component in the specification of the cryptosystem.

Definition 3.29. *A symmetric-key cryptosystem is specified by*

$$\Gamma_{sym} = (\mathcal{P}, \mathcal{C}, \mathcal{K}, \mathcal{G}_{gen}, \mathcal{E}, \mathcal{D}, n) \tag{3.22}$$

The set of plaintexts in this cryptosystem is $\mathcal{P} \subseteq \mathfrak{B}^*$. *The set of ciphertexts is* $\mathcal{C}$, *and the key space is* $\mathcal{K}$. *Its security parameter is* $n \in \mathbb{P}$.

(a) $\mathcal{G}_{gen}$ *is the key-generation algorithm. The input and output of this randomized algorithm are* 1^n *and* $k \in \mathcal{K}$ *respectively. Thus* $k \overset{R}{\leftarrow} \mathcal{G}_{gen}(1^n)$. *If* $\mathcal{K} \subseteq \mathfrak{B}^*$ *then we assume that* $|k| \geq n$.

(b) $\mathcal{E}$ *is the set of encryption mappings or algorithms which are indexed by the key* $k \in \mathcal{K}$. *An encryption algorithm* $E_k(\cdot)$ *takes the plaintext* $x \in \mathcal{P}$ *as input and maps it into a ciphertext* $c \in \mathcal{C}$. *The encryption algorithm can possibly be randomized. Therefore* $c \overset{R}{\leftarrow} E_k(x)$.

(c) $\mathcal{D}$ *is the set of decryption mappings or algorithms which are indexed by the key* $k \in \mathcal{K}$. *A decryption algorithm* $D_k(\cdot)$ *takes the ciphertext* $c \in \mathcal{C}$ *as input and maps it into the plaintext* $x \in \mathcal{P}$. *The decryption algorithm is deterministic. Thus,* $D_k(c) = x$. □

If we assume that $\mathcal{P} \subseteq \mathfrak{B}^*$, then the computational complexity of the encryption algorithm for $x \in \mathcal{P}$, and $k \in \mathcal{K}$ depends upon $(|k| + |x|)$. However, for simplicity we shall assume that it depends only upon $|k|$, as $|x|$ can generally be bounded.

Security of the Symmetric-Key Cryptosystem

We next examine intuitively the security of the cryptographic scheme Γ_{sym} in presence of an *eavesdropping adversary* $\mathcal{A}_{advrs}$. It is assumed that the adversary $\mathcal{A}_{advrs}$ uses only PPT-complexity algorithms to decipher the ciphertext. It is also assumed that the eavesdropper was able to obtain only a single ciphertext. Security is defined by using the notion of *indistinguishability encryptions*. As per Kerckhoff's assumption (principle) the adversary $\mathcal{A}_{advrs}$ is aware of the cryptographic scheme that he or she is eavesdropping. Furthermore, the adversary is also aware of the security parameter $n \in \mathbb{P}$.

For a given cryptosystem Γ_{sym}, consider an experiment $X_{exper}(\cdot, \cdot)$ which is a function of the adversary $\mathcal{A}_{advers}$ and the security parameter n. As we shall see, the output of this experiment is either 0 or 1.

In the experiment $X_{exper}(\mathcal{A}_{advers}, n)$, the adversary outputs two different plaintexts x_0 and x_1 of equal length. These two plaintexts have equal length because of theoretical requirement in the analysis. Note, however that the cryptographic scheme Γ_{sym} can encrypt plaintexts of arbitrary length.

A different entity generates a random key $k \in \mathcal{K}$. This entity also selects a random bit b from the set $\mathfrak{B}$. It then creates a ciphertext $c \overset{R}{\leftarrow} E_k(x_b)$, where $E_k(\cdot)$ is a PPT-complexity encryption algorithm which uses the key $k \in \mathcal{K}$. This ciphertext is given to the adversary. Upon seeing the ciphertext c, the adversary selects a bit $b' \in \mathfrak{B}$. The output of the experiment $X_{exper}(\mathcal{A}_{advers}, n)$ is equal to 1 if $b = b'$, and 0 otherwise. Therefore, if $X_{exper}(\mathcal{A}_{advers}, n) = 1$, then the adversary $\mathcal{A}_{advrs}$ has succeeded.

The symmetric-key cryptographic scheme Γ_{sym} is said to have indistinguishable encryptions in the presence of an eavesdropping adversary which uses PPT-complexity deciphering algorithm, if the probability of the event

$$X_{exper}(\mathcal{A}_{advers}, n) = 1$$

is at most negligibly greater than $1/2$. Observe that it is incorrect to specify this bounding probability to be $1/2$, as this can simply be achieved by selecting the bit b' randomly. This notion of security via indistinguishable encryptions is made precise in the following definition.

Definition 3.30. *A symmetric-key cryptosystem is specified by*

$$\Gamma_{sym} = (\mathcal{P}, \mathcal{C}, \mathcal{K}, \mathcal{G}_{gen}, \mathcal{E}, \mathcal{D}, n) \tag{3.23a}$$

The security of Γ_{sym} in the presence of an eavesdropping adversary $\mathcal{A}_{advrs}$ is specified by considering an indistinguishability experiment.

(a) *The adversary $\mathcal{A}_{advrs}$ is aware of the cryptographic scheme Γ_{sym}, the security parameter $n \in \mathbb{P}$, and intercepts only a single ciphertext. In addition, the adversary uses only PPT-complexity algorithm while attempting to decipher the ciphertext.*
(b) *For the given cryptosystem Γ_{sym}, the indistinguishability experiment is $X_{exper}(\cdot, \cdot)$. It is a function of the adversary $\mathcal{A}_{advrs}$ and the security parameter n.*
    (i) *Using the security parameter $n \in \mathbb{P}$, the adversary outputs two different plaintexts x_0 and x_1 of equal length.*
    (ii) *A different entity generates a random key $k \in \mathcal{K}$ by using $\mathcal{G}_{gen}(1^n)$. This entity also selects a random bit b from the set $\mathfrak{B} = \{0, 1\}$. It then creates a ciphertext $c \overset{R}{\leftarrow} E_k(x_b)$, where $E_k(\cdot)$ is a PPT-complexity encryption algorithm which uses the key $k \in \mathcal{K}$. The entity gives the ciphertext c to the adversary $\mathcal{A}_{advrs}$.*
    (iii) *Upon seeing the ciphertext c, the adversary $\mathcal{A}_{advrs}$ selects a bit $b' \in \mathfrak{B}$.*
    (iv) *The output of the experiment $X_{exper}(\mathcal{A}_{advers}, n) = 1$ if $b = b'$, and 0 otherwise. Therefore, if $X_{exper}(\mathcal{A}_{advers}, n) = 1$, then the adversary $\mathcal{A}_{advrs}$ has succeeded.*
(c) *The cryptographic scheme Γ_{sym} is said to have indistinguishable encryptions in the presence of adversary $\mathcal{A}_{advrs}$ if the probability of the event*

$$X_{exper}(\mathcal{A}_{advers}, n) = 1 \tag{3.23b}$$

is at most negligibly greater than $1/2$. *That is,*

$$P(X_{exper}(\mathcal{A}_{advers}, n) = 1) \leq \frac{1}{2} + \epsilon(n) \tag{3.23c}$$

where $\epsilon(\cdot)$ *is a negligible function of the security parameter* n. *The probability* $P(\cdot)$ *is taken over: all probabilistic choices available to the adversary, all random choices in generating the key* k, *the random bit* b, *and in the encrypting algorithm.*

The cryptographic scheme Γ_{sym} *has indistinguishable encryptions in presence of an eavesdropper, if the above conditions are true for each such adversary which uses deciphering algorithm with PPT-complexity.* □

This definition implies that the adversary $\mathcal{A}_{advers}$ is successful in guessing the correctly encrypted message with a probability which is negligibly larger than a simple guess, where the simple guess is correct with probability $1/2$.

3.16.3 Security of an Asymmetric-Key Cryptosystem

The benefits and several examples of asymmetric-key cryptosystem have been discussed earlier in the chapter. These are also called public-key cryptosystems. In these schemes, a key is a two-tuple. That is, each key has two components: a private key and a public key. The strength of asymmetric-key (or public-key) cryptosystems is based upon the concept of trapdoor one-way function. These functions are computationally easy to evaluate, but hard to invert. Such functions have been informally discussed, defined, and used earlier in the chapter. We define such functions precisely in this subsection. We also examine the security of such cryptosystems. As in the case of symmetric-key cryptosystem, the selected key is random, and the encrypting algorithm can possibly be probabilistic.

One-Way Function

A one-way function is easily computable in polynomial-time. However, it is hard to invert it by an adversary which uses a PPT-complexity algorithm.

Definition 3.31. *One-way function. Let* $\mathfrak{B} = \{0, 1\}$ *be an alphabet, and* $\mathfrak{B}^*$ *be the corresponding dictionary of words. Consider a function* $f : \mathfrak{B}^* \to \mathfrak{B}^*$. *The function* $f(\cdot)$ *is one-way if it is easy to compute, and hard to invert. Easy computation and hard inversion are:*

(a) *Easy computation: A polynomial-time complexity algorithm* A_f *exists which can compute* $f(x), \forall\, x \in \mathfrak{B}^*$.
(b) *Hard inversion: This is defined via an inversion experiment* $Y_{exper}(\cdot, \cdot, \cdot)$. *The experiment is a function of:* A_{inv} *which is a PPT-complexity algorithm, the function* $f(\cdot)$, *and* $n \in \mathbb{P}$ *the security parameter of the asymmetric-key cryptographic scheme.*
   (i) *Select* $x \xleftarrow{R} \mathfrak{B}^n$, *and compute* $y = f(x) \in \mathfrak{B}^*$.
   (ii) *The input to the algorithm* A_{inv} *is* 1^n *and* y. *Its output is* $x' \in \mathfrak{B}^*$.

(iii) *The output of the experiment* $Y_{exper}(A_{inv}, f, n) = 1$ *if* $f(x') = y$, *and* 0 *otherwise. Note that, in this experiment it is not necessary to find* x. *It is sufficient to find* x' *such that* $f(x') = f(x) = y$.

(iv) *For every PPT-complexity algorithm* A_{inv}, $\exists$ *a negligible function* $\epsilon(\cdot)$ *of the security parameter* n, *such that*

$$P(Y_{exper}(A_{inv}, f, n) = 1) \le \epsilon(n) \tag{3.24}$$

Note that the probability is taken over different values of $x \in \mathfrak{B}^n$. □

Asymmetric-Key Cryptosystem

The definition of asymmetric-key cryptosystem Γ_{asym} is similar to the definition of symmetric-key cryptosystem.

Definition 3.32. *An asymmetric-key cryptosystem is specified by*

$$\Gamma_{asym} = (\mathcal{P}, \mathcal{C}, \mathcal{K}, \mathcal{G}_{gen}, \mathcal{E}, \mathcal{D}, n) \tag{3.25}$$

The set of plaintexts in this cryptosystem is $\mathcal{P} \subseteq \mathfrak{B}^*$. *The set of ciphertexts is* $\mathcal{C}$, *and the key space is* $\mathcal{K}$. *Its security parameter is* $n \in \mathbb{P}$.

(a) $\mathcal{G}_{gen}$ *is the key-generation algorithm. The input and output of this randomized algorithm are* 1^n *and* $k \in \mathcal{K}$ *respectively. Thus* $k \overset{R}{\leftarrow} \mathcal{G}_{gen}(1^n)$. *Each key* $k = (k_{priv}, k_{pub})$ *is a two-tuple, where* k_{priv} *is the private key and* k_{pub} *is the public key. If* $\mathcal{K} \subseteq \mathfrak{B}^* \times \mathfrak{B}^*$ *then we assume that* $|k_{priv}| \ge n$, *and* $|k_{pub}| \ge n$. *The public key that is used for encryption is available to the public. The private key is used for decryption and is kept private by the person who receives the ciphertext.*

(b) $\mathcal{E}$ *is the set of encryption mappings or algorithms which are indexed by the public key* k_{pub}. *An encryption algorithm* $E_{k_{pub}}(\cdot)$ *takes the plaintext* $x \in \mathcal{P}$ *as input and maps it into a ciphertext* $c \in \mathcal{C}$. *The encryption algorithm can possibly be randomized. Therefore* $c \overset{R}{\leftarrow} E_{k_{pub}}(x)$.

(c) $\mathcal{D}$ *is the set of decryption mappings or algorithms which are indexed by the private key* k_{priv}. *A decryption algorithm* $D_{k_{priv}}(\cdot)$ *takes the ciphertext* $c \in \mathcal{C}$ *as input and maps it into the plaintext* $x \in \mathcal{P}$. *That is,* $D_{k_{priv}}(c) = x$. *The decryption algorithm is deterministic.* □

The characterization of security of a public-key cryptosystem based upon indistinguishable encryptions in the presence of adversary is similar to that of symmetric-key cryptosystems with minor modifications. The adversary $\mathcal{A}_{advrs}$ in the asymmetric-key cryptosystem has access to the public key k_{pub}, but not the corresponding private key k_{priv} of the receiver.

Reference Notes

Readable introduction to cryptography is given in the books by Bruen, and Forcinito (2005), Hankerson, Hoffman, Leonard, Lindner, Phelps, Rodger, and Wall (2000), Hardy, and Walker (2003),

Haykin (2001), and Smart (2016). The textbook by Katz, and Lindell (2008) provides a modern and rigorous introduction to cryptography. A scholarly discussion on the foundations of cryptography can be found in Goldreich (2001). Trappe, and Washington (2006) provide an accessible account of the Rijndael algorithm.

Encyclopedic, yet readable and extremely useful references to this subject are the books by Menezes, Oorschot, and Vanstone (1996), and Schneier (1996). There are several books which deal exclusively with cryptography. The authors of these books are Buchmann (2001), Garrett (2001), Mollin (2006), Stinson (2006), and Wagstaff (2003). Discussion on information-theoretic cryptography can also be found in Wells (1999), and Welsh (1988). A mathematical description of public-key cryptography can be found in Galbraith (2012).

Problems

1. Let $(\mathcal{P}, \mathcal{C}, \mathcal{K}, \mathcal{E}, \mathcal{D})$ be a cryptosystem with perfect secrecy. Prove that $|\mathcal{P}| \leq |\mathcal{C}| \leq |\mathcal{K}|$.
   Hint: See Stinson (2006). Assume that $P(Y = y) > 0$ for all values of $y \in \mathcal{C}$. If $P(Y = y) = 0$, then this $y \in \mathcal{C}$ can be removed from $\mathcal{C}$. Consequently, for a specified value of $x \in \mathcal{P}$, $P(Y = y \mid X = x) = P(Y = y) > 0$. Therefore for each $y \in \mathcal{C}$, there must exist at least a single key $z \in \mathcal{K}$, such that $E_z(x) = y$. This implies that $|\mathcal{C}| \leq |\mathcal{K}|$.
   In any cryptosystem, each encoding is injective. Therefore $|\mathcal{P}| \leq |\mathcal{C}|$.
   Thus $|\mathcal{P}| \leq |\mathcal{C}| \leq |\mathcal{K}|$.
2. Prove that the one-time pad cryptosystem provides perfect secrecy if the random variable corresponding to the keyword is uniformly distributed.
   Hint: See Katz, and Lindell (2008). Let $(\mathcal{P}, \mathcal{C}, \mathcal{K}, \mathcal{E}, \mathcal{D})$ be a one-time pad cryptosystem. The random variables X, Y, and Z take values in the sets $\mathcal{P}, \mathcal{C}$, and $\mathcal{K}$ respectively. The random variable Z has a uniform distribution. Let the ciphering scheme encrypt m bits at a time, where $m \in \mathbb{P}$. Observe that

$$\begin{aligned}P(Y = y \mid X = x) &= P((X + Z) \equiv y \pmod 2 \mid X = x) \\ &= P((x + Z) \equiv y \pmod 2) \\ &= P(Z \equiv (x + y) \pmod 2) = 2^{-m}\end{aligned}$$

   Also

$$\begin{aligned}P(Y = y) &= \sum_{x \in X} P(Y = y \mid X = x) P(X = x) \\ &= 2^{-m} \sum_{x \in X} P(X = x) = 2^{-m}\end{aligned}$$

   Therefore $P(Y = y \mid X = x) = P(Y = y)$. This result implies perfect secrecy.
3. Let $(\mathcal{P}, \mathcal{C}, \mathcal{K}, \mathcal{E}, \mathcal{D})$ be a cryptosystem, such that $|\mathcal{P}| = |\mathcal{C}| = |\mathcal{K}| = m$. The random variables X, Y and Z take values in the sets $\mathcal{P}$, $\mathcal{C}$, and $\mathcal{K}$ respectively. Assume that the random variables X and Z are independent of each other. This cryptosystem has perfect secrecy iff every key is used with probability m^{-1}, and for every value of $x \in \mathcal{P}$ and every value of $y \in \mathcal{C}$, there is a unique key $z \in \mathcal{K}$, such that $E_z(x) = y$. Prove this result.

Hint: See Stinson (2006). Under the assumption of perfect secrecy of the cryptosystem, for each $x \in \mathcal{P}$ and each $y \in \mathcal{C}$ there exists at least a single key $z \in \mathcal{K}$, such that $E_z(x) = y$. This yields $|\mathcal{C}| = |\{E_z(x) \mid z \in \mathcal{K}\}| \le |\mathcal{K}|$.
But by hypothesis $|\mathcal{C}| = |\mathcal{K}|$, therefore $|\{E_z(x) \mid z \in \mathcal{K}\}| = |\mathcal{K}|$. This implies that there do not exist two different keys z_1 and z_2 which yield the same value y of ciphertext for the same value of the plaintext x. Therefore, for any $x \in \mathcal{P}$ and any $y \in \mathcal{C}$ there is precisely one key $z \in \mathcal{K}$ such that $E_z(x) = y$.
For a given $y \in \mathcal{C}$, let $E_{z_i}(x_i) = y$ for $1 \le i \le m$. Then for $x_i \in \mathcal{P}$

$$\begin{aligned} P(X = x_i \mid Y = y) &= \frac{P(X = x_i) P(Y = y \mid X = x_i)}{P(Y = y)} \\ &= \frac{P(X = x_i) P(Z = z_i)}{P(Y = y)} \end{aligned}$$

But because of perfect secrecy $P(X = x_i \mid Y = y) = P(X = x_i)$, the above equation reduces to

$$P(X = x_i) = \frac{P(X = x_i) P(Z = z_i)}{P(Y = y)}$$

Therefore $P(Y = y) = P(Z = z_i)$ for $1 \le i \le m$. This implies that $P(Z = z_i) = m^{-1}$ for $1 \le i \le m$. The proof of the converse is left to the reader.

4. Let X, Y and Z be the random variables which take values in the sets $\mathcal{P}$, $\mathcal{C}$, and $\mathcal{K}$ respectively. Prove that for a cryptosystem with perfect secrecy,

$$H(X \mid Y) = H(X), \quad \text{and } H(Y \mid X) = H(Y)$$

5. Let X, Y and Z be the random variables which take values in the sets $\mathcal{P}$, $\mathcal{C}$, and $\mathcal{K}$ respectively. Prove the following results.
   (a) $H(Z \mid Y) = H(X) + H(Z) - H(Y)$
   (b) $H(Y \mid X) = H(X,Y,Z) - H(X) - H(Z \mid X,Y)$
   (c) $H(Y \mid X) = H(Z) - H(Z \mid X,Y)$

   Hint: See Stinson (2006), and Trappe, and Washington (2006).
   (a) Use of chain rule of entropies yields

   $$\begin{aligned} H(X,Y,Z) &= H(X \mid Y,Z) + H(Z \mid Y) + H(Y) \\ H(X,Y,Z) &= H(Y \mid X,Z) + H(X \mid Z) + H(Z) \end{aligned}$$

   Observe that $H(X \mid Y,Z) = 0$, since X is completely determined given Y and Z. Similarly $H(Y \mid X,Z) = 0$. The result follows by using the observation $H(X \mid Z) = H(X)$. The final result is the key-equivocation principle.
   (b) The result follows by noting that

   $$H(X,Y,Z) = H(Z \mid X,Y) + H(Y \mid X) + H(X)$$

   (c) Consider $H(X,Y,Z) = H(Y \mid X,Z) + H(X \mid Z) + H(Z)$. The result follows by observing that $H(Y \mid X,Z) = 0$ and $H(X \mid Z) = H(X)$; and using the result in part (b).

6. Prove that RSA's signature verification process is correct.

7. Prove that the polynomial $\xi(x) = \left(x^8 + x^4 + x^3 + x + 1\right)$ is irreducible over GF_2.
Hint: See Hardy, and Walker (2003). Find an element of order 255 modulo the polynomial $\xi(x)$. Try the element $(\alpha + 1)$. The number $255 = 3 \times 5 \times 17$. It can be shown that

$$\begin{aligned}(\alpha+1)^{17} &= (\alpha+1)^{16}(\alpha+1) \\ &= \left(\alpha^{16}+1\right)(\alpha+1)\end{aligned}$$

Thus

$$\begin{aligned}(\alpha+1)^{17} \pmod{\xi(\alpha)} &\equiv \left(\alpha^{17}+\alpha^{16}+\alpha+1\right) \pmod{\xi(\alpha)} \\ &\equiv \left(\alpha^{7}+\alpha^{6}+\alpha^{5}+1\right) \pmod{\xi(\alpha)}\end{aligned}$$

Next compute

$$(\alpha+1)^{34} = (\alpha+1)^{17}(\alpha+1)^{17}$$

Thus

$$\begin{aligned}(\alpha+1)^{34} \pmod{\xi(\alpha)} &\equiv \left(\alpha^{14}+\alpha^{12}+\alpha^{10}+1\right) \pmod{\xi(\alpha)} \\ &\equiv \left(\alpha^{6}+\alpha^{4}+\alpha^{3}+\alpha^{2}\right) \pmod{\xi(\alpha)}\end{aligned}$$

Similarly

$$\begin{aligned}&(\alpha+1)^{51} \pmod{\xi(\alpha)} \\ &\equiv \alpha^2\left(\alpha^4+\alpha^2+\alpha+1\right)\left(\alpha^7+\alpha^6+\alpha^5+1\right) \pmod{\xi(\alpha)} \\ &\equiv \alpha^2\left(\alpha^{11}+\alpha^{10}+\alpha^7+\alpha^5+\alpha^4+\alpha^2+\alpha+1\right) \pmod{\xi(\alpha)} \\ &\equiv \alpha^2(\alpha+1) \pmod{\xi(\alpha)}\end{aligned}$$

Also

$$(\alpha+1)^{255} = (\alpha+1)^{51}(\alpha+1)^{51}(\alpha+1)^{51}(\alpha+1)^{51}(\alpha+1)^{51}$$

This yields

$$\begin{aligned}&(\alpha+1)^{255} \\ &\equiv \alpha^{10}(\alpha+1)^5 \pmod{\xi(\alpha)} \\ &\equiv \left(\alpha^{15}+\alpha^{14}+\alpha^{11}+\alpha^{10}\right) \pmod{\xi(\alpha)} \\ &\equiv 1 \pmod{\xi(\alpha)}\end{aligned}$$

Thus $(\alpha + 1)$ has order 255 modulo $\xi(\alpha)$. Therefore $\xi(x)$ is an irreducible polynomial. The steps in reaching this conclusion are indeed tedious. The arduous reader might discover a more systematic procedure to establish if a polynomial over GF_2 of any degree $n \in \mathbb{P}$ is irreducible.

References

1. Apostol, T. M., 1976. *Introduction to Analytic Number Theory*, Springer-Verlag, Berlin, Germany.

2. Baldoni, M. W., Ciliberto, C., and Cattaneo, G. M. P., 2009. *Elementary Number Theory, Cryptography, and Codes*, Springer-Verlag, Berlin, Germany.
3. Brown, E., 2000. "Three Fermat Trails to Elliptic Curves," The College Mathematics Journal, Vol. 31, No. 3, pp. 162-172.
4. Bruen, A. A., and Forcinito, M. A., 2005. *Cryptography, Information Theory, and Error-Correction*, John Wiley & Sons, Inc., New York, New York.
5. Buchmann, J. A., 2001. *Introduction to Cryptography*, Springer-Verlag, Berlin, Germany.
6. Cohen, H., and Frey, G., 2006. *Handbook of Elliptic and Hyperelliptic Curve Cryptography*, CRC Press, New York.
7. Diffie, W., and Hellman, M. E., 1976. "New Directions in Cryptography," IEEE Transactions on Information Theory, Vol. 22, No. 6, pp. 644-654.
8. El Gamal, T., 1985. "A Public Key Cryptosystem and a Signature Scheme Based on Discrete Logarithm," IEEE Transactions on Information Theory, Vol. 31, No. 4, pp. 469-472.
9. Forouzan, B. A., 2006. *Cryptography and Network Security*, McGraw-Hill Book Company, New York, New York.
10. Galbraith, S. D., 2012. *Mathematics of Public Key Cryptography*, Cambridge University Press, Cambridge, Great Britain.
11. Garrett, P., 2001. *Making, Breaking Codes: An Introduction to Cryptology*, Prentice-Hall, Upper Saddle River, New Jersey.
12. Gentry, C., 2009. "A Fully Homomorphic Encryption Scheme," Ph. D. dissertation, Stanford University, Stanford, California.
13. Goldreich, O., 2001. *Foundations of Cryptography: Basic Tools*, Cambridge University Press, Cambridge, Great Britain.
14. Goldwasser, S., and Micali, S., 1984. "Probabilistic Encryption," Journal of Computer and Systems Science, Vol. 28, No. 2, pp. 270-299.
15. Hankerson, D., Hoffman, D. G., Leonard, D. A., Lindner, C. C., Phelps, K. T., Rodger, C. A., and Wall, J. R., 2000. *Coding Theory and Cryptography, The Essentials*, Second Edition, Marcel Dekker, Inc., New York, New York.
16. Hardy, D. W., and Walker, C. L., 2003. *Applied Algebra, Codes, Ciphers, and Discrete Algorithms,* Prentice-Hall, Upper Saddle River, New Jersey.
17. Hasse, H., 1978. *Number Theory*, Springer-Verlag, Berlin, Germany.
18. Haykin, S., 2001. *Communication Systems*, Fourth Edition, John Wiley & Sons, Inc., New York, New York.
19. Hoffstein, J., Pipher, J., and Silverman, J. H., 2008. *An Introduction to Mathematical Cryptography*, Springer-Verlag, Berlin, Germany.
20. Katz, J., and Lindell, Y., 2008. *Introduction to Modern Cryptography*, Chapman and Hall/CRC Press, New York, New York.
21. Kaufman, C., Perlman, R., and Speciner, M., 2002. *Network Security, Private Communication in a Public World*, Prentice-Hall, Upper Saddle River, New Jersey.
22. Koblitz, N., 1989. "Hyperelliptic Cryptosystems," Journal of Cryptology, Vol. 1, Issue 3, pp. 139-150.
23. Koblitz, N., 1994. *A Course in Number Theory and Cryptography*, Second Edition, Springer-Verlag, Berlin, Germany.
24. Koblitz, N., 1999. *Algebraic Aspects of Cryptography*, Springer-Verlag, Berlin, Germany.
25. Kumanduri, R., and Romero, C., 1998. *Number Theory with Computer Applications*, Prentice-Hall, Englewood Cliffs, New Jersey.
26. Menezes, A., Oorschot, P. von, and Vanstone, S., 1996. *Handbook of Applied Cryptography*, CRC Press, New York.

27. Mollin, R. A., 1998. *Fundamental Number Theory with Applications*, Chapman and Hall/CRC Press: New York, New York.
28. Mollin, R. A., 1999. *Algebraic Number Theory*, Chapman and Hall/CRC Press: New York, New York.
29. Mollin, R. A., 2006. *An Introduction to Cryptography*, Second Edition, Chapman and Hall/CRC Press: New York, New York.
30. Mollin, R. A., 2005. *Codes: The Guide to Secrecy from Ancient to Modern Times*, Chapman and Hall/CRC Press: New York, New York.
31. Niederreiter, H., and Xing, C. 2009. *Algebraic Geometry in Coding Theory and Cryptography*, Princeton University Press, Princeton, New Jersey.
32. Prasolov, V., and Solovyev, Y. 1997. *Elliptic Functions and Elliptic Integrals*, American Mathematical Society, Providence, Rhode Island.
33. Riesel, H., 1987. *Prime Numbers and Computer Methods for Factorization*, Birkhauser, Boston, Massachusetts.
34. Rivest, R., Adleman, L., and Dertouzos, M., 1978. "On Data Banks and Privacy Homomorphisms," Foundations of Secure Computation, Vol. 4, Issue 11, pp. 169-180.
35. Rosen, K. H., Editor-in-Chief, 2000. *Handbook of Discrete and Combinatorial Mathematics,* CRC Press: Boca Raton, Florida.
36. Schmitt, S., Zimmer, H. G., 2003. *Elliptic Curves, A Computational Approach,* Walter De Gruyter, New York, New York.
37. Schneier, B., 1996. *Applied Cryptography*, Second Edition, John Wiley & Sons, Inc., New York, New York.
38. Silverman, J. H., and Tate, J., 1992. *Rational Points on Elliptic Curves*, Springer-Verlag, Berlin, Germany.
39. Smart, N. P., 2016. *Cryptography Made Simple*, Springer-Verlag, Berlin, Germany.
40. Stallings, W., 2006. *Cryptography and Network Security, Principles and Practices*, Fourth Edition, Prentice-Hall, Upper Saddle River, New Jersey.
41. Stinson, D. R., 2006. *Cryptography, Theory and Practice*, Third Edition, CRC Press: New York, New York.
42. Trappe, W., and Washington, L. C., 2006. *Introduction to Cryptology and Coding Theory,* Second Edition, Prentice-Hall, Upper Saddle River, New Jersey.
43. Wagstaff Jr., S. S., 2003. *Cryptanalysis of Number Theoretic Ciphers,* CRC Press: New York, New York.
44. Washington, L. C., 2003. *Elliptic Curves, Number Theory and Cryptography*, Chapman and Hall/CRC Press, New York, New York.
45. Wells, R. B., 1999. *Applied Coding and Information Theory for Engineers*, Prentice-Hall, Englewood Cliffs, New Jersey.
46. Welsh, D., 1988. *Codes and Cryptography*, Oxford University Press, Oxford, Great Britain.

CHAPTER 4

Internet Traffic

$$f_X(x) = \frac{1}{\sqrt{2\pi}\sigma} \exp\left\{-\frac{1}{2}\left(\frac{x-\mu}{\sigma}\right)^2\right\}$$

$$x \in \mathbb{R}$$

Gaussian probability density function

Johann Carl Friedrich Gauss. Gauss was born on 30 April, 1777 in Brunswick, Duchy of Brunswick (now Germany). He is often regarded as the greatest mathematician who ever lived. Gauss was a child prodigy. It is said that, he was able to read, and perform arithmetical calculations by the age of three.

Gauss earned his doctorate from the University of Helmstedt in 1799. He proved the Fundamental Theorem of Algebra in his dissertation. This theorem states that every nonzero single-variable polynomial with complex coefficients of degree *n*, has exactly *n* complex roots, possibly repeated. At the age of 24 Gauss published a treatise on number theory, *Disquistiones Arithmeticae*, (1801).

Besides number theory, Gauss made fundamental contribution to differential geometry, complex analysis, astronomy, geodesy, statistics, mechanics, and magnetism. The Gaussian elimination procedure of systematically solving linear equations was also devised by Gauss. He died on 23 February, 1855 in Göttingen, Hanover (now Germany).

4.1 Introduction

The Internet has experienced phenomenal growth in the past several years. A major contribution to its growth has been the ubiquitous presence of the World Wide Web (sometimes simply abbreviated as the Web). Therefore, an understanding of the network traffic is crucial in order to properly design and build a computer network and offer applications like World Wide Web, electronic mail, information retrieval, and social networks. The following topics are discussed in this chapter: self-similarity in the Local Area Network (LAN) and Internet traffic, power-law probability distributions, modeling of highly variable measurements, Internet traffic modeling, and explanation of self-similarity in Internet traffic. Sections are also devoted to the asymptotic analysis of R/S statistic, transform analysis of fractional Brownian motion process, and Internet traffic modeling via chaotic maps.

Network traffic can be considered to be a time series representing bytes or packets per unit time which flow on a communication channel. Recall that packets are units of information, not necessarily of equal length (measured in bytes), which flow on a telecommunication channel.

Till a few years back, voice traffic was predominant on networks, however modern day networks carry voice, data, image, and video. Voice traffic has been thoroughly understood by engineers. It is generally Poissonian, with exponentially distributed call lengths. Multiplexing (merging) several voice traffic streams results in a Poisson process. This traffic was traditionally carried on circuit-switched networks. On the other hand, Internet is a packet-switched data network. Merging several data sources does not result in a Poissonian stream of packets. Therefore the design techniques used in modeling circuit-switched networks do not carry over easily into the packet-switched world. Consequently new design principles must be conceived to model packet-switched networks.

Some of the hurdles in the design of packet networks are: burstiness, heterogeneity, and statistical nonstationarity of the data. The ever changing topology of the network also provides a challenge. Furthermore, diverse applications like video on demand and voice over IP (VoIP) have different performance requirements and constraints. All these characterizations have different impact on different layers of the Internet.

The difference between the telephony (voice) and Internet traffic is next characterized more precisely.

Telephony traffic

The telephony traffic is described by Poissonian arrival streams, and exponentially distributed hold times (call length duration). Consider graphically representing the telephony traffic by the number of call arrivals in a time interval of some fixed length, say Δ. Assume that the number of call arrivals in successive time intervals of length Δ is plotted at time instants: $t = \Delta, 2\Delta, 3\Delta, \ldots$. Another plot of the number of call arrivals in successive time intervals of length Δ', is plotted at time instants: $t = \Delta', 2\Delta', 3\Delta', \ldots$, where $\Delta' > \Delta$. Then it is the characteristic of the telephony traffic that the second plot appears to be visually *smoother* than the first plot. The two plots are said to have different time-scales.

Internet traffic

We represent the Internet traffic by the number of packet (or byte) arrivals in a time interval of length, say Δ. Assume that the number of packet arrivals in successive time intervals of length Δ is plotted at time instants: $t = \Delta, 2\Delta, 3\Delta, \ldots$. Another plot of the number of packet arrivals in successive time intervals of length Δ' is plotted at time instants: $t = \Delta', 2\Delta', 3\Delta', \ldots$, where $\Delta' \neq \Delta$. Then the first and the second plots appear to be visually *similar*. This is irrespective of whether $\Delta' > \Delta$ or $\Delta' < \Delta$.

Furthermore, if $\Delta' > \Delta$, then the second graph is not smoother than the first plot, as in the case of telephony traffic. This result appears to be counter intuitive. Nevertheless, the two plots appear to be similar. Therefore it can be concluded that the Internet traffic contains similar features at all time scales. Consequently, the Internet traffic is said to be *self-similar*. The phrase self-similar was coined by Benoit Mandelbrot (1924-2010) and his coworkers.

An important facet of packet traffic is worth noting. While smoothing (averaging) removes small-scale fluctuations in the packet traffic, the large-scale features remain unaltered. This feature can be modeled by *long-range dependence.* In addition, the Internet traffic appears to be *bursty* across an extremely wide range of time scales. This burstiness can be described by :

(a) Power-law probability density function like Pareto, and generalized Zipf distribution. These are examples of heavy-tailed distributions. A distribution is said to be heavy-tailed if its tail decays more slowly, than for example an exponential and Gaussian distribution. Random variables with heavy-tailed distributions can have extremely large (infinite) variance, which in turn implies extreme variability.
(b) Self-similarity measures. Self-similar traffic is also referred to as fractal traffic. Fractals model Internet traffic behavior over time scales from a few fractions of a second onward. Internet traffic behavior below a few fractions of a second can be modeled by multifractals.
(c) Special measures to characterize burstiness.

Heavy-tailed distributions can be found in: time consumed by different processes in the central processing unit (CPU) of a computer, sizes of files in a file system, Web item sizes, inter-key stroke times when a person types. Size of bursts of data while transferring files exhibit heavy-tailed behavior. Furthermore, duration of bursts or idle periods of individual Ethernet Local Area Network (LAN) connections exhibit heavy-tailed behavior. It has been determined that the popularity of documents at Web sites follow a generalized Zipf distribution. Some sources of burstiness in the Internet traffic are listed below. Traffic burstiness occurs due to both the self-similar sources and the self-similar behavior of the users. Some of the causes of burstiness are:

(a) The traffic stream generated during Web browsing sessions is self-similar. Observe that during a Web browsing session, the traffic flows from several different geographical regions to a single requesting-point.
(b) The heavy-tailed behavior of the requested files is a major cause of self-similarity in the network traffic. This is because, the file-size distribution which is heavy-tailed can give rise to extremely large file size requests with very small probability.
(c) People's surfing behavior. Individual Internet connections range from extremely small to extremely large periods. These connections in turn request extremely low to super-high traffic-rates.
(d) Inherent self-similar characteristics of video-traffic.
(e) Retransmissions of lost packets contribute to the burstiness. Furthermore, end to end congestion control in the network dramatically decreases the rate at which data is transmitted.

This chapter relies heavily upon the methods and techniques of applied analysis, probability theory, stochastic processes, and chaos theory. Note that, $P(\cdot)$ is the probability function in this chapter.

4.2 Self-Similarity in the LAN and Internet Traffic

Self-similarity in the LAN and Internet traffic is explained in this section. Note that computers or workstations are interconnected via a LAN, which is generally an Ethernet LAN. The LAN is then connected via a special processor (called a *router*, or a *gateway*) to the Wide Area Network (WAN). A conglomeration of networks like WANs form the Internet. Recall that a network of networks is the Internet, and the World Wide Web is an application implemented upon it.

4.2.1 Self-Similarity in the LAN Traffic

Self-similarity in the traffic on Ethernet LANs has been observed in several experiments. This self-similarity phenomenon is explained as follows. Ethernet sources can be described by simple "on" and "off" models. During the "on" period, packets are generated at regular intervals, while during the "off" interval no packets are generated. Also assume that the lengths of the "on" interval or "off" interval have high-variability or infinite variance. Infinite variances can be modeled by using the Pareto distribution. Superposition of many such "on" and "off" sources (sometimes called packet-trains) with precisely alternating "on" and "off" periods generate self-similar or long-range dependent packet traffic in the limit.

Note that infinite variance implies extreme variability or variability at all time scales. These also capture the observed phenomenon of nonnegligible probability that the "on" and "off" periods can last for very long times. The burstiness of packet-trains can be quantified by the Hurst parameter H, named after the civil engineer (hydrologist) Harold Edwin Hurst (1900-1978). Hurst encountered long-range dependence in his studies on measuring the long-term storage of water flowing in the river Nile.

The Hurst parameter is a measure of the burstiness of the traffic. The theoretical range of values of the Hurst parameter is zero to unity. However, its observed value on the LANs has been found to range typically from 0.75 to 0.95.

The infinite variance syndrome is sometimes called the *Noah effect*. This is in reference to the Biblical story of Noah and the *Big Flood*. Also the long-range dependence and the self-similarity of the packet traffic is called the *Joseph effect*. This is in reference to the Biblical figure who predicted the "seven fat years and seven lean years." These Biblical names were first used in this context by Benoit Mandelbrot, a pioneer in the study of fractals. The Old Testament fables of Joseph and Noah demonstrate how persistence and the presence of outliers were observed in the water levels in rivers. A phenomenon which unfortunately affected the production and price of agricultural commodities during the good (or bad) old days.

4.2.2 Self-Similarity in the Internet Traffic

The prevalence of heavy-tailed probability distributions in the Internet traffic is explored in this subsection. These distributions are present in: files requested by the users, files transmitted through the network, transmission durations of files, and files stored on the servers.

Popularity of the documents on the Web can be described by a generalized Zipf distribution. As in the case of the LAN traffic, the behavior of WAN traffic can be explained by considering a superposition of several on- and off-periods. A Web session can be considered to be on, when it is either sending or receiving data. It is in the off-state if there is no activity by the user. If these durations have heavy-tailed distributions, then the net traffic on the Web will be self-similar in nature.

The off-durations can be modeled by Pareto distribution, with parameter $\alpha \approx 1.5$. The corresponding value of α, for the on-period lies in the interval $(1, 2)$ Therefore the self-similarity of network traffic can be explained by the presence of heavy-tailed distribution of the file transmission times (or equivalently file-size). Important properties of power-law distributions like Pareto and Zipf are explained in the next section.

4.3 Power-Law Probability Distributions

Power-law probability distributions and self-similar processes are inter-related. Power-law distributions are of the form $f(x) = kx^{-\alpha}$, for $x \geq x_0$ and $\alpha > 0$. The Pareto and generalized Zipf probability distributions are examples of power-law distributions. Pareto distribution is a continuous-valued distribution. In contrast the Zipf distribution is a discrete-valued distribution. The tails of such distributions asymptotically decay hyperbolically. These are also examples of a larger class of distributions called heavy-tailed distributions. This is in contrast to light-tailed distributions like exponential and Gaussian which have an exponentially decreasing tail. For proper parametric values,

Pareto and Zipf distributions can have infinite variance. The popularity of a Web page follows the Zipf distribution. Generalized form of the classical Zipf distribution are also used to model the Web proxy-cache.

Definition 4.1. *Random variable X has a power-law-tailed distribution if the probability $P(X > x) \sim cx^{-\alpha}$, as $x \to \infty$, where $c \in \mathbb{R}^+$, and $\alpha \in (0, 2)$. The parameter α is called the tail-index or shape parameter, or exponent, and c is called the scale coefficient.* □

Note that the power-law-tailed distribution has a variance which tends towards infinity.

4.3.1 Pareto Distribution

A random variable X is said to have a Pareto distribution, if the probability density function of X is given by

$$f_X(x) = \begin{cases} 0, & x \le x_0 \\ \dfrac{\alpha}{x_0}\left(\dfrac{x_0}{x}\right)^{\alpha+1}, & x > x_0 \end{cases} \tag{4.1}$$

Observe that $x_0 \in \mathbb{R}^+$ is called the location parameter, and $\alpha \in \mathbb{R}^+$ is called the shape parameter. Expressions for its mean, variance, and cumulative distribution function are given in the chapter on probability theory. Its complementary cumulative distribution is $F_X^c(x) = (x_0/x)^\alpha$, for $x > x_0$.

For large values of x, the Pareto density function decays much more slowly than the exponential density function. If $\alpha \in (0, 1]$, the random variable X has infinite mean. However if $\alpha \in (1, 2]$, the random variable X has finite mean, but infinite variance. Therefore as α decreases, a significant portion of the probability mass is concentrated in the tail of the distribution. Consequently, the density function has heavier tail, as the value of α decreases. Some important properties of the Pareto distribution are listed below.

(a) Let $m > 0$, and $x > x_0$ then

$$P(X \ge mx) = m^{-\alpha} P(X \ge x)$$

This result implies that Pareto distribution is scaling.

(b) Let $y > y_0 > x_0$, then

$$P(X > y \mid X > y_0) = \left(\frac{y_0}{y}\right)^\alpha$$

Notice that this conditional distribution is also Paretian.

4.3.2 Zipf Distribution

A random variable X has a Zipf distribution, and it takes values in the set $\{1, 2, \ldots, n\}$. Its distribution is given by

$$p_X(x) = \frac{\kappa}{x}, \quad 1 \le x \le n \tag{4.2a}$$

$$\kappa^{-1} = \sum_{x=1}^{n} \frac{1}{x} \tag{4.2b}$$

Note that

$$\sum_{x=1}^{n} p_X(x) = 1 \tag{4.2c}$$

and κ is the normalizing constant. Like the Pareto, this distribution is also scaling. This distribution is a candidate to model the popularity of a Web-page. The popularity of the xth most popular Web-page is proportional to x^{-1}. A generalized form of the Zipf distribution is used to model the Web proxy-cache. In this model, $p_X(x)$ is proportional to $x^{-\alpha}$, where $\alpha \in \mathbb{R}^+$.

4.4 Modeling Highly Variable Measurements

Measurements are said to be highly variable, if the observations take values over several orders of magnitude. Usually in such measurements, a majority of values are small. However some measurements take extremely large values with nonnegligible probabilities. Therefore highly variable measured values imply very large standard deviation. Highly variable measurements in the Internet are ubiquitous.

Lognormal, Pareto, Weibull, and generalized Zipf's probability distributions can be used to model highly variable measurements. A common property of these probability distributions, is that these are heavy-tailed or fat-tailed for appropriate parameter values. Another class of heavy-tailed probability distributions correspond to the α-stable random variables. These distributions exhibit the heavy-tailed property for proper range of values of α.

Different classes of heavy-tailed probability distributions: long-tailed, subexponential, regularly varying distributions can be used as per mathematical convenience. Certain invariance properties of distributions with power-law tails are also discussed in this section. The random variables described in this section generally take nonnegative values.

4.4.1 Heavy-Tailed Distributions

Heavy-tailed distributions are probability distributions with tails heavier than the exponential distribution. Some examples of heavy-tailed distributions are: Cauchy, lognormal, Pareto, and Weibull distributions. In the following discussion, the cumulative distribution function, and the complementary cumulative distribution function of the random variable X, are denoted by $F_X(\cdot)$ and $F_X^c(\cdot)$ respectively. The cumulative distribution function is sometimes simply referred to as the distribution function for brevity.

Definition 4.2. *The random variable X is heavy-tailed if for each $\epsilon > 0$,*

$$\lim_{x\to\infty} \frac{F_X^c(x)}{e^{-\epsilon x}} \to \infty \tag{4.3}$$

□

Thus, if a random variable is heavy-tailed, then the corresponding distribution $F_X(\cdot)$ has a heavy-tail. If $r \in \mathbb{R}^+$, then $F_X(x) = (1 - x^{-r})$, where $x \geq 1$ is an example of heavy-tailed distribution. An important subclass of heavy-tailed distributions is the class of long-tailed distribution.

Definition 4.3. *The random variable X is long-tailed if $F_X^c(x) > 0$, and for all $y \in \mathbb{R}^+$,*

$$\lim_{x\to\infty} P(X > x+y \mid X > x) = \lim_{x\to\infty} \frac{F_X^c(x+y)}{F_X^c(x)} = 1 \tag{4.4}$$

□

The above definition, implies that if the random variable X exceeds a large value, then it is most likely to exceed any larger value also. All random variables with long-tailed distributions are heavy-tailed distributions. However, the converse is not true. Note that the exponentially distributed random variable X with parameter $\lambda \in \mathbb{R}^+$ is not long-tailed, as

$$\frac{F_X^c(x+y)}{F_X^c(x)} = e^{-\lambda y}, \quad y \in \mathbb{R}^+$$

A subset of random variables with long-tailed distributions is the class of subexponentially distributed random variables.

Definition 4.4. *Let X and Y be independent, and identically distributed random variables. The random variable X, or its cumulative distribution function $F_X(\cdot)$ is said to be subexponential if $F_X^c(x) > 0$, and*

$$\lim_{x\to\infty} \frac{P(X+Y > x)}{P(X > x)} = 2 \tag{4.5}$$

□

The following observations are immediate from the above definition of a subexponential random variable or distribution.

Observation 4.1. Let $X_1, X_2, \ldots, X_n$ be a sequence of $n \in \mathbb{P}\backslash\{1\}$ independent, and identically distributed random variables. Denote a generic such random variable by X. The random variable X, or its cumulative distribution function $F_X(\cdot)$ is subexponential if one of the following equivalent conditions holds:

(a)

$$\lim_{x\to\infty} \frac{P(X_1 + X_2 + \ldots + X_n > x)}{P(X > x)} = n \tag{4.6a}$$

(b)

$$\lim_{x\to\infty} \frac{P(X_1 + X_2 + \ldots + X_n > x)}{P(\max(X_1, X_2, \ldots, X_n) > x)} = 1 \tag{4.6b}$$

Part (a) can be established by induction. Part (b) of the observation implies that the sum of random variables is likely to get large, if one of the random variables gets large. It can be shown, that if the above limits are true for some $n \geq 2$, then these are true for all $n \geq 2$. □

Pareto, lognormal, and Weibull distributions are subexponential for appropriate parameter values. A subclass of subexponential distributions are distributions with regularly varying tail.

Definition 4.5. *The random variable X, or its cumulative distribution function $F_X(\cdot)$ is said to be regularly varying with tail index $\alpha \in \mathbb{R}_0^+$, if $F_X^c(x) > 0$, and for all $t \in \mathbb{R}^+$,*

$$\lim_{x\to\infty} \frac{F_X^c(tx)}{F_X^c(x)} = t^{-\alpha} \tag{4.7}$$

The distribution function has slowly varying tail if $\alpha = 0$. □

For a regularly varying distribution function, $F_X^c(x) = x^{-\alpha} L_s(x)$, where $L_s(\cdot)$ is some slowly varying function. Some examples of $L_s(\cdot)$ are constants, functions converging to a constant, Pareto tails, logarithms, or iterated logarithms.

4.4.2 Properties of Power-Law-Tailed Distributions

Invariance properties of random variables with power-law-tailed distributions like: aggregation, maximization, and weighted mixture are next described. For simplicity in discussion, assume that the random variables take nonnegative values. Recall that a random variable X has a power-law-tailed distribution if $P(X > x) \sim cx^{-\alpha}$, as $x \to \infty$, where $c \in \mathbb{R}^+$, and $\alpha \in (0, 2)$. The parameter α is called the tail index or shape parameter, and c is called the scale coefficient.

Aggregation

As per Lévy's central limit theorem: if $X_1, X_2, \ldots, X_n$ is a sequence of independent, and identically distributed random variables, $\{a_n\}$ is a sequence of real numbers, and

$$S_n = \frac{(X_1 + X_2 + \cdots + X_n)}{n^{1/\alpha}} + a_n, \quad \alpha \in (0, 2]$$

Then as $n \to \infty$, random variable X is the limit in distribution of S_n if and only if X is stable. In this case the random variable X is said to be α-stable. The case $\alpha = 2$, corresponds to the Gaussian central limit theorem. However, if $\alpha \in (0, 2)$ the α-stable random variable X has a probability distribution with heavy-tail.

Therefore if $V_1, V_2, \ldots, V_n$ are independent, and identically distributed α-stable random variables with heavy-tails, and V is a similarly distributed random variable, there exists a real number b_n such that

$$V_1 + V_2 + \cdots + V_n = n^{1/\alpha} V + b_n, \quad \alpha \in (0, 2)$$

Maximization

Let $X_1, X_2, \ldots, X_n$ be a sequence of n independent random variables, such that

$$P(X_i > x) \sim c_i x^{-\alpha}; \quad c_i \in \mathbb{R}^+, \quad \alpha \in (0, 2), \quad 1 \le i \le n$$

Define M_k to be the k-th successive maxima for $1 \le k \le n$. That is,

$$M_k = \max(X_1, X_2, \ldots, X_k)$$

It can be shown that

$$P(M_k > x) \sim c_{M_k} x^{-\alpha}, \quad \text{where} \quad c_{M_k} = \sum_{i=1}^{k} c_i$$

This result implies that the distribution of k-th successive maxima of random variables with power-law tails also has a power-law tail.

Weighted Mixture

Let $X_1, X_2, \ldots, X_n$ be a sequence of n independent random variables, such that

$$P(X_i > x) \sim c_i x^{-\alpha}; \quad c_i \in \mathbb{R}^+, \quad \alpha \in (0,2), \quad 1 \le i \le n$$

Also let N_n be a weighted mixture of the X_i's. That is, $N_n = X_i$ with probability p_i, where $p_i \in [0,1]$, for $1 \le i \le n$ and $\sum_{i=1}^{n} p_i = 1$. It can be shown that

$$P(N_n > x) = \sum_{i=1}^{n} p_i P(X_i > x) \sim c_{N_n} x^{-\alpha}, \text{ where } c_{N_n} = \sum_{i=1}^{n} p_i c_i$$

Thus the probability distribution of a weighted mixture of distributions with power-law tails with index α, also has a power-law tail with index α, but with a different scale coefficient than that of the individual X_i's.

4.5 Internet Traffic Modeling

In this section, some parsimonious models of the Internet traffic are developed. The Internet traffic is characterized by long-range dependence and slow-decaying variances. This can be described in terms of self-similar processes and power-law distributions. Self-similar processes are described mathematically in this section. Some special measures to describe burstiness of the Internet traffic are also defined.

Special measures to describe burstiness of the data are: peak-to-mean-ratio, index of dispersion for counts (IDC), squared coefficient of variation of the process, and the Hurst parameter. For example, peak-to-mean ratio of packet arrivals is the ratio of peak arrival rate of the packets to its mean arrival rate. This measure requires the nontrivial task of determining the length of the averaging interval. The IDC is given by the variance of the number of packet arrivals divided by its mean value computed over a specified time interval. Squared coefficient of variation of a random variable is the ratio of its variance and its squared-mean. Computation of this parameter depends upon the sample size. Techniques to determine the Hurst parameter are also described. The following topics are discussed in the rest of this section: description of self-similar processes, fractional Brownian motion, second order self-similarity, Hurst's law, and estimation of the Hurst parameter.

4.5.1 Self-Similar Processes

A self-similar process is a stochastic process which is invariant in distribution under suitable scaling of the time parameter $t \in T$. In the following definition, the indexing set T of the stochastic process is equal to either $\mathbb{R}$, or $\mathbb{R}_0^+$, or $\mathbb{R}^+$.

Definition 4.6. *A stochastic process $\{Y(t), t \in T\}$ with continuous time parameter t, is said to be statistically self-similar with self-similarity parameter $H > 0$, if for any $u > 0$, the finite-dimensional distribution of the process $\{u^{-H} Y(ut), t \in T\}$ is identical to the finite-dimensional* *distribution of the process* $\{Y(t), t \in T\}$ *. That is, for any* $n \in \mathbb{P}$*, and* $t_1, t_2, \ldots, t_n \in T$*, the distribution of* $(Y(t_1), Y(t_2), \ldots, Y(t_n))$ *is identical to that of*

$$\left(u^{-H}Y(ut_1), u^{-H}Y(ut_2), \ldots, u^{-H}Y(ut_n)\right) \tag{4.8}$$

□

The parameter H in the above definition is called the Hurst parameter or index of self-similarity. The Hurst parameter H is used to quantify the degree of self-similarity (or burstiness) of the Internet traffic. Let $\mathcal{E}(\cdot)$ denote the expectation operator. For any $s, t \in T$, the above definition yields for any $u \in \mathbb{R}^+$:

(a) $\mathcal{E}(Y(t)) = u^{-H}\mathcal{E}(Y(ut))$
(b) $Var(Y(t)) = u^{-2H}Var(Y(ut))$
(c) $Cov(Y(s), Y(t)) = u^{-2H}Cov(Y(us), Y(ut))$

Other definitions of self-similarity are common, including second order self-similarity, which states that the first and second moments scale as in the above remarks. The Brownian motion process (BMP) is an example of self-similar process. The BMP has been discussed in the chapter on stochastic processes.

4.5.2 Fractional Brownian Motion

Fractional Brownian motion process (FBMP) is a generalization of the Brownian motion process. It is occasionally used in analysing self-similar traffic. Fractional Brownian motion stochastic process is one of several processes to model self-similarity. Following is the classical definition of FBMP for $t \in \mathbb{R}$.

Definition 4.7. (*FBMP*) *The random process* $\{B_H(t), t \in \mathbb{R}\}$ *is a fractional Brownian motion process with Hurst parameter* H *if:*

(a) *The process is Gaussian and self-similar with stationary increments.*
(b) $0 < H < 1$.
(c) $\sigma^2 = Var(B_H(1))$.

If $\sigma = 1$*, the FBMP is called a standard FBMP.* □

Some observations about FBMP are listed below. Also define a function $a(t)$ for $t \in \mathbb{R}$ as

$$a(t) = \begin{cases} -1, & t < 0 \\ 0, & t = 0 \\ 1, & t > 0 \end{cases}$$

Observations 4.2. Observations about FBMP.

1. For $H = 0.5$, the FBMP is a BMP.
2. $B_H(0) = 0$ almost surely. This follows from the observation that $B_H(0) = B_H(u0)$ has the same distribution as $u^H B_H(0)$, where $u > 0$.

3. For a fixed $t \in \mathbb{R}$, $B_H(-t)$ is equal to $-B_H(t)$ in distribution. This statement is true because: $B_H(0) = 0$ and $B_H(-t) = (B_H(-t) - B_H(0))$ is equivalent to $(B_H(0) - B_H(t)) = -B_H(t)$ in distribution because of the stationarity of the increments.
4. $\mathcal{E}(B_H(t)) = 0$ for any $t \in \mathbb{R}$. Self-similarity of the process implies $\mathcal{E}(B_H(2t)) = 2^H \mathcal{E}(B_H(t))$.
   Note that $\mathcal{E}(B_H(2t) - B_H(t)) = \mathcal{E}(B_H(t) - B_H(0))$ follows from the stationarity of the increments.
   Therefore $\mathcal{E}(B_H(2t)) = 2\mathcal{E}(B_H(t))$.
   Thus $2^H \mathcal{E}(B_H(t)) = 2\mathcal{E}(B_H(t))$. Since $H \neq 1, \mathcal{E}(B_H(t)) = 0$.
5. $\mathcal{E}(B_H^2(t)) = \sigma^2 |t|^{2H}$.
   Since $B_H(-t)$ is equal to $-B_H(t)$ in distribution

$$\begin{aligned}\mathcal{E}\left(B_H^2(t)\right) &= \mathcal{E}\left(B_H^2(a(t)\cdot|t|)\right) = |t|^{2H}\,\mathcal{E}\left(B_H^2(a(t))\right)\\ &= |t|^{2H}\,\mathcal{E}\left(B_H^2(1)\right) = \sigma^2 |t|^{2H}\end{aligned} \tag{4.9}$$

   where we used the relationship, $B_H(-1) = -B_H(1)$ in distribution.
6. $Var(B_H(t) - B_H(s)) = \sigma^2 |t-s|^{2H}$. This observation is a consequence of the property of stationary-increments.
7. $Cov(B_H(t), B_H(s)) = c_H(t,s) = \sigma^2 \left\{|s|^{2H} + |t|^{2H} - |t-s|^{2H}\right\}/2$.
   As $\mathcal{E}\left(\{B_H(s) - B_H(t)\}^2\right) = \mathcal{E}\left(B_H^2(s)\right) + \mathcal{E}\left(B_H^2(t)\right) - 2\mathcal{E}(B_H(s)B_H(t))$, the result follows. Therefore it can be observed that the FBMP is nonstationary, as $c_H(t,s)$ is not a function of $|t-s|$.
8. Check that the self-similarity parameter H satisfies $H < 1$.

$$\begin{aligned}\mathcal{E}(|B_H(2)|) &= \mathcal{E}(|B_H(2) - B_H(1) + B_H(1)|)\\ &\le \mathcal{E}(|B_H(2) - B_H(1)|) + \mathcal{E}(|B_H(1)|)\\ &= 2\mathcal{E}(|B_H(1)|)\end{aligned}$$

   Also $\mathcal{E}(|B_H(2)|) = 2^H \mathcal{E}(|B_H(1)|)$. Therefore $2^H \le 2$, which implies $H \le 1$.
9. The autocovariance of the increments of the FBMP in two nonoverlapping intervals (t_1, t_2) and (t_3, t_4), where $t_1 < t_2 < t_3 < t_4$ is given by

$$\begin{aligned}&\mathcal{E}\left[\{B_H(t_4) - B_H(t_3)\}\{B_H(t_2) - B_H(t_1)\}\right]\\ &= \frac{\sigma^2}{2}\left[|t_4 - t_1|^{2H} + |t_3 - t_2|^{2H} - |t_4 - t_2|^{2H} - |t_3 - t_1|^{2H}\right]\end{aligned} \tag{4.10a}$$

   For a BMP

$$\mathcal{E}\left[\{B_H(t_4) - B_H(t_3)\}\{B_H(t_2) - B_H(t_1)\}\right] = 0, \quad H = 0.5 \tag{4.10b}$$

10. FBMP exhibits the interesting phenomenon of long-run correlation. Consider the autocovariance between the increment of this process from 0 to t and the increment of the process from t to $2t$. This autocovariance is

$$\begin{aligned}&\mathcal{E}\left[\{B_H(t) - B_H(0)\}\{B_H(2t) - B_H(t)\}\right]\\ &= \sigma^2 |t|^{2H}\left\{2^{(2H-1)} - 1\right\}\end{aligned}$$

At $H = 0.5$, the above autocovariance is equal to zero, while for $H > 0.5$, the autocovariance is nonzero. This value increases, as H increases. This persistent behavior is in direct contrast to what is generally assumed about stochastic phenomenon. It is generally assumed incorrectly, that the random phenomenon become uncorrelated as time becomes large.

11. The increment process of the FBMP is $\{X(t), t \in \mathbb{R}\}$, where

$$X(t) = B_H(t) - B_H(t - \Delta) \tag{4.11a}$$

and $\Delta > 0$. Note that the increment process is stationary by definition, and $\mathcal{E}(X(t)) = 0$. Also $Var(X(t)) = \sigma^2 \Delta^{2H}$. The autocovariance of this increment process is

$$Cov(X(t), X(t-\tau)) = \frac{\sigma^2}{2}\left[|\tau + \Delta|^{2H} + |\tau - \Delta|^{2H} - 2|\tau|^{2H}\right] \tag{4.11b}$$

Observe that $Cov(X(t), X(t-\tau))$ is independent of t. It depends only on τ and Δ. Therefore for notational convenience, let

$$r(\tau, \Delta) \triangleq Cov(X(t), X(t-\tau)) \tag{4.11c}$$

Then

$$r(\tau, \Delta) \simeq \sigma^2 \Delta^2 H(2H-1)|\tau|^{2H-2}, \qquad \Delta \ll |\tau| \tag{4.11d}$$

This process is generally termed *fractional Gaussian noise process*. □

The following theorem characterizes FBMP.

Theorem 4.1. *If the random process* $\{Y(t), t \in \mathbb{R}\}$:

(a) *Is a Gaussian process withe zero mean, and* $Y(0) = 0$.
(b) $\mathcal{E}(Y^2(t)) = \sigma^2 |t|^{2H}$ *for some* $\sigma > 0$ *and* $0 < H < 1$.
(c) *The process has stationary increments.*

Then the process is fractional Brownian motion.

Proof. The process $Y(t), t \in \mathbb{R}$ is Gaussian with zero mean, and from the hypothesis of the theorem, the covariance function of the process is of the form

$$c(t,s) = \frac{\sigma^2}{2}\left[|s|^{2H} + |t|^{2H} - |t-s|^{2H}\right]$$

which is the covariance of a FBMP. Furthermore, its finite-dimensional distributions are uniquely determined by its covariance function. □

Representation of Fractional Brownian Motion

A representation of standard FBMP in the time domain is obtained.

Definition 4.8. *Let* $\chi^+ = \max(0, \chi)$, *and* $\chi^- = \max(0, -\chi)$, *where* $\chi \in \mathbb{R}$. □

This definition yields

$$|\chi| = \chi^+ + \chi^-$$

The following theorem gives an explicit representation of FBMP. The FBMP is represented as a weighted average of BMP over the infinite past.

Theorem 4.2. *Let* $\{B(t), t \in \mathbb{R}\}$ *be the standard Brownian motion process. For notational convenience define*

$$\zeta = (H - 1/2), \quad \text{where} \quad 0 < H < 1 \tag{4.12a}$$

Also define a weight function $w_H(\cdot, \cdot)$ *as*

$$w_H(t, u) = \left\{(t-u)^+\right\}^\zeta - \left\{(-u)^+\right\}^\zeta, \qquad t, u \in \mathbb{R} \tag{4.12b}$$

The standard fractional Brownian motion process $B_H(t), t \in \mathbb{R}$ *can be represented as*

$$Y(t) = \frac{1}{C_H} \int_{-\infty}^{\infty} w_H(t, u)\, dB(u), \qquad t \in \mathbb{R} \tag{4.12c}$$

where

$$C_H^2 = \int_0^\infty \left((1+u)^\zeta - u^\zeta\right)^2 du + \frac{1}{2H} \tag{4.12d}$$

$$= \frac{\{\Gamma(H + 1/2)\}^2}{\Gamma(2H+1)\sin(\pi H)} \tag{4.12e}$$

and $\Gamma(\cdot)$ *is the gamma function. When* $H = 1/2, C_{1/2} = 1$*, and* $Y(t)$ *is equal to* $\int_0^t dB(u)$ *for* $t \geq 0$*, and* $-\int_t^0 dB(u)$ *for* $t < 0$*.*

Proof. Note that the weighting function is well-defined for $0 < H < 1$. In order to prove that the random process $\{Y(t), t \in \mathbb{R}\}$ is a standard FBMP, it has to be shown that $Y(t), t \in \mathbb{R}$ is a Gaussian process withe zero mean. In addition, the relationship $\mathcal{E}\left(Y^2(t)\right) = |t|^{2H}$ for $0 < H < 1$ has to be established. Furthermore, the stationarity of the increments also has to be established.

The statement is clearly true if $H = 1/2$. Next suppose that $H \neq 1/2$. Observe that $\mathcal{E}(Y(t)) = 0$ because the process $B(t), t \in \mathbb{R}$ is a BMP. For $t > 0$, we have

$$w_H(t, u) = \begin{cases} 0, & t \leq u \\ (t-u)^\zeta, & 0 \leq u < t \\ (t-u)^\zeta - (-u)^\zeta, & u < 0 \end{cases}$$

Since $Y(t)$ is expressed as a Wiener integral, we have

$$\begin{aligned} \mathcal{E}\left(Y(t)^2\right) &= C_H^{-2}\left\{\int_{-\infty}^0 \left((t-u)^\zeta - (-u)^\zeta\right)^2 du + \int_0^t (t-u)^{2\zeta} du\right\} \\ &= \frac{t^{2H}}{C_H^2}\left\{\int_{-\infty}^0 \left((1-u)^\zeta - (-u)^\zeta\right)^2 du + \int_0^1 (1-u)^{2\zeta} du\right\} = t^{2H} \end{aligned}$$

It can be similarly shown that

$$\mathcal{E}\left(Y(t)^2\right) = |t|^{2H}, \quad \text{for } t < 0$$

For any $s, t \in \mathbb{R}$

$$\begin{aligned}
&\mathcal{E}\left((Y(t)-Y(s))^2\right) \\
&= C_H^{-2}\int_{-\infty}^{\infty}\left(\left\{(t-u)^+\right\}^{\varsigma}-\left\{(s-u)^+\right\}^{\varsigma}\right)^2 du \\
&= C_H^{-2}\int_{-\infty}^{\infty}\left(\left\{(t-s-u)^+\right\}^{\varsigma}-\left\{(-u)^+\right\}^{\varsigma}\right)^2 du \\
&= \mathcal{E}\left(Y^2(t-s)\right)
\end{aligned}$$

Consequently for any $b_k, t_k \in \mathbb{R}$, for $1 \le k \le n$, and any $h \in \mathbb{R}$,

$$\mathcal{E}\left(\left\{\sum_{k=1}^{n} b_k\left(Y(t_k+h)-Y(h)\right)\right\}^2\right) = \mathcal{E}\left(\left\{\sum_{k=1}^{n} b_k Y(t_k)\right\}^2\right)$$

Stationarity of the increment process follows because BMP has stationary increments. Furthermore, $\mathcal{E}\left(Y(1)^2\right) = 1$, that is $B_H(t)$ is standard fractional Brownian motion process. See the problem section for an explicit evaluation of C_H^2. □

4.5.3 Second Order Self-Similarity

Experimental data is sampled at regular intervals. Consequently self-similar discrete-time process is studied in this subsection. Let $\{X(i), i \in \mathbb{Z}\}$ be a discrete-time stochastic process. For the purpose of modeling, it is required that the process $X(\cdot)$ be stationary. More explicitly, it is assumed that this process is second order stationary. Assume that $\mathcal{E}(X(i)) = \mu, Var(X(i)) = \sigma^2$, and $Cov(X(i), X(j)) = Cov(X(i+k), X(j+k))$ for any $i, j, k \in \mathbb{Z}$.

Also let $\{Y(i), i \in \mathbb{Z}\}$ be a discrete time stochastic process, where

$$X(i) = (Y(i) - Y(i-1)), \quad \forall\, i \in \mathbb{Z}$$

Thus $X(\cdot)$ is an *increment process* of $Y(\cdot)$. In addition, assume that the process $Y(\cdot)$ is self-similar, that is $Y(i)$ is equal to $n^{-H}Y(ni)$ in distribution where $n \in \mathbb{P}$ and the Hurst parameter $H \in (0,1)$. This assumption implies that $Y(0) = 0$, and $\mathcal{E}(Y(i)) = 0$ for any $i \in \mathbb{Z}$. This in turn implies $\mathcal{E}(X(i)) = \mu = 0$ for any $i \in \mathbb{Z}$. Thus $Var(Y(1)) = \sigma^2$. Next define

$$\left\{X^{(m)}(k), k \in \mathbb{Z}, m \ge 2\right\}$$

as an aggregate process of $\{X(i), i \in \mathbb{Z}\}$, where

$$X^{(m)}(k) = \frac{1}{m}\sum_{j=m(k-1)+1}^{mk} X(j)$$

In other words, $X(j)$ is split into nonoverlapping blocks of size m, and their values are averaged. The parameter m is called the aggregation level. The variable k is used to index these blocks. Thus

$$\begin{aligned}
X^{(m)}(k) &= \frac{1}{m}\left(Y(mk) - Y(m(k-1))\right) \\
&= m^{H-1}\left(Y(k) - Y(k-1)\right), \quad \text{in distribution} \\
&= m^{H-1}X(k)
\end{aligned}$$

Therefore the equality of processes $X(\cdot)$ and $m^{1-H}X^{(m)}(\cdot)$ in distribution appears to be plausible. That is, the process $X(\cdot)$ is said to be *exactly self-similar* to the process $X^{(m)}(\cdot)$ for any $m \in \mathbb{P}$.

Different properties of self-similar processes are next studied. We study the phenomenon of the slow decay of variance of the aggregated process $X^{(m)}(\cdot)$. In addition, the concept of the long-range dependence of the process $X^{(m)}(\cdot)$ is also quantified. It is essentially concluded that the autocovariance of the aggregated process also decays slowly.

Slowly Decaying Variance

Since $Var(X(k)) = \sigma^2$, we have

$$\mathcal{E}\left(X^{(m)}(k)\right) = 0$$

$$Var\left(X^{(m)}(k)\right) = \frac{\sigma^2}{m^\beta}, \qquad \beta = 2(1-H)$$

Observe that this variance is proportional to $m^{-\beta}$. The value of the parameter β for classical models lies in the interval $(1,2)$. Therefore as $m \to \infty$, the variance in the classical models decay much faster than the variance of the aggregated process when $\beta \in (0,1)$. Note that if $\beta \in (0,1)$ then $H \in (1/2,1)$, and if $\beta \in (1,2)$ then $H \in (0,1/2)$.

Long-Range Dependence

The autocovariance of the process $X^{(m)}(\cdot)$ differs from the autocovariance of the process $X(\cdot)$ by a constant scale factor. Therefore we only study the autocovariance of the process $X(\cdot)$. Let the autocovariance of the process $X(\cdot)$ be $\gamma(n) = Cov(X(i), X(i-n))$. The process $Y(\cdot)$ is self-similar, and it is to be modeled by a FBMP. Furthermore, the process $X(\cdot)$ is its incremental process, then $\gamma(n)$ should be

$$\gamma(n) = \frac{\sigma^2}{2}\left(|n+1|^{2H} + |n-1|^{2H} - 2|n|^{2H}\right), \qquad \forall n \in \mathbb{Z}$$

If $H = 1/2$, then $\gamma(n) = 0$, that is the process $X(\cdot)$ is uncorrelated. However, if $H \approx 1$, then $\gamma(n) \simeq \sigma^2$ for all values of $n \in \mathbb{Z}$. Furthermore,

$$\gamma(n) \simeq \sigma^2 H(2H-1)|n|^{2H-2}, \qquad |n| \gg 1$$

If $H \in (1/2,1)$, the process $X(\cdot)$ exhibits a positive correlation. The corresponding value of β lies in the interval $(0,1)$, and $\gamma(n)$ is proportional to $n^{-\beta}$. Therefore $\sum_{n=-\infty}^{\infty}\gamma(n)$ tends to infinity if $\beta \in (0,1)$. Such processes are called *long-range dependent.* Alternately, the process is said to have *strong dependence,* or that it has *long memory,* or that it is a $1/f$ *noise*.

The process $X(\cdot)$ is said to be *short-range dependent* if the autocovariance function is summable, that is if $\sum_{n=-\infty}^{\infty}|\gamma(n)| < \infty$. This occurs if $\beta \in (1,2)$, that is $H \in (0,1/2)$. Thus classical models of stochastic processes capture short-range dependence and not long-range dependence.

Generally, classical models have autocorrelation of the form $\gamma(n) = \sigma^2\rho^{|n|}$, where $\rho \in (0,1)$ and $n \in \mathbb{Z}$. Thus $\sum_{n=-\infty}^{\infty}|\gamma(n)| < \infty$. This is in contrast to long-range dependent processes which have hyperbolically decaying autocorrelation functions.

In summary, it can be concluded that the most significant feature of long-range dependent process is that the correlation of its aggregated process tends asymptotically towards zero much more slowly, than the correlation of the traditional (classical) aggregated processes.

4.5.4 Self-Similarity and the Hurst's Law

Self-similar processes also provide an explanation of the *Hurst's law* (or Hurst's effect). Let $\{X_i \mid 1 \le i \le n\}$ be a set of random observations, and $\overline{X}(n)$ and $S^2(n)$ be its sample mean and sample variance respectively.

$$\overline{X}(n) = \frac{1}{n}\sum_{i=1}^{n} X_i, \text{ and } S^2(n) = \frac{1}{(n-1)}\sum_{i=1}^{n}\left(X_i - \overline{X}(n)\right)^2, \quad \text{for } n > 1$$

Define

$$w_k = (X_1 + X_2 + \ldots + X_k), \qquad 1 \le k \le n \tag{4.13a}$$

$$W_k = (X_1 + X_2 + \ldots + X_k) - k\overline{X}(n), \qquad 1 \le k \le n \tag{4.13b}$$

The *range* $r(n)$ of the variables $X_1, X_2, \ldots, X_n$ is defined as

$$r(n) = \max(0, w_1, w_2, \ldots, w_n) - \min(0, w_1, w_2, \ldots, w_n) \tag{4.13c}$$

The corresponding *adjusted range* $R(n)$ of the variables $X_1, X_2, \ldots, X_n$ is defined as

$$R(n) = \max(0, W_1, W_2, \ldots, W_n) - \min(0, W_1, W_2, \ldots, W_n) \tag{4.13d}$$

The *rescaled adjusted range statistic* (or *R/S statistic*) is given by

$$\frac{R(n)}{S(n)} = \frac{1}{S(n)}\left[\max(0, W_1, W_2, \ldots, W_n) - \min(0, W_1, W_2, \ldots, W_n)\right] \tag{4.13e}$$

Hurst found that several naturally occurring time series are well-represented by the relation

$$\mathcal{E}\left(\frac{R(n)}{S(n)}\right) \sim an^H, \text{ as } n \to \infty \tag{4.13f}$$

where a is a finite positive constant and H is the Hurst parameter. The parameter H is equal to 0.5 for a phenomenon which can be modeled by a short-range dependent process. While $H \in (0.5, 1.0)$ for self-similar processes found on the Internet. This behavior is called the *Hurst effect* or *Hurst phenomenon*.

The reader should also refer to some elementary results on probabilistic properties of range of a sequence of random variables in the chapter on probability theory.

4.5.5 Estimation of the Hurst Parameter

The Hurst parameter H can be regarded as a measure of self-similarity of a given set of data points. The range of values of H for long-range dependence is $(0.5, 1.0)$. There are several techniques to determine the Hurst parameter. These are variance-time plots, rescaled range analysis, wavelet analysis, periodograms, and Whittle estimator. Only two techniques to estimate the Hurst parameter are next outlined. The first technique depends upon computing the variance of the aggregated process $X^{(m)}(\cdot)$, and the other depends upon the R/S statistic.

A set of observations $\{x_i \mid 1 \le i \le n\}$ is given. Their sample mean and sample variance are $\overline{x}_n$ and $s^2(n)$ respectively. Let the estimated value of the Hurst parameter be $\widehat{H}$. The goal is to determine the value of $\widehat{H}$.

Variance-time plots: It is known from the last subsection that

$$Var\left(X^{(m)}(k)\right) = bm^{-\beta}; \quad b > 0, \;\; 0 < \beta < 2, \;\; H = (1 - \beta/2)$$

This yields

$$\log\left(Var\left(X^{(m)}(k)\right)\right) = -\beta \log m + \log b \tag{4.14}$$

This equation suggests a procedure to determine the value of H, given a set of data points. An algorithm to estimate Hurst parameter via variance-time plot is outlined.

Algorithm 4.1. *Estimation of Hurst Parameter via Variance-Time Plot.*

Input: $x_i, i = 1, 2, 3, \ldots$.
Output: Estimate of Hurst parameter, $\widehat{H}$.
begin
Step 1: For different values of m, estimate the mean and variance of $X^{(m)}(k)$. These mean and variance computations are each based upon several sample values of $X^{(m)}(\cdot)$.
Step 2: Plot logarithm of m and the corresponding logarithm of the variance, for different values of m.
Step 3: For large values of m, find a best-fit straight line. For a self-similar process, the slope is equal to $-\widehat{\beta}$, which is an estimate of $-\beta$. Then $\widehat{H} = \left(1 - \widehat{\beta}/2\right)$.
end (*end of Hurst parameter estimation via variance-time plot*)

R/S Statistic: This technique to estimate the Hurst parameter depends upon the relationship

$$\mathcal{E}\left(\frac{R(n)}{S(n)}\right) \sim an^{H}, \quad a > 0, \; n \to \infty$$

This yields

$$\log\left(\mathcal{E}\left(\frac{R(n)}{S(n)}\right)\right) \sim H \log n + \log a, \quad n \to \infty$$

In the following algorithm, $x_i, i = 1, 2, 3, \ldots$ is a set of the data points, and $\overline{x}_n$ and $s^2(n)$ are the sample mean and sample variance respectively. An algorithm to estimate Hurst parameter via R/S statistic is outlined.

Algorithm 4.2. *Estimation of Hurst Parameter via R/S Statistic.*

Input: $x_i, i = 1, 2, 3, \ldots$.
Output: Estimate of the Hurst parameter, $\widehat{H}$.
begin

Step 1: For different values of n :
Compute $\overline{x}_n$ and $s^2(n)$ for the given set of data points.
Compute $y_k = (x_1 + x_2 + \ldots + x_k) - k\overline{x}_n$, for $1 \le k \le n$.
Compute

$$z(n) = \frac{1}{s(n)} [\max(0, y_1, y_2, \ldots, y_n) - \min(0, y_1, y_2, \ldots, y_n)]$$

Step 2: The values of n and the corresponding $z(n)$ are plotted on a log-log scale. This plot is often referred to as a *Pox plot.*
Step 3: For large values of n, find a best-fit straight line.
For a self-similar process, the slope is equal to $\widehat{H}$.
end (*end of Hurst parameter estimation via R/S statistic*)

A quantitative explanation of self-similarity is provided in the next section.

4.6 Explanation of Self-Similarity in the Internet Traffic

A framework for the study of self-similarity in Internet traffic is provided in this section. Relationship between heavy-tails and self-similarity is elucidated. It is demonstrated that the superposition of several strictly alternating independent, and identically distributed "on" and "off" sources, each exhibiting Noah effect, asymptotically produces self-similar traffic on large time scales. During the "on" interval, the source outputs a *packet train.* Evidently, during the "off" interval, there is no output from the source. The Noah effect is characterized by the infinite variance syndrome. Therefore heavy-tailed distributions are used to model the "on" and "off" durations. The parameter α, which describes the heaviness of the tail is linked to the Hurst parameter H.

It is established that the cumulative packet traffic can be modeled as a fractional Brownian motion process. Equivalently its increment process, the number of packets per unit time, can be modeled by a fractional Gaussian noise process. This FBMP $B_H(t)$ has a zero mean, stationary increments, and covariance function

$$Cov(B_H(s), B_H(t)) = \frac{\sigma^2}{2} \left\{ |s|^{2H} + |t|^{2H} - |t - s|^{2H} \right\}$$

where $1/2 < H < 1$, and σ^2 is the variance of the number of packets in unit time.

4.6.1 Notation and Assumptions

Let $\{W(t), t \in \mathbb{R}_0^+\}$ be a two-state stationary stochastic process. The variable $W(t)$ takes two values: 0 and 1. When $W(t) = 1$, there is a packet at time t, and when $W(t) = 0$, there is no packet. All contiguous values of t for which $W(t) = 1$ constitutes an "on" interval, while all contiguous values of t for which $W(t) = 0$ form an "off" interval. The "on" and "off" intervals are assumed to alternate. The lengths of "on" intervals are independent, and identically distributed random variables. Similarly, the lengths of "off" intervals are independent, and identically distributed random variables. Furthermore, the random "on" and "off" intervals are stochastically independent of each other. A successive pair of "on" and "off" intervals constitute an interrenewal period.

Assume that there are M independent, and identically distributed sources. The source m is described by an alternating renewal process

$$\left\{W^{(m)}(t), t \in \mathbb{R}_0^+\right\}, \quad 1 \leq m \leq M \tag{4.15a}$$

The net packet count at time t due to M sources is equal to $\sum_{m=1}^{M} W^{(m)}(t)$. Define the aggregated cumulative packet count in the interval $[0, t]$ by

$$\widetilde{W}_M(t) = \int_0^t \sum_{m=1}^{M} W^{(m)}(u)\, du, \quad t \in \mathbb{R}_0^+ \tag{4.15b}$$

Let the random variables A and B specify the "on" and "off" intervals. Also let the probability density function, cumulative distribution function, and the complementary cumulative distribution function of the random variable A be $f_1(t), F_1(t)$, and $F_1^c(t)$ respectively for $t \in \mathbb{R}_0^+$. Similarly the probability density function, cumulative distribution function, and the complementary cumulative distribution function of the random variable B be $f_2(t), F_2(t)$, and $F_2^c(t)$ respectively for $t \in \mathbb{R}_0^+$. Also let $\mathcal{E}(A) = \mu_1, \mathcal{E}(B) = \mu_2, Var(A) = \sigma_1^2$, and $Var(B) = \sigma_2^2$. Let the Laplace transform of $f_1(t)$ and $f_2(t)$ be $\widehat{f}_1(s)$ and $\widehat{f}_2(s)$ respectively, where $s \in \mathbb{C}$.

The final result is obtained via the central limit theorem. Therefore the process $\int_0^t W(u)\, du$, $t \in \mathbb{R}_0^+$ is first characterized. This integral is the single source cumulative packet count process in the interval $[0, t]$.

4.6.2 Single Source Characterization

As $t \to \infty$, $\mathcal{E}(W(t)) = \mu_1/(\mu_1 + \mu_2)$. Also since the process $W(t)$, $t \in \mathbb{R}_0^+$ is stationary, $\mathcal{E}(W(t)) = \mu_1/(\mu_1 + \mu_2)$ for all $t \in \mathbb{R}_0^+$. This yields

$$\mathcal{E}\left(\int_0^t W(u)\, du\right) = \frac{\mu_1 t}{(\mu_1 + \mu_2)}$$

Use of stationarity yields

$$\mathcal{E}\left(\int_0^t \int_0^t W(u) W(v)\, dudv\right) = 2 \int_0^t \int_0^v \mathcal{E}(W(u) W(0))\, dudv$$

Define $V(t) = Var\left(\int_0^t W(u)\, du\right)$. Therefore

$$V(t) = 2 \int_0^t \int_0^v \gamma(u)\, dudv$$
$$\gamma(u) = \mathcal{E}(W(u) W(0)) - \{\mathcal{E}(W(0))\}^2$$

$\mathcal{E}(W(t) W(0))$ is determined as follows.

$$\begin{aligned}\mathcal{E}(W(t) W(0)) &= P(\text{the process is in "on" state at time 0 and } t) \\ &= \pi_{11}(t) \frac{\mu_1}{(\mu_1 + \mu_2)}\end{aligned}$$

where

$$\pi_{11}(t) = P\,(\text{process is in "on" state at time } t \mid \text{process is in "on" state at time } 0)$$

As $t \to \infty, \pi_{11}(t) \to \mu_1/(\mu_1+\mu_2)$ and $\gamma(t) \to 0$. Let the Laplace transform of $\pi_{11}(t), \gamma(t)$, and $V(t)$ be $\widehat{\pi}_{11}(s), \widehat{\gamma}(s)$, and $\widehat{V}(s)$ respectively, where $s \in \mathbb{C}$. Note that $\widehat{V}(s) = 2\widehat{\gamma}(s)/s^2$. Use of the equilibrium renewal theory of alternating processes yields

$$\widehat{\pi}_{11}(s) = \frac{1}{s} - \frac{\left(1-\widehat{f}_1(s)\right)\left(1-\widehat{f}_2(s)\right)}{\mu_1 s^2\left\{1-\widehat{f}_1(s)\widehat{f}_2(s)\right\}}$$

Consequently

$$\begin{aligned}\widehat{\gamma}(s) &= \frac{\mu_1}{(\mu_1+\mu_2)}\left[\widehat{\pi}_{11}(s) - \frac{\mu_1}{s(\mu_1+\mu_2)}\right] \\ &= \frac{\mu_1\mu_2}{(\mu_1+\mu_2)^2 s} - \frac{\left(1-\widehat{f}_1(s)\right)\left(1-\widehat{f}_2(s)\right)}{(\mu_1+\mu_2)s^2\left\{1-\widehat{f}_1(s)\widehat{f}_2(s)\right\}}\end{aligned}$$

The analysis at this stage is split into four parts:

Case 1: Both the random variables A and B have finite variance.

Case 2: Both the random variables A and B have infinite variance.

Case 3: The random variable A has infinite variance and the random variable B has finite variance.

Case 4: The random variable A has finite variance and the random variable B has infinite variance.

Case 1: The random variables A and B have finite variance. If the first and second moments of the random variables A and B are finite, then $\widehat{f}_j(s) \simeq 1 - \mu_j s + \left(\sigma_j^2 + \mu_j^2\right)s^2/2$ as $s \to 0$ for $j = 1, 2$. Simplification results in

$$\begin{aligned}\widehat{\gamma}(s) &\simeq \frac{\left(\mu_2^2\sigma_1^2 + \mu_1^2\sigma_2^2\right)}{2(\mu_1+\mu_2)^3}, \quad \text{as } s \to 0 \\ \widehat{V}(s) &\simeq \frac{\left(\mu_2^2\sigma_1^2 + \mu_1^2\sigma_2^2\right)}{s^2(\mu_1+\mu_2)^3}, \quad \text{as } s \to 0\end{aligned}$$

Inverting $\widehat{V}(s)$ yields an expression for $V(t)$ as t approaches ∞. Thus

$$V(t) \simeq \frac{\left(\mu_2^2\sigma_1^2 + \mu_1^2\sigma_2^2\right)t}{(\mu_1+\mu_2)^3}, \quad \text{as } t \to \infty$$

Therefore if the first and second moments of the random variables A and B are finite, then $V(t)$ is of the form $\sigma_{\lim}^2 t$ as $t \to \infty$, where $\sigma_{\lim}^2$ is a constant.

Case 2: The random variables A and B have infinite variance. The asymptotic value of variance $V(t)$, when the variances of the random variables A and B approach infinity is determined. Using renewal theory, we have

$$\pi_{11}(t) = \int_t^{\infty} \frac{F_1^c(x)}{\mu_1} dx + \int_0^t m_{12}(x) F_1^c(t-x)\, dx$$

where $m_{12}(t)$ is the probability density that the end of an "off" interval occurs at time t, given that the process was in "on" state at time 0. Define

$$G_j(t) = \int_t^{\infty} \frac{F_j^c(x)}{\mu_j} dx, \quad j = 1, 2$$

Let the Laplace transform of $G_j(t)$ be $\widehat{G}_j(s)$, then

$$\widehat{G}_j(s) = \frac{1}{s} - \frac{\left(1 - \widehat{f}_j(s)\right)}{\mu_j s^2}, \qquad j = 1, 2$$

$$\widehat{f}_j(s) = 1 - \mu_j s + \mu_j s^2 \widehat{G}_j(s), \quad j = 1, 2$$

Substituting this expressions for $\widehat{f}_j(s)$ in the equation for $\widehat{\gamma}(s)$ yields

$$\widehat{\gamma}(s) = \frac{\mu_1\mu_2}{(\mu_1+\mu_2)^2 s} - \frac{\mu_1\mu_2}{(\mu_1+\mu_2)} \frac{\left(1 - s\widehat{G}_1(s)\right)\left(1 - s\widehat{G}_2(s)\right)}{\left\{1 - \prod_{j=1}^{2}\left(1 - \mu_j s + \mu_j s^2 \widehat{G}_j(s)\right)\right\}}$$

As $s \to 0$

$$\widehat{\gamma}(s) \simeq \frac{\left(\mu_2^2\mu_1\widehat{G}_1(s) + \mu_1^2\mu_2\widehat{G}_2(s) - \mu_1^2\mu_2^2\right)}{(\mu_1+\mu_2)^3}$$

Expressions for $\widehat{G}_j(s)$'s as $s \to 0$ are next determined. As $t \to \infty$, assume that

$$F_j^c(t) \simeq l_j t^{-\alpha_j} L_j(t), \quad 1 < \alpha_j < 2, \quad j = 1, 2$$

where $l_j > 0$ is a constant and $L_j(t) > 0$ is a slowly varying function at infinity, that is $\lim_{x\to\infty} L_j(tx)/L_j(x) = 1$ for any $t > 0$. Note that $1 < \alpha_j < 2$ for $j = 1, 2$ is a requisite for infinite variance. Then as $t \to \infty$ and $s \to 0$

$$G_j(t) \simeq \frac{l_j}{\mu_j(\alpha_j - 1)} t^{-\alpha_j+1} L_j(t), \quad j = 1, 2$$

$$\mu_j \widehat{G}_j(s) \simeq \frac{l_j \Gamma(2-\alpha_j)}{(\alpha_j - 1)} s^{\alpha_j - 2} L_j\left(\frac{1}{s}\right), \quad j = 1, 2$$

where $\Gamma(\cdot)$ is the gamma function. Then defining $a_j = l_j\Gamma(2-\alpha_j)/(\alpha_j - 1)$ for $j = 1, 2$ yields

$$\widehat{V}(s) \simeq \frac{2}{(\mu_1+\mu_2)^3}\left\{\mu_2^2 a_1 s^{\alpha_1-4} L_1\left(\frac{1}{s}\right) + \mu_1^2 a_2 s^{\alpha_2-4} L_2\left(\frac{1}{s}\right)\right\}, \quad \text{as } s \to 0$$

$$V(t) \simeq \frac{2}{(\mu_1+\mu_2)^3} t^3 \left\{\frac{\mu_2^2 a_1}{\Gamma(4-\alpha_1)} \frac{L_1(t)}{t^{\alpha_1}} + \frac{\mu_1^2 a_2}{\Gamma(4-\alpha_2)} \frac{L_2(t)}{t^{\alpha_2}}\right\}, \quad \text{as } t \to \infty$$

Define

$$b = \lim_{t\to\infty}\left(\frac{L_1(t)}{t^{\alpha_1}}\right)\left(\frac{t^{\alpha_2}}{L_2(t)}\right)$$

Consider the following subcases, where $\alpha_{\min} = \min(\alpha_1, \alpha_2)$.

(a) Let $\alpha_1 = \alpha_2$, then $0 < b < \infty$ and

$$V(t) \simeq \frac{2}{(\mu_1+\mu_2)^3} \frac{\{\mu_2^2 a_1 b + \mu_1^2 a_2\}}{\Gamma(4-\alpha_{\min})} t^{3-\alpha_{\min}} L_2(t), \quad \text{as } t \to \infty$$

(b) Let $\alpha_1 < \alpha_2$, then $b \to \infty$ and

$$V(t) \simeq \frac{2}{(\mu_1+\mu_2)^3} \frac{\mu_2^2 a_1}{\Gamma(4-\alpha_{\min})} t^{3-\alpha_{\min}} L_1(t), \quad \text{as } t \to \infty$$

(c) Let $\alpha_1 > \alpha_2$, then $b \to 0$ and

$$V(t) \simeq \frac{2}{(\mu_1+\mu_2)^3} \frac{\mu_1^2 a_2}{\Gamma(4-\alpha_{\min})} t^{3-\alpha_{\min}} L_2(t), \quad \text{as } t \to \infty$$

In all these subcases, $V(t)$ is of the form $\sigma_{\lim}^2 L(t) t^{2H}$ as $t \to \infty$, where $2H = (3-\alpha_{\min})$, and $\sigma_{\lim}^2$ is a constant.

Case 3: The random variable A has infinite variance and the random variable B has finite variance. In this case the heavy-tail dominates $V(t)$ as $t \to \infty$. Therefore

$$V(t) \simeq \frac{2}{(\mu_1+\mu_2)^3} \frac{\mu_2^2 a_1}{\Gamma(4-\alpha_1)} t^{3-\alpha_1} L_1(t), \quad \text{as } t \to \infty$$

In this case $\alpha_{\min} = \alpha_1$.

Case 4: The random variable A has finite variance and the random variable B has infinite variance. In this case the heavy-tail dominates $V(t)$ as $t \to \infty$. Therefore

$$V(t) \simeq \frac{2}{(\mu_1+\mu_2)^3} \frac{\mu_1^2 a_2}{\Gamma(4-\alpha_2)} t^{3-\alpha_2} L_2(t), \quad \text{as } t \to \infty$$

In this case $\alpha_{\min} = \alpha_2$.

The above discussion can be summarized in the following lemma.

Lemma 4.1. *The variance $V(t)$ of a process with alternating "on" and "off" intervals, which have heavy-tailed distributions is of the form $\sigma_{\lim}^2 L(t) t^{2H}$ as $t \to \infty$, where the values $\sigma_{\lim}$, $L(t)$, and H are chosen appropriately.* □

4.6.3 Multiple Source Characterization

The cumulative packet count at time t due to M independent, and identically distributed sources is equal to $\sum_{m=1}^{M} W^{(m)}(t)$. For large values of M, use of the central limit theorem yields

$$\lim_{M\to\infty} M^{-1/2} \sum_{m=1}^{M} \left\{ W^{(m)}(t) - \mathcal{E}\left(W^{(m)}(t)\right) \right\} = G(t), \quad t \in \mathbb{R}_0^+$$

where $\{G(t), t \in \mathbb{R}_0^+\}$ is a stationary Gaussian process, since the $W^{(m)}(t)$'s are stationary. This is a zero mean process with covariance function $\gamma(t), t \in \mathbb{R}_0^+$. Using the relationship $V(t) \simeq \sigma_{\lim}^2 L(t) t^{2H}$ as $t \to \infty$ results in

$$\lim_{\mathcal{T}\to\infty}\left(\mathcal{T}^{2H}L(\mathcal{T})\right)^{-1/2}\int_0^{\mathcal{T}t}G(u)\,du = B_H(t), \quad t\in\mathbb{R}_0^+$$

This limit is Gaussian with zero mean and stationary increments because the integral of $G(t)$ should possess these properties. Furthermore, the variance of this limit must be $\sigma_{\text{lim}}^2 t^{2H}$ for a specified value of t. These are the requirements for a fractional Brownian motion process. Therefore $\{B_H(t), t\in\mathbb{R}_0^+\}$ is indeed a FBMP. The results of this section are summarized in the following theorem.

Theorem 4.3. *Let μ_1 and μ_2 be the average lengths of "on" and "off" durations of the packet generation processes respectively. The positive parameters α_1 and α_2 describe the tails of the complementary cumulative distribution functions of the "on" and "off" durations respectively. Let the number of identical packet-generation sources be M. For very large values of M and $\mathcal{T}$, the aggregate cumulative packet process $\left\{\widetilde{W}_M(\mathcal{T}t), t\in\mathbb{R}_0^+\right\}$ behaves statistically as*

$$\mathcal{T}M\frac{\mu_1}{(\mu_1+\mu_2)}t+\mathcal{T}^H\sqrt{L(\mathcal{T})M}B_H(t), \quad t\in\mathbb{R}_0^+ \tag{4.16a}$$

where

$$\alpha_{\min}=\min(\alpha_1,\alpha_2), \ \ \alpha_{\min}\in(1,2), \ \ H=(3-\alpha_{\min})/2, \ \ L(\mathcal{T})>0 \tag{4.16b}$$

is a slowly varying function at infinity, and

$$Var(B_H(t))=\sigma_{\text{lim}}^2 t^{2H} \tag{4.16c}$$

The parameter σ_{lim} is as described in the above discussion. □

This fundamental result is originally due to M. S. Taqqu, W. Willinger, and R. Sherman. This theorem states that the main contribution to the packet generation process, for very large values of M and $\mathcal{T}$ is provided by $\mathcal{T}M\mu_1/(\mu_1+\mu_2)$. The fluctuations are provided by the time-dependent term $\mathcal{T}^H\sqrt{L(\mathcal{T})M}B_H(t)$.

Example 4.1. Consider the following cases.

(a) The first and second moments of the random variables A and B are finite, then

$$\sigma_{\text{lim}}^2=\frac{\left(\mu_2^2\sigma_1^2+\mu_1^2\sigma_2^2\right)}{(\mu_1+\mu_2)^3}$$

(b) The random variables A and B have identical heavy-tailed distribution. Let $\mu_1=\mu_2=\mu, \alpha_1=\alpha_2=\alpha$, where $1<\alpha<2$, $L_1(t)=L_2(t)$, $l_1=l_2=1$, and $H=(3-\alpha)/2$. Then

$$\sigma_{\text{lim}}^2=\frac{1}{2\mu(\alpha-1)(2-\alpha)(3-\alpha)}$$

(c) The random variable A is heavy-tailed, the random variable B has finite variance, and $H=(3-\alpha_1)/2$. Then

$$\sigma_{\text{lim}}^2=\frac{2\mu_2^2 l_1}{(\mu_1+\mu_2)^3(\alpha_1-1)(2-\alpha_1)(3-\alpha_1)}$$

□

4.7 Asymptotic Analysis of the R/S Statistic

It has been mentioned in an earlier section, that the purpose of collecting the R/S statistic of a set of data points is to determine the Hurst parameter of the process. In this section, certain asymptotic results about rescaled adjusted range statistic (or R/S statistic) are obtained. This is done via the concept of *exchangeable* random variables. Consider a sequence of random variables $X_1, X_2, \ldots, X_n$. The random variables $X_1, X_2, \ldots, X_n$ are called *exchangeable* if any permutation $X_{\widetilde{\sigma}_1}, X_{\widetilde{\sigma}_2}, \ldots, X_{\widetilde{\sigma}_n}$, of $X_1, X_2, \ldots, X_n$ has a joint probability distribution which is identical to that of $X_1, X_2, \ldots, X_n$. For example, a set of independent, and identically distributed random variables are exchangeable. Also if the random variables $X_1, X_2, \ldots, X_n$ have a joint normal distribution with $\mathcal{E}(X_i) = \mu, Var(X_i) = \sigma^2$, for $i = 1, 2, \ldots, n$, and $Cov(X_i, X_j) = \rho\sigma^2$ for $i \neq j$, $i, j = 1, 2, \ldots, n$, then $X_1, X_2, \ldots, X_n$ are exchangeable.

Let $X_1, X_2, \ldots, X_n$ be a random sample, and

$$\overline{X}(n) = \frac{1}{n}\sum_{j=1}^{n} X_j$$

Define

$$U_i = \left(X_i - \overline{X}(n)\right), \quad \text{for } i = 1, 2, \ldots, n$$

and

$$w_k = \sum_{j=1}^{k} X_j, \; W_k = \sum_{j=1}^{k} U_j, \quad 1 \leq k \leq n$$

The variables $r(n)$ and $R(n)$ have been defined in an earlier section as the range and adjusted range of the random variables $X_1, X_2, \ldots, X_n$ respectively. These are

$$r(n) = \max(0, w_1, w_2, \ldots, w_n) - \min(0, w_1, w_2, \ldots, w_n)$$

and

$$R(n) = \max(0, W_1, W_2, \ldots, W_n) - \min(0, W_1, W_2, \ldots, W_n)$$

Our goal is to compute the expectations $\mathcal{E}(r(n))$ and $\mathcal{E}(R(n))$. This is achieved by using an extremely curious combinatorial result. This combinatorial result is related to permutation of a sequence of real numbers, and a certain maximum value related to these numbers.

Theorem 4.4. *Let the set S be equal to $\{a_i \mid a_i \in \mathbb{R}, 1 \leq i \leq n\}$. Also, let a permutation of the elements of this set be $\widetilde{\sigma}$, and*

$$U_k(\widetilde{\sigma}) = \sum_{i=1}^{k} a_{\widetilde{\sigma}_i}, \quad 1 \leq k \leq n \tag{4.17a}$$

Define the number of positive terms in the sequence $U_1(\widetilde{\sigma}), U_2(\widetilde{\sigma}), \ldots, U_n(\widetilde{\sigma})$ to be equal to $N(\widetilde{\sigma})$. Then

$$\sum_{\widetilde{\sigma}} \max(0, U_1(\widetilde{\sigma}), U_2(\widetilde{\sigma}), \ldots, U_n(\widetilde{\sigma})) = \sum_{\widetilde{\sigma}} N(\widetilde{\sigma})\, a_{\widetilde{\sigma}_1} \tag{4.17b}$$

where the summation extends over all the $n!$ permutations of the elements of the set S.

Proof. See the problem section. □

Expectation of $r(n)$

The expectation of $r(n)$ is established in a sequence of lemmas. In these lemmas, the functions χ^+ and χ^- are used. As defined earlier, if $\chi \in \mathbb{R}$ then $\chi^+ = \max(0, \chi)$, and $\chi^- = \max(0, -\chi)$.

Lemma 4.2. *Let the sequence of random variables $X_1, X_2, \ldots, X_n$ be exchangeable, and $w_k = \sum_{j=1}^{k} X_j, 1 \leq k \leq n$.*

Then $\mathcal{E}(\max(0, w_1, w_2, \ldots, w_n)) = \mathcal{E}(\sum_{k=1}^{n} X_1 \theta(w_k))$, where $\theta(x) = 1$ if $x > 0$, and $\theta(x) = 0$ otherwise.

Proof. We first provide a motivation for the result. Let $\widetilde{\sigma}$ be a permutation of the numbers $1, 2, \ldots, n$ such that

$$\widetilde{\sigma} = \begin{pmatrix} 1, & 2, & \ldots, & n \\ \widetilde{\sigma}_1, & \widetilde{\sigma}_2, & \ldots, & \widetilde{\sigma}_n \end{pmatrix}$$

Define $Y_k(\widetilde{\sigma}) = \sum_{j=1}^{k} X_{\widetilde{\sigma}_j}$, for $1 \leq k \leq n$, then the stated relationship is equivalent to

$$\mathcal{E}(\max(0, Y_1(\widetilde{\sigma}), Y_2(\widetilde{\sigma}), \ldots, Y_n(\widetilde{\sigma}))) = \mathcal{E}\left(\sum_{k=1}^{n} X_{\widetilde{\sigma}_1} \theta(Y_k(\widetilde{\sigma}))\right)$$

Since both sides are the same for all permutations $\widetilde{\sigma}$, the above equation is equivalent to

$$\sum_{\widetilde{\sigma}} \mathcal{E}(\max(0, Y_1(\widetilde{\sigma}), Y_2(\widetilde{\sigma}), \ldots, Y_n(\widetilde{\sigma}))) = \sum_{\widetilde{\sigma}} \mathcal{E}\left(\sum_{k=1}^{n} X_{\widetilde{\sigma}_1} \theta(Y_k(\widetilde{\sigma}))\right)$$

or

$$\mathcal{E}\left(\sum_{\widetilde{\sigma}} \max(0, Y_1(\widetilde{\sigma}), Y_2(\widetilde{\sigma}), \ldots, Y_n(\widetilde{\sigma}))\right) = \mathcal{E}\left(\sum_{\widetilde{\sigma}} \sum_{k=1}^{n} X_{\widetilde{\sigma}_1} \theta(Y_k(\widetilde{\sigma}))\right)$$

But a much stronger result has been stated in the last theorem. It is

$$\sum_{\widetilde{\sigma}} \max(0, Y_1(\widetilde{\sigma}), Y_2(\widetilde{\sigma}), \ldots, Y_n(\widetilde{\sigma})) = \sum_{\widetilde{\sigma}} \sum_{k=1}^{n} X_{\widetilde{\sigma}_1} \theta(Y_k(\widetilde{\sigma}))$$

Thus the statement in the lemma is true because the elements of the sequence of random variables $X_1, X_2, \ldots, X_n$ are exchangeable. □

Lemma 4.3. $\mathcal{E}(\max(0, w_1, w_2, \ldots, w_n)) = \sum_{k=1}^{n} k^{-1} \mathcal{E}(w_k^+)$.

Proof. From the preceding lemma

$$\begin{aligned} &\mathcal{E}(\max(0, w_1, w_2, \ldots, w_n)) \\ &= \mathcal{E}\left(\sum_{k=1}^{n} X_1 \theta(w_k)\right) = \sum_{k=1}^{n} k^{-1} \mathcal{E}\left(\sum_{j=1}^{k} X_j \theta(w_k)\right) \\ &= \sum_{k=1}^{n} k^{-1} \mathcal{E}(w_k \theta(w_k)) = \sum_{k=1}^{n} k^{-1} \mathcal{E}(w_k^+) \end{aligned}$$

□

Lemma 4.4. $\mathcal{E}(\min(0, w_1, w_2, \ldots, w_n)) = -\sum_{k=1}^{n} k^{-1}\mathcal{E}(w_k^-)$.
Proof. Notice that

$$\min(0, w_1, w_2, \ldots, w_n) = -\max(0, -w_1, -w_2, \ldots, -w_n)$$

Therefore

$$\begin{aligned}\mathcal{E}(\min(0, w_1, w_2, \ldots, w_n)) &= -\mathcal{E}(\max(0, -w_1, -w_2, \ldots, -w_n)) \\ &= -\sum_{k=1}^{n} k^{-1}\mathcal{E}(w_k^-)\end{aligned}$$

where we used the result of the last lemma. □

Lemma 4.5. *Let* $\overline{X}(k) = k^{-1}\sum_{i=1}^{k} X_i$, *for* $k = 1, 2, \ldots n$.
Then $\mathcal{E}(r(n)) = \sum_{k=1}^{n} k^{-1}\mathcal{E}(|w_k|) = \sum_{k=1}^{n} \mathcal{E}(|\overline{X}(k)|)$.
Proof. The proof is left to the reader. □

The above results are valid for any set of exchangeable random variables $X_1, X_2, \ldots, X_n$.

Expectation of $R(n)$

The expected value of the adjusted range is next obtained.

Lemma 4.6. *If* $X_1, X_2, \ldots, X_n$ *is a set of exchangeable random variables, and* $\overline{X}(n) = \sum_{j=1}^{n} X_j/n$, $U_j = (X_j - \overline{X}(n))$, *for* $j = 1, 2, \ldots, n$; *then the random variables*

$$U_1, U_2, \ldots, U_n$$

are a set of exchangeable random variables.

Proof. In order to prove this result, it is sufficient to show that the characteristic function of $U_1, U_2, \ldots, U_n$ is the same as the characteristic function of $U_{\widetilde{\sigma}_1}, U_{\widetilde{\sigma}_2}, \ldots, U_{\widetilde{\sigma}_n}$. The set of random variables $U_{\widetilde{\sigma}_1}, U_{\widetilde{\sigma}_2}, \ldots, U_{\widetilde{\sigma}_n}$ is an arbitrary permutation of the random variables $U_1, U_2, \ldots, U_n$.

Let the characteristic function of $X_1, X_2, \ldots, X_n$ and $U_1, U_2, \ldots, U_n$ be

$$\mathcal{C}_{X_1,X_2,\ldots,X_n}(t_1, t_2, \ldots, t_n), \text{ and } \mathcal{C}_{U_1,U_2,\ldots,U_n}(t_1, t_2, \ldots, t_n)$$

respectively, where $t_j \in \mathbb{R}$ for $1 \leq j \leq n$. Also let $i = \sqrt{-1}$.

$$\begin{aligned}&\mathcal{C}_{U_1,U_2,\ldots,U_n}(t_1, t_2, \ldots, t_n) \\ &= \mathcal{E}\left(\exp\left\{i\sum_{j=1}^{n} t_j U_j\right\}\right) \\ &= \mathcal{E}\left(\exp\left\{i\sum_{j=1}^{n} t_j\left(X_j - \overline{X}(n)\right)\right\}\right) \\ &= \mathcal{E}\left(\exp\left\{i\sum_{j=1}^{n}\left(t_j - \overline{t}(n)\right)X_j\right\}\right) \\ &= \mathcal{C}_{X_1,X_2,\ldots,X_n}\left(t_1 - \overline{t}(n), t_2 - \overline{t}(n), \ldots, t_n - \overline{t}(n)\right)\end{aligned}$$

where $\overline{t}(n) = \sum_{m=1}^{n} t_m/n$. As $X_1, X_2, \ldots, X_n$ are exchangeable random variables,

$$\begin{aligned}
&\mathcal{C}_{X_1,X_2,\ldots,X_n}\left(t_1 - \overline{t}(n), t_2 - \overline{t}(n), \ldots, t_n - \overline{t}(n)\right) \\
&= \mathcal{C}_{X_{\widetilde{\sigma}_1},X_{\widetilde{\sigma}_2},\ldots,X_{\widetilde{\sigma}_n}}\left(t_1 - \overline{t}(n), t_2 - \overline{t}(n), \ldots, t_n - \overline{t}(n)\right) \\
&= \mathcal{E}\left(\exp\left\{i\sum_{j=1}^{n}\left(t_j - \overline{t}(n)\right)X_{\widetilde{\sigma}_j}\right\}\right) \\
&= \mathcal{E}\left(\exp\left\{i\sum_{j=1}^{n} t_j\left(X_{\widetilde{\sigma}_j} - \overline{X}(n)\right)\right\}\right) \\
&= \mathcal{E}\left(\exp\left\{i\sum_{j=1}^{n} t_j U_{\widetilde{\sigma}_j}\right\}\right) = \mathcal{C}_{U_{\widetilde{\sigma}_1},U_{\widetilde{\sigma}_2},\ldots,U_{\widetilde{\sigma}_n}}\left(t_1, t_2, \ldots, t_n\right)
\end{aligned}$$

Therefore

$$\mathcal{C}_{U_1,U_2,\ldots,U_n}\left(t_1, t_2, \ldots, t_n\right) = \mathcal{C}_{U_{\widetilde{\sigma}_1},U_{\widetilde{\sigma}_2},\ldots,U_{\widetilde{\sigma}_n}}\left(t_1, t_2, \ldots, t_n\right)$$

□

The next lemma follows from the above lemmas.

Lemma 4.7. *Let* $\overline{U}(k) = \sum_{j=1}^{k} U_j/k$, *for* $k = 1, 2, \ldots, n$.
Then $\mathcal{E}(R(n)) = \sum_{k=1}^{n} k^{-1}\mathcal{E}(|W_k|) = \sum_{k=1}^{n}\mathcal{E}\left(\left|\overline{U}(k)\right|\right)$.
Proof. The proof is left to the reader. □

Example 4.2. Assume that $X_1, X_2, \ldots, X_n$ are exchangeable normally distributed random variables. Let $\mathcal{E}(X_k) = \mu$, $Var(X_k) = \sigma^2$, for $k = 1, 2, \ldots, n$. Also $Cov(X_i, X_j) = \sigma^2\rho$, for $i \neq j$ and $i, j = 1, 2, \ldots, n$.

$\mathcal{E}(W_k) = 0$, and $Var(W_k) = (k/n)(n-k)\sigma^2(1-\rho)$. It is known that if Z is a normally distributed random variable with zero mean and variance $Var(Z)$, then $\mathcal{E}(|Z|) = 2\sqrt{Var(Z)/2\pi}$. Then

$$\begin{aligned}
\mathcal{E}(R(n)) &= \sum_{k=1}^{n}\frac{\mathcal{E}(|W_k|)}{k} = \sum_{k=1}^{n}\frac{2\left\{Var(W_k)\right\}^{1/2}}{k\sqrt{2\pi}} \\
&= 2\sigma\left\{\frac{(1-\rho)}{2\pi n}\right\}^{1/2}\sum_{k=1}^{n}\left\{\frac{n-k}{k}\right\}^{1/2}
\end{aligned}$$

For large value of n, $\mathcal{E}(R(n))$ can be approximated as

$$\begin{aligned}
\mathcal{E}(R(n)) &\sim 2\sigma\left\{\frac{(1-\rho)}{2\pi n}\right\}^{1/2}\int_0^n\left\{\frac{n-x}{x}\right\}^{1/2}dx \\
&= 2\sigma\left\{\frac{(1-\rho)\,n}{2\pi}\right\}^{1/2}\int_0^1\left\{\frac{1-y}{y}\right\}^{1/2}dy
\end{aligned}$$

This yields $\mathcal{E}(R(n)) \sim \sigma\left\{\pi n(1-\rho)/2\right\}^{1/2}$. If $\rho = 0$ then $\mathcal{E}(R(n)) \sim \sigma\left\{\pi n/2\right\}^{1/2}$. Therefore $\mathcal{E}(R(n))$ is proportional to $\sigma n^{1/2}$. □

In the above example, it was established that if $X_1, X_2, \ldots, X_n$ are exchangeable normally distributed random variables, then for large values of n, $\mathcal{E}(R(n))$ is proportional to $n^{1/2}$. Similarly, it is possible to expect that for the sequence of exchangeable random variables $X_1, X_2, \ldots, X_n$ with some other distributions, $\log \mathcal{E}(R(n))$ is a linear function of $\log n$, as $n \to \infty$.

4.8 Transform Analysis of Fractional Brownian Motion

Fractional Brownian motion can also be studied via different transform techniques. It is well-known that transforms help us see patterns. In this section we study power spectrum of the increment of FBMP. In addition, the Wigner-Ville transform, continuous wavelet transform, and discrete wavelet transform techniques are also applied to examine FBMP.

4.8.1 Power Spectrum of the Increment of a FBMP

The power spectrum of a stationary random process is the Fourier transform of its covariance function. As per the definition of the FBMP, its increments are stationary. This implies that the power spectrum of the increment of the FBMP can be computed.

Let $\{B_H(t), t \in \mathbb{R}\}$ be a FBMP, its increment process is $\{X(t), t \in \mathbb{R}\}$, where $X(t) = (B_H(t) - B_H(t-\Delta))$, and $\Delta > 0$. This increment process is stationary by definition of the FBMP. The autocovariance of this increment process is

$$Cov(X(t), X(t-\tau)) = \frac{\sigma^2}{2}\left[|\tau + \Delta|^{2H} + |\tau - \Delta|^{2H} - 2|\tau|^{2H}\right] \tag{4.18a}$$

Observe that $Cov(X(t), X(t-\tau))$ is independent of t, and depends only on τ and Δ. Denote it by $r(\tau, \Delta)$.Then $r(\tau, \Delta) \simeq \sigma^2 \Delta^2 H (2H-1)|\tau|^{2H-2}$, $\Delta \ll |\tau|$. The Fourier transform of

$$r(\tau, \Delta) = \frac{\sigma^2}{2}\left[|\tau + \Delta|^{2H} + |\tau - \Delta|^{2H} - 2|\tau|^{2H}\right] \tag{4.18b}$$

is next computed. In order to compute the power spectrum of the process, the Fourier transform of functions of the type $|t|^{\gamma-1}, t \in \mathbb{R}, \gamma > 0$ has to be determined. The Fourier transform of this function is obtained via its Laplace transform. The Laplace transform of $t^{\gamma-1}$, $\gamma > 0$, and $t \in \mathbb{R}_0^+$ is

$$\int_0^\infty t^{\gamma-1} e^{-st} dt = \frac{\Gamma(\gamma)}{s^\gamma}, \quad s \in \mathbb{C}\backslash\{0\} \tag{4.19a}$$

where $\Gamma(\cdot)$ is the gamma function. The Fourier transform of the function $|t|^{\gamma-1}, t \in \mathbb{R}, \gamma > 0$ can now be established by substituting $s = i\omega$ in the Laplace transform of this function. Thus

$$\Im\left(|t|^{\gamma-1}\right) = \frac{2\Gamma(\gamma)}{|\omega|^\gamma}\cos\left(\frac{\pi}{2}\gamma\right), \quad \gamma > 0, \ \omega \neq 0 \tag{4.19b}$$

where $\Im(\cdot)$ is the Fourier transform operator. The power spectrum of the process can now be computed. It is the Fourier transform of $r(t, \Delta)$. Thus

$$r(t, \Delta) \leftrightarrow \frac{4\sigma^2 \Gamma(2H+1)}{|\omega|^{2H+1}} \sin(\pi H) \sin^2\left(\frac{\omega\Delta}{2}\right), \quad \omega \neq 0 \tag{4.20}$$

where the symbol $\leftrightarrow$ denotes a Fourier transform pair. The proof is left to the reader. Observe that, as $\omega\Delta \to 0$, the Fourier transform of $r(t, \Delta)$ is proportional to $|\omega|^{-(2H-1)}$.

4.8.2 Wigner-Ville Transform Technique

The FBMP is next analysed using Wigner-Ville's transform. The Wigner-Ville transform helps in finding a time-dependent spectrum of the FBMP. The Wigner-Ville transform of $B_H(t)$ is

$$\widetilde{W}_{B_H}(\tau,\omega)=\int_{-\infty}^{\infty}B_H\left(\tau+\frac{t}{2}\right)\overline{B_H\left(\tau-\frac{t}{2}\right)}e^{-i\omega t}dt, \qquad \tau,\omega\in\mathbb{R} \tag{4.21a}$$

It can be shown that

$$\begin{aligned}&\mathcal{E}\left(\widetilde{W}_{B_H}(\tau,\omega)\right)\\&=\frac{\sigma^2\Gamma(2H+1)}{|\omega|^{2H+1}}\sin(\pi H)\left[1-2^{(1-2H)}\cos(2\omega\tau)\right]\end{aligned} \tag{4.21b}$$

where $\tau\in\mathbb{R}$, and $\omega\in\mathbb{R}\backslash\{0\}$. Note that $\mathcal{E}\left(\widetilde{W}_{B_H}(\tau,\omega)\right)\geq 0$, if and only if $H\in[0.5,1)$. A better insight into this spectrum can be obtained by averaging this time-dependent spectrum over time. Let this averaging interval be $\mathcal{T}$. Then

$$S_{B_H}(\omega,\mathcal{T})=\frac{1}{\mathcal{T}}\int_0^{\mathcal{T}}\mathcal{E}\left(\widetilde{W}_{B_H}(\tau,\omega)\right)d\tau \tag{4.22a}$$

Use of this definition yields

$$\begin{aligned}&S_{B_H}(\omega,\mathcal{T})\\&=\frac{\sigma^2\Gamma(2H+1)}{|\omega|^{2H+1}}\sin(\pi H)\left[1-2^{(1-2H)}\frac{\sin(2\omega\mathcal{T})}{2\omega\mathcal{T}}\right]\end{aligned} \tag{4.22b}$$

where $\omega\in\mathbb{R}\backslash\{0\}$.

Observations 4.3. Following observations can be made about the time averaged spectrum.

1. If the value of $\mathcal{T}$ is chosen such that $2\omega\mathcal{T}=\pi n$, where $n\in\mathbb{Z}\backslash\{0\}$, then

$$S_{B_H}\left(\omega,\frac{\pi n}{2\omega}\right)=\frac{\sigma^2\Gamma(2H+1)}{|\omega|^{2H+1}}\sin(\pi H), \qquad \forall\, n\in\mathbb{Z}\backslash\{0\}$$

   where $\omega\in\mathbb{R}\backslash\{0\}$. From this relationship, the value of the parameter H can be evaluated.
2. If for a given frequency ω_0, the value of $\mathcal{T}$ is chosen such that $\mathcal{T}\gg 1/\omega_0$, then

$$S_{B_H}(\omega_0,\mathcal{T})\simeq\frac{\sigma^2\Gamma(2H+1)}{|\omega_0|^{2H+1}}\sin(\pi H)$$

□

4.8.3 Continuous Wavelet Transform Technique

The continuous wavelet transform technique is next applied to examine FBMP. The continuous wavelet transform of function $f(t), t\in\mathbb{R}$, is defined as

$$W_f(\psi, a, b) = \int_{-\infty}^{\infty} f(t)\, \overline{\psi_{a,b}(t)} dt \tag{4.23a}$$

where

$$\psi_{a,b}(t) = \frac{1}{\sqrt{|a|}} \psi\left(\frac{t-b}{a}\right), \quad a, b \in \mathbb{R},\ a \neq 0 \tag{4.23b}$$

$$\int_{-\infty}^{\infty} \psi(t)\, dt = 0 \tag{4.23c}$$

and

$$C_\psi = \int_{-\infty}^{\infty} \frac{|\Psi(\omega)|^2}{|\omega|} d\omega < \infty \tag{4.23d}$$

$$\psi(t) \leftrightarrow \Psi(\omega) \tag{4.23e}$$

Define

$$f_{B_H}(s, t, a) = \mathcal{E}\left(W_{B_H}(\psi, a, s)\, W_{B_H}(\psi, a, t)\right), \qquad s, t, a \in \mathbb{R},\ a \neq 0$$

For simplicity, assume the wavelet function to be real-valued. Thus

$$f_{B_H}(s, t, a) = \frac{1}{|a|} \int_{-\infty}^{\infty} \int_{-\infty}^{\infty} c_H(x, y)\, \psi\left(\frac{x-s}{a}\right) \psi\left(\frac{y-t}{a}\right) dxdy$$

Since $c_H(x, y) = \sigma^2 \left\{ |x|^{2H} + |y|^{2H} - |x-y|^{2H} \right\} /2$

$$f_{B_H}(s, t, a) = -\frac{\sigma^2}{2\,|a|} \int_{-\infty}^{\infty} \int_{-\infty}^{\infty} |x-y|^{2H}\, \psi\left(\frac{x-s}{a}\right) \psi\left(\frac{y-t}{a}\right) dxdy$$

Substituting $(x-y)/a = u$ and $(x-s)/a = v$ in the above equation yields

$$f_{B_H}(s, t, a) = -\frac{\sigma^2}{2} |a|^{2H+1} \int_{-\infty}^{\infty} \int_{-\infty}^{\infty} |u|^{2H}\, \psi(v)\, \psi\left(v - u + \frac{s-t}{a}\right) dudv$$

Note that $f_{B_H}(s, t, a)$ is a function of $(s-t)$. Define

$$\lambda(\tau, a) = f_{B_H}\left(t + \frac{\tau}{2}, t - \frac{\tau}{2}, a\right), \qquad t, \tau \in \mathbb{R}$$

Therefore

$$\lambda(\tau, a) = -\frac{\sigma^2}{2} |a|^{2H+1} \int_{-\infty}^{\infty} \int_{-\infty}^{\infty} |u|^{2H}\, \psi(v)\, \psi\left(v - u + \frac{\tau}{a}\right) dudv$$

The function $\lambda(\cdot, \cdot)$ is amenable to Fourier analysis. Let the Fourier transform of $\lambda(\tau, a)$ be $\Lambda(\omega, a)$. The transform pair

$$\psi\left(v - u + \frac{t}{a}\right) \leftrightarrow |a|\, e^{-i\omega a(u-v)} \Psi(a\omega)$$

implies

$$\Lambda(\omega, a) = \frac{|a|\, \sigma^2 \Gamma(2H+1)}{|\omega|^{2H+1}} \sin(\pi H)\, |\Psi(a\omega)|^2, \quad \omega \neq 0$$

This discussion is summarized in the following lemma.

Lemma 4.8. *Let $\{B_H(t), t \in \mathbb{R}\}$ be a FBMP, and $W_{B_H}(\cdot,\cdot,\cdot)$ be its wavelet transform. The mother wavelet function $\psi(\cdot)$ is assumed to be real-valued. The scale parameter of the wavelet transform is $a \in \mathbb{R}\backslash\{0\}$. If*

$$\lambda(\tau, a) = \mathcal{E}\left(W_{B_H}\left(\psi, a, t+\frac{\tau}{2}\right) W_{B_H}\left(\psi, a, t-\frac{\tau}{2}\right)\right) \tag{4.24a}$$

and $\Lambda(\omega, a)$ is its Fourier transform, then for $a, \tau \in \mathbb{R}$, $a \neq 0$

$$\lambda(\tau, a) = -\frac{\sigma^2}{2}|a|^{2H+1} \int_{-\infty}^{\infty}\int_{-\infty}^{\infty} |u|^{2H} \psi(v)\psi\left(v-u+\frac{\tau}{a}\right) dudv \tag{4.24b}$$

$$\Lambda(\omega, a) = \frac{|a|\,\sigma^2\Gamma(2H+1)}{|\omega|^{2H+1}} \sin(\pi H)\left|\Psi(a\omega)\right|^2, \qquad \omega \neq 0 \tag{4.24c}$$

□

4.8.4 Discrete Wavelet Transform Technique

Some results on discrete wavelet transform of a FBMP are derived in this subsection. Let $\psi(\cdot)$ be the mother wavelet function, which satisfies the admissibility condition. Admissibility condition is a requirement for the original function to be recovered from its wavelet transform. Assume the mother wavelet function to be real-valued for simplicity. The discrete wavelet transform of a function $f(t), t \in \mathbb{R}$ is defined as

$$f(t) = \sum_{m,n\in\mathbb{Z}} d(m,n)\,\psi_{m,n}(t) \tag{4.25a}$$

$$\psi_{m,n}(t) = 2^{\frac{m}{2}}\psi(2^m t - n), \qquad \forall\, m,n \in \mathbb{Z} \tag{4.25b}$$

$$d(m,n) = \int_{-\infty}^{\infty} f(t)\,\psi_{m,n}(t)\,dt, \qquad \forall\, m,n \in \mathbb{Z} \tag{4.25c}$$

If the function $f(\cdot)$ is a FBMP, $\mathcal{E}(d(m,j)\,d(m,k))$, for $m, j, k \in \mathbb{Z}$ is evaluated. Then

$$\begin{aligned}&\mathcal{E}(d(m,j)\,d(m,k)) \\ &= 2^m \int_{-\infty}^{\infty}\int_{-\infty}^{\infty} c_H(x,y)\,\psi(2^m x - j)\,\psi(2^m y - k)\,dxdy\end{aligned}$$

Note that

$$c_H(x,y) = \frac{\sigma^2}{2}\left\{|x|^{2H} + |y|^{2H} - |x-y|^{2H}\right\}$$

Use of the admissibility condition of wavelets yields

$$\begin{aligned}&\mathcal{E}(d(m,j)\,d(m,k)) \\ &= -\sigma^2 2^{m-1} \int_{-\infty}^{\infty}\int_{-\infty}^{\infty} |x-y|^{2H}\,\psi(2^m x - j)\,\psi(2^m y - k)\,dxdy\end{aligned}$$

Substitution of $u = 2^m(x-y)$, and $v = (2^m x - j)$ in the above equation yields

$$\mathcal{E}\left(d\left(m,j\right)d\left(m,k\right)\right)$$
$$= -\frac{\sigma^2}{2^{m(2H+1)+1}}\int_{-\infty}^{\infty}\int_{-\infty}^{\infty}|u|^{2H}\psi\left(v\right)\psi\left(v-u+j-k\right)dudv$$

Observe that

$$\int_{-\infty}^{\infty}|u|^{2H}\psi\left(v-u+j-k\right)du$$
$$= -\frac{\Gamma\left(2H+1\right)}{\pi}\sin\left(\pi H\right)\int_{-\infty}^{\infty}\frac{\Psi\left(\omega\right)}{|\omega|^{2H+1}}e^{i\omega(v+j-k)}d\omega$$

Therefore

$$\mathcal{E}\left(d\left(m,j\right)d\left(m,k\right)\right)$$
$$= \frac{\sigma^2\Gamma\left(2H+1\right)\sin\left(\pi H\right)}{\left(2\pi\right)2^{m(2H+1)}}\int_{-\infty}^{\infty}\frac{|\Psi\left(\omega\right)|^2}{|\omega|^{2H+1}}e^{i\omega(j-k)}d\omega$$

The variance coefficient $Var\left(d\left(m,j\right)\right)$ is obtained by letting $j=k$ in the above equation. Note that higher the value of m, smaller is the covariance. The above result is summarized in the following lemma.

Lemma 4.9. *Let $\{f(t) \mid t \in \mathbb{R}\}$ be a FBMP. Its discrete wavelet transform is given by*

$$f(t) = \sum_{m,n\in\mathbb{Z}} d\left(m,n\right)\psi_{m,n}\left(t\right)$$

The mother wavelet function $\psi\left(\cdot\right)$ is assumed to be real-valued. Then

$$\mathcal{E}\left(d\left(m,j\right)d\left(m,k\right)\right)$$
$$= \frac{\sigma^2\Gamma\left(2H+1\right)\sin\left(\pi H\right)}{\left(2\pi\right)2^{m(2H+1)}}\int_{-\infty}^{\infty}\frac{|\Psi\left(\omega\right)|^2}{|\omega|^{2H+1}}e^{i\omega(j-k)}d\omega \tag{4.26a}$$

$$Var\left(d\left(m,j\right)\right)$$
$$= \frac{\sigma^2\Gamma\left(2H+1\right)\sin\left(\pi H\right)}{\left(2\pi\right)2^{m(2H+1)}}\int_{-\infty}^{\infty}\frac{|\Psi\left(\omega\right)|^2}{|\omega|^{2H+1}}d\omega \tag{4.26b}$$

where $m,j,k \in \mathbb{Z}$ □

4.9 Internet Traffic Modeling via Chaotic Maps

It is possible to model the Internet traffic via chaotic maps. Such maps in essence are nonlinear. Therefore, the chaotic maps are a viable alternative to the stochastic modeling of the Internet traffic.

Dynamical systems in nature are inherently nonlinear. Among the several man-made phenomena, Internet is essentially nonlinear in several different respects. In order to study a nonlinear system, it is fruitful to linearize the nonlinear equations for a restricted range of values, and then examine the behavior of such complex systems. This approach to the analysis of nonlinear systems yields useful results only on rare occasions. The application of chaotic maps to study packet generation process appears to be one such example. This paradigm of modeling the Internet traffic is an alternative to the use of stochastic processes. The value of nonlinear models lies in their simplicity. However, the simplicity and parsimony in the nonlinear modeling approach comes at the price of analytical tractability. The basics of chaos theory have been discussed in the chapter on optimization, stability theory, and chaos theory.

Sources of traffic in the Internet alternate between "on" and "off" states. During the "on" state packets are emitted by the source, while during the "off" state no packets are generated. This phenomenon can be modeled as an iterated system, described by functions that take initial values as their argument and yield the next initial value as its result. These are of the form $x_{n+1} = f(x_n)$, $n \in \mathbb{N}$, where x_0 is the initial value. Note that such systems are completely deterministic. Consider a one-dimensional nonlinear map in which the evolution of the state x_n for any $n \in \mathbb{N}$ is specified by

$$x_{n+1} = \begin{cases} f_1(x_n), & \text{if } \ 0 \le x_n < d \\ f_2(x_n), & \text{if } \ d \le x_n \le 1 \end{cases} \tag{4.27a}$$

where $d \in (0,1)$. A condition for this map to be chaotic is that it satisfy the SIC (sensitivity to initial conditions) property. This property implies that any small change in initial value of the orbit may result in a significant change in the eventual behavior of the orbit. Using this model, a packet generation process can be modeled by assuming that the traffic source is either in a passive or active state at time n. If the value of x_n is below the *threshold* value d, then the source is in a passive state, and it is dormant. However, if the value of $x_n \in [d,1]$ then the source is active, and it generates a single packet. The packet arrival process is described by the *indicator* variable y_n for $n \in \mathbb{N}$ where

$$y_n = \begin{cases} 0, & \text{if } \ 0 \le x_n < d \\ 1, & \text{if } \ d \le x_n \le 1 \end{cases} \tag{4.27b}$$

In this scheme, x_n and y_n are called hidden and observed variables respectively. The task is to find proper functions $f_1(\cdot)$ and $f_2(\cdot)$ which satisfy the SIC property of chaotic maps and match the statistical properties of the packet generation process. Observe that this simple model can be modified in several ways. For example the map might have more than two segments, and the threshold values for the sequences x_n and y_n can be different.

Of particular interest is the *sojourn* or *dwell time* of a state. If a trajectory spends exactly n iterations in the same state (either 0 or 1) before leaving that state, then n is a sojourn time of the trajectory in that state. Recall that

$$f^1(x) \triangleq f(x), f^2(x) \triangleq f(f(x)), f^3(x) \triangleq f(f(f(x))), \ldots$$

and so on. Following is a formal definition of the sojourn time.

Definition 4.9. *Let $V = [0,1]$ be a set, and $f : V \to V$ be a chaotic map. The interval V is partitioned into r parts by real numbers $a_i \in V, 0 \le i \le r$ such that $0 = a_0 < a_1 < \cdots < a_r = 1$. Also let:*

(a) $S_i = \{x \mid a_i \le x < a_{i+1}\}$ *for some* $i \in \mathbb{Z}_r$.
(b) $z \notin S_i, f^{n+1}(z) \notin S_i$, *for some* $z \in V$ *and* $n \in \mathbb{P}$.
(c) $f^k(z) \in S_i$ *for all values of* $k \in \{1, 2, \ldots, n\}$.

Then n is a sojourn time in interval S_i. □

Based upon the above concepts, different chaotic map models can be developed. One such model is the Bernoulli map.

Bernoulli Map

Bernoulli maps are piecewise-linear maps. These maps are generalization of the doubling map, and consist of two linear segments. A Bernoulli map is defined as

$$x_{n+1} = \begin{cases} x_n/d, & \text{if } \ 0 \leq x_n < d \\ (x_n - d) / (1 - d), & \text{if } \ d \leq x_n \leq 1 \end{cases} \tag{4.28}$$

Therefore, if $x_n \in [d, 1]$ then y_n is set to 1, and a packet is generated. It is convenient to define $\phi \triangleq (1 - d)$. It can be shown that the invariant density $\rho(x)$ of this map is equal to unity for $x \in [0, 1]$. Using this model, it can be inferred that a packet is generated with probability ϕ.

Let $P(\cdot)$ denote the probability of an event, and $\mathcal{E}(\cdot)$ and $Var(\cdot)$ be the expectation and variance of a random variable respectively. Assume that the indicator variable y_n is a random variable. In addition, a sequence of indicator random variables are also assumed to be independent of each other. Then following properties can be inferred immediately.

Observations 4.4. Observations about Bernoulli map.

1. The value of the slope in each linear segment is greater than unity, so the map is expansive.
2. $\mathcal{E}(y_n) = \int_d^1 \rho(x)\, dx = \phi$.
3. $\mathcal{E}(y_n^2) = \int_d^1 \rho(x)\, dx = \phi$, and $Var(y_n) = \phi(1 - \phi)$.
4. The active and passive periods of this process are geometrically distributed. That is:
   (a) $P(\text{length of active period } = k) = \phi^k (1 - \phi), \forall\, k \in \mathbb{N}$.
   (b) $P(\text{length of passive period } = k) = d^k (1 - d), \forall\, k \in \mathbb{N}$. □

The above set of observations imply that the Bernoulli maps, model generation of packets with geometrically distributed packet burst lengths.

4.9.1 Nonlinear Maps

A powerful and generalized version of intermittency map is next developed. It has been experimentally determined that packet burst lengths in the Internet have a power-law distribution. In order to model this phenomenon, nonlinearity is introduced into the chaotic maps. A possible candidate is the *single intermittency map*. These are called intermittency maps because, they have been used to model a phenomenon called intermittency in turbulence. Intermittent phenomenon in turbulence has alternating periods of "bursts" followed by "regular" phases. The single intermittent map can be used to model packet sources which are characterized by heavy-tailed "off," and light-tailed "on" distributions (or vice-versa).

The following single intermittency map is a generalization of the Bernoulli map. This map has two segments, of which one is linear, and the other is nonlinear. A single intermittency map is defined by

$$x_{n+1} = \begin{cases} \epsilon + x_n + c x_n^m, & \text{if} \quad 0 \le x_n < d \\ (x_n - d) / (1 - d), & \text{if} \quad d \le x_n \le 1 \end{cases} \tag{4.29a}$$

$$c = \frac{1 - \epsilon - d}{d^m} \tag{4.29b}$$

where $0 \le \epsilon \ll d, \ 0 < d < 1$, and $m > 1$. This map exhibits heavy-tail in the "off" state, while in the "on" state it has a light tail. The purpose of ϵ in the above definition, is to limit the maximum duration of the "off" state. A double intermittency map is

$$x_{n+1} = \begin{cases} \epsilon_1 + x_n + c_1 x_n^{m_1}, & \text{if} \ \ 0 \le x_n < d \\ x_n - \epsilon_2 - c_2 (1 - x_n)^{m_2}, & \text{if} \ \ d \le x_n \le 1 \end{cases} \tag{4.30a}$$

$$c_1 = \frac{1 - \epsilon_1 - d}{d^{m_1}}, \ \text{and} \ c_2 = \frac{d - \epsilon_2}{(1 - d)^{m_2}} \tag{4.30b}$$

where $0 \le \epsilon_1, \epsilon_2 \ll d, \ 0 < d < 1$, and $m_1, m_2 > 1$. This map exhibits heavy-tail in both "off" and "on" states. This is a requirement for modeling packet traffic in the Internet, because both the "on" and "off" periods of packet generation processes in networks have heavy-tailed distributions. The next family of maps is a generalized version of the double intermittency maps. In these maps, nonlinearity is present in both the segments of the chaotic mapping. These are called *renormalized-group double intermittency maps.* Such maps are so named because these are based upon renormalized-group technique used in theoretical physics.

Definition 4.10. *The renormalized-group double intermittency map is*

$$x_{n+1} = \begin{cases} f_1 (x_n), & \textit{if} \quad 0 \le x_n < d \\ f_2 (x_n), & \textit{if} \quad d \le x_n \le 1 \end{cases} \tag{4.31a}$$

$$f_1 (x) = \epsilon_1 + g_1 (x), \quad f_2 (x) = 1 - \epsilon_2 - g_2 (1 - x) \tag{4.31b}$$

$$g_i (x) = \frac{x}{\left\{1 - c_i x^{(m_i - 1)}\right\}^{1/(m_i - 1)}}, \quad i = 1, 2 \tag{4.31c}$$

$$c_1 = \frac{(1 - \epsilon_1)^{(m_1 - 1)} - d^{(m_1 - 1)}}{(1 - \epsilon_1)^{(m_1 - 1)} d^{(m_1 - 1)}} \tag{4.31d}$$

$$c_2 = \frac{(1 - \epsilon_2)^{(m_2 - 1)} - (1 - d)^{(m_2 - 1)}}{(1 - \epsilon_2)^{(m_2 - 1)} (1 - d)^{(m_2 - 1)}} \tag{4.31e}$$

where $0 \le \epsilon_1, \epsilon_2 \ll d; 0 < d < 1;$ *and* $m_1, m_2 > 1$. *The indicator variable* $y_n = 0$, *if* $0 \le x_n < d$, *and* $y_n = 1$, *if* $d \le x_n \le 1$.

Note that $f_1 (0) = \epsilon_1, f_1 (d) = 1, f_2 (d) = 0$ *and* $f_2 (1) = (1 - \epsilon_2)$. □

Observe in the above definition that identical functional form is used in both the "on" and "off" states, except for reflection and translation. It is shown subsequently, that if $\epsilon_1 = \epsilon_2 = 0$, the functional form of $f_1 (\cdot)$ and $f_2 (\cdot)$ remain invariant under composition. This is in contrast to a typical nonlinear function which quickly becomes complex under successive iterations. The nonlinear function $g (\cdot)$ is next studied. It is used in both the segments of the map. Let

$$g(x) = \frac{x}{\{1 - qx^{(m-1)}\}^{1/(m-1)}} \tag{4.32a}$$

$$q = \frac{(1 - d^{m-1})}{d^{m-1}} \tag{4.32b}$$

where $x \in [0, d)$, $m > 1$. Also $g(0) = 0$ and $g(d) = 1$.

Genesis of the Nonlinear Function $g(\cdot)$

Renormalization-group technique is used in determining the origin of the functional form of the nonlinear function $g(\cdot)$. It uses the doubling transformation, and a functional form of fixed point. Let $\mathcal{D}$ be an operator (map), and $f^*(\cdot)$ be a function. Then $f^*(\cdot)$ is a functional-fixed point of the operator $\mathcal{D}$, if $\mathcal{D}f^*(x) = f^*(x)$ for all values of of x in its domain. Next consider a function

$$f(x) = x + ux^m; \qquad x \in [0, 1], \ m > 1$$

Then $f^2(x) = x + ux^m + u\{x + ux^m\}^m$. As $x \to 0$, $f^2(x) \to x + ux^m = f(x)$. Similarly, it can be shown that $f^n(x) \to f(x)$ as $x \to 0$. Observe that $f(0) = 0$, and $f'(0) = 1$, where $f'(x)$ denotes the first derivative of $f(x)$ with respect to x. In light of this example, a fair question to ask would be: Does there exists an operator $\mathcal{D}$ such that its repeated application to a function of the type given in the above equation would yield a functional-fixed point $f^*(\cdot)$ of $\mathcal{D}$?

Of special interest to us is an operator called the doubling operator. It is defined as

$$\mathcal{D}f^*(x) = \alpha f^*\left(f^*\left(\frac{x}{\alpha}\right)\right)$$

where the term α preserves functional self-similarity. Thus a doubling transformation is a functional composition, and a rescaling of the independent, and dependent variables. If a repeated application of the doubling operator results in $f^*(x)$, then $\mathcal{D}f^*(x) = f^*(x)$. That is

$$\alpha f^*\left(f^*\left(\frac{x}{\alpha}\right)\right) = f^*(x)$$

Assume that $x \in [0, 1]$, and the boundary conditions are $f^*(0) = 0$, and $f^{*\prime}(0) = 1$, where $f^{*\prime}(x)$ denotes the first derivative of $f^*(x)$ with respect to x. An important step in the analysis is to assume the following recursive equation.

$$J(f(x)) = J(x) - b$$

where b is a free parameter. The above equation implies $f(x) = J^{-1}(J(x) - b)$. Thus

$$J(f(f(x))) = J(f(x)) - b = J(x) - 2b$$
$$\frac{1}{2}J(f(f(x))) = \frac{1}{2}J(x) - b$$

Since $\alpha f(f(x)) = f(\alpha x)$, $J(\alpha f(f(x))) = J(f(\alpha x)) = \{J(\alpha x) - b\}$. That is

$$J(\alpha f(f(x))) = J(\alpha x) - b$$

Compare the last two equations. In order to solve the functional-fixed point equation, $J(\cdot)$ must observe the relationship

$$\frac{1}{2}J^*(x) = J^*(\alpha x)$$

This relationship is satisfied, if $J^*(x) = x^{-(m-1)}$, and $\alpha = 2^{1/(m-1)}$. Also, since $J^*(f^*(x)) = \{J^*(x) - b\}$

$$\begin{aligned} f^*(x) &= J^{*-1}(J^*(x) - b) \\ &= \left\{x^{-(m-1)} - b\right\}^{-1/(m-1)} \\ &= \frac{x}{\{1 - bx^{(m-1)}\}^{1/(m-1)}} \end{aligned}$$

A binomial series expansion, near $x = 0$ yields $f^*(x) \simeq x + \eta x^m$, where $\eta = b/(m-1)$. Therefore $u = b/(m-1)$. Substituting $b = q$, in the above equation results in the function $g(x), x \in [0, d)$. In addition, q is adjusted so that $g(d) = 1$.

Observations 4.5. Some properties of the function $g(x), x \in [0, d)$, where $d \in (0, 1)$ are listed below.

1. $g(0) = 0$, and $g(d) = 1$.
2. $g(x) = x/d$ as $m \to 1$.
3. A binomial series expansion of $g(x)$ yields $g(x) \simeq x + \eta x^m$, where $\eta = q/(m-1)$ and $m > 1$.
4. Denote the first derivative of $g(x)$ with respect to x by $g'(x)$.

$$g'(x) = \frac{1}{\{1 - qx^{(m-1)}\}^{m/(m-1)}}$$

Thus $g'(0) = 1$, and $g'(x) > 1$ if $x > 0$. Therefore $g(x)$ is expansive everywhere except at the origin.

5. The nth iterate of $g(x)$ has the same functional form as $g(x)$. That is

$$g^n(x) = \frac{x}{\{1 - nqx^{(m-1)}\}^{1/(m-1)}}, \quad n \in \mathbb{P}$$

This observation implies that the sequence $x, g(x), g^2(x), \ldots, g^n(x)$ is invariant under composition. Therefore $g(x)$ exhibits so-called *functional self-similarity*. Since $x \in [0, d)$, the sequence elements have to be less than d. If $g(x)$ is defined as $h(x, q)$, then

$$h^n(x, q) = \varrho h\left(x/\varrho, nq\varrho^{m-1}\right)$$

6. Also

$$g^{-n}(x) = \frac{x}{\{1 + nqx^{(m-1)}\}^{1/(m-1)}}, \quad n \in \mathbb{P}$$

□

The renormalization-group double intermittency map is a generalization of the two-segment Bernoulli map, and the single and double intermittency maps.

Examples 4.3. In the following examples, the functions $g_1(\cdot)$ and $g(\cdot)$ are as defined in the renormalized-group intermittency map.

1. The double intermittency map is obtained by using the Taylor's series expansion for both $g_1(x)$ and $g_2(x)$ near $x = 0$.
2. The single intermittency map is obtained by using the Taylor's series expansion for both $g_1(x)$ and $g_2(x)$ near $x = 0$. Also let $m_1 = m$, $m_2 \to 1$, and $\epsilon_1 = \epsilon$, $\epsilon_2 = 0$, then $f_1(x) \simeq \epsilon + x + cx^m$ for $x \to 0$, and $f_2(x) = (x-d)/(1-d)$.
3. The Bernoulli map is obtained by using the Taylor's series expansion for both $g_1(x)$ and $g_2(x)$ near $x = 0$. Also let $m_1, m_2 \to 1$, and $\epsilon_1 = \epsilon_2 = 0$, then $f_1(x) = x/d$, and $f_2(x) = (x-d)/(1-d)$. □

4.9.2 Sojourn Time Probability

A useful measure to characterize a chaotic map is the time an orbit spends in a particular state. This is equal to the sojourn time of the packet source. It is the length of continuous sequence of 1's (0's) that would be output by the indicator variable y_n when the map is in state 1 (0). If the nonlinear segments have the form $g(x)$, then it is shown in this subsection, that the sojourn times of the orbit have a power-law distribution.

Observation 4.6. Let $V = [0,1]$ be a set, and $f : V \to V$ be an expansive chaotic map. The interval V is partitioned into r parts by real numbers $a_i \in V$, $0 \le i \le r$ such that $0 = a_0 < a_1 < \cdots < a_r = 1$. Also let $x_{-1} \notin (a_{i-1}, a_i)$, $x_0 = f(x_{-1}) \in (a_{i-1}, a_i)$. The sojourn time in a partition is K, if for $k < K$, $f^k(x_0) \in (a_{i-1}, a_i)$, and $f^K(x_0) \notin (a_{i-1}, a_i)$.

(a) Assume that the function $f_i(x)$, the restriction of $f(x)$ to the interval (a_{i-1}, a_i) can be translated and/or reflected to the nonlinear function $g(\cdot)$.

$$g(x) = \frac{x}{\left\{1 - qx^{(m-1)}\right\}^{1/(m-1)}}$$

$$q = \frac{\left(1 - d^{m-1}\right)}{d^{m-1}}$$

where $x \in [0, d)$, $m > 1$, and $g(d) = 1$.

(b) Let $\vartheta(x_0)$ be the reinjection probability density of the orbit into the ith interval (a_{i-1}, a_i). That is, $\vartheta(x_0)$ is the probability density of $x_0 \in (a_{i-1}, a_i)$. Assume that it is slowly varying over this interval. In this case $\vartheta(x_0) \sim \vartheta(0)$.

Then as $k \to \infty$, the complementary cumulative distribution function of K, specified by $P(K > k)$ is proportional to $k^{-1/(m-1)}$. Consequently, the probability $P(K = k)$ is proportional to $k^{-m/(m-1)}$. Thus the sojourn time of the orbit has a power-law distribution.

Observe that, as per Taqqu-Willinger-Sherman theorem, the Hurst's parameter $H = (3 - \alpha)/2$, where $\alpha \in (1, 2)$, and α is the shape parameter of the power-law distribution. Letting $\alpha = 1/(m-1)$, results in

$$H = \frac{(3m-4)}{(2m-2)}, \quad \text{for} \quad m \in (3/2, 2)$$

□

Thus, this observation relates the Hurst's parameter H, the chaotic map parameter m, and the shape parameter α of the power-law distribution of the chaotic orbit's sojourn time. See the problem section for a proof of this observation. This observation implies that chaos-theoretic techniques can be used to generate power-law distributions.

4.9.3 Asymptotic Spectral Analysis

In this subsection, the power spectrum of the chaotic map $x_{n+1} = f(x_n), \forall n \in \mathbb{N}$, is determined. In this equation, $f(x) \to x + ux^m$ as $x \to 0$, for $x \in [0, d]$. The function $f(\cdot)$ is arbitrary beyond $x = d$, but with an additional requirement that this segment of the map produce a random reinjection in the interval $(x_0, x_0 + dx_0)$, where $x_0 \in [0, d]$ with a probability $\vartheta(x_0)\, dx_0$.

The power spectrum of this map is subsequently related to the Hurst parameter, which characterizes the self-similar packet traffic. Let the correlation of the sequence $\{x_n \mid n \in \mathbb{N}\}$ be $\{C(j) \mid j \in \mathbb{N}\}$, and the corresponding power spectrum be $\widetilde{F}(\omega)$. Then

$$C(j) = \lim_{N\to\infty} \frac{1}{N} \sum_{n=0}^{N} x_{n+j} x_n, \quad \forall j \in \mathbb{N} \tag{4.33a}$$

$$\widetilde{F}(\omega) = \lim_{N\to\infty} \frac{1}{N} \sum_{j=0}^{N} \cos(j\omega)\, C(j) \tag{4.33b}$$

The correlation function $C(\cdot)$ is next expressed in terms of the probability $P(K = k) \triangleq P(k)$. Recall that $P(K = k)$ is the probability that the sojourn time K in the segment on which $f(\cdot)$ is defined, is k units. For simplicity, assume that k is continuous. The probability $P(K = k)$ is related to $\vartheta(x_0)$ via the relationship

$$\vartheta(x_0)\, dx_0 = \vartheta(x_0(k)) \left| \frac{dx_0}{dk} \right| dk \triangleq P(K = k)\, dk$$

The correlation function is approximated as follows. Assume that a burst of packets is represented by a single pulse. Further assume that x_n is nearly equal to zero in the region other than $[0, d]$, and equal to a single pulse of height one in the chaotic (bursty) region. Under this approximation, $C(j)$ is proportional to the conditional property of finding a pulse at time j units, when it is specified that a pulse occurred at time zero. Thus

$$\begin{aligned}
C(1) &= P(1) \\
C(2) &= P(2) + P(1)^2 = P(2) + C(1)\, P(1) \\
&\vdots \\
C(j) &= P(j) + C(1)\, P(j-1) + \ldots + C(j-1)\, P(1)
\end{aligned}$$

Define $P(0) = 0$ and $C(0) = 1$, then

$$C(j) = \sum_{k=0}^{j} C(j-k)\, P(k) + \delta_{j0}, \quad \forall j \in \mathbb{N} \tag{4.34}$$

where δ_{j0} is equal to 1 if $j = 0$, and equal to 0 otherwise. Since only the behavior of the correlation function, as $j \to \infty$, is of interest, we pass on to continuous time variables. In this limit let $C(j) \to c(\cdot)$. Similarly as $k \to \infty$, let $P(k) \to p(\cdot)$. Therefore the above relationship in continuous time domain is

$$c(t) = \int_0^t c(t-\tau)\, p(\tau)\, d\tau + \delta(t) \tag{4.35}$$

The first term on the right-hand side of the above equation is a convolution integral, and the second term is Dirac's delta function. Let the single-sided Laplace transform of $c(t)$ and $p(t)$ be $\widehat{c}(s)$ and $\widehat{p}(s)$ respectively, where $s = (\sigma + i\omega) \in \mathbb{C}$ and $i = \sqrt{-1}$. Thus

$$\widehat{c}(s) = \frac{1}{(1 - \widehat{p}(s))} \tag{4.36}$$

Let the spectral density of the corresponding continuous time correlation function $c(t)$ be $\widetilde{f}(\omega)$. Thus

$$\widetilde{f}(\omega) = \int_0^\infty c(t) \cos(\omega t)\, dt \tag{4.37}$$

Use of the definition of the Laplace transform, and the above equation results in

$$\widetilde{f}(\omega) = \frac{1}{2} \{\widehat{c}(i\omega) + \widehat{c}(-i\omega)\} \tag{4.38}$$

Also $\int_1^\infty p(t)\, dt = 1$ yields $\vartheta(0) \int_1^\infty t^{-m/(m-1)} dt = 1$. Furthermore,

$$\int_1^\infty t^{-m/(m-1)} dt = (m-1), \quad m > 1$$

Therefore

$$p(t) = \frac{1}{(m-1)} t^{-m/(m-1)}, \quad t \in [1, \infty), \; m > 1 \tag{4.39}$$

Let $\mathcal{E}(\cdot)$ be the expectation operator, then the average value of the transit time is $\mathcal{E}(K) = (2-m)^{-1}$ for $1 < m < 2$. The expected value $\mathcal{E}(K)$ is infinite (does not exist) for $m \geq 2$. The Laplace transform $\widehat{p}(s)$ of the probability density function $p(t)$ is

$$\widehat{p}(s) = \frac{1}{(m-1)} \int_1^\infty e^{-st} t^{-m/(m-1)} dt \tag{4.40}$$

Note that $\int_1^\infty e^{-st} t^{-m/(m-1)} dt$ in the above equation is an exponential integral, and its properties are well-known. As mentioned in the last subsection, the values of m in the interval $(3/2, 2)$ are of interest. It can be shown (see the problem section) that as $\omega \to 0$:

(a) The power spectrum $\widetilde{f}(\omega)$ is proportional to $\omega^{-(2m-3)/(m-1)}$ for $m \in (3/2, 2)$.
(b) The power spectrum $\widetilde{f}(\omega)$ is constant for $m \in (1, 3/2)$.
(c) The power spectrum $\widetilde{f}(\omega)$ is proportional to $|\ln \omega|$ for $m = 3/2$.

The negative of the slope of the complementary cumulative distribution of the sojourn time is equal to $\alpha = 1/(m-1)$. Therefore, if $m \in (3/2, 2)$, then $\alpha \in (1, 2)$. This implies that for this set of values of m, the output sequence of the nonlinear map has a finite mean, and infinite variance. Also the corresponding Hurst parameter

$$H = \frac{(3m-4)}{(2m-2)}, \quad H \in (1/2, 1)$$

Therefore

$$m = \frac{(4-2H)}{(3-2H)}, \quad \text{and} \quad \frac{(2m-3)}{(m-1)} = (2H-1)$$

That is, for $m \in (3/2, 2)$, $\widetilde{f}(\omega)$ is proportional to $\omega^{-(2H-1)}$. However, if both the segments of the chaotic map are nonlinear, as in the renormalized-group double intermittency map, select $m = \max(m_1, m_2)$.

Reference Notes

Internet's self-similar traffic can be modeled via stochastic techniques. Another fruitful approach to model Internet traffic is via nonlinear dynamical methods. Basics of both approaches are introduced in this chapter.

Self-similarity was first observed in Ethernet traffic in the trail-blazing paper by Leland, Taqqu, Willinger, and Wilson (1994). For a pioneering work on the evidence of self-similarity in the World Wide Web data transfers, refer to the paper by Crovella, and Bestavros (1997).

The section on modeling of highly variable measurements is based upon Athreya, and Ney (1972), Mandelbrot (1997), Embrechts, Klüppelberg, and Mikosch (1998), Sigman (1999), and Willinger, Alderson, and Li (2004).

A paper by Hurst (1951) is a classic work on the use of R/S statistic. The curious combinatorial result in the section on R/S statistic is from Kac (1954). This result is associated with a galaxy of mathematicians including K. L. Chung, F. J. Dyson, G. A. Hunt, and M. Kac.

A paper by Mandelbrot, and Van Ness (1968) is a readable account of fractional Brownian motion processes. The mathematician, Mandelbrot is generally regarded as the father of "fractals."

The theorem in the section on explanation of self-similarity in Internet traffic is due to Taqqu, Willinger, and Sherman (1997). See also Whitt (2002) for an advanced introduction to this subject. Asymptotic analysis of R/S statistic follows the work of Boes, and Salas-La Cruz (1973). For a clear and lucid description of self-similarity, the reader can refer to the books by Beran (1994), Doukhan, Oppenheim, and Taqqu (2003), Embrechts, and Maejima (2002), and Samorodnitsky, and Taqqu (1994). The spectral representation of FBMP is due to Flandrin (1989). Wornell (1995) wrote another scholarly textbook on the application of wavelets to fractals.

The discussion on chaotic maps and its use in modeling packet generation is based upon the work of Samuel (1999). Important and fundamental results on application of chaotic maps can be found in the textbook by Schuster (1995).

Problems

1. Let $X_1, X_2, \ldots, X_n$ be a sequence of $n \in \mathbb{P}\backslash\{1\}$ independent, and identically distributed random variables. Denote a generic such random variable by X. The random variable X, or its cumulative distribution function $F_X(\cdot)$ is subexponential if one of the following equivalent conditions holds:

   (a)
   $$\lim_{x\to\infty}\frac{P(X_1+X_2+\ldots+X_n>x)}{P(X>x)}=n$$

   (b)
   $$\lim_{x\to\infty}\frac{P(X_1+X_2+\ldots+X_n>x)}{P(\max(X_1,X_2,\ldots,X_n)>x)}=1$$

Prove the above result.

Hint: There are two parts in the problem.

(a) See Athreya, and Ney (1972). The result can be established by induction.

(b) See Embrechts, Klüppelberg, and Mikosch (1998). Note that

$$\begin{aligned}&P(\max(X_1, X_2, \ldots, X_n) > x)\\&= 1 - \{F_X(x)\}^n\\&= F_X^c(x)\left\{1 + F_X(x) + \ldots + \{F_X(x)\}^{n-1}\right\}\\&\sim nF_X^c(x)\end{aligned}$$

The result follows by using part (a) of the problem.

2. Prove that BMP is a self-similar process.

Hint: Observe, that $B(t), t \in \mathbb{R}$ is a Gaussian process. Therefore, it is sufficient to note its expected value, and covariance function. Let $u > 0$, then $\mathcal{E}(B(ut)) = \mathcal{E}(B(ut) - B(0)) = 0 = u^{1/2}\mathcal{E}(B(t))$. Also it can be established that its covariance function

$$\begin{aligned}c(t, s) &= Cov(B(s), B(t))\\&= u^{-1}c(ut, us)\end{aligned}$$

3. $\{B_H(t), t \in \mathbb{R}\}$ is a fractional Brownian motion process. Prove that the covariance of increments of the FBMP in two nonoverlapping intervals (t_1, t_2) and (t_3, t_4), where $t_1 < t_2 < t_3 < t_4$ is given by

$$\begin{aligned}&\mathcal{E}[\{B_H(t_4) - B_H(t_3)\}\{B_H(t_2) - B_H(t_1)\}]\\&= \frac{\sigma^2}{2}\left[|t_4 - t_1|^{2H} + |t_3 - t_2|^{2H} - |t_4 - t_2|^{2H} - |t_3 - t_1|^{2H}\right]\end{aligned}$$

4. $\{B_H(t), t \in \mathbb{R}\}$ is a fractional Brownian motion process. Prove that

$$\begin{aligned}&\mathcal{E}[\{B_H(t+h) - B_H(h)\}\{B_H(s+h) - B_H(h)\}]\\&= \mathcal{E}[B_H(t)B_H(s)]\end{aligned}$$

Hint:

$$\mathcal{E}[\{B_H(t+h) - B_H(h)\}\{B_H(s+h) - B_H(h)\}]$$

$$\begin{aligned}&= \mathcal{E}[B_H(t+h)B_H(s+h)] - \mathcal{E}[B_H(t+h)B_H(h)]\\&\quad - \mathcal{E}[B_H(s+h)B_H(h)] + \mathcal{E}\left[B_H^2(h)\right]\\&= \frac{\sigma^2}{2}\left\{|t+h|^{2H} + |s+h|^{2H} - |t-s|^{2H}\right\}\\&\quad - \frac{\sigma^2}{2}\left\{|t+h|^{2H} + |h|^{2H} - |t|^{2H}\right\}\\&\quad + \frac{\sigma^2}{2}\left[-\left\{|s+h|^{2H} + |h|^{2H} - |s|^{2H}\right\} + 2|h|^{2H}\right]\\&= \frac{\sigma^2}{2}\left[|s|^{2H} + |t|^{2H} - |t-s|^{2H}\right] = \mathcal{E}[B_H(t)B_H(s)]\end{aligned}$$

5. In the integral representation of the fractional Brownian motion, prove that

$$C_H^2 = \frac{\{\Gamma(H+1/2)\}^2}{\Gamma(2H+1)\sin(\pi H)}, \quad H \in (0,1)$$

Hint: See Samorodnitsky, and Taqqu (1994); and Taqqu in Doukhan, Oppenheim, and Taqqu (2003). Define $(H+1/2) = b$, where $H \in (0,1)$. Therefore $b \in (1/2, 3/2)$. Let the Fourier transform of the weight function $w_H(t,u)$, where $t, u \in \mathbb{R}$ be $\widetilde{w}_H(t,\omega)$. This is

$$\widetilde{w}_H(t,\omega) = \Gamma(b)\frac{\left(e^{-i\omega t}-1\right)}{(-i\omega)^b}$$

where $i = \sqrt{-1}$. Using Parseval's theorem, we obtain

$$\begin{aligned} C_H^2 &= \int_{-\infty}^{\infty} |w_H(1,u)|^2 \, du = \frac{1}{2\pi}\int_{-\infty}^{\infty} |\widetilde{w}_H(1,\omega)|^2 \, d\omega \\ &= \frac{4\{\Gamma(b)\}^2}{\pi}\int_0^{\infty} \frac{\sin^2(\omega/2)}{\omega^{2b}} d\omega \end{aligned}$$

At $b = 1$, use of the Fourier transform of the function $sinc(t)$, $t \in \mathbb{R}$ and Parseval's theorem yields

$$\int_0^{\infty} \frac{\sin^2(\omega/2)}{\omega^2} d\omega = \frac{\pi}{4}$$

Therefore at $H = 1/2$ we have $C_H^2 = 1$. However, if

$$b \in (1/2, 1) \cup (1, 3/2)$$

the result GR3.823 in Gradshteyn and Ryzhik (1980), p. 447 yields

$$\int_0^{\infty} \frac{\sin^2(\omega/2)}{\omega^{2b}} d\omega = -\frac{\Gamma(1-2b)\sin(\pi b)}{2}$$

The above result can also be derived by using the artifice of integration by parts, and the Laplace transform of $t^{\gamma-1}$, $\gamma > 0$, and $t \in \mathbb{R}_0^+$. It is

$$\int_0^{\infty} t^{\gamma-1} e^{-st} dt = \frac{\Gamma(\gamma)}{s^{\gamma}}, \quad s \in \mathbb{C}\backslash\{0\}$$

Use of the relationships $\Gamma(z)\Gamma(1-z) = \pi/\sin(\pi z)$ where $z \in \mathbb{C}\backslash\mathbb{Z}$, and $z\Gamma(z) = \Gamma(z+1)$ yields the stated result.

6. Let $\{X(i), i \in \mathbb{Z}\}$ be a discrete-time stationary process. Also let

$$\begin{aligned} \mathcal{E}(X(i)) = \mu, \ \ Var(X(i)) = \sigma^2 \\ Cov(X(i), X(i+j)) = \sigma^2 r(|j|), \quad \forall\, i, j \in \mathbb{Z} \end{aligned}$$

Let $\left\{X^{(m)}(k), k \in \mathbb{Z}, m \geq 2\right\}$ be an aggregate process of $\{X(i), i \in \mathbb{Z}\}$, where

$$X^{(m)}(k) = \frac{1}{m}\sum_{j=m(k-1)+1}^{mk} X(j), \quad \forall\, k \in \mathbb{Z}$$

Then $\mathcal{E}\left(X^{(m)}(k)\right) = \mu$. Also let

$$Var\left(X^{(m)}(k)\right) \triangleq \sigma_m^2(k)$$

and

$$Cov\left(X^{(m)}(k), X^{(m)}(k+j)\right) \triangleq \sigma_m^2(k)\, r^{(m)}(|j|), \quad \forall\, k, j \in \mathbb{Z}$$

(a) If

$$b_0 = 0, \text{ and } b_m = \frac{m^2 \sigma_m^2(1)}{\sigma^2}, \quad \forall\, m \in \mathbb{P}$$

then prove that

$$\sigma_m^2(k) = \frac{\sigma^2}{m^2}\left\{ m + 2\sum_{j=1}^{m-1}(m-j)\, r(j) \right\}, \quad \forall\, k \in \mathbb{Z}$$

$$r(m) = \frac{1}{2}\left(b_{m+1} - 2b_m + b_{m-1}\right), \quad \forall\, m \in \mathbb{P}$$

Observe that, $\sigma_m^2(k)$ is independent of k.

(b) If

$$a_0 = 0, \text{ and } a_j = \frac{j^2 \sigma_{jm}^2(1)}{\sigma_m^2(1)}, \quad \forall\, j \in \mathbb{P}$$

then prove that

$$r^{(m)}(j) = \frac{1}{2}\left(a_{j+1} - 2a_j + a_{j-1}\right), \quad \forall\, j \in \mathbb{P}$$

(c) Recall that $Var\left(X^{(m)}(k)\right) = \sigma_m^2(k) = \sigma^2/m^\beta$, where $0 < \beta < 2$. Therefore, if $\sigma_j^2(1) = \sigma^2/j^\xi$, prove that $b_j = a_j = j^{2-\xi}$, and consequently $r(j) = r^{(m)}(j)$ for any $j \in \mathbb{P}$. This implies $\xi = \beta$.

Hint: Use the following results

(a) Establish the relationship

$$b_m = m + 2\sum_{j=1}^{m-1}(m-j)\, r(j), \quad m \geq 2$$

(b) Result follows from analogy with part (a) of the problem.

(c) Use the definitions of b_j and a_j for $j \in \mathbb{P}$.

7. Let the set S be equal to $\{a_i \mid a_i \in \mathbb{R}, 1 \leq i \leq n\}$. Also, let a permutation of the elements of this set be $\widetilde{\sigma}$, and

$$U_k(\widetilde{\sigma}) = \sum_{i=1}^{k} a_{\widetilde{\sigma}_i}, \quad 1 \leq k \leq n$$

Define the number of positive terms in the sequence

$$U_1(\widetilde{\sigma}), U_2(\widetilde{\sigma}), \ldots, U_n(\widetilde{\sigma})$$

to be equal to $N(\widetilde{\sigma})$. Show that

$$\sum_{\widetilde{\sigma}} \max\left(0, U_1(\widetilde{\sigma}), U_2(\widetilde{\sigma}), \ldots, U_n(\widetilde{\sigma})\right) = \sum_{\widetilde{\sigma}} N(\widetilde{\sigma})\, a_{\widetilde{\sigma}_1}$$

where the summation extends over all the $n!$ permutations of the elements of the set S.
Hint: See Kac (1954). Define

$$\theta(x) = \begin{cases} 1, & x > 0 \\ 0, & x \le 0 \end{cases}$$

and $V_k(\widetilde{\sigma}) = \sum_{i=2}^{k} a_{\widetilde{\sigma}_i}$, $2 \le k \le n$. It can be observed that

$$\max(0, a_{\widetilde{\sigma}_1}) = a_{\widetilde{\sigma}_1}\theta(a_{\widetilde{\sigma}_1})$$

and for $k \ge 2$

$$\begin{aligned}
&\max(0, U_1(\widetilde{\sigma}), U_2(\widetilde{\sigma}), \ldots, U_k(\widetilde{\sigma})) \\
&- \max(0, U_1(\widetilde{\sigma}), U_2(\widetilde{\sigma}), \ldots, U_{k-1}(\widetilde{\sigma})) \\
&= \theta(U_k(\widetilde{\sigma}))\{a_{\widetilde{\sigma}_1} + \max(0, V_2(\widetilde{\sigma}), V_3(\widetilde{\sigma}), \ldots, V_k(\widetilde{\sigma})) \\
&- \max(0, U_1(\widetilde{\sigma}), U_2(\widetilde{\sigma}), \ldots, U_{k-1}(\widetilde{\sigma}))\}
\end{aligned}$$

Sum the above equations over all permutations ($\widetilde{\sigma}$'s) for a specified first k terms. Then sum over $\binom{n}{k}$ such sets. These operations yield

$$\begin{aligned}
&\sum_{\widetilde{\sigma}} \{\max(0, U_1(\widetilde{\sigma}), U_2(\widetilde{\sigma}), \ldots, U_k(\widetilde{\sigma})) \\
&- \max(0, U_1(\widetilde{\sigma}), U_2(\widetilde{\sigma}), \ldots, U_{k-1}(\widetilde{\sigma}))\} \\
&= \sum_{\widetilde{\sigma}} a_{\widetilde{\sigma}_1}\theta(U_k(\widetilde{\sigma}))
\end{aligned}$$

Finally, sum over all k's, where $k = 1, 2, \ldots, n$.

$$\sum_{\widetilde{\sigma}} \max(0, U_1(\widetilde{\sigma}), U_2(\widetilde{\sigma}), \ldots, U_n(\widetilde{\sigma})) = \sum_{\widetilde{\sigma}} a_{\widetilde{\sigma}_1} \sum_{k=1}^{n} \theta(U_k(\widetilde{\sigma}))$$

The result follows by noting that $\sum_{k=1}^{n} \theta(U_k(\widetilde{\sigma})) = N(\widetilde{\sigma})$.

8. Find the Fourier transform of the autocovariance of the increment of the FBMP. That is, if

$$r(\tau, \Delta) = \frac{\sigma^2}{2}\left[|\tau + \Delta|^{2H} + |\tau - \Delta|^{2H} - 2|\tau|^{2H}\right]$$

prove that

$$r(t, \Delta) \leftrightarrow \frac{4\sigma^2 \Gamma(2H+1)}{|\omega|^{2H+1}} \sin(\pi H) \sin^2\left(\frac{\omega \Delta}{2}\right), \quad \omega \in \mathbb{R}\backslash\{0\}$$

9. For a FBMP, using Wigner-Ville's analysis establish

$$\mathcal{E}\left(\widetilde{W}_{B_H}(\tau, \omega)\right) = \frac{\sigma^2 \Gamma(2H+1)}{|\omega|^{2H+1}} \sin(\pi H)\left[1 - 2^{(1-2H)} \cos(2\omega\tau)\right]$$

where $\tau \in \mathbb{R}$, and $\omega \in \mathbb{R}\backslash\{0\}$.
Hint: Note that

$$\widetilde{W}_{B_H}(\tau, \omega) = \int_{-\infty}^{\infty} B_H\left(\tau + \frac{t}{2}\right) \overline{B_H\left(\tau - \frac{t}{2}\right)} e^{-i\omega t} dt, \quad \tau, \omega \in \mathbb{R}$$

Then

$$\begin{aligned}
\mathcal{E}\left(\widetilde{W}_{B_H}(\tau,\omega)\right) &= \int_{-\infty}^{\infty} \mathcal{E}\left(B_H\left(\tau+\frac{t}{2}\right)\overline{B_H\left(\tau-\frac{t}{2}\right)}\right)e^{-i\omega t}dt \\
&= \int_{-\infty}^{\infty} c_H\left(\tau+\frac{t}{2},\tau-\frac{t}{2}\right)e^{-i\omega t}dt \\
&= \frac{\sigma^2}{2}\int_{-\infty}^{\infty}\left[\left|\tau+\frac{t}{2}\right|^{2H}+\left|\tau-\frac{t}{2}\right|^{2H}-|t|^{2H}\right]e^{-i\omega t}dt
\end{aligned}$$

10. Compute the probability distribution function of the sojourn time of orbit of a chaotic nonlinear map.
Hint: See Samuel (1999). Observe that 0 is a fixed point of the map $g(x)$. For $K > k$, and $m > 1$, the reinjection point $x_0 \in (0, y)$, where

$$y = \lim_{x\to d} g^{-k}(x) \to \frac{d}{\left\{1 + kqd^{(m-1)}\right\}^{1/(m-1)}}$$

Assume k to be continuous in this analysis. In the relationship

$$g^k(x) = \frac{x}{\left\{1 - kqx^{(m-1)}\right\}^{1/(m-1)}}$$

let $x \to x_0$ and $g^k(x_0) \to d$. This gives the number of iterations necessary to cross the threshold d. That is

$$k \simeq \frac{1}{q}\left\{\frac{1}{x_0^{m-1}} - \frac{1}{d^{m-1}}\right\}$$

Observe that

$$\vartheta(x_0)\,dx_0 = \vartheta(x_0(k))\left|\frac{dx_0}{dk}\right|dk \triangleq P(K=k)\,dk$$

Since $\vartheta(x_0)$, the reinjection probability density function of the orbit, is slowly varying in the interval, let it be equal to $\vartheta(0)$. Therefore $P(K=k) \simeq \vartheta(0)\,|dx_0/dk|$. This implies that, as $k \to \infty$, $P(K>k)$ and $P(K=k)$ are proportional to $k^{-1/(m-1)}$ and $k^{-m/(m-1)}$ respectively.

11. Determine asymptotic expressions for $\widetilde{f}(\omega)$, the power spectrum of the output sequence of the renormalized-group chaotic map.
Hint: Three cases are discussed. These correspond to $m \in (3/2, 2)$, $m \in (1, 3/2)$, and $m = 3/2$.
Case (a): $m \in (3/2, 2)$. The Laplace transform $\widehat{p}(s)$ of the probability density function $p(t)$ is

$$\widehat{p}(s) = \frac{1}{(m-1)}s^{\varphi-1}\Gamma(1-\varphi, s)$$

where $\varphi = m/(m-1)$, and $\Gamma(\cdot,\cdot)$ is the upper incomplete gamma function. Thus

$$\widehat{p}(s) = \frac{\Gamma(1-\varphi)}{(m-1)}s^{\varphi-1} - \sum_{n\in\mathbb{N}}\frac{(-s)^n}{(mn-n-1)\,n!}$$

Note that the above representation is valid, provided $(mn-n-1) \neq 0$, for $n \in \mathbb{N}$. This condition is true for any $m \in (3/2, 2)$. Therefore

$$\widehat{c}(s) = \frac{1}{(1-\widehat{p}(s))} = \left\{ -\frac{\Gamma(1-\varphi)}{(m-1)} s^{\varphi-1} + \sum_{n\in\mathbb{P}} \frac{(-s)^n}{(mn-n-1)\, n!} \right\}^{-1}$$

Define

$$a(s) = -\frac{\Gamma(1-\varphi)}{(m-1)} s^{\varphi-1}, \text{ and } b(s) = \sum_{n\in\mathbb{P}} \frac{(-s)^n}{(mn-n-1)\, n!}$$

Then the power spectrum $\widetilde{f}(\omega)$ is

$$\widetilde{f}(\omega) = \frac{1}{2}\{\widehat{c}(i\omega) + \widehat{c}(-i\omega)\} = \frac{1}{2} \frac{\{a(i\omega) + b(i\omega) + a(-i\omega) + b(-i\omega)\}}{\{a(i\omega) + b(i\omega)\}\{a(-i\omega) + b(-i\omega)\}}$$

The power spectrum can be obtained by computing

$$a(i\omega) + a(-i\omega) = -\frac{\Gamma(1-\varphi)}{(m-1)} \omega^{\varphi-1} 2\cos\left\{\frac{\pi}{2}(\varphi-1)\right\}$$

$$b(i\omega) + b(-i\omega) = 2\sum_{n\in\mathbb{P}} \frac{(-1)^n \omega^{2n}}{(2mn-2n-1)(2n)!}$$

$$a(i\omega)\, a(-i\omega) = \left\{\frac{\Gamma(1-\varphi)}{(m-1)} \omega^{\varphi-1}\right\}^2$$

$$b(i\omega)\, b(-i\omega) = \left\{\sum_{n\in\mathbb{P}} \frac{(-i\omega)^n}{(mn-n-1)\, n!}\right\}\left\{\sum_{n\in\mathbb{P}} \frac{(i\omega)^n}{(mn-n-1)\, n!}\right\}$$

and

$$a(i\omega)\, b(-i\omega) + a(-i\omega)\, b(i\omega)$$
$$= -2\frac{\Gamma(1-\varphi)}{(m-1)} \sum_{n\in\mathbb{P}} \frac{(-1)^n \omega^{n+\varphi-1}}{(mn-n-1)\, n!} \cos\left\{\frac{\pi}{2}(n+\varphi-1)\right\}$$

The above expressions are evaluated as $\omega \to 0$. The numerator of $\widetilde{f}(\omega)$ is proportional to $\omega^{\varphi-1}$, and the denominator is proportional to ω^2. Consequently $\widetilde{f}(\omega)$ is proportional to $\omega^{\varphi-3}$, as $\omega \to 0$.

Case (b): $m \in (1, 3/2)$. The Laplace transform $\widehat{p}(s)$ of the probability density function $p(t)$ is

$$\widehat{p}(s) = \frac{1}{(m-1)} s^{\varphi-1} \Gamma(1-\varphi, s)$$

where $\varphi = m/(m-1)$, and $\Gamma(\cdot,\cdot)$ is the upper incomplete gamma function. Thus

$$\widehat{p}(s) = \frac{\Gamma(1-\varphi)}{(m-1)} s^{\varphi-1} - \sum_{n\in\mathbb{N}} \frac{(-s)^n}{(mn-n-1)\, n!}$$

Note that the above representation is valid, provided $(mn-n-1) \neq 0$, for $n \in \mathbb{N}$. Therefore, the above series representation is valid for $m \in (1, 3/2)$ where $m \neq (n+1)/n$ for $n = 3, 4, 5, \ldots$. In this subcase expand $\widetilde{f}(\omega)$ as in Case (a). The terms in the numerator and denominator of $\widetilde{f}(\omega)$ are each proportional to ω^2. Therefore $\widetilde{f}(\omega)$ is approximately constant as $\omega \to 0$.

Next consider the subcase, where $m = (n+1)/n$ for $n = 3, 4, 5, \ldots$. For these values of m

$$\widehat{p}(s) = \frac{1}{(m-1)} \int_1^\infty e^{-st} t^{-m/(m-1)} dt = n \int_1^\infty e^{-st} t^{-(n+1)} dt$$

Observe that $\int_1^\infty e^{-st} t^{-(n+1)} dt = E_{n+1}(s)$, where $E_{n+1}(\cdot)$ is the exponential integral. Expand e^{-st} for small values of s, and evaluate $\widetilde{f}(\omega)$ as $\omega \to 0$. It turns out that in this subcase $\widetilde{f}(\omega)$ is also approximately constant as $\omega \to 0$.
Summarizing, $\widetilde{f}(\omega)$ is approximately constant as $\omega \to 0$ for all values of $m \in (1, 3/2)$.
Case (c): $m = 3/2$. The Laplace transform of the sojourn time density function is given by

$$\widehat{p}(s) = 2 \int_1^\infty e^{-st} t^{-3} dt$$

Note that $\int_1^\infty e^{-st} t^{-3} dt$ is equal to $E_3(s)$, an exponential integral. Therefore

$$\widehat{p}(s) = e^{-s}(1-s) - s^2 \left\{ \gamma + \ln s + \sum_{n \in \mathbb{P}} \frac{(-1)^n s^n}{nn!} \right\}, \quad |\arg(s)| < \pi$$

where γ is Euler's constant. As $\omega \to 0$, it can be shown that $\widetilde{f}(\omega)$ is proportional to $|\ln \omega|$.

References

1. Abramowitz, M., and Stegun, I. A., 1965. *Handbook of Mathematical Functions,* Dover Publications, Inc., New York, New York.
2. Adler, R. J., Feldman, R. E., and Taqqu, M. S., Editors, 1998. *A Practical Guide to Heavy Tails, Statistical Techniques and Applications*, Birkhauser, Boston, Massachusetts.
3. Anis, A. A., and Lloyd, E. H., 1976. "The Expected Value of the Adjusted Rescaled Hurst Range of Independent Normal Summands," Biometrika, Vol. 63, No. 1, pp. 111-117.
4. Athreya, K. B., and Ney, P. E. 1972. *Branching Processes,* Springer-Verlag, Berlin, Germany.
5. Beran, J., 1994. *Statistics for Long-Memory Processes*, Chapman and Hall/CRC Press, New York, New York.
6. Boes, D. C., and Salas-La Cruz, J. D., 1973. "On the Expected Range and Expected Adjusted Range of Partial Sums of Exchangeable Random Variables," Journal of Applied Probability, Vol. 10, Issue 3, pp. 671-677.
7. Crovella, M. E., and Bestavros, A., 1997. "Self-Similarity in World Wide Web Traffic: Evidence and Possible Causes," IEEE/ACM Transactions on Networking, Vol. 5, No. 6, pp. 835-846.
8. Crovella, M. E., and Krishnamurthy, B., 2006. *Internet Measurement*: *Infrastructure, Traffic, and Applications*, John Wiley & Sons, Inc., New York, New York.
9. Daubechies, I., 1992. *Ten Lectures on Wavelets,* Society for Industrial and Applied Mathematics, Philadelphia.
10. Doukhan, P., Oppenheim, G., and Taqqu, M. S., Editors, 2003. *Theory and Applications of Long Range Dependence,* Birkhauser, Boston, Massachusetts.

11. Embrechts, P., Klüppelberg, C., and Mikosch, T., 1997. *Modeling Extremal Events for Insurance and Finance*, Springer-Verlag, Berlin, Germany.
12. Embrechts, P., and Maejima, M., 2002. *Selfsimilar Processes*, Princeton University Press, Princeton, New Jersey.
13. Erramilli, A., Roughan, M., Veitch, D., and Willinger, W., 2002. "Self-Similar Traffic and Network Dynamics," Proc. IEEE, Vol. 90, No. 5, pp. 800-819.
14. Erramilli, A., Singh, R. P., and Pruthi, P., 1995. "An Application of Deterministic Chaotic Maps to Model Packet Traffic," Queueing Systems, Vol. 20, Issues 1-2, pp. 171-206.
15. Flandrin, P., 1989. "On the Spectrum of Fractional Brownian Motion," IEEE Transactions on Information Theory, Vol. 35, No. 1, pp. 197-199.
16. Hurst, H. E., 1951. "Long-Term Storage Capacity of Reservoirs," Transactions of the American Society of Civil Engineers, Vol. 116, Issue 1, pp. 770-808.
17. Kac, M., 1954. "Toeplitz Matrices, Translation Kernels and a Related Problem in Probability Theory," Duke Mathematics Journal, Vol. 21, No. 3, pp. 501-509.
18. Leland, W. E., Taqqu, M. S., Willinger, W., and Wilson, D. V. 1994. "On the Self-Similar Nature of Ethernet-Traffic (Extended Version)," IEEE/ACM Transactions on Networking, Vol. 2, No. 1, pp. 1-15.
19. Mandelbrot, B. B., 1997. *Fractals and Scaling in Finance*: *Discontinuity, Concentration, and Risk*, Springer-Verlag, Berlin, Germany.
20. Mandelbrot, B. B., and Van Ness, J. W., 1968. "Fractional Brownian Motions, Fractional Noises and Applications," Society of Industrial and Applied Mathematics Review, Vol. 10, No. 4, pp. 422-437.
21. Nikias, C. L., and Shao, M., 1995. *Signal Processing with Alpha-Stable Distributions, and Applications*, John Wiley & Sons, Inc., New York, New York.
22. Pruthi, P., 1995. "An Application of Chaotic Maps to Packet Traffic Modeling," Ph. D. dissertation, KTH, Stockholm, Sweden.
23. Resnick, S. I., 2007. *Heavy-Tail Phenomena Probabilistic and Statistical Modeling,* Springer-Verlag, Berlin, Germany.
24. Samorodnitsky, G. and Taqqu, M. S. 1994. *Stable Non-Gaussian Random Processes*, Chapman and Hall, New York, New York.
25. Samuel, L. G., 1999. "The Application of Nonlinear Dynamics to Teletraffic Modeling," Ph. D. dissertation, University of London, England.
26. Schuster, H. G., 1995. *Deterministic Chaos*: *An Introduction*, Third Edition, John Wiley & Sons, Inc., New York, New York.
27. Sigman, K., 1999. "Appendix: A Primer on Heavy-Tailed Distributions," Queueing Systems, Vol. 33, Issues 1-3, pp. 261-275.
28. Spitzer, F., 1956. "A Combinatorial Lemma and its Applications to the Probability Theory," Transaction of the American Mathematics Society, Vol. 82, No. 2, pp. 323-339.
29. Taqqu, M. S., Willinger, W., and Sherman, R., 1997. "Proof of a Fundamental Result in Self-Similar Traffic Modeling," Computer Communication Review, Vol. 27, Issue 2, pp. 5-23.
30. Willinger, W., Alderson, D., and Li, L., 2004. "A Pragmatic Approach to Dealing with High-Variability in Network Measurements", IMC'04, Taormina, Sicily, Italy, pp. 88-100.
31. Whitt, W., 2002. *Stochastic-Process Limits*, Springer-Verlag, Berlin, Germany.
32. Wornell, G., 1995. *Signal Processing with Fractals*: *A Wavelet Based Approach*, Prentice-Hall, Englewood Cliffs, New Jersey.

CHAPTER 5

Dynamics, Control, and Management of Internet Congestion

$$\mathbb{D} = \{z \in \mathbb{C} \mid |z| < 1\}$$
$$\partial\mathbb{D} = \{z \in \mathbb{C} \mid |z| = 1\}$$

Poincaré disc and its boundary

Jules Henri Poincaré. Henri Poincaré was born on 29 April, 1854 in Nancy, Lorraine, France. He received his doctorate in mathematics in 1879 from École des Mines. Poincaré did fundamental work in several branches of mathematics. He is said to be the originator of algebraic topology and the theory of analytic functions of several complex variables. Poincaré also contributed to number theory, geometry, celestial mechanics, the three-body problem, and electromagnetic wave theory. His work was also a forerunner to the special theory of relativity, which was propounded by Albert Einstein in 1905. Among his philosophical writings, *Science and Hypothesis*, (1901) stands out. Henri Poincaré died on 17 July, 1912 in Paris, France.

5.1 Introduction

The subject of this chapter is the study of dynamics, control and management of Internet congestion. The topics discussed in this chapter are: congestion control and resource allocation, network basics, and description of Transmission Control and Internet protocols. A simple periodic TCP model, queue management, a mean-field model for multiple TCP connections, traffic rate and congestion control in networks, and its stability are also discussed. Traffic management technique via packet scheduling and regulation, and deterministic network calculus are also outlined. Certain aspects of multiple access of a communicating medium are also described. Techniques to model wireless networks via the use of stochastic geometry are also provided.

Congestion is the state of a network, in which performance degradation takes place due to saturation of resources. Internet congestion occurs when the cumulative traffic demand on the network resources exceeds its availability. Consequences of this phenomenon of congestion include decrease in network throughput, and increase in delivery time of packets. Severe congestion is also referred to as "congestion collapse."

A basic feature of the Internet architecture is the connectionless end-to-end packet service. This service uses the Internet Protocol. A connectionless design offers flexibility and robustness. This advantage comes with several challenges. One is a requirement of careful design to provide acceptable service under heavy traffic conditions. If the dynamics of packet forwarding algorithm is not carefully streamlined, severe congestion might result. This phenomenon has been alternately referred to as "Internet meltdown."

Congestion in a network has to be managed at different time-scales. For example, traffic can be managed via provisioning, technology, and pricing strategy, if the time-scale for congestion management is of the order of few months to years. If the time-scale is of the order of several minutes to weeks, then it can be managed via capacity management, pricing strategy for different flows, and route-control. Note that capacity is the rate at which data flows. It can be specified either in units of bits or packets per unit time. Traffic control over time scales from fraction of a second to few minutes can be managed via: routing algorithms, active queue management, and feedback mechanism. In all these techniques to manage congestion, the key phrase is *optimization of network resources* over different time-scales. In this chapter, only congestion management over short time-scales is addressed.

The science and art of congestion management in a network deals with allocation of resources in a network such that the network can operate at an acceptable performance level when the demand is more than or equal to the capacity of certain network resources. If the traffic load on the network increases, the throughput of the network might decrease. Since congestion is caused by the demand being greater than the available resources, the congestion is not simply relieved by increasing the resources. Besides increasing the resources, the dynamics of the traffic flow also has to be managed. This can be accomplished via proper protocol design and appropriate queue management. Following are some myths about congestion control.

(a) Congestion is caused by slow processors. Therefore the problem of congestion might be alleviated by increasing the speed of the processors.
(b) Speed of the links is slow. Higher speed links might decrease congestion.
(c) Buffers space is low. Increasing buffer space might provide a relief from excessive traffic demand.
(d) A combination of the above "remedies" might cause the congestion to disappear.

Congestion problem cannot simply be solved by increasing the speed of the processors or links; or increasing the buffer space. These static solutions might help, but are not the final and ultimate "answer." Congestion in a network also has to be managed via smart dynamics. This is done via *packet scheduling algorithms* and *queue management*. These two schemes are different but complementary. The scheduling algorithms decide which packet should be sent by the host computer. These are basically used to manage allocation of capacity among different flows. Therefore scheduling algorithms are generally managed from the edges of the network. Queue management algorithms manage the length of packet queues. These algorithms decide, which, when, and where packets are either retained or dropped appropriately.

A common theme recurring in the Internet congestion control is that of feedback. In a *feedback system*, the present output also depends upon the past inputs. Therefore by introducing suitable inputs in a feedback system in the present, the output of the system in the immediate future is influenced. This in a nutshell, albeit imprecise, is the definition of a feedback system. Internet congestion control is accomplished via feedback. The congestion control processes in the current Internet represent a feedback system of an extremely large size.

The current Internet is based upon the best-effort principle. This principle implies that all packets are treated equally. The network does not provide any guarantees about packet losses and delays. This paradigm is successful for data packets. However, if other application flows have more stringent requirements about packet loss and delay, then clearly best-effort principle is not sufficient. A possible solution to this requirement is to build a separate network for each and every application. However, this option does not appear to be economically feasible. Therefore modern communication networks are evolving towards an integrated-services network. In order to achieve the goal of providing different performance goals for different traffic streams, the network must be capable of providing *differentiation*. This is made possible by the use of clever packet *scheduling* techniques. Thus, all is not doom and gloom in the midst of Internet chaos, anarchy, and congestion. It is still possible to manage successful transmission of different types of traffic flows, albeit with some difficulty.

The topics discussed in this chapter rely upon applied analysis, theory of matrices and determinants, probability theory, stochastic processes, optimization techniques, and stability theory. Unless stated explicitly, it is assumed in this chapter that, if for example $A_i, i = 1, 2, \ldots$ is a sequence of random variables with identical distribution, then a generic random variable with this distribution is denoted by A. The expectation of the random variable A is denoted by $\mathcal{E}(A)$, where $\mathcal{E}(\cdot)$ is the expectation operator.

The interplay between congestion control and resource allocation is demonstrated in the next section.

5.2 Congestion Control and Resource Allocation

Congestion control and resource allocation in a network are intrinsically related to each other. This is demonstrated via the following two simple, yet evocative examples. These examples are also the genesis of advanced congestion control algorithms. In these examples, a packet is a unit of information.

Example 5.1. Consider a single link of capacity c units. It is fed by two independent sources at the rate of $r_1(t)$ and $r_2(t)$ units at time t. The transmission link provides feedback to the sources indicating whether the net arrival rate of packets from the two sources

$$r(t) = \{r_1(t) + r_2(t)\}$$

is less than or equal to c, which is the link capacity. Using this information, the two sources adjust their rates via the following relationships.

$$\frac{dr_i(t)}{dt} = \begin{cases} 1, & \text{if } r(t) \le c \\ -\zeta r_i(t), & \text{if } r(t) > c \end{cases}, \quad i = 1, 2$$

where ζ is a positive constant. The behavior of this system is next examined at steady-state, that is as $t \to \infty$. Define

$$s(t) = \{r_1(t) - r_2(t)\},$$

then

$$\frac{ds(t)}{dt} = \begin{cases} 0, & \text{if } r(t) \le c \\ -\zeta s(t), & \text{if } r(t) > c \end{cases}$$

In light of the above two sets of equations, consider the following cases.

Case $r(t) \le c$: The source rates $r_1(t)$ and $r_2(t)$ increase, but $s(t)$ remains constant.

Case $r(t) > c$: Define $Y(t) = \{s(t)\}^2$, then

$$\frac{dY(t)}{dt} = -2\zeta Y(t)$$

Since $\zeta > 0$, $dY(t)/dt < 0$, unless $Y(t) = 0$.

From the above two cases, it can be concluded that as $t \to \infty$, we have

$$\{r_1(t) + r_2(t)\} \to c,$$

and $Y(t) \to 0$. This in turn implies that at steady-state, each source rate tends towards $c/2$. The following lessons can be learned from this example.

Evidently, the feedback mechanism tends to increase the source rates, if $r(t) \leq c$. However if $r(t) > c$, the source rates decrease. Thus the feedback mechanism provides congestion control. In addition, at steady-state the congestion control algorithm drives the system towards a fair resource allocation. That is, as $t \to \infty$ the link capacity is shared equally among the two sources. Also observe that the link is completely utilized at steady-state. □

Example 5.2. A network has a single link of capacity c units. This link is fed by two independent sources. The source rates are $r_1(t)$ and $r_2(t)$ units at time t. However the packets from these two sources arrive at the link after a delay of δ_1 and δ_2 time units respectively. The net arrival rate of packets from the two sources is equal to $r(t)$, where

$$r(t) = \{r_1(t-\delta_1) + r_2(t-\delta_2)\}$$

The link lets the sources know, if the cumulative rate $r(t)$ is less than c. However, this information arrives from the link to the sources, after a delay of τ_1 and τ_2 time units respectively. Using this information the sources adjust their rates as per the following relationships.

$$\frac{dr_i(t)}{dt} = \begin{cases} 1, & \text{if } r(t-\tau_i) \leq c \\ -\zeta r_i(t), & \text{if } r(t-\tau_i) > c \end{cases}, \quad i = 1, 2$$

where ζ is a positive constant. The above equation at time $(t+\tau_i)$ can be expressed as

$$\frac{dr_i(t+\tau_i)}{dt} = \begin{cases} 1, & \text{if } r(t) \leq c \\ -\zeta r_i(t+\tau_i), & \text{if } r(t) > c \end{cases}, \quad i = 1, 2$$

The behavior of this system is next examined at steady-state, that is as $t \to \infty$. Define

$$s(t) = \{r_1(t+\tau_1) - r_2(t+\tau_2)\},$$

then

$$\frac{ds(t)}{dt} = \begin{cases} 0, & \text{if } r(t) \leq c \\ -\zeta s(t), & \text{if } r(t) > c \end{cases}$$

As in the last example, note the following two cases.

Case $r(t) \leq c$: The source rates $r_1(t+\tau_1)$ and $r_2(t+\tau_2)$ increase, but $s(t)$ remains constant.

Case $r(t) > c$: Define $Y(t) = \{s(t)\}^2$, then

$$\frac{dY(t)}{dt} = -2\zeta Y(t)$$

Since $\zeta > 0$, $dY(t)/dt < 0$, unless $Y(t) = 0$.

From the above two cases, it can be concluded that as $t \to \infty$, then $Y(t) \to 0$. In addition, as $t \to \infty$, then

$$\{r_1(t+\tau_1) + r_2(t+\tau_2)\} \to c$$

This observation implies that, in the presence of delays, and at steady-state, the rate allocation is not fair at all instants of time. However, it is fair at time-shifted values of the source rates. This in turn implies that when the source rates are averaged over a sufficient length of time, the capacity allocation is fair.

It appears that the algorithm outlined in this example, eases congestion on a single link. However, this algorithm is not easily extensible to an arbitrary general network. □

The important lesson to be learned from these two examples is that a feedback mechanism might possibly ease congestion. In the next section network basics are outlined.

5.3 Network Basics

The Internet is a collection of computers which are interconnected by transmission links. Its purpose is to efficiently and reliably transfer information between source and destination computers. The proper functioning of the Internet is based upon a set of rules called protocols. It can perhaps safely be asserted that the Transmission Control and the Internet Protocols are at the heart of the spectacular success of the Internet. TCP and IP are the acronyms of the Transmission Control Protocol and the Internet Protocol respectively. These protocols can most conveniently be described in terms of Open Systems Interconnection's (OSI) seven layer architecture (model) of the Internet.

Recall that the layer one is the physical layer. This layer is governed by a collection of protocols that are responsible for transmission of bits (a bit is a 0 or a 1) over a physical medium. The second layer is the data link layer. Several bits form a frame. The suite of data link protocols ensure that a frame is transmitted from one end of a link to the other end reliably. This layer can also include error detection and correction of bits in a frame.

The third layer of the OSI architecture is the network layer. IP belongs to the network layer. Several computers are connected by different links to form a network. These computers must be able to determine the routing of packets, so that the packets reach their designated destinations. The function of determining the paths the packets must follow is called *routing*. Routing of packets is accomplished by appending unique network addresses of the source and destination computers to the frames. The unit of information at the network layer is called the *packet*. The computers which perform the function of routing are called the *routers*. Therefore, in summary the network layer performs the packet delivery function.

The network layer does not guarantee that the packets are delivered to the destination computer reliably. The packets on their way from the source to the destination computer might be corrupted by noise or might get lost due to the finite buffer size at intermediate and destination computers (nodes).

The fourth layer of the OSI architecture is called the transport layer. It is responsible for introducing reliability to the network layer. The protocol at this layer ensures that the lost packets are detected, and might possibly be retransmitted. TCP and UDP (User Datagram Protocol) are the two predominant transport layer protocols. The TCP is used by file transfer applications where reliable in-sequence delivery of packets from the source to the destination is desired. The UDP is used by delay-sensitive applications where a certain level of packet-loss is tolerable. In this protocol there is no retransmission of packets. A possible application of the UDP is video transmission.

The fifth layer is the session layer. It supervises the dialog between the source and destination computers. This layer sets up a connection between the source and destination computers prior to an exchange of information between them. The sixth layer is the presentation layer. This layer is responsible for encryption, compression, and syntax conversion. The seventh layer is the application layer. It provides services such as file transfer, remote login, directory service, terminal emulation, World Wide Web, electronic mail, and remote job execution.

Observe that OSI is a seven-layered architecture of the Internet. There are other layered architectures which explain the design of the Internet. For example, there are four- and five-layered architectures. Each layered architecture can be mapped into the other layered architecture. Let us consider the four-layered architecture, which is also called the TCP/IP reference model. The functionality of layers one and two of the OSI model map to the functionality of layer one of the TCP/IP reference model. Similarly, the functionality of layers three and four of the OSI model map to the functionality of layers two and three of the TCP/IP reference model respectively. And, the functionality of layers five, six, and seven of the OSI model map to the functionality of layer four of the TCP/IP reference model. Finally, it should be noted that each layered architecture has its own set of advantages, and disadvantages.

5.4 The Transmission Control and Internet Protocols

In order to gain a basic understanding of the dynamics and control of Internet congestion, it is necessary to understand the fundamentals of TCP, UDP, and IP.

5.4.1 Transmission Control Protocol

The TCP functions at the transport layer. The performance of the TCP can be evaluated by first gaining a thorough understanding of its basics. The following features of TCP are explained in this subsection:

- TCP mechanism
- Acknowledgment and time-out of segments
- Flow control
- Congestion avoidance algorithm.

There are several flavors of TCP. Some of these are, TCP-Tahoe, Reno, NewReno, FACK, SACK, and Vegas. The different versions of the TCP are described in the open literature. Only the basics of the protocol are outlined.

TCP Mechanism

The TCP is a connection-oriented, full-duplex, streaming, and reliable service mechanism. In this connection-oriented protocol, two application processes on two different computers initially establish a TCP connection before they begin to send data to each other. This protocol is also a full-duplex service. That is, it supports data flow in both directions of the transmission links. This implies that once a TCP connection has been set up between two application processes on two different computers, either process can send data to the other over the same connection simultaneously. Therefore, after a connection is established between two processes, these applications can stream data to one another. However, note that the transfer of information at the TCP layer is in packet mode. TCP makes sure that every single byte is delivered to its destination in order, and without any duplication. This is possible by an acknowledgment scheme. In this scheme, unacknowledged data is retransmitted to the receiving computer. Therefore, as per this scheme, it is possible to have multiple retransmissions of the same segment of data. Consequently, this mode of data transfer is preferred by applications such as the World Wide Web, electronic mail, and file transfer.

The fundamental reason that the TCP congestion control algorithms work is that the TCP source adjusts its sending rate according to the rate of packet drops in the network. The assumption is that the packet drops in the network are indications of congestion in the network. Features of TCP relevant to congestion and its control are next described. These features are: acknowledgment and time-out of segments, flow control, and congestion control.

Acknowledgment and Time-Out of Segments

It is possible for the packets to get corrupted in transit. Some of these data segments might possibly be corrected by the use of error correcting code. However, it is also quite possible that a segment might not have been correctable. In this case, the segment is discarded. Therefore, in order to achieve reliable transmission, the mechanism of *acknowledgment* and *time-out* is used. TCP requires the destination (receiving) computer to confirm the correct delivery of data. The receiving computer sends an acknowledgment for each received segment to the transmitting computer. This acknowledgment segment is also called an ACK. Acknowledgment can also be cumulative. That is, the receiving computer can convey to the transmitting computer, that contiguous segments up to a certain point in time have been received successfully. The acknowledgment information can be sent in a separate packet, or it can piggyback on a data segment flowing from the destination computer to the source computer. Time-out implies that the source node retransmits the segment, if its acknowledgment has not yet been received till a fixed duration of time.

Flow Control

Flow of data that is transmitted from the source host to the destination host needs to be controlled. If this flow control is not instituted, the receiving computer might possibly be flooded by incoming data stream. Therefore, buffer overflow at the receiving host can occur in the absence of flow control. TCP uses the concept of sliding window to implement flow control.

The size of window at the transmitting host controls the number of bytes in transit. These are the transmitted but not yet acknowledged bytes. If all the data in a window is in transit, the source host stops transmitting any more data segments and waits for acknowledgment from the destination host. Once an acknowledgment arrives, the source can transmit new data segments which are not more than the number of acknowledged bytes.

The window size at the source computer can be either fixed or dynamic. Depending upon the buffer space available in the receiving host, the receiver advertises its window size to the source host. This is accomplished by the receiver transmitting this information in one of its data segments to the source host. This way, the window size is adjusted dynamically if necessary.

In summary, flow control means regulating the flow of senders data such that the buffers at the destination computers do not overflow. Therefore flow control addresses the problem of data flow between the source and the destination computers.

Congestion Avoidance Algorithm

Flow control does not address the problem of buffer overflow in routers which are lying between the source and destination computers. Therefore congestion control deals with the interaction between the host computer and the network. Congestion control implies slowing down the rate at which data enters the network, when packets are getting dropped in the network. Congestion control enables each source to determine the number of packets it can submit safely without causing congestion. However, determining this number is a nontrivial task. One way for congestion control in the Internet was suggested by V. Jacobson. He suggested a mechanism which operates in the host computer to cause the TCP connections to slow down during congestion. Thus his technique of congestion control is called a "congestion avoidance algorithm." It is a scheduling type of congestion control algorithm. That is, packets are scheduled for transmission over the network, before congestion occurs. In this sense it is a proactive algorithm.

The aim of congestion avoidance algorithm is to adjust transmission window of the source computer such that buffer overflow does not occur not only at the receiver, but also at the intermediate routers. Note that routers are intermediate nodes, which route the data to the destination. The daunting task is to determine the available buffer space in the network routers. To achieve this goal, the source computer uses a parameter called *CongestionWindow*.

The basic principle of TCP is the retransmission mechanism. If a data segment is lost, then it is retransmitted. To achieve this goal, TCP maintains a *retransmission timer* for each transmitted segment. Its value is set for a duration called the retransmission time-out period. If an acknowledgment is received before the time-out period, the timer is cleared; otherwise the timer expires. In the later case, the data segment is retransmitted.

TCP assumes that the network is congested whenever a retransmission timer at the source computer has expired and adjusts the *CongestionWindow* parameter by using three algorithms. These are the slow start, congestion avoidance, and multiplicative decrease mechanisms. See Figure 5.1.

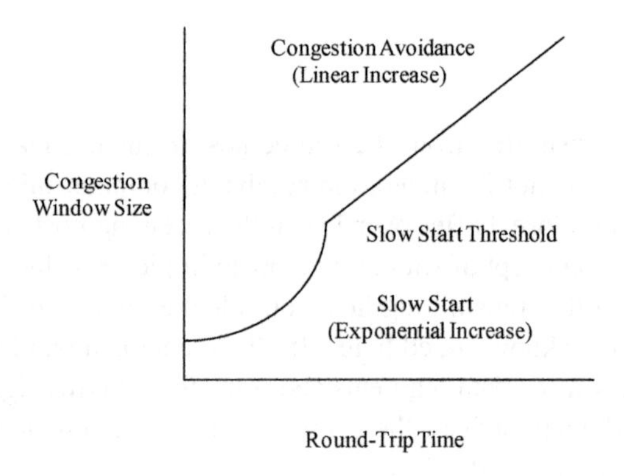

Figure 5.1. Round-trip time versus congestion window size.

Slow Start

The slow-start mechanism is used to increase the size of the congestion window quickly. This has the effect of determining the available buffer space in the network. In the beginning the *CongestionWindow* is set to one segment. This value is incremented each time a segment is acknowledged. For example, when the acknowledgment of the first segment is obtained, the value of *CongestionWindow* becomes two. Next the source, sends two segments. When the acknowledgment of these two segments is achieved, the value of the *CongestionWindow* becomes four and so on. Remember that multiple segments can be acknowledged by a single acknowledgment segment from the destination host to the source host.

Congestion Avoidance

Using the slow-start mechanism, it can be observed that the value of congestion window size practically gets doubled after each round of acknowledgment. Therefore this leads to an exponential growth of window size, which in turn might lead to congestion. To overcome this problem, the congestion avoidance algorithm is used. This algorithm triggers a linear increase of the *CongestionWindow* after it reaches a threshold. This threshold value is dynamically adjusted via a variable called *ssthresh.* Once the *CongestionWindow* reaches *ssthresh,* the value of the *CongestionWindow* is incremented by *1/CongestionWindow* each time an acknowledgment is received. Thus the congestion window size is effectively incremented by one for every round-trip time (RTT). Note that the time between sending a packet and receiving the acknowledgment is called round-trip time. In summary, the size of the congestion window grows exponentially in the slow-start phase, but grows linearly in the congestion avoidance phase.

Multiplicative Decrease

The parameter *ssthresh* determines the transition from the slow-start phase to the congestion avoidance phase. Multiplicative decrease algorithm determines the value of the threshold *ssthresh.* As per this algorithm, the TCP sets *ssthresh* to half of current *CongestionWindow* value each time a time-out occurs. The minimum allowable value of the threshold *ssthresh* is two segments. Consequently, successive time-outs indicate severe network congestion, and these result in a multiplicative decrease in the sending rate of the segments.

Therefore the value of the *CongestionWindow* parameter increases additively during the congestion avoidance phase, but there is a multiplicative decrease in the value of *ssthresh* during network congestion.

5.4.2 User Datagram Protocol

User Datagram Protocol is a much simpler version of TCP. UDP is also a protocol at the transport layer. This protocol is connectionless, unreliable, and datagram-oriented. It is assumed that the source does not provide a stream of data. However, the application provides segmented data to the UDP for transportation as an independent datagram. Lack of reliability implies that, there is no acknowledgment and retransmission scheme in this protocol. Furthermore, there is no flow and congestion control. Consequently, the destination host does not know if a datagram has been lost due to errors or buffer overflow.

The strength of this protocol lies in its simplicity. It is therefore used in applications such as network management, routing table update, real-time multimedia, and multicasting.

5.4.3 Internet Protocol

The IP is an important protocol, in the Internet communication network. The unit of information transfer in this protocol is an IP datagram (or packet). This protocol is a connectionless unreliable datagram model of communication. Connectionless service implies that each packet travels through the network independent of any other packet. Packets from the same message might travel to their destination via different paths.

The connectionless and unreliable mode of transmission of packets is often referred to as a *best-effort delivery* service. The term best-effort implies that IP does not provide any type of tracking (flow control) or error control (except for header checksum) of the datagram.

The IP is a network layer protocol, which is used by both TCP and UDP. This protocol encapsulates the higher layer protocols, such as the TCP segments, and the UDP datagrams. Thus the TCP segments, and the UDP datagrams are the payloads for the IP datagram payload. An IP header is appended to the TCP segment or the UDP datagram to form an IP datagram. Since this protocol is a network layer protocol, it forwards the IP datagram to the next hop router towards the destination computer.

5.5 A Simple Periodic TCP Model

A simple, approximate, and periodic model of TCP is considered in this section. The model is simple because it does not take into consideration the complete dynamics of TCP. Therefore its discussion is only intended to provide a glimpse into the performance of this protocol. This model assumes steady-state behavior of the protocol. Therefore transient stage of the protocol like slow start does not appear in the model. This is justifiable because our goal is to model the long term performance of the protocol. Consequently, only the congestion avoidance phase of the protocol is considered.

Let W be the maximum window size (in units of packets). The window size of a source-destination pair is the maximum number of packets that can be transmitted without acknowledgment. The source TCP maintains a variable called *AdvertisedWindow* to track the current window size. This value is used in flow control. Furthermore, the receiver window size depends upon the available buffer space.

The window size, which is controlled by the TCP source, depends upon the level of congestion. The time interval between the transmission of the packet by the TCP source, and the receipt of its acknowledgment from the receiver is called the round-trip time. Denote the round-trip time by R, and assume that it is a constant. In general, this value is probabilistic, and depends upon the level of congestion in the network. This assumption that the round-trip time is constant, implies that the path between the source and destination has sufficient capacity and low traffic so that there are no significant queueing delays. It is also assumed that the time required to send all the packets in a window is smaller than the round-trip time.

In the congestion avoidance phase of the algorithm, window size increases linearly if packets are not lost in the network, and decreases multiplicatively if a packet is lost. More specifically, the window size increases by a single packet for each round-trip time, and decreases by half in case of a packet loss. Denote the size of the window by $w(t)$, where $t \in \mathbb{R}_0^+$ is the time parameter. Note that the round-trip time begins with the back-to-back transmission of $w(t)$ packets, which is the size of the TCP congestion window at time t. After $w(t)$ packets have been transmitted, no other packets are sent till an acknowledgment is received for one of these $w(t)$ packets.

For simplicity, denote this window size by simply w. If the current window size is w, then this window size increases by $1/w$ after each acknowledgment of a packet. Thus the new window size is $(w + 1/w)$ after the acknowledgment of a single packet. Therefore the window size increases by unity after w packet acknowledgments. In general, all the packets in a window are transmitted before, any acknowledgment is received for the packet. Thus the window size effectively increases by unity for each round-trip time.

The packet-loss process is a measure of traffic load or congestion in the network. The loss of a packet can be triggered by any node along the path of the TCP connection between the source and destination nodes. Let p be the nonzero packet-loss probability. Furthermore, the model assumes that the packet-loss probability is constant. Therefore, the model implies that approximately $1/p$ packets are delivered correctly to the receiver for each lost packet. As already mentioned, the window size is halved in the event of a packet loss.

Also assume that this packet loss is light to moderate. The analysis of the model also presumes that there is sufficient data at the transmitting node to send. In addition, the receiver window is assumed to be not too small. At steady-state, the window size achieves its maximum value W by the time a packet is lost. Therefore, when a packet is lost, the window size decreases to $W/2$.

Thus the window size function appears like a sawtooth waveform. This essentially follows from the fact that, the window size increases linearly and decreases by half multiplicatively when a packet is lost. Thus the minimum and maximum values of the window size $w(\cdot)$ in this model are $W/2$ and W respectively. This increase in window size from $W/2$ to W occurs in time T, where

$$T = \frac{W}{2}R$$

This is because each of the $(W - W/2)$ window increases occurs in a single round-trip time. Consequently the window size $w(\cdot)$ is periodic sawtooth with period T. See Figure 5.2.

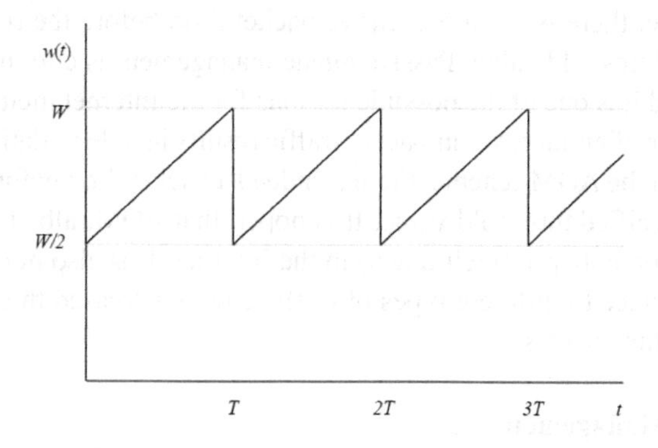

Figure 5.2. Periodic window size of a TCP flow versus time at steady-state.

In this figure, the window size $w(t)$ is shown to take a continuum of values for simplicity. However, it only takes discrete values in its implementation. At steady-state, the window size $w(t)$ at time $t \in \mathbb{R}_0^+$ is

$$w(t) = \frac{W}{2} + \frac{(t - nT)}{R}; \quad nT < t \leq (n+1)T, \ \forall n \in \mathbb{N}$$

Recall that at the end of each period, a single packet is lost with probability p. Therefore, on the average $1/p$ packets are acknowledged by the receiver. Thus the number of packets transmitted successfully during each period is

$$\frac{T}{R}\left\{\frac{1}{2}\left(\frac{W}{2} + W\right)\right\} = \frac{1}{p}$$

Substituting $T = WR/2$ in the above equation yields

$$W = \left(\frac{8}{3p}\right)^{1/2}$$

The average sending rate of the TCP source $X(p)$ is given by

$$X(p) = \frac{1/p}{T} = \frac{1}{R}\left(\frac{3}{2p}\right)^{1/2} \quad \text{packets per unit time} \tag{5.1}$$

The above relationship is known as the inverse square-root p law. This law demonstrates, that the throughput of a TCP rate is inversely proportional to the round-trip time R, and the square root of the average packet-loss probability p.

5.6 Queue Management

Queue management is the set of schemes (algorithms) that control the length of packet-queues at the routers by dropping packets in order to manage packet-traffic. Queue management algorithms can again be divided into two classes. The two classes of algorithms are: passive queue management (PQM), and active queue management (AQM).

In the PQM scheme, there is not a preventive packet drop before the router-buffer gets full or reaches a prespecified threshold value. Passive queue management is currently widely prevalent in the Internet routers, and it is one of the possible reasons for the Internet melt down. A melt down in network is said to occur, if an increase in packet traffic results in a degradation of its throughput.

In direct contrast, in the AQM scheme, there is indeed a packet drop before the router buffer gets full or reaches a prespecified threshold value. It is hoped that AQM, albeit difficult to implement, might prevent congestion collapse (melt down) in the Internet. It is also hoped that AQM provides improved quality of service to different types of traffic due to increased throughput and a decrease in the delivery time of the packets.

5.6.1 Passive Queue Management

As already mentioned, passive queue management technique does not employ preventive packet drop for newly arriving packets until the router-buffer reaches a prespecified value. When the number of buffered packets reaches this value, subsequently arriving packets are dropped. Observe that this scheme does not warn the packet sources to back-off, so that they might decrease their traffic level and avoid network congestion. Some examples of PQM are tail-drop, drop-from-front, and push-out.

- *Tail-Drop Scheme*: In the tail-drop scheme of PQM, packets drop from the tail of the queue. If the queue is full, or it has reached a prespecified threshold, newly arriving packets are dropped. Nothing happens to the packets which are already in the buffer.
- *Drop-From Front Scheme*: In the drop-from front discipline, if the queue is full, or it has reached a prespecified threshold, the newly arriving packet is accepted in the queue, but the packet which is resident at the head of the queue is discarded. In this scheme, the data source "sees" the packet loss earlier than that in the tail-drop scheme.
- *Push-Out Scheme*: In the push-out discipline, if the queue is full, or it has reached a prespecified threshold, the newly arriving packet is accepted in the queue, but the packet which is resident at the tail of the queue is discarded. This technique of packet buffering does not appear to be very efficient.

The limitations of the PQM should be evident. A larger buffer in the router results in increased throughput, but this is at the expense of increased average delay per packet. More specifically, the problems with PQM are: lock-out and full queues.

- *Lock-out*: In some traffic scenarios tail-drop allows a single connection or a limited number of flows to monopolize the buffer space of the router. This prevents other traffic flows from getting space in the queue. Therefore the lock-out phenomenon might cause unequal distribution of resources.
- *Full Queues*: Note that in the tail-drop scheme packets are not dropped till the queue is full. Therefore, this results in long queueing delays.

Due to these drawbacks, new Internet routers are not designed with PQM.

5.6.2 Active Queue Management

In order to overcome the disadvantages of PQM, AQM is recommended for new Internet routers. The goals of AQM are:

(a) Decrease the number of packets dropped in routers. This is professed to improve the throughput.
(b) Maintain small queue lengths. This results in smaller queueing delay. This is beneficial to the traffic flows which are delay-sensitive. However small queue lengths might result in low link utilization. This limitation poses a challenge in the AQM design.
(c) Avoid lock-out behavior by splitting the capacity fairly among competing traffic flows.

These goals are accomplished by AQM via:

(a) Preventive random packet-drops even before the router-buffer is full.
(b) Packets are dropped at random with increasing intensity when the traffic rates increase.

An example of AQM is a random early detection scheme. It is hoped that this queue management scheme overcome the shortcomings of the PQM techniques. This scheme is discussed in the next subsection.

5.6.3 Random Early Detection

Random Early Detection (RED) is an example of active queue management. It was proposed by V. Jacobson and S. Floyd. This AQM scheme is intended to alleviate some of the congestion control problems. In order to manage congestion, the router which uses the RED technique of AQM, discards arriving packets as a function of the estimated queue length. A fluid-flow model of RED is initially developed in this subsection. This is followed by a model of the RED scheme. Slow start behavior of the TCP is not considered in this model, because it typically exists for only a few round-trip times.

A Fluid-Flow Model of RED

The interaction between TCP flows and RED mode of active queue management is modeled. The router dynamics is first described. This is followed by a description of the packet-discard model. The queue-length, estimated queue-length, and TCP-window dynamics are subsequently determined.

Router Dynamics: Consider a router with a single link of capacity C_ℓ packets per unit time. It is assumed that this router manages N number of TCP flows. This number N is assumed to be a constant in the model. These flows are labeled $i = 1, 2, \ldots, N$. Let $W_i(t)$ and $R_i(t)$ denote the

TCP window size and packet round-trip time associated with flow i at time $t \in \mathbb{R}_0^+$. Assume that a_i is the fixed propagation time associated with flow i. Recall that propagation delay is the speed-of-light latency for a single bit over the medium. Also note that speed of light is different for different media. It is equal to the distance between the source and destination, divided by the speed of the light in the physical medium. Also define $X_i(t)$ to be the instantaneous throughput of flow i. Thus

$$R_i(t) = a_i + \frac{Q(t - R_i(t))}{C_\ell}, \text{ and } X_i(t) = \frac{W_i(t)}{R_i(t)}, \quad t \in \mathbb{R}_0^+, \; 1 \le i \le N \tag{5.2}$$

where $Q(t)$ is the queue length (packets) of the router at time t. Let the maximum value of the queue length (buffer size) and the window size be $Q_{\max}$ and $W_{\max}$ respectively. Therefore $Q(t) \in [0, Q_{\max}]$, and $W_i(t) \in [0, W_{\max}]$. The window size and the queue length are assumed to be stochastic.

Packet-Discard Model: RED's packet-discard model depends upon an estimate of average queue length at the router. Let $x(t)$ be an estimate of the average queue length at the router at time t. This estimate is governed by a parameter α which lies in the interval $(0, 1)$. Assume that the queue length is sampled every $\delta_{samp} \in \mathbb{R}^+$ units of time. This estimate is specified by

$$x(t) = (1 - \alpha) x(t - \delta_{samp}) + \alpha Q(t - \delta_{samp}), \qquad 0 < \alpha < 1 \tag{5.3}$$

A router which uses RED accepts all packets until the estimated queue length reaches a value $\sigma_{\min}$, after which it drops a packet with a probability which increases linearly with an increase in estimate of the queue length. When the estimated queue length reaches a value $\sigma_{\max}$, all arriving packets are dropped. Denote the packet loss probability by $p(x(t))$. Then

$$p(x) = \begin{cases} 0, & 0 \le x < \sigma_{\min} \\ (x - \sigma_{\min}) L_{RED}, & \sigma_{\min} \le x \le \sigma_{\max} \\ 1, & \sigma_{\max} < x \end{cases} \tag{5.4a}$$

where

$$L_{RED} = \frac{p_{\max}}{(\sigma_{\max} - \sigma_{\min})} \tag{5.4b}$$

In the above equations, explicit dependence of $x(\cdot)$ on time is not shown for clarity. Therefore the performance of this packet-dropping scheme depends upon the choice of four parameters $\alpha, \sigma_{\min}, \sigma_{\max}$, and $p_{\max}$. The basic idea behind this scheme is that the router detects congestion early by computing an estimate of the average queue length and then drops a packet with probability $p(x)$. The value of this probability, in turn depends upon a judicious choice of these four parameters.

Queue-Length Dynamics: A differential equation which connects the expected values of the queue length, the window size, and round-trip time is given below. The derivative of the queue length with respect to time is

$$\frac{dQ(t)}{dt} = \begin{cases} \max\{0, g(t)\}, & Q(t) = 0 \\ g(t), & Q(t) > 0 \end{cases}$$

where

$$g(t) = -C_\ell + \sum_{i=1}^{N} \frac{W_i(t)}{R_i(t)}$$

In the above equation, the term $-C_\ell$ contributes to the decrease in queue length. The term $\sum_{i=1}^{N} W_i(t)/R_i(t)$ contributes to the increase in queue length due the arrival of packets from the N TCP flows. If the queue at this router is a bottleneck, then $Q(t) > 0$ with probability close to unity. Consequently

$$\frac{dQ(t)}{dt} \simeq g(t)$$

Taking expectations on both sides of the above equation and using the approximation

$$\mathcal{E}\left\{\frac{W_i(t)}{R_i(t)}\right\} \simeq \frac{\mathcal{E}(W_i(t))}{\mathcal{E}(R_i(t))}$$

yields

$$\frac{d\mathcal{E}(Q(t))}{dt} \simeq \sum_{i=1}^{N} \frac{\mathcal{E}(W_i(t))}{\mathcal{E}(R_i(t))} - C_\ell \tag{5.5}$$

This approximation is again used in the discussion of window dynamics.

Estimated Queue-Length Dynamics: Next consider a discretized version of the queue-length estimate, which is

$$x((k+1)\delta_{samp}) = (1-\alpha)x(k\delta_{samp}) + \alpha Q(k\delta_{samp}), \quad k \in \mathbb{N}$$

This equation is converted into a differential equation as follows. Assume that

$$\frac{dx(t)}{dt} = Jx(t) + MQ(t)$$

where J and M are constants. Then

$$x(t_{k+1}) = e^{J(t_{k+1}-t_k)}x(t_k) + M\int_{t_k}^{t_{k+1}} e^{J(t_{k+1}-\tau)}Q(\tau)\,d\tau$$

In the right-hand side of the above equation $Q(\tau)$ can be approximated by $Q(t_k)$. Let $t_{k+1} = (k+1)\delta_{samp}$ and $t_k = k\delta_{samp}$ and comparison of the above equation with the discretized equation of the queue-length estimate yields

$$J = \frac{1}{\delta_{samp}} \ln(1-\alpha)$$
$$M = -J$$

Thus

$$\frac{dx(t)}{dt} = \frac{1}{\delta_{samp}} \ln(1-\alpha)\{x(t) - Q(t)\}$$

Taking expectation of both sides yields

$$\frac{d\mathcal{E}(x(t))}{dt} = \frac{1}{\delta_{samp}} \ln(1-\alpha)\{\mathcal{E}(x(t)) - \mathcal{E}(Q(t))\} \tag{5.6}$$

TCP Window Dynamics: It is assumed in this model that packet losses of flow i are described by a Poisson process $\{N_i(t), t \in \mathbb{R}_0^+\}$ with time-varying rate parameter $\lambda_i(t) > 0$, where $N_i(t)$ denotes the total number of packets lost by flow i till time t. Furthermore, t is the point in time when the flow detects losses. This is different from when the actual dropping of packet occurs at the queue. Assume that loss of packets due to time-outs do not occur. Therefore using the additive-increase multiplicative-decrease behavior of TCP yields

$$dW_i(t) = \frac{dt}{R_i(t)} - \frac{W_i(t)}{2} dN_i(t)$$

The first term in the above equation represents the additive-increase component, in which the window size increases by unity for every round-trip time. The second term represents the multiplicative decrease component, which halves the window size when a packet loss occurs. Taking expectation on both sides of the above equation yields

$$d\mathcal{E}(W_i(t)) \simeq \mathcal{E}\left\{\frac{dt}{R_i(t)}\right\} - \frac{\mathcal{E}(W_i(t))\lambda_i(t)}{2} dt$$

where the approximation $\mathcal{E}(W_i(t)\, dN_i(t)) \simeq \mathcal{E}(W_i(t))\,\mathcal{E}(dN_i(t))$ has been used. In the above equation, $\lambda_i(t)$ is the loss indication received by the source. It reaches the source after a delay of τ_i time units, which is approximately one round-trip after a packet has been dropped at the queue. Therefore

$$\lambda_i(t) \simeq p(\mathcal{E}(x(t-\tau_i)))\,\mathcal{E}(X_i(t-\tau_i))$$

where $X_i(\cdot)$ is the instantaneous throughput of flow i. Use of the approximations

$$\mathcal{E}\left\{\frac{1}{R_i(t)}\right\} \simeq \frac{1}{\mathcal{E}(R_i(t))}$$

$$\mathcal{E}(X_i(t-\tau_i)) = \mathcal{E}\left\{\frac{W_i(t-\tau_i)}{R_i(t-\tau_i)}\right\} \simeq \frac{\mathcal{E}(W_i(t-\tau_i))}{\mathcal{E}(R_i(t-\tau_i))}$$

yields

$$\frac{d\mathcal{E}(W_i(t))}{dt} \simeq \frac{1}{\mathcal{E}(R_i(t))} - \frac{\mathcal{E}(W_i(t))\,\mathcal{E}(W_i(t-\tau_i))}{2\mathcal{E}(R_i(t-\tau_i))} p(\mathcal{E}(x(t-\tau_i))), \quad i = 1, 2, \ldots, N \tag{5.7}$$

The above N equations, and the expressions for $d\mathcal{E}(Q(t))/dt$ and $d\mathcal{E}(x(t))/dt$ form $(N+2)$ coupled equations in $(N+2)$ unknowns. These unknowns are $\mathcal{E}(Q(t))$, $\mathcal{E}(x(t))$, and $\mathcal{E}(W_i(t))$, $1 \le i \le N$. Recall that τ_i is approximately equal to the round-trip time, and $R_i(\cdot)$ is a function of $Q(\cdot)$. Therefore the unknowns can be determined numerically. Also, observe that these equations are nonlinear.

Analysis of RED

A dynamic and nonlinear model of interaction between TCP flows and RED scheme has been developed thus far in this subsection, by using a fluid-flow approach. In this model, the TCP time-out mechanism was ignored. This nonlinear model is studied further via linearization. For notational convenience let

$$\mathcal{E}(W_i(t)) \triangleq W(t), \quad 1 \le i \le N \tag{5.8a}$$

$$\mathcal{E}(Q(t)) \triangleq \theta(t) \tag{5.8b}$$

$$\mathcal{E}(x(t)) \triangleq \upsilon(t) \tag{5.8c}$$

$$\mathcal{E}(R_i(t)) \triangleq R(t) = a_{\varpi} + \frac{\theta(t-R(t))}{C_\ell}, \quad 1 \le i \le N \tag{5.8d}$$

$$p(\mathcal{E}(x(t))) \triangleq \varrho(t) \tag{5.8e}$$

where, it is assumed for simplicity, that a_{ϖ} is the propagation time for each of the N TCP sessions. Note that the expected window size $\mathcal{E}(W_i(t))$ and the average round-trip times are independent of i. Also observe that $\varrho(t)$ is simply the packet drop probability at time t. Denote the first derivative of $W(t)$ and $\theta(t)$ with respect to time as

$$\frac{dW(t)}{dt} \triangleq \dot{W}(t) \tag{5.9a}$$

$$\frac{d\theta(t)}{dt} \triangleq \dot{\theta}(t) \tag{5.9b}$$

Nonlinear Model: In the above notation, the dynamical equations describing TCP and RED scheme for $t \in \mathbb{R}_0^+$ are

$$\dot{W}(t) = \frac{1}{R(t)} - \frac{W(t)W(t-R(t))}{2R(t-R(t))}\varrho(t-R(t)) \tag{5.10a}$$

$$\dot{\theta}(t) = \frac{W(t)}{R(t)}N - C_\ell \tag{5.10b}$$

where $R(t) = \{a_{\varpi} + \theta(t-R(t))/C_\ell\}$.

Linearization: The purpose of linearization of the above nonlinear equations is to study the RED active queue management scheme. When $\dot{W}(t) = \dot{\theta}(t) = 0$, denote the variables $W(t)$, $\theta(t)$, $R(t)$, and $\varrho(t)$ by W_0, θ_0, R_0, and ϱ_0 respectively. Thus

$$W_0^2\varrho_0 = 2, \quad W_0 = \frac{R_0 C_\ell}{N}, \quad \text{and } R_0 = a_{\varpi} + \frac{\theta_0}{C_\ell} \tag{5.11}$$

For typical networks $W_0 \gg 1$. Let

$$W(t) \triangleq W_0 + \delta W(t), \; \theta(t) \triangleq \theta_0 + \delta\theta(t), \text{ and } \varrho(t) \triangleq \varrho_0 + \delta\varrho(t)$$

Assume that the terms which depend upon $(t - R(t))$ are approximated by $(t - R_0)$. Thus the expression for $\dot{W}(t)$ can be written as

$$\dot{W}(t) = \frac{1}{R(t)} - \frac{W(t)W(t-R_0)}{2R(t-R_0)}\varrho(t-R_0)$$

The nonlinear model is linearized as

$$\delta\dot{W}(t) = -\lambda_0\delta W(t) - \frac{R_0 C_\ell^2}{2N^2}\delta\varrho(t-R_0) \tag{5.12a}$$

$$\delta\dot{\theta}(t) = \frac{N}{R_0}\delta W(t) - \frac{1}{R_0}\delta\theta(t) \tag{5.12b}$$

where $\lambda_0 = 2/(W_0 R_0)$. The eigenvalues of this dynamics are $-\lambda_0$ and $-1/R_0$. Observe that all the network parameters are positive quantities. Consequently the negative eigenvalues imply that the states of the nonlinear dynamics are locally asymptotically stable. See the problem section for the derivation of the above equations.

Analysis of RED Scheme via Laplace Transform

Let the single-sided Laplace transform of $\delta W(t)$, $\delta\theta(t)$, and $\delta\varrho(t)$ be $\widehat{\delta W}(s)$, $\widehat{\delta\theta}(s)$, and $\widehat{\delta\varrho}(s)$ respectively, where $s \in \mathbb{C}$. Then

$$\widehat{\delta W}(s) = -\frac{\left(R_0 C_\ell^2 / 2N^2\right) e^{-sR_0}}{(s+\lambda_0)} \widehat{\delta \varrho}(s)$$

$$\widehat{\delta \theta}(s) = \frac{N/R_0}{(s+1/R_0)} \widehat{\delta W}(s)$$

Therefore

$$\widehat{\delta \theta}(s) = -\frac{\left(C_\ell^2 / 2N\right) e^{-sR_0}}{(s+1/R_0)(s+\lambda_0)} \widehat{\delta \varrho}(s)$$

Let the single-sided Laplace transform of $\theta(t)$ and $\upsilon(t)$ be $\widehat{\theta}(s)$ and $\widehat{\upsilon}(s)$ respectively, where $s \in \mathbb{C}$. Let $\dot{\upsilon}(t)$ be the first derivative of $\upsilon(t)$ with respect to time t. Define

$$K \triangleq -\frac{\{\ln(1-\alpha)\}}{\delta_{samp}}$$

The relationship

$$\dot{\upsilon}(t) = -K\{\upsilon(t) - \theta(t)\}$$

yields

$$\widehat{\upsilon}(s) = \frac{K}{(s+K)} \widehat{\theta}(s)$$

Use of the RED dynamics results in

$$\delta\varrho(t) = L_{RED}\delta\upsilon(t)$$

$$\widehat{\delta\varrho}(s) = L_{RED}\frac{K}{(s+K)} \widehat{\delta\theta}(s)$$

Combining the two equations in $\widehat{\delta\varrho}(s)$ and $\widehat{\delta\theta}(s)$ yields

$$L(s) + 1 = 0 \tag{5.13a}$$

where $L(s)$ is

$$L(s) = \frac{K L_{RED}\left(C_\ell^2/2N\right) e^{-sR_0}}{(s+K)(s+1/R_0)(s+\lambda_0)} \tag{5.13b}$$

The function $L(\cdot)$ is called the loop transfer-function in classical control theory. Note that control theory is the art and science of controlling the states of a dynamical system. The function $L(\cdot)$ has three poles, and a time delay. The three poles of $L(s)$ are in the left half s-plane, consequently a system specified by $L(s)$ is asymptotically stable. Let $s = (s_r + i\omega) \in \mathbb{C}$, where $i = \sqrt{-1}$. The variable s_r is the real part of s, and ω is its imaginary component. The variable ω is called (angular) frequency in control-theory literature. Observe that $|L(i\omega)|$ is a monotonically decreasing function of ω. If $|L(0)| \geq 1$, then for an appropriate choice of parameters, there exists a unique frequency ω_g such that $|L(i\omega_g)| = 1$. This frequency ω_g is called the unity-gain crossover frequency. The condition $|L(0)| \geq 1$ implies

$$\frac{L_{RED}(R_0 C_\ell)^3}{(2N)^2} \geq 1 \tag{5.14a}$$

Select ω_g such that

$$\omega_g \ll \min\left\{\lambda_0, \frac{1}{R_0}\right\} \tag{5.14b}$$

where $\lambda_0 = 2/(W_0 R_0) = 2N/\left(R_0^2 C_\ell\right)$, then

$$L(i\omega_g) \simeq \frac{K L_{RED} \left(C_\ell^2/2N\right) e^{-i\omega_g R_0}}{(\lambda_0/R_0)(i\omega_g + K)} = \frac{K L_{RED} (R_0 C_\ell)^3}{(2N)^2} \frac{e^{-i\omega_g R_0}}{(i\omega_g + K)}$$

Since the loop transfer-function $L(\cdot)$ is Nyquist-stable for the RED scheme, the parameters ω_g, L_{RED}, and K are selected such that $|L(i\omega_g)| = 1$. That is

$$\left| \frac{K L_{RED} (R_0 C_\ell)^3}{(2N)^2 (i\omega_g + K)} \right| = 1 \tag{5.14c}$$

In addition, its phase margin is required to be high. That is,

$$\left(\pi - \omega_g R_0 - \tan^{-1} \omega_g/K\right)$$

should be positive and high. These requirements provide guidelines for selecting parameters in the RED scheme. The reader should be aware that high-frequency effects were not considered in this analysis.

Analysis of Modified RED Schemes

Two modified RED schemes are analysed. In these schemes let the parameter K tend to infinity, that is the value of the parameter α used in averaging the queue length is unity. This implies that the packet-dropping scheme uses only the current value of the queue length to determine if a packet has to be dropped or if it enters the queue.

Scheme 1: In this scheme, let the loop transfer-function $L(\cdot)$ is specified by

$$L(s) = \frac{K_{s1} \left(C_\ell^2/2N\right) e^{-sR_0}}{(s + 1/R_0)(s + \lambda_0)}$$

where K_{s1} is a constant. Observe that this loop transfer-function is stable. The loop's unity-gain crossover frequency is selected to be the geometric mean of the frequencies $1/R_0$ and λ_0. Thus

$$\omega_g = \sqrt{\frac{\lambda_0}{R_0}} = \sqrt{\frac{2N}{R_0^3 C_\ell}} \tag{5.15}$$

The constant K_{s1} is selected such that $|L(i\omega_g)| = 1$. Since $\lambda_0 = 2/(W_0 R_0)$, and typically $W_0 > 2$, therefore $\lambda_0 < 1/R_0$. This implies that $\omega_g R_0 < 1$. Note that if $L(i\omega_g) = |L(i\omega_g)| e^{i\eta}$, then it can be shown that $\eta = -\pi/2 - \omega_g R_0$. Thus $(\pi - |\eta|)$ is positive, and lower-bounded by $(\pi/2 - 1)$, which is positive. Therefore, the design parameters should be selected such that the phase margin $(\pi/2 - \omega_g R_0)$ is as high as possible.

Scheme 2: In this scheme, the loop transfer-function is designed such that

$$L(s) = \frac{K_{s2} \left(C_\ell^2/2N\right) e^{-sR_0}}{(s + 1/R_0)(s + \lambda_0)} \left\{ \frac{(s + \lambda_0)}{\lambda_0 s} \right\}$$

The corresponding loop's unity-gain crossover frequency is selected to be

$$\omega_g = \frac{\varkappa}{R_0} \tag{5.16a}$$

where $\varkappa$ is determined from the positivity requirement of the phase margin. The constant K_{s2} is selected such that $|L(i\omega_g)| = 1$. This yields

$$\frac{K_{s2}\left(C_\ell^2/2N\right)}{|i\omega_g + 1/R_0|\,\lambda_0\omega_g} = 1$$

Note that if $L(i\omega_g) = |L(i\omega_g)|\,e^{i\eta}$, it can be shown that

$$\eta = -\frac{\pi}{2} - \omega_g R_0 - \tan^{-1}(\omega_g R_0) = -\frac{\pi}{2} - \varkappa - \tan^{-1}\varkappa \tag{5.16b}$$

Let the phase margin be equal to $\left(\pi/2 - \varkappa - \tan^{-1}\varkappa\right)$. This value is positive, if $\varkappa \in (0, 0.86)$. A mean-field model of multiple TCP/IP connections is discussed in the next section.

5.7 A Mean-Field Model of Multiple TCP/IP Connections

A mean-field model of multiple TCP connections at steady-state is studied in this section. This model includes the RED scheme, the packet tail-drop mechanism, and their combination. An expression for the window-size probability density function is obtained. The queue-length dynamics for each user connection is also described. The assumptions of the model are initially outlined.

We consider the TCP Reno in the congestion avoidance phase. Each connection in this scheme implements a window flow control. The window size limits the number of packets which are transmitted without an acknowledgment from the receiver. Thus the window size per user (connection) places an upper limit on the number of transmitted packets during a single round-trip time (RTT). The window size of a connection increases by a single packet per round-trip time, provided no packet loss occurs. For simplicity, we assume that no time-outs occur in the model. However, this feature can be included in the model. When a packet loss occurs due to active queue management or the tail-drop mechanism, the window size is halved. This model also assumes that the round-trip times of packet acknowledgment are not too high. This is a requirement for the analysis of a stable system.

We assume that the window size of each connection is written in each packet from which it originates. Furthermore, the slow-start phase of the user connection, and the negotiated maximum window size is also neglected. Scheduling of packets in the router queue is first in first out (FIFO). It is also assumed that the N users (flows) are downloading a file of infinite size. That is, the buffer is assumed to have continuous data (fluid approximation). This implies that the source router is a bottleneck.

Let the window size of connection i at time t be equal $W_i(t)$, where $i = 1, 2, \ldots, N$. The window size is equal to the number of packets transmitted during a single round-trip time. The round-trip time between the source and the receiver is assumed to be larger than the time required to transmit all the packets in the window. The link rate of the router is assumed to be C_ℓ packets per unit time, and the link rate allocated to each user is assumed to be C_N packets per unit time, where $C_\ell = NC_N$. We initially develop stochastic differential equations to describe window-size evolution. If N is very large, it turns out that the stochastic differential equations become deterministic.

Let the buffer space allocated to each flow be B_{bsp} units of packets. The average queue size per flow at time t is $Q_N(t)$ in units of packets. The packet drop probability due to a full ith flow-buffer and/or the RED scheme is denoted by $F(Q_N(t))$. The tail-drop probability $F(q) = 0$ if $q < B_{bsp}$, and $F(q) = 1$ if $q \geq B_{bsp}$.

The packet drop probabilities under the RED scheme have been described in the last section. Its parameters where $\sigma_{\min}, \sigma_{\max}$, and $p_{\max}$. The RED scheme was implemented on the basis of all the N cumulative flows. However, we modify it by implementing the RED scheme on a per-connection basis. We make the packet drop probability $F(Q_N(t))$ equal to zero for $Q_N(t)$ below $\sigma_{\min}/N$. The drop probability increases to $p_{\max}$ at $Q_N(t) = \sigma_{\max}/N$, and equal to unity if $Q_N(t) = B_{bsp}$. Packet loss during transmission of packets is not considered in this model.

Denote the round-trip time of connection i at time t be $R_i(t)$. The round-trip time of the packet is the sum of propagation delay from the source to the destination and back, plus the queueing delay at the source router. The propagation delay of each connection i is assumed to be a constant. Let it be equal to a_{ϖ} units of time. The instantaneous throughput of a connection at time t is its window size divided by the RTT at time t.

The window-size evolution is described next. This is followed by a derivation of the differential equation to describe the window queue-size. Subsequently, a partial differential equation of the window-size probability density function is also obtained.

5.7.1 Window-Size Evolution

The window-size reduction at connection i is assumed to occur according to a nonhomogeneous Poisson process with stochastic intensity

$$\frac{W_i(t - R_i(t))}{R_i(t - R_i(t))} F(Q_N(t - R_i(t))) \tag{5.17a}$$

where $W_i(t) = 0$ for $t < 0$.

The above expression implies that the packet loss rate is proportional to the transmission rate $(W_i(t - R_i(t)) / R_i(t - R_i(t)))$ a single RTT in the past, multiplied by the packet drop probability $F(Q_N(t - R_i(t)))$ one RTT in the past. Therefore the packet loss from the N connections can be described by N mutually independent time-varying (nonhomogeneous) Poisson processes. These processes are denoted by $\{N_i(t), t \in \mathbb{R}_0^+\}$, where $i = 1, 2, \ldots, N$. The expectation of $N_i(t)$ is

$$\Lambda_i(t) = \int_0^t \frac{W_i(y - R_i(y))}{R_i(y - R_i(y))} F(Q_N(y - R_i(y)))\, dy \tag{5.17b}$$

We next describe the evolution of the window size by making a fluid approximation of the queue. If a packet loss does not occur, then the window size increases linearly at rate $1/R_i(t)$. However, if there is an indication of packet drop, then the packet transmitted at time $(t - R_i(t))$ from connection i is lost. In this case, the window size is halved. Therefore the evolution of the window size is described by the following equation.

$$dW_i(t) = \frac{dt}{R_i(t)} - \frac{W_i(t)}{2} dN_i(t) \tag{5.17c}$$

where $W_i(0), i = 1, 2, \ldots, N$ is known. Note that in the above expression, we ignored the possibility of the event that in the RTT before time t, the connection i window entered fast retransmit/fast recovery phase due to loss of packets in the preceding RTT. This is a plausible assumption, if the packet loss rate is small.

The round-trip time of each connection i is

$$R(t) = a_{\varpi} + \frac{Q_N(t - R(t))}{C_N} \tag{5.17d}$$

The above stochastic dynamical system is next approximated by a fluid-flow model. We develop differential equations for the queue size and window-size probability density function for time $t \in \mathbb{R}_0^+$, as $N \to \infty$. As the total number of TCP connections $N \to \infty$ let: $Q_N(t) \to q(t)$, and $R(t) \to r(t)$ for $t \in \mathbb{R}_0^+$. Also define $F(q(t)) = k(t)$. Let the continuous probability density of the window size w at time t be $p(t, w)$. We further assume that $q(t), r(t), k(t)$ and $p(t, w)$ are continuous functions of t. The functions $q(\cdot)$ and $p(\cdot, \cdot)$ are also assumed to be differentiable in t. Thus

$$r(t) = a_{\varpi} + \frac{q(t - r(t))}{C_N}$$

In the next two subsections we develop differential equations for the queue size $q(t)$ and the window probability density $p(t, w)$.

5.7.2 Differential Equation of the Queue Size

If there are no packet losses, the connection i fills the buffer at the rate $W_i(t)/R(t)$. The proportion of lost packets is $F(Q_N(t))$. In addition, the buffer is emptied at the rate of C_N packets per unit time. If $Q_N(t) > 0$, the rate of change of buffer occupancy due to connection i is $G_i(t)$, where

$$G_i(t) = \frac{W_i(t)}{R(t)} \{1 - F(Q_N(t))\} - C_N$$

Then the net rate of change of buffer occupancy is

$$\begin{aligned} N\frac{dQ_N(t)}{dt} &= \sum_{i=1}^{N} G_i(t) + \left\{ \sum_{i=1}^{N} G_i(t) \right\}^{-} I_{Q_N(t)=0} \\ &= \sum_{i=1}^{N} \frac{W_i(t)}{R(t)} \{1 - F(Q_N(t))\} - NC_N \\ &\quad + \left\{ \sum_{i=1}^{N} \frac{W_i(t)}{R(t)} \{1 - F(Q_N(t))\} - NC_N \right\}^{-} I_{Q_N(t)=0} \end{aligned}$$

where $I_{Q_N(t)=0}$ is the indicator function of $Q_N(t) = 0$. The second term in the above equation ensures that the queue size does not become negative. The notation $x^- \triangleq \max(0, -x)$, is used in the above equation. Thus

$$\begin{aligned} \frac{dQ_N(t)}{dt} &= \frac{1}{N} \sum_{i=1}^{N} \frac{W_i(t)}{R(t)} \{1 - F(Q_N(t))\} - C_N \\ &\quad + \left\{ \frac{1}{N} \sum_{i=1}^{N} \frac{W_i(t)}{R(t)} \{1 - F(Q_N(t))\} - C_N \right\}^{-} I_{Q_N(t)=0} \end{aligned}$$

Letting $N \to \infty$, and integrating, we obtain

$$q(t) - q(0) = \int_{y=0}^{t} \left\{ \int_w \frac{w}{r(y)} p(y, w)\, dw \{1 - k(y)\} - C_N + \left\{ \int_w \frac{w}{r(y)} p(y, w)\, dw \{1 - k(0)\} - C_N \right\}^- I_{Q_N(t)=0} \right\} dy \tag{5.18}$$

Thus we have the following result.

Observation 5.1. The differential equation for queue size $q(t), t \in \mathbb{R}_0^+$ as the number of TCP connections $N \to \infty$ is

$$\frac{dq(t)}{dt} = \int_w \frac{w}{r(t)} p(t, w)\, dw \{1 - k(t)\} - C_N - \left\{ \int_w \frac{w}{r(t)} p(t, w)\, dw \{1 - k(0)\} - C_N \right\}^- I_{Q_N(t)=0} \tag{5.19}$$

where $r(t) = a_{\varpi} + q(t - r(t))/C_N$. Furthermore, if $F(q(t)) = 1$ when $q(t) > \sigma_{\max}/N$, then $k(t)$ is determined by $\int_w wp(t, w)\, dw \{1 - k(t)\}/r(t) = C_N$. Note that the value $\sigma_{\max}/N$ is used in implementing the RED scheme. □

It remains to determine the probability density function of the window size $p(\cdot, \cdot)$.

5.7.3 Window-Size Probability Density Function

The window-size dynamics is more complicated to describe than the queue-size dynamics. The window-size probability density function is described in the following observation.

Observation 5.2. As the number of TCP connections $N \to \infty$, the differential equation for window-size probability density $p(t, w)$ at time $t \in \mathbb{R}_0^+$ is

$$\frac{\partial p(t, w)}{\partial t} = \left\{ p(t, 2w) \frac{2w}{r(t - r(t))} 2 - p(t, w) \frac{w}{r(t - r(t))} \right\} k(t - r(t)) - \frac{1}{r(t)} \frac{\partial p(t, w)}{\partial w} \tag{5.20}$$

where $r(t) = a_{\varpi} + q(t - r(t))/C_N$. It is assumed that the initial window probability density function is continuous. The above equation is called the Fokker-Planck equation of the TCP Reno window probability density function. □

The derivation of this observation is given in the problem section.

5.7.4 Analysis of Stable System

If the round-trip time is sufficiently small, the differential equation of the queue size, and the partial differential equation of the window-size probability density function further simplify. In this case,

$q(t)$ tends to a constant. This in turn implies that the round-trip time $r(t)$, and the packet loss probability $F(q(t))$ also tend to constants r and k respectively. Therefore, if the system is stable, define $f(w) = p(t,w)$. Thus

$$C_N = \frac{(1-k)}{r} \int_w w f(w)\, dw$$

$$\frac{df(w)}{dw} = k\{4wf(2w) - wf(w)\}$$

The relationship $C_N = (1-k)\int_w wf(w)\,dw/r$ is actually Little's law. It says that at stead-state, the packet throughput is equal to the product of the fraction of packets which are not lost, and the ratio of the average window size and the round-trip time. The next set of observations give an explicit expression for the window-size probability density function for a stable system. This explicit expression for $f(w)$ is evaluated by using the Mellin transform.

Observations 5.3. Related results, and unique solution of the differential equation

$$\frac{df(w)}{dw} = k\{4wf(2w) - wf(w)\} \tag{5.21}$$

is specified, where $\int_w f(w)\,dw = 1$.

1. Let the Mellin transform of $f(w)$ be $F_M(s) = \int_0^\infty w^{s-1} f(w)\,dw$, $\operatorname{Re}(s) \geq 1$. Thus

$$F_M(s) = \left(\frac{2}{k}\right)^{(s-1)/2} \frac{\Gamma(s/2)}{\sqrt{\pi}} \prod_{j\in\mathbb{N}} \frac{\left(1-2^{-2j-s}\right)}{\left(1-2^{-2j-1}\right)} \tag{5.22}$$

2. Let

$$f(w) = \sum_{i\in\mathbb{N}} b_i \exp\left(-k4^i \frac{w^2}{2}\right) \tag{5.23a}$$

then

$$\Phi_{ML} = \prod_{j\in\mathbb{N}} \left(1-2^{-2j-1}\right) = \sum_{i\in\mathbb{N}} \frac{2^i}{\prod_{j=1}^{i}(1-4^j)} = 0.41942\ldots \tag{5.23b}$$

$$b_0 = \left(\frac{2k}{\pi}\right)^{1/2} \frac{1}{\Phi_{ML}} \tag{5.23c}$$

$$b_i = b_{i-1}\frac{4}{1-4^i} = b_0 \frac{4^i}{\prod_{j=1}^{i}(1-4^j)}, \quad \forall\, i \in \mathbb{P} \tag{5.23d}$$

3. The average value of the window size is equal to $\mathcal{E}(W) = F_M(2)$. Therefore, an approximate value of $\mathcal{E}(W)$ is

$$\sqrt{\frac{2}{k\pi}}1.641632\ldots = \frac{1.3098\ldots}{\sqrt{k}}$$

This result is in consonance with the result derived in an earlier section. □

These observations are established in the problem section.

5.8 Traffic Rate, Congestion Management, and Stability

This section describes issues related to traffic rate, congestion management, and stability in networks. It is based upon the fundamental work of Frank Kelly and his collaborators. Different congestion management (control) strategies raise the issue of stability of traffic flow. Useful insight can be gained in such issues by modeling the complete network via a system of differential equations. Traffic-rate control, stability, and fairness in capacity management is studied in this section by considering all the nodes and links of the network. Consider a network specified by an undirected graph $G = (V, E)$, where the vertex set V represents the routers, and the edge set E represents the transmission links. The number of edges in the graph is $|E|$, and this edge set is $\{1, 2, \ldots, |E|\}$. Associated with each edge $e \in E$ is its transmission capacity $c_e \in \mathbb{R}^+$. Its units are in bits per unit time (bit rate).

Denote the set of users who access this network by $\mathfrak{R}$. Therefore the number of users in this set is $|\mathfrak{R}|$. Let this user set be $\{1, 2, \ldots, |\mathfrak{R}|\}$. Associated with each user $r \in \mathfrak{R}$ is a traffic demand $\rho_r \in \mathbb{R}^+$ in units of bits per unit time. This traffic demand has to be transmitted between a source-destination pair. The actual allocation of this demand in the network is $\beta_r \in \mathbb{R}^+$ where $\beta_r \leq \rho_r$. For simplicity assume that user r's allocation β_r, follows a single path (route) π_r in the network. Recall that a path can be specified as an ordered sequence of links. If a link $e \in E$ belongs to a path π_r, then it is denoted by $e \in \pi_r$. Denote the set of these routes by $\mathcal{P} = \{\pi_1, \pi_2, \ldots, \pi_{|\mathfrak{R}|}\}$, where $|\mathcal{P}| = |\mathfrak{R}|$. Since there is a one-to-one correspondence between a user and its traffic route, the users and the corresponding routes are indexed by r in rest of this section. The set of links associated with each route is specified by a $|E| \times |\mathfrak{R}|$ matrix $A = [a_{er}]$, where $a_{er} = 1$ if $e \in \pi_r$. That is, $a_{er} = 1$ if e lies on route $r \in \mathfrak{R}$, and $a_{er} = 0$ otherwise.

Associated with each user r is its utility $U_r(\beta_r)$. That is, utility of the user r is a function of β_r. Assume that this utility function is an increasing, strictly concave, and continuously differentiable function of β_r. Traffic that leads to such utility function is called *elastic traffic*. Let $U_r'(\beta_r)$ denote the first derivative of $U_r(\beta_r)$ with respect to β_r. Also $U_r'(\beta_r)$ should be nonnegative for all values of $\beta_r \in \mathbb{R}^+$in order to find the optimum value of β_r. In addition assume that the utilities are additive, so that the net utility of all the users in the set $\mathfrak{R}$ is $\sum_{r \in \mathfrak{R}} U_r(\beta_r)$. Denote the set of utilities by $U = \{U_r(\cdot) \mid r \in \mathfrak{R}\}$.

Further assume that the user $r \in \mathfrak{R}$ with a traffic allocation of β_r is charged by the network, an amount w_r per unit time. Also assume that w_r is directly proportional to β_r. Therefore let $w_r = \lambda_r \beta_r$, where $\lambda_r \in \mathbb{R}^+$ is the charge per unit flow for the user r. For convenience in notation, the following vectors are defined

$$\beta \triangleq \left(\beta_1, \beta_2, \ldots, \beta_{|\mathfrak{R}|}\right)$$

$$U(\beta) \triangleq \left(U_1(\beta_1), U_2(\beta_2), \ldots, U_{|\mathfrak{R}|}\left(\beta_{|\mathfrak{R}|}\right)\right)$$

$$U'(\beta) \triangleq \left(U_1'(\beta_1), U_2'(\beta_2), \ldots, U_{|\mathfrak{R}|}'\left(\beta_{|\mathfrak{R}|}\right)\right)$$

$$\lambda \triangleq \left(\lambda_1, \lambda_2, \ldots, \lambda_{|\mathfrak{R}|}\right)$$

$$w \triangleq \left(w_1, w_2, \ldots, w_{|\mathfrak{R}|}\right)$$

$$C \triangleq \left(c_1, c_2, \ldots, c_{|E|}\right)$$

$$\mu \triangleq \left(\mu_1, \mu_2, \ldots, \mu_{|E|}\right)$$

The interpretation of the vector μ is given in the next subsection. In a misuse of notation, the above symbols (for vectors) are also used to represent corresponding column vectors. Also note that a traffic unit need not necessarily be a bit. The flow or rate vector β is said to be *feasible* if it satisfies the constraint $A\beta \leq C$, and $\beta \geq 0$.

5.8.1 Traffic Rate Control and Fairness

Interaction between traffic rate control and fairness is studied in this subsection. Assume that the *network* represents the physical resources, like the nodes and the transmission links; while the *system* includes the complete set of *users* as well as the *network*. It is shown that the formulation of an overall system optimization problem can be decomposed into problems of optimization at the user's level and at the network level. This is useful in practice, because in a very large network, it is difficult to know the utility function and determine optimal traffic allocation for each user.

It turns out that in this model, each user's utility value, network resources, and the overall system resources can be optimized separately. Define $\mathfrak{B}(w)$ to be the set of users for which their charges per unit time are positive. That is, $\mathfrak{B}(w) \triangleq \{r \in \mathfrak{R} \mid w_r > 0\}$. The aforementioned optimization problems are explicitly stated below.

(a) *Single User Optimization Problem*: A user $r \in \mathfrak{R}$, is specified by its utility function $U_r(\cdot)$, and $\lambda_r \in \mathbb{R}^+$. It aims to maximize $\{U_r(w_r/\lambda_r) - w_r\}$, over $w_r \geq 0$. That is, the user $r \in \mathfrak{R}$, selects w_r amount per unit time to pay for a flow of w_r/λ_r units per unit time. The utility of flow w_r/λ_r is $U_r(w_r/\lambda_r)$.

(b) *Network Optimization Problem*: The network is specified by A, C, and w. It aims to maximize $\sum_{r\in\mathfrak{B}(w)} w_r \ln \beta_r$, given that $A\beta \leq C$, and $\beta \geq 0$. That is, the network maximizes an aggregate of weighted logarithmic utility functions, where the w_r's are selected by the users. Observe that the network does not require a knowledge of the users' utility functions, but is aware of their charge (amount) per unit time. In addition, it uses the $\ln(\cdot)$ function in its objective function. As we shall study subsequently, this leads to the concept of *weighted proportionally fair* criteria of optimization.

(c) *System Optimization Problem*: The system is specified by U, A, and C. It aims to maximize $\sum_{r\in\mathfrak{R}} U_r(\beta_r)$, given that $A\beta \leq C$, and $\beta \geq 0$. That is, the system maximizes aggregate utility of all users, subject to the capacity constraints.

These optimization scenarios can be solved independently via Lagrangian techniques. In what follows, it will be demonstrated that these apparently distinct constrained optimization problems are interrelated. Consequently, the network optimization problem can be reduced to a distributed implementation of single user control. In addition, the Lagrangian dual of the network optimization problem is also determined. Its relationship with the above three problems is also indicated. These problems and their *solutions* are explicitly stated below. See the problem section for the complete solution of these four optimization problems.

Single User Optimization Problem

The optimization problem for a single user $r \in \mathfrak{R}$ is $USER_r[U_r; \lambda_r]$, where $\lambda_r \in \mathbb{R}^+$. This optimization problem is:

$$\max U_r\left(\frac{w_r}{\lambda_r}\right) - w_r \tag{5.24a}$$

over $w_r \geq 0$.

The variable w_r can be found from the set of conditions, labeled $\mathcal{C}_{USER}$.

$$\mathcal{C}_{USER}: \ w_r \geq 0, \ \lambda_r \geq U_r'\left(\frac{w_r}{\lambda_r}\right), \ w_r\left\{U_r'\left(\frac{w_r}{\lambda_r}\right) - \lambda_r\right\} = 0 \tag{5.24b}$$

In this optimization problem w_r is said to solve $USER_r\left[U_r; \lambda_r\right]$ for $r \in \mathfrak{R}$. □

Network Optimization Problem

The network optimization problem is specified by $NETWORK\left[A, C; w\right]$. This optimization problem is:

$$\max \sum_{r \in \mathfrak{B}(w)} w_r \ln \beta_r \tag{5.25a}$$

subject to $A\beta \leq C$, and $\beta \geq 0$.

The vector β can be found from the set of conditions, labeled $\mathcal{C}_{NET}$.

$$\begin{aligned} \mathcal{C}_{NET}: \quad & A\beta \leq C, \quad \beta \geq 0, \ \mu^T(C - A\beta) = 0, \ \mu \geq 0; \\ & \lambda_r = \textstyle\sum_{e \in \pi_r} \mu_e, \text{ and } w_r = \lambda_r \beta_r, \quad \text{for } r \in \mathfrak{R} \end{aligned} \tag{5.25b}$$

For a fixed set of users and arbitrary vector w, the rate allocation vector β solves the problem $NETWORK\left[A, C; w\right]$. In the network optimization problem the scalars μ_e can be interpreted as the *implied cost* of unit flow through link e. The μ_e's are also referred to as the *shadow price* per unit of flow through link e. □

System Optimization Problem

The system optimization problem is specified by $SYSTEM\left[U, A, C\right]$. The optimization problem is:

$$\max \sum_{r \in \mathfrak{R}} U_r(\beta_r) \tag{5.26a}$$

subject to $A\beta \leq C$, and $\beta \geq 0$.

The vector β can be found from the set of conditions, labeled $\mathcal{C}_{SYSTEM}$.

$$\begin{aligned} \mathcal{C}_{SYSTEM}: \quad & A\beta \leq C, \quad \beta \geq 0, \ \widetilde{\mu}^T(C - A\beta) = 0, \ \widetilde{\mu} \geq 0 \\ & U'(\beta) \leq \widetilde{\lambda}, \ \beta^T\left(U'(\beta) - \widetilde{\lambda}\right) = 0 \\ & \widetilde{\mu} = \left(\widetilde{\mu}_1, \widetilde{\mu}_2, \ldots, \widetilde{\mu}_{|E|}\right) \\ & \widetilde{\lambda}_r \triangleq \textstyle\sum_{e \in \pi_r} \widetilde{\mu}_e, \ r \in \mathfrak{R}, \ \widetilde{\lambda} \triangleq \left(\widetilde{\lambda}_1, \widetilde{\lambda}_2, \ldots, \widetilde{\lambda}_{|\mathfrak{R}|}\right) \end{aligned} \tag{5.26b}$$

□

Lagrangian Dual of the Network Optimization Problem

The Lagrangian dual of the network optimization problem $NETWORK\left[A, C; w\right]$ is denoted by $DUAL\left[A, C; w\right]$. Let

$$\omega(\mu) = -\sum_{r\in\mathfrak{B}(w)} w_r \ln\left\{\sum_{e\in\pi_r} \mu_e\right\} + \sum_{e\in E} \mu_e c_e - \sum_{r\in\mathfrak{B}(w)} w_r (1 - \ln w_r) \tag{5.27a}$$

The dual is:

$$\min_{\mu} \omega(\mu) \tag{5.27b}$$

over $\mu \geq 0$. □

Observe that the last term of $\omega(\mu)$ does not affect the value of μ, when $\omega(\mu)$ is minimized. However this term which is independent of μ, affects the value of the objective function. Equivalently, the dual problem is

$$\max_{\mu}\left\{\sum_{r\in\mathfrak{B}(w)} w_r \ln\left(\sum_{e\in\pi_r} \mu_e\right) - \sum_{e\in E} \mu_e c_e\right\}$$

over $\mu \geq 0$.

The next theorem establishes a relationship between the above seemingly different optimization problems. This theorem demonstrates that the $SYSTEM[U, A, C]$ optimization problem is decomposed into:

$$NETWORK[A, C; w], \text{ and } USER_r[U_r; \lambda_r], \ r \in \mathfrak{R}$$

subproblems. It justifies an earlier claim, that the overall system optimization problem can be decomposed into problems of optimization at the user's level and at the network level.

Theorem 5.1. (*System decomposition theorem*) *There exist vectors w, λ, and β such that*:

(a) *w_r is obtained by solving $USER_r[U_r; \lambda_r]$, for $r \in \mathfrak{R}$.*
(b) *The vector μ is obtained from $DUAL[A, C; w]$.*
(c) *The traffic allocation vector β is obtained from $NETWORK[A, C; w]$.*
(d) *$w_r = \beta_r \lambda_r$, where $\lambda_r = \sum_{e\in\pi_r} \mu_e$, $r \in \mathfrak{R}$.*

The vector β solves $SYSTEM[U, A, C]$.
Proof. See the problem section. □

User's Utility Functions

The concept of fairness among different traffic flows is next examined by using the formulation of the above optimization problems. Users' utility functions can be used to determine the specified fairness objective. A convenient class of utility functions $U_r(\cdot)$, for user $r \in \mathfrak{R}$ is

$$U_r(\beta_r) = \begin{cases} \dfrac{w_r \beta_r^{(1-\varphi)}}{(1-\varphi)}, & \varphi \in \mathbb{R}_0^+ \backslash \{1\} \\ w_r \ln \beta_r, & \varphi = 1 \end{cases} \tag{5.28a}$$

where $w_r \in \mathbb{R}_0^+$, and the parameter φ is a measure of fairness in allocation of capacity (demand) to different users. This utility function is nondecreasing and concave function of β_r. Note that, if $U_r'(\beta_r)$ is the first derivative of $U_r(\beta_r)$ with respect to β_r, then

$$U_r'(\beta_r) = \frac{w_r}{\beta_r^{\varphi}}, \quad \beta_r \in \mathbb{R}^+, \text{ and } \varphi \in \mathbb{R}_0^+ \tag{5.28b}$$

For simplicity, assume that the utility functions of all $|\mathfrak{R}|$ users have the same value of φ. The goal is to provide optimal capacity sharing by maximizing the sum of utilities of all the users, which is $\sum_{r\in\mathfrak{R}} U_r(\beta_r)$. Let β be the optimal traffic allocation vector, and $\widetilde{\beta}$ be another feasible flow vector. Let the optimal flow β_r be perturbed by $\delta\beta_r$, such that $\widetilde{\beta}_r = (\beta_r + \delta\beta_r)$. Therefore, the net change in this objective function at β is

$$\sum_{r\in\mathfrak{R}} w_r \frac{\delta\beta_r}{\beta_r^{\varphi}} = \sum_{r\in\mathfrak{R}} w_r \left\{ \frac{\widetilde{\beta}_r - \beta_r}{\beta_r^{\varphi}} \right\} \tag{5.29a}$$

Since the objective function $\sum_{r\in\mathfrak{R}} U_r(\beta_r)$, where $\beta_r \in \mathbb{R}^+$, and $r \in \mathfrak{R}$ is concave, and β is the optimal traffic allocation vector, the above expression is nonpositive. That is

$$\sum_{r\in\mathfrak{R}} w_r \left\{ \frac{\widetilde{\beta}_r - \beta_r}{\beta_r^{\varphi}} \right\} \le 0, \quad \beta_r \in \mathbb{R}^+, \text{ and } \varphi \in \mathbb{R}_0^+ \tag{5.29b}$$

The above inequality is called the *φ-fairness* property of the flow allocation. We next consider some special cases. If $\varphi = 0$, then overall throughput is maximized. Choice of $\varphi = 1$ results in the weighted proportional fairness. Optimization with $\varphi = 2$ yields *minimum potential delay fairness*. If $\varphi \to \infty$ and $w_r = 1$ for each $r \in \mathfrak{R}$, then we have the *max-min fairness* criteria. Observe that small values of the parameter φ emphasize network utilization at the expense of individual flows. However, larger values of the parameter φ favour users with lower values of β_r. We next elaborate upon these cases.

Case 1: If $\varphi = 0$, then the goal is to maximize the overall throughput $\sum_{r\in\mathfrak{R}} w_r\beta_r$, provided the link capacity constraints are satisfied. This is perhaps not a fair solution.

Case 2: The choice of $\varphi = 1$ implies implies maximization of $\sum_{r\in\mathfrak{R}} w_r \ln \beta_r$, where $\beta_r \in \mathbb{R}^+$. This is the weighted proportional fairness criteria. The weighted proportionally fair criteria provides a compromise between fairness and optimality. Let β be the optimal traffic allocation vector for this problem, and $\widetilde{\beta}$ be another feasible flow vector. If the optimal flow β_r is perturbed by $\delta\beta_r$, such that $\widetilde{\beta}_r = (\beta_r + \delta\beta_r)$, then the net change in this objective function at β is

$$\sum_{r\in\mathfrak{R}} w_r \frac{\delta\beta_r}{\beta_r} = \sum_{r\in\mathfrak{R}} w_r \left\{ \frac{\widetilde{\beta}_r - \beta_r}{\beta_r} \right\} \le 0 \tag{5.29c}$$

Thus if the traffic allocation vector deviates from the optimal vector β to another feasible vector $\widetilde{\beta}$, then the weighted sum of proportional changes in each user's rate is nonpositive. Therefore the traffic allocation corresponding to the utility $U_r(\beta_r) = w_r \ln \beta_r$ is called weighted proportionally fair. If w_r is equal to unity for each $r \in \mathfrak{R}$, then the allocation is simply called *proportionally fair*.

Case 3: $\varphi = 2$ implies maximization of $-\sum_{r\in\mathfrak{R}} w_r/\beta_r$, which is same as minimization of $\sum_{r\in\mathfrak{R}} w_r/\beta_r$. This is *minimum potential delay fairness*. If w_r is interpreted as the file size, then w_r/β_r is equal to the transfer time of the file. Thus the goal of the optimization problem in this case is to minimize the sum of file transfer times (or delays).

Case 4: $\varphi \to \infty$ and $w_r = 1$ for each $r \in \mathfrak{R}$ yields the max-min fairness criteria. The corresponding value of the optimized objective function is $\max_\beta \min_{r\in\mathfrak{R}} \beta_r$. Max-min fairness is a popular form of fairness criteria. A traffic-rate allocation is said to be max-min fair if a user's rate cannot be incremented without decreasing the rate of another user who is receiving a smaller rate.

This definition of max-min fairness implies that absolute priority is given to users with smaller rates. However, max-min fairness comes at the cost of suboptimality of the network resources. The max-min fairness criteria is specified more formally as follows. The flow vector β is said to be max-min fair if β is feasible, and in addition a flow β_k cannot be made larger while maintaining feasibility without decreasing β_i for some flow $i \neq k$ such that $\beta_i \leq \beta_k$.

Alternately, let β be a max-min fair flow vector, and $\widetilde{\beta}$ be another feasible vector which satisfies the capacity constraints. Then if $\beta_k < \widetilde{\beta}_k$ for some $k \in \mathfrak{R}$, then there exists $i \neq k$, such that $\widetilde{\beta}_i < \beta_i \leq \beta_k$.

The fact that $\varphi \to \infty$ indeed leads to the max-min fairness criteria is established in the problem section.

Implementation of the NETWORK and DUAL Algorithms

The optimization problems $NETWORK\left[A, C; w\right]$ and $DUAL\left[A, C; w\right]$ appear to be mathematically tractable in principle. However, the size of a network might possibly pose a significant challenge in its implementation on a centralized processor. An alternative would be to design decentralized algorithms. However, decentralized systems pose the problem of stochastic perturbations, and time lags. Nevertheless such models provide significant insights into the behavior of the networks.

5.8.2 Congestion Control via Feedback and Its Stability

Congestion control in a complete network, via feedback is studied in this subsection. Traffic flow in a network is modeled via a set of differential equations, and its stability is evaluated by using Lyapunov's techniques. Since the model which is about to be described is dynamical, some variables introduced at the beginning of this section are functions of time t. For simplicity in presentation, it is assumed that user r's traffic flow rate is $\beta_r\left(t\right) \in \mathbb{R}_0^+$ at time t, and its path is π_r, where $r \in \mathfrak{R}$. Two algorithms are discussed. These are the primal and the dual algorithms.

Primal Algorithm

Consider a link $e \in E$ in the network. As the link e becomes more heavily loaded, the network incurs increasing cost. The increase in load might result in an increase in delay or loss of packets, or the cost might be reflected in terms of additional equipment or resources that have to be allocated or purchased. Let $\mathcal{C}_e\left(y\right)$ be the rate at which cost is incurred at link e, where y is the total flow through it. Assume that $\mathcal{C}_e\left(y\right)$ is differentiable, and let

$$\frac{d\mathcal{C}_e\left(y\right)}{dy} = p_e\left(y\right), \quad e \in E \tag{5.30a}$$

where $p_e\left(y\right)$ is the price charged by the link e per unit flow. The function $p_e\left(\cdot\right)$ is assumed to be positive, continuous, strictly increasing function of its argument, and upper bounded by unity for each $e \in E$. Therefore $\mathcal{C}_e\left(\cdot\right)$ is a strictly convex function, and

$$\mathcal{C}_e\left(y\right) = \int_0^y p_e\left(x\right) dx, \quad e \in E \tag{5.30b}$$

Note that the total flow y through link e at time t is $\xi_e\left(t\right)$ where

$$y = \xi_e(t) = \sum_{r \in \mathfrak{R}} \sum_{e \in \pi_r} \beta_r(t) \tag{5.30c}$$

Also define

$$\mu_e(t) = p_e(\xi_e(t)), \quad e \in E \tag{5.30d}$$

We are now ready to consider a system of differential equations

$$\frac{d\beta_r(t)}{dt} = \kappa_r \left\{ w_r - \beta_r(t) \sum_{e \in \pi_r} \mu_e(t) \right\}, \quad r \in \mathfrak{R} \tag{5.30e}$$

These equations have the following interpretation expressed in the language of feedback systems theory. Suppose that the link e marks a fraction $p_e(y)$ of packets with a feedback signal when the total flow through the link e is y units. The flow (user) r interprets this feedback signal as an indication of congestion, and thus requires a reduction in the flow $\beta_r(t)$. The above equation has two components on its right-hand side. The first component represents a steady increase at rate proportional to $w_r > 0$. The second component represents a steady decrease at a rate proportional to the strength of the cumulative stream of congestion indication signals received by flow r. Furthermore, the parameter w_r represents the user's *willingness-to-pay,* and the parameter $\kappa_r > 0$ depends upon the round-trip time of the flow (between the source-destination pair). The Lyapunov function for the primal system of differential equations is established in the following theorem.

Theorem 5.2. (*Global stability of primal algorithm*) *If* $w_r \in \mathbb{R}^+$ *for each* $r \in \mathfrak{R}$, *then* $\mathcal{U}(\cdot)$ *is a Lyapunov function for the primal set of differential equations, where*

$$\mathcal{U}(\beta(t)) = \sum_{r \in \mathfrak{R}} w_r \ln \beta_r(t) - \sum_{e \in E} C_e(\xi_e(t)) \tag{5.31}$$

The unique value β *which maximizes* $\mathcal{U}(\beta(t))$ *is a stable point of the system to which all orbits (trajectories) of the dynamical system converge.*

Proof. See the problem section. □

The traffic-allocation on route $r \in \mathfrak{R}$ at the stable point is

$$\beta_r = \frac{w_r}{\sum_{e \in \pi_r} \mu_e} \tag{5.32}$$

The above relationship was obtained independently in the optimization of the primal problem $NETWORK[A, C; w]$. Recall that in the above equation, the variable μ_e is the shadow price per unit flow through link e. Also observe that the first term in $\mathcal{U}(\cdot)$ is the objective function of the $NETWORK[A, C; w]$ problem. Therefore the functions $p_e(\cdot), e \in E$ may be chosen such that the maximization of the Lyapunov function $\mathcal{U}(\cdot)$ arbitrarily closely approximates the optimization problem $NETWORK[A, C; w]$. In this spirit, the primal algorithm is a relaxation of the network problem. A dual of the above primal algorithm is next outlined.

Dual Algorithm

In the primal algorithm, traffic rates change gradually, and shadow prices are specified in terms of traffic rates. In the dual algorithm the shadow prices change gradually, but rates are given as functions of shadow prices. Let

$$\frac{d\mu_e(t)}{dt} = \kappa_e \left\{ \sum_{\substack{e \in \pi_r \\ r \in \mathfrak{R}}} \beta_r(t) - q_e(\mu_e(t)) \right\}, \quad e \in E \tag{5.33a}$$

where

$$\beta_r(t) = \frac{w_r}{\sum_{e \in \pi_r} \mu_e(t)}, \quad r \in \mathfrak{R} \tag{5.33b}$$

and $\kappa_e > 0$. Also $q_e(\eta)$ defined over $\eta \geq 0$, is a continuous and strictly increasing function of η, where $q_e(0) = 0$. It is the flow through link e which generates a price of η units per unit flow at link $e \in E$. Therefore $p_e(q_e(\eta)) = \eta$. Define

$$\mathcal{D}_e(\eta) = \int_0^{\eta} q_e(v)\, dv, \quad e \in E \tag{5.33c}$$

As in the case of the primal algorithm, the Lyapunov function for the dual system of differential equations can be established. It is stated in the following theorem.

Theorem 5.3. (*Global stability of dual algorithm*) *If $w_r \in \mathbb{R}^+$ for each $r \in \mathfrak{R}$, then $\mathcal{V}(\cdot)$ is a Lyapunov function for the dual set of differential equations, where*

$$\mathcal{V}(\mu(t)) = \sum_{r \in \mathfrak{R}} w_r \ln \left\{ \sum_{e \in \pi_r} \mu_e(t) \right\} - \sum_{e \in E} \mathcal{D}_e(\mu_e(t)) \tag{5.34}$$

The unique value μ which maximizes $\mathcal{V}(\mu(t))$ is a stable point of the system to which all orbits of the dynamical system converge.

Proof. See the problem section. □

At the stable point, the traffic-allocation β_r on route r is identical to that obtained for the primal algorithm. Also observe that the functions $q_e(\cdot), e \in E$ may be chosen such that the maximization of the Lyapunov function $\mathcal{V}(\cdot)$ arbitrarily closely approximates the optimization problem $DUAL[A, C; w]$. In this spirit, the dual algorithm is a relaxation of the $DUAL[A, C; w]$ problem.

User Interaction and Extension of the Algorithms

Next consider the stability of systems in which the users are able to vary w_r over time in response to the congestion experienced by them. Assume that the user's model $w_r(t)$ as

$$w_r(t) = \beta_r(t)\, U_r'(\beta_r(t)), \quad r \in \mathfrak{R} \tag{5.35a}$$

For this user the charge per unit flow $\lambda_r(t) = w_r(t)/\beta_r(t)$, and the corresponding user optimization problem is $USER_r[U_r; \lambda_r(t)]$. The modified differential equations for the primal and dual algorithms are obtained by replacing w_r by $w_r(t)$. It can be shown that both modified algorithms have a unique stable (equilibrium) point, which is identical for both algorithms. It can also be shown that the stable point maximizes the Lyapunov function $\mathcal{W}(\cdot)$, of the corresponding primal set of differential equations, where

$$\mathcal{W}(\beta(t)) = \sum_{r \in \mathfrak{R}} U_r(\beta_r(t)) - \sum_{e \in E} \mathcal{C}_e(\xi_e(t)) \tag{5.35b}$$

Also the first term in $\mathcal{W}(\cdot)$ is the objective function of $SYSTEM[U,A,C]$. Therefore the functions $p_e(\cdot), e \in E$ may be chosen such that the maximization of the Lyapunov function $\mathcal{W}(\cdot)$ arbitrarily closely approximates the optimization problem $SYSTEM[U,A,C]$. Therefore the time-varying version of the primal algorithm in this sense, is also a relaxation of the $SYSTEM[U,A,C]$ problem. It is also possible to determine similarly the Lyapunov function of the corresponding dual set of differential equations.

The dynamical system of equations model only macroscopic characteristics of traffic flow through the network. Furthermore, the set of differential equations which modeled different algorithms, did not consider the stochastic perturbations and time lags in the network. The rate of convergence of the algorithms to their unique equilibrium points were also not investigated. The mathematical model of this subsection was also not extended to the TCP algorithm. These issues and extension of these algorithms can be found in the literature.

5.8.3 Max-Min Fair Algorithms

We explicitly describe the max-min fair algorithm for capacity allocation in case of a single link, and also for a network. The max-min fairness concept is initially described for a single link. This is followed by the max-min fair algorithm for a network. The max-min fair allocation of capacity resource in a network consists of maximizing the network's use of user sessions with the minimum allocation.

Max-Min Fair Flow Allocation in a Single Link

The set of flows which have to access a link of capacity C_ℓ units is denoted by $\mathfrak{R}$. Let this set be $\{1,2,\ldots,|\mathfrak{R}|\}$. Also let the $|\mathfrak{R}|$ flow rates be $\rho_1,\rho_2,\ldots,\rho_{|\mathfrak{R}|}$. If $\sum_{i\in\mathfrak{R}}\rho_i \le C_\ell$, then no special flow allocation procedure is required. However, if $\sum_{i\in\mathfrak{R}}\rho_i > C_\ell$, the ith flow is allocated only β_i units of capacity, so that $0\le\beta_i\le\rho_i$ for $1\le i\le|\mathfrak{R}|$ and $\sum_{i\in\mathfrak{R}}\beta_i\le C_\ell$. A flow vector $\beta=\left(\beta_1,\beta_2,\ldots,\beta_{|\mathfrak{R}|}\right)$ which satisfies these constraints is said to be *feasible*. A max-min fair flow allocation vector is next defined.

Definition 5.1. *A flow vector* $\beta=\left(\beta_1,\beta_2,\ldots,\beta_{|\mathfrak{R}|}\right)$ *is said to be max-min fair if* $\beta_i\le\rho_i$, $\forall\, i\in\{1,2,\ldots,|\mathfrak{R}|\}$; *and for any* $i\in\{1,2,\ldots,|\mathfrak{R}|\}$, *the rate* β_i *cannot be made larger while maintaining feasibility without decreasing* β_j *for some flow* $j\ne i$ *such that* $\beta_j\le\beta_i$. □

An algorithm for max-min rate allocation in a single link is next outlined. The goal is to maximize the allocation of the link capacity to flows with minimum capacity (rate).

Algorithm 5.1. *Max-Min Fair Flow Allocation Algorithm in a Single Link.*

Input: A link with capacity C_ℓ, and flow rates $\rho_i, 1\le i\le|\mathfrak{R}|$, where $\rho_i\le\rho_{i+1}$ for each $i\in\{1,2,\ldots,|\mathfrak{R}|-1\}$. Assume that $\sum_{i\in\mathfrak{R}}\rho_i > C_\ell$.
Output: Max-min fair rates $\beta_1,\beta_2,\ldots,\beta_{|\mathfrak{R}|}$ such that $\sum_{i\in\mathfrak{R}}\beta_i\le C_\ell$, and $\beta_i\le\rho_i$ for each $i\in\{1,2,\ldots,|\mathfrak{R}|\}$.
begin

(*initialization*)
$\widetilde{W} \leftarrow \{1, 2, \ldots, |\mathfrak{R}|\}$
$m \leftarrow |\mathfrak{R}|$
$\widetilde{R} \leftarrow C_\ell$
($\widetilde{W}$ *denotes remaining flows.*)
(m *denotes the remaining number of flows.*)
($\widetilde{R}$ *denotes the remaining capacity.*)
while $m > 0$ **do**
begin
find the smallest flow $i \in \widetilde{W}$.
if $\rho_i < \widetilde{R}/m$ **then** $\beta_i \leftarrow \rho_i$ **else** $\beta_i \leftarrow \widetilde{R}/m$
$m \leftarrow (m - 1)$
$\widetilde{R} \leftarrow \widetilde{R} - \beta_i$
$\widetilde{W} \leftarrow \widetilde{W} \backslash \{i\}$
end (*end of while-loop*)
end (*end of single link max-min fair flow allocation algorithm*)

It is left to the reader to establish the equivalence between the definition of max-min fair flow allocation and the above algorithm. The above ideas are next extended analogously to max-min fair flow allocation in a network.

Max-Min Fair Flow Allocation in a Network

Assume that a telecommunication network is represented by a directed and connected graph $G = (V, E)$, where V is the set of vertices (nodes), and E is the set of edges (links). The sets V and E represent the routers and the transmission links respectively. The transmission carrying capacity of each link $e \in E$ is $c_e \in \mathbb{R}^+$, where the set of links E is equal to $\{1, 2, \ldots, |E|\}$.

The goal of this network is to carry traffic between specified node pairs. A node pair consists of a source node, and a destination node. It is assumed that a user session consists of transmitting traffic between a source-destination node pair via a single fixed path in the network. Denote the set of user sessions (demands, flows, traffic flows) by $\mathfrak{R}$, and let this set be represented by $\{1, 2, \ldots, |\mathfrak{R}|\}$. Let $\beta_i \in \mathbb{R}^+$ be the allowable traffic in the network for the session $i \in \mathfrak{R}$, when the max-min fair flow algorithm is used. A max-min fair traffic flow allocation vector $\beta = \left(\beta_1, \beta_2, \ldots, \beta_{|\mathfrak{R}|}\right)$ is next defined.

Definitions 5.2. *A telecommunication network is represented by a directed and connected graph* $G = (V, E)$, *where* V *is the vertex set,* E *is the edge set, and* $c_e \in \mathbb{R}^+$ *is the link capacity of the link* $e \in E$. *Let the set of user sessions be* $\mathfrak{R}$. *This set is represented by* $\{1, 2, \ldots, |\mathfrak{R}|\}$.

1. *Let* $\beta_i \in \mathbb{R}^+$ *be the allocated traffic flow for session* i, *where* $i \in \mathfrak{R}$ *and* $|\mathfrak{R}|$ *is the total number of sessions. The allocated flow on link* e *of the network is* F_e, *where*

$$F_e = \sum_{i \in \mathfrak{R}} \delta_{ei} \beta_i \tag{5.36}$$

where δ_{ei} *is equal to unity, if the flow* i *uses link* $e \in E$*, otherwise it is equal to zero. If* $F_e \leq c_e$*,* $\forall\ e \in \{1, 2, \ldots, |E|\}$*, the vector* $\beta = \left(\beta_1, \beta_2, \ldots, \beta_{|\mathfrak{R}|}\right)$ *is said to be feasible.*

2. *A flow vector* $\beta = \left(\beta_1, \beta_2, \ldots, \beta_{|\mathfrak{R}|}\right)$ *is said to be max-min fair if* β_i *is feasible* $\forall\ i \in \mathfrak{R}$*. Furthermore, the flow* β_i *cannot be made larger while maintaining feasibility without decreasing* β_j *for some flow* $j \neq i$ *such that* $\beta_j \leq \beta_i$*.*
3. *For a given feasible flow vector* β*, a link* $e \in E$ *is a bottleneck (saturated) link for a session* $i \in \mathfrak{R}$*, if it uses the link* e*, such that* $F_e = c_e$ *and* $\beta_{i'} \leq \beta_i$ *for all user sessions* i' *which pass through this link* e*.*
4. *The number of flows passing through the link* e *is denoted by* n_e*.* □

An algorithm for max-min fair flow allocation on a network is next outlined. The algorithm starts with an all-zero flow vector. It then increases the flows on all paths together until the flow on a single (or more than one) link equals to its capacity. A saturated link is a bottleneck link for at least a single user session which uses it. In the next step of the algorithm, all user sessions not using the bottleneck links are incremented equally until the next set of links get saturated. Observe that the flows which used saturated link (or links) in an earlier iteration might also use the newly saturated link (or links). Each of these newly saturated links is a bottleneck link for at least a single session which pass through it.

During each iteration of the while-loop of the algorithm outlined below, an equal increment is added to all sessions which do not pass through a saturated link. Therefore all updated sessions during an iteration have the same traffic flow value. A feasible flow vector β is max-min fair if and only if each user session has associated with it a saturated link with respect to β. This observation should be evident by the construction of the algorithm. The algorithm terminates, when each flow passes through at least a single saturated link.

Algorithm 5.2. *Max-Min Fair Flow Allocation Algorithm in a Network.*

Input: A telecommunication network represented by a directed and connected graph $G = (V, E)$. Also $c_e \in \mathbb{R}^+$ is the capacity of the link $e \in E$. The set of user sessions $\mathfrak{R}$ is $\{1, 2, \ldots, |\mathfrak{R}|\}$.
Each user session has a fixed path associated with it.
Output: Max-min fair flow allocation vector $\beta = \left(\beta_1, \beta_2, \ldots, \beta_{|\mathfrak{R}|}\right)$
begin

(*initialization*)
(F_e *is the allocated flow on link* e.)
(L *is the set of unsaturated links.*)
(ϑ *is the set of user sessions not passing through any any saturated link at the beginning of the iteration.*)
(n_e *is the number of sessions passing through link* e.)
($\Delta\beta$ *is the flow increment added to all user sessions in the set* ϑ)
$\beta \leftarrow (0, 0, \ldots, 0)$
$F_e \leftarrow 0,\ \forall\ e \in E$
$L \leftarrow E$
$\vartheta \leftarrow \mathfrak{R}$

while $\vartheta \neq \varnothing$ **do**
begin
(*compute the number of flows passing through each link* $e \in L$)
$n_e \leftarrow \sum_{s \in \vartheta} \delta_{es}, \ \forall \, e \in L$
(*from all unsaturated links, compute smallest possible value of increment* $\Delta\beta$)
$\Delta\beta \leftarrow \min_{e \in L} (c_e - F_e) / n_e$
(*increment* $\Delta\beta$ *is added to all sessions not yet passing through a saturated link*)
$\beta_s \leftarrow (\beta_s + \Delta\beta), \ \forall \, s \in \vartheta$
(*all sessions in* ϑ *have the same flow*)
$F_e \leftarrow \sum_{s \in \vartheta} \delta_{es} \beta_s, \ \forall \, e \in L$
(*next remove all saturated links, which are all edges with* $c_e = F_e$)
$L \leftarrow L \backslash \{e \mid c_e = F_e, e \in L\}$
(*find the set of all flows which use unsaturated links*)
$\vartheta \leftarrow \{s \mid \textit{all flows } s, \textit{ which use unsaturated links}\}$
end (*end of while-loop*)
end (*end of network max-min fair flow allocation algorithm*)

It is possible to generalize the above algorithms. These generalizations can be found in the literature.

5.9 Traffic Management via Packet Scheduling and Regulation

This section is concerned with the control and management of congestion in integrated networks via traffic scheduling and regulation. Packet networks have traditionally served data streams well. However such data networks were not designed to provide performance guarantees to traffic flows with real-time constraints, such as voice and video. This goal is achievable via traffic differentiation. This in turn can be accomplished by packet scheduling mechanisms. Packet scheduling mechanism can be implemented in routers and switches. For each packet which arrives at the router or switch, the packet scheduler determines the instant of time at which it is to be forwarded to the next router or switch. In absence of such scheduling mechanism, the packet service follows the discipline of first-in first-out. The requirements for an acceptable packet scheduling scheme are: *flexibility*, *fairness* and *mathematical tractability*. Flexibility means that the network should be able to accept traffic streams with different performance requirements. Fairness implies that a select few classes of flows should not degrade the service of other classes. Mathematical tractability is a requirement for providing performance guarantees.

The packet scheduling mechanism which is described in this section is called generalized processor sharing (GPS). Before the GPS scheduling scheme is defined, the processor sharing (PS) model is described. Let the service rate of the processor be $\mathcal{R}$. In the PS scheduling model, if there are $n(t)$ customers (packets) in the queue at time t, then the server applies $\mathcal{R}/n(t)$ of its resources to each of the $n(t)$ customers. Thus there is actually no queue, and the rate at which each customer receives service changes each time a new arrival joins the system or a packet exits when its service requirement is over.

Consider a modification of the PS discipline. Assume that there is more than a single class of customers. Packets of different classes queue up in separate queues, and the server applies equal share of its resources to the customers at the *head* of nonempty queues. This modified PS scheme is called head-of-the-line PS (HOLPS) packet scheduling scheme. Generalized processor sharing is a further modification of HOLPS, where the service offered to each of the head-of-line packets is in proportion to the weight associated with each queue. This is a flow-based multiplexing discipline that is analyzable, efficient, fair and flexible. A packetized version of the GPS is called the packetized generalized processor sharing (PGPS) scheme. Some researchers also call it weighted fair queueing (WFQ).

Traffic regulation implies regulation of traffic in order to achieve the desired service guarantees for different traffic classes. This can be achieved via the concept of leaky bucket. The generalized processor sharing scheme, when combined with leaky bucket admission control, yields to quantifiable worst-case performance guarantees on throughput and queueing delay.

5.9.1 Generalized Processor Sharing

In this subsection, a generalized processor sharing which uses a traffic scheduling discipline is discussed. This scheme is simple, fluid-based, mathematically tractable, isolates traffic flows, and provides service differentiation. Other traffic differentiation schemes like priority scheduling might result in starvation of low priority flows.

As already mentioned, GPS algorithm is fluid-based. This means that the incoming traffic stream is infinitely divisible, that is bits are permitted to be fractions. There are techniques to overcome this idealization and obtain a realistic scheduling policy. The GPS scheme is initially studied at a single-node (server) level. The node is modeled as an infinite buffer space in which session traffic waits (if necessary), and there is a single link of capacity $\mathcal{R}$ bits per unit time.

The GPS policy is a *work-conserving* scheduling discipline. This means that the link is busy when traffic is waiting to be served in the node. Assume that there are N traffic flows, and define $\mathcal{I} = \{1, 2, \ldots, N\}$ to be a set of flows (sessions). Associated with each flow $i \in \mathcal{I}$ is a weight $\phi_i \in \mathbb{R}^+$. A session is said to be backlogged at time t if there is traffic of that session waiting in the queue to be serviced. Let $\mathcal{B}(t)$ be the set of backlogged sessions at time t. Then the service rate $\mathcal{R}_i(t)$ received by ith session at time t is

$$\mathcal{R}_i(t) = \begin{cases} \dfrac{\phi_i}{\sum_{j \in \mathcal{B}(t)} \phi_j} \mathcal{R}, & \text{if } i \in \mathcal{B}(t) \\ 0, & \text{otherwise} \end{cases} \tag{5.37}$$

It should be evident that the ith traffic stream's delay is decreased by increasing the value of its weight ϕ_i. Let $\mathcal{S}_i(\tau, t)$ be the amount of session i traffic served in the time interval $(\tau, t]$. If the session i is continuously backlogged in the interval $(\tau, t]$, it can be shown that

$$\mathcal{S}_i(\tau, t) \geq (t - \tau) g_i, \quad g_i = \frac{\phi_i}{\sum_{j=1}^{N} \phi_j} \mathcal{R} \tag{5.38}$$

See the problem section for a proof of the above assertion. It further implies that a GPS node *guarantees* for any session i that is continuously backlogged in the interval $(\tau, t]$

$$\frac{S_i(\tau,t)}{S_j(\tau,t)} \geq \frac{\phi_i}{\phi_j}, \quad 1 \leq j \leq N \tag{5.39}$$

where session j has nonzero flow in the interval $(\tau, t]$. The above relationship is sometimes taken as a definition of the GPS scheduling discipline. In summary, this scheme has two attractive characteristics:

(a) Each backlogged flow is guaranteed a minimum service rate, thus providing fairness among all flows.
(b) If a flow is dormant, then its allocated service rate is redistributed among the backlogged flows in proportion to their weights.

Observe that the GPS scheduling policy is idealized, because it assumes that packets are infinitely divisible. GPS also assumes that the node can serve multiple sessions simultaneously. This idealization is addressed in the packetized generalized processor sharing scheme.

5.9.2 Packetized Generalized Processor Sharing Scheme

As any packet stream is not infinitely divisible, the GPS scheduling scheme has to be approximated so that the packet flows from different sessions are scheduled efficiently. In this subsection, a simple packet-by-packet transmission scheme is discussed. This scheme closely approximates the GPS discipline. The service discipline, which is described next, is called the packetized generalized processor sharing scheme.

Assume that *a packet has arrived if and only if its last bit has arrived.* Let d_k be the time at which packet $\widetilde{p}_k$ departs the node under the GPS scheme, where $k = 1, 2, \ldots$. An extremely good approximation of GPS is a work-conserving packet scheduling scheme that serves packets in increasing order of d_k. However, this approximation poses a problem. In this proposed scheme, when a server becomes free at time τ, the next packet to leave the server under GPS *might not have arrived* at time τ. Consequently, when the PGPS server becomes free, it selects the first packet that would complete service in the GPS scheme if *no additional packets were to arrive at the node after time* τ. The time-penalty, a packet has to pay when it departs the node under PGPS relative to GPS, is next evaluated.

Lemma 5.1. *Let $\widetilde{p}$ and $\widetilde{\pi}$ be packets in a GPS system at time instant τ, and packet $\widetilde{p}$ completes service before packet $\widetilde{\pi}$, if there are no arrivals after τ. Then packet $\widetilde{p}$ also completes service before packet $\widetilde{\pi}$ for any sequence of packet arrivals after time τ.*

Proof. See the problem section. □

The above lemma implies that, when compared to the GPS discipline, the only packets delayed more in the PGPS discipline are those that arrive too late to be transmitted in their GPS order. Based upon the above discussion, the following set of useful bounds can be obtained.

Observations 5.4. Let $\mathcal{I} = \{1, 2, \ldots, N\}$ be the set of sessions.

1. Let d_k and $\widehat{d}_k$ be the departure instants of packet $\widetilde{p}_k$ under the GPS and PGPS scheme respectively. Let $L_{\max}$ be the maximum packet length (in bits), and $\mathcal{R}$ be the rate of the server. Then for all packets $\widetilde{p}_k$:

$$\widehat{d}_k - d_k \leq \frac{L_{\max}}{\mathcal{R}} \tag{5.40a}$$

2. Let $\mathcal{S}_i(\tau, t)$ and $\widehat{\mathcal{S}}_i(\tau, t)$ be the amount of session i traffic (measured in bits and not packets) served in the time interval $[\tau, t]$, under the GPS and PGPS schemes respectively. Then for all times τ and each session $i \in \mathcal{I}$:
$$\mathcal{S}_i(0, \tau) - \widehat{\mathcal{S}}_i(0, \tau) \le L_{\max} \tag{5.40b}$$
3. Let $\mathcal{Q}_i(\tau)$ and $\widehat{\mathcal{Q}}_i(\tau)$ be the backlog (measured in units of traffic) of session i in the time interval $[0, \tau]$, under the GPS and PGPS schemes respectively. Then for all times τ and each session $i \in \mathcal{I}$:
$$\widehat{\mathcal{Q}}_i(\tau) - \mathcal{Q}_i(\tau) \le L_{\max} \tag{5.40c}$$
This result is immediate from observation 2. □

The proofs of some of these observations are provided in the problem section. These proofs are based upon the concept of *busy period.* Busy period is defined as the time during which packets are backlogged.

5.9.3 Leaky Bucket for Traffic Regulation

A buffered leaky bucket is a mechanism to regulate traffic which enters the network. This scheme limits packet input to the network to a specified burst size and average transmission rate. The leaky bucket mechanism has a token bucket into which tokens are fed at a rate of $\widetilde{\rho}$ tokens per unit time. The bucket can hold up to $\widetilde{\sigma}$ bits worth of tokens. Data is permitted to leave the server only if there are matching tokens available in the bucket. When data leaves the server, the corresponding number of tokens is decreased in the bucket (hence the name leaky bucket). However, if data arrives and there are no matching tokens, then the data is buffered in the source buffer. Note that there is no bound on the number of packets that can be buffered. Therefore, over a time interval of length t units, the number of packets output by the server is bounded by $(\widetilde{\sigma}+\widetilde{\rho}t)$. Consequently this scheme provides an attractive strategy for admission control.

Sometimes a leaky bucket is further qualified by a parameter, called peak rate c (tokens per unit time). Note that $\widetilde{\rho}$ is the average sustainable rate, and c is the peak rate, where $\widetilde{\rho} \le c$. Therefore, a leaky bucket is identified by the triple $(\widetilde{\sigma}, \widetilde{\rho}, c)$. A configuration of a leaky bucket is shown in Figure 5.3.

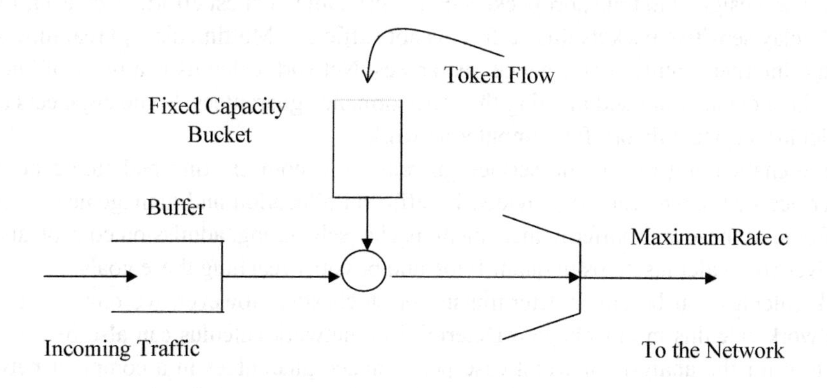

Figure 5.3. Leaky bucket $(\widetilde{\sigma}, \widetilde{\rho}, c)$, with capacity $\widetilde{\sigma}$, and token flow rate $\widetilde{\rho}$.

Each session $i \in \mathcal{I}$ is associated with the triple $(\widetilde{\sigma}_i, \widetilde{\rho}_i, c_i)$. The following parameters are further introduced to describe the flow i.

(a) Let $\mathcal{A}_i(\tau, t)$ be the amount of session i flow that leaves the leaky bucket and enters the network in time interval $(\tau, t]$.

$$\mathcal{A}_i(\tau, t) \le \min\{(t-\tau)c_i, \widetilde{\sigma}_i + \widetilde{\rho}_i(t-\tau)\}, \; \forall\, t \ge \tau \ge 0 \tag{5.41a}$$

(b) Assume that the session starts out with a full bucket of tokens. The total number of tokens accepted in the time interval $(0, t]$ is $\mathcal{K}_i(t)$. This number neither includes the full bucket of tokens that the session i begins with, nor does it include arriving tokens that find the bucket full. Then the number of tokens $\mathcal{T}_i(t)$ in the token bucket at time t is

$$\mathcal{T}_i(t) = \widetilde{\sigma}_i + \mathcal{K}_i(t) - \mathcal{A}_i(0, t) \tag{5.41b}$$

Note that $\mathcal{T}_i(0) = \widetilde{\sigma}_i$.

The following observations should be immediate.

Observations 5.5. Let $\mathcal{I} = \{1, 2, \ldots, N\}$ be the set of sessions, and $i \in \mathcal{I}$.

1. We have
$$\{\mathcal{K}_i(t) - \mathcal{K}_i(\tau)\} \le \widetilde{\rho}_i(t-\tau), \quad \forall\, \tau \le t$$
2. Use of the last observation, and the definition of $\mathcal{T}_i(t)$ yields,
$$\mathcal{A}_i(\tau, t) \le \mathcal{T}_i(\tau) - \mathcal{T}_i(t) + \widetilde{\rho}_i(t-\tau), \quad \forall\, \tau \le t$$

□

5.10 Network Calculus

The Internet was designed to transport packets on the principle of "best effort." However, for transportation of delay sensitive packets this "effort" is not sufficient. Multimedia and real time applications over the Internet require performance guarantees. Network calculus is a body of knowledge which helps in understanding and meeting these performance guarantees. Some engineers also call network calculus a system theory for computer networks.

It is quite challenging to provide service guarantees in both existing and future networking systems. Service guarantees can be provided by efficient allocation and management of network resources. Some of these are: buffer management, packet scheduling, admission control, and traffic regulation. Network calculus helps in quantifying and possibly reaching these goals.

Network calculus can be either deterministic or stochastic. However, we only discuss deterministic network calculus in this chapter. Deterministic network calculus can also be viewed as a theoretical tool for the analysis of worst case performance guarantees in a computer networking system. We initially summarize some mathematical preliminaries needed to develop network calculus. This is followed by elucidation of certain basic concepts. Subsequently bounds on buffers size, delay, and output flow of a networking element are obtained.

5.10.1 Mathematical Preliminaries

The results of this section can be understood by initially introducing some special functions and operations. We introduce wide-sense increasing, and subadditive functions. The min-plus convolution and deconvolution operations of two functions are also defined.

Definition 5.3. *A function $f(\cdot)$ is wide-sense increasing iff $f(s) \leq f(t)$ for all $s \leq t$. The range of these functions is the set of extended nonnegative real numbers $\mathbb{R}_0^+ \cup \{\infty\}$. The parameter t can be either discrete or continuous.*

(a) *If the parameter t is discrete, then $\{f(t), t \in \mathbb{Z}\}$ is a sequence.*
(b) *If the parameter t is continuous, then the function $f(t), t \in \mathbb{R}$ is left-continuous.*

Denote the set of such functions by $\mathcal{G}$. □

Wide-sense increasing function can count the volume of data arriving or departing to/from some network element. Thus these functions are nonnegative, nondecreasing, and can possibly take infinite value.

Definitions 5.4. *Basic definitions.*

1. *A function $f \in \mathcal{G}$ and $f(t) = 0$ for all $t < 0$. Denote the set of such functions by $\mathcal{F}$.*
2. *The functions $f, g \in \mathcal{F}$, and $f(t) \leq g(t)$ for all $t \geq 0$. This inequality is denoted as $f \leq g$.*
3. *Point-wise sum and minimum of functions f and g for $t \in \mathbb{R}$. The notation:*
   (a) $(f + g)(t) \triangleq f(t) + g(t)$.
   (b) $\min(f, g)(t) \triangleq \min(f(t), g(t))$.
4. *Let $f \in \mathcal{F}$, then f is a subadditive function if and only if $f(t+s) \leq f(t) + f(s)$ for all $s, t \geq 0$.* □

Observations 5.6. Some elementary facts about subadditive functions.

1. A concave function which passes through the origin is subadditive. However, a subadditive function is not necessarily concave.
2. If f and g are subadditive, then so is $(f + g)$. □

An operation called min-plus convolution is defined below. This operation provides a compact way to study network calculus.

Definition 5.5. *Let f and g be either functions or sequences in $\mathcal{F}$. The min-plus convolution of f and g is*

$$(f \circledast g)(t) = \inf_{0 \leq \tau \leq t} \{f(\tau) + g(t - \tau)\} \tag{5.42}$$

and $(f \circledast g)(t) = 0$ for $t < 0$. □

Note that, if necessary the limits in the inf operation can be 0_- and t_+ to include jumps in the functions $f(\cdot)$ and $g(\cdot)$ at $t = 0$ and t. Observe that the min-plus convolution operation is analogous to the classical convolution operation of classical analysis.

Analogous to the delta-function in classical analysis, we have an impulse function $\delta_T(t), t \geq 0$ for the min-plus convolution operator, where $T \geq 0$. For $\delta_T(\cdot)$ to be the impulse function for this operator, we should have for any $f \in \mathcal{F}$, $(f \circledast \delta_T)(t) = f(t-T)$. This impulse function is defined below.

Definition 5.6. *For* $T \geq 0$, *the impulse function* $\delta_T(t), t \geq 0$ *is*

$$\delta_T(t) = \begin{cases} 0, & 0 \leq t \leq T \\ \infty, & t > T \end{cases} \tag{5.43}$$

At $T = 0$, *we denote* $\delta_0(t) \triangleq \delta(t)$. □

The following set of observations about min-plus convolution are immediate from the above definitions.

Observation 5.7. Let the functions $f, g, h \in \mathcal{F}$. Then the convolution operator $\circledast$ satisfies the following properties.

(a) Closure: $(f \circledast g) \in \mathcal{F}$.
(b) The identity element is $\delta(t)$: $(f \circledast \delta) = f$.
(c) Associativity: $((f \circledast g) \circledast h) = (f \circledast (g \circledast h))$.
(d) Commutativity: $(f \circledast g) = (g \circledast f)$.
(e) Distributive over min operation: $(f \circledast \min(g, h)) = \min((f \circledast g), (f \circledast h))$.
(f) Addition of a constant. For any $K_{const} \in \mathbb{R}_0^+$, $((f + K_{const}) \circledast g) = (f \circledast g) + K_{const}$.
(g) Subadditivity. If f and g are subadditive functions, then so is $(f \circledast g)$. □

Another useful operation is the dual of the min-plus convolution operation. It is called the min-plus deconvolution operation.

Definition 5.7. *Let* f *and* g *be either functions or sequences in* $\mathcal{F}$. *The min-plus deconvolution of* f *and* g *is*

$$(f \oslash g)(t) = \sup_{\tau \geq 0} \{f(t+\tau) - g(\tau)\} \tag{5.44}$$

□

Note that if both $f(t)$ and $g(t)$ are infinite for some t, then $(f \oslash g)(t)$ is not defined. Furthermore, $(f \oslash g)(t)$ does not always belong to the set $\mathcal{F}$. Thus, it is possible that $(f \oslash g)(t)$ is not necessarily zero for $t \leq 0$. Some properties of the min-plus deconvolution operation are listed below.

Observation 5.8. Let the functions $f, g, h \in \mathcal{F}$.

(a) If $f \leq g$, then $(f \oslash h) \leq (g \oslash h)$, and $(h \oslash f) \geq (h \oslash g)$.
(b) Composition of $\oslash$: $((f \oslash g) \oslash h) = (f \oslash (g \circledast h))$.
(c) Composition of $\oslash$ and $\circledast$: $((f \circledast g) \oslash g) \leq (f \circledast (g \oslash g))$.
(d) Duality between $\oslash$ and $\circledast$: $(f \oslash g) \leq h$ if and only if $f \leq (g \circledast h)$. This result is the reason for the phrase "deconvolution" operation.
(e) Self-deconvolution: $(f \oslash f)$ is a subadditive function of $\mathcal{F}$. Also, $(f \oslash f)(0) = 0$. □

The proofs of some of these results are provided in the problem section.

5.10.2 Basic Concepts

We first develop some basic concepts about data (traffic) flow. Data flow can be described in terms of the cumulative function $F_{cum}(\cdot)$, where $F_{cum}(t)$ is the total number of bits observed on the data flow in the time interval $[0,t]$. It is assumed that $F_{cum}(0)=0$. More conveniently, we assume that $F_{cum} \in \mathcal{F}$. In practice, t and $F_{cum}(t)$ take discrete values, however we permit these variables to take continuous values for mathematical convenience. Such mathematical descriptions are called *fluid models*.

Consider a networking system $\mathcal{N}$ which is modeled as a blackbox. It receives an input at its ingress point, and produces an output at its egress point after a certain delay. The input and output data flows are described by their cumulative flows in the interval $[0,t]$. Denote the cumulative input and output data flows in the interval $[0,t]$ by $F_{in}(t)$ and $F_{out}(t)$ respectively, where $F_{in}, F_{out} \in \mathcal{F}$. We also assume in this section that data is not lost. Such systems are called *lossless systems*.

The system $\mathcal{N}$ can be described in terms of its backlog, virtual delay, and arrival and service curves. Based upon this description important bounds are derived in the next subsection for lossless systems with service guarantees.

Definitions 5.8. *We define backlog, virtual delay, and arrival and service curves. The cumulative input and output data flows of a networking system $\mathcal{N}$ in the interval $[0,t]$ are $F_{in}(t)$ and $F_{out}(t)$ respectively, where $F_{in}, F_{out} \in \mathcal{F}$.*

1. *If the system $\mathcal{N}$ is lossless*:
   (a) *Backlog at time t is $(F_{in}(t) - F_{out}(t))$.*
   (b) *Virtual delay at time t is $d(t) = \inf\{\tau \ge 0 \mid F_{in}(t) \le F_{out}(t+\tau)\}$.*
2. *The arrival curve $\widetilde{A}(\cdot)$. Let $\widetilde{A} \in \mathcal{F}$. The flow F_{in} is constrained by $\widetilde{A}$ if and only if for all $s \le t$*:
$$F_{in}(t) - F_{in}(s) \le \widetilde{A}(t-s) \tag{5.45a}$$
   The arrival curve $\widetilde{A}(t)$ is also called an envelope for the flow $F_{in}(t)$. Thus
$$F_{in}(t) \le \inf_{0\le s\le t}\left\{\widetilde{A}(t-s) + F_{in}(s)\right\} = \left(\widetilde{A} \circledast F_{in}\right)(t), \quad \forall\, t \ge 0 \tag{5.45b}$$
3. *The service curve $\widetilde{S}(\cdot)$. Let $\widetilde{S} \in \mathcal{F}$. The service curve of the data flow is $\widetilde{S}$ if and only if for all $t \ge 0$, there exists t_0, such that $t \ge t_0 \ge 0$, and*
$$F_{out}(t) - F_{in}(t_0) \ge \widetilde{S}(t-t_0) \tag{5.46a}$$
   Thus
$$F_{out}(t) \ge \inf_{0\le s\le t}\left\{\widetilde{S}(t-s) + F_{in}(s)\right\} = \left(\widetilde{S} \circledast F_{in}\right)(t) \tag{5.46b}$$

□

Backlog is equal to the data (bits) held inside the networking system $\mathcal{N}$. If the system is a single buffer, then the backlog is indeed equal to the queue length. However, if the system $\mathcal{N}$ has a complex networking topology, and $F_{in}(t)$ and $F_{out}(t)$ are observable, then the backlog is simply equal to the number of bits in transit from the ingress point to the egress point. If we plot $F_{in}(t)$, and $F_{out}(t)$ versus t, then backlog is the vertical deviation between the two curves.

The virtual delay of a bit which arrives at time t is the delay experienced by this bit, due to the servicing of all the bits which arrived before it. The definition of the virtual delay function implies that if the function $F_{out}(\cdot)$ is continuous, then $F_{out}(t + d(t)) = F_{in}(t)$. If we plot $F_{in}(t)$, and $F_{out}(t)$ versus t, then the delay is the horizontal deviation between the two curves.

Knowledge of the arrival and service curves facilitate in providing appropriate resources in the system $\mathcal{N}$. Thus, the arrival curve puts an upper constraint on the input data flow. In order to provide an acceptable level of performance to the data flow, the system $\mathcal{N}$ needs to offer guarantees. This is done in the form of service curve. The data flow receives service which is is equal to at least $\widetilde{S}(s)$ in all intervals $[0, s]$ included in the busy period beginning at time 0. Furthermore $\widetilde{S}(0) = 0$. The above concepts are illustrated via the following examples.

Examples 5.3. Elementary examples.

1. Consider a system $\mathcal{N}$ fed by data from a link of speed r bits per unit time. Then the arrival curve of the system $\mathcal{N}$ is $\widetilde{A}(t) = rt$, where $t \geq 0$.
2. A node provides a guaranteed node delay of T time units, where $T > 0$. The bits in this node are served in a first-in first-out order. Thus $(F_{in} \circledast \delta_T)(t) = F_{in}(t - T)$, where $t \geq 0$. Therefore the service curve of this node is $\widetilde{S} = \delta_T$.
3. Consider two networking elements $\mathcal{N}_1$ and $\mathcal{N}_2$ in tandem (series). Let the service curves of these elements be $\widetilde{S}_1$ and $\widetilde{S}_2$ respectively. It can be shown that the service curve of the composite system is $\widetilde{S} = \widetilde{S}_1 \circledast \widetilde{S}_2$.
4. Some routers offer a service curve of the form $\widetilde{S}(t) = r(t - T)^+$, where $r, T > 0$, and $t \geq 0$. Consider two such routers that are connected in tandem, with parameters $r_i, T_i > 0$, where $i = 1, 2$. The service curve of the two routers in tandem is of the form $\widetilde{S}(t) = r(t - T)^+$, where $r = \min(r_1, r_2)$, $T = (T_1 + T_2)$, and $t \geq 0$.
5. Leaky bucket mechanism is a technique to regulate the burst size and the peak rate of a link. Assume that the rate of flow of tokens in the bucket is $\widetilde{\rho}$ tokens per unit time (sustainable rate of the leaky bucket). The capacity of the bucket is $\widetilde{\sigma}$ tokens. Data leaves the server only if there are matching tokens available in the bucket. If data arrives and there are no matching tokens in the bucket, then the data is buffered in the source buffer. Observe that there is no bound on the number of packets that can be buffered. Thus, over a time interval of length t units, the number of packets output by the server is bounded by $(\widetilde{\sigma}+\widetilde{\rho}t)$. Assume that the bucket is empty at time 0. Therefore a data link which is regulated by this mechanism provides an arrival curve $\widetilde{A}(t) = (\widetilde{\sigma}+\widetilde{\rho}t)$, for $t > 0$, and 0 otherwise. □

5.10.3 Fundamental Bounds

Based upon the mathematical preliminaries and basic concepts, three fundamental bounds of deterministic network calculus are developed. These are backlog bound, delay bound, and description of output flow.

Observations 5.9. The cumulative input and output data flows of a networking system $\mathcal{N}$ in the interval $[0, t]$ are $F_{in}(t)$ and $F_{out}(t)$ respectively, where $F_{in}, F_{out} \in \mathcal{F}$. Its arrival and service curves are $\widetilde{A}$ and $\widetilde{S}$ respectively.

1. *Backlog bound*: The backlog $(F_{in}(t) - F_{out}(t))$ at time $t \geq 0$ is upper-bounded as

$$(F_{in}(t) - F_{out}(t)) \leq \sup_{s \geq 0} \left\{ \widetilde{A}(s) - \widetilde{S}(s) \right\} = \left(\widetilde{A} \oslash \widetilde{S} \right)(0) \tag{5.47}$$

2. *Delay bound*: The delay bound is obtained by considering the virtual delay of a hypothetical system, in which the cumulative input and output data flows for all $t \geq 0$ are $\widetilde{A}(t)$, and $\widetilde{S}(t)$ respectively. We assume that such a system exists, which implies $\widetilde{A}(t) \leq \widetilde{S}(t), \forall\, t \geq 0$. Denote this virtual delay by $\Delta(s)$. That is,

$$\Delta(s) = \inf\left\{\tau \geq 0 \mid \widetilde{A}(s) \leq \widetilde{S}(s+\tau)\right\} \tag{5.48}$$

   Denote the supremum of $\Delta(s)$ by $H\left(\widetilde{A}, \widetilde{S}\right)$.

   The virtual delay $d(t)$ at time t is upper-bounded as $d(t) \leq H\left(\widetilde{A}, \widetilde{S}\right)$ for all $t \geq 0$.

3. *Output flow*: The output flow is constrained (limited) by the arrival curve $\left(\widetilde{A} \oslash \widetilde{S}\right)(t)$. □

These observations are proved in the problem section.

Examples 5.4. The use of backlog and delay bounds is illustrated in the following examples.

1. A flow into a node is described by the four parameters $L_{\max}, r_{pk}, b$, and r_{sus}. The parameter $L_{\max}$ is the maximum packet size in bits, r_{pk} is the peak rate in bits per unit time, b is the burst tolerance, and r_{sus} is the sustainable rate. Assume that $r_{pk} > r_{sus}$ and $b > L_{\max}$.
   The node offers a service curve of the form

$$\widetilde{S}(t) = \mathcal{R}(t-T)^{+}, \quad \text{where } \mathcal{R}, T > 0, \text{ and } t \geq 0$$

   Thus the node provides a maximum latency of T time units. Assume that $\mathcal{R} \geq r_{sus}$. The arrival curve for this node is

$$\widetilde{A}(t) = \min\{L_{\max} + r_{pk}t, b + r_{sus}t\}, \text{ for } t > 0$$

   The required buffer size can be determined from the vertical deviation between the two curves $\widetilde{A}(t)$, and $\widetilde{S}(t)$, where $t \geq 0$. The buffer size for the flow is bounded by

$$Q_{\max} = \sup_{t \geq 0}\left\{\widetilde{A}(t) - \widetilde{S}(t)\right\}$$

   If

$$t^{*} = \frac{b - L_{\max}}{r_{pk} - r_{sus}}$$

   we have

$$Q_{\max} = \begin{cases} b + r_{sus}T, & t^{*} \leq T \\ L_{\max} + r_{pk}T + (t^{*} - T)(r_{pk} - \mathcal{R})^{+}, & t^{*} > T \end{cases}$$

   The maximum delay $D_{\max}$ for the flow can be determined from the horizontal deviation between the two curves $\widetilde{A}(t)$, and $\widetilde{S}(t)$, where $t \geq 0$. It is

$$D_{\max} = T + \frac{L_{\max} + t^{*}(r_{pk} - \mathcal{R})^{+}}{\mathcal{R}}$$

2. Consider two routers with service curves

$$\widetilde{S}_i(t) = r_i(t - T_i)^{+}, \quad \text{where } r_i, T_i > 0,\ i = 1, 2, \text{ and } t \geq 0$$

These two routers are connected in tandem. The arrival curve to this tandem system is specified as:

$$\widetilde{A}(0) = 0$$

$$\widetilde{A}(t) = b + r_{arr}t, \text{ for } t > 0, \text{ where } b, r_{arr} > 0$$

Assume that $r_{arr} < r_i$, where $i = 1, 2$.
It has been shown in an earlier example that the service curve of the two routers in tandem is

$$\widetilde{S}(t) = r(t - T)^+, \text{ where } r = \min(r_1, r_2), \; T = (T_1 + T_2), \text{ and } t \geq 0$$

The delay bound $D_{\max}$ of the composite system is then found to be $D_{\max} = (b/r + T)$. □

5.11 Multiple Access of Communicating Medium

Protocols that specify access of a communication medium simultaneously by several users are described in this section. Note that the terms user, station, and node are used synonymously in this section. Consider a set of users who share a common medium (channel) for communication. Such scenario might occur in the case of satellite communication, ring networks, shared buses, wireless communication, and so on. Proper access of the shared medium by different users is called medium access control.

Different techniques for medium access control are initially described. This is followed by a simplified throughput performance evaluation of a wired Ethernet local area network (LAN). This is repeated for a wireless Ethernet LAN.

5.11.1 Medium Access Control Techniques

Medium access control occurs in the data link layer of the Open Systems Interconnection's seven-layered model. The functionality of the data link layer can be divided into two sublayers. The lower sublayer performs the function of resolving access to the shared medium, and the upper sublayer implements the data link layer functionality.

The schemes for sharing a communication medium properly by several users are called access protocols. These involve efficient and effective sharing of a medium by different users. In broad terms, there are two types of such protocols. These are:

- Static medium access control.
- Dynamic medium access control.

In static and dynamic medium controls, the resources allocated to each user are fixed and variable respectively. The dynamic medium access control mechanism, can again be divided into two types. These are:

- Controlled access protocols.
- Random access protocols.

These are discussed in the rest of this section.

Static Medium Access Control

The static medium controls are also called channelization protocols. These are essentially channel resource partitioning schemes. These are useful, if each node has sufficient data to be transferred. Some of the static medium access control schemes are:

- Frequency-division multiple access (FDMA).
- Time-division multiple access (TDMA).
- Wavelength division multiplexing (WDM).
- Code-division multiple access (CDMA).

In FDMA, the frequency spectrum of the accessible medium (communication channel) is split equally among all nodes across the communication link. Thus the communication link dedicates a fixed bandwidth to each node for the period of the connection. Therefore simultaneous transmission is possible in this protocol for each node. In this scheme, unused communication subchannels are idle. Also observe that this is an analog scheme.

In a time-division multiplexing (TDM) scheme, time is divided into frames of fixed duration. Each frame-duration is further divided into a fixed number of time slots. Then, each node is allocated a fixed time slot within each frame-period. Thus a node is able to use the complete bandwidth of the medium for the duration of the time slot. However, it is possible for a time slot to be idle. Thus, this scheme does not use the channel resources efficiently. Observe that the TDM scheme is digital. This type of TDM is called synchronous TDM. We shall see later that there is also an asynchronous TDM.

WDM is used on links which use fiber optic technology. This scheme is similar to FDMA. Each user is allocated a pair of wavelengths, one wavelength for transmitting and another for receiving. Further, different users have different pairs of wavelengths. Observe that this is an analog scheme.

In the CDMA scheme, the communication channel carries bit streams from different users simultaneously. Multiple access to the communication channel is possible, if the nodes use different codes. The underlying theory of CDMA is based upon the theory of orthogonal codes.

The dynamic medium access control mechanisms: controlled access protocols, and random access protocols are next outlined. Only a qualitative explanation of these protocols is initially provided.

Controlled Access Protocols

Controlled access protocols utilize the channel bandwidth more efficiently than static access protocols. Some of the controlled access schemes are:

- Asynchronous time-division multiplexing.
- Reservation protocol.
- Polling protocol.
- Token passing protocol.

There are two types of TDMA schemes. These are the synchronous and asynchronous TDMA. Synchronous TDMA scheme was described earlier. It does not use the communication medium efficiently. This loss of efficiency can be averted by using asynchronous time-division multiplexing.

In the asynchronous TDMA, time slots are dynamically allocated to improve efficient utilization of the communication medium. A node is allocated a time slot, only if it has data to transmit. Thus, a time slot does not go unused, as in the case of synchronous TDMA. Note that in synchronous time-division multiplexing scheme, a time slot carries only data. On the other hand, in asynchronous time-division multiplexing scheme a time slot has to carry both the data and the addressing information. The address is that of the destination node. Asynchronous time-division multiplexing is also called statistical time-division multiplexing.

In the reservation scheme, a node is required to make a reservation prior to transmitting data. Time is split into intervals. In each interval, a reservation frame is initially sent. Assume that there are n number of nodes. In this case, the reservation frame has precisely n number of mini time slots. Each mini time slot corresponds to a node. If a node has sufficient data to send, it indicates its intention in its corresponding mini time slot. The nodes that made reservations, then transmit their data frames subsequent to the transmission of the reservation frame.

In the polling protocol, one of the nodes is designated as a master node. The master node then polls each other node in a round-robin fashion, and requests them to transmit up to only a certain maximum number of frames. The master node can also determine whether a node has finished transmitting by sensing a lack of signal on the communication channel. The master node continues this process by polling all other nodes in a cyclic fashion. A disadvantage of this scheme is the overhead due to the polling process itself. Further, if the master node fails, then the overall scheme fails.

The master node is absent in the token passing protocol. The token passing protocol organizes the nodes of the network into a logical ring. That is, each node has a predecessor node and a successor node. A small, special-purpose frame called the token is used in this scheme. The token frame moves from a node to another node in a circular fashion. Upon receiving the token from its predecessor node, the node sends the token immediately to its successor node if it has no frames to send. However, if it has frames to transmit, then it is allowed to transmit up to a certain maximum number of frames. It subsequently transfers the token to its successor node.

This decentralized scheme appears to be efficient. However, failure of a single node disrupts transmission of frames on the communication channel. Fiber distributed data interface (FDDI) is a popular token ring protocol.

Random Access Protocols

Stations generally use communication medium only randomly. Further, not all nodes want to transmit data at all times. Consequently, the number of active nodes is also not fixed. Thus static medium control protocols, and controlled access protocols are generally not suitable for high throughput under such conditions. Random access protocols are contention-based. These later protocols provide higher throughput for random access of the communication medium. Some of the random access protocols are:

- ALOHA.
- Carrier-sense multiple access (CSMA).
- Carrier-sense multiple access with collision detection (CSMA/CD).
- Carrier-sense multiple access with collision avoidance (CSMA/CA).

We discuss these four random access protocols in order.

ALOHA

ALOHA is a random multiple access technique in which a node can transmit data at all times. As its name implies, it was developed at the University of Hawaii in the early 1970's. It was originally designed for a wireless LAN. However the ALOHA scheme can be implemented on any shared medium.

In this scheme, each source of data can send a frame at all times. This introduces the possibility of collision of frames on the communication channel from different sources. This is partially alleviated as follows. If a frame successfully reaches its destination (receiver node), then the receiver sends an acknowledgement frame to the transmitting node.

If the transmitting node does not receives an acknowledgement frame within a certain fixed period of time, then it assumes that the transmission of the frame was not successful. After this fixed period of time, the transmitting node resends the frame after a random period of time. The reason for introducing randomness is as follows. Remember that if a collision occurs, then frames from other sources are also colliding. Therefore, if all sources involved in frame collisions try to resend their corresponding frames after a fixed period of time, then there can be repeated collisions. In order to avoid this scenario, each transmitting node resends the frame after a random interval of time.

CSMA

In the CSMA protocol, a node senses the medium to detect the absence or presence of another frame (from a different source) on the communication channel. It uses a carrier-sensing mechanism for this process.

If it detects the presence of another frame on the communication channel, it waits for the transmission of this later frame to finish. However, if the medium is sensed to be idle, then it can transmit a frame. This in turn improves the performance of the protocol. Thus, in this scheme multiple nodes can transmit on the same communication channel, albeit not simultaneously. This is the main idea behind the CSMA scheme.

Observe that in this scheme, there is still a possibility of collision. When a node transmits a frame to another node, it takes a finite amount of time for the leading edge of the first bit of the frame to reach another node because of the finite speed of electromagnetic waves in the communication medium. This delay is called the *propagation delay*. During this delay, it is quite possible for another node to sense the communication medium, and find it to be idle. In this case the other node might try to transmit a frame on the medium. This can result in a collision.

CSMA with collision detection, and CSMA with collision avoidance are modifications of the CSMA protocol.

CSMA/CD

The CSMA/CD protocol is generally used in the wired Ethernet LAN. This protocol is an improvement over the CSMA protocol. In this scheme, nodes have the capability to detect collision between frames on the communication channel. Before sending a frame on the communication channel, the node senses the medium for the presence of frame from another source. This sensing is done by monitoring the energy level of electromagnetic signal on the channel.

(a) If the node senses, that the channel is busy, then it waits a random period of time, before it attempts to send the frame.
(b) However if it senses that the channel is idle, then it begins its transmission of the frame. The node continues to monitor the medium even after it has sent the frame. It does so till the transmission of the frame is successful. However, if a collision was detected during this transmission, then the node suspends its transmission immediately. It subsequently waits a random period of time before attempting a retransmission of the frame. This is the random backoff strategy.

A simplified mathematical model of the CSMA/CD protocol is described later in the section.

CSMA/CA

The CSMA/CA protocol is used in wireless Ethernet LANs. The CSMA in the CSMA/CA protocol means that a node senses the communication medium before transmitting a frame. We next determine the reason for not using the CSMA/CD protocol in wireless communication.

The CSMA/CD protocol assumes that the nodes can detect collision due to simultaneous transmission of frames originating from different nodes. This detection is not easy in wireless communication. This is so, because due to attenuation, the strength of the received signal can be few orders of magnitude smaller than that of the transmitted signal. Therefore designing a collision detection hardware will be extremely expensive. Thus the protocol that is used in wireless LANs is CSMA with collision awareness.

Consequently, in wireless LANs the communication channels are generally half-duplex. That is, these channels cannot transmit and receive signals at the same time on a single frequency. In a wireless LAN, if a node detects a collision it does not suspend its transmission immediately. Actually it transmits its frame in its entirety. Therefore, its medium access protocol should be designed carefully. As we shall see, the CSMA/CA protocol does try to avoid collisions, albeit not completely. In addition to frame collisions in a wireless network, there are two more problems. These are the *hidden*, and *exposed node* problems.

Hidden Node Problem: Consider two nodes N_1 and N_2. These two nodes are located such that they are not within the receiving range of each other. Also consider another node N_0 which is within the receiving range of both nodes N_1 and N_2. If the two nodes N_1 and N_2 simultaneously transmit frames to node N_0, then there will be a collision at node N_0. Note that this is possible, because N_1 and N_2 do not know of each other's transmissions. Therefore, they can transmit frames simultaneously. Thus nodes N_1 and N_2 are said to be *hidden* from each other. Consequently hidden nodes reduce the effective transmission capacity of the wireless LAN. See Figure 5.4.

$$N_1 \Longrightarrow N_0 \Longleftarrow N_2$$

Figure 5.4. Nodes N_1 and N_2 are hidden from each other.
The arrows indicate direction of transmission of frames.

Exposed Node Problem: Consider three nodes N_1, N_2, and N_3. These three nodes are located such that:

(a) Node N_1 is within the receiving range of only node N_2. However nodes N_1 and N_3 cannot communicate directly with each other.

(b) Node N_2 can communicate with both nodes N_1 and N_3.
(c) Node N_3 is within the receiving range of only node N_2. However nodes N_3 and N_1 cannot communicate directly with each other.

Also consider another node N_0 which is within the receiving range of only node N_3. Node N_3 wishes to transmit frames to node N_0. Further, node N_2 is in the process of transmitting frames to node N_1. Node N_3 senses this later transmission, and does not transmit frames to node N_0, even though it can do so without any collision. This is called the *exposed node* problem. This problem forces a node to be unnecessarily more conservative, effectively decreasing the net throughput of the communication channel. In this configuration of nodes, the node N_3 is said to be *exposed*. See Figure 5.5.

$$N_1 \Longleftarrow N_2 \Longrightarrow N_3 \Longrightarrow N_0$$

Figure 5.5. Node N_3 is exposed.
The arrows indicate direction of transmission of frames.

The CSMA/CA protocol attempts to avoid collisions by introducing certain timing parameters. These are Short Inter-Frame Spacing, Distributed Coordination Function Inter-Frame Space, and Extended Inter-Frame Space.

(a) *Short Inter-Frame Space* (*SIFS*): SIFS is the delay introduced by the CSMA/CA protocol at the receiving node. It occurs between the reception of a frame, and the transmission of the acknowledgement (ACK) frame. Actually, it is the time required by a node to switch from receiving mode to transmitting mode. This later transformation is required to send the ACK frame to the source node.
(b) *Distributed Coordination Function Inter-Frame Space* (*DIFS*): In order to transmit a frame, a node must initially wait for the communication channel to be in the idle state for at least DIFS length of time. In order to use this delay, the previous frame should also have been received correctly.
(c) *Extended Inter-Frame Space* (*EIFS*): If the previous frame was received incorrectly, then this implies collisions on the channel. Therefore the transmitting node must sense the channel to be in idle state for at least EIFS length of time, before transmitting a frame.

Note that $SIFS < DIFS < EIFS$. The precise values of these three variables depend upon the type of the physical layer. The CSMA/CA protocol works as follows:

Sending Node: Sending node mechanism.

1. Sender node senses the medium. If it is idle for DIFS length of time, then it transmits the entire frame to the receiving node. It is assumed that the previous frame was transmitted correctly by the node.
2. If the sending node senses that the channel is busy during the DIFS interval, then it waits till the channel becomes idle. At this point it waits an additional DIFS interval, and generates a random backoff delay.
   It selects a random integer $x \in [0, CW]$, where CW is a positive integer called the contention window.

While $x > 0$, the node senses the medium for one slot time. If the medium is idle throughout the time slot, then x is decremented by unity. That is, $x \leftarrow (x - 1)$. This is repeated till x is equal to zero.
Thus, when the backoff counter reaches zero, the sending node transmits the frame.
The timer is frozen during the backoff process, if a transmission is sensed on the medium. In this case, the whole process of waiting for the channel to be idle is repeated again (which includes a new DIFS interval followed by a different backoff period).
3. After transmission of a frame, if the sending node does not receive the ACK acknowledgement frame from the recipient node, due to either errors or collision after a certain period of time, the protocol reactivates the backoff algorithm after the channel remains idle for an EIFS length of time.

Receiving Node: If the received frame arrives without errors, then an ACK acknowledgement frame is sent to the sending node, after a delay of SIFS.

Even after introducing the above timing variables, it is still possible to have frame collisions. The CSMA/CA protocol has certain other optional features. These features allow the nodes to reserve the medium for a certain duration of time. This is facilitated by introducing two control frames. These are: Request to Send (RTS), and Clear to Send (CTS) frames. These two frames are of short duration. This minimizes the risk of frame collision.

Assume that a sender node wants to transmit a data frame to a recipient node. The sender node initially sends an RTS frame to the recipient node. The RTS frame contains the duration of the time period for which the use of the medium has been requested. After an SIFS delay, the recipient node answers back by broadcasting a CTS frame. The CTS frame also has information about the duration of the time period for which the reservation was requested. The CTS frame performs two functions. It gives permission to the sender node to transmit the data frame. It also instructs other nodes to not send any frames for the duration of the reservation period.

This feature is useful when there are multiple frame collisions. If this is not the case, then it is not necessary to use this feature. A simplified mathematical model of the CSMA/CA protocol is described later in the section.

5.11.2 CSMA/CD Model

A simplified mathematical model of the CSMA/CD protocol is described in this subsection. In brief, the wired Ethernet LAN, which is a single bus uses this protocol. It has the following features.

(a) Each node listens for the presence of carrier signal on the bus (medium). A node transmits a frame, only when there is no energy on the bus.
(b) During the transmission of a frame, the transmitting node listens for collision. If a collision is detected, it aborts the transmission and reschedules it.

The random backoff strategy is ignored in this model. Also assume that two-way communication is possible on the LAN bus. The important variables used in the model are:

$\tau =$ end-to-end propagation delay on the wired LAN bus (channel).
$m =$ single-frame transmission time.
$a = \tau / m$. Assume that $a \ll 1$.
$p =$ probability with which a node transmits in time interval 2τ if it senses an idle channel.

$\nu =$ probability that only one node transmits, and is successful in the time interval 2τ. This is actually the no-collision probability.

$t_v =$ average time to successfully transmit a single frame. It is the virtual transmission time.

$J_{avg} =$ average number of retransmissions that are required to resolve collision.

$n =$ number of nodes that are involved in possible transmission. Assume that $n \gg 1$.

$\lambda_{\max} =$ maximum throughput (frames per unit time).

$\lambda =$ average number of frames per unit time that are transmitted over the bus from *all* nodes.

$\rho =$ traffic intensity.

Analysis

The throughput performance of a wired Ethernet LAN is next described. Let A be a node located at one end of the LAN bus, and B be another node located at the other end of the LAN bus. Thus the propagation delay that a frame transmission undergoes between nodes A and B is τ units of time. Assume that a collision occurs between frames transmitted from these two nodes. In the worst case it takes nodes A and B, 2τ units of time to detect the collision, and immediately turn-off the transmission. Thus the virtual transmission time is

$$t_v = (m + \tau + 2\tau J_{avg})$$

where m is the single-frame transmission time, τ is the propagation delay, and $2\tau J_{avg}$ is the average time required to resolve collisions once detected. The value J_{avg} is next determined. The probability that it takes $k \in \mathbb{P}$ intervals, each of length 2τ, to resolve collisions is p_k, where

$$p_k = \nu (1 - \nu)^{k-1}, \quad k \in \mathbb{P}$$

Observe that the number of retransmissions required to resolve collisions has a geometric distribution. The average number of such retransmissions is equal to J_{avg}, where

$$J_{avg} = \sum_{k \in \mathbb{P}} k p_k = \frac{1}{\nu}$$

Our next goal is to determine ν. Assume that n number of nodes are involved in possible transmission of frames. The probability that a node wants to transmit in the interval 2τ is p. Then the probability that precisely a single node out of n nodes transmits successfully in the interval 2τ is

$$\nu = np (1 - p)^{n-1}$$

For large values of n, the maximized value of ν obtained by varying p, is $\nu_{\max} = e^{-1}$. Therefore

$$\begin{aligned} t_v &= \{m + \tau (1 + 2e)\} \\ &= m \{1 + a (1 + 2e)\} \end{aligned}$$

Thus the maximum throughput $\lambda_{\max}$ in units of number of frames per unit time is

$$\lambda_{\max} = \frac{1}{t_v} \text{ frames per unit time}$$

The traffic intensity ρ is

$$\rho = \lambda m < \frac{1}{\{1 + a (1 + 2e)\}} \tag{5.49}$$

For example, if $a = 0.1$, then $\rho < 0.6$.

5.11.3 CSMA/CA Model

A simplified mathematical model of the CSMA/CA protocol is described in this subsection. In brief, the wireless Ethernet LAN which uses this protocol has the following features.

(a) Each node listens for the presence of carrier signal in the communication medium before transmitting the frame.
(b) The protocol attempts to avoid frame collisions by fine tuning the timing parameters in the nodes which use this protocol.

The assumptions made in the model are initially listed.

(a) All nodes are within the transmission range of each other. This implies that there can only be a single successful transmission at a time. Thus there are no hidden and exposed nodes.
(b) All nodes have frames to transmit at all times. This is often called the assumption of *infinite backlog*.
(c) All frames have the same length L.
(d) Backoff time intervals of all nodes have an exponential distribution with parameter $\beta > 0$. That is the length of mean backoff interval of each node is β^{-1} units of time.

The variables used in the model are:

$n =$ number of nodes, each of which is infinitely backlogged with frames.
$L =$ length of each frame.
$r =$ the constant data rate of each node.
$\delta =$ length of the slot time. Remember, that backoff interval is usually specified in units of slots.
$\beta =$ the parameter of the backoff interval, which has an exponential distribution.
$T_h =$ the frame-transmission overhead time. It includes SIFs, DIFSs, and EIFs. It also includes times for the transmission of the ACK frame; and also that of RTS and CTS frames if used. It also includes times for transferring all other frame overheads.
$T_c =$ the time elapsed during a collision before the next backoff period. It also includes other time intervals that occur during a frame collision.
$\gamma =$ the probability that a frame transmission suffers a collision.
$t_v =$ average time to successfully transmit a single frame. It is the virtual transmission time.
$t_p =$ average time to transmit the actual data content of a frame.
$\rho =$ normalized throughput of the wireless Ethernet LAN.

Analysis

It is shown in the problem section, that the probability a frame incurs a collision is

$$\gamma = 1 - e^{-(n-1)\beta\delta}$$

The time required to successfully transmit a single frame is made up of three stochastically independent components. The first component is the time spent by a frame in the backoff state, which is ready to be transmitted. The second component, is the actual time spent by a frame, and the corresponding overhead time intervals, on the communication channel. The third component is the time spent by the frame transmission during frame collision.

Thus in this simplified model

$$t_v = \frac{1}{n\beta} + (1-\gamma)\left\{\frac{L}{r} + T_h\right\} + \gamma T_c$$

In the above expression for t_v, the first term on the right-hand side, $1/(n\beta)$ corresponds to the average time spent by a frame in the backoff state, which is ready to be transmitted. The second term, $(1-\gamma)\{L/r + T_h\}$ is the actual average time spent by a frame, and the corresponding overhead time intervals, on the communication channel. The third term, γT_c is the average time spent by the frame transmission during frame collisions. Further

$$t_p = (1-\gamma)\frac{L}{r}$$

Therefore the normalized throughput is

$$\rho = \frac{t_p}{t_v} = \frac{(1-\gamma)\dfrac{L}{r}}{\dfrac{1}{n\beta} + (1-\gamma)\left\{\dfrac{L}{r} + T_h\right\} + \gamma T_c} \tag{5.50a}$$

Observe in the above expression for ρ that: as $\beta \to 0$, $\rho \to 0$; and as $\beta \to \infty$, $\rho \to 0$. Therefore, it is possible to maximize ρ by varying the value of β. It can be shown that ρ is maximized for

$$\beta = \frac{1}{2nT_c}\left\{\sqrt{1 + \frac{4nT_c}{(n-1)\delta}} - 1\right\}, \quad \text{where} \quad n \geq 2 \tag{5.50b}$$

5.12 Stochastic Geometry of Wireless Networks

Wireless networks are ubiquitous. These networks consist of source-destination pairs which transmit information over communication channels without wires. Absence of wires makes them more attractive than wired networks. Economics also plays a key role in their adaptation. Moreover, wireless networks represent critical Internet enabling technology.

Some performance aspects of wireless networks are studied in this section from a stochastic-theoretic perspective. The goal is to evaluate performance characteristics of the network like: signal-to-interference-ratio, outage probability, and transmission capacity.

Elements of wireless technology are initially described in this section. Subsequently mathematical preliminaries used in the probabilistic-characterization of a wireless network are specified. This is followed by the development of an interference model. Outage probability and transmission capacity of certain wireless networks are also evaluated.

5.12.1 Basics of Wireless Technology

Basics of wireless networks are initially described in this subsection. Certain relevant characteristics of propagation of electromagnetic energy are also elucidated. The effects of neighboring transmitters at a receiver are also described. This is followed by some examples of wireless networks. Some performance parameters of wireless network are also listed. Finally, a system model for a wireless network is developed.

Propagation of Electromagnetic Energy

Transmission of information takes place via electromagnetic waves in a wireless network. Electromagnetic waves travel from a transmitter antenna to the corresponding receiver-site antenna. The antennas convert electrical current into electromagnetic waves, and vice-versa. Some common modes of electromagnetic wave propagation are: line-of sight propagation, reflection, diffraction, and scattering.

Line of sight propagation implies direct radiation of electromagnetic energy between transmitter and receiver antennas which are visible to each other. In this mode of transmission, the power received at the receiver antenna is inversely proportional to the square of the transmission distance.

Reflection of an electromagnetic wave occurs at the interface of two dissimilar media. In this case, a fraction of the incident energy (wave) returns to its media of origination. The attenuation of the electromagnetic energy in this case depends upon the frequency, angle of incidence, and the type of the media. This effect is generally perceived indoors.

Diffraction occurs at the corners or sharp edges. This is in contrast to the line-of-sight transmission. Electromagnetic waves bend over hills, and travel around buildings and get into the shadowed regions. The effect of diffraction on the electromagnetic wave is much more severe than either line-of-sight propagation or reflection. Moreover, it is much more dominant outdoors than indoors. This effect is called shadowing.

Scattering occurs when electromagnetic waves impinge irregular objects. Instead of traveling in a straight line, the electromagnetic energy is radiated in all directions. This impacts the strength of electromagnetic energy which are far from the scatterer.

Sometimes, the transmitted signal arrives at the receiver by following more than a single path. The superposition of these multipath signals at the receiver can either be constructive or destructive. This results in variation in received signal strengths. This effect is called multipath fading. Fading varies with radio frequency, geographic area, and of course time. This is generally modeled as a random process.

In addition to the above effects; refraction, absorption, and polarization impact the strength of electromagnetic energy at the receiver.

Influence of Neighboring Transmitters

The performance of a wireless network depends upon the strength of the received signal at the receiver, the ambient noise, and interference from neighboring transmitting stations. A wireless network has several concurrent transmissions which are essentially using the same media. Therefore a receiver might receive unwanted electromagnetic energy from some of its active neighboring transmitters. This gives rise to the phenomenon of interference. The net interference at a receiver, in turn depends upon the seemingly random spatial configuration of other transmitters. This important effect is quantified by the signal-to-interference-plus-noise-ratio (SINR) measured at the receiver.

Some Examples of Wireless Networks

Some examples of wireless networks are: ad hoc, cellular, cognitive, femtocell, mesh, sensor, and ultra-wideband networks.

- Ad hoc wireless network is a decentralized type of network. This network has wireless links. It is ad hoc because each node is willing and capable of forwarding data destined for other nodes.
- Cellular systems segment a region into several geographical areas. A transmitter or radio relay antenna is located in each area. This antenna is responsible for forwarding calls from one area to the next area.
- Cognitive wireless networks use techniques borrowed from machine learning and knowledge representation.
- Femtocell networks help cell service-providers in offering service at the edges of a cell and possibly indoors.
- Wireless mesh networks are networks whose topology is in the form of a mesh. Ad hoc networks can have a mesh topology.
- Wireless sensor networks are networks of devices which track environmental conditions like temperature, pressure, wind, and pressure. These devices cooperate with each other and pass the collected information to a centrally located device.
- Ultra-wideband networks use low energy level, short-range and high bandwidth communication. A possible application is sensor data collection.

Parameters of Interest

Some possible parameters of interest in understanding the working of a wireless network are:

- *Signal-to-interference-plus-noise-ratio* (*SINR*): The interference at a specific wireless station, which is caused by neighboring wireless stations depends upon the location of these other stations. The spatial locations of these other stations are best described via the use of stochastic geometry. Thus communication channel noise, along with interference are limiting factors in performance of such networks. In contrast, classical techniques in communication engineering are simply concerned with signal-to-noise-ratio (SNR).
- *Outage Probability*: This is the probability that the SINR at a reference receiver is below a specified threshold required for proper reception. In order to understand the effect of only interference, the random noise is generally assumed to be zero.
- *Transmission capacity*: The transmission capacity of the network is defined as the total number of successful transmissions which occur in the network per unit area, which satisfy certain specific constraints on the outage probability.

System Model

Wireless networks can be analyzed by using tools from stochastic geometry and random graph theory. Mathematical techniques are introduced in the next subsection to develop a signal-to-interference-plus-noise-ratio model of a large wireless ad hoc network. The network is assumed to be *uncoordinated.* That is, there is absolutely no coordination between different transmitters regarding transmission decisions. All transmitting nodes are assumed to share the same narrow band communication channel. Transmitter/receiver mobility is not included in the model. Further, each transmitter has an associated receiver located at a fixed distance from it.

It is also assumed that the transmitting stations are independently and uniformly located over the network arena. Under this assumption, it is reasonable to model the location of the transmitters via a Poisson point process (PPP). The model results are obtained for a typical (average) network configuration.

Communication channel strength is assumed to be completely determined by path-loss and fading. Path-loss is the decay in the amplitude of the signal, as it travels in space. As mentioned earlier, fading of a signal also occurs due to randomized environmental conditions.

The goal is to develop a stochastic-geometry based model of a wireless ad hoc network. We also evaluate interference, outage probability of any given link, and transmission capacity of the network.

5.12.2 Preliminaries on Stochastic Modeling

Certain mathematical preliminaries in stochastic modeling of wireless ad hoc networks are described. Poisson point process is initially described. This is used in modeling the location of ad hoc network-stations in space. Distribution of Euclidean distance in infinite random networks is also obtained. This is useful in mapping homogeneous PPP in $\mathbb{R}^m$ to nonhomogeneous Poisson process in $\mathbb{R}$. Finally, use of Slivnyak's theorem is described.

Poisson Point Process

A Poisson point process is used to model the location of transmitters in the wireless network. Further, the distance between each transmitter-receiver pair is assumed to be a constant. A Poisson point process Φ is a countable set of random variables X_i's, which take values in the set S_P. The set S_P is typically the Euclidean space $\mathbb{R}^m$, where $m \in \mathbb{P}$. Let $\widetilde{S}_P$ be the Borel σ-algebra of the set S_P. Thus,

$$\Phi = \{X_n \mid X_n, n \in \mathbb{P}\}$$

is a PPP where:

(a) For every set $B \in \widetilde{S}_P$, the number of points $N(B)$ in it is

$$N(B) = \sum_{n \in \mathbb{P}} \zeta_{X_n}(B) \tag{5.51a}$$

where $\zeta_x(B)$ is the Dirac measure. That is, $\zeta_x(B) = I_B(x)$, where $I_B(\cdot)$ is the indicator function.

The random variable $N(B)$ has a Poisson distribution with mean equal to $\Lambda(B) \in \mathbb{R}^+$. That is

$$P(N(B) = k) = \exp(-\Lambda(B)) \frac{(\Lambda(B))^k}{k!}, \quad k \in \mathbb{N} \tag{5.51b}$$

Also, $\Lambda(\cdot)$ is called the intensity measure of the point process.

(b) If $B_1, B_2, \ldots, B_m$ are disjoint sets in $\widetilde{S}_P$, then $N(B_1), N(B_2), \ldots, N(B_m)$ are mutually independent random variables.

A PPP can either be homogeneous or nonhomogeneous. In a homogeneous PPP, the mean of the random variable $N(B)$ equals to $\Lambda(B) = \lambda |B|$, where $\lambda \in \mathbb{R}^+$, and $|B|$ denotes the hypervolume of the set B. In a nonhomogeneous PPP, $\Lambda(B) = \int_B \lambda(u)\, du$, and $\lambda(\cdot)$ is a positive function. The function $\lambda(\cdot)$ is also called the intensity or rate function of the PPP.

Other relevant details in describing point processes are described in the chapter on stochastic processes.

Euclidean Distance Distribution in Infinite Random Networks

Probability distribution of distance between a point in a PPP and its nth nearest neighbor is determined, where $n \in \mathbb{P}$.

Observations 5.10. Consider a homogeneous PPP in m-dimensional space $\mathbb{R}^m$, with intensity $\lambda > 0$, where $m = 1, 2, 3, \ldots$. Also let $R_n, n \in \mathbb{P}$ be the random variable which denotes the Euclidean distance from an arbitrarily selected point in $\mathbb{R}^m$ to its nth nearest neighbor in Φ.

Let the volume of the m-dimensional hypersphere $\mathcal{H}_r$ of radius $r \in \mathbb{R}^+$ be $V_m(r)$, where $m = 1, 2, 3, \ldots$. Thus

$$V_m(r) = c_m r^m \tag{5.52a}$$

where

$$c_m = \frac{\pi^{m/2}}{\Gamma(m/2+1)} \tag{5.52b}$$

and $\Gamma(\cdot)$ is the gamma function. Observe that $c_1 = 2, c_2 = \pi$, and $c_3 = 4\pi/3$. Also $\|\cdot\|$ is the Euclidean distance metric defined on points in space $\mathbb{R}^m$.

1. The cumulative distribution function of random variable R_n is $F_{R_n}(\cdot)$. We have

$$F_{R_n}(r) = 1 - \frac{\Gamma(n, \lambda c_m r^m)}{\Gamma(n)}, \quad r > 0 \tag{5.53a}$$

where $\Gamma(\cdot,\cdot)$ is the incomplete gamma function. In particular

$$F_{R_1}(r) = 1 - \exp(-\lambda c_m r^m), \quad r > 0$$

The probability density function of the random variable R_n is $f_{R_n}(\cdot)$. We have

$$f_{R_n}(r) = \frac{nm}{r} \exp(-\lambda V_m(r)) \frac{(\lambda V_m(r))^n}{n!}, \quad r > 0 \tag{5.53b}$$

2. Let $y \in \mathbb{R}^m$, and $X_i \in \mathbb{R}^m$ be points of the homogeneous Poisson process which are ordered according to their Euclidean distance from the point y. The random variables $R_i^m = \|y - X_i\|^m \triangleq Z_i, i \in \mathbb{P}$ are distributed as a one-dimensional Poisson process of intensity λc_m. In other words, Z_1, and $(Z_i - Z_{i-1})$, where $i > 1$ are exponentially distributed, each with parameter λc_m, and $\mathcal{E}(Z_i) = i/(\lambda c_m)$.
3. The random variables

$$R_i = \|y - X_i\|, \quad i \in \mathbb{P} \tag{5.54}$$

are distributed as a one-dimensional nonhomogeneous Poisson process with intensity function $\lambda c_m m r^{m-1}$. □

The above observations are established in the problem section.

Slivnyak's Theorem

We evaluate the strength of the received signal at a specific receiver. This receiver is also called the reference receiver. As per Slivnyak's theorem, if a point x is in an homogeneous PPP Φ, then the probability law of $\Phi - \{x\}$ conditioned on the fact that $x \in \Phi$ is the same as the probability law of simply Φ.

As the network is infinitely large and spatially homogeneous, the laws of PPP are not affected, if an additional transmitter/receiver pair is added to it. Furthermore, the performance of this transmitter/receiver pair is representative of a typical transmitter/receiver pair in the network. That is, the performance of this pair exhibits the average node-performance in the network. This is true as per Slivnyak's theorem. Thus the distribution of point process in space $\mathbb{R}^2$ is unaffected if the reference receiver is placed at the origin, and the associated reference transmitter at a fixed distance away from the receiver.

5.12.3 Interference Model

Interference at the reference receiver due to unwanted electromagnetic signals from neighboring transmitters is initially characterized. This is followed by evaluating the effect of interference from a nearby transmitter in the absence of fading. Effect of interference, in presence of fading is subsequently evaluated.

Signal-to-Interference-plus-Noise-Ratio

Let $x \in \mathbb{R}^m$, and $y \in \mathbb{R}^m$ be the spatial locations of the transmitter and receiver respectively. Possible values of m are $1, 2$, and 3. Each transmitter is assigned a receiver. The fixed distance between each such transmitter and receiver pair is $\varrho \in \mathbb{R}^+$. Also let $P_t(x)$ and $P_r(y)$ be the transmitted and received powers respectively. The power received at the receiver decays with distance as

$$P_r(y) = P_t(x) A_p H_{xy} \|x - y\|^{-\alpha}, \quad x \neq y \tag{5.55}$$

where A_p is a propagation constant, H_{xy} is a random variable which accounts for multipath fading and shadowing, $\|\cdot\|$ is the Euclidean norm. Also, $\alpha = 2b$ is the path-loss exponent, where b is the *amplitude loss exponent.* It can vary from 0.8 to 4. The value $b = 1$ corresponds to free-space propagation. The popular probabilistic fading models are the so-called Rayleigh and Nakagami fading. These use the exponential and gamma probability distributions respectively. For simplicity, the propagation constant A_p is assumed to be unity.

A wireless network has several concurrent transmissions which are essentially using the same media. Therefore the receiver might receive unwanted electromagnetic energy from some of its active neighboring transmitters. This gives rise to the phenomenon of interference. This important effect is quantified via the signal-to-interference-plus-noise-ratio (SINR). It is

$$SINR = \frac{S}{W_{ns} + I}$$

where S is the received signal power at the receiver, W_{ns} is the communication channel noise power, and I is the interference power. Noise is essentially unwanted signal due to spurious sources.

Definition 5.9. *Certain useful ratios involving signal, interference, and noise. These are: Signal-to-Interference-plus-Noise-Ratio (SINR), and Signal-to-Interference-Ratio (SIR).*

1. *Signal-to-interference-plus-noise-ratio is*

$$SINR = \frac{S}{W_{ns} + I} \tag{5.56a}$$

where S is the received signal power at the reference receiver, W_{ns} is the communication channel noise power, and I is the interference power. The communication channel noise power W_{ns}, and the interference I are stochastically independent of each other.

2. *Signal-to-interference-ratio is*

$$SIR = \frac{S}{I} \tag{5.56b}$$

□

The location of all interfering transmitters is modeled as a homogeneous Poisson point process $\varPhi_t = \{X_n \mid X_n, n \in \mathbb{P}\}$ of intensity $\lambda > 0$, where the random variable X_n takes values in the Euclidean space $\mathbb{R}^m$. The random variable X_n specifies the location of a transmitter.

The total interference measured at a point $x \in \mathbb{R}^m$ is $I(x)$ where

$$I(x) = \sum_{Y \in \varPhi_t} P_Y H_Y \ell(x - Y) \tag{5.57}$$

In the above expression: for each $Y \in \varPhi_t$, assume that $P_Y = P_t$ is the constant transmitted power, H_Y is the random variable that characterizes the combined effect of shadowing and fading, and $\ell(\cdot)$ is the path loss function. For simplicity, we assume that the transmitter power P_t equals unity.

The H_Y's are assumed to be independent and identically distributed as random variable H, with mean $\mathcal{E}(H) = 1$. It is also assumed that the Poisson point process $\varPhi_t$ is stochastically independent of the random variables H_Y's.

The path loss function is $\ell : \mathbb{R}^m \to \mathbb{R}^+$. We model $\ell(\cdot)$ as a power law: $\ell(x_i) = k_0 \|x_i\|^{-\alpha}$, where $\alpha > 0$ is the path-loss exponent, and $\|x_i\| \in \mathbb{R}^+$ for each $x_i \neq 0$. Also assume $k_0 = 1$ for simplicity.

As per Slivnyak's theorem, evaluation of $I(x)$ is independent of the location where interference is measured. More specifically, it is irrelevant, whether x is in $\varPhi_t$ or not, provided its contribution to $I(\cdot)$ is not included if $\varPhi_t$ is conditioned on having a point at x. This is true, because the network is assumed to be a homogeneous Poisson point process. Interference from a nearby transmitter without fading is next evaluated.

Interference from a Nearby Transmitter without Fading

The probability distribution of interference in the absence of fading from the n-th nearest neighboring transmitter is determined. This is accomplished by assuming H_Y to be equal to unity. Let I_n be the interference at the origin from the n-th nearest neighboring transmitter, where $n \in \mathbb{P}$. Let the random variable R_n specify the distance between the origin and the n-th nearest neighboring transmitter. Therefore $I_n = R_n^{-\alpha}$. The results obtained in determining the probability distribution of distance between a point in a PPP, and its nth nearest neighbor are used. These were developed in the last subsection.

Initially consider the nearest transmitter. That is $n = 1$. Then

$$\begin{aligned} P(I_1 \leq w) &= P\left(R_1^{-\alpha} \leq w\right) \\ &= P\left(R_1 \geq w^{-1/\alpha}\right) \\ &= \exp\left(-\lambda c_m w^{-\delta}\right), \quad w \in \mathbb{R}^+ \end{aligned} \tag{5.58a}$$

where $\delta = m/\alpha$. Therefore.

$$P(I_1 > w) \sim \lambda c_m w^{-\delta}, \quad w \to \infty$$

This is a heavy-tailed distribution. Also note that the probability density function $f_{I_1}(\cdot)$ of random variable I_1 is

$$f_{I_1}(w) = \lambda c_m \delta w^{-\delta-1} \exp\left(-\lambda c_m w^{-\delta}\right), \quad w \in \mathbb{R}^+$$

and

$$f_{I_1}(w) \sim \lambda c_m \delta w^{-\delta-1}, \quad w \to \infty$$

The mean of the random variable I_1 is

$$\mathcal{E}(I_1) = (\lambda c_m)^{1/\delta} \Gamma\left(1 - \delta^{-1}\right), \quad \delta > 1$$

If $\delta \leq 1$, then $\mathcal{E}(I_1)$ does not exist. That is, it is infinite. This result is a consequence of the heavy-tailed distribution of the random variable I_1 for small value of δ. It can be shown that $\mathcal{E}\left(I_1^k\right)$, for $k \in \mathbb{P}$ exists for values of δ which satisfy $k < \delta$.

The above results are next extended for any value of $n > 1$. We have

$$P(I_n \leq w) = P\left(R_n^{-\alpha} \leq w\right) = P\left(R_n \geq w^{-1/\alpha}\right) = \frac{\Gamma\left(n, \lambda c_m w^{-\delta}\right)}{\Gamma(n)}$$

where $\Gamma(\cdot,\cdot)$ is the incomplete gamma function. Thus

$$P(I_n \leq w) = \exp\left(-\lambda c_m w^{-\delta}\right) \sum_{k=0}^{n-1} \frac{\left(\lambda c_m w^{-\delta}\right)^k}{k!} \tag{5.58b}$$

and

$$P(I_n > w) = \exp\left(-\lambda c_m w^{-\delta}\right) \sum_{k \geq n} \frac{\left(\lambda c_m w^{-\delta}\right)^k}{k!}$$

As $w \to \infty$, the dominant term in the above series occurs at $k = n$. Therefore

$$P(I_n > w) \sim \frac{1}{n!} (\lambda c_m)^n w^{-n\delta}, \quad w \to \infty$$

Also, $\mathcal{E}\left(I_n^k\right)$, for $k \in \mathbb{P}$ exists for values of δ which satisfy $k < n\delta$. Interference in presence of fading is next evaluated.

Interference Evaluation in Presence of Fading

It can be observed from the expression for the interference $I(\cdot)$, that its value is dependent upon the seemingly randomized location of the transmitters. Moreover, these transmitters can also be mobile. Stochastic geometry can be used to model the locations of the interfering transmitters.

The location of transmitters is modeled as a homogeneous PPP

$$\Phi_t = \{X_n \mid X_n, n \in \mathbb{P}\}$$

of intensity $\lambda > 0$, where the random variable X_n takes values in the Euclidean space $\mathbb{R}^m$. The value of the random variable X_n specifies the location of a transmitter.

Let R_n, be the distances of points in the homogeneous PPP Φ_t from an arbitrary origin o in increasing order, where $n = 1, 2, 3, \ldots$. Then

$$\Phi = \{R_i = \|X_i\| \mid i \in \mathbb{P}\}$$

is a nonhomogeneous Poisson process defined on $\mathbb{R}_0^+$ with intensity measure $\mu(r) = \lambda c_m m r^{m-1}$, where $r > 0$, and $c_m = \pi^{m/2}/\Gamma(m/2+1)$. This conclusion follows by assuming Euclidean distance distribution in infinite random networks. Observe that interference can be modeled as a shot noise process. Description of this process is outlined in the chapter on stochastic processes. The interference measured at the origin o is

$$I \triangleq I(o) = \sum_{R \in \Phi} H_R R^{-\alpha} \tag{5.59a}$$

As mentioned earlier, the H_R's are assumed to be independent and identically distributed as random variable H, with mean $\mathcal{E}(H) = 1$. Furthermore, the random variables H_R's are independent of the point process Φ. The Laplace functional of the interference function $I(\cdot)$ is

$$\begin{aligned} \widehat{f}_I(s) &= \mathcal{E}(\exp(-sI)) \\ &= \exp\left\{-\int_0^\infty \mathcal{E}\{1 - \exp(-sHr^{-\alpha})\}\mu(r)\,dr\right\}, \; s \in \mathbb{C} \end{aligned} \tag{5.59b}$$

where the expectation in the integrand is with respect to the random variable H. The above expression for $\widehat{f}_I(s)$ simplifies to

$$\widehat{f}_I(s) = \exp\left(-\lambda c_m \mathcal{E}\left(H^\delta\right)\Gamma(1-\delta)\,s^\delta\right) \tag{5.59c}$$

where $\delta = m/\alpha$, $\delta \in (0,1)$, and $\Gamma(\cdot)$ is the gamma function. See the problem section for a derivation of the above expression. If the transmitter power P_t is not unity, then s is simply replaced by $P_t s$.

For $\alpha \leq m$, we have I tending towards infinity. This is true, because in this case, in an infinite network, the sum of interference from several far away transmitters at the receiver tends towards infinity. If the size of the network is finite, then the interference I will indeed be finite.

We next compute the average value of the interference. Note that $\mathcal{E}(I)$ does not exist for $\delta < 1$, because the first derivative of $\widehat{f}_I(s)$ with respect to s evaluated at s equal to zero does not exist. This is true because of singularity of the path-loss expression at the origin. It can be shown similarly that all higher moments of I do not exist. All moments of I exist, if a bounded path-loss expression is used.

For Rayleigh fading, the random variable H has an exponential distribution with unity as its average value. Therefore $\mathcal{E}\left(H^\delta\right) = \Gamma(1+\delta)$. Using the result

$$\Gamma(1+\delta)\,\Gamma(1-\delta) = \frac{\pi\delta}{\sin(\pi\delta)}, \quad \text{for} \quad \delta \in (0,1)$$

we obtain

$$\widehat{f}_I(s) = \exp\left(-\lambda c_m s^\delta \frac{\pi\delta}{\sin \pi\delta}\right)$$

The above expression is the Laplace transform of a Lévy stable distribution. If the dimension $m = 2$ and $\alpha = 4$, then $c_2 = \pi$, and $\delta = 1/2$. Then

$$\widehat{f}_I(s) = \exp\left(-\frac{1}{2}\pi^2 \lambda s^{1/2}\right)$$

The corresponding interference probability density function $f_I(\cdot)$ is

$$f_I(x) = \frac{\lambda}{4}\left(\frac{\pi}{x}\right)^{3/2} \exp\left(-\frac{\pi^4\lambda^2}{16x}\right), \quad x \in \mathbb{R}_0^+$$

It is shown in the problem section that the single-sided Laplace transform of $f_I(x), x \in \mathbb{R}_0^+$ is indeed $\widehat{f_I}(s)$. The cumulative distribution function of the random variable I is

$$F_I(x) = 1 - \operatorname{erf}\left(\frac{\pi^2\lambda}{4\sqrt{x}}\right), \quad x \in \mathbb{R}_0^+$$

where $\operatorname{erf}(x) = 2\int_0^x e^{-t^2}dt/\sqrt{\pi}$ is the error function.

5.12.4 A Combination Model for Path Loss and Fading

A point process model is described in this subsection to simultaneously reflect the effect of both path loss and fading. Let

$$\Phi_t = \{X_n \mid X_n, n \in \mathbb{P}\}$$

be a homogeneous PPP with unit intensity, where the random variable X_n takes values in the space $\mathbb{R}^2$. Also let

$$\widetilde{\Phi} = \{Y_n \mid Y_n = \|X_n\|^\alpha, X_n \in \Phi_t, n \in \mathbb{P}\}$$

be a path loss point process, which excludes the effect of fading, and $\|\cdot\|$ is the Euclidean metric in $\mathbb{R}^2$.

The fading is modeled via a sequence of independent and identically distributed random variables H_n, $n \in \mathbb{P}$. Each of these random variables is distributed as continuous random variable H, with cumulative distribution function $F_H(\cdot)$, and probability density function $f_H(\cdot)$.

A path loss process with fading is

$$\Xi = \left\{\xi_n \mid \xi_n = Y_n/H_n, Y_n \in \widetilde{\Phi}, n \in \mathbb{P}\right\}$$

In this model, if the nth transmitter transmits unit power, then the power received at the origin is ξ_n^{-1}. We next invoke the mapping theorem to establish that both $\widetilde{\Phi}$ and Ξ are PPPs. Observe that the process $\widetilde{\Phi}$ is a PPP, as it is obtained from the PPP Φ_t via deterministic mapping. Moreover, it is a one-dimensional Poisson process. Similarly, the process Ξ is a PPP, as it is obtained from $\widetilde{\Phi}$ via an independent displacement. We next determine the intensity functions of the PPP's $\widetilde{\Phi}$ and Ξ.

As the process Φ_t is a stationary PPP with unit intensity, the average number of points in the ball $b(o, x)$ is equal to $c_2x^2 = \pi x^2$, where $x \in \mathbb{R}^+$. The single dimension PPP $\widetilde{\Phi}$ has the same number of points in the interval $[0, y)$, where $y = x^\alpha$. Therefore the average number of points in $\widetilde{\Phi}$ in the interval $[0, y)$ is equal to $\pi x^2 = \pi y^{2/\alpha} = \pi y^\delta$, where $\delta = 2/\alpha$. Therefore the intensity function of the process $\widetilde{\Phi}$ is equal to $\widetilde{\mu}(y) = \pi\delta y^{\delta-1}$, where $y > 0$.

The displacement theorem is used to determine the intensity function of the PPP Ξ. Note that $\xi \in \Xi$, where $\xi = Y/H$, and $Y \in \widetilde{\Phi}$. Therefore

$$P(y/H \le x) = P(y/x \le H) = 1 - F_H(y/x)$$

The displacement kernel is

$$\rho(y,x) = \frac{d}{dx}\{1 - F_H(y/x)\} = \frac{y}{x^2} f_H(y/x)$$

As per the displacement theorem, the corresponding intensity function $\mu(x)$, for $x > 0$ is

$$\begin{aligned}\mu(x) &= \int_0^{\infty} \widetilde{\mu}(y)\,\rho(y,x)\,dy \\ &= \int_0^{\infty} \pi\delta y^{\delta-1} \frac{y}{x^2} f_H(y/x)\,dy \\ &= \pi\delta x^{\delta-1}\mathcal{E}\left(H^{\delta}\right)\end{aligned}$$

Assume that the random variable H has a gamma distribution, with parameter $k \in \mathbb{P}$, and unit mean. The probability density function of this random variable is

$$f_H(y) = \frac{k\,(ky)^{k-1}\,e^{-ky}}{\Gamma(k)}, \quad y \in \mathbb{R}_0^+$$

where $\Gamma(\cdot)$ is the gamma function. This probability density function of the fading random variable H, is also referred to as the Nakagami-k (power) fading model. The mean value of random variable H is unity. This is the probability density function of an exponentially distributed random variable with unit parameter, if $k = 1$. It can be shown that

$$\mathcal{E}\left(H^{\delta}\right) = \frac{\Gamma(k+\delta)}{k^{\delta}\Gamma(k)}$$

Therefore, for the Nakagami-k fading model

$$\mu(x) = \pi\delta x^{\delta-1}\mathcal{E}\left(H^{\delta}\right) = \pi\delta x^{\delta-1}\frac{\Gamma(k+\delta)}{k^{\delta}\Gamma(k)}, \quad x \in \mathbb{R}^+$$

Typically $\alpha > 2$, consequently $\delta \in (0,1)$, and $\mathcal{E}\left(H^{\delta}\right) < 1$. This is true because

$$\frac{\Gamma(k+\delta)}{k^{\delta}\Gamma(k)} < 1, \quad \text{for } k \in \mathbb{P} \quad \text{and} \quad \delta \in (0,1)$$

This fact is established in the problem section. Also as $k \to \infty$, $F_H(x) \to 1$ for all values of $x \geq 1$, and $\mathcal{E}\left(H^{\delta}\right) \to 1$. This implies that the affect of fading is very small for large values of k.

Assume that the node situated at the origin can successfully decode signals form neighboring transmitters after undergoing path loss (and possibly fading) smaller than some threshold. The above analysis demonstrates that fading will typically decrease the number of such neighbors. These results are summarized in the following observation.

Observation 5.11. Let

$$\Phi_t = \{X_n \mid X_n, n \in \mathbb{P}\} \tag{5.60a}$$

be a homogeneous PPP with unit intensity, where the random variable X_n takes values in the space $\mathbb{R}^2$. The X_n's specify the location of nodes in $\mathbb{R}^2$. Also let

$$\widetilde{\Phi} = \{Y_n \mid Y_n = \|X_n\|^{\alpha}, X_n \in \Phi_t, n \in \mathbb{P}\} \tag{5.60b}$$

The fading is modeled via a sequence of independent and identically distributed random variables H_n, $n \in \mathbb{P}$. Each of these random variables is distributed as continuous random variable H, with cumulative distribution function $F_H(\cdot)$, and probability density function $f_H(\cdot)$. A path loss process with fading is

$$\Xi = \left\{ \xi_n \mid \xi_n = Y_n / H_n, Y_n \in \widetilde{\Phi}, n \in \mathbb{P} \right\} \tag{5.60c}$$

Then

(a) $\widetilde{\Phi}$ is a nonhomogeneous one-dimensional Poisson process with intensity function $\widetilde{\mu}(y) = \pi\delta y^{\delta-1}$, where $y \in \mathbb{R}^+$. This process is ordered.
(b) Ξ is a nonhomogeneous one-dimensional Poisson process with intensity function $\mu(x) = \pi\delta x^{\delta-1}\mathcal{E}\left(H^\delta\right)$, where $x \in \mathbb{R}^+$. If Nakagami-k (power) fading model is assumed

$$\mu(x) = \pi\delta x^{\delta-1}\frac{\Gamma(k+\delta)}{k^\delta \Gamma(k)}, \quad x \in \mathbb{R}^+ \tag{5.60d}$$

where $k \in \mathbb{P}$. This process is no longer ordered in general.
Typically $\alpha > 2$, consequently $\delta \in (0,1)$, and $\mathcal{E}\left(H^\delta\right) < 1$. Also as $k \to \infty$, $F_H(x) \to 1$ for all values of $x \geq 1$, and $\mathcal{E}\left(H^\delta\right) \to 1$. Note that the intensities of the Poisson point processes $\widetilde{\Phi}$ and Ξ are proportional to each other. □

These results are used subsequently to evaluate transmission capacity of the network in presence of both path loss and fading.

5.12.5 Outage Probability

Link outage in a wireless network is said to occur if the transmitted information fails to reach the destination (receiver) node. We assume that this occurs if $SINR$ is less than some threshold value $\theta \in \mathbb{R}^+$. That is, $SINR < \theta$. The transmission success probability, and outage probability are formally defined.

Definition 5.10. *Transmission success and outage probabilities.*

1. *For some fixed $\theta \in \mathbb{R}^+$, the transmission success probability is $p_s(\theta)$, where*

$$p_s(\theta) = P(SINR > \theta) \tag{5.61}$$

and $SINR$ is the signal-to-interference-plus-noise-ratio.
2. *The link outage probability is $(1 - p_s(\theta))$.* □

Recall that each transmitter-receiver distance is $\varrho \in \mathbb{R}^+$. Further assume that the fading of the received signal has Rayleigh distribution. Therefore the signal power S at the receiver has exponential probability distribution with mean $\varrho^{-\alpha}$. Initially assume that $\varrho = 1$. Recall that signal-to-interference-plus-noise-ratio is $SINR = S/(W_{ns} + I)$ where S is the desired signal power at the reference receiver, W_{ns} is the communication channel noise power, and I is the interference power. Furthermore, the communication channel noise power W_{ns}, and the interference I are stochastically independent of each other. Also, signal-to-interference-ratio is $SIR = S/I$. If the communication channel noise power W_{ns} is equal to zero, then

$$p_s(\theta) = P(SIR > \theta) = P(S > I\theta) = \mathcal{E}(\exp(-\theta I))$$

where the expectation is over the random variable I. Observe that the complementary cumulative distribution of SIR is the outage probability. Also note that an expression for $\mathcal{E}(\exp(-sI))$ has already been obtained. Therefore

$$p_s(\theta) = \mathcal{E}(\exp(-\theta I)) = \widehat{f}_I(\theta) = \exp\left(-\lambda c_m \theta^\delta \frac{\pi\delta}{\sin \pi\delta}\right)$$

where $\delta = m/\alpha$, and $\delta \in (0,1)$. As the length of the link between the transmitter and and corresponding receiver is ϱ, then

$$p_s(\theta) = \mathcal{E}(\exp(-\theta\varrho^\alpha I)) = \exp\left(-\lambda c_m \theta^\delta \varrho^m \frac{\pi\delta}{\sin \pi\delta}\right) \tag{5.62}$$

Let the probability density function of SIR be $f_{SIR}(\cdot)$. It is obtained by taking the derivative of $(1 - p_s(\theta))$ with respect to θ.

Observation 5.12. The probability density function of SIR in a Poisson network with Rayleigh fading is

$$f_{SIR}(x) = c\delta x^{\delta-1} \exp\left(-cx^\delta\right), \quad x \in \mathbb{R}^+ \tag{5.63a}$$

where $c = \lambda c_m \varrho^m (\pi\delta) / \sin(\pi\delta)$, and $\delta = m/\alpha$, $\delta \in (0,1)$. This is a Weibull distribution, with mean value equal to

$$\mathcal{E}(SIR) = c^{-1/\delta}\Gamma(1 + 1/\delta) \tag{5.63b}$$

□

It is also convenient to include the effect of noise and interference simultaneously. Denote the success probability in the presence of noise and interference by $p_s^{W_{ns}+I}(\theta)$, and only in the presence of interference by $p_s^I(\theta)$. We have

$$\begin{aligned} p_s^{W_{ns}+I}(\theta) &= P(S > \theta(W_{ns} + I)) \\ &= \mathcal{E}(\exp(-\theta\varrho^\alpha (W_{ns} + I))) \\ &= \mathcal{E}(\exp(-\theta\varrho^\alpha W_{ns}))\, p_s^I(\theta) \end{aligned}$$

where $p_s^I(\theta) = p_s(\theta)$. If the transmitted power is P_t, then substitute W_{ns}/P_t for W_{ns} in the above expression. Observe that the interference-only success probability $p_s^I(\theta)$ is not dependent upon the transmitter power P_t.

5.12.6 Transmission Capacity

The transmission capacity of a wireless network is equal to the spatial intensity of successful transmissions, subject to a specific allowable outage probability. Upper-bounding the outage probability serves as a quality of service measure of the wireless network. We are interested in the transmission capacity of Poisson bipolar wireless networks. In such networks, the transmitters are spatially distributed as per a Poisson point process, and the receiver associated with a transmitter is at a fixed distance from it.

Definition 5.11. *Poisson bipolar network. A Poisson bipolar network consists of transmitters* $\{X_n \mid n \in \mathbb{P}\}$ *which are distributed as per a homogeneous PPP with intensity* $\lambda > 0$.

The random variables $X_n, n \in \mathbb{P}$ take values in the space $\mathbb{R}^m$. The receiver corresponding to the transmitter X_n is located at $Y_n \in \mathbb{R}^m$; and for some $\varrho \in \mathbb{R}^+$, $\|X_n - Y_n\| = \varrho$, $\forall$ $n \in \mathbb{P}$. □

As per the displacement theorem, the receiver set $\{Y_n \mid n \in \mathbb{P}\}$ also forms a homogeneous PPP of intensity λ.

Definition 5.12. *Transmission capacity of a bipolar network. Consider a Poissonian bipolar network of intensity $\lambda > 0$. The reference receiver is located at the origin in $\mathbb{R}^m$, and its corresponding transmitter is located at a distance $\varrho \in \mathbb{R}^+$ from it. Let the probability of successful reception at this receiver in presence of interference from other transmitters in the network be $f_s(\lambda) \triangleq p_s(\theta)$. Also let the target outage probability be $\varepsilon \in (0,1)$. The transmission capacity of the network $\lambda_{tc}(\varepsilon)$ is*

$$\lambda_{tc}(\varepsilon) = (1-\varepsilon)\max\{\lambda \mid f_s(\lambda) \geq (1-\varepsilon)\} \tag{5.64}$$

□

Observe that the success probability $f_s(\lambda)$ is a strictly monotonically decreasing function of λ. Therefore

$$\lambda_{tc}(\varepsilon) = (1-\varepsilon) f_s^{-1}(1-\varepsilon) \tag{5.65}$$

The computation of transmission capacity is next evaluated for some illustrative cases.

Example 5.5. Consider a two-dimensional network with Rayleigh fading. That is, $m = 2, c_2 = \pi$, and $\delta = 2/\alpha$. From an earlier derivation

$$f_s(\lambda) = p_s(\theta) = \exp\left(-\pi\lambda\theta^\delta\varrho^2\frac{\pi\delta}{\sin\pi\delta}\right)$$

Define $\gamma = \pi^2\theta^\delta\varrho^2\delta/\sin(\pi\delta)$. This results in

$$\lambda_{tc}(\varepsilon) = -\frac{1}{\gamma}(1-\varepsilon)\ln(1-\varepsilon)$$

As $\ln(1-\varepsilon) = -\sum_{n\in\mathbb{P}}\varepsilon^n/n$, we have

$$\begin{aligned}\lambda_{tc}(\varepsilon) &= \frac{1}{\gamma}\left(\varepsilon - \frac{\varepsilon^2}{2}\right) + O(\varepsilon^3)\\ &= \frac{\varepsilon}{\gamma} + O(\varepsilon^2)\end{aligned}$$

This approximation is true for only small values of ε. For larger values of ε, the transmission capacity $\lambda_{tc}(\cdot)$ is not a monotonically increasing function of ε. □

Example 5.6. Transmission capacity of a two-dimensional network without fading is evaluated. That is, $m = 2, c_2 = \pi, \alpha = 4$, and $H = 1$ (no fading). Therefore $\delta = 1/2$. From an earlier derivation

$$\widehat{f}_I(s) = \exp\left(-\lambda c_m \Gamma(1-\delta)s^\delta\right) = \exp\left(-\lambda\pi^{3/2}s^{1/2}\right)$$

where we used the result $\Gamma(1/2) = \sqrt{\pi}$. The probability density function of the random variable I is

$$f_I(x) = \frac{\pi\lambda}{2x^{3/2}} \exp\left(-\frac{\pi^3\lambda^2}{4x}\right), \quad \lambda > 0, \text{ and } x \in \mathbb{R}_0^+$$

The cumulative distribution function of the random variable I is

$$P(I \le x) = F_I(x) = 1 - \operatorname{erf}\left(\frac{\pi^{3/2}\lambda}{2\sqrt{x}}\right), \quad x \in \mathbb{R}_0^+$$

where $\operatorname{erf}(x) = 2\int_0^x e^{-t^2} dt/\sqrt{\pi}$ is the error function. For a specified signal power S at the receiver, we have $S = \varrho^{-4}$

$$\begin{aligned} p_s(\theta) &= P\left(\varrho^{-4} > I\theta\right) = P\left(I < 1/\left(\theta\varrho^4\right)\right) \\ &= 1 - \operatorname{erf}\left(\frac{\pi^{3/2}\lambda\varrho^2\sqrt{\theta}}{2}\right) \end{aligned}$$

As $p_s(\theta) = f_s(\lambda)$, and $\lambda_{tc}(\varepsilon) = (1-\varepsilon) f_s^{-1}(1-\varepsilon)$, we have

$$\lambda_{tc}(\varepsilon) = \frac{2(1-\varepsilon)}{\pi^{3/2}\varrho^2\sqrt{\theta}} \operatorname{erf}^{-1}(\varepsilon)$$

For small values of ε, we have

$$\lambda_{tc}(\varepsilon) \sim \frac{\varepsilon}{\pi\varrho^2\sqrt{\theta}} + O\left(\varepsilon^2\right)$$

□

Example 5.7. A general fading network model is considered in this example. A bound on the transmission capacity is obtained by considering only the effect of dominating transmitters. The set of such transmitters Φ_{dom} is specified as

$$\Phi_{dom} = \left\{X \mid X \in \Phi_t,\ H\varrho^{-\alpha}/\left(H_X \|X\|^{-\alpha}\right) < \theta\right\}$$

where H is the fading-coefficient random variable of the reference transmitter-receiver link, and H_X is the fading-coefficient random variable of the interfering transmitter at location $X \in \mathbb{R}^2$. Recall that Ξ is a single dimensional nonhomogeneous Poisson process with intensity function $\mu(\cdot)$, where

$$\mu(x) = \pi\lambda\delta x^{\delta-1}\mathcal{E}\left(H^\delta\right), \quad x \in \mathbb{R}^+$$

where $\delta = 2/\alpha$. For a given H, the probability that there are zero number of dominant points of the process Ξ within a distance $\varrho^\alpha\theta/H$ of the origin is

$$\begin{aligned} P\left(\Phi_{dom} = \varnothing \mid H\right) &= \exp\left(-\int_0^{\varrho^\alpha\theta/H} \mu(x)\, dx\right) \\ &= \exp\left(-\pi\lambda\mathcal{E}\left(H^\delta\right)\varrho^2\theta^\delta H^{-\delta}\right) \end{aligned}$$

Consequently

$$p_s(\theta) < \mathcal{E}\left(\exp\left(-\pi\lambda\mathcal{E}\left(H^\delta\right)\varrho^2\theta^\delta H^{-\delta}\right)\right)$$

where the outer expectation operator on the right-hand side of the above inequality is with respect to the random variable H. Use of Jensen's inequality results in

$$\mathcal{E}\left(\exp\left(-\pi\lambda\mathcal{E}\left(H^{\delta}\right)\varrho^{2}\theta^{\delta}H^{-\delta}\right)\right)\geq\exp\left(-\pi\lambda\varrho^{2}\theta^{\delta}\mathcal{E}\left(H^{\delta}\right)\mathcal{E}\left(H^{-\delta}\right)\right)$$

From the last two inequalities, it is hoped that the following approximation is true. That is,

$$p_{s}\left(\theta\right)\simeq\exp\left(-\pi\lambda\varrho^{2}\theta^{\delta}\mathcal{E}\left(H^{\delta}\right)\mathcal{E}\left(H^{-\delta}\right)\right)$$

As $p_s(\theta) = f_s(\lambda)$, and $\lambda_{tc}(\varepsilon) = (1-\varepsilon) f_s^{-1}(1-\varepsilon)$, we have

$$\lambda_{tc}\left(\varepsilon\right)\simeq\frac{-\left(1-\varepsilon\right)\ln\left(1-\varepsilon\right)}{\pi\varrho^{2}\theta^{\delta}\mathcal{E}\left(H^{\delta}\right)\mathcal{E}\left(H^{-\delta}\right)}\sim\frac{\varepsilon}{\pi\varrho^{2}\theta^{\delta}\mathcal{E}\left(H^{\delta}\right)\mathcal{E}\left(H^{-\delta}\right)}$$

Note that the above approximate result turns out to be identical to the corresponding expression for Rayleigh fading.

In presence of fading, $\mathcal{E}\left(H^{\delta}\right)\mathcal{E}\left(H^{-\delta}\right) > 1$. Consequently, $\lambda_{tc}(\varepsilon)$ is negatively impacted by fading. □

Reference Notes

For a comprehensive introduction to the mathematics of Internet congestion control, the reader can refer to the masterful monograph by Srikant (2004). The subject of Internet congestion has been studied and surveyed by several groups. Some of these are: Gevros, Crowcroft, Kirstein, and Bhatti (2001), Low, Paganini, and Doyle (2002), Ryu, Rump, and Qiao (2003), and Varma (2015).

Jain (1990) develops a good case for intelligent congestion management. His observations are still valid today. The two examples given in the section on congestion control and resource allocation are from the monograph by Srikant (2004). He attributes these results to the seminal papers by Chiu, and Jain (1989); and Bonomi, Mitra, and Seery (1995). Basic facts about TCP can be found in Hassan, and Jain (2004). TCP congestion control was first proposed by Jacobson (1988), and the simple periodic TCP model is developed in Mathis, Semke, Mahdavi, and Ott (1997). A mean-field model of multiple TCP/IP connections is due to Baccelli, McDonald, and Reynier (2002).

The collective wisdom on queue management and congestion avoidance in the Internet has been crystallized in the RFC 2309 by the luminaries: Braden, Clark, Crowcroft, Davie, Deering, Estrin, Floyd, Jacobson, Minshall, Partridge, Peterson, Ramakrishnan, Shenker, Wroclawski, and Zhang (1998). The random-early-detection technique of active queue management was first proposed by Floyd, and Jacobson in 1993, who are among some of the earliest workers in the field of Internet congestion management. The analysis of the RED algorithm discussed in this chapter is based upon the work of Misra, Gong, and Towsley (2000), and Hollot, Misra, Towsley, and Gong (2001a & b, 2002).

The section on rate and congestion control in networks, and stability is based upon the pioneering work of Frank Kelly and his colleagues. Details can be found in Kelly (1997), Kelly, Maulloo, and Tan (1998), and Kelly (2001). Description of a user's utility function was adapted from Mo, and Walrand (2000). The reader should refer to the textbook by Courcoubetis, and Weber (2003), for an enlightening discussion on control of congestion via pricing. The simple and elegant max-min fair algorithms for capacity allocation are from Bertsekas, and Gallager (1992).

The section on traffic management via packet scheduling is based upon the landmark papers by Parekh, and Gallager (1993, 1994). They proposed the GPS scheme. A packetized version of the GPS is called the packetized generalized processor sharing (PGPS) scheme. It was first proposed by Demers, Keshav, and Shenker (1990). However, they called it weighted fair queueing (WFQ).

The ground work for deterministic network calculus was laid down in a pair of seminal papers by Cruz (1991a and b). Significant results in this area were also obtained by Agrawal, Cruz, Okino, and Rajan (1999), Chang (2000), and Le Boudec, and Thiran (2004). Le Boudec, and Thiran's masterful monograph is a definitive work on network calculus.

The CSMA/CD LAN model is based upon Schwartz (1987); and the CSMA/CA LAN model is based upon Kumar, Manjunath, and Kuri (2004).

The section on stochastic geometry of wireless networks is based upon the scholarly work of Baccelli, and Blaszczyszyn (2009a & b, Volumes I and II); and Haenggi, Andrews, Baccelli, Dousse, and Franceschetti (2009). Haenggi (2013) provides an introduction to stochastic geometry of wireless networks. Another useful introduction to stochastic geometry is by Stoyan, Kendall, and Mecke (1995).

Problems

1. Find the solution of the differential equation
$$\frac{dx(t)}{dt} = J(t)\,x(t) + M(t)\,Q(t), \quad t \in \mathbb{R}$$
The initial value $x(t_0)$ is given. Also assume sufficient continuity conditions about $J(t)$ and $M(t)$ so that the solution exists.
Hint: Prove that
$$x(t) = x(t_0)\,e^{f(t)} + e^{f(t)} \int_{t_0}^{t} M(\tau)\,Q(\tau)\,e^{-f(\tau)} d\tau$$
where $f(t) = \int_{t_0}^{t} J(y)\,dy$.
2. Linearize the nonlinear equations which model the interaction between TCP flows and the RED scheme.
Hint: See Hollot, Misra, Towsley, and Gong, (2002). The nonlinear dynamical equations describing the TCP and RED scheme for $t \in \mathbb{R}_0^+$ are
$$\dot{W}(t) = \frac{1}{R(t)} - \frac{W(t)\,W(t-R(t))}{2R(t-R(t))}\varrho(t-R(t))$$
$$\dot{\theta}(t) = \frac{W(t)}{R(t)}N - C_\ell$$
where $R(t) = \{a_\varpi + \theta(t-R(t))/C_\ell\}$. When $\dot{W}(t) = \dot{\theta}(t) = 0$, denote the variables $W(t), \theta(t), R(t)$, and $\varrho(t)$ by W_0, θ_0, R_0, and ϱ_0 respectively. Thus

$$W_0^2 \varrho_0 = 2, \;\; W_0 = \frac{R_0 C_\ell}{N}, \;\; \text{and } R_0 = a_\varpi + \frac{\theta_0}{C_\ell}$$

Let

$$\begin{aligned} W(t) &\triangleq W_0 + \delta W(t) \\ \theta(t) &\triangleq \theta_0 + \delta\theta(t) \\ \varrho(t) &\triangleq \varrho_0 + \delta\varrho(t) \end{aligned}$$

Assume that the terms which depend upon $(t - R(t))$ are approximated by $(t - R_0)$. Thus the expression for $\dot{W}(t)$ can be written as

$$\dot{W}(t) = \frac{1}{R(t)} - \frac{W(t)\, W(t - R_0)}{2R(t - R_0)} \varrho(t - R_0)$$

The nonlinear model is linearized as

$$\begin{aligned} \delta\dot{W}(t) &= -\frac{N}{R_0^2 C_\ell} \{\delta W(t) + \delta W(t - R_0)\} \\ &\quad - \frac{1}{R_0^2 C_\ell} \{\delta\theta(t - R_0) - \delta\theta(t - 2R_0)\} \\ &\quad - \frac{R_0 C_\ell^2}{2N^2} \delta\varrho(t - R_0) \\ \delta\dot{\theta}(t) &= \frac{N}{R_0} \delta W(t) - \frac{1}{R_0} \delta\theta(t - R_0) \end{aligned}$$

For typical networks $W_0 \gg 1$. This leads to a simplification in the expression for $\delta\dot{W}(t)$ and $\delta\dot{\theta}(t)$. Thus

$$\begin{aligned} \delta\dot{W}(t) &= -\lambda_0 \delta W(t) - \frac{R_0 C_\ell^2}{2N^2} \delta\varrho(t - R_0) \\ \delta\dot{\theta}(t) &= \frac{N}{R_0} \delta W(t) - \frac{1}{R_0} \delta\theta(t) \end{aligned}$$

where $\lambda_0 = 2/(W_0 R_0)$.

3. Determine the partial differential equation which describes the probability density $p(t, w)$ of the window size w at time t, when the number of TCP connections tends to infinity.
Hint: See Baccelli, McDonald, and Reynier (2002). The window dynamics of stream i connection is described by

$$dW_i(t) = \frac{dt}{R(t)} - \frac{W_i(t)}{2} dN_i(t), \;\; \text{where } i \in \{1, 2, \ldots, N\}$$

Thus

$$\begin{aligned} &\frac{1}{N} \sum_{i=1}^{N} (W_i(t) - W_i(0)) \\ &= \frac{1}{N} \sum_{i=1}^{N} \int_0^t \left\{ \frac{1}{R(y)} dy + \left(\frac{W_i(y)}{2} - W_i(y) \right) dN_i(y) \right\} \end{aligned}$$

In order to overcome singular behavior associated with the TCP connections disappearing from the system, a bounded function $g(\cdot)$ is introduced. It is a function of the window size w. The function $g(\cdot)$ takes nonnegative values. In addition, it has bounded derivatives, and $g(0) = 0$. Therefore on both sides of the above equation, replace each value of window size w by $g(w)$. Thus we obtain

$$\begin{aligned}&\frac{1}{N}\sum_{i=1}^{N}\{g(W_i(t)) - g(W_i(0))\}\\&= \frac{1}{N}\sum_{i=1}^{N}\int_0^t \left\{\frac{dg(W_i(y))}{dw}\frac{1}{R(y)}dy\right.\\&\quad \left. + \left(g\left(\frac{W_i(y)}{2}\right) - g(W_i(y))\right)dN_i(y)\right\}\end{aligned}$$

That is

$$\begin{aligned}&\frac{1}{N}\sum_{i=1}^{N}\{g(W_i(t)) - g(W_i(0))\}\\&= \frac{1}{N}\sum_{i=1}^{N}\int_0^t \left\{\frac{dg(W_i(y))}{dw}\frac{1}{R(y)}dy\right.\\&\quad \left. + \left(g\left(\frac{W_i(y)}{2}\right) - g(W_i(y))\right)\frac{W_i(y-R(y))}{R(y-R(y))}F(Q_N(y-R(y)))\,dy\right\}\end{aligned}$$

We also make the approximation

$$\mathcal{E}(W_i(y-R(y)) \mid W_i(y), Q_N(y-R(y))) \simeq W_i(y)$$

If $N \to \infty$, we have $R(\cdot) \to r(\cdot)$, and $F(Q_N(\cdot)) \to k(\cdot)$. Furthermore, the above equation can be written as

$$\begin{aligned}&\int_w \{p(t,w) - p(0,w)\}g(w)\,dw\\&= \int_{y=0}^{t}\int_w \frac{1}{r(y)}\frac{dg(w)}{dw}p(y,w)\,dwdy\\&\quad + \int_{y=0}^{t}\int_w\int_v g(w)\,2vp(y-r(y),v;y,2w)\frac{k(y-r(y))}{r(y-r(y))}dvdwdy\\&\quad - \int_{y=0}^{t}\int_w\int_v g(w)\,vp(y-r(y),v;y,w)\frac{k(y-r(y))}{r(y-r(y))}dvdwdy\\&= -\int_w g(w)\int_{y=0}^{t}\frac{1}{r(y)}\frac{\partial p(y,w)}{\partial w}dydw\\&\quad + \int_w g(w)\int_{y=0}^{t}\{4wp(y,2w) - wp(y,w)\}\frac{k(y-r(y))}{r(y-r(y))}dydw\end{aligned}$$

The result stated in the observation is obtained from the above equation.

4. Find a solution of the differential equation

$$\frac{df(w)}{dw} = k\{4wf(2w) - wf(w)\}, \quad w \in \mathbb{R}_0^+$$

where $\int_w f(w)\,dw = 1$.
Hint: Let the Mellin transform of $f(w)$ be $F_M(s) = \int_0^\infty w^{s-1} f(w)\,dw$, where $\text{Re}(s) \geq 1$. The differential equation yields

$$sF_M(s) = k\left(1 - 2^{-s}\right)F_M(s+2)$$

Let $F_M(s) \triangleq v(s)(2/k)^{s/2}\,\Gamma(s/2)$. Substituting this expression in the above equation yields

$$v(s) = \left(1 - 2^{-s}\right)v(s+2)$$

The above equality is satisfied by $v(s) = \alpha\prod_{j\in\mathbb{N}}\left(1 - 2^{-s-2j}\right)$, where α is a constant. Therefore

$$F_M(s) \triangleq \alpha\left(\frac{2}{k}\right)^{s/2}\Gamma(s/2)\prod_{j\in\mathbb{N}}\left(1 - 2^{-s-2j}\right)$$

The value of α is obtained by noting that $F_M(1) = 1$. This is true, because $f(\cdot)$ is a probability density function. Therefore

$$F_M(s) = \left(\frac{2}{k}\right)^{(s-1)/2}\frac{\Gamma(s/2)}{\sqrt{\pi}}\prod_{j\in\mathbb{N}}\frac{\left(1 - 2^{-2j-s}\right)}{\left(1 - 2^{-2j-1}\right)}$$

The next step is to invert $F_M(s)$. Represent $f(w)$ as

$$f(w) = \sum_{i\in\mathbb{N}} b_i \exp\left(-k4^i\frac{w^2}{2}\right)$$

Recall that the Mellin transform of $\exp\left(-bw^2\right)$ is $b^{-s/2}\Gamma(s/2)/2$, where $b \in \mathbb{R}^+$, and $\text{Re}(s) > 0$. Therefore

$$F_M(s) = \left(\frac{2}{k}\right)^{s/2}\frac{\Gamma(s/2)}{2}\sum_{i\in\mathbb{N}} b_i 2^{-is}$$

Comparison of the two expressions for $F_M(s)$ yields

$$\sum_{i\in\mathbb{N}} b_i 2^{-is} = \left(\frac{2k}{\pi}\right)^{1/2}\prod_{j\in\mathbb{N}}\frac{\left(1 - 2^{-2j-s}\right)}{\left(1 - 2^{-2j-1}\right)}$$

The final result, which is stated as an observation follows by comparison of terms on both sides of the above equation.

5. A single user $r \in \mathfrak{R}$ is specified by $USER_r[U_r; \lambda_r]$ where $\lambda_r \in \mathbb{R}^+$. Solve for w_r:

$$\max U_r\left(\frac{w_r}{\lambda_r}\right) - w_r$$

over $w_r \geq 0$.
Hint: See Kelly (1997). Let the Lagrangian be $\mathcal{L}_{USER}(w_r)$, where

$$\mathcal{L}_{USER}(w_r) = \{U_r(w_r/\lambda_r) - w_r\}$$

Then

$$\frac{\partial \mathcal{L}_{USER}(w_r)}{\partial w_r} = \frac{1}{\lambda_r} U_r'\left(\frac{w_r}{\lambda_r}\right) - 1$$

where $U_r'(\cdot)$ is the first derivative of $U_r(\cdot)$.

(a) If $w_r > 0$ then $\partial \mathcal{L}_{USER}(w_r)/\partial w_r = 0$. This implies $U_r'(w_r/\lambda_r) = \lambda_r$.
(b) If $w_r = 0$ then $\partial \mathcal{L}_{USER}(w_r)/\partial w_r \leq 0$. This implies $U_r'(w_r/\lambda_r) \leq \lambda_r$.

These in turn imply $w_r \partial \mathcal{L}_{USER}(w_r)/\partial w_r = 0$, that is

$$w_r \{U_r'(w_r/\lambda_r) - \lambda_r\} = 0$$

Summarizing, a w_r with the following conditions solves $USER_r[U_r; \lambda_r]$.

$$w_r \geq 0, \ \ \lambda_r \geq U_r'\left(\frac{w_r}{\lambda_r}\right), \ \ w_r\left\{U_r'\left(\frac{w_r}{\lambda_r}\right) - \lambda_r\right\} = 0$$

6. The network is specified by $NETWORK[A, C; w]$. Solve for β:

$$\max \sum_{r \in \mathfrak{B}(w)} w_r \ln \beta_r$$

subject to $A\beta \leq C$, and $\beta \geq 0$. Recall that $\mathfrak{B}(w) \triangleq \{r \in \mathfrak{R} \mid w_r > 0\}$.
Hint: See Kelly (1997). Let $C = A\beta + z$, where $z = (z_1, z_2, \ldots, z_{|E|})$. The nonnegative vector elements of z, are called the *slack variables*. Define $\mu = (\mu_1, \mu_2, \ldots, \mu_{|E|})$. The Lagrangian of this optimization problem is

$$\begin{aligned}\mathcal{L}_{NET}(\beta, z, \mu) &= \sum_{r \in \mathfrak{B}(w)} w_r \ln \beta_r + \mu^T (C - A\beta - z) \\ &= \sum_{r \in \mathfrak{B}(w)} w_r \ln \beta_r + \sum_{e \in E} \mu_e (c_e - z_e) - \sum_{r \in \mathfrak{R}} \beta_r \sum_{e \in \pi_r} \mu_e\end{aligned}$$

We maximize $\mathcal{L}_{NET}(\beta, z, \mu)$ such that $\beta, z \geq 0$.
Next compute partial derivatives of $\mathcal{L}_{NET}(\beta, z, \mu)$ with respect to β_r and z_e. Therefore

$$\frac{\partial \mathcal{L}_{NET}(\beta, z, \mu)}{\partial \beta_r} = \begin{cases} \dfrac{w_r}{\beta_r} - \sum_{e \in \pi_r} \mu_e, & r \in \mathfrak{B}(w) \\ -\sum_{e \in \pi_r} \mu_e, & r \notin \mathfrak{B}(w) \end{cases}$$

$$\frac{\partial \mathcal{L}_{NET}(\beta, z, \mu)}{\partial z_e} = -\mu_e$$

(a) Consider the partial derivative of the Lagrangian with respect to β_r.
  (i) If $r \in \mathfrak{B}(w)$ then $w_r > 0$. Also $w_r = \lambda_r \beta_r$. Therefore $\beta_r > 0$.
  Thus $\partial \mathcal{L}_{NET}(\beta, z, \mu)/\partial \beta_r = 0$. This implies $\beta_r = w_r / \sum_{e \in \pi_r} \mu_e$.
  (ii) If $r \notin \mathfrak{B}(w)$ then $w_r = 0$. Also $w_r = \lambda_r \beta_r$, and $\lambda_r > 0$ imply $\beta_r = 0$.
  This in turn implies $\partial \mathcal{L}_{NET}(\beta, z, \mu)/\partial \beta_r \leq 0$.

(b) Consider the partial derivative of the Lagrangian with respect to z_e.
  (i) If $z_e > 0$ then $\partial \mathcal{L}_{NET}(\beta, z, \mu)/\partial z_e = 0$. This implies $\mu_e = 0$.

(ii) If $z_e = 0$ then $\partial \mathcal{L}_{NET}(\beta, z, \mu)/\partial z_e \leq 0$. This implies $\mu_e \geq 0$.

Case (a) implies $w_r = \beta_r \sum_{e\in\pi_r} \mu_e$. However $w_r = \beta_r \lambda_r$, which implies $\lambda_r = \sum_{e\in\pi_r} \mu_e$. Similarly, Case (b) implies $\mu_e z_e = 0$ and $\mu_e \geq 0$, that is $\mu^T z = 0$ and $\mu \geq 0 \Rightarrow \mu^T(C - A\beta) = 0$, and $\mu \geq 0$.

In summary, a β with the following conditions

$$A\beta \leq C, \quad \beta \geq 0, \quad \mu^T(C - A\beta) = 0, \quad \mu \geq 0;$$
$$\lambda_r = \sum_{e\in\pi_r} \mu_e, \text{ and } w_r = \lambda_r \beta_r, \quad \text{for } r \in \mathfrak{R}$$

solves $NETWORK[A, C; w]$.

7. The system is specified by $SYSTEM[U, A, C]$. Solve for β:

$$\max \sum_{r\in\mathfrak{R}} U_r(\beta_r)$$

subject to $A\beta \leq C$, and $\beta \geq 0$.

Hint: See Kelly (1997). Let $C = A\beta + z$, where $z = (z_1, z_2, \ldots, z_{|E|})$ and $z \geq 0$. Also define $\widetilde{\mu} = (\widetilde{\mu}_1, \widetilde{\mu}_2, \ldots, \widetilde{\mu}_{|E|})$. The Lagrangian of this optimization problem is

$$\begin{aligned}\mathcal{L}_{SYS}(\beta, z, \widetilde{\mu}) &= \sum_{r\in\mathfrak{R}} U_r(\beta_r) + \widetilde{\mu}^T(C - A\beta - z) \\ &= \sum_{r\in\mathfrak{R}} U_r(\beta_r) + \sum_{e\in E} \widetilde{\mu}_e(c_e - z_e) - \sum_{r\in\mathfrak{R}} \beta_r \sum_{e\in\pi_r} \widetilde{\mu}_e\end{aligned}$$

We maximize $\mathcal{L}_{SYS}(\beta, z, \widetilde{\mu})$ such that $\beta, z \geq 0$.

Next compute partial derivatives of $\mathcal{L}_{SYS}(\beta, z, \widetilde{\mu})$ with respect to β_r and z_e. Therefore

$$\frac{\partial \mathcal{L}_{SYS}(\beta, z, \widetilde{\mu})}{\partial \beta_r} = U_r'(\beta_r) - \sum_{e\in\pi_r} \widetilde{\mu}_e$$
$$\frac{\partial \mathcal{L}_{SYS}(\beta, z, \widetilde{\mu})}{\partial z_e} = -\widetilde{\mu}_e$$

(a) Consider the partial derivative of the Lagrangian with respect to β_r.
   (i) If $\beta_r > 0$ then $\partial \mathcal{L}_{SYS}(\beta, z, \widetilde{\mu})/\partial \beta_r = 0$. This implies $U_r'(\beta_r) = \sum_{e\in\pi_r} \widetilde{\mu}_e$.
   (ii) If $\beta_r = 0$ then $\partial \mathcal{L}_{SYS}(\beta, z, \widetilde{\mu})/\partial \beta_r \leq 0$. This implies $U_r'(\beta_r) \leq \sum_{e\in\pi_r} \widetilde{\mu}_e$.

(b) Consider the partial derivative of the Lagrangian with respect to z_e.
   (i) If $z_e > 0$ then $\partial \mathcal{L}_{SYS}(\beta, z, \widetilde{\mu})/\partial z_e = 0$. This implies $\widetilde{\mu}_e = 0$.
   (ii) If $z_e = 0$ then $\partial \mathcal{L}_{SYS}(\beta, z, \widetilde{\mu})/\partial z_e \leq 0$. This implies $\widetilde{\mu}_e \geq 0$.

Case (a) implies $U_r'(\beta_r) \leq \sum_{e\in\pi_r} \widetilde{\mu}_e$ and $\{U_r'(\beta_r) - \sum_{e\in\pi_r} \widetilde{\mu}_e\} \beta_r = 0$, and Case (b) implies $\widetilde{\mu}_e z_e = 0$ and $\widetilde{\mu}_e \geq 0$. That is, $\widetilde{\mu}^T z = 0$ and $\widetilde{\mu} \geq 0 \Rightarrow \widetilde{\mu}^T(C - A\beta) = 0$, and $\widetilde{\mu} \geq 0$.

Define $\widetilde{\lambda}_r \triangleq \sum_{e\in\pi_r} \widetilde{\mu}_e$, and $\widetilde{\lambda} \triangleq (\widetilde{\lambda}_1, \widetilde{\lambda}_2, \ldots, \widetilde{\lambda}_{|\mathfrak{R}|})$.

In summary, a β with the following conditions solves $SYSTEM[U, A, C]$.

$$A\beta \leq C, \quad \beta \geq 0, \quad \widetilde{\mu}^T(C - A\beta) = 0, \quad \widetilde{\mu} \geq 0$$
$$U'(\beta) \leq \widetilde{\lambda}, \quad \beta^T\left(U'(\beta) - \widetilde{\lambda}\right) = 0$$

8. Determine the Lagrangian dual of $NETWORK[A, C; w]$ (the network problem). Denote this problem by $DUAL[A, C; w]$.
Hint: The vector β solves $NETWORK[A, C; w]$ such that

$$\sum_{r\in\mathfrak{B}(w)} w_r \ln \beta_r$$

is maximized subject to $A\beta \le C$, and $\beta \ge 0$. Note that $\mathfrak{B}(w) \triangleq \{r \in \mathfrak{R} \mid w_r > 0\}$. Let $\mu = \left(\mu_1, \mu_2, \ldots, \mu_{|E|}\right)$. The Lagrangian of the network optimization problem is

$$\begin{aligned}\mathcal{L}_N(\beta,\mu) &= \sum_{r\in\mathfrak{B}(w)} w_r \ln \beta_r + \mu^T (C - A\beta) \\ &= \sum_{r\in\mathfrak{B}(w)} w_r \ln \beta_r + \sum_{e\in E} \mu_e c_e - \sum_{r\in\mathfrak{R}} \beta_r \sum_{e\in\pi_r} \mu_e\end{aligned}$$

Next compute partial derivatives of $\mathcal{L}_N(\beta,\mu)$ with respect to β_r.

$$\frac{\partial \mathcal{L}_N(\beta,\mu)}{\partial \beta_r} = \begin{cases} \dfrac{w_r}{\beta_r} - \sum_{e\in\pi_r} \mu_e, & r \in \mathfrak{B}(w) \\ -\sum_{e\in\pi_r} \mu_e, & r \notin \mathfrak{B}(w) \end{cases}$$

Consider the partial derivative of the Lagrangian with respect to β_r.
(a) If $r \in \mathfrak{B}(w)$ then $w_r > 0$. Since $w_r = \lambda_r \beta_r$, it is inferred that $\beta_r > 0$. Thus $\partial \mathcal{L}_N(\beta,\mu)/\partial \beta_r = 0$. This implies $\beta_r = w_r / \sum_{e\in\pi_r} \mu_e$.
(b) If $r \notin \mathfrak{B}(w)$ then $w_r = 0$. Also $w_r = \lambda_r \beta_r$, and $\lambda_r > 0$ imply $\beta_r = 0$. This in turn implies $\partial \mathcal{L}_N(\beta,\mu)/\partial \beta_r \le 0$.
Substituting for β_r in the Lagrangian yields

$$\begin{aligned}\omega(\mu) &= \max_{\beta\ge 0} \mathcal{L}_N(\beta,\mu) \\ &= \sum_{r\in\mathfrak{B}(w)} w_r \ln \left\{ \frac{w_r}{\sum_{e\in\pi_r} \mu_e} \right\} + \sum_{e\in E} \mu_e c_e - \sum_{r\in\mathfrak{B}(w)} w_r \\ &= -\sum_{r\in\mathfrak{B}(w)} w_r \ln \left\{ \sum_{e\in\pi_r} \mu_e \right\} + \sum_{e\in E} \mu_e c_e - \sum_{r\in\mathfrak{B}(w)} w_r (1 - \ln w_r)\end{aligned}$$

Therefore the dual problem is $\min_\mu \omega(\mu)$ over $\mu \ge 0$. Note that the last term of $\omega(\mu)$ does not affect the value of μ, when $\omega(\mu)$ is minimized. However this term which is independent of μ, affects the value of the objective function. It can be shown that solving the dual yields $A\beta \le C$, $\beta \ge 0$, $\mu^T (C - A\beta) = 0$, $\mu \ge 0$.

9. Prove the system decomposition theorem.
Hint: See Kelly (1997). Lagrangian duality solves for μ via $DUAL[A, C; w]$ if and only if there exists β such that (β, μ) satisfies the conditions specified in $\mathcal{C}_{NET}$. However the tuple $(\beta, \widetilde{\mu})$ which satisfies $\mathcal{C}_{SYSTEM}$ identifies a solution to the conditions specified in $\mathcal{C}_{NET}$, if the relationship $\widetilde{\mu} = \mu$ is permitted. This implies $\widetilde{\lambda} = \lambda$. Since β solves $NETWORK[A, C; w]$, the vector w is defined. Finally, the conditions $\mathcal{C}_{SYSTEM}$ together with the definition of w imply conditions in $\mathcal{C}_{USER}$. This proves the existence of the vectors w, λ, and β.

Conversely, assume that the vectors w, λ, and β exist such that they satisfy the conditions (a), (b), (c), and (d) specified in the statement of the theorem. Using Lagrangian duality it is known that there exists $\widehat{\beta}$ such that the triple

$$\left(\widehat{\beta}, \lambda, \mu\right)$$

satisfies the conditions specified in $\mathcal{C}_{NET}$. It is next demonstrated that $\widehat{\beta} = \beta$. Condition (d) yields $w_r = \beta_r \lambda_r$, for $r \in \mathfrak{R}$, and $w_r = \widehat{\beta}_r \lambda_r$ by condition in $\mathcal{C}_{NET}$. Since $\lambda_r > 0$, it follows that $\widehat{\beta} = \beta$ for each $r \in \mathfrak{R}$. Therefore (β, λ, μ) satisfies conditions in $\mathcal{C}_{NET}$. Also, by (a) w_r and λ_r satisfy conditions in $\mathcal{C}_{USER}$ for each $r \in \mathfrak{R}$. Therefore (β, λ, μ) satisfies conditions in $\mathcal{C}_{SYSTEM}$. Consequently, it can be concluded that β solves $SYSTEM\left[U, A, C\right]$, and β is uniquely determined. Furthermore, since $\lambda_r \in \mathbb{R}^+$ for each $r \in \mathfrak{R}$, the vector w is also uniquely determined.

10. A network is specified by an undirected graph $G = (V, E)$, where the vertex set V represents the routers, and the edge set E represents the transmission links. Associated with each edge $e \in E$ is its transmission capacity $c_e \in \mathbb{R}^+$. Denote the set of users who access this network by $\mathfrak{R}$. Let this user set be $\{1, 2, \ldots, |\mathfrak{R}|\}$. Associated with each user $r \in \mathfrak{R}$ is a traffic demand $\rho_r \in \mathbb{R}^+$. This traffic demand has to be transmitted between a source-destination pair. The actual allocation of this demand in the network is $\beta_r \in \mathbb{R}^+$ where $\beta_r \leq \rho_r$. For simplicity assume that each user allocation β_r, follows a single route π_r in the network. The set of links associated with each route is specified by an $|E| \times |\mathfrak{R}|$ matrix $A = [a_{er}]$, where $a_{er} = 1$ if $e \in \pi_r$. That is, $a_{er} = 1$ if e lies on route $r \in \mathfrak{R}$, and $a_{er} = 0$ otherwise. Associated with each user $r \in \mathfrak{R}$ is its utility $U_r(\beta_r)$. It is

$$U_r(\beta_r) = \lim_{\varphi \to \infty} \frac{\beta_r^{(1-\varphi)}}{(1-\varphi)}$$

Establish that the maximization of the sum of utility of all the users $\sum_{r \in \mathfrak{R}} U_r(\beta_r)$ yields the max-min fair allocation of capacity to the users.

Hint: We essentially follow the sketch of the proof given in Srikant (2004). Let $\beta(\varphi)$ be the optimal solution to the optimization problem

$$\max \sum_{r \in \mathfrak{R}} U_r(\beta_r), \quad \text{subject to} \quad A\beta \leq C, \quad \text{and} \quad \beta \geq 0$$

Let $\widetilde{\beta}$ be another flow vector, which satisfies the link constraint $A\widetilde{\beta} \leq C$ and $\widetilde{\beta} \geq 0$. Assume that $\widetilde{\beta} \neq \beta(\varphi)$, and that there exists $k \in \mathfrak{R}$ such that $\beta_k(\varphi) < \widetilde{\beta}_k$ for all sufficiently large values of φ. As the flow vector $\beta(\varphi)$ is optimal, we have

$$\sum_{r \in \mathfrak{R}} \frac{\widetilde{\beta}_r - \beta_r(\varphi)}{\beta_r^{\varphi}(\varphi)} < 0$$

This implies

$$\frac{\widetilde{\beta}_k - \beta_k(\varphi)}{\beta_k^{\varphi}(\varphi)} < -\sum_{r \in \mathfrak{R} \setminus \{k\}} \frac{\widetilde{\beta}_r - \beta_r(\varphi)}{\beta_r^{\varphi}(\varphi)}$$

As $\widetilde{\beta}_k - \beta_k(\varphi) > 0$, we have

$$1 < \sum_{r \in \mathfrak{R} \setminus \{k\}} -\left\{\frac{\widetilde{\beta}_r - \beta_r(\varphi)}{\widetilde{\beta}_k - \beta_k(\varphi)}\right\} \left\{\frac{\beta_k(\varphi)}{\beta_r(\varphi)}\right\}^{\varphi}$$

$$\le \sum_{\substack{r \in \mathfrak{R} \backslash \{k\} \\ \text{only positive terms}}} -\left\{ \frac{\widetilde{\beta}_r - \beta_r(\varphi)}{\widetilde{\beta}_k - \beta_k(\varphi)} \right\} \left\{ \frac{\beta_k(\varphi)}{\beta_r(\varphi)} \right\}^{\varphi}$$

As $\varphi \to \infty$, the above inequality is true provided there exists an $i \in \mathfrak{R} \backslash \{k\}$ such that $\beta_i(\infty) \le \beta_k(\infty)$, otherwise all terms on the right-hand side will tend to zero. Furthermore, the assumption $\beta_k(\infty) < \widetilde{\beta}_k$ implies $\widetilde{\beta}_i < \beta_i(\infty)$. Thus, if we assume that $\beta_k(\infty) < \widetilde{\beta}_k$, then there exists $i \in \mathfrak{R}$ such that $\widetilde{\beta}_i < \beta_i(\infty) \le \beta_k(\infty)$. These inequalities are in conformance with the max-min fairness criteria of capacity allocation.

11. Prove the theorem about the global stability of the primal algorithm.
Hint: See Kelly, Maulloo, and Tan (1998). The Lyapunov function of the system of differential equations describing the primal algorithm of congestion control is determined. Since $w_r \in \mathbb{R}^+$ is a constant for each $r \in \mathfrak{R}$, and $\mathcal{C}_e(\cdot)$ is a strictly convex function for each $e \in E$, the function $\mathcal{U}(\cdot)$ is strictly concave on the positive orthant. Also its maximum is within the interior. Thus the maximizing value of $\beta(\cdot)$ is unique. Next take the partial derivatives of $\mathcal{U}(\beta(t))$ with respect to β_r.

$$\frac{\partial \mathcal{U}(\beta(t))}{\partial \beta_r} = \frac{w_r}{\beta_r(t)} - \sum_{e \in \pi_r} \mu_e(t), \quad \forall\, r \in \mathfrak{R}$$

Equating these derivatives to zero yields the maximum of $\mathcal{U}(\beta(t))$. Also

$$\frac{d\mathcal{U}(\beta(t))}{dt} = \sum_{r \in \mathfrak{R}} \frac{\partial \mathcal{U}(\beta(t))}{\partial \beta_r} \frac{d\beta_r(t)}{dt}$$

$$= \sum_{r \in \mathfrak{R}} \frac{\kappa_r}{\beta_r(t)} \left\{ w_r - \beta_r(t) \sum_{e \in \pi_r} \mu_e(t) \right\}^2$$

The above result implies that $\mathcal{U}(\cdot)$ is strictly increasing with time t, for all values of $\beta(t)$, except its unique value β which maximizes the value of the function $\mathcal{U}(\cdot)$. Therefore $\mathcal{U}(\cdot)$ is a Lyapunov function for the primal system of differential equations.

12. Prove the theorem about global stability of the dual algorithm.
Hint: See Kelly, Maulloo, and Tan (1998). The Lyapunov function of the system of differential equations describing the dual algorithm of congestion control is determined. Since $w_r \in \mathbb{R}^+$, for each $r \in \mathfrak{R}$ is a constant, and $\mathcal{D}_e(\cdot)$ is a strictly convex function for each $e \in E$, the function $\mathcal{V}(\cdot)$ is strictly concave on the positive orthant. Also its maximum is within the interior. Thus the maximizing value of $\mu(\cdot)$ is unique. Next take the partial derivatives of $\mathcal{V}(\mu(t))$ with respect to μ_e.

$$\frac{\partial \mathcal{V}(\mu(t))}{\partial \mu_e} = \sum_{\substack{e \in \pi_r \\ r \in \mathfrak{R}}} \frac{w_r}{\sum_{k \in \pi_r} \mu_k(t)} - q_e(\mu_e(t)), \quad \forall\, e \in E$$

Equating these derivatives to zero yields the maximum of $\mathcal{V}(\mu(t))$. Also

$$\frac{d\mathcal{V}(\mu(t))}{dt} = \sum_{e \in E} \frac{\partial \mathcal{V}(\mu(t))}{\partial \mu_e} \frac{d\mu_e(t)}{dt}$$

$$= \sum_{e \in E} \kappa_e \left\{ \sum_{\substack{e \in \pi_r \\ r \in \mathfrak{R}}} \beta_r(t) - q_e(\mu_e(t)) \right\}^2$$

The above result implies that $\mathcal{V}(\mu(t))$ is strictly increasing with time t, for all values of $\mu(t)$, except its unique value which maximizes the value of the function $\mathcal{V}(\cdot)$. Therefore $\mathcal{V}(\cdot)$ is a Lyapunov function for the dual system of differential equations.

13. Let $\mathcal{I} = \{1, 2, \ldots, N\}$ be the set of sessions, in a GPS server. A session $i \in \mathcal{I}$ is continuously backlogged in the interval $(\tau, t]$. Prove that the amount of session i's traffic $\mathcal{S}_i(\tau, t) \geq (t - \tau) g_i$, where $g_i = \phi_i \mathcal{R} / \sum_{j=1}^{N} \phi_j$.
Hint: See Parekh, and Gallager (1993). Let $\mathcal{B}(t)$ be the set of backlogged sessions at time t. Then the service rate $\mathcal{R}_i(t)$ received by ith session at time t is
$$\mathcal{R}_i(t) = \begin{cases} \dfrac{\phi_i}{\sum_{j \in \mathcal{B}(t)} \phi_j} \mathcal{R}, & \text{if } i \in \mathcal{B}(t) \\ 0, & \text{otherwise} \end{cases}$$
Then
$$\begin{aligned} \mathcal{S}_i(\tau, t) &= \int_{\tau}^{t} \mathcal{R}_i(u)\, du \\ &= \int_{\tau}^{t} \frac{\phi_i}{\sum_{j \in \mathcal{B}(u)} \phi_j} \mathcal{R} du \\ &\geq \int_{\tau}^{t} \frac{\phi_i}{\sum_{j=1}^{N} \phi_j} \mathcal{R} du \\ &= (t - \tau) g_i \end{aligned}$$

14. Let $\widetilde{p}$ and $\widetilde{\pi}$ be packets in a GPS system at time instant τ, and packet $\widetilde{p}$ completes service before packet $\widetilde{\pi}$, if there are no arrivals after τ. Prove that packet $\widetilde{p}$ also completes service before packet $\widetilde{\pi}$ for any sequence of packet arrivals after time τ.
Hint: See Parekh, and Gallager (1993). The flows to which $\widetilde{p}$ and $\widetilde{\pi}$ belong are both backlogged from time τ till packet $\widetilde{p}$ exits the node. By the definition of the GPS scheme, the ratio of the service received by these different flows is independent of any future arrivals of packets. The proof follows by using the definition of GPS scheme.

15. Let d_k and $\widehat{d}_k$ be the departure instants of packet $\widetilde{p}_k$ under the GPS and PGPS scheme respectively. Let $L_{\max}$ be the maximum packet length, and $\mathcal{R}$ be the rate of the server. Prove that for all packets $\widetilde{p}_k$
$$\widehat{d}_k - d_k \leq \frac{L_{\max}}{\mathcal{R}}$$
Hint: See Parekh, and Gallager (1993); and Kumar, Manjunath, and Kuri (2004). Recall that both GPS and PGPS are work-conserving disciplines, therefore their busy periods coincide. That is, a GPS server is in a busy period if and only if the PGPS server is in a busy period. Therefore it is sufficient to establish the result for each busy period. During the busy period, same packets are served by both schemes. However departure instants of the packets in the two schemes may be different. For simplicity assume that the busy period begins at time zero. Index the packets by their departure order under PGPS. The packet $\widetilde{p}_1$ is the first to depart under PGPS in this busy period, and packet $\widetilde{p}_k$ is the kth packet to depart, where $k = 1, 2, \ldots$. Let the length of the kth packet be L_k. Then the packet departure instants under the PGPS discipline are
$$\widehat{d}_1 = \frac{L_1}{\mathcal{R}}, \widehat{d}_2 = \frac{L_1 + L_2}{\mathcal{R}}, \ldots, \widehat{d}_k = \frac{\sum_{i=1}^{k} L_i}{\mathcal{R}}$$

The departure instant d_k, of packet $\widetilde{p}_k$ under the GPS scheme is next examined. There are two cases.

Case (a): In the GPS scheme, all the packets $\widetilde{p}_1, \widetilde{p}_2, \ldots, \widetilde{p}_{k-1}$ depart strictly before d_k. Consequently

$$d_k \geq \frac{\sum_{i=1}^{k} L_i}{\mathcal{R}} = \widehat{d}_k$$

This satisfies the stated condition in the problem.

Case (b): There is an index m such that $d_k < d_m$, where $1 \leq m \leq (k-1)$. In addition, m is the largest such index. Consequently

$$d_{m+1} \leq d_k, d_{m+2} \leq d_k, \ldots, d_{k-1} \leq d_k$$

That is, the packet $\widetilde{p}_m$, departs after the packets $\widetilde{p}_{m+1}, \widetilde{p}_{m+2}, \ldots, \widetilde{p}_k$ under the GPS scheme. (Actually, the packet $\widetilde{p}_m$ can depart with packet $\widetilde{p}_k$ under GPS). Next consider $\widehat{s}_m$, the instant at which the service of the packet $\widetilde{p}_m$ commenced under the PGPS scheme. The packets $\widetilde{p}_{m+1}, \widetilde{p}_{m+2}, \ldots, \widetilde{p}_k$ cannot be present at time instant $\widehat{s}_m$ in the PGPS scheme, otherwise one of them would have been selected for service instead of the packet $\widetilde{p}_m$. Therefore

$$d_k \geq \widehat{s}_m + \frac{\sum_{i=m+1}^{k} L_i}{\mathcal{R}}$$

Observe that

$$\widehat{d}_k = \widehat{s}_m + \frac{\sum_{i=m}^{k} L_i}{\mathcal{R}}$$

Thus

$$d_k \geq \widehat{d}_k - \frac{L_m}{\mathcal{R}}$$

The stated result follows, since $L_{\max}$ is the largest packet size.

16. Let $\mathcal{I} = \{1, 2, \ldots, N\}$ be the set of sessions, and $\mathcal{S}_i(\tau, t)$ and $\widehat{\mathcal{S}}_i(\tau, t)$ be the amount of session $i \in \mathcal{I}$ traffic (measured in bits and not packets) served in the time interval $[\tau, t]$, under the GPS and PGPS schemes respectively. For all times τ and session i, prove that

$$\mathcal{S}_i(0, \tau) - \widehat{\mathcal{S}}_i(0, \tau) \leq L_{\max}, \quad i \in \mathcal{I}$$

Hint: See Parekh, and Gallager (1993); and Kumar, Manjunath, and Kuri (2004). Note that when a session i packet is being transmitted the slope of $\widehat{\mathcal{S}}_i(0, t)$ is equal to $\mathcal{R}$, and when there is no transmission, this slope is equal to zero. The slope of $\mathcal{S}_i(0, \tau)$ follows similar law. Thus the difference $\left\{\mathcal{S}_i(0, t) - \widehat{\mathcal{S}}_i(0, t)\right\}$ reaches its maximal value when session i packets begin transmission under the PGPS discipline. Let t be some such time, and the corresponding length of the packet be L bits. This packet finishes its transmission at $t + L/\mathcal{R}$. Also let τ be the time at which this packet completes transmission under the GPS scheme. Note that session i packets are served in the same order in both the PGPS and GPS schemes. Therefore

$$\mathcal{S}_i(0, \tau) = \widehat{\mathcal{S}}_i\left(0, t + \frac{L}{\mathcal{R}}\right)$$

From the last problem

$$\tau \geq \left(t + \frac{L}{\mathcal{R}}\right) - \frac{L_{\max}}{\mathcal{R}}$$

The above two relationships imply

$$\mathcal{S}_i\left(0, t + \frac{L - L_{\max}}{\mathcal{R}}\right) \leq \widehat{\mathcal{S}}_i\left(0, t + \frac{L}{\mathcal{R}}\right)$$
$$= \widehat{\mathcal{S}}_i(0, t) + L$$

The last step uses the observation, that the slope of $\mathcal{S}_i(0, t)$ is at most $\mathcal{R}$. The result follows.

17. Let the functions $f, g, h \in \mathcal{F}$. Prove the following results.
    (a) Composition of $\oslash$ and $\circledast$: $((f \circledast g) \oslash g) \leq (f \circledast (g \oslash g))$.
    (b) Duality between $\oslash$ and $\circledast$: $(f \oslash g) \leq h$ if and only if $f \leq (g \circledast h)$.
    (c) Self-deconvolution: $(f \oslash f)$ is a subadditive function of $\mathcal{F}$. Furthermore, $(f \oslash f)(0) = 0$.

    Hint: The proofs follow the careful and crafty work of Le Boudec and Thiran (2004). These involve meticulous use of the definitions of convolution and deconvolution operators

    (a) Composition of $\oslash$ and $\circledast$:
    Observe that

$$\begin{aligned}
((f \circledast g) \oslash g)(t) &= \sup_{\tau \geq 0} \{(f \circledast g)(t + \tau) - g(\tau)\} \\
&= \sup_{\tau \geq 0} \left\{ \inf_{0 \leq v \leq t+\tau} \{f(t + \tau - v) + g(v)\} - g(\tau) \right\} \\
&= \sup_{\tau \geq 0} \left\{ \inf_{-\tau \leq w \leq t} \{f(t - w) + g(w + \tau)\} - g(\tau) \right\} \\
&\leq \sup_{\tau \geq 0} \left\{ \inf_{0 \leq w \leq t} \{f(t - w) + g(w + \tau)\} - g(\tau) \right\} \\
&\leq \sup_{\tau \geq 0} \inf_{0 \leq w \leq t} \left\{ f(t - w) + \sup_{y \geq 0} \{g(w + y) - g(y)\} \right\} \\
&= \inf_{0 \leq w \leq t} \left\{ f(t - w) + \sup_{y \geq 0} \{g(w + y) - g(y)\} \right\} \\
&= \inf_{0 \leq w \leq t} \{f(t - w) + (g \oslash g)(w)\} \\
&= (f \circledast (g \oslash g))(t)
\end{aligned}$$

    (b) Duality between $\oslash$ and $\circledast$:
    We first establish: $(f \oslash g) \leq h \Rightarrow f \leq (g \circledast h)$. Assume that $(f \oslash g)(v) \leq h(v)$ for all $v \geq 0$. For any $v, y \geq 0$ we have

$$f(v + y) - g(y) \leq \sup_{\tau \geq 0} \{f(v + \tau) - g(\tau)\} = (f \oslash g)(v) \leq h(v)$$

    This implies

$$f(v + y) \leq g(y) + h(v)$$

    Substitute $t = (v + y)$ in the above inequality. It results in

$$f(t) \leq g(t - v) + h(v)$$

    Also

$$f(t) \leq \inf_{0 \leq v \leq t} \{g(t - v) + h(v)\} = (g \circledast h)(t), \quad \forall\, t \geq 0$$

We next establish: $(f \oslash g) \leq h \Leftarrow f \leq (g \circledast h)$. Assume that $f(y) \leq (g \circledast h)(y)$ for $y \geq 0$. As $g, h \in \mathcal{F}$, then for any $t \in \mathbb{R}$ we have

$$\begin{aligned} f(y) &\leq \inf_{0 \leq v \leq y} \{g(y-v) + h(v)\} \\ &= \inf_{v \in \mathbb{R}} \{g(y-v) + h(v)\} \\ &\leq g(y-t) + h(t) \end{aligned}$$

Substitute $\tau = (y - t)$ in the above inequality. This yields

$$f(t+\tau) - g(\tau) \leq h(t), \quad \forall \tau \in \mathbb{R}$$

Thus

$$\sup_{\tau \in \mathbb{R}} \{f(t+\tau) - g(\tau)\} \leq h(t)$$

As $g(\tau) = 0$ for $\tau < 0$, we have $\sup_{\tau<0} \{f(t+\tau) - g(\tau)\} = f(t)$. Thus the above inequality is

$$\sup_{\tau \geq 0} \{f(t+\tau) - g(\tau)\} \leq h(t)$$

for all t.

(c) Self-deconvolution:
It follows from the definition of deconvolution operation that $(f \oslash f)(0) = 0$ and $(f \oslash f) \in \mathcal{F}$. Also

$$\begin{aligned} &(f \oslash f)(u) + (f \oslash f)(v) \\ &= \sup_{y \geq 0} \{f(u+y) - f(y)\} + \sup_{\tau \geq 0} \{f(v+\tau) - f(\tau)\} \\ &= \sup_{x \geq -v} \{f(u+v+x) - f(v+x)\} + \sup_{\tau \geq 0} \{f(v+\tau) - f(\tau)\} \\ &\geq \sup_{x \geq 0} \left\{ \sup_{\tau \geq 0} \{f(u+v+x) - f(v+x) + f(v+\tau) - f(\tau)\} \right\} \\ &\geq \sup_{x \geq 0} \{f(u+v+x) - f(v+x) + f(v+x) - f(x)\} \\ &= (f \oslash f)(u+v) \end{aligned}$$

18. Establish the three fundamental bounds of deterministic network calculus. These are the backlog and delay bounds, and a bound on the output flow.
Hint: We essentially follow the proofs given in the monograph by Le Boudec and Thiran (2004).
(a) Backlog bound: For any $t \geq 0$ there exists an s such that

$$\{F_{in}(t) - F_{in}(t-s)\} \leq \widetilde{A}(s), \text{ and } \{F_{out}(t) - F_{in}(t-s)\} \geq \widetilde{S}(s)$$

Subtraction yields $\{F_{in}(t) - F_{out}(t)\} \leq \left\{\widetilde{A}(s) - \widetilde{S}(s)\right\}$
(b) Delay bound: Let $t \geq 0$ be fixed. The definition of virtual delay implies for all $\tau < d(t)$

$$F_{in}(t) > F_{out}(t+\tau)$$

Use of the service curve definition at time $(t+\tau)$ implies that there is some s_0 such that

$$F_{out}(t+\tau) - F_{in}(t+\tau-s_0) \geq \widetilde{S}(s_0)$$

Combining the last two inequalities yield

$$F_{in}(t) > F_{in}(t+\tau-s_0) + \widetilde{S}(s_0)$$

The last inequality implies $t+\tau-s_0 < t$. Thus we have

$$\widetilde{A}(s_0-\tau) \geq \{F_{in}(t) - F_{in}(t+\tau-s_0)\} > \widetilde{S}(s_0)$$

Therefore

$$\tau \leq \Delta(s_0-\tau) \leq H\left(\widetilde{A}, \widetilde{S}\right)$$

This result is true for all $\tau < d(t)$, thus $d(t) \leq H\left(\widetilde{A}, \widetilde{S}\right)$ for all $t \geq 0$.

(c) Output flow: We consider

$$\{F_{out}(t) - F_{out}(t-s)\}, \quad \text{where} \quad 0 \leq (t-s) \leq t$$

Next use the definition of service curve applied at time $(t-s)$. Assume that there exists $v \geq 0$ such that $(t-s-v) \geq 0$ and

$$F_{out}(t-s) - F_{in}(t-s-v) \geq \widetilde{S}(v)$$

Therefore

$$F_{out}(t) - F_{out}(t-s) \leq F_{out}(t) - F_{in}(t-s-v) - \widetilde{S}(v)$$

Next, we use the inequality $F_{out}(t) \leq F_{in}(t)$. Thus

$$\begin{aligned} F_{out}(t) - F_{out}(t-s) &\leq F_{in}(t) - F_{in}(t-s-v) - \widetilde{S}(v) \\ &\leq \widetilde{A}(s+v) - \widetilde{S}(v) \end{aligned}$$

Thus $\{F_{out}(t) - F_{out}(t-s)\}$ is bounded by $\left(\widetilde{A} \oslash \widetilde{S}\right)(s)$.

19. In the analysis of the CSMA/CA protocol, it is assumed that the backoff time interval of each of the n nodes was exponentially distributed with parameter $\beta \in \mathbb{R}^+$. Further, the backoff time interval of each node is independent of the backoff time interval of all other nodes. Prove that a frame transmitted from a node suffers a collision with a frame transmitted from any of the other $(n-1)$ nodes, in a time interval δ (single slot time) with probability $\gamma = \left(1 - e^{-(n-1)\beta\delta}\right)$.
Hint: Let $T_1, T_2, \ldots, T_{n-1}$ be independent exponentially distributed random variables each with parameter β. Each of these random variables represent the backoff time intervals of the $(n-1)$ nodes. Let

$$U = \min\{T_1, T_2, \ldots, T_{n-1}\}$$

For any $\delta \geq 0$, we have

$$\gamma = P(U \leq \delta)$$

$$\begin{aligned} P(U > \delta) &= P(T_1 > \delta, T_2 > \delta, \ldots, T_{n-1} > \delta) \\ &= P(T_1 > \delta) P(T_2 > \delta) \ldots P(T_{n-1} > \delta) \\ &= e^{-(n-1)\beta\delta} \end{aligned}$$

The result follows.

20. Consider a homogeneous Poisson point process in m-dimensional space $\mathbb{R}^m$, with intensity $\lambda > 0$, where $m = 1, 2, 3, \ldots$. Also let $R_n, n \in \mathbb{P}$ be the random variable which denotes the Euclidean distance from an arbitrarily selected point in $\mathbb{R}^m$ to its nth nearest neighbor in Φ. The volume of the m-dimensional hypersphere $\mathcal{H}_r$ of radius $r \in \mathbb{R}^+$ be $V_m(r)$, where $m = 1, 2, 3, \ldots$. We have

$$V_m(r) = c_m r^m$$

and

$$c_m = \frac{\pi^{m/2}}{\Gamma(m/2+1)}$$

and $\Gamma(\cdot)$ is the gamma function. Also $\|\cdot\|$ is the Euclidean distance metric defined on points in space $\mathbb{R}^m$. Prove the following results.

(a) The cumulative distribution function of random variable R_n is $F_{R_n}(\cdot)$. We have

$$F_{R_n}(r) = 1 - \frac{\Gamma(n, \lambda c_m r^m)}{\Gamma(n)}, \quad r > 0$$

where $\Gamma(\cdot, \cdot)$ is the incomplete gamma function. The probability density function of the random variable R_n is $f_{R_n}(\cdot)$. We have

$$f_{R_n}(r) = \frac{nm}{r} \exp(-\lambda V_m(r)) \frac{(\lambda V_m(r))^n}{n!}, \quad r > 0$$

(b) Let $y \in \mathbb{R}^m$ and $X_i \in \mathbb{R}^m$ be points of the homogeneous Poisson process which are ordered according to their Euclidean distance from the point y. The random variables $Z_i = \|y - X_i\|^m, i \in \mathbb{P}$ are distributed as a one-dimensional Poisson process of intensity λc_m. In other words, Z_1, and $(Z_i - Z_{i-1})$, where $i > 1$ are exponentially distributed, each with parameter λc_m, and $\mathcal{E}(Z_i) = i/(\lambda c_m)$.

(c) The random variables $R_i = \|y - X_i\|, i \in \mathbb{P}$ are distributed as a one-dimensional nonhomogeneous Poisson process. Its intensity function is $\mu(r) = \lambda c_m m r^{m-1}$.

Hint: See Haenggi (2005).

(a) As per the definition of PPP, the probability of having k points in an m-dimensional hypersphere of radius $r \in \mathbb{R}^+$ is

$$P(N(V_m(r)) = k) = \exp(-\lambda V_m(r)) \frac{(\lambda V_m(r))^k}{k!}, \quad k \in \mathbb{N}$$

The cumulative distribution function of random variable R_n is $F_{R_n}(\cdot)$ is

$$F_{R_n}(r) = \sum_{k=n}^{\infty} P(N(V_m(r)) = k), \quad r > 0$$

The cumulative distribution function $F_{R_n}(\cdot)$ for $r > 0$ can also be expressed as

$$\begin{aligned} F_{R_n}(r) &= 1 - \sum_{k=0}^{n-1} P(N(V_m(r)) = k) \\ &= 1 - \exp(-\lambda c_m r^m) \sum_{k=0}^{n-1} \frac{(\lambda c_m r^m)^k}{k!} \\ &= 1 - \frac{\Gamma(n, \lambda c_m r^m)}{\Gamma(n)} \end{aligned}$$

where $\Gamma(\cdot,\cdot)$ is the incomplete gamma function. The probability density function of the random variable R_n is

$$f_{R_n}(r) = \frac{dF_{R_n}(r)}{dr}$$

After some algebraic churn we obtain

$$f_{R_n}(r) = \frac{nm}{r}\exp\left(-\lambda V_m(r)\right)\frac{\left(\lambda V_m(r)\right)^n}{n!}, \quad r > 0$$

(b) From part (a) of the problem, the cumulative distribution function of the random variable R_n^m is

$$F_{R_n^m}(r) = 1 - \frac{\Gamma(n, \lambda c_m r)}{\Gamma(n)}, \quad r > 0$$

and its probability density function is

$$f_{R_n^m}(r) = \frac{\lambda c_m \left(\lambda c_m r\right)^{n-1} e^{-\lambda c_m r}}{\Gamma(n)}, \quad r > 0$$

This is the probability density function of an Erlang distributed random variable. Its mean is $\mathcal{E}(R_n^m) = n/(\lambda c_m)$. The result follows by recalling that an Erlang distributed random variable is obtained by adding independent and identically distributed exponential random variables.

(c) From part (b) of the problem, the random variables $Z_i = \|y - X_i\|^m$, $i \in \mathbb{P}$ are distributed as a one-dimensional Poisson process of intensity λc_m. As r^m, for $r > 0$ is monotonic on r, $R_i = Z_i^{1/m}$ is also a one dimensional Poisson process. Denote the intensity function of the process R_i by $\mu(\cdot)$. The interval $(0, r] \subset \mathbb{R}$ corresponds to an m- dimensional hypersphere $b(o, r)$ of radius r, centered at the origin o in space $\mathbb{R}^m$. Therefore

$$\int_0^r \mu(w)\, dw = \int_{b(o,r)} \lambda dy = \lambda V_m(r) = \lambda c_m r^m$$

This implies $\mu(r) = \lambda c_m m r^{m-1}$.

21. Show that for $s \in \mathbb{C}$,

$$\int_0^\infty \left\{1 - \exp\left(-s P_t H r^{-\alpha}\right)\right\} \mu(r)\, dr = \lambda c_m P_t^\delta H^\delta \Gamma(1-\delta)\, s^\delta$$

where $\mu(r) = \lambda c_m m r^{m-1}$, $r > 0$, $c_m = \pi^{m/2}/\Gamma(m/2+1)$, $m \in \mathbb{P}$; P_t is the constant transmitter power of each interfering station; H is the random variable that characterizes the combined effect of shadowing and fading; $\alpha > 0$ is the path-loss exponent, and $\delta = m/\alpha$. Also $\delta \in (0,1)$. This integral is used in evaluating the Laplace functional of the interference function I.

Hint: See Haenggi (2013). Substitute $x = s P_t H r^{-\alpha}$ in the integral to obtain

$$\lambda c_m \delta P_t^\delta H^\delta s^\delta \int_0^\infty \left(1 - e^{-x}\right) x^{-\delta-1} dx$$

Next use integration by parts to obtain the stated result.

22. Establish the following results.

(a) This result is attributed to P. Laplace. For $a > 0$, and $b \geq 0$

$$\int_0^\infty \exp\left(-a^2x^2 - b^2x^{-2}\right) dx = \frac{\sqrt{\pi}}{2a} \exp\left(-2ab\right)$$

(b) Let

$$f(t) = \frac{1}{\sqrt{t}} \exp\left(-\frac{k}{t}\right), \quad k \geq 0, \text{ and } t \in \mathbb{R}_0^+$$

The single-sided Laplace transform of $f(t)$, is

$$\widehat{f}(s) = \mathcal{L}\left[f(t)\right] = \int_0^\infty f(t)e^{-st}dt = \sqrt{\frac{\pi}{s}} \exp\left(-2\sqrt{ks}\right), \quad \operatorname{Re}(s) > 0$$

(c) Let

$$g(t) = \frac{1}{t^{3/2}} \exp\left(-\frac{k}{t}\right), \quad k > 0, \text{ and } t \in \mathbb{R}_0^+$$

The single-sided Laplace transform of $g(t)$, is

$$\begin{aligned}
\widehat{g}(s) &= \mathcal{L}\left[g(t)\right] \\
&= \int_0^\infty g(t)e^{-st}dt \\
&= \sqrt{\frac{\pi}{k}} \exp\left(-2\sqrt{ks}\right), \quad \operatorname{Re}(s) > 0
\end{aligned}$$

(a) Hint: See Albano, Tewodros, Beyerstedt, and Moll (2011). Using the method of completing square, we have

$$\begin{aligned}
I &= \int_0^\infty \exp\left(-a^2x^2 - b^2x^{-2}\right) dx \\
&= \exp\left(-2ab\right) \int_0^\infty \exp\left(-\left(ax - bx^{-1}\right)^2\right) dx
\end{aligned}$$

Substituting $t = b/(ax)$ in the above integral yields

$$I = \exp\left(-2ab\right) \frac{b}{a} \int_0^\infty \exp\left(-\left(at - bt^{-1}\right)^2\right) \frac{dt}{t^2}$$

Adding the two expressions for I and dividing by 2 results in

$$I = \frac{\exp\left(-2ab\right)}{2a} \int_0^\infty \exp\left(-\left(ax - bx^{-1}\right)^2\right) \left(a + \frac{b}{x^2}\right) dx$$

Substitute $\left(ax - bx^{-1}\right) = u/\sqrt{2}$ in the above integral. It yields

$$\begin{aligned}
I &= \frac{\exp\left(-2ab\right)}{2^{3/2}a} \int_{-\infty}^\infty \exp\left(-u^2/2\right) du \\
&= \frac{\sqrt{\pi}}{2a} \exp\left(-2ab\right)
\end{aligned}$$

(b) Hint: See Andrews, and Shivamoggi (1988). Let

$$\begin{aligned}\widehat{f}(s) &= \mathcal{L}[f(t)] \\ &= \int_0^\infty \frac{1}{\sqrt{t}} \exp\left(-st - kt^{-1}\right) dt\end{aligned}$$

Substitute $t = x^2$ in the above integral. It results in

$$\widehat{f}(s) = 2\int_0^\infty \exp\left(-sx^2 - kx^{-2}\right) dx$$

Using the result in part (a) of the problem yields the stated result.

(c) Hint: See Andrews, and Shivamoggi (1988). Differentiate formally both sides of the result in part (b) with respect to k.

23. Show that:

(a) The Laplace transform of

$$f_I(x) = \frac{\lambda}{4}\left(\frac{\pi}{x}\right)^{3/2} \exp\left(-\frac{\pi^4\lambda^2}{16x}\right), \quad \lambda > 0, \text{ and } x \in \mathbb{R}_0^+$$

is

$$\widehat{f}_I(s) = \exp\left(-\frac{1}{2}\pi^2\lambda s^{1/2}\right)$$

(b) The Laplace transform of

$$f_I(x) = \frac{\pi\lambda}{2x^{3/2}} \exp\left(-\frac{\pi^3\lambda^2}{4x}\right), \quad \lambda > 0, \text{ and } x \in \mathbb{R}_0^+$$

is

$$\widehat{f}_I(s) = \exp\left(-\lambda\pi^{3/2}s^{1/2}\right)$$

Hint: In the last problem part (c):

(a) Substitute $2\sqrt{k} = \pi^2\lambda/2$.

(b) Substitute $2\sqrt{k} = \pi^{3/2}\lambda$.

24. Prove that, for $k \in \mathbb{P}$ and $\delta \in (0,1)$

$$\frac{\Gamma(k+\delta)}{k^\delta \Gamma(k)} < 1$$

Hint: See Wendel (1948). The stated inequality is established by using the Hölder's inequality. It is

$$\int_0^\infty |f(t)\, g(t)|\, dt \le \left\{\int_0^\infty |f(t)|^p\, dt\right\}^{1/p} \left\{\int_0^\infty |g(t)|^q\, dt\right\}^{1/q}$$

where

$$\left(p^{-1} + q^{-1}\right) = 1, \text{ and } p, q > 1$$

As $\delta \in (0,1)$, we let $p = 1/\delta$, and $q = 1/(1-\delta)$. Also let

$$\begin{aligned}f(t) &= \left(e^{-t}t^k\right)^\delta \\ g(t) &= \left(e^{-t}t^{k-1}\right)^{(1-\delta)}\end{aligned}$$

Therefore

$$\int_0^\infty e^{-t} t^{k+\delta-1} dt < \left\{ \int_0^\infty e^{-t} t^k dt \right\}^\delta \left\{ \int_0^\infty e^{-t} t^{k-1} dt \right\}^{1-\delta}$$

which is

$$\Gamma(k+\delta) < \{\Gamma(k+1)\}^\delta \{\Gamma(k)\}^{1-\delta}$$

As $\Gamma(k+1) = k\Gamma(k)$, we have $\Gamma(k+\delta) < k^\delta \Gamma(k)$.

References

1. Abramowitz, M., and Stegun, I. A., 1965. *Handbook of Mathematical Functions*, Dover Publications, Inc., New York, New York.
2. Agrawal, R., Cruz, R. L., Okino, C., and Rajan, R., 1999. "Performance Bounds for Flow Control Protocols," IEEE/ACM Transactions on Networking, Vol. 7, No. 3, pp. 310-323.
3. Albano, M., Tewodros, A., Beyerstedt, E., and Moll, V. H., 2011. "The Integrals in Gradshteyn and Ryzhik. Part 19: The Error Function," Scientia, Series A: Mathematical Sciences, Vol. 21, pp. 25-42.
4. Andrews, L. C., and Shivamoggi, B. K., 1988. *Integral Transforms for Engineers and Applied Mathematicians*, Macmillan Publishing Company, New York, New York.
5. Baccelli, F., and Blaszczyszyn, 2009a. "Stochastic Geometry and Wireless Networks," Volume I - Theory, NOW, Foundations and Trends in Networking, Vol. 3, Nos. 3-4, pp. 249-449.
6. Baccelli, F., and Blaszczyszyn, 2009b. "Stochastic Geometry and Wireless Networks," Volume II - Applications, NOW, Foundations and Trends in Networking, Vol. 4, Nos. 1-2, pp. 1-312.
7. Baccelli, F., McDonald, D. R., and Reynier, J., 2002. "A Mean-Field Model for Multiple TCP Connections Through a Buffer Implementing RED," Performance Evaluation, Vol. 49, Issues 1-4, pp. 77-97.
8. Bertsekas, D., and Gallager, R. 1992. *Data Networks*, Second Edition, Prentice-Hall, Upper Saddle River, New Jersey.
9. Bonomi, F., Mitra, D., and Seery, J. B., 1995. "Adaptive Algorithms for Feedback-Based Flow Control in High Speed Wide Area ATM Networks," IEEE Journal on Selected Areas in Communications, Vol. 13, No. 7, pp. 1267-1283.
10. Braden, B., Clark, D., Crowcroft, J., Davie, B., Deering, S., Estrin, D., Floyd, S., Jacobson, V., Minshall, G., Partridge, C., Peterson, L., Ramakrishnan, K., S. Shenker, Wroclawski, J., and Zhang, L., 1998. "Recommendations on Queue Management and Congestion Avoidance in the Internet," RFC 2309, IETF.
11. Chang, C.-S., 2000. *Performance Guarantees in Telecommunication Networks,* Springer-Verlag, Berlin, Germany.
12. Chiu, D. M., and Jain, R., 1989. "Analysis of the Increase and Decrease Algorithms for Congestion Avoidance in Computer Networks," Computer Networks and ISDN Systems, Vol. 17, Issue 1, pp. 1-14.
13. Courcoubetis, C., and Weber, R., 2003. *Pricing Communication Networks, Economics, Technology and Modeling*, Second Edition, John Wiley & Sons, Inc., New York, New York.
14. Crowcroft, J., 2002. "Towards a Field Theory for Networks," Computer Laboratory, University of Cambridge.
15. Cruz, R. L., 1991a. "A Calculus for Network Delay, Part I: Network Elements in Isolation," IEEE Trans. on Information Theory, Vol. IT-37, No. 1, pp. 114-131.

16. Cruz, R. L., 1991b. "A Calculus for Network Delay, Part II: Network Analysis," IEEE Trans. on Information Theory, Vol. IT-37, No. 1, pp. 132-141.
17. Demers, A., Keshav, S., and Shenker, S., 1990. "Analysis and Simulation of a Fair Queueing Algorithm," Internetworking: Research and Experience, Vol. 1, No. 1, pp. 3-26.
18. Floyd, S., and Jacobson, V., 1993. "Random Early Detection Gateways for Congestion Avoidance," IEEE/ACM Transactions on Networking, Vol. 1 , No. 4, pp. 397-413.
19. Franceschetti, M., and Meester, R., 2007. *Random Networks, for Communication from Statistical Physics to Information Systems*, Cambridge University Press, New York, New York.
20. Gevros, P., Crowcroft, J., Kirstein, P., and Bhatti, S., 2001. "Congestion Control Mechanisms and the Best Effort Service Model," IEEE Network, Vol. 15, Issue 3, pp. 16-26.
21. Haenggi, M., 2005. "On Distances in Uniformly Random Networks," IEEE Trans. on Information Theory, Vol. IT-51, No. 10, pp. 3584-3586.
22. Haenggi, M., 2013. *Stochastic Geometry for Wireless Networks*, Cambridge University Press, New York, New York.
23. Haenggi, M., Andrews, J. G., Baccelli, F., Dousse, O., and Franceschetti, M., December 2009. "Stochastic Geometry and Random Graphs for the Analysis and Design of Wireless Networks," IEEE Journal on Selected Areas in Communications, Vol. 27, No. 7, pp. 1029-1046.
24. Haenggi, M., and Ganti, R. K., 2008. "Interference in Large Wireless Networks", NOW, Foundations and Trends in Networking, Vol. 3, No. 2, pp. 127-248.
25. Hassan, M., and Jain, R., 2004. *High Performance TCP/IP Networking, Concepts, Issues, and Solutions*, Prentice-Hall, Upper Saddle River, New Jersey.
26. Hollot, C. V., Misra, V., Towsley, and Gong, W. B., 2001a. "A Control Theoretic Analysis of RED," IEEE INFOCOM, 1510-1519.
27. Hollot, C. V., Misra, V., Towsley, and Gong, W. B., 2001b. "On Designing Improved Controllers for AQM Routers Supporting TCP Flows," IEEE INFOCOM, pp. 1726-1734.
28. Hollot, C. V., Misra, V., Towsley, and Gong, W. B., 2002. "Analysis and Design of Controllers for AQM Routers Supporting TCP Flows," IEEE Transactions on Automatic Control, Vol. 47, Issue 6, pp. 945-959.
29. Jacobson, V., 1988. "Congestion Avoidance and Control," ACM SIGCOMM Computer Communication Review, Vol. 18, Issue 4, pp. 314-329.
30. Jain, R., 1990. "Congestion Control in Computer Networks: Issues and Trends," IEEE Network Magazine, Vol. 4, Issue 3, pp. 24-30.
31. Johari, R., 2000. "Mathematical Modeling and Control of Internet Congestion," SIAM News, Vol. 33, No. 2.
32. Kelly, F. P., with an addendum by Kelly, F. P. and Johari, R., 1997. "Charging and Rate Control for Elastic Traffic," European Transactions on Telecommunications, Vol. 8, Issue 1, pp. 33-37.
33. Kelly, F. P., Maulloo, A. K., and Tan, D. K. H., 1998. "Rate Control in Communication Networks: Shadow Prices, Proportional Fairness and Stability," Journal of the Operational Research Society, Vol. 49, No. 3, pp. 237-252.
34. Kelly, F. P., 2001. "Mathematical Modeling of the Internet," in *Mathematics Unlimited - 2001 and Beyond.* Editors: Engquist, B., and Schmid, W., Springer-Verlag, Berlin, Germany, pp. 685-702.
35. Keshav, S., 2012. *Mathematical Foundations of Computer Networking,* Addison-Wesley Publishing Company, New York, New York.
36. Kumar, A., Manjunath, D., and Kuri, J., 2004. *Communication Networking, An Analytical Approach,* Morgan Kaufmann Publishers, San Francisco, California.
37. Le Boudec, J.-Y., and Thiran, P., 2004. *Network Calculus: A Theory of Deterministic Queueing Systems for the Internet,* LNCS 2050, Springer-Verlag, Berlin, Germany.

38. Low, S. H., and Lapsley, D. E., 1999. "Optimization Flow Control, I: Basic Algorithm and Convergence," IEEE/ACM Transactions on Networking, Vol. 7, No. 6, pp. 861-874.
39. Low, S. H., Paganini, F., and Doyle, J. C., 2002. "Internet Congestion Control," IEEE Control Systems Magazine, Vol. 22, Issue 1, pp. 28-43.
40. Low, S. H., 2017. *Analytical Methods for Network Congestion Control*, Morgan and Claypool Publishers, San Rafael, California.
41. Mathis, M., Semke, J., Mahdavi, J., and Ott, T. 1997. "The Macroscopic Behavior of the TCP Congestion Avoidance Algorithm," ACM Computer Communication Review, Vol. 27, No. 3, pp. 67-82.
42. Misra, V., Gong, W., and Towsley, D., 2000. "Fluid-Based Analysis of a Network of AQM Routers Supporting TCP Flows with an Application to RED," Proceedings of ACM SIGCOMM, Stockholm, Sweden, pp. 151-160.
43. Mo, J., and Walrand, J., 2000. "Fair End-to-End Window-Based Congestion Control," IEEE/ACM Transactions on Networking, Vol. 8, No. 5, pp. 556-567.
44. Padhye, J., Firoiu, V., Towsley, D., and Kurose, J., 2000. "Modeling TCP Throughput: A Simple Model and its Empirical Validation," IEEE/ACM Transactions on Networking, Vol. 8, No. 2, pp. 133-145.
45. Parekh, A. K., and Gallager, R. G., 1993. "A Generalized Processor Sharing Approach to Flow Control in Integrated Services Networks: The Single-Node Case," IEEE/ACM Transactions on Networking, Vol. 1, No. 3, pp. 344-357.
46. Parekh, A. K., and Gallager, R. G., 1994. "A Generalized Processor Sharing Approach to Flow Control in Integrated Services Networks: The Multiple Node Case," IEEE/ACM Transactions on Networking, Vol. 2, No. 2, pp. 137-150.
47. Ryu, S., Rump, C., and Qiao, C., Third Quarter 2003. "Advances in Internet Congestion Control," IEEE Communications Surveys & Tutorials, Vol. 5, Issue 1, pp. 28-39.
48. Schwartz, M., 1987. *Telecommunication Networks*: *Protocols, Modeling and Analysis*, Addison-Wesley Publishing Company, Reading, Massachusetts.
49. Srikant, R., 2004. *The Mathematics of Internet Congestion Control*, Birkhauser, Boston, Massachusetts.
50. Stoyan, D., Kendall, W. S., and Mecke, J., 1995. *Stochastic Geometry and its Applications*, Second Edition, John Wiley & Sons, Inc., New York, New York.
51. Van Mieghem, P., 2006. *Performance Analysis of Communications Networks and Systems*, Cambridge University Press, Cambridge, Great Britain.
52. Varma, S., 2015. *Internet Congestion Control,* Morgan Kaufmann Publishers, San Francisco, California.
53. Vaze, R., 2015. *Random Wireless Networks, an Information Theoretic Approach*, Cambridge University Press, New Delhi, India.
54. Veres, A., and Boda, M., 2000. "The Chaotic Nature of TCP Congestion Control," Proc. IEEE INFOCOM'2000, pp. 1715-1723.
55. Weber, S., Andrews, J. G., Jindal, N., December 2010. "An Overview of the Transmission Capacity of Wireless Networks," IEEE Transactions on Communications, Vol. Com-58, No. 12, pp. 3593-3604.
56. Wendel, J. G., 1948. "Note on the Gamma Function," Amer. Math. Monthly, Vol. 55, Issue 9, pp. 563-564.

CHAPTER 6

Queueing Theory

$$p_X(x) = e^{-\lambda}\frac{\lambda^x}{x!}$$

$$\forall x \in \mathbb{N}, \ \lambda > 0$$

Poisson distribution

Siméon Denis Poisson. Siméon Poisson was born on 21 June, 1781 in Pithiviers, France. Parents of Siméon Poisson had first hoped that he would become a surgeon, but strong mathematical talent led him in 1798 to enter École Polytechnique, where he was a student of P. S. Laplace (1749-1827). Poisson later became a professor at this institute. He made contributions to celestial mechanics, potential theory, differential equations, probability theory, and several other branches of applied mathematics.

Poisson's oft repeated view of mathematics: *Life is good for only two things, discovering mathematics, and teaching mathematics.* Poisson died on 25 April, 1840 in Sceaux (near Paris), France.

6.1 Introduction

Queueing theory is a very versatile scientific discipline. It can be used to model computer systems, telecommunication networks, and several other applications. The Internet is a prime example of a network of queues. The following topics are discussed in this chapter. Queue-theoretic terminology and notation is initially introduced. This is followed by the statement and proof of the ubiquitous Little's law. Limiting (steady-state) probabilities of a queueing system are next described. Birth-and-death process based queueing systems, non-Poisson queues, priority queueing systems, and Markovian queueing networks are also discussed. Some important queueing bounds and approximations are also stated. Finally, applications of large deviation techniques are discussed.

Human beings also encounter queueing situations frequently. For example customers arrive at a bank and form a waiting line to be eventually served by a teller. In this scenario clients of the bank wishing to have a transaction are the customers, and the bank-tellers are the servers. In the case of a packet network, the packets are the customers, and the router is the server. In the rest of this chapter, the term customer is used to indicate a single unit which requires service.

A queue is basically a line waiting to be served. In addition to waiting space, a queueing system has two parties. One party consists of the servers, who provide the service requested by the customer, and the other party is the set of customers, who request service from the servers. Therefore in a queueing system, a server may be either busy servicing a customer, or it may be idle. When a customer enters a queueing system, it gets serviced by an idle server if available. However, if all the servers in this system are busy, it joins the waiting line along with customers who might already be present in the queue. The customers in the waiting line get serviced as the servers become free. Further, the customer in the waiting line to be serviced first depends upon the servicing discipline.

A single queueing center is primarily characterized by its *capacity* C, and the *arrival rate* R at which the customers arrive at the facility to be serviced. The capacity C is the maximum service rate at which the customers can get processed at the queueing center.

The flow of customers at the queueing center is *steady* (or stable) in the long run, if the capacity of the queueing center is larger than the customer arrival rate. This occurs, if $R < C$. On the other hand, if the capacity of the queueing center is less than the customer arrival rate, then all mayhem (congestion) may occur in general. This occur if $R > C$ and results in an *unsteady* (or unstable) flow of customers. In this chapter, we only analyze queueing systems for which $R < C$. Next consider the case of a network of queues, in which the queueing centers are connected in some configuration. For stable flow of customers in the network, it is still required that $R < C$ on each channel of the network.

Basics of applied analysis, matrices and determinants, probability theory, stochastic processes, optimization techniques, and Internet traffic are required to study this chapter.

6.2 Terminology and Notation

As per Leonard Kleinrock, the word *queueing*, is perhaps the only word in English language with five consecutive vowels in it. This subject of queueing theory merits special attention, because in queueing scenarios intuitive answers are often misleading. The purpose of queueing theory is to evaluate the response time of the customers and the efficiency of the servers. These are evaluated as a function of: the customer source population, the arrival pattern of the customer, the service time characteristics, number of servers, the capacity of the queueing system, and the service (or queue) discipline. These parameters are next described in detail.

(a) *The Customer Population*: This is the total number of customers which can request service. This number can be either finite or infinite. The customer population influences the performance of a queueing system.

(b) *The Arrival Pattern of the Customer*: Assume that the queueing system is being observed from time $t_0 \geq 0$, and the customers arrive at instants $t_1, t_2, \ldots$ such that $0 \leq t_0 \leq t_1 \leq t_2 \leq \ldots$. The variables $a_k = t_k - t_{k-1}$, where $k = 1, 2, \ldots$, are called *interarrival times* of the customers. Let a_k be an instance of the random variable A_k, where $k \in \mathbb{P}$. It is assumed in this chapter that these interarrival times have identical probability distribution. Furthermore, these are independent of each other. The random variable A is used to denote any such arbitrary interarrival time. Thus the probability distribution of the random variable A essentially characterizes the arrival pattern of a customer.
It should be noted that the customers might either arrive in a single stream or in a bulk (simultaneous arrivals). If they arrive in bulk, then the bulk interarrival time distribution and bulk-size distribution have to be specified.

(c) *The Service Time Characteristics*: The service times of different customers have identical probability distribution. If the service is rendered in bulk, then this bulk-size distribution also has to be specified. Also the service times of customers are assumed to be independent of each other.

(d) *Number of Servers*: The performance of a queueing system depends upon the number of servers. If there is a single server, then the queueing system is said to be a single-server system. However, if there are multiple servers, then the queueing system is said to be a multiserver system. It is also possible to have an infinite server system. This later queueing system is possible, if all arriving customers find a server immediately.

(e) *The Capacity of the Queueing System*: The capacity of a queueing system is the maximum number of customers which can be held in it. This number can be either finite or infinite. It includes the number of places available for the customers to queue (wait) and also get service. That is, capacity is total number of waiting positions plus number of servers.

(f) *The Service Discipline*: The service or queue discipline is the scheme for selecting the next customer from the waiting line to be offered service. The most popular service discipline is the *first-come, first-served* (FCFS) queueing scheme. This scheme is also called the *first-in, first-out* (FIFO) discipline. Another example of service discipline is *last-come, first-served* (LCFS) (also called *last-in, first-out* (LIFO)). *Service in random-order* (SIRO) is another servicing discipline. In this scheme each customer waiting in the queue has equal probability of being selected for service. In the *priority service* (PRI) discipline customers are split into several priority classes. Then preferential service is offered by classes.

The above six parameters of a queue are generally specified in a special notation. It is: $A/B/c/K/N_{pop}/Z$:

(a) A describes the customer interarrival time distribution.
(b) B describes the customer service time distribution.
(c) c is equal to the number of servers.
(d) K is the capacity of the queueing system.
(e) N_{pop} is the total number of customers (population) which can request service.
(f) Z is the queue service discipline.

If the capacity of the queueing system is infinite, the customer population is infinite, and the service discipline is FCFS; then a shorter notation $A/B/c$ is used. That is, $A/B/c/\infty/\infty/FCFS$ is abbreviated as $A/B/c$. The following symbols are generally used for A and B.

(a) M for exponential probability distribution (M for memoryless).
(b) D for deterministic probability distribution.
(c) G for general probability distribution.

For example $G/M/c$ queue has a general customer interarrival time distribution, its customer service time distribution is exponential, and there are c number of servers. Furthermore, the capacity of the queueing system is infinite, the customer population is infinite, and the service discipline is FCFS.

It is generally difficult to obtain the transient performance of a queueing system. Therefore results are only specified for the performance of queueing systems at steady-state (equilibrium), provided they exist. The conditions for the existence of steady-state results are discussed within the context of each queueing system.

The interarrival and service time processes are also assumed to be stochastically independent of each other.

The commonly used symbols in this chapter are next listed. In this description, $\mathcal{E}(\cdot)$ and $Var(\cdot)$ are the expectation and variance operators respectively.

Notation Used in Describing Queues

$c =$ number of servers with identical characteristics.
$A =$ random variable which represents the customer interarrival time.
$B =$ random variable which represents the customer service time.
$\Lambda_s =$ random variable which represents the number of customers receiving service at equilibrium.
$\Lambda_q =$ random variable which represents the number of customers in the queue (waiting) at equilibrium.

$\Lambda =$ random variable which represents the number of customers in the queueing system at equilibrium.
$\Omega_q =$ random variable which represents the total time a customer waits in the queue at equilibrium.
$\Omega =$ random variable which represents the total time a customer spends in the queueing system at equilibrium.
$L_s =$ average number of customers receiving service at steady-state $= \mathcal{E}(\Lambda_s)$.
$L_q =$ average number of customers in the queue (waiting) at steady-state $= \mathcal{E}(\Lambda_q)$.
$L =$ average number of customers in the queueing system at steady-state $= \mathcal{E}(\Lambda)$.
$W_s =$ average time a customer spends receiving service $= \mathcal{E}(B)$.
$W_q =$ average time a customer waits in the queue at steady-state, before it receives service $= \mathcal{E}(\Omega_q)$.
$W =$ average time a customer spends in the queueing system at steady-state $= \mathcal{E}(\Omega)$.
$\lambda =$ arrival rate of customers to the queueing system $= 1/\mathcal{E}(A)$, where $\mathcal{E}(A) > 0$.
$\mu =$ service rate per server $= 1/\mathcal{E}(B)$, where $\mathcal{E}(B) > 0$.
$\sigma_A =$ standard deviation of the random variable A.
$\sigma_B =$ standard deviation of the random variable B.
$C_A =$ coefficient of variation of the random variable A.
$C_B =$ coefficient of variation of the random variable B.
$\rho =$ average utilization of a single server.
$\zeta = \lambda/\mu =$ average number of customer arrivals during mean service time of a server.
$p_n =$ probability that there are n customers in the queueing system at steady-state $= P(\Lambda = n.)$, where $n \in \mathbb{N}$.
$t =$ time variable. Generally $t \in \mathbb{R}_0^+$.
$f_A(\cdot) =$ probability density function of the interarrival time of the customers.
$F_A(\cdot) =$ cumulative distribution function of the interarrival time of the customers.
$f_B(\cdot) =$ probability density function of the service time of the customers.
$F_B(\cdot) =$ cumulative distribution function of the service time of the customers.
$f_{\Omega_q}(\cdot) =$ probability density function of the time a customer waits in the queueing system at equilibrium.
$F_{\Omega_q}(\cdot) =$ cumulative distribution function of the time a customer waits in the queueing system at equilibrium.
$f_\Omega(\cdot) =$ probability density function of the total time a customer spends in the queueing system at equilibrium.
$F_\Omega(\cdot) =$ cumulative distribution function of the total time a customer spends in the queueing system at equilibrium.

It is assumed in the rest of the chapter that

$$\mathcal{E}(A), \mathcal{E}(B) \in \mathbb{R}^+$$

That is, the average customer interarrival and service times are assumed to be finite and positive. The following relationships are immediately observed.

$$\Lambda = \Lambda_q + \Lambda_s, \quad L = L_q + L_s \tag{6.1a}$$

$$\Omega = \Omega_q + B, \quad W = W_q + W_s \tag{6.1b}$$

It should also be noted that steady-state results are valid only if $\rho < 1$. In the next section it is also demonstrated that at steady-state $L = \lambda W$, and $L_q = \lambda W_q$. Also notice that $L_s = \lambda W_s$. These are simple, yet extremely useful results in queueing theory.

6.3 Little's Law

Little's law is perhaps one of the most fundamental results of queueing theory. It states that in a queueing system at steady-state

$$L = \lambda W$$

J. D. C. Little provided the first mathematical proof of this relationship in 1961, even though this relationship was known much earlier. This law is analogous to the *law of conservation* found in many physical systems. Little's law does not depend explicitly upon: customer interarrival time distribution, customer service time distribution, the number of servers in the queueing system, and the queueing discipline.

A similar relationship is obtained connecting L_q, W_q, and λ, which is $L_q = W_q \lambda$. The relationship $L_s = \lambda W_s$ is the application of Little's law at the server level. Little's law is next stated formally.

Theorem 6.1. *Let $\alpha(t)$ be the number of customers who arrive in the time interval $[0, t]$. The average customer arrival rate λ is defined by*

$$\lambda = \lim_{t \to \infty} \frac{\alpha(t)}{t} \tag{6.2a}$$

Also let the number of customers present in the queueing system at time τ be $N(\tau)$. Define

$$L = \lim_{t \to \infty} \frac{1}{t} \int_0^t N(\tau)\, d\tau \tag{6.2b}$$

If W_j is the time spent by the jth customer in the queueing system, then define the average residence time of a customer in this system by

$$W = \lim_{n \to \infty} \frac{1}{n} \sum_{j=1}^{n} W_j \tag{6.2c}$$

If the values λ and W exist (finite) then

$$L = \lambda W \tag{6.2d}$$

Sketch of the proof. As per the hypothesis of the theorem:
$\alpha(t) =$ total number of customer arrivals during the interval $[0, t]$.
$N(t) =$ total number of customers in the system at time t.
Next define:
$\delta(t) =$ total number of customer departures during the interval $[0, t]$.
$\omega(t) =$ total time spent in the system by all customers who arrive during the interval $[0, t]$.
$\lambda(t) =$ mean arrival rate of customers who arrive during the interval $[0, t]$.
$W(t) =$ mean residence (in the queueing system) time of customers who arrive during the interval $[0, t]$.
$L(t) =$ mean number of customers who arrive during the interval $[0, t]$.
Assume that the system is empty at time $t = 0$. Use of the above definitions yields

$$N(t) = \alpha(t) - \delta(t)$$
$$\omega(t) = \int_0^t N(\tau)\, d\tau$$

The mean values for $t > 0$ are

$$\lambda(t) = \frac{\alpha(t)}{t}, \quad W(t) = \frac{\omega(t)}{\alpha(t)}, \quad L(t) = \frac{\omega(t)}{t}$$

From which it follows that

$$L(t) = \lambda(t)\, W(t)$$

By hypothesis of the theorem the limits $\lim_{t\to\infty} \lambda(t)$, and $\lim_{t\to\infty} W(t)$ exist and are finite. Actually $\lim_{t\to\infty} \lambda(t) = \lambda$, and it can be shown that

$$\lim_{t\to\infty} W(t) = \lim_{t\to\infty} \frac{1}{\alpha(t)} \sum_{j=1}^{\alpha(t)} W_j = \lim_{n\to\infty} \frac{1}{n} \sum_{j=1}^{n} W_j = W$$

Consequently the limit of $L(t)$ as $t \to \infty$ also exists, and is given by $\lim_{t\to\infty} L(t) = L$. Thus the above limits satisfy the equation $L = \lambda W$. □

6.4 Steady-State Probabilities

Some characteristics of limiting probabilities of a queueing system are described. As defined in the last section, $N(t)$ is equal to the total number of customers in the queueing system at time t. Also, $\alpha(t)$ and $\delta(t)$ are the total number of customer arrivals and departures respectively in the time interval $[0, t]$. The following variables are initially defined.

$n =$ number of customers in a queueing system, $n \in \mathbb{N}$. This variable can be used to specify the state of the queueing system.

$\alpha_n(t) =$ total number of customer arrivals which occur in state n of the queueing system during the time interval $[0, t]$.

$\delta_n(t) =$ total number of customer departures which occur in state n of the queueing system during the time interval $[0, t]$.

$a_n(t) =$ arrival point probability that n customers are present in the queueing system when a customer arrives at time t.

$d_n(t) =$ departure point probability that n customers are left in the queueing system after a customer departs at time t.

$P(N(t) = n) =$ probability that the number of customers $N(t)$ in the system at time t is equal to n.

Limiting (steady-state or equilibrium) probabilities, provided they exist are defined below. That is, as $t \to \infty$.

$p_n =$ long run probability that n customers are present in the queueing system.

$a_n =$ arrival point probability that n customers are present in the queueing system when a customer arrives.

$d_n =$ departure point probability that n customers are left in the queueing system after a customer departs.

$$p_n = \lim_{t \to \infty} P(N(t) = n), \quad a_n = \lim_{t \to \infty} \frac{\alpha_n(t)}{\alpha(t)}, \quad d_n = \lim_{t \to \infty} \frac{\delta_n(t)}{\delta(t)} \tag{6.3}$$

Lemma 6.1. *In a queueing system in which customers arrive and depart one at a time $a_n = d_n$ for all $n \in \mathbb{N}$, at equilibrium.*

Proof. Since arrivals and departures occur singly $|\alpha_n(t) - \delta_n(t)| \leq 1$ for large values of t. Also

$$N(t) = (N(0) + \alpha(t) - \delta(t))$$

Therefore

$$\frac{\delta_n(t)}{\delta(t)} = \frac{\alpha_n(t) + \delta_n(t) - \alpha_n(t)}{\alpha(t) + N(0) - N(t)}$$

Noting that $N(0)$ is finite, and $N(t)$ is finite because of the assumption of the existence of equilibrium

$$\begin{aligned} d_n &= \lim_{t \to \infty} \frac{\delta_n(t)}{\delta(t)} = \lim_{t \to \infty} \frac{\alpha_n(t) + \delta_n(t) - \alpha_n(t)}{\alpha(t) + N(0) - N(t)} \\ &= \lim_{t \to \infty} \frac{\alpha_n(t)}{\alpha(t)} = a_n \end{aligned}$$

□

Next consider a queueing system in which customer arrival process is Poissonian. In this queueing system, customer arrivals always see time averages. That is, a customer arrival from a Poisson stream finds the same state distribution as a *random* observer from outside the queueing system. This law is succinctly stated as: *Poisson arrivals see time averages.* It is often abbreviated as PASTA. This assertion is proved in the next lemma.

Lemma 6.2. *In a queueing system in which customers arrive according to a Poisson process, then*

$$a_n(t) = d_n(t) = P(N(t) = n), \quad \forall\, n \in \mathbb{N}, \ t \geq 0 \tag{6.4}$$

Proof.

$$\begin{aligned} & a_n(t) \\ &= \lim_{\Delta t \to 0} P(n \text{ in system at } t \mid \text{arrival occurs in } (t, t + \Delta t)) \\ &= \lim_{\Delta t \to 0} \frac{P(n \text{ in system at } t \text{ and arrival occurs in } (t, t + \Delta t))}{P(\text{arrival occurs in } (t, t + \Delta t))} \\ &= \lim_{\Delta t \to 0} \left\{ \frac{P(\text{arrival occurs in } (t, t + \Delta t) \mid n \text{ in system at } t)}{P(\text{arrival occurs in } (t, t + \Delta t))} \right\} \\ & \quad \times P(n \text{ in system at } t) \\ &= P(n \text{ in system at } t) = P(N(t) = n) \end{aligned}$$

where the last step follows from the property of Poisson process that the probability of an arrival is independent of the number of customers already present in the queueing system. Note that $P(N(t) = n) = d_n(t)$, consequently $a_n(t) = d_n(t)$, $\forall\, n \in \mathbb{N}$, $t \geq 0$.

More specifically at steady-state $a_n = p_n$ for any $n \in \mathbb{N}$. □

6.5 Birth-and-Death Process Based Queueing Systems

In the birth-and-death process based queueing systems, the interarrival times of customers and the server-processing times have exponential distributions. Therefore the properties of an exponentially distributed random variable are summarized below. This is followed by a general description of a birth-and-death based queueing system. Several examples of such queueing systems are also given. Note that such queueing systems are also called Poisson queues.

6.5.1 Exponentially Distributed Random Variable

Some properties of an exponentially distributed random variable are summarized below.

Observations 6.1. A random variable T is an exponentially distributed random variable. Its probability density function is given by

$$f_T(t) = \begin{cases} 0, & t \in (-\infty, 0) \\ \lambda e^{-\lambda t}, & t \in [0, \infty) \end{cases}$$

where $\lambda \in \mathbb{R}^+$.

1. Its mean $\mathcal{E}(T) = 1/\lambda$, and variance $Var(T) = 1/\lambda^2$.
2. Its cumulative distribution function is given by
$$F_T(t) = \begin{cases} 0, & t \in (-\infty, 0) \\ \left(1 - e^{-\lambda t}\right), & t \in [0, \infty) \end{cases}$$
3. The probability density function $f_T(\cdot)$ is a strictly decreasing function of t for $t > 0$. Therefore
$$P(0 \le T \le t) > P(u \le T \le t + u), \quad u > 0$$
4. A random variable T is said to have the memoryless property if
$$P(T \ge t + u \mid T \ge u) = P(T \ge t), \quad \forall\, t, u \in [0, \infty)$$
An exponentially distributed random variable is the only continuous random variable, that possesses the memoryless property.
5. The minimum of several independent, and exponentially distributed random variables is also an exponentially distributed random variable.
6. If $t > 0$, then for small values of Δt
$$P(T \le t + \Delta t \mid T > t) = \left(1 - e^{-\lambda \Delta t}\right) \simeq \lambda \Delta t$$
7. Relationship between exponential distribution and the Poisson process.
   (a) A Poisson process is a counting random process $\{N(t), t \in \mathbb{R}_0^+\}$. In this process, $N(t)$ counts the total number of events which have occurred up to time t, such that
   $$P(N(t) = n) = e^{-\lambda t} \frac{(\lambda t)^n}{n!}, \quad \forall\, n \in \mathbb{N}, \ \lambda \in \mathbb{R}^+$$
   Note that $P(N(t) = 0) = e^{-\lambda t}$ = probability that the first event of the Poisson process occurs after time t. Furthermore, since $\mathcal{E}(N(t)) = \lambda t$, the expected number of events per unit time is λ. Therefore λ is the *mean rate* at which the events occur.

(b) The time interval between two successive occurrences of a Poisson process are called *interarrival times*. Interarrival times have an exponential distribution with parameter λ. Furthermore, these interarrival times are stochastically independent of each other.

(c) Aggregation of several independent streams of Poisson processes is a Poisson process. Similarly, the independent disaggregation of a Poisson process can produce several independent streams of Poisson processes. □

All the above observations are useful in describing Poissonian streams of customers who enter a queueing system. For example, if the customer interarrival times at a queue have an exponential distribution with parameter λ, and these times are mutually independent, then the number of customers who arrive in the elapsed time t have a Poisson distribution with parameter λt.

6.5.2 Birth-and-Death Process Description of a Queue

The queues described in this section assume that the respective queueing systems are governed by birth-and-death processes. The input and output stream of customers to the queueing system are the arrivals and departures respectively. The *birth* refers to the *arrival* of a new customer, and the *death* refers to the *departure* of a customer. It has been mentioned in the chapter on stochastic processes, that the birth-and-death process is an example of a continuous-time Markov chain. The steady-state results of this process are summarized below.

Let $\{N(t), t \in \mathbb{R}_0^+\}$ be a birth-and-death process which describes a queueing system. In this process, $N(t)$ is equal to the number of customers in the queueing system at time t. The state space of this process is the set of integers $\mathbb{N}$. If $N(t) = j$, the birth and death rates of the process at this time are equal to λ_j and μ_j respectively, for $j \in \mathbb{N}$. Trivially, death is not possible if $N(t) = 0$, that is $\mu_0 = 0$. In this general formulation of birth-and-death process $\lambda_j \in \mathbb{R}^+$ for all $j \in \mathbb{N}$, and $\mu_j \in \mathbb{R}^+$ for all $j \in \mathbb{P}$. As mentioned at the beginning of the chapter, each customer interarrival time is independent of the other, each service time is independent of the other; and the customer arrival process and the service times are stochastically independent of each other.

In addition, define p_j to be the probability that j customers are present in the queueing system at steady-state, where $j \in \mathbb{N}$. Then the birth-and-death balance equations are:

$$0 = \lambda_{j-1}p_{j-1} - (\lambda_j + \mu_j)\,p_j + \mu_{j+1}p_{j+1}, \quad \forall j \in \mathbb{P}$$
$$0 = -\lambda_0 p_0 + \mu_1 p_1$$

Define

$$\alpha_0 = 1$$
$$\alpha_j = \frac{\lambda_0\lambda_1\ldots\lambda_{j-1}}{\mu_1\mu_2\ldots\mu_j}, \quad \forall j \in \mathbb{P}$$

If $\sum_{j\in\mathbb{N}} \alpha_j < \infty$, then

$$p_j = \frac{\alpha_j}{\sum_{j\in\mathbb{N}} \alpha_j}, \quad \forall j \in \mathbb{N}$$

It can be shown that the condition $\sum_{j\in\mathbb{N}} \alpha_j < \infty$ is sufficient for the birth-and-death process to reach equilibrium (steady-state). In this case, the birth-and-death process is also said to be *ergodic*. This formulation is next extended to several different queueing systems. These are:

(a) $M/M/\infty$ queue.

(b) $M/M/c/c$ queue.
(c) $M/M/1$ queue.
(d) $M/M/1/K$ queue.
(e) $M/M/c$ queue.
(f) $M/M/1/K/K$ queue.

The results of these queueing systems are considered at equilibrium. In these formulations the Erlang distribution, named in honor of the Danish engineer Agner Krarup Erlang (1878-1929), is used. He is generally regarded as the founder of queueing theory. Let $S_1, S_2, \ldots, S_k$ be k independent exponentially distributed random variables, each with parameter $\mu \in \mathbb{R}^+$. Then the random variable $Y_k = \sum_{j=1}^{k} S_j$ has an Erlang distribution $f_{Y_k}(t)$ given by

$$f_{Y_k}(t) = \begin{cases} 0, & t \in (-\infty, 0) \\ \dfrac{\mu (\mu t)^{k-1} e^{-\mu t}}{(k-1)!}, & t \in [0, \infty) \end{cases}$$

Its mean and variance are

$$\mathcal{E}(Y_k) = \frac{k}{\mu}, \quad \text{and} \quad Var(Y_k) = \frac{k}{\mu^2}$$

respectively. If $k = 1$, then the random variable Y_1 is exponentially distributed. The reader should note that this distribution is related to the gamma distribution in an obvious manner.

6.5.3 $M/M/\infty$ Queue

An $M/M/\infty$ queue has independent, and exponentially distributed customer interarrival times, each with parameter $\lambda \in \mathbb{R}^+$. That is, the arriving process is Poissonian with parameter λ. The service times of the servers are independent, and exponentially distributed random variables, each with parameter $\mu \in \mathbb{R}^+$. Furthermore, there are an infinite number of servers, and the source population is infinite. Therefore, any arriving customer receives service immediately. This queue is modeled as a birth-and-death process with

$$\lambda_n = \lambda, \quad \forall\, n \in \mathbb{N}$$
$$\mu_n = n\mu, \quad \forall\, n \in \mathbb{P}$$

The probability p_n, that there are n customers in the queueing system at steady-state is given by

$$p_n = \frac{\zeta^n}{n!} e^{-\zeta}, \quad \forall\, n \in \mathbb{N}$$

where $\zeta = \lambda/\mu$. Therefore the number of customers in this queue has a Poisson distribution with parameter ζ. For this queueing system

$$L = L_s = \zeta, \;\; L_q = 0$$
$$W = W_s = \mu^{-1}, \;\; W_q = 0$$

The variance of the number of customers in the queue is equal to ζ. It can also be established that if the queueing system is $M/G/\infty$, the forms of the above equations do not change. Thus the distribution of the number of customers in this queue is also Poissonian.

6.5.4 $M/M/c/c$ Queue

An $M/M/c/c$ is a queueing system in which a customer is lost, if it does not find a free server immediately upon its arrival. That is, the customer is not allowed to wait. This queueing system has $c \in \mathbb{P}$ servers, and its capacity is also c. Consequently, this system is called a c-channel *loss system*. In this queueing scheme the arrival process is Poissonian with parameter $\lambda \in \mathbb{R}^+$. The service times of the servers are independent, and have identical exponential distribution with parameter $\mu \in \mathbb{R}^+$. The customer population is also assumed to be infinite. This queueing system is modeled as a birth-and-death process. Its parameters are:

$$\lambda_n = \begin{cases} \lambda, & n = 0, 1, 2, \ldots, c-1 \\ 0, & n \geq c \end{cases}$$

$$\mu_n = \begin{cases} n\mu, & n = 1, 2, \ldots, c \\ 0, & n > c \end{cases}$$

Using the birth-and-death equations, it can be shown that the steady-state probability p_n for $n = 0, 1, 2, \ldots, c$ is given by

$$p_n = p_0 \frac{\zeta^n}{n!}, \quad n = 0, 1, 2, \ldots, c$$

where $\zeta = \lambda/\mu$. Also $p_n = 0$ for $n > c$. Use of the relationship $\sum_{n=0}^{c} p_n = 1$ yields

$$p_n = \frac{\dfrac{\zeta^n}{n!}}{\sum_{j=0}^{c} \dfrac{\zeta^j}{j!}}, \quad n = 0, 1, 2, \ldots, c$$

This probability distribution is known as the truncated Poisson distribution. This queueing system was first investigated by Erlang in the year 1917. The probability p_c is called the blocking probability. It is the probability that all c servers are busy, so that an arriving customer is blocked from entering the queueing system. The probability p_c is given a special name in honor of this pioneering engineer: *Erlang's $\mathcal{B}$ function* or *Erlang's loss formula.* That is

$$p_c = \mathcal{B}(c, \zeta) = \frac{\dfrac{\zeta^c}{c!}}{\sum_{j=0}^{c} \dfrac{\zeta^j}{j!}} \tag{6.5}$$

The average utilization of a single server in this queueing system has been defined as ρ. Therefore the average utilization of all c servers is equal to ρc.

$$\rho c = \sum_{n=1}^{c} n p_n = \zeta (1 - p_c)$$

In telephony literature, ζ is called the *offered traffic*, and ρc the *carried traffic*. The Erlang's $\mathcal{B}$ formula is generally used to calculate the blocking probability of an incoming telephone call on a group of c trunks.

The average arrival rate to the queue is given by $\overline{\lambda}$. It is

$$\overline{\lambda} = \sum_{n=0}^{c} \lambda_n p_n = \sum_{n=0}^{c-1} \lambda p_n = \lambda (1 - p_c)$$

Also

$$L = \sum_{n=0}^{c} n p_n = \zeta (1 - p_c)$$

Thus $L = \rho c$. There is no queueing, therefore L_q and W_q are each equal to zero. Also the average wait time of the customers who enter the system is $W = L/\overline{\lambda} = 1/\mu$. Furthermore, the distribution of this time is exponentially distributed with parameter μ. The Erlang's $\mathcal{B}$ formula can also be computed recursively as follows.

$$\mathcal{B}(0, \zeta) = 1$$

$$\mathcal{B}(n, \zeta) = \frac{\zeta \mathcal{B}(n-1, \zeta)}{n + \zeta \mathcal{B}(n-1, \zeta)}, \quad \forall\, n \in \mathbb{P}$$

All the above formulae of this queueing system are also valid for the $M/G/c/c$ queueing system. This implies that only the mean value of the service time is significant for the $M/G/c/c$ queueing system. However, the expression for the cumulative distribution function $P(\Omega \leq t), t \in \mathbb{R}_0^+$ is different for the $M/M/c/c$ and $M/G/c/c$ queueing systems.

6.5.5 $M/M/1$ Queue

An $M/M/1$ queueing system is a single server queue with infinite waiting space. The arrival process is Poissonian with parameter $\lambda \in \mathbb{R}^+$, and the service time has an exponential distribution with parameter $\mu \in \mathbb{R}^+$. The customer population is infinite, and the service discipline is FCFS. In this queueing scheme

$$\lambda_n = \lambda, \quad \forall\, n \in \mathbb{N}$$

$$\mu_n = \mu, \quad \forall\, n \in \mathbb{P}$$

This queueing process is ergodic if $\zeta < 1$, where $\zeta = \lambda/\mu$. This is a necessary condition for the existence of steady-state results. Using the birth-and-death equations, it can be shown that the steady-state probability p_n is given by

$$p_n = (1 - \zeta) \zeta^n, \quad \forall\, n \in \mathbb{N}$$

That is, the random variable Λ representing the total number of customers in the queueing system has a geometric distribution. The server utilization ρ is given by

$$\rho = \sum_{n=1}^{\infty} p_n = (1 - p_0) = \zeta$$

Therefore the probability that the server is busy is equal to $(1 - p_0) = \zeta$. The following expressions for this queue can be computed immediately.

$$L = \mathcal{E}(\Lambda) = \sum_{n=0}^{\infty} n p_n = \frac{\zeta}{(1 - \zeta)}$$

$$Var(\Lambda) = \frac{\zeta}{(1 - \zeta)^2}$$

Use of Little's law yields average waiting time in the queueing system as

$$W = \frac{1}{\mu(1-\zeta)}$$

Consequently

$$W_q = W - \frac{1}{\mu} = \frac{\zeta}{\mu(1-\zeta)}$$

$$L_q = \frac{\zeta^2}{(1-\zeta)}$$

The distribution of the random variable Ω is next determined. This is computed by first observing that if a customer arrives when there are already n customers in the queueing system, then its wait time in the system is equal to the sum of $(n+1)$ independent exponentially distributed random variables, each with parameter μ. Therefore this wait time is a random variable Y_{n+1}, which has an Erlang distribution. Therefore, for $t \in \mathbb{R}_0^+$ the cumulative distribution function of the random variable Ω is given by

$$\begin{aligned} F_\Omega(t) &= P(\Omega \le t) = \sum_{n=0}^{\infty} p_n P(Y_{n+1} \le t) \\ &= (1-\zeta) \sum_{n=0}^{\infty} \zeta^n \int_0^t \frac{\mu(\mu x)^n e^{-\mu x}}{n!} dx \\ &= \left(1 - e^{-t/W}\right) \end{aligned}$$

Thus

$$F_\Omega(t) = \left(1 - e^{-t/W}\right), \quad t \in \mathbb{R}_0^+$$

This derivation shows that Ω is an exponentially distributed random variable with parameter $1/W$. Therefore $Var(\Omega) = W^2$. It can be similarly established that the cumulative distribution function of the random variable Ω_q is given by $F_{\Omega_q}(t) = \left(1 - \zeta e^{-t/W}\right), t \in \mathbb{R}_0^+$. Its proof is left as an exercise to the reader. The customer departure process of this queueing system is next examined.

Departure Process

The interval between successive customer departures from an $M/M/1$ queue at steady-state is characterized. It is actually established that the departure process is Poissonian, with parameter λ. Denote the random variable corresponding to the interdeparture interval by E. Let its probability density function be $f_E(t), t \in \mathbb{R}_0^+$. The Laplace transform of $f_E(t)$ is denoted by $\widehat{f}_E(s)$, where $s \in \mathbb{C}$. Also let the Laplace transforms of the customer interarrival time and service time probability density functions be $\widehat{f}_A(s)$ and $\widehat{f}_B(s)$ respectively. Consider the following two events:

(a) Assume that a customer C_1 is in service, and customer C_2 is waiting in line to be serviced next. When the customer C_1 departs, the customer C_2 enters service, and departs after its service is completed. Therefore in this case, the interdeparture time of customers is given by the service time of customer C_2. Its probability density function is equal to $f_B(t)$, for $t \in \mathbb{R}_0^+$. Therefore the Laplace transform of this interdeparture time is equal to $\widehat{f}_B(s)$.

(b) Assume that the customer C_1 is in service, and it departs the queueing system with zero customers waiting for service. Then the interdeparture interval time is equal to the sum of an interarrival time of a customer and its service time. Furthermore, these two intervals are independent of each other. Thus the Laplace transform of this interdeparture time is equal to $\widehat{f}_A(s)\,\widehat{f}_B(s)$.

The probability of the first event is ζ, and the probability of the second event is equal to $(1-\zeta)$. Therefore

$$\widehat{f}_E(s) = \zeta \widehat{f}_B(s) + (1-\zeta)\,\widehat{f}_A(s)\,\widehat{f}_B(s)$$

Substituting

$$\widehat{f}_A(s) = \frac{\lambda}{(\lambda+s)}, \quad \text{and} \quad \widehat{f}_B(s) = \frac{\mu}{(\mu+s)}$$

yields $\widehat{f}_E(s) = \widehat{f}_A(s)$. This result implies that the customer interdeparture times at steady-state are exponentially distributed with parameter λ. A little reflection also shows that the interdeparture times are also independent of each other. Therefore the departure process of an $M/M/1$ queue is Poissonian with parameter λ. This property is useful in studying Markovian queueing networks.

As stated earlier, steady-state condition for the queue exists if $\lambda < \mu$. However, if $\lambda \geq \mu$ it can be shown that as $t \to \infty$, the output process of an $M/M/1$ queue is also Poissonian with parameter μ. In summary, the output process of an $M/M/1$ queue is Poissonian with parameter $\min(\lambda, \mu)$.

6.5.6 $M/M/1/K$ Queue

An $M/M/1/K$ is a single server queue with finite waiting space. The capacity of this queue is K, that is it can hold at most K customers in the queueing system including the customer in service. The arrival process in this queue is Poissonian with parameter $\lambda \in \mathbb{R}^+$, and the service time has an exponential distribution with parameter $\mu \in \mathbb{R}^+$. The customer population is infinite, and the service discipline is FCFS. In this queueing scheme

$$\lambda_n = \begin{cases} \lambda, & n = 0, 1, 2, \ldots, K-1 \\ 0, & n \geq K \end{cases}$$

$$\mu_n = \begin{cases} \mu, & n = 1, 2, \ldots, K \\ 0, & n > K \end{cases}$$

Using the birth-and-death equations, it can be shown that the steady-state probability p_n for $n = 0, 1, 2, \ldots, K$ is given by

$$p_n = p_0 \zeta^n, \quad n = 1, 2, \ldots, K$$

where $\zeta = \lambda/\mu$. Use of the relationship $\sum_{n=0}^{K} p_n = 1$ results in

$$p_n = \begin{cases} \dfrac{(1-\zeta)\,\zeta^n}{\left(1-\zeta^{K+1}\right)}, & \zeta \neq 1 \\[2ex] \dfrac{1}{(K+1)}, & \zeta = 1 \end{cases}$$

where $n = 0, 1, 2, \ldots, K$. Also $p_n = 0$ for $n > K$. Note that if $\zeta < 1$ these probabilities approach the steady-state probabilities of an $M/M/1$ queue as $K \to \infty$. The average queue length at steady-state is

$$L = \sum_{n=0}^{K} n p_n$$

This yields

$$L = \begin{cases} \dfrac{\zeta}{(1-\zeta)} - \dfrac{(K+1)\,\zeta^{K+1}}{\left(1-\zeta^{K+1}\right)}, & \zeta \neq 1 \\ \dfrac{K}{2}, & \zeta = 1 \end{cases}$$

Similarly

$$\begin{aligned} L_q &= \sum_{n=1}^{K} (n-1)\, p_n \\ &= L - (1 - p_0) \end{aligned}$$

Let $\overline{\lambda}$ be the average arrival rate of customers into the queueing system. Then

$$\overline{\lambda} = \sum_{n=0}^{(K-1)} \lambda p_n = \lambda (1 - p_K)$$

The average wait times of the customers who enter the queueing system is obtained by using Little's law.

$$\begin{aligned} W &= \frac{L}{\overline{\lambda}} \\ W_q &= \frac{L_q}{\overline{\lambda}} \\ &= W - \frac{1}{\mu} \end{aligned}$$

The server utilization ρ is given by

$$\rho = \sum_{n=1}^{K} p_n = (1 - p_0)$$

The derivation of cumulative distribution functions of the random variables Ω and Ω_q is complicated. It requires the computation of the probability that an arriving customer who enters the system finds n customers already present in the system. Denote this probability by q_n, where $n = 0, 1, 2, \ldots, (K-1)$. It is next proved that $q_n = p_n / (1 - p_K)$.

$$\begin{aligned} q_n &= P\,(n \text{ customers in the system} \mid \text{arrival about to occur}) \\ &= \frac{P\,(\text{arrival about to occur} \mid n \text{ customers in the system})\, p_n}{\sum_{n=0}^{K} P\,(\text{arrival about to occur} \mid n \text{ customers in the system})\, p_n} \\ &= \frac{\lambda p_n}{\sum_{n=0}^{K-1} \lambda p_n} = \frac{p_n}{(1 - p_K)} \end{aligned}$$

Define Y_n to be a random variable equal to the sum of n independent exponentially distributed random variables, each with parameter μ. This random variable has Erlang distribution. Then for $t \in \mathbb{R}_0^+$

$$\begin{aligned} F_{\Omega}(t) &= P(\Omega \le t) = \sum_{n=0}^{K-1} q_n P(Y_{n+1} \le t) \\ &= \sum_{n=0}^{K-1} q_n \int_0^t \frac{\mu (\mu x)^n e^{-\mu x}}{n!} dx \\ &= 1 - e^{-\mu t} \sum_{n=0}^{K-1} q_n \sum_{j=0}^{n} \frac{(\mu t)^j}{j!} \end{aligned}$$

The following result was used in the above derivation.

$$\int_0^t \frac{x^n e^{-x}}{n!} dx = 1 - e^{-t} \sum_{j=0}^{n} \frac{t^j}{j!}, \quad \forall\, n \in \mathbb{N}$$

Similarly for $t \in \mathbb{R}_0^+$

$$\begin{aligned} F_{\Omega_q}(t) &= P(\Omega_q \le t) = P(\Omega_q = 0) + P(0 < \Omega_q \le t) \\ &= q_0 + \sum_{n=1}^{K-1} q_n P(Y_n \le t) \\ &= q_0 + \sum_{n=1}^{K-1} q_n \int_0^t \frac{\mu (\mu x)^{n-1} e^{-\mu x}}{(n-1)!} dx \\ &= 1 - e^{-\mu t} \sum_{n=0}^{K-2} q_{n+1} \sum_{j=0}^{n} \frac{(\mu t)^j}{j!} \end{aligned}$$

The reader can verify that all the expressions derived in this subsection reduce to those derived for the $M/M/1$ queue, if K tends to infinity.

6.5.7 $M/M/c$ Queue

An $M/M/c$ is a multiserver queue with infinite waiting space and infinite customer population. The number of servers is equal to $c \in \mathbb{P}$, and the arrival process is Poissonian with parameter $\lambda \in \mathbb{R}^+$. The service times of the servers are independent, and have identical exponential distribution with parameter $\mu \in \mathbb{R}^+$. The service discipline is FCFS. In this queueing scheme

$$\begin{aligned} \lambda_n &= \lambda, \quad \forall\, n \in \mathbb{N} \\ \mu_n &= \begin{cases} n\mu, & n = 1, 2, \ldots, c \\ c\mu, & n > c \end{cases} \end{aligned}$$

Define $\rho = \lambda / (c\mu)$ and $\zeta = \lambda / \mu$. This queueing process is ergodic if $\rho < 1$. Using the birth-and-death equations, it can be shown that the steady-state probability p_n, $\forall\, n \in \mathbb{N}$ is given by

$$p_0 = \left\{ \sum_{n=0}^{c-1} \frac{\zeta^n}{n!} + \frac{\zeta^c}{c!\,(1-\rho)} \right\}^{-1}$$

$$p_n = \begin{cases} \dfrac{\zeta^n}{n!} p_0, & n = 1, 2, \ldots, c \\ \rho^{n-c} p_c, & n > c \end{cases}$$

The probability that an arriving customer joins the waiting queue is called the *Erlang's C function* or *Erlang's delay formula.* It is denoted by $\mathcal{C}(c, \zeta)$. Therefore

$$\mathcal{C}(c, \zeta) = \sum_{n=c}^{\infty} p_n$$

Thus

$$\mathcal{C}(c, \zeta) = \frac{\dfrac{\zeta^c}{c!}}{\dfrac{\zeta^c}{c!} + (1 - \rho) \sum_{n=0}^{c-1} \dfrac{\zeta^n}{n!}} \tag{6.6}$$

It can also be shown that

$$(1 - \rho) \mathcal{C}(c, \zeta) = \frac{\zeta^c}{c!} p_0$$

The number of customers waiting in the queue L_q is

$$L_q = \sum_{n=c}^{\infty} (n - c) p_n = \frac{\rho}{(1 - \rho)^2} \frac{\zeta^c}{c!} p_0 = \frac{\rho}{(1 - \rho)} \mathcal{C}(c, \zeta)$$

The average waiting time in the queue W_q is obtained by using the Little's law. Therefore

$$W_q = \frac{L_q}{\lambda} = \frac{\mathcal{C}(c, \zeta)}{c\mu (1 - \rho)}$$

The net average waiting time in the queueing system is $W = (W_q + W_s) = \left(W_q + \mu^{-1}\right)$, and $L = \lambda W$. The average number of customers receiving service at steady-state $L_s = \mathcal{E}(\Lambda_s)$.

$$L_s = \sum_{n=0}^{c-1} n p_n + \sum_{n=c}^{\infty} c p_n = \zeta = c\rho$$

Consequently, the average utilization of a single server is $L_s / c = \rho$. Similarly, it can be shown that the expected number of idle servers in the queueing system is equal to $(c - L_s) = c(1 - \rho)$.

The cumulative distribution function $F_{\Omega_q}(t), t \in \mathbb{R}_0^+$ is next computed. Define Z_n to be a random variable equal to the sum of n independent exponentially distributed random variables, each with parameter $c\mu$. This random variable has Erlang distribution. Note that

$$P(\Omega_q = 0) = \sum_{n=0}^{c-1} p_n = 1 - \mathcal{C}(c, \zeta)$$

$$\begin{aligned} P(0 < \Omega_q \leq t) &= \sum_{n=c}^{\infty} p_n P(Z_{n-c+1} \leq t) = \sum_{k=0}^{\infty} p_{k+c} P(Z_{k+1} \leq t) \\ &= \sum_{k=0}^{\infty} p_c \rho^k \int_0^t \frac{\mu c (\mu c x)^k e^{-\mu c x}}{k!} dx = \mathcal{C}(c, \zeta) \left\{1 - e^{-\mu(c - \zeta)t}\right\} \\ P(\Omega_q \leq t) &= P(\Omega_q = 0) + P(0 < \Omega_q \leq t) = 1 - \mathcal{C}(c, \zeta) e^{-\mu(c - \zeta)t} \end{aligned}$$

Thus

$$F_{\Omega_q}(t) = 1 - \mathcal{C}(c, \zeta)\, e^{-\mu t(c-\zeta)}, \quad t \in \mathbb{R}_0^+$$

As in the case of an $M/M/1$ queue it can be established that as $t \to \infty$ and $\rho < 1$, the output process of the $M/M/c$ queue is also Poissonian with parameter λ. That is, the interdeparture times of customers have an exponential distribution with parameter λ. However, if $\rho \geq 1$ it can be shown that as $t \to \infty$, the output process of the $M/M/c$ queue is also Poissonian with parameter $c\mu$. In summary, the output process of an $M/M/c$ queue is Poissonian with parameter $\min(\lambda, c\mu)$.

6.5.8 $M/M/1/K/K$ Queue

An $M/M/1/K/K$ queue is a single server queueing system with finite population K, and queueing capacity also equal to K. The service time of the server has an exponential distribution with parameter $\mu \in \mathbb{R}^+$. The service discipline is FCFS. In this scheme, each customer is in one of two states. These states are the *queueing-mode* and the *think-mode.*

(a) In the queueing mode, the customer is getting service or waiting in the queue to get service. Assume that there are n customers in this state. The service discipline in the queue is FCFS. After getting service, the customer enters the think mode. The possible values of n are $0, 1, 2, \ldots, K$.
(b) In the think mode, the customer spends time which has an exponential distribution with parameter $\lambda \in \mathbb{R}^+$. After thinking, the customer enters the queue mode. The number of customers in this mode is $(K - n)$, where n is the number of customers in the queueing mode.

The classical computer time-sharing system can be modeled by this queueing scheme. This model is also referred to as the machine repairman problem in the classical literature in operations research. This system is modeled as a birth-and-death process.

$$\lambda_n = \begin{cases} (K - n)\lambda, & n = 0, 1, 2, \ldots, K \\ 0, & n > K \end{cases}$$

$$\mu_n = \begin{cases} \mu, & n = 1, 2, \ldots, K \\ 0, & n = 0, \text{ and } n > K \end{cases}$$

The steady-state probabilities p_n are given by

$$p_0 = \left\{ \sum_{n=0}^{K} \frac{K!}{(K-n)!} \zeta^n \right\}^{-1}$$

$$p_n = \frac{K!}{(K-n)!} \zeta^n p_0, \quad n = 1, 2, \ldots, K$$

where $\zeta = \lambda/\mu$. The expression for p_0 can be written in terms of Erlang's $\mathcal{B}$ formula. The expression p_0 is equal to $\mathcal{B}(K, \zeta^{-1})$, where $\mathcal{B}(\cdot, \cdot)$ is Erlang's $\mathcal{B}$ function. The server utilization ρ is given by

$$\rho = \sum_{n=1}^{K} p_n = (1 - p_0)$$

The average number of customers in the queueing system is equal to L.

$$L = \sum_{n=0}^{K} n p_n = K - \frac{\rho}{\zeta}$$

The average number of customers waiting in the queueing system is equal to L_q.

$$L_q = \sum_{n=1}^{K} (n-1)\, p_n = L - \rho$$

The average arrival rate to the queue is given by $\overline{\lambda}$. It is

$$\overline{\lambda} = \sum_{n=0}^{K} \lambda_n p_n$$

$$= \sum_{n=0}^{K} \lambda (K-n)\, p_n = \mu\rho$$

The average customer resident time in the queueing system W, and average wait time in the queue W_q can be computed by using the Little's law.

$$W = \frac{L}{\overline{\lambda}}, \text{ and } W_q = \frac{L_q}{\overline{\lambda}}$$

which yields as expected

$$W = W_q + \frac{1}{\mu}$$

Use of the above relationships results in

$$K = \overline{\lambda}\left(W + \frac{1}{\lambda}\right)$$

This last relationship is the application of Little's law to the entire system. The asymptotic behavior of the normalized customer resident time μW, is next examined. Note that:

(a) If $K = 1$, then $\mu W = 1$.
(b) If K is very high and $\rho \approx 1$, then μW tends towards $\left(K - \zeta^{-1}\right)$. That is, under this condition μW is a linear function of K.

6.6 Non-Poisson Queues

Queueing systems in which the customer interarrival and service time distributions are not simultaneously exponentially distributed are considered in this section. Such queues are called non-Poisson queues. The following queueing systems are discussed in this section:

(a) $D/D/1$
(b) $M/G/1$
(c) $G/M/1$
(d) $G/M/c$
(e) $G/G/1$

In each of the $M/G/1$, $G/M/1$, $G/M/c$, and $G/G/1$ queueing systems: Customer interarrival times are independent of each other, customer service times are independent of each other; and the customer arrival process and the service processing times are stochastically independent of each other.

6.6.1 $D/D/1$ Queue

The $D/D/1$ is a single server queueing system in which the interarrival times of customers are equi-spaced in time, and the service time of each customer is equal to a constant. The waiting space for the customer is assumed to be infinite, and the customer population is infinite. Furthermore, the service discipline is FCFS.

Let a and b be customer interarrival and service times respectively. Both times are assumed to be positive. The following three cases are considered.

(a) The service time b is smaller than the interarrival time a.
(b) The service time b is equal to the interarrival time a.
(c) The service time b is greater than the interarrival time a.

As defined earlier, the number of customers in the queueing system at time t is equal to $N(t)$. Assume that the queueing system is initially empty, that is $N(0) = 0$. Also *assume* that the first arrival occurs at time $t = 0_+$, and the $(n+1)$th arrival occurs at time $(na)_+$, for any $n \in \mathbb{N}$.

Case 1: $b < a$. When the first customer arrives, the server is busy till time b, and idle from time b till time a. The second customer arrives at time a, and the sever is busy till time $(a+b)$, and idle from time $(a+b)$ till time $2a$. The third customer arrives at time $2a$. This process repeats itself. Therefore

$$N(t) = \begin{cases} 1, & na < t \le na + b \\ 0, & na + b < t \le (n+1)a \end{cases}$$

for any $n \in \mathbb{N}$. The utilization of this server ρ, is equal to b/a.

Case 2: $b = a$. The server is always busy in this scenario. However, only a single customer is present in the queueing system at all times. That is

$$N(t) = 1, \quad \forall\, t \in \mathbb{R}^+$$

The utilization of this server ρ, is equal to 1.

Case 3: $b > a$. The service time is longer than the interarrival time. Therefore the server is always busy, and the queue length grows as time increases. The net number of customer arrivals in time t is equal to $\alpha(t)$, and the net number of serviced customers in time t is equal to $\delta(t)$. Note that the nth customer's service is finished at time $(nb)_+$, for any $n \in \mathbb{P}$. Thus

$$\begin{aligned} &\alpha(t) = (n+1), && na < t \le (n+1)a, \quad n \in \mathbb{N} \\ &\delta(t) = n, && nb < t \le (n+1)b, \quad n \in \mathbb{N} \\ &N(t) = \alpha(t) - \delta(t), && \forall\, t \in \mathbb{R}^+ \end{aligned}$$

The utilization of this server ρ, is equal to 1. Thus as $t \to \infty$, the number of customers in the queueing system $N(t)$ tends towards infinity.

6.6.2 $M/G/1$ Queue

An $M/G/1$ queue is a single server queue with Poissonian arrivals and general service time distribution. That is, the customer interarrival times have an exponential distribution with parameter $\lambda \in \mathbb{R}^+$. The customer population and the waiting space are each assumed to be infinite. Furthermore, the service discipline is FCFS.

Therefore the queueing process is not a Markov process. However imbedded within this non-Markov process is a Markov chain, which yields to manageable analysis. Let $N(t)$ be the total number of customers in the system at time t. The state space of this process is discrete, which is $\mathbb{N}$. Also let $t_1, t_2, t_3, \ldots$ be successive instants of completion of service times. Then $N(t_n)$ is the number of customers left behind by the nth customer. Define:

N_n = number of customers $N(t_n)$ in the queueing system after the nth customer departs.
B_n = service time of the nth customer.
V_n = number of customers who arrived during the service time B_n, of the nth customer.

Then for all $n \in \mathbb{P}$

$$N_{n+1} = \begin{cases} N_n - 1 + V_{n+1}, & N_n \geq 1 \\ V_{n+1}, & N_n = 0 \end{cases}$$

The sequence $\{N_n, n \in \mathbb{N}\}$ is a discrete state Markov chain. It is actually an imbedded Markov chain of the process $\{N(t), t \in \mathbb{R}_0^+\}$. The service times are independent of each other, therefore we denote these by the random variable B. The finite mean value of this random variable is $\mathcal{E}(B) = 1/\mu$. Also the random variable V_n depends only on B, and not on the queue size. Therefore, it is denoted simply by V in subsequent discussion. Let the probability density function of the random variable B be $f_B(t), t \in \mathbb{R}_0^+$. Then

$$P(V = v) = \int_0^\infty P(V = v \mid B = t) f_B(t)\, dt$$

$$P(V = v \mid B = t) = e^{-\lambda t} \frac{(\lambda t)^v}{v!}, \quad \forall v \in \mathbb{N}$$

Thus

$$P(N_{n+1} = j \mid N_n = i) = P(V = j - i + 1), \quad i \geq 1,\ j \geq (i-1)$$

Let $(j - i + 1) = m$, and $i \geq 1$. Thus

$$P(N_{n+1} = j \mid N_n = i) = \begin{cases} k_m, & m \geq 0 \\ 0, & m < 0 \end{cases}$$

$$k_m = \int_0^\infty e^{-\lambda t} \frac{(\lambda t)^m}{m!} f_B(t)\, dt, \quad \forall m \in \mathbb{N}$$

where $k_m = P(m \text{ arrivals occur during a service time } B = t)$.

It can also be noted that $P(N_{n+1} = j \mid N_n = 0) = P(N_{n+1} = j \mid N_n = 1)$. Therefore the imbedded process $\{N_n, n \in \mathbb{N}\}$ is Markovian, since the transition probabilities depend only on the indices (i, j). Furthermore, since the state space is discrete, the process is a Markov chain. Let the transition probabilities of the Markov chain be p_{ij} where $i, j \in \mathbb{N}$, and the transition matrix is $\mathcal{P} = [p_{ij}]$. Also

$$\mathcal{P} = \begin{bmatrix} k_0 & k_1 & k_2 & \cdots \\ k_0 & k_1 & k_2 & \cdots \\ 0 & k_0 & k_1 & \cdots \\ 0 & 0 & k_0 & \cdots \\ \vdots & \vdots & \vdots & \ddots \end{bmatrix}$$

As $k_m > 0$ for all $m \in \mathbb{N}$ each state is reachable from any other state. Thus this Markov chain is irreducible (and aperiodic). It can also be shown that the Markov chain is positive-recurrent when $\rho = \lambda \mathcal{E}(B) < 1$. Therefore the Markov chain is ergodic if $\rho < 1$. This is a requirement for the existence of the steady-state probabilities $\psi_n, n \in \mathbb{N}$, where ψ_n is the steady-state probability of n customers in the system after a customer departs. If

$$\psi = \left[\psi_0 \; \psi_1 \; \psi_2 \cdots\right]$$

then

$$\psi = \psi\mathcal{P}, \text{ and } \sum_{n\in\mathbb{N}} \psi_n = 1$$

Thus

$$\psi_i = \psi_0 k_i + \sum_{j=1}^{i+1} \psi_j k_{i-j+1}, \qquad \forall\, i \in \mathbb{N}$$

For $|z| \leq 1$, define the generating functions $\Psi(z)$ and $K(z)$ as

$$\Psi(z) = \sum_{i\in\mathbb{N}} \psi_i z^i$$

$$K(z) = \sum_{i\in\mathbb{N}} k_i z^i$$

Let the Laplace transform of the service time probability density function $f_B(t), t \in \mathbb{R}_0^+$ be defined as $\widehat{f}_B(s)$, where $s \in \mathbb{C}$. Thus

$$K(z) = \widehat{f}_B(\lambda - \lambda z)$$

$$\Psi(z) = \frac{\psi_0 (1-z) K(z)}{K(z) - z}$$

Also $\Psi(1) = 1$, and $K(1) = 1$. Denote the first derivative of $K(z)$ with respect to z by $K'(z)$, then $K'(1) = \rho$. Substituting $z = 1$ in the above equation, and using L'Hospital's rule yields $\psi_0 = (1-\rho)$. Therefore

$$\Psi(z) = \frac{(1-\rho)(1-z) K(z)}{K(z) - z} \tag{6.7}$$

This relationship is known as the *Pollaczek-Khinchin (P-K) formula,* after mathematicians Felix Pollaczek (1892-1981) and A. Y. Khinchin (1894-1959). Let p_n be the steady-state probability of n customers in the queueing system at an arbitrary point in time. The relationship between ψ_n and p_n is next established. It is shown that $\psi_n = p_n, \forall\, n \in \mathbb{N}$.

In this queueing system customers arrive and depart singly. Therefore at steady-state, the departure probabilities ψ_n are equal to the arrival probabilities a_n, $\forall\, n \in \mathbb{N}$. Furthermore, since the arrival process is Poissonian, the PASTA law is applicable. That is, at steady-state $p_n = a_n$, $\forall\, n \in \mathbb{N}$. Thus $\psi_n = p_n$, $\forall\, n \in \mathbb{N}$. Let the generating function of the probabilities p_n's be $P(z)$. Then $P(z) = \Psi(z)$. Therefore

$$P(z) = \frac{(1-\rho)(1-z)\widehat{f}_B(\lambda - \lambda z)}{\widehat{f}_B(\lambda - \lambda z) - z}$$

If the first derivative of $P(z)$ with respect to z is denoted by $P'(z)$, then $P'(1) = L$. Differentiating $P(z)$ with respect to z, and using L'Hospital's rule twice results in

$$L = \rho + \frac{K''(1)}{2(1-\rho)}$$

where $K''(z)$ is the second derivative of $K(z)$ with respect to z. Also

$$\begin{aligned}K''(1) &= \lambda^2 \int_0^\infty t^2 f_B(t)\,dt \\ &= \lambda^2 \left(\frac{1}{\mu^2} + \sigma_B^2\right) \\ &= \rho^2 \left(1 + C_B^2\right)\end{aligned}$$

where σ_B^2 is the variance of the service time, and C_B^2 is the squared coefficient of variation of the service time. Thus

$$L = \rho + \frac{\rho^2 \left(1 + C_B^2\right)}{2(1-\rho)} \tag{6.8a}$$

The average number of customers waiting in the queue L_q, can be obtained from the relationship $L = (L_q + \rho)$. The corresponding average time W, a customer spends in the queueing system, and the average waiting time W_q, at equilibrium can be obtained via Little's law. For example

$$W_q = \frac{\rho\left(1 + C_B^2\right)}{2\mu(1-\rho)} \tag{6.8b}$$

An alternate technique to derive an expression for L, the average number of customers at steady-state in the ergodic queue is given below.

Expected Value Analysis

It is also possible to derive an expression for the average queue length via a mean-value technique. This technique is again due to Pollaczek and Khinchin. The expression for N_{n+1} can be written as

$$N_{n+1} = N_n - \Delta_n + V_{n+1}$$
$$\Delta_n = \begin{cases} 1, & N_n \geq 1 \\ 0, & N_n = 0 \end{cases}$$

Therefore

$$\mathcal{E}(N_{n+1}) = \mathcal{E}(N_n) - \mathcal{E}(\Delta_n) + \mathcal{E}(V_{n+1})$$

At steady-state, that is as $n \to \infty$, $\mathcal{E}(N_n) = \mathcal{E}(N_{n+1}) = L$. Consequently

$$\mathcal{E}(\Delta_n) = \mathcal{E}(V_{n+1})$$

However

$$\begin{aligned}\mathcal{E}(V) &= \int_0^\infty \mathcal{E}(V \mid B = t)\, f_B(t)\,dt \\ &= \int_0^\infty \lambda t f_B(t)\,dt = \lambda \mathcal{E}(B) = \rho\end{aligned}$$

Therefore

$$\mathcal{E}(\Delta_n) = \mathcal{E}(V) = \rho$$

Squaring the expressions for N_{n+1} yields

$$N_{n+1}^2 = N_n^2 + \Delta_n^2 + V_{n+1}^2 - 2N_n\Delta_n + 2N_nV_{n+1} - 2\Delta_nV_{n+1}$$

Taking expectations and noting that as $n \to \infty$, $\mathcal{E}\left(N_n^2\right) = \mathcal{E}\left(N_{n+1}^2\right)$ yields

$$0 = \mathcal{E}\left(\Delta_n^2\right) + \mathcal{E}\left(V_{n+1}^2\right) - 2\mathcal{E}\left(N_n\Delta_n\right) + 2\mathcal{E}\left(N_nV_{n+1}\right) - 2\mathcal{E}\left(\Delta_nV_{n+1}\right)$$

Note that $\Delta_n^2 = \Delta_n$, and $N_n\Delta_n = N_n$. Also N_n is independent of the random variable V_{n+1}. This implies that Δ_n is independent of the random variable V_{n+1}. Thus

$$0 = \rho + \mathcal{E}\left(V_{n+1}^2\right) - 2L + 2\mathcal{E}\left(N_n\right)\mathcal{E}\left(V_{n+1}\right) - 2\mathcal{E}\left(\Delta_n\right)\mathcal{E}\left(V_{n+1}\right)$$

Simplification results in

$$L = \frac{\rho - 2\rho^2 + \mathcal{E}\left(V^2\right)}{2\left(1-\rho\right)}$$

Note that $P\left(V = v\right) = k_v$, $\forall\ v \in \mathbb{N}$. Therefore $\mathcal{E}\left(V^2\right)$ is evaluated as follows. As $K\left(z\right) = \widehat{f}_B\left(\lambda - \lambda z\right)$, $K'\left(1\right) = \rho$, and $K''\left(1\right) = \lambda^2\mathcal{E}\left(B^2\right)$, we have

$$\begin{aligned}\mathcal{E}\left(V^2\right) &= K''\left(1\right) + K'\left(1\right) \\ &= \lambda^2\mathcal{E}\left(B^2\right) + \rho\end{aligned}$$

Therefore

$$\begin{aligned}L &= \rho + \frac{\lambda^2\mathcal{E}\left(B^2\right)}{2\left(1-\rho\right)} \\ &= \rho + \frac{\rho^2\left(1 + C_B^2\right)}{2\left(1-\rho\right)}\end{aligned}$$

Not surprisingly, this expression for L is identical to the expression derived via the generating function technique.

Examples 6.1. Some basic examples.

1. Let the service time be exponentially distributed with parameter μ. That is, the queue is an $M/M/1$ queue. For this service time distribution, $C_B^2 = 1$. Thus

$$\begin{aligned}L &= \frac{\rho}{\left(1-\rho\right)} \\ L_q &= \frac{\rho^2}{\left(1-\rho\right)}\end{aligned}$$

Identical expressions were obtained by using the birth-and-death model for an $M/M/1$ queue.

2. For an $M/D/1$ queue the service time is deterministic, therefore $C_B^2 = 0$. Consequently

$$\begin{aligned}L &= \rho + \frac{\rho^2}{2\left(1-\rho\right)} \\ L_q &= \frac{\rho^2}{2\left(1-\rho\right)}\end{aligned}$$

Observe that due to zero variation in service time, L_q for an $M/D/1$ queue is equal to half of the L_q value of an $M/M/1$ queue. □

6.6.3 $G/M/1$ Queue

The $G/M/1$ queue is a single server queuing system in which the interarrival times of the customers have a general distribution. These interarrival intervals are also independent of each other. The service times of the customers are independent, and each exponentially distributed with parameter $\mu \in \mathbb{R}^+$. Furthermore, the customer service times, and the interarrival times are independent of each other. In this queueing system, the customer population, and the waiting space are assumed to be infinite. The service discipline is FCFS. The random variable representing the interarrival times is A. Let the number of customers in the system at time t be $N(t)$, and at $t = t_0 = 0$ be $N(0) = 0$. The customers arrive at instants $t_1, t_2, \ldots$. That is, the nth customer arrives at the instant t_n. Define:

$N_n =$ number of customers $N(t_n)$ in the queueing system immediately before the nth arrival.
$\Xi_{n+1} =$ number of customers served in the time interval $[t_n, t_{n+1})$.

Then

$$N_{n+1} = N_n + 1 - \Xi_{n+1}, \text{ if } N_n \geq 0, \; \Xi_{n+1} \leq (N_n + 1)$$

This queue can be analysed using the imbedded Markov chain approach as in the case of an $M/G/1$ queue. Actually the sequence $\{N_n, n \in \mathbb{N}\}$ is a discrete state Markov chain. It is an imbedded Markov chain of the process $\{N(t), t \in \mathbb{R}_0^+\}$. As the interarrival times of the customers have identical distribution and the service time is exponential, the random variable Ξ_n can be replaced by Ξ. Let the probability that m customers are served in the period Ξ has a probability g_m, where

$$\begin{aligned} g_m &= P(\Xi = m) \\ &= \int_0^\infty e^{-\mu t} \frac{(\mu t)^m}{m!} f_A(t)\, dt, \quad \forall\, m \in \mathbb{N} \end{aligned}$$

The generating function of the sequence $g_m, m \in \mathbb{N}$ is $G(z)$. It is specified by

$$\begin{aligned} G(z) &= \sum_{m=0}^{\infty} g_m z^m, \qquad |z| \leq 1 \\ &= \widehat{f}_A(\mu - \mu z) \end{aligned}$$

where $\widehat{f}_A(s), s \in \mathbb{C}$ is the Laplace transform of the interarrival time probability density function $f_A(t), t \in \mathbb{R}_0^+$. The transition probabilities of the Markov chain are given by

$$P(N_{n+1} = j \mid N_n = i) = p_{ij}, \quad \forall\, i, j \in \mathbb{N}$$

where for any $i \in \mathbb{N}$

$$p_{ij} = \begin{cases} g_{i+1-j}, & 1 \leq j \leq (i+1) \\ 0, & (i+1) < j \\ h_i, & j = 0 \end{cases}$$

$$p_{i0} = 1 - \sum_{k=0}^{i} g_k \triangleq h_i$$

The transition probability matrix $\mathcal{P} = [p_{ij}]$ is

$$\mathcal{P} = \begin{bmatrix} h_0 & g_0 & 0 & 0 & 0 & \cdots \\ h_1 & g_1 & g_0 & 0 & 0 & \cdots \\ h_2 & g_2 & g_1 & g_0 & 0 & \cdots \\ \vdots & \vdots & \vdots & \vdots & \vdots & \ddots \end{bmatrix}$$

Let $\mathcal{E}(A) = 1/\lambda$, and $\rho = \lambda/\mu$, then the Markov chain is ergodic if $\rho < 1$. If the Markov chain is ergodic then the equilibrium probability distribution exists. This distribution is

$$\varphi_n = \lim_{k \to \infty} P(N_k = n), \qquad \forall\, n \in \mathbb{N}$$

In the above equation, φ_n is the probability of n customers in the system just prior to a customer arrival at steady-state. If

$$\varphi = \left[\varphi_0 \, \varphi_1 \, \varphi_2 \cdots\right]$$

then

$$\varphi = \varphi \mathcal{P}, \quad \text{and} \quad \sum_{n=0}^{\infty} \varphi_n = 1$$

Thus

$$\varphi_i = \sum_{j=i-1}^{\infty} \varphi_j g_{j+1-i}, \qquad \forall\, i \in \mathbb{P}$$

That is,

$$\varphi_i = \sum_{j=i-1}^{\infty} \varphi_j \int_0^{\infty} e^{-\mu t} \frac{(\mu t)^{j+1-i}}{(j+1-i)!} f_A(t)\, dt, \qquad \forall\, i \in \mathbb{P}$$

Next try a solution of the form $\varphi_i = \varpi \sigma^i$, where ϖ is a positive constant, and substitute in the above equation. This yields

$$\sigma = \widehat{f}_A(\mu - \mu\sigma) \tag{6.9}$$

Therefore σ is a zero of the equation

$$\sigma - \widehat{f}_A(\mu - \mu\sigma) = 0$$

Use of the relationship $\sum_{i=0}^{\infty} \varphi_i = 1$, results in $\varpi = (1 - \sigma)$ provided $\sigma < 1$. Therefore

$$\varphi_i = (1 - \sigma)\, \sigma^i, \qquad \forall\, i \in \mathbb{N}$$

That is, the equilibrium distribution is geometric. If $\rho < 1$ it can be shown that σ is a unique zero inside the unit circle $|z| = 1$, and $0 < \sigma < 1$. This discussion is summarized in the following lemma.

Lemma 6.3. *In a $G/M/1$ queue, the distribution of the number of customers found in the queueing system by a customer arrival has a geometric distribution at equilibrium.* □

Example 6.2. Consider an $M/M/1$ queue. The arrival process is Poissonian in this queue, and

$$\widehat{f}_A(s) = \frac{\lambda}{s + \lambda}$$

It can indeed be verified that $\sigma = \rho$, if $\rho < 1$. □

Steady-State Probabilities at any Time

Recall that p_j is the long run probability that j customers are present in the queueing system. Actually p_j is equal to the proportion of time that there are j customers in the queueing system as $t \to \infty$. Thus p_j's are time averages. That is, if

$$p_{ij}(t) = P(N(t) = j \mid N(0) = i), \quad t \in \mathbb{R}_0^+$$

then

$$\lim_{t \to \infty} p_{ij}(t) = p_j, \quad \forall j \in \mathbb{N}$$

Similarly a_j has been defined as the arrival point probability that j customers are present in the queueing system when a customer arrives. Therefore a_j is equal to φ_j for any $j \in \mathbb{N}$. However the probabilities p_j and a_j are not equal in general for this queueing system. Relationship between these two probabilities is next established.

At steady-state, the rate at which the number of customers in the queueing system change from n to $(n-1)$ is equal to the rate at which the number of customers change from $(n-1)$ to n. These two scenarios are described below.

(a) The rate at which the number of customers in the system change from n to $(n-1)$ is equal to the fraction of time the system has n customers, which is p_n, multiplied by the service rate μ. This rate is equal to μp_n, where $n \in \mathbb{P}$.
(b) The rate at which the number of customers in the system change from $(n-1)$ to n is equal to the fraction of the time an arriving customer finds $(n-1)$ customers, which is φ_{n-1}, times the arrival rate λ. This rate is equal to $\lambda\varphi_{n-1}$, where $n \in \mathbb{P}$.

Equating the two rates results in

$$\mu p_n = \lambda \varphi_{n-1}, \quad \forall n \in \mathbb{P}$$

Therefore

$$p_n = \rho(1-\sigma)\sigma^{n-1}, \quad \forall n \in \mathbb{P}$$

Using the relationship $\sum_{n=0}^{\infty} p_n = 1$ yields

$$p_0 = (1-\rho)$$

Waiting Time Distribution

The equilibrium probability distribution $\varphi_n, n \in \mathbb{N}$ is geometric and the service time distribution is exponential in a $G/M/1$ queue. This is similar to an $M/M/1$ queueing system. Therefore the waiting time distributions of an $M/M/1$ and $G/M/1$ queue have the same form. Consequently the waiting time distribution $P(\Omega_q \le t)$ of the $G/M/1$ queue is given by

$$F_{\Omega_q}(t) = P(\Omega_q \le t) = \left\{1 - \sigma e^{-\mu(1-\sigma)t}\right\}, \quad t \in \mathbb{R}_0^+ \tag{6.10a}$$

This leads to

$$W_q = \frac{\sigma}{\mu(1-\sigma)} \tag{6.10b}$$

$$L_q = \frac{\rho\sigma}{(1-\sigma)} \tag{6.10c}$$

A Lagrange Series Solution of σ

Classical textbooks in queueing theory propose iterative techniques to compute σ. However, σ can be evaluated via a Lagrange's series expansion. We describe this analytical technique via examples. A queue in which the customer interarrival times have gamma distribution is described below. Customer interarrival times with bounded Pareto distribution are considered in the problem section. As per Lagrange, if $z = a + \xi\widetilde{\varphi}(z)$, $z = a$ when $\xi = 0$, and $\widetilde{\varphi}(z)$ is analytic inside and on a circle C containing $z = a$, then z has the following series expansion

$$z = a + \sum_{n\in\mathbb{P}} \frac{\xi^n}{n!} \frac{d^{n-1}}{da^{n-1}} [\widetilde{\varphi}(a)]^n$$

Substituting $z = \mu(1-\sigma)$ in $\sigma - \widehat{f}_A(\mu - \mu\sigma) = 0$, we obtain

$$\sigma = 1 - \frac{z}{\mu}. \text{ and } z = \mu - \mu\widehat{f}_A(z)$$

Observe that z can be evaluated using Lagrange series. In Lagrange's series expansion substitute $a = \mu$, $\xi = -\mu$, and $\widetilde{\varphi}(z) = \widehat{f}_A(z)$. This yields

$$z = \mu - \mu \sum_{n\in\mathbb{P}} \frac{(-\mu)^{n-1}}{n!} \frac{d^{n-1}}{d\mu^{n-1}} \left[\widehat{f}_A(\mu)\right]^n \tag{6.11}$$

In the above series expansion, z can be evaluated by using Faà di Bruno's formula. Recall that Faà di Bruno's formula is used for calculating the nth derivative of a composite function. Let

$$\frac{d^j}{d\mu^j}\widehat{f}_A(\mu) \triangleq \widehat{f}_A^{(j)}(\mu), \quad \forall\, j \in \mathbb{N}$$

then for $n \geq 2$

$$\frac{d^{n-1}}{d\mu^{n-1}} \left[\widehat{f}_A(\mu)\right]^n =$$

$$\sum_{\substack{\kappa=1 \\ \kappa_1,\kappa_2,\ldots,\kappa_{n-1}\geq 0 \\ \kappa_1+\kappa_2+\ldots+\kappa_{n-1}=\kappa \\ \kappa_1+2\kappa_2+\ldots+(n-1)\kappa_{n-1}=(n-1)}}^{(n-1)} \left\{ \frac{(n-1)!}{\kappa_1!\kappa_2!\ldots\kappa_{n-1}!} \frac{n!}{(n-\kappa)!} \left[\widehat{f}_A(\mu)\right]^{n-\kappa} \prod_{j=1}^{(n-1)} \left\{ \frac{\widehat{f}_A^{(j)}(\mu)}{j!} \right\}^{\kappa_j} \right\}$$

The above equation requires constrained summation over integer partitions of κ. This requires computer storage of partitioned sets of values of κ or generation of these partitions at the time of computation. However the following procedure is suggested for simplicity. For convenience in notation let

$$\widetilde{d}_n \triangleq \frac{(-\mu)^{n-1}}{n!} \frac{d^{n-1}}{d\mu^{n-1}} \left[\widehat{f}_A(\mu)\right]^n, \quad \forall\, n \in \mathbb{P}$$

Then

$$z = \mu - \mu \sum_{n \in \mathbb{P}} \widetilde{d}_n, \quad \text{and} \quad \sigma = \sum_{n \in \mathbb{P}} \widetilde{d}_n$$

In actual computation of σ, only a finite number of terms $(J+1)$ can be used in the series specified by the above equation. Therefore

$$\sigma \simeq \sum_{n=1}^{(J+1)} \widetilde{d}_n$$

Thus $\widetilde{d}_n$ for $1 \le n \le (J+1)$ is computed recursively as follows. Define

$$\begin{aligned}
b_j &\triangleq \frac{(-\mu)^j}{j!} \frac{d^j}{d\mu^j} \widehat{f}_A(\mu), \qquad 0 \le j \le J \\
c_{m,j} &\triangleq \frac{(-\mu)^j}{j!} \frac{d^j}{d\mu^j} \left[\widehat{f}_A(\mu) \right]^m, \qquad 0 \le j \le J, \quad 1 \le m \le (J+1) \\
c_{1,j} &= b_j, \qquad 0 \le j \le J \\
c_{m,j} &= \sum_{k=0}^{j} b_k c_{m-1,j-k}, \qquad 0 \le j \le J, \quad 2 \le m \le (J+1) \\
\widetilde{d}_n &= \frac{c_{n,n-1}}{n}, \qquad 1 \le n \le (J+1)
\end{aligned}$$

The above equations required in the computation of $\widetilde{d}_n$, for $1 \le n \le (J+1)$ can be condensed in an algorithmic form as follows. In this algorithm $v_m, w_m, 0 \le m \le J$ are intermediate variables.

Using $\widetilde{d}_n$, $1 \le n \le (J+1)$, σ can be computed. Note that larger values of ρ require larger values of J in the computation of σ.

Algorithm 6.1. *Computation of* $\widetilde{d}_n$*, for* $1 \le n \le (J+1)$ *in* $G/M/1$ *queue.*

Input: $\widehat{f}_A(s)$, μ, and J.
Output: $\widetilde{d}_n$, for $1 \le n \le (J+1)$
(*initialization*)
Compute $b_j, 0 \le j \le J$.
Let $v_0 \leftarrow 1$ and $v_m \leftarrow 0$, $1 \le m \le J$.
begin
  for $n = 1$ **to** J **do**
  begin
    $w_m \leftarrow \sum_{l=0}^{m} b_l v_{m-l}, \quad 0 \le m \le J$
    $\widetilde{d}_n \leftarrow w_{n-1}/n$
    $v_m \leftarrow w_m, \quad 0 \le m \le J$
  end (*end of n for-loop*)
  (*last step*)
  $\widetilde{d}_{J+1} \leftarrow \sum_{l=0}^{J} b_l v_{J-l} / (J+1)$
end (*end of computation of* $\widetilde{d}_n$*'s*)

The use of the Lagrange-series technique is next demonstrated via an example.

Example 6.3. This example assumes that the interarrival times have a gamma distribution. The probability density function $f_A(\cdot)$ of a random variable A with gamma distribution is given by

$$f_A(x) = \begin{cases} 0, & x < 0 \\ \dfrac{r\lambda (r\lambda x)^{r-1} e^{-r\lambda x}}{\Gamma(r)}, & x \geq 0 \end{cases}$$

$$\Gamma(r) = \int_0^{\infty} x^{r-1} e^{-x} dx$$

where $\Gamma(\cdot)$ is the gamma function and $r, \lambda \in \mathbb{R}^+$. If $r = 1$, then A is an exponentially distributed random variable. The average value is $\mathcal{E}(A) = \lambda^{-1}$ and the variance of the random variable A is $(r\lambda^2)^{-1}$. The Laplace transform of $f_A(x)$ is given by $\widehat{f}_A(s)$.

$$\widehat{f}_A(s) = \left(\frac{r\lambda}{s + r\lambda}\right)^r$$

The $\widetilde{d}_n$ values for $n \in \mathbb{P}$ can be computed directly. These are given by

$$\widetilde{d}_n = \frac{1}{(1 + r\rho)^{n-1}} \left(\frac{r\rho}{1 + r\rho}\right)^{nr} \frac{\Gamma(nr + n - 1)}{\Gamma(n+1)\Gamma(nr)}, \qquad \forall n \in \mathbb{P}$$

As $\widetilde{d}_n$'s are relatively simple, these can be computed via the following recursion.

$$\widetilde{d}_1 = \left(\frac{r\rho}{1 + r\rho}\right)^r$$

$$\widetilde{d}_2 = \frac{\widetilde{d}_1^2 r}{(1 + r\rho)}$$

$$\widetilde{d}_n = \varsigma_{n-1} \widetilde{d}_{n-1}, \qquad n \geq 3$$

$$\varsigma_{n-1} = \frac{\widetilde{d}_1}{(1 + r\rho)} \frac{(nr + n - 2)}{n} \prod_{k=0}^{n-3} \left(\frac{nr + k}{nr - r + k}\right), \qquad n \geq 3$$

After computing a finite number of $\widetilde{d}_n$'s, σ can be computed. The other queue performance parameters follow easily. A brief check of these equations can be made as follows. Note that if $r = 1$, then the arrival process is Poissonian. In this case

$$\widetilde{d}_n = \frac{1}{n}\binom{2n-2}{n-1} \frac{\rho^n}{(1+\rho)^{2n-1}}, \qquad \forall n \in \mathbb{P}$$

Thus

$$\sigma = \frac{\rho}{(1+\rho)} \sum_{n \in \mathbb{N}} \frac{1}{(n+1)} \binom{2n}{n} \frac{\rho^n}{(1+\rho)^{2n}}$$

The above equation can be simplified by observing that the generating function of the Catalan numbers

$$\frac{\binom{2n}{n}}{(n+1)}, \quad n \in \mathbb{N}$$

is given by

$$\sum_{n \in \mathbb{N}} \frac{1}{(n+1)} \binom{2n}{n} t^n = \frac{1 - \sqrt{1-4t}}{2t}$$

Catalan numbers are named after the Belgian mathematician Charles Eugène Catalan (1814-1894). The above equations yield $\sigma = \rho$ as expected. Using the above equations, σ is plotted as a function of ρ in Figure 6.1.

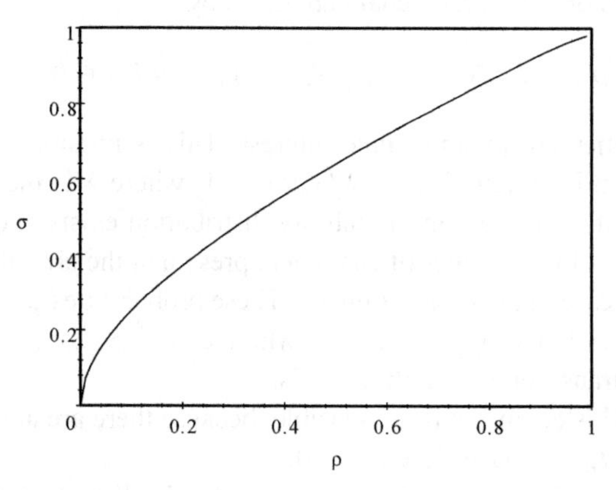

Figure 6.1. Interarrival times with gamma distribution, $r = 0.5$, ρ versus σ.

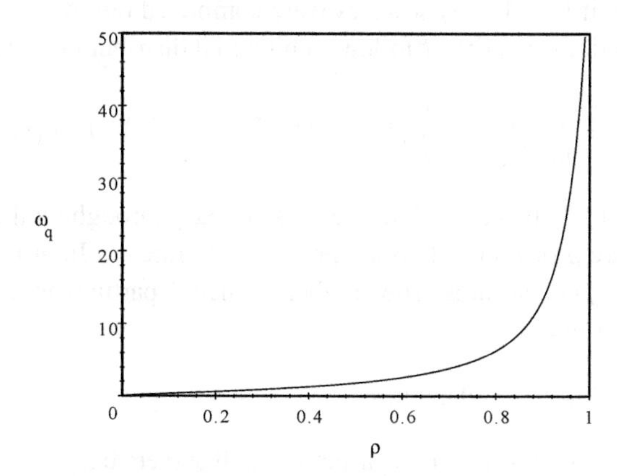

Figure 6.2. Interarrival times with gamma distribution, $r = 0.5$, ρ versus ω_q.

Let the normalized average waiting time ω_q be equal to $\mu\mathcal{E}(\Omega_q)$. Figure 6.2 has a plot of ρ versus ω_q. The value of r used in these figures is equal to 0.5. □

6.6.4 $G/M/c$ Queue

The $G/M/c$ queueing system is similar to the $G/M/1$ queue, except that there are $c \in \mathbb{P}$ number of servers, each with identical, independent, and exponentially distributed service time distributions with parameter $\mu \in \mathbb{R}^+$. The customer interarrival times have a general distribution. The cumulative distribution function of the interarrival times is given by $F_A(t)$ for $t \in \mathbb{R}_0^+$. As in the case of the $G/M/1$ queue, let N_n be equal to the number of customers in the queueing system immediately before the nth customer arrival. The nth customer arrives at time instant t_n. Also Ξ_{n+1} is equal to the number of customers served in the interval $[t_n, t_{n+1})$. Then

$$N_{n+1} = N_n + 1 - \Xi_{n+1}, \text{ if } N_n \geq 0, \ \Xi_{n+1} \leq (N_n + 1)$$

It can be inferred that $\{N_n, n \in \mathbb{N}\}$ is a discrete state Markov chain. The random variable Ξ_n is independent of n, however the net service rate during the interval $[t_n, t_{n+1})$ is state-dependent. The transition probabilities of the Markov chain are given by

$$p_{ij} = P(N_{n+1} = j \mid N_n = i), \qquad \forall\, i, j \in \mathbb{N}$$

Steady-state probabilities are of immediate interest. This is possible if the Markov chain is ergodic. The Markov chain is ergodic if $\rho = \lambda/(\mu c) < 1$, where λ is the average arrival rate of the customers. Therefore the equilibrium probability distribution exists under this condition. It is the probability distribution of the number of customers present at the arrival instant, as $t \to \infty$. Its definition is similar to the case of the $G/M/1$ queue. These probabilities $\varphi_i, i \in \mathbb{N}$ satisfy the linear system of equations $\varphi = \varphi\mathcal{P}$, and $\sum_{n=0}^{\infty} \varphi_n = 1$, where $\varphi = [\varphi_0\ \varphi_1\ \varphi_2\ \cdots]$ and $\mathcal{P} = [p_{ij}]$. It remains to determine the transition probabilities p_{ij}'s.

Case 1: $(i+1) < j$. This condition is not feasible, because there are at most $(i+1)$ customers present in the interval $[t_n, t_{n+1})$. Therefore $p_{ij} = 0$.

Case 2: $j \leq (i+1) \leq c$. No customers are waiting, that is all customers are with the servers. Furthermore, the newly arrived customer enters service. Therefore, the next arriving customer finds j customers in the system if $(i+1-j)$ services were completed out of $(i+1)$ services. Also the number of service completions in a time t follows a binomial distribution. Thus

$$p_{ij} = \int_0^\infty \binom{i+1}{j} \left(1 - e^{-\mu t}\right)^{i+1-j} \left(e^{-\mu t}\right)^j dF_A(t)$$

Case 3: $c \leq j \leq (i+1)$. In this case all the servers are busy throughout the interarrival interval. Consequently the departure process is a Poisson process with rate $c\mu$. In order to go from state i to j, there has to be $(i+1-j)$ departures. The number of such departures in time t follow a Poisson distribution with parameter $c\mu t$.

$$\begin{aligned} p_{ij} &= P(i+1-j \text{ departures}) \\ &= \int_0^\infty P(i+1-j \text{ departures} \mid t, \text{all servers busy})\, dF_A(t) \\ &= \int_0^\infty \frac{(c\mu t)^{i+1-j}}{(i+1-j)!} e^{-c\mu t} dF_A(t) \end{aligned}$$

For convenience in notation define

$$\theta_n = \int_0^\infty \frac{(c\mu t)^n}{n!} e^{-c\mu t} dF_A(t), \quad \forall\, n \in \mathbb{N}$$

Therefore

$$p_{ij} = \theta_{i+1-j}$$

Case 4: $j < c < (i+1)$. The queueing system initially has $i \geq c$ customers. During the interarrival time A, some of the servers start becoming idle until only j servers are busy.

Let U be the time interval during which $(i - c + 1)$ customers have been served when all the c servers are busy. Thus $0 < U < A$. Let $H(u)$ and $h(u)$ be the cumulative distribution function and probability density function respectively of the random variable U. Therefore the variable U is the sum of $(i - c + 1)$ independent exponentially distributed random variables, each with a parameter $c\mu$. Consequently

$$h(u) = \frac{(c\mu)(c\mu u)^{i-c}}{(i-c)!} e^{-c\mu u}, \qquad u \in \mathbb{R}_0^+$$

Also, there are $(c - j)$ service completions in the time interval $(A - U)$. Therefore use of the binomial distribution results in

$$\begin{aligned} p_{ij} &= \int_0^\infty \int_0^t \binom{c}{c-j} \left\{1 - e^{-\mu(t-u)}\right\}^{c-j} \left\{e^{-\mu(t-u)}\right\}^j dH(u)\, dF_A(t) \\ &= \binom{c}{j} \frac{(c\mu)^{i-c+1}}{(i-c)!} \int_0^\infty \int_0^t \left\{1 - e^{-\mu(t-u)}\right\}^{c-j} e^{-\mu(t-u)j} u^{i-c} e^{-c\mu u} du dF_A(t) \end{aligned}$$

The matrix equation relating the stationary (steady-state) probabilities is $\varphi = \varphi \mathcal{P}$. That is

$$\varphi_j = \sum_{i \in \mathbb{N}} \varphi_i p_{ij}, \qquad \forall\, j \in \mathbb{N}$$

For $j \geq c$,

$$\varphi_j = \sum_{i=0}^{j-2} \varphi_i 0 + \sum_{i=j-1}^{\infty} \varphi_i \theta_{i+1-j} = \sum_{k \in \mathbb{N}} \varphi_{j+k-1} \theta_k$$

As in the case of the $G/M/1$ queue, try a solution of the form $\varphi_j = \varrho \sigma^j, j \geq c$, and ϱ is a positive constant. Substituting it in the above equation yields $\sigma = \widehat{f}_A(c\mu - c\mu\sigma)$. Therefore σ is a zero of the equation

$$\sigma - \widehat{f}_A(c\mu - c\mu\sigma) = 0$$

If $\rho < 1$, it can be shown that $0 < \sigma < 1$, and σ is unique. The constant ϱ and the probabilities φ_j for $0 \leq j \leq (c-1)$ can be determined recursively from the equations

$$\varphi_j = \sum_{i \in \mathbb{N}} \varphi_i p_{ij}, \qquad 0 \leq j \leq (c-1)$$

and the relationship $\sum_{i \in \mathbb{N}} \varphi_i = 1$. Also

$$\varrho = \frac{(1-\sigma)\left\{1 - \sum_{i=0}^{c-1} \varphi_i\right\}}{\sigma^c}$$

Waiting Time Distribution

The waiting time distribution can be obtained as in the case of the $G/M/1$ queue.

$$P(\Omega_q = 0) = \sum_{i=0}^{c-1} \varphi_i = 1 - \frac{\varrho \sigma^c}{(1-\sigma)}$$

Also

$$F_{\Omega_q}(t) = P(\Omega_q \le t) = \left\{1 - \frac{\varrho\sigma^c}{(1-\sigma)} e^{-\mu c(1-\sigma)t}\right\}, \quad t \in \mathbb{R}_0^+ \tag{6.12a}$$

Thus

$$W_q = \frac{\varrho\sigma^c}{c\mu(1-\sigma)^2} \tag{6.12b}$$

6.6.5 $G/G/1$ Queue

A $G/G/1$ queue is a single server queueing system in which customer interarrival times A, and service times B, have general distributions. The service discipline in this scheme is FCFS. Also, the service and interarrival times are independent of each other. Furthermore, the waiting space for the arriving customers and the customer population are each infinite. Let $\lambda = 1/\mathcal{E}(A)$, $\mu = 1/\mathcal{E}(B)$, and $\rho = \lambda/\mu < 1$. The condition $\rho < 1$ is required for the queue to be ergodic. The distribution of the waiting time of a customer in this queue at steady-state is based upon a Weiner-Hopf type of integral equation.

Let the waiting times of the nth and $(n+1)$th customer be $\Omega_q^{(n)}$ and $\Omega_q^{(n+1)}$ respectively. Also let the time between the arrival instants of these two customers be $A^{(n)}$, and the service time of the nth customer be $B^{(n)}$. Then

$$\Omega_q^{(n+1)} = \begin{cases} \Omega_q^{(n)} + B^{(n)} - A^{(n)}, & \Omega_q^{(n)} + B^{(n)} - A^{(n)} > 0 \\ 0, & \Omega_q^{(n)} + B^{(n)} - A^{(n)} \le 0 \end{cases}$$

Define $C^{(n)} = \left(B^{(n)} - A^{(n)}\right)$, then the cumulative distribution function of the wait time of the $(n+1)$th customer is given by $P\left(\Omega_q^{(n+1)} \le t\right), t \in \mathbb{R}_0^+$. Thus

$$\begin{aligned} P\left(\Omega_q^{(n+1)} \le t\right) &= P\left(\Omega_q^{(n+1)} = 0\right) + P\left(0 < \Omega_q^{(n+1)} \le t\right) \\ &= P\left(\Omega_q^{(n)} + C^{(n)} \le 0\right) + P\left(0 < \Omega_q^{(n)} + C^{(n)} \le t\right) \\ &= P\left(\Omega_q^{(n)} + C^{(n)} \le t\right) \end{aligned}$$

Let the cumulative distribution function and probability density function of the random variable $C^{(n)}$ be $F_{C^{(n)}}(t)$, and $f_{C^{(n)}}(t)$ respectively, where $t \in \mathbb{R}$. Then

$$\begin{aligned} F_{C^{(n)}}(t) &= P\left(B^{(n)} - A^{(n)} \le t\right) \\ &= \int_0^\infty P\left(B^{(n)} \le t + x \mid A^{(n)} = x\right) f_A(x)\, dx \end{aligned}$$

Since service time and interarrival times are independent of each other

$$\begin{aligned} F_{C^{(n)}}(t) &= \int_0^\infty F_B(t+x) f_A(x)\, dx \\ &= \int_t^\infty F_B(x) f_A(x-t)\, dx \end{aligned}$$

Observe that $F_{C^{(n)}}(t)$ is independent of n. Therefore define $F_{C^{(n)}}(t) \triangleq F_C(t)$, and denote the random variable $(B - A)$ by C. Also for $t \in \mathbb{R}_0^+$

$$
\begin{aligned}
F_{\Omega_q^{(n+1)}}(t) &= P\left(\Omega_q^{(n+1)} \le t\right) = P\left(\Omega_q^{(n)} + C^{(n)} \le t\right) \\
&= \int_{0_-}^{\infty} P\left(C^{(n)} \le t - w \mid \Omega_q^{(n)} = w\right) f_{\Omega_q^{(n)}}(w)\, dw
\end{aligned}
$$

Since $\Omega_q^{(n)}$ is independent of $A^{(n)}$ and $B^{(n)}$, it is indeed independent of $C^{(n)}$. Consequently,

$$F_{\Omega_q^{(n+1)}}(t) = \int_{0_-}^{\infty} F_{C^{(n)}}(t - w)\, f_{\Omega_q^{(n)}}(w)\, dw$$

It can be shown that if the queue is ergodic $\lim_{n\to\infty} F_{\Omega_q^{(n)}}(t) = F_{\Omega_q}(t)$. This yields

$$F_{\Omega_q}(t) = \int_{0_-}^{\infty} F_C(t - w)\, f_{\Omega_q}(w)\, dw, \qquad t \ge 0$$

Also $F_{\Omega_q}(t) = 0$ for $t < 0$. Summarizing, we have

$$F_{\Omega_q}(t) = \begin{cases} \int_{0_-}^{\infty} F_C(t - w)\, f_{\Omega_q}(w)\, dw, & t \ge 0 \\ 0, & t < 0 \end{cases} \tag{6.13a}$$

This Wiener-Hopf type integral equation is originally due to D. V. Lindley. Other representations of $F_{\Omega_q}(t)$ are:

$$F_{\Omega_q}(t) = \begin{cases} \int_{0_-}^{\infty} F_{\Omega_q}(w)\, f_C(t - w)\, dw, & t \ge 0 \\ 0, & t < 0 \end{cases} \tag{6.13b}$$

and

$$F_{\Omega_q}(t) = \begin{cases} \int_{-\infty}^{t} F_{\Omega_q}(t - w)\, f_C(w)\, dw, & t \ge 0 \\ 0, & t < 0 \end{cases} \tag{6.13c}$$

These integral equations appear to be almost like the convolution integral, but they are not. Notice that $F_{\Omega_q}(t) = 0$, for negative values of t. The solution of this integral equation via a so-called spectral approach is next outlined. Define

$$F_{\Omega_q -}(t) = \begin{cases} 0, & t \ge 0 \\ \int_{-\infty}^{t} F_{\Omega_q}(t - w)\, f_C(w)\, dw, & t < 0 \end{cases}$$

Therefore

$$F_{\Omega_q}(t) + F_{\Omega_q -}(t) = \int_{-\infty}^{t} F_{\Omega_q}(t - w)\, f_C(w)\, dw, \qquad t \in \mathbb{R}$$

Next define the two-sided (bilateral) Laplace transforms of $F_{\Omega_q}(t)$ and $F_{\Omega_q -}(t)$ as $\widetilde{F}_{\Omega_q}(s)$ and $\widetilde{F}_{\Omega_q -}(s)$ respectively, where $s \in \mathbb{C}$. As the probability density functions of the random variables A and B are defined for $t \in \mathbb{R}_0^+$, let the single-sided Laplace transforms of $f_A(t)$ and $f_B(t)$ be $\widehat{f}_A(s)$ and $\widehat{f}_B(s)$ respectively, where $s \in \mathbb{C}$. Also since $C = (B - A)$ the probability density function $f_C(t)$ of the random variable C is defined for all real values of t. Let the two-sided Laplace transform of $f_C(t)$ be $\widetilde{f}_C(s)$. Then

$$\widetilde{f}_C(s) = \widehat{f}_A(-s)\, \widehat{f}_B(s)$$

The two-sided Laplace transform of $\int_{-\infty}^{t} F_{\Omega_q}(t - w)\, f_C(w)\, dw$, for $t \in \mathbb{R}$ is next evaluated.

$$\begin{aligned}
&\int_{-\infty}^{\infty}\int_{-\infty}^{t} F_{\Omega_q}(t-w)\, f_C(w)\, e^{-st} dw dt \\
&= \int_{-\infty}^{\infty}\int_{-\infty}^{\infty} e^{-s(t-w)} F_{\Omega_q}(t-w)\, f_C(w)\, e^{-sw} dw dt \\
&= \int_{-\infty}^{\infty} f_C(w)\, e^{-sw} \int_{-\infty}^{\infty} e^{-s(t-w)} F_{\Omega_q}(t-w)\, dt dw \\
&= \widetilde{f}_C(s)\, \widetilde{F}_{\Omega_q}(s)
\end{aligned}$$

where the relationship $F_{\Omega_q}(t-w) = 0$ for $w \geq t$ was used. Thus

$$\widetilde{F}_{\Omega_q}(s) + \widetilde{F}_{\Omega_q -}(s) = \widetilde{F}_{\Omega_q}(s)\, \widehat{f}_A(-s)\, \widehat{f}_B(s)$$

Denote the single-sided Laplace transform of the waiting time cumulative distribution function $F_{\Omega_q}(t)$ by $\widehat{F}_{\Omega_q}(s)$, where $s \in \mathbb{C}$. Use of the relationship $\widetilde{F}_{\Omega_q}(s) = \widehat{F}_{\Omega_q}(s)$, and the above equation yields

$$\widehat{F}_{\Omega_q}(s) = \frac{\widetilde{F}_{\Omega_q -}(s)}{\left\{\widehat{f}_A(-s)\, \widehat{f}_B(s) - 1\right\}} \tag{6.14}$$

Thus $\widehat{F}_{\Omega_q}(s)$ is obtained from $\widetilde{F}_{\Omega_q -}(s)$, and the single-sided Laplace transforms of the probability density functions of the random variables A and B. The determination of $\widetilde{F}_{\Omega_q -}(s)$ often requires techniques from the complex variable theory. Therefore the function $\widehat{F}_{\Omega_q}(s)$ is determined via a technique called the spectral factorization.

Spectral Factorization

In this scheme, the expression $\left\{\widehat{f}_A(-s)\, \widehat{f}_B(s) - 1\right\}$ is represented as a rational function of s. Therefore let

$$\left\{\widehat{f}_A(-s)\, \widehat{f}_B(s) - 1\right\} = \frac{\Theta(s)}{\Theta_-(s)} \tag{6.15}$$

where:

(a) For $\operatorname{Re}(s) > 0$, select $\Theta(s)$ such that it is an analytic function of s with no zeros in this half-plane.
(b) For $\operatorname{Re}(s) < K_c$ and $K_c > 0$, select $\Theta_-(s)$ such that it is an analytic function of s with no zeros in this half-plane.
(c) Also

$$\lim_{|s|\to\infty} \frac{\Theta(s)}{s} = 1, \qquad \operatorname{Re}(s) > 0$$

$$\lim_{|s|\to\infty} \frac{\Theta_-(s)}{s} = -1, \quad \operatorname{Re}(s) < K_c$$

Use of the above equations results in

$$\widehat{F}_{\Omega_q}(s)\, \Theta(s) = \widetilde{F}_{\Omega_q -}(s)\, \Theta_-(s), \qquad 0 < \operatorname{Re}(s) < K_c$$

Observe that $\widehat{F}_{\Omega_q}(s)\, \Theta(s)$ and $\widetilde{F}_{\Omega_q -}(s)\, \Theta_-(s)$ are both analytic and bounded in the common strip. Using complex analysis it can be established that

$$\widehat{F}_{\Omega_q}(s)\,\Theta(s) = \widetilde{F}_{\Omega_q-}(s)\,\Theta_-(s) = K_r$$

where K_r is a constant. Thus

$$\widehat{F}_{\Omega_q}(s) = \frac{K_r}{\Theta(s)}$$

This equation implies that a knowledge of K_r and $\Theta(s)$ is sufficient to determine $\widehat{F}_{\Omega_q}(s)$. The constant K_r can be determined via any of the following three relationships.

$$K_r = \lim_{s \to 0} \frac{\Theta(s)}{s} = \lim_{s \to 0} \frac{d\Theta(s)}{ds} = F_{\Omega_q}(0_+) \tag{6.16}$$

See the problem section for a proof of these relationships. The relationship $K_r = F_{\Omega_q}(0_+)$ implies that the constant K_r is equal to the probability that an arriving customer does not queue. In general, this constant K_r is not always equal to $(1-\rho)$. Actually $(1-\rho)$ is the fraction of time the server is idle. However, these expressions are equal for an $M/G/1$ queueing system.

Finally, the cumulative distribution function $F_{\Omega_q}(\cdot)$ can be determined by inverting $\widehat{F}_{\Omega_q}(s)$. This distribution function can also be obtained by working with $\Theta_-(s)$ instead of $\Theta(s)$.

Example 6.4. The use of spectral factorization technique is demonstrated for the $M/M/1$ queue. We determine $F_{\Omega_q}(t)$, the waiting time cumulative distribution function of the customers in this queue.

In this queue $f_A(t) = \lambda e^{-\lambda t}$ and $f_B(t) = \mu e^{-\mu t}$ for $t \in \mathbb{R}_0^+$. Therefore $\widehat{f}_A(s) = \lambda/(s+\lambda)$ and $\widehat{f}_B(s) = \mu/(s+\mu)$. Let $\lambda/\mu = \rho < 1$. Observe that $C = (B - A)$ and $\widetilde{f}_C(s) = \widehat{f}_A(-s)\,\widehat{f}_B(s)$ lead to

$$\frac{\Theta(s)}{\Theta_-(s)} = \left\{\widehat{f}_A(-s)\,\widehat{f}_B(s) - 1\right\} = \frac{s(s+\mu-\lambda)}{(s+\mu)(\lambda - s)}$$

The zeros of this rational function are at $s = 0$ and $(\lambda - \mu)$. The zeros of the denominator, which are generally called poles, are at $s = \lambda$ and $-\mu$. It is required that $\Theta(s)$ be analytic and have no zeros in the half-plane $\operatorname{Re}(s) > 0$. Also $\Theta_-(s)$ must be analytic and have no zeros in the half-plane $\operatorname{Re}(s) < K_c$, where $K_c > 0$. This is true if $K_c = \lambda$. All these conditions are satisfied if

$$\Theta(s) = \frac{s(s+\mu-\lambda)}{(s+\mu)}$$

$$\Theta_-(s) = (\lambda - s)$$

The constant K_r is evaluated

$$K_r = \lim_{s \to 0} \frac{\Theta(s)}{s} = (1-\rho)$$

Therefore

$$\widehat{F}_{\Omega_q}(s) = \frac{K_r}{\Theta(s)} = \frac{(1-\rho)(s+\mu)}{s(s+\mu-\lambda)}$$

Inversion of $\widehat{F}_{\Omega_q}(s)$ yields

$$F_{\Omega_q}(t) = 1 - \rho e^{-\mu(1-\rho)t}, \quad t \in \mathbb{R}_0^+$$

This indeed is the cumulative distribution function of the waiting time of a customer in an ergodic $M/M/1$ queue. □

Mean Waiting Time

The mean waiting time W_q, of a $G/G/1$ queue at steady-state is next obtained. Note that

$$\Omega_q^{(n+1)} = \max\left(0, \Omega_q^{(n)} + C^{(n)}\right)$$

where $C^{(n)} = \left(B^{(n)} - A^{(n)}\right)$. Also define $X^{(n)}$ as

$$X^{(n)} = \min\left(0, \Omega_q^{(n)} + C^{(n)}\right)$$

therefore

$$\Omega_q^{(n+1)} + X^{(n)} = \Omega_q^{(n)} + C^{(n)}$$

Note that $\mathcal{E}\left(\Omega_q^{(n+1)}\right) = \mathcal{E}\left(\Omega_q^{(n)}\right)$ for an ergodic queue. Also as $n \to \infty$, $X^{(n)} \to X$, and $C^{(n)} \to C$. Taking expectation on both sides of the above equation yields

$$\mathcal{E}(X) = \mathcal{E}(C) = \left(\frac{1}{\mu} - \frac{1}{\lambda}\right)$$

Using the definitions of $\Omega_q^{(n+1)}$ and $X^{(n)}$, it can be concluded that

$$\Omega_q^{(n+1)} X^{(n)} = 0$$

Squaring both sides of the relationship $\left(\Omega_q^{(n+1)} + X^{(n)}\right) = \left(\Omega_q^{(n)} + C^{(n)}\right)$ yields

$$\left(\Omega_q^{(n+1)}\right)^2 + 2\Omega_q^{(n+1)} X^{(n)} + \left(X^{(n)}\right)^2 = \left(\Omega_q^{(n)}\right)^2 + 2\Omega_q^{(n)} C^{(n)} + \left(C^{(n)}\right)^2$$

Letting $n \to \infty$, and taking expectations on both sides of the above equation results in

$$\mathcal{E}\left(X^2\right) = 2W_q \mathcal{E}(C) + \mathcal{E}\left(C^2\right)$$
$$W_q = \frac{\mathcal{E}\left(X^2\right) - \mathcal{E}\left(C^2\right)}{2\mathcal{E}(C)} = \frac{\mathcal{E}\left(X^2\right) - \mathcal{E}\left(C^2\right)}{2\mathcal{E}(X)}$$

where use was made of the observation that $\Omega_q^{(n)}$ and $C^{(n)}$ are independent of each other, and $\mathcal{E}(\Omega_q) = W_q$. The above average value of the waiting time can also be expressed via the moments of the idle period of the queue. The idle period I of the queue is said to occur only when the system remains idle for a nonzero interval of time. Define $I^{(n)}$ as the idle time (if any) preceding the $(n+1)$th arrival. Then

$$I^{(n)} = -\min\left(0, \Omega_q^{(n)} + C^{(n)}\right)$$
$$\Omega_q^{(n+1)} - I^{(n)} = \Omega_q^{(n)} + C^{(n)}$$

Therefore $I^{(n)} = -X^{(n)}$. Thus either $I^{(n)} = 0$ or $I^{(n)} > 0$. Further, if $I^{(n)} > 0$ then $I^{(n)}$ equals the idle period. As $n \to \infty$, $I^{(n)} \to I$ in an ergodic queue. Denote the arrival-point probabilities by $\{\varphi_n\}$. Thus

$$\mathcal{E}\left(I^{(n)}\right) = P\,(\text{queueing system is found empty by an arrival})\,\mathcal{E}\,(\text{idle time})$$

$$\mathcal{E}\,(X) = -\mathcal{E}\left(I^{(n)}\right) = -\varphi_0 \mathcal{E}\,(I)$$

Thus $\mathcal{E}\,(C) = -\varphi_0 \mathcal{E}\,(I)$. Also

$$\begin{aligned}\mathcal{E}\left(X^2\right) &= P\,(\text{queueing system is found empty by an arrival})\,\mathcal{E}\left(I^2\right) \\ &= \varphi_0 \mathcal{E}\left(I^2\right)\end{aligned}$$

Finally

$$W_q = -\frac{\mathcal{E}\,(I^2)}{2\mathcal{E}\,(I)} - \frac{\mathcal{E}\,(C^2)}{2\mathcal{E}\,(C)} \tag{6.17}$$

The above analysis is further clarified via an example.

Example 6.5. Consider an ergodic $M/G/1$ queueing system. In this queueing system, the arrivals are Poissonian with parameter λ. Also

$$\mathcal{E}\,(I) = \frac{1}{\lambda}, \text{ and } \mathcal{E}\left(I^2\right) = \frac{2}{\lambda^2}$$

It can be verified that

$$W_q = \frac{\rho\left(1 + C_B^2\right)}{2\mu\,(1-\rho)}$$

where $\rho < 1$, and C_B is the coefficient of variation of the service time. The condition $\rho < 1$ ensures ergodicity of the queueing system. □

6.7 Priority Queueing Systems

In priority based queueing systems, all customers are not necessarily treated equal, some customers get preferential treatment. This type of queueing discipline in which some customers get preferential treatment is called a *priority queueing system*. In direct contrast, the FCFS queueing discipline has no priorities. A customer which arrives first is served first in a FCFS scheme.

In a priority queueing scheme, arriving customers are assigned to different classes. Assume that there are n classes. Classes are numbered from 1 through n. The customers in class 1 have the highest priority, and those in class 2 have the next highest priority, and so on. The customers in class n have the lowest priority. Therefore, customers of class i are given preferential treatment over customers of class j, if $1 \leq i < j \leq n$. Furthermore, customers which belong to the same class are served in a FCFS manner. Note that the higher the priority, lower is its number. There are two important types of priority queueing disciplines. These are: *nonpreemptive*, and *preemptive* queueing schemes.

These schemes are elucidated as follows. Assume that a customer of class j is in service, and the newly arrived customer belongs to class i such that $i < j$.

In a nonpreemptive queueing scheme, a newly arrived customer which belongs to class i, is required to wait till the customer in service which belongs to class j completes its service. Only after the completion of service of this class j customer, that the customer of class i enters service, provided no other customer of class higher than i is waiting for service.

In a preemptive queueing system, the arriving customer of class i interrupts the service of the class j customer which was getting service. The preemptive queueing scheme can be further classified into: *preemptive-resume*, and *preemptive-repeat* queueing schemes.

In preemptive-resume queueing scheme the customer whose service was interrupted resumes its service at the point at which it was interrupted, when it reenters service. In preemptive-repeat queueing scheme the customer whose service was interrupted repeats the service from the beginning, when it reenters service.

Assume that all customers who enter a queueing system exit only after their service is completed. Such queueing schemes are called *work conserving*. Furthermore, in a work conserving queueing system, the server is not forced to be idle when customers are waiting to be served. The following two priority queueing schemes are discussed in this chapter: a single server queueing system with nonpreemptive priorities and general service time distributions, and a multiserver queueing system with nonpreemptive priorities and exponentially distributed service times. In each queueing scheme, we shall assume that customer interarrival time is independent of all others, each service time is independent of all others; and the customer arrival processes and the service times are stochastically independent of each other.

6.7.1 Single Server Nonpreemptive Priorities

Consider a single server queueing system with infinite waiting space, and n classes of customers. The highest priority customers belong to class 1, and the lowest priority customers belong to class n. The service discipline of this queueing system is nonpreemptive. Furthermore, the customer population of each class, and waiting space for each class of customers are assumed to be infinite. Customer arrivals within each class are described by a Poisson process. For example the Poisson parameter of class m customer is $\lambda_m \in \mathbb{R}^+, 1 \leq m \leq n$. Let the service time random variable of class m customer be B_m, and the corresponding cumulative distribution function be $F_{B_m}(t), t \in \mathbb{R}_0^+$ for $1 \leq m \leq n$. Also let $\mathcal{E}(B_m) = \mu_m^{-1} \in \mathbb{R}^+$. The service discipline is FCFS within each priority class. The steady-state average waiting time W_{q_m}, of class m-customers is determined, where $1 \leq m \leq n$.

Define

$$\lambda = \sum_{m=1}^{n} \lambda_m$$

and the combined service time cumulative distribution function is specified by

$$F_B(t) = \frac{1}{\lambda} \sum_{m=1}^{n} \lambda_m F_{B_m}(t), \qquad t \in \mathbb{R}_0^+$$

The corresponding probability density function is defined by $f_B(t)$ for $t \in \mathbb{R}_0^+$. Also define

$$\rho_m = \frac{\lambda_m}{\mu_m}, \qquad 1 \leq m \leq n$$

$$\widetilde{\sigma}_0 = 0, \text{ and } \widetilde{\sigma}_m = \sum_{i=1}^{m} \rho_i, \qquad 1 \leq m \leq n$$

This queue is ergodic if $\rho = \widetilde{\sigma}_n < 1$. Consider a customer of class m which enters the queueing system at time t_a and enters service at time t_b. Its waiting time is equal to $T = (t_b - t_a)$. The time T has the following components.

(a) At time t_a there can be either one or zero number of customers in service. Let the time required to complete the service of the customer in service be T_0.
(b) Also assume that there are N_i number of customers of class i, $1 \leq i \leq m$ which have to be serviced before this newly arrived customer. These customers are already present in the queueing system at time t_a. Let the time required to complete service of N_i customers of class i be equal to T_i for $1 \leq i \leq m$.
(c) During the interval T, assume that N_i' number of customers of class i, $1 \leq i \leq (m-1)$ have arrived. Let the time required to complete service of N_i' customers of class i be equal to T_i' for $1 \leq i \leq (m-1)$.

Then

$$T = T_0 + \sum_{i=1}^{m} T_i + \sum_{i=1}^{m-1} T_i'$$

Taking expectations on both sides results in

$$\mathcal{E}(T) = \mathcal{E}(T_0) + \sum_{i=1}^{m} \mathcal{E}(T_i) + \sum_{i=1}^{m-1} \mathcal{E}(T_i')$$

Observe that:

(a) $\mathcal{E}(T) =$ average waiting time of customers of class m, which is W_{q_m}.
(b) For $1 \leq i \leq m$, $\mathcal{E}(T_i) = \mathcal{E}(B_i N_i)$. Since the service time B_i, and the number of customers N_i are independent of each other, $\mathcal{E}(T_i) = \mathcal{E}(B_i)\mathcal{E}(N_i)$. But $\mathcal{E}(B_i) = \mu_i^{-1}$ and $\mathcal{E}(N_i) = \lambda_i W_{q_i}$. Therefore $\mathcal{E}(T_i) = \rho_i W_{q_i}$.
(c) For $1 \leq i \leq (m-1)$, $\mathcal{E}(T_i') = \mathcal{E}(B_i N_i')$. Similarly the service time B_i, and the number of customers N_i' are independent of each other, $\mathcal{E}(T_i') = \mathcal{E}(B_i)\mathcal{E}(N_i')$. Note that $\mathcal{E}(N_i')$ is equal to $\lambda_i W_{q_m}$. Therefore $\mathcal{E}(T_i') = \rho_i W_{q_m}$.
(d) The component of $\mathcal{E}(T_0)$ which is due to a customer of class i in service is equal to

$$\lambda_i \mathcal{E}\left(\int_0^{B_i} (B_i - x)\, dx\right) = \frac{\lambda_i \mathcal{E}(B_i^2)}{2}$$

Thus

$$\mathcal{E}(T_0) = \sum_{i=1}^{n} \frac{\lambda_i \mathcal{E}(B_i^2)}{2} = \frac{\lambda}{2}\int_0^{\infty} t^2 f_B(t)\, dt$$

Use of these results yield

$$W_{q_m} = \frac{\mathcal{E}(T_0) + \sum_{i=1}^{m} \rho_i W_{q_i}}{(1 - \widetilde{\sigma}_{m-1})}, \qquad 1 \leq m \leq n$$

This simplifies to

$$W_{q_m} = \frac{\mathcal{E}(T_0)}{(1 - \widetilde{\sigma}_{m-1})(1 - \widetilde{\sigma}_m)}, \qquad 1 \leq m \leq n \tag{6.18a}$$

Note that if $\rho < 1$, then $\widetilde{\sigma}_m < 1$ for $1 \leq m \leq n$. If the total average waiting time of a customer in the queue is denoted by W_q, then

$$W_q = \frac{1}{\lambda} \sum_{m=1}^{n} \lambda_m W_{q_m} \tag{6.18b}$$

The corresponding average queue lengths can be determined by applying Little's law to each priority stream.

6.7.2 Multiserver Nonpreemptive Priorities

Consider a multiserver queueing system with nonpreemptive priority scheme. In this queueing system there are c servers, and n classes of customers. The service discipline of the customers is nonpreemptive, and the servicing scheme is FCFS within each class of customers. The class m customer stream is described by a Poisson process with parameter $\lambda_m \in \mathbb{R}^+$, for $1 \leq m \leq n$. The c servers have identical exponentially distributed service time distributions with parameter $\mu \in \mathbb{R}^+$. Define

$$\lambda = \sum_{m=1}^{n} \lambda_m, \quad \zeta = \frac{\lambda}{\mu}$$

$$\rho_m = \frac{\lambda_m}{c\mu}, \quad 1 \leq m \leq n$$

$$\widetilde{\sigma}_0 = 0, \quad \widetilde{\sigma}_m = \sum_{i=1}^{m} \rho_i, \quad 1 \leq m \leq n$$

Assume that the queue is ergodic, that is $\widetilde{\sigma}_n = \rho = \lambda/(c\mu) < 1$. Let the steady-state average waiting time of customers which belong to class m be equal to W_{q_m}, for $1 \leq m \leq n$. Then proceeding as in the last subsection, it can be shown that

$$W_{q_m} = \frac{\mathcal{E}(T_0)}{(1-\widetilde{\sigma}_{m-1})(1-\widetilde{\sigma}_m)}, \quad 1 \leq m \leq n \tag{6.19a}$$

$$\mathcal{E}(T_0) = \frac{\mathcal{C}(c,\zeta)}{c\mu} \tag{6.19b}$$

In the above equation $\mathcal{C}(\cdot,\cdot)$ is the Erlang's $\mathcal{C}$ function which has been defined in the subsection on $M/M/c$ queueing system.

6.8 Queueing Networks

Stochastic models of a network of queues are discussed in this section. A queueing network is made up of several queues, in which customers move from one queueing center to another according to some random or fixed order, and then possibly exit the network from different queueing centers. There is a special class of queueing networks which are amenable to analysis. These are the Markovian queueing networks. There are two types of Markovian queueing networks: *open* and *closed queueing networks*.

(a) In an open queueing network, customers are allowed to enter the queueing network from outside. Furthermore, customers are also allowed to leave the queueing network.
(b) In a closed queueing network, the finite customer population of the queueing network remains fixed. External customers are not allowed to enter the network, and customers from within the network are not allowed to leave the network.

The queueing centers in a queueing network are also sometimes referred to as the nodes of the network.

6.8.1 Open Queueing Networks

An opening queueing network is made up of several interconnected queueing centers. Customers can enter the queueing centers, either from outside the network or from other queueing centers. Customers are also permitted to depart from the network. The customer population is assumed to be infinite. An open queueing network is specified as follows.

The queueing network has N queueing centers (nodes), each with infinite waiting space. The node i of the network has $c_i \in \mathbb{P}$ servers. Each of these servers at a given node i has an exponentially distributed service time with parameter $\mu_i \in \mathbb{R}^+$. The ith node of the network also receives customers from outside the network. These arrive according to a Poisson process with parameter $\gamma_i \in \mathbb{R}^+$. Upon completion of service at node i, a customer moves to node j with probability p_{ij}. This customer is in fact an internal arrival at node j. Furthermore, a customer upon completion of service at node i, leaves the queueing network with probability $\left(1 - \sum_{j=1}^{N} p_{ij}\right)$.

Furthermore, the external customer arrival processes and all service times at all queueing centers are stochastically independent of each other. A queueing network of this type was first analysed by J. R. Jackson. The results of his analysis are noted in the following observation. This result is valid at steady-state of the queueing network.

Observation 6.2. Assume that an open network of queues is in equilibrium.

(a) The average arrival rate at node i is λ_i, where

$$\lambda_i = \gamma_i + \sum_{j=1}^{N} \lambda_j p_{ji}, \qquad 1 \le i \le N$$

(b) Let the steady-state probability that there are $n_i \in \mathbb{N}$ customers at node i for $1 \le i \le N$ be $p(n_1, n_2, \ldots, n_N)$. If $\lambda_i < c_i \mu_i$ for $1 \le i \le N$, then

$$p(n_1, n_2, \ldots, n_N) = p_1(n_1)\, p_2(n_2) \ldots p_N(n_N)$$

where $p_i(n_i)$ is the steady-state probability that there are n_i customers at the ith node. It is specified by the probability that there are n_i customers in an $M/M/c_i$ queueing system, where the arrival rate is λ_i and the service rate is μ_i at each of the c_i servers. That is, each node i behaves as if it were an *isolated* $M/M/c_i$ queueing system. □

An intuitive explanation of this interesting behavior of such networks is next provided. Observe that the customer departure process of an ergodic $M/M/c$ queue is Poissonian with parameter λ, where λ is the parameter of its input Poisson process. Furthermore, the aggregate of several independent Poissonian streams of customers is a Poisson process. Also independent disaggregation of a Poisson process results in several independent Poissonian streams.

Thus, the input stream of customers at node i in this queueing network mimics a Poisson process with parameter λ_i. Furthermore, the output of this queue behaves as a Poisson process with parameter λ_i, because the node behaves as an $M/M/c_i$ queueing system. Also, the stream of customers which move form node i to node j with probability p_{ij} mimics a Poissonian stream with parameter $\lambda_i p_{ij}$.

Actually, it can be demonstrated that in queueing networks with feedback, the arrival processes at different nodes may not be Poissonian in general. Nevertheless, Jackson's powerful result stands true.

It can also be established that at steady-state, that is under equilibrium, the total external input flow rate of customers is equal to the total external output flow rate of customers. A formal proof of these results can be obtained by setting up the so-called flow-balance equations.

6.8.2 Closed Queueing Networks

A closed queueing network is another example of a Markovian network. In this queueing network the total customer population within the network is a finite constant. Customers are neither permitted to enter the network nor depart from it. A closed queueing network is specified as follows.

The total number of customers within the network is equal to $K \in \mathbb{P}$. Also, the queueing network has N queueing centers, each with sufficient waiting space. That is, all K customers can be resident at any single queueing center simultaneously. Furthermore, the node i of the network has $c_i \in \mathbb{P}$ servers. Each of these servers at a given node has an exponentially distributed service time with parameter μ_i. Upon completion of service at node i, a customer moves to node j with probability p_{ij}. Also, the service times at all queueing centers are stochastically independent of each other. A queueing network of this type was first analysed by W. J. Gordon and G. P. Newell. The results of their analysis are summarized in the next observation.

Observation 6.3. Assume that a closed network of queues is in equilibrium.

(a) According to the law of total probability,

$$\sum_{j=1}^{N} p_{ij} = 1, \qquad 1 \leq i \leq N$$

(b) Let $\mathcal{P} = [p_{ij}]$ be a Markov transition probability matrix, which is assumed to be irreducible. Let $p = \begin{bmatrix} p_1 & p_2 & \cdots & p_N \end{bmatrix}$ denote the equilibrium (stationary) probabilities of this Markov chain. Then $p\mathcal{P} = p$, that is

$$p_j = \sum_{i=1}^{N} p_i p_{ij}, \;\; 1 \leq j \leq N, \;\; \text{and} \;\; \sum_{i=1}^{N} p_i = 1$$

(c) In this closed queueing network, let $n(t) = (n_1(t), n_2(t), \ldots, n_N(t))$, where $n_i(t)$ is the number of customers at queueing center i at time $t \in \mathbb{R}_0^+$, and $\sum_{i=1}^{N} n_i(t) = K$. The process $\{n(t), t \in \mathbb{R}_0^+\}$ is an irreducible continuous-time Markov chain with finite state space. Consequently this Markov chain possesses an equilibrium distribution.

(d) Denote the average arrival rate, or equivalently the average service completion rate at node j by $\lambda_K(j)$ for $1 \leq j \leq N$. Then

$$\lambda_K(j) = \sum_{i=1}^{N} \lambda_K(i) \, p_{ij}$$

Therefore $\lambda_K(j) = \lambda_K p_j$, for $1 \leq j \leq N$, where $\lambda_K = \sum_{j=1}^{N} \lambda_K(j)$. That is, λ_K is the average service completion rate for the entire queueing network. It is called the *system throughput*.

(e) Let the steady-state probability that there are n_i customers at node i for $1 \leq i \leq N$ be $p(n_1, n_2, \ldots, n_N)$, where $\sum_{i=1}^{N} n_i = K$. Then

$$p(n_1, n_2, \ldots, n_N) = \frac{1}{\mathcal{G}(K)} \prod_{i=1}^{N} \frac{\{\zeta_K(i)\}^{n_i}}{d_i(n_i)}$$

$$\zeta_K(i) = \frac{\lambda_K(i)}{\mu_i}, \qquad 1 \leq i \leq N$$

$$d_i(n_i) = \begin{cases} n_i! & n_i \leq c_i \\ c_i! c_i^{n_i - c_i} & n_i > c_i \end{cases}, \qquad 1 \leq i \leq N$$

$$\mathcal{G}(K) = \sum_{\widehat{n} \in \mathcal{N}} \prod_{i=1}^{N} \frac{\{\zeta_K(i)\}^{n_i}}{d_i(n_i)}$$

where $\widehat{n} = (n_1, n_2, \ldots, n_N)$ and $\mathcal{N}$ is the set of all vectors $\widehat{n}$ such that $\sum_{i=1}^{N} n_i = K$. The variable $\zeta_K(i)$ is the utilization of server i, for $1 \leq i \leq N$. Note that the cardinality of the set $\mathcal{N}$ is equal to

$$|\mathcal{N}| = \binom{N+K-1}{K}$$

The above binomial coefficient gives the total number of ways of distributing K customers among N nodes. □

As $\lambda_K(i)$ is proportional to p_i, for $1 \leq i \leq N$, an alternative representation of the probabilities $p(n_1, n_2, \ldots, n_N)$ is

$$p(n_1, n_2, \ldots, n_N) = \frac{1}{\mathcal{H}(K)} \prod_{i=1}^{N} \frac{1}{d_i(n_i)} \left(\frac{p_i}{\mu_i}\right)^{n_i}$$

$$\mathcal{H}(K) = \sum_{\widehat{n} \in \mathcal{N}} \prod_{i=1}^{N} \frac{1}{d_i(n_i)} \left(\frac{p_i}{\mu_i}\right)^{n_i}$$

As in the case of open queueing networks, these equations can be checked by setting up the flow-balance equations. Of special interest is the queueing network in which each queueing center has a single server, that is $c_i = 1$ for $1 \leq i \leq N$. Recursive equations are next developed to evaluate queueing networks of this type. This analysis is called the *mean value analysis* (MVA).

Mean Value Analysis

The mean value analysis technique is a recursive scheme to evaluate a closed queueing network. This analysis assumes that each queueing center of the network has a single server. The MVA technique essentially depends upon Little's law and the *arrival theorem.* This theorem states that at steady-state, a customer in a closed queueing network, when it moves from one queueing center to another, observes the mean state of the network with itself removed.

Consider a closed queueing network with N single server queueing centers, and a customer population of K customers. The service times at all queueing centers are exponentially distributed and are stochastically independent of each other. Let a customer move from center v to center j after its completion of service. The probability of the state of the queueing system as observed by this customer is determined. The probability of the state, that this customer observes when there are n_k customers at centers $k = 1, 2, \ldots, N$; such that $\sum_{i=1}^{N} n_i = (K-1)$ is next computed. Thus

$$\begin{aligned}
&P\left(\text{the customer sees } n_k \text{ customers at server } k, \text{ where } 1 \le k \le N \mid \right. \\
&\left. \text{and customer moves from node } \upsilon \text{ to } j\right) \\
&= \frac{\begin{array}{c} P\left(\text{the network state is } (n_1, n_2, \ldots, n_\upsilon + 1, \ldots, n_j, \ldots, n_N),\right. \\ \left.\text{and customer moves from node } \upsilon \text{ to } j\right) \end{array}}{P\left(\text{customer moves from node } \upsilon \text{ to } j\right)} \\
&= \frac{p\left(n_1, n_2, \ldots, n_\upsilon + 1, \ldots, n_j, \ldots, n_N\right) \mu_\upsilon p_{\upsilon j}}{\sum_{\widehat{n}} p\left(n_1, n_2, \ldots, n_\upsilon + 1, \ldots, n_N\right) \mu_\upsilon p_{\upsilon j}}, \quad \sum_{i=1}^{N} n_i = (K-1) \\
&= K_d \prod_{i=1}^{N} \left(\frac{p_i}{\mu_i}\right)^{n_i}, \quad \sum_{i=1}^{N} n_i = (K-1)
\end{aligned}$$

where $\widehat{n} = (n_1, n_2, \ldots, n_N)$ and K_d is a constant which is independent of $n_1, n_2, \ldots, n_N$. As the above expression is the probability mass function defined on the set of vectors $\widehat{n}$, such that $\sum_{i=1}^{N} n_i = (K-1)$, it follows from Gordon and Newell's result that

$$\begin{aligned}
&P\left(\text{the customer sees } n_k \text{ customers at server } k, \text{ where } 1 \le k \le N \mid \right. \\
&\left. \text{customer moves from node } \upsilon \text{ to } j\right) \\
&= p\left(n_1, n_2, \ldots, n_N\right), \text{ and } \sum_{i=1}^{N} n_i = (K-1)
\end{aligned}$$

The above result is true for all values of υ. It is formalized in the following arrival theorem.

Theorem 6.2. *A closed queueing network has N single server queueing centers. The service time of each server is exponentially distributed. Furthermore, the service times at all queueing centers are stochastically independent of each other. Also, the customer population in this queueing network is K. At steady-state, a customer which arrives at a queueing center j sees the queueing network as if it has a stationary (steady-state) distribution in the same queueing system with $(K-1)$ customers.* □

In order to develop a recursive mean value analysis of the queueing system at steady-state, the following variables are defined for $1 \le j \le N$.

$\lambda_K(j)$ = average customer arrival or completion rate at server j.
$W_K(j)$ = average time a customer spends at server j.
$L_K(j)$ = average number of customers at server j.
As per Little's law

$$L_K(j) = \lambda_K(j) W_K(j), \quad 1 \le j \le N$$

Also

$$\sum_{j=1}^{N} L_K(j) = K$$

Average time $W_K(j)$, a customer spends in the queueing system at server j is next derived. As per the arrival theorem, this average time of a customer who moves to server j is:

$$W_K(j) = \frac{\{1 + \mathcal{E}\,(\text{number of customers at server } j \text{ as seen by an arrival})\}}{\mu_j} = \frac{\{1 + L_{K-1}(j)\}}{\mu_j}$$

Note that

$$L_{K-1}(j) = \lambda_{K-1}(j)\, W_{K-1}(j) = \lambda_{K-1} p_j W_{K-1}(j)$$

$$(K-1) = \sum_{j=1}^{N} L_{K-1}(j) = \lambda_{K-1} \sum_{j=1}^{N} p_j W_{K-1}(j)$$

Therefore

$$\lambda_{K-1} = \frac{(K-1)}{\sum_{j=1}^{N} p_j W_{K-1}(j)}$$

Thus

$$W_K(j) = \frac{1}{\mu_j} + \frac{(K-1)\, p_j W_{K-1}(j)}{\mu_j \sum_{j=1}^{N} p_j W_{K-1}(j)}$$

These values can be computed recursively by noting that $W_1(j) = \mu_j^{-1}$, for $1 \le j \le N$. Also note that, the server utilization is $\zeta_K(j) = \lambda_K(j)/\mu_j$, for $1 \le j \le N$. This remarkably simple recursive algorithm is called the MVA algorithm.

Algorithm 6.2. *Mean Value Analysis* (*MVA*).

Input: $N, K, \mathcal{P}$, and μ_j for $1 \le j \le N$.
Output: The vector p, and λ_K; and
$\lambda_K(j), W_K(j), L_K(j)$, and $\zeta_K(j)$ for $1 \le j \le N$.
begin

Step 1: Compute the vector p via the equation $p\mathcal{P} = p$, where $\sum_{j=1}^{N} p_j = 1$.
Step 2: (*initialization*) Let $L_0(j) = 0$, for $1 \le j \le N$.
Step 3: **for** $1 \le k \le K$ **do**

$$W_k(j) \leftarrow \frac{\{1 + L_{k-1}(j)\}}{\mu_j}, \qquad 1 \le j \le N$$

$$\lambda_k \leftarrow \frac{k}{\sum_{j=1}^{N} p_j W_k(j)}$$

$$\lambda_k(j) \leftarrow \lambda_k p_j, \qquad 1 \le j \le N$$

$$L_k(j) \leftarrow \lambda_k(j)\, W_k(j), \qquad 1 \le j \le N$$

Step 4:

$$\zeta_K(j) \leftarrow \frac{\lambda_K(j)}{\mu_j}, \qquad 1 \le j \le N$$

end (*end of MVA algorithm*)

The bottleneck-server in the queueing network is the server which has the largest utilization. Therefore the bottleneck server is $\arg\max_{1 \le j \le N} \{\zeta_K(j)\}$.

The MVA is an admirable algorithm for evaluating a closed queueing network, with a single server at each queueing center, independent routing of customers, and independent exponentially distributed service times. However, it is also possible to develop computationally efficient algorithms to evaluate closed queueing networks, with possibly multiple servers at each queueing center, independent routing of customers, and independent exponentially distributed service times

6.9 Bounds and Approximations

Sometimes it is useful to obtain bounds and approximations of certain queueing scenarios which would otherwise be analytically intractable. Moreover, such bounds and approximations are important from a practical perspective. Heavy traffic approximations, approximations for multiserver queues, bounds for $G/G/1$ queue, and diffusion approximation are discussed in this section.

In this section, the probability density function, the mean, and variance of the customer interarrival times are given by $f_A(t)$, $1/\lambda$ and σ_A^2 respectively, where $t \in \mathbb{R}_0^+$. Similarly, the probability density function, the mean, and variance of the customer service times are given by $f_B(t)$, $1/\mu$ and σ_B^2 respectively, where $t \in \mathbb{R}_0^+$. In addition, the queues are also assumed to be ergodic.

6.9.1 Heavy Traffic Approximations

Heavy traffic approximations for the $G/G/1$ and $G/M/c$ queues are considered in this subsection. Laplace-transform theoretic techniques are used to obtain these approximations.

$G/G/1$ Heavy Traffic Approximation

The probability density functions of the interarrival and service times of the $G/G/1$ queue are specified by $f_A(t)$ and $f_B(t)$ respectively, where $t \in \mathbb{R}_0^+$. The single-sided Laplace transforms of these functions are $\widehat{f}_A(s)$ and $\widehat{f}_B(s)$ respectively, where $s \in \mathbb{C}$. The assumption of heavy traffic implies that the utilization of the queue $\rho \approx 1$. Under these conditions $\widehat{f}_A(s)$ and $\widehat{f}_B(s)$ are expanded near the origin $(s \to 0)$.

$$\widehat{f}_A(s) = 1 + s\widehat{f}_A^{(1)}(0) + \frac{s^2}{2}\widehat{f}_A^{(2)}(0) + o\left(s^2\right)$$

$$\widehat{f}_B(s) = 1 + s\widehat{f}_B^{(1)}(0) + \frac{s^2}{2}\widehat{f}_B^{(2)}(0) + o\left(s^2\right)$$

where $\widehat{f}_A^{(1)}(s)$ and $\widehat{f}_A^{(2)}(s)$ are the first and second derivative of $\widehat{f}_A(s)$ with respect to s. Similarly, $\widehat{f}_B^{(1)}(s)$ and $\widehat{f}_B^{(2)}(s)$ are the first and second derivative of $\widehat{f}_B(s)$ with respect to s. Note that

$$\widehat{f}_A^{(1)}(0) = -\mathcal{E}(A) = -\frac{1}{\lambda}, \quad \widehat{f}_A^{(2)}(0) = \mathcal{E}\left(A^2\right) = \sigma_A^2 + \frac{1}{\lambda^2}$$

$$\widehat{f}_B^{(1)}(0) = -\mathcal{E}(B) = -\frac{1}{\mu}, \quad \widehat{f}_B^{(2)}(0) = \mathcal{E}\left(B^2\right) = \sigma_B^2 + \frac{1}{\mu^2}$$

Recall that the expression $\left\{\widehat{f}_A(-s)\,\widehat{f}_B(s) - 1\right\}$ occurs in the analysis of the $G/G/1$ queue. Substituting the above values in this expression yields

$$\widehat{f}_A(-s)\,\widehat{f}_B(s) - 1 = s\left[\frac{(1-\rho)}{\lambda} + \frac{s}{2}\left\{\sigma_A^2 + \sigma_B^2 + \frac{(1-\rho)^2}{\lambda^2}\right\}\right] + o\left(s^2\right)$$

As $\rho = \lambda/\mu \approx 1$, we have

$$\widehat{f}_A(-s)\,\widehat{f}_B(s) - 1 \simeq K_h s\,(s - s_0)$$

where

$$K_h = \frac{\left(\sigma_A^2 + \sigma_B^2\right)}{2}$$
$$s_0 = -\frac{2(1-\rho)}{\lambda\left(\sigma_A^2 + \sigma_B^2\right)}$$

Thus use of the analysis of the $G/G/1$ queue results in

$$\widetilde{F}_{\Omega_q-}(s) = \left\{\widehat{f}_A(-s)\,\widehat{f}_B(s) - 1\right\}\widehat{F}_{\Omega_q}(s)$$
$$\simeq K_h s\,(s - s_0)\,\widehat{F}_{\Omega_q}(s)$$

Recall that $s\widehat{F}_{\Omega_q}(s) = \widehat{f}_{\Omega_q}(s)$, therefore $s\widehat{F}_{\Omega_q}(s) \to 1$ as $s \to 0$. Consequently for small values of s

$$\widetilde{F}_{\Omega_q-}(s) \simeq -K_h s_0$$

Thus

$$s\,(s - s_0)\,\widehat{F}_{\Omega_q}(s) \simeq -s_0$$

Inverting the Laplace transform $\widehat{F}_{\Omega_q}(s)$ yields

$$F_{\Omega_q}(t) \simeq \left(1 - e^{s_0 t}\right), \qquad t \in \mathbb{R}_0^+$$

Therefore

$$W_q \simeq -\frac{1}{s_0} = \frac{\lambda\left(\sigma_A^2 + \sigma_B^2\right)}{2(1-\rho)} \tag{6.20}$$

These results are summarized in the following theorem.

Theorem 6.3. *Under the assumption of heavy traffic* ($\rho \approx 1$), *the steady-state waiting time distribution of a* $G/G/1$ *queue is approximated by an exponential distribution with parameter* $2(1-\rho)/\left\{\lambda\left(\sigma_A^2 + \sigma_B^2\right)\right\}$. □

This analysis is due to J. F. C. Kingman.

$G/M/c$ Heavy Traffic Approximation

Under the assumption of heavy traffic, $P(\Omega_q = 0) \approx 0$. That is, $\varrho\sigma^c \simeq (1-\sigma)$. Also if the queue is heavily utilized, then $W = \mathcal{E}(\Omega) \simeq \mathcal{E}(\Omega_q)$. Thus, the average time a customer spends in the queueing system is approximated by

$$W \simeq \frac{1}{c\mu(1-\sigma)}$$

Recall that σ is obtained from the relationship

$$\sigma = \widehat{f}_A(c\mu - c\mu\sigma)$$

Substituting $\xi = (c\mu - c\mu\sigma)$ in the above equation yields

$$\sigma = \widehat{f}_A(\xi) = 1 - \frac{\xi}{c\mu}$$

Next expand $\widehat{f}_A(\xi)$ in a power series in ξ.

$$\begin{aligned}\widehat{f}_A(\xi) &= 1 + \xi \widehat{f}_A^{(1)}(0) + \frac{\xi^2}{2}\widehat{f}_A^{(2)}(0) + o(\xi^2) \\ &= 1 - \frac{\xi}{\lambda} + \frac{\xi^2}{2}\left(\sigma_A^2 + \frac{1}{\lambda^2}\right) + o(\xi^2)\end{aligned}$$

Thus

$$1 - \frac{\xi}{c\mu} \simeq 1 - \frac{\xi}{\lambda} + \frac{\xi^2}{2}\left(\sigma_A^2 + \frac{1}{\lambda^2}\right)$$

Simplification results in

$$\xi \simeq \frac{2(1-\rho)}{\lambda\left(\sigma_A^2 + \frac{1}{\lambda^2}\right)}$$

As $\xi = c\mu(1-\sigma)$, the average wait time W is approximated as

$$W \simeq \frac{\lambda\left(\sigma_A^2 + \frac{1}{\lambda^2}\right)}{2(1-\rho)}$$

Also since the service time is exponentially distributed $\sigma_B^2 = \mu^{-2}$. Furthermore, $\rho \approx 1$, therefore $\lambda \simeq c\mu$. Using these observations, the average time W a customer spends in the queue is approximated as

$$W \simeq \frac{\lambda\left(\sigma_A^2 + \sigma_B^2/c^2\right)}{2(1-\rho)} \tag{6.21}$$

The above expression has inspired Kingman to make a conjecture about the average time a customer spends in the queueing system under heavy traffic conditions in a $G/G/c$ queue. As per his conjecture, this time is given by the above expression. He further stated that this net queue-resident time of a customer under heavy traffic is exponentially distributed. His conjecture has been confirmed by J. Köllerström.

6.9.2 Approximations for Multiserver Queues

It can be observed that the average waiting time of a customer in an $M/G/1$ queue is $\left(1 + C_B^2\right)/2$ times the average waiting time of a customer in an $M/M/1$ queue. That is

$$W_q|_{M/G/1} = \frac{\left(1 + C_B^2\right)}{2} W_q|_{M/M/1}$$

where C_B is the coefficient of variation of the service time. Using this result, it has been suggested that

$$W_q|_{M/G/c} \simeq \frac{(1+C_B^2)}{2} W_q|_{M/M/c} = \frac{\mathcal{C}(c,\zeta)(1+C_B^2)}{2c\mu(1-\rho)}$$

where $\mathcal{C}(\cdot,\cdot)$ is Erlang's $\mathcal{C}$ function. If the coefficient of variation of the customer interarrival time interval is C_A, then as $\rho \to 1$

$$W_q|_{G/G/1} \simeq \frac{(C_A^2 + C_B^2)}{2} W_q|_{M/M/1}$$

This heavy traffic approximation of the $G/G/1$ queue suggests that as $\rho \to 1$

$$W_q|_{G/G/c} \simeq \frac{(C_A^2 + C_B^2)}{2} W_q|_{M/M/c}$$

Bounds for $G/G/1$ are discussed in the next subsection.

6.9.3 $G/G/1$ Queue Bounds

Upper and lower bounds for average waiting time of an ergodic $G/G/1$ queue at steady-state are derived. Recall that A and B are customer interarrival and service time random variables respectively, and $C = (B - A)$. It has been shown in the discussion on $G/G/1$ queue that the average waiting time W_q in this queue depends upon the first and second moments of the length of the idle period I, and the random variable C. Note that $\mathcal{E}(C) = -\lambda^{-1}(1-\rho)$, and $\mathcal{E}(I) = -\mathcal{E}(C)/\varphi_0$, and $\varphi_0 \in (0,1]$, therefore

$$\mathcal{E}(I) \geq -\mathcal{E}(C) = \frac{(1-\rho)}{\lambda}$$

An upper bound for W_q is obtained by noting that the variance of a random variable is always greater than or equal to zero. Using these observations, it can be proved that

$$W_q \leq \frac{\lambda(\sigma_A^2 + \sigma_B^2)}{2(1-\rho)}$$

See the problem section for a proof of this inequality. A possible lower bound for W_q is next obtained. If a lower bound for $\mathcal{E}(X^2)$ is determined, then a lower bound for W_q can be obtained. See the section on $G/G/1$ queue for a definition of the variable X. This lower bound is determined by using the concept of *stochastically smaller* random variables. A random variable Z_1 is said to be stochastically smaller than the random variable Z_2 if and only if $P(Z_1 > z) \leq P(Z_2 > z)$ for all $z \in \mathbb{R}$. Thus the moment $\mathcal{E}(Z_1^k) \leq \mathcal{E}(Z_2^k)$ for any positive value of k.

It is next established that the random variable $-X$ is stochastically smaller than the random variable A. Recall from the discussion on $G/G/1$ queue that

$$\begin{aligned} -X &= -\min(0, \Omega_q + C) \\ &= \max(0, -\Omega_q - C) = \max(0, A - B - \Omega_q) \end{aligned}$$

Since A, B, and Ω_q are all nonnegative random variables, it must follow that $-X$ is stochastically smaller than the random variable A. This result implies that $\mathcal{E}\left((-X)^2\right) = \mathcal{E}(X^2) \leq \mathcal{E}(A^2)$. Therefore

$$W_q = \frac{\mathcal{E}(X^2) - \mathcal{E}(C^2)}{2\mathcal{E}(C)} = \frac{\lambda\{\mathcal{E}(C^2) - \mathcal{E}(X^2)\}}{2(1-\rho)}$$
$$\geq \frac{\lambda\{\mathcal{E}(C^2) - \mathcal{E}(A^2)\}}{2(1-\rho)}$$

Finally

$$W_q \geq \frac{\lambda^2\sigma_B^2 + \rho(\rho-2)}{2\lambda(1-\rho)} \tag{6.22}$$

This lower bound is useful provided the numerator of the bound is positive. This numerator is positive iff $\sigma_B^2 > (2-\rho)/(\lambda\mu)$.

6.9.4 Diffusion Approximation

Diffusion approximation technique is applied to the $G/G/1$ queue to obtain asymptotic average queue length under heavy traffic conditions. Recall that $\alpha(t)$ and $\delta(t)$ are the total number of customer arrivals and departures respectively during the time interval $[0,t]$. Also, $N(t)$ is equal to the total number of customers in the queueing system at time t. Asymptotic expressions for the mean and variance of the arrival and departure processes are initially determined. Let the arrival and departure process be $\{\alpha(t), t \in \mathbb{R}_0^+\}$ and $\{\delta(t), t \in \mathbb{R}_0^+\}$ respectively. Then

$$N(t) = N(0) + \alpha(t) - \delta(t), \quad t \in \mathbb{R}_0^+$$

In this discussion, it is assumed that $N(t)$ does not become zero. If this is the case, then the departure process becomes approximately independent of the arrival process.

Customers are assumed to arrive at instants $t_1, t_2, \ldots$ such that $0 \leq t_1 \leq t_2 \ldots$. The interarrival times of the customers are a_k, $k = 1, 2, \ldots$. The nth customer arrives at time t_n, where

$$t_n = \sum_{k=1}^{n} a_k, \quad \forall\, n \in \mathbb{P}$$

Let the random variables corresponding the arrival instant and interarrival time of the nth customer be T_n and A_n respectively. It is assumed that these interarrival times are independent, and identically distributed random variables. The mean and variance of the interarrival times are $1/\lambda$ and σ_A^2 respectively. Therefore $\mathcal{E}(T_n) = n/\lambda$, and $Var(T_n) = n\sigma_A^2$. As the A_k's are independent, and identically distributed random variables, the central limit theorem can be used to determine the asymptotic distribution of T_n. Therefore for large values of n,

$$P\left(\frac{T_n - n/\lambda}{\sqrt{n}\sigma_A} \leq x\right) = \Phi(x)$$

where $\Phi(\cdot)$ is the cumulative distribution function of standard normal random variable. Define

$$t = \sqrt{n}\sigma_A x + \frac{n}{\lambda}$$

Therefore $P(T_n \leq t) = \Phi(x)$. Also for very large values n, $t \sim n/\lambda$, consequently

$$n \sim \lambda t - x\lambda\sigma_A\sqrt{\lambda t}$$

It can be observed by using a result from renewal theory that

$$P(\alpha(t) \geq n) = P(T_n \leq t)$$

Therefore

$$P(\alpha(t) \geq n) = \Phi(x)$$

That is

$$P\left(\alpha(t) \geq \lambda t - x\lambda\sigma_A\sqrt{\lambda t}\right) = \Phi(x)$$

$$P\left(\frac{\alpha(t) - \lambda t}{\lambda\sigma_A\sqrt{\lambda t}} \geq -x\right) = \Phi(x) = 1 - \Phi(-x)$$

This yields

$$P\left(\frac{\alpha(t) - \lambda t}{\lambda\sigma_A\sqrt{\lambda t}} \leq -x\right) = \Phi(-x)$$

and

$$P\left(\frac{\alpha(t) - \lambda t}{\lambda\sigma_A\sqrt{\lambda t}} \leq x\right) = \Phi(x)$$

The last equation implies that the asymptotic distribution of $\alpha(t)$ is Gaussian, with

$$\mathcal{E}(\alpha(t)) \simeq \lambda t$$
$$Var(\alpha(t)) \simeq \lambda^3\sigma_A^2 t$$

It can be similarly proved that the asymptotic distribution of $\delta(t)$ is Gaussian, with

$$\mathcal{E}(\delta(t)) \simeq \mu t$$
$$Var(\delta(t)) \simeq \mu^3\sigma_B^2 t$$

Define $\widetilde{N}(t) = (N(t) - N(0))$, and since $\rho = \lambda/\mu = \mathcal{E}(B)/\mathcal{E}(A)$, using the above results, it follows that for heavy traffic and as $t \to \infty$

$$\mathcal{E}\left(\widetilde{N}(t)\right) \simeq (\lambda - \mu)t = \mu(\rho - 1)t$$
$$Var\left(\widetilde{N}(t)\right) \simeq \left(\lambda^3\sigma_A^2 + \mu^3\sigma_B^2\right)t$$

The above results are summarized in the following theorem.

Theorem 6.4. *In a $G/G/1$ queue, let $N(t)$ be the total number of customers at time t, where $t \in \mathbb{R}_0^+$. Let λ and μ be the customer arrival rate and service rate respectively. Similarly, let σ_A^2, and σ_B^2 be variances of the customer interarrival time and service time respectively. Then under heavy traffic, and as $t \to \infty$, the random variable $\widetilde{N}(t) = (N(t) - N(0))$ has a Gaussian distribution with mean $\mu(\rho - 1)t$ and variance $\left(\lambda^3\sigma_A^2 + \mu^3\sigma_B^2\right)t$.* □

Observations 6.4. Some related observations.

1. The asymptotic mean and variance expressions for the processes $\alpha(t)$ and $\delta(t)$ for $t \in \mathbb{R}_0^+$ as obtained above, also follow from the renewal theory.

2. The process $\{N(t), t \in \mathbb{R}_0^+\}$ can be approximated by a diffusion process with infinitesimal mean m and variance σ^2, where

$$m = \lim_{\Delta t \to 0} \frac{\mathcal{E}(N(t+\Delta t) - N(t))}{\Delta t} \simeq \mu(\rho - 1)$$

$$\sigma^2 = \lim_{\Delta t \to 0} \frac{Var(N(t+\Delta t) - N(t))}{\Delta t} \simeq (\lambda^3\sigma_A^2 + \mu^3\sigma_B^2)$$

Assume that these results are valid for large values of t, and that the queue has heavy traffic.
3. Observe that both the mean and the variance of the number of customers in the queueing system grow linearly with t. Furthermore, the expressions for $\mathcal{E}\left(\widetilde{N}(t)\right)$ and m appear to be reasonable for $\rho > 1$. However for $\rho < 1$, this expression takes negative value. This anomaly is taken care of next by inserting a "reflecting boundary" at the origin for $N(t)$. □

Also assume that $N(0) = x_0$. For the stochastic process $\{N(t), t \in \mathbb{R}_0^+\}$, this boundary acts as a reflection barrier. Then as per an observation of the continuous-time Markov process of the diffusion type, as $t \to \infty$, $N(t)$ is exponentially distributed with parameter $-2m/\sigma^2$, where $m < 0$. The reader should refer to the chapter on stochastic processes, for a justification of this observation. Thus the average number of customers in the queueing system for large values of t and under the conditions of heavy traffic is given by

$$L \simeq -\frac{\sigma^2}{2m} = \frac{(\lambda^3\sigma_A^2 + \mu^3\sigma_B^2)}{2\mu(1-\rho)} \tag{6.23a}$$

$$= \frac{\mu^2(\rho^3\sigma_A^2 + \sigma_B^2)}{2(1-\rho)} \tag{6.23b}$$

Example 6.6. For an $M/G/1$ queue, $\sigma_A^2 = \lambda^{-2}$. Therefore the diffusion approximation for this queue is

$$L \simeq \frac{(\rho^3 + \lambda^2\sigma_B^2)}{2\rho^2(1-\rho)}$$

The exact value of L via the Pollaczek-Khinchin formula is

$$L = \rho + \frac{\rho^2 + \lambda^2\sigma_B^2}{2(1-\rho)}$$

$$= \frac{(2\rho - \rho^2) + \lambda^2\sigma_B^2}{2(1-\rho)}$$

It can be observed that the diffusion approximation is good when $\rho \approx 1$. □

6.10 Application of Large Deviation Techniques

This section is concerned with the application of large deviation techniques to some aspects of resource management in queues. We develop the theory of effective bandwidths, and apply the theory of large deviations to study a single server queue with large buffer space. Elements of large deviation theory have been discussed in the chapter on probability theory.

6.10.1 Theory of Effective Bandwidths

The theory of effective bandwidths is simultaneously a simple, robust, and efficient technique for resource management of queueing systems. Effective bandwidth is a stochastic technique to describe the flow of traffic in a network. It is based upon the principle of large deviation. It turns out that when traffic streams from independent sources are multiplexed together, the effective bandwidth of each traffic stream is independent of other streams. In addition, the effective bandwidth of a mixture of several independent traffic streams is equal to the sum of their individual effective bandwidths. Thus effective bandwidth possesses the additive property. We initially develop motivation for the concept of effective bandwidth of a traffic stream. This is followed by its formal definition, properties, and examples.

Motivation

We study a queueing scenario in which the probability that the queue length exceeds a specified value of buffer size is determined. This study in turn provides a motivation for the definition of effective bandwidth of a stochastic traffic stream.

A single server queue with infinite buffer space is studied in discrete time. Let the amount of data in the queue just after the tth instant be $\widetilde{Q}_t$, where $t \in \mathbb{Z}$. Also the amount of data which arrives in the interval $(t-1, t)$ is $\widetilde{A}_t \in \mathbb{R}_0^+$. The server processes $\widetilde{C}_t \in \mathbb{R}^+$ amount of work at time t. Therefore

$$\widetilde{Q}_t = \left(\widetilde{Q}_{t-1} + \widetilde{A}_t - \widetilde{C}_t\right)^+, \quad t \in \mathbb{Z}$$

where we used the notation $(x)^+ = \max(x, 0)$. Also assume that the amounts of data which arrives in different time slots are independent, and identically distributed random variables. A representative of these random variables is denoted by $\widetilde{A}$. Let the cumulant (logarithm of the moment generating function) of the random variable $\widetilde{A}$ be $\Upsilon(\cdot)$. Also assume that the amounts of transmitted (serviced) work, the $\widetilde{C}_t$'s are independent of each other. In addition, the $\widetilde{A}_t$ and $\widetilde{C}_t$ sequences are stochastically independent of each other. Initially assume that $\widetilde{C}_t = \widetilde{C} \in \mathbb{R}^+$ for all $t \in \mathbb{Z}$. Thus

$$\widetilde{Q}_t = \left(\widetilde{Q}_{t-1} + \widetilde{A}_t - \widetilde{C}\right)^+, \quad t \in \mathbb{Z}$$

This recursive relationship can be solved by imposing a boundary condition. We evaluate $\widetilde{Q}_0$ by assuming that $\widetilde{Q}_{-T} = 0$, where $T \in \mathbb{P}$. This value of $\widetilde{Q}_0$ is denoted by $\widetilde{Q}_0^{-T}$. The following result is obtained by successive use of the above recursion. Its proof is left to the reader.

Lemma 6.4. *Let $\widetilde{S}_t, t \in \mathbb{N}$ be the cumulative arrival process. Define $\widetilde{S}_0 = 0$, and $\widetilde{S}_t = \sum_{j=0}^{t-1} \widetilde{A}_{-j}, t \in \mathbb{P}$. If $\widetilde{Q}_{-T} = 0$, and $T \in \mathbb{P}$, then*

$$\widetilde{Q}_0^{-T} = \max_{0 \le t \le T} \left\{\widetilde{S}_t - \widetilde{C}t\right\} \tag{6.24}$$

□

A bound on the probability that the queue length exceeds m is determined as follows. Denote this probability by $P\left(\widetilde{Q}_0^{-T} > m\right)$. Thus

$$P\left(\widetilde{Q}_0^{-T} > m\right) = P\left(\max_{0 \le t \le T} \left\{\widetilde{S}_t - \widetilde{C}t\right\} > m\right)$$

$$= P\left(\cup_{0\leq t\leq T}\left\{\widetilde{S}_t - \widetilde{C}t\right\} > m\right)$$
$$\leq \sum_{t=0}^{T} P\left(\widetilde{S}_t > m + \widetilde{C}t\right)$$

In the above analysis, the second equality follows because

$$\max_{0\leq t\leq T}\left\{\widetilde{S}_t - \widetilde{C}t\right\} > m$$

if and only if there exists some t such that $\left\{\widetilde{S}_t - \widetilde{C}t\right\} > m$. The last inequality follows from the observation that the probability of a union of events is upper-bounded by the sum of their probabilities. In the next step Chernoff's bound is used. For $\theta > 0$ we have

$$P\left(\widetilde{Q}_0^{-T} > m\right) \leq \sum_{t=0}^{T} \mathcal{E}\left(e^{\theta\left(\widetilde{S}_t - m - \widetilde{C}t\right)}\right)$$

$$= e^{-\theta m}\sum_{t=0}^{T} \mathcal{E}\left(e^{\theta\left(\widetilde{S}_t - \widetilde{C}t\right)}\right)$$
$$= e^{-\theta m}\sum_{t=0}^{T} e^{-t\left(\widetilde{C}\theta - \Upsilon(\theta)\right)}$$

where $\mathcal{E}(\cdot)$ is the expectation operator. Next assume that $T \to \infty$. This implies that the queue was empty at time $-\infty$. Also assume that $\Upsilon(\theta)/\theta < \widetilde{C}$. If $\widetilde{C} = (\Upsilon(\theta)/\theta + \epsilon)$ for some $\epsilon > 0$, then the summation on the right-hand side of the above inequality is equal to a constant, say K_b. If

$$\lim_{T\to\infty} P\left(\widetilde{Q}_0^{-T} > m\right) \triangleq P\left(\widetilde{Q} > m\right)$$

then

$$P\left(\widetilde{Q} > m\right) \leq K_b e^{-\theta m}, \quad \theta > 0$$

The above inequality implies that at steady-state, if $\Upsilon(\theta)/\theta < \widetilde{C}$, the probability that the buffer exceeds m is upper-bounded exponentially with a decay parameter θ. Therefore, if it is required that $P\left(\widetilde{Q} > m\right)$ be below a specific value (specified by the decay parameter θ), then the intensity of the traffic stream should be such that $\Upsilon(\theta)/\theta$ is less than the server processing capacity $\widetilde{C}$. The quantity $\Upsilon(\theta)/\theta$ is called the *effective bandwidth* of the traffic source with decay rate θ. Note that θ has the units of data^{-1} and $\Upsilon(\theta)$ has the units of time^{-1}.

In the continuous-time case, if $\widetilde{A}(t)$ is the amount of work that arrives from a source in the interval $[0, t]$, then define $\Upsilon_t(\theta)$ as

$$\Upsilon_t(\theta) = \frac{1}{t}\ln \mathcal{E}\left(e^{\theta \widetilde{A}(t)}\right), \quad t \in \mathbb{R}^+$$

and its effective bandwidth is $\Upsilon_t(\theta)/\theta$. The above discussion leads to the following definitions of effective bandwidth of sources with stationary increments, which operate either in discrete or continuous time.

Definitions 6.1. *Effective bandwidth for both discrete and continuous times are stated.*

1. *Consider a discrete time process*
$$\left\{\widetilde{A}(t), t \in \mathbb{P}\right\}$$
with stationary increments. The traffic which arrives in time interval (increment) i is $\widetilde{A}_i$ where $i \in \mathbb{P}$, and $\widetilde{A}(t) = \sum_{i=1}^{t} \widetilde{A}_i$. The effective bandwidth of the source is
$$\vartheta(\theta, t) = \frac{1}{\theta t} \ln \mathcal{E}\left(e^{\theta \widetilde{A}(t)}\right); \quad \theta \in \mathbb{R}^+, \quad t \in \mathbb{P} \tag{6.25a}$$
2. *Consider a continuous-time process*
$$\left\{\widetilde{A}(t), t \in \mathbb{R}_0^+\right\}$$
with stationary increments. The traffic which arrives in time interval $[0, t]$ is $\widetilde{A}(t)$. *The effective bandwidth of the source is*
$$\vartheta(\theta, t) = \frac{1}{\theta t} \ln \mathcal{E}\left(e^{\theta \widetilde{A}(t)}\right); \quad \theta, t \in \mathbb{R}^+ \tag{6.25b}$$

□

Properties of Effective Bandwidth

Some useful properties of effective bandwidth of discrete-time processes are listed below. Assume that the processes have stationary increments. Similar statements are true about continuous-time processes.

1. Let $\vartheta(\theta, t)$ be the effective bandwidth of a source. Then θ has units of the unit of data^{-1} (for example bit^{-1}), and t has the units of time. Similarly, $\vartheta(\theta, t)$ has units of data per unit time (for example bits per second).
2. If the process $\left\{\widetilde{A}(t), t \in \mathbb{P}\right\}$ has independent increments, then $\vartheta(\theta, t)$ does not depend upon t.
3. Effective bandwidths are additive. That is, if a traffic stream is sum of finite number of independent traffic streams, then the effective bandwidth of the merged traffic streams is equal to the sum of the effective bandwidths of individual traffic streams.
Let $\widetilde{A}_i(t)$ be a traffic stream, with effective bandwidth $\vartheta_i(\theta, t)$, where $1 \le i \le n$ and $n \in \mathbb{P}$. Also the traffic stream $\widetilde{A}_i(t)$ is independent of traffic stream $\widetilde{A}_j(t)$ for $i \neq j$, and $1 \le i, j \le n$. If the effective bandwidth of $\widetilde{A}(t) = \sum_{i=1}^{n} \widetilde{A}_i(t)$ is $\vartheta(\theta, t)$ then
$$\vartheta(\theta, t) = \sum_{i=1}^{n} \vartheta_i(\theta, t)$$
4. If the parameter t is fixed, then the effective bandwidth $\vartheta(\theta, t)$ is increasing in θ, and lies between the mean and peak of the arrival rate measured over an interval of length t. That is
$$\frac{\mathcal{E}\left(\widetilde{A}(t)\right)}{t} \le \vartheta(\theta, t) \le \frac{\mathcal{A}(t)}{t}$$
where $\mathcal{A}(t) = \sup\left\{x \mid P\left(\widetilde{A}(t) > x\right) > 0\right\}$, and $\mathcal{A}(t)$ can possibly be infinite. □

Some of the above observations are established in the problem section.

Examples 6.7. Some useful examples.

1. *Bernoulli process*: An arrival stream is an independent sequence of Bernoulli random variables, each with parameter p. Its cumulant is given by $\Upsilon(\theta) = \ln\left(pe^{\theta} + 1 - p\right)$. Therefore its effective bandwidth is

$$\vartheta(\theta, t) = \vartheta(\theta, 1) = \frac{\Upsilon(\theta)}{\theta} = \frac{\ln\left(pe^{\theta} + 1 - p\right)}{\theta}$$

Observe that:

(a) $\lim_{\theta \to 0} \vartheta(\theta, t) = p =$ the mean arrival rate of the input stream.
(b) $\lim_{\theta \to \infty} \vartheta(\theta, t) = 1$.

2. *Poisson process*: Consider a Poisson process with rate $\lambda \in \mathbb{R}^{+}$. This process has independent increments, and the number of arrivals in interval $[0, t]$ has a Poisson distribution with parameter λt. Recall that the cumulant of a random variable with Poisson distribution with parameter λ is $\lambda\left(e^{\theta} - 1\right)$. Therefore the effective bandwidth of this process is given by

$$\vartheta(\theta, t) = \frac{1}{\theta t} \ln \mathcal{E}\left(e^{\theta \widetilde{A}(t)}\right) = \frac{\lambda t\left(e^{\theta} - 1\right)}{\theta t} = \frac{\lambda\left(e^{\theta} - 1\right)}{\theta}.$$

As expected, the effective bandwidth of the Poissonian stream does not depend upon t. Also
(a) $\lim_{\theta \to 0} \vartheta(\theta, t) = \lambda =$ the mean arrival rate of the input stream.
(b) $\lim_{\theta \to \infty} \vartheta(\theta, t)$ does not exist.

3. *Gaussian sources*: Consider a continuous-time process such that its cumulative arrival process $\widetilde{A}(t), t \in \mathbb{R}_0^{+}$ is $\widetilde{A}(t) = \left\{\lambda t + \widetilde{B}(t)\right\}$.

(a) Assume that $\widetilde{B}(t), t \in \mathbb{R}_0^{+}$ is a Brownian motion process. This process has independent increments, $\widetilde{B}(0) = 0$, and is normally distributed with zero mean and variance $\sigma^2 t$. Observe that $\widetilde{B}(t)$ can take both positive and negative values, because it is normally distributed. This is not possible in an arrival process. However, it can be an approximation when the arrival rate is much higher than the perturbations. Therefore

$$\vartheta(\theta, t) = \frac{1}{\theta t} \ln \mathcal{E}\left(e^{\theta \widetilde{A}(t)}\right) = \frac{1}{\theta t}\left\{\lambda \theta t + \frac{\sigma^2 \theta^2 t}{2}\right\} = \lambda + \frac{\sigma^2 \theta}{2}$$

(b) Assume that $\widetilde{B}(t), t \in \mathbb{R}_0^{+}$ is a fractional Brownian motion process, with Hurst parameter H, where $H \in (0, 1)$. The parameter $H = 1/2$ results in the Brownian motion process. This process is normally distributed with zero mean and variance $\sigma^2 t^{2H}$. Therefore

$$\vartheta(\theta, t) = \frac{1}{\theta t} \ln \mathcal{E}\left(e^{\theta \widetilde{A}(t)}\right) = \frac{1}{\theta t}\left\{\lambda \theta t + \frac{\sigma^2 \theta^2 t^{2H}}{2}\right\} = \lambda + \frac{\sigma^2 \theta}{2} t^{2H-1}$$

□

Effective Bandwidth of Multiplexed Traffic Streams

Consider a discrete-time, bufferless, single server queueing system with processing capacity $\widetilde{C} \in \mathbb{R}^{+}$. The queue serves $\widetilde{J}$ types of sources. Also, there are $n_j \in \mathbb{N}$ number of sources of type j, where $1 \leq j \leq \widetilde{J}$. The net traffic which arrives from a source i of type j in a single time slot (interval) is $\widetilde{A}_{ji}$, where $1 \leq i \leq n_j$, and $1 \leq j \leq \widetilde{J}$. All these different traffic streams are stochastically independent processes. The distributions of these processes may depend upon j but not on the parameter i. The cumulant of the traffic which arrives from a source of type j is denoted by $\Upsilon_j(\theta)$, where $\theta > 0$ for $1 \leq j \leq \widetilde{J}$. The effective bandwidth of source of type j is $\vartheta_j(\theta) = \Upsilon_j(\theta)/\theta$ for $1 \leq j \leq \widetilde{J}$.

Denote the net traffic which arrives into the queueing system in a time interval by $\widetilde{A}$, and the effective bandwidth of the net traffic by $\vartheta(\theta)$. Therefore

$$\widetilde{A} = \sum_{j=1}^{\widetilde{J}} \sum_{i=1}^{n_j} \widetilde{A}_{ji}$$

$$\vartheta(\theta) = \sum_{j=1}^{\widetilde{J}} n_j \vartheta_j(\theta)$$

Use of Chernoff's bound yields

$$\ln P\left(\widetilde{A} \geq \widetilde{C}\right) \leq \ln \mathcal{E}\left(e^{\theta\left(\widetilde{A}-\widetilde{C}\right)}\right) = \theta\left(\vartheta(\theta) - \widetilde{C}\right)$$

In this model, it is also required that

$$\ln P\left(\widetilde{A} \geq \widetilde{C}\right) \leq -\gamma_0, \text{ where } \gamma_0 \in \mathbb{R}^+$$

This constraint is satisfied if the vector $n = \left(n_1, n_2, \ldots, n_{\widetilde{J}}\right)$ belongs to the set

$$\mathcal{N}_e = \left\{n \mid \inf_{\theta}\left(\theta\left(\vartheta(\theta) - \widetilde{C}\right)\right) \leq -\gamma_0\right\}$$

If it is further assumed that all the sources possess stationary increments, then the above discussion provides a reasonable framework for trade-off between flows of different types.

6.10.2 Queue with Large Buffer

We evaluate the asymptotic performance of a single server queue with infinite buffer space. For simplicity, this queue is studied in discrete time. The queue is serviced at a constant rate $\widetilde{C} \in \mathbb{R}^+$. Let the amount of data in the queue just after the tth instant be $\widetilde{Q}_t$, where $t \in \mathbb{Z}$. Also the amount of data which arrives in the interval $(t-1, t)$ is $\widetilde{A}_t \in \mathbb{R}_0^+$. Assume that the amounts of data which arrive in each time slot are independent, and identically distributed random variables. A representative of these random variables is denoted by $\widetilde{A}$, and its cumulant is $\Upsilon(\cdot)$. For simplicity, assume that this cumulant is finite. It has been shown in an earlier subsection that, if $\widetilde{S}_0 = 0$ and $\widetilde{S}_t = \sum_{j=0}^{t-1} \widetilde{A}_{-j}, t \in \mathbb{P}$, then the value of $\widetilde{Q}$ at time 0 is given by

$$\widetilde{Q} = \sup_{t \geq 0}\left\{\widetilde{S}_t - \widetilde{C}t\right\}$$

where it is assumed that the queue was empty at time $-\infty$. The main result of this subsection is based upon the following set of observations. These observations are established in the problem section.

Observations 6.5. Let

$$\mathcal{E}\left(\widetilde{A}\right) < \widetilde{C}, \ y \in \mathbb{R}, \ q > 0,$$
$$\ell(x) = \sup_{\theta \in \mathbb{R}} (\theta x - \Upsilon(\theta))$$
$$\widetilde{Q} = \sup_{t \geq 0} \left\{ \widetilde{S}_t - \widetilde{C}t \right\}$$

1.
$$\limsup_{y \to \infty} \frac{1}{y} \ln P\left(\widetilde{Q}/y > q\right) \leq -q \sup\left\{\theta > 0 \mid \Upsilon(\theta) < \theta\widetilde{C}\right\} \tag{6.26a}$$

2.
$$\liminf_{y \to \infty} \frac{1}{y} \ln P\left(\widetilde{Q}/y > q\right) \geq -\inf_{t \in \mathbb{R}^+} \left\{ t\ell\left(\widetilde{C} + q/t\right)\right\} \tag{6.26b}$$

3. If $\mathcal{I}(q) = \inf_{t \in \mathbb{R}^+} \left\{ t\ell\left(\widetilde{C} + q/t\right)\right\}$, then

$$\mathcal{I}(q) = \inf_{t \in \mathbb{R}^+} \sup_{\theta \geq 0} \left\{ \theta\left(q + \widetilde{C}t\right) - t\Upsilon(\theta)\right\} = q \sup\left\{\theta > 0 \mid \Upsilon(\theta) < \theta\widetilde{C}\right\} \tag{6.26c}$$

□

The above observation can be condensed in the following theorem. It is generally called the large deviation principle for queue size.

Theorem 6.5. *Let* $\mathcal{E}\left(\widetilde{A}\right) < \widetilde{C}, y \in \mathbb{R}, q > 0, \ell(x) = \sup_{\theta \in \mathbb{R}} (\theta x - \Upsilon(\theta))$, *and*

$$\widetilde{Q} = \sup_{t \geq 0} \left\{ \widetilde{S}_t - \widetilde{C}t \right\}$$

Then
$$\lim_{y \to \infty} \frac{1}{y} \ln P\left(\widetilde{Q}/y > q\right) = -\mathcal{I}(q) \tag{6.27}$$

where $\mathcal{I}(q) = \inf_{t \in \mathbb{R}^+} \left\{ t\ell\left(\widetilde{C} + q/t\right)\right\}$.

Proof. The proof follows immediately from the above set of observations. □

Definition 6.2. *The optimizing values of* θ *and* t *in the expression for* $\mathcal{I}(q)$ *are denoted by* θ_c *and* t_c *respectively. These are called the critical point of the queueing system.* □

Thus

$$\begin{aligned}
\mathcal{I}(q) &= \inf_{t \in \mathbb{R}^+} \sup_{\theta \geq 0} \left\{ \theta\left(q + \widetilde{C}t\right) - t\Upsilon(\theta)\right\} \\
&= \inf_{t \in \mathbb{R}^+} \sup_{\theta \geq 0} \left\{ \theta\left(q + \widetilde{C}t\right) - \theta t\vartheta(\theta, t)\right\} \\
&= \theta_c\left(q + \widetilde{C}t_c\right) - \theta_c t_c \vartheta(\theta_c, t_c)
\end{aligned}$$

The optimizing value t_c can be interpreted as the possible length of the busy period prior to buffer overflow. That is, the buffer fills up to level qy in time $t_c y$ units. If the optimizing value θ_c is small, then the effective bandwidth is closer to its mean value. A large value of θ_c means that the effective bandwidth is closer to its maximum (peak) value.

Usually in practice, single server queues have a finite waiting space. Therefore, there is a finite probability that the arriving traffic overflows. The probability of buffer overflow $P\left(\text{overflow}\right)$, due to a finite buffer of size qy is related to $\mathcal{I}\left(\cdot\right)$ as $\ln P\left(\text{overflow}\right) \sim -y\mathcal{I}\left(q\right)$. A more refined estimate of the buffer overflow probability can be obtained by using the Bahadur-Rao theorem. As $y \to \infty$

$$P\left(\text{overflow}\right) \sim \frac{1}{\sqrt{2\pi y}\theta_c\sigma_c} \exp\left(-y\mathcal{I}\left(q\right)\right)$$

$$\sigma_c^2 = \frac{\partial^2}{\partial\theta^2} \ln \mathcal{E}\left(e^{\theta\widetilde{A}(t)}\right)$$

where σ_c^2 is evaluated at $\theta = \theta_c$, and $t = t_c$ (critical point).

Corollary 6.1. *Consider a single server queue, in which the service time is also a random variable $\widetilde{C}_t$. The independent sequence of random variables $\widetilde{A}_t$'s and $\widetilde{C}_t$'s are also independent of each other. Then in the above theorem, replace the random variable $\widetilde{A}_t$ by $\left(\widetilde{A}_t - \widetilde{C}_t\right)$, and $\widetilde{C}$ by 0. If $\Upsilon_{\widetilde{A}}\left(\cdot\right)$ and $\Upsilon_{\widetilde{C}}\left(\cdot\right)$ are cumulants of $\widetilde{A}_t$ and $\widetilde{C}_t$ respectively, then*

$$\Upsilon\left(\theta\right) = \left\{\Upsilon_{\widetilde{A}}\left(\theta\right) + \Upsilon_{\widetilde{C}}\left(-\theta\right)\right\} \tag{6.28a}$$

$$\ell\left(x\right) = \inf_{y}\left\{\ell_{\widetilde{A}}\left(y\right) + \ell_{\widetilde{C}}\left(y - x\right)\right\} \tag{6.28b}$$

where $\ell\left(\cdot\right), \ell_{\widetilde{A}}\left(\cdot\right)$, and $\ell_{\widetilde{C}}\left(\cdot\right)$ are large deviation rate functions of $\left(\widetilde{A}_t - \widetilde{C}_t\right)$, $\widetilde{A}_t$, and $\widetilde{C}_t$ respectively. □

Example 6.8. Consider a continuous-time process such that its cumulative arrival process, $\widetilde{A}\left(t\right), t \in \mathbb{R}_0^+$ is

$$\widetilde{A}\left(t\right) = \left\{\lambda t + \widetilde{B}\left(t\right)\right\}$$

In this process $\widetilde{B}\left(t\right), t \in \mathbb{R}_0^+$ is a fractional Brownian motion process, with Hurst parameter H, where $H \in (0, 1)$. This process is normally distributed with zero mean and variance $\sigma^2 t^{2H}$. Its effective bandwidth is given by

$$\vartheta\left(\theta, t\right) = \lambda + \frac{\sigma^2\theta}{2} t^{2H-1}$$

Assume that $\lambda < \widetilde{C}$ for the queue to be in equilibrium. The critical point of this queueing system is at t_c and θ_c where

$$t_c = \frac{H}{(1-H)} \frac{q}{\left(\widetilde{C} - \lambda\right)} > 0$$

$$\theta_c = \frac{q + t_c\left(\widetilde{C} - \lambda\right)}{\sigma^2 t_c^{2H}} > 0$$

Using the relationship

$$\mathcal{I}\left(q\right) = \theta_c\left(q + \widetilde{C}t_c\right) - \theta_c t_c \vartheta\left(\theta_c, t_c\right)$$

we obtain

$$\mathcal{I}(q) = \frac{1}{2}\left\{\frac{q^{(1-H)}\left(\widetilde{C}-\lambda\right)^{H}}{\sigma H^{H}(1-H)^{(1-H)}}\right\}^{2}$$

□

The analysis discussed in this subsection can also be extended to model a queue shared by several independent sources.

Reference Notes

Study of queueing theory is useful in evaluating response time metric and buffer resources. Moreover, the Internet can be regarded as a giant queueing network.

Fundamentals of queueing theory are covered in several books. These include books by Allen (1990), Bolch, Greiner, de Meer, and Trivedi (1998), Cooper (1981), Gross, and Harris (1985), Kleinrock (1975), Medhi (2003), and Ross (2010). The books by Jain (1991), Kant (1992), and Lavenberg (1983) cover the application of queueing theory to computer performance modeling. The Lagrange-series solution of the $G/M/1$ queue is due to Bhatnagar (1998).

The theory of effective bandwidths has been developed by several researchers. The notes in this chapter are based upon the publications of Hui (1990), Kelly (1996), Courcoubetis, Siris, and Stamoulis (1999), Ganesh, O'Connell, and Wischik (2004), Kumar, Manjunath, and Kuri (2004), and Likhanov, and Mazumdar (1999). Rabinovitch's master's thesis (2000) provides a comprehensive introduction to the subject of effective bandwidth.

Problems

1. Prove that the minimum of several independent and exponentially distributed random variables is an exponentially distributed random variable.
   Hint: Let $T_1, T_2, \ldots, T_n$ be independent exponentially distributed random variables with positive parameters $\lambda_1, \lambda_2, \ldots, \lambda_n$ respectively. Let

   $$U = \min\{T_1, T_2, \ldots, T_n\}$$

   For any $t \geq 0$,

   $$\begin{aligned} P(U > t) &= P(T_1 > t, T_2 > t, \ldots, T_n > t) \\ &= P(T_1 > t)\, P(T_2 > t) \ldots P(T_n > t) \\ &= e^{-\lambda_1 t} e^{-\lambda_2 t} \ldots e^{-\lambda_n t} = e^{-\sum_{i=1}^{n} \lambda_i t} \end{aligned}$$

   Thus U is exponentially distributed with parameter $\lambda = \sum_{i=1}^{n} \lambda_i$.

2. Prove that for an $M/M/1$ queue $P(\Omega_q \le t) = \left(1 - \zeta e^{-t/W}\right)$, for $t \ge 0$.
   Hint:

$$
\begin{aligned}
P(\Omega_q \le t) &= P(\Omega_q = 0) + P(0 < \Omega_q \le t) \\
P(\Omega_q = 0) &= p_0 = (1 - \zeta) \\
P(0 < \Omega_q \le t) &= \sum_{n=1}^{\infty} p_n P(Y_n \le t) = \zeta \left(1 - e^{-t/W}\right)
\end{aligned}
$$

   where $W^{-1} = \mu(1 - \zeta)$, and Y_n is equal to the sum of n independent exponentially distributed random variables, each with parameter μ.
3. Prove that the Erlang's $\mathcal{B}$ and $\mathcal{C}$ functions are related by

$$
\begin{aligned}
\mathcal{C}(c,\zeta) &= \frac{\mathcal{B}(c,\zeta)}{\{1 - \rho(1 - \mathcal{B}(c,\zeta))\}} \\
\mathcal{C}(c,\zeta) &> \mathcal{B}(c,\zeta)
\end{aligned}
$$

   Hint: The inequality follows by observing that $\rho \in (0,1)$ and $\mathcal{B}(c,\zeta) < 1$.
4. Evaluate $F_{\Omega}(t)$ for an $M/M/c$ queueing system.
5. Derive the following expressions for an $M/M/c/K$ queueing system.
   (a) The average queue length L.
   (b) The cumulative distribution function $P(\Omega \le t)$, for $t \in \mathbb{R}_0^+$.
   (c) The cumulative distribution function $P(\Omega_q \le t)$, for $t \in \mathbb{R}_0^+$.
6. Evaluate the performance of a $G/M/1$ queue with Paretian interarrival times.
   Hint: See Bhatnagar (1998). Basic properties of a random variable A with bounded Pareto distribution are first described. Let the probability density function of A be $f_A(\cdot)$. Then

$$
f_A(x) = \begin{cases} \widetilde{K}\beta a_l^{\beta} x^{-(\beta+1)}, & a_l \le x \le a_u \\ 0, & \text{otherwise} \end{cases}
$$

$$
\widetilde{K} = \frac{1}{(1 - \nu^{\beta})}
$$

   In the above equation, $\beta > 0$ is the shape parameter, a_l and a_u are the lower and upper location parameters such that, $0 < a_l < a_u < \infty$, and $\nu = a_l/a_u$. Note that $0 < \nu < 1$. The range of valid values of β is determined by the value of ν and the requirement that $\widetilde{K}$ be positive. The jth moment of the random variable A is $\mathcal{E}\left(A^j\right)$, $j = 1, 2, 3, \ldots$.

$$
\mathcal{E}\left(A^j\right) = \begin{cases} a_l^j \dfrac{\beta}{(\beta - j)} \dfrac{\left(1 - \nu^{(\beta - j)}\right)}{(1 - \nu^{\beta})}, & \beta \ne j \\ a_l^{\beta} \dfrac{\beta}{(\nu^{\beta} - 1)} \ln \nu, & \beta = j \end{cases}
$$

   The moment $\mathcal{E}\left(A^j\right)$, exists for appropriate values of j, β, and ν. If a_u tends to infinity, then the random variable A has the classical Pareto distribution. In this case the first and second moments exist only for restricted range of values of β. The Laplace transform of $f_A(x)$ is given by $\widehat{f}_A(s)$.

$$\widehat{f}_A(s) = \widetilde{K} \beta a_l^\beta s^\beta B_{\beta+1}(s)$$
$$B_w(s) = \int_{a_l s}^{a_u s} e^{-x} x^{-w} dx$$

Let $\widehat{f}_A^{(m)}(s)$ be the mth derivative of $\widehat{f}_A(s)$ with respect to s. Then

$$\widehat{f}_A^{(m)}(s) = (-1)^m \widetilde{K} \beta a_l^\beta s^{(\beta-m)} B_{\beta+1-m}(s), \quad m = 1, 2, 3, \ldots$$

Then as per the algorithm outlined, $\widehat{f}_A^{(m)}(\mu), 1 \leq m \leq (J+1)$ have to be computed efficiently. This is done via computing $B_w(\mu)$ for positive and negative values of w. Let $\alpha_l = a_l\mu$, and $\alpha_u = a_u\mu$. Then

$$B_0(\mu) = \left(e^{-\alpha_l} - e^{-\alpha_u}\right)$$
$$B_1(\mu) = E_1(\alpha_l) - E_1(\alpha_u)$$

where $E_1(x)$ is the exponential integral.

$$E_1(x) = \int_x^\infty \frac{e^{-t}}{t} dt, \quad x > 0$$

The following two representations of $E_1(x)$ are used in the computations.

$$E_1(x) = -\gamma - \ln x - \sum_{j\geq 1} \frac{(-x)^j}{jj!}$$
$$E_1(x) = \cfrac{e^{-x}}{x + \cfrac{1}{1 + \cfrac{1}{x + \cfrac{2}{1 + \cfrac{2}{x + \ldots}}}}}$$

where Euler's constant $\gamma = 0.5772156649\ldots$. The series representation of $E_1(x)$ is used when the value of x is small ($0 < x \leq 1$). The continued fraction representation of $E_1(x)$ is used to compute $E_1(x)$ for larger values of x (that is for $x > 1$). In actual computation of $E_1(x)$ via continued fractions, a simple recursive procedure can be used to compute the continuants of the continued fraction. Using the expression for $B_w(s)$ results in

$$B_w(\mu) = -wB_{w+1}(\mu) + \left(\alpha_l^{-w} e^{-\alpha_l} - \alpha_u^{-w} e^{-\alpha_u}\right)$$

The above equation is also useful in the following form.

$$B_{w+1}(\mu) = -\frac{1}{w}\left(B_w(\mu) + \alpha_u^{-w} e^{-\alpha_u} - \alpha_l^{-w} e^{-\alpha_l}\right), \quad w \neq 0$$

Another representation for the evaluation of $B_w(\mu)$ for noninteger values of w is necessary. The lower incomplete gamma function $\gamma(\cdot, \cdot)$ is

$$\gamma(b, x) = \int_0^x e^{-t} t^{b-1} dt, \quad b > 0$$
$$\gamma(b, x) = x^b e^{-x} \sum_{n\geq 0} \frac{\Gamma(b) x^n}{\Gamma(b+n+1)}, \quad b > 0$$

Then

$$B_w(\mu) = \gamma(1-w, \alpha_u) - \gamma(1-w, \alpha_l), \qquad w < 1$$

Steps to compute $B_w(\mu)$ for all possible values of w (positive or negative) are next outlined. Note that $w = (\beta + 1 - m)$, where $0 \le m \le J$. J is the order of the highest derivative. If β is an integer, then w is an integer. However, if β has nonintegral values, then w has nonintegral values.

(a) Since $\beta > 0$, let β be a real number with integral values. Then $\widehat{f}_A(\mu)$ can be computed by computing $B_1(\mu), B_2(\mu), \ldots B_{\beta+1}(\mu)$ recursively. If $J \le (\beta+1)$, then the $B_w(\mu)$'s required in the computations of the derivatives $\widehat{f}_A^{(m)}(\mu)$'s are already computed while computing $B_{\beta+1}(\mu)$. However if $J > (\beta+1)$, then w can possibly take negative values, then $B_w(\mu)$'s can also be computed recursively.

(b) If β is a real number with a nonintegral value, then first compute $B_{w_{frac}}(\mu)$, where $w_{frac} = (\beta - \lfloor\beta\rfloor) < 1$, via series expansion of $\gamma(b, x)$. Then $B_{\beta+1}(\mu)$ required in the computation of $\widehat{f}_A(\mu)$ can be computed by using $B_{w_{frac}}(\mu)$ as an initial value in the recursion. If $J < \beta$, then the $B_w(\mu)$'s required in the computations of the derivatives $\widehat{f}_A^{(m)}(\mu)$'s are already computed while computing $B_{\beta+1}(\mu)$. However if $J > \beta$, then w can possibly take negative values. In this case, the $B_w(\mu)$'s can also be generated and using $B_{w_{frac}}(\mu)$ as an initial value in the recursion.

Finally, the b_j, $0 \le j \le J$ required in the computation of σ are

$$b_j = \frac{\widetilde{K}\beta\alpha_l^{\beta}}{j!} B_{\beta+1-j}(\mu), \qquad 0 \le j \le J$$

These can be computed recursively as follows.

$$b_0 = \widetilde{K}\beta\alpha_l^{\beta} B_{\beta+1}(\mu),$$

$$b_j = \frac{1}{j}\frac{B_{\beta+1-j}(\mu)}{B_{\beta+2-j}(\mu)} b_{j-1}, \quad 1 \le j \le J$$

Note that, for large values of α_l and $-w$, $B_w(\mu)$ is very large. However, in order to compute b_j for $1 \le j \le J$, only the ratio of $B_{\beta+1-j}(\mu)$ and $B_{\beta+2-j}(\mu)$ has to be computed. Therefore the following scheme is adapted to compute the b_j's. Define

$$D_{\beta+1-j}(\mu) = \frac{B_{\beta+1-j}(\mu)}{B_{\beta+2-j}(\mu)}$$

$$b_j = \frac{1}{j} D_{\beta+1-j}(\mu) b_{j-1}, \quad 1 \le j \le J$$

Therefore

$$D_{\beta+1-j}(\mu) = -(\beta+1-j) + C_{\beta+1-j}(\mu), \ 1 \le j \le J$$

where

$$C_\beta(\mu) = \frac{\left(1 - e^{G_\beta(\mu)}\right)}{\alpha_l^{\beta} e^{\alpha_l} B_{\beta+1}(\mu)}$$

$$C_{\beta-j}(\mu) = \frac{\alpha_l\left(1 - e^{G_{\beta-j}(\mu)}\right)}{D_{\beta+1-j}(\mu)\left(1 - e^{G_{\beta+1-j}(\mu)}\right)} C_{\beta+1-j}(\mu), \quad 1 \le j \le (J-1)$$

$$G_{\beta-j}(\mu) = (\alpha_l - \alpha_u + (\beta - j)\ln\nu), \qquad 0 \le j \le (J-1)$$

After computing b_j, for $0 \leq j \leq J$, σ is evaluated via the outlined algorithm. This algorithm is used to plot ρ versus σ in Figure 6.3. Figure 6.4 has a plot of ρ versus $\omega_q = \mu \mathcal{E}(\Omega_q)$. In these plots, $a_l = 1$, $a_u = 10$, and $\beta = 2.0$.

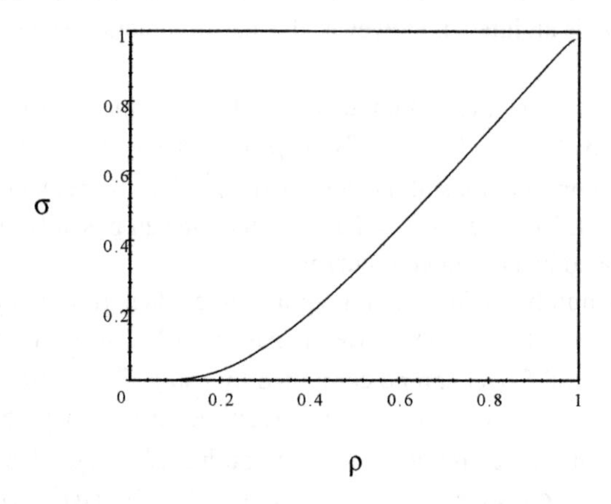

Figure 6.3. Interarrival times with bounded Pareto distribution. ρ versus σ.

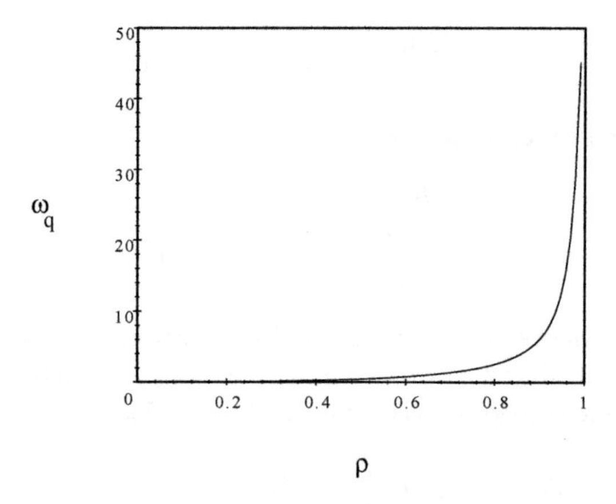

Figure 6.4. Interarrival times with bounded Pareto distribution. ρ versus ω_q.

7. The random variables A and B are defined for only nonnegative values. Therefore their respective probability density functions $f_A(t)$ and $f_B(t)$ are defined for $t \geq 0$. Let the single-sided Laplace transform of these density functions be $\widehat{f}_A(s)$ and $\widehat{f}_B(s)$ respectively. The probability density function of the random variable $C = (B - A)$ is $f_C(t)$ for $t \in \mathbb{R}$. Let the two-sided Laplace transform of $f_C(t)$ be $\widetilde{f}_C(s)$. Prove that $\widetilde{f}_C(s) = \widehat{f}_A(-s)\widehat{f}_B(s)$.
   Hint:
$$f_C(t) = \int_0^{\infty} f_B(t+x) f_A(x)\, dx$$
   Then
$$\widetilde{f}_C(s) = \int_{-\infty}^{\infty} f_C(t) e^{-st} dt$$

$$= \int_{-\infty}^{\infty} \int_{0}^{\infty} f_B(t+x) f_A(x) dx e^{-st} dt$$
$$= \int_{0}^{\infty} f_A(x) e^{sx} \int_{-\infty}^{\infty} f_B(t+x) e^{-st-sx} dt dx$$

Substituting $(t+x)=u$, results in

$$\widetilde{f_C}(s) = \int_{0}^{\infty} f_A(x) e^{sx} \int_{-\infty}^{\infty} f_B(u) e^{-su} du dx$$

Since $f_B(t)=0$ for $t<0$ the result follows.

8. In a $G/G/1$ queue prove that

$$K_r = \lim_{s\to 0} \frac{\Theta(s)}{s} = \lim_{s\to 0} \frac{d\Theta(s)}{ds} = F_{\Omega_q}(0_+)$$

Hint: See Kleinrock (1975). Observe that $s\widehat{F}_{\Omega_q}(s) = \widehat{f}_{\Omega_q}(s)$ implies $\lim_{s\to 0} s\widehat{F}_{\Omega_q}(s) = 1$. Therefore $\lim_{s\to 0} sK_r/\Theta(s) = 1$. This in turn yields

$$K_r = \lim_{s\to 0} \frac{\Theta(s)}{s}$$

Using the above equation, the Taylor's series expansion of the function $\Theta(s)$ about $s=0$, and noticing that $\Theta(0)=0$ by construction (because $\left\{\widehat{f}_A(0)\widehat{f}_B(0)-1\right\}=0$) results in

$$K_r = \lim_{s\to 0} \frac{d\Theta(s)}{ds}$$

Finally, $K_r = \widehat{F}_{\Omega_q}(s)\Theta(s)$, and the result $\lim_{|s|\to\infty} \Theta(s)/s = 1$ yields

$$K_r = \lim_{s\to\infty} s\widehat{F}_{\Omega_q}(s) = \lim_{s\to\infty} s\int_{0_-}^{\infty} e^{-sx} F_{\Omega_q}(x) dx$$
$$= \lim_{s\to\infty} \int_{0_-}^{\infty} e^{-y} F_{\Omega_q}\left(\frac{y}{s}\right) dy = F_{\Omega_q}(0_+)$$

9. Establish an upper bound for the average waiting time of a customer in a $G/G/1$ queue at steady-state. Prove that

$$W_q \le \frac{\lambda\left(\sigma_A^2+\sigma_B^2\right)}{2(1-\rho)}$$

Hint: Note that

$$W_q = -\frac{\mathcal{E}(I^2)}{2\mathcal{E}(I)} - \frac{\mathcal{E}(C^2)}{2\mathcal{E}(C)} = -\frac{Var(I)+\{\mathcal{E}(I)\}^2}{2\mathcal{E}(I)} - \frac{\mathcal{E}(C^2)}{2\mathcal{E}(C)}$$
$$\le -\frac{\{\mathcal{E}(I)\}^2}{2\mathcal{E}(I)} - \frac{Var(C)+\{\mathcal{E}(C)\}^2}{2\mathcal{E}(C)} = -\frac{\mathcal{E}(I)}{2} - \frac{Var(C)}{2\mathcal{E}(C)} - \frac{\mathcal{E}(C)}{2}$$

Using the observation that $\{\mathcal{E}(I)+\mathcal{E}(C)\} = \mathcal{E}(I)(1-\varphi_0) \ge 0$ we have

$$W_q \le -\frac{Var(C)}{2\mathcal{E}(C)}$$

The result follows by noting that $C=(B-A)$, and $\mathcal{E}(C) = \mathcal{E}(B)-\mathcal{E}(A) = -\lambda^{-1}(1-\rho)$.

10. Prove the following properties of effective bandwidth of discrete-time processes with stationary increments.
    (a) If the process $\left\{\widetilde{A}(t), t \in \mathbb{P}\right\}$ has independent increments, then prove that effective bandwidth $\vartheta(\theta, t)$ does not depend upon t.
    (b) Let $\widetilde{A}_i(t)$ be a traffic stream, with effective bandwidth $\vartheta_i(\theta, t)$, where $1 \le i \le n$ and $n \in \mathbb{P}$. Also the traffic stream $\widetilde{A}_i(t)$ is independent of traffic stream $\widetilde{A}_j(t)$ for $i \neq j$, and $1 \le i, j \le n$. If the effective bandwidth of $\widetilde{A}(t) = \sum_{i=1}^n \widetilde{A}_i(t)$ is $\vartheta(\theta, t)$ then

$$\vartheta(\theta, t) = \sum_{i=1}^{n} \vartheta_i(\theta, t)$$

    (c) If the parameter t is fixed, then the effective bandwidth $\vartheta(\theta, t)$ lies between the mean and peak of the arrival rate measured over an interval of length t.

All of the above properties essentially follow from the definition of effective bandwidth.

(a) Hint: See Rabinovitch (2000).

$$\begin{aligned}\vartheta(\theta, t) &= \frac{1}{\theta t} \ln \mathcal{E}\left(e^{\theta \widetilde{A}(t)}\right) = \frac{1}{\theta t} \ln \mathcal{E}\left(e^{\theta \sum_{i=1}^{t} \widetilde{A}_i}\right) \\ &= \frac{1}{\theta t} \ln \mathcal{E}\left(\prod_{i=1}^{t} e^{\theta \widetilde{A}_i}\right) = \frac{1}{\theta t} \ln \prod_{i=1}^{t} \mathcal{E}\left(e^{\theta \widetilde{A}_i}\right)\end{aligned}$$

The last equality is due to independent increments.

$$\vartheta(\theta, t) = \frac{1}{\theta t} \sum_{i=1}^{t} \ln \mathcal{E}\left(e^{\theta \widetilde{A}_i}\right) = \frac{t}{\theta t} \ln \mathcal{E}\left(e^{\theta \widetilde{A}_1}\right)$$

The last equality is due to stationarity of increments. Finally

$$\vartheta(\theta, t) = \frac{1}{\theta} \ln \mathcal{E}\left(e^{\theta \widetilde{A}_1}\right) = \vartheta(\theta, 1)$$

Thus $\vartheta(\theta, t)$ is independent of t.

(b) Hint: See Rabinovitch (2000).

$$\begin{aligned}\vartheta(\theta, t) &= \frac{1}{\theta t} \ln \mathcal{E}\left(e^{\theta \widetilde{A}(t)}\right) = \frac{1}{\theta t} \ln \mathcal{E}\left(e^{\theta \sum_{i=1}^{n} \widetilde{A}_i(t)}\right) \\ &= \frac{1}{\theta t} \ln \mathcal{E}\left(\prod_{i=1}^{n} e^{\theta \widetilde{A}_i(t)}\right) = \frac{1}{\theta t} \ln\left(\prod_{i=1}^{n} \mathcal{E}\left(e^{\theta \widetilde{A}_i(t)}\right)\right) \\ &= \frac{1}{\theta t} \sum_{i=1}^{n} \ln\left(\mathcal{E}\left(e^{\theta \widetilde{A}_i(t)}\right)\right) = \sum_{i=1}^{n} \vartheta_i(\theta, t)\end{aligned}$$

(c) Hint: See Kumar, Manjunath, and Kuri (2004).
Using Jensen's inequality, we obtain

$$\vartheta(\theta, t) = \frac{1}{\theta t} \ln \mathcal{E}\left(e^{\theta \widetilde{A}(t)}\right) \ge \frac{1}{\theta t} \ln e^{\theta \mathcal{E}\left(\widetilde{A}(t)\right)} = \frac{\mathcal{E}\left(\widetilde{A}(t)\right)}{t}$$

Let R be the peak rate of the source. Therefore $\widetilde{A}(t) \le tR$. For $\theta > 0$

$$\vartheta(\theta, t) = \frac{1}{\theta t} \ln \mathcal{E}\left(e^{\theta \widetilde{A}(t)}\right) \le \frac{1}{\theta t} \ln \mathcal{E}\left(e^{\theta t R}\right) = R$$

11. Prove the following observations about the application of large deviation technique to a single server queue with large buffer.
Let $\mathcal{E}\left(\widetilde{A}\right) < \widetilde{C}$, $y \in \mathbb{R}$, $q > 0$, $\ell(x) = \sup_{\theta \in \mathbb{R}} (\theta x - \Upsilon(\theta))$, and $\widetilde{Q} = \sup_{t \geq 0} \left\{\widetilde{S}_t - \widetilde{C}t\right\}$.
(a) $\limsup_{y \to \infty} \frac{1}{y} \ln P\left(\widetilde{Q}/y > q\right) \leq -q \sup\left\{\theta > 0 \mid \Upsilon(\theta) < \theta\widetilde{C}\right\}$.
(b) $\liminf_{y \to \infty} \frac{1}{y} \ln P\left(\widetilde{Q}/y > q\right) \geq -\inf_{t \in \mathbb{R}^+} \left\{t\ell\left(\widetilde{C} + q/t\right)\right\}$.
(c) If $\mathcal{I}(q) = \inf_{t \in \mathbb{R}^+} \left\{t\ell\left(\widetilde{C} + q/t\right)\right\}$, then

$$\mathcal{I}(q) = \inf_{t \in \mathbb{R}^+} \sup_{\theta \geq 0} \left\{\theta\left(q + \widetilde{C}t\right) - t\Upsilon(\theta)\right\} = q \sup\left\{\theta > 0 \mid \Upsilon(\theta) < \theta\widetilde{C}\right\}$$

Hint: We essentially follow the elegant proof given in Ganesh, O'Connell, and Wischik (2004).
(a) This result can be proved by using Chernoff's bound.

$$P\left(\widetilde{Q} > yq\right) = P\left(\sup_{t \geq 0} \left\{\widetilde{S}_t - \widetilde{C}t\right\} > yq\right) \leq \sum_{t \geq 0} P\left(\widetilde{S}_t - \widetilde{C}t > yq\right)$$

Use of Chernoff's bound results in

$$P\left(\widetilde{Q} > yq\right) \leq e^{-\theta yq} \sum_{t \geq 0} e^{-t\left(\widetilde{C}\theta - \Upsilon(\theta)\right)}, \quad \theta > 0$$

If $\Upsilon(\theta) < \theta\widetilde{C}$, we obtain

$$\limsup_{y \to \infty} \frac{1}{y} \ln P\left(\widetilde{Q} > yq\right) \leq -q\theta$$

In the next step, taking the supremum over all $\theta > 0$ yields the stated result.
(b) Let $t \in \mathbb{R}^+$ be a fixed number. Also if $x \in \mathbb{R}$, then $\lceil x \rceil \in \mathbb{Z}$ is defined as the smallest integer greater than or equal to x. Observe that $\widetilde{Q} > yq$ if and only if there exists u such that

$$\left(\widetilde{S}_u - \widetilde{C}u\right) > yq$$

Thus

$$P\left(\widetilde{Q} > yq\right) = P\left(\widetilde{S}_u - \widetilde{C}u > yq\right) \geq P\left(\widetilde{S}_{\lceil yt \rceil} - \widetilde{C}\lceil yt \rceil > yq\right)$$

Consequently

$$\begin{aligned}
\liminf_{y \to \infty} \frac{1}{y} \ln P\left(\widetilde{Q} > yq\right) &\geq \liminf_{y \to \infty} \frac{1}{y} \ln P\left(\widetilde{S}_{\lceil yt \rceil} - \widetilde{C}\lceil yt \rceil > yq\right) \\
&\geq \liminf_{y \to \infty} \frac{t}{\lceil yt \rceil} \ln P\left(\widetilde{S}_{\lceil yt \rceil} - \widetilde{C}\lceil yt \rceil > \frac{\lceil yt \rceil}{t} q\right)
\end{aligned}$$

The last inequality follows from the observation that $y \leq \lceil yt \rceil / t$. Let $n \triangleq \lceil yt \rceil$, then

$$\begin{aligned}
\liminf_{y \to \infty} \frac{1}{y} \ln P\left(\widetilde{Q} > yq\right) &\geq t \liminf_{n \to \infty} \frac{1}{n} \ln P\left(\widetilde{S}_n - \widetilde{C}n > nq/t\right) \\
&= t \liminf_{n \to \infty} \frac{1}{n} \ln P\left(\widetilde{S}_n/n > \widetilde{C} + q/t\right) \\
&\geq -t\ell\left(\widetilde{C} + q/t\right)
\end{aligned}$$

The last inequality follows from Cramer's theorem of large deviation. As t was selected to be an arbitrary positive real number, we have

$$\liminf_{y\to\infty}\frac{1}{y}\ln P\left(\widetilde{Q}>yq\right)\geq-\inf_{t\in\mathbb{R}^+}\left\{t\ell\left(\widetilde{C}+q/t\right)\right\}$$

(c) We initially prove that $\mathcal{I}(q)=\inf_{t\in\mathbb{R}^+}\sup_{\theta\geq0}\left\{\theta\left(q+\widetilde{C}t\right)-t\Upsilon(\theta)\right\}$.

$$\begin{aligned}\mathcal{I}(q)&=\inf_{t\in\mathbb{R}^+}\left\{t\ell\left(\widetilde{C}+q/t\right)\right\}=\inf_{t\in\mathbb{R}^+}\left\{t\sup_{\theta\in\mathbb{R}}\left(\theta\left(\widetilde{C}+q/t\right)-\Upsilon(\theta)\right)\right\}\\&=\inf_{t\in\mathbb{R}^+}\sup_{\theta\in\mathbb{R}}\left\{\theta\left(q+\widetilde{C}t\right)-t\Upsilon(\theta)\right\}\end{aligned}$$

As

$$\mathcal{E}\left(\widetilde{A}\right)<\widetilde{C}\Rightarrow\mathcal{E}\left(\widetilde{A}\right)<\frac{\left(q+\widetilde{C}t\right)}{t}$$

and using the property of the large deviation rate function, the supremum can be restricted to be over $\theta\geq0$. This yields the required result.
It is next proved that

$$\inf_{t\in\mathbb{R}^+}\sup_{\theta\geq0}\left\{\theta\left(q+\widetilde{C}t\right)-t\Upsilon(\theta)\right\}=q\sup\left\{\theta>0\mid\Upsilon(\theta)<\theta\widetilde{C}\right\}$$

This is proved in two parts.
Part (i): We prove that

$$\inf_{t\in\mathbb{R}^+}\sup_{\theta\geq0}\left\{\theta\left(q+\widetilde{C}t\right)-t\Upsilon(\theta)\right\}\geq q\sup\left\{\theta>0\mid\Upsilon(\theta)<\theta\widetilde{C}\right\}$$

Observe that for $t\in\mathbb{R}^+,\theta>0$, and $\Upsilon(\theta)<\theta\widetilde{C}$,

$$\theta\left(q+\widetilde{C}t\right)-t\Upsilon(\theta)=\theta q+t\left(\theta\widetilde{C}-\Upsilon(\theta)\right)\geq\theta q$$

Therefore taking the supremum over θ yields

$$\sup_{\theta>0,\Upsilon(\theta)<\theta\widetilde{C}}\left\{\theta\left(q+\widetilde{C}t\right)-t\Upsilon(\theta)\right\}\geq q\sup\left\{\theta>0\mid\Upsilon(\theta)<\theta\widetilde{C}\right\}$$

As

$$\sup_{\theta\geq0}\left\{\theta\left(q+\widetilde{C}t\right)-t\Upsilon(\theta)\right\}\geq\sup_{\theta>0,\Upsilon(\theta)<\theta\widetilde{C}}\left\{\theta\left(q+\widetilde{C}t\right)-t\Upsilon(\theta)\right\}$$

we have

$$\sup_{\theta\geq0}\left\{\theta\left(q+\widetilde{C}t\right)-t\Upsilon(\theta)\right\}\geq q\sup\left\{\theta>0\mid\Upsilon(\theta)<\theta\widetilde{C}\right\}$$

Note that the right-hand side of the above inequality is independent of t. Therefore the desired inequality follows by taking the infimum over $t\in\mathbb{R}^+$.
Part (ii): We prove that

$$\inf_{t\in\mathbb{R}^+}\sup_{\theta\geq0}\left\{\theta\left(q+\widetilde{C}t\right)-t\Upsilon(\theta)\right\}\leq q\sup\left\{\theta>0\mid\Upsilon(\theta)<\theta\widetilde{C}\right\}$$

Assume that

$$\theta_c = \sup\left\{\theta > 0 \mid \Upsilon(\theta) < \theta\widetilde{C}\right\}$$

If $\theta_c = \infty$, then the inequality is immediate. Therefore assume that $\theta_c < \infty$. It is known that the cumulant $\Upsilon(\theta)$ is convex, finite, continuous and differentiable everywhere. This implies $\Upsilon(\theta_c) = \theta_c\widetilde{C}$ and $\Upsilon'(\theta_c) > \widetilde{C}$, where $\Upsilon'(\cdot)$ is the first derivative of $\Upsilon(\cdot)$. Again using the fact that $\Upsilon(\theta)$ is convex in θ, we have

$$\Upsilon(\theta) \geq \theta_c\widetilde{C} + \Upsilon'(\theta_c)(\theta - \theta_c)$$

Therefore

$$\begin{aligned}
&\inf_{t\in\mathbb{R}^+}\sup_{\theta\geq 0}\left\{\theta\left(q+\widetilde{C}t\right) - t\Upsilon(\theta)\right\} \\
&\leq \inf_{t\in\mathbb{R}^+}\sup_{\theta\geq 0}\left\{\theta\left(q+\widetilde{C}t\right) - t\left(\theta_c\widetilde{C} + \Upsilon'(\theta_c)(\theta-\theta_c)\right)\right\} \\
&= \inf_{t\in\mathbb{R}^+}\sup_{\theta\geq 0}\left\{\theta\left(q - t\left(\Upsilon'(\theta_c) - \widetilde{C}\right)\right) + \theta_c t\left(\Upsilon'(\theta_c) - \widetilde{C}\right)\right\}
\end{aligned}$$

Next perform the optimization over $\theta \geq 0$. This yields

$$\begin{aligned}
&\inf_{t\in\mathbb{R}^+}\sup_{\theta\geq 0}\left\{\theta\left(q+\widetilde{C}t\right) - t\Upsilon(\theta)\right\} \\
&\leq \inf_{t\in\mathbb{R}^+}\begin{cases}\infty, & \text{if } t < q/\left(\Upsilon'(\theta_c) - \widetilde{C}\right) \\ \theta_c t\left(\Upsilon'(\theta_c) - \widetilde{C}\right), & \text{if } t \geq q/\left(\Upsilon'(\theta_c) - \widetilde{C}\right)\end{cases} \\
&= \theta_c q = q\sup\left\{\theta > 0 \mid \Upsilon(\theta) < \theta\widetilde{C}\right\}
\end{aligned}$$

This proves part (ii). Parts (i) and (ii) together imply

$$\inf_{t\in\mathbb{R}^+}\sup_{\theta\geq 0}\left\{\theta\left(q+\widetilde{C}t\right) - t\Upsilon(\theta)\right\} = q\sup\left\{\theta > 0 \mid \Upsilon(\theta) < \theta\widetilde{C}\right\}$$

References

1. Allen, A., 1990. *Probability, Statistics, and Queueing Theory*, Second Edition, Academic Press, New York, New York.
2. Bhatnagar, N., 1998. "A Lagrange Series Solution of a $G/M/1$ Queue," unpublished.
3. Bolch, G., Greiner, S., De Meer, H., and Trivedi, K. S., 1998. *Queueing Networks and Markov Chains*, John Wiley & Sons, Inc., New York, New York.
4. Bucklew, J. A., 1990. *Large Deviation Techniques in Decision, Simulation, and Estimation,* John Wiley & Sons, Inc., New York, New York.
5. Cooper, R. B., 1981. *Introduction to Queueing Theory*, Second Edition, North-Holland, New York, New York.

6. Courcoubetis, C., Siris, V. A., and Stamoulis, G. D., 1999. "Application of the Many Source Asymptotic and Effective Bandwidths to Traffic Engineering," Telecommunication Systems, Vol. 12, Issues 2-3, pp. 167-191.
7. Ganesh, A., O'Connell, N., and Wischik, D., 2004. *Big Queues,* Vol. 1838 of Lecture Notes in Mathematics, Springer-Verlag, Berlin, Germany.
8. Gross, D., and Harris, C. M., 1985. *Fundamentals of Queueing Theory,* Second Edition, John Wiley & Sons, Inc., New York, New York.
9. Hollander, F. D., 2000. *Large Deviations,* American Mathematical Society, Providence, Rhode Island.
10. Hui, J. Y. N., 1990. *Switching and Traffic Theory for Integrated Broadband Networks*, Kluwer Academic Publishers, Norwell, Massachusetts.
11. Jain, R., 1991. *The Art of Computer Systems Performance Analysis*, John Wiley & Sons, Inc., New York, New York.
12. Kant, K., 1992. *Introduction to Computer System Performance Evaluation,* McGraw-Hill Book Company, New York, New York.
13. Kelly, F. P., 1996. "Notes on Effective Bandwidth," in *Stochastic Networks*: *Theory and Applications*. Editors: F. P. Kelly, S. Zachary, and I. B. Ziedins, Royal Statistical Society Lecture Notes Series, 4. Oxford University Press, pp. 141-168.
14. Kleinrock, L., 1975. *Queueing Systems, Volumes I and II,* John Wiley & Sons, Inc., New York, New York.
15. Kumar, A., Manjunath, D., and Kuri, J., 2004. *Communication Networking, An Analytical Approach,* Morgan Kaufmann Publishers, San Francisco, California.
16. Lavenberg, S. S., Editor, 1983. *Computer Performance Modeling Handbook,* Academic Press, New York, New York.
17. Likhanov, N., and Mazumdar, R. R., 1999. "Cell Loss Asymptotics for Buffers Fed with a Large Number of Independent Stationary Sources," Journal of Applied Probability, Vol. 36, Issue 1, pp. 86-96.
18. Mandjes, M., 2009. *Large Deviations for Gaussian Queues,* John Wiley & Sons, Inc., New York, New York.
19. Medhi, J., 1994. *Stochastic Processes*, Second Edition, John Wiley & Sons, Inc., New York, New York.
20. Medhi, J., 2003. *Stochastic Models in Queueing Theory*, Second Edition, John Wiley & Sons, Inc., New York, New York.
21. Rabinovitch, P., 2000. "Statistical Estimation of Effective Bandwidth," Master of Information and System Sciences thesis, Carleton University.
22. Ross, S. M., 2010. *Introduction to Probability Models,* Tenth Edition, Academic Press, New York, New York.
23. Shwartz, A., and Weiss, A., 1995. *Large Deviations for Performance Analysis,* Chapman and Hall/CRC Press, New York, New York.
24. Takács, L., 1962. *Introduction to the Theory of Queues,* Oxford University Press, Oxford, Great Britain.

CHAPTER 7

Stochastic Structure of the Internet and the World Wide Web

$$\lim_{x \to \infty} \frac{\pi(x)}{\frac{x}{\ln x}} = 1$$

Prime number theorem

Paul Erdös. Paul Erdös was born on 26 March, 1913 in Budapest, Hungary. He received a Ph.D. in mathematics from Eötvös University when he was only twenty one years old. Erdös is well-known for his work on combinatorics, number theory, graph theory, and several other branches of mathematics.

In the year 1845, a French mathematician J. Bertrand conjectured that: There exists at least one prime number between integers n and 2n, where n is greater than unity. Pafnuty Chebyshev (1821-1894) proved Bertrand's conjecture in 1850. However, Erdös gave an elementary proof of this conjecture, when he was only eighteen years old. Erdös, along with Atle Selberg also provided an elementary proof of the well-known prime number theorem (PNT).

Erdös did not have a permanent home. He traveled extensively throughout the world, and produced approximately 1500 papers with about 500 coauthors. He died on 20 September, 1996 in Warsaw, Poland.

7.1 Introduction

The Internet, as the name says is a collection of interconnected networks. It can also be called a network of networks. The World Wide Web (WWW) is a hypermedia information-application available on the Internet. The structure of the Internet and the World Wide Web is studied in this chapter from a stochastic perspective. The World Wide Web, sometimes simply called the Web, and the Internet can be modeled abstractly as random graphs or networks. Some important metrics to characterize random graphs are first discussed in this chapter. This is followed by experimental evidence to support the premise that the Internet and the Web have a stochastic structure. Some important metrics are also revisited. Classical random graph theory, study of random graphs via z-transforms, the $\mathcal{K}$-core of a random graph, graph spectra, small-world networks, and network evolution models are also discussed. Stochastic network from the perspective of hyperbolic geometry is also examined. Search techniques on networks, virus epidemics and immunization in the Internet, error and attack tolerance of networks, and congestion in a stochastic network are also briefly described. Finally, a section on exponential random graph models is included.

Recall that a graph is a two-tuple mathematical structure. The two-tuples are: a set of vertices, also called nodes; and a set of edges (links or arcs) which connect some of these vertices. The nodes in the Internet are the computers (hosts) and the routers, and the edges are the cables and wires that physically connect them. The nodes of the WWW are the Web documents, connected via directed hyperlinks. As the number of nodes in such graphs is very large, the properties of such graphs can be studied from a probabilistic perspective.

The phrase "random graphs" is restricted to the study of classical random graphs, in several textbooks and scientific literature related to the study of stochastic networks. These graphs were first studied in considerable detail by P. Erdös and A. Rényi (1921-1970). In contrast, this phrase is used in this chapter to refer to any type of random graph, including the classical random graphs.

The subject of this chapter is actually a subset of a field called network science. Network science is the study of complex networks in general. Some examples of complex networks are: computer networks, biological networks, social networks, cognitive and semantic networks, and telecommunication networks.

The mathematical prerequisites for studying this chapter are applied analysis, theory of matrices, graph theory, basics of probability theory, stochastic processes, optimization, and geometry. We denote the expectation operator by $\mathcal{E}(\cdot)$. The notation $a_n \asymp b_n$ implies $a_n = \Theta(b_n)$.

7.2 Metrics to Characterize Random Graphs

Let $G = (V, E)$ be a graph, where V is the set of vertices, and E is the set of edges. In general, this graph is directed. Let the set of vertices V be equal to $\{1, 2, \ldots, n\}$. In graph theory, $|E|$ and $|V|$ are called the size and order of the graph respectively. If the graph G represents a network, we shall sometimes call the parameter n as the "size" of the network.

Assume that $|V| = n$ is very large, that is $n \to \infty$. Since the graph is large, it is convenient to study the properties of such graphs in a probabilistic framework. Such graphs are termed random graphs. In addition, note that a random graph is actually a statistical ensemble of random graphs, which consists of multiple realizations. Some convenient metrics to characterize a random graph are: degree distribution, clustering measures (coefficient), average path length, diameter of a graph, heterogeneity of networks, graph spectra, and betweenness. All these interrelated measures are like fingerprints of the network.

Degree Distribution: Consider a random graph. Let ν be a random variable which represents the degree of a vertex. Among several possible candidates for the distribution of ν are: the Poisson distribution, power-law distribution, or any other distribution defined on a discrete state-space which takes nonnegative values.

Random graphs with Poissonian degree distribution were first studied in depth by P. Erdös and A. Rényi. Networks with power-law degree distribution are called *scale-free*. The phrase "scale-free" refers to any functional form $g(x)$ that remains unaltered to within a multiplicative factor under rescaling of the independent variable x. That is, $g(ax) = h(a) g(x)$, where a is a nonzero constant, and $h(\cdot)$ is some function of a. Recall that the Zipf distribution has this form. Its continuous analog is the Pareto distribution. The Paretian probability density function, and its complementary cumulative distribution function also have this scale-free form.

Clustering Measures: *Clustering coefficient of a vertex* measures the "density" of connections in the vicinity of the vertex. *Clustering coefficient of a graph* is the average of all individual vertex clustering coefficients. The concept of clustering is only valid for undirected graphs.

Average Path Length of a Graph: Let all the edges of a connected graph (network) be of unit length. Also let the shortest-path length between the nodes i and j be d_{ij}. If this graph is random, the average (characteristic) path length ℓ of the graph is the average value of d_{ij}'s, which are also averaged over all possible realizations of the randomness. In brief, the parameter ℓ determines the average separation of pairs of vertices. In a fully connected network, the value of ℓ is equal to one.

Diameter of a Graph: The diameter of a graph is the largest distance between any pair of vertices of a connected graph. It is assumed in this chapter that the length of an edge is equal to unity, and the distance between any pair of vertices is equal to the length of the shortest path between them. Some authors refer to the average path length ℓ as the diameter of the graph.

Heterogeneity of Networks: The heterogeneity of a network $\varkappa$ indicates a characterization of the level of its heterogeneity. If the random variable ν represents the degree of a node of a random graph, then

$$\varkappa = \frac{\mathcal{E}\left(\nu^2\right)}{\mathcal{E}\left(\nu\right)} \tag{7.1}$$

where $\mathcal{E}(\cdot)$ is the expectation operator. A network is said to be homogeneous if $\varkappa \simeq \mathcal{E}(\nu)$, and heterogeneous if $\varkappa \gg \mathcal{E}(\nu)$. For a large scale-free real-world random graph $\varkappa \to \infty$.

Graph Spectra: A graph G with n vertices can be represented by its adjacency matrix. The size of this matrix is n. The eigenvalues of this adjacency matrix represent the spectrum of the graph G. The graph spectra has been defined in the chapter on graph theory.

Betweenness: The *betweenness* or *betweenness centrality* (BC) of a vertex m is a measure of the total number of shortest paths (sometimes called geodesics) between all possible pairs of vertices of a graph that pass through this vertex. This measure is an indicator of importance of a vertex in the flow of network traffic through it. The assumption here is that traffic is routed in the network via the shortest path.

The betweenness of an edge e is similarly defined. It measures the total number of shortest paths between all possible pairs of vertices of a graph that pass through this edge.

Convenient computational procedures are described in the later part of this chapter to determine the above parameters.

7.3 Experimental Evidence

The foundation of any body of knowledge is the set of experimental facts which support it. Therefore, in this section, we discuss some experimental evidence related to the description of:

(a) The World Wide Web, and
(b) The Internet

It has been found that if the World Wide Web and the Internet are represented as graphs, then the degree distribution of the graph generally follows a power-law. Therefore the degree distribution of these graphs is of the form $\asymp k^{-\gamma}, \gamma > 0$, where k denotes the degree of a node. The parameter γ is called the exponent of the distribution. A fruitful technique to study the Internet topology is by the method of $\mathcal{K}$-shell decomposition. The map of the Internet is studied experimentally at the autonomous system (routing domain) level. This approach appears to be robust, and provides a peek at the structure of the Internet and its functional requirements.

World Wide Web

The WWW can be considered to be a graph, in which the nodes are the documents (Web pages) and the edges are the hyperlinks that provide pointers from one document to another. These documents are written in HyperText Markup Language (HTML). The WWW graph is considered to be directed, as its edges (links) are directed. Consequently, this network has two kinds of degree distributions. These are the distributions of the outgoing and the incoming edges. It has been found experimentally that the tails of both these distributions follow a power-law. Thus the probability distribution of the outgoing and incoming edges are of the form: $\asymp k^{-\gamma_{out}}$, and $\asymp k^{-\gamma_{in}}$ respectively for large values of k. The range of values of γ_{out} is from 2.3 to 2.8, and the value of γ_{in} is about 2.1.

Internet

The Internet is a conglomeration of networks of computers and several other telecommunication devices which are interconnected via transmission links. Note that the Web documents are accessed through the Internet. The Internet graph can be studied at two levels. These are at the router level, and at the autonomous system level. An autonomous system is a subnetwork of several routers. These subnetworks are formed for the purpose of administration and hierarchically aggregating routing information.

(a) Router level: In this classification, the routers are the nodes and the transmission links between the routers are the edges.
(b) Autonomous system level: An autonomous system or domain represents a node. And an edge is any physical link between two domains.

It has been found that the probability distribution of the degree of nodes also follows a power-law in both the above two cases. Denote the power-law exponent in the case of the router level classification by γ^{rt}, and the exponent at the autonomous system level classification by γ^{as}. Experimental range of values of γ^{rt} is from 2.3 to 2.5. These values are identical for both the in-degree and the out-degree. Similarly, the experimental range of values of γ^{as} is from 2.1 to 2.2. This range is identical for both the in-degree and the out-degree. More refined measurements suggest that perhaps the tail of the node distribution is heavy-tailed (possibly Weibullian).

Besides these two types of networks, there are a multitude of complex networks, which can be represented abstractly by a graph. The probability distribution of degree of graph in many such cases exhibits a power-law behavior. For example, consider a phone-call network. In this network, the nodes are the telephone numbers, and every completed call is an edge which is directed from the caller to the receiver. The value of power-law exponent in this graph is equal to 2.1. This value is the same for both the in-degree and the out-degree distributions.

Another classical example is the acquaintance network which was made famous in the "small-world" experiments of social psychologist Stanley Milgram. These experiments gave birth to the phrase "six degrees of separation." These experiments determined the distribution of path lengths in an acquaintance network by requesting the participants to pass a letter to one of their friends in an attempt to ultimately get it to an assigned target individual. In the course of the experiment several letters were lost, but about a fourth of the letters did indeed reach their target. These letters reached their assigned target by passing through the hands of about six people on average. Hence the phrase: six degrees of separation. This phrase actually did not appear in Milgram's paper. The first apparent user of this phrase was J. Guare. Perhaps, the number six may not be an accurate estimate, but the idea upon which it is based is correct. Therefore, as per the experiment of Milgram, the "small-world effect" implies that any two persons can establish contact with each other by going through only a small chain of intermediate acquaintances.

Some examples of complex-statistical networks which exhibit this power-law behavior are summarized below without much elaboration. These are:

(a) Biological networks: cellular networks, ecological and food webs, genetic regulatory networks, networks of metabolic reactions, protein networks, and structure of neural networks.
(b) Information networks: citation networks, networks in linguistics, and science collaboration networks.

(c) Physical/technological networks: electric power grid networks in United States of America, electronic circuits, Internet, phone call networks, train routes, and the WWW.
(d) Social networks: acquaintance networks, electronic mail messages, movie actor collaboration network, sometimes called the Hollywood network, and the web of human sexual contacts.

7.3.1 Macroscopic Structure of the World Wide Web

It has been experimentally determined that the WWW has five main components. See Figure 7.1.

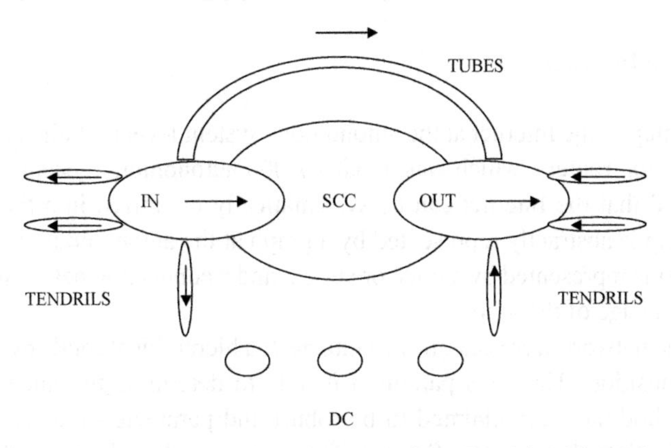

Figure 7.1. Macroscopic structure of the World Wide Web.

The components of WWW are:

(a) Giant strongly connected component: This is the central core of the Web in which each vertex (page) is reachable from another along directed links. This component is termed SCC (strongly connected component).
(b) In component: This component consists of pages that can reach pages in the SCC, but not vice-versa. This component is termed IN. For example, these can be new pages, which have not yet been accessed.
(c) Out component: This component consists of pages that are reachable from the SCC, but not vice-versa. This component is termed OUT. For example, these pages might belong to a company Web-site that consists of only internal links.
(d) TENDRILS: These are pages which are neither accessible from SCC, nor can they access the pages in SCC. These TENDRIL pages are the set of nodes that are reachable from certain pages of IN. These can also access portions of OUT, without actually passing through the SCC. It is certainly feasible to have a TENDRIL hanging off from the IN component to be connected into a TENDRIL leading into the OUT component, thus forming a TUBE. Therefore a TUBE is a pipe with portions touching the IN and OUT components, without actually touching the SCC.
(e) Disconnected component: Sets of Web pages which do not belong to any of the above four major components are called the disconnected components (DC).

7.3.2 A $\mathcal{K}$-shell Decomposition of the Internet Topology

A useful technique to study Internet topology is via $\mathcal{K}$-shell decomposition. The Internet map can be abstractly represented by a graph, which in turn can be studied by its $\mathcal{K}$-shell decomposition.

The concept of $\mathcal{K}$-shell of a graph can be explained in terms of the $\mathcal{K}$-core of the graph. A $\mathcal{K}$-core of a graph is obtained by pruning all vertices (with their incident edges) of degree less than $\mathcal{K}$, till all vertices have a degree greater than or equal to $\mathcal{K}$.

The $\mathcal{K}$-core decomposition establishes a hierarchy in the network. It is able to identify the heart or the nucleus of the network. In addition, it is useful in studying the cooperative processes within the network, and in visualization of the complex network. This in turn is useful in examining the topology and traffic of the network. The $\mathcal{K}$-core decomposition is also an artifice to probe the centrality and connectivity properties of the large scale network. Note that connectivity in a graph is related to robustness of the network to attacks and faults; and routing efficiency.

Structure of the Internet

A topological map of the Internet at the autonomous system level is studied in this subsection. A study of over a million routers, which was about $20,000$ autonomous systems (subnetworks), and $70,000$ links, found that the Internet can be systematically organized into three groups. Assume that the Internet map is abstractly represented by a graph at the autonomous system (AS) level. An autonomous system is represented by a node or vertex, and a connection between a pair of the nodes is represented by an edge of the graph.

The nodes in the network were categorized into nested hierarchical shells by using the technique of $\mathcal{K}$-shell decomposition. Thus this paradigm is able to determine the nucleus of the network. Moreover, this method has been claimed to be robust and parameter-free. The nodes in the AS-graph were classified into three groups (layers). There are only three layers in the network because the Internet AS graph has a small diameter. See Figure 7.2. The three groups are:

(a) A set of highly connected nodes called the nucleus. These nodes form the core of the network.
(b) The nodes which are not in the nucleus form the crust of the network. The set of nodes that belong to the largest connected component of this crust form the middle layer. Thus the middle layer of the network forms a set of peer-connected nodes.
(c) The remaining nodes of the crust form the peripheral layer of nodes. These nodes which might belong to smaller clusters of the crust or they might simply be leaves, are sparsely connected. The nodes in this layer connect only through the nucleus, and are like tendrils hanging from the nucleus of the network. This is in analogy with the tendrils of a jellyfish.

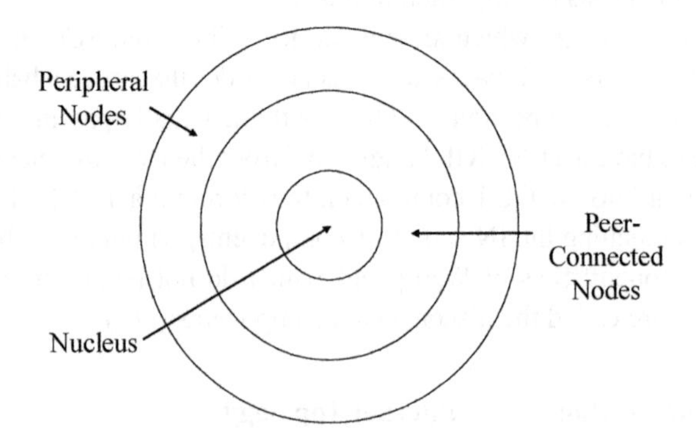

Figure 7.2 Classification of Internet Nodes.

Properties of Nodes in the Three Layers

Structure of the Nucleus: The nucleus is made up of about 100 nodes. The nodes in the nucleus are separated by one or two hops.

Structure of the Middle Layer: The middle layer is made up of about $15,000$ nodes, which is approximately 70 percent of the network. These set of nodes are peer-connected. Thus these nodes do not require the set of nodes in the nucleus for communication. This is a significant observation, because the nodes in this layer do not load up the nodes in the nucleus for communication. It takes three or four hops for the two nodes within this layer to communicate with each other.

Structure of the Peripheral Layer: The peripheral layer is made up of approximately $5,000$ nodes, which is approximately 30 percent of the network. This layer has several small cluster of nodes, and the nodal degree of nodes in this layer is small. These nodes depend upon the nodes in the nucleus for communication. Therefore, if two nodes in this layer want to communicate with each other, the communication requires two or three hops. For example, a source node requires one hop to the nucleus node, possibly one more hop within the nucleus, and finally one more hop back to the target node in the peripheral layer.

In general, if the nodes in the nucleus are used for communication between any pair of nodes in the AS-graph, then approximately four hops are necessary for communication. If the nodes in the nucleus are removed, then approximately seven or eight hops are necessary for communication between any pair of nodes in the network. However, this later choice might relieve the nodes in the nucleus from congestion. Thus connection via the nucleus decreases path length significantly. The condition for the existence of a $\mathcal{K}$-core of a random graph is studied later in the chapter.

Certain mathematical characterizations of a network are defined and discussed in more detail in the next section.

7.4 Some Important Metrics Revisited

Some metrics of a random graph like degree distribution, clustering coefficient, and betweenness are next discussed in more detail.

7.4.1 Degree Distribution

Let ν be a random variable which represents the vertex degree, in a random graph,. The degree distribution is specified by the probability $P(\nu = k) \triangleq p_\nu(k)$, where $k = 0, 1, 2, \ldots, K$ and $\sum_{k=0}^{K} p_\nu(k) = 1$. Denote the average degree by $\overline{k} \triangleq \mathcal{E}(\nu) = \sum_{k=0}^{K} k p_\nu(k)$, where $\mathcal{E}(\cdot)$ is the expectation operator. It is possible for K to be extremely large (infinite).

Examples 7.1. Some illustrative examples.

1. The random variable ν has a *Poisson distribution* with parameter $\overline{k}$. The moments of this distribution are finite. It is shown in a subsequent section that the degree distribution of a classical random graph asymptotically has a Poisson distribution. Classical random graphs are studied in more detail in the next section.
2. The random variable ν has a *discrete exponential distribution*, that is $p_\nu(k)$ is proportional to $e^{-k/\kappa}$, where κ is a positive constant, and $k \in \mathbb{N}$. The moments of this distribution are also finite.

3. The random variable ν has a *power-law distribution*, that is $p_\nu(k)$ is proportional to $k^{-\gamma}$, where $k \in \mathbb{P}$. The exponent of this distribution is said to be γ, which is a positive real number. However, note that the exponent of k in this distribution is actually negative. This distribution is sometimes called the generalized Zipf distribution. It is named after the late Harvard psychologist George Kingsley Zipf, who was a pioneer in the use of this distribution. Networks with this distribution can be scale-free. Assume that the graph is connected. Also assume that the value of ν ranges from k_0 to k_c, where $k_0 \ll k_c$, and $\gamma > 0$. Then

$$p_\nu(k) = \mathfrak{B}k^{-\gamma}, \qquad k_0 \le k \le k_c \tag{7.2a}$$

$$\mathfrak{B}^{-1} \simeq \begin{cases} \ln(k_c/k_0), & \gamma = 1 \\ k_0^{-(\gamma-1)}\left(1-(k_0/k_c)^{(\gamma-1)}\right)/(\gamma-1), & \gamma \ne 1 \end{cases} \tag{7.2b}$$

In the above equation, the normalizing constant $\mathfrak{B}$ is positive. Continuum approximation was used to evaluate $\mathfrak{B}^{-1}$. The m-th moment of the random variable ν, $\mathcal{E}(\nu^m), m \in \mathbb{P}$ is next evaluated.

$$\mathcal{E}(\nu^m) \simeq \begin{cases} \mathfrak{B}\ln(k_c/k_0), & \gamma = (m+1) \\ \mathfrak{B}k_0^{m-\gamma+1}\left(1-(k_0/k_c)^{(\gamma-m-1)}\right)/(\gamma-m-1), & \gamma \ne (m+1) \end{cases} \tag{7.3}$$

Two cases are considered below, in which the order (number of vertices) of the network is either infinite or finite.

(a) *Infinite network*: Assume that the order of the network is infinite and $k_c \to \infty$. The value $\mathfrak{B}$ exists, provided $\gamma > 1$. The average value of the degree exists if $\gamma > 2$. Thus

$$\mathfrak{B} \simeq (\gamma-1)k_0^{(\gamma-1)}, \qquad \gamma > 1$$

$$\mathcal{E}(\nu^m) \simeq \frac{(\gamma-1)}{(\gamma-m-1)}k_0^m, \quad \gamma > (m+1)$$

(b) *Finite network*: Consider the network which is finite, yet very large. Let the order of the network be n. Then $1 \le k_0 < k_c < n$. Two subcases are evaluated, by assuming a relationship between k_0, k_c, and n. These yield nearly identical results for large values of n.

(i) Select

$$\left(\frac{k_0}{k_c}\right)^{(\gamma-1)} \simeq \frac{1}{n}$$

Since $k_c < n$, the above relationship is possible if $\gamma > 1$. Thus the above equation yields $k_c \simeq k_0 n^{1/(\gamma-1)}$. Consequently

$$\mathfrak{B} \simeq k_0^{(\gamma-1)}\frac{(\gamma-1)}{(1-1/n)}, \quad \gamma > 1$$

The m-th moment of the random variable ν, is $\mathcal{E}(\nu^m), m \in \mathbb{P}$. It is:

- If $\gamma = (m+1)$, then

$$\mathcal{E}(\nu^m) \simeq \frac{\mathfrak{B}\ln n}{(\gamma-1)}$$

- If $\gamma \neq (m+1)$, then

$$\mathcal{E}(\nu^m) \simeq \frac{\mathfrak{B} k_0^{m-\gamma+1}\left(1 - n^{-(\gamma-m-1)/(\gamma-1)}\right)}{(\gamma - m - 1)}$$

Finally, the expression for the moment can be further approximated if $\gamma > 1$ and $(m+1-\gamma) \neq 0$.

- If $(m+1-\gamma) < 0$, then

$$\mathcal{E}(\nu^m) \simeq \frac{(\gamma-1)}{(\gamma-m-1)} k_0^m$$

- If $(m+1-\gamma) > 0$, then

$$\mathcal{E}(\nu^m) \simeq \frac{(\gamma-1)}{(m-\gamma+1)} k_0^m n^{(m-\gamma+1)/(\gamma-1)}$$

Note that the exponent of n in the expression for $\mathcal{E}(\nu^m)$ is a linear function of m, which is the order of the moment. A distribution with this kind of $\mathcal{E}(\nu^m)$ dependence is said to be a *fractal distribution.* Furthermore, the coefficient of linear dependence d_f is called the *fractal dimension.* In this example,

$$d_f = \frac{1}{(\gamma-1)}$$

(ii) An estimate of k_c could also have been obtained from the relationship

$$\int_{k_c}^{\infty} p_\nu(k)\, dk \simeq 1/n$$

Under this approximation, the normalizing constant $\mathfrak{B}$ exists provided $\gamma > 1$. Thus

$$\mathfrak{B} \simeq \frac{(\gamma-1)}{n} k_c^{(\gamma-1)}$$

Therefore

$$\left(\frac{k_0}{k_c}\right)^{(\gamma-1)} \simeq \frac{1}{n}$$

This result reduces to the last subcase for large values of n.

4. The *truncated power-law distribution* characterized by an exponent γ, and an exponential cut-off parameter κ. It is specified by

$$p_\nu(k) = C k^{-\gamma} e^{-k/\kappa}, \quad \forall\, k \in \mathbb{P} \tag{7.4a}$$

where γ and κ are positive constants, and C is the normalization constant. It is given by

$$C^{-1} = Li_\gamma\left(e^{-1/\kappa}\right) \tag{7.4b}$$

where $Li_\gamma(x)$ is the γ-th polylogarithm of x. □

7.4.2 Clustering Coefficient

The clustering coefficient of a graph is the probability that the two nearest neighbors of a randomly chosen vertex are also nearest neighbors of one another. Assume that the graph $G = (V, E)$ is simple, labeled, and undirected. Consider a vertex $i \in V$, with degree d_i. If the first neighbors of the node i were part of a clique, there would be $d_i (d_i - 1)/2 = \binom{d_i}{2}$ edges between them. The clustering coefficient C_i of vertex $i \in V$ is defined as the ratio of the number of edges y_i between its d_i neighboring vertices and $\binom{d_i}{2}$. Thus

$$C_i = \frac{y_i}{\binom{d_i}{2}}, \qquad i \in V \tag{7.5a}$$

The denominator is the maximum possible number of edges between neighbors of vertex i. Therefore $0 \le C_i \le 1$. Note that the d_i neighbors of the vertex i induce a subgraph G_i of graph G. Then y_i is equal to the number of edges in the graph G_i. Let C_G be the clustering coefficient of the graph, then

$$C_G = \frac{1}{n} \sum_{i \in V} C_i \tag{7.5b}$$

where $|V| = n$. An alternative definition of the clustering coefficient of a graph is C'_G, where

$$C'_G = \frac{\sum_{i \in V} \binom{d_i}{2} C_i}{\sum_{i \in V} \binom{d_i}{2}} \tag{7.5c}$$

This alternative definition of clustering coefficient has the following interpretation. The clustering coefficient of a graph C'_G is the ratio of: number of pairs (i, j), and (i, k) of adjacent edges for which (j, k) is an edge; and the number of pairs (i, j), and (i, k) of adjacent edges. Thus

$$C'_G = \frac{\text{three times the number of triangles}}{\text{number of pairs of adjacent edges}} \tag{7.5d}$$

In this interpretation, a triangle is made up of three interconnected nodes. The presence of the factor three in the numerator accounts for the fact that each triangle contributes to three connected triples of vertices. This definition is also applicable to multigraphs. Use of the above definition yields the following lemma.

Lemma 7.1. *Let $G = (V, E)$ be an undirected, simple, and labeled graph, where $V = \{1, 2, \ldots, n\}$. Let the adjacency matrix of this graph be A, and d_i be the degree of the vertex $i, 1 \le i \le n$. Also, let N_Δ be equal to the number of triangles in the graph G, and N_α be equal to the number of pairs of adjacent edges.*

1. *Some related observations made in the chapter on graph theory are summarized below.*
   (a) *The matrix A is symmetric.*
   (b) $d_i = \sum_{j=1}^{n} a_{ij}$, *for* $1 \le i \le n$.
   (c) $2|E| = \sum_{i=1}^{n} d_i = \sum_{1 \le i,j \le n} a_{ij} = tr(A^2)$, *where $tr(\cdot)$ is the trace operator of a matrix.*

2. *Let* $A^2 = \left[a_{ij}^{(2)}\right]$, *then*

$$N_\alpha = \sum_{i=1}^{n} \binom{d_i}{2} = \frac{1}{2}\sum_{i=1}^{n} d_i \left(d_i - 1\right) = \frac{1}{2} \sum_{\substack{1 \le i,j \le n \\ i \ne j}} a_{ij}^{(2)} \tag{7.6a}$$

3. *The number of triangles (loops of length three) in the graph G is*

$$N_\Delta = \frac{1}{6} tr\left(A^3\right) \tag{7.6b}$$

4. *The clustering coefficient* C'_G *is*

$$C'_G = \frac{3N_\Delta}{N_\alpha} \tag{7.6c}$$

Proof. The proof is left to the reader. □

If a graph is random, then it is sometimes convenient to define the clustering coefficient of a vertex as the ratio of the mean number of edges between the neighbors of a vertex and the mean number of possible edges between the neighbors of the vertex. The clustering coefficient of a graph is similarly defined. This redefinition does not substantially alter the physical significance of this parameter.

Examples 7.2. Following examples compute clustering coefficient of a graph in specific cases.

1. Consider an undirected graph $G = (V, E)$, which is a double star. Let the set of vertices V be equal to $\{1, 2, \ldots, n\}$, where $n \ge 3$. In this graph, the vertices 1 and 2 are adjacent to each other and to all other $(n-2)$ vertices. There are no other edges. The degrees of the vertices in this graph are: $d_1 = d_2 = (n-1)$, and $d_i = 2$ for $3 \le i \le n$. Then $C_1 = C_2 = 2/(n-1)$, and $C_i = 1$ for $3 \le i \le n$. Thus

$$C_G = \frac{1}{n}\left\{\frac{4}{(n-1)} + n - 2\right\}$$

where $C_G \approx 1$ for large values of n. Also

$$C'_G = \frac{\frac{2}{(n-1)}\binom{n-1}{2} + \frac{2}{(n-1)}\binom{n-1}{2} + (n-2)}{\binom{n-1}{2} + \binom{n-1}{2} + (n-2)} = \frac{3}{n}$$

2. It has been defined in the chapter on graph theory that a ring-lattice $R_{n,2k}$ is a graph with n vertices in which each vertex is connected to its first $2k$ neighbors (k on either side). It is demonstrated that the clustering coefficient of a ring-lattice graph $R_{n,2k}$ is

$$C_G = \frac{3(k-1)}{2(2k-1)}$$

where $1 \le k \le (n-1)/2$. In this graph, the degree of each node is $2k$ and the number of edges between these neighbors is given by

$$y = \underbrace{(k-1) + (k-1) + \cdots + (k-1)}_{k \text{ terms}} + (k-1) + (k-2) + \cdots + 2 + 1$$
$$= \frac{3k}{2}(k-1)$$

Taking the ratio of y and $\binom{2k}{2}$ yields the value of the clustering coefficient C_G. For this graph $C_G = C'_G$. □

7.4.3 Betweenness

Consider a graph $G = (V, E)$, where V is the set of vertices, and E is the set of edges. A distance metric is also defined between each pair of vertices, so that shortest paths between a pair of vertices are well-defined. This graph can be either directed or undirected. Let the total number of shortest paths between vertices i and j be $B(i,j) > 0$, and $B(i,m,j)$ of these paths pass through the vertex $m \in V$, where the vertex m is also allowed to be either vertex i or j. Then the ratio $B(i,m,j)/B(i,j)$ is a measure of importance of vertex m in the flow of traffic between vertices i and j. Thus $\beta_v(m)$, the betweenness measure of vertex m is defined as

$$\beta_v(m) = \sum_{\substack{i,j \in V \\ i \neq j, B(i,j) > 0}} \frac{B(i,m,j)}{B(i,j)}, \quad \forall\, m \in V \tag{7.7a}$$

Thus higher the value of $\beta_v(m)$, higher is the importance of the node m in the flow of network traffic. It would be fair to conjecture that higher the value of the betweenness of a vertex, higher is its nodal degree.

The concept of the betweenness of a vertex can also be extended to the betweenness of an edge $e \in E$. The betweenness $\beta_\varepsilon(e)$, of an edge e is defined as

$$\beta_\varepsilon(e) = \sum_{\substack{i,j \in V \\ i \neq j, B(i,j) > 0}} \frac{D(i,e,j)}{B(i,j)}, \quad \forall\, e \in E \tag{7.7b}$$

where $D(i,e,j)$ is the total number of shortest paths between vertices i and j that pass through the edge e.

It can be shown that for random trees with power-law degree distribution, the probability distribution of both the vertex and edge betweenness also follows a power-law. See the problem section for the justification of the assertion about edge betweenness. The probability distribution of these betweenness measures has also been observed to have a power-law distribution for real networks with loops, provided the degree correlations are absent.

7.5 Classical Random Graph Theory

Classical probabilistic aspects of graph theory are outlined in this section. Study of classical random graph theory enables us to gain a perspective on the structure of the Internet and the Web graphs. A random graph on n vertices is a graph in which each edge is likely to occur with some probability. P. Erdös and A. Rényi were pioneers in the study of this discipline. Assume that all the graphs discussed in this section are undirected, simple, and labeled. Two types of random graphs can be considered. These are defined below.

Definitions 7.1. *Models A and B of a random graph.*

1. *Model A: A random graph $G(n, p)$ has n labeled vertices, and the probability of the existence of an edge between any pair of vertices is p, where $0 \leq p \leq 1$. Furthermore, the existences of different edges are mutually independent events.*
2. *Model B: A random graph $G(n, e)$ has n labeled vertices, and exactly e number of edges. Each such graph occurs with probability $1/\binom{N_{cl}}{e}$ where $N_{cl} = \binom{n}{2}$.* □

A Model A random graph is also called an uncorrelated or Bernoulli random graph. Each edge in this random graph occurs with a fixed probability, and the occurrences of these events are independent random variables. A Bernoulli random graph has at most N_{cl} edges.

The Model A is easier to work with, but the Model B is easier to define. However the results of Model A can be easily extended to Model B. If p is defined to be e/N_{cl}, then the corresponding results for Model B can generally be obtained. Consequently the properties of Model A random graphs are studied.

Almost every graph is said to have a property $\mathcal{P}$, if the probability that a random graph has property $\mathcal{P}$ approaches 1 as $n \to \infty$.

Assume that the vertices of the graph are labeled $\{1, 2, \ldots, n\}$. The total number of labeled graphs in the probability space of Model A is $2^{N_{cl}}$. In this model, a graph with e edges occurs with a probability

$$p^e (1-p)^{N_{cl}-e}, \quad \text{where } 0 \leq e \leq N_{cl}$$

Degree Distribution

Consider a random graph which belongs to Model A. Assume that the connection probability between two vertices is $p \in (0, 1)$. The average degree of a vertex is first evaluated. The average number of edges in the graph is $N_{cl}p$, and the average number of ends of edges is equal to twice this value. So the average degree of a vertex is

$$\overline{k} = \frac{n(n-1)p}{n} = (n-1)p \sim np$$

The degree ν of vertex i follows a binomial distribution with parameters $(n-1)$ and p. Therefore

$$p_\nu(k) = \binom{n-1}{k} p^k (1-p)^{n-1-k}, \quad 0 \leq k \leq (n-1)$$

If $n \to \infty, p \to 0$, and $(n-1)p \to \lambda$, where $\lambda \in \mathbb{R}^+$, then

$$p_\nu(k) \to \frac{\lambda^k}{k!} e^{-\lambda}, \quad \forall\, k \in \mathbb{N}$$

Therefore for large n and finite np, the degree distribution of a classical random graph has a Poisson distribution. That is,

$$\mathcal{E}(\nu) = \overline{k} = \lambda \simeq np$$

Clustering Coefficient

The clustering coefficient of a random graph which belongs to Model A is evaluated as follows. Assume that $n \to \infty$. The probability that two of the nodes are connected is equal to the probability that two randomly selected nodes are connected. Thus the clustering coefficient of the classical random graph is equal to $C_G = p \simeq \overline{k}/n$.

Note that for a fixed value of $\overline{k}$ the clustering coefficient tends to zero in the limit as the order (number of vertices) of the graph becomes infinitely large.

Average Path Length

A heuristic derivation of the average path length ℓ is given, where it is assumed that the network is connected. In this network, the number of neighbors at a distance unity from any vertex $i \in V$ is $\overline{k}$. If the edges are random and omitting the effect of cycles, the number of neighboring vertices at a distance d can be approximated by $\overline{k}^d$. Thus, if $d = \ell$, then $\overline{k}^\ell = n$, the number of nodes in the graph. This yields

$$\ell = \frac{\ln n}{\ln \overline{k}}$$

Note that for this graph $\ell/n \to 0$ as $n \to \infty$. This result can also be established rigorously.

Connectivity Properties

It is instructive to examine the connectivity properties of a random graph G, as the value of p is increased. For example, if a graph has at least $(n-1)$ edges, then the graph can possibly have isolated vertices. However if the number of edges in the random graph is less than $(n-1)$, the graph is certain to have an isolated vertex.

Some connectivity properties of n-vertex random graphs which belong to Model A are listed below. In the following observations a component of a graph is defined to be a maximal connected subgraph. Thus, a component of a graph has a subset of vertices of the graph, each of which is reachable from the others by following a sequence of edges.

Observations 7.1. Assume that the graph has n vertices, and the probability of an edge between any pair of vertices is p. Also the occurrence of different edges are mutually independent events, and $n \to \infty$.

1. If the probability p is a nonzero constant, then almost every graph $G(n,p)$ is connected.
2. Let P_n be the probability that a graph $G(n,p)$ is connected, and $q = (1-p)$. Then:
   (a) $P_1 = 1$, and

$$P_n = 1 - \sum_{k=1}^{n-1} \binom{n-1}{k-1} q^{k(n-k)} P_k, \quad \text{for} \quad n = 2, 3, 4, \ldots$$

   (b) Also $P_n \simeq (1 - nq^{n-1})$, for large n and $q < 1$. □

The above set of observations is established in the problem section. Next consider a random undirected graph with a fixed and large number of vertices. The connectivity properties of such random graphs are examined, by varying the number of their edges. This is accomplished by varying the value of p. The graph is said to evolve if the value p is increased from zero to unity. The *evolution of graphs* as the value of p is increased is next studied. The order of the connected components of the random graph and related properties are studied.

Observations 7.2. Observations about Model A random graphs.

1. If $p \in o\left(n^{-1}\right)$, then almost every graph has no cycles.
2. Let $p = 2c/n$, where c is a constant.
   (a) If $0 < c < 1/2$, then the largest component of graph has order $\simeq \ln n$. The graph may possibly have cycles.
   (b) If $c = 1/2$, then the largest component of graph has order $\simeq n^{2/3}$. The graph may possibly have cycles.
   (c) If $c > 1/2$, then the largest component of graph has order $\widehat{p}n$, where $\widehat{p}$ is a constant.
3. Let $p = c(\ln n)/n$, where c is a constant.
   (a) If $0 < c < 1$, then almost every graph $G(n,p)$ is disconnected.
   (b) If $c > 1$, then almost every graph $G(n,p)$ is connected. □

Based upon the above set of observations, the evolution of Model A random graphs can be described, as the value of p changes from zero to unity. To make the discussion meaningful, the connection probability p is made a function of n, as $n \to \infty$. Initially, there are a set of n isolated vertices. This stage corresponds to a value of p equal to zero. Edges are added to the vertices, as the value of p increases. Finally, a completely connected graph is obtained with $n(n-1)/2$ edges. This final stage corresponds to a connection probability of p equal to unity. For values of p which lie between 0 and 1, the graph connectedness increases as p increases. The purpose of the random graph theory is to determine a value of p at which a specific property of a graph arises. The main contribution of Erdös and Rényi was that several important properties of random graphs appear quite suddenly as the value of p changes. Attention is restricted to the connectivity property of random graphs in this chapter.

In the beginning p is equal to zero. This stage has n isolated vertices. As the value of p is increased approximately to $o\left(n^{-1}\right)$, then almost every graph has no cycles. In the next stage of evolution let $p = 2c/n$, where c is a nonnegative constant. If $0 < c < 1/2$, then the largest component of graph has order $\ln n$. Such graphs may possibly have cycles.

However if $c = 1/2$, the structure of the graph changes radically. At this value, the graphs are said to exhibit the phenomenon of *double jump*, because the structure of the graph is substantially different for $c < 1/2$, $c = 1/2$, and $c > 1/2$. At $c = 1/2$, the largest component of graph has order $n^{2/3}$. This graph may possibly have cycles. The order of the largest component jumps from $\ln n$ to $n^{2/3}$. If the value of c is such that it is greater than $1/2$, then the largest component of graph has order $\widehat{p}n$, where $\widehat{p}$ is a constant. The number of vertices which are outside this largest component approaches $o(n)$. The value of the constant $\widehat{p}$ is established in the next section.

The value of p at $c = 1/2$ is $p = 1/n$. This value of p is said to be the *phase-transition point*, because at this point a *giant component* of the graph emerges.

Consider $p = c(\ln n)/n$, where c is a constant. Observe that if $0 < c < 1$, then almost every graph $G(n,p)$ is disconnected. However if $c > 1$, then almost every graph $G(n,p)$ is connected. The value $(\ln n)/n$ is said to be a *threshold probability function* for the disappearance of isolated vertices.

Evolution of Model B random graphs can be similarly described.

7.6 Study of Random Graphs via z-Transforms

The z-transform of the distribution function of a discrete random variable is a convenient construct to study the characteristics of a random graph. This formalism facilitates the study of random graphs with any degree distribution. For simplicity, it is assumed that the random graphs are simple, undirected, and labeled. Furthermore, the number of vertices n is assumed to be very large (infinite). For a random graph G with arbitrary degree distribution: concept of phase transition, average path length, clustering coefficient of the graph, distribution of component sizes below the phase transition point, and distribution of component sizes above the phase transition point are studied. These results are finally illustrated via examples.

7.6.1 Phase Transition

The random variable of most interest is the degree of a vertex in a random graph. The discrete probability distribution function of this random variable ν is specified by $p_\nu(k)$, $\forall\ k \in \mathbb{N}$. Its z-transform is defined by $\mathcal{G}_0(z)$, where $z \in \mathbb{C}$. Therefore

$$\mathcal{G}_0(z) = \sum_{k\in\mathbb{N}} p_\nu(k) z^k \tag{7.8a}$$

$$p_\nu(k) = \frac{1}{k!}\frac{d^k}{dz^k}\mathcal{G}_0(z)\bigg|_{z=0}, \qquad \forall\, k \in \mathbb{N} \tag{7.8b}$$

Note that $\mathcal{G}_0(1) = 1$, and $\mathcal{G}_0(z)$ is absolutely convergent for all $|z| < 1$. Consequently, there are no singularities in this region. Also the average degree of a vertex ν is given by

$$\mathcal{E}(\nu) = \sum_{j\in\mathbb{P}} j p_\nu(j) = \mathcal{G}_0'(1) \tag{7.8c}$$

where $\mathcal{E}(\cdot)$ is the expectation operator, and $\mathcal{G}_0'(z)$ is the first derivative of $\mathcal{G}_0(z)$ with respect to z. Thus the average number of neighbors of a randomly chosen vertex is $\mathcal{E}(\nu) = \overline{k}$. In order to study the clustering phenomenon of a random network, the degree distribution of the first neighbors of a randomly selected node also has to be studied. This can be achieved in the following way. A randomly selected edge starts from a vertex i and reaches a node with degree $(k+1)$, with probability proportional to $(k+1)\,p_\nu(k+1)$. The number of edges which emerge from this node is equal to k. These are the number of edges which contribute to the second neighbors of the vertex i. This count of edges excluded the edge from which we arrived. Denote this probability by $q_\nu(k)$. Thus

$$q_\nu(k) = \frac{(k+1)\,p_\nu(k+1)}{\sum_{j\in\mathbb{P}} j p_\nu(j)}, \qquad \forall\, k \in \mathbb{N} \tag{7.9a}$$

The probability that any of these k outgoing edges connect back to the original vertex i, or any of its other immediate neighbors is inversely proportional to n, which is the order of the network. Therefore this probability is assumed to be negligible as $n \to \infty$. Let the z-transform of the probability distribution $q_\nu(k)$, $\forall\ k \in \mathbb{N}$ be $\mathcal{G}_1(z)$, that is

$$\mathcal{G}_1(z) = \sum_{k\in\mathbb{N}} q_\nu(k) z^k \tag{7.9b}$$

It follows that

$$\mathcal{G}_1(z) = \frac{1}{\xi_1}\mathcal{G}_0'(z) \tag{7.9c}$$

$$\xi_1 \triangleq \mathcal{G}_0'(1) = \overline{k} \tag{7.9d}$$

Therefore, the average number of vertices two steps away from vertex i via one of its neighbors is equal to $\mathcal{G}_1'(1)$, where $\mathcal{G}_1'(z)$ is equal to the first derivative of $\mathcal{G}_1(z)$ with respect to z. Also

$$\mathcal{G}_1'(1) = \sum_{k\in\mathbb{N}} kq_\nu(k) = \frac{\sum_{k\in\mathbb{P}}(k-1)kp_\nu(k)}{\xi_1} = \frac{(\mu_2 - \overline{k})}{\xi_1} = \frac{\xi_2}{\xi_1} \tag{7.9e}$$

$$\xi_2 \triangleq (\mu_2 - \overline{k}) \tag{7.9f}$$

where μ_2 is the second moment of the random variable ν. Therefore, $\mathcal{G}_1'(1)$ times the mean degree of vertex i, which is just ξ_1, yields the mean number of second neighbors of a vertex. This is simply equal to ξ_2. This line of reasoning can be extended iteratively to further neighbors. The average number of edges from each second neighbor, other than the one we arrived along is also equal to $\mathcal{G}_1'(1)$. Therefore the average number of neighbors ξ_m at distance m (each edge has a length of unity) is given by

$$\xi_m = \frac{\xi_2}{\xi_1}\xi_{m-1}, \quad m \geq 2$$

Thus

$$\xi_m = \left(\frac{\xi_2}{\xi_1}\right)^{m-1}\xi_1 \tag{7.10}$$

Depending upon the ratio ξ_2/ξ_1 the value ξ_m can either converge or diverge as m becomes large. Therefore:

(a) If $\xi_2 < \xi_1$, then the average total number of neighbors of vertex i at all distances m is finite. Consequently there can be no giant component in the random graph, as $n \to \infty$.
(b) If $\xi_2 > \xi_1$, then the average total number of neighbors of vertex i at all distances m diverges. Thus there is a giant component in the random graph, as $n \to \infty$.

Consequently the phase transition occurs at $\xi_2 = \xi_1$. That is when

$$(\mu_2 - 2\overline{k}) = 0 \tag{7.11}$$

This relationship was first derived by M. Molloy and B. Reed for any general vertex distribution.

7.6.2 Average Path Length

The average path length ℓ between a pair of vertices in a random network is a typical distance through the network. As mentioned earlier, assume that the length of each edge is unity. If $\xi_2 < \xi_1$, then a majority of vertices are not connected to one another. Therefore the average path length measure is not very meaningful. However, if $\xi_2 > \xi_1$ a giant component is present in the random graph. In this giant component all vertices are connected to one another via some edges. If $\xi_2/\xi_1 \gg 1$, then $\xi_\ell \simeq n$. Thus

$$n \simeq \xi_\ell = \left(\frac{\xi_2}{\xi_1}\right)^{\ell-1}\xi_1$$

This yields

$$\ell = \frac{\ln(n/\xi_1)}{\ln(\xi_2/\xi_1)} + 1 \tag{7.12}$$

Observe that the value ℓ grows logarithmically with the graph order n. This phenomenon is called the *small-world effect.*

7.6.3 Clustering Coefficient of the Random Graph

Random networks exhibit clustering of the nodes. Consider a specific vertex $l \in V$, where V is the set of vertices in the graph G. Its neighboring vertex i has k_i edges emerging from it other than the edge attached to the vertex l. Furthermore, the distribution of these edges is given by $q_\nu(k)$, $\forall\, k \in \mathbb{N}$. This vertex i is connected to another vertex $j \in V$ with probability $k_i k_j / (n\xi_1)$. Note that the distribution of the edges which emerge from vertex j, except the edge which is attached to vertex i is given by $q_\nu(k)$, $\forall\, k \in \mathbb{N}$. The average of this probability is indeed the clustering coefficient C_G of the graph.

$$C_G = \frac{1}{n\xi_1}\left\{\sum_{k\in\mathbb{N}} k q_\nu(k)\right\}^2 = \frac{1}{n\xi_1}\left(\frac{\xi_2}{\xi_1}\right)^2 = \frac{\xi_1}{n}\left(\frac{\mu_2 - \xi_1}{\xi_1^2}\right)^2$$

Thus

$$C_G = \frac{\xi_1}{n}\left\{C_v^2 + \frac{\xi_1 - 1}{\xi_1}\right\}^2 \tag{7.13}$$

where C_v^2 is the squared coefficient of variation of the random variable ν. In the above equation, it was essentially assumed that the correlations between the vertices were absent.

7.6.4 Distribution of Component Sizes

We study the distribution of component sizes of the random graph, both below and above the phase transition point. Recall that a component of a graph is defined to be a maximal connected subgraph, and the component size is the number of vertices in it.

Distribution of Component Sizes below the Phase Transition Point

The distribution of the sizes of the connected components of an undirected random graph below the phase transition point is initially examined. In this regime giant components of the graph are absent. Furthermore, there is no significant clustering. In this case, the clustering coefficient C_G actually tends to zero, as $n \to \infty$. The probability that any two vertices i and j with degrees k_i and k_j respectively, are connected is given by $k_i k_j / (n\xi_1)$, which tends to zero as $n \to \infty$. This in turn implies that any finite component of connected vertices of the random graph has no closed loops in it. That is, all finite components of the graph, in limit are tree-like.

Denote the z-transform of the probability distribution of cluster sizes (number of vertices) by $\mathcal{H}_1(z)$. Select an edge at random in the random graph and reach a vertex from which $k \in \mathbb{P}$ more edges emanate, and then to every other vertex reachable from it, and so on. We designate this edge as type (a). It is also possible to follow a randomly chosen edge and find only a single vertex and no more edges leaving it. This is called a type (b) edge.

Each edge of type (a) then leads to another cluster whose size distribution can be described by the z-transform $\mathcal{H}_1(z)$. Note that the number of k edges which emanate from the randomly chosen vertex, excluding the one from which we arrived, has a probability distribution $q_\nu(k)$, $\forall\, k \in \mathbb{N}$. Thus

$$\mathcal{H}_1(z) = z \sum_{k\in\mathbb{N}} q_\nu(k) \{\mathcal{H}_1(z)\}^k = z\mathcal{G}_1(\mathcal{H}_1(z)) \tag{7.14}$$

In the above equation, the factor z corresponds to the single vertex at the end of the randomly selected edge. The distribution of the sizes of clusters to which a randomly chosen vertex belongs is next determined. The number of edges which leave such vertex has a distribution $p_\nu(k)$, $\forall\, k \in \mathbb{N}$. The z-transform of cluster sizes, from each of the k edges is $\mathcal{H}_1(z)$. Therefore the z-transform of the cluster sizes obtained by selecting a vertex at random is $\mathcal{H}_0(z)$. It is

$$\mathcal{H}_0(z) = z \sum_{k\in\mathbb{N}} p_\nu(k) \{\mathcal{H}_1(z)\}^k = z\mathcal{G}_0(\mathcal{H}_1(z)) \tag{7.15}$$

The above equations for $\mathcal{H}_0(z)$ and $\mathcal{H}_1(z)$ are not easily solvable. Even for the case of Poissonian degree distribution, these equations do not have a closed-form solution. However, first few terms in the expansion can be found by iteration. We next evaluate $\mathcal{H}_0'(1)$ and $\mathcal{H}_1'(1)$, where $\mathcal{H}_0'(z)$ and $\mathcal{H}_1'(z)$ are the first derivative of $\mathcal{H}_0(z)$ and $\mathcal{H}_1(z)$ with respect to z respectively. Note that $\mathcal{H}_0(1) = \mathcal{H}_1(1) = 1$. Taking the derivative of $\mathcal{H}_0(z)$ with respect to z yields

$$\mathcal{H}_0'(z) = \sum_{k\in\mathbb{N}} p_\nu(k) \{\mathcal{H}_1(z)\}^k + z \sum_{k\in\mathbb{P}} k p_\nu(k) \{\mathcal{H}_1(z)\}^{k-1} \mathcal{H}_1'(z)$$

$$\mathcal{H}_0'(1) = \sum_{k\in\mathbb{N}} p_\nu(k) + \sum_{k\in\mathbb{P}} k p_\nu(k) \mathcal{H}_1'(1) = 1 + \mathcal{G}_0'(1)\mathcal{H}_1'(1)$$

Similarly

$$\mathcal{H}_1'(1) = 1 + \mathcal{G}_1'(1)\mathcal{H}_1'(1)$$

$$\mathcal{H}_1'(1) = \frac{1}{1 - \mathcal{G}_1'(1)} \tag{7.16a}$$

Therefore

$$\mathcal{H}_0'(1) = 1 + \frac{\mathcal{G}_0'(1)}{1 - \mathcal{G}_1'(1)} \tag{7.16b}$$

Recall that $\mathcal{G}_0'(1) = \overline{k} = \xi_1$ and $\mathcal{G}_1'(1) = \xi_2/\xi_1$. Denote the mean graph component size below phase transition by φ_{bl}. It is equal to $\mathcal{H}_0'(1)$.

$$\varphi_{bl} = 1 + \frac{\xi_1^2}{\xi_1 - \xi_2} \tag{7.17}$$

Note that at $\xi_1 = \xi_2$, the expression for the mean component size diverges. This in turn implies the formation of the giant component. This condition for phase transition has also been derived in an earlier subsection.

Distribution of Component Sizes above the Phase Transition Point

A giant component exists above the phase transition point. Consequently as $n \to \infty$, the order of the giant component also becomes infinite. Define $\mathcal{H}_0(z)$ and $\mathcal{H}_1(z)$ to be the z-transforms of the distribution of component sizes *excluding* the giant component, above the transition point. Therefore the relationships: $\mathcal{H}_1(z) = z\mathcal{G}_1(\mathcal{H}_1(z))$, and $\mathcal{H}_0(z) = z\mathcal{G}_0(\mathcal{H}_1(z))$ are still valid. However $\mathcal{H}_0(1) \neq 1$, and $\mathcal{H}_1(1) \neq 1$. Note that $\mathcal{H}_0(1)$ is equal to the fraction of vertices which are not in the giant component. Therefore $\widehat{\rho} = (1 - \mathcal{H}_0(1))$ is equal to the fraction of the total number of vertices which belong to the giant component. The fraction $\widehat{\rho}$ can be determined from the following relationships. Let $v = \mathcal{H}_1(1)$, then

$$\mathcal{H}_0(1) = \mathcal{G}_0(v), \quad \widehat{\rho} = (1 - \mathcal{G}_0(v)), \quad v = \mathcal{G}_1(v)$$

The value $\widehat{\rho}$ can be determined iteratively in general. The average size of the nongiant components φ_{ab} is

$$\varphi_{ab} = \frac{\mathcal{H}_0'(1)}{\mathcal{H}_0(1)}$$

This value can be determined by differentiating the z-transform functions with respect to z. Thus

$$\begin{aligned}
\mathcal{H}_1'(1) &= \frac{\mathcal{G}_1(\mathcal{H}_1(1))}{\{1 - \mathcal{G}_1'(\mathcal{H}_1(1))\}} \\
\mathcal{H}_0'(1) &= \mathcal{G}_0(\mathcal{H}_1(1)) + \mathcal{G}_0'(\mathcal{H}_1(1))\,\mathcal{H}_1'(1) \\
&= \mathcal{H}_0(1) + \frac{\mathcal{G}_0'(\mathcal{H}_1(1))\,\mathcal{G}_1(\mathcal{H}_1(1))}{\{1 - \mathcal{G}_1'(\mathcal{H}_1(1))\}}
\end{aligned}$$

Therefore

$$\begin{aligned}
\frac{\mathcal{H}_0'(1)}{\mathcal{H}_0(1)} &= 1 + \frac{\mathcal{G}_0'(\mathcal{H}_1(1))\,\mathcal{G}_1(\mathcal{H}_1(1))}{\{1 - \mathcal{G}_1'(\mathcal{H}_1(1))\}\,\mathcal{H}_0(1)} \\
&= 1 + \frac{v\mathcal{G}_0'(v)}{\{1 - \mathcal{G}_1'(v)\}\,\mathcal{G}_0(v)}
\end{aligned}$$

Note that

$$\mathcal{G}_0'(v) = \sum_{k=1}^{\infty} k p_\nu(k)\, v^{k-1} = \xi_1 \sum_{k=1}^{\infty} q_\nu(k-1)\, v^{k-1} = \xi_1 \mathcal{G}_1(v) = \xi_1 v$$

Finally

$$\varphi_{ab} = 1 + \frac{\xi_1 v^2}{\{1 - \mathcal{G}_1'(v)\}(1 - \widehat{\rho})} \tag{7.18}$$

If $\widehat{\rho} = 0$, then $v = 1$ and the expression for φ_{ab} reduces to the one derived for the average component size below the transition.

7.6.5 Examples

The concepts discussed in this section are further clarified via several examples.

Example 7.3. The degree distribution of a random graph has a Poisson distribution with parameter λ. Therefore the z-transform of this distribution is $\mathcal{G}_0(z) = e^{\lambda(z-1)}$. The expected value of the degree is equal to $\mathcal{E}(\nu) = \overline{k} = \xi_1 = \lambda$. Also

$$\mathcal{G}_1(z) = \frac{1}{\xi_1}\mathcal{G}_0'(z) = e^{\lambda(z-1)} = \mathcal{G}_0(z)$$

That is, the distribution of outgoing edges at a node is the same, irrespective of whether we reached there by selecting a vertex at random or by following a randomly chosen edge. For the Poisson distributed random variable, the second moment is equal to $\mu_2 = \left(\lambda + \lambda^2\right)$. Then $\xi_2 = \left(\mu_2 - \overline{k}\right) = \lambda^2$. Therefore the average distance ℓ is given by

$$\ell = \frac{\ln\left(n/\xi_1\right)}{\ln\left(\xi_2/\xi_1\right)} + 1 = \frac{\ln n}{\ln \lambda}$$

The phase transition occurs at $\xi_2 = \xi_1$, that is when $\lambda = 1$. This corresponds to a Model A graph of order $n = 1/p$. The mean graph component size below phase transition φ_{bl} is

$$\varphi_{bl} = \frac{1}{(1 - \lambda)}$$

The average size of the nongiant components φ_{ab} is

$$\varphi_{ab} = \frac{1}{\{1 - \lambda\left(1 - \widehat{\rho}\right)\}}$$

where

$$\widehat{\rho} = 1 - e^{-\lambda\widehat{\rho}}$$

Recall that $\widehat{\rho} < 1$ is the size of the giant component as a fraction of the total graph order n. Therefore an expression for $\widehat{\rho}$ can be obtained explicitly via Lagrange's series expansion. Thus

$$\widehat{\rho} = 1 - \frac{1}{\lambda}\sum_{j\in\mathbb{P}} \frac{j^{j-1}}{j!}\left(\lambda e^{-\lambda}\right)^j$$

The coefficient of variation of a Poisson distributed random variable is equal to $C_v = 1/\sqrt{\lambda}$. Therefore the clustering coefficient of the random graph is

$$C_G = \frac{\lambda}{n}$$

If $\lambda = np$, then $C_G = p$. □

Example 7.4. If the vertex distribution has a truncated power-law distribution then

$$\mathcal{G}_0(z) = \frac{Li_\gamma\left(ze^{-1/\kappa}\right)}{Li_\gamma\left(e^{-1/\kappa}\right)}$$

$$\mathcal{G}_0'(1) = \frac{Li_{\gamma-1}\left(e^{-1/\kappa}\right)}{Li_\gamma\left(e^{-1/\kappa}\right)} = \xi_1$$

$$\mathcal{G}_1(z) = \frac{Li_{\gamma-1}\left(ze^{-1/\kappa}\right)}{zLi_{\gamma-1}\left(e^{-1/\kappa}\right)}$$

where $Li_\gamma(\cdot)$ is the γ-th polylogarithm function. Denoting the second derivative of $\mathcal{G}_0(z)$ with respect to z by $\mathcal{G}_0''(z)$ yields

$$\xi_2 = \mathcal{G}_0''(1) = \frac{\left\{Li_{\gamma-2}\left(e^{-1/\kappa}\right) - Li_{\gamma-1}\left(e^{-1/\kappa}\right)\right\}}{Li_\gamma\left(e^{-1/\kappa}\right)}$$

The average vertex to vertex distance ℓ is obtained by using the value ξ_1 and ξ_2.

$$\begin{aligned}\ell &= \frac{\ln(n/\xi_1)}{\ln(\xi_2/\xi_1)} + 1 \\ &= \frac{\ln n + \ln\left\{Li_\gamma\left(e^{-1/\kappa}\right)/Li_{\gamma-1}\left(e^{-1/\kappa}\right)\right\}}{\ln\left\{Li_{\gamma-2}\left(e^{-1/\kappa}\right)/Li_{\gamma-1}\left(e^{-1/\kappa}\right) - 1\right\}} + 1\end{aligned}$$

As $\kappa \to \infty$

$$\begin{aligned}\mathcal{G}_0(z) &= \frac{Li_\gamma(z)}{\zeta(\gamma)}, \quad \gamma > 1 \\ \mathcal{G}_0'(1) &= \frac{\zeta(\gamma-1)}{\zeta(\gamma)} = \xi_1, \quad \gamma > 2 \\ \mathcal{G}_1(z) &= \frac{Li_{\gamma-1}(z)}{z\zeta(\gamma-1)}, \quad \gamma > 2 \\ \xi_2 = \mathcal{G}_0''(1) &= \frac{\{\zeta(\gamma-2) - \zeta(\gamma-1)\}}{\zeta(\gamma)}, \quad \gamma > 3\end{aligned}$$

where $\zeta(\cdot)$ is Riemann's zeta function. Also the corresponding average vertex to vertex distance ℓ is

$$\ell = \frac{\ln n + \ln\{\zeta(\gamma)/\zeta(\gamma-1)\}}{\ln\{\zeta(\gamma-2)/\zeta(\gamma-1) - 1\}} + 1, \quad \gamma > 3$$

The phase transition occurs at $\xi_2 = \xi_1$. This occurs if $\zeta(\gamma-2) = 2\zeta(\gamma-1)$. This gives a critical value of γ, which is $\gamma_c = 3.4788\ldots$. Therefore a giant component exists below this value of γ, and there is no giant component above this value. Use of the relationship $v = \mathcal{G}_1(v)$, yields

$$v = \frac{Li_{\gamma-1}(v)}{v\zeta(\gamma-1)}$$

where $2 < \gamma < \gamma_c$. The size of the giant component can be calculated using the above equation for $\gamma \in (2, \gamma_c)$. However, if $\gamma \le 2$, $v = 0$ also satisfies the above equation. Consequently $\widehat{\rho} = (1 - \mathcal{G}_0(0)) \to 1$. This implies that the giant component is the entire graph in this regime of γ. □

Example 7.5. This example is about clustering coefficient of a graph.

(a) If the degree distribution of the graph is Poissonian with parameter $\lambda = np$. It has already been shown that $C_G = p$. This can also be verified by using the general expression for C_G.
(b) Next, consider a graph with power-law degree distribution. Let the exponent of the power-law be $2 < \gamma < 3$; and $k_c \simeq n^{1/(\gamma-1)}$. Then $\xi_1 = \overline{k} \approx 1$, and $\xi_2 = \left(\mu_2 - \overline{k}\right) \simeq \mu_2 \simeq k_c^{(3-\gamma)} \simeq n^{(3-\gamma)/(\gamma-1)}$. Thus

$$C_G \simeq \frac{\xi_2^2}{n} \simeq n^{(7-3\gamma)/(\gamma-1)}$$

Since $C_G \le 1$, the above approximation is valid provided $7/3 < \gamma < 3$. □

Example 7.6. Recall that the heterogeneity parameter $\varkappa$ of a random graph is the ratio of the second and first moment of the random variable ν.

(a) The heterogeneity parameter $\varkappa$ of a random graph, whose degree distribution is Poissonian with parameter λ, is equal to $(\lambda + 1)$.

(b) For a random graph with power-law degree distribution, with exponent γ, the heterogeneity parameter $\varkappa$ is given by:

$$\varkappa \simeq \begin{cases} k_0 \left(k_c/k_0\right)^{(3-\gamma)}, & 2<\gamma<3 \\ \{k_0 \ln n\}/2, & \gamma=3 \end{cases}$$

where, n is the order of the network. □

7.7 The $\mathcal{K}$-core of a Random Graph

The $\mathcal{K}$-core of an undirected finite graph is its maximal (largest) subgraph with smallest vertex-degree at least $\mathcal{K}$. The question we seek to address in this section is whether a nonnull $\mathcal{K}$-core exists in a random graph $G(n,p)$, where the number of vertices $n \to \infty$, and the edge existence probability is $p \in (0,1)$. The discussion in this section is presumed to be nontechnical.

It is known that if $n \to \infty$, and $p \to 0$, then $np \to \lambda \in \mathbb{R}^+$, and the degree distribution of the random graph is Poissonian with parameter λ. Assume that the vertex-degree distribution of the graph is $P(\nu = k) \triangleq p_\nu(k), \forall\, k \in \mathbb{N}$. Therefore $p_\nu(k) = e^{-\lambda}\lambda^k/k!$, $\forall\, k \in \mathbb{N}$, and the average value of the vertex degree is $\mathcal{E}(\nu) = \lambda$, where $\mathcal{E}(\cdot)$ is the expectation operator.

The $\mathcal{K}$-core of a random graph is generated as per the following algorithm. In each step of the algorithm, a single vertex of degree smaller than $\mathcal{K}$ and the attached edges are removed from the graph. Thus after T steps, there are $(n-T)$ vertices left in the graph, where $T \in \{0,1,2,\ldots,n\}$. Define time $t \triangleq T/n$ and $\Delta t \triangleq 1/n$. Observe that $t \in [0,1]$. The introduction of the variable t is important, because, if $n \to \infty$, then $\Delta t \to 0$, and t takes all values in the interval $[0,1]$. Furthermore, this allows us to convert difference equations into differential equations.

Let the vertex-degree random variable at time t be $\nu(t)$. In this notation, $\nu(0)$ is simply equal to ν. The vertex-degree distribution at time t is $P(\nu(t) = k) \triangleq p_{\nu(t)}(k), \forall\, k \in \mathbb{N}$. Also assume that the average value of the vertex degree at time t is $\overline{k}(t) = \mathcal{E}(\nu(t))$. In this notation $\overline{k}(0) = \lambda$. Denote the total number of vertices after the Tth step of the algorithm by $N(T)$. Thus $N(T) = n(1-t)$. Also let $N_k(T)$ be the number of vertices, each with k neighbors at time t. Thus

$$p_{\nu(t)}(k) = \frac{N_k(T)}{N(T)}, \quad \forall\, k \in \mathbb{N}, \quad t \in [0,1] \tag{7.19a}$$

An expression for $N_k(T+1)$ can be written in terms of $N_k(T)$, the degree distribution $p_{\nu(t)}(k), \forall\, k \in \mathbb{N}$ of the graph at time t, and an indicator function $I_k(t)$ for vertices of degree at most $(\mathcal{K}-1)$ at time t. Thus

$$I_k(t) = \begin{cases} 1, & \text{if } k < \mathcal{K} \\ 0, & \text{if } k \geq \mathcal{K} \end{cases} \tag{7.19b}$$

The average value of this indicator function at time t, is $\mathcal{E}(I_k(t)) = \sum_k I_k(t)\, p_{\nu(t)}(k) = \sum_{k=0}^{(\mathcal{K}-1)} p_{\nu(t)}(k)$. Thus the expected change of $N_k(\cdot)$ in the $(T+1)$st step is

$$\begin{aligned} &N_k(T+1) - N_k(T) \\ &= -\frac{I_k(t)\, p_{\nu(t)}(k)}{\mathcal{E}(I_k(t))} + \frac{\mathcal{E}(kI_k(t))}{\mathcal{E}(I_k(t))}\left[-\frac{kp_{\nu(t)}(k)}{\overline{k}(t)} + \frac{(k+1)\, p_{\nu(t)}(k+1)}{\overline{k}(t)}\right] \end{aligned} \tag{7.19c}$$

for all $k \in \mathbb{N}$. There are two terms in the right-hand side of the above difference equation. The first term is due to the removal of the vertex in the $(T+1)$st step. Let the degree of the current vertex be k. The first term is present if k is less than $\mathcal{K}$, and absent if k is greater than or equal to $\mathcal{K}$. That is, a current vertex with a degree higher than $k < \mathcal{K}$ is not removed. We ensure this by the use of the indicator function $I_k(t)$.

The second term is due to the contribution of the neighbors of the removed vertex. The degree of these vertices decrements by one. There are $\mathcal{E}(kI_k(t))/\mathcal{E}(I_k(t))$ neighbors, each of them has a degree k with probability $q_{\nu(t)}(k) = kp_{\nu(t)}(k)/\overline{k}(t)$. The negative term $-kp_{\nu(t)}(k)/\overline{k}(t)$ is due to the vertices with degree k, and the positive term $(k+1)p_{\nu(t)}(k+1)/\overline{k}(t)$ is due to the vertices with degree $(k+1)$, before the removal of their incident edges. As

$$N_k(T) = N(T)p_{\nu(t)}(k) = \frac{(1-t)p_{\nu(t)}(k)}{\Delta t}$$

we obtain

$$\begin{aligned} N_k(T+1) - N_k(T) &= \frac{(1-t-\Delta t)p_{\nu(t+\Delta t)}(k) - (1-t)p_{\nu(t)}(k)}{\Delta t} \\ &= \frac{d}{dt}\{(1-t)p_{\nu(t)}(k)\}, \quad \text{as } n \to \infty \end{aligned}$$

Thus

$$\begin{aligned} &\frac{d}{dt}\{(1-t)p_{\nu(t)}(k)\} \\ &= -\frac{I_k(t)p_{\nu(t)}(k)}{\mathcal{E}(I_k(t))} + \frac{\mathcal{E}(kI_k(t))}{\mathcal{E}(I_k(t))}\left[-\frac{kp_{\nu(t)}(k)}{\overline{k}(t)} + \frac{(k+1)p_{\nu(t)}(k+1)}{\overline{k}(t)}\right] \end{aligned} \tag{7.20}$$

for all $k \in \mathbb{N}$. The above infinite set of equations are analysed in several steps.

Step 1: If the above set of equations are summed over all values of k, we simply obtain the consistency relationship $\frac{d}{dt}(1-t) = -1$.

Step 2: However, if the above equation is multiplied on both sides by k and summed over all values of k yields

$$\begin{aligned} &\frac{d}{dt}\{(1-t)\mathcal{E}(\nu(t))\} \\ &= \frac{\mathcal{E}(kI_k(t))}{\mathcal{E}(I_k(t))}\left[-1 - \frac{\mathcal{E}\left(\nu(t)^2\right)}{\overline{k}(t)} + \frac{\mathcal{E}(\nu(t)(\nu(t)-1))}{\overline{k}(t)}\right] \\ &= -2\frac{\mathcal{E}(kI_k(t))}{\mathcal{E}(I_k(t))} \end{aligned}$$

Let $\widetilde{m}(t) \triangleq \{(1-t)\mathcal{E}(\nu(t))\}$, then

$$\frac{d\widetilde{m}(t)}{dt} = -2\frac{\mathcal{E}(kI_k(t))}{\mathcal{E}(I_k(t))}$$

where $\widetilde{m}(0) = \lambda$.

Step 3: A crucial simplification of the infinite set of difference-differential equations is obtained by noting that the algorithm never directly removes a vertex of degree $k \geq \mathcal{K}$. The properties of these vertices are possibly modified by the random removal of their incident edges. Therefore, the degree distribution of these vertices is still Poissonian, with parameter $\widetilde{\beta}(t)$, where $\widetilde{\beta}(0) = \lambda$. Thus

$$(1-t)\, p_{\nu(t)}(k) = \frac{N_k(T)}{n} = e^{-\widetilde{\beta}(t)} \frac{\widetilde{\beta}(t)^k}{k!}, \quad \forall\, k \geq \mathcal{K} \tag{7.21}$$

Substitution of this observation in the equation

$$\frac{d}{dt}\{(1-t)\, p_{\nu(t)}(k)\} = \frac{\mathcal{E}(kI_k(t))}{\mathcal{E}(I_k(t))}\left[-\frac{kp_{\nu(t)}(k)}{\overline{k}(t)} + \frac{(k+1)\, p_{\nu(t)}(k+1)}{\overline{k}(t)}\right], \quad \forall\, k \geq \mathcal{K}$$

yields

$$\frac{d}{dt}\widetilde{\beta}(t) = -\frac{\widetilde{\beta}(t)}{\widetilde{m}(t)} \frac{\mathcal{E}(kI_k(t))}{\mathcal{E}(I_k(t))}$$

The above result is independent of $\mathcal{K}$. Also, $d\widetilde{m}(t)/dt = -2\mathcal{E}(kI_k(t))/\mathcal{E}(I_k(t))$, via Step 2. Thus

$$\frac{2}{\widetilde{\beta}(t)} \frac{d}{dt}\widetilde{\beta}(t) = \frac{1}{\widetilde{m}(t)} \frac{d}{dt}\widetilde{m}(t)$$

Using the initial conditions $\widetilde{\beta}(0) = \widetilde{m}(0) = \lambda$, we obtain

$$\widetilde{m}(t) = \frac{\widetilde{\beta}(t)^2}{\lambda} \tag{7.22}$$

and

$$\frac{d}{dt}\widetilde{\beta}(t) = -\frac{\lambda}{\widetilde{\beta}(t)} \frac{\mathcal{E}(kI_k(t))}{\mathcal{E}(I_k(t))}$$

Step 4: In this step $\mathcal{E}(kI_k(t))$ and $\mathcal{E}(I_k(t))$ are eliminated from the above equation. Define

$$\widetilde{E}_0(x) \triangleq 1, \quad x \in \mathbb{R}^+ \tag{7.23a}$$

$$\widetilde{E}_{\mathcal{K}}(x) \triangleq 1 - \sum_{k=0}^{(\mathcal{K}-1)} e^{-x} \frac{x^k}{k!}, \quad x \in \mathbb{R}^+ \tag{7.23b}$$

Then

$$1 = \sum_{k \in \mathbb{N}} p_{\nu(t)}(k) = \sum_{k \in \mathbb{N}} I_k(t)\, p_{\nu(t)}(k) + \sum_{k=\mathcal{K}}^{\infty} \frac{1}{(1-t)} e^{-\widetilde{\beta}(t)} \frac{\widetilde{\beta}(t)^k}{k!}$$

Thus

$$1 = \mathcal{E}(I_k(t)) + \frac{1}{(1-t)} \widetilde{E}_{\mathcal{K}}\left(\widetilde{\beta}(t)\right)$$

Similarly, using the relationship $\overline{k}(t) = \sum_{k \in \mathbb{N}} kp_{\nu(t)}(k)$, we obtain

$$\overline{k}(t) = \mathcal{E}(kI_k(t)) + \frac{\widetilde{\beta}(t)}{(1-t)} \widetilde{E}_{\mathcal{K}-1}\left(\widetilde{\beta}(t)\right)$$

Step 5: Use of results from Steps 3 and 4 yields

$$\frac{d}{dt}\widetilde{\beta}(t) = -\frac{\widetilde{\beta}(t) - \lambda \widetilde{E}_{\mathcal{K}-1}\left(\widetilde{\beta}(t)\right)}{1 - t - \widetilde{E}_{\mathcal{K}}\left(\widetilde{\beta}(t)\right)} \tag{7.24}$$

The above equation is nonlinear, and therefore difficult to solve for an arbitrary value of $\mathcal{K}$.

Step 6: Nevertheless significant properties of the $\mathcal{K}$-core of the graph can be determined by finding the halting point of the algorithm. The algorithm stops at time t_f when the $\mathcal{K}$-core of the graph has been determined. It terminates at time $t_f \in (0,1)$, when the degree of the remaining vertices is at least greater than or equal to $\mathcal{K}$. Just after the time instant t_f, the remaining graph has $n(1-t_f)$ vertices, $I_k(t_f) = 0$, $\mathcal{E}(I_k(t_f)) = \mathcal{E}(kI_k(t_f)) = 0$, and $\widetilde{\beta}_f \triangleq \widetilde{\beta}(t_f) > 0$. Using results in Steps 3 and 4 we obtain

$$1 - t_f = \widetilde{E}_{\mathcal{K}}\left(\widetilde{\beta}_f\right) \tag{7.25a}$$

$$\frac{\widetilde{\beta}_f}{\lambda} = \widetilde{E}_{\mathcal{K}-1}\left(\widetilde{\beta}_f\right) \tag{7.25b}$$

Observe that $\widetilde{\beta}_f$ is determined from the last equation. If there is more than a single positive solution for $\widetilde{\beta}_f$, then the largest one, but smaller than λ is selected. The halting time t_f is determined from the relationship

$$t_f = 1 - \widetilde{E}_{\mathcal{K}}\left(\widetilde{\beta}_f\right)$$

The number of vertices in the $\mathcal{K}$-core of the graph, if it exists, is equal to $n_{\mathcal{K}}(\lambda) \triangleq n(1-t_f)$. The corresponding number of edges in the $\mathcal{K}$-core of the graph is equal to $e_{\mathcal{K}}(\lambda)$.

$$e_{\mathcal{K}}(\lambda) = \frac{\overline{k}(t_f)\, n_{\mathcal{K}}(\lambda)}{2} = \frac{1}{2\lambda} n \widetilde{\beta}_f^2$$

Conditions for the Existence of $\widetilde{\beta}_f$

The smallest value of λ above which nonempty $\mathcal{K}$-core of the graph $G(n,p)$ exists is called its threshold. This threshold value is denoted by $\lambda_{\mathcal{K}}$. Therefore $\widetilde{\beta}_f \in \mathbb{R}^+$ exists if λ is greater than $\lambda_{\mathcal{K}}$.

Case $\mathcal{K} = 1$: The 1-core of the graph is determined by simply removing vertices of zero degree. Thus $n_1(\lambda) = n(1 - p_\nu(0)) = n\left(1 - e^{-\lambda}\right)$.

Alternately, observe that the 1-core of the graph always exists for any $\lambda \in \mathbb{R}^+$, and $t_f = e^{-\lambda}$. Thus the number of vertices in the 1-core of the graph is equal to $n_1(\lambda) = n\left(1 - e^{-\lambda}\right)$.

Case $\mathcal{K} = 2$: Define $\varphi(x) = x/\widetilde{E}_{\mathcal{K}-1}(x)$. We need to determine, if there exists a value of x which satisfies $\lambda = \varphi(x)$. Observe that $\varphi(x) = x/\left(1 - e^{-x}\right)$ is strictly increasing from 1 to ∞ for $0 < x < \infty$. Therefore the threshold value of λ is $\lambda_2 = 1$. That is, if $\lambda > \lambda_2$, then there is exactly a single positive solution $x = \widetilde{\beta}_f$, of the equation $\lambda = \varphi(x)$. Furthermore, $\varphi(x) < \lambda$ for $x \in \left(0, \widetilde{\beta}_f\right)$.

Case $\mathcal{K} \geq 3$: We need to determine whether $\lambda = x/\widetilde{E}_{\mathcal{K}-1}(x)$ is possible. The following lemma is initially established.

Lemma 7.2. *Let $\mathcal{K} \geq 3$, and $\varphi(x) \triangleq x/\widetilde{E}_{\mathcal{K}-1}(x)$. Denote the first derivative of $\varphi(x)$, with respect x by $\varphi'(x)$. Then $\varphi(x) \to \infty$, as $x \to 0$ or $x \to \infty$. Also $\varphi(x)$ has a unique minimum value at $x_0 > 0$. That is, $\varphi'(x) < 0$ for $0 < x < x_0$ and $\varphi'(x) > 0$ for $x > x_0$.*

Proof. Observe that $\varphi(x) > 0$, and it is differentiable for all values of $x \in \mathbb{R}^+$. It also follows from the definition of $\varphi(x)$ that $\varphi(x) \to \infty$, as $x \to 0$ or $x \to \infty$. Therefore, $\varphi(x)$ has a minimum value at some $x_0 > 0$. Denote the first derivative of $\widetilde{E}_{\mathcal{K}-1}(x)$, with respect x by $\widetilde{E}'_{\mathcal{K}-1}(x)$. It can be shown that

$$\widetilde{E}'_{\mathcal{K}-1}(x) = e^{-x}\frac{x^{(\mathcal{K}-2)}}{(\mathcal{K}-2)!}$$

Therefore

$$\frac{\widetilde{E}_{\mathcal{K}-1}(x)}{x\widetilde{E}'_{\mathcal{K}-1}(x)} = (\mathcal{K}-2)!\sum_{j\in\mathbb{N}}\frac{x^j}{(j+\mathcal{K}-1)!}$$

is an increasing function of $x \in \mathbb{R}^+$. Thus

$$x\frac{d}{dx}\ln\varphi(x) = 1 - x\frac{\widetilde{E}'_{\mathcal{K}-1}(x)}{\widetilde{E}_{\mathcal{K}-1}(x)}$$

is increasing in $x \in \mathbb{R}^+$. As $\ln\varphi(x)$ attains a minimum value at x_0, we have

$$\left.\frac{d}{dx}\ln\varphi(x)\right|_{x=x_0} = 0$$

Therefore, since $\frac{d}{dx}\ln\varphi(x)$ is an increasing function of $x \in \mathbb{R}^+$, we have $\frac{d}{dx}\ln\varphi(x) < 0$ for $x < x_0$ and $\frac{d}{dx}\ln\varphi(x) > 0$ for $x > x_0$. □

Define $\lambda_{\mathcal{K}} = \varphi(x_0)$, where x_0 is the value of $x \in \mathbb{R}^+$ at which $\varphi(x) \triangleq x/\widetilde{E}_{\mathcal{K}-1}(x)$ is minimum.

If $\lambda > \lambda_{\mathcal{K}}$, two values of x satisfy $\lambda = \varphi(x)$. Let these values be $x^-(\lambda)$ and $x^+(\lambda)$, such that $0 < x^-(\lambda) < x_0 < x^+(\lambda)$. Furthermore, $\varphi(x) < \lambda$ for $x \in (x^-(\lambda), x^+(\lambda))$. As mentioned earlier, of the two values of x which satisfy $\lambda = \varphi(x)$, the larger of the two values, which is smaller than λ is chosen for $\widetilde{\beta}_f$.The above cases are summarized in the next set of observations.

Observations 7.3. Let $G(n, p)$ be a random graph, where the number of vertices $n \to \infty$, and the edge existence probability is $p \in (0, 1)$. Assume that, as $n \to \infty$ and $p \to 0$, we have $np \to \lambda \in \mathbb{R}^+$. The average vertex degree in this graph is equal to λ, and the average number of edges is equal to $n\lambda/2$.

1. The 1-core of the graph always exists. The average number of vertices and edges in the 1-core are equal to $n_1(\lambda) = n(1 - e^{-\lambda})$, and $e_1(\lambda) = n\lambda/2$ respectively.
2. Consider two cases if $\mathcal{K} \geq 2$.
   (a) Let $\mathcal{K} = 2$. For the 2-core of the graph to exist, the threshold value of λ is equal to $\lambda_2 = 1$. That is, if $\lambda > \lambda_2$, then there is exactly a single positive solution $x = \widetilde{\beta}_f$, of the equation $\lambda = x/(1 - e^{-x})$.
   (b) Let $\mathcal{K} \geq 3$, and $x_0 \in \mathbb{R}^+$ be the unique minimizing value of $\varphi(x) \triangleq x/\widetilde{E}_{\mathcal{K}-1}(x)$, where $x \in \mathbb{R}^+$ and $\widetilde{E}_{\mathcal{K}}(x) = 1 - \sum_{k=0}^{(\mathcal{K}-1)} e^{-x}\frac{x^k}{k!}$. The value x_0 is a zero if

$$1 - \sum_{k=0}^{(\mathcal{K}-2)} e^{-x} \frac{x^k}{k!} - e^{-x} \frac{x^{(\mathcal{K}-1)}}{(\mathcal{K}-2)!} = 0 \tag{7.26a}$$

and $\lambda_{\mathcal{K}} = \varphi(x_0)$.
Therefore if $\lambda > \lambda_{\mathcal{K}}$ the equation $\varphi(x) = \lambda$ is satisfied by two values of x. The larger of the two values, which is smaller than than λ is chosen to be $\widetilde{\beta}_f$.

If the $\mathcal{K}$-core of the graph exists, then the number of vertices and edges in it are

$$n_{\mathcal{K}}(\lambda) = n\widetilde{E}_{\mathcal{K}}\left(\widetilde{\beta}_f\right), \text{ and } e_{\mathcal{K}}(\lambda) = \frac{n\widetilde{\beta}_f^2}{2\lambda}, \quad \mathcal{K} \geq 2 \tag{7.26b}$$

respectively. □

The threshold value $\lambda_{\mathcal{K}}$ is computed for some elementary cases in the following examples.

Examples 7.7. Some illustrative examples.

1. Let $\mathcal{K} = 3$, then x_0 is a zero of

$$1 - \sum_{k=0}^{1} e^{-x} \frac{x^k}{k!} - e^{-x} x^2 = 0$$

The above equation yields $x_0 = 1.7933$. Also $\lambda_3 = \varphi(x_0)$, where

$$\varphi(x) = \frac{x}{1 - \sum_{k=0}^{1} e^{-x} \frac{x^k}{k!}}$$

Thus $\lambda_3 = 3.3509$. Fraction of vertices in the 3-core of the graph at $\lambda = \lambda_3$ is $\widetilde{E}_3(x_0)$. It is

$$\widetilde{E}_3(x_0) = 1 - \sum_{k=0}^{2} e^{-x_0} \frac{x_0^k}{k!} = 0.26759$$

2. Let $\mathcal{K} = 4$, then x_0 is a zero of

$$1 - \sum_{k=0}^{2} e^{-x} \frac{x^k}{k!} - e^{-x} \frac{x^3}{2!} = 0$$

The above equation yields $x_0 = 3.3836$. Also $\lambda_4 = \varphi(x_0)$, where

$$\varphi(x) = \frac{x}{1 - \sum_{k=0}^{2} e^{-x} \frac{x^k}{k!}}$$

Thus $\lambda_4 = 5.1493$. Fraction of vertices in the 4-core of the graph at $\lambda = \lambda_4$ is $\widetilde{E}_4(x_0)$. It is

$$\widetilde{E}_4(x_0) = 1 - \sum_{k=0}^{3} e^{-x_0} \frac{x_0^k}{k!} = 0.43805$$

3. Let $\mathcal{K} = 5$, then x_0 is a zero of

$$1 - \sum_{k=0}^{3} e^{-x} \frac{x^k}{k!} - e^{-x} \frac{x^4}{3!} = 0$$

The above equation yields $x_0 = 4.8813$. Also $\lambda_5 = \varphi(x_0)$, where

$$\varphi(x) = \frac{x}{1 - \sum_{k=0}^{3} e^{-x} \frac{x^k}{k!}}$$

Thus $\lambda_5 = 6.7995$. Fraction of vertices in the 5-core of the graph at $\lambda = \lambda_5$ is $\widetilde{E}_5(x_0)$. It is

$$\widetilde{E}_5(x_0) = 1 - \sum_{k=0}^{4} e^{-x_0} \frac{x_0^k}{k!} = 0.53844$$

□

7.8 Graph Spectra

Computation of graph spectra is another useful technique to study a random graph. Let G be an undirected, simple, and labeled graph with n vertices. This graph is completely specified by its adjacency matrix A. If the eigenvalues of this adjacency matrix are $\lambda_1, \lambda_2, \ldots, \lambda_n$, then the spectral density $\rho(\lambda)$ of the graph G is

$$\rho(\lambda) = \frac{1}{n} \sum_{i=1}^{n} \delta(\lambda - \lambda_i) \tag{7.27}$$

In the above equation, $\delta(\cdot)$ is the Dirac's delta function. As $n \to \infty$, $\rho(\cdot)$ approaches a continuous function. The characteristics of graph spectra are studied for two types of random graphs. These are: a random graph, in which as $n \to \infty$, the degree distribution follows the Poisson law; and a random graph, in which the degree distribution follows a power-law.

7.8.1 Graph with Poissonian Degree Distribution

As we studied in the chapter on matrices and determinants, if each element of an infinite random matrix is a Gaussian distributed random variable with zero mean and unit variance, and these elements are uncorrelated; the spectral density of this matrix obeys a semicircle law. This law was discovered by the physicist Eugene Wigner.

Let $B = [B_{ij}]$ be an $n \times n$ real, symmetric, and random matrix. The matrix elements $B_{ij}, 1 \le i \le j \le n$ are stochastically independent Gaussian random variables, where $B_{ii} \sim \mathcal{N}(0, 2\sigma^2), 1 \le i \le n$, and $B_{ij} \sim \mathcal{N}(0, \sigma^2), 1 \le j < i \le n$. It can be shown that as $n \to \infty$ the density of eigenvalues of the matrix $B/\sqrt{n}$ converges to the semicircular distribution:

$$\rho(\lambda) = \begin{cases} \dfrac{1}{2\pi\sigma^2}\sqrt{4\sigma^2 - \lambda^2}, & \text{if } |\lambda| < 2\sigma \\ 0, & \text{otherwise} \end{cases} \tag{7.28}$$

Many of the assumptions made in deriving the semicircle law do not exist for the adjacency matrix of an uncorrelated random graph. The off-diagonal elements of the adjacency matrix A, of a $G(n,p)$ graph are equal to either 1 or 0, where $0 < p < 1$. That is, the degree distribution of this graph is Poissonian. The off-diagonal entries 1 and 0 of this symmetric matrix A occur with probabilities p and $(1-p)$ respectively.

Thus these matrix elements are independently distributed Bernoulli random variables with parameter p. Since $p \neq 0$, the mean of these nondiagonal elements is nonzero. It can be shown that, if the matrix A is normalized by $\sqrt{n}$, and as $n \to \infty$, its spectral density has a semicircular form. Furthermore, its eigenvalues lie in the interval $(-2\sigma, 2\sigma)$, where $\sigma^2 = p(1-p)$.

Also note that in this random graph, the average value of the degree of a node is equal to $(n-1)p$, and the average number of edges is equal to $n(n-1)p/2$. The later expression is an indication of connectivity of the n nodes.

7.8.2 Graph with Power-Law Degree Distribution

The spectra of graphs with power-law degree distribution is studied in this subsection. It is demonstrated that the largest eigenvalues of such graphs also follow a power-law probability distribution. Besides, there is empirical evidence that the larger eigenvalues of the adjacency matrix of such graphs also observe a power-law.

Let $G = (V, E)$ be an undirected, simple, and labeled graph, where V is the set of vertices, and E is the set of edges. In this graph, the number of vertices is equal to $|V| = n$. The eigenvalues of its adjacency matrix are $\lambda_1 \leq \lambda_2 \leq \ldots \leq \lambda_n$. Also, let the maximum degree of a vertex be $d_{\max}$, then it has been established in the chapter on graph theory that

$$|\lambda_i| \leq \min\left\{d_{\max}, \sqrt{2|E|}\right\}, \quad 1 \leq i \leq n$$

Also recall that the eigenvalues of a star graph S_n for $n \geq 3$ are: 0 and $\pm\sqrt{n-1}$. The multiplicity of the eigenvalue 0 is $(n-2)$. In addition to these facts, use of the interlacing eigenvalue theorem yields the following lemma.

Lemma 7.3. *Let G be an undirected and simple graph. It is a union of the following sets of graphs:*

(a) *Vertex disjoint stars $S_{(d_i+1)}$, with (d_i+1) set of nodes, for $1 \leq i \leq c$. Also assume that $d_i \leq d_{i+1}$, for $1 \leq i \leq (c-1)$.*

(b) *Vertex disjoint graph components $G_j, 1 \leq j \leq l$, each with maximum degree of a vertex equal to $d_{\max}(G_j)$ and edge sets E_{G_j}, such that their maximum eigenvalues are less than those of any of the star graphs. That is*

$$\min\left\{d_{\max}(G_j), \sqrt{2|E_{G_j}|}\right\} = o(d_1), \quad 1 \leq j \leq l \tag{7.29a}$$

Furthermore, these graph components are disjoint from all the stars $S_{(d_i+1)}$, $1 \leq i \leq c$.

(c) *A graph H, whose edge set is E, and maximum vertex degree is d such that*

$$\min\left\{d, \sqrt{2|E|}\right\} = o(d_1) \tag{7.29b}$$

In addition, this graph is permitted to have arbitrary intersection with the star graphs $S_{(d_i+1)}$'s and the graphs G_j's.

Under the above hypothesis, the largest eigenvalues of the graph G satisfy

$$\sqrt{d_i}\,(1 - o(1)) \leq \lambda_i \leq \sqrt{d_i}\,(1 + o(1)), \quad 1 \leq i \leq c \tag{7.29c}$$

Proof. The proof is left to the reader. □

The graph G in the above lemma has been constructed such that its dominant eigenvalues are those corresponding to the eigenvalues of their star subcomponents. Also the graph H can be any sparse graph. It can be connected, disconnected, or a tree. Assume that the order n of this graph G tends to infinity. Let the vertex degrees in this graph range from k_0 to k_c, and its degree distribution be $p_\nu(k), k_0 \leq k \leq k_c$. Also let $\int_{k_c}^{\infty} p_\nu(k)\, dk \simeq 1/n$. Further assume that the highest degree sequence of this graph $d_1, d_2, \ldots, d_c$ have a power-law distribution with exponent $\gamma > 1$. In the notation of power-law distribution: $d_c = k_c \simeq k_0 n^{1/(\gamma-1)}$. Therefore, the spectral radius of the graph, or the maximum eigenvalue of the graph G described in the lemma is upper bounded by $\sqrt{d_c} \simeq \sqrt{k_0} n^{1/\{2(\gamma-1)\}}$.

7.9 Small-World Networks

Small-world networks are graphs which have small values of average path length ℓ, and high values of the clustering coefficient C_G. This is in direct contrast to classical random networks (which have Poissonian degree distribution). The classical random networks have low values of both the average path length and the clustering coefficient. Both the Internet and the Web graph exhibit the small-world effect. D. J. Watts and S. H. Strogatz did pioneering work in the study of such graphs. Watts and Strogatz constructed a graph which exhibits the small-world effect.

A network (graph) which demonstrates this effect is built as follows. Initially start with a ring-lattice $R_{n,2k}$ where $1 \leq k \leq (n-1)/2$. In this undirected graph of n nodes, every node is connected to $2k$ immediate neighbors (k on either side), and the distance between adjacent nodes is assumed to be unity. For the graph to be sparse assume that $n \gg 2k \gg \ln n \gg 1$. In the next step of this construction, the graph is randomized. Each edge of $R_{n,2k}$ is randomly rewired with probability p such that self-loops and duplicate edges are avoided. This procedure introduces npk long-range edges. If $p = 0$, the graph is $R_{n,2k}$, however if $p = 1$, the graph is completely random. For intermediate values of p, the randomized graph exhibits small-world effect of varying degrees.

Several baroque small-world models can be constructed. For simplicity, a small-world model originally developed by A. Barrat and M. Weigt is studied in this section. It is a variant of the small-world model proposed by Watts and Strogatz. In this model, n nodes are arranged in a ring configuration and each node is connected to its $2k$ nearest neighbors (k on either side). Let the vertex set of this graph be $V = \{1, 2, \ldots, n\}$. Note that the degree of each node is equal to $2k$, and the total number of edges in this network is equal to kn. This deterministic network (graph) is randomized as follows. Some edges are rewired at random with probability p. These edges tend to decrease the average path length ℓ, between these nodes. Consequently these edges are also called shortcuts. At each node, the rewiring scheme is designed as follows.

(a) Links to k neighboring nodes are left untouched.
(b) Of the remaining k edges to the neighboring nodes, some are connected to other nodes with probability p, and the remaining are left untouched with probability $(1 - p)$.

In this randomized graph, the minimum degree of each node is equal to k, the total number of edges in the graph remains unchanged at kn, and the graph remains connected. Its degree distribution, clustering coefficient, and the average path length ℓ averaged over all pairs of vertices is next studied.

Degree Distribution

Let the random variable representing the degree of node i be ϕ_i, for $1 \le i \le n$. Let

$$\phi_i = k + n_i, \quad n_i \ge 0$$
$$n_i = n_i^{(1)} + n_i^{(2)}$$

where $n_i^{(1)} \le k$ links are left in place at node i, each with probability $(1-p)$, and $n_i^{(2)}$ links have been reconnected "towards" node i, each with probability p/n. The probability distributions of the random variables $n_i, n_i^{(1)}$, and $n_i^{(2)}$ are readily obtained.

$$P\left(n_i^{(1)} = r\right) = \binom{k}{r} (1-p)^r p^{k-r}, \quad 0 \le r \le k$$
$$P\left(n_i^{(2)} = r\right) \simeq \frac{(kp)^r}{r!} e^{-kp}, \quad \forall\, r \in \mathbb{N}, \quad \text{for large } n$$

These distributions yield

$$P(\phi_i = r) = \sum_{j=0}^{\min(r-k,k)} \binom{k}{j} (1-p)^j p^{k-j} \frac{(kp)^{r-k-j}}{(r-k-j)!} e^{-kp}, \quad k \le r$$

Note that

$$\mathcal{E}\left(n_i^{(1)}\right) = k(1-p), \quad Var\left(n_i^{(1)}\right) = k(1-p)p$$
$$\mathcal{E}\left(n_i^{(2)}\right) = Var\left(n_i^{(2)}\right) = kp$$
$$\mathcal{E}(\phi_i) = 2k, \quad Var(\phi_i) = k(2-p)p$$

The probability mass function $P(\phi_i = r), r \in \mathbb{N}$ has a pronounced peak at a value of r equal to $2k$. Furthermore, this function decays exponentially for large values of k.

Clustering Coefficient

The clustering coefficient of a network is a useful parameter for describing small-world networks. Small-world networks have a relatively high clustering coefficient. Denote the clustering coefficient of the random network by $\widetilde{C}_G(p)$. Recall that the clustering coefficient of a ring-lattice graph $R_{n,2k}$ is given by $\widetilde{C}_G(0) = 3(k-1)/\{2(2k-1)\}$. Therefore this is the value of the clustering coefficient of the network when $p = 0$. If $p > 0$, two neighbors of vertex i which were connected at $p = 0$ are still neighbors of i and linked together with probability $(1-p)^3$, up to terms of order n^{-1}. Therefore $\widetilde{C}_G(p) \simeq \widetilde{C}_G(0)(1-p)^3$.

If $p \approx 1$ and $n \gg 1$ the graph is random and the clustering coefficient is approximately equal to $\mathcal{E}(\phi_i)/n = 2k/n$.

Average Path Length

The average path length ℓ, is determined as a function of n, p and k. It has been established in the problem section that for $n \gg 1$,

$$\ell \simeq \begin{cases} n/(4k), & p = 0 \\ \ln(n)/\ln(2k - 0.5), & p \approx 1, \quad k \gg 1 \end{cases}$$

The average number of rewired (reconnected) links is equal to $n_c = npk$. Recall that these rewired links are also referred to as shortcuts, because they indeed shorten the distance between a randomly selected pair of nodes. Therefore, if there is less than a single shortcut on the average, then ℓ is dominated by connections on the regular lattice. In this case ℓ increases linearly with n. However, if $n \to \infty$ and a large value of p is held constant, ℓ becomes greater than unity and starts to scale as $\ln n$.

The transition between these two regimes takes place at some intermediate value of n equal to n_b. Let $n_b = p^{-\tau}, \tau \geq 0$. It is initially demonstrated that $\tau \geq 1$, and eventually established that $\tau = 1$ for all values of k.

Let $n = n_b$, and substituting $p = n_b^{-1/\tau}$, the average number of rewired links is obtained as $n_c = n_b^{1-1/\tau}k$. Therefore, if $(1 - 1/\tau) < 0$ and $n_b \to \infty$ the unacceptable observation that $n_c \to 0$ is obtained. Consequently $(1 - 1/\tau) \geq 0$, that is $\tau \geq 1$.

The analysis is next split in two parts. The technique used in this analysis is called *renormalization*. Initially assume that $k = 1$ and

$$\ell = nf\left(\frac{n}{n_b}\right)$$

$$f(x) \simeq \begin{cases} 1/4, & x \ll 1 \\ (\ln x)/x, & x \gg 1 \end{cases}$$

The value of τ in the expression $n_b = p^{-\tau}$ is next determined. Assume that the number of vertices n is an even number. The graph is modified as follows. Merge a node with its neighbor to form a single node. Links between the merged nodes exist, provided there was a prior direct path between the two pairs of nodes. Therefore, the number of shortcuts is preserved. Let n' be the number of nodes in the modified graph, and p' its new rewiring probability. Then

$$n' = \frac{n}{2}, \quad p' = 2p$$

The above equations are called the renormalized equations. Thus the modified length of the average path is halved for paths which run around the circumference of the ring, and remains the same along shortcuts. Therefore, if n' is large, and the value of p' is small, the length along the shortcuts is relatively small, and so can be neglected. Denote the average path length in the modified graph by ℓ'. Thus $\ell' = \ell/2$.

If $n \gg 1$ and $p \ll 1$, then $\ell' = n'f(n'p'^{\tau})$. Use of the above relationships results in $\ell = nf(2^{\tau-1}np^{\tau})$. However, it is known that $\ell = nf(np^{\tau})$. That is, $2^{\tau-1} = 1$, which in turn implies that $\tau = 1$. Therefore it can be inferred that

$$\ell = nf(np), \quad \text{if } n \gg 1, \ p \ll 1, \text{ and } k = 1$$

Next consider the case when $k > 1$. A slightly different renormalization transformation is used. Assume that the number of vertices n is divisible by k. To examine this case, merge k adjacent nodes to form a single node, with links between the newly formed nodes as in the last case. Again, the number of shortcuts is preserved in this transformation. The corresponding renormalization equations are

$$n' = \frac{n}{k}, \;\; p' = k^2 p, \;\; k' = 1, \;\; \ell' = \ell$$

The average distance ℓ' is not affected if $n \to \infty$ and $p \ll 1$. This occurs because, the number of vertices along the path connecting two nodes is reduced by a factor k, but the vertices that can be traversed in a single step are also reduced by an identical factor, and the two effects cancel out. Since $k' = 1, n' \gg 1$, and $p' \ll 1$ the results from the last case are applicable. Thus

$$\ell' = n' f\left(n'p'\right)$$

Use of the renormalization equations results in

$$\ell = \frac{n}{k} f\left(npk\right), \;\; n \gg k, \;\; \text{and} \;\; k^2 p \ll 1$$

Thus ℓ is estimated with an accuracy of $\pm k$. Also, if $p \simeq k^{-2}$, the network exhibits random behavior. Furthermore, n_b is proportional to p^{-1}. This in turn implies that $\tau = 1$. The above discussion is summarized in the following lemma.

Lemma 7.4. *Consider a ring-lattice $R_{n,2k}$. This graph has n nodes, and every node is connected to $2k$ immediate neighbors (k on either side), and the distance between adjacent nodes is assumed to be unity. This deterministic network is randomized by rewiring some edges with probability p. At each node, links to k neighboring nodes are left untouched. Of the remaining k links to the neighboring nodes, some are connected to other nodes with probability p, and the remaining are left untouched with probability $(1 - p)$.*

(a) *The average number of rewired links is equal to $n_c = npk$.*
(b) *If $p = 0$, then $\ell \simeq n/4k$.*
(c) *If $p \approx 1$, and $k \gg 1$, then $\ell \simeq \ln(n) / \ln(2k - 0.5)$.*
(d) *Let $n \gg k$, $k^2 p \ll 1$, and*

$$f(x) \simeq \begin{cases} 1/4, & x \ll 1 \\ (\ln x)/x, & x \gg 1 \end{cases} \tag{7.30}$$

then the average path length $\ell = (n/k)\, f(n_c)$. □

7.10 Network Evolution Models

Network evolution models are discussed in this section. Experimental results indicate that several large networks are scale-free. That is, the tails of their degree distribution follow a power-law. A simple and elegant model of evolution of a scale-free network was first introduced by Albert-László Barabási and Réka Albert (BA). This model, its variant, and its generalized version are discussed in this section.

Scale-free networks can also be generated by assigning vertex weights. In this modeling approach, a link (edge) exists between a pair of vertices with probability which is a function of the weights of the two vertices. This explanation could also be a possible reason for the prevalence of scale-free networks.

A popular model of Web graph evolution is the "copying model." The essential idea of the copying model is that a new Web page is generally made by copying an old one, and introducing some hyperlinks from the already existing Web pages.

7.10.1 Barabási and Albert Model

Most real world networks *grow* from a small set of nodes and links to a larger network. This occurs by initially starting with a small network, and continuously adding new nodes and links over time. This is how the WWW grows in time, which is via continuous addition of new Web pages and their associated hyperlinks. In addition to this growth, the new nodes exhibit a phenomenon called *preferential attachment*. This is the process by which a new node attaches to an existing node depending upon the nodal degree of the older node. Higher the nodal degree of the older node, higher are the chances of the new node attaching itself to this old node. This mode of nodal linkage is justified because, Web page with higher nodal degrees are presumably more popular. That is, a new Web page has a higher probability of attaching itself to Web pages with higher degree. The BA model is successful in creating a network in which the nodal degree of the associated graph exhibits a power-law distribution. Furthermore, this model also exhibits small-world properties. All this is accomplished by the use of simple dynamical principles. This model is next developed.

This network is modeled by an undirected and connected graph. Assume that at time t equal to 0, the network has $n(0)$ nodes. Furthermore, the average nodal degree in this graph is $\overline{k}(0)$. Thus the number of edges in this graph is equal to $n(0)\overline{k}(0)/2$. At each time step (of duration one unit), nodes and edges are added according to the following rules of network growth and preferential attachment.

(a) *Growth*: At every time step a new vertex is added with m number of edges, where $m \leq n(0)$. These m number of edges are connected to m different older vertices.
(b) *Preferential attachment*: A new edge is connected to an old node i with a probability $\wp(\cdot)$ proportional to its degree $k_i(t)$ at time $t > 0$.

$$\wp(k_i(t)) = \frac{k_i(t)}{\sum_j k_j(t)} \tag{7.31a}$$

After t time steps, the number of nodes and edges in the network is equal to $n(t) = (n(0) + t)$ and mt respectively. It is next established that this BA evolutionary paradigm creates a scale-free network. A continuum approach is used to develop this model. That is, the degree $k_i(t)$ of vertex i is assumed to be a continuous function of time. Note that the rate at which $k_i(t)$ changes with time is proportional to $\wp(k_i(t))$. Thus

$$\frac{\partial k_i(t)}{\partial t} = m\wp(k_i(t)) \simeq \frac{mk_i(t)}{\{2mt + n(0)\overline{k}(0)\}} \tag{7.31b}$$

Assuming that the network is large, that is for large value of t

$$\frac{\partial k_i(t)}{\partial t} = \frac{k_i(t)}{2t}, \qquad k_i(t_i) = m$$

Observe that the node i at its introduction time t_i has $k_i(t_i) = m$. Therefore

$$k_i(t) = m\left(\frac{t}{t_i}\right)^{1/2} \tag{7.31c}$$

Using this equation it can be observed that smaller the value of t_i, larger is the value of $k_i(t)$. That is, older vertices have higher degrees. Therefore for large values of t older vertices have very high values of degree. This is the so-called "rich-gets-richer" phenomenon which is prevalent in real networks. Using the above equation, the probability that a vertex has a degree $k_i(t)$ smaller than k, that is $P(k_i(t) < k)$ can be written as

$$P(k_i(t) < k) = P\left(t_i > \frac{m^2 t}{k^2}\right)$$

Since the nodes are added at equal time intervals to the evolving network, the t_i value has a constant probability density function. Thus

$$P(t_i) = \frac{1}{n(0) + t}$$

Use of the above two equations yields

$$P\left(t_i > \frac{m^2 t}{k^2}\right) = 1 - \left(\frac{m}{k}\right)^2 \frac{t}{(n(0) + t)}$$

That is

$$P(k_i(t) < k) = 1 - \left(\frac{m}{k}\right)^2 \frac{t}{(n(0) + t)}$$

Then as $t \to \infty$

$$P(k) = p_\nu(k) \simeq 2m^2 k^{-3} \tag{7.31d}$$

where ν is the random variable, which represents the nodal degree. Thus, the degree distribution is asymptotically independent of time and the network reaches *stationary scale-free state,* despite its continuous growth (as $n(t) = (n(0) + t)$). This equation also demonstrates that a preferential attachment growth generates a network with power-law degree distribution. The BA model also asserts that the network has vertices of very high degrees, with a nonnegligible probability. It can also be verified that in this model

$$1 = \int_m^\infty p_\nu(k)\, dk$$

$$\overline{k} = \int_m^\infty k p_\nu(k)\, dk = 2m \tag{7.31e}$$

where $\mathcal{E}(\nu) = \overline{k} \triangleq \overline{k}(t)$ as $t \to \infty$. Furthermore, the variance of the nodal degree ν does not exist in this model. The BA model discussed in this subsection is one of several possible *linear preference* models. This model is linear because the preferential attachment probability $\wp(k_i(t))$ is linear in $k_i(t)$. A variant of the BA model is outlined in the next subsection.

If the vertex attachment probability $\wp(k_i(t))$ is proportional to $\{k_i(t)\}^{\alpha_e}$, where $\alpha_e \neq 1$ then the network evolution model is said to have a nonlinear preference. It has been determined that the networks which evolve under these assumptions do not create a scale-free graph, as the order of the network approaches infinity.

7.10.2 A Variant of the Barabási and Albert Model

The BA model is a special type of preference model that produces power-law-tailed degree distribution. The exponent of the power-law degree distribution produced by this model is equal to 3. However the exponent γ of the power-law degree distribution of the Internet and Web graphs lies between 2 and 3, that is $\gamma \in (2, 3)$. In this subsection, a simple variant of the BA model is considered. It allows the exponent γ to take values in the interval $(2, 3)$. This model is due to S. N. Dorogovtsev and J. F. F. Mendes. It also obeys the law of preferential attachment. Assume that the network graph is undirected. The evolution of the network is next described.

(a) A new vertex is created at each time step, where the length of each time step is assumed to be a single unit. This vertex is attached preferentially to a single older vertex via an edge. The probability that the new vertex attaches itself to an older vertex i at time t is proportional to its degree $k_i(t)$. This is equivalent to the BA model with the parameter $m = 1$.
(b) During this time step, c_e new links are created between two unconnected nodes i and j with a probability proportional to the product of their degrees. If $c_e > 0$, the network is developing, and if $c_e < 0$, then the network is decaying. And lastly, if $c_e = 0$, the equivalent of a BA model is obtained.

Initially assume that $c_e \geq 0$. Use of continuum theory, and for large values of t

$$\frac{\partial k_i(t)}{\partial t} = \frac{k_i(t)}{\int_0^t k_j(t)\, du_j} + 2c_e \frac{k_i(t)\left\{\int_0^t k_j(t)\, du_j - k_i(t)\right\}}{\left\{\int_0^t k_j(t)\, du_j\right\}^2 - \int_0^t k_j^2(t)\, du_j}, \quad k_i(t_i) = 1 \tag{7.32}$$

In the above equation the summation over all vertices $\sum_j k_j(t)$ has been approximated by an integral over all introduction times u_j of vertex j. The first term on the right-hand side of the above equation represents the contribution due to linear preferential attachment. And the second term represents the contribution due to c_e edges. As per the evolution rules of this model, every vertex can be at either end of the newly created edge. Thus the probability of vertex i becoming an end of the new link is proportional to both $k_i(t)$, and the summation $\sum_{j,j\neq i} k_j(t)$. The denominator of the second term is the normalization factor. It is the integral approximation of $\sum_{i,j;i\neq j} k_i(t)\, k_j(t)$. For large values of t

$$\frac{\partial k_i(t)}{\partial t} = (1 + 2c_e)\frac{k_i(t)}{\int_0^t k_j(t)\, du_j}, \quad k_i(t_i) = 1$$

Note that $\int_0^t k_j(t)\, du_j = 2(1 + c_e)\, t$. Defining

$$\widetilde{c}_e = \frac{(1 + 2c_e)}{2(1 + c_e)}$$

yields

$$k_i(t) = \left(\frac{t}{t_i}\right)^{\widetilde{c}_e}$$

Proceeding as in the BA model, as $t \to \infty$ we have

$$P(k) \simeq \frac{1}{\widetilde{c}_e} k^{-\gamma}, \quad \widetilde{c}_e(\gamma - 1) = 1$$

Thus

$$\gamma = 2 + \frac{1}{(1+2c_e)}$$

Observe that if $c_e = 0$, then $\widetilde{c}_e = 1/2$, and $\gamma = 3$, which is the value of the exponent in the BA model. If $c_e \to \infty$ then $\widetilde{c}_e \to 1$, and $\gamma \to 2$.

If $c_e < 0$, then in addition to the creation of a new node at every time step, $-c_e$ edges are deleted at random from the network. The above expressions are also valid for negative values of c_e. However, the limiting value of c_e is -1. This is true, because the rate of removal of edges in the network cannot be higher than the rate of addition of new nodes and edges.

7.10.3 A Generalization of Barabási and Albert Model

A generalization of Barabási and Albert model of a growing network is outlined. The evolving network has directed edges. In this network evolution model, at each time step $t = 1, 2, 3, \ldots$ a new node is added to the network. Along with this node, $m \in \mathbb{P}$ new directed edges go out of unspecified vertices, possibly also from outside of the network. The new m edges are incident upon vertices of the old network and also possibly on the new node. Thus new edges may also appear between old nodes. The following analysis assumes that $t \gg 1$.

The node created at time $s \in \mathbb{P}$ is labeled as simply s. Also let $q(s,t)$ be equal to the number of incoming links (in-degree) at node s at time t, where $s \leq t$. The probability that a new link points to a node s is proportional to $(q(s,t) + A)$, where the parameter $A > 0$ characterizes additional attractiveness of a node. The in-degree distribution of node s at time t is $p(q,s,t)$. As a mathematical convenience, we let $A \triangleq ma$, where $a > 0$.

More specifically, if each newly created vertex is the source of all m edges, then the degree of node s at time t is equal to $k(s,t)$, where

$$k(s,t) = (q(s,t) + m)$$

This is the case of the Barabási and Albert model, if in addition we also have $A = m$. That is, $a = 1$.

The general evolution model is discussed in the rest of the subsection. The total number of edges in the network at time $(t+1)$ is equal to $m(t+1)$. Thus the probability that a new edge is incident upon vertex s equals

$$\frac{(q(s,t) + A)}{mt + At}$$

Note that

$$\sum_{1 \leq s \leq t} (q(s,t) + A) = mt + At = (1+a)\,mt, \quad t \in \mathbb{P}$$

Probability that the vertex s receives exactly l edges of the m new edges at time $(t+1)$ is $P_s(q,l,m)$, where $l \in [0,m]$. It is

$$P_s(q,l,m) = \binom{m}{l} \left\{ \frac{q+am}{(1+a)\,mt} \right\}^l \left\{ 1 - \frac{q+am}{(1+a)\,mt} \right\}^{m-l}, \quad l \in [0,m]$$

Therefore the in-degree distribution of an individual vertex of a very large network ($t \gg 1$) is specified by

$$p(q,s,t+1) = \sum_{l=0}^{m} P_s(q-l,l,m)\,p(q-l,s,t), \quad s \leq t$$

In the construction of this network, vertices are born without incoming edges. Therefore, $p(q,s,s) = \delta_{q0}$, where δ_{q0} is the Kronecker symbol. That is, for integers i and j, δ_{ij} is equal to one if $i = j$, and zero otherwise. For $t \gg 1$, we obtain the finite-difference equation

$$p(q,s,t+1) = \left\{1 - \frac{(q+am)}{(1+a)t}\right\} p(q,s,t) + \frac{(q-1+am)}{(1+a)t} p(q-1,s,t) + O\left(\frac{p}{t^2}\right)$$

Approximating the finite difference, as a derivative as $t \gg 1$, we obtain

$$(1+a)t\frac{\partial p(q,s,t)}{\partial t} = (q-1+am)p(q-1,s,t) - (q+am)p(q,s,t)$$

Using the above equation, expressions for

$$p(0,s,t), p(1,s,t), p(2,s,t), \ldots$$

can be determined explicitly. The following result follows formally via induction.

$$p(q,s,t) = \frac{\Gamma(q+ma)}{\Gamma(ma)\,q!}\left(\frac{s}{t}\right)^{am/(1+a)}\left\{1 - \left(\frac{s}{t}\right)^{1/(1+a)}\right\}^q, \quad s \le t,\ q \in \mathbb{N}$$

where $\Gamma(\cdot)$ is the gamma function. The average in-degree connectivity is given by $\overline{q}(s,t)$, where

$$\overline{q}(s,t) = \sum_{q \in \mathbb{N}} q p(q,s,t) = am\left\{\left(\frac{s}{t}\right)^{-1/(1+a)} - 1\right\}, \quad s \le t$$

Therefore, for fixed time t, the average connectivity of a node $s \ll t$ is proportional to $s^{-\beta}$, where $\beta = 1/(a+1)$.

Evaluation of Degree Distribution in the Network

We determine the degree distribution of the network as $t \gg 1$. This degree distribution is $P(q,t), q \in \mathbb{N}$ where

$$P(q,t) = \frac{1}{t}\sum_{s=1}^{t} p(q,s,t)$$

Use of the finite-difference equation results in

$$\begin{aligned}&(t+1)P(q,t+1) - p(q,t+1,t+1)\\&= \left\{t - \frac{(q+am)}{(1+a)}\right\}P(q,t) + \frac{(q-1+am)}{(1+a)}P(q-1,t) + O\left(\frac{P}{t}\right)\end{aligned}$$

Approximating the finite difference, as a derivative as $t \gg 1$, we obtain

$$\begin{aligned}&(1+a)t\frac{\partial P(q,t)}{\partial t} + (1+a)P(q,t)\\&+ (q+am)P(q,t) - (q-1+am)P(q-1,t)\\&= (1+a)\delta_{q0}\end{aligned}$$

Assume that the limit $P(q) = \lim_{t\to\infty} P(q,t)$ exists. The stationary degree distribution is specified by

$$(1+a)\,P(q)+(q+am)\,P(q)-(q-1+am)\,P(q-1)=(1+a)\,\delta_{q0}$$

The above recursive equation yields

$$P(0)=\frac{(1+a)}{1+a(m+1)}$$
$$P(q)=\frac{(q-1+am)}{\{q+1+a(m+1)\}}P(q-1),\quad q\in\mathbb{P}$$

Thus

$$P(q)=(1+a)\frac{\Gamma(q+ma)}{\Gamma(ma)}\frac{\Gamma[a(m+1)+1]}{\Gamma[q+a(m+1)+2]},\quad q\in\mathbb{N}$$

where $\Gamma(\cdot)$ is the gamma function. Using Stirling's approximation for $\Gamma(r)$, as $r\to\infty$, we obtain

$$P(q)\simeq(1+a)\frac{\Gamma[a(m+1)+1]}{\Gamma(ma)}(q+ma)^{-(2+a)},\quad \text{as } (q+ma)\gg 1$$

That is,

$$P(q)\sim(q+ma)^{-(2+a)},\quad \text{as } (q+ma)\gg 1 \tag{7.33}$$

Therefore, the tail of the degree distribution has a power-law with exponent $\gamma=(2+a)$. As $a>0$, we have $\gamma>2$. The results of this subsection are also used in describing the popularity-similarity model of network evolution. An equivalent model is developed in a subsequent section via hyperbolic geometry.

7.10.4 Intrinsic Vertex Weight Model

Power-law vertex degree distribution of a graph can also be obtained by assigning weights to vertices. Assume that the network is modeled by an undirected simple graph. Let the vertex set of the graph be $V=\{1,2,\ldots,n\}$, where the total number of vertices n tends to infinity. Recall that ν is a random variable, which represents the degree of a vertex. Also $p_\nu(k)$ is the probability that the degree of a vertex is k. For scale-free networks, $p_\nu(k)=\mathfrak{B}k^{-\gamma}, k_0\le k\le k_c<n$, where $\mathfrak{B}$ is the normalizing constant. In the rest of this subsection, upper limit of k is assumed to be n instead of $(n-1)$, as $n\to\infty$.

This model assigns a weight to each vertex. It is a measure of the importance of the vertex. The weight of each vertex is a nonnegative real number, and is randomized for analytical tractability. Let Ω be a random variable, which represents the vertex weight. This random variable takes nonnegative real values. The probability density function and the cumulative distribution function of the random variable Ω are $\widetilde{f}_\Omega(\omega)$, and $\widetilde{F}_\Omega(\omega), \omega\in\mathbb{R}_0^+$ respectively. Furthermore, weights of different vertices are assumed to be independent of each other.

Also let $\widetilde{p}(\omega_i,\omega_j)$ be the probability with which a link is formed between vertices $i,j\in V$, where $i\neq j$. This is also a symmetric function of its arguments. Let $k(\omega)$ be the degree of a vertex with weight ω. Then

$$k(\omega)=n\int_0^\infty \widetilde{p}(\omega,x)\,\widetilde{f}_\Omega(x)\,dx,\quad \omega\in\mathbb{R}_0^+ \tag{7.34}$$

In the rest of this subsection, it is assumed that the degree k of a vertex takes a continuum of values. In addition, $k(\cdot)$ is assumed to be an invertible function of ω. We next consider two types of the probability function $\widetilde{p}(\cdot,\cdot)$. These are the:

(a) Gravity model
(b) Threshold model

Gravity Model

In the gravity model, $\widetilde{\rho}(\omega_i, \omega_j) = \omega_i \omega_j / \omega_M^2$; $i, j \in V, i \neq j$, and ω_M is the largest value of ω in the network. This model is a generalization of the Erdös-Rényi model, where an edge exists between any pair of vertices with a fixed probability. Thus

$$k(\omega) = \frac{n\omega}{\omega_M^2} \mathcal{E}(\Omega) \tag{7.35a}$$

where $\mathcal{E}(\Omega)$ is the expected value of the random variable Ω. Therefore

$$p_\nu(k) = \frac{d\omega}{dk} \widetilde{f}_\Omega(\omega) \tag{7.35b}$$

$$= \frac{\omega_M^2}{n\mathcal{E}(\Omega)} \widetilde{f}_\Omega \left(\frac{\omega_M^2}{n\mathcal{E}(\Omega)} k \right) \tag{7.35c}$$

If it is assumed that $\widetilde{f}_\Omega(\cdot)$ follows a power-law, then it should be immediately evident that so does the degree distribution $p_\nu(\cdot)$. Therefore, if the intrinsic vertex weights follow a power-law distribution, then the gravity model results in a scale-free network.

Threshold Model

It is established in the threshold model that the degree distribution $p_\nu(\cdot)$ follows a power-law, for some probability density functions $\widetilde{f}_\Omega(\cdot)$ which do not follow a power-law. Let

$$\widetilde{\rho}(\omega_i, \omega_j) = \begin{cases} 1, & \omega_i + \omega_j \geq \theta_{th} \\ 0, & \text{otherwise} \end{cases} \tag{7.36}$$

where $i, j \in V$, and $i \neq j$; and the threshold parameter $\theta_{th} \in \mathbb{R}_0^+$. Therefore, an edge is formed between the two vertices i and j, if $(\omega_i + \omega_j) \geq \theta_{th}$. A threshold value of zero results in a complete graph, which is not very interesting. Therefore assume that $\theta_{th} > 0$. Thus

$$k(\omega) = n \int_{\theta_{th}-\omega}^{\infty} \widetilde{f}_\Omega(x)\, dx \tag{7.37a}$$

$$= n \left\{ 1 - \widetilde{F}_\Omega(\theta_{th} - \omega) \right\}, \quad \omega \in \mathbb{R}_0^+ \tag{7.37b}$$

where $0 \leq k(\omega) < n$. Assume that there is a one-to-one correspondence between k and ω. Let $\widetilde{F}_\Omega^{-1}(\cdot)$ be the inverse of the cumulative distribution function $\widetilde{F}_\Omega(\cdot)$. Then

$$p_\nu(k) = \frac{d\omega}{dk} \widetilde{f}_\Omega(\omega) \tag{7.37c}$$

$$= \frac{\widetilde{f}_\Omega \left(\theta_{th} - \widetilde{F}_\Omega^{-1}(1 - k/n) \right)}{n \widetilde{f}_\Omega \left(\widetilde{F}_\Omega^{-1}(1 - k/n) \right)} \tag{7.37d}$$

We next consider two cases. In the first case, Ω has an exponential distribution. In the second case, Ω has a Pareto distribution.

Case 1: Exponential distribution: Let the random variable Ω be exponentially distributed, with parameter $\widetilde{\lambda} > 0$. Its probability density function is given by

$$\widetilde{f}_{\Omega}(\omega) = \begin{cases} 0, & x \in (-\infty, 0) \\ \widetilde{\lambda} e^{-\widetilde{\lambda}\omega}, & \omega \in [0, \infty) \end{cases}$$

Therefore

$$p_{\nu}(k) = \frac{e^{-\widetilde{\lambda}\omega}}{k} = \frac{n}{k^2} e^{-\widetilde{\lambda}\theta_{th}}, \quad ne^{-\widetilde{\lambda}\theta_{th}} \le k < n \tag{7.38}$$

In the above equation, smallest value of k occurs as $\omega \to 0$. Therefore the smallest value of k is $ne^{-\widetilde{\lambda}\theta_{th}}$. Thus the above expression for $p_{\nu}(k)$ is valid for $ne^{-\widetilde{\lambda}\theta_{th}} \le k < n$. Notice that $p_{\nu}(k) \propto k^{-2}$. This result implies that even non power-law distributions of the random variable Ω, can result in scale-free networks.

Case 2: Pareto distribution: The random variable Ω has Pareto distribution with parameters $\omega_0, \widetilde{\sigma} \in \mathbb{R}^+$. Its probability density function is given by

$$\widetilde{f}_{\Omega}(\omega) = \begin{cases} 0, & \omega \le \omega_0 \\ \dfrac{\widetilde{\sigma}}{\omega_0} \left(\dfrac{\omega_0}{\omega} \right)^{\widetilde{\sigma}+1}, & \omega > \omega_0 \end{cases}$$

where ω_0 and $\widetilde{\sigma}$ are its parameters. The cumulative distribution function of Ω is given by

$$\widetilde{F}_{\Omega}(\omega) = \begin{cases} 0, & \omega \le \omega_0 \\ 1 - \left(\dfrac{\omega_0}{\omega} \right)^{\widetilde{\sigma}}, & \omega > \omega_0 \end{cases}$$

The inverse of the cumulative distribution function of Ω is given by

$$\widetilde{F}_{\Omega}^{-1}(x) = \frac{\omega_0}{(1-x)^{1/\widetilde{\sigma}}}, \quad 0 \le x < 1$$

Therefore, if $(\theta_{th} - \omega_0) \ge \omega$

$$k = n \left\{ \frac{\omega_0}{\theta_{th} - \omega} \right\}^{\widetilde{\sigma}}$$

$$n \left\{ \frac{\omega_0}{\theta_{th} - \omega_0} \right\}^{\widetilde{\sigma}} \le k < n$$

Nontrivial networks form if $\theta_{th} > 2\omega_0$. Observe that the smallest value of k occurs as $\omega \to \omega_0$. Also

$$\omega = \theta_{th} - \left(\frac{n}{k} \right)^{1/\widetilde{\sigma}} \omega_0$$

Therefore

$$p_{\nu}(k) = \frac{n^{1/\widetilde{\sigma}}}{\left\{ k^{1/\widetilde{\sigma}} \theta_{th} / \omega_0 - n^{1/\widetilde{\sigma}} \right\}^{\widetilde{\sigma}+1}}$$

If $n (\omega_0 / \theta_{th})^{\widetilde{\sigma}} \ll k < n$, then $p_{\nu}(k)$ can be approximated by

$$p_{\nu}(k) \simeq \left(\frac{\omega_0}{\theta_{th}} \right)^{\widetilde{\sigma}+1} n^{1/\widetilde{\sigma}} k^{-(\widetilde{\sigma}+1)/\widetilde{\sigma}} \tag{7.39}$$

Therefore if the random variable Ω has a Pareto distribution, the underlying graph has a power-law degree distribution with parameter

$$\gamma = \frac{(\widetilde{\sigma} + 1)}{\widetilde{\sigma}} > 1$$

There are several other mathematical models of evolution of scale-free networks. These models are not discussed due to space limitation.

7.10.5 Web Graph Evolution Model

We describe the evolution of the WWW via a copying model. Copying model or process is a mechanism in which the WWW grows by addition of Web pages and hyperlinks. The WWW is modeled as a digraph, in which the Web page is a vertex, and a hyperlink is a directed edge.

In this copying model, a Web page is added at each time step; and the hyperlinks are added to it either randomly or by copying some out-going links (edges) from other Web pages. Copying of an existing link means that the head of the link is *copied* as head of the new edge.

In this model the network begins with an initial digraph. Subsequently, at each time step, a new vertex v is added. This new vertex has on the average $m \in \mathbb{P}$ out-going edges. The new edges are created as follows.

- With probability $\xi \in (0, 1)$, the model adds m out-going links from the new node v to m vertices selected uniformly and randomly from the existing set of vertices.
- Further, with probability $(1 - \xi)$ another existing vertex u is randomly and uniformly selected from existing set of vertices and then out-going links from the vertex u are copied to vertex v. That is, if there are edges, (u, w_i) , $1 \leq i \leq m$, then the edges (v, w_i) , $1 \leq i \leq m$ are created. Note that the head of the edge (u, w_i) is w_i. It makes the head of the new edge (v, w_i) for $1 \leq i \leq m$.
  - If the vertex u has more than m out-going edges, then the m links are selected at random.
  - However, if the selected vertex u has less than m out-going edges, then all of its links are copied, and another existing vertex u' is selected uniformly and randomly from the existing set of vertices and its out-going links are copied to the vertex v. This process is repeated till a total of m out-going links are created from the new vertex v.

This process is indeed intuitive because, if a new Web page is created, the author of the Web page would copy certain number of out-going links to an existing Web page. In addition, with a certain probability the newly created vertex will have an affinity or *preference* for the out-going links of an already existing vertex. Therefore, the degree distribution of the digraph in this model has asymptotically a power-law form if the WWW is viewed as a directed graph. A simplified form of such model is described in the problem section.

A similar process can be described to model removal of vertices and edges. In a general time-evolving digraph model, nodes and edges are either created or destroyed at each time step. Thus for large digraphs, the model can be described by four discrete-time stochastic processes. These are the node and edge *creation* processes, and the node and edge *deletion* processes. Each such process is a function of the time step, and the current state of the digraph. It can be shown that both the in-degree and out-degree distribution of the time-evolved random digraph asymptotically have power-law distributions, with different parameters.

In the above model, the digraph grows by the addition of a single vertex at each time step. Therefore this is called a *linear growth* model. As the Web is growing at an exponential pace, it is also possible to model this scenario by the creation of a fraction of vertices of the current digraph at each time step. This leads to an *exponential growth* model.

7.11 Hyperbolic Geometry of Networks

A geometric framework is developed to study certain topological properties of a large complex network. It is tacitly assumed that hyperbolic geometry underlies these complex networks. A consequence of this assumption is the heterogeneous degree distribution of the underlying graph of the network. Another consequence of this assumption is the emergence of strong clustering of nodes of the network graph.

Hyperbolic geometry has been discussed in a separate chapter on geometry. Nevertheless, immediately useful properties of hyperbolic geometry are initially summarized. Like Euclidean geometry, hyperbolic geometry is a logically consistent geometry. Subsequently, a hyperbolic geometry based framework is developed to model the structure and dynamics of a large network. A network model based upon the concepts of popularity and similarity is also developed.

7.11.1 Relevant Hyperbolic Geometry

Similar to Euclidean geometry, hyperbolic geometry can also be specified by a set of five axioms. The first four axioms are similar to the Euclidean geometry. However, Playfair's axiom of Euclidean geometry has to be replaced by a different axiom. The replaced axiom is: for every line ℓ and for every point A not on the line ℓ; there exist at least two lines that pass through the point A, and are also parallel to line ℓ.

There are several models of hyperbolic geometry. Some of these are the upper half-plane model, Poincaré disc model, hyperboloid model, and Beltrami-Klein model. We list some features of only the Poincaré disc model.

- Lines in the Poincaré disc model of hyperbolic geometry are geodesics. These are shortest paths between two distinct points on the disc. The geodesics are disc diameters and arcs of Euclidean circles which intersect the boundary perpendicularly.
- The sum of interior angles of a triangle is less than π.
- Let r_e and r_h be the Euclidean and hyperbolic distance respectively, of a point from the center of the disc. We have
$$r_e = \tanh \frac{r_h}{2}$$
- Euclidean space is flat, whereas hyperbolic space has a negative curvature equal to $-\zeta^2 < 0$, where $\zeta > 0$. The circumference $L(r)$ and area $A(r)$ of the hyperbolic circle of radius $r > 0$ are:
$$L(r) = \left(\frac{2\pi}{\zeta}\right) \sinh \zeta r, \quad \text{and} \quad A(r) = \left(\frac{2\pi}{\zeta^2}\right) (\cosh(\zeta r) - 1)$$
Observe that for large values of ζr, both circumference and area grow exponentially as $e^{\zeta r}$.
- Let the hyperbolic distance between two points with polar coordinates (r, θ) and (r', θ') be x. Note that r, r', and x are hyperbolic lengths. These values are related via the "hyperbolic cosine rule." It is:
$$\cosh \zeta x = \cosh \zeta r \cosh \zeta r' - \sinh \zeta r \sinh \zeta r' \cos \Delta\theta$$

where $\Delta\theta = \{\pi - |\pi - |\theta - \theta'||\}$ is the angle between the points (r, θ) and (r', θ').

– If $\zeta \to 0$, we obtain the law of cosines for a triangle in Euclidean geometry, which is

$$x^2 \simeq r^2 + r'^2 - 2rr' \cos \Delta\theta$$

– For sufficiently large ζr, $\zeta r'$, and $\Delta\theta > \sqrt{e^{-2\zeta r} + e^{-2\zeta r'}}$ the hyperbolic distance x is

$$x \simeq r + r' + \frac{2}{\zeta} \ln \sin \frac{\Delta\theta}{2} \simeq r + r' + \frac{2}{\zeta} \ln \frac{\Delta\theta}{2}$$

There is a close analogy between hyperbolic spaces and b-ary trees. The circumference of a hyperbolic circle of radius r is analogous to the number of nodes in a b-ary tree which are r hops away from the root. Similarly, the area of a hyperbolic circle of radius r is analogous to the number of nodes in a b-ary tree which are at most r hops away from the root. In both cases, the number of nodes grow approximately as b^r. If we let $\zeta = \ln b$, the number of nodes grow as $e^{\zeta r}$. This expression is equal to the growth rate of circumference and area of a circle of radius r in hyperbolic space. This comparison suggests that the b-ary trees are the discretized versions of hyperbolic spaces. This in turn implies that the metric structure of trees and hyperbolic spaces are similar. Therefore trees, possibly with infinite number of nodes, permit nearly isometric embeddings into hyperbolic spaces. Consequently, it is possible to tessellate the hyperbolic space (Poincaré disc) via certain special class of trees. This is in contrast to Euclidean spaces, where trees do not embed in general.

7.11.2 Modeling Complex Networks

It is generally assumed in complex networks that their nodes can be classified into groups. It is also quite plausible, that there exists a hierarchy of classification of nodes. That is, nodes can be divided into several groups. Each group of nodes is further split into subgroups, and the subgroups are further split into several smaller subsubgroups, and so on. The relationships between such groups of nodes can be represented approximately by a graph-theoretic tree-like structure. It is not the claim that complex networks form a tree-like structure. Nevertheless, it is quite possible that the network is in a tree-form.

Our goal is an explanation of the emergence of a hierarchical network of nodes via hyperbolic geometry. At the birth of the network, classification of the nodes is not clear. However, as more nodes join the network, the new nodes connect to pre-existing nodes which are similar to it. This results in a hierarchy of network nodes. As the number of nodes becomes very large, the hierarchy of nodes can be modeled via hyperbolic geometry.

As an example, consider the evolution of the Internet. Initially, a pair of computers were connected. As the network evolved, computers which were administered by the same entity were connected together. This led to the emergence of autonomous systems (ASes). These ASes can be further classified based upon their size, type of customers, physical location, and several other factors. Thus a hierarchy of nodes develops as such networks evolve. Similarly, Web pages also connect to other analogous Web pages via hyperlinks. Eventually a hierarchy of Web pages develops.

Uniform Node Density, and Curvature $K = -1$

We next describe the evolution of network topology via hyperbolic geometry. Consider the upper half-plane model $\mathbb{H} = \{(v, w) \in \mathbb{R}^2 \mid w > 0\}$ of the hyperbolic geometry. Initially assume that the curvature of the surface $K = -1$. In this model, $n \gg 1$ nodes are spread uniformly on the hyperbolic plane over a disk of radius $R \gg 1$. In this model, the parameter R represents abstractly, the depth of the invisible tree-like hierarchy of the nodes. It turns out that R is an increasing function of n. Thus hierarchical nature of the network deepens with increasing number of nodes. The hypothesis of uniform density of nodes implies that a node is assigned an angular coordinate $\theta \in [0, 2\pi)$ with uniform density $\beta(\theta) = 1/(2\pi)$. The density $\rho(r), r \in [0, R]$ of the radial coordinate of the node is

$$\rho(r) = \frac{L(r)}{A(R)} = \frac{\sinh r}{(\cosh R - 1)} \simeq e^{r-R} \sim e^{r}$$

In order to form a network, the connection probability $p(\cdot)$ between two nodes has to be specified. This connection probability is specified as a function of the hyperbolic distance x between the two nodes. Let

$$p(x) = u(R - x), \quad x \in [0, R)$$

where $u(\cdot)$ is the step function. The step function $u(t)$ is equal to unity for $t > 0$, and equal to zero for $t < 0$. The above expression implies that a pair of nodes are connected, if and only if the hyperbolic distance between them is less than R. Our goal is to determine the degree distribution of the network graph $P(\nu = k) \triangleq p_{\nu}(k)$, where $k = 0, 1, 2, \ldots$.

Let $\nu(r)$ be a random variable which represents the degree of a vertex (node) at radial distance r, and $p_{\nu(r)}(k(r))$ be the probability that the degree of a node is equal to $k(r)$. Let the corresponding expected value of the degree of nodes at radial distance r be $\mathcal{E}(\nu(r)) = \overline{k}(r)$. The value $\overline{k}(r)$ is next determined.

A disc of radius R, centered at O contains all n nodes in the network. These nodes are distributed uniformly (in a probabilistic sense) within the disc. A second disc, also of radius R, is centered at O'. A single node is located at the center O'. Let the hyperbolic distance between O and O' be r. See Figure 7.3.

If $r > R$, then the node at O' is not connected to any node in the disc with center O. However, if $r \le R$, then the node at O' is connected to all nodes which lie within both discs. In the later case denote the intersecting region and area by $S(r)$ and $\psi(r)$ respectively.

Then $\overline{k}(r) = \delta\psi(r)$, where δ is equal to the node density. Note that $\delta = n/A(R) = n/\{2\pi(\cosh R - 1)\}$. It remains to determine $\psi(r)$. The infinitesimal area $d\psi$ in polar coordinates (y, θ) is $d\psi = \sinh y \, dy d\theta$. Therefore the area $\psi(r)$ of the intersecting region $S(r)$ is

$$\begin{aligned} \psi(r) &= \iint_{S(r)} d\psi = 2\int_0^{R-r} \sinh y \int_0^{\pi} d\theta dy + 2\int_{R-r}^{R} \sinh y \int_0^{\theta_y} d\theta dy \\ &= 2\pi\left[\cosh(R - r) - 1\right] + 2\int_{R-r}^{R} \theta_y \sinh y dy \end{aligned}$$

where θ_y is obtained by using the hyperbolic law of cosines

$$\cosh R = \cosh r \cosh y - \sinh r \sinh y \cos\theta_y, \quad \theta_y \in [0, \pi]$$

For large values of R, r, and y, we use the approximation $\theta_y \simeq 2e^{(R-r-y)/2}$. Therefore

$$\overline{k}(r) = n\left\{\frac{4}{\pi}e^{-r/2} - \left(\frac{4}{\pi} - 1\right)e^{-r}\right\} \simeq \frac{4}{\pi}ne^{-r/2}, \quad r \in [0, R]$$

Using the above expression for $\overline{k}(r)$, average nodal degree in the network graph can be determined. It is

$$\overline{k} = \int_0^R \rho(r)\,\overline{k}(r)\,dr = \frac{8}{\pi} n e^{-R/2}$$

The above expression yields $R = 2\ln\left(8n/\left(\pi\overline{k}\right)\right)$. Therefore $R \sim \ln n$. That is, the radius of the network is proportional to logarithm of the number of nodes in the network. Thus, R mimics the height of a balanced tree with n nodes. This analysis reinforces our hypothesis of hierarchical nature of the network. Also observe that

$$\overline{k}(r) = \frac{\overline{k}}{2} e^{(R-r)/2} \sim e^{-r/2}$$

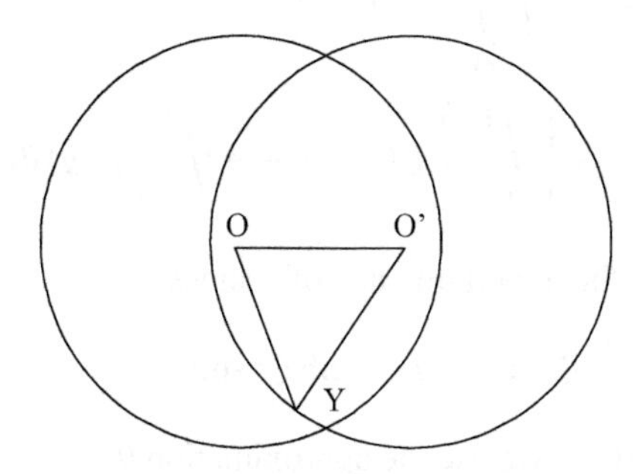

Figure 7.3. The two overlapping circles each have radius R. The centers of the two circles are O and O'. The hyperbolic distance between the centers of the two circles is r. The overlapping region is denoted by $S(r)$. In the triangle, $O'OY$, the hyperbolic lengths of lines OO', OY, and $O'Y$ are r, y, and R respectively. The angle $O'OY$ is equal to θ_y.

Finally, the degree distribution of the network graph $P(\nu = k)$, where $k = 0, 1, 2, \ldots$ can be determined by using the concept of "hidden variable." The radial coordinate r is used as a hidden variable. A node with radial coordinate r is assumed to have a Poissonian distribution $g(\cdot \mid r)$ with parameter $\overline{k}(r)$. Thus

$$P(k) = \int_0^R g(k \mid r)\,\rho(r)\,dr$$

$$g(k \mid r) = e^{-\overline{k}(r)} \frac{\overline{k}(r)^k}{k!}, \quad k = 0, 1, 2, \ldots$$

The use of Poissonian distribution is justified for a sparse network. The above expressions yield

$$P(k) = 2\left(\frac{\overline{k}}{2}\right)^2 \frac{\Gamma\left(k-2, \overline{k}/2\right)}{k!} \sim k^{-3} \tag{7.40}$$

where $\Gamma(\cdot,\cdot)$ is the incomplete gamma function. This result implies that the nodal degree distribution indeed follows a power law.

Quasi-uniform Node Density and Arbitrary Negative Curvature

The model developed earlier is generalized. Assume that the radius of the disc $R \gg 1$. Also assume that the radial density of nodes $\rho(r)$ is

$$\rho(r) = \alpha \frac{\sinh \alpha r}{(\cosh \alpha R - 1)} \simeq \alpha e^{\alpha(r-R)} \sim e^{\alpha r}, \quad r \in [0, R]$$

where $\alpha > 0$. Thus the radial density $\rho(r)$ is quasi-uniform. This corresponds to an average branching factor of $b = e^{\alpha}$ in tree-like hierarchy. Further assume that the curvature K of the hyperbolic space is allowed any negative value. That is, $K = -\zeta^2 < 0$, where $\zeta > 0$. Observe that if we let $\alpha = \zeta$, the radial node density is uniform.

As noted earlier, the expected value of the degree of nodes at radial distance r is equal to $\overline{k}(r)$. It is

$$\begin{aligned}
\overline{k}(r) &= \frac{n}{2\pi} \int\int_{S(r)} \rho(y)\, dy d\theta \\
&= n \left\{ \int_0^{R-r} \rho(y)\, dy + \frac{1}{\pi} \int_{R-r}^{R} \rho(y)\, \theta_y dy \right\}
\end{aligned}$$

where θ_y is obtained by using the hyperbolic law of cosines

$$\cosh \zeta R = \cosh \zeta r \cosh \zeta y - \sinh \zeta r \sinh \zeta y \cos \theta_y, \quad \theta_y \in [0, \pi]$$

For large values of $\zeta R, \zeta r$, and ζy, we use the approximation $\theta_y \simeq 2e^{\zeta(R-r-y)/2}$. Defining

$$\xi \triangleq \frac{\alpha}{\zeta} \left(\frac{\alpha}{\zeta} - \frac{1}{2} \right)^{-1}$$

we obtain

$$\begin{aligned}
\overline{k}(r) &= n \left\{ \frac{2}{\pi} \xi e^{-\zeta r/2} - \left(\frac{2}{\pi} \xi - 1 \right) e^{-\alpha r} \right\} \\
&= \begin{cases} n(2\xi/\pi)\, e^{-\zeta r/2}, & \text{if } \alpha > \zeta/2 \\ n(1 + \zeta r/\pi)\, e^{-\zeta r/2}, & \text{if } \alpha \to \zeta/2 \\ n(1 - 2\xi/\pi)\, e^{-\alpha r}, & \text{if } \alpha < \zeta/2 \end{cases}
\end{aligned}$$

Using the above expression for $\overline{k}(r)$, average nodal degree in the network graph can be determined in the three cases. It is

$$\begin{aligned}
\overline{k} &= \int_0^R \rho(r)\, \overline{k}(r)\, dr \\
&= \begin{cases} n(2/\pi)\, \xi^2 e^{-\zeta R/2}, & \text{if } \alpha > \zeta/2 \\ n(\zeta/2)\, R \{1 + (\zeta/(2\pi))\, R\}\, e^{-\zeta R/2}, & \text{if } \alpha \to \zeta/2 \\ n\alpha R (1 - (2\xi/\pi))\, e^{-\alpha R}, & \text{if } \alpha < \zeta/2 \end{cases}
\end{aligned}$$

The degree distribution of the network graph $P(\nu = k)$, where $k = 0, 1, 2, \ldots$ is determined again by using the concept of hidden variable. The radial coordinate r is the hidden variable. A node with radial coordinate r is assumed to have a Poissonian distribution $g(\cdot \mid r)$ with parameter $\overline{k}(r)$. For $\alpha/\zeta > 0$, the nodal degree distribution scales as

$$P(k) \sim k^{-\gamma}, \quad \text{where} \quad \gamma = \begin{cases} 2\dfrac{\alpha}{\zeta} + 1, & \text{if } \alpha/\zeta \geq 1/2 \\ 2, & \text{if } \alpha/\zeta < 1/2 \end{cases} \tag{7.41}$$

Thus the proposed network model for $\alpha/\zeta \geq 1/2$ generates a scale-free network with a power-law distribution of the nodal degree with exponent $\gamma = (2\alpha/\zeta + 1) \geq 2$. Observe that if $\alpha = \zeta$, we have $\gamma = 3$. This result corresponds to the preferential attachment model.

7.11.3 Popularity-Similarity Network Dynamics

It is demonstrated in this subsection that the concepts of both popularity and similarity are driving forces behind the evolution of networks. Dynamics of preferential attachment helped explain the emergence of scale-free networks earlier in the chapter. The principle of preferential attachment incorporates the principle of popularity in the evolution of a complex network. It is shown in this subsection that both popularity and similarity simultaneously shape the structure of a network. In this later model, as the network evolves, new connections prefer nodes which optimize certain objective functions between popularity and similarity traits of network nodes.

Simply stated, new nodes form a link with popular nodes, and as well as nodes which are similar to them. As the network evolves, older nodes have a higher chance of becoming popular. Assume that a new node is born at time instants $t = 1, 2, 3, \ldots$ and so on. The node born at time $t \in \mathbb{P}$ is its label. The concept of similarity is modeled by placing the nodes on a circle. The angular distance between the nodes models their similarity distance. A convenient approach to model the trade-off between popularity and similarity is to determine new connections by optimizing the product of popularity and similarity measures. We next describe the network evolution.

- Initially the network is empty.
- A time instant $t \in \mathbb{P}$, node t is born at coordinates (r_t, θ_t) on a circle. The radial distance $r_t \in \mathbb{R}^+$, and θ_t is uniformly distributed on the interval $[0, 2\pi)$. Furthermore $r_t = \ln t$.
- The new node t connects to $m \in \mathbb{P}$ number of existing nodes $s < t$. Furthermore, these m nodes are hyperbolically closest to the node t. The value m is related to the average nodal degree $\overline{k}$ as $\overline{k} = 2m$, and θ_{st} is the angular distance between nodes s and t.
- At time instants $t \leq m$, the node t connects to all existing nodes.
- In this model, it appears that older nodes have a higher chance of becoming more popular. However, it might happen that the popularity of older nodes might also fade. To model this phenomenon, we let the radial coordinate of each existing node s, where $s < t$ change to $r_s(t) = \beta r_s + (1 - \beta) r_t$, where $\beta \in [0, 1]$.
  For $\beta = 1$, the nodes do not move. However for $\beta = 0$, the nodes drift away. In the later case all nodes lie on the circle of radius r_t.

The above formulation has an intriguing geometric interpretation. Consider a hyperbolic space of curvature $-4 = -\zeta^2$. Changing the value of the curvature via any other positive value of ζ will simply rescale the results. That is, the general conclusions are not affected qualitatively. Consider two points (r_s, θ_s) and (r_t, θ_t) in a hyperbolic space. The hyperbolic distance and the angle between these two points is x_{st} and θ_{st} respectively. We have

$$\begin{aligned} \cosh \zeta x_{st} &= \cosh \zeta r_s \cosh \zeta r_t - \sinh \zeta r_s \sinh \zeta r_t \cos \theta_{st} \\ \theta_{st} &= \{\pi - |\pi - |\theta_s - \theta_t||\} \end{aligned}$$

For $\zeta = 2$, and large values of r_s and r_t we have

$$x_{st} \simeq r_s + r_t + \ln\left(\frac{\theta_{st}}{2}\right)$$

If we let $r_s = \ln s$ and $r_t = \ln t$, then $x_{st} = \ln(st\theta_{st}/2)$. Therefore the set of nodes $s < t$ which minimize either x_{st} or $st\theta_{st}$ (or simply $s\theta_{st}$) for fixed t are identical. This is true because the logarithm is a monotonically increasing function. Therefore, the hyperbolic distance is a suitable single metric for representing the attractiveness attributes. It represents simultaneously the radial popularity and angular similarity of nodes in a single metric. It is next demonstrated that this network growth model produces the same quantitative results generated by the generalized Barabási and Albert network growth model. Thus there is an equivalence between the generalized preferential attachment model of Barabási and Albert and the popularity-similarity model.

Let R_t be the radius of a hyperbolic disc centered at the new node t. The probability that there is a connection between nodes s and t is

$$\begin{aligned} P(x_{st} \leq R_t) &= P\left(\theta_{st} \leq 2e^{-(r_s(t)+r_t-R_t)}\right) \\ &\simeq \frac{2}{\pi}e^{-(r_s(t)+r_t-R_t)} \end{aligned}$$

The average number of nodes which lie within this hyperbolic disc of radius R_t is

$$\begin{aligned} \mu_{R_t} &= \int_1^t P(x_{it} \leq R_t)\, di \\ &= \frac{2}{\pi}e^{-(r_t-R_t)} \int_1^t e^{-r_i(t)} di \\ &= \frac{2}{\pi}e^{-(r_t-R_t)} \frac{1}{(1-\beta)} \left\{1 - e^{-(1-\beta)r_t}\right\} \end{aligned}$$

where the relationship $r_i(t) = \beta r_i + (1-\beta) r_t$, $r_i = \ln i$, for $\beta \in [0,1)$ was used. The above result implies

$$R_t = r_t - \ln\left\{\frac{2\left(1 - e^{-(1-\beta)r_t}\right)}{\pi\mu_{R_t}(1-\beta)}\right\}$$

Thus, R_t is the radius of a hyperbolic disc centered at node t, which on the average contains hyperbolically closest μ_{R_t} existing nodes. As per the specification of the model, the node t is connected to m nodes. Therefore $\mu_{R_t} = m$. Consequently

$$P(x_{st} \leq R_t) = \frac{m(1-\beta)}{\left(1 - e^{-(1-\beta)r_t}\right)} e^{-r_s(t)}$$

Observe that $P(x_{st} \leq R_t)$ is also equal to the probability that an existing node created at time instant s is connected to the node created at time instant t, if the new node t on an average connects to the hyperbolically closest m existing nodes. We denote this probability by $\Pi(r_s(t))$. That is,

$$\Pi(r_s(t)) = P(x_{st} \leq R_t)$$

As

$$\int_1^t e^{-r_i(t)} di = \frac{1}{(1-\beta)} \left\{1 - e^{-(1-\beta)r_t}\right\}$$

the expression for $\Pi(r_s(t))$ can also be written as

$$\begin{aligned}\Pi\left(r_s\left(t\right)\right) &= m\frac{e^{-r_s(t)}}{\int_1^t e^{-r_i(t)}di} \\ &= m\frac{e^{-(\beta r_s+(1-\beta)r_t)}}{\int_1^t e^{-(\beta r_i+(1-\beta)r_t)}di} \\ &= m\frac{\left(\frac{s}{t}\right)^{-\beta}}{\int_1^t \left(\frac{i}{t}\right)^{-\beta} di}\end{aligned}$$

We now recall results from the generalized Barabási and Albert model of a growing network.

Results from the Generalized Barabási and Albert Network Evolution Model

At each time instant $t \in \mathbb{P}$, a new node node is created. This node is labeled t. We also let it connect to m existing nodes $s < t$ via out-going links from it, where $m \in \mathbb{P}$ and $s \in \mathbb{P}$. The in-degree of a node created at time s at some later time t is $q\left(s,t\right)$. Therefore the degree of a vertex s at time t is $k\left(s,t\right) = \left(m + q\left(s,t\right)\right)$, as the out-degree of each node is m.

The probability that a new link points to a node s is proportional to $\left(q\left(s,t\right) + A\right)$, where the parameter $A > 0$ characterizes additional attractiveness of a node. For mathematical convenience, we let $A \triangleq ma$, where $a > 0$. Note that $\sum_{1\leq s\leq t}\left(q\left(s,t\right) + A\right) = \left(m + A\right)t$. Also the average in-degree connectivity of a node s at time t is given by $\overline{q}\left(s,t\right)$, where $s \leq t$. It is

$$\overline{q}\left(s,t\right) = A\left\{\left(\frac{s}{t}\right)^{-1/(1+a)} - 1\right\}, \quad s \leq t$$

Similarly, let the average connectivity of a node s be given by $\overline{k}\left(s,t\right)$ for $s \leq t$. Therefore

$$\begin{aligned}\overline{k}\left(s,t\right) &= m + \overline{q}\left(s,t\right) \\ &= m + A\left\{\left(\frac{s}{t}\right)^{-1/(1+a)} - 1\right\}, \quad s \leq t\end{aligned}$$

For large values of t, that is $t \gg 1$, an existing node s with degree $k\left(s,t\right)$ attracts a link from the newly created node t with probability $\Pi\left(k\left(s,t\right)\right)$, where

$$\begin{aligned}\Pi\left(k\left(s,t\right)\right) &= m\frac{\left(q\left(s,t\right) + A\right)}{\left(m + A\right)t} \\ &= m\frac{\left(k\left(s,t\right) - m + A\right)}{\left(m + A\right)t}, \quad s \leq t\end{aligned}$$

Similarly

$$\Pi\left(\overline{k}\left(s,t\right)\right) = m\frac{\left(\overline{k}\left(s,t\right) - m + A\right)}{\sum_{i=1}^t\left(\overline{k}\left(i,t\right) - m + A\right)}, \quad s \leq t$$

The denominator in the above expression is

$$\sum_{i=1}^{t}\left(\overline{k}\left(i,t\right) - m + A\right) \simeq \int_1^t \left\{\overline{k}\left(i,t\right) - m + A\right\} di$$

Therefore for $t \gg 1$

$$\Pi\left(\overline{k}\left(s,t\right)\right)=m\frac{\left(\dfrac{s}{t}\right)^{-1/(1+a)}}{\displaystyle\int_1^t\left(\dfrac{i}{t}\right)^{-1/(1+a)}di},\quad s\leq t$$

Also the tail of the degree distribution is $P\left(q\right)\sim\left(q+ma\right)^{-\gamma}$, as $\left(q+ma\right)\gg 1$, where $\gamma=\left(2+a\right)$.

Application of the Generalized Barabási and Albert Network Evolution Model

If β is set equal to $1/\left(1+a\right)$, then

$$\Pi\left(r_s\left(t\right)\right)=\Pi\left(\overline{k}\left(s,t\right)\right)$$

This result implies that the probability a node $s<t$ attracts a link from a new node t, is identical in both the popularity-similarity and the generalized Barabási and Albert network evolution models. Consequently the degree distribution in the two models should be identical. Therefore, the tail of the degree distribution has a power-law with exponent γ, where

$$\gamma=\left(2+a\right)=1+\frac{1}{\beta},\quad \beta\in\left(0,1\right)$$

More specifically, $\gamma\in\left(2,3\right)$ if $\beta\in\left(0.5,1\right)$.

7.12 Search Techniques on Networks

It has often been posited that one of the goals of the Internet is organization of stored human knowledge. Actually, this goal of the Internet was anticipated by Vannevar Bush (1890-1974) several years before the invention of the Internet. Bush was a prominent engineer and visionary of the last century. He published a thought provoking article *As We May Think* in the July 1945 issue of Atlantic Monthly. In this article he envisioned that new forms of encyclopedia would appear in near future. However, a prelude to this noble goal is the requirement that it is possible to search efficiently for information on the Internet.

The documents available on the Internet have made it almost an inexhaustible source of information. These documents in particular reside on the Web pages. A significant activity of human beings in the twenty first century has been to search for useful information on the Internet. Thus it is imperative that efficient and useful search processes be studied and discovered. The software processes which perform these searches are called Web search engines. From an algorithmic viewpoint, the important elements of a search process are crawling, indexing, and page ranking.

(a) *Crawling* is the process of downloading "interesting and important" documents from the Web, and transforming and storing them on the search engines.
(b) *Indexing* is the process of maintaining a list of important words, and a corresponding list of their occurrences in the Web pages. For example, the names of the authors, keywords used in a technical paper can be termed important, and are thus indexed. When a user submits a query to a search engine in the form of several words, a logical intersection of Web pages of their occurrence is obtained from the *index*.

(c) The list of documents produced as a result of a query might be extremely long. The process of assigning the interesting and important label to a document as a result of a query is called *page ranking*. Thus page ranking is the process of bestowing importance to a document.

The following topics are discussed in this section: application of singular value decomposition (SVD) techniques to search engines, detailed network search, and random and guided search on the network.

7.12.1 Application of SVD Techniques

The search engine catalogs exhaustively the complete document collection in the SVD search technique. Thus the search engine creates a map of the data found on it. This technique is suitable when it is expensive to search in real time. More specifically, the technique uses a method called *latent semantic indexing* (LSI). It is used in the fields of text mining and information retrieval. The search engine is built by first creating a dictionary of key terms which exist in the documents. Let these terms be $\tau_1, \tau_2, \ldots, \tau_m$. These terms can be either single words, or phrases made up of more than a single word. Let the list of documents or pages be $\widehat{D}_1, \widehat{D}_2, \ldots, \widehat{D}_n$. The search engine searches for the occurrences of the terms in the dictionary, and determines their frequency of occurrence. For example, in document $\widehat{D}_j$ the frequency of occurrence of term τ_i is denoted by a_{ij}. Thus associated with the document $\widehat{D}_j, 1 \leq j \leq n$ is a column vector a_j of size m, where its ith element is equal to $a_{ij}, 1 \leq i \leq m$. The vector a_j associated with the document $\widehat{D}_j$ is called the *document vector,* and the process of generating these columns is called *indexing.*

Let $\widehat{A}$ be an $m \times n$ matrix, where

$$\widehat{A} = \begin{bmatrix} a_1 \; a_2 \; \cdots \; a_n \end{bmatrix} = [a_{ij}] \tag{7.42}$$

This matrix is called the *term-by-document matrix.* A query is submitted by a user of the search engine to look for documents containing the terms specified in the query. The terms in the query are represented by a *query vector* where

$$q = \begin{bmatrix} q_1 \; q_2 \; \cdots \; q_m \end{bmatrix}^T \tag{7.43a}$$

$$q_i = \begin{cases} 1, & \text{if the term } \tau_i \text{ appears in the query} \\ 0, & \text{otherwise} \end{cases} \tag{7.43b}$$

In order to measure the proximity of the query vector to a document vector, compute

$$\cos\theta_j = \frac{q^T a_j}{\|q\|_2 \, \|a_j\|_2}, \qquad 1 \leq j \leq n$$

where $\|\cdot\|_2$ is the Euclidean norm. If $|\cos\theta_j| \geq \Upsilon$ for a prespecified threshold value Υ, then the document $\widehat{D}_j$ is considered to be useful and is indicated so to the user who submitted the query. The value Υ is generally determined via experimentation. In practice, the matrix $\widehat{A}$ is sparse and its columns are normalized to have unit length. Similarly, the length of the query vector is normalized to unity. In this case $\cos\theta_j = q^T a_j$, for $1 \leq j \leq n$.

It should be noted that the index creation process is generally not very precise due to imperfections in the use of vocabulary. Consequently the information obtained by evaluating $|\cos\theta_j|$'s for $1 \leq j \leq n$ is not accurate. Therefore a SVD technique is used to filter out "noise" in the collected information.

The SVD of the matrix $\widehat{A}$ is $\widehat{U}\Sigma\widehat{V}^T$ where $\widehat{U}$ and $\widehat{V}$ are orthogonal matrices of order m and n respectively. Also let the rank of the matrix $\widehat{A}$ be $r_{\widehat{A}}$. Furthermore, all the terms in the matrix $\Sigma = [\sigma_{ij}]$ are all zeros except $r_{\widehat{A}}$ number of positive σ_{ii}'s. For simplicity assume that $\widehat{p} = \min(m,n)$, and define $\sigma_{ii} \triangleq \sigma_i$ for $1 \le i \le r_{\widehat{A}} \le \widehat{p}$, where $\sigma_i > 0$ for $1 \le i \le r_{\widehat{A}}$. Define $\widehat{U} = \begin{bmatrix} u_1 \ u_2 \cdots \ u_m \end{bmatrix}$, and $\widehat{V} = \begin{bmatrix} v_1 \ v_2 \cdots \ v_n \end{bmatrix}$, then

$$\widehat{A} = \sum_{i=1}^{r_{\widehat{A}}} \sigma_i u_i v_i^T$$

In practice, only k values of the σ_i's are considered, where $k \le r_{\widehat{A}}$. Let e_j be a unit column vector of size n and length 1, where all its elements are zero, except its jth element, which is equal to unity. Then $a_j = \widehat{A}e_j$ for $1 \le j \le n$. Thus

$$\cos\theta_j = \frac{q^T \widehat{A} e_j}{\|q\|_2 \left\| \widehat{A} e_j \right\|_2}, \qquad 1 \le j \le n$$

The computational effort is minimized via the following considerations. Define

$$\mathcal{S} = \Sigma\widehat{V}^T = \begin{bmatrix} s_1 \ s_2 \cdots \ s_n \end{bmatrix}$$

Then $s_j = \mathcal{S}e_j$, $\widehat{A}e_j = \widehat{U}s_j$, and $\left\|\widehat{A}e_j\right\|_2 = \|s_j\|_2$. Thus

$$\cos\theta_j = \frac{q^T \widehat{U} s_j}{\|q\|_2 \|s_j\|_2}, \qquad 1 \le j \le n \tag{7.44}$$

In this technique, $\widehat{U}$ and $\mathcal{S}$ need to be computed only once. Moreover, the entire SVD is not required. In addition, k is chosen to be significantly less than $r_{\widehat{A}}$, which is the rank of the matrix $\widehat{A}$.

The storage and computational requirements of this technique are significant. Furthermore, the discriminating power of this scheme has its limitations. Consequently, processing an enormous collection of Web pages is beyond the scope of this algorithm. Thus this scheme is limited to a relatively smaller collection of documents. Examples of such collections are medical and legal documents. These types of documents are generally controlled and nonlinked collections. This is in contrast to collection of documents on the Web, which are linked and extremely large in size. This is the subject of next subsection.

7.12.2 Detailed Network Search

It is indeed a challenging task to retrieve information from the World Wide Web, which is quite mammoth and heterogeneous. In order to retrieve "useful" information from the World Wide Web, the Web pages can be ranked based upon some criteria. The graph structure of the Web can be used to determine this ranking. In order to do this, two schemes are discussed. These are:

(a) Page-Brin's scheme due to Lawrence Page and Sergey Brin.
(b) Kleinberg's scheme due to Jon Kleinberg.

It is assumed in this discussion that the graph associated with the Web does not have isolated components.

Page-Brin's Scheme

Remember that the Web can be modeled as a directed graph in which the Web pages are the nodes, and the hyperlinks connecting them are the edges. The forward links of a Web page are the out-edges, and its back-links are the in-edges of this graph.

Page ranking is a technique for assigning significance to a page. The page-rank of a page is considered to be high if the sum of the page-ranks of its back-links is high. This intuitive notion is justified if a page either has several predecessor pages connected via back-links, or it has very few highly ranked predecessor pages connected via back-links. The page-rank measure on a directed graph is next defined recursively.

Definition 7.2. *Let $G = (V, E)$ be a directed graph, in which $V = \{1, 2, \ldots, n\}$ is the vertex set, and each vertex $i \in V$ represents a Web page, in which an in-edge of the graph represents a back-link, and an out-edge represents a forward link. The page-rank r_i of a vertex $i \in V$ is a real number. Also let F_i be the set of pages to which the vertex i points to, and B_i be the set of pages which point to vertex i. If the out-degree of vertex $i \in V$ is $d_i^+ = |F_i| \in \mathbb{P}$, then*

$$r_i = \sum_{j \in B_i} \frac{r_j}{d_j^+}, \quad \forall\, i \in V \tag{7.45}$$

The page-rank vector R is $\begin{bmatrix} r_1 \; r_2 \; \cdots \; r_n \end{bmatrix}^T$. □

This definition of page-rank can be restated in terms of the adjacency matrix $A = [a_{ij}]$ of the graph G. A matrix $W = [w_{ij}]$ is defined via the adjacency matrix A as:

$$w_{ij} = \begin{cases} 1/d_i^+, & \text{if } \; a_{ij} = 1 \\ 0, & \text{if } \; a_{ij} = 0 \end{cases} \tag{7.46}$$

Note that the sum of the elements of each row of the matrix W is equal to either 0 or 1. Therefore

$$R = W^T R \tag{7.47}$$

Notice that R is an eigenvector of matrix W^T, and the corresponding eigenvalue is unity. The page-rank vector R can be determined by solving the set of equations $\left(W^T - I\right) R = 0$ and $R^T e = 1$, where

$$e = \begin{bmatrix} 1 \, 1 \cdots \, 1 \end{bmatrix}^T \tag{7.48}$$

is an all-1 vector, and I is the identity matrix of size n. Another useful way to find R is via the following scheme.

Initially assume that W is a primitive stochastic matrix. That is, each element of this matrix is nonnegative, the sum of the elements of each row is equal to unity, and the digraph associated with the matrix is connected. Then using Perron-Frobenius theory, it is known that the spectral radius of the matrix W is $\varrho(W) = 1$, and that an eigenvector of W^T corresponding to unit eigenvalue is nonnegative.

The page-rank vector R can be computed iteratively. Define $R^{(0)} = e/n$, and $R^{(k)} = W^T R^{(k-1)}$ for $k = 1, 2, \ldots$. It can be shown that $R = \lim_{k \to \infty} R^{(k)}$. From the Perron-Frobenius theory it can be inferred that $\lim_{k \to \infty} W^k = eR^T$. Observe that $R^{(k)} = \left\{W^T\right\}^k R^{(0)}$, then

$$\lim_{k\to\infty} R^{(k)} = \lim_{k\to\infty} \left\{W^T\right\}^k R^{(0)}$$
$$= \left\{eR^T\right\}^T \frac{e}{n} = R$$

It should be noted that the above formulation can also be interpreted as an account of a random walk on the Web graph. The rank of a page i can be interpreted as the asymptotic probability that a random surfer is browsing this page. Furthermore, W is the transition matrix of the underlying Markov process. However, the W matrix associated with a real Web graph may not be strictly stochastic, because this matrix can have a row of all zero elements. That is, the out-degree of the vertex corresponding to this row is equal to zero. To overcome this problem, this row can be modified by replacing all of its elements by $1/n$'s. Denote the modified W matrix by W'. This operation is next formally defined. Let the out-degree of node $i \in V$ be denoted by d_i^+. A matrix $D = [d_{ij}]$ is next introduced, where

$$d_{ij} = \begin{cases} 1/n, & \text{if } d_i^+ = 0 \\ 0, & \text{otherwise} \end{cases} \tag{7.49a}$$

and $1 \le i, j \le n$. Then

$$W' = W + D \tag{7.49b}$$

However the matrix W' may not be irreducible, which is a requirement for the application of Perron-Frobenius theory. In order to force irreducibility, a new matrix W'' is defined.

$$W'' = \alpha W' + (1 - \alpha)\frac{ee^T}{n} \tag{7.49c}$$

where $0 < \alpha < 1$. Generally, the value of α is chosen to be closer to unity. This matrix W'' turns out to be both stochastic and primitive. Recall that the primitive property of a matrix implies irreducibility. Thus each state of the modified Markov process is reachable from every other state of the process. This guarantees that a unique page-rank vector R exists. This vector is computed via equations

$$R^T = R^T W'', \quad R^T e = 1 \tag{7.49d}$$

The Page-Brin's algorithm and its refinements have been implemented commercially, and its initial impact on the commercial front has been of googolian proportions. The above ideas are illustrated via the following toy example.

Example 7.8. Let $G = (V, E)$ be a directed graph with 3 vertices. Its adjacency matrix A is given by

$$A = \begin{bmatrix} 0 & 1 & 0 \\ 0 & 0 & 0 \\ 1 & 1 & 0 \end{bmatrix}$$

The W, W', and W'' matrices, assuming $\alpha = 0.91$ are

$$W = \begin{bmatrix} 0 & 1 & 0 \\ 0 & 0 & 0 \\ 1/2 & 1/2 & 0 \end{bmatrix}$$

$$W' = \begin{bmatrix} 0 & 1 & 0 \\ 1/3 & 1/3 & 1/3 \\ 1/2 & 1/2 & 0 \end{bmatrix}$$

$$W'' = \begin{bmatrix} 0.03 & 0.94 & 0.03 \\ 1/3 & 1/3 & 1/3 \\ 0.485 & 0.485 & 0.03 \end{bmatrix}$$

Notice that the matrix W'' is stochastic and primitive. Since $R^T = R^T W''$ and $R^T e = 1$, the page-rank vector R is

$$R = \begin{bmatrix} 0.2779874 & 0.5309560 & 0.1910566 \end{bmatrix}^T$$

Note that $r_1 + r_2 + r_3 = 1$. □

Query Processing. Once the page-rank vector is determined, an actual query is processed as follows. To answer a customer query, an inverted file storage of document contents is required. In concept, an inverted file storage is similar to the index in the back of a book. Each row in this table contains two columns. In the first column is a list of terms. In the second column is a list of all documents that use that term.

Index: Inverted file storage of document contents

term 1 $\rightarrow$ doc 1, doc 3, doc 8, doc 11
term 2 $\rightarrow$ doc 3, doc 5, doc 17
term 3 $\rightarrow$ doc 8, doc 17
⋮ ⋮ ⋮

Assume that the query is for terms 1 and 3. Then the relevancy set of documents is

$$\{1, 3, 8, 11, 17\}$$

In the next step, the page-ranks of these documents r_1, r_3, r_8, r_{11}, and r_{17} are looked up in the page-rank vector R. These five page-ranks are then sorted in decreasing order. For example, assume that $r_8 \geq r_{17} \geq r_3 \geq r_1 \geq r_{11}$. In this case the document 8 is the most important, followed by the document 17, and then document 3, and so on.

Kleinberg's Scheme

Kleinberg's scheme is more complicated than Page-Brin's scheme. This scheme is also called HITS (Hypertext Induced Topic Search). The HITS method of information retrieval introduces the concept of *authorities* and *hubs*. An authority is a document with multiple inlinks, and hub document has multiple outlinks. The HITS algorithm assumes that good authorities are pointed to by good hubs, and good hubs point to good authorities. Web pages with high authority scores are likely to contain relevant information. A Web page with a high hub score may not have directly relevant information, but it indicates where to find useful information. It is definitely possible for a document to be both an authority and a hub. For each document (page), this algorithm computes both an authority score (measure or weight), and a hub score. Thus a document is considered authoritative, if its authority score is high.

Definition 7.3. *Let $G = (V, E)$ be a directed graph, in which $V = \{1, 2, \ldots, n\}$ is the vertex set, and each vertex $i \in V$ represents a Web page. Let $e_{ij} \in E$ represent a directed edge from vertex*

i to vertex j. Each vertex $i \in V$ has two measures: x_i and y_i. The authority score of a page i is x_i and the hub score is y_i. Then

$$x_i = \sum_{\{j | e_{ji} \in E\}} y_j, \quad 1 \leq i \leq n \tag{7.50a}$$

$$y_i = \sum_{\{j | e_{ij} \in E\}} x_j, \quad 1 \leq i \leq n \tag{7.50b}$$

The authority and the hub vectors of the Web graph are x and y respectively, where

$$x = \begin{bmatrix} x_1 \; x_2 \; \cdots \; x_n \end{bmatrix}^T, \; \textit{and} \; y = \begin{bmatrix} y_1 \; y_2 \; \cdots \; y_n \end{bmatrix}^T \tag{7.50c}$$

□

Let $A = [a_{ij}]$ be the adjacency matrix of the graph G. The above definition yields

$$x_i = \sum_{j=1}^{n} a_{ji} y_j, \text{ and } y_i = \sum_{j=1}^{n} a_{ij} x_j, \quad 1 \leq i \leq n$$

That is

$$x = A^T y, \text{ and } y = Ax \tag{7.51}$$

Also note that

$$x = A^T A x, \text{ and } y = A A^T y \tag{7.52}$$

Thus the matrix $A^T A$ determines the authority scores. Therefore it is called the *authority matrix*. Similarly, the matrix AA^T which determines the hub scores is called the *hub matrix*. Furthermore, $A^T A$ and AA^T are symmetric, positive semidefinite, and nonnegative matrices. Thus x is a dominant right-hand eigenvector of matrix $A^T A$, and y is a dominant right-hand eigenvector of matrix AA^T.

Computation of Authority and Hub Vectors. The authority and hub vectors are computed iteratively via the following set of recursive equations.

$$\begin{aligned} y^{(0)} &= e \\ x^{(k)} &= A^T y^{(k-1)}, \text{ and } y^{(k)} = A x^{(k)}, \quad k = 1, 2, \ldots \\ \lim_{k \to \infty} x^{(k)} &= x, \text{ and } \lim_{k \to \infty} y^{(k)} = y \end{aligned}$$

where e is a column vector of all unit elements. Note that each iterative step can also be expressed as

$$x^{(k)} = A^T A x^{(k-1)}, \text{ and } y^{(k)} = A A^T y^{(k-1)}$$

After each iteration $x^{(k)}$ and $y^{(k)}$ are normalized. Let the normalization measure be $\eta(\cdot)$. That is, after computation of $x^{(k)}$'s and $y^{(k)}$'s in each iteration, these vectors are normalized as

$$x^{(k)} \leftarrow \frac{x^{(k)}}{\eta\left(x^{(k)}\right)}, \text{ and } y^{(k)} \leftarrow \frac{y^{(k)}}{\eta\left(y^{(k)}\right)}$$

A possible but preferred normalizing measure is to take $\eta\left(x^{(k)}\right)$ to be the signed component of maximal magnitude. In this case, $\eta\left(x^{(k)}\right)$ converges to the dominant eigenvalue, and $x^{(k)}$ converges to a corresponding normalized eigenvector x as $k \to \infty$. The vector y is similarly computed.

7.12.3 Random and Guided Search on the Network

Consider a search scenario in a network, where search is conducted for information retrieval from some Web pages. Assume that the graph representing this network is undirected and uncorrelated, and that during search, same vertex is not visited more than once. The search cost is defined as the number of steps until approximately the entire graph is revealed. Two search strategies are discussed in this subsection: random search strategy, and guided search strategy.

Random Search Strategy

In the random search strategy, search starts at a randomly selected vertex, and at each successive step, jump is made to a randomly selected nearest neighbor of the vertex. This strategy is evidently not clever. The number of steps Ξ_s, needed to find some target vertex, provided that the search started from a randomly selected vertex, is next determined. Recall that the discrete probability distribution function of the random variable ν, the degree of a node in the network, is specified by $p_\nu(k), \forall k \in \mathbb{N}$. Define the average degree of a randomly chosen end-vertex of a randomly selected edge as $\widetilde{\mu}$. Therefore

$$\widetilde{\mu} = \sum_{k \in \mathbb{P}} k \frac{k p_\nu(k)}{\xi_1}$$

where ξ_1 is the average nodal degree. Thus $\widetilde{\mu} = \mu_2/\xi_1$, where μ_2 is the second moment of the random variable ν. If it is assumed that the network is scale-free with $2 < \gamma < 3$, then $\mu_2 \sim n^{(3-\gamma)/(\gamma-1)}$, where n is the order of the network. Therefore the required average number of steps is given by $\Xi_s = n/\widetilde{\mu}$. That is

$$\Xi_s = \frac{n}{\widetilde{\mu}} \sim n^{2(\gamma-2)/(\gamma-1)}$$

Note that $\widetilde{\mu} < n/\xi_1$ for $2 < \gamma < 3$. Therefore the average number of search steps can be reduced by following a guided search technique which is next discussed. In guided search strategy, each vertex "knows" about its nearest neighbors. Thus only this "local information" is used in the search.

Guided Search Strategy

In guided search strategy, it is assumed that vertices are knowledgeable about their neighboring vertices, and also sometimes about their neighbors' neighbors. We consider a search strategy in which vertices are knowledgeable about their neighbors' neighbors. A generating function formalism is used to study this local search in random graphs. The z-transform of the distribution of the degree of a node is given by $\mathcal{G}_0(z) = \sum_{k \in \mathbb{N}} p_\nu(k) z^k$. Therefore $\mathcal{E}(\nu) = \mathcal{G}_0'(1)$ = average degree of a randomly chosen vertex, where $\mathcal{G}_0'(z)$ is the first derivative of $\mathcal{G}_0(z)$ with respect to z. A randomly selected edge arrives at a vertex with probability proportional to the degree of the vertex, that is $k p_\nu(k)$. Therefore, the z-transform of this normalized probability is

$$\frac{\sum_{k \in \mathbb{N}} k p_\nu(k) z^k}{\sum_{k \in \mathbb{N}} k p_\nu(k)} = z \frac{\mathcal{G}_0'(z)}{\mathcal{G}_0'(1)}$$

Therefore the z-transform of the probability distribution function of the number of outgoing edges from the vertex we arrived at by first following a random edge, but *excluding* the edge we just arrived from is $\mathcal{G}_1(z)$.

$$\mathcal{G}_1(z) = \frac{\mathcal{G}_0'(z)}{\mathcal{G}_0'(1)}$$

As local search algorithm is of interest, it is reasonable to assume that the vertices are knowledgeable about their second neighbors. Therefore, the distribution of the second neighbors is next computed. It is assumed that the probability that a second neighbor connects to any of the first neighbors or to one another is approximately proportional to n^{-1} as $n \to \infty$. This probability is therefore ignored. The distribution of the second neighbors of the original randomly chosen vertex is specified by $\mathcal{J}_A(z)$. Let the average number of such second neighbors be θ_{2A}. Then

$$\mathcal{J}_A(z) = \sum_{k \in \mathbb{N}} p_\nu(k) \{\mathcal{G}_1(z)\}^k = \mathcal{G}_0(\mathcal{G}_1(z))$$

$$\theta_{2A} = \mathcal{G}_0'(1)\,\mathcal{G}_1'(1) \tag{7.53a}$$

where $\mathcal{G}_1'(z)$ is the first derivative of $\mathcal{G}_1(z)$ with respect to z. However, if the original node was not selected at random, but arrived at by following a random edge, then the distribution of the number of second neighbors is specified by $\mathcal{J}_B(z)$. Let the average number of such second neighbors be θ_{2B}. Then

$$\mathcal{J}_B(z) = \mathcal{G}_1(\mathcal{G}_1(z))$$

$$\theta_{2B} = \{\mathcal{G}_1'(1)\}^2 \tag{7.53b}$$

Assuming that the random walk along edges proceeds node to node, the cost of the search is defined as the number of steps until the entire graph is approximately searched. Let n be the order of the network. If the original node was arrived at by following a random edge, then the number of steps Ψ_s in the search is given by

$$\Psi_s \sim \frac{n}{\theta_{2B}} \tag{7.53c}$$

This search strategy is sometimes termed a random walk search.

Examples 7.9. The above results are elucidated via the following examples.

1. Consider a graph with Poissonian degree distribution. In this graph $\mathcal{G}_1(z) = \mathcal{G}_0(z) = e^{\lambda(z-1)}$. Thus $\mathcal{G}_0'(1) = \mathcal{G}_1'(1) = \lambda$, and $\theta_{2A} = \theta_{2B} = \lambda^2$. Using these values the number of steps in the guided search is given by $\Psi_s \sim n/\lambda^2$.
   This result demonstrates that the search technique in Poissonian graph has a search cost that scales linearly with the order of the network.
2. Assume that the degree distribution of the graph follows a power-law. Let the exponent of the power-law be $2 < \gamma < 3$; and $k_c \simeq n^{1/(\gamma-1)}$. Then $\xi_1 = \overline{k} \approx 1$, and $\xi_2 = (\mu_2 - \overline{k}) \simeq \mu_2 \simeq k_c^{(3-\gamma)} \simeq n^{(3-\gamma)/(\gamma-1)}$. Thus $\mathcal{G}_1'(1) = \xi_2/\xi_1 \simeq n^{(3-\gamma)/(\gamma-1)}$. Therefore $\theta_{2A} \simeq n^{(3-\gamma)/(\gamma-1)}$, and $\theta_{2B} \simeq n^{2(3-\gamma)/(\gamma-1)}$. Finally, $\Psi_s \sim n^{(3\gamma-7)/(\gamma-1)}$, where $\gamma \in (7/3, 3)$.
   This result demonstrates that the search technique in power-law graph has a search cost that scales sublinearly with the order of the network. □

An alternate guided search strategy utilizes high degree nodes intentionally. This search technique can be very efficient, as it utilizes information about the degree of its neighboring nodes. It assumes that each vertex of the network is aware of the degree of its neighboring nodes, and also the information stored in them. The intuition behind this algorithm is as follows. If every vertex has information about its immediate neighbors, the obvious way to examine the largest number of vertices at each step is to inspect the most connected vertex, and iterate the process. This is evidently an intelligent technique for doing search.

The algorithm is executed as follows. It starts from a single source vertex on a network. The query is transmitted to all its neighbors. If a neighbor replies: "Yes, I have it," then the search is over, or the reply is, "No, I do not have it, but my degree is k." If the inquiring vertex receives a negative reply from all of its neighbors, the source vertex selects a vertex with the highest value of the degree k, and passes the responsibility of further querying to this node. This process is repeated iteratively. Thus the hope is that the required query is answered positively on the average in the smallest number of inquiries.

7.13 Virus Epidemics and Immunization in the Internet

One of the primary purposes of the Internet is data exchange. However, this basic function of the Internet can also be used to cause it harm. This is done in the form of computer virus (strain). Computer viruses are small programs that can replicate themselves by infecting other clean programs. These newly infected computer programs, when executed on another computer in turn infect other computer programs on the other computer. Thus this insidious cycle of infection propagates through the network, as a virus. Hence the use of the term computer virus. Scale-free networks, of which the Internet is potentially a prime example are susceptible to the persistence and diffusion of virus. Propagation of the virus physically occurs via routers and servers, and can quickly reach epidemic proportions.

There are three main classes of computer virus. The first strain is the *file virus*. These are viruses that infect application programs. The second type of strain is the *boot-sector virus*. These viruses infect the boot-sector of the hard drives. Boot-sector of a disk contains a small program responsible for loading the operating system of the computer. The third type of strain is the *macro virus*. These viruses are independent of the platform's hardware and software. Macro viruses infect data files created by word processors and spreadsheets. Such viruses are coded using *macro* instructions and appended to the documents. A very hostile type of cyber-organism is the *worm*. These viruses spread via electronic mail.

The effect of virus in a network can be modeled by representing individuals as the nodes of a graph, and the interactions of the individuals by links of the graph along which the virus spreads.

7.13.1 A Model of Spread of Epidemic

A preliminary model of the spread of virus in an undirected graph (network) is developed. The modeling technique borrows heavily from the epidemiological models developed for understanding the spread of diseases in large populations. We consider both:

(a) Homogeneous networks.
(b) Heterogeneous networks.

Recall that the heterogeneity parameter of a network has been defined to be equal to

$$\varkappa = \frac{\mathcal{E}\left(\nu^{2}\right)}{\mathcal{E}\left(\nu\right)}$$

where ν is a random variable representing the degree of a node in the undirected graph. A network is said to be homogeneous if its heterogeneity parameter $\varkappa \simeq \mathcal{E}(\nu)$, and heterogeneous if $\varkappa \gg \mathcal{E}(\nu)$. For a large scale-free real-world random graph $\varkappa \to \infty$.

Homogeneous Networks

In a homogeneous network $\mathcal{E}(\nu^2) \simeq \{\mathcal{E}(\nu)\}^2$, where ν is random variable representing the degree of a node in the undirected graph. Thus the heterogeneity of a homogeneous network $\varkappa \simeq \mathcal{E}(\nu)$. This model also assumes that new nodes and edges are not added to the network during the spread of virus. Furthermore, all nodes are considered to be equivalent.

In this model, a vertex of the graph can be in two states: infected or susceptible. That is, a node is either infected with a virus or it is susceptible to get infected by it. At each time step, a susceptible node is infected with probability υ_e if it is connected to at least a single infected node. It is possible for the infected node to get cured and become again susceptible with probability p_e. Denote the effective rate of spread of virus by λ_e. Then

$$\lambda_e = \frac{\upsilon_e}{p_e} \tag{7.54}$$

Assume that p_e is equal to unity in the rest of this section. It only introduces a redefinition of the time scale of virus (disease) propagation. In summary, individual nodes alternate between the states of susceptibility and infection. Also denote the density of infected nodes present in the network at time t by $\chi(t)$. The density correlations among different nodes are ignored. Let $\overline{k}$ be the average nodal degree, then for $\chi(t) \ll 1$

$$\frac{d\chi(t)}{dt} = -\chi(t) + \lambda_e \overline{k} \chi(t) \{1 - \chi(t)\} \tag{7.55}$$

The first term on the right-hand side of the above equation represents the infected nodes becoming healthy with unit rate. The second term is the average rate of density of nodes which are newly infected. This second term is equal to the product of infection spreading rate λ_e, average number of links emanating from each node, the probability $(1 - \chi(t))$ that a given link points to a healthy node, and $\chi(t)$. Use of the factor $\overline{k}$ in the second term implies that each node has approximately the same number of links. This in turn implies that the network is homogeneous. The last factor $\chi(t)$ implies that the rate of change of density of infection is proportional to the density of infected individuals $\chi(t)$. This assumption is called the *homogeneous mixing hypothesis*. At steady-state, that is as $t \to \infty$, we have $d\chi(t)/dt \to 0$ and denote $\lim_{t\to\infty} \chi(t)$ by χ. Thus

$$\chi\{-1 + \lambda_e \overline{k}(1 - \chi)\} = 0,$$

Define the epidemic threshold $\Lambda_e = \overline{k}^{-1}$. As $\chi \geq 0$ we have

$$\chi = \begin{cases} 0, & \lambda_e < \Lambda_e \\ (1 - \Lambda_e/\lambda_e), & \lambda_e \geq \Lambda_e \end{cases} \tag{7.56}$$

The variable χ is generally referred to as the *prevalence* of the network. The above equation yields an important conclusion about the spread of virus in a homogeneous network: that is the presence of a *nonzero* value of the threshold Λ_e of the spread rate. For spread rates above this threshold, the infection spreads and becomes persistent. Below this threshold, the infection decays to a zero value at steady-state.

Heterogeneous Networks

A heterogeneous network has been defined as a network for which the heterogeneity parameter $\varkappa \gg \mathcal{E}(\nu)$. Such networks are generally scale-free. Therefore epidemics in scale-free networks are next discussed. Denote the density of infected nodes in the network with degree (connectivity) k at time t by $\chi_k(t)$. Thus $\chi_k(t)$ is the probability that at time t, a node with k links is infected. Also let $\Im(t) \triangleq \{\chi_{k'}(t)\}$ be the set of densities of such infected nodes. Then for $\chi_k(t) \ll 1$

$$\frac{d\chi_k(t)}{dt} = -\chi_k(t) + \lambda_e k(1 - \chi_k(t)) X_k(\Im(t)) \tag{7.57}$$

The second term on the right-hand side of the above equation is equal to the product of infection spreading rate λ_e, actual number of connections k, the density $(1 - \chi_k(t))$ of healthy vertices with k links that might possibly get infected via a neighboring node, and $X_k(\Im(t))$. The term $X_k(\Im(t))$ is the probability that an edge emanating from a vertex of degree k points to an infected vertex. Observe that $X_k(\cdot)$ is a function of the set of densities of infected vertices $\Im(t)$ (including vertices of degree k).

This formulation is further simplified by assuming that nodal degrees are uncorrelated. That is, the probability that a vertex i of degree k is connected to a vertex j of degree k', is independent of the degree k of the originating vertex i. Under these conditions, $X_k(\cdot)$ is independent of k. Therefore denote $X_k(\cdot)$ by $X(\cdot)$. At steady-state, let $\lim_{t\to\infty} \chi_k(t) \triangleq \chi_k$, and $\lim_{t\to\infty} X(\Im(t)) \triangleq X(\lambda_e)$, then

$$\chi_k = \frac{\lambda_e k X(\lambda_e)}{1 + \lambda_e k X(\lambda_e)} \tag{7.58a}$$

Therefore, higher the value of k, higher is the value of χ_k. Notice that χ_k's are functions of λ_e. Also recall that the probability that an edge points to a vertex of degree k is equal to $kp_\nu(k)/\overline{k}$, $\forall$ $k \in \mathbb{P}$. Thus

$$X(\lambda_e) = \frac{1}{\overline{k}} \sum_{k\in\mathbb{P}} kp_\nu(k)\chi_k \tag{7.58b}$$

Define χ to be the average value of χ_k's. That is

$$\chi = \sum_{k\in\mathbb{P}} p_\nu(k)\chi_k \tag{7.58c}$$

Use of the above equations yields

$$X(\lambda_e) = \frac{1}{\overline{k}} \sum_{k\in\mathbb{P}} kp_\nu(k) \frac{\lambda_e k X(\lambda_e)}{1 + \lambda_e k X(\lambda_e)}$$

Observe that $X(\lambda_e) = 0$ is a solution of the above equation. A value of $X(\lambda_e)$ in the interval $(0, 1]$ can also be obtained by considering the intersection of curves $y_1(X) = X$ and $y_2(X)$, where

$$y_2(X) = \frac{1}{\overline{k}} \sum_{k\in\mathbb{P}} kp_\nu(k) \frac{\lambda_e k X}{1 + \lambda_e k X}$$

In the above equation, the dependence of $X(\cdot)$ on λ_e was dropped for simplicity. The function $y_2(\cdot)$ is monotonically increasing function of X in the interval $(0, 1]$. Furthermore, $y_2(0) = 0$ and $y_2(1) < 1$. Consequently, in order to find a value of $X(\lambda_e)$ in the interval $(0, 1]$,

$$\left.\frac{dy_2(X)}{dX}\right|_{X=0} \geq 1$$

However

$$\left.\frac{dy_2(X)}{dX}\right|_{X=0} = \lambda_e \frac{\mathcal{E}(\nu^2)}{\overline{k}}$$

Define

$$\Lambda_e = \frac{\overline{k}}{\mathcal{E}(\nu^2)} \tag{7.59}$$

Thus $\Lambda_e = \varkappa^{-1}$, where $\varkappa$ is the heterogeneity of the network. Therefore a nonzero value of $X(\lambda_e)$ is obtained in the interval $(0, 1]$ provided $\lambda_e \geq \Lambda_e$. For scale-free networks, with degree exponent γ in the interval $(2, 3]$, the heterogeneity of a network $\varkappa \to \infty$. That is, $\Lambda_e \to 0$. This conclusion implies that for scale-free networks with degree exponent γ in the interval $(2, 3]$, a *null epidemic threshold* exists. Consequently, as per this model, scale-free networks are ideal candidates for the spread of computer viruses.

Continuous k approximation of $X(\lambda_e)$ and χ are next computed for a scale-free network, for $\chi \ll 1$ and consequently $X(\lambda_e) \ll 1$. It is also assumed that the parameter γ is in the range $(2, 3]$. In this range of values, the degree distribution of the undirected graph is given by

$$p_\nu(k) \simeq (\gamma - 1)\, k_0^{\gamma-1} k^{-\gamma}, \qquad k_0 \leq k < \infty$$

In this range of values $\overline{k} = \mathcal{E}(\nu) = k_0(\gamma - 1)/(\gamma - 2)$. The approximations for $X(\lambda_e)$ and χ are

$$X(\lambda_e) \simeq \frac{1}{\overline{k}} \int_{k_0}^{\infty} k p_\nu(k) \frac{\lambda_e k X(\lambda_e)}{1 + \lambda_e k X(\lambda_e)} dk$$

$$\chi \simeq \int_{k_0}^{\infty} p_\nu(k) \frac{\lambda_e k X(\lambda_e)}{1 + \lambda_e k X(\lambda_e)} dk$$

Define $\omega_e = k_0 \lambda_e$. Then after some algebraic manipulations, for $2 < \gamma < 3$

$$X(\lambda_e) \simeq \omega_e^{(\gamma-2)/(3-\gamma)}$$

$$\chi \simeq \frac{(\gamma - 1)}{(\gamma - 2)} \omega_e X(\lambda_e)$$

Thus

$$\chi \simeq \frac{(\gamma - 1)}{(\gamma - 2)} \omega_e^{1/(3-\gamma)}$$

The above analysis reflects the fact that there is an absence of epidemic threshold in a scale-free network for values of γ in the interval $(2, 3)$. Also there is a nonzero prevalence χ for all values of ω_e. That is, the infection can proliferate even for small values of the spreading rate λ_e. Similarly for $\gamma = 3$,

$$X(\lambda_e) \simeq \frac{1}{\omega_e} \left\{ e^{1/\omega_e} - 1 \right\}^{-1}$$

$$\chi \simeq 2e^{-1/\omega_e}$$

Again there is an absence of epidemic threshold in a scale-free network for $\gamma = 3$. The infection can proliferate even for small values of the spreading rate λ_e.

7.13.2 Immunization Process

It has been discussed in the last subsection, that for large scale-free networks, for values of the degree exponent $\gamma \in (2,3]$, there is no threshold value of disease spread rate below which viruses disappear at steady-state. Thus scale-free networks are susceptible to the spread and persistence of computer viruses. Two immunization strategies to overcome this persistence are studied. These are:

(a) Uniform immunization.
(b) Targeted immunization.

Uniform Immunization

Uniform immunization is a simple scheme to introduce a certain fraction of immune vertices spread uniformly in the network. Let the density of immune vertices be f_e. It is called the *immunity* factor. In this scheme the virus spread rate λ_e, is effectively decreased by a factor f_e. Therefore the new virus spread rate is $\lambda_e (1 - f_e)$. Consequently the probability of finding and infecting a susceptible and nonimmune vertex is equal to $\lambda_e (1 - f_e)$.

For homogeneous networks define $\digamma_e$ as the critical immunization value above which the density of infected vertices at steady-state is equal to zero. This value depends upon λ_e. Thus

$$\digamma_e = 1 - \frac{\Lambda_e}{\lambda_e} \tag{7.60}$$

The prevalence factor χ_f for the homogeneous network at steady-state in presence of immunization is given by

$$\chi_f = \begin{cases} 0, & f_e > \digamma_e \\ (\digamma_e - f_e) / (1 - f_e), & f_e \le \digamma_e \end{cases} \tag{7.61}$$

Therefore uniform immunization levels which are larger than $\digamma_e$, completely protect homogeneous networks, and consequently large epidemics are not possible. It should be the goal of immunization software that this network operates above this threshold. However if the network is scale-free, the critical immunization value $\digamma_e$ can be obtained from the equation

$$\lambda_e (1 - \digamma_e) = \Lambda_e = \frac{1}{\varkappa} \tag{7.62}$$

For scale-free networks $\varkappa \to \infty$, therefore $\digamma_e = 1$. That is, only a complete immunization of the scale-free network ensures a virus-free network. Therefore uniform immunization procedures are not effective in controlling the spread of viruses in such networks.

Targeted Immunization

The heterogeneity property of scale-free networks can be utilized to control the spread of viruses in such networks. A *targeted* immunization procedure can be applied to the network by immunizing the nodes with higher nodal degrees. It is the most highly connected nodes which spread the virus faster. Therefore it is reasonable and strategic to immunize these nodes against the virus. This policy also appears to be the most optimal.

Let the range of degree distribution of the graph underlying the network be from k_0 through k_c. A fraction of the nodes with the highest degrees is immunized in this network. The nodes with degree k greater than $g_e k_c$ are immunized, where $g_e \in (0,1)$. The virus cannot propagate along the edges which emanate from these set of nodes. Therefore, the removal of nodes and edges from the network which do not participate in the virus propagation induces a new connectivity pattern. Denote the random variable corresponding to this modified degree distribution by ν_{g_e}. Then the nodes with degrees in the range $k_0 \le k \le g_e k_c$ help propagate the virus. This in turn determines a new value of propagate rate threshold Λ_{e_g}, where

$$\Lambda_{e_g} = \frac{\mathcal{E}(\nu_{g_e})}{\mathcal{E}(\nu_{g_e}^2)} \tag{7.63}$$

Therefore the effective network is protected from the virus propagation if $\lambda_e \le \Lambda_{e_g}$.

7.14 Error and Attack Tolerance of Networks

The topological aspects of a network's tolerance to errors and attacks are studied in this section. Note again that in this section a network is modeled as a simple, undirected, and labeled graph. Removal of either an edge or a node can possibly alter the connectivity of a network. Removal of a node also implies the removal of all its incident edges as well. Consequently a node removal has a potentially more deleterious impact on the network than a single edge removal.

Removal of nodes with low degrees has practically little effect on the connectivity of the network. For example removal of vertices with degree zero or one does not effect the connectivity of vertices with higher degrees. In contrast, removal of vertices of high degrees can have a damaging effect upon the connectivity of the network. Vertices can fail at random, or can be attacked and removed by an external agent.

Consider a network in which the degree distribution of the underlying graph is specified by $p_\nu(k), \forall k \in \mathbb{N}$. It is assumed in this model that a vertex is either present or absent (nonfunctional). Let $b(k)$ be the probability that a vertex is present. It is specified as a function of degree k of a vertex. Define a z-transform $\mathcal{F}_0(z)$ as

$$\mathcal{F}_0(z) = \sum_{k \in \mathbb{N}} p_\nu(k)\, b(k)\, z^k \tag{7.64a}$$

Observe that $p_\nu(k)\, b(k)$ is the probability that a vertex of degree k is present in the network. Note that $\mathcal{F}_0(1) \ne 1$. The value $\mathcal{F}_0(1)$ is equal to the fraction of all vertices that are present. Also define $\mathcal{F}_1(z)$ as

$$\mathcal{F}_1(z) = \frac{\sum_{k \in \mathbb{P}} k p_\nu(k)\, b(k)\, z^{k-1}}{\sum_{k \in \mathbb{P}} k p_\nu(k)} = \frac{\mathcal{F}_0'(z)}{\xi_1} \tag{7.64b}$$

where $\mathcal{F}_0'(z)$ is the first derivative of $\mathcal{F}_0(z)$ with respect z, and ξ_1 is the average nodal degree.

Using these z-transforms, we define the z-transform of the distribution of the size of connected components of present vertices which are accessible from a randomly chosen vertex as $\mathcal{I}_0(z)$. Also define $\mathcal{I}_1(z)$ as the z-transform of the distribution of the size of connected components of present vertices which are accessible from a randomly chosen edge.

$$\mathcal{I}_0(z) = 1 - \mathcal{F}_0(1) + z\mathcal{F}_0(\mathcal{I}_1(z)) \tag{7.65a}$$

$$\mathcal{I}_1(z) = 1 - \mathcal{F}_1(1) + z\mathcal{F}_1(\mathcal{I}_1(z)) \tag{7.65b}$$

where $\mathcal{I}_0(1) = \mathcal{I}_1(1) = 1$. The above equations are similar to the relationships $\mathcal{H}_0(z) = z\mathcal{G}_0(\mathcal{H}_1(z))$, and $\mathcal{H}_1(z) = z\mathcal{G}_1(\mathcal{H}_1(z))$, which were derived in an earlier section. The models for errors in the network and attack tolerance are next developed below.

7.14.1 Error Tolerance

Assume that the vertices fail at random. That is, the probability $b(k)$ is equal to a constant b. In this case $\mathcal{F}_0(z) = b\mathcal{G}_0(z)$ and $\mathcal{F}_1(z) = b\mathcal{G}_1(z)$. Recall that $\mathcal{G}_0(z) = \sum_{k\in\mathbb{N}} p_\nu(k) z^k$, and $\mathcal{G}_1(z) = \mathcal{G}_0'(z)/\mathcal{G}_0'(1)$. Thus

$$\mathcal{I}_0(z) = 1 - b + bz\mathcal{G}_0(\mathcal{I}_1(z)) \tag{7.66a}$$

$$\mathcal{I}_1(z) = 1 - b + bz\mathcal{G}_1(\mathcal{I}_1(z)) \tag{7.66b}$$

Let $\mathcal{I}_0'(z)$ be the first derivative of $\mathcal{I}_0(z)$ with respect to z. Similarly $\mathcal{I}_1'(z)$ is the first derivative of $\mathcal{I}_1(z)$ with respect to z. The average size of components of present vertices is given by $\mathcal{I}_0'(1)$. Therefore

$$\mathcal{I}_0'(1) = b\{1 + \mathcal{G}_0'(1)\mathcal{I}_1'(1)\}$$
$$\mathcal{I}_1'(1) = b\{1 + \mathcal{G}_1'(1)\mathcal{I}_1'(1)\}$$

Thus

$$\mathcal{I}_0'(1) = b\left\{1 + \frac{b\mathcal{G}_0'(1)}{1 - b\mathcal{G}_1'(1)}\right\}$$

Phase transition occurs at a critical value of b, denoted by b_c. It is

$$\begin{aligned} b_c &= \frac{1}{\mathcal{G}_1'(1)} = \frac{\xi_1}{\xi_2} = \frac{\xi_1}{(\mu_2 - \xi_1)} \\ &= \frac{1}{\left(\dfrac{\mu_2}{\xi_1} - 1\right)} \end{aligned} \tag{7.67}$$

Recall that $\xi_2 \triangleq (\mu_2 - \xi_1)$, and first and second moments of the vertex-degree random variable in the original network are ξ_1 and μ_2 respectively. Thus, if a fraction $b < b_c$ of the vertices are present in the random network, then the giant component is absent. That is, connectivity is lost from the network.

Examples 7.10. In these examples, denote the fraction of nodes which are removed at the critical condition by f_c. That is, $f_c = (1 - b_c)$. Also the number of nodes in the network is equal to n.

1. Random graphs: The degree distribution is Poissonian as $n \to \infty$. Thus $\xi_1 = np$, and $\mu_2 = \left\{(np)^2 + np\right\}$. Therefore

$$b_c = \frac{1}{np}, \text{ and } f_c = \left(1 - \frac{1}{np}\right)$$

   Consider the case of a random network in which $np = 1$. Recall that this is the condition at which a giant component forms in the original network. Under this condition, $b_c = 1$, and $f_c = 0$. This implies that any number of node removals leads to network fragmentation.

2. Scale-free networks: These networks have a power-law degree distribution. Consider the following two cases as $n \to \infty$.
   (a) If $1 < \gamma < 2$, then
$$\frac{\mu_2}{\xi_1} \simeq \frac{(2-\gamma)}{(3-\gamma)} k_0 n^{1/(\gamma-1)}$$
   (b) If $\gamma = 2$, then
$$\frac{\mu_2}{\xi_1} \simeq k_0 \frac{n}{\ln n}$$
   (c) If $2 < \gamma < 3$, then
$$\frac{\mu_2}{\xi_1} \simeq \frac{(\gamma-2)}{(3-\gamma)} k_0 n^{(3-\gamma)/(\gamma-1)}$$
   If $\gamma \in (1,3)$, the ratio μ_2/ξ_1 diverges when $n \to \infty$. This implies $f_c \to 1$. This is generally the case of actual networks. Thus scale-free random networks with extremely large number of vertices do not fragment under random failures.
3. Truncated power-law networks: For networks with truncated power-law distribution
$$\frac{\mu_2}{\xi_1} = \frac{Li_{\gamma-2}\left(e^{-1/\kappa}\right)}{Li_{\gamma-1}\left(e^{-1/\kappa}\right)}$$
$$b_c = \left\{\frac{Li_{\gamma-2}\left(e^{-1/\kappa}\right)}{Li_{\gamma-1}\left(e^{-1/\kappa}\right)} - 1\right\}^{-1}$$
   Observe that the expression for b_c is valid for both $\gamma > 3$ and also for $2 < \gamma < 3$. This is in contrast to the scale-free networks. This is due to the presence of the exponential cutoff parameter κ. If $\kappa \to \infty$ then
$$b_c = \left\{\frac{\zeta(\gamma-2)}{\zeta(\gamma-1)} - 1\right\}^{-1}$$
   where Riemann's zeta function $\zeta(x)$ is defined for $x > 1$. Therefore the above expression is valid for $\gamma > 3$. Actually $b_c \to 0$ for all $\gamma \leq 3$. Therefore for these range of values of γ, networks in which $n \to \infty$, infinite cluster exists even for infinitesimal occupation probabilities. That is, if failing vertices are selected at random, there always exists a giant component in the network. Therefore the scale-free network is extremely robust with respect to this property. □

7.14.2 Attack Tolerance

Vertices with larger degrees are removed in the attack mode of vertex failure. As mentioned earlier, $b(k)$ is the probability that a vertex of degree k is present. Vertices with degrees higher than $k_{\max}$ are removed, in an attack on the network. Thus, in the intentional attack mode

$$b(k) = \begin{cases} 1, & k \leq k_{\max} \\ 0, & k > k_{\max} \end{cases} \tag{7.68a}$$

$$\mathcal{F}_0(z) = \sum_{k=1}^{k_{\max}} p_\nu(k) z^k \tag{7.68b}$$

$$\mathcal{F}_1(z) = \frac{\sum_{k=1}^{k_{\max}} k p_\nu(k) z^{k-1}}{\sum_{k\in\mathbb{P}} k p_\nu(k)} = \frac{\mathcal{F}_0'(z)}{\xi_1} \tag{7.68c}$$

Also

$$\mathcal{F}_1'(z) = \frac{\mathcal{F}_0''(z)}{\xi_1} = \frac{1}{\xi_1} \sum_{k=1}^{k_{\max}} k(k-1) p_\nu(k) z^{k-2}$$

Phase transition in the modified network occurs at $\mathcal{F}_1'(1) = 1$. Denote the value of $k_{\max}$ at phase transition by $\widetilde{k}_{\max}$. That is

$$\sum_{k=1}^{\widetilde{k}_{\max}} k(k-1) p_\nu(k) = \xi_1 = \sum_{k \in \mathbb{P}} k p_\nu(k) \tag{7.69}$$

Let the degree distribution follow power-law. That is, let $p_\nu(k) = \mathfrak{B} k^{-\gamma}$, $\forall\, k \in \mathbb{P}$, and $\mathfrak{B} = 1/\zeta(\gamma)$. Then the above equation yields

$$H_{\widetilde{k}_{\max}}^{(\gamma-2)} - H_{\widetilde{k}_{\max}}^{(\gamma-1)} = \zeta(\gamma - 1) \tag{7.70}$$

where $H_m^{(r)}$ is the m-th harmonic number of order r.

$$H_m^{(r)} = \sum_{k=1}^{m} \frac{1}{k^r} \tag{7.71}$$

The fraction of vertices removed at transition (critical point) is given by f_c.

$$f_c = 1 - \frac{H_{\widetilde{k}_{\max}}^{(\gamma)}}{\zeta(\gamma)} \tag{7.72}$$

The relationship $\left\{ H_{\widetilde{k}_{\max}}^{(\gamma-2)} - H_{\widetilde{k}_{\max}}^{(\gamma-1)} \right\} = \zeta(\gamma - 1)$ can be used to determine numerically the value of $\widetilde{k}_{\max}$ for $\gamma > 2$. This value of $\widetilde{k}_{\max}$ is then substituted in the above equation to find the value of f_c. It can be shown that only very small values of f_c are necessary to destroy the connectedness of the giant component. Recall that original networks with $\gamma < 2$ do not have a finite mean. Furthermore, $f_c = 0$ for $\gamma > 3.4788\ldots$, where 3.4788 is the solution of $\zeta(\gamma - 2) = 2\zeta(\gamma - 1)$. In this case the network does not have a giant component before the intentional attack.

In order to gain a better insight on attack tolerance of the network, consider the degree distribution $p_\nu(k) = \mathfrak{B} k^{-\gamma}$, $k_0 \le k \le k_c, 1 < \gamma$, where $\mathfrak{B}$ is the normalizing constant. Let the fraction of nodes with degree higher than $k_{\max}$ be equal to g. Assuming $\int_{k_c}^{\infty} p_\nu(k)\, dk \simeq 1/n$ results in

$$g = \sum_{k=k_{\max}+1}^{k_c} p_\nu(k) \simeq \sum_{k=k_{\max}+1}^{\infty} p_\nu(k) - \frac{1}{n} \tag{7.73a}$$

As $n \to \infty$ for very large networks, then

$$k_{\max} \simeq k_0 \left(g + \frac{1}{n} \right)^{1/(1-\gamma)} \simeq k_0 g^{1/(1-\gamma)} \tag{7.73b}$$

The removal of these nodes with high degree results in the removal of a fraction $\widetilde{g}$ of edges from the remaining nodes. Thus

$$\widetilde{g}\xi_1 = \sum_{k=k_{\max}+1}^{k_c} k p_\nu(k)$$

Finally

$$\widetilde{g} \simeq \frac{\ln (ng)}{\ln n}, \quad \gamma = 2 \tag{7.74a}$$

and

$$\widetilde{g} \simeq g^{(\gamma-2)/(\gamma-1)}, \quad \gamma > 2 \tag{7.74b}$$

Therefore it can be observed that very small values of g can result in the removal of a very large number of edges as $n \to \infty$.

The analysis of this subsection suggests that large networks with power-law degree distribution, are extremely sensitive to the removal of vertices with highest degrees.

7.15 Congestion in a Stochastic Network

Congestion is the state of the network in which the time it takes to transmit a packet dramatically increases over its normal transmission time. In this state, the net packet-throughput collapses precipitously. A simple field-theoretic model of network congestion is developed in this section. The network is modeled as an undirected graph $G = (V, E)$, where V is the set of hosts and routers; and E is the set of transmission links which connect the elements of the set V. It is assumed that hosts are vertex elements which create packets, and routers simply route packets between source and destination pairs. A fraction $p_h \in (0, 1)$ of the $|V|$ vertices represent hosts. Therefore the network has $p_h |V|$ average number of hosts. It is also assumed that the time for a packet to travel from one node to its neighbor is unit time. Therefore the average transit time of a packet between a pair of source and destination nodes, in a lightly loaded network is the average path length ℓ, of the network. The following additional variables are used in developing the network congestion model.

(a) The average number packets generated per unit time per host is equal to L, and its value at steady-state is denoted by $\widetilde{L}$.
(b) Total number of packets in the network at time t is equal to $\mathcal{N}(t)$, and its value at steady-state is denoted by $\widetilde{\mathcal{N}}$.
(c) Average time spent by a packet in the system is equal to $\mathcal{T}(t)$, and its value at steady-state is denoted by $\widetilde{\mathcal{T}}$.
(d) Critical load (packets per unit time) at which congestion sets in is denoted by $\widetilde{L}_c$.

A technique based upon Little's law of queueing theory is used. Little's law states that: the average number of packets in a queueing system at steady-state is equal to the product of the average arrival rate of the packets and the average time spent by the packets in the system. Thus

$$\frac{d\mathcal{N}(t)}{dt} = p_h |V| L - \frac{\mathcal{N}(t)}{\mathcal{T}(t)} \tag{7.75}$$

In the above equation, the first term $p_h |V| L$, on the right-hand side is the average arrival rate to the queues from $p_h |V|$ hosts. The second term $\mathcal{N}(t) / \mathcal{T}(t)$, represents the number of packets delivered per unit time. Assumption of steady-state implies $d\mathcal{N}(t)/dt = 0$. Thus

$$p_h |V| \widetilde{L} - \frac{\widetilde{\mathcal{N}}}{\widetilde{\mathcal{T}}} = 0 \tag{7.76}$$

Two types of networks are analysed. These are a network with equally busy nodes, and a scale-free network.

7.15.1 Network with Equally-Busy Nodes

Consider a network in which all nodes are equally busy. That is, the average queue size at each node at steady-state $\widetilde{Q}$ is equal to $\widetilde{\mathcal{N}} / |V|$. Also

$$\widetilde{\mathcal{T}} = \ell\left(1 + \widetilde{Q}\right) = \ell\left\{1 + \frac{\widetilde{\mathcal{N}}}{|V|}\right\} \tag{7.77a}$$

where it is known that the time for a packet to travel from one node to its neighbor is unit time. Use of the above two equations yields

$$\widetilde{\mathcal{N}} = \frac{\ell p_h \widetilde{L}}{1 - \ell p_h \widetilde{L}} |V| \tag{7.77b}$$

At congestion point $\widetilde{\mathcal{N}} \to \infty$. This occurs when the critical packet generation rate, $\widetilde{L}_c$ is

$$\widetilde{L}_c = \frac{1}{\ell p_h} \tag{7.77c}$$

Therefore the critical value of the packet generation rate is inversely proportional to the average path length of the network.

7.15.2 Scale-Free Networks

In scale-free networks, some nodes are more important than others. The important nodes have high nodal-degree. Consequently these high-degree nodes get congested first, and subsequently, the network gets congested. Therefore, a convenient measure to quantify node usage is the betweenness measure $\beta_v(m)$, of a vertex $m \in V$. Recall that the betweenness of a vertex is a relative measure of the amount of traffic flowing through it. Let $\widetilde{B}_v(m)$ be the normalized betweenness of the node m in the network, where

$$\widetilde{B}_v(m) = \frac{\beta_v(m)}{\sum_{w \in V} \beta_v(w)}, \quad m \in V \tag{7.78}$$

Therefore the average queue size $\widetilde{Q}(m)$ at vertex $m \in V$ at steady-state is approximated as

$$\widetilde{Q}(m) \simeq \widetilde{B}_v(m) \widetilde{\mathcal{N}}, \quad m \in V \tag{7.79a}$$

Let the set of nodes in the path between the source-node $\widetilde{s}$ and the destination-node $\widetilde{d}$ be denoted by $\varsigma_{\widetilde{s}\widetilde{d}}$. If the transit time of a packet from the source-node $\widetilde{s}$ to the destination-node $\widetilde{d}$, at steady-state is $\tau_{\widetilde{s}\widetilde{d}}$, then

$$\tau_{\widetilde{s}\widetilde{d}} \simeq \sum_{m \in \varsigma_{\widetilde{s}\widetilde{d}}} \widetilde{Q}(m); \;\; \widetilde{s} \neq \widetilde{d}, \;\; \widetilde{s}, \widetilde{d} \in V \tag{7.79b}$$

where we used the assumption that the time for a packet to travel from one node to its neighbor is unit time. The total delay $\widetilde{\mathcal{T}}$, at steady can be approximated by

$$\widetilde{\mathcal{T}} = \ell + \widetilde{\mathcal{D}} \widetilde{\mathcal{N}} \tag{7.80a}$$

$$\widetilde{\mathcal{D}} = \frac{1}{|V|(|V| - 1)} \sum_{\substack{\widetilde{s}, \widetilde{d} \in V \\ \widetilde{s} \neq \widetilde{d}}} \sum_{m \in \varsigma_{\widetilde{s}\widetilde{d}}} \widetilde{B}_v(m) \tag{7.80b}$$

Therefore, at steady-state

$$\widetilde{\mathcal{N}} = \frac{\ell p_h |V| \widetilde{L}}{1 - p_h |V| \widetilde{\mathcal{D}} \widetilde{L}} \tag{7.81a}$$

Note that, at the congestion point $\widetilde{\mathcal{N}} \to \infty$. The corresponding critical packet generation rate is $\widetilde{L}_c$, where

$$\widetilde{L}_c = \frac{1}{p_h |V| \widetilde{\mathcal{D}}} \tag{7.81b}$$

Thus the above expression for $\widetilde{L}_c$ is a reasonable approximation for packet rate at the onset of congestion. Note that in a network in which all the nodes are equally loaded, the normalized betweenness of a vertex $m \in V$ is equal to

$$\widetilde{B}_v (m) = \frac{1}{|V|} \tag{7.82a}$$

Consequently

$$\widetilde{\mathcal{D}} = \frac{\ell}{|V|} \tag{7.82b}$$

This in turn implies

$$\widetilde{L}_c = \frac{1}{\ell p_h} \tag{7.82c}$$

as expected.

7.16 Exponential Random Graph Models

Exponential random graph models are used to analyse empirical data about social and other forms of networks. Such graph models are studied in this section. The reason for the phrase "exponential" will become clear shortly. We assume that all graphs in this section are simple. In general, these graphs can either be undirected or directed.

Let $\mathcal{G}$ be an ensemble of such graphs. Denote a random graph variable which takes values in this set by $\widetilde{G}$. Let G be an instance of the random graph variable $\widetilde{G}$. The probability of occurrence of the graph G is $P\left(\widetilde{G} = G\right) \triangleq P(G)$. Thus $\sum_{G \in \mathcal{G}} P(G) = 1$.

The network is modeled by a set of observables (measures, features). Let this set of measures be $\{f_i \mid 1 \le i \le r\}$. For example, these measures can be number of edges in the graph, number of vertices, number of triangles (a subgraph of cycle-length 3), and so on. Let $f_i(G)$ be the value f_i, in graph G, where $1 \le i \le r$. If $\mathcal{E}(\cdot)$ is the expectation operator, the expected value of the random variable $f_i\left(\widetilde{G}\right)$ is

$$\mathcal{E}\left(f_i\left(\widetilde{G}\right)\right) = \sum_{G \in \mathcal{G}} f_i(G) P(G), \quad 1 \le i \le r$$

The measured value of $\mathcal{E}\left(f_i\left(\widetilde{G}\right)\right)$ is $\overline{f}_i \in \mathbb{R}$, for $1 \le i \le r$. We let

$$\sum_{G \in \mathcal{G}} f_i(G) P(G) = \overline{f}_i, \quad 1 \le i \le r$$

The probabilities $P(G)$, $G \in \mathcal{G}$ are determined by maximizing the entropy of the random graph variable $\widetilde{G}$, subject to certain constraints.

$$\underset{P(G),G\in\mathcal{G}}{\textit{Maximize}} \; -\sum_{G\in\mathcal{G}} P(G)\ln P(G)$$

Subject to:

$$\sum_{G\in\mathcal{G}} P(G) = 1$$

$$\sum_{G\in\mathcal{G}} f_i(G)\,P(G) = \overline{f}_i, \quad 1 \le i \le r$$

and $P(G) \ge 0$, $\forall\, G \in \mathcal{G}$. Using the method of Lagrange multipliers, it can be shown that

$$P(G) = \frac{e^{\mathcal{H}(G)}}{Z(\beta)}, \quad \mathcal{H}(G) = \sum_{i=1}^{r} \beta_i f_i(G), \quad G \in \mathcal{G} \tag{7.83a}$$

where $Z(\beta)$ is a normalizing constant, and the β_i's for $1 \le i \le r$ are called ensemble parameters. These are the unknowns which have to be determined from the collected data. As $\sum_{G\in\mathcal{G}} P(G) = 1$, we obtain

$$Z(\beta) = \sum_{G\in\mathcal{G}} e^{\mathcal{H}(G)} \tag{7.83b}$$

See the problem section for a derivation of the above form of the $P(G)$'s. As the probability $P(G)$ is proportional to $\exp(\mathcal{H}(G))$, we call such scheme exponential random graph model. The function $\mathcal{H}(\cdot)$ is often called the graph Hamiltonian. This is named after William R. Hamilton (1805-1865). Let w be another quantity of interest in the model. The expected value of this quantity in the ensemble is given by

$$\mathcal{E}\left(w\left(\widetilde{G}\right)\right) = \sum_{G\in\mathcal{G}} w(G)\,P(G)$$

Computation of any $\mathcal{E}\left(w\left(\widetilde{G}\right)\right)$ is generally hard in this model. However, it can be shown that

$$\mathcal{E}\left(f_i\left(\widetilde{G}\right)\right) = \sum_{G\in\mathcal{G}} f_i(G)\,P(G) = \frac{\partial \ln Z(\beta)}{\partial \beta_i}, \quad 1 \le i \le r \tag{7.84}$$

The above methodology is illustrated via the following examples.

Examples 7.11. Let the set of vertices in a simple graph G be $V = \{1, 2, 3, \ldots, n\}$. The adjacency matrix of the graph G is $A(G) = [a_{ij}(G)]$, where

$$a_{ij}(G) = \begin{cases} 1, & \text{if there is an edge between vertices } i \text{ and } j \\ 0, & \text{otherwise} \end{cases}$$

1. Classical random graph. Let the number of edges in the undirected graph G be $m(G)$, where

$$m(G) = \sum_{1\le i<j\le n} a_{ij}(G)$$

Also, let the Hamiltonian of this graph be $\mathcal{H}(G) = \beta m(G)$, where β is a real number. Therefore, as $Z(\beta) = \sum_{G\in\mathcal{G}} e^{\mathcal{H}(G)}$, we have

$$\begin{aligned}
Z(\beta) &= \sum_{G\in\mathcal{G}} \exp(\beta m(G)) = \sum_{G\in\mathcal{G}} \exp\left\{\beta \sum_{1\le i<j\le n} a_{ij}(G)\right\} \\
&= \sum_{G\in\mathcal{G}} \prod_{1\le i<j\le n} \exp\{\beta a_{ij}(G)\} = \prod_{1\le i<j\le n} \sum_{\sigma=0,1} \exp\{\beta\sigma\} \\
&= \left(1+e^{\beta}\right)^{\binom{n}{2}}
\end{aligned}$$

Therefore $\ln Z(\beta) = \binom{n}{2} \ln\left(1+e^{\beta}\right)$. The average number of edges in the random graph is equal to $\mathcal{E}\left(f\left(\widetilde{G}\right)\right)$, where

$$\mathcal{E}\left(f\left(\widetilde{G}\right)\right) = \frac{\partial \ln Z(\beta)}{\partial\beta} = \binom{n}{2}\frac{e^{\beta}}{\left(1+e^{\beta}\right)}$$

Let $\mathcal{E}\left(f\left(\widetilde{G}\right)\right) = \overline{m}$, then

$$\beta = \ln\left\{\frac{\overline{m}}{\binom{n}{2} - \overline{m}}\right\}$$

We next compare this model with the Bernoulli random graph (classical random graph model A). In the Bernoulli random graph, the probability of an edge between a pair of vertices is p. Therefore the average number of edges in the Bernoulli random graph is $p\binom{n}{2}$. This expression is equal to $\overline{m}$ provided

$$p = \frac{e^{\beta}}{\left(1+e^{\beta}\right)}$$

Also note that for $G \in \mathcal{G}$

$$P(G) = \frac{e^{\mathcal{H}(G)}}{Z(\beta)} = \frac{e^{\beta m(G)}}{\left(1+e^{\beta}\right)^{\binom{n}{2}}} = \left\{\frac{e^{\beta}}{1+e^{\beta}}\right\}^{m(G)} \left\{\frac{1}{1+e^{\beta}}\right\}^{\binom{n}{2}-m(G)}$$

Therefore, if we substitute $p = e^{\beta}/\left(1+e^{\beta}\right)$ in the above expression, we obtain

$$P(G) = p^{m(G)}(1-p)^{\binom{n}{2}-m(G)}$$

Thus the Bernoulli random graph, and the exponential random graph model are equivalent if $p = e^{\beta}/\left(1+e^{\beta}\right)$.

2. Generalized undirected random graph. Let the graph G be undirected, and

$$\beta = \left\{\beta_{ij} \mid 1 \le i < j \le n\right\}$$

be the set of ensemble parameters. The Hamiltonian of the graph is

$$\mathcal{H}(G) = \sum_{1\le i<j\le n} \beta_{ij} a_{ij}(G)$$

Therefore, as $Z(\beta) = \sum_{G\in\mathcal{G}} e^{\mathcal{H}(G)}$, we have

$$\begin{aligned}
Z(\beta) &= \sum_{G\in\mathcal{G}} \prod_{1\le i<j\le n} \exp\left\{\beta_{ij} a_{ij}(G)\right\} \\
&= \prod_{1\le i<j\le n} \sum_{\sigma=0,1} \exp\left\{\beta_{ij}\sigma\right\} = \prod_{1\le i<j\le n} \left\{1+\exp\left(\beta_{ij}\right)\right\}
\end{aligned}$$

Thus

$$\ln Z(\beta) = \sum_{1 \le i < j \le n} \ln \{1 + \exp(\beta_{ij})\}$$

The probability of occurrence of an edge between vertices i and j is p_{ij}, where

$$p_{ij} = \frac{\partial \ln Z(\beta)}{\partial \beta_{ij}} = \frac{\exp(\beta_{ij})}{\{1 + \exp(\beta_{ij})\}}, \quad i \ne j$$

If the graph is sparse, then $p_{ij} \simeq \exp(\beta_{ij})$, for $i \ne j$, and $1 \le i, j \le n$.

3. Generalized directed random graph. Consider a directed graph G. Let the in-degree of the vertex j be $k_j^{in}(G)$, where $k_j^{in}(G) = \sum_{i=1, i \ne j}^{n} a_{ij}(G)$, $1 \le j \le n$. Similarly, let the out-degree of the vertex i be $k_i^{out}(G)$, where $k_i^{out}(G) = \sum_{j=1, i \ne j}^{n} a_{ij}(G)$, $1 \le i \le n$.
Let $\{\beta_j^{in} \mid 1 \le j \le n\}$ and $\{\beta_i^{out} \mid 1 \le i \le n\}$ be the ensemble parameters, and β be the union of these two sets. The Hamiltonian of the graph is

$$\begin{aligned}
\mathcal{H}(G) &= \sum_{j=1}^{n} \beta_j^{in} k_j^{in}(G) + \sum_{i=1}^{n} \beta_i^{out} k_i^{out}(G) \\
&= \sum_{\substack{1 \le i,j \le n \\ i \ne j}} \left(\beta_j^{in} + \beta_i^{out}\right) a_{ij}(G)
\end{aligned}$$

Therefore, as $Z(\beta) = \sum_{G \in \mathcal{G}} e^{\mathcal{H}(G)}$, we have

$$\begin{aligned}
Z(\beta) &= \sum_{G \in \mathcal{G}} \prod_{\substack{1 \le i,j \le n \\ i \ne j}} \exp\left\{\left(\beta_j^{in} + \beta_i^{out}\right) a_{ij}(G)\right\} \\
&= \prod_{\substack{1 \le i,j \le n \\ i \ne j}} \sum_{\sigma = 0,1} \exp\left\{\left(\beta_j^{in} + \beta_i^{out}\right) \sigma\right\} \\
&= \prod_{\substack{1 \le i,j \le n \\ i \ne j}} \left\{1 + \exp\left(\beta_j^{in} + \beta_i^{out}\right)\right\}
\end{aligned}$$

Thus

$$\ln Z(\beta) = \sum_{\substack{1 \le i,j \le n \\ i \ne j}} \ln \left\{1 + \exp\left(\beta_j^{in} + \beta_i^{out}\right)\right\}$$

The average in-degree of vertex j in the random graph is equal to $\mathcal{E}\left(k_j^{in}\left(\widetilde{G}\right)\right)$, $1 \le j \le n$. Similarly, the average out-degree of vertex i in the random graph is equal to $\mathcal{E}\left(k_i^{out}\left(\widetilde{G}\right)\right)$, $1 \le i \le n$. These are

$$\mathcal{E}\left(k_j^{in}\left(\widetilde{G}\right)\right) = \frac{\partial \ln Z(\beta)}{\partial \beta_j^{in}} = \sum_{\substack{i=1 \\ i \ne j}}^{n} \frac{\exp\left(\beta_j^{in} + \beta_i^{out}\right)}{\left\{1 + \exp\left(\beta_j^{in} + \beta_i^{out}\right)\right\}}, \quad 1 \le j \le n$$

$$\mathcal{E}\left(k_i^{out}\left(\widetilde{G}\right)\right) = \frac{\partial \ln Z(\beta)}{\partial \beta_i^{out}} = \sum_{\substack{j=1 \\ i \ne j}}^{n} \frac{\exp\left(\beta_j^{in} + \beta_i^{out}\right)}{\left\{1 + \exp\left(\beta_j^{in} + \beta_i^{out}\right)\right\}}, \quad 1 \le i \le n$$

It can be observed that

$$\sum_{j=1}^{n} \mathcal{E}\left(k_j^{in}\left(\widetilde{G}\right)\right) = \sum_{i=1}^{n} \mathcal{E}\left(k_i^{out}\left(\widetilde{G}\right)\right)$$

as expected. □

Reference Notes

The textbooks by Bonato (2008), Bornholdt, and Schuster (editors, 2003), Caldarelli (2007), Cohen, and Havlin (2010), Dorogovtsev, and Mendes (2003), Estrada (2012), Newman (2010), and Pastor-Satorras, and Vespignani (2004) provide an accessible introduction to the general subject of random networks. Surveys of these types of networks are discussed in the very comprehensive and eminently readable papers by Albert, and Barabási (2002), Dorogovtsev, and Mendes (2002), Newman (2003), Chakrabarti, and Faloutsos (2006), Dorogovtsev, Goltsev, and Mendes (2007), Haddadi, Rio, Iannaccone, Moore, and Mortier (2008), Chung (2010), and Spencer (2010). Popular yet illuminating accounts of the Internet science have been given in Watts (1999), and Barabási (2002).

The numerical values stated in the section on experimental evidence are from the afore mentioned survey papers, and the pioneering work of Faloutsos, Faloutsos, and Faloutsos (1999). A macroscopic structure of the World Wide Web was determined in the measurements of Broder, Kumar, Maghoul, Raghavan, Rajagopalan, Stata, Tomkins, and Wiener (2000). Internet topology data comparison is also provided in Huffaker, Fomenkov, and claffy (2012). A model of the Internet topology using the $\mathcal{K}$-shell decomposition was studied by Carmi, Havlin, Kirkpatrick, Shavitt, and Shir (2007); and Alvarez-Hamelin, Dall'Asta, Barrat, and Vespignani (2006).

Classical random graph theory was studied at length by Erdös, and Rényi. The discussion of classical random graph in this chapter is based upon the lucid books by Frank, and Frisch (1971), and Palmer (1985). Other references on this subject are the books by Bollobás (1985), Kolchin (1999), Sachkov (1997), West (2001), and Bonato (2008). The section on the z-transform formalism is due to Newman (2003). Actually random graphs were studied about ten years earlier than Erdös, and Rényi in Solomonoff, and Rapoport (1951).

For erudite and historical discussion of power-laws, see Mitzenmacher (2002), and Newman (2005). The results on spectra of a graph are based upon the work of Farkas, Derényi, Barabási, and Vicsek (2001), Mihail, and Papadimitriou (2003), Chung, Lu, and Vu (2003), and Van Mieghem (2011).

The discussion on small-world networks is based upon the seminal works of Watts, and Strogatz (1998), Barrat, and Weigt (2000), Newman, and Watts (1999), and Watts (1999). The pioneering Barabási-Albert model of graphical evolution was first proposed by them in the year 1999. The hyperbolic geometry based complex network model was formulated by Krioukov, Papadopoulos, Kitsak, Vahdat, and Boguñá (2010). The intrinsic vertex weight models are based upon the works of Caldarelli, Capocci, Rios, and Muñoz (2002), and Masuda, Miwa, and Konno (2004). The copying model of evolution of a graph was proposed by Kumar, Raghavan, Rajagopalan, Sivakumar, Tomkins, and Upfal (2000). An all encompassing generalized Web graph model to date is due to

Cooper, and Frieze (2003). For a detailed and precise discussion of clustering coefficient, see the paper by Bollobás, and Riordan in the book by Bornholdt, and Schuster (editors, 2003). The subject of popularity versus similarity in growing networks is discussed in detail in Papadopoulos, Kitsak, Serrano, Boguñá, and Krioukov (2012).

The network search technique based upon SVD is from Berry, and Browne (1999), and Meyer (2000). Guided search on the network was investigated by Adamic, Lukose, Puniyani, and Huberman (2001). The Page-Brin, and Kleinberg's algorithms are from the papers by Brin, and Page (1998), Kleinberg (1999), and Langville, and Meyer (2003). There is another search technique, called the Stochastic Approach for Link Structure Analysis (SALSA). It was developed by Lempel, and Moran (2000). This was not discussed in this chapter due to space limitation. Premier work on epidemics and immunization in the Internet has been done by Pastor-Satorras, and Vespignani (2004). The section on congestion in a stochastic network is based upon the work of Arrowsmith, Mondragón, and Woolf (2005). The section on exponential random graph models is based upon the work of Park, and Newman (2004), and Newman (2010).

A branch of physics called *percolation theory* has also been used to explain complex network phenomena. However this technique has not been discussed due to space limitation. For a critique of some of the models discussed in this chapter, the reader is referred to the papers by Mitzenmacher (2004), Alderson (2008), and Willinger, Alderson, and Doyle (2009). In light of these papers, it is quite possible that the experimental results and some models discussed in this chapter might need further refinement.

Finally, since the Internet is an ever changing phenomenon, we have not yet completely fathomed its structure. The different models and their ramifications have to be interpreted carefully. For example, it would simply be serendipitous to recover all the "important" features of a graph (network) from its degree distribution. It is hoped that the different models discussed in this chapter will provide a stepping stone for further inquiry into the structure of the Internet and the World Wide Web.

Problems

1. The edge connecting probability p of a random graph $G(n,p)$ is a nonzero constant. Prove that almost every graph $G(n,p)$ is connected.
   Hint: See West (2001). Partition the set of vertices V into two disjoint sets V_1 and V_2, such that V_2 is the complement of the set V_1. The graph G is disconnected, if there are no edges between the vertices of these two sets. Let $|V_1| = k$, then $|V_2| = (n-k)$ where $1 \leq k \leq (n-1)$. Therefore there can be a total of $k(n-k)$ edges between these two sets of vertices. The probability that these edges do not exist is given by $(1-p)^{k(n-k)}$. Occurrence of edges in the sets V_1 and V_2 is not relevant. Let p_{D_n} be the probability that the graph G is disconnected. Then
   $$p_{D_n} \leq \frac{1}{2}\sum_{k=1}^{n-1}\binom{n}{k}(1-p)^{k(n-k)}$$
   The above inequality is symmetric in k and $(n-k)$. Therefore

$$p_{D_n} \leq \sum_{k=1}^{\lfloor n/2 \rfloor} \binom{n}{k} (1-p)^{k(n-k)}$$

Note that $\binom{n}{k} < n^k$ and $(1-p)^{(n-k)} \leq (1-p)^{n/2}$ for $k \leq n/2$. Consequently

$$p_{D_n} < \sum_{k=1}^{\lfloor n/2 \rfloor} \left\{ n (1-p)^{n/2} \right\}^k$$

Observe that $x \triangleq n(1-p)^{n/2} < 1$ for sufficiently large values of n. Therefore for sufficiently large values of n, the upper bound on p_{D_n} is a geometric series which converges. Thus $p_{D_n} < x/(1-x)$. If p is a constant, then $x \to 0$ and $p_{D_n} \to 0$ as $n \to \infty$.

2. Let the probability that a graph $G(n,p)$ is connected be equal to P_n. Also let $q = (1-p)$. Prove that:
   (a) $P_1 = 1$, and $P_n = 1 - \sum_{k=1}^{n-1} \binom{n-1}{k-1} q^{k(n-k)} P_k$, for $n = 2, 3, 4, \ldots$.
   (b) $P_n \simeq (1 - nq^{n-1})$, for large n and $q < 1$.

   Hint: See Ross (2010).

   (a) A recursive formula for P_n is initially derived. Partition the set of nodes into two components. The two nonintersecting vertex sets V_1 and V_2 form the two components. Assume that there are k nodes in the set V_1 and $(n-k)$ nodes in the set V_2. These two components are not connected. Without loss of generality assume that the node 1 is in set V_1. The set V_1 is formed by selecting $(k-1)$ nodes out of the remaining $(n-1)$ nodes.
   Furthermore, there can be a total of $k(n-k)$ edges between the set of vertices V_1 and V_2. The probability that all these edges are absent is equal to $q^{k(n-k)}$. Also assume that the nodes which belong to the set V_1 are connected. The probability of this event is equal to P_k. Therefore the combined probability that there is no edge which connects a node in V_1 to a node in V_2; and that the set of nodes in V_1 are connected is given by $q^{k(n-k)} P_k$. Therefore

$$\begin{aligned} &P(\text{node 1 is a part of a vertex set } V_1 \text{ of cardinality } k) \\ &= \binom{n-1}{k-1} q^{k(n-k)} P_k, \quad k = 1, 2, \ldots, n \end{aligned}$$

   Hence use of the law of total probability yields

$$\sum_{k=1}^{n} \binom{n-1}{k-1} q^{k(n-k)} P_k = 1$$

   That is

$$P_n = 1 - \sum_{k=1}^{n-1} \binom{n-1}{k-1} q^{k(n-k)} P_k, \qquad n = 2, 3, \ldots$$

   (b) Therefore the P_n's can be determined recursively by using the above relationship and noting that $P_1 = 1$.
   Thus $P_2 = p$, $P_3 = \{1 - q^2(2p+1)\}$, and so on. For sufficiently large values of n, and as $q < 1$, and $P_k \leq 1$

$$1 - P_n \leq \sum_{k=1}^{n-1} \binom{n-1}{k-1} q^{k(n-k)}$$

Observe that $k\,(n-k)$ is a concave function of k, where

$$k\,(n-k) \geq \begin{cases} \{(n-2)\,k+n\}\,/2, & \text{if } 1 \leq k \leq n/2 \\ \{(n-2)\,(n-k)+n\}\,/2, & \text{if } n/2 \leq k \leq (n-1) \end{cases}$$

Combining the above two sets of inequalities yields

$$1 - P_n \lesssim nq^{n-1}$$

A lower bound for $(1-P_n)$ is next obtained. Let Q_i be the event that the vertex i is isolated from all other $(n-1)$ vertices for $1 \leq i \leq n$. Therefore, if any event Q_i occurs, the graph is disconnected. Then

$$1 - P_n = P\,(\text{graph is disconnected}) \geq P\left(\bigcup_{i=1}^{n} Q_i\right)$$

Use of the inclusion-exclusion bounds results in

$$P\left(\bigcup_{i=1}^{n} Q_i\right) \geq \sum_{i=1}^{n} P\,(Q_i) - \sum_{1\leq i<j\leq n} P\,(Q_i Q_j)$$

Note that $P\,(Q_i) = q^{n-1}$ and $P\,(Q_i Q_j) = q^{2n-3}$ for $i \neq j$. Therefore

$$1 - P_n \geq nq^{n-1} - \binom{n}{2} q^{2n-3}$$

Since $\binom{n}{2} q^{2n-3} / \left(nq^{n-1}\right) \to 0$ for large value of n, $1 - P_n \gtrsim nq^{n-1}$. Combining the two bounds yields $P_n \simeq \left(1 - nq^{n-1}\right)$, for large n and $q < 1$.

3. Prove that, if $p = o\left(n^{-1}\right)$, then almost every Model A random graph has no cycles.
Hint: See Palmer (1985) and West (2001). Let $\widetilde{X}$ be a random variable which denotes the number of cycles in G. A cycle of length k occurs with probability p^k, and there are $\binom{n}{k}(k-1)!/2$ possible cycles for $3 \leq k \leq n$. Then

$$\mathcal{E}\left(\widetilde{X}\right) = \sum_{k=3}^{n} \binom{n}{k} \frac{(k-1)!}{2} p^k \leq \sum_{k=3}^{n} \frac{(pn)^k}{2k}$$

Therefore, if $pn \to 0$, then $\mathcal{E}\left(\widetilde{X}\right) \to 0$.

4. Prove that if $p = n^{-1}$ then the largest component of a Model A graph has an order $\simeq n^{2/3}$.
Hint: See Palmer (1985). Let Y_m be a random variable which indicates the number of components of G that are trees of order at least m. Then

$$\mathcal{E}\,(Y_m) = \sum_{k=m}^{n} \binom{n}{k} k^{k-2} p^{k-1} (1-p)^{\binom{k}{2}-(k-1)+k(n-k)}$$

Note that

$$\binom{n}{k} = O\,(1)\,(n)_k \frac{e^k}{k^{k+1/2}}$$

$$(1-p)^{\binom{k}{2}-(k-1)+k(n-k)} = O\,(1)\left(\frac{e^{0.5k/n}}{e}\right)^k$$

Therefore

$$\mathcal{E}(Y_m) = O(1)\, n \sum_{k=m}^{n} \frac{(n)_k}{n^k} k^{-5/2} e^{0.5k^2/n}$$

Also

$$\frac{(n)_k}{n^k} = O(1)\, e^{-0.5k^2/n}$$

Thus

$$\mathcal{E}(Y_m) = O(1)\, n \sum_{k=m}^{n} k^{-5/2}$$

The sum $\sum_{k=m}^{n} k^{-5/2}$ can be approximated by an integral. Therefore

$$\sum_{k=m}^{n} k^{-5/2} \simeq \frac{2}{3}\left(\frac{1}{m^{3/2}} - \frac{1}{n^{3/2}}\right)$$

Let $m = \omega_n n^{2/3}$, where $\omega_n \to \infty$ arbitrarily slowly as $n \to \infty$. Consequently

$$\mathcal{E}(Y_m) = O(1)\left(\frac{1}{\omega_n^{3/2}} - \frac{1}{n^{1/2}}\right)$$

Therefore $\mathcal{E}(Y_m) \to 0$ for $m = \omega_n n^{2/3}$. Thus use of Markov's inequality results in $P(Y_m \geq 1) \to 0$, for $m = \omega_n n^{2/3}$, where $\omega_n \to \infty$ arbitrarily slowly as $n \to \infty$.
It can be similarly established that $\mathcal{E}(Y_m^2) \sim \mathcal{E}(Y_m)^2$ for $m = n^{2/3}/\omega_n$. Use of the second moment method results in $P(Y_m = 0) \to 0$, that is $P(Y_m \geq 1) \to 1$, for $m = n^{2/3}/\omega_n$, where $\omega_n \to \infty$ arbitrarily slowly as $n \to \infty$.
Therefore almost every graph which belongs to Model A with $p = n^{-1}$ has a biggest tree of order about $n^{2/3}$. Furthermore, there are no other components of order greater than or equal to $n^{2/3}\omega_n$.

5. In Model A graph, define Y_m to be a random variable which indicates the number of components of G that are trees of order at least m. Also let ω_n be a number which tends to ∞ arbitrarily slowly as $n \to \infty$. Let $p = 2c/n$, $e^{-\alpha} = 2ce^{1-2c}$, and $c \neq 1/2$. Show that

$$P(Y_m \geq 1) \to 0 \text{ for } m = \frac{1}{\alpha}\left(\ln n - \frac{5}{2}\ln\ln n\right) + \omega_n$$

but

$$P(Y_m \geq 1) \to 1 \text{ for } m = \frac{1}{\alpha}\left(\ln n - \frac{5}{2}\ln\ln n\right) - \omega_n$$

Stated alternately, this problem asserts that in almost every graph the largest component that is a tree, has an order $\ln n$.
Hint: See Palmer (1985). As in the last problem

$$\mathcal{E}(Y_m) = \sum_{k=m}^{n} \binom{n}{k} k^{k-2} p^{k-1} (1-p)^{\binom{k}{2}-(k-1)+k(n-k)}$$

Note that

$$\binom{n}{k} = O(1)\,(n)_k \frac{e^k}{k^{k+1/2}}$$

$$(1-p)^{\binom{k}{2}-(k-1)+k(n-k)} = O(1)\left(\frac{e^{ck/n}}{e^{2c}}\right)^k$$

Therefore

$$\mathcal{E}(Y_m) = O(1)\, n \sum_{k=m}^{n} \frac{(n)_k}{n^k} k^{-5/2} \left\{ (e^c)^{k/n} \left(2ce^{1-2c}\right) \right\}^k$$

For $0 < c < 1/2$, use the approximation

$$\frac{(n)_k}{n^k} = O(1)\, e^{-0.5k^2/n}$$

Thus

$$\frac{(n)_k}{n^k} e^{ck^2/n} = O(1) \exp\left\{ \frac{k^2}{n}\left(c - \frac{1}{2}\right) \right\} = O(1)$$

Consequently

$$\mathcal{E}(Y_m) = O(1)\, n \sum_{k=m}^{n} k^{-5/2} \left(2ce^{1-2c}\right)^k$$

Observe that $0 < c < 1/2$ implies $0 < 2ce^{1-2c} < 1$. As $e^{-\alpha} = 2ce^{1-2c}$, we have

$$\mathcal{E}(Y_m) = O(1) \frac{n}{m^{5/2}} e^{-\alpha m}$$

Letting $m = \left(\ln n - \frac{5}{2}\ln\ln n\right)/\alpha + \omega_n$, where $\omega_n \to \infty$ arbitrarily slowly as $n \to \infty$, it can be shown that $\mathcal{E}(Y_m) = O(1)\, e^{-\alpha\omega_n} = o(1)$. The result follows by using Markov's inequality. This establishes the first part of the problem for $0 < c < 1/2$.

For $c > 1/2$, we establish the result in Steps 0, 1, and 2. Step 0 is a preliminary step. Its result is used in Step 2.

Step 0: Define $f(x, c)$ as

$$f(x,c) = -\alpha + cx - \sum_{i \in \mathbb{P}} \frac{x^i}{i(i+1)}$$

Note that

$$\sum_{i \in \mathbb{P}} \frac{x^i}{i(i+1)} = \frac{(1-x)\ln(1-x) + x}{x}$$

As $e^{-\alpha} = 2ce^{1-2c}$, we have

$$f(x,c) = \ln(2c) - 2c + cx - \frac{(1-x)\ln(1-x)}{x}$$

The critical point of $f(x, c)$ is at $x = 0$, and $c = 1/2$. This is obtained from the equations $\partial f(x,c)/\partial x = 0$, and $\partial f(x,c)/\partial c = 0$. It can also be shown that the maximum value of $f(x, c)$ occurs at this point, and it is equal to $f(0, 1/2) = 0$.

Therefore for some $\epsilon > 0$, we have $\exp\{f(x,c)\} \le (1-\epsilon)$ for all $c > 1/2$, and all $x \in (0,1)$.

Step 1: The summation on the right-hand side of the expression for $\mathcal{E}(Y_m)$ is split into two parts, where

$$\mathcal{E}(Y_m) = O(1)\, n \sum_{k=m}^{n} \frac{(n)_k}{n^k} k^{-5/2} \left\{ (e^c)^{k/n} e^{-\alpha} \right\}^k$$

The first part of the summation is from $k = m$ to a value of k such that $k^2 = O(n)$. It can be shown, as in the last problem that this lower portion of the summation goes to zero.
Step 2: The upper part of the summation on the right-hand side of the expression for $\mathcal{E}(Y_m)$ is over values of k in the interval $[k_u\sqrt{n}, n]$ where k_u is some positive constant. Denote this sum by $F_u(n)$, and replace the term $k^{-5/2}$ by its smallest value. Thus

$$F_u(n) = O(1)\, n^{-1/4} \sum_{k\in\left[k_u\sqrt{n},n\right]} \frac{(n)_k}{n^k}\left\{(e^c)^{k/n}\, e^{-\alpha}\right\}^k$$

Next, we use the following upper bound for $(n)_k / n^k$.

$$\frac{(n)_k}{n^k} \leq \exp\left\{-\frac{(k-1)^2}{n}\sum_{i\in\mathbb{P}}\frac{1}{i(i+1)}\left(\frac{k-1}{n}\right)^{i-1}\right\}$$

Thus

$$F_u(n) = O(1)\, n^{-1/4} \sum_{k\in\left[k_u\sqrt{n},n\right]} \left\{\exp\left[f\left(\frac{k-1}{n}, c\right)\right]\right\}^{k-1}$$

Step 0 implies that the summation in the above equation converges. Therefore $\mathcal{E}(Y_m) = o(1)$. Note that the proof of this result for $c > 1/2$ is also true for all $c > 0$, and $c \neq 1/2$.
Use of the second-moment method yields the stated result.

6. Assume that a Model A random graph has n vertices, and the probability of an edge between any pair of vertices is p. Also the occurrences of different edges are mutually independent events, and $n \to \infty$. Let $p = c(\ln n)/n$. Prove that:
   (a) If $0 < c < 1$, then almost every graph $G(n,p)$ is disconnected.
   (b) If $c > 1$, then almost every graph $G(n,p)$ is connected.
   Hint: See Palmer (1985), and West (2001).
   Let Q be a random variable which denotes the number of isolated vertices. Also let Q_i indicate whether the vertex i is isolated. The indicator variable Q_i takes a value of 1 if it is isolated, and 0 otherwise, for $1 \leq i \leq n$. Therefore $\mathcal{E}(Q) = \sum_{i=1}^n \mathcal{E}(Q_i) = n(1-p)^{n-1}$. Note that

$$\begin{aligned}(1-p)^n &= \exp\{\ln(1-p)^n\} = \exp\{n\ln(1-p)\}\\ &= \exp\left\{n\left(-p-\frac{p^2}{2}-\frac{p^3}{3}-\ldots\right)\right\}\\ &= \exp(-np)\exp\left\{-np^2\left(\frac{1}{2}+\frac{p}{3}+\ldots\right)\right\}\end{aligned}$$

   Therefore $\mathcal{E}(Q) \sim ne^{-np}$ if $np^2 \to 0$, that is $p \in o(1/\sqrt{n})$. If $p = c(\ln n)/n$, then $\mathcal{E}(Q) \sim n^{(1-c)}$, where c may be a function of n.
   (a) If $0 < c < 1$, and $n \to \infty$; then $\mathcal{E}(Q) \to \infty$. By using the second moment method it has to be demonstrated that $\mathcal{E}(Q^2) \sim \{\mathcal{E}(Q)\}^2$. Note that $Q_i^2 = Q_i$, then

$$\mathcal{E}(Q^2) = \sum_{1\leq i\leq n}\mathcal{E}(Q_i^2) + \sum_{\substack{1\leq i,j\leq n\\ i\neq j}}\mathcal{E}(Q_iQ_j) = \mathcal{E}(Q) + n(n-1)\,\mathcal{E}(Q_iQ_j)$$

   Observe that Q_iQ_j has a value equal to 1, only when the vertices i and j are simultaneously isolated, which disallows $\{2(n-2)+1\}$ edges. Therefore $\mathcal{E}(Q_iQ_j) = (1-p)^{2n-3} \sim e^{-2np}$. Consequently

$$\mathcal{E}\left(Q^2\right) \sim \mathcal{E}(Q) + n(n-1)e^{-2np} \sim \mathcal{E}(Q) + \{\mathcal{E}(Q)\}^2$$

That is

$$\frac{\mathcal{E}\left(Q^2\right)}{\{\mathcal{E}(Q)\}^2} \sim \frac{1}{\mathcal{E}(Q)} + 1$$

As $\mathcal{E}(Q) \to \infty$, we have $\mathcal{E}\left(Q^2\right) \sim \{\mathcal{E}(Q)\}^2$, and $\mathcal{E}\left(Q^2\right) \to \infty$. That is, $P(Q=0) \to 0$, which implies that almost every graph is disconnected.

(b) If $c > 1$ and independent of n, and $n \to \infty$ it can be concluded that $\mathcal{E}(Q) \sim n^{(1-c)} \to 0$. Use of Markov's inequality results in $P(Q=0) \to 1$. That is, almost every graph is connected.

7. Obtain the z-transform of the number of neighbors which are m steps away from a randomly chosen vertex. Denote it by $\mathcal{G}^{(m)}(z)$. Using this formalism obtain an estimate of the typical length ℓ, of the shortest path between two randomly chosen vertices of a connected graph. Assume that the distance between adjacent pair of nodes is equal to unity.
Hint: See Newman, Strogatz, and Watts (2001). Recall that $\mathcal{G}_0(z)$ is the z-transform of the vertex distribution, and $\mathcal{G}_1(z)$ is the z-transform of distribution of the outgoing edges of a vertex reached by a chosen edge. Similarly, the z-transform of number of neighbors which are m steps away from a randomly chosen vertex can be stated recursively as

$$\mathcal{G}^{(m)}(z) = \begin{cases} \mathcal{G}_0(z), & m = 1 \\ \mathcal{G}^{(m-1)}(\mathcal{G}_1(z)), & m \geq 2 \end{cases}$$

Define ξ_m to be the average number of mth nearest neighbors. It can be computed by evaluating the first derivative of $\mathcal{G}^{(m)}(z)$ with respect to z, at $z = 1$. Therefore

$$\begin{aligned} \xi_m &\triangleq \mathcal{G}^{(m)\prime}(1) = \mathcal{G}_1'(1)\,\xi_{m-1}, \quad m \geq 2 \\ &= \left(\frac{\xi_2}{\xi_1}\right)^{m-1} \xi_1 \end{aligned}$$

Average path length ℓ is attained when the total number of neighbors of a vertex which are out to distance ℓ is equal to n the total number of vertices in the graph. That is

$$n = 1 + \sum_{m=1}^{\ell} \xi_m$$

Substituting the expression for ξ_m in the above expression yields

$$\ell = \frac{\ln\left\{\xi_1^2 + (n-1)(\xi_2 - \xi_1)\right\} - \ln \xi_1^2}{\ln(\xi_2/\xi_1)}$$

If $n \gg \xi_1$ and $\xi_2 \gg \xi_1$

$$\ell \simeq \frac{\ln(n/\xi_1)}{\ln(\xi_2/\xi_1)} + 1$$

Note that these results are only approximate, and additionally assume that the graph is connected.

8. If $\widehat{\rho} < 1$, find the solution of the equation

$$\widehat{\rho} = 1 - e^{-\lambda \widehat{\rho}}$$

Hint: Use of Lagrange's series expansion results in

$$\widehat{\rho} = 1 - \frac{1}{\lambda} \sum_{j \in \mathbb{P}} \frac{j^{j-1}}{j!} \left(\lambda e^{-\lambda}\right)^j$$

9. The degree distribution of a random graph is

$$p_\nu(k) = \left(1 - e^{-1/\kappa}\right) e^{-k/\kappa}, \quad \forall\, k \in \mathbb{N}, \ \kappa > 0$$

Prove that

$$\mathcal{G}_0(z) = \frac{\left(1 - e^{-1/\kappa}\right)}{\left(1 - ze^{-1/\kappa}\right)}$$
$$\mathcal{G}_1(z) = \{\mathcal{G}_0(z)\}^2$$

10. Obtain asymptotes for cluster size distribution using the z-transform $\mathcal{H}_0(z)$.
Hint: See Newman, Strogatz, and Watts (2001). The z-transform of cluster sizes ψ_s is $\mathcal{H}_0(z)$. Near the phase transition point, the tail of the distribution is approximated by

$$\psi_s \sim s^{-\alpha} e^{-s/s^*}$$

Calculate α and s^* as follows. The parameter s^* is related to the radius of convergence $|z^*|$ of the z-transform $\mathcal{H}_0(z)$. That is, $s^* = 1/\ln|z^*|$. The value $|z^*|$ is equal in absolute value of the position of z^* of the singularity of $\mathcal{H}_0(z)$ which is closest to the origin. Observe that $\mathcal{H}_0(z) = z\mathcal{G}_0(\mathcal{H}_1(z))$. Therefore the singularity is caused by either a singularity in $\mathcal{H}_1(z)$ or $\mathcal{G}_0(z)$. However, the singularity in $\mathcal{G}_0(z)$ is possibly outside the unit circle. And there is a singularity in $\mathcal{H}_1(z)$ as $z \to 1$ (phase transition). Consequently, the singularity in $\mathcal{H}_0(z)$ which is sufficiently close to the phase transition, and nearest to the origin, is also a singularity in $\mathcal{H}_1(z)$.
Let $\mathcal{H}_1(z) = w$, then $z = \mathcal{H}_1^{-1}(w)$. Using the relationship $\mathcal{H}_1(z) = z\mathcal{G}_1(\mathcal{H}_1(z))$ yields

$$w = z\mathcal{G}_1(w) = \mathcal{H}_1^{-1}(w)\,\mathcal{G}_1(w)$$
$$z = \frac{w}{\mathcal{G}_1(w)}$$

The value of w which corresponds to the singularity is denoted by w^*. At this value, the first derivative of $\mathcal{H}_1^{-1}(w)$ is equal to zero. Setting $dz/dw = 0$ yields

$$\mathcal{G}_1(w^*) = w^* \mathcal{G}_1'(w^*)$$

It is quite possible for the above equation to not have a finite solution. In this case, ψ_s may not be of the form $s^{-\alpha} e^{-s/s^*}$. Recall that at the phase transition point $\mathcal{G}_1(1) = \mathcal{G}_1'(1) = 1$. At this point $w^* = z^* = 1$, and $s^* \to \infty$. In the next step expand $\mathcal{H}_1^{-1}(w)$ about $w = 1$. Let $w = (1 + \epsilon)$, then

$$\mathcal{H}_1^{-1}(1+\epsilon) = 1 + \frac{1}{2}\epsilon^2 \left.\frac{d^2 z}{dw^2}\right|_{z=w=1} + O\left(\epsilon^3\right)$$
$$= 1 - \frac{1}{2}\epsilon^2 \mathcal{G}_1''(1) + O\left(\epsilon^3\right)$$

In general $\mathcal{G}_1''(1) \neq 0$. Therefore, if $\mathcal{H}_1^{-1}(1+\epsilon) = z$, then $\mathcal{H}_1(z) = (1+\epsilon)$ and $z = \left\{1 - \frac{1}{2}\epsilon^2 \mathcal{G}_1''(1) + O\left(\epsilon^3\right)\right\}$. Thus

$$\epsilon \simeq \left\{\frac{2}{\mathcal{G}_1''(1)}(1-z)\right\}^{1/2}$$
$$\mathcal{H}_1(z) = (1+\epsilon) \simeq 1 + \left\{\frac{2}{\mathcal{G}_1''(1)}\right\}^{1/2}(1-z)^{\beta}, \quad \beta = \frac{1}{2}$$

$\mathcal{H}_0(z)$ is also of similar form. The relationship between the exponents α and β is next established. Use of the relationship $\psi_s \sim s^{-\alpha} e^{-s/s^*}$ results in

$$\mathcal{H}_0(z) = \sum_{s=0}^{m-1} \psi_s z^s + C \sum_{s=m}^{\infty} s^{-\alpha} e^{-s/s^*} z^s + \epsilon(m)$$

In the above equation, C is a constant, and $\epsilon(m)$ is an error term, which is much smaller than the second term. In this equation, the first term is a finite polynomial, and has no singularities on the finite plane. However, singularities lie in the second term.
As

$$\beta = \lim_{z \to 1_-} \left\{1 + (z-1)\frac{\mathcal{H}_0''(z)}{\mathcal{H}_0'(z)}\right\}$$

We have

$$\beta = \lim_{m\to\infty} \lim_{z\to 1_-} \left\{\frac{1}{z} + \frac{(z-1)}{z}\frac{\sum_{s=m}^{\infty} s^{2-\alpha} z^{s-1}}{\sum_{s=m}^{\infty} s^{1-\alpha} z^{s-1}}\right\}$$

Approximating the sums by integrals as $m \to \infty$ results in

$$\beta = \lim_{m\to\infty} \lim_{z\to 1_-} \left\{\frac{1}{z} + \frac{(1-z)}{z \ln z}\frac{\Gamma(3-\alpha, -m\ln z)}{\Gamma(2-\alpha, -m\ln z)}\right\}$$

where $\Gamma(\cdot,\cdot)$ is the upper incomplete gamma function. Taking the limits in the proper order yields $\beta = (\alpha - 1)$. That is

$$\alpha = \beta + 1 = \frac{3}{2}$$

provided $\mathcal{G}_1''(1) \neq 0$. Therefore the tail of the cluster distribution near the phase transition is proportional to $n^{-3/2}$. Thus if the component to which a randomly chosen vertex belongs has to be determined, the exponent of n is $-3/2$, otherwise it can be established that the exponent is $-5/2$.

11. The ring-lattice graph $R_{n,2k}$ has n vertices, and the degree of each vertex is $2k$. The n vertices lie on a ring, and each vertex is connected to k vertices on either side. Find the average path length of the $R_{n,2k}$ graph, where $k \geq 1$, and the length of each edge is unity.
Hint: Assume that $n = (2kr + 2)$ and $r \geq 1$. Since the graph has symmetry, the average path length between any pair of nodes is the same as the average path length from a single node to all other nodes. Select a single vertex. From this node, there are $2k$ nodes of length j, where

$1 \leq j \leq r$; and a single node at a distance of $(r+1)$ units. The sum of these paths S is given by

$$S = 2k \left\{ \frac{r(r+1)}{2} \right\} + (r+1) = \frac{n}{2}(r+1)$$

Therefore the average path length ℓ is given by

$$\ell = \frac{S}{(n-1)} = \frac{n(n+2k-2)}{4k(n-1)} \simeq \frac{n}{4k}$$

12. Find the average path length of a randomized ring lattice graph $R_{n,2k}$, where $k \geq 1$, and the length of each edge is unity. The randomization is done as per the Barrat-Weigt's scheme, and $n \gg 1$.
Hint: The average path length is given by

$$\ell = \frac{\ln(n/\xi_1)}{\ln(\xi_2/\xi_1)} + 1$$

where $\xi_1 = 2k$ = average degree, $\xi_2 = (\mu_2 - \xi_1)$, μ_2 = second moment of the degree, and $\xi_2/\xi_1 \gg 1$. If $p \approx 1$, use of these relationships results in $\xi_2/\xi_1 \simeq (2k - 0.5)$. Therefore, if $k \gg 1$ and $p \approx 1$, yields $\ell \simeq \ln(n) / \ln(2k - 0.5)$.
13. Power-laws also occur in optimization scenarios. Let N be the total number of mutually independent abstract events, and $r_i \in (0, 1]$ be the resource allocated to event i, where $1 \leq i \leq N$. The total available resource is equal to $\widetilde{R}$. Also let $l_i \in \mathbb{R}_0^+$ be the cost and p_i be the probability of initiation of event i, where $1 \leq i \leq N$. Trivially assume that all p_i's are positive. The expected cost of all the N events is equal to J. Thus $\sum_{i=1}^{N} p_i = 1$, and

$$J = \sum_{i=1}^{N} p_i l_i, \quad \text{and} \quad \sum_{i=1}^{N} r_i \leq \widetilde{R}$$

Assume that $l_i \triangleq f(\widetilde{\alpha}, r_i), \widetilde{\alpha} \in \mathbb{R}_0^+, 1 \leq i \leq N$, and c is a constant, where:
(a) If $\widetilde{\alpha} = 0$, then $f(\widetilde{\alpha}, r_i) = -c \ln r_i$
(b) If $\widetilde{\alpha} > 0$, then

$$f(\widetilde{\alpha}, r_i) = \frac{c}{\widetilde{\alpha}} \left(r_i^{-\widetilde{\alpha}} - 1 \right)$$

Minimize J, subject to given constraint on resources.
Hint: See Doyle, and Carlson (2000). Observe that $f(\widetilde{\alpha}, 1) = 0$. Denote the first derivative of $f(\widetilde{\alpha}, r_i)$ with respect to r_i by $f'(\widetilde{\alpha}, r_i)$. Thus $f'(\widetilde{\alpha}, r_i) = -c r_i^{-(\widetilde{\alpha}+1)}$ for all $\widetilde{\alpha} \in \mathbb{R}_0^+$. The value J is minimized by setting up the Lagrangian

$$\mathcal{L}(r_1, r_2, \ldots, r_N, \lambda) = \sum_{i=1}^{N} p_i f(\widetilde{\alpha}, r_i) + \lambda \left\{ \sum_{i=1}^{N} r_i - \widetilde{R} \right\}$$

where $\lambda \in \mathbb{R}$. Taking the derivative of $\mathcal{L}(r_1, r_2, \ldots, r_N, \lambda)$ with respect to λ and r_i yields

$$\frac{\partial \mathcal{L}(r_1, r_2, \ldots, r_N, \lambda)}{\partial \lambda} = \sum_{i=1}^{N} r_i - \widetilde{R}$$

$$\frac{\partial \mathcal{L}(r_1, r_2, \ldots, r_N, \lambda)}{\partial r_i} = p_i f'(\widetilde{\alpha}, r_i) + \lambda, \quad 1 \leq i \leq N$$

Next equate the derivatives to zero, and denote the minimized value of J by $J_{\min}$.

(a) $\widetilde{\alpha}=0$: In this case $\sum_{i=1}^{N} r_i = \widetilde{R}$, and $r_i = cp_i/\lambda$. Thus

$$r_i = p_i\widetilde{R}, \text{ and } l_i = -c\ln\left(p_i\widetilde{R}\right), \quad 1 \le i \le N$$

$$J_{\min} = -c\sum_{i=1}^{N} p_i \ln\left(p_i\widetilde{R}\right)$$

The above equation also has an information-theoretic interpretation.

(b) $\widetilde{\alpha}>0$: In this case $\sum_{i=1}^{N} r_i = \widetilde{R}$, and $r_i = (cp_i/\lambda)^{1/(\widetilde{\alpha}+1)}$. Let $A=\sum_{i=1}^{N} {p_i}^{1/(\widetilde{\alpha}+1)}$, then

$$r_i = \frac{\widetilde{R}}{A}{p_i}^{1/(\widetilde{\alpha}+1)}, \quad 1 \le i \le N$$

$$l_i = \frac{c}{\widetilde{\alpha}}\left\{\left(\frac{A}{\widetilde{R}}\right)^{\widetilde{\alpha}} {p_i}^{-\widetilde{\alpha}/(\widetilde{\alpha}+1)} - 1\right\}, \quad 1 \le i \le N$$

$$J_{\min} = \frac{c}{\widetilde{\alpha}}\left\{\frac{A^{\widetilde{\alpha}+1}}{\widetilde{R}^{\widetilde{\alpha}}} - 1\right\}$$

This case can be applied to the management of a Web site. A Web site layout can be considered to be a partition of a one-dimensional document into files of length l_i, which are accessed with frequency p_i, where $1 \le i \le N$. The goal is to minimize $J=\sum_{i=1}^{N} p_i l_i$, which is proportional to the average delay, a user might experience in downloading files from this Web site. The Web management and navigation become challenging with a plethora of tiny files. In addition, the hyperlinks connecting them consume precious space. Consequently there has to be a balance between Web site management and fast downloads. In this scenario, it turns out that $l_i \propto r_i^{-\widetilde{\alpha}}$, where $\widetilde{\alpha} \in (1/2, 1]$.

14. Consider a simplified copying digraph-evolution model of the WWW, in which a vertex represents a Web page, and a directed edge represents a hyperlink. At the beginning of its evolution, the directed graph has a single vertex with a link to itself. At each time step $t \in \mathbb{P}$, a new vertex v is created with out-degree 1, and in-degree 0. In this time step, an edge is created as follows. A vertex u is chosen uniformly at random from vertices created at an earlier time step.
    - With probability $\xi \in (0,1)$, an edge (v,u) is created.
    - With probability $(1-\xi)$ an edge (v,w) is created, where (u,w) is the out-going edge from the vertex u. That is, the new vertex v copies the out-going link of vertex u to itself.

    In this model, vertices and edges are never deleted. Show that this evolving directed graph has a power-law (Zipfian) in-degree distribution.

    Hint: See Mitzenmacher (2002); and Blum, Chan, and Rwebangira (2006). Let $X_i(t)$ be the number of vertices with in-degree $i \in \mathbb{N}$ when there are $t \in \mathbb{P}$ vertices in the digraph. For $i \in \mathbb{P}$, $X_i(t)$ changes to $X_i(t+1)$ at time step $(t+1)$ as follows. The positive change in $X_i(t)$ due to creation of the edge (v,u) is

$$\frac{\xi X_{i-1}(t)}{t} - \frac{\xi X_i(t)}{t}$$

    where the first term represents the probability that u is selected at random, and has an in-degree of $(i-1)$. The second term represents the probability that u is selected at random, and has an in-degree of i.

    The positive change in $X_i(t)$ due to creation of the edge (v,w) is

$$\frac{(1-\xi)(i-1)X_{i-1}(t)}{t} - \frac{(1-\xi)iX_i(t)}{t}$$

where the first term represents the probability that w is selected randomly as an outgoing neighbor of a randomly selected vertex, and has an in-degree of $(i-1)$. The second term represents the probability that w is thus selected at random, and has an in-degree of i. Therefore

$$X_i(t+1) = X_i(t) + \frac{\xi}{t}\{X_{i-1}(t) - X_i(t)\}$$
$$+\frac{(1-\xi)}{t}\{(i-1)X_{i-1}(t) - iX_i(t)\}, \quad i \in \mathbb{P}$$

Similarly

$$X_0(t+1) = X_0(t) + 1 - \frac{\xi}{t}X_0(t)$$

In the above equation, 1 appears on the right-hand side because each newly created vertex has an in-degree 0. Further, $\xi X_0(t)/t$ represents decrease in the fraction of vertices with in-degree 0. Substituting $X_i(t) = c_i t$, $i \in \mathbb{N}$ in the above equations yields

$$c_i\{1+\xi+(1-\xi)i\} = c_{i-1}\{\xi+(1-\xi)(i-1)\}$$
$$c_0 = 1 - \xi c_0$$

Therefore

$$\frac{c_i}{c_{i-1}} = 1 - \frac{(2-\xi)}{\{1+\xi+(1-\xi)i\}} \sim 1 - \frac{(2-\xi)}{(1-\xi)}\left\{\frac{1}{i}\right\}, \quad i \in \mathbb{P}$$

It is known that $\Pi_{i=1}^{n}(1+\lambda/i) = \Theta(n^{\lambda})$. Therefore for large values of t, we have

$$c_i = c_0 \prod_{j=1}^{i}\left\{1 - \frac{\beta}{j}\right\} = \Theta\left(i^{-\beta}\right) \sim Ci^{-\beta}, \quad i \in \mathbb{P}$$

where $c_0 = 1/(1+\xi)$, $\beta = (2-\xi)/(1-\xi)$, and C is an appropriate positive constant. Note that for $\xi \in (0,1)$, we have $\beta \in (2,\infty)$.

15. Let $G = (V, E)$ be a directed graph with 4 vertices. Use the Page-Brin's algorithm, to find the page-rank vector R of the graph specified by its adjacency matrix A.

$$A = \begin{bmatrix} 0 & 1 & 0 & 0 \\ 0 & 0 & 0 & 0 \\ 0 & 1 & 0 & 1 \\ 0 & 1 & 0 & 0 \end{bmatrix}$$

Assume that $\alpha = 0.9$.
Hint: The W, W', and W'' matrices, assuming $\alpha = 0.9$ are

$$W = \begin{bmatrix} 0 & 1 & 0 & 0 \\ 0 & 0 & 0 & 0 \\ 0 & 1/2 & 0 & 1/2 \\ 0 & 1 & 0 & 0 \end{bmatrix}, \quad W' = \begin{bmatrix} 0 & 1 & 0 & 0 \\ 1/4 & 1/4 & 1/4 & 1/4 \\ 0 & 1/2 & 0 & 1/2 \\ 0 & 1 & 0 & 0 \end{bmatrix}$$

$$W'' = \begin{bmatrix} 0.025 & 0.925 & 0.025 & 0.025 \\ 0.250 & 0.250 & 0.250 & 0.250 \\ 0.025 & 0.475 & 0.025 & 0.475 \\ 0.025 & 0.925 & 0.025 & 0.025 \end{bmatrix}$$

The matrix W'' is stochastic and primitive. The vector R can be computed from the equations $R^T = R^T W''$ and $R^T e = 1$. Computation of the page-rank vector R is left to the reader.

16. Compare and contrast the similarity between the LSI (latent semantic indexing) and HITS (Kleinberg's) algorithms of searching on the Web.
Hint: See Chakrabarti (2003).

17. Find the probability distribution of $\beta_\varepsilon(e)$, the edge betweenness of an undirected random tree graph. It is given that the nodal degrees of this tree graph has a power-law distribution.
Hint: See Goh, Oh, Jeong, Kahng, and Kim (2002). Consider a growing Barabási-Albert (BA) type of tree with $m = 1$, where a new vertex attaches (at time t) to an already existing vertex i, with probability proportional to $k_i(t) / \sum_j k_j(t)$, where $k_j(t)$ is the degree of node j at time t. The network has $n(t) = (t+1)$ vertices and $E(t) = t$ edges at time t. Note that this construction produces a tree, because $m = 1$. Furthermore, since the growth follows a BA type of evolution, the nodal degrees of the tree have a power-law distribution.
Each edge e of a tree divides the vertices of the tree into two groups which are attached to either sides of the edge. Let the size of the tree cluster on the descendent side be $\widetilde{m}$ vertices, and the size of the tree cluster on the ancestral side be the remaining $(t+1-\widetilde{m})$ vertices at time t. Assume that the edge e was born at time s. Let $P_s(\widetilde{m}, t)$ be the probability that the edge e born at time s bridges these two clusters. Then the edge betweenness g of the edge e is given by

$$g = \beta_\varepsilon(e) = 2\widetilde{m}(t+1-\widetilde{m})$$

The rate equation of this process is

$$P_s(\widetilde{m}, t+1) = r_1(\widetilde{m}, t) P_s(\widetilde{m}, t) + r_2(\widetilde{m}-1, t) P_s(\widetilde{m}-1, t)$$

where $r_1(\widetilde{m}, t)$ is the probability that a new vertex attaches to the $(t+1-\widetilde{m})$ cluster of nodes on the ancestor side, and $r_2(\widetilde{m}-1, t)$ is the probability with $(\widetilde{m}-1)$ vertices on the descendent side. Also

$$r_1(\widetilde{m}, t) \simeq \frac{(t-\widetilde{m}) + 1/2}{t}$$
$$r_2(\widetilde{m}, t) = 1 - r_1(\widetilde{m}, t)$$

As the betweenness of edge g is independent of its birth time s, define $P(\widetilde{m}, t)$ as

$$P(\widetilde{m}, t) = \frac{1}{t}\sum_{s=1}^{t} P_s(\widetilde{m}, t)$$

Thus $P(\widetilde{m}, t)$ is the probability that a given edge connects two clusters of sizes $\widetilde{m}$ and $(t+1-\widetilde{m})$ vertices averaged over its birth time. Note that the edge betweenness centrality (BC) equal to g is unaltered due to this averaging process. Thus the modified rate equation, as $t \to \infty$ is:

$$(t+1) P(\widetilde{m}, t+1) = r_1(\widetilde{m}, t) t P(\widetilde{m}, t) + r_2(\widetilde{m}-1, t) t P(\widetilde{m}-1, t)$$

Define $P(\widetilde{m}, t) = \widetilde{\mathcal{P}}(\widetilde{m}/t)$, and $x = \widetilde{m}/t$. This yields

$$P(\widetilde{m}, t) = \widetilde{\mathcal{P}}(x), \quad P(\widetilde{m}, t+1) \simeq \widetilde{\mathcal{P}}(x), \quad P(\widetilde{m}-1, t) = \widetilde{\mathcal{P}}(x - 1/t)$$
$$r_1(\widetilde{m}, t) \simeq (1 - x + 1/(2t)), \quad r_2(\widetilde{m}-1, t) \simeq (x - 1/(2t))$$

Approximate

$$\widetilde{\mathcal{P}}(x-1/t) \simeq \widetilde{\mathcal{P}}(x) - \frac{1}{t}\frac{d\widetilde{\mathcal{P}}(x)}{dx}$$

Substituting these relationships in the above process equation and letting $t \to \infty$ results in $d\widetilde{\mathcal{P}}(x)/dx$ proportional to $-\widetilde{\mathcal{P}}(x)/x$. Thus $\widetilde{\mathcal{P}}(x)$ is proportional to x^{-2}. However $g = 2\widetilde{m}(t+1-\widetilde{m}) \simeq 2xt^2$ for finite values of $\widetilde{m}$ and large values of t. Therefore the probability distribution of the edge BC, $P_{\varepsilon}(g)$ is proportional to g^{-2}. Summarizing, the edge BC of a random tree graph with AB type of evolutionary model has a power-law distribution with the exponent parameter equal to 2.

18. In the exponential random graph model, determine the probability $P(G), G \in \mathcal{G}$ by maximizing the entropy of the random graph variable $\widetilde{G}$, subject to the stated constraints.
Hint: See Newman (2010). Initially ignore the nonnegativity constraint of the probabilities, which is $P(G) \geq 0, \forall\, G \in \mathcal{G}$. Let the entropy of the random graph variable $\widetilde{G}$ be S, where

$$S = -\sum_{G \in \mathcal{G}} P(G) \ln P(G)$$

In this definition of entropy, the base of the entropy is chosen to be the Euler's number e, instead of 2. However, this does not affect the final result. Let $\alpha \in \mathbb{R}$, and $\beta = (\beta_1, \beta_2, \ldots, \beta_r) \in \mathbb{R}^r$ be the Lagrange multipliers. The Lagrangian is

$$\begin{aligned}\mathcal{L}(P(G), G \in \mathcal{G}; \alpha; \beta) &= S + \alpha\left\{\sum_{G \in \mathcal{G}} P(G) - 1\right\} \\ &\quad + \sum_{i=1}^{r} \beta_i \left\{\sum_{G \in \mathcal{G}} f_i(G) P(G) - \overline{f}_i\right\}\end{aligned}$$

Taking partial derivatives of $\mathcal{L}(P(G), G \in \mathcal{G}; \alpha; \beta)$ with respect to $P(G)$, $G \in \mathcal{G}$, α, and $\beta_i, 1 \leq i \leq r$ yields

$$\frac{\partial \mathcal{L}}{\partial P(G)} = -\ln P(G) - 1 + \alpha + \sum_{i=1}^{r} \beta_i f_i(G), \quad G \in \mathcal{G}$$

$$\frac{\partial \mathcal{L}}{\partial \alpha} = \sum_{G \in \mathcal{G}} P(G) - 1$$

$$\frac{\partial \mathcal{L}}{\partial \beta_i} = \sum_{G \in \mathcal{G}} f_i(G) P(G) - \overline{f}_i, \quad 1 \leq i \leq r$$

Next define $\mathcal{H}(G) = \sum_{i=1}^{r} \beta_i f_i(G)$ and $\ln Z(\beta) = (1-\alpha)$, and equate to zero the above partial derivatives. It yields the stated result. Observe that the objective function (entropy) is a concave function of the probabilities, and the constraints are convex functions. Therefore the entropy is maximized. Also, note that in the stated results, it fortunately turns out that $P(G) \geq 0, \forall\, G \in \mathcal{G}$ indeed.

References

1. Abramowitz, M., and Stegun, I. A., 1965. *Handbook of Mathematical Functions*, Dover Publications, Inc., New York, New York.
2. Adamic, L. A., Lukose, R. M., Puniyani, A. R., and Huberman, B. A., 2001. "Search in Power-Law Networks," Phys. Rev. E, Vol. 64, 046135, Issue 4.
3. Adler, R. J., Feldman, R. E., and Taqqu, M. S., Editors, 1998. *A Practical Guide to Heavy Tails, Statistical Techniques and Applications*, Birkhauser, Boston, Massachusetts.
4. Aiello, W., Broder, A., Janssen, J., and Milios, E., Editors, 2008. Algorithms and Models for the Web-Graph: Fourth International Workshop, WAW 2006, Banff, Canada, November 30 - December 1, LNCS 4936, Springer-Verlag, Berlin, Germany.
5. Albert, R., and Barabási, A.-L., 2002. "Statistical Mechanics of Complex Networks," Rev. Mod. Phys. Vol. 74, pp. 67-97.
6. Alderson, D. A., 2008. "Catching the Network Science Bug: Insight and Opportunity for the Operations Researcher," Operations Research, Vol. 56, No. 5, pp. 1047-1065.
7. Alon, N., and Spencer, J. H., 2000. *The Probabilistic Method*, Second Edition, John Wiley & Sons, Inc., New York, New York.
8. Alvarez-Hamelin, J. I., Dall'Asta, L., Barrat, A., and Vespignani, A., 2006. "K-core Decomposition of Internet Graphs: Hierarchies, Self-Similarity and Measurement Biases," DELIS-TR 356.
9. Arrowsmith, D. K., Mondragón, R. J., and Woolf, M., 2005. "Data Traffic, Topology and Congestion," in *Complex Dynamics in Communication Networks*. Editors: Kocarev, L., and Vattay, G., Springer-Verlag, Berlin, Germany, pp. 127-157.
10. Barabási, A.-L., 2002. *Linked*: *The New Science of Networks*, Perseus Press, New York, New York.
11. Barabási, A.-L., 2010. *Bursts*: *The Hidden Pattern Behind Everything We Do*, Penguin Group, New York, New York.
12. Barabási, A.-L., and Albert, R., 1999. "Emergence of Scaling in Random Networks," Science. Vol. 286, Issue 5439, pp. 509-512.
13. Barrat, A., and Weigt, M., 2000. "On the Properties of Small-World Network Models," Eur. Phys. J. B. Vol. 13, pp. 547-560.
14. Barrat, A., Barthélemy, M., and Vespignani, A., 2008. *Dynamical Processes on Complex Networks*, Cambridge University Press, Cambridge, Great Britain.
15. Berry, M. W., and Browne M., 1999. *Understanding Search Engines, Mathematical Modeling and Text Retrieval*, Society of Industrial and Applied Mathematics, Philadelphia, Pennsylvania.
16. Blum, A., Chan, T-H. H., and Rwebangira, M. R., 2006. "A Random-Surfer Web-Graph Model," Proceedings of the eighth Workshop on Algorithm Engineering and Experiments and the third Workshop on Analytic Algorithmics and Combinatorics, pp. 238-246.
17. Bollobás, B., 1985. *Random Graphs*, Second Edition, Academic Press, New York, New York.
18. Bonato, A. 2008. *A Course on the Web Graph*, American Mathematical Society, Providence, Rhode Island.
19. Bornholdt, S., and Schuster, H. G., Editors, 2003. *Handbook of Graphs and Networks*, Wiley-VCH GmbH & Co. KGaA, Weinheim, Germany.
20. Brin, S., and Page, L., 1998. "The Anatomy of a Large-Scale Hypertextual Web Search Engine," Computer Networks and ISDN Systems, Vol. 30, Issues 1-7, pp. 107-117.

21. Broder, A., Kumar, R., Maghoul, F., Raghavan, P., Rajagopalan, S., Stata, R., Tomkins, A., and Wiener, J., 2000. "Graph Structure in the Web," The International Journal of Computer and Telecommunication Networks, Vol. 33, Issues 1-6, pp. 309-320.
22. Bush, V., 1945. "As We May Think," Atlantic Monthly, Vol. 176, Issue 1, pp. 101-108.
23. Caldarelli, G., 2007. *Scale-Free Networks: Complex Webs in Nature and Technology*, Oxford University Press, Oxford, Great Britain.
24. Caldarelli, G., and Vespignani, A., 2007. *Large Scale Structure and Dynamics of Complex Networks*, World Scientific, River Edge, New Jersey.
25. Caldarelli, G., Capocci, A., Rios, P. D. L., and Muñoz, M. A., 2002. "Scale-Free Networks from Varying Vertex Intrinsic Fitness," Phys. Rev. Letters, Vol. 89, No. 25.
26. Carmi, S., Havlin, S., Kirkpatrick, S., Shavitt, Y., and Shir, E., 2007. "A Model of Internet Topology Using k-shell Decomposition," Proc. National Academy of Sciences, Vol. 104, No. 27, pp. 11150-11154.
27. Chakrabarti. D., and Faloutsos, C., 2006. "Graph Mining: Laws, Generators, and Algorithms," ACM Computing Surveys, Vol. 38, No. 1, Article 2.
28. Chakrabarti. D., and Faloutsos, C., 2012. *Graph Mining: Laws, Tools, and Case Studies*, Morgan and Claypool Publishers, San Rafael, California.
29. Chiang, M., Low, S. H., Calderbank, A. R. , and Doyle, J. C., 2007. "Layering as Optimization Decomposition: A Mathematical Theory of Network Architectures," Proc. of the IEEE, Vol. 95, No. 1, pp. 255-312.
30. Chung, F., 2010. "Graph Theory in the Information Age," Notices of the American Mathematical Society, Vol. 57, No. 6, pp. 726-732.
31. Chung, F., Lu, L., and Vu, V., 2003. "Eigenvalues of Random Power Law Graphs," Combinatorica, Vol. 7, pp. 21-33.
32. Chung, F., and Lu, L. 2006. *Complex Graphs and Networks,* American Mathematical Society, Providence, Rhode Island.
33. Cohen, R., and Havlin, S., 2010. *Complex Networks: Structure, Robustness and Function,* Cambridge University Press, Cambridge, Great Britain.
34. Cooper, C., and Frieze, A., 2003. "A General Model of Web graphs," Random Structures and Algorithms, Vol. 22, No. 3, pp. 311-335.
35. Csermely, P., 2009. *Weak Links: Stabilizers of Complex Systems from Proteins to Social Networks*, Second Edition, Springer-Verlag, Berlin, Germany.
36. Dorogovtsev, S. N., Mendes, J. F. F., and Samukhin, A. N., 2000. "Structure of Growing Networks with Preferential Linking," Phys. Rev. Lett., Vol. 85, Issue 21, pp. 4633-4636.
37. Dorogovtsev, S. N., and Mendes, J. F. F., 2002. "Evolution of Networks," Advances in Physics, Vol. 51, Issue 4, pp. 1079-1187.
38. Dorogovtsev, S. N., and Mendes, J. F. F., 2003. *Evolution of Networks*, Oxford University Press, Oxford, Great Britain.
39. Dorogovtsev, S. N., Goltsev, A. V., and Mendes, J. F. F., 2007. "Critical Phenomena in Complex Networks," Vol. 80, Issue 4, pp. 1275-1335.
40. Doyle, J., and Carlson, J. M., 2000. "Power Laws, Highly Optimized Tolerance, and Generalized Source Coding," Physical Review Letters, Vol. 84, No. 24, pp. 5656-5659.
41. Durrett, R., 2007. *Random Graph Dynamics*, Cambridge University Press, Cambridge, Great Britain.
42. Erdős, P., and Spencer, J., 1974. *Probabilistic Methods in Combinatorics*, Academic Press, New York, New York.
43. Estrada, E., 2012. *The Structure of Complex Networks: Theory and Applications*, Oxford University Press, Oxford, Great Britain.
44. Faloutsos, M., Faloutsos, P., and Faloutsos, C., 1999. "On Power-Law Relationships of the Internet Topology," Proc. ACM SIGCOMM, Computing Communication Review, Vol. 29, pp. 251-262.

45. Farkas, I. J., Derényi, I., Barabási, A. L., and Vicsek, T., 2001. "Spectra of "Real-World" Graphs: Beyond the Semi-Circle Law," Phys. Rev. E Vol. 64 026704, Issue 2 11, pp. 267041-2670412.
46. Franceschetti, M., and Meester, R., 2007. *Random Networks for Communication: From Statistical Physics to Information Systems,* Cambridge University Press, Cambridge, Great Britain.
47. Frank, H., and Frisch, I. T., 1971. *Communication, Transmission, and Transportation Network,* Addison-Wesley Publishing Company, New York, New York.
48. Goh, K., Oh, E., Jeong, H., Kahng, B., and Kim, D., 2002. "Classification of Scale-Free Networks," Proc. National Academy of Sciences, Vol. 99, No. 20, pp. 12583-12588.
49. Goldenberg, A., Zheng, A. X., Fienberg, S. E., and Airoldi, E. M., 2009. "A Survey of Statistical Network Models," NOW, Foundation and Trends in Machine Learning, Vol. 2, No. 2, pp. 129-233.
50. Guare, J., 1990. *Six Degree of Separation: A Play,* Vintage Press: New York, New York.
51. Haddadi, H., Rio, M., Iannaccone, G., Moore, A., and Mortier, R., Second Quarter 2008. "Network Topologies: Inference, Modeling, and Generation," IEEE Communications Surveys & Tutorials, Vol. 10, Issue 2, pp. 48-69.
52. Hartmann, A. K., and Weigt, M., 2005. *Phase Transitions in Combinatorial Optimization Problems,* Wiley-VCH GmbH & Co. KGaA, Weinheim, Germany.
53. Huffaker, B., Fomenkov, M., and claffy, kc, 2012. "Internet Topology Data Comparison." Technical Report, Cooperative Association for Internet Data Analysis (CAIDA), University of California, San Diego.
54. Janson, S., Knuth, D. E., Łuczak, T., and Pittel, B., 1993. "The Birth of the Giant Component," Random Structures and Algorithms, Vol. 4, pp. 233-358.
55. Janson, S., and Łuczak, M. J., 2006. "A Simple Solution to the k-core Problem," Random Structures and Algorithms, Vol. 30, Issues 1-2, pp. 50-62.
56. Janson, S., Łuczak, T., and Ruciński, A., 2000. *Random Graphs*, John Wiley & Sons, Inc., New York, New York.
57. Katz, L., 1953. "A New Status Index Derived from Sociometric Data Analysis," Psychometrika, Vol. 18, Issue 1, pp. 39-43.
58. Kleinberg, J., 1999. "Authoritative Sources in a Hyperlinked Environment," Journal of the ACM, Vol. 46, Issue 5, pp. 604-632.
59. Kolaczyk, E. D., 2009. *Statistical Analysis of Network Data: Methods and Models*, Springer-Verlag, Berlin, Germany.
60. Kolchin, V. F., 1999. *Random Graphs*, Cambridge University Press, Cambridge, Great Britain.
61. Krioukov, D., Papadopoulos, F., Kitsak, M., Vahdat, A., and Boguñá, M., 2010. "Hyperbolic Geometry of Complex Networks," Phys. Rev. E, Vol. 82, 36106, Issue 3.
62. Kumar, R., Raghavan, P., Rajagopalan, S., Sivakumar, D., Tomkins, A., and Upfal, E., 2000. "Stochastic Models for the Web Graph," Proceedings of the 41st IEEE Symposium of Foundations of Computer Science, pp. 57-65.
63. Langville, A. N., and Meyer, C. D., 2003. "A Survey of Eigenvector Methods of Web Information Retrieval," Department of Mathematics, North Carolina State University, Raleigh, North Carolina.
64. Langville, A. N., and Meyer, C. D., 2006. *Google's PageRank and Beyond: The Science of Search Engine Rankings*, Princeton University Press, Princeton, New Jersey.
65. Lempel, R., and Moran, S., 2000. "The Stochastic Approach for Link-Structure Analysis (SALSA) and the TKC Effect," Computer Networks, Vol. 33, pp. 387-401.
66. Lewis, T. G., 2009. *Network Science: Theory and Practice*, John Wiley & Sons, Inc., New York, New York.
67. Masuda, N., Miwa, H., and Konno, N., 2004. "Analysis of Scale-Free Networks Based on a Threshold Graph with Intrinsic Vertex Weights," Phys. Rev. E, Vol. 70, 036124, Issue 3.

68. Meyer, C., 2000. *Matrix Analysis and Applied Linear Algebra*, Society of Industrial and Applied Mathematics, Philadelphia, Pennsylvania.
69. Mihail, M., and Papadimitriou, C. H., 2003. "On the Eigenvalue Power Law," preprint.
70. Milgram, S., 1967. "The Small World Problem," Psychology Today, Vol. 1, No. 1, pp. 61-67.
71. Mitzenmacher, M., 2002. "A Brief History of Generative Models for Power Law and Lognormal Distributions." Technical Report, Harvard University, Cambridge, Massachusetts.
72. Mitzenmacher, M., 2004. "Editorial: The Future of Power Law Research." Internet Mathematics, Vol. 2, No. 4, pp. 525-534.
73. Molloy, M., and Reed, B., 1995. "A Critical Point for Random Graphs with a given Degree Sequence." Random Structures and Algorithms, Vol. 6, Issues 2-3, pp. 161-180.
74. Newman, M. E. J., 2003. "The Structure and Function of Complex Networks," Society of Industrial and Applied Mathematics Review, Vol. 45, No. 2, pp. 167-256.
75. Newman, M. E. J., 2005. "Power Laws, Pareto Distributions and Zipf's Law," Contemporary Physics, Vol. 46, Issue 5, pp. 323-351.
76. Newman, M. E. J., 2010. *Networks: An Introduction*, Oxford University Press, Oxford, Great Britain.
77. Newman, M. E. J., Strogatz, S. H., and Watts, D. J., 2001. "Random Graphs with Arbitrary Degree Distributions and their Applications," Phys. Rev. E 64, 026118, Issue 2.
78. Newman, M. E. J., and Watts, D. J., 1999. "Renormalization Group Analysis of the Small-World Network Model," Phys. Lett. A Vol. 263, Issues 4-6, pp. 341-346.
79. Palmer, E. M., 1985. *Graphical Evolution*, John Wiley & Sons, Inc., New York, New York.
80. Papadopoulos, F., Kitsak, M., Serrano, M. A., Boguñá, M., and Krioukov, D., 2012. "Popularity versus Similarity in Growing Networks," Nature, Vol. 489, Issue 7417, pp. 537-540.
81. Park, J., and Newman, M. E. J., 2004. "The Statistical Mechanics of Networks," Phys. Rev. E, Vol. 70 066117, Issue 6.
82. Pastor-Satorras, R., and Vespignani, A., 2004. *Evolution and Structure of the Internet*, Cambridge University Press, Cambridge, Great Britain.
83. Pittel, B., Spencer, J., and Wormald, N., 1996. "Sudden Emergence of a Giant k-core in a Random Graph," J. Combin. Theor. Vol. 67, Issue 1, pp. 111-151.
84. Rosen, K. H., Editor-in-Chief, 2018. *Handbook of Discrete and Combinatorial Mathematics,* Second Edition, CRC Press: Boca Raton, Florida.
85. Ross, S. M., 2010. *Introduction to Probability Models*, Tenth Edition, Academic Press, New York, New York.
86. Sachkov, V. N., 1997. *Probabilistic Methods in Combinatorial Analysis,* Cambridge University Press, Cambridge, Great Britain.
87. Serrano, M. A., and Boguñá, M., 2005. "Weighted Configuration Model," Science Of Complex Networks: From Biology to the Internet and WWW: CNET 2004. AIP Conference Proceedings, Vol. 776, pp. 101-107.
88. Siganos, G., Faloutsos, M., Faloutsos, P., and Faloutsos, C., 2003. "Power Laws and the AS-Level Internet Topology," IEEE/ACM Transactions on Networking, Vol. 11, No. 4, pp. 514-524.
89. Sigman, K., 1999. "Appendix: A Primer on Heavy-Tailed Distributions," Queueing Systems, Vol. 33, Issues 1-3, pp. 261-275.
90. Solomonoff, R., and Rapoport, A., 1951. "Connectivity of Random Nets," Bulletin of Mathematical Biophysics, Vol. 13, Issue 2, pp. 107-117.
91. Spencer, J., 2010. "The Giant Component: The Golden Anniversary," Notices of the American Mathematical Society, Vol. 57, No. 6, pp. 720-724.
92. Szabó, G., Alava, M., and Kertész, J., 2002. "Shortest Paths and Load Scaling in Scale-Free Trees," Phys. Rev. E. Vol. 66, pp. 026101/1-8.

93. Van Mieghem, P., 2011. *Graph Spectra for Complex Networks*, Cambridge University Press, Cambridge, Great Britain.
94. Wasserman, S., Faust, K., 1994. *Social Network Analysis,* Cambridge University Press, Cambridge, Great Britain.
95. Watts, D. J., 1999. *Small Worlds,* Princeton University Press, Princeton, New Jersey.
96. Watts, D. J., and Strogatz, S. H., 1998. "Collective Dynamics of 'Small-World' Networks," Nature, Vol. 393, pp. 440-442.
97. West, D. B., 2001. *Introduction to Graph Theory*, Second Edition, Prentice-Hall, Upper Saddle River, New Jersey.
98. Whittle, P., 2007. *Networks*: *Optimization and Evolution*, Cambridge University Press, Cambridge, Great Britain.
99. Willinger, W., Alderson, D., and Doyle, J. C. 2009. "Mathematics and the Internet: A Source of Enormous Confusion and Great Potential," Notices of the American Mathematical Society, Vol. 56, No. 5, pp. 586-599.
100. Zipf, G. K., 1949. *Human Behavior and the Principle of Least Effort*, Hafner Publishing Company, New York, New York.

CHAPTER 8

Graph-Theoretic Algorithms

$\mathcal{M} = (\Gamma, Q, \delta, \varrho)$

Deterministic Turing machine

Alan Mathison Turing. Alan Turing was born on 23 June, 1912 in London, England. He studied mathematics at King's College, Cambridge. In the year 1935, Alan Turing was elected a fellow of King's College, for a dissertation on the central limit theorem.

In the year 1936 he published his celebrated paper: *On computable numbers, with an application to the Entscheidungsproblem,* in the Proceedings of London Mathematical Society. The Turing machine, as we know it today was introduced in this paper. This paper contains ideas which are at the foundation of computer science and mathematics. A significant hallmark of this work was that Turing machines laid the mathematical foundation for modern computer technology. The Turing machine can also be used as a mathematical model for describing algorithms.
During the second world war Turing played a significant role in decoding the messages sent by the Enigma machines of the German navy. He died on 7 June, 1954 in Wilmslow, Cheshire, England, under very unfortunate circumstances.

8.1 Introduction

Some useful graph-theoretic algorithms are discussed in this chapter. These algorithms are used in routing of packets, and also in the design and analysis of least cost reliable networks. They can also be used for designing networks to maximize throughput. A graph is also sometimes referred to as a network in this chapter. The shortest path algorithms, special path algorithms, algebraic path techniques, geographic routing in hyperbolic space, minimum spanning tree algorithms, connectivity of a graph, network reliability, maximum flows and minimum cost flows, network coding, and design of communication networks are discussed in this chapter.

The phrases "traffic" and "flow" are used synonymously in this chapter. A *flow* is the amount of some commodity which travels along an edge per unit time. For example, it can be packets per unit time, or simply the bit rate required to transmit information. A basic knowledge of applied analysis, matrices, graph theory, probability theory, and geometry are required to study this chapter.

A graph $G = (V, E)$, where the set of vertices (nodes) is V and the set of edges (links or arcs) is E. The set of vertices V is represented by either $\{1, 2, \ldots, n\}$ or $\{v_1, v_2, \ldots, v_n\}$. The choice of representation of the vertex set V depends upon ease of notation in the topic of discussion. Similarly, an edge of the graph G is represented by (i, j), or (v_i, v_j), or e_{ij}, or simply by e.

8.2 Shortest Path Algorithms

Shortest path algorithms are developed for a digraph $G = (V, E)$, where the vertex set is V and the edge set is E. Also $|V| = n$, and $|E| = m$. The graph G is assumed to be connected and simple. A *length* (or *weight* or *cost*) metric is specified for each arc $(v_i, v_j) \in E$. It is denoted by c_{ij}, where

$c_{ij} \in \mathbb{R}_0^+$. If $(v_i, v_j) \notin E$ and $i \neq j$, then $c_{ij} \to \infty$. Furthermore, $c_{ii} = 0, 1 \leq i \leq n$. Recall from the chapter on graph theory that a path between a pair of nodes is a sequence of arcs which connect the two nodes, without any repetitive intermediate nodes. The length of a directed path $\mathcal{P}$ is given by $\sum_{(v_i,v_j)\in\mathcal{P}} c_{ij}$. The length of path from the vertex v_s to the vertex v_d is the *shortest path* if the length of the path between these two vertices is the shortest. Note that this path is not necessarily unique. That is, it is possible to have other paths with the same cost, between the specified pair of nodes. The following shortest path algorithms are discussed in this section. The Bellman-Ford's shortest path method is described. This is followed by an algorithm which computes the shortest path from a single node to all other nodes of the network. A popular and efficient algorithm to do this is due to E. W. Dijkstra. An efficient algorithm for the shortest path computation between all pairs of nodes due to R. W. Floyd and S. Warshall is also described.

8.2.1 Bellman-Ford Shortest Path Algorithm

The Bellman-Ford shortest path method computes the shortest path between a source node and all other nodes of a graph. Without loss of any generality, assume that the source node is v_1. Define the length of the shortest path from node v_1 to node $v_j \in V$ to be $d_j, 1 \leq j \leq n$. For the scalar d_j to be the shortest path, $d_j \leq (d_i + c_{ij})$ for all $(v_i, v_j) \in E$. This statement is often called the *shortest path optimality condition.* Furthermore, these shortest path lengths satisfy the following equations.

$$d_1 = 0 \tag{8.1a}$$

$$d_j = \min_{\substack{1 \leq q \leq n \\ q \neq j}} \{d_q + c_{qj}\}, \quad 2 \leq j \leq n \tag{8.1b}$$

These nonlinear and implicit relationships are called Bellman's equations, which are named after the mathematician Richard E. Bellman (1920-1984). These equations are easily solvable, if the graph G is acyclic. Recall that a digraph is acyclic iff the vertices are numbered such that there exists an arc from v_i to v_j only if $i < j$. Bellman's equations for an acyclic graph are

$$d_1 = 0 \tag{8.2a}$$

$$d_j = \min_{1 \leq q < j} \{d_q + c_{qj}\}, \quad 2 \leq j \leq n \tag{8.2b}$$

These equations can be solved by backward substitution. Observe that d_1 is known, d_2 depends solely on d_1, d_3 depends on d_1 and d_2, and so on. In general, d_j depends on $d_1, d_2, \ldots, d_{j-1}$ for $2 \leq j \leq n$. The computational complexity of these operations is $O\left(n^2\right)$. Bellman's equations can be solved in general for any digraph by the method of successive approximation. This iterative method is due to Bellman and Ford. In this algorithm,

$$d_j^{(k)}, \; 1 \leq k \leq (n-1)$$

are the successive approximations of d_j for $1 \leq j \leq n$. The shortest distance $d_j = d_j^{(n-1)}$ for $1 \leq j \leq n$. The successive approximations are

$$d_1^{(k)} = 0, \;\; 1 \leq k \leq (n-1)$$

$$d_j^{(1)} = c_{1j}, \;\; 2 \leq j \leq n$$

$$d_j^{(k+1)} = \min \left\{ d_j^{(k)}, \min_{\substack{1 \leq q \leq n \\ q \neq j}} \left\{ d_q^{(k)} + c_{qj} \right\} \right\}, \; 2 \leq j \leq n, \;\; k = 1, 2, \ldots, (n-2)$$

Observe that

$$d_j^{(1)} \geq d_j^{(2)} \geq d_j^{(3)} \geq \ldots \geq d_j^{(n-1)} = d_j, \quad 1 \leq j \leq n$$

The computational complexity of this procedure is $O\left(n^3\right)$.

Next consider the case of a directed graph G in which the c_{ij}'s can take negative values. It is possible to have negative cycles in this graph. A negative cycle is a directed cycle of negative length. If the network graph does not contain any negative cycles, then the distances d_i's are well-defined, and $d_j \leq d_i + c_{ij}$ for all arcs (v_i, v_j) in the graph G. In this case, there exists a shortest path from node v_1 (source node) to each node v_j with no repeated nodes.

8.2.2 Dijkstra's Shortest Path Algorithm

Let G be a simple digraph. Given a source node $v_s \in V$, Dijkstra's algorithm computes the shortest path between the node v_s and the nodes in the set $V - \{v_s\}$. The output of this algorithm is specified in terms of a shortest path tree T_{spt}. The tree T_{spt} is rooted at the node v_s. In this tree, a path from node v_s to node v_t is the shortest path between these two nodes. Furthermore, the arcs in this tree are directed away from the node v_s. The tree can be represented by a *predecessor vector* $\mathcal{T} = (\tau_1, \tau_2, \ldots, \tau_n)$ as follows. The vector element τ_i is the predecessor-node (or simply predecessor) of node v_i in the tree, where $1 \leq i \leq n$.

Let the path $\mathcal{P}_t$ from node v_s to node v_t be specified by $(v_s, v_1, v_2, \ldots, v_t)$. In this example, the predecessor of node v_1 is v_s. Denote the predecessor of v_1 by τ_1, that is $\tau_1 = v_s$. Similarly the predecessor of node v_2, denoted by τ_2, is equal to v_1, and so on. Predecessor of the source node v_s is denoted by $\tau_s = \varnothing$ (null-indicator). This information is compactly represented by the predecessor vector $\mathcal{T}$.

Dijkstra's algorithm is a *greedy* algorithm. This implies that, selecting a path *locally* under this greedy criteria turns out to be the optimal result *globally* for all pairs of vertices. Consequently, if $\mathcal{P}_t = (v_s, v_1, v_2, \ldots, v_t)$ is a shortest path from v_s to v_t then $\mathcal{P}_k = (v_s, v_1, v_2, \ldots, v_k)$ is a shortest path from v_s to v_k for $1 \leq k \leq t$.

Associated with each node $v_j \in V$ in this tree is a scalar quantity d_j. It is the length of the directed path from node v_s to node v_j. Note that the distance d_s of the node v_s from itself is equal to 0. These distances have to satisfy the shortest path optimality conditions.

Dijkstra's algorithm is an iterative process. The number of iterations in this algorithm is equal to $O(n)$. In this process a list of vertices L is maintained, which have not yet been chosen to generate the final shortest path tree. In each iteration of the algorithm, two operations are performed. These are node selection and distance update. In the node selection process, a node $v_i \in L$ is chosen such that d_i is the smallest. If there is a tie, select arbitrarily. Recall from the chapter on graph theory that A_{v_i} is the adjacency set of vertex v_i. It is the set of all arcs emanating from vertex v_i. The distance update operation performs the updating of the distances d_j where $(v_i, v_j) \in A_{v_i}$ and $v_j \in L$. The distances are output in a vector $\mathcal{D} = (d_1, d_2, \ldots, d_n)$. The computational complexity of this algorithm is $O\left(n^2\right)$. The correctness of Dijkstra's algorithm is established in the following theorem.

Theorem 8.1. *Let $G = (V, E)$, be a simple digraph with $c_{ij} \in \mathbb{R}_0^+$ for all arcs $(v_i, v_j) \in E$, and the source node is v_s. Let v_d be a vertex in the set $V - \{v_s\}$, and it is reachable from the vertex v_s. Then Dijkstra's algorithm finds a shortest path from the vertex v_s to vertex v_d.*

Sketch of the proof. The sketch of the proof is essentially by induction. Assume that after first k iterations, the rooted shortest path tree T_{spt} has k vertices which are nearest to the vertex v_s. Then as per the greedy rule, the rooted tree T_{spt} selects the $(k+1)$th vertex which is nearest to the vertex v_s. Therefore after n iterations, the shortest path between the vertices v_s and v_d has definitely been determined. □

Algorithm 8.1. *Dijkstra's Shortest Path Algorithm.*

Input: Digraph $G = (V, E)$, $c_{ij} \in \mathbb{R}_0^+$ for all arcs $(v_i, v_j) \in E$, and the source node v_s.
Output: The rooted shortest path tree T_{spt} specified by the predecessor vector $\mathcal{T}$, and the distance vector $\mathcal{D}$.
begin
  (*initialization*)
  $d_s \leftarrow 0$
  $d_j \leftarrow \infty$, for all nodes $v_j \in V - \{v_s\}$
  $\tau_s \leftarrow \varnothing$
  $L \leftarrow V$
  while $L \neq \varnothing$ **do**
  ($\varnothing$ *is the empty set*)
  begin
    (*select vertex*)
    find $v_i \in L$ **such that** $d_i = \min\{d_j \mid v_j \in L\}$
    $L \leftarrow L - \{v_i\}$
    if $L = \varnothing$, **stop** (*the computation is complete*)
    (*update distance*)
    for all $((v_i, v_j) \in A_{v_i}) \wedge (v_j \in L)$ **do**
    begin
      if $(d_i + c_{ij}) < d_j$ **then**
      begin
        $d_j \leftarrow (d_i + c_{ij})$
        $\tau_j \leftarrow v_i$
      end (*end of if statement*)
    end (*end of for-loop*)
  end (*end of while-loop*)
end (*end of Dijkstra's algorithm*)

Example 8.1. The shortest path tree, and the shortest distance vector for the digraph $G = (V, E)$ is determined. The graph is specified as follows.

$$V = \{1, 2, 3, 4, 5, 6\}$$

$$E = \{(1,2), (1,3), (2,3), (2,4), (3,4), (3,5), (4,6), (5,3), (5,4), (5,6), (6,4)\}$$

See Figure 8.1.

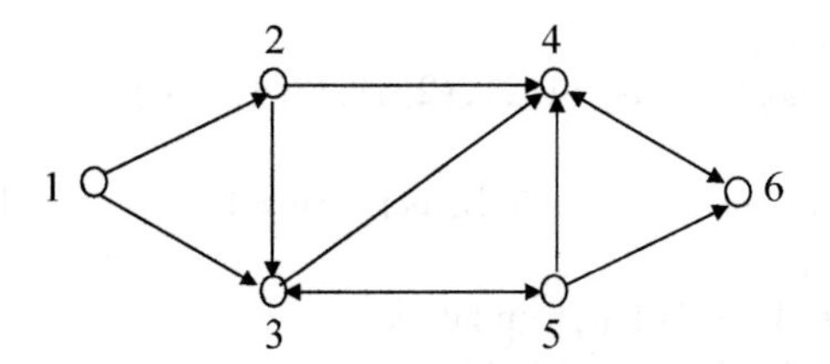

Figure 8.1. Shortest path computation for the graph $G = (V, E)$.

The lengths of the arcs in the set E are:

$$c_{12} = 3, c_{13} = 5, c_{23} = 1, c_{24} = 2, c_{34} = 6, c_{35} = 2,$$
$$c_{46} = 5, c_{53} = 7, c_{54} = 3, c_{56} = 1, \text{ and } c_{64} = 4$$

Assume that the source node is vertex 1. The adjacency sets of the vertices are listed for convenience. These are: $A_1 = \{(1,2),(1,3)\}$, $A_2 = \{(2,3),(2,4)\}$, $A_3 = \{(3,4),(3,5)\}$, $A_4 = \{(4,6)\}$, $A_5 = \{(5,3),(5,4),(5,6)\}$, and $A_6 = \{(6,4)\}$.

The distance and predecessor vectors to be determined are

$$\mathcal{D} = (d_1, d_2, d_3, d_4, d_5, d_6) \text{ and } \mathcal{T} = (\tau_1, \tau_2, \tau_3, \tau_4, \tau_5, \tau_6)$$

respectively. As per Dijkstra's algorithm, these are obtained via following steps.

Initialization:

$\mathcal{D} \leftarrow (0, \infty, \infty, \infty, \infty, \infty)$, $\tau_1 \leftarrow \varnothing$, $L \leftarrow \{1,2,3,4,5,6\}$

Iteration 1:

Select vertex. $\min(d_1, d_2, d_3, d_4, d_5, d_6) = d_1 = 0$.

Select vertex 1. $L \leftarrow L - \{1\}$.

At this stage $L = \{2,3,4,5,6\}$.

Update distance. Since $A_1 = \{(1,2),(1,3)\}$ update d_2 and d_3.

$d_2 \leftarrow \min(\infty, 0+3)$, that is $d_2 \leftarrow 3$. Also $\tau_2 \leftarrow 1$.

$d_3 \leftarrow \min(\infty, 0+5)$, that is $d_3 \leftarrow 5$. Also $\tau_3 \leftarrow 1$.

Thus $\mathcal{D} = (0, 3, 5, \infty, \infty, \infty)$.

The arc set of the generated tree is $\{(1,2),(1,3)\}$.

Iteration 2:

Select vertex. $\min(d_2, d_3, d_4, d_5, d_6) = d_2 = 3$.

Select vertex 2. $L \leftarrow L - \{2\}$.

At this stage $L = \{3,4,5,6\}$.

Update distance. Since $A_2 = \{(2,3),(2,4)\}$ update d_3 and d_4.

$d_3 \leftarrow \min(5, 3+1)$, that is $d_3 \leftarrow 4$. Also $\tau_3 \leftarrow 2$.

$d_4 \leftarrow \min(\infty, 3+2)$, that is $d_4 \leftarrow 5$. Also $\tau_4 \leftarrow 2$.

Thus $\mathcal{D} = (0, 3, 4, 5, \infty, \infty)$.

The arc set of the generated tree is $\{(1,2),(2,3),(2,4)\}$.

Iteration 3:

Select vertex. $\min(d_3, d_4, d_5, d_6) = d_3 = 4$. Select vertex 3. $L \leftarrow L - \{3\}$.

At this stage $L = \{4,5,6\}$.

Update distance. Since $A_3 = \{(3,4),(3,5)\}$ update d_4 and d_5.

$d_4 \leftarrow \min(5, 4+6)$, d_4 and τ_4 remain unchanged.

$d_5 \leftarrow \min(\infty, 4+2)$, that is $d_5 \leftarrow 6$. Also $\tau_5 \leftarrow 3$.

Thus $\mathcal{D} = (0, 3, 4, 5, 6, \infty)$.
The arc set of the generated tree is $\{(1,2), (2,3), (2,4), (3,5)\}$.
Iteration 4:
Select vertex. $\min(d_4, d_5, d_6) = d_4 = 5$. Select vertex 4. $L \leftarrow L - \{4\}$.
At this stage $L = \{5, 6\}$.
Update distance. Since $A_4 = \{(4,6)\}$ update d_6.
$d_6 \leftarrow \min(\infty, 5+5)$, that is $d_6 \leftarrow 10$. Also $\tau_6 \leftarrow 4$.
Thus $\mathcal{D} = (0, 3, 4, 5, 6, 10)$.
The arc set of the generated tree is $\{(1,2), (2,3), (2,4), (3,5), (4,6)\}$.
Iteration 5:
Select vertex. $\min(d_5, d_6) = d_5 = 6$. Select vertex 5. $L \leftarrow L - \{5\}$.
At this stage $L = \{6\}$.
Update distance. Since $A_5 = \{(5,3), (5,4), (5,6)\}$ update d_6.
$d_6 \leftarrow \min(10, 6+1)$, that is $d_6 \leftarrow 7$. Also $\tau_6 \leftarrow 5$.
Thus $\mathcal{D} = (0, 3, 4, 5, 6, 7)$.
The arc set of the generated tree is $\{(1,2), (2,3), (2,4), (3,5), (5,6)\}$.
Iteration 6:
Select vertex. $\min(d_6) = d_6 = 7$. Select vertex 6. $L \leftarrow L - \{6\}$.
At this stage $L = \varnothing$.
Stop since $L = \varnothing$.
End of Dijkstra's algorithm.
The final distance and predecessor vectors are

$$\mathcal{D} = (0, 3, 4, 5, 6, 7) \text{ and } \mathcal{T} = (\varnothing, 1, 2, 2, 3, 5)$$

respectively. The arc set of the shortest path tree is

$$\{(1,2), (2,3), (2,4), (3,5), (5,6)\}$$

□

8.2.3 All-Pairs Shortest Path Algorithms

Two algorithms for finding shortest paths between all pairs of nodes are developed in this subsection. Let $G = (V, E)$ be a simple digraph, where $V = \{1, 2, \ldots, n\}$. Associated with each arc in the graph is a cost (length). That is, the cost of the arc connecting node i to node j is c_{ij}. These costs can be conveniently represented in a matrix $C = [c_{ij}]$ called the cost matrix. This matrix is of size n. For simplicity assume that $c_{ij} \in \mathbb{R}_0^+$, if $(i,j) \in E$, for $1 \le i, j \le n$. Furthermore, $c_{ij} \to \infty$, if arc $(i,j) \notin E$ and $i \ne j$ for $1 \le i, j \le n$; and $c_{ii} = 0$ for $1 \le i \le n$.

The goal is to compute shortest paths from each of the n nodes to the other $(n-1)$ nodes. Let d_{ij} be the length of the shortest path from node i to node j. Evidently $d_{ii} = 0$. Let D be an $n \times n$ matrix, where $D = [d_{ij}]$. Also define an $n \times n$ matrix

$$D^{(w)} = \left[d_{ij}^{(w)}\right]$$

where $d_{ij}^{(w)}$ is the length of the shortest path from node i node j with no more than w arcs between the two nodes i and j. Observe that $d_{ij}^{(1)} = c_{ij}$ and $d_{ij}^{(n-1)} = d_{ij}$ for $1 \le i, j \le n$. The d_{ij}'s can be computed from the following recursive relationship

$$d_{ij}^{(1)} = c_{ij} \tag{8.3a}$$

$$d_{ij}^{(w)} = \min_{1 \le k \le n} \left\{ d_{ik}^{(w-1)} + c_{kj} \right\}, \quad 2 \le w \le (n-1) \tag{8.3b}$$

Observe that the minimization operation in the above recursion is similar to matrix multiplication. If in matrix multiplication, the scalar multiplication is replaced by scalar addition, and scalar addition operation is replaced by minimization, then the above recursion can be stated in terms of a modified type of matrix multiplication. Denote this new matrix multiplication operation by $\circledast$. If $X = [x_{ij}], Y = [y_{ij}]$, and $Z = [z_{ij}]$, are square matrices of the same size n, then

$$X = Y \circledast Z$$

$$x_{ij} = \min_{1 \le k \le n} \{ y_{ik} + z_{kj} \}, \quad 1 \le i, j \le n$$

Therefore the shortest path computation can be stated in terms of this modified matrix operation. It is

$$D^{(1)} = C \tag{8.4a}$$

$$D^{(w)} = D^{(w-1)} \circledast C, \quad 2 \le w \le (n-1) \tag{8.4b}$$

$$D^{(n-1)} = D \tag{8.4c}$$

It can be observed that $D^{(2)} = C \circledast C$, $D^{(3)} = (C \circledast C) \circledast C$, and so on. This procedure is outlined in the following algorithm.

Algorithm 8.2. *All-Pairs Shortest Path Algorithm via Matrix Multiplication.*

Input: Digraph $G = (V, E)$, and the cost matrix $C = [c_{ij}]$, where $c_{ij} \in \mathbb{R}_0^+$ for $1 \le i, j \le n$.
Output: The shortest-path distance matrix $D = D^{(n-1)}$.
begin
    $D^{(1)} = C$
    $D^{(w)} = D^{(w-1)} \circledast C$ for $2 \le w \le (n-1)$
end (*end of modified matrix multiplication algorithm*)

The complexity of these matrix operations can be computed as follows. Each modified matrix multiplication can be performed in $\Theta(n^3)$ operations and there are $(n-2)$ such operations. Therefore the computational complexity of determining the matrix D is $\Theta(n^4)$.

It is possible to perform the $(n-2)$ modified matrix multiplication operations in $\Theta(\log n)$ steps, by the technique of repeated squaring. Therefore the computational complexity of determining the matrix D can be improved to $\Theta(n^3 \log n)$.

Example 8.2. The shortest path lengths between all pairs of a digraph $G = (V, E)$ are determined, where

$$V = \{1, 2, 3, 4\}$$

$$E = \{(1,2),(1,4),(2,3),(2,4),(3,2),(3,4),(4,1),(4,3)\}$$

The cost matrix is given by

$$C = \begin{bmatrix} 0 & 3 & \infty & 2 \\ \infty & 0 & 1 & 4 \\ \infty & 5 & 0 & 7 \\ 9 & \infty & 6 & 0 \end{bmatrix}$$

Then

$$D^{(1)} = C$$
$$D^{(2)} = C \circledast C = \begin{bmatrix} 0 & 3 & 4 & 2 \\ 13 & 0 & 1 & 4 \\ 16 & 5 & 0 & 7 \\ 9 & 11 & 6 & 0 \end{bmatrix}$$
$$D = D^{(2)} \circledast D^{(2)} = D^{(2)}$$

In this example D turns out to be equal to $D^{(2)}$. □

An all-pairs shortest path algorithm due to R. W. Floyd and S. Warshall is more efficient than the above algorithm. It has a computational complexity of $\Theta(n^3)$ steps. The basic idea behind this algorithm is as follows. Again, assume that all the elements of the cost-matrix are nonnegative.

Define $d_{ij}^{(w)}$ to be the length of the shortest path between vertices i and j, such that all intermediate vertices on the path (if any) are in the set $\{1, 2, \ldots, w\}$, where $1 \le w \le n$.

A shortest path between the vertices i and j which uses vertices from the set $\{1, 2, \ldots, w\}$ as internal nodes either passes through or does not pass through the node w.

(a) If the shortest path passes through the node w, then $d_{ij}^{(w)} = d_{iw}^{(w-1)} + d_{wj}^{(w-1)}$.
(b) If the shortest path does not pass through the node w, then $d_{ij}^{(w)} = d_{ij}^{(w-1)}$.

Consequently a recursive definition of the shortest paths can be given as follows. Thus, $\forall\, i, j \in V$

$$d_{ij}^{(0)} = c_{ij}$$
$$d_{ij}^{(w)} = \min\left\{d_{ij}^{(w-1)}, d_{iw}^{(w-1)} + d_{wj}^{(w-1)}\right\}, \quad 1 \le w \le n$$
$$d_{ij} = d_{ij}^{(n)}$$

The shortest paths can be kept track of via a square matrix Φ of size n. Let $\Phi = [\varphi_{ij}]$, where the element φ_{ij} is generated as follows. Initialize φ_{ij}'s as

$$\varphi_{ij} \leftarrow \begin{cases} j, & c_{ij} < \infty \\ \varnothing, & c_{ij} \to \infty \end{cases}$$

While updating $d_{ij}^{(w)}$'s let

$$\varphi_{ij} \leftarrow \begin{cases} \varphi_{iw}, & d_{ij}^{(w-1)} > \left(d_{iw}^{(w-1)} + d_{wj}^{(w-1)}\right) \\ \varphi_{ij}, & d_{ij}^{(w-1)} \le \left(d_{iw}^{(w-1)} + d_{wj}^{(w-1)}\right) \end{cases}$$

The shortest path can then be extracted from φ_{ij}'s. Observe that φ_{ij} is the first vertex after vertex i in a shortest path from vertex i to vertex j. If $\varphi_{ij} = \varnothing$ at the end of the algorithm, then a path does not exist from the node i to node j.

At the end of this algorithm, the shortest path between the vertices i and j is recovered from the matrix Φ. Let this shortest path be given by the sequence of vertices

$$(i, i_1, i_2, i_3, \ldots, i_q, j)$$

Then,

$$i_1 = \varphi_{ij},\ i_2 = \varphi_{i_1 j},\ i_3 = \varphi_{i_2 j},\ \ldots,\ j = \varphi_{i_q j}$$

Therefore the shortest path from vertex i to vertex j is obtained by examining the elements of the jth column of the matrix Φ.

Algorithm 8.3. *Floyd-Warshall's Algorithm.*

Input: Digraph $G = (V, E)$, and the cost matrix $C = [c_{ij}]$, where $c_{ij} \in \mathbb{R}_0^+$ for $1 \leq i, j \leq n$.
Output: The shortest-path distance matrix D and the path matrix Φ.
begin
  (*initialize*)
  $d_{ij} \leftarrow c_{ij}, \forall\, i, j \in V$
  $$\varphi_{ij} = \begin{cases} j, & \text{if } c_{ij} < \infty \\ \varnothing, & \text{if } c_{ij} \to \infty \end{cases}, \ \forall\, i, j \in V$$
  for $w = 1$ **to** n **do**
  begin
    for $i = 1$ **to** n **do**
    begin
      for $j = 1$ **to** n **do**
      begin
        if $d_{ij} > (d_{iw} + d_{wj})$
        begin
          $d_{ij} \leftarrow (d_{iw} + d_{wj})$
          $\varphi_{ij} \leftarrow \varphi_{iw}$
        end (*end of if statement*)
      end (*end of j for-loop*)
    end (*end of i for-loop*)
  end (*end of w for-loop*)
end (*end of Floyd-Warshall's algorithm*)

Certain special path algorithms are described in the next section. These are extensions of the shortest path algorithms.

8.3 Special Path Algorithms

Special path computation algorithms are discussed in this section. These are: computation of K shortest paths between a pair of vertices, routing algorithm with restrictions, routing algorithm with constraints, and disjoint paths algorithm. All these algorithms are useful in the reliable design of networks.

8.3.1 *K* Shortest Paths Algorithm

It is sometimes necessary to determine K number of shortest paths between a pair of nodes in a network. If additional constraints are placed upon these paths, then such constrained paths can be selected from these $K \in \mathbb{P}$ shortest paths. A simple, and connected digraph $G = (V, E)$; the cost matrix C, which specifies the cost of each arc; and a pair of vertices v_s and $v_t \in V$ are specified. The source node is v_s, and v_t is the target or destination node. It is required to find $K \in \mathbb{P}$ shortest paths between these two nodes, provided they exist. These paths are not permitted to contain repeated nodes. Such paths are called *elementary paths.* Let $|V| = n$ and $C = [c_{ij}]$, where $c_{ij} \in \mathbb{R}_0^+, 1 \leq i, j \leq n$. Without any loss of generality, denote v_s by v_1. If $K = 1$, then Dijkstra's algorithm can be used. However, if K is greater than 1, then the algorithm outlined below should be used. It was discovered by J. Y. Yen.

Let $V = \{v_1, v_2, \ldots, v_n\}$. Define the kth shortest path between the vertices $v_s = v_1$ and v_t as $\pi^k = \left(v_1^k, v_2^k, \ldots, v_{p_k}^k, v_t\right)$, where $v_1 \triangleq v_1^k$ and $1 \leq k \leq K$. It is assumed that this path is elementary. The length of the path π^{k-1} is less than or equal to the length of the path π^k, for $2 \leq k \leq K$. Also, π_i^k is defined to be the *deviation* of path π^k from the path π^{k-1} at vertex v_i. This implies that the path π^k follows the path of π^{k-1} from v_1 up till the vertex v_i, and then deviates to a vertex v_{i+1} which is different from all the v_{i+1}th vertices of the paths π^w, for $1 \leq w \leq (k-1)$. The complete π^k path is generated by computing shortest subpath from v_i^{k-1} to v_t. This subpath does not pass through vertices $v_1^{k-1}, v_2^{k-1}, \ldots, v_i^{k-1}$. Therefore

$$\pi^k = \left(v_1^{k-1}, v_2^{k-1}, \ldots, v_i^{k-1}, v_{i+1}^k, v_{i+2}^k, \ldots, v_{p_k}^k, v_t\right)$$

Define the initial subpath as

$$\left(v_1^{k-1}, v_2^{k-1}, \ldots, v_i^{k-1}\right) = \left(v_1^k, v_2^k, \ldots, v_i^k\right) \triangleq \rho_i^k$$

It is called the *root* of the path π^k. The second subpath

$$\left(v_i^k, v_{i+1}^k, v_{i+2}^k, \ldots, v_{p_k}^k, v_t\right)$$

of π^k, is called its *spur*. It is denoted by ψ_i^k. The algorithm generates in order the paths: π^1, $\pi^2, \ldots, \pi^{K-1}$, and finally π^K. The algorithm assumes that at least K paths exist between the pair of vertices v_1 and v_t, and that $K > 1$. If this is not the case, then the algorithm can be suitably modified. The list of K shortest paths in order of nondecreasing length is $\Pi_\ell = \left\{\pi^1, \pi^2, \ldots, \pi^K\right\}$. It is also assumed that the ties between paths of equal length are resolved by following some lexicographic ordering. The computational complexity of this algorithm is $O\left(Kn^3\right)$, and its storage requirement is of order $O(Kn)$.

Algorithm 8.4. *K Shortest Path Algorithm.*

Input: Digraph $G = (V, E)$, cost matrix C, the nodes v_s and v_t.
The number of shortest paths to be determined is equal to $K \in \mathbb{P} \backslash \{1\}$.
Output: The list of shortest paths $\Pi_\ell = \{\pi^1, \pi^2, \ldots, \pi^K\}$.
begin
(*initialization*)
($\mathcal{L}$ *is a temporary list of paths from which the* π^k*'s are selected*)
$\mathcal{L} \leftarrow \varnothing$ and $\Pi_\ell \leftarrow \varnothing$
Using Dijkstra's algorithm, find the shortest path in graph G,
with source node v_s and destination node v_t.
Let the number of shortest paths (of equal length) be ξ.
if $\xi \geq K$ **then**
begin
enter any K shortest paths in Π_ℓ
stop (*the algorithm terminates*)
end
else
(*the number of shortest paths* $\xi < K$)
begin
Select any shortest path π^1 and enter it in Π_ℓ.
That is, $\Pi_\ell \leftarrow \Pi_\ell \cup \pi^1$
Enter remaining $(\xi - 1)$ paths in $\mathcal{L}$.
go to label J
end
label J
$k \leftarrow 2$
while $k \leq K$ **do**
begin
(*find all the deviations* π_i^k *of the* $(k-1)$*th shortest path* $\pi^{k-1}, 1 \leq i \leq p_{k-1}$)
for $i = 1$ **to** p_{k-1} **do**
begin
for $w = 1$ **to** $(k-1)$ **do**
begin
Determine if the sequence of first i vertices in the path
π^{k-1} coincides with the first i vertices of the path π^w.
If true, then set the cost of the arc $\left(v_i^{k-1}, v_{i+1}^w\right)$
equal to ∞
end (*end of* w *for-loop*)
Use Dijkstra's algorithm to find the shortest path,
from v_i^{k-1} to v_t. Exclude from consideration the set of
vertices $\{v_1^{k-1}, v_2^{k-1}, \ldots, v_i^{k-1}\}$ in this shortest path
computation. If there is more than a single shortest
path, select any one, and denote it by ψ_i^k.

Next form π_i^k by concatenating $\rho_i^k = \left(v_1^{k-1}, v_2^{k-1}, \ldots, v_i^{k-1}\right)$ with ψ_i^k, and then insert π_i^k into the list $\mathcal{L}$.
Replace the cost of the arc elements $\left(v_i^{k-1}, v_{i+1}^w\right)$'s set to ∞, with original value.
end (*end of i for-loop*)
Select the shortest path in the list $\mathcal{L}$. Denote it by π^k.
$\Pi_\ell \leftarrow \Pi_\ell \cup \pi^k$, and $\mathcal{L} \leftarrow \mathcal{L} - \pi^k$
$k \leftarrow k+1$
end (*end of while-loop*)
end (*end of K shortest path algorithm*)

Validation of the Algorithm

The path π^k should be a deviation at some vertex v_i, $i \geq 1$ from any one of the paths π^w, for $1 \leq w \leq (k-1)$. Therefore the algorithm determines all shortest deviations from all π^w's. Furthermore, due to the iterative feature of the algorithm, all the shortest deviations of paths π^w, for $1 \leq w \leq (k-2)$ exist in the list $\mathcal{L}$. Therefore, only the shortest deviations from π^{k-1} need to be considered. Finally, π^k path is the shortest path in the list $\mathcal{L}$.

Also, if the sequence of first i vertices in the path π^{k-1}, coincides with the first i vertices of the paths π^w for $1 \leq w \leq (k-1)$, then the cost of the arc $\left(v_i^{k-1}, v_{i+1}^w\right)$ is set equal to ∞. This precludes the regeneration of π^w as a deviation of π^{k-1} at the vertex v_i. It can also be observed that at the beginning of the iteration k, the number of paths in the list $\mathcal{L}$ need not be larger than $\{K-(k-1)\}$. Furthermore, the k-loop contributes a factor K, the i-loop contributes a factor n, and each iteration of the i-loop contributes a factor n^2 (computational complexity of the Dijkstra's algorithm) to the overall computational complexity of the algorithm, which is $O\left(Kn^3\right)$.

8.3.2 Routing with Transit-Times and Constraints

The classical shortest path routing algorithms due to Bellman-Ford, Dijkstra, and Floyd-Warshall address the problem of minimizing a single metric associated with the arcs in the graph. Sometimes, more than a single metric is associated with an arc of the graph. The problem in this scenario is to find paths between a pair of vertices which satisfy multiple constraints associated with different metrics. Two such formulations are addressed. These are: shortest path routing in a graph with transit-times, and constrained routing problem.

Routing in a Graph with Transit-Times

Consider a restricted shortest path problem in which each arc of the graph is associated with a cost metric, and an integer-valued transit-time metric. It is the aim of this restricted shortest path algorithm to find a shortest path from a source node to a destination node, provided the transit time of the path is bounded by a certain fixed value.

The problem is first formulated precisely. Let $G=(V,E)$, be a digraph where $|V|=n$, and $|E|=m$. This graph is assumed to be connected and simple. Cost and time metrics are defined for each arc $(v_i, v_j) \in E$. Denote these by c_{ij} and t_{ij} respectively. It is assumed that $c_{ij} \in \mathbb{R}_0^+$ and $t_{ij} \in \mathbb{P}$ if $(v_i, v_j) \in E$. If $(v_i, v_j) \notin E$ and $i \neq j$, then $c_{ij} = t_{ij} \to \infty$. Furthermore, $c_{ii} = t_{ii} = 0$, for $1 \leq i \leq n$. The length of a directed path $\mathcal{P}$ is given by $\sum_{(v_i,v_j)\in\mathcal{P}} c_{ij}$. The corresponding net transit-time of the path is given by $\sum_{(v_i,v_j)\in\mathcal{P}} t_{ij}$. Also assume that the source node is v_1. The goal is to find a shortest path from the origin v_1 to a destination node such that the total transit-time is at most T_{time} time units, where $T_{time} \in \mathbb{P}$. For simplicity, we shall assume that the nodes are reachable from the origin in this duration of time.

Define the length of the shortest path from node v_1 to node $v_j \in V$ to be

$$d_j(t), 1 \leq j \leq n$$

with the assumption that the path requires no more than t time units of net transit-time. These shortest path lengths satisfy the following generalized Bellman equations.

$$d_j(t) \to \infty, \;\; t < 0, \;\; 1 \leq j \leq n \tag{8.5a}$$

$$d_1(t) = 0, \;\; t = 0, 1, 2, \ldots, T_{time} \tag{8.5b}$$

$$d_j(0) \to \infty, \;\; 2 \leq j \leq n \tag{8.5c}$$

$$d_j(t) = \min\left\{ d_j(t-1), \min_{\substack{1\leq k\leq n \\ k\neq j}} \{d_k(t - t_{kj}) + c_{kj}\} \right\},$$

$$2 \leq j \leq n; \;\; t = 1, 2, \ldots, T_{time} \tag{8.5d}$$

The above equations can be solved recursively, if it is assumed that the underlying graph is acyclic. In an acyclic graph, the vertices are numbered such that $(v_i, v_j) \in E$ implies $i < j$. The recursive technique determines $d_j(T_{time})$ for $2 \leq j \leq n$. The computational complexity of this algorithm for acyclic graphs is $O(n^2 T_{time})$. Extension to general graphs is also straightforward.

Constrained Routing Problem

Routing algorithms with two constraints are next studied. In this problem two metrics are assigned to each arc of the graph. The problem is first formulated. Let $G = (V, E)$, be a digraph where $|V| = n$, and $|E| = m$. This graph is assumed to be connected and simple. Associated with each arc $(v_i, v_j) \in E$ is a cost (length) metric and a delay (time) metric. Denote these by c_{ij} and t_{ij} respectively. It is assumed that $c_{ij}, t_{ij} \in \mathbb{P}$, if $(v_i, v_j) \in E$. However, if $(v_i, v_j) \notin E$ and $i \neq j$, then $c_{ij} = t_{ij} \to \infty$. Furthermore, $c_{ii} = t_{ii} = 0$, for $1 \leq i \leq n$. The cost of a directed path $\mathcal{P}$ is given by $\mathcal{C}(\mathcal{P}) = \sum_{(v_i,v_j)\in\mathcal{P}} c_{ij}$. The corresponding delay associated with this path is given by $\mathfrak{D}(\mathcal{P}) = \sum_{(v_i,v_j)\in\mathcal{P}} t_{ij}$. Also assume that the source node is v_s and the destination (target) node is v_d. The goal is to find a path $\mathcal{P}$ from the origin v_s to a destination node v_d, such that the cost of this path is bounded by $\varPsi$, and the delay of the path is bounded by $\varOmega$. That is, $\mathcal{C}(\mathcal{P}) \leq \varPsi$ and $\mathfrak{D}(\mathcal{P}) \leq \varOmega$, where $\varPsi, \varOmega \in \mathbb{P}$. Let

$$c_{\max} = \max\{c_{ij} \mid (v_i, v_j) \in E, 1 \leq i, j \leq n\} \tag{8.6a}$$

$$t_{\max} = \max\{t_{ij} \mid (v_i, v_j) \in E, 1 \leq i, j \leq n\} \tag{8.6b}$$

and define a path $\mathcal{P}$ to be *feasible*, if $\mathcal{C}(\mathcal{P}) \leq \varPsi$ and $\mathfrak{D}(\mathcal{P}) \leq \varOmega$. Observe that an elementary path $\mathcal{P}$ has at most $(n-1)$ arcs. If $(n-1)c_{\max} \leq \varPsi$, then there exists a feasible path from vertices v_s to v_d if and only if there exists a path $\mathcal{P}$ from v_s to v_d such that $\mathfrak{D}(\mathcal{P}) \leq \varOmega$. Similarly, if $(n-1)t_{\max} \leq \varOmega$, then there exists a feasible path from vertices v_s to v_d if and only if there exists a path $\mathcal{P}$ from v_s to v_d such that $\mathcal{C}(\mathcal{P}) \leq \varPsi$. If $(n-1)c_{\max} \leq \varPsi$ and $(n-1)t_{\max} \leq \varOmega$, then any path between vertices v_s and v_d is feasible. The interesting case, when both $(n-1)\, c_{\max} > \Psi$ and $(n-1)\, t_{\max} > \Omega$ is addressed in the rest of this subsection.

Further assume that the *greatest common divisor* (gcd) of c_{ij} and Ψ is 1, for all edges $(v_i, v_j) \in E$. Similarly, the greatest common divisor of t_{ij} and Ω is assumed to be 1, for all edges $(v_i, v_j) \in E$. The reason for these assumptions will become clear shortly.

For each vertex $v \in V$, a set of cost-delay pairs is computed. Denote this set by F_v. It consists of two-tuples of type (c, t) such that c is the cost (or length) and t is the total delay (or transit-time), of the path from the vertex v to the target node v_d. The size of the set F_v is kept to a minimum by excluding two-tuples which are not feasible, and which are redundant. For example, the two-tuple (c, t) is said to be infeasible, if either $c > \Psi$ or $t > \Omega$. The two-tuple (c, t) is said to be redundant if there exists another two-tuple (c', t') such that $c' \le c$ and $t' \le t$. Furthermore, a cost-delay set F_v is said to be minimal if it has neither infeasible nor redundant two-tuples. Following are some useful observations about minimal sets.

(a) If the set F_v is minimal, then $|F_v| \le \min\{\Psi, \Omega\}$.
(b) For every set F_v there is a unique greatest minimal subset $F'_v \subseteq F_v$. Denote the greatest minimal set of F_v by $Min\,(F_v)$.
(c) Let $F_v = \{(c_1, t_1), (c_2, t_2), \ldots, (c_q, t_q)\}$ be a minimal set associated with the vertex v. If in this minimal set $c_1 \le c_2 \le \ldots \le c_q$, then $t_1 \ge t_2 \ge \ldots \ge t_q$. If this were not the case, then there would be at least a single redundant element in the set F_v. Such (ordered) representation of a minimal set is considered to be canonical.
(d) Let F_{v_1} and F_{v_2} be two minimal sets, such that these sets are in their canonical form. Their union $F_{v_1} \cup F_{v_2}$ is always feasible, but possibly redundant. The minimal set $Min\,(F_{v_1} \cup F_{v_2})$ can be obtained in $O\,(\min\{\Psi, \Omega\})$ steps.
(e) The translation of a minimal set F_v by (c, t) is defined as follows.

$$F_v + (c, t) = \{(c_i + c, t_i + t) \mid (c_i, t_i) \in F_v\} \tag{8.7}$$

The resulting set is nonredundant, however it might possibly contain infeasible two-tuples. A minimal set $Min\,(F_v + (c, t))$ can be obtained by omitting the infeasible two-tuples. The computational complexity of these operations is $O\,(\min\{\Psi, \Omega\})$.

Note that the assumption of $\gcd(c_{ij}, \Psi) = \gcd(t_{ij}, \Omega) = 1$, for all edges $(v_i, v_j) \in E$ helps in obtaining unique minimal subsets of F_v, for any $v \in V$.

Outline of the Algorithm

The algorithm to determine the constrained path is essentially iterative. At each iteration, the cost-delay set of each node is updated by using the cost and transit-time values of all its successor vertices. The algorithm terminates, when the cost-delay sets of all the nodes do not get updated in an iteration. Denote the cost-delay set of vertex $v \in V$, at iteration j by F_v^j, for $0 \le j \le n$, where the target node is v_d. The algorithm determines for a vertex $v \in V - \{v_d\}$ the path to the node v_d (if it exists), its cost, and delay.

Let $\omega_1, \omega_2, \ldots, \omega_k$ be the successor nodes of the node v. Denote the cost and delay of the arc (v, ω_i) by c_i and t_i respectively for $1 \le i \le k$. Using this notation, the cost-delay set of vertex v at iteration $(j+1)$ is given by

$$F_v^{j+1} = Min\left(F_v^j \cup \bigcup_{i=1}^{k} \left(F_{\omega_i}^j + (c_i, t_i)\right)\right) \tag{8.8}$$

The following pseudocode assumes that $(n-1)\, c_{\max} > \Psi$ and $(n-1)\, t_{\max} > \Omega$. It further assumes that the procedures to compute $Min\,(F_{v_1} \cup F_{v_2})$ for $v_1, v_2 \in V$ and $Min\,(F_v + (c,t))$ for $v \in V$ are available.

Observe that the algorithm terminates after n iterations. Actually, the algorithm can be stopped when no updating occurs for all the vertices in an iteration. A feasible path exists from the vertex $v \in V - \{v_d\}$ to the target node v_d iff $F_v^n \neq \varnothing$. Furthermore, if $(c,t) \in F_v^n$ then there exists a path $\mathcal{P}$ from vertex v to the vertex v_d such that $\mathcal{C}\,(\mathcal{P}) = c$ and $\mathfrak{D}\,(\mathcal{P}) = t$.

Algorithm 8.5. *Constrained Routing Algorithm.*

Input: Digraph $G = (V, E)\,, c_{ij}, t_{ij} \in \mathbb{P}$, for all arcs $(v_i, v_j) \in E$.
The target node is v_d, $|V| = n$, and $|E| = m$.
Cost and transit-time bounds: $\Psi, \Omega \in \mathbb{P}$.
Also $\gcd\,(c_{ij}, \Psi) = \gcd\,(t_{ij}, \Omega) = 1$, for all arcs $(v_i, v_j) \in E$.
Output: The cost-delay set of each vertex F_v, for $v \in V - \{v_d\}$.
begin
  (*initialization*)
  $F_{v_d}^j \leftarrow \{(0,0)\}$, for $j = 0, 1, 2, \ldots, n$
  $F_v^0 \leftarrow \varnothing, \forall\, v \in V - \{v_d\}$
  for $j = 0$ **to** $(n-1)$ **do**
  begin
    for all $v \in V - \{v_d\}$ **do**
    begin
      $(\omega_1, \omega_2, \ldots, \omega_k$,*are the successor nodes of the node* $v)$
      (*where* k *depends upon* v)
      (*the cost of the arc* (v, ω_i) *is* c_i *for* $1 \le i \le k$)
      (*the delay of the arc* (v, ω_i) *is* t_i *for* $1 \le i \le k$)
      $F_v^{j+1} \leftarrow Min\left(F_v^j \cup \bigcup_{i=1}^k \left(F_{\omega_i}^j + (c_i, t_i)\right)\right)$
    end (*end of for-loop*)
  end (*end of* j *for-loop*)
  $F_v \leftarrow F_v^n, \forall\, v \in V$
end (*end of constrained routing algorithm*)

The path information can be included in the above algorithm as follows. Assume that the cost-delay two-tuple $(c,t) \in F_v$ was obtained from:

(a) The arc (v, w_1) with cost c' and delay t'.
(b) The two-tuple $(c_1, t_1) \in F_{w_1}$.
(c) The operation $Min\,(F_v \cup (F_{w_1} + (c', t')))$.

Therefore the path can be uniquely identified, by associating with the two-tuple $(c,t) \in F_v$, the arc (v, w_1) and a pointer to the two-tuple $(c_1, t_1) \in F_{w_1}$. Subsequently look for the path from w_1 with cost delay (c_1, t_1), and so on. See Figure 8.2.

The computational complexity of this algorithm is determined as follows. The cost-delay set of each node is obtained after at most n iterations. Furthermore, at each iteration, there are at most m updates. Each such update has a translation and merge (union) operation, which have a computational complexity of $O(\min(\Psi, \Omega))$ operations. Therefore the computational complexity of this algorithm is $O(mn \min(\Psi, \Omega))$.

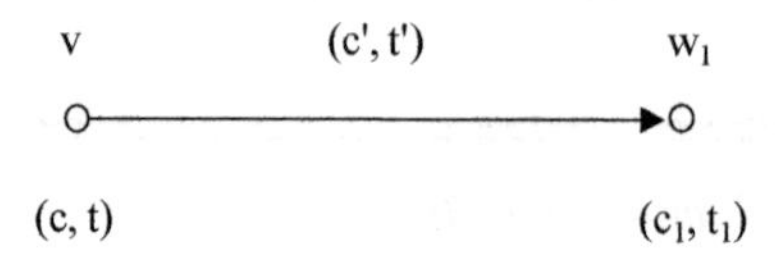

Figure 8.2. Path determination.

8.3.3 Disjoint Paths Algorithm

A communication network which can carry information (traffic) even in the presence of failures of its nodes or edges is called a survivable network. In order for a network to be survivable, it is necessary to have a disjoint pair of paths between pairs of nodes. Evidently, a network with two disjoint paths is more reliable than a network with a single path between a pair of nodes. For example, if a single link fails in one path, then communication between the two end nodes is not disrupted because of the presence of the second path.

The disjoint paths between a pair of vertices can be either edge-disjoint or node-disjoint, or both edge-disjoint and node-disjoint. In edge-disjoint paths between a pair of vertices, the goal is to find a pair of shortest paths without any common edges. Similarly, in node-disjoint paths between a pair of vertices, the paths do not have any common vertices, except the source and destination vertices.

A technique to compute edge-disjoint paths between a pair of vertices is outlined in this subsection. It is originally due to J. W. Suurballe. Let $G = (V, E)$, be a digraph where $|V| = n$. A length (cost) metric $c_{ij} \in \mathbb{R}_0^+$ is defined for each arc $(v_i, v_j) \in E$. Also assume that this graph is connected and simple. The source and target (destination) nodes are v_s and v_t respectively. The aim of the algorithm is to find a pair of edge-disjoint paths from v_s to v_t of minimum total length. That is, the sum of the lengths of the two edge-disjoint paths should be minimal. As the graph G is connected, then all vertices of the graph are reachable from the vertex v_s. Further assume that there exist at least two paths between v_s and v_t such that there is no common edge between them.

The disjoint path algorithm is developed in terms of a minimum cost flow problem in a capacitated network. This problem has been addressed in the field of operations research. Therefore it is first described briefly.

Formulation in Terms of a Minimum Cost Flow Problem in a Capacitated Network

Consider a digraph $G = (V, E)$, and two metrics which are associated with each arc in E. These metrics are the length metric c_{ij} and the capacity metric u_{ij} for each arc $(v_i, v_j) \in E$. These metrics are both nonnegative. Also associated with each vertex $v_i \in V$ is a source or sink of traffic specified by $b(i)$. If $b(i) > 0$, then the vertex v_i is a supplier of traffic. However, if $b(i) < 0$, then v_i is a receiver of traffic. For conservation of traffic flow on the network it is required that $\sum_{i=1}^{n} b(i) = 0$, where $n = |V|$ is the total number of vertices in the graph. The goal is to find the traffic flow x_{ij} on each arc $(v_i, v_j) \in E$ such that traffic conservation is preserved and traffic capacity constraints are observed. That is

$$\sum_{j=1}^{n} x_{ij} - \sum_{j=1}^{n} x_{ji} = b(i), \quad 0 \le x_{ij} \le u_{ij}, 1 \le i, j \le n$$

The traffic flows are specified in an $n \times n$ matrix $x = [x_{ij}]$. The cost of flow of this traffic is $z(x) = \sum_{i=1}^{n} \sum_{j=1}^{n} x_{ij} c_{ij}$. Minimum cost requirement for flow of this traffic implies that this cost has to be minimized.

The minimum cost flow problem is transformed into the edge-disjoint path problem as follows. Let the arc capacities be $u_{ij} = 1$, for $1 \le i, j \le n$, $b(s) = 2$, and $b(t) = -2$, and all other $b(i)$'s are each equal to zero. That is, in this unit-capacity network two units of traffic are supplied by the source node v_s, and two units of traffic have to be received at the destination node v_t. Observe that, since two units of traffic have to be transmitted from the source node to the destination node in a unit capacity network, the traffic flows in disjoint paths, and each path carries one unit of traffic. Note that the flows along cycles do not affect the flow value.

Cost Transformation

The next step in the development of the disjoint-path algorithm is the modification of the cost matrix C. In the discussion which follows, it is assumed that if an arc $(v_i, v_j) \in E$ then $(v_j, v_i) \notin E$. However, if both $(v_i, v_j) \in E$ and $(v_j, v_i) \in E$, then the digraph G can be modified by a node-splitting transformation so that there is only a single arc between any pair of nodes. Technique to implement this mapping is discussed in the chapter on graph theory.

This minimum cost flow problem can be addressed by using Dijkstra's algorithm twice. Initially compute the shortest path tree T_{spt} of the digraph G rooted at the vertex v_s. Denote the shortest path from the vertex v_s to vertex v_t by $\mathcal{P}_1$. Let the distance vector be $\mathcal{D}_1 = (d_1, d_2, \ldots, d_n)$, where d_i is the shortest distance from the source vertex v_s to vertex v_i for $1 \le i \le n$. Next perform the following two steps. These are the cost transformation and the graph modification steps.

Transform the costs associated with each arc to $c'_{ij} = (c_{ij} + d_i - d_j)$. Denote the corresponding cost matrix by C', that is $C' = [c'_{ij}]$. This procedure is called the *cost transformation.* We establish the following facts about this cost transformation.

Let a path $\mathcal{P}$ between the vertices v_s and v_m be $(v_s, v_1, v_2, \ldots, v_k, v_m)$. The length of this path using the cost matrix C is $\sum_{e \in \mathcal{P}} c_e$. Similarly, let the length of this path using the cost matrix C' be $\sum_{e \in \mathcal{P}} c'_e$. Then it can be shown that

$$\sum_{e \in \mathcal{P}} c'_e = \sum_{e \in \mathcal{P}} c_e - d_m$$

Thus the length of any given path $\mathcal{P}$ from the source vertex to an arbitrary destination vertex v_m when the cost matrix is C', is less than the length the corresponding path length with cost matrix C by a fixed amount equal to the shortest path length from the source vertex v_s to the destination vertex v_m.

This fact implies that the ranking a pair of paths with identical source and destination vertices by their net length remains unchanged by this cost transformation. Consequently, if the shortest path pairs problem for the transformed length is solved with cost matrix C', a solution of the problem with the original cost matrix C is obtained. This observation is summarized in the following statement.

Fact 1: The ranking of paths by path length between the source vertex v_s and a destination vertex v_m remains unchanged by this cost transformation.

The next important fact follows from the property of the d_i's. The distance d_i is the shortest distance from the source vertex v_s to the vertex v_i for $1 \leq i \leq n$.

Fact 2: Observe that if (v_i, v_j) is an edge in the shortest path tree T_{spt}, then $c'_{ij} = 0$. For all other edges of the digraph, c'_{ij} is nonnegative.

Therefore it can be concluded from this later fact that the cost transformation effectively removes the arcs in the path $\mathcal{P}_1$, because they are saturated due to their unit capacity.

Also observe that if

$$z(x) = \sum_{i=1}^{n}\sum_{j=1}^{n} x_{ij}c_{ij}, \text{ and } z'(x) = \sum_{i=1}^{n}\sum_{j=1}^{n} x_{ij}c'_{ij}$$

then

$$\begin{aligned} z(x) - z'(x) &= \sum_{i=1}^{n}\sum_{j=1}^{n} x_{ij}(d_j - d_i) \\ &= \sum_{i=1}^{n}\sum_{j=1}^{n} x_{ji}d_i - \sum_{i=1}^{n}\sum_{j=1}^{n} x_{ij}d_i \\ &= \sum_{i=1}^{n} d_i \sum_{j=1}^{n} (x_{ji} - x_{ij}) \\ &= -\sum_{i=1}^{n} d_i b(i) = 2(d_t - d_s) = 2d_t \end{aligned}$$

Fact 3: As $2d_t$ is a constant, minimizing $z(x)$ to obtain minimum cost flow of traffic is equivalent to minimizing $z'(x)$.

In the next step, reverse the orientation of all the arcs in the path $\mathcal{P}_1$ to obtain a new digraph G_r. The cost of these reversed arcs is zero, and their capacity is unity. The matrix C' is again suitably modified to C''. This new graph is called the *residual digraph* (*network*). An optimal solution obtained by using the cost matrix C'', and the residual digraph G_r, can be readily transformed into an optimal solution in the original graph G. Therefore, if another unit of traffic flow is sent from the vertex v_s to vertex v_t along the shortest path, a pair of shortest edge-disjoint paths is determined as per the theory of minimum cost flow in a capacitated network.

Next find a shortest path $\mathcal{P}_2$ from the vertex v_s to vertex v_t in the residual digraph G_r using the cost matrix C''. Let the set of arcs in the path $\mathcal{P}_1$ be A_1. Let A_2 be the set of arcs in the path $\mathcal{P}_2$, with the reversed arcs in their original orientation. The edge-disjoint pair of shortest paths $\mathcal{P}'$ and $\mathcal{P}''$ between the vertices v_s and v_t is obtained by deleting all arcs which belong to the set $A_1 \cap A_2$ from the set $A_1 \cup A_2$. The above discussion is condensed in the following edge-disjoint paths algorithm.

Algorithm 8.6. *Edge-Disjoint Paths Algorithm.*

Input: Digraph $G = (V, E)$, where $|V| = n$.
The cost matrix $C = [c_{ij}]$ has nonnegative elements.
The source and target nodes are v_s and v_t respectively.
Assumption: There are at least two edge-disjoint paths between v_s and v_t. If $(v_i, v_j) \in E$ then $(v_j, v_i) \notin E$, for $1 \leq i, j \leq n$. This limitation can

be overcome by a suitable transformation.
Output: The edge-disjoint pair of shortest paths $\mathcal{P}'$ and $\mathcal{P}''$ between the vertices v_s and v_t. The sum of the lengths of the two edge-disjoint paths is minimal.
begin

Step 1: Determine the shortest path tree T_{spt}, of the digraph G rooted at the vertex v_s. Compute the shortest path $\mathcal{P}_1$ from vertex v_s to vertex v_t. Let the distance vector corresponding to the tree T_{spt} be $\mathcal{D}_1 = (d_1, d_2, \ldots, d_n)$.

Step 2: Perform the cost transformation:
$c'_{ij} \leftarrow (c_{ij} + d_i - d_j)$, for $1 \leq i, j \leq n$.
After this cost transformation, all the arcs which belong to the tree T_{spt} have zero cost. Note that $C' = [c'_{ij}]$.

Step 3: Modify the graph G, by reversing the orientation of all the arcs in the path $\mathcal{P}_1$ to obtain a new graph G_r.
This new graph is called the residual graph.
Again modify the cost matrix C' to reflect the modified orientation. Let it be C''.

Step 4: Using the residual graph G_r and the cost matrix C'', compute the shortest path from v_s to v_t and denote it by $\mathcal{P}_2$.

Step 5: Let the set of arcs in the path $\mathcal{P}_1$ be A_1. Let A_2 be the set of arcs in the path $\mathcal{P}_2$, with the reversed arcs in their original orientation.
The set of all the arcs in the paths $\mathcal{P}'$ and $\mathcal{P}''$ is obtained by deleting all arcs which belong to $A_1 \cap A_2$ from $A_1 \cup A_2$.

end (*end of edge-disjoint paths algorithm*)

See the problem section to establish the correctness of this algorithm. The computational complexity of this algorithm is the same as that of the Dijkstra's algorithm, which is $O(n^2)$. A complete mathematical justification of this algorithm can be provided by using the theory of minimum cost flow in a capacitated network. This theory is developed later in this chapter.

Example 8.3. A pair of edge-disjoint shortest paths for the digraph $G = (V, E)$ is determined. The digraph is specified as

$$V = \{1, 2, 3, 4, 5, 6\}$$

$$E = \{(1,2), (1,3), (2,3), (2,4), (3,4), (3,5), (4,6), (5,4), (5,6), (6,1)\}$$

See Figure 8.3. The cost matrix is

$$C = \begin{bmatrix} 0 & 2 & 1 & \infty & \infty & \infty \\ \infty & 0 & 4 & 1 & \infty & \infty \\ \infty & \infty & 0 & 1 & 3 & \infty \\ \infty & \infty & \infty & 0 & \infty & 2 \\ \infty & \infty & \infty & 3 & 0 & 2 \\ 4 & \infty & \infty & \infty & \infty & 0 \end{bmatrix}$$

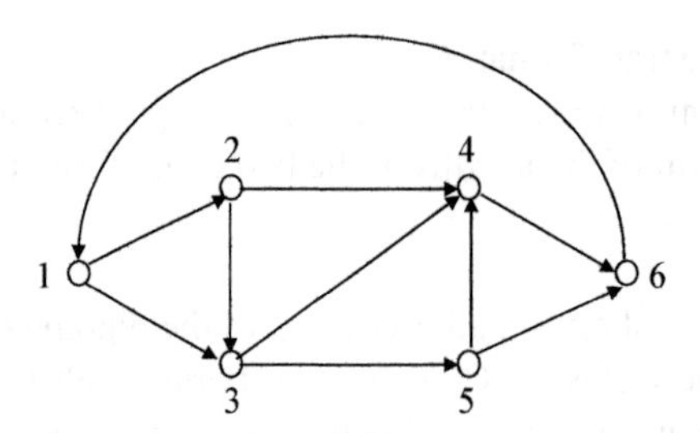

Figure 8.3. Edge-disjoint shortest path computation for the graph $G = (V, E)$.

Assume that the source and destination node numbers are 1 and 6 respectively. The steps in the computation are:

Step 1: Determine the shortest path tree for the digraph G rooted at the vertex 1. Compute the shortest path from node 1 to node 6 and denote it by $\mathcal{P}_1$. Let the distance and precedence vectors be $\mathcal{D}_1$ and $\mathcal{T}_1$ respectively. Use of Dijkstra's algorithm yields

$$\mathcal{P}_1 = (1, 3, 4, 6), \; \mathcal{D}_1 = (d_1, d_2, d_3, d_4, d_5, d_6) = (0, 2, 1, 2, 4, 4),$$

and $\mathcal{T}_1 = (\varnothing, 1, 1, 3, 3, 4)$. The cost of the path $\mathcal{P}_1$ is equal to 4.

Step 2: Perform the cost transformation: $c'_{ij} \leftarrow (c_{ij} + d_i - d_j)$, for $1 \leq i, j \leq 6$. Determine $C' = \left[c'_{ij}\right]$.

Step 3: Modify the graph G, by reversing the orientation of all the arcs in the path $\mathcal{P}_1$ to obtain a new reduced graph G_r. The vertex set of this graph is same as that of graph G, but the arc set is

$$E_r = \{(1, 2), (2, 3), (2, 4), (3, 1), (3, 5), (4, 3), (5, 4), (5, 6), (6, 1), (6, 4)\}$$

Also

$$C'' = \begin{bmatrix} 0 & 0 & \infty & \infty & \infty & \infty \\ \infty & 0 & 5 & 1 & \infty & \infty \\ 0 & \infty & 0 & \infty & 0 & \infty \\ \infty & \infty & 0 & 0 & \infty & \infty \\ \infty & \infty & \infty & 5 & 0 & 2 \\ 8 & \infty & \infty & 0 & \infty & 0 \end{bmatrix}$$

Step 4: Compute the shortest path from node 1 to node 6 in the residual graph G_r with the cost matrix C''. Denote this shortest path by $\mathcal{P}_2$. Let the corresponding distance and precedence vectors be $\mathcal{D}_2$ and $\mathcal{T}_2$ respectively. Use of Dijkstra's algorithm results in

$$\mathcal{P}_2 = (1, 2, 4, 3, 5, 6), \; \mathcal{D}_2 = (0, 0, 1, 1, 1, 3),$$

and $\mathcal{T}_2 = (\varnothing, 1, 4, 2, 3, 5)$. The cost of this path is equal to 3.

Step 5: The set of arcs in $\mathcal{P}_1$ is $A_1 = \{(1, 3), (3, 4), (4, 6)\}$. Similarly, the set of arcs in $\mathcal{P}_2$, with the reversed arcs in their original orientation is

$$A_2 = \{(1, 2), (2, 4), (3, 4), (3, 5), (5, 6)\}$$

Observe the orientation of the arcs in the set A_2. The set of arcs in $A_1 \cap A_2$ is $\{(3, 4)\}$, and

$$A_1 \cup A_2 = \{(1, 2), (1, 3), (2, 4), (3, 4), (3, 5), (4, 6), (5, 6)\}$$

After discarding the arc $(3, 4)$, from the set $A_1 \cup A_2$, the set of arcs in the paths $\mathcal{P}'$ and $\mathcal{P}''$ are obtained. The arcs in these paths are

$$\{(1,2),(1,3),(2,4),(3,5),(4,6),(5,6)\}$$

From this set, $\mathcal{P}' = (1,2,4,6)$ and $\mathcal{P}'' = (1,3,5,6)$ are determined. The cost of these paths are 5 and 6 respectively. This solution also happens to be vertex-disjoint. Also note that the paths $\mathcal{P}'$ and $\mathcal{P}''$ are each more expensive than the shortest path $\mathcal{P}_1 = (1,3,4,6)$, the cost of which is equal to 4. □

A procedure to compute a pair of edge-disjoint paths in a digraph has been discussed in this subsection. This procedure can be extended to compute a pair of vertex-disjoint paths from vertex v_s to vertex v_t.

In vertex-disjoint path computation, the two disjoint paths follow different nodes, except at the source and destination nodes. This can be facilitated by splitting each vertex $v \in V$ into two vertices v' and v'' and connecting them by an arc (v'', v') of zero cost and infinite capacity. Furthermore, the original arc (v_i, v_j) is replaced by an arc $(v_{i'}, v_{j''})$. Computation of edge-disjoint paths in the modified graph from vertex v_s' to vertex v_t'' gives vertex-disjoint paths in the original graph.

8.4 Algebraic Path Techniques

Basics of algebraic path techniques and related applications are discussed in this section. Algebraic path techniques are generalization of the shortest path problem defined on digraphs. Use of algebraic structures like semiring, dioid, and prebimonoid are examined.

8.4.1 Semirings and Dioids

Motivation for the use of semirings and dioids is developed initially by describing the algebraic path problem. This is followed by the definitions of monoid and semiring, and justification for the use of semirings. An ordered semiring called dioid is also defined. In addition, application of such operations to the elements of vectors and matrices is also described. These are all building blocks for describing algebraic path techniques.

Algebraic Path Problem

Consider a digraph $G = (V, E)$, in which the vertex set is V, the edge set is E, and $|V| = n$. The digraph G is assumed to be connected and simple. A length, or weight, or cost metric is specified for each arc $(v_i, v_j) \in E$, where $v_i, v_j \in V$. It is denoted by c_{ij}, where $c_{ij} \in \mathbb{R}_0^+$. If $(v_i, v_j) \notin E$ and $i \neq j$, then $c_{ij} \to \infty$; and $c_{ii} = 0$, for $1 \leq i \leq n$. Recall that a path between a pair of nodes is a sequence of arcs which connect the two nodes without any repetitive intermediate nodes. The length of a directed path $\mathcal{P}$ is given by $w(\mathcal{P}) = \sum_{(v_i,v_j)\in\mathcal{P}} c_{ij}$. Let the set of all paths between vertices v_s and v_t be $\mathcal{P}^{st}$.

The path from the vertex v_s to the vertex v_t is the *shortest path* if the length of the path between these two vertices is the shortest. This path is not necessarily unique. Let d_{st} be the length of the shortest path from node v_s to node v_t. Therefore

$$w(\mathcal{P}) = \sum_{(v_i,v_j)\in\mathcal{P}} c_{ij} = c_{si_1} + c_{i_1 i_2} + \ldots + c_{i_k t}$$

$$d_{st} = \min_{\mathcal{P} \in \mathcal{P}^{st}} w(\mathcal{P})$$

If in the above equations, the min and $+$ operations are replaced by the binary operators (relations) $\oplus$ and $\otimes$ respectively, we obtain

$$d_{st} = \bigoplus_{\mathcal{P} \in \mathcal{P}^{st}} w(\mathcal{P}) = \bigoplus_{\mathcal{P} \in \mathcal{P}^{st}} \{c_{si_1} \otimes c_{i_1 i_2} \otimes \cdots \otimes c_{i_k t}\}$$

It is expected that the binary operators $\oplus$ and $\otimes$ defined on some set observe special properties. If this underlying set, and the binary operators form a semiring, the problem is called an *algebraic path problem*. These informal notions are next made precise.

Monoids and Semirings

Monoid and semiring algebraic structures help in specifying the algebraic path problem formally.

Definition 8.1. *A monoid $\mathcal{M} = (M, \star)$ is a nonempty set M, and a binary operation $\star$ defined on M, such that the following properties are satisfied.*

(a) *Associative property: The monoid is associative. That is, $a \star (b \star c) = (a \star b) \star c$ for all $a, b, c \in M$.*
(b) *Identity property: The set M has an element e, called the identity or the neutral element of $\mathcal{M}$, such that $e \star a = a \star e = a$ for all $a \in M$.* □

Definition 8.2. *Let $\mathcal{M} = (M, \star)$ be a monoid. The monoid $\mathcal{M}$ is commutative if it also satisfies the property: $a \star b = b \star a$ for all $a, b \in M$.* □

A semiring is next defined.

Definition 8.3. *A semiring $\mathcal{S} = (S, \oplus, \otimes)$ is a nonempty set S, and two binary operations, $\oplus$ (addition) and $\otimes$ (multiplication), defined on S. Further:*

(a) *The algebraic structure $(S, \oplus)$ is a commutative monoid, with neutral element $\overline{0}$. Sometimes $\overline{0}$ is called the zero of the semiring.*
(b) *The algebraic structure $(S, \otimes)$ is a monoid, with neutral element $\overline{1}$. Sometimes $\overline{1}$ is called the unit element of the semiring. If the context is clear, $a \otimes b$ is simply denoted by ab for all $a, b \in S$.*
(c) *The multiplication operation $\otimes$ is distributive over the addition operation $\oplus$. That is:*

$$a \otimes (b \oplus c) = (a \otimes b) \oplus (a \otimes c) \tag{8.9a}$$

$$(b \oplus c) \otimes a = (b \otimes a) \oplus (c \otimes a) \tag{8.9b}$$

for all $a, b \in S$.
(d) *Also $\overline{0} \otimes a = \overline{0}$, and $a \otimes \overline{0} = \overline{0}$, $\forall\ a \in S$. Therefore $\overline{0}$ is called the annihilator for $\otimes$.* □

Sometimes the semiring $\mathcal{S} = (S, \oplus, \otimes)$, is simply referred to as the semiring S. Observe that semirings are different than rings. In semirings, the addition operation $\oplus$ forms only a commutative monoid, and not necessarily a commutative group. Thus, elements in semirings do not have an additive inverse. Further, all rings are semirings.

Motivation for Use of Semirings

The axioms of semirings are a natural requirement for any path problem. Observe that the operation $\oplus$ must be commutative and associative, because the summation operator $\bigoplus_{\mathcal{P}\in\mathcal{P}^{st}}$ in the expression

$$d_{st} = \bigoplus_{\mathcal{P}\in\mathcal{P}^{st}} \{c_{si_1} \otimes c_{i_1 i_2} \otimes \ldots \otimes c_{i_k t}\}$$

must be independent of the order of the operands. Further, the $\otimes$ operator which computes the length of the path from the length of its arcs must be associative. Note that $\overline{1}$ is the length of an empty path, and $c_{ij} \in S$ for $1 \le i, j \le n$. Presence of distributivity of $\otimes$ over $\oplus$ helps in efficient path computations.

Ordered Semirings

Special types of semirings are of interest. Recall from the chapter on number theory that a binary relation is an order (or partial order) if it is reflexive, antisymmetric, and transitive. More specifically, let $\preceq$ be a binary relation defined on a nonempty set S. It satisfies reflexive property: if $a \in S$, then $a \preceq a$; it satisfies antisymmetric property: if $a \preceq b$ and $b \preceq a$, then $a = b$; and it satisfies transitive property: if $a \preceq b$ and $b \preceq c$, then $a \preceq c$.

Definitions 8.4. *Order, total order, and dioid.*

1. *A relation $\preceq$ on a nonempty set S is an order, if it satisfies the properties of reflexivity, antisymmetry, and transitivity.*
2. *An ordered relation $\preceq$ on a nonempty set S is a total order, if $\forall\ a, b \in S$ either $a \preceq b$ or $b \preceq a$. This implies that all elements of the set S can be ordered (compared).*
3. *Let $\mathcal{S} = (S, \oplus, \otimes)$ be a semiring. The semiring $\mathcal{S}$ is called a dioid, if it is canonically ordered. That is, if the relation $\preceq$ is an order on S, and $\forall\ a, b \in S$:*

$$a \preceq b \Leftrightarrow \exists\ c \in S \text{ such that } a \oplus c = b \tag{8.10}$$

□

Definition 8.5. *The binary operator $\oplus$ defined on a set S is idempotent, if $a \oplus a = a$ for all $a \in S$.* □

Observations 8.1. Let $\mathcal{S} = (S, \oplus, \otimes)$ be a semiring.

1. If the binary operator $\oplus$ is an idempotent, then the set S is canonically ordered. This is true because the antisymmetry property of the order relation $\preceq$ can be inferred.
2. Let $\mathcal{S}$ be a dioid with order relation $\preceq$.
   (a) As $\overline{0}$ is the additive identity, $\overline{0} \preceq a$ for all $a \in S$.
   (b) The direction of order $\preceq$ is maintained if addition and multiplication is performed on both sides with the same element. That is, $\forall\ a, b \in S$, where $a \preceq b$; and $\forall\ x \in S$

$$a \oplus x \preceq b \oplus x, \quad \text{and } a \otimes x \preceq b \otimes x$$

(c) The direction of order $\preceq$ is maintained if inequalities are added and multiplied. That is, $\forall$ $a, b, c, d \in S$, if $a \preceq b$; and $c \preceq d$, then

$$a \oplus c \preceq b \oplus d, \text{ and } a \otimes c \preceq b \otimes d$$

□

Vector and Matrix Representation

The notation for semiring addition and multiplication operations with vector and matrix elements is next introduced. Let $\mathcal{S} = (S, \oplus, \otimes)$ be a semiring; and $A = [a_{ij}]$, and $B = [b_{ij}]$, where $a_{ij}, b_{ij} \in S$. Then

$$A \oplus B \triangleq [a_{ij} \oplus b_{ij}], \text{ and } AB \triangleq \left[\bigoplus_k a_{ik} \otimes b_{kj}\right]$$

8.4.2 Applications

Some applications of the algebraic framework discussed in the last subsection are outlined. Bellman-Ford's shortest path algorithm, path enumeration in a digraph, and certain optimal path algorithms are studied. The set of nonnegative integers including infinity is denoted by $\mathbb{N}^* = \mathbb{N} \cup \{+\infty\}$. Similarly, the set of nonnegative real numbers including infinity is denoted by $\mathbb{R}_0^{+*} = \mathbb{R}_0^+ \cup \{+\infty\}$.

Bellman-Ford Shortest Path Algorithm

Bellman-Ford shortest path algorithm has been described earlier in the chapter. It determines the shortest path between a source node, and all other nodes in the digraph. This algorithm is described again by using the notation developed in the last subsection. In the digraph $G = (V, E)$, a length metric is defined for each edge $(v_i, v_j) \in E$, where $v_i, v_j \in V$. This length metric is denoted by c_{ij}, where $c_{ij} \in \mathbb{R}_0^{+*}$. If $(v_i, v_j) \notin E$ and $i \neq j$, then $c_{ij} = \infty$, and $c_{ii} = 0$, for $1 \leq i \leq n$.

Without loss of any generality, assume that the source node is v_1. Define the length of the shortest path from node v_1 to node $v_j \in V$ to be $d_j, 1 \leq j \leq n$. These shortest path lengths satisfy the following equations.

$$\begin{aligned} d_1 &= 0 \\ d_j &= \min_{\substack{1 \leq q \leq n \\ q \neq j}} \{d_q + c_{qj}\}, \quad 2 \leq j \leq n \end{aligned}$$

The above set of equations can be written in matrix form by using the operators from the semiring $(\mathbb{R}_0^{+*}, \min, +)$. In this semiring $\overline{0} = \infty$, and $\overline{1} = 0$. This semiring is also a dioid. Let $C = [c_{ij}]$, $d^T = \begin{bmatrix} d_1 \ d_2 \ \cdots \ d_n \end{bmatrix}$, and $u^T = \begin{bmatrix} u_1 \ u_2 \ \cdots \ u_n \end{bmatrix}$, where $u_1 = \overline{1}$, and $u_i = \overline{0}$, $2 \leq i \leq n$. The above set of equations can be restated as

$$d_1 = \overline{1}, \text{ and } d^T = d^T C \oplus u^T$$

The d vector can be determined iteratively as:

$$\begin{aligned} \left(d^{(1)}\right)^T &= \begin{bmatrix} 0 \ c_{12} \ \cdots \ c_{1n} \end{bmatrix} \\ \left(d^{(k+1)}\right)^T &= \left(d^{(k)}\right)^T C \oplus u^T, \quad k = 1, 2, \ldots, (n-2) \end{aligned}$$

The d vector always exists because the edge costs are nonnegative. Note that $\left(d^{(n-1)}\right)^T = d^T$. The above formulation can be generalized to determine shortest path between every pair of nodes in the digraph G. Let $D = [d_{ij}]$, where d_{ij} is the shortest path distance between nodes v_i and v_j. Also let $U = [u_{ij}]$ where $u_{ii} = \overline{1}$ for $1 \leq i \leq n$, and $u_{ij} = \overline{0}$, for $i \neq j$, and $1 \leq i, j \leq n$. The matrix form of Bellman-Ford equations is

$$D = DC \oplus U$$

The matrix D can be determined iteratively.

Path Enumeration in a Digraph

The general problem of enumerating all possible paths between a pair of vertices can be conveniently addressed via an algebraic formulation. In the digraph $G = (V, E)$, let $V = \{v_1, v_2, \ldots, v_n\}$. A general path enumeration is possible in a dioid $\mathcal{S} = (S, \oplus, \otimes)$.

The set S of the dioid is defined as follows. Denote the set of a sequence of ordered vertices of length $k \in \{1, 2, \ldots, n\}$ by $V^{(k)}$. The elements of the set S are the subsets of: $V \cup V^{(2)} \cup \ldots \cup V^{(n)}$ and the empty set $\varnothing$. Thus the set S is the power set of ordered vertex sequences.

The addition $\oplus$ of two sets of vertex-sequences is simply defined as their set union. The zero element $\overline{0}$ of the dioid is the empty set $\varnothing$. The multiplication operation $\otimes$ performed on two sets of sequences, is the set where concatenation is possible between a sequence from the first set, and a sequence from the second set. For example, if $S_1, S_2 \in S$, then

$$S_1 \otimes S_2 = \{s_1 s_2 \mid s_1 \in S_1, s_2 \in S_2, \text{ also the last vertex in the sequence } s_1 \\ \text{is identical to the first vertex in the sequence } s_2\}$$

Further, in the concatenation $s_1 s_2$, the common vertex is not duplicated. Therefore, the unit element $\overline{1}$ is the set of all elements of the set V. It is also possible to list all paths with a specific property P_{prop}. An example would be enumeration of elementary paths in the digraph.

Optimal Paths or Quality of Service Routing

An example of optimal path computation is the determination of the shortest path in a digraph. The Bellman-Ford shortest path algorithm was discussed earlier in this subsection. In this algorithm, shortest path was determined between a pair of nodes in the digraph. It is possible to have other criteria for optimal path routing. This can be in the form of Quality of Service routing. In such schemes, each arc of the digraph is associated with a measure of quality like: bandwidth, delay, and loss. The goal in such routing schemes would be to compute optimal routing path from source node to destination node as per the specified metric. It is also possible to associate each edge with more than a single metric, with a prespecified priority among them. These are illustrated in the following examples which describe computation of: path with smallest number of hops between a pair of nodes, path with largest capacity, most reliable path, and a multicriteria path-computing scheme in a digraph.

Examples 8.4. Some optimal path algorithms in a simple, and connected digraph $G = (V, E)$ are described, where V is the vertex set, E is the edge set, and $|V| = n$.

1. *Path with smallest number of hops in a digraph.* In the digraph, let each edge weight be equal to unity. Apply any shortest path algorithm like Bellman-Ford shortest path algorithm to de-

termine a path with smallest number of hops between a pair of nodes. The dioid of interest is $\mathcal{S} = (S, \oplus, \otimes) = (\mathbb{N}^*, \min, +)$, where $\overline{0} = \infty$, and $\overline{1} = 0$.

2. *Largest capacity path (widest path) in a digraph.* The edge weights are assigned bandwidth/capacity (nonnegative real numbers). The capacity of a path is equal to the bandwidth of a link (on the path) with the smallest capacity. This bandwidth is often termed the bottleneck capacity of the path. The goal is to determine a path from a source node to destination node with the largest capacity. The dioid of interest is $\mathcal{S} = (S, \oplus, \otimes) = \left(\mathbb{R}_0^{+*}, \max, \min\right)$, where $\overline{0} = 0$, and $\overline{1} = \infty$.
3. *Path of maximum reliability in a digraph.* The edge weight is the probability of successful transmission of information (packets) along the edge. Assume that the probability of successful transmission of information along each edge is independent of such probabilities associated with every other edge. The goal is to determine a path from a source node to the destination node that has the highest reliability (probability). The dioid of interest is $\mathcal{S} = (S, \oplus, \otimes) = ([0, 1], \max, \times)$, where $\overline{0} = 0$, and $\overline{1} = 1$.
4. *Multicriteria path in a digraph.* Let each edge weight be a two-tuple. For each pair of vertices v_i and v_j in the digraph, the transit time of packets is t_{ij}, and the cost of the link is c_{ij}. The goal is to minimize the cost of transmission between a pair of nodes. However, if several paths have equal cost, then the next criterion is the smallest transit time. A preference relation $\preceq$ on the set of two-tuples (c, t) is defined as:

$$(c_1, t_1) \preceq (c_2, t_2) \Leftrightarrow (c_1 < c_2) \;\text{ or }\; (c_1 = c_2 \;\text{ and }\; t_1 \leq t_2)$$

The semiring of interest is $\mathcal{S} = (S, \oplus, \otimes)$, where $S = \mathbb{R}_0^{+*} \times \mathbb{R}_0^{+*}$, and

$$(c_1, t_1) \oplus (c_2, t_2) = \begin{cases} (c_1, t_1), & c_1 < c_2 \\ (c_2, t_2), & c_2 < c_1 \\ (c_1, \min(t_1, t_2)), & c_1 = c_2 \end{cases}$$

$$(c_1, t_1) \otimes (c_2, t_2) = (c_1 + c_2, t_1 + t_2)$$

Note that $\overline{0} = (\infty, \infty)$, and $\overline{1} = (0, 0)$. This semiring is also a dioid. It is also possible to have more than two criteria per link. Some candidates for different criteria are: bandwidth, cost, delay, loss, jitter, and so on. □

8.4.3 Prebimonoids

An algebraic structure called prebimonoid is used in describing interdomain routing in the Internet. The prebimonoid is first formally defined.

Definition 8.6. *A prebimonoid $\mathcal{S}_{pre} = (S_{pre}, \oplus, \otimes)$ is an algebraic structure, where S_{pre} is a nonempty set, and $(S_{pre}, \oplus)$ is a commutative monoid with neutral element $\overline{0}$. The binary operator $\otimes$ is not associative, and its neutral element is $\overline{1}$; and $\overline{0}$ is the annihilator for $\otimes$.* □

If the associativity of the operator $\otimes$, and the distribution of $\otimes$ over $\oplus$, is imposed upon the prebimonoid algebraic structure, we obtain a semiring. The canonical order $\preceq$ of prebimonoid is defined as in semirings.

Consider the digraph $G = (V, E)$, in which the vertex set is V, the edge set is E, and $|V| = n$. The digraph G is assumed to be connected and simple. A length (or weight or cost) metric is specified for each arc $(v_i, v_j) \in E$. It is denoted by c_{ij}, where $c_{ij} \in \mathbb{R}_0^{+*}$ for $1 \le i, j \le n$. Recall, that a path between a pair of nodes is a sequence of arcs which connect the two nodes without any repetitive intermediate nodes. Let the directed path $\mathcal{P}$ between vertices v_s and v_t be $v_s, v_{i_1}, v_{i_2}, \ldots v_{i_k}, v_t$. As associativity of the binary operator $\otimes$ is not defined, the weight of this path can be determined by a proper specification of sequence of successive $\otimes$-multiplications. The left-weight of this path $\mathcal{P}$ is defined as

$$w_L(\mathcal{P}) = c_{si_1} \otimes (c_{i_1 i_2} \otimes (\cdots \otimes c_{i_k t} \cdots))$$

This $\otimes$-multiplication is performed from right to left. Let the set of all paths between vertices v_s and v_t be $\mathcal{P}^{st}$. The weight of the empty path is $\overline{1}$. The prebimonoid of interest is

$$\mathcal{S}_{pre} = (S_{pre}, \oplus, \otimes) = \left(\mathbb{R}_0^{+*}, \min, +\right)$$

where $\overline{0} = \infty$, and $\overline{1} = 0$. An optimal path from vertex v_s to vertex v_t, if it exists, with $\preceq$-minimum left-weight d_{st} is

$$d_{st} = \bigoplus_{\mathcal{P} \in \mathcal{P}^{st}} w_L(\mathcal{P})$$

Similar definitions can be made for right-weights. The above formulation can be generalized to determine optimal path between every pair of nodes in the digraph G. Let $D_L = [d_{ij}]$, where d_{ij} is possibly an optimal path distance between vertices v_i and v_j. Also let $U = [u_{ij}]$, where $u_{ii} = \overline{1}$ for $1 \le i \le n$, and $u_{ij} = \overline{0}$, for $i \ne j$, and $1 \le i, j \le n$. The matrix form of Bellman-Ford type of optimal path formulation is

$$D_L = CD_L \oplus U$$

where $C = [c_{ij}]$. The matrix D_L can be determined iteratively. However, it may not always yield optimal solution because of lack of distributivity of $\otimes$ over $\oplus$.

Interdomain Routing

Internet is made up of several autonomous systems (ASes) or domains. An AS is a collection of routers which are under control of a single administrative entity. Connectivity among ASes can be specified via an AS-graph. In this graph, the set of vertices is the ASes, and the set of edges represent physical communication links between ASes. Intradomain routing, that is routing within an AS, generally uses shortest path algorithms. However, interdomain routing among ASes is policy based. In the AS graph (network), each network link has a label, and each network path has a signature.

ASes exchange routing information (routes) among themselves. Consequently it is quite possible for an AS to have in its collection of imported routes, more than a single route to reach a destination AS. Therefore, an AS has to decide the route it has to select in order to reach its destination AS. In contrast to the shortest path algorithms, the selection of routes between pairs of ASes is policy based.

The mechanism which governs routing between ASes is the Border Gateway Protocol (BGP). We consider only an extremely simplified version of BGP. In this version of BGP, an AS can be either a customer, or a provider, or a peer of another AS. These are actually business relationships. That is, an AS A can be customer of AS B; and AS B is a provider for AS A. Thus, AS A pays AS B for providing connectivity to its users. Peer ASes do not charge each other for providing services to each other and their customers. Note that an AS can be in many different relationships simultaneously. It can be a provider to several ASes, a customer of others, and a peer of others. The following guidelines for interdomain routing have been proposed.

- *Exporting to a customer*: During an exchange of routing information with a customer; an AS can export its routes, as well as the routes it learned from its providers and peers. Thus, an AS *provides* transit services to its customers.
- *Exporting to a provider*: During an exchange of routing information with a provider; an AS can export its routes, as well as the route it learned from its customers. However, it cannot export routes it learned from other providers or peers. Thus, an AS *does not provide* transit services to its provider.
- *Exporting to a peer*: During an exchange of routing information with a peer; an AS can export its routes, as well as the routes it learned from its customers. However, it cannot export routes that it learned from other providers or other peers. Thus, an AS *does not provide* transit services to its peers
- Routes learned from customers are preferred to routes learned from either peers or providers. Routes learned from peers are preferred to routes learned from providers.
- Preferences among the set of customer routes are not specified. The same is true for preferences among sets of peer and provider routes.
- The subgraph of the AS-graph induced by customer-provider arcs should not have directed cycles.

The above routing constraints ensure that an AS is not (indirectly) its own provider. The routing constraints also ensure robust convergence of the algorithm, without any coordination among the ASes. Also note that the above guidelines do not guarantee optimality of the paths. The preferences among routes can be modeled by a prebimonoid $\mathcal{S}_{pre} = (S_{pre}, \oplus, \otimes)$. In this prebimonoid, the set $S_{pre} = \{c, r, p, \overline{0}, \overline{1}\}$; element c stands for a customer route and customer link, element r stands for a peer route and peer link, and element p stands for a provider route and provider link. The element $\overline{0}$ represents forbidden path. These are the paths which cannot be exported. The element $\overline{1}$ represents trivial path composed of a single node. The canonical order $\preceq$ among elements of the set S_{pre} is defined as $a \preceq b \Leftrightarrow a \oplus b = b, \forall\, a, b \in S_{pre}$.

Thus preferences among allowed paths are encoded by the $\oplus$ operator; and the allowed paths are encoded by the operator $\otimes$. The operations $\oplus$ and $\otimes$ are specified in Table 8.1.

For example, observe that in the table for $\oplus$ operation, the relationship $r \oplus c = c \oplus r = c$ implies that $r \preceq c$. That is, customer routes are preferred over peer routes.

In the $\otimes$-table, the first set of operands appear in the first column, and the second set of operands appear in the first row. The first operand represents a *label* assigned to a link of the AS network. The second operand represents a *signature* assigned to a path in the AS network. The result of the $\otimes$ operation is a signature assigned to a path in the AS network.

In the table for $\otimes$ operation, the relationship $c \otimes r = \overline{0}$ implies that a peer path cannot be extended by a customer link. Similarly, $r \otimes c = r$ implies that an AS can export to a peer a route learned from a customer, thereby becoming a peer route.

$\oplus$	$\overline{1}$	c	r	p	$\overline{0}$
$\overline{1}$	$\overline{1}$	$\overline{1}$	$\overline{1}$	$\overline{1}$	$\overline{1}$
c	$\overline{1}$	c	c	c	c
r	$\overline{1}$	c	r	r	r
p	$\overline{1}$	c	r	p	p
$\overline{0}$	$\overline{1}$	c	r	p	$\overline{0}$

$\otimes$	$\overline{1}$	c	r	p	$\overline{0}$
$\overline{1}$	$\overline{1}$	c	r	p	$\overline{0}$
c	c	c	$\overline{0}$	$\overline{0}$	$\overline{0}$
r	r	r	$\overline{0}$	$\overline{0}$	$\overline{0}$
p	p	p	p	p	$\overline{0}$
$\overline{0}$	$\overline{0}$	$\overline{0}$	$\overline{0}$	$\overline{0}$	$\overline{0}$

Table 8.1. Operations $\oplus$ and $\otimes$.

8.5 Geographic Routing in Hyperbolic Space

Geographic routing of a message from a source s to a target (destination) t consists of routing it simply based upon the physical coordinates of s, t and the neighboring nodes of s. The geographic routing scheme is *greedy*. That is, the nodes forward the message to a neighboring node which is closest to the destination. This scheme appears to be simple and efficient if the number of nodes in the network is very large. However, it has drawbacks. There is no guarantee that greedy routing provides shortest paths between a pair of nodes. Furthermore, greedy routing might fail, if the message reaches an intermediate node which is closest to the destination than all of its neighbors. This node is said to have a local minimum. Messages reaching this node do not reach the specified target node.

A possible scheme to overcome this dilemma is to assign the location of nodes of the network virtual coordinates in a suitable metric space, and route successfully yet greedily in this virtual (latent) space. Such mappings are sometimes called *greedy embeddings*. The virtual space might possibly have higher dimension, and use a non-Euclidean metric. A possible candidate for the virtual space is a hyperbolic space.

Robert Kleinberg established in a landmark paper that any finite and connected graph has a greedy embedding in a hyperbolic space. He demonstrated, that a graph G can be embedded in hyperbolic space by generating a spanning tree of the graph G. Let this spanning tree's maximum nodal degree be δ_{max}. In this scheme, the spanning tree of the graph G is made a subtree of an infinite complete δ_{max}-regular tree. Recall that in a complete δ_{max}-regular tree (typically infinite) each internal vertex has exactly δ_{max} children (branches), and all leaves are at the same depth. It is possible to route along the edges of the tree. However, it is hoped that the non-tree edges of the graph will be used by the greedy routing scheme to produce possibly shorter paths between source-destination pairs. Kleinberg's important result is stated as a theorem.

Theorem 8.2. *Every connected and finite graph has a greedy embedding in a hyperbolic space.* □

The rest of this section is devoted to establishing the truth of this result. We initially discuss preliminaries of greedy embeddings. This is followed by a summary of immediately useful facts about hyperbolic geometry. Finally a greedy embedding of the network graph in a hyperbolic space is demonstrated.

8.5.1 Greedy Embeddings

The communication network is represented by an undirected graph $G = (V, E)$. The set of vertices V in the graph G represent the routers in the network. The set of edges E represent the links between the nodes of the network. It is also assumed that the graph G is connected. Let the number of vertices in the network be $|V| = n$. The neighbors of a vertex $v \in V$ are the set of vertices $N_v = \{w \mid w \in V, (v, w) \in E\}$.

The initial goal is to assign vertices in the set V virtual coordinates in a metric space (X, d). The distance metric d is used for greedy forwarding of a message (or packet). Furthermore, the mapping of the vertices in the set V to elements in the set X has to be a greedy embedding. This is elaborated upon in the rest of this subsection.

Definition 8.7. *Let a network be represented by an undirected and connected graph $G = (V, E)$. The set of neighboring vertices of vertex $v \in V$ is N_v. A greedy embedding of vertices in the graph G in the metric space (X, d) is a mapping $f : V \to X$. In this map, for each two-tuple $(s, t) \in V \times V$, where $s \neq t$; there exists a $u \in N_s$ such that*

$$d(f(u), f(t)) < d(f(s), f(t)) \tag{8.11}$$

□

Thus a greedy embedding is a map in which at each vertex there always exists at least a single neighboring vertex which is closer to the target node. This mapping ensures that the message does not encounter a local minima, and reaches its target node via greedy forwarding.

Observations 8.2. Basic observations about greedy embeddings.

1. Let H be a subgraph of an undirected graph G. The vertex sets of the two graphs are identical. Then every greedy embedding of H is also a greedy embedding of G.
2. Let $T = (V, E)$, and $T_s = (V_s, E_s)$ be two tree graphs, where T_s is a subtree of T. That is, $V_s \subseteq V$ and $E_s \subseteq E$. Then every greedy embedding of T in a metric space (X, d) restricts to a greedy embedding of T_s. This result is established in the problem section. □

As mentioned earlier, a spanning tree of the network graph is determined in this scheme. Let this spanning tree's maximum nodal degree be δ_{max}. *Assume* that it is possible to have a greedy embedding of an infinite complete δ_{max}-regular tree into the hyperbolic plane for each $\delta_{max} \geq 3$. Further, every finite tree with maximum degree δ_{max}, is a subtree of an infinite complete δ_{max}-regular tree. Using the last set of observations, it can be inferred that every finite tree has a greedy embedding into the hyperbolic plane. This will complete the proof of the theorem, provided the assumption is justified.

Therefore, in order to establish the theorem, we have to demonstrate a greedy embedding of an infinite δ_{max}-regular tree into the hyperbolic plane for each $\delta_{max} \geq 3$.

8.5.2 Applicable Hyperbolic Geometry

Elements of hyperbolic geometry were described in the chapter on geometry. Two well-known models of hyperbolic geometry are the upper half-plane model and the Poincaré disc model. We summarize certain immediately useful facts about hyperbolic geometry.

Upper Half-Plane Model and Poincaré Disc Model of Hyperbolic Geometry

1. The first fundamental form:
   (a) In the upper half-plane model of hyperbolic geometry, the first fundamental form is:

$$ds^2 = \frac{dv^2 + dw^2}{w^2}, \quad (v, w) \in \mathbb{R}^2, w > 0$$

   The corresponding set of points is denoted by

$$\mathbb{H} = \left\{ (v, w) \in \mathbb{R}^2 \mid w > 0 \right\}$$

(b) In the Poincaré disc model of hyperbolic geometry, the first fundamental form is:

$$ds^2 = \frac{4\left(dv^2 + dw^2\right)}{\left(1 - v^2 - w^2\right)^2}, \quad (v, w) \in \mathbb{R}^2, \left(v^2 + w^2\right) < 1$$

The corresponding set of points is denoted by

$$\mathbb{D} = \left\{(v, w) \in \mathbb{R}^2 \mid \left(v^2 + w^2\right) < 1\right\}$$

2. The boundary points in both models are also called *ideal points*:
   (a) The boundary of the points in $\mathbb{H}$ is the set of points on the real line, and the point at infinity. It is denoted by $\partial\mathbb{H}$.
   (b) The boundary of the points in $\mathbb{D}$ is the set of points on the circumference of a circle of unit radius and center $(0, 0)$. It is denoted by $\partial\mathbb{D}$.
3. Geodesics in the two models.
   (a) A geodesic in the upper half-plane model is an arc of a semicircle with center on the real axis. It meets $\partial\mathbb{H}$ in two ideal points p, q. Note that a straight line is regarded as a degenerate case of a circle. In this case, a vertical line is an arc of a circle which intersects $\partial\mathbb{H}$ at two points. One point is on the real line, and the other point is at infinity.
   (b) The geodesics in the Poincaré disc model are disc diameters and arcs of Euclidean circles which intersect the boundary $\partial\mathbb{D}$ perpendicularly.
4. Let p, q, and r be ideal points. The geodesics joining p to q, q to r, and r to p enclose a region in $\mathbb{H}$ (or $\mathbb{D}$). This region is a curvilinear triangle with vertices p, q, and r. It is also called an *ideal triangle*.
5. The mapping

$$Q(z) = -\frac{iz + 1}{iz - 1}, \quad z \in \mathbb{C}\backslash\{-i\}$$

is an isometry from the upper half-plane model, to the Poincaré disc model.

Relationship to Moebius Transformations

In addition to the above basic descriptions of the hyperbolic geometries, we also consider certain mappings called Moebius transformations. These mappings are of the form

$$f(z) = \frac{az + b}{cz + d}, \quad \text{where} \quad z \in \mathbb{C}^*;\ a, b, c, d \in \mathbb{C};\ \text{and}\ (ad - bc) \neq 0$$

where $\mathbb{C}^* = \mathbb{C} \cup \{+\infty\}$. This transformation can be specified as a matrix of size 2.

$$\begin{bmatrix} a & b \\ c & d \end{bmatrix}, \quad a, b, c, d \in \mathbb{C}$$

In the following summary, $PSL(2, \mathbb{R})$ is the group of real matrices of size 2, with determinant equal to unity, modulo $\{\pm I\}$; where I is an identity matrix of size 2.

1. The set of all Moebius transformations of $\mathbb{H}$ is denoted by $\mathcal{M}(\mathbb{H})$. The set $\mathcal{M}(\mathbb{H})$ forms a group under composition of Moebius transformations.
   If the coefficients a, b, c, and d are normalized so that $(ad - bc) = 1$, then the group $\mathcal{M}(\mathbb{H})$ can be identified with $PSL(2, \mathbb{R})$, the projective special linear group.

2. The set of all Moebius transformations of $\mathbb{D}$ is denoted by $\mathcal{M}(\mathbb{D})$. The set $\mathcal{M}(\mathbb{D})$ forms a group under composition of the Moebius transformation.
   If the coefficients a and b are normalized so that $\left(|a|^2 - |b|^2\right) = 1$, then the group $\mathcal{M}(\mathbb{D})$ can be identified with $PSL(2, \mathbb{C})$, the projective special linear group.

Modular Group $PSL(2, \mathbb{Z})$

Recall that the modular group $PSL(2, \mathbb{Z})$ is a subgroup of $PSL(2, \mathbb{R})$. This group consists of all transformations specified by matrices with integer values. We state some results related to this group. Some of these results are established in the problem section.

1. The special linear group $SL(2, \mathbb{Z})$ is generated by the matrices

$$A = \begin{bmatrix} 0 & -1 \\ 1 & 0 \end{bmatrix}$$

$$W = \begin{bmatrix} 1 & 1 \\ 0 & 1 \end{bmatrix}$$

2. The special linear group $SL(2, \mathbb{Z})$ is also generated by the matrices

$$A = \begin{bmatrix} 0 & -1 \\ 1 & 0 \end{bmatrix}$$

$$B = \begin{bmatrix} 0 & 1 \\ -1 & 1 \end{bmatrix}$$

   This result follows from the last observation by noting that $B = A^{-1}W^{-1}$.
3. Let I be an identity matrix of size 2. The modular group $PSL(2, \mathbb{Z}) = SL(2, \mathbb{Z}) / \{\pm I\}$ is generated by the matrices A and B. Therefore, $A^2 = B^3 = -I$. That is, the matrices A and B are elements in $PSL(2, \mathbb{Z})$ of order 2 and 3 respectively. Note that I is identified with $-I$ in this group.
4. The map $Q(\cdot)$ is an isometry, where $Q(z) = -(iz + 1) / (iz - 1)$, for $z \in \mathbb{C} \backslash \{-i\}$. It maps points in the half-plane to points in the Poincaré disc. Therefore, if F is any isometry of $\mathbb{H}$, then isometries of the Poincaré disc model are $Q \circ F \circ Q^{-1}(\cdot)$.
   From this fact, it follows that the matrix transformations A and B are represented in the Poincaré disc model by the mappings

$$A : z \mapsto -z$$

$$B : z \mapsto \frac{(1 + 2i) z + 1}{z + (1 - 2i)}$$

5. In the Poincaré disc model, consider an ideal triangle $\varDelta$ with vertices at $-1, 1, i$. Under the mapping A, these vertices are transformed to $1, -1, -i$ respectively. Similarly, under the mapping B, these vertices are transformed to $1, i, -1$ respectively. Thus, in the mapping B the triangle $\triangle$ is preserved, but the vertices are cyclically permuted.
   Therefore, as g ranges over all elements of $PSL(2, \mathbb{Z})$, the ideal triangles $g(\varDelta)$ tessellate the Poincaré disc.

The dual of this tessellation is an infinite complete 3-regular tree T. Let $u = (2 - \sqrt{3})\, i$, and $v = -u$. Also let, $e = (u, v)$. The vertex set of the tree T is $\{g(u) \mid g \in PSL(2, \mathbb{Z})\}$, and the edge set of T is $\{g(e) \mid g \in PSL(2, \mathbb{Z})\}$. The group $PSL(2, \mathbb{Z})$ acts simultaneously vertex-transitively and edge-transitively on the tree T. Actually, $PSL(2, \mathbb{Z})$ is isomorphic to the group of automorphisms of the tree T which maintain the cyclic ordering of edges around each vertex.

8.5.3 Greedy Embedding in Hyperbolic Space

The infinite tessellation of the Poincaré disc, and the dual of this via an infinite complete 3-regular tree T was demonstrated in the last subsection. Both these tessellations are invariant under the action of the discrete isometry group $PSL(2, \mathbb{Z})$. The following set of observations are used in the computation of virtual coordinates of the greedy embedding.

Observations 8.3. Some basic observations.

1. There exists an embedding of an infinite complete 3-regular tree T in the Poincaré disc $\mathbb{D}$ which is greedy.
2. This observation generalizes the above result to infinite complete δ-regular trees, where $\delta > 3$. There exists an embedding of an infinite complete δ-regular tree T in the Poincaré disc $\mathbb{D}$ which is greedy, where $\delta > 3$. □

The above observations are established in the problem section.

Computation of Virtual Coordinates of the Greedy Embedding

We describe a distributed algorithm to determine the virtual coordinates of the greedy embedding of a connected undirected graph in the Poincaré disc $\mathbb{D}$. A procedure is outlined to determine virtual coordinates of each vertex in the connected and undirected graph G. The virtual coordinate of vertex $s \in V$ is $f(s)$ in the Poincaré disc. In this scheme, a spanning tree T of the graph G is initially determined. Let the maximum degree of a node in the tree be $\delta_{max} \triangleq \delta$. Assume that $\delta \geq 3$.

Consider the interior P of an ideal polygon. Its vertices are the δ-th roots of unity. Observe that P is preserved by the mapping $\rho(z) = e^{2\pi i/\delta} z$. It generates a cyclic subgroup of order δ in the hyperbolic isometry group $PSL(2, \mathbb{R})$. Also note that $PSL(2, \mathbb{R})$ acts transitively on clock-wise ordered triples of ideal points. Therefore, it is possible to pick any side of the ideal polygon ∂P and find a hyperbolic isometry σ which maps its end points to 1 and -1, and maps the mid-point of the arc (side of the ideal polygon ∂P) connecting these endpoints to $-i$. The following two hyperbolic isometries are used in this scheme.

$$A : z \mapsto -z,$$
$$B : z \mapsto \sigma\left(\rho\left(\sigma^{-1}(z)\right)\right)$$

1. Initially a spanning tree T of the undirected graph G is determined. The maximum degree of a node in this tree graph is determined. Let it be $\delta_{max} \triangleq \delta$. It is assumed that $\delta \geq 3$. The tree T is a greedily embedded in the complete infinite δ-regular tree in the Poincaré disc.

2. A root node (distinguished node) r is selected arbitrarily in the tree T. This tree is converted into an arborescence, as follows. Each node $w \neq r$ in the tree has a parent $p(w)$ node (predecessor node), and the arcs $(p(w), w)$ form an arborescence rooted at node r.
   Let the degree of the node $w \neq r$ be d_w. As the graph is an arborescence, its in-degree $d_w^- = 1$, and out-degree $d_w^+ = (d_w - 1)$. The parent of the node w is numbered 0, and the successor nodes of w are numbered (indexed) $1, 2, \ldots, (d_w - 1)$ in some arbitrary order.
3. A Moebius transformation μ_w for each vertex $w \in V$ is determined.
   (a) The Moebius transformation at root node r is $\mu_r = A$. Let $u = \sigma(0)$, $v = -u$, and the virtual coordinate of node r is $f(r) = \mu_r^{-1}(v)$.
   (b) Each node w computes the Moebius transformation required for each of its child node w'. This Moebius transformation is equal to $\mu_{w'} = B^i \circ A \circ \mu_w$, where i is the index of the child node. Note that, $i \in [1, (d_w - 1)]$. The node w transmits to its child node w' the coefficients of the Moebius transformation $\mu_{w'}$.
   (c) Consider the node $w \neq r$: The Moebius transformation μ_w maps $f(p(w))$ to u and $f(w)$ to v. Thus the virtual coordinate of node w is $f(w) = \mu_w^{-1}(v)$.

In order to prove the correctness of the above algorithm, we need to establish that for each arc $(p(w), w)$ in the tree T, the function $f(\cdot)$ maps $p(w)$ and w to adjacent nodes in the complete infinite δ-regular tree T in the Poincaré disc. Observe that, as μ_w is an automorphism of T, it is sufficient to prove that $\mu_w(f(p(w)))$ and $\mu_w(f(w))$ are adjacent nodes in the tree T. Note that $\mu_w(f(w)) = \mu_w(\mu_w^{-1}(v)) = v$ as per definition. Therefore, it is sufficient to prove that $\mu_w(f(p(w))) = u$. Observe that $A \circ \mu_{p(w)}$ maps $f(p(w))$ to $A(v) = -v = u$. Furthermore, u is a fixed point of B, therefore $\mu_w = B^i \circ A \circ \mu_w$ maps $f(p(w))$ to u. This completes the proof of correctness of the distributed algorithm.

For this algorithm to function properly, the topology of the network has to be stationary in time.

8.6 Minimum Spanning Tree Algorithms

Let $G = (V, E)$ be an undirected, simple, and connected graph, where V is the set of vertices, and E is the set of edges. Let the cardinality of these sets be $|V| = n$ and $|E| = m$. Also each edge of the graph G is associated with a cost (or length). Let the cost of the edge $(v_i, v_j) \in E$ be $c_{ij} \in \mathbb{R}_0^+$. The costs are specified in a cost matrix C, where $c_{ij} \in \mathbb{R}_0^+$ for all edges $(v_i, v_j) \in E$, $c_{ij} \to \infty$ for all $(v_i, v_j) \notin E$ and $i \neq j$, and $c_{ii} = 0$, for $1 \leq i, j \leq n$.

A connected subgraph of a graph G is a spanning tree of the graph, if it contains (spans) all the vertices. The number of edges in this tree-subgraph is equal to $(n - 1)$. Denote this subgraph by G_T. Furthermore, each pair of vertices in a tree have a unique path connecting them. A spanning tree of a graph is not unique. That is, there can be several spanning trees of a graph.

A spanning tree of graph G is a *minimum spanning tree* (MST) of the graph, if the sum of the cost of its edges is minimum. Let the edge set of this tree be E_T, then $G_T = (V, E_T)$. Denote the cost of the MST by $\mathcal{C}(E_T)$. Note also that the MST of a graph is not unique. That is, different spanning trees of a graph can have identical cost which is minimum.

The goal of a MST algorithm is to minimize the cost of the spanning tree. Therefore, the cost of the tree is called the objective function. Algorithms to generate MST's are *greedy*. A greedy algorithm tries to improve upon its objective function in successive iterations. Each iteration makes a choice which *locally* optimizes the objective function. Furthermore, this is done without any concern for *global* optimality. In the case of generating MST, this local optimization turns out to be serendipitous. That is, the local greedy approach also turns out to be the best globally.

Two algorithms to generate MST of a graph are discussed below. These are the Jarník's and Kruskal's algorithms, discovered in 1930 and 1956 respectively. These algorithms are so named after the mathematicians V. Jarník and J. B. Kruskal. An algorithm similar to Jarník's algorithm was discovered independently by R. C. Prim in 1957. These two algorithms depend upon the following result.

Theorem 8.3. *Let $G = (V, E)$ be an undirected, simple, and connected graph. The cost of an edge $(v_i, v_j) \in E$ is $c_{ij} \in \mathbb{R}_0^+$. During the generation of the MST, a forest of trees $G_1, G_2, \ldots, G_r$ has been generated, where $G_j = (V_j, E_j), V_j \subset V$, and $E_j \subset E$, for $1 \leq j \leq r$. Also the set of edges $\widetilde{E} = \bigcup_{j=1}^r E_j$ belong to the MST of G. Consider a least cost edge $(v_a, v_b) \in E \backslash \widetilde{E}$ such that $v_a \in V_i$ and $v_b \notin V_i$. Then the edge (v_a, v_b) is an edge which belongs to the MST of the graph G, and can be added to the edge set $\widetilde{E}$, thereby creating a new forest of the graph G.*

Proof.

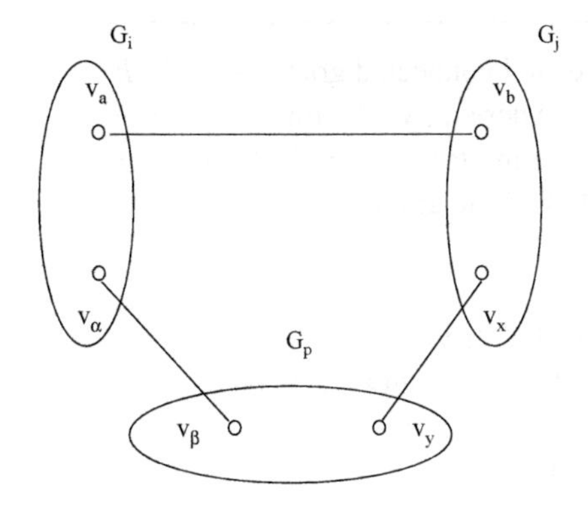

Figure 8.4. Pictorial demonstration of the proof of MST generation.

The proof is by contradiction. As per the hypothesis, the least cost edge in $E \backslash \widetilde{E}$ is (v_a, v_b), where $v_a \in V_i$ and $v_b \in V_j$ such that $i \neq j$. Assume that this edge does not belong to the MST of the graph G. Instead, let the MST include the edge $(v_\alpha, v_\beta) \in E \backslash \widetilde{E}$ such that $v_\alpha \in V_i$ and $v_\beta \in V_p$ where $p \neq i$, however p can be equal to j. In order to generate a MST, let the edge (v_x, v_y) connect the graphs G_j and G_p, where $v_x \in G_j$ and $v_y \in G_p$. See Figure 8.4.

Let the cost of the MST that includes the edge (v_α, v_β) but not the edge (v_a, v_b) be C_0. Addition of the edge (v_a, v_b) to the tree introduces a cycle, that can be broken by removing the edge (v_α, v_β). Observe that the connectivity of the MST is not affected by this operation, however the cost of the new spanning tree is $(C_0 + c_{ab} - c_{\alpha\beta})$. According to the hypothesis, $c_{ab} < c_{\alpha\beta}$, therefore $(C_0 + c_{ab} - c_{\alpha\beta}) < C_0$. This contradicts the assumption, that C_0 is the cost of the initial spanning tree. Note that if $c_{ab} = c_{\alpha\beta}$, then the MST is not unique. This completes the proof. □

The following corollary is immediate.

Corollary 8.1. *The least cost edge incident to any vertex of the graph G belongs to the MST.* □

Two algorithms to generate MSTs are next outlined. These differ only in the way they select the subtrees $G_j = (V_j, E_j)$ to which an edge $e \in E$ is added.

8.6.1 Jarník's Algorithm

Jarník's MST algorithm is next outlined. Observe that in each iteration of this algorithm, a single tree is generated. Each iteration requires $O(n)$ operations, and the number of iterations $(n-1)$ is of order $O(n)$. Thus, the computational complexity of this algorithm is $O(n^2)$. This algorithm is validated by the use of the above theorem. The algorithm initially starts with a tree $G_1 = (V_1, E_1)$, where $V_1 = \{v_1\}$ and $E_1 = \varnothing$. The algorithm is justified by $(n-1)$ successive applications of the above theorem.

Algorithm 8.7. *Jarník's MST Algorithm.*

Input: Undirected, simple, and connected graph $G = (V, E)$.
The cost matrix $C = [c_{ij}]$, where $c_{ij} \in \mathbb{R}_0^+$ for $1 \le i, j \le n$.
Output: The minimum spanning tree $G_T = (V, E_T)$.
The cost $\mathcal{C}(E_T)$ of the MST of the graph G.
begin
  (*initialization*)
  Select any arbitrary vertex $v_1 \in V$.
  Let $Q \leftarrow \{v_1\}, R \leftarrow (V - Q)$, and $E_T \leftarrow \varnothing$.
  $\mathcal{C}(E_T) \leftarrow 0$
  for $i = 1$ **to** $(n-1)$ **do**
  begin
    select an edge (v_j, v_k) such that $c_{jk} = \min\{c_{ab} \mid v_a \in Q, v_b \in R\}$
    (*if there is a tie in minimization pick arbitrarily*)
    $Q \leftarrow Q \cup \{v_k\}$
    $R \leftarrow (V - Q)$
    $E_T \leftarrow E_T \cup (v_j, v_k)$
    $\mathcal{C}(E_T) \leftarrow \mathcal{C}(E_T) + c_{jk}$
  end (*end of i for-loop*)
end (*end of Jarník's algorithm*)

Example 8.5. The MST of the undirected graph $G = (V, E)$ is determined. See Figure 8.5.

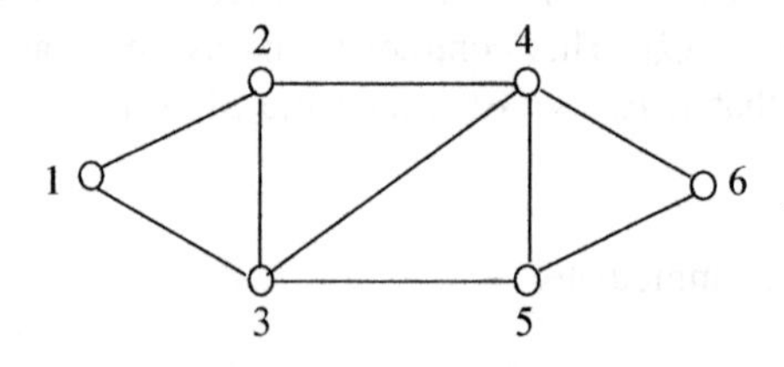

Figure 8.5. The undirected graph $G = (V, E)$ for MST generation.

The graph is specified as follows: $V = \{1, 2, 3, 4, 5, 6\}$, and

$$E = \{(1,2), (1,3), (2,3), (2,4), (3,4), (3,5), (4,5), (4,6), (5,6)\}$$

The lengths of the arcs in the set E are $c_{12} = 3$, $c_{13} = 5$, $c_{23} = 1$, $c_{24} = 5$, $c_{34} = 6$, $c_{35} = 2$, $c_{45} = 3$, $c_{46} = 4$, and $c_{56} = 1$. Since $|V| = 6$, the number of edges in the MST is equal to 5.

Initialization: $Q \leftarrow \{1\}, R \leftarrow \{2, 3, 4, 5, 6\}$, and $E_T \leftarrow \varnothing$. Also let $\mathcal{C}(E_T) \leftarrow 0$.

Iteration 1: Select least cost edge $= (1, 2), c_{12} = 3$. Then $Q = \{1, 2\}$, $R = \{3, 4, 5, 6\}$, and $E_T = \{(1, 2)\}$. Also $\mathcal{C}(E_T) \leftarrow \mathcal{C}(E_T) + c_{12}$. That is, $\mathcal{C}(E_T) = 3$.

Iteration 2: Select least cost edge $= (2, 3), c_{23} = 1$. Then $Q = \{1, 2, 3\}$, $R = \{4, 5, 6\}$, and $E_T = \{(1, 2), (2, 3)\}$. Also $\mathcal{C}(E_T) \leftarrow \mathcal{C}(E_T) + c_{23}$. That is, $\mathcal{C}(E_T) = 4$.

Iteration 3: Select least cost edge $= (3, 5), c_{35} = 2$. Then $Q = \{1, 2, 3, 5\}$, $R = \{4, 6\}$, and $E_T = \{(1, 2), (2, 3), (3, 5)\}$. Also $\mathcal{C}(E_T) \leftarrow \mathcal{C}(E_T) + c_{35}$. That is, $\mathcal{C}(E_T) = 6$.

Iteration 4: Select least cost edge $= (5, 6), c_{56} = 1$. Then $Q = \{1, 2, 3, 5, 6\}$, $R = \{4\}$, and $E_T = \{(1, 2), (2, 3), (3, 5), (5, 6)\}$. Also $\mathcal{C}(E_T) \leftarrow \mathcal{C}(E_T) + c_{56}$. That is, $\mathcal{C}(E_T) = 7$.

Iteration 5: Select least cost edge $= (4, 5), c_{45} = 3$. Then $Q = \{1, 2, 3, 4, 5, 6\}$, $R = \varnothing$, and $E_T = \{(1, 2), (2, 3), (3, 5), (5, 6), (4, 5)\}$. Also $\mathcal{C}(E_T) \leftarrow \mathcal{C}(E_T) + c_{45}$. That is, $\mathcal{C}(E_T) = 10$.

End of Jarník's MST algorithm.

The edge set of the MST is

$$E_T = \{(1,2), (2,3), (3,5), (5,6), (4,5)\}$$

and its cost is equal to 10. Observe that $|E_T| = 5$. □

8.6.2 Kruskal's Algorithm

Kruskal's algorithm generates a MST of a graph by initially sorting edges in increasing order of costs. The algorithm starts with a forest of n subtrees. Each subtree has a single node and no edges. In the next steps, edges are added to the forest of trees so that they do not create a cycle. Note that an edge becomes a candidate for MST if it is of the least cost among the edges not already a part of the MST. The algorithm is terminated once a tree with $(n - 1)$ edges is created. Recall that a graph with n vertices has a spanning tree with $(n - 1)$ edges.

This algorithm was independently discovered by J. B. Kruskal (1956), and by H. Loberman and A. Weinberger (1957). Initialization phase of the algorithm requires a sort of the m edges by their cost. A simple sort of the m edges has a computational complexity of $O(m^2)$. More refined algorithms have a computational complexity of $O(m \log m)$ operations. The formal algorithm for generating the MST via Kruskal's algorithm is given below.

Algorithm 8.8. *Kruskal's MST Algorithm.*

Input: Undirected, simple, and connected graph $G = (V, E)$.
The cost matrix $C = [c_{ij}]$, where $c_{ij} \in \mathbb{R}_0^+$ for $1 \leq i, j \leq n$.
Output: The minimum spanning tree $G_T = (V, E_T)$.
The cost $\mathcal{C}(E_T)$ of the MST of the graph G.
begin
(*initialization*)

Sort the m edges in increasing order of cost.
Edges with identical value are ranked arbitrarily.
Let $E_T \leftarrow \varnothing$, and $\mathcal{C}(E_T) \leftarrow 0$
while $|E_T| \leq (n-1)$ **do**
begin
Examine the first unexamined edge in the ascending-ordered list of edges. If this edge e does not form a cycle with edges in E_T, then $E_T \leftarrow E_T \cup e$ and $\mathcal{C}(E_T) \leftarrow \mathcal{C}(E_T) + c_e$, where c_e is the cost of the edge e.
end (*end of while-loop*)
end (*end of Kruskal's algorithm*)

Observe that in the main part of the algorithm, the algorithm may generate a forest of trees. In order to determine whether adding an edge $(v_i, v_j) \in E$ introduces a cycle, we only need to test whether the vertices v_i and v_j belong to the same tree. The complexity of this cycle detection step is $O(n)$. Therefore, the complexity of the MST algorithm, after the initial sorting is $O(mn)$. As $m \geq (n-1)$ in a connected graph, the overall complexity of the Kruskal's algorithm is $O(m^2)$. Jarník's algorithm generally outperforms Kruskal's algorithm on dense graphs. Kruskal's MST algorithm is illustrated below via an example.

Example 8.6. The MST of the graph of the preceding example (see Figure 8.5) is determined.
Initialization: First sort the edge list in order of increasing costs. The list is

$$\{(2,3),(5,6),(3,5),(1,2),(4,5),(4,6),(1,3),(2,4),(3,4)\}$$

Let $E_T \leftarrow \varnothing$, and $\mathcal{C}(E_T) \leftarrow 0$.

Iteration 1: Examine the edge $(2,3)$. Evidently it does not form a circuit with edges in E_T. Let $E_T = \{(2,3)\}$, and $\mathcal{C}(E_T) \leftarrow \mathcal{C}(E_T) + c_{23}$. That is, $\mathcal{C}(E_T) = 1$. Also $|E_T| = 1$.

Iteration 2: Examine the edge $(5,6)$. Evidently it does not form a circuit with edges in E_T. Let $E_T = \{(2,3),(5,6)\}$, and $\mathcal{C}(E_T) \leftarrow \mathcal{C}(E_T) + c_{56}$. That is, $\mathcal{C}(E_T) = 2$. Also $|E_T| = 2$.

Iteration 3: Examine the edge $(3,5)$. Evidently it does not form a circuit with edges in E_T. Let $E_T = \{(2,3),(5,6),(3,5)\}$, and $\mathcal{C}(E_T) \leftarrow \mathcal{C}(E_T) + c_{35}$. That is, $\mathcal{C}(E_T) = 4$. Also $|E_T| = 3$.

Iteration 4: Examine the edge $(1,2)$. Evidently it does not form a circuit with edges in E_T. Let $E_T = \{(2,3),(5,6),(3,5),(1,2)\}$, and $\mathcal{C}(E_T) \leftarrow \mathcal{C}(E_T) + c_{12}$. That is, $\mathcal{C}(E_T) = 7$. Also $|E_T| = 4$.

Iteration 5: Examine the edge $(4,5)$. Evidently it does not form a circuit with edges in E_T. Let $E_T = \{(2,3),(5,6),(3,5),(1,2),(4,5)\}$, and $\mathcal{C}(E_T) \leftarrow \mathcal{C}(E_T) + c_{45}$. That is, $\mathcal{C}(E_T) = 10$. Also $|E_T| = 5$.

As $|E_T| = 5$, E_T is the edge set of the MST with a cost $\mathcal{C}(E_T) = 10$.

End of Kruskal's MST algorithm.

The edge set of the MST is

$$E_T = \{(1,2),(2,3),(3,5),(5,6),(4,5)\}$$

and its cost is equal to 10. As expected this result is identical to that obtained by using Jarník's algorithm. As minimum spanning trees are not unique, this coincidence is not true in general. □

8.7 Connectivity of a Graph

The purpose of a network is to provide connectivity. Therefore, it is often necessary to check the connectivity of a network. An undirected graph $G = (V, E)$ is connected if and only if there is a path between every pair of vertices, otherwise it is disconnected. The connectivity of a graph can be described by its *connectedness matrix*. If $|V| = n$, then this matrix is a square matrix of size n. The (i, j)th element of this matrix is equal to 1, if vertex v_i is connected to vertex v_j, irrespective of the number of arcs that are required in getting there. Let the number of disconnected components of a simple graph $G = (V, E)$ be equal to d. Then

$$G = \bigcup_{i=1}^{d} G_i, \text{ where } G_i = (V_i, E_i), \; 1 \leq i \leq d$$

$$V = \bigcup_{i=1}^{d} V_i, \text{ and } E = \bigcup_{i=1}^{d} E_i.$$

$$V_i \cap V_j = \varnothing, \; E_i \cap E_j = \varnothing, \text{ for all } i \neq j, \; 1 \leq i, j \leq d$$

Evidently, the graph is connected if d is equal to 1. The basic description of the algorithm to find the connectedness of a graph is as follows. The algorithm starts with the adjacency matrix A of the graph. The basis of the algorithm is the *fusion* of adjacent vertices. It initially starts with any vertex in the graph. The algorithm then fuses all the vertices which are adjacent to it into a single vertex. The fused vertex is again fused with its adjacent vertices. This process is repeated till it cannot fuse any more vertices. That is, a connected component of graph is fused into a single fused-vertex. At the end of this fusion operation, the algorithm checks to see if all the vertices of the graph G are fused. The algorithm terminates if all the vertices have been fused, otherwise it starts a new vertex-fusion process. If the number of fused vertices d is equal to 1 at the end of the algorithm, then the graph G is connected, otherwise it is disconnected.

The fusion of the vertex v_j to vertex v_i is achieved by logical addition of the jth row to the ith row, and the jth column to the ith column of the adjacency matrix A. Denote the logical addition (or) operator by $\vee$, where

$$0 \vee 0 = 0, 0 \vee 1 = 1, 1 \vee 0 = 1, \text{ and } 1 \vee 1 = 1$$

After this fusion operation the jth column and jth row are discarded from the matrix. Generally, the discarded rows and columns in the original matrix are marked as such. This tagging ensures that the corresponding elements of the adjacency matrix are not considered in the subsequent fusion operations.

Algorithm 8.9. *Connectedness Algorithm.*

Input: An undirected, and simple graph $G = (V, E)$. It is specified by its adjacency matrix A.
Output: The vertex sets $V_i, 1 \leq i \leq d$, of the disconnected components.
begin

(*initialize*)
$Y \leftarrow A$ (*initialize* Y *by the adjacency matrix* A. $Y = [y_{ij}]$)
$d \leftarrow 0$
$i \leftarrow 1$
label J
$d \leftarrow d+1$
$V_d \leftarrow \{v_i\}$
for $j = 1$ **to** n **do**
begin
  if $((y_{ij} = 1) \wedge (i < j))$ **then**
  begin
    Logically add row j to row i, and column j to column i of matrix Y. Discard row j and column j of matrix Y.
    The phrase $disc$ (short for discard) is entered in the elements of these columns and rows of the matrix Y.
    $V_d \leftarrow V_d \cup \{v_j\}$
  end (*end of if statement*)
end (*end of* j *for-loop*)
Find a row $k > i$ which has not been discarded.
if such a row has been found $i \leftarrow k$, **go to label** J
else stop
end (*end of connectedness algorithm*)

If $d = 1$ at the end of the algorithm, then the graph G is connected. However, if $d > 1$ at the end of the algorithm, then the graph is disconnected. The computational complexity of this connectedness algorithm is $O(n^2)$. The steps in this algorithm are clarified by the following example.

Example 8.7. Consider a graph $G = (V, E)$ specified by its adjacency matrix A. In this graph $V = \{1, 2, 3, 4, 5\}$. See Figure 8.6. We determine the connectivity of the graph G.

$$A = \begin{bmatrix} 0 & 1 & 0 & 1 & 0 \\ 1 & 0 & 0 & 1 & 0 \\ 0 & 0 & 0 & 0 & 1 \\ 1 & 1 & 0 & 0 & 0 \\ 0 & 0 & 1 & 0 & 0 \end{bmatrix}$$

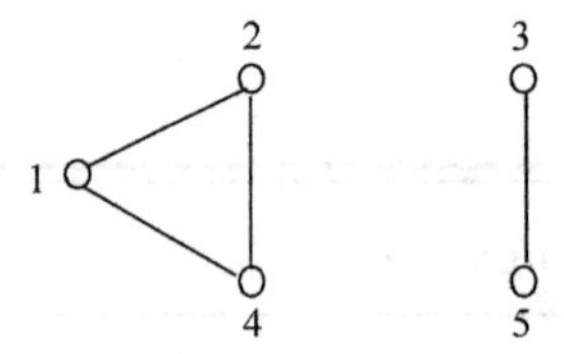

Figure 8.6. The undirected graph $G = (V, E)$ for testing connectivity.

Initialization: $Y \leftarrow A$, $d \leftarrow 0$, $i \leftarrow 1$.

Step 1: $d \leftarrow 1$, $V_1 \leftarrow \{1\}$. Note that $y_{12} = 1$. Therefore logically add row 2 to row 1, and column 2 to column 1 of the matrix Y. Discard row 2 and column 2 of matrix Y. Also set $V_1 = \{1, 2\}$. At the end of this step, the Y matrix is:

$$Y = \begin{bmatrix} 1 & disc & 0 & 1 & 0 \\ disc & disc & disc & disc & disc \\ 0 & disc & 0 & 0 & 1 \\ 1 & disc & 0 & 0 & 0 \\ 0 & disc & 1 & 0 & 0 \end{bmatrix}$$

Step 2: Observe that $y_{14} = 1$. Therefore logically add row 4 to row 1, and column 4 to column 1 of the matrix Y. Discard row 4 and column 4 of the matrix Y. Let $V_1 = \{1, 2, 4\}$. At the end of this step, the Y matrix is:

$$Y = \begin{bmatrix} 1 & disc & 0 & disc & 0 \\ disc & disc & disc & disc & disc \\ 0 & disc & 0 & disc & 1 \\ disc & disc & disc & disc & disc \\ 0 & disc & 1 & disc & 0 \end{bmatrix}$$

There are no more 1's in the first row of matrix Y. Therefore the first component of the graph G has been determined.

Step 3: The row 3 has not yet been discarded. Set $i \leftarrow 3$. Also $d \leftarrow 2$, $V_2 \leftarrow \{3\}$. Note that $y_{35} = 1$. Therefore logically add row 5 to row 3, and column 5 to column 3 of the matrix Y. Discard row 5 and column 5 of the matrix Y. Also let $V_2 = \{3, 5\}$. At the end of this step, the Y matrix is:

$$Y = \begin{bmatrix} 1 & disc & 0 & disc & disc \\ disc & disc & disc & disc & disc \\ 0 & disc & 1 & disc & disc \\ disc & disc & disc & disc & disc \\ disc & disc & disc & disc & disc \end{bmatrix}$$

The algorithm terminates, since all the rows have been examined.

End of connectedness algorithm.

At the end of this algorithm, the set of vertices of the connected components are $V_1 = \{1, 2, 4\}$ and $V_2 = \{3, 5\}$. Therefore the graph G is disconnected, and has two components. □

The connectivity of an undirected and simple graph G can also be determined by computing the eigenvalues of its corresponding Laplacian matrix. Since this Laplacian matrix is real and positive semidefinite, its eigenvalues are real and nonnegative numbers. Furthermore, the number of connected components of the graph G is equal to the multiplicity of the eigenvalue zero of its Laplacian matrix. The theoretical details of this approach are provided in the chapter on graph theory. A probabilistic approach to the study of network connectivity is via the evaluation of its reliability. This is studied next.

8.8 Network Reliability

The reliability of a system is defined as the probability that it functions at a random instant of time. Very large systems like telecommunication networks can be modeled as graphs. The vertices of this graph represent the centers where communication intelligence resides, and the edges represent the transmission links. A telecommunication network is actually composed of two parts in general: the backbone or mesh network, and the access networks. The access networks are geographically localized, while the mesh network provides interconnection between the different access networks. The local access networks are generally trees, while the backbone networks are meshes.

The nodes and links of a telecommunication network are failure-prone. We therefore determine the probability that the access network remains connected. In addition, the probability that a mesh network remains connected is also determined. The probability of network-connectedness is defined as its reliability in this chapter.

8.8.1 Access Network Reliability

For simplicity, assume that access networks are described by tree graphs. Also assume that these graphs are undirected, simple, and connected. Let this graph be $G = (V, E)$, where $|V| = n$, and $|E| = (n-1)$. Recall that the cardinality of an edge set of a spanning tree is $|E| = (|V| - 1)$. Let the operational probability of a link $e_i \in E$ be p_i for $1 \le i \le (n-1)$, and the operational probability of a vertex v_j be n_j for $1 \le j \le n$. Also assume that all the probabilistic events are stochastically independent of each other. If either a single link or a single node is not operational, the tree network gets disconnected. Denote the reliability of the tree network by $\mathcal{R}_{rel}(G)$. It can be computed as

$$\mathcal{R}_{rel}(G) = \prod_{j=1}^{n} n_j \prod_{i=1}^{(n-1)} p_i \tag{8.12}$$

The access network is generally modeled as a rooted tree. In a rooted tree, a distinguished node is termed the root of the tree. This node is generally a part of the backbone node. An algorithm to find the expected number of nodes which are able to communicate with the root of the tree is next developed. Denote this expectation by $\mathcal{N}_c(G)$.

Without loss of generality, assume that the root of the tree is the vertex v_1. Let the probability, with which a vertex v_j communicates with vertex v_1 be λ_j, for $1 \le j \le n$. Note that $\lambda_1 = n_1$, and $\mathcal{N}_c(G) = \sum_{j=1}^{n} \lambda_j$. The graph in this algorithm is specified by M_{v_i}, for $1 \le i \le n$, where M_{v_i} is the list-of-neighbors representation of the graph G, where the root node is v_1.

Algorithm 8.10. *Computation of the Expected Number of Nodes which are able to Communicate with the Root Node.*

Input: Undirected tree graph $G = (V, E)$, $|V| = n$, where $n \ge 2$.
The graph is specified by M_{v_i}, $1 \le i \le n$, and the root node is v_1.
The operating probability of the edges and nodes are equal to p_i,
for $1 \le i \le (n-1)$, and n_j, for $1 \le j \le n$ respectively.
Output: Expected number of nodes which can communicate with the root node: $\mathcal{N}_c(G)$.
begin

(*initialization*)
$\lambda_1 = n_1$
$B \leftarrow \{v_1\}$

while $B \neq \varnothing$ **do**
begin
$F \leftarrow \varnothing$
(*the set* F *is an intermediate variable*)
for all $v_k \in B$ **do**
begin
for all $v_l \in M_{v_k}$ **do**
begin
if $M_{v_l} \neq \varnothing$ **do**
begin
$M_{v_l} \leftarrow M_{v_l} - \{v_k\}$
(*remove the predecessor-node* v_k *of* v_l *from* M_{v_l})
$\lambda_l \leftarrow \lambda_k p_l n_l$
(p_l *is the operating probability of link* (v_k, v_l))
$F \leftarrow F \cup \{v_l\}$
end (*end of if statement*)
end (*end of for-loop*)
end (*end of for-loop*)
$B \leftarrow F$
end (*end of while-loop*)
$\mathcal{N}_c(G) = \sum_{j=1}^{n} \lambda_j$
end (*end of algorithm*)

The computational complexity of this algorithm depends upon the number of nodes in the tree. The proof of the correctness of this algorithm is left as an exercise to the reader. This algorithm is illustrated via the following example.

Example 8.8. A tree graph $G = (V, E)$ is specified by

$$V = \{1, 2, 3, 4, 5, 6, 7, 8, 9, 10, 11, 12\}$$

and the edge set E can be determined by the list-of-neighbors representation of the graph G. In this graph

$$M_1 = \{2, 3, 5\}, \ M_2 = \{1, 4, 6\}, \ M_3 = \{1, 7, 8\}, \ M_4 = \{2\},$$
$$M_5 = \{1, 9, 10, 11\}, \ M_6 = \{2\}, \ M_7 = \{3, 12\}, \ M_8 = \{3\},$$
$$M_9 = M_{10} = M_{11} = \{5\}, \text{ and } M_{12} = \{7\}$$

Let the link operating probabilities be $p_i = \alpha$, for $1 \leq i \leq 11$, and the node operating probabilities are $n_j = \beta$, for $1 \leq j \leq 12$. See Figure 8.7. We determine the expected number of nodes which are able to communicate with the root node. The algorithm proceeds by examining nodes at each level of the tree. The results at the end of each iteration are stated.

Initialization: $\lambda_1 = \beta$, and $B = \{1\}$.
Iteration 1: $\lambda_2 = \lambda_3 = \lambda_5 = \alpha\beta^2$, and $B = \{2, 3, 5\}$.
Iteration 2: $\lambda_4 = \lambda_6 = \lambda_7 = \lambda_8 = \lambda_9 = \lambda_{10} = \lambda_{11} = \alpha^2\beta^3$,

and $B = \{4, 6, 7, 8, 9, 10, 11\}$.
Iteration 3: $\lambda_{12} = \alpha^3\beta^4$, and $B = \{12\}$.
Iteration 4: $B = \varnothing$.
Finally, $\mathcal{N}_c(G) = \sum_{j=1}^{12} \lambda_j = \beta\left(1 + 3\alpha\beta + 7\alpha^2\beta^2 + \alpha^3\beta^3\right)$.

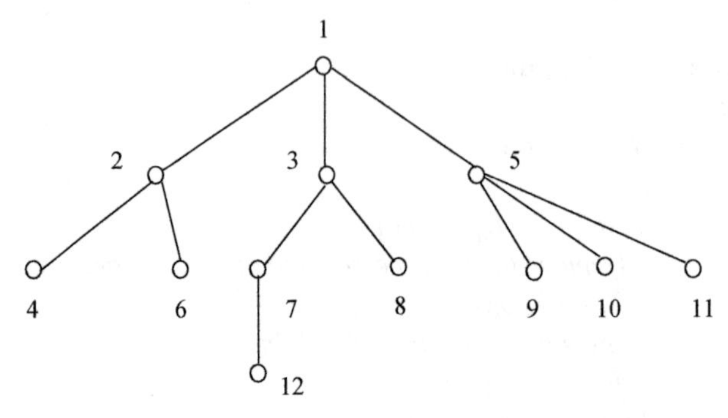

Figure 8.7. A tree graph.

□

8.8.2 Mesh Network Reliability

Telecommunication mesh network is modeled as an undirected, simple, and connected graph $G = (V, E)$. The connectedness in this context implies internode reachability. It is assumed in this model that a link $e_i \in E$ works fault-free with a probability p_i, and that all the nodes are working perfectly. Further assume that the operating probabilities of different links are independent of each other. Given this model, the probability that the entire network is connected is evaluated. In contrast to the graph-theoretic connectedness discussed in a prior section, this connectedness has a probabilistic interpretation.

This probability of connectedness depends upon the different spanning trees of the graph G. If at least a single spanning tree of the network is functioning, then the network is connected. The probability of at least a single functioning spanning tree is called the network reliability. *Spanning tree redundancy* and the *network reliability* are next defined more precisely.

Definition 8.8. *In the undirected, simple, and connected graph* $G = (V, E)$, $|V| = n$ *and* $|E| = m$. *Each link (edge) of the network is in two states. It is either operating correctly or it is in failed mode. The reliability of a link is the probability* p_i *that the edge* $e_i \in E$ *is functioning properly, and its unreliability is* $q_i = (1 - p_i)$, *where* $1 \leq i \leq m$. □

Let the number of the spanning trees of the graph G be $\eta(G)$. Denote the spanning trees of this network by $\widetilde{\tau}_1, \widetilde{\tau}_2, \ldots, \widetilde{\tau}_{\eta(G)}$. Also define an event T_i associated with the tree $\widetilde{\tau}_i$, for $1 \leq i \leq \eta(G)$. The event T_i occurs if the tree is operating correctly. The number of edges in a spanning tree of graph G is equal to $(n-1)$. Let the edges of the tree $\widetilde{\tau}_i$ be $\left\{e_{i_1}, e_{i_2}, \ldots, e_{i_{(n-1)}}\right\}$. Denote this edge set of the tree $\widetilde{\tau}_i$ by E_i, where $1 \leq i \leq \eta(G)$. The probability of event T_i is denoted by $P(T_i)$, where

$$P(T_i) = \prod_{k=1}^{(n-1)} p_{i_k}, \quad 1 \leq i \leq \eta(G) \tag{8.13}$$

Thus $P(T_i)$ is the probability that the spanning tree T_i is operating correctly.

Definitions 8.9. *Spanning tree redundancy, and network reliability are defined below.*

1. *Spanning tree redundancy* $\mathcal{R}_{red}(G)$ *of the network is defined as the expected number of spanning trees which are operating correctly. That is*

$$\mathcal{R}_{red}(G) = \sum_{i=1}^{\eta(G)} P(T_i) \tag{8.14}$$

2. *Network reliability* $\mathcal{R}_{rel}(G)$ *is the probability that all the nodes of the network are able to communicate with each other.* □

The redundancy and reliability of some very simple networks are computed in the following examples.

Examples 8.9. It is assumed in these examples, that the nodes operate without failure.

1. Compute the redundancy and reliability of a tree graph G with n nodes. Assume that the probability of reliability of every link is p.
   A tree becomes disconnected if any one of the $(n-1)$ edge fails. Therefore $\mathcal{R}_{red}(G) = \mathcal{R}_{rel}(G) = p^{n-1}$.
2. Compute the reliability of a ring network R_n, with n nodes. In a ring network, the degree of each node is 2. The probability that a link does not fail is given by p. Define $q = (1-p)$.
   This ring network has n trees. It remains connected, if either all the links are operating, or at most a single link fails. Therefore

$$\mathcal{R}_{red}(G) = np^{n-1}$$
$$\mathcal{R}_{rel}(G) = \left(p^n + nqp^{n-1}\right)$$

□

Procedure to determine the reliability of a mesh communication network is next outlined. It is assumed that the network is modeled by an undirected graph. A technique to determine the spanning trees of an undirected graph has been outlined in the chapter on graph theory.

Reliability Computation

There are several general techniques to compute network reliability. Some of these are: state-space enumeration technique, use of inclusion-exclusion principle, method of disjoint-products, matroid-theoretic technique, and the method of factoring. State-space enumeration is a basic technique for computing reliability. The second technique uses the inclusion-exclusion principle of probability theory. However its computational complexity is exponential. The disjoint-products technique requires addition of only $\eta(G)$ probabilities. The matroid-theory based algorithm to compute network reliability is both efficient and elegant. The method of factoring is used when the network has a special structure. It uses the state of an individual link. This later method is not discussed. The methods based upon inclusion-exclusion principle, disjoint-products, and matroid theory are next outlined.

Inclusion-Exclusion Principle

As per the method of inclusion-exclusion principle, the reliability of the network $\mathcal{R}_{rel}(G)$ with $\eta(G)$ number of spanning trees is given by

$$\begin{aligned}\mathcal{R}_{rel}(G) &= P\left(T_1 \cup T_2 \cup \ldots \cup T_{\eta(G)}\right) \\ &= \sum_{1\le i\le \eta(G)} P(T_i) - \sum_{1\le i<j\le \eta(G)} P(T_i \cap T_j) \\ &\quad + \sum_{1\le i<j<k\le \eta(G)} P(T_i \cap T_j \cap T_k) \\ &\quad - \sum_{1\le i<j<k<l\le \eta(G)} P(T_i \cap T_j \cap T_k \cap T_l) + \ldots \\ &\quad + (-1)^{\eta(G)+1} P\left(T_1 \cap T_2 \cap \ldots \cap T_{\eta(G)}\right) \end{aligned} \tag{8.15}$$

This technique requires summation of $\left(2^{\eta(G)} - 1\right)$ terms. Consequently even for moderate values of $\eta(G)$, this expression for number of term becomes very large quickly.

Disjoint-Products Principle

The network reliability $\mathcal{R}_{rel}(G)$ using disjoint-products principle is

$$\begin{aligned}\mathcal{R}_{rel}(G) &= P\left(T_1 \cup T_2 \cup \ldots \cup T_{\eta(G)}\right) \\ &= P(T_1) + P(T_1^c \cap T_2) + P(T_1^c \cap T_2^c \cap T_3) + \ldots \\ &\quad + P\left(T_1^c \cap T_2^c \cap \ldots \cap T_{\eta(G)-1}^c \cap T_{\eta(G)}\right) \end{aligned} \tag{8.16}$$

This equation has only $\eta(G)$ terms, however it requires logical (Boolean) simplification of the terms of type $T_1^c \cap T_2^c \cap \ldots \cap T_{k-1}^c \cap T_k$.

Matroid-Theoretic Algorithm to Compute Network Reliability

An algorithm to compute network reliability by using the disjoint-products principle and matroid theory is described. Let the edge set of the graph be $E = \{e_1, e_2, \ldots, e_m\}$, where $|E| = m$. The algorithm is implemented by initially listing the spanning trees in a lexicographic order. This is the order in which a list of strings appears in a dictionary-order. Consider an example, in which the edge set of a spanning tree $\widetilde{T}_i$ is

$$E_i = \left\{e_{i_1}, e_{i_2}, \ldots, e_{i_{(n-1)}}\right\}$$

These edges are listed lexicographically if $i_1 < i_2 < \ldots < i_{(n-1)}$.

It turns out that the disjoint-products technique has $\eta(G)$ number of terms. The reason that the lexicographic ordering works is because the spanning trees so ordered form a matroid. This algorithm uses the set difference operation. Let H_1 and H_2 be two sets, then recall that the set difference is the set

$$\begin{aligned}(H_1 - H_2) &= \{x \mid (x \in H_1) \wedge (x \notin H_2)\} \\ &= H_1 \cap H_2^c\end{aligned}$$

Furthermore, each term in the representation of $\mathcal{R}_{rel}(G)$ is a product of only p_i's and q_i's. The proof of the correctness of this algorithm can been developed by using concepts from matroid theory. The computational complexity of this algorithm depends upon the number of spanning trees.

Algorithm 8.11. *Matroid-Theoretic Network Reliability Computation Algorithm.*

Input: Undirected graph $G = (V, E)$, $E = \{e_1, e_2, \ldots, e_m\}$, and $|V| = n$.
Operational probability of edge e_i is p_i, where $1 \leq i \leq m$.
Output: Network reliability $\mathcal{R}_{rel}(G)$.
begin
  (*initialization*)
  Determine the $\eta(G)$ spanning trees. List these trees in lexicographic order. Let these trees be $\widetilde{T}_1, \widetilde{T}_2, \ldots, \widetilde{T}_{\eta(G)}$.
  The edge set of spanning tree $\widetilde{T}_i$ is
  $E_i = \left\{e_{i_1}, e_{i_2}, \ldots, e_{i_{(n-1)}}\right\}$, where $1 \leq i \leq \eta(G)$.
  for $i = 1$ **to** $\eta(G)$ **do**
  begin
    $B_i \leftarrow \varnothing$
    if $(i > 1)$ **then**
    begin
      for $j = 1$ **to** $(i - 1)$ **do**
      begin
        $\delta \leftarrow \min\{s \mid e_s \in (E_j - E_i)\}$
        $B_i \leftarrow B_i \cup \{e_\delta\}$
        (*note that* $(E_j - E_i)$ *is the set difference operation*)
      end (*end of j for-loop*)
    end (*end of if statement*)
    $r_i \leftarrow \prod_{e_k \in E_i} p_k \prod_{e_k \in B_i} q_k$
  end (*end of i for-loop*)
  $\mathcal{R}_{rel}(G) \leftarrow \sum_{i=1}^{\eta(G)} r_i$
end (*end of network reliability computation algorithm*)

The above algorithm is clarified in the following example.

Example 8.10. The spanning tree redundancy and reliability of a network specified by the graph $G = (V, E)$ is determined. In this graph,

$$V = \{1, 2, 3, 4\},$$

$$E = \{(1,2), (1,3), (1,4), (2,3)\},$$

$|V| = 4$, and $|E| = 4$. Its adjacency matrix is

$$A = \begin{bmatrix} 0 & 1 & 1 & 1 \\ 1 & 0 & 1 & 0 \\ 1 & 1 & 0 & 0 \\ 1 & 0 & 0 & 0 \end{bmatrix}$$

See Figure 8.8.

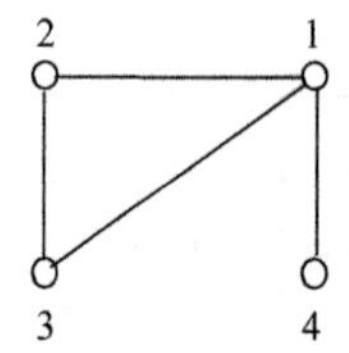

Figure 8.8. Reliability computation of graph $G = (V, E)$.

For ease in notation, define

$$e_1 = (1,2), e_2 = (2,3), e_3 = (1,3), \text{ and } e_4 = (1,4)$$

The probability of an operational link e_i is equal to $P(e_i) \triangleq p_i, 1 \le i \le 4$.

Using the matrix-tree theorem, the number of spanning trees of the graph can be determined to be equal to 3. The spanning tree polynomial of this graph is

$$\wp(G) = (e_1 e_2 e_4 + e_1 e_3 e_4 + e_2 e_3 e_4)$$

Let the edge set of tree $\widetilde{\mathcal{T}}_1$ be $E_1 = \{e_1, e_2, e_4\}$. Similarly the edge set of tree $\widetilde{\mathcal{T}}_2$ is $E_2 = \{e_1, e_3, e_4\}$, and that of tree $\widetilde{\mathcal{T}}_3$ is $E_3 = \{e_2, e_3, e_4\}$. The spanning tree redundancy is computed to be

$$\begin{aligned} \mathcal{R}_{red}(G) &= \sum_{i=1}^{3} P(T_i) \\ &= (p_1 p_2 p_4 + p_1 p_3 p_4 + p_2 p_3 p_4) \end{aligned}$$

The network reliability is next computed by the inclusion-exclusion principle, the disjoint-products method, and also by the matroid-theoretic algorithm.

Inclusion-Exclusion Principle: Using the inclusion-exclusion principle, the network reliability $\mathcal{R}_{rel}(G)$ is given by

$$\begin{aligned} \mathcal{R}_{rel}(G) &= P(T_1 \cup T_2 \cup T_3) \\ &= P(T_1) + P(T_2) + P(T_3) \\ &\quad - P(T_1 \cap T_2) - P(T_1 \cap T_3) - P(T_2 \cap T_3) \\ &\quad + P(T_1 \cap T_2 \cap T_3) \\ &= p_1 p_2 p_4 + p_1 p_3 p_4 + p_2 p_3 p_4 - 2 p_1 p_2 p_3 p_4 \end{aligned}$$

Observe that for the event $T_1 \cap T_2$ to occur both the trees $\widetilde{\mathcal{T}}_1$ and $\widetilde{\mathcal{T}}_2$ should be operational. That is, the edges in their corresponding edge sets, $E_1 = \{e_1, e_2, e_4\}$ and $E_2 = \{e_1, e_3, e_4\}$ should be operational (functioning) simultaneously. Therefore the event $T_1 \cap T_2$ occurs, if the edges in the set $E_1 \cup E_2 = \{e_1, e_2, e_3, e_4\}$ are functioning. The probability of this event to occur is therefore equal to $p_1 p_2 p_3 p_4$. The probability of events $T_1 \cap T_3$, $T_2 \cap T_3$, and $T_1 \cap T_2 \cap T_3$ is similarly computed.

Disjoint-Products Principle: Use of disjoint-products directly gives

$$\begin{aligned}\mathcal{R}_{rel}(G) &= P(T_1 \cup T_2 \cup T_3) \\ &= P(T_1) + P(T_1^c \cap T_2) + P(T_1^c \cap T_2^c \cap T_3)\end{aligned}$$

For ease in notation denote the event $\{e_i \text{ operates}\}$ by ς_i, and the event $\{e_i \text{ fails}\}$ by ς_i^c. Therefore the event $T_1^c \cap T_2$ is $(\varsigma_1 \cap \varsigma_2 \cap \varsigma_4)^c \cap (\varsigma_1 \cap \varsigma_3 \cap \varsigma_4) = (\varsigma_1^c \cup \varsigma_2^c \cup \varsigma_4^c) \cap (\varsigma_1 \cap \varsigma_3 \cap \varsigma_4) = (\varsigma_1 \cap \varsigma_2^c \cap \varsigma_3 \cap \varsigma_4)$. The probability of this event is given by $P(T_1^c \cap T_2) = p_1 q_2 p_3 p_4$. Similarly $P(T_1^c \cap T_2^c \cap T_3) = q_1 p_2 p_3 p_4$. Then

$$\begin{aligned}\mathcal{R}_{rel}(G) &= (p_1 p_2 p_4 + p_1 q_2 p_3 p_4 + q_1 p_2 p_3 p_4) \\ &= (p_1 p_2 p_4 + p_1 p_3 p_4 + p_2 p_3 p_4 - 2 p_1 p_2 p_3 p_4)\end{aligned}$$

Matroid-Theoretic Algorithm: Different steps in the network reliability computation are explicitly outlined.

Initialization: The edge sets of the three spanning trees in lexicographic order are

$$E_1 = \{e_1, e_2, e_4\}, \; E_2 = \{e_1, e_3, e_4\}, \text{ and } E_3 = \{e_2, e_3, e_4\}$$

Begin i for-loop
Iteration 1: At the end of the iteration $i = 1$, $B_1 = \varnothing$, and $r_1 = p_1 p_2 p_4$.
Iteration 2: At the end of the iteration $i = 2$, $B_2 = \{e_2\}$, and $r_2 = p_1 q_2 p_3 p_4$.
Iteration 3: At the end of the iteration $i = 3$, $B_3 = \{e_1\}$, and $r_3 = q_1 p_2 p_3 p_4$.
End of i loop.

$$\mathcal{R}_{rel}(G) = \sum_{i=1}^{3} r_i = (p_1 p_2 p_4 + p_1 q_2 p_3 p_4 + q_1 p_2 p_3 p_4)$$

End of network reliability computation algorithm.

If we assume $p_i = (1 - \epsilon)$, for $1 \le i \le 4$, where $0 < \epsilon \ll 1$, then it can be shown that $\mathcal{R}_{red}(G) \simeq 3(1 - 3\epsilon)$, and $\mathcal{R}_{rel}(G) \simeq (1 - \epsilon)$. □

Node Failures

It has been assumed in the reliability model of the network that the nodes are operating without any faults. It is not yet known how to extrapolate the above undirected graph model to include both node and arc failures simultaneously. However, observe that if any single node fails, then the network gets disconnected.

It appears that the complete reliability of the network can possibly be obtained by multiplying: the reliability of the network, assuming only arc-failures; and the product of the node operational probabilities. This rule incorrectly assumes that the operational probabilities of links and nodes are stochastically independent of each other.

8.9 Maximum Flows and Minimum Cost Flows

Each link (edge) in a telecommunication network is generally associated with both cost and speed (capacity). This gives rise to the following two types of problems.

(a) Assume that a network is specified by only link capacity metric. A network of this type is called a capacitated network. It is interesting to determine the *maximum traffic flow* that is possible in this network.
(b) A network is specified by both link cost and link capacity metrics, and certain traffic flow requirements between source and destination nodes. A least cost scheme for transfer of traffic has to be determined. This gives rise to *minimum cost flow* problem.

In the maximum traffic flow problem, a network with nonnegative link capacities is specified. The goal of the problem is to determine maximum flow that is possible between the source node and the sink (destination) node.

Minimum cost flow algorithms are also discussed in this section. In addition to the capacity constraints in a network, each link is assigned a cost. The purpose of this algorithm is to determine the least cost transportation of goods (network traffic) such that the demands at specified vertices are met by supplies available at the source vertices. This formulation of the problem is a generalization of the maximum flow and the shortest path problems.

Abstractions of such problems are studied in this section. More specifically, the problems associated with maximum flows, and minimum cost flows are examined. In summary, traffic *flows* in a network are studied.

It is implicitly assumed in this section that all the digraphs are simple and connected. The theoretical details developed in this section can also be used to justify the correctness of the edge-disjoint shortest paths algorithm. The edge-disjoint shortest paths algorithm was developed earlier in the chapter.

8.9.1 Maximum Flows

The maximum flow problem determines the maximum possible flow in a capacitated network. The network variables are defined below.

Definition 8.10. *Basic definitions.*

(a) *Let* $G = (V, E)$ *be a directed graph, where* V *is the vertex (node) set,* E *is the set of edges,* $|V| = n$, $|E| = m$, *and* $V = \{1, 2, \ldots, n\}$.
(b) *Associated with this graph is a mapping* $u : E \to \mathbb{R}_0^+$, *that is for each edge* $e \in E$ *there is a value* $u(e) \geq 0$ *called the capacity of the edge* e. *If an arc is denoted by* (i, j), *where* i *and* j *are the vertices, then its capacity is specified by* u_{ij}. *For simplicity assume that the capacities are finite. Notice that arcs with zero capacity are also allowed. The mapping* u *is generally specified by an* $n \times n$ *matrix denoted by* U, *where* $U = [u_{ij}]$.
(c) *The graph does not contain parallel arcs. That is, two or more arcs with the same head and tail nodes are not permissible. Also assume that the graph is connected.*
(d) *In this graph there are two special vertices called the source node* s *and the sink (target, destination) node* t, *such that* t *is accessible from* s.
(e) *The flow network or a capacitated network is specified by* $N = (G, u, s, t)$. □

The maximum flow scheme routes maximum flow of traffic from the source to the sink node. Some more notation and assumptions are listed below.

Definitions 8.11. *Let* $N = (G, u, s, t)$ *be a capacitated network.*

1. *A feasible flow on N is a mapping $x : E \to \mathbb{R}_0^+$ which satisfies the following conditions:*
   (a) *Capacity constraints.* $0 \leq x_{ij} \leq u_{ij}$ *for all arcs* $(i, j) \in E$.
   (b) *Flow conservation.* $\sum_{\{j|(i,j)\in E\}} x_{ij} = \sum_{\{j|(j,i)\in E\}} x_{ji}$ *for all* $i \in V - \{s, t\}$.
   In this statement, the left-hand side expression $\sum_{\{j|(i,j)\in E\}} x_{ij}$ *represents the total flow out of node* i, *and the right-hand side expression* $\sum_{\{j|(j,i)\in E\}} x_{ji}$ *represents the total flow into the node* i.

   The mapping x is generally specified by an $n \times n$ matrix denoted by x, such that $x = [x_{ij}]$.
2. *The value of flow x is $v = \sum_{\{j|(s,j)\in E\}} x_{sj}$. It is equal to the total flow leaving the source node.*
3. *A maximum flow on N is a flow for which v has a maximum value.* □

Observation 8.4. The capacity constraint implies that each edge can carry a nonnegative flow bounded from above by the capacity of the edge. The flow conservation constraint means that flows are preserved at all vertices except at the source and sink nodes. That is, the amount which flows into a node (other than the source and sink nodes) equals the amount that flows out of it. Thus for any $i \in V$

$$\sum_{j\in V} x_{ji} - \sum_{j\in V} x_{ij} = \begin{cases} -v, \; i = s \\ 0, \; i \neq s, t \\ v, \; i = t \end{cases}$$

□

The maximum flow problem in a capacitated network is next stated formally. *In a capacitated network $N = (G, u, s, t)$, maximize the value of v, which is the flow between the source node s and the sink node t, and determine the corresponding flow x.* In order to develop the theory further, the concept of cut of a set is introduced.

Definitions 8.12. *Let $N = (G, u, s, t)$ be a capacitated network, where $G = (V, E)$ is a directed graph.*

1. *The arc $(i, j) \in E$ is saturated in flow x if $x_{ij} = u_{ij}$.*
2. *A cut $\langle S, T \rangle$ is a partition of the vertex set V into two subsets S and T such that $T = (V - S)$. The arcs which are directed from S to T are called forward arcs, and the arcs which are directed from T to S are called backward arcs. The sets of forward and backward arcs are denoted by (S, T) and (T, S) respectively.*
3. *The cut $\langle S, T \rangle$ is called an s-t cut if $s \in S$ and $t \in T$. The capacity of the s-t cut $\langle S, T \rangle$ is denoted by $u \langle S, T \rangle = \sum_{(i,j)\in(S,T)} u_{ij}$.*
4. *The flow across an s-t cut is called an s-t flow.*
5. *A minimum cut is an s-t cut which has a minimum capacity.* □

Theorem 8.4. *The value of any s-t flow is less than or equal to the capacity of any s-t cut in the network.*

Proof. Let x be a flow in the network and $\langle S, T \rangle$ be an s-t cut. If $i \in S$, then

$$\left\{ \sum_{j\in V} x_{ij} - \sum_{j\in V} x_{ji} \right\}$$

equals v if $i = s$, and it is equal to zero if $i \neq s, t$. Therefore

$$v = \sum_{i \in S} \left\{ \sum_{j \in V} x_{ij} - \sum_{j \in V} x_{ji} \right\}$$
$$= \sum_{i \in S} \sum_{j \in S} (x_{ij} - x_{ji}) + \sum_{i \in S} \sum_{j \in T} (x_{ij} - x_{ji})$$

The first set of summations add to zero. Thus

$$v = \sum_{i \in S} \sum_{j \in T} (x_{ij} - x_{ji})$$

This equation implies that the value v of the flow is equal to the net flow through the s-t cut $\langle S, T \rangle$. However $x_{ji} \geq 0$ and $x_{ij} \leq u_{ij}$, consequently

$$v \leq \sum_{(i,j) \in (S,T)} u_{ij} = u \langle S, T \rangle \tag{8.17}$$

□

The result of the above theorem should be intuitively evident. Any s-t flow must pass through any s-t cut in the network. Thus the value of any s-t flow cannot exceed the capacity of an s-t cut with the smallest value. Alternately, let x be a feasible flow between a source node s and a target node t, with value v. If v happens to be equal to the capacity of some s-t cut $\langle S, T \rangle$ in the network, then x is a maximum flow and $\langle S, T \rangle$ is a minimum cut.

Corollary 8.2. *Let x be a feasible flow between a source node s and a target node t, with value v. Assume that $\langle S, T \rangle$ is any s-t cut in the network, and the net flow across this cut is equal to v. Then*

$$v = \sum_{(i,j) \in (S,T)} x_{ij} - \sum_{(i,j) \in (T,S)} x_{ij} \tag{8.18}$$

Proof. See the problem section. □

Example 8.11. Consider a graph $G = (V, E)$, where $V = \{1, 2, 3, 4, 5, 6\}$ and the edges are specified by the adjacency matrix A. The capacities of the arcs are specified by the capacity matrix U.

$$A = \begin{bmatrix} 0&1&0&1&0&0 \\ 0&0&1&0&1&0 \\ 0&0&0&0&0&1 \\ 0&1&0&0&1&0 \\ 0&0&1&0&0&1 \\ 0&0&0&0&0&0 \end{bmatrix}, \quad U = \begin{bmatrix} 0&5&0&7&0&0 \\ 0&0&6&0&3&0 \\ 0&0&0&0&0&8 \\ 0&4&0&0&5&0 \\ 0&0&5&0&0&6 \\ 0&0&0&0&0&0 \end{bmatrix}$$

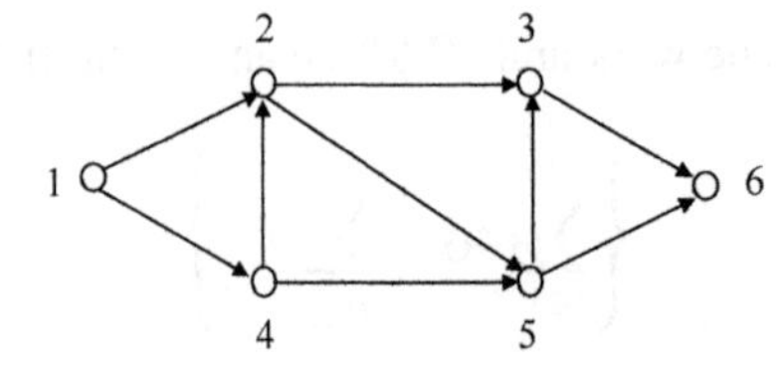

Figure 8.9. The graph $G = (V, E)$.

See Figure 8.9. Let the source node be $s = 1$, and the sink node be $t = 6$. A possible feasible flow x is specified by:

$$x_{12} = 3,\ x_{14} = 4,\ x_{23} = 4,\ x_{25} = 2,\ x_{36} = 5,$$
$$x_{42} = 3,\ x_{45} = 1,\ x_{53} = 1,\ \text{and } x_{56} = 2$$

The value of this flow is $v = 7$. Note that $v = x_{12} + x_{14} = x_{36} + x_{56} = 7$. Also, none of the arcs is saturated, because all arc flows are less than their respective capacities. Observe that flow conservation is maintained at all the nodes. The flow across the cut $\langle S, T \rangle$ where $S = \{1, 2\}$ and $T = \{3, 4, 5, 6\}$ is next evaluated. This flow is equal to $x_{14}+x_{23}+x_{25}-x_{42} = 4+4+2-3 = 7 = v$. The capacity of this cut is computed to be

$$u \langle S, T \rangle = u_{14} + u_{23} + u_{25} = 7 + 6 + 3 = 16$$

Therefore the value of the flow in the network is bounded from above by 16. The capacity of the cut $\langle S', T' \rangle$ where $S' = \{1\}$ and $T' = \{2, 3, 4, 5, 6\}$ is computed to be

$$u \langle S', T' \rangle = u_{12} + u_{14} = 5 + 7 = 12$$

This cut capacity provides an improved value for the upper bound on the flow. Consider another feasible flow x'.

$$x'_{12} = 5,\ x'_{14} = 7,\ x'_{23} = 6,\ x'_{25} = 2,\ x'_{36} = 6,$$
$$x'_{42} = 3,\ x'_{45} = 4,\ x'_{53} = 0,\ \text{and } x'_{56} = 6$$

The value of this flow is $v' = 12$, which is equal to the capacity $u \langle S', T' \rangle$. We subsequently show that if the value of a flow is equal to capacity of some cut, then that flow is maximal, and the corresponding cut is minimum. Therefore, $\langle S', T' \rangle$ is the minimum cut and the value of the maximal flow is equal to 12. □

It is next proved that the maximum value of a flow in a capacitated network equals the minimum capacity of all s-t cuts. An algorithm for determining maximum flows is also developed. These are based upon the concept of *augmenting path* and *residual network*.

The concept of a residual network of a graph G is based upon the following simple idea. If the edge $(i, j) \in E$ of capacity u_{ij} carries a flow of $x_{ij} \in \mathbb{R}^+$ units, then it can potentially carry another $(u_{ij} - x_{ij})$ units of flow on this edge. Also observe that the edge (i, j) can carry an additional x_{ji} units, which is the flow from node j to node i. Thus the residual network $G(x)$ with respect to the flow x is defined as follows. Replace *each* edge (i, j) in the graph G by *two* edges (i, j) and (j, i). Let the capacity of these two edges be $r_{ij} = (u_{ij} - x_{ij})$ and $r_{ji} = x_{ij}$ respectively.

However, if the initial network specified by the graph G has edges (i, j) and (j, i) then the residual graph will have two parallel edges from node j to node i. This restriction can be overcome by a suitable transformation of the original graph.

Definitions 8.13. *Assume that $N = (G, u, s, t)$ is a capacitated network.*

1. *For a given flow x, the residual capacity r_{ij} of an arc $(i, j) \in E$ is the maximum additional flow which can be sent from vertex i to vertex j using the arcs (i, j) and (j, i). The residual capacity $r_{ij} = (u_{ij} - x_{ij}) + x_{ji}$, where $x_{ij} \in \mathbb{R}^+$. Observe that $(u_{ij} - x_{ij})$ represents the unused capacity of the arc (i, j); and x_{ji} is the value of flow on arc (j, i) that can be cancelled to increase the flow from vertex i to vertex j. The capacities in the residual network are specified in an $n \times n$ matrix r.*

2. *The residual network $G(x)$ for a specified value of flow x are those arcs in G which have a positive residual capacity.*
3. *An augmenting path in $G(x)$ is a directed path from vertex s to vertex t in the residual network $G(x)$.*
4. *The capacity of a directed path is the smallest value of arc capacity which lies on the path.* □

Lemma 8.1. *A flow x in a network $N = (G, u, s, t)$ has a value v. The extra s-t flow that is possible in this network is less than or equal to the residual capacity of any s-t cut.*

Proof. Let x' be a flow of value $(v + \Delta v)$, where $\Delta v \geq 0$. Therefore

$$v + \Delta v \leq \sum_{(i,j)\in(S,T)} u_{ij}$$

Use of the result

$$v = \sum_{(i,j)\in(S,T)} x_{ij} - \sum_{(i,j)\in(T,S)} x_{ij}$$

yields

$$\begin{aligned}\Delta v &\leq \sum_{(i,j)\in(S,T)} (u_{ij} - x_{ij}) + \sum_{(i,j)\in(T,S)} x_{ij} \\ &= \sum_{(i,j)\in(S,T)} (u_{ij} - x_{ij} + x_{ji}) = \sum_{(i,j)\in(S,T)} r_{ij}\end{aligned}$$

Therefore

$$\Delta v \leq \sum_{(i,j)\in(S,T)} r_{ij}$$

□

Thus for any network flow x of value v, the additional flow that is possible between the source and sink nodes is less than or equal to the residual capacity of any s-t cut. Using the above definitions, properties of augmenting paths can be established. The next theorem is called the *augmenting path theorem.*

Theorem 8.5. *Augmenting path theorem. A flow x in a flow network $N = (G, u, s, t)$ is maximal if and only if there are no augmenting path from s to t.*

Proof. It is evident that if there is an augmenting path, then the flow is not maximal. Conversely, assume that the flow x does not permit an augmenting path in the network. Let S be the set of all nodes i, including the source node s, for which there is an augmenting path from s to node i. Furthermore, let $T = V - S$. Using the definitions of augmenting path, and vertex sets S and T; it follows that $x_{ij} = u_{ij}$ and $x_{ji} = 0$ for all values of $i \in S$ and $j \in T$. Therefore

$$\begin{aligned}v &= \sum_{i\in S}\sum_{j\in T} (x_{ij} - x_{ji}) \\ &= \sum_{i\in S}\sum_{j\in T} u_{ij} = u\langle S, T\rangle\end{aligned}$$

which is the capacity of the cut $\langle S, T\rangle$. But $v \leq u\langle S, T\rangle$ from theorem 8.3, therefore the flow is maximal. □

Thus a flow x is a maximum flow iff the residual network specified by $G(x)$ has no augmenting path. The next set of theorems are useful in developing a useful and intuitive maximal flow algorithm. The integral property of the flow is proved in the next theorem. It is called the *integral flow theorem.*

Theorem 8.6. *Integral flow theorem. Let $N = (G, u, s, t)$ be a flow network in which all capacities are integers. There exists a maximal flow on N such that the value of flow on any arc is integral.*

Proof. As per the hypothesis of the theorem, all capacities are integers. Initially assume that the flow x'_{ij} on arc (i, j) is equal to zero for all $i, j \in V$. Let the corresponding value of the flow be v'. If this flow is not maximal, then an augmenting path exists such that there is an integral flow $v'' > v'$. If v'' is not maximal, then the flow network admits an augmenting path, and so on. This process can be continued iteratively till the flow does not permit any more augmenting paths. Note that in this process the value of the flow increases from the previous step by at least unity. Consequently at the end of this process the value of the maximal flow is integral, and there is no augmenting path. □

The celebrated *max-flow min-cut theorem* was proved by L. R. Ford and D. R. Fulkerson, and also independently by P. Elias, A. Feinstein and C. E. Shannon in the year 1956.

Theorem 8.7. (*Max-flow min-cut theorem*) *Consider a flow network $N = (G, u, s, t)$. The maximum value of an s-t flow on the network N is equal to the minimum capacity of an s-t cut.*

Proof. The inequality $v \leq u\langle S, T\rangle$, the augmenting path and integral flow theorems together imply this theorem if the capacities are all integers.

However, if the capacities are rational numbers, then it can be reduced to the integer case by multiplying all the capacities by the lowest common multiple of the denominators. This is the case when all the capacities can be represented on a computer. The case of real-valued capacities can be addressed by using a continuity argument and observing that v is a continuous function of the x_{ij}'s.

Furthermore, the existence of a cut with minimum capacity is guaranteed because there are only a finite number of s-t cuts. □

A Max-Flow Min-Cut Algorithm

An augmenting path algorithm for determining maximal flow in a network is outlined below. It is based upon the above theoretical details.

Algorithm 8.12. *Outline of Max-Flow Min-Cut Algorithm.*

Input: A flow network $N = (G, u, s, t)$. The capacity matrix is U.
Output: Maximum flow x.
begin

$x \leftarrow 0$
$r \leftarrow U$
while ($G(x)$ contains a directed path between vertices s and t) **do**
begin

find an augmenting path $\mathcal{P}$ between vertices s and t in $G(x)$
$\theta \leftarrow \min\{r_{ij} \mid (i,j) \in \mathcal{P}\}$
augment the flow along the path $\mathcal{P}$ by θ units
update the residual network $G(x)$
end (*end of while-loop*)
find the maximal flow from the final residual network $G(x)$
end (*end of the outline of max-flow min-cut algorithm*)

This algorithm requires a technique to determine an augmenting path $\mathcal{P}$ in the residual network $G(x)$. Furthermore, it is not clear if it terminates in a finite number of steps. The above algorithm has to be further developed to meet this goal. This is done via a *labeling procedure*. The purpose of this labeling procedure is to determine via a search algorithm an augmenting path which starts at node s and terminates at node t.

In order to reach the target node t from the source node s, the algorithm determines via a *fanning out* process, all the nodes which are reachable from the source node s along a directed path in the residual network $G(x)$. In this process, the algorithm splits the nodes in the network into two parts: *labeled* and *unlabeled nodes*. The labeled nodes of the network $G(x)$ are a set of nodes which have been reached from s during the fanning out process. Thus, there is a direct path from s to all the labeled nodes. The unlabeled nodes are the nodes which have not yet been reached from the node s during the fanning out process.

The algorithm essentially selects a labeled node in the network $G(x)$, and scans its adjacency list of vertices for unlabeled nodes. It then labels these additional nodes. Iteration of this process ultimately results in labeling the sink node t. When this occurs, the algorithm transfers maximum possible flow along this directed path from node s to node t. Based upon this path and its flow, the algorithm determines a new residual network. The algorithm unlabels all the nodes, and repeats the process. The algorithm eventually terminates when it has scanned all the labeled nodes and it was not successful in labeling the sink node t.

Recall that the vertex i is the predecessor of vertex j if the arc $(i,j) \in E$. This is represented in the algorithm as predecessor$(j) = i$.

Algorithm 8.13. *Max-Flow Min-Cut Labeling Algorithm.*

Input: A flow network $N = (G, u, s, t)$. The capacity matrix is U.
Output: Maximum flow x.
begin
$x \leftarrow 0$
$r \leftarrow U$
label node t
while (node t is labeled) **do**
begin
unlabel all nodes in the network
label node s
$L \leftarrow \{s\}$

```
        predecessor(j) ← ∅ for all j ∈ V
        while ((L ≠ ∅) ∧ (t is unlabeled)) do
        begin
            (remove a node i from the set L)
            L ← L − {i}
            for (all arcs (i, j) in residual network G(x)
            emanating from node i) do
            begin
                if (r_ij > 0 and node j is unlabeled) then
                begin
                    predecessor(j) ← i
                    label node j
                    L ← L ∪ {j}
                end (end of if statement)
            end (end of for-loop)
        end (end of while-loop)
        if (t is labeled) then
        begin
            (augmentation of the flow)
            determine the augmenting path P in G(x) from s to t
                by using the predecessor table to trace back the path
                from the target node t to the source node s
            θ ← min {r_ij | (i, j) ∈ P}
            augment the flow along the path P by θ units
            update G(x)
        end (end of if statement)
    end (end of while-loop)
    find the maximal flow from the final residual network G(x)
end (end of max-flow min-cut labeling algorithm)
```

The correctness of the algorithm is next established. At the end of each iteration, the above algorithm either terminates or finds an augmenting path. If the algorithm terminates, it has to be proved that a maximum flow x has been determined. At the end of the algorithm, denote the labeled nodes as S and the unlabeled nodes as T, which is equal to $(V - S)$. Observe that $s \in S$ and $t \in T$. The algorithm fails to label any node in T from any node in S. This implies $r_{ij} = 0$ for all pairs $(i, j) \in S \times T$. Also $r_{ij} = (u_{ij} - x_{ij}) + x_{ji}$; and $(u_{ij} - x_{ij}) \geq 0$ and $x_{ji} \geq 0$ implies $u_{ij} = x_{ij}$ for all arcs $(i, j) \in S \times T$ and $x_{ij} = 0$ for all $(i, j) \in T \times S$. Therefore

$$v = \sum_{(i,j)\in(S,T)} x_{ij} - \sum_{(i,j)\in(T,S)} x_{ij} = \sum_{(i,j)\in(S,T)} u_{ij} = u\langle S, T\rangle$$

That is, the value of the flow v is equal to the capacity of the cut $\langle S, T\rangle$. But it is known that $v \leq u\langle S, T\rangle$ for all cuts, therefore v is the maximum flow, and $\langle S, T\rangle$ is the minimum cut. This algorithm can also be regarded as a constructive proof of the max-flow min-cut theorem with integer capacities. It should be noted that the algorithm can sometimes run forever for nonintegral values of the capacities. Specifically, the labeling algorithm may not terminate if the capacities are irrational numbers. However the max-flow min-cut algorithm is also true for irrational capacities. Several other superior algorithms exist in literature to determine maximum flow in a network. These are not discussed in this chapter.

Complexity of the Labeling Algorithm

Observe that in each path augmentation step, the algorithm scans any edge or node at most a single time. Therefore each path augmentation step requires $O(m)$ steps, where m is equal to the number of edges in the graph. The maximum flow is bounded by $nu_{\max}$, where n is equal to the number of vertices in the graph, and $u_{\max} = \max_{(i,j)\in E} \{u_{ij}\}$. If the network capacities are integers, the algorithm increments the value of the flow by at least a single unit. Therefore the maximum number of augmentations is $nu_{\max}$. Thus the computational complexity of the algorithm for integral values of the edge capacities is $O(nmu_{\max})$. The above labeling algorithm is illustrated below via a simple example.

Example 8.12. Consider a graph $G = (V, E)$, where $V = \{1, 2, 3, 4\}$ and the edges are specified by the adjacency matrix A. The capacities of the arcs are specified in the capacity matrix U. The source and sink nodes are 1 and 3 respectively. Maximum flow in the graph G is determined.

See Figure 8.10. This network has four cuts. Their capacities are listed in Table 8.2. It can be observed that the cut with the minimum capacity is $\langle S, T\rangle$, where $S = \{1, 4\}$, $T = \{2, 3\}$, and its capacity is $u\langle S, T\rangle = 9$. Therefore the value of the maximum flow is equal to 9. This can indeed be checked by following the algorithm. The iteration numbers are with respect to the outer while-loop of the labeling algorithm.

$$A = \begin{bmatrix} 0 & 1 & 0 & 1 \\ 0 & 0 & 1 & 1 \\ 0 & 0 & 0 & 0 \\ 0 & 0 & 1 & 0 \end{bmatrix}, \quad U = \begin{bmatrix} 0 & 4 & 0 & 7 \\ 0 & 0 & 8 & 3 \\ 0 & 0 & 0 & 0 \\ 0 & 0 & 5 & 0 \end{bmatrix}$$

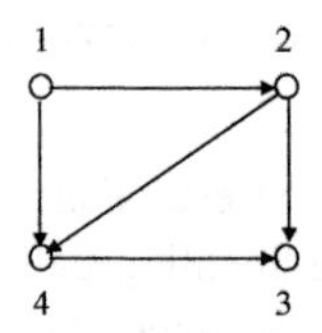

Figure 8.10. Maximum flow computation for the graph $G = (V, E)$.

S	T	$u\langle S, T\rangle$
$\{1\}$	$\{2, 3, 4\}$	11
$\{1, 2\}$	$\{3, 4\}$	18
$\{1, 4\}$	$\{2, 3\}$	9
$\{1, 2, 4\}$	$\{3\}$	13

Table 8.2. s-t cuts and their capacities.

Initialization: $x \leftarrow 0, r \leftarrow U$, and label the sink node 3.

Iteration 1: The path $\mathcal{P} = (1, 2, 3)$. Also $\theta = \min\{r_{12}, r_{23}\} = \min\{4, 8\} = 4$. Therefore $x_{12} = x_{23} = 4$. The values $x_{24} = x_{14} = x_{43} = 0$ remain unchanged. The capacities of the residual network are $r_{12} = 0$, $r_{21} = 4$, $r_{23} = 4$, and $r_{32} = 4$. The capacities $r_{24} = 3$, $r_{14} = 7$, and $r_{43} = 5$ remain unchanged.

Iteration 2: The path $\mathcal{P} = (1, 4, 3)$. Also $\theta = \min\{r_{14}, r_{43}\} = \min\{7, 5\} = 5$. Therefore $x_{14} = x_{43} = 5$. The values $x_{12} = x_{23} = 4$, and $x_{24} = 0$ remain unchanged. The capacities of the residual network are $r_{14} = 2$, $r_{41} = 5$, $r_{43} = 0$, and $r_{34} = 5$. The capacities $r_{12} = 0$, $r_{21} = 4$, $r_{23} = 4$, $r_{32} = 4$, and $r_{24} = 3$ remain unchanged.

Iteration 3: No complete path exists from the source node 1 to sink node 3. Therefore the algorithm terminates. Thus the flow x is

$$x_{12} = 4,\ x_{23} = 4,\ x_{24} = 0,\ x_{14} = 5,\ \text{and}\ x_{43} = 5$$

The value of the maximal flow $v = (x_{12} + x_{14}) = (x_{23} + x_{43}) = 9$. As noted earlier, the value $v = 9$ is also equal to the capacity of the cut $\langle S, T\rangle$, where

$$S = \{1, 4\},\ T = \{2, 3\},\ \text{and}\ \langle S, T\rangle = 9$$

This cut also happens to be a minimum capacity cut. □

We next extend the above results to include networks with multiple sources and sinks, networks with both arc and vertex capacities, and flows with nonnegative lower and upper bounds.

Extensions of the Single-Source and Single-Sink Maximum Flow Problem

1. *A network with multiple sources and sinks*: Assume that a graph has n_s sources and n_t sinks. The problem is to find maximum flow from all sources to all sinks. This problem can be converted to a single source and a single sink problem by adding a master source node and a master sink node.
   A direct edge is added from the master source node to each source node. The capacity of each of these links is set to infinity, or equal to the finite value if the supply at that source is limited. Also a direct edge is added from each of the original sink nodes to the master sink node. The capacity of each of these links is set to infinity, or equal to the finite value if the demand at the sink is limited.
2. *Graphs with both arc and vertex capacities*: Consider a single source and a single sink graph in which, the arcs and the nodes have finite capacities. The net flow which can enter a node $i \in V$ is equal to w_i units. That is, the capacity of node $i \in V$ is w_i for all $1 \le i \le n$. This problem can be converted to a capacitated single-source single-sink maximum flow problem.
   Replace each interior node i (nodes other than the source and sink nodes) by a pair of nodes i' and i'', and an arc (i'', i'). The nodes i' and i'' are the called the *out-node* and the *in-node* respectively. The capacity of the arc (i'', i') is equal to w_i. Furthermore, the capacity of the arc (i', j'') is equal to u_{ij} in the modified network.
3. *Flows with nonnegative lower and upper bounds*: The problem with nonnegative lower and upper bound can also be studied using the techniques discussed in this section. Observe that the maximum flow problem always has a feasible solution with zero lower bounds. In this case, the zero flow is feasible. However if there is a restriction on the lowest value of flow on the arcs, then it is not always possible to obtain a feasible solution. This problem is not discussed any further. It is addressed extensively in books on network flows and optimization.

Combinatorial application of max-flow and min-cut theorem is next outlined.

Combinatorial Application of Max-Flow and Min-Cut Theorem

Some problems in network connectivity can be studied using the max-flow min-cut theorem. Two directed paths from the source node s to the sink node t are said to be *arc-disjoint* if they do not have any arc in common. *Node-disjoint* paths are defined similarly. Two paths are node-disjoint if they do not have any common nodes, except the nodes s and t. The following observations can be made about these applications of max-flow min-cut theorem.

Observations 8.5. Some basic observations.

1. The maximum number of arc-disjoint paths from the source node s to the sink node t, is equal to the minimum number of arcs required to be removed from the network to disconnect all paths from the node s to the node t.
   This result can be established by assigning a capacity of 1 to each arc in the network, and using the max-flow min-cut theorem.
2. Similarly, the maximum number of node-disjoint paths from the source node s to the sink node t, is equal to the minimum number of nodes, excluding nodes s and t, whose removal is sufficient to disconnect all paths from the node s to the node t. □

See the problem section for proofs of the above observations.

8.9.2 Minimum Cost Flows

The minimum cost flow problem determines the smallest cost required to transport information over a capacitated network from some nodes to other nodes of the network. This problem is also referred to as the least cost routing of commodities (goods, information, or network traffic) in a capacitated network. Thus this problem is a generalization of the maximum flow problem and the shortest path problem. The network parameters are defined below.

Definition 8.14. *Basic definitions.*

(a) *Let $G = (V, E)$ be a directed graph, where V is the vertex (node) set, E is the set of edges, $|V| = n$, $|E| = m$, and $V = \{1, 2, \ldots, n\}$.*
(b) *Associated with this graph is a mapping $u : E \rightarrow \mathbb{N}$. That is, for each edge $e \in E$ there is a value $u(e) \geq 0$ called the capacity of the edge e. If an edge is denoted by (i, j), where i and j are vertices, then its capacity is specified by u_{ij}. For simplicity assume that the capacities are finite and nonnegative integers. Notice that edges with zero capacity are also allowed. The mapping u is generally specified by an $n \times n$ matrix also denoted by U, where $U = [u_{ij}]$.*
(c) *Associated with this graph is another mapping $c : E \rightarrow \mathbb{N}$. That is, for each edge $e \in E$ there is a value $c(e) \geq 0$ called the cost of the edge e. The cost of the edge (i, j), is given by c_{ij}. The mapping c is generally specified by an $n \times n$ matrix C, where $C = [c_{ij}]$.*
   If $H \subseteq E$, then the total cost of edges in the set H is denoted by $\mathcal{C}(H)$. Also if $\mathcal{P}$ is a path in the graph G, then the total cost of edges in this path is denoted by $\mathcal{C}(\mathcal{P})$.
(d) *The digraph does not contain parallel arcs. That is, two or more arcs with the same head and tail nodes are not permissible. Also assume that the digraph is connected.*
(e) *There is a mapping $b : V \rightarrow \mathbb{Z}$. That is, for each node (vertex) $i \in V$ there is a number $b(i) \in \mathbb{Z}$. If $b(i) > 0$, then the node i is a supply (source) node. If $b(i) < 0$, then the node i is a demand (sink) node. The mapping $b(\cdot)$ is generally specified by a vector of size n. Let $b = (b(1), b(2), \ldots, b(n))$.*

(f) *The flow network or a capacitated network is specified by* $N = (G, u, c, b)$. □

It is also assumed that there exists at least a single directed path between each node pair, which can carry an extremely "large" flow. This condition facilitates the determination of a minimum cost flow. Some more notation and assumptions are listed below.

Definitions 8.15. *Additional definitions.*

1. *A feasible flow on capacitated network* N *is a mapping* $x : E \rightarrow \mathbb{N}$ *which satisfies the following conditions*:
   (a) *Capacity constraints*: $0 \leq x_{ij} \leq u_{ij}$ *for all arcs* $(i, j) \in E$.
   (b) *Flow conservation*: $\sum_{\{j|(i,j)\in E\}} x_{ij} - \sum_{\{j|(j,i)\in E\}} x_{ji} = b(i)$ *for all* $i \in V$.
   (c) *The supplies and demands at the nodes satisfy* $\sum_{i\in V} b(i) = 0$.

   The mapping x *is generally specified by an* $n \times n$ *matrix* x, *such that* $x = [x_{ij}]$
2. *The cost of the feasible flow* x *is given by* $z(x) = \sum_{(i,j)\in E} c_{ij} x_{ij}$.
3. *A minimum (optimal) cost flow is a feasible flow which has a minimum cost.* □

The minimum cost flow problem is next stated formally. *Minimize the cost* $z(x)$ *of flow* x *in the capacitated network* $N = (G, u, c, b)$, *and determine the flow* x.

The feasibility of the minimum cost flow problem can be determined by solving a maximum flow problem. Add two nodes, s^* and t^* to the network. For each node i for which $b(i) > 0$, introduce an arc (s^*, i) with capacity $b(i)$. Also for each node i for which $b(i) < 0$, introduce an arc (i, t^*) with capacity $-b(i)$. Solve this modified network problem for maximum flow. If the solution is such that all the arcs (s^*, i) are saturated, then the minimum cost flow problem has a feasible solution. Before this problem is addressed, additional terminology is introduced.

Definitions 8.16. *Pseudoflow, residual network, vertex potential function, cost of a path, and negative cycle are defined.*

1. *A pseudoflow* x *is a function which assigns a nonnegative flow value to each edge* $e \in E$ *of the graph. It satisfies the arc capacity and nonnegativity constraints, but possibly not the flow conservation constraint. Its matrix representation is* $x = [x_{ij}]$.
2. *The residual network and the residual capacity are defined as follows. For a flow or a pseudoflow* x, *the residual network* $G(x)$ *has for each arc* $(i, j) \in E$: *two arcs* (i, j) *and* (j, i) *such that*:
   (a) *The cost of the arc* (i, j) *is* c_{ij} *and residual capacity* $r_{ij} = (u_{ij} - x_{ij})$.
   (b) *The cost of the arc* (j, i) *is* $c_{ji} = -c_{ij}$ *and residual capacity* $r_{ji} = x_{ij}$.

   The residual network $G(x)$ *has arcs of only positive residual capacity.*
3. $\pi : V \rightarrow \mathbb{Z}$ *is called the vertex potential function. For each* $i \in V$, $\pi(i)$ *is called the potential of vertex* i. *It is related to the flow conservation constraint at vertex* i. *For a specified set of vertex potentials, the reduced cost of arc* (i, j) *in the residual network* $G(x)$ *is* $c_{ij}^{\pi} = (c_{ij} - \pi(i) + \pi(j))$. *The* $\pi(i)$*'s are also represented by a vector* $\pi = (\pi(1), \pi(2), \ldots, \pi(n))$.
4. *The cost of a path* $\mathcal{P}$ *in* $G(x)$ *is* $C(\mathcal{P}) = \sum_{(i,j)\in\mathcal{P}} c_{ij}$. *Its reduced cost is equal to* $C^{\pi}(\mathcal{P}) = \sum_{(i,j)\in\mathcal{P}} c_{ij}^{\pi}$.
5. *A negative cycle* W *in a residual network* $G(x)$ *is a directed cycle for which its cost is negative. Let* $C(W)$ *denote the cost of the negative cycle. Then* $C(W) < 0$. □

The concept of node (vertex) potential is useful in describing an algorithm for minimum cost flow in a capacitated network.

Typically algorithms assign traffic flows incrementally. Therefore, it is useful to revisit the concept of residual network. It acts as a "remaining flow network," and determines the optimal value of the incremental flow. Let the cost and capacity of the edge (i,j) be c_{ij} and u_{ij} respectively. If the edge $(i,j) \in E$ of capacity u_{ij} carries a flow of x_{ij} units, then it can carry another $(u_{ij} - x_{ij})$ units of flow on this edge. Also observe that it can carry an additional x_{ij} units of flow from node j to node i, which is equivalent to canceling the existing flow on the edge (i,j). Thus the residual network $G(x)$ with respect to the flow x is defined as follows. Replace *each* edge (i,j) in the graph G by *two* edges (i,j) and (j,i). Let the capacity of these two edges be $r_{ij} = (u_{ij} - x_{ij})$ and $r_{ji} = x_{ij}$ respectively. Next observe that sending flow on the link (i,j) increases the cost by c_{ij} per unit of flow. Also sending flow from node j to i on the same link decreases the traffic flow cost. Therefore $c_{ji} = -c_{ij}$.

Observations 8.6. Let $\pi(\cdot)$ be a vertex potential function.

1. If $\mathcal{P}$ is a path from i to j in the residual network $G(x)$, then
$$\mathcal{C}^{\pi}(\mathcal{P}) = (\mathcal{C}(\mathcal{P}) - \pi(i) + \pi(j)) \tag{8.19}$$
2. If W is a directed cycle in the residual network $G(x)$, then $\mathcal{C}^{\pi}(W) = \mathcal{C}(W)$.
3. $z(x) = \sum_{(i,j)\in E} c_{ij}x_{ij} = \sum_{(i,j)\in E} c^{\pi}_{ij}x_{ij} + \sum_{i\in V} \pi(i)\, b(i)$. Therefore x minimizes $\sum_{(i,j)\in E} c_{ij}x_{ij}$ if and only if it minimizes $\sum_{(i,j)\in E} c^{\pi}_{ij}x_{ij}$.
4. A flow $x \geq 0$ in a network G is feasible if and only if a flow $x' \geq 0$ in a residual network $G(\widetilde{x})$ is feasible, where:
If $\widetilde{x}_{ij} \leq x_{ij}$, then $x'_{ij} = (x_{ij} - \widetilde{x}_{ij})$, and $x'_{ji} = 0$.
If $\widetilde{x}_{ij} \geq x_{ij}$, then $x'_{ij} = 0$, and $x'_{ji} = (\widetilde{x}_{ij} - x_{ij})$.
Also $c'_{ij}x'_{ij} = c_{ij}(x_{ij} - \widetilde{x}_{ij})$, where c'_{ij} is the cost in the residual network. This observation provides a technique to translate an optimal solution in the residual network, to an optimal solution in the original network. □

See the problem section for a proof of the last observation. In order to study minimum cost flow in a capacitated network, the flows in a directed graph have to be further characterized in terms of its paths and cycles.

Characterization of Flows in a Directed Graph

We characterize flows in a directed graph in terms of its paths and cycles. Consider a directed graph $G = (V,E)$, where V is the set of vertices, E is the set of edges, $|V| = n$, and $|E| = m$. Recall that a directed cycle is a directed walk with identical first and last vertices, and no repeated intermediate vertices. Let $\mathcal{P}$ be a directed path between a pair of nodes, and W be a directed cycle in the graph. Denote the collection of all directed paths $\mathcal{P}$, and all directed cycles W in the graph by $\widetilde{\mathcal{P}}$ and $\widetilde{W}$ respectively.

The traffic flow on arc $(i,j) \in E$ is $x_{ij} \in \mathbb{R}^{+}_{0}$. The flow x_{ij} depends upon the decision variables $g_1(\mathcal{P})$'s and $g_2(W)$'s. Therefore it can be represented as

$$x_{ij} = \sum_{\mathcal{P}\in\widetilde{\mathcal{P}}} \delta_{ij,\mathcal{P}}\, g_1(\mathcal{P}) + \sum_{W\in\widetilde{W}} \delta_{ij,W}\, g_2(W)$$

where $\delta_{ij,\mathcal{P}}$ is equal to 1 if the arc (i, j) is in the path $\mathcal{P}$, otherwise it is equal to 0. Similarly $\delta_{ij,W}$ is equal to 1 if the arc (i, j) is in the cycle W, otherwise it is equal to 0. The decision variables $g_1(\cdot)$ and $g_2(\cdot)$ depend upon $\mathcal{P}$ and W respectively. The above equation demonstrates that the flow along a path or a cycle has a representation in terms of nonnegative arc flows, and that every nonnegative arc flow can be represented as a combination of path and cycle flows.

Definition 8.17. *A circulation in a directed graph G is a flow in the network in which the flow is conserved at each vertex.* □

Some properties of flows in a graph are listed below.

Observations 8.7. Consider a directed graph $G = (V, E)$, where V is the set of vertices, E is the set of edges, $|V| = n$, and $|E| = m$.

1. A circulation in the graph can be represented as cycle flow with at most m directed cycles.
2. Consider a capacitated network $N = (G, u, c, b)$ with feasible flows x' and x''. Then the flow x'' is equal to the flow x' plus the flow on at most m directed cycles in the residual network $G(x')$. In addition, the cost of the flow x'' is equal to the cost of the flow x' plus the cost of the flow on these m augmenting cycles. □

See the problem section for proofs of the above set of observations. The last observation is useful in proving a key result in minimum cost flow of traffic in a capacitated network.

Characterization of Minimum Cost Flows

There are several equivalent conditions for characterizing minimum cost flow in a capacitated network. These are: negative cycle, reduced cost, and complementary slackness optimality conditions. The next theorem characterizes the *negative cycle optimality conditions*. This condition states that a feasible flow x is a minimum cost flow if and only if the residual network $G(x)$ has no negative directed cycles.

Theorem 8.8. *In a capacitated network $N = (G, u, c, b)$, a feasible flow x is a minimum cost flow if and only if the residual network $G(x)$ has no negative directed cycles.*

Proof. Assume that x is a feasible flow and the residual network $G(x)$ has a negative cycle. Then a feasible solution with a lower cost is possible by augmenting positive flow along the cycle. This implies that if x is a minimum cost flow, then $G(x)$ cannot have a negative cycle.

Conversely, assume that x' is a feasible flow and the residual network $G(x')$ has no negative cycle. Also consider x'' to be an optimal flow such that $x' \neq x''$. Therefore $z(x'') \leq z(x')$, because x'' is an optimal flow. The flow $y = (x'' - x')$ can be expressed as a sum of at most m flow augmenting cycles with respect to the flow x'. The sum of the cost of these cycles is equal to $(z(x'') - z(x'))$. Furthermore, the lengths of all cycles in $G(x')$ are nonnegative. This implies $(z(x'') - z(x')) \geq 0$. Thus $z(x'') = z(x')$, and x' is also an optimal flow. Consequently x' is a least cost flow with no negative cycles. □

The optimality condition can also be obtained in terms of the reduced cost c_{ij}^{π}'s. The *reduced cost optimality conditions* are

$$c_{ij}^{\pi} = (c_{ij} - \pi(i) + \pi(j)) \geq 0, \quad \forall \text{ arcs } (i, j) \in E \tag{8.20}$$

In the following discussion, the words *length, distance* and *cost* are sometimes used synonymously. This is because the algorithm to find the minimum cost flow uses the shortest-path algorithm multiple times.

Theorem 8.9. *In a capacitated network* $N = (G, u, c, b)$, *a feasible flow* x *is a minimum cost flow if and only if for some set of node potentials* $\{\pi(i) \mid 1 \le i \le n\}$, *the reduced costs satisfy:* $c_{ij}^{\pi} \ge 0$ *for all arcs* (i, j) *in* $G(x)$.

Proof. Suppose that $c_{ij}^{\pi} \ge 0$ for all arcs (i, j) in $G(x)$, then for every directed cycle W, $\sum_{(i,j)\in W} c_{ij}^{\pi} \ge 0$. This implies $\mathcal{C}(W) = \mathcal{C}^{\pi}(W) \ge 0$. Thus there are no negative cycles in the residual network $G(x)$. Therefore by the preceding theorem, x is a minimum cost flow in the network N.

Conversely, suppose that x is a feasible least cost flow in the network $G(x)$. Let d_i denote the shortest path distance from node 1 to node i, for $1 \le i \le n$. If the network N does not contain any negative cycles, then the distances d_i's are well-defined, and $d_j \le d_i + c_{ij}$ for all arcs (i, j) in $G(x)$. That is, $\{c_{ij} - (-d_i) + (-d_j)\} \ge 0$. Define $\pi(i) = -d_i$ for $1 \le i \le n$, then $c_{ij}^{\pi} \ge 0$ for all arcs (i, j) in $G(x)$. Thus the feasible solution x satisfies the reduced cost optimality conditions. □

Finally, the *complementary slackness optimality conditions* are established. These are so named because these conditions are related to a similar technique used in the theory of linear programming.

Theorem 8.10. *In a capacitated network* $N = (G, u, c, b)$, *a feasible flow* x *is a minimum cost flow if and only if for some set of node potentials* $\{\pi(i) \mid 1 \le i \le n\}$, *for each arc* $(i, j) \in E$:

$$\textit{if } c_{ij}^{\pi} > 0, \ \textit{then } x_{ij} = 0 \tag{8.21a}$$

$$\textit{if } c_{ij}^{\pi} < 0, \ \textit{then } x_{ij} = u_{ij} \tag{8.21b}$$

$$\textit{if } 0 < x_{ij} < u_{ij}, \ \textit{then } c_{ij}^{\pi} = 0 \tag{8.21c}$$

Proof. We initially demonstrate that the reduced cost optimality conditions imply the above results. Assume that the flow x satisfies the reduced cost optimality conditions. For any arc $(i, j) \in E$:

(a) If $c_{ij}^{\pi} > 0$ then the residual network $G(x)$ does not have the arc (j, i) because $c_{ji}^{\pi} = -c_{ij}^{\pi} < 0$, which contradicts the condition that all reduced costs in the residual network are nonnegative. Thus $x_{ij} = 0$.
(b) If $c_{ij}^{\pi} < 0$ then the residual network $G(x)$ does not have the arc (i, j) because it contradicts the nonnegativity condition of the reduced costs in the residual network. Therefore $x_{ij} = u_{ij}$.
(c) If $0 < x_{ij} < u_{ij}$ then the residual network $G(x)$ has both the arcs (i, j) and (j, i). Also from the reduced cost optimality condition $c_{ij}^{\pi} \ge 0$ and $c_{ji}^{\pi} \ge 0$. However $c_{ji}^{\pi} = -c_{ij}^{\pi}$. Therefore $c_{ij}^{\pi} = c_{ji}^{\pi} = 0$.

Therefore it can be inferred that if the flow x and the node potentials $\pi(\cdot)$ satisfy the reduced cost optimality conditions, then the complementary slackness conditions stated in the theorem are true.

The converse statement: If the flow x and the node potentials $\pi(\cdot)$ satisfy the complementary slackness conditions, then these also satisfy the reduced cost optimality conditions. See the problem section for a proof of this converse. □

Algorithm to Determine the Minimum Cost Flow

A minimum cost flow algorithm based upon successive computation of shortest paths is described. This goal is achieved via the concept of pseudoflows. Recall that a pseudoflow satisfies the arc capacity and nonnegativity constraints but possibly not the flow conservation constraint.

Definitions 8.18. *In a capacitated network $N = (G, u, c, b)$:*

1. *For any pseudoflow x, the imbalance of vertex $i \in V$ is given by $f(i)$.*

$$f(i) = b(i) + \sum_{\{j \mid (j,i) \in E\}} x_{ji} - \sum_{\{j \mid (i,j) \in E\}} x_{ij}, \quad \forall\, i \in V \tag{8.22}$$

2. *For any vertex $i \in V$:*
   (a) *If $f(i) > 0$, then $f(i)$ is referred to as the excess of the vertex i. The vertex i is called an excess vertex.*
   (b) *Similarly if $f(i) < 0$, then $-f(i)$ is referred to as the deficit of the vertex $i \in V$. The vertex i is called the deficit vertex.*
   (c) *If $f(i) = 0$, then the vertex i is called a balanced vertex.*
3. *Let V_e and V_d be the set of excess and deficit vertices.* □

Observations 8.8. The following observations readily follow from the above definitions.

1. The conservation conditions hold. That is

$$\sum_{i \in V} f(i) = \sum_{i \in V} b(i) = 0, \quad \text{and} \quad \sum_{i \in V_e} f(i) = -\sum_{i \in V_d} f(i)$$

2. If a flow network contains an excess node, then it also contains a deficit node. □

A residual network for a pseudoflow can also be defined. The reduced cost optimality conditions and pseudoflows can be used to describe a minimum cost flow algorithm based upon the computation of successive shortest paths. The theoretical basis of this algorithm is the following set of results.

Lemma 8.2. *Let $N = (G, u, c, b)$ be a capacitated network. Also x is a pseudoflow or a flow which satisfies the reduced cost optimality conditions with respect to a node potential $\pi(\cdot)$. Let c_{ij}^{π} be the length of the arc (i, j) in the residual network $G(x)$. Denote the shortest path distances from an arbitrary node s to all other nodes in the residual network $G(x)$ by $\mathcal{D} = (d_1, d_2, \ldots, d_n)$. Then:*

(a) *The pseudoflow x also satisfies the reduced cost optimality conditions if the node potential vector is $\pi' = (\pi - \mathcal{D})$.*
(b) *The reduced cost $c_{ij}^{\pi'} = 0$, if the arc (i, j) is in a shortest path from vertex s to every other vertex.*

Proof. Note that $c_{ij}^{\pi} \geq 0$ for all arcs (i, j) in the residual network $G(x)$. This follows from the hypothesis that x satisfies the reduced cost optimality conditions with respect to the potentials $\pi(\cdot)$. Also since $\mathcal{D}$ is the shortest path distances vector with c_{ij}^{π} as arc-lengths, the vector elements d_i's satisfy the shortest path optimality conditions. These are

$$d_j \leq d_i + c_{ij}^{\pi}, \ \forall \text{ arcs } (i,j) \text{ in } G(x)$$

Substituting $c_{ij}^{\pi} = (c_{ij} - \pi(i) + \pi(j))$ in the above relationship yields

$$c_{ij} - (\pi(i) - d_i) + (\pi(j) - d_j) \geq 0$$

The left-hand side in the above inequality is equal to $c_{ij}^{\pi'}$. Thus $c_{ij}^{\pi'} \geq 0$. This proves the first part of the lemma.

In the shortest path from the node s to another node k, if arc (i,j) is in this path, then

$$d_j = \left(d_i + c_{ij}^{\pi}\right)$$

Substituting $c_{ij}^{\pi} = (c_{ij} - \pi(i) + \pi(j))$ and the above result in the expression for $c_{ij}^{\pi'}$ yields $c_{ij}^{\pi'} = 0$. This proves the second part of the lemma. □

Lemma 8.3. *Let $N = (G,u,c,b)$ be a capacitated network. Also x is a pseudoflow or a flow which satisfies the reduced cost optimality conditions with respect to a node potential $\pi(\cdot)$. Another flow x' is obtained from the flow x by sending flow on a shortest path from node s to some other node k, then the flow x' also satisfies the reduced cost optimality conditions.*

Proof. Let $\mathcal{D}$ be the shortest path distances vector from a node s to all other nodes in the network $G(x)$ with costs c_{ij}^{π}'s. Let the potential π' be equal to $(\pi - \mathcal{D})$. Then from the preceding lemma, $c_{ij}^{\pi'} = 0$ for all arcs (i,j) in the shortest path $\mathcal{P}$ from node s to node k. If the flow is augmented on this arc (i,j) then this flow might possibly modify flow on the arc (j,i) of the residual network. However, $c_{ij}^{\pi'} = 0$ for all arcs $(i,j) \in \mathcal{P}$, consequently $c_{ji}^{\pi'} = 0$. Therefore the arc (j,i) also satisfies the reduced cost optimality conditions. □

The mathematical machinery developed thus far can be used to describe an algorithm to compute a minimum cost flow in a capacitated network. The algorithm depends upon the successive use of the shortest path algorithm. The minimum cost flow algorithm begins with a pseudoflow $x = 0$. The corresponding residual network is $G(x) = G$, and the corresponding cost of the arc (i,j) in the network $G(x)$ equal to c_{ij}^{π} is initialized to c_{ij}. Also initialize the flow imbalance $f(i)$ by $b(i)$ for all $i \in V$. The r_{ij}'s (residual capacity) are initialized to u_{ij}'s for all $1 \leq i,j \leq n$. The set of excess and deficit vertices, V_e and V_d are also determined.

In the next step, an excess vertex v_e and a deficit vertex v_d are selected. A shortest path $\mathcal{P}$ is computed from the vertex v_e to v_d in the residual network $G(x)$, where the cost of the arcs are c_{ij}^{π}'s. The flow along the path $\mathcal{P}$ is augmented by θ, where the value θ is minimum of $f(v_e), -f(v_d)$, and $\min\{r_{ij} \mid (i,j) \in \mathcal{P}\}$. Thus the flow x is updated by this value. Subsequently $\pi, f(\cdot), G(x), V_e, V_d$, and c_{ij}^{π}'s are updated.

If the flow along a path $\mathcal{P}$ is augmented, then the residual capacities of the arcs in $\mathcal{P}$ are decreased by θ, and the arcs which have a reverse direction to those in $\mathcal{P}$ have their capacities increased by θ. Furthermore, $f(v_e)$ is decreased by θ, and $f(v_d)$ is increased by θ.

Observe that if any node is imbalanced, then the sets V_e and V_d are nonempty. If this is the case, then the shortest path computation procedure is repeated. Therefore the algorithm computes successive shortest paths as long as nodes are imbalanced. Notice that the shortest path computation is performed in a network with nonnegative arc-lengths.

This minimum cost flow algorithm appears to be intuitively the simplest. However several other algorithms exist in literature which are computationally superior.

Algorithm 8.14. *Minimum Cost Flow Algorithm.*

Input: A flow network $N = (G, u, c, b)$.
Output: Minimum cost flow x.
begin
    (*initialization*)
    $x \leftarrow 0$
    $G(x) \leftarrow G$
    $\pi \leftarrow (0, 0, \ldots 0)$
    $f(i) \leftarrow b(i), \forall\, i \in V$
    $V_e \leftarrow \{i \mid f(i) > 0, i \in V\}$
    $V_d \leftarrow \{i \mid f(i) < 0, i \in V\}$
    while $V_e \neq \varnothing$ **do**
    begin
        Step 1: Select a vertex $v_e \in V_e$, and a vertex $v_d \in V_d$ such that
            v_d is reachable from v_e in $G(x)$.
            If there is no such v_d then **stop**
        Step 2: The costs in $G(x)$ are c_{ij}^{π}'s. Use these costs to find a
            shortest path $\mathcal{P}$ in $G(x)$ from vertex v_e to vertex v_d.
            Determine the corresponding distance vector $\mathcal{D}$.
        Step 3: $\theta \leftarrow \min\{f(v_e), -f(v_d), \min\{r_{ij} \mid (i,j) \in \mathcal{P}\}\}$
            Augment the flow along the path $\mathcal{P}$ by θ units, that is
            the flow x is updated.
        Step 4: Update $G(x), \pi, f(\cdot), V_e, V_d$, r_{ij}'s, and c_{ij}^{π}'s.
    end (*end of while-loop*)
    Determine the optimal flow from the residual network $G(x)$.
end (*end of minimum cost flow algorithm*)

Complexity of the Minimum Cost Flow Algorithm

Let $b_{\max} = \max_{i \in V}\{|b(i)|\}$, the algorithm requires at most $nb_{\max}$ iterations, where there is a shortest path computation in each iteration. Let $c_{\max} = \max_{(i,j)\in E}\{c_{ij}\}$, then $\max_{(i,j)\in E}\left\{c_{ij}^{\pi}\right\} = nc_{\max}$. If $\widetilde{S}(n, m, c_{\max})$ is the number of steps required to perform shortest path calculations in a network with nonnegative arc-lengths and cost matrix $C = [c_{ij}]$, then the overall complexity of the minimum cost algorithm is

$$O\left(nb_{\max}\widetilde{S}(n, m, nc_{\max})\right)$$

A candidate to solve the shortest path problem is the Dijkstra's algorithm.

Example 8.13. The minimum cost flow in a capacitated network with 4 vertices is computed. See Figure 8.11. The network is described by a directed graph G, which in turn is described by the adjacency matrix A, the capacity matrix U, and the cost matrix C. The supply and demand flow is specified by the vector b, where $b = (6, 0, 0, -6)$. Note that $\sum_{i=1}^{4} b(i) = 0$.

The minimum cost flow x, and the corresponding cost $z(x)$ are determined.

$$A = \begin{bmatrix} 0 & 1 & 1 & 0 \\ 0 & 0 & 0 & 1 \\ 0 & 1 & 0 & 1 \\ 0 & 0 & 0 & 0 \end{bmatrix}, \quad U = \begin{bmatrix} 0 & 4 & 5 & 0 \\ 0 & 0 & 0 & 6 \\ 0 & 3 & 0 & 8 \\ 0 & 0 & 0 & 0 \end{bmatrix}, \quad C = \begin{bmatrix} 0 & 2 & 3 & \infty \\ \infty & 0 & \infty & 1 \\ \infty & 2 & 0 & 4 \\ \infty & \infty & \infty & 0 \end{bmatrix}$$

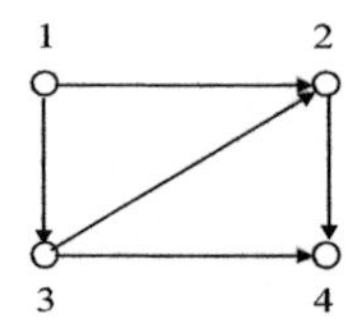

Figure 8.11. Minimum cost flow computation for the graph $G = (V, E)$.

Initialization: $x \leftarrow 0, \pi \leftarrow 0, f \leftarrow b, r_{ij} \leftarrow u_{ij}$ for $1 \leq i, j \leq 4$; $V_e = \{1\}$, and $V_d = \{4\}$.
Iteration 1: $v_e = 1, v_d = 4, \mathcal{D} = (0, 2, 3, 3), \mathcal{P} = (1, 2, 4)$,
$\pi \leftarrow \pi - \mathcal{D}$, therefore $\pi = (0, -2, -3, -3)$.
$\theta = \min\{f(1), -f(4), \min\{r_{12}, r_{24}\}\} = \min\{6, 6, 4, 6\} = 4$. The flows are

$$x_{12} \leftarrow \theta \Rightarrow x_{12} = 4$$
$$x_{24} \leftarrow \theta \Rightarrow x_{24} = 4$$

The reduced costs are

$$c_{12}^{\pi} \leftarrow c_{12} - \pi(1) + \pi(2) = 0$$
$$c_{13}^{\pi} \leftarrow c_{13} - \pi(1) + \pi(3) = 0$$
$$c_{24}^{\pi} \leftarrow c_{24} - \pi(2) + \pi(4) = 0$$
$$c_{32}^{\pi} \leftarrow c_{32} - \pi(3) + \pi(2) = 3$$
$$c_{34}^{\pi} \leftarrow c_{34} - \pi(3) + \pi(4) = 4$$

Also the residual capacities and other reduced costs are

$$r_{12} \leftarrow (r_{12} - \theta) \Rightarrow r_{12} = (4 - 4) = 0$$
$$r_{21} \leftarrow \theta \Rightarrow r_{21} = 4;\ c_{21}^{\pi} = -c_{12}^{\pi} = 0$$
$$r_{24} \leftarrow (r_{24} - \theta) \Rightarrow r_{24} = (6 - 4) = 2$$
$$r_{42} \leftarrow \theta \Rightarrow r_{42} = 4;\ c_{42}^{\pi} = -c_{24}^{\pi} = 0$$

The updated values of the supply and demand flow are

$$f(1) \leftarrow (f(1) - \theta) = 2$$
$$f(4) \leftarrow f(4) + \theta = -2$$

Also $V_e = \{1\}$, and $V_d = \{4\}$.

Iteration 2: Note that the residual network consists of edges with positive residual capacity. Therefore arc $(1, 2)$ is not present in the residual network. Thus $v_e = 1, v_d = 4, \mathcal{D} = (0, 3, 0, 3), \mathcal{P} = (1, 3, 2, 4)$,

$\pi \leftarrow \pi - \mathcal{D}$, therefore $\pi = (0, -5, -3, -6)$.

$\theta = \min \{f(1), -f(4), \min \{r_{13}, r_{32}, r_{24}\}\} = \min \{2, 2, 5, 3, 2\} = 2$. The flows are

$$x_{13} \leftarrow \theta \Rightarrow x_{13} = 2$$
$$x_{32} \leftarrow \theta \Rightarrow x_{32} = 2$$
$$x_{24} \leftarrow (x_{24} + \theta) \Rightarrow x_{24} = 6$$

The reduced costs are

$$c_{13}^{\pi} \leftarrow c_{13} - \pi(1) + \pi(3) = 0$$
$$c_{24}^{\pi} \leftarrow c_{24} - \pi(2) + \pi(4) = 0$$
$$c_{32}^{\pi} \leftarrow c_{32} - \pi(3) + \pi(2) = 0$$
$$c_{34}^{\pi} \leftarrow c_{34} - \pi(3) + \pi(4) = 1$$

Also the residual capacities and other reduced costs are

$$r_{13} \leftarrow (r_{13} - \theta) \Rightarrow r_{13} = (5 - 2) = 3$$
$$r_{31} \leftarrow \theta = 2 \Rightarrow r_{31} = 2;\ c_{31}^{\pi} = -c_{13}^{\pi} = 0$$
$$r_{32} \leftarrow (r_{32} - \theta) \Rightarrow r_{32} = (3 - 2) = 1$$
$$r_{23} \leftarrow \theta = 2 \Rightarrow r_{23} = 2;\ c_{23}^{\pi} = -c_{32}^{\pi} = 0$$
$$r_{24} \leftarrow (r_{24} - \theta) \Rightarrow r_{24} = (2 - 2) = 0$$
$$r_{42} \leftarrow (r_{42} + \theta) \Rightarrow r_{42} = (4 + 2) = 6;\ c_{42}^{\pi} = -c_{24}^{\pi} = 0$$

The updated values of the supply and demand flow are

$$f(1) \leftarrow (f(1) - \theta) = 0$$
$$f(4) \leftarrow f(4) + \theta = 0$$

Thus $V_e = V_d = \varnothing$.

Last Step: The minimum cost flow matrix x is

$$x = \begin{bmatrix} 0 & 4 & 2 & 0 \\ 0 & 0 & 0 & 6 \\ 0 & 2 & 0 & 0 \\ 0 & 0 & 0 & 0 \end{bmatrix}$$

The minimum cost of the flow is $z(x) = \sum_{1 \leq i,j \leq 4} c_{ij} x_{ij} = 24$. □

8.10 Network Coding

Network coding is a relatively new technique to implement information flow in a communication network. It promises to improve network performance when compared to traditional techniques. Traditional computer communication networks transmit information between a source-destination node pair in units of information called packets. Packets reach their destination by traversing intermediate nodes of the network. The form of the packet created at the source node essentially remains unchanged until it reaches its destination node. The purpose of an intermediate node is to receive a packet from an incoming edge, and forward it to an outgoing edge. This type of information transmission is therefore called *store-and-forward,* or simply *routing.* In contrast, the network coding technique permits the intermediate nodes to *mix* the incoming packets, generate new packets, and subsequently transmit them on outgoing edges. It is assumed that the destination node is able to reconstruct the packets that were originally sent from the source node. It is further assumed that there is sufficient computational power in the nodes to provide this mixing of packets. This paradigm offers to improve throughput, reliability, and robustness of the network. It is useful in applications in which network bandwidth (capacity) is a critical resource.

Some of the possible applications of this technique are wireless communication, network monitoring and management, distributed storage, security, and quantum networks. The basics of network coding are next demonstrated via examples. In these examples the exclusive-or operator $\oplus$ is used. The $\oplus$ operation is actually modulo 2 addition operation. If $\mathbb{Z}_2 = \{0, 1\}$, then $0 \oplus 0 = 0$, $0 \oplus 1 = 1$, $1 \oplus 0 = 1$, and $1 \oplus 1 = 0$. The basic idea behind network coding is illustrated in the following examples.

Example 8.14. Consider a multidigraph shown in the Figure 8.12 (a). The vertex set of this multidigraph is V, where $V = \{v_i \mid 0 \leq i \leq 6\}$. Notice that there are two edges between the vertices v_3 and v_4. The information carrying capacity of each edge in this multidigraph is unity. For simplicity, assume that this capacity is equal to a single bit per unit time. Further, the source node in the network is v_0, and the terminal nodes are represented by v_5 and v_6.

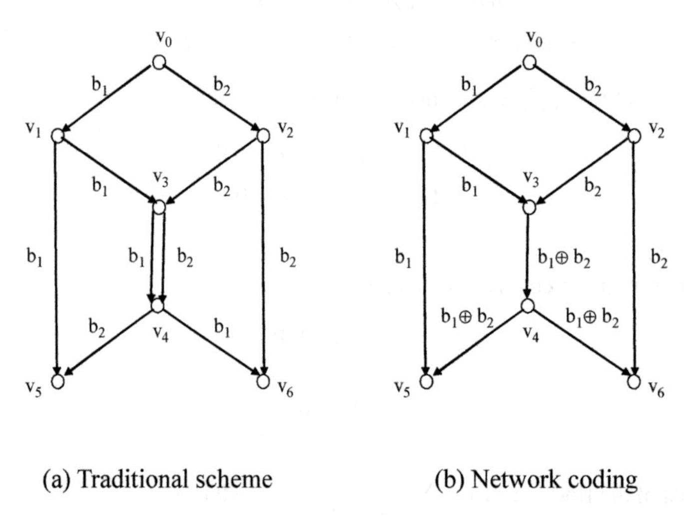

(a) Traditional scheme (b) Network coding

Figure 8.12. Comparison of traditional routing and network coding schemes.

The goal of the network is for the source node v_0 to multicast information to the two destination nodes v_5 and v_6. It is also required that the network maximize the rate at which the source node transmits information to the destination nodes. As each destination node has two incoming edges, each of unit capacity, the destination nodes can receive information at a maximum rate of two bits per unit time. Bits b_1 and b_2 are transmitted from the source node v_0 to the destination nodes v_5 and v_6. In the initial scenario, each intermediate node in the network operates in a store-and-forward mode. As mentioned earlier, this is the routing mode of transmission of information. The flow of specific bits on different edges is shown in Figure 8.12 (a). This bit-flow pattern on different edges demonstrates that optimal routing of bits has been achieved for this network.

Next consider the network represented by a digraph in Figure 8.12 (b). In this digraph, each edge has a unit capacity. Furthermore, there is only a single edge between vertices v_3 and v_4. The reader can convince herself that the traditional routing of information does not yield a transmission rate of two bits per unit time at the destination nodes. Optimal information transmission rate between the source node and the target nodes can yet be achieved by using the exclusive-or operation $\oplus$ performed at node v_3.

Note that, if $(b_1 \oplus b_2)$ is transmitted on the edge (v_3, v_4), and subsequently on the edges (v_4, v_5) and (v_4, v_6), then the destination node v_5 can recover bit b_2 via the operation $b_1 \oplus (b_1 \oplus b_2)$, where it received the bit b_1 via the edge (v_1, v_5). Similarly, the destination node v_6 can receive the bits b_1 and b_2. This scheme of mixing streams of information is called network coding. It has been tacitly assumed in this scheme that the computation time of $(b_1 \oplus b_2)$ at the node v_3 is negligible. The network configurations shown in Figure 8.12 are often called butterfly networks. □

Example 8.15. Consider a wireless network shown in Figure 8.13 (a), in which the base station is represented by the vertex v_0. Vertices v_1 and v_2 exchange information via the vertex v_0. It is assumed in this wireless network that a node cannot simultaneously transmit and receive. It is also assumed that a node cannot receive information from more than a single neighboring node.

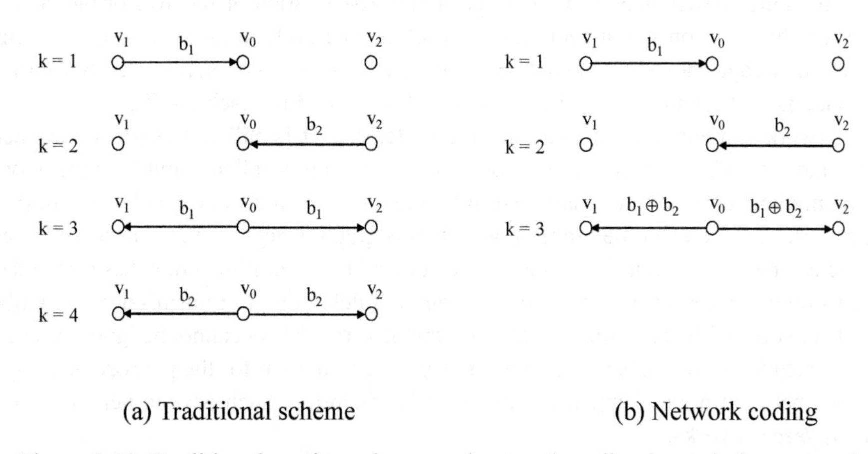

(a) Traditional scheme (b) Network coding

Figure 8.13. Traditional routing scheme and network coding in a wireless network.

In the traditional method of exchange of information, vertices v_1 and v_2 transmit bits b_1 and b_2 respectively to vertex v_0 in successive time slots $k = 1$, and 2. The vertex v_0 subsequently broadcasts bits b_1 and b_2 sequentially to the vertices v_1 and v_2 in time slots, $k = 3$, and 4 respectively. These operations occur in a total of four time slots.

In the network coding scheme shown in Figure 8.13 (b), the base station v_0 can speed up the process by broadcasting $(b_1 \oplus b_2)$ to v_1 and v_2 in time slot $k = 3$. The vertex v_1 can recover bit b_2 via the operation $b_1 \oplus (b_1 \oplus b_2)$. The vertex v_2 can similarly recover the bit b_1. These operations occur in three time slots, $k = 1, 2$, and 3. Thus the network coding scheme improves the efficiency of transmission of information by the base station. This model also assumes that the exclusive-or operation at the nodes takes negligible time. □

A graph-theoretic communication network model, linear network coding scheme, and its properties are described in the remainder of this section. This is followed by a discussion of the throughput-advantage due to network coding.

8.10.1 A Communication Network Model

We develop an elementary communication network model for describing multicast of information from a source node to a set of receiver nodes by using graph-theoretic techniques. Principles of network coding scheme are demonstrated subsequently via this model. Next, we have: description of a communication network model by using graph theory, description of source coding used in the network, communication channel capacities, and flow of symbols in the network.

Description of the Network

The network is represented by a multidigraph $G = (V, E)$, where V is the set of nodes (vertices) in the network, and E is the set of edges. An edge is also called a channel. As the graph G is a multidigraph, parallel edges between any pair of nodes are permitted.

The set of input and output edges of a node $v \in V$ are denoted by $\Gamma_I (v)$ and $\Gamma_O (v)$ respectively. Also assume that there are no self-loops in the multidigraph. That is,

$$\Gamma_I (v) \cap \Gamma_O (v) = \varnothing, \quad \forall\, v \in V$$

The set of source nodes is $S \subset V$. The set of sink (destination or receiver or target) nodes is $\mathcal{T}_{rcvr} \subseteq V \backslash S$. Information is transmitted from each source node to all sink nodes. Assume that there are no input edges at each node $s \in S$. That is, $\Gamma_I (s) = \varnothing$ for each $s \in S$. Similarly, there are no output edges at each node $t \in \mathcal{T}_{rcvr}$. That is, $\Gamma_O (t) = \varnothing$ for each $t \in \mathcal{T}_{rcvr}$.

If the network contains only a single source node, then it is called a single-source network. However, if the network contains multiple source nodes, then it is called a multi-source network.

The communication network is assumed to be modeled by an acyclic graph. This implies that there are no directed cycles in the multidigraph. If this graph is acyclic, then successive messages at each node can be processed independent of each other. However, if the graph has directed cycles, then the encoding (processing) delay at each node, and delay due to transmission of symbols on the channel convolve with each other. In this case the incurred delays cannot be ignored. Therefore, if the communication network is modeled as an acyclic graph, then for the purpose of analysis, all transmissions between nodes happen instantaneously. Therefore such acyclic networks are often termed *delay-free* networks.

Source Coding

The alphabet used in communication is $\mathcal{X}$, where $|\mathcal{X}| \geq 2$. It is the set of source symbols. The alphabet is usually the set of elements of a finite field $\mathbb{F}_q$, where $|\mathcal{X}| = q \in \mathbb{P} \backslash \{1\}$. This finite field is also denoted by GF_q. For example, if $q = 2$, then

$$\mathbb{F}_2 = GF_2 = \mathbb{Z}_2$$

In this field, a symbol is simply a bit. The field $\mathbb{F}_q$ is also often called a base field. The source emits information in codeword (block code) of length $m \in \mathbb{P}$.

Channel Capacity

As the graph G is a multidigraph, the capacity of each edge is unity, and multiple edges between a pair of nodes essentially simulate the effect of capacity of an edge between a pair of nodes.

Flows in Networks

If there is only a single source node and a single sink node in the network modeled as a digraph, then the value of the maximum flow is equal to the capacity of a min-cut between these two nodes. This is the celebrated max-flow min-cut theorem established in the last section. This flow can be achieved via the traditional routing (store and forward) scheme.

Next consider another scenario in which information is multicast at a rate ω_m from a single source node $s \in V$ to a set of sink nodes $\mathcal{T}_{rcvr} \subseteq V \backslash \{s\}$, where $|\mathcal{T}_{rcvr}| \geq 2$. Multicast of information from the source node s to the sink nodes in the set $\mathcal{T}_{rcvr}$ implies that the information from the node s is delivered to all the nodes in the set $\mathcal{T}_{rcvr}$ simultaneously. Let the value of the maximum flow from the node s to node $t_i \in \mathcal{T}_{rcvr}$ be $maxflow\,(t_i)$. This value is equal to the minimum value of a s-t_i cut. Therefore

$$\omega_m \leq maxflow\,(t_i)\,, \quad \forall\, t_i \in \mathcal{T}_{rcvr}$$

As the information is *multicast* from the source node s to each sink node in the set $\mathcal{T}_{rcvr}$, then

$$\omega_m \leq \min_{t_i \in \mathcal{T}_{rcvr}} maxflow\,(t_i)$$

In the store and forward mode of communication, it is not quite evident that this maximum flow-bound is achievable for multicast transmission. It is the thesis of the network coding paradigm of transmitting information, that this bound is certainly achievable (with equality) for multicast transmission for networks which use network coding techniques. Moreover, linear network coding can be used to achieve this goal for a communication network which can be modeled as an acyclic graph, provided the number of symbols in the alphabet $\mathcal{X}$ is sufficiently large. These assertions are established in the rest of this section.

8.10.2 Linear Network Coding

We develop a basis to demonstrate the benefits of network coding in this subsection. Linear network coding is achieved by judiciously combining linearly the symbols emitted by the source node, at each non-source node of the network. This results in optimal multicast transmission of symbols from a source node to sink nodes. The communication network is first described formally.

Definition 8.19. *Let $\mathcal{N} = (G, s, \mathcal{T}_{rcvr}, \mathcal{X}, m)$ represent a communication network.*

(a) *The network is represented by a multidigraph $G = (V, E)$, where V is the set of nodes (vertices) in the network, and E is the set of edges (channels), where, $|V|$ and $|E|$ are finite.*
(b) *The set of input and output edges of a node $v \in V$ are denoted by $\Gamma_I\,(v)$ and $\Gamma_O\,(v)$ respectively. There are no self-loops in the graph, and the graph G is acyclic. That is, there are no directed cycles in the network.*
(c) *A unique source node $s \in V$ transmits symbols from the alphabet $\mathcal{X}$, to a set of sink (destination, or receiver, or target) nodes $\mathcal{T}_{rcvr} \subseteq V \backslash \{s\}$.*
Typically, $\mathcal{X}$ is the set of elements of a finite field $\mathbb{F}_q = GF_q$, where $|\mathcal{X}| = q \in \mathbb{P} \backslash \{1\}$. This field is often called the base field.

(d) *The number of output edges at the source node s is equal to $m \in \mathbb{P}$. That is, $|\Gamma_O(s)| = m$. For ease in discussion, we shall specify that the source node s has m imaginary input edges. This set of edges is denoted by $\Gamma_I(s)$.*
Also $\Gamma_O(v) = \varnothing$ for each $v \in \mathcal{T}_{rcvr}$.
(e) *Assume that each node $v \in V \backslash \{s\}$ has at least a single input channel.*
(f) *As G is a multidigraph, multiple edges are permitted between each pair of nodes. The capacity of each edge is unity. That is, one symbol from the alphabet $\mathcal{X}$ can be transmitted over each edge per unit time.*
The multiplicity of edges between each pair of nodes is used to represent the capacity of direct transmission between them. □

The source node $s \in V \backslash \mathcal{T}_{rcvr}$ generates messages. A single message consists of m symbols drawn from the alphabet $\mathcal{X}$. Therefore, a message can be represented as an m-dimensional vector in the vector space $\mathbb{F}_q^{(m)}$. Let $y \in \mathcal{X}^m$ be a message, where $y^T = \begin{bmatrix} y_1 & y_2 & \cdots & y_m \end{bmatrix}$ is a row vector of size m. As $|\Gamma_O(s)| = m$, the source node s transmits the m components of the vector y on its m different output channels simultaneously. Encoding of the incoming symbols at each intermediate node in transit can be described via a mapping. The transformed symbols are then sent on the output channels of the non-sink nodes. This mapping, which may not necessarily be linear, provides a local description of the network.

Definition 8.20. *Local mapping of a network code. Consider a communication network $\mathcal{N} = (G, s, \mathcal{T}_{rcvr}, \mathcal{X}, m)$, where $G = (V, E)$. For each $e \in E$, the local encoding mapping is*

$$\widetilde{k}_e : \mathcal{X}^{|\Gamma_I(v)|} \to \mathcal{X} \tag{8.23}$$

where $e \in \Gamma_O(v)$, and $v \in V \backslash \mathcal{T}_{rcvr}$. The corresponding network code is said to be m-dimensional. □

The encoding mapping for each edge $e \in E$, can also be specified in terms of the map of the message $y \in \mathcal{X}^m$ created at the source node s. This is called a global mapping of the network code.

Definition 8.21. *Global mapping of a network code. Consider a communication network $\mathcal{N} = (G, s, \mathcal{T}_{rcvr}, \mathcal{X}, m)$, where $G = (V, E)$. The global encoding mapping for each edge $e \in E$ is*

$$\widetilde{f}_e : \mathcal{X}^m \to \mathcal{X} \tag{8.24}$$

where $\mathcal{X}^m$ is the set of messages which can be generated by the source node $s \in V \backslash \mathcal{T}_{rcvr}$. For the imaginary $e \in \Gamma_I(s)$, the mappings $\widetilde{f}_e(\cdot)$'s are the projections from the space $\mathcal{X}^m$ to the respective m different coordinates. □

It can be observed that for each node $v \in V \backslash \mathcal{T}_{rcvr}$, each channel $e \in \Gamma_O(v)$, and $y \in \mathcal{X}^m$; $\widetilde{f}_e(y)$ is uniquely determined by

$$\left\{ \widetilde{f}_d(y) \mid d \in \Gamma_I(v), y \in \mathcal{X}^m \right\}$$

via the local encoding mapping $\widetilde{k}_e(\cdot)$.

We next consider local and global *linear* network coding mappings. If the local encoding mapping $\widetilde{k}_e(\cdot)$ at vertex $v \in V \backslash \mathcal{T}_{rcvr}$ and edge $e \in \Gamma_O(v)$ is linear, then $\widetilde{k}_e(z) = z^T k_e$, where $z \in \mathcal{X}^{|\Gamma_I(v)|}$ is a column vector which represents the symbols received at the node v, and $k_e \in \mathcal{X}^{|\Gamma_I(v)|}$ is a column vector. Similarly, if the global encoding mapping $\widetilde{f}_e(\cdot)$ at $e \in E$ is linear, then $\widetilde{f}_e(y) = y^T f_e$, where f_e is a column vector of size m, and $f_e \in \mathcal{X}^m$. Note that $y \in \mathcal{X}^m$ is the message generated by the source node $s \in V \backslash \mathcal{T}_{rcvr}$.

If in the acyclic communication network $\mathcal{N} = (G, s, \mathcal{T}_{rcvr}, \mathcal{X}, m)$, m-dimensional network coding is used, then linear local mappings imply linear global mappings. Conversely, linear global mappings also imply linear local mappings. The local and global linear mappings of network code are next described.

Definitions 8.22. *Let $\mathcal{N} = (G, s, \mathcal{T}_{rcvr}, \mathcal{X}, m)$, represent a communication network, where $G = (V, E)$.*

1. *Local linear mapping of a network code. Each node $v \in V \backslash \mathcal{T}_{rcvr}$ performs a linear mapping. It is specified by a $|\Gamma_I(v)| \times |\Gamma_O(v)|$ matrix*
$$K_v = [k_{de}] \tag{8.25a}$$
where $k_{de} \in \mathcal{X}, d \in \Gamma_I(v)$, and $e \in \Gamma_O(v)$. Note that k_{de} is called the local encoding kernel for the pair (d, e) of adjacent channels, and K_v is called the local encoding kernel at node v. A pair of channels $(d, e) \in E \times E$ is called an adjacent pair if there exists a node $v \in V$ such that $d \in \Gamma_I(v)$ and $e \in \Gamma_O(v)$.
The matrix K_v implicitly assumes ordering among both the input and output channels.
2. *Global linear mapping of a network code. For each node $v \in V \backslash \mathcal{T}_{rcvr}, d \in \Gamma_I(v)$, $e \in \Gamma_O(v)$, and $k_{de} \in \mathcal{X}$, there exist column vectors $f_d, f_e \in \mathcal{X}^m$ such that*
$$f_e = \sum_{d \in \Gamma_I(v)} k_{de} f_d, \quad e \in \Gamma_O(v) \tag{8.25b}$$
For the node $v = s$, the f_d's for the m imaginary channels $d \in \Gamma_I(s)$ are the standard basis of the vector space $\mathbb{F}_q^{(m)}$.
The vector f_e is called the global encoding kernel for channel $e \in E$. □

In order to implement linear network coding, and recover the transmitted message $y \in \mathcal{X}^m$ at each sink node $t \in \mathcal{T}_{rcvr}$, it is necessary that the global encoding kernels $f_e, e \in \Gamma_I(t)$ be known.

Example 8.16. The terminology developed in this subsection is clarified via the butterfly network shown in Figure 8.12 (b). The set of vertices in this network is $V = \{v_i \mid 0 \le i \le 6\}$. In this network, the source node is $s = v_0$, and the set of sink nodes is $\mathcal{T}_{rcvr} = \{v_5, v_6\}$. The alphabet is $\mathcal{X} = \{0, 1\}$, and the base field $\mathbb{Z}_2$. The message vector to be multicast from the source node s to the sink nodes is $y^T = \begin{bmatrix} b_1 & b_2 \end{bmatrix}$, where $b_1, b_2 \in \mathcal{X}$. Let $(b_1 + b_2) \equiv c \pmod 2$. The local encoding mappings are:

$$\widetilde{k}_{(s,v_1)}(b_1, b_2) = b_1, \ \widetilde{k}_{(s,v_2)}(b_1, b_2) = b_2,$$
$$\widetilde{k}_{(v_1,v_3)}(b_1) = \widetilde{k}_{(v_1,v_5)}(b_1) = b_1,$$
$$\widetilde{k}_{(v_2,v_3)}(b_2) = \widetilde{k}_{(v_2,v_6)}(b_2) = b_2,$$
$$\widetilde{k}_{(v_3,v_4)}(b_1, b_2) = c,$$
$$\widetilde{k}_{(v_4,v_5)}(c) = \widetilde{k}_{(v_4,v_6)}(c) = c$$

Let the imaginary input channels (edges) at the source node s be (o, s) and $(o, s)'$ respectively. The global encoding mappings are:

$$\begin{aligned}
\widetilde{f}_e(y) &= b_1, \quad \text{for } e = (o,s), (s,v_1), (v_1,v_3), (v_1,v_5) \\
\widetilde{f}_e(y) &= b_2, \quad \text{for } e = (o,s)', (s,v_2), (v_2,v_3), (v_2,v_6) \\
\widetilde{f}_e(y) &= c, \quad \text{for } e = (v_3,v_4), (v_4,v_5), (v_4,v_6)
\end{aligned}$$

The network code specified for the butterfly network is actually linear. Let the alphabetical order among the channels be

$$(o,s), (o,s)', (s,v_1,), \ldots, (v_4,v_6)$$

The local encoding kernels are:

$$K_s = \begin{bmatrix} 1 & 0 \\ 0 & 1 \end{bmatrix}, \quad K_{v_1} = K_{v_2} = K_{v_4} = \begin{bmatrix} 1 & 1 \end{bmatrix}, \quad K_{v_3} = \begin{bmatrix} 1 \\ 1 \end{bmatrix}$$

The global encoding kernels are:

$$\begin{aligned}
f_e &= \begin{bmatrix} 1 & 0 \end{bmatrix}^T, \quad \text{for } e = (o,s), (s,v_1), (v_1,v_3), (v_1,v_5) \\
f_e &= \begin{bmatrix} 0 & 1 \end{bmatrix}^T, \quad \text{for } e = (o,s)', (s,v_2), (v_2,v_3), (v_2,v_6) \\
f_e &= \begin{bmatrix} 1 & 1 \end{bmatrix}^T, \quad \text{for } e = (v_3,v_4), (v_4,v_5), (v_4,v_6)
\end{aligned}$$

□

8.10.3 Properties of Linear Network Coding

Certain important properties of linear network coding are described in this subsection. We initially introduce notation about maximum flow rates between a source node and non-source node(s).

Definitions 8.23. *Let $\mathcal{N} = (G, s, \mathcal{T}_{rcvr}, \mathcal{X}, m)$, represent a communication network, where $G = (V, E)$ is a multidigraph, and the capacity of each edge is unity.*

1. *The value of maximum flow between the source node s, and a node $t \in V \backslash \{s\}$ is denoted by $maxflow\,(t)$.*
2. *The value of maximum flow between the source node s, and a set of non-source nodes $\mathcal{T} \subseteq V \backslash \{s\}$ is denoted by $maxflow\,(\mathcal{T})$.* □

The $maxflow\,(t)$ can be determined via the max-flow min-cut theorem. It is equal to the capacity of the minimum cut between the s and $t \in V \backslash \{s\}$ node.

The value $maxflow\,(\mathcal{T})$ is determined as follows. Assume that $\mathcal{T}$ is nonempty. Insert a new node $v_{\mathcal{T}}$ in the graph G. Also insert $maxflow\,(t)$ number of edges between the vertices t and $v_{\mathcal{T}}$, for each $t \in \mathcal{T}$. The capacity of each newly formed edge is unity. Then $maxflow\,(v_{\mathcal{T}}) = maxflow\,(\mathcal{T})$.

Recall that the span of a set of vectors $\widetilde{R}$ is the set of all possible finite linear combinations of the vectors in $\widetilde{R}$. The span of a set of vectors in the set $\widetilde{R}$ is denoted by $L\left(\widetilde{R}\right)$.

Definitions 8.24. *Let $\mathcal{N} = (G, s, \mathcal{T}_{rcvr}, \mathcal{X}, m)$, represent a communication network, where $G = (V, E)$. The network uses linear network coding. The global encoding kernel for channel $e \in E$ is f_e.*

1. *For a node $t \in V$, the linear span of a set of vectors $\{f_e \mid e \in \Gamma_I(t)\}$ is*

$$L_t = L(\{f_e \mid e \in \Gamma_I(t)\}) \tag{8.26a}$$

2. *For a set of nodes $\mathcal{T} \subseteq V$, let*

$$L_{\mathcal{T}} = L(\cup_{t \in \mathcal{T}} L_t) \tag{8.26b}$$

3. *For a collection of channels $\xi \subseteq E$, the linear span of a set of vectors $\{f_e \mid e \in \xi\}$ is*

$$L_{\xi} = L(\{f_e \mid e \in \xi\}) \tag{8.26c}$$

Also $L_{\varnothing} = \{0\}$, where 0 is a zero column m-vector. □

The next theorem establishes the maximum-flow bound for linear network coding.

Theorem 8.11. (*Maximum-flow bound*) *Let $\mathcal{N} = (G, s, \mathcal{T}_{rcvr}, \mathcal{X}, m)$ be the communication network which uses linear network coding. For a set of non-source nodes $\mathcal{T} \subseteq V \backslash \{s\}$, we have*

$$\dim(L_{\mathcal{T}}) \leq \min\{m, maxflow(\mathcal{T})\} \tag{8.27}$$

Proof. See the problem section. □

Definition 8.25. *Multicast with linear network coding. Consider a communication network $\mathcal{N} = (G, s, \mathcal{T}_{rcvr}, \mathcal{X}, m)$, which uses linear network coding. Let $\dim(L_t) = m$, and $maxflow(t) \geq m$ for every non-source node $t \in \mathcal{T} \subseteq V \backslash \{s\}$. Then linear multicast transmission from the source node s to the set of non-source nodes $\mathcal{T}$ is said to occur.* □

Let ξ be a set of channels. It might possibly include imaginary channels at the input of the source node s. Let

$$F_{\xi} = \begin{bmatrix} f_{e_1} & f_{e_2} & \cdots & f_{e_{|\xi|}} \end{bmatrix}$$

be an $m \times |\xi|$ matrix. If $y \in \mathcal{X}^m$ is transmitted at the source node s, then at the vertex t, the symbol $y^T f_e$ is received for each $e \in \Gamma_I(t)$. These symbols are actually elements of the row vector $y^T F_{\Gamma_I(t)}$.

Therefore the elements of the row vector y^T can be determined uniquely at the node t, if and only if the rank of the matrix $F_{\Gamma_I(t)}$ is equal to m. That is, $\dim(L_t) = m$.

Next consider the maximum-flow bound theorem. The node $t \in V \backslash \{s\}$ can decode the message y if and only if $maxflow(t) \geq m$. However, if $maxflow(t) < m$ there are no guarantees.

Existence of Linear Network Coding for Multicasting

The existence of linear network coding scheme for the communication network

$$\mathcal{N} = (G, s, \mathcal{T}_{rcvr}, \mathcal{X}, m)$$

depends upon the value of $m \in \mathbb{P}$, the network topology specified by the multidigraph $G = (V, E)$, and the choice of the base field $\mathbb{F}_q$. Note again that $\mathcal{X}$ is the set of elements of a finite field $\mathbb{F}_q$, where $|\mathcal{X}| = q \in \mathbb{P} \backslash \{1\}$.

Assume that the capacity of each min-cut to each sink node $t \in \mathcal{T}_{rcvr}$ is at least h. In this scheme, nodes in the network linearly combine the incoming information symbols which belong to the alphabet $\mathcal{X}$. We shall prove that there exists a multicast transmission scheme over a sufficiently large finite field $\mathbb{F}_q$ which can transmit information simultaneously from the source node $s \in V$ to all nodes in the set $\mathcal{T}_{rcvr} \subseteq V \backslash \{s\}$ at a rate equal to m, provided $m \leq h$. In order to prove this result, the following result, called sparse zeros lemma is necessary.

Lemma 8.4. *Sparse zeros lemma. Let $g(z_1, z_2, \ldots, z_\eta) \in \mathbb{F}_q[z_1, z_2, \ldots, z_\eta]$ be a nonzero multivariate polynomial with degree in each variable z_i at most $\delta \in \mathbb{N}$. The elements in the field $\mathbb{F}_q$ belong to the alphabet $\mathcal{X}$, and $|\mathcal{X}| = q$ is finite. If $q > \delta$, then there exist $a_1, a_2, \ldots, a_\eta \in \mathcal{X}$ such that $g(a_1, a_2, \ldots, a_\eta) \neq 0$.*

Proof. See the problem section. □

Theorem 8.12. *A communication network $\mathcal{N} = (G, s, \mathcal{T}_{rcvr}, \mathcal{X}, m)$ uses linear network coding and base field $\mathbb{F}_q$. In this communication network $m \leq \min_{t_i \in \mathcal{T}_{rcvr}} maxflow(t_i)$. Multicast transmission from the source node s to the set of sink nodes in the set $\mathcal{T}_{rcvr}$ is possible at a rate m, provided $|\mathcal{X}| = q$ is sufficiently large.*

Proof. See the problem section. □

8.10.4 Throughput Improvement

Information throughput improvement due to network coding is examined in this subsection. Information throughput is defined as the number of symbols that are transmitted per unit time from the source to the sink nodes in a given network. The set of symbols belong to the alphabet $\mathcal{X}$. For a given network, we determine the ratio of the throughput with and without network coding. This evaluation requires the use of combinatorial optimization techniques.

Assume that the communication network is represented by an undirected graph $G = (V, E)$. Associated with this graph is a mapping $u : E \to \mathbb{R}_0^+$. That is, for each edge $e \in E$ there is a value $u(e) \geq 0$ called the capacity of the edge e. The capacity $u(e)$ is also denoted by u_e for each $e \in E$. We investigate multicast transmission between a source node $s \in V$, and a set of receiving (sink) nodes $\mathcal{T}_{rcvr} \subseteq V \backslash \{s\}$, where $|\mathcal{T}_{rcvr}| \geq 2$.

Throughput in the network with and without network coding is denoted by $\mathcal{Y}_{NC}(G, u)$ and $\mathcal{Y}(G, u)$ respectively. The network coding throughput-advantage defined as

$$\frac{\mathcal{Y}_{NC}(G, u)}{\mathcal{Y}(G, u)}$$

is studied in the rest of this subsection. It is shown that the network coding throughput-advantage can be assessed in terms of Steiner and multicast trees of the graph G.

Trees

Steiner tree is named in honor of the mathematician Jakob Steiner (1796-1863). Such trees were foreseen earlier by other mathematicians like Pierre de Fermat (1601-1665), and Johann Carl Friedrich Gauss (1777-1855).

A Steiner tree in graph theory is a generalization of the minimum spanning tree. Consider an undirected graph $G = (V, E)$ in which a nonnegative weight is assigned to each edge. Let $R \subseteq V$ be a subset of vertices. The least cost connected subgraph spanning R is called a Steiner tree. The spanning subgraph is a tree because it is least cost. Use of vertices in the set $V \backslash R$ is allowed in the construction of the Steiner tree. If $R = V$, then the Steiner tree is simply a minimum spanning tree of the graph G. A formal definition of a Steiner tree is given below.

Definition 8.26. *Consider a graph $G = (V, E)$, and a distinguished set of nodes $R \subseteq V$. A nonnegative cost is assigned to each edge in the graph G. A Steiner tree is a minimum-cost spanning tree $T = (V', E')$ where $R \subseteq V' \subseteq V$, and $E' \subseteq E$.*

Vertices in the set R are called the terminal nodes, and those in the set $V \backslash R$ are called the Steiner vertices. □

Note that a Steiner tree is not unique. Determination of a minimum cost Steiner tree is known to be computationally expensive. This is actually the celebrated *Steiner tree problem.*

A multicast tree in the undirected graph $G = (V, E)$ is a tree which spans the source node and the receiver nodes. Such trees can also include vertices in the set $V - s \cup \mathcal{T}_{rcvr}$. Thus all Steiner trees which span the vertices in the set $s \cup \mathcal{T}_{rcvr}$ are multicast trees, but not vice versa. Actually, all multicast trees are candidates for Steiner trees. Terminology associated with multicast trees relevant to network coding is introduced in the following definition.

Definition 8.27. *Let a communication network be represented by an undirected graph $G = (V, E)$. Also, let $s \in V$ be a source node, and $\mathcal{T}_{rcvr} \subseteq V \backslash \{s\}$ be the set of receiver nodes. The capacity of an edge is specified by the mapping $u : E \to \mathbb{R}_0^+$. The capacity of edge e is $u(e) \triangleq u_e$, $\forall\, e \in E$. Let $R = s \cup \mathcal{T}_{rcvr}$.*

(a) *A tree $T = (V', E')$ is a multicast tree which spans the vertices in the set R, where $R \subseteq V' \subseteq V$, and $E' \subseteq E$.*
(b) *Let Υ be the set of multicast trees associated with the graph G and the set R. Packing of multicast trees is an assignment $x_\tau \in \mathbb{R}_0^+$, for each $\tau \in \Upsilon$ such that*

$$\sum_{\tau \in \Upsilon,\ e \text{ is in tree } \tau} x_\tau \le u_e, \quad \forall\, e \in E \tag{8.28}$$

The value of the packing is $\sum_{\tau \in \Upsilon} x_\tau$.
(c) *The maximum value of the packing, $\sum_{\tau \in \Upsilon} x_\tau$ is called the fractional multicast tree packing number.* □

The word "fractional" in the phrase *fractional multicast tree packing number* is used because the assignments x_τ's are allowed to take nonnegative and possibly nonintegral values.

Throughput without Network Coding - Fractional Multicast Tree Packing

Model for throughput computation in a multicast network without network coding can be formulated in terms of multicast trees in the undirected graph G. Let Υ be the set of all multicast trees in the graph G which span vertices in the set $\{s\} \cup \mathcal{T}_{rcvr}$. In such trees, there is a single path between the source node s and each target node in the set $\mathcal{T}_{rcvr}$. Thus, the multicast tree spans the source vertex and all the sink vertices. Associated with each tree $\tau \in \Upsilon$ is a flow variable x_τ. It is equal to the flow rate from the source node s to the sink nodes which uses the tree τ. The flow x_τ is allowed to take nonnegative fractional values. That is, $x_\tau \in \mathbb{R}_0^+$.

Maximum throughput $\mathcal{Y}(G, u)$ from the source node to the sink nodes is the fractional multicast tree packing number for the graph G. It is achieved by solving the following linear programming problem. We state both the primal and dual of the linear program associated with the *fractional multicast tree packing*. Feasibility of the problem is also assumed.

Input: Undirected graph $G = (V, E)$, capacity function $u : E \to \mathbb{R}_0^+$, where $u(e) \triangleq u_e$, $\forall e \in E$. The source node is $s \in V$, and sink nodes $\mathcal{T}_{rcvr} \subseteq V \backslash \{s\}$. Let $\mathcal{T}_{rcvr} = \{t_1, t_2, \ldots, t_\kappa\}$, where $|\mathcal{T}_{rcvr}| = \kappa$, and $\kappa \geq 2$.

The set of multicast trees spanning the nodes $\{s\} \cup \mathcal{T}_{rcvr}$ is Υ.

Output: Maximum throughput between the source node and the set of sink nodes without network coding: $\mathcal{Y}(G, u)$.

Primal:
Objective: *Maximize* $\sum_{\tau\in\Upsilon} x_\tau$
Subject to:

$$\sum_{\tau\in\Upsilon,\ e \text{ is in tree } \tau} x_\tau \leq u_e, \quad \forall e \in E$$

$$x_\tau \geq 0, \quad \forall \tau \in \Upsilon$$

The maximized value of $\sum_{\tau\in\Upsilon} x_\tau$ is $\mathcal{Y}(G, u)$.

Dual: The decision variables in this dual are $y_e, e \in E$.
Objective: *Minimize* $\sum_{e\in E} u_e y_e$
Subject to:

$$\sum_{e \text{ is in tree } \tau} y_e \geq 1, \quad \forall \tau \in \Upsilon$$

$$y_e \geq 0, \quad \forall e \in E$$

The minimized value of $\sum_{e\in E} u_e y_e$ is $\mathcal{Y}(G, u)$, by the strong duality theorem of linear programming.

In the dual program, the variable y_e can be viewed as cost of edge $e \in E$. The objective of the dual program is to minimize $\sum_{e\in E} u_e y_e$. The corresponding constraint implies that the cost of *each* multicast tree is at least unity.

The above primal and dual problems are computationally expensive to solve, as the set of all multicast trees Υ can be exponentially large in the number of vertices $|V|$ and the number of sink nodes $|\mathcal{T}_{rcvr}|$.

Throughput with Network Coding

The flows in the communication networks which use network coding, are called *conceptual flows*. These flows are called conceptual, because such flows are only conceptual and not physical. In this scheme, different conceptual flows coexist in the network *without* contending for the edge capacities. This is in contrast to the traditional *commodity flows* in networks. In the commodity flow model of information flow, data is stored and forwarded and not coded. In the commodity flow model, different flows compete for the capacity of a shared edge.

The undirected graph $G = (V, E)$ is converted into a directed graph $G_D = (V, E_D)$ as follows. Each undirected edge $e \in E$ with end nodes $v \in V$ and $w \in V$ is converted into two directed edges $a_1 = (v, w) \in E_D$ and $a_2 = (w, v) \in E_D$. Further,

$$u_e = u_{a_1} + u_{a_2}, \quad \forall e \in E$$

Also, the edge capacities in the directed network are nonnegative. That is, $u_a \in \mathbb{R}_0^+$, $\forall\, a \in E_D$. Thus the number of directed edges in the digraph G_D is equal to twice the number of edges in the undirected graph G. That is, $|E_D| = 2\,|E|$.

The conceptual flow rate to a sink node $t_i \in \mathcal{T}_{rcvr}$ is denoted by $x^i \in \mathbb{R}_0^+$. The corresponding flow rate on an edge $a \in E_D$ is $x^i\,(a) \in \mathbb{R}_0^+$. The total incoming flow rate x^i at vertex $v \in V$ is denoted by $x_{in}^i\,(v) \in \mathbb{R}_0^+$. Similarly, the total outgoing flow rate x^i at vertex $v \in V$ is denoted by $x_{out}^i\,(v) \in \mathbb{R}_0^+$. The multicast target flow rate is $x^* \in \mathbb{R}_0^+$. The goal (objective) is to maximize x^*.

Each conceptual flow rate should also satisfy the capacity constraint. Further, for each flow, conservation of flow rate should be preserved at all non-source and non-receiver nodes. The incoming flow rates at the source node, and the outgoing flow rate at each sink node are all zero. Also the flow rate at each sink node is equal to at least the target flow rate. The conceptual flow problem is stated as an optimization problem.

Input: Undirected graph $G = (V, E)$, capacity function $u : E \to \mathbb{R}_0^+$, where $u\,(e) \triangleq u_e$, $\forall$ $e \in E$. The source node is $s \in V$, and sink nodes $\mathcal{T}_{rcvr} \subseteq V \backslash \{s\}$. Let $\mathcal{T}_{rcvr} = \{t_1, t_2, \ldots, t_\kappa\}$, where $|\mathcal{T}_{rcvr}| = \kappa$, and $\kappa \geq 2$.

The undirected graph G is transformed into a directed graph $G_D = (V, E_D)$. Each edge $e \in E$ is converted into two directed arcs $a_1 \in E_D$ and $a_2 \in E_D$ of opposite orientations, each with nonnegative capacities.

Output: Maximized value of target flow rate x^*, which is $\mathcal{Y}_{NC}\,(G, u)$.

Objective: Maximize x^*

Subject to orientation and conceptual flow constraints:

(a) *Orientation constraints*:

$$\begin{aligned} u_a &\in \mathbb{R}_0^+, \quad \forall\, a \in E_D \\ u_{a_1} + u_{a_2} &= u_e, \quad a_1, a_2 \in E_D, \quad \forall\, e \in E \end{aligned}$$

(b) *Conceptual flow constraints*: $\forall\, i \in [1, 2, \ldots, \kappa]$

$$\begin{aligned} 0 \leq x^i\,(a) &\leq u_a, \quad \forall\, a \in E_D \\ x_{in}^i\,(v) &= x_{out}^i\,(v), \quad \forall\, v \in V - \{s, t_i\} \\ x_{in}^i\,(s) &= 0 \\ x_{out}^i\,(t_i) &= 0 \end{aligned}$$

(c) *Rate constraints at receiver nodes*: $x^* \leq x_{in}^i\,(t_i)\,, \forall\, i \in [1, 2, \ldots, \kappa]$.

The optimum value of the objective function in this optimization problem is denoted by $\mathcal{Y}_{NC}\,(G, u)$.

The maximum value of the ratio $\mathcal{Y}_{NC}\,(G, u)\,/\mathcal{Y}\,(G, u)$, obtained by varying u the capacities of the arcs, can be expressed alternately by considering a Steiner tree problem.

Steiner Tree Problem

Weights are assigned to edges in the undirected graph $G = (V, E)$ via the mapping $w : E \rightarrow \mathbb{R}_0^+$. That is, for each edge $e \in E$ there is a value $w(e) \geq 0$ called the weight of the edge e. The weight $w(e)$ is also denoted by $w_e \ \forall \ e \in E$. We determine a Steiner tree in the graph G in which the set of terminating nodes is $s \cup \mathcal{T}_{rcvr}$. Consider the following optimization problem. It is framed as a bidirected cut integer program. This is an optimization problem in which the decision variables are integers.

Input: Undirected graph $G = (V, E)$, weight function $w : E \rightarrow \mathbb{R}_0^+$, source node $s \in V$, and sink nodes $\mathcal{T}_{rcvr} \subseteq V \backslash \{s\}$. Let $\mathcal{T}_{rcvr} = \{t_1, t_2, \ldots, t_\kappa\}$, where $|\mathcal{T}_{rcvr}| = \kappa$, and $\kappa \geq 2$. The terminal nodes of the Steiner tree are $s \cup \mathcal{T}_{rcvr}$.

The undirected graph $G = (V, E)$ is converted into a directed graph $G_D = (V, E_D)$ by replacing each edge $e \in E$ by two directed arcs $a_1, a_2 \in E_D$ of opposite orientations, and the weight of each of these arcs is equal to w_e.

Comment: A *separating* or *valid* set of vertices is a set $C \subset V$, such that the source node $s \in C$, and $V \backslash C$ contains at least a single sink node. The set of edges from C to $V \backslash C$ is $\delta(C) = \{(u, v) \in E_D \mid u \in C, v \notin C\}$.

Output: The minimized value of $\sum_{a \in E_D} w_a u_a$ is $OPT(G_D, w)$. The weight of the Steiner tree is $OPT(G, w) = OPT(G_D, w)$, where the set of terminating nodes is $s \cup \mathcal{T}_{rcvr}$.

The decision variable $u_a = 1$ if $a \in E_D$ is in the Steiner tree, and $u_a = 0$ otherwise.

Objective: *Minimize* $\sum_{a \in E_D} w_a u_a$

Subject to:

$$\sum_{a \in \delta(C)} u_a \geq 1, \quad \forall \, C$$

$$u_a \in \{0, 1\}, \quad \forall \, a \in E_D$$

The inequality $\sum_{a \in \delta(C)} u_a \geq 1, \forall \, C$ ensures that the source node s is connected to each node in the set $\mathcal{T}_{rcvr}$. If the weights in the above problem are interpreted as capacities, these constraints imply that the capacity of the minimum cut between the source node s and each receiver in the set $\mathcal{T}_{rcvr}$ is at least unity. Let the minimized value of $\sum_{a \in E_D} w_a u_a$ be $OPT(G_D, w)$. Also, let the optimal solution to the above problem have arcs in the set E'_D, where $u_a = 1$ for all $a \in E'_D$. Let $G'_D = (V, E'_D)$ be the graph induced by E'_D. If the orientation of the arcs in the set E'_D is removed, we obtain a Steiner tree of the undirected graph G with a weight $OPT(G, w) = OPT(G_D, w)$.

We next show that this construction indeed produces a Steiner tree of the undirected graph G. Let C' be the set of all vertices that can be reached from s in G'. If the set C' does not include all the sink nodes in the set $\mathcal{T}_{rcvr}$, then the constraint $\sum_{a \in \delta(C)} u_a \geq 1, \forall \, C$ is violated. This inequality ensures that there is a path between the source node s and each sink node in the set $\mathcal{T}_{rcvr}$. Further, the minimization of the arc weights implies that the underlying graph is indeed a tree.

Determining $OPT(G_D, w)$ with integral values of the decision variables is computationally very expensive. The problem becomes relatively easier to solve, if the integral constraints are relaxed. By relaxing the integrality constraint $u_a \in \{0, 1\}, \forall \, a \in E_D$; and letting u_a take values in the interval $[0, 1]$ for each $a \in E_D$, we can obtain a minimum value of $\sum_{a \in E_D} w_a u_a$. This constraint can be further relaxed by letting $u_a \geq 0, \forall \, a \in E_D$. This later constraint is justified by noticing that if a solution is feasible with $u_a \geq 1$, then it remains feasible by setting $u_a = 1$. Let the corresponding optimum value be $OPT_{LP}(G, w)$. The subscript LP implies that the original problem with integer decision variables has been converted to a linear programming (LP) problem. Evidently $OPT_{LP}(G, w) \leq OPT(G, w)$. The maximum value of $OPT(G, w) / OPT_{LP}(G, w)$ over all possible values of the weights is called the *integrality gap* of this bidirected cut relaxation. Observe that the integrality gap is invariant under multiplicative scaling of the weights. Similarly, the network coding throughput-advantage $\mathcal{Y}_{NC}(G, u) / \mathcal{Y}(G, u)$ is invariant under multiplicative scaling of the capacities. We next determine the relationship between these two ratios.

Throughput Advantage

We show that the maximum network coding throughput-advantage of a communication system modeled by an undirected graph G is identical to to the integrality gap of the bidirected cut relaxation for the Steiner tree optimization problem.

Theorem 8.13. *Throughput-advantage due to network coding. Consider a communication network represented by an undirected graph $G = (V, E)$. The capacities of the edges are specified by the capacity function $u : E \to \mathbb{R}_0^+$. Information is multicast from the source node $s \in V$, to a set of sink nodes $\mathcal{T}_{rcvr} \subseteq V \backslash \{s\}$, where $|\mathcal{T}_{rcvr}| \geq 2$. Maximum throughput from the source node s to the set of sink nodes $\mathcal{T}_{rcvr}$ without network coding is $\mathcal{Y}(G, u)$.*

The undirected graph G is transformed into a directed graph $G_D = (V, E_D)$. Each edge $e \in E$ is converted into two directed arcs $a_1, a_2 \in E_D$, of opposite orientations, each with nonnegative capacities. If the capacity of edge e is $u(e) \triangleq u_e$, $\forall\, e \in E$; then the sum of the capacities of the arcs a_1 and a_2 is also equal to u_e. Maximum multicast throughput from the source node s to the set of sink nodes $\mathcal{T}_{rcvr}$ with network coding in this model is $\mathcal{Y}_{NC}(G, u)$.

Weights are assigned to edges in the undirected graph $G = (V, E)$ via the mapping $w : E \to \mathbb{R}_0^+$. If the weight of edge e is $w(e) \triangleq w_e$, then the weight of each arc a_1 and a_2 is also equal to w_e, $\forall\, e \in E$.

The weight of a Steiner tree in the graph G is $OPT(G, w)$. Its set of terminating nodes is the set $s \cup \mathcal{T}_{rcvr}$. The optimum value of the corresponding bidirected cut relaxation is $OPT_{LP}(G, w)$. Then

$$\max_u \frac{\mathcal{Y}_{NC}(G, u)}{\mathcal{Y}(G, u)} = \max_w \frac{OPT(G, w)}{OPT_{LP}(G, w)} \tag{8.29}$$

Proof. See the problem section. □

8.11 Design of Communication Networks

Some problems related to the design of communication networks are described in this section. The goal of network design is optimal or near-optimal determination of: traffic flow, traffic routing, capacity of links and nodes in the network. The network design should be both cost effective, and meet user demands. Design of a communication network is often a blend of mathematical techniques and art. Several topics developed in this chapter can be conveniently incorporated in the design of a communication network.

Network design problems use graph theory extensively, and are generally computationally expensive. Therefore approximations and heuristics are used in addressing such problems. A heuristic solution to a problem provides a feasible solution, but it may not be optimal. A communication network in general has two parts. These are: the backbone network, and the local access networks.

The backbone network connects routers (switching centers) via high speed links. The routers are generally geographically distributed. This network essentially carries the high volume traffic. Therefore it has to be a highly reliable network. Consequently this scenario can be conveniently modeled by a graph, where the routers and links are the vertices and edges of the graph respectively.

The other type of network, which is the access network, transports the customer traffic to the routers. Therefore this network can ideally be modeled by a tree graph. Several design issues and subproblems in the design of these networks are discussed briefly in this section. These are:

- General network design problem
- Uncapacitated network design problem
- Capacitated network design problem.
- Traffic capacity assignment in a network
- Design of survivable networks
- Traffic routing problem.

Heuristics for the solution of these problems can be found in open literature. Sometimes it is possible to address these design issues within the framework of a linear programming or an integer programming problem.

Network Design Problem

In its most general setup, the network is modeled by a directed graph $G = (V, E)$. Let $|V| = n$ and $|E| = m$. In this model, the set of vertices V is the set of routers, and the set of edges E is the set of transmission links. The nodes are geographically distributed. These are the locations at which traffic originates and terminates. The traffic flow between a pair of nodes can be specified by an $n \times n$ matrix. The edges of the graph represent a fiber-optic cable, or a satellite link, or any other form of the medium which can be used for transmission of information. Since information can be transmitted in either direction between a pair of routers, the graph G is directed. Furthermore, there is a cost associated with transmitting information between a node-pair. The purpose of the network design problem is to establish location of these routers, interconnect them with transmission links at a minimum cost, and satisfy traffic constraints.

Note that both the router and a transmission link have a finite traffic carrying capacity. Relaxation of these constraints results in an *uncapacitated* network. Otherwise it is a *capacitated* network. In some design problems it is mathematically convenient to relax the constraint of finite capacity.

Uncapacitated and Capacitated Network Design

In an uncapacitated network design model, the edges and the nodes are assumed to have unlimited capacity. This problem is of practical significance, because for some applications the capacity of modern transmission links like fiber-optic cables is more than sufficient. Furthermore, if the capacity constraints are relaxed, the problem and algorithm focus is on reducing installation and traffic flow costs.

Capacitated network design problem can be classified into capacitated minimum spanning tree problem and capacitated concentrator location problem. The capacitated minimum spanning tree problem and the capacitated concentrator location problem are briefly discussed.

Capacitated Minimum Spanning Tree Problem: The capacitated minimum spanning tree problem occurs in local access network design. The end user (customer) is connected to the concentrator (router) at a minimum cost via transmission links of finite capacity. The customers or other traffic sources are the nodes in the network, and the transmission links are the edges in the tree graph. If the transmission links have infinite capacity, the design problem reduces to finding a minimum spanning tree, hence the term capacitated minimum spanning tree. Heuristics are generally employed to address the capacitated minimum spanning tree problem. A well-known and effective heuristic is due to L. R. Esau and K. C. Williams.

Capacitated Concentrator Location Problem: In this problem, concentrators of finite capacity are to be set at certain locations so that the traffic demands of customers are met. This problem occurs in the design of both the local access network and the backbone network. The goal is to select the concentrator locations from a list of candidate geographical sites, and assign users to each concentrator site. The network connecting each set of users to a concentrator is a tree.

Traffic Capacity Assignment

The traffic capacity assignment problem is to install a combination of low and high capacity cables for each edge in either a directed or undirected graph G, such that all traffic demands are satisfied. Furthermore, the traffic flow on each edge is at most equal to the capacity of the edge. Finally, the total cost is required to be minimized.

Design of Survivable Networks

The goal of network design is to meet customer's traffic demand, say p percentage of the time in a cost effective manner. Realistically the value p is not equal to one hundred percent in general. This is because of the node and or link failures in the network. Furthermore, if the traffic constraints are met by a sparse network, then the network is vulnerable to failures. If this is the case, redundancy should be built into the network to provide alternate paths of communication in the network. However providing redundancy is expensive. Therefore there is a trade-off between providing a specified quality of service and cost.

In the design of a survivable network, it is generally assumed that the probability of simultaneous failure of two or more physical components (nodes or edges) is very small. Therefore in such designs it is generally assumed that only single failures occur.

Traffic Routing Problem

Several algorithms were discussed in earlier sections on traffic routing. The problem we seek to address here is called the *traveling salesman problem* (TSP). In this problem, a traveling salesman starts at a home town and visits a certain number of cities and returns to the home city. The travel should be such that the total distance covered is the smallest. This is a difficult optimization problem. It has resisted the attempts of mathematicians to provide an effective algorithm to address this problem, when the number of cities is large. Some of the possible applications of this problem in telecommunication engineering are: sequencing of jobs, cluster analysis, and drilling of printed circuit boards.

Reference Notes

Lawler's book (1976) is prime source of graph-theoretic algorithms. Presentation of the algorithm to compute K shortest paths follows Christofides (1975). The constrained routing algorithm was developed by J. M. Jaffe (1984) and later refined by Puri, and Tripakis (2002). The disjoint path algorithm is generally credited to Suurballe (1974), Suurballe, and Tarjan (1984), and Bhandari (1999).

The section on algebraic path techniques is based upon scholarly books by Gondran, and Minoux (1984), and Baras, and Theodorakopoulo (2010). The works of Rote (1990), Gao, and Rexford (2001), Sobrinho (2005), and Sobrinho, and Griffin (2010) are certainly inspirational. Kleinberg (2007) proposed geographic routing using hyperbolic space. The section on geographic routing is essentially based upon the pioneering work of Kleinberg.

The theorem used in validating the minimum spanning tree algorithms is from Eiselet, and Sandblom (2000). The algorithm to determine the connectivity of a graph is from the book by Claiborne (1990).

The notes in this chapter benefitted from the superb monographs on reliability theory by Colbourn (1986), and Shier (1991). Kershenbaum (1993) provides a lucid presentation of reliability of access networks. The papers by Ball (1980), and Ball, and Provan (1988) are very illuminating. The inspiration for the section on mesh-network reliability was a paper by Shier (1996).

The books by Frank, and Frisch (1971), and Ahuja, Magnanti, and Orlin (1993) are prolific sources of references. The discussion and algorithms on max-flow min-cut theorem and the minimum cost flow are based on the later book.

The paper by Ahlswede, Cai, Li, and Yeung (2000) is ground-breaking work on network coding. The basics of network coding can be found in Fragouli, and Soljanin (2007), Ho, and Lun (2008), and Yeung (2008, 2010). An elegant algebraic introduction to network coding can be found in Koetter, and Médard (2003). The theorem on throughput-advantage due to network coding is due to Agarwal, and Charikar (2004).

A comprehensive and scholarly introduction to routing algorithms used in the Internet can be found in the book by Medhi and Ramasamy (2017). The reader can refer to the textbook by Grover (2003), and Pióro, and Medhi (2004) for the heuristics and approximations used in the design of communication networks. Also refer to the landmark paper authored by Chiang, Low, Calderbank, and Doyle (2007).

A handbook of discrete combinatorial mathematics by Rosen (2018) is a comprehensive and useful source of precise definitions, algorithms, and theorems.

Problems

1. Let G be a completely connected undirected simple graph with $(n+2)$ nodes. Prove that the total number of different paths, without repeated nodes, between a pair of nodes is given by

$$S_n = n! \sum_{r=0}^{n} \frac{1}{r!}$$

For large values of n, an approximation of $S_n \sim \sqrt{2\pi n} n^n e^{-n+1}$.
Hint: See Lawler (1976). Lawler attributes this problem to V. Klee. The number of different paths with r intermediate nodes is $\binom{n}{r} r!$ for $0 \leq r \leq n$.

2. A communication network can be modeled as a digraph $G = (V, E)$, where the vertices represent the routers, and the arcs represent the communication links. Let $|V| = n$. The probability that the arc (link) (v_i, v_j) is functioning is $p_{ij}, 1 \leq i, j \leq n$. Assume that the links operate stochastically independent of each other. Find the most reliable path between a pair of nodes.
Hint: Define $c_{ij} = -\ln p_{ij}$, if $p_{ij} > 0$, otherwise $c_{ij} \to \infty$ for $1 \leq i, j \leq n$. Compute the shortest path with the c_{ij}'s as "length."
3. Let $G = (V, E)$ be a digraph with a cost matrix $C = [c_{ij}]$. Develop an algorithm for determining the second shortest path between any pair of vertices.
Hint: See Claiborne (1990). Using Dijkstra's algorithm, determine the shortest path between the specified pair of vertices v_s and v_t. Let the sequence of arcs in this path be $\{a_1, a_2, \ldots, a_w\}$. Obtain a graph by removing the arc a_1 from G. Apply Dijkstra's algorithm to the modified graph with edge set $E \backslash \{a_1\}$, and determine the shortest path between the vertices v_s and v_t. Repeat this procedure by removing only the arc a_2, and determine the shortest path, by applying Dijkstra's algorithm to the modified graph with edge set $G \backslash \{a_2\}$. This procedure is repeated for all the w arcs. The second shortest path between the vertices v_s and v_t is the shortest of the w such generated paths. Note that the removal of an arc implies that the corresponding arc cost is ∞. This scheme may not necessarily be the most efficient, but it is conceptually simple.
4. Develop an algorithm to compute the shortest path between a pair of vertices in a digraph with transit-times.
5. Justify completely the correctness of the edge-disjoint paths algorithm.
Hint: Use the theory of minimum cost flow in a capacitated network.
6. Write a pseudocode to develop the vertex-disjoint shortest paths algorithm.
7. Consider a simple directed graph $G = (V, E)$. Assume that the graph is connected. Write pseudocode:
   (a) To detect a cycle in the graph.
   (b) Determine all the cycles in the graph.

   In each case determine the computational complexity of the algorithm.
8. Let $T = (V, E)$, and $T_s = (V_s, E_s)$ be two tree graphs, where T_s is a subtree of T. That is, $V_s \subseteq V$ and $E_s \subseteq E$. Prove that every greedy embedding of T in a metric space (X, d) restricts to a greedy embedding of T_s.
Hint: See Kleinberg (2007). Let the mapping $f : V \to X$ be a greedy embedding. Therefore, for any two distinct nodes $s, t \in V_s$ there exists in the tree T, a path $s = s_0, s_1, s_2, \ldots, s_k = t$ such that the sequence $d(f(s_i, t))$ is monotonically decreasing as i goes from 0 to k. This path must lie in the subtree T_s, as T_s is a subtree of T and $s, t \in V_s$. Therefore vertex $s \in V_s$ has a neighbor $s_1 \in V_s$ which satisfies $d(f(s_1), f(t)) < d(f(s), f(t))$.
9. Show that the special linear group $SL(2, \mathbb{Z})$ is generated by the matrices

$$A = \begin{bmatrix} 0 & -1 \\ 1 & 0 \end{bmatrix}, \text{ and } W = \begin{bmatrix} 1 & 1 \\ 0 & 1 \end{bmatrix}$$

Hint: See Rankin (1977), and Conrad (2015). Let H be the group generated by the matrices A and W. Note that it is a subgroup of $SL(2, \mathbb{Z})$. Also observe that $\det A = \det W = 1$, $A^2 = -I$, and $A^{-1} = A^3$. Also

$$W^n = \begin{bmatrix} 1 & n \\ 0 & 1 \end{bmatrix}, \; n \in \mathbb{Z}; \text{ and } AW = \begin{bmatrix} 0 & -1 \\ 1 & 1 \end{bmatrix}$$

and $(AW)^3 = -I$. Let $Y \in SL(2, \mathbb{Z})$, where

$$Y = \begin{bmatrix} a & b \\ c & d \end{bmatrix}, \text{ and } \det Y = 1$$

Therefore

$$AY = \begin{bmatrix} -c & -d \\ a & b \end{bmatrix}, \text{ and } W^n Y = \begin{bmatrix} a + nc & b + nd \\ c & d \end{bmatrix}$$

Assume that the matrix element $c \neq 0$.

Step 1: If $|a| \geq |c|$, let $a = (cq + r)$, where $0 \leq r < |c|$. Then

$$W^{-q} Y = \begin{bmatrix} r & b - qd \\ c & d \end{bmatrix}$$

Observe in the above matrix that $r < |c|$. Next, premultiply $(W^{-q} Y)$ by A. This results in

$$A\left(W^{-q} Y\right) = \begin{bmatrix} -c & -d \\ r & b - qd \end{bmatrix}$$

Note that the determinant of the above matrix is equal to unity, as $(ad - bc) = 1$. The above process is repeated, if the lower left entry in the above matrix is nonzero. After few steps, we obtain a matrix with determinant equal to unity of the form

$$\begin{bmatrix} \pm 1 & m \\ 0 & \pm 1 \end{bmatrix} = W^m \text{ or } -W^{-m}, \text{ for some } m \in \mathbb{Z}$$

Therefore, we have $hY = \pm W^m$ for some $h \in H$ and $m \in \mathbb{Z}$. Observe that $W^m \in H$, and $-I = A^2 \in H$. Consequently, $Y = \pm h^{-1} W^m \in H$.

Step 2: However, if $|a| < |c|$, then premultiply matrix Y with matrix A. Proceed next as in Step 1.

10. Let

$$A = \begin{bmatrix} 0 & -1 \\ 1 & 0 \end{bmatrix}, \text{ and } B = \begin{bmatrix} 0 & 1 \\ -1 & 1 \end{bmatrix}$$

The map $Q(\cdot)$ is an isometry, where $Q(z) = -(iz + 1)/(iz - 1)$, and $z \in \mathbb{C} \backslash \{-i\}$. It maps points in the half-plane to points in the Poincaré disc. Therefore, if F is any isometry of $\mathbb{H}$, then isometries of the Poincaré disc model are $Q \circ F \circ Q^{-1}(\cdot)$.

Show that the matrix transformations A and B are represented in the Poincaré disc model by the mappings:

(a) The map corresponding to matrix A is: $A : z \mapsto -z$

(b) The map corresponding to matrix B is:

$$B : z \mapsto \frac{(1+2i)\,z+1}{z+(1-2i)}$$

Hint: Let $w = Q(z) = -(iz+1)/(iz-1)$. Therefore

$$z = \frac{i(1-w)}{(1+w)}$$

This implies

$$Q^{-1}(z) = \frac{i(1-z)}{(1+z)}$$

(a) The Moebius transformation corresponding to the matrix A is $A(z) = -1/z$. Therefore

$$A\left(Q^{-1}(z)\right) = \frac{i(1+z)}{(1-z)}$$

Thus

$$Q\left(A\left(Q^{-1}(z)\right)\right) = -z$$

(b) The Moebius transformation corresponding to the matrix B is $B(z) = 1/(-z+1)$. Therefore

$$B\left(Q^{-1}(z)\right) = \frac{(z+1)}{(1+i)\,z+(1-i)}$$

Thus

$$Q\left(B\left(Q^{-1}(z)\right)\right) = \frac{(1+2i)\,z+1}{z+(1-2i)}$$

11. Prove that there exists an embedding of an infinite complete 3-regular tree T in the Poincaré disc $\mathbb{D}$ which is greedy.
Hint: See Kleinberg (2007). Consider an ideal triangle Δ with vertices at $-1, 1, i$. As g ranges over all elements of $PSL(2,\mathbb{Z})$, the ideal triangles $g(\Delta)$ tessellate the Poincaré disc $\mathbb{D}$. The dual of this tessellation is an infinite complete 3-regular tree T. Thus the infinite tiling of $\mathbb{D}$ and the infinite complete 3-regular tree T are invariant under the action of the modular group $PSL(2,\mathbb{Z})$. As per our notation, the distance function in the Poincaré disc is $d_{\mathbb{D}}(\cdot,\cdot)$.
In order to establish that an infinite complete 3-regular tree T is a greedy embedding, it is sufficient to prove that for every path $s = s_0, s_1, s_2, \ldots, s_k = t$ in tree T, the inequality $d_{\mathbb{D}}(s_0, t) > d_{\mathbb{D}}(s_1, t)$ is satisfied. As the action of the group $PSL(2,\mathbb{Z})$ on T is edge-transitive, we may assume without loss of generality that $s_0 = u$ and $s_1 = v$, where $u = (2-\sqrt{3})\,i$, and $v = -u$. If the edge (u, v) is deleted from the tree T, then the tree T is split into two subtrees. One subtree has all of its nodes in the upper half-plane, and the other in the lower half-plane. As v lies on the path from u to t, we infer that t lies in the lower half-plane. Instead of proving $d_{\mathbb{D}}(u,t) > d_{\mathbb{D}}(v,t)$, a more general result is established. We prove that for any z in the lower half of the Poincaré disc $\mathbb{D}$, $d_{\mathbb{D}}(u,z) > d_{\mathbb{D}}(v,z)$. See Figure 8.14. Let S be a geodesic arc from u to z, and let r be the intersection of S with the real line. Therefore

$$d_{\mathbb{D}}(u,z) = d_{\mathbb{D}}(u,r) + d_{\mathbb{D}}(r,z)$$

Observe that the geodesic arc from v to z lies entirely in the lower half of the Poincaré disc $\mathbb{D}$. Therefore

$$d_{\mathbb{D}}(v,z) < d_{\mathbb{D}}(v,r) + d_{\mathbb{D}}(r,z)$$

Next note that $d_{\mathbb{D}}(u,r) = d_{\mathbb{D}}(v,r)$, as complex conjugation is an isometry of Poincaré disc $\mathbb{D}$. Consequently $d_{\mathbb{D}}(v,z) < d_{\mathbb{D}}(u,r) + d_{\mathbb{D}}(r,z) = d_{\mathbb{D}}(u,z)$.

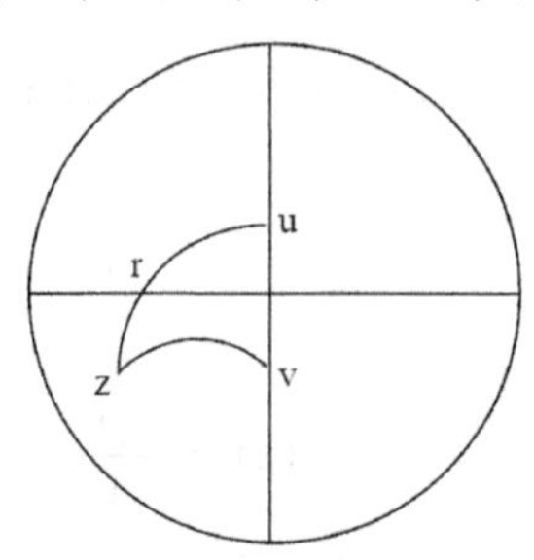

Figure 8.14. Greedy embedding of an infinite complete 3-regular tree T in the Poincaré disc $\mathbb{D}$.

12. Prove that there exists an embedding of an infinite complete δ-regular tree T in the Poincaré disc $\mathbb{D}$ which is greedy, where $\delta > 3$.
Hint: See Kleinberg (2007). Existence of greedy embedding of an infinite complete δ-regular tree T in the Poincaré disc $\mathbb{D}$, where $\delta > 3$, is demonstrated. In this setting, the group Γ analogous to the modular group $PSL(2,\mathbb{Z})$ is defined implicitly. The actual definition of Γ depends upon the geometry. Consider the interior P of an ideal polygon. Its vertices are the δ-th roots of unity. Observe that P is preserved by the mapping $\rho(z) = e^{2\pi i/\delta} z$. It generates a cyclic subgroup of order δ in the hyperbolic isometry group $PSL(2,\mathbb{R})$. Observe that $PSL(2,\mathbb{R})$ acts transitively on clock-wise ordered triples of ideal points. It is possible to pick any side of the ideal polygon ∂P and find a hyperbolic isometry σ which maps its end points to 1 and -1, and maps the mid-point of the arc (side of the ideal polygon ∂P) connecting these endpoints to $-i$. Next consider the hyperbolic isometries

$$A : z \mapsto -z, \quad \text{and} \quad B : z \mapsto \sigma\left(\rho\left(\sigma^{-1}(z)\right)\right)$$

The isometry A rotates the disk by a half-turn. Isometry B preserves the set P, however the vertices of the polygon are cyclically permuted. The isometries A and B generate a discrete subgroup $\Gamma \subset PSL(2,\mathbb{R})$ so that the set $\{g(P) \mid g \in \Gamma\}$ specifies a tiling of the Poincaré disc $\mathbb{D}$, by the ideal δ-gons. Furthermore, the dual of this tessellation is an infinite complete δ-regular tree T. The nodes of this tree are all the points in the set $\{g(\sigma(0)) \mid g \in \Gamma\}$. Observe that the complex numbers $u = \sigma(0)$, $v = -u$ are purely imaginary, and each edge of the tree T is the image of the edge (u,v) under an appropriate transformation $g \in \Gamma$.
The proof that T is indeed greedily embedded in $\mathbb{D}$ is similar to the proof for the case of the infinite complete 3-regular tree. The action of group Γ on T is edge-transitive. Therefore, it is sufficient to consider a path in the tree with first hop (u,v) and whose destination (target) node t lies in the lower half-plane. The proof that $d_{\mathbb{D}}(u,z) > d_{\mathbb{D}}(v,z)$ for each z in the lower half-plane is identical.
13. Let x be a feasible flow between a source node s and a target node t, with value v. Assume that $\langle S,T\rangle$ is any s-t cut in the network, and the net flow across this cut is equal to v. Prove that

$$v = \sum_{(i,j)\in(S,T)} x_{ij} - \sum_{(i,j)\in(T,S)} x_{ij}$$

Hint: See Ahuja, Magnanti, and Orlin (1993).

14. Prove the following statements.
    (a) The maximum number of arc-disjoint paths from the source node s to the sink node t is equal to the minimum number of arcs required to be removed from the network which disconnects all paths from the node s to the node t.
    (b) The maximum number of node-disjoint paths from the source node s to the sink node t, is equal to the minimum number of nodes, excluding nodes s and t, whose removal is sufficient to disconnect all paths from the node s to the node t.

    Hint: See Ahuja, Magnanti, and Orlin (1993).
15. Prove that a flow $x \geq 0$ in a network G is feasible if and only if a flow $x' \geq 0$ in a residual network $G(\widetilde{x})$ is feasible, where:
    If $\widetilde{x}_{ij} \leq x_{ij}$, then $x'_{ij} = (x_{ij} - \widetilde{x}_{ij})$, and $x'_{ji} = 0$.
    If $\widetilde{x}_{ij} \geq x_{ij}$, then $x'_{ij} = 0$, and $x'_{ji} = (\widetilde{x}_{ij} - x_{ij})$.
    Also show that $c'_{ij} x'_{ij} = c_{ij}(x_{ij} - \widetilde{x}_{ij})$, where c'_{ij} is the cost in the residual network.
    Hint: See Ahuja, Magnanti, and Orlin (1993). Note that $r_{ij} = (u_{ij} - \widetilde{x}_{ij})$, and $r_{ji} = \widetilde{x}_{ij}$.
    Assume that $\widetilde{x}_{ij} \leq x_{ij} \leq u_{ij}$. In this case let $x'_{ij} = (x_{ij} - \widetilde{x}_{ij})$, and $x'_{ji} = 0$. Observe that $x'_{ij} \leq r_{ij}$. Therefore the flow x'_{ij} is feasible.
    Next assume that $\widetilde{x}_{ij} \geq x_{ij}$. In this case let $x'_{ij} = 0$, and $x'_{ji} = (\widetilde{x}_{ij} - x_{ij})$. Observe that $0 \leq x'_{ji} \leq \widetilde{x}_{ij} = r_{ji}$. Therefore the flow x'_{ij} is feasible.
    The relationship between the costs is next established. In the residual network $c'_{ij} = c_{ij}$, and $c'_{ji} = -c_{ij}$.
    Assume that $\widetilde{x}_{ij} \leq x_{ij} \leq u_{ij}$. Therefore $\left(c'_{ij} x'_{ij} + c'_{ji} x'_{ji}\right) = c_{ij}(x_{ij} - \widetilde{x}_{ij})$.
    Next assume that $\widetilde{x}_{ij} \geq x_{ij}$. Therefore $\left(c'_{ij} x'_{ij} + c'_{ji} x'_{ji}\right) = c_{ij}(x_{ij} - \widetilde{x}_{ij})$.
16. Consider a directed graph $G = (V, E)$, where V is the set of vertices, E is the set of edges, $|V| = n$, and $|E| = m$. Prove the following statements about characterization of flows in the directed graph.
    (a) A circulation in the graph can be represented as cycle flow with at most m directed cycles.
    (b) Consider a capacitated network $N = (G, u, c, b)$ with feasible flows x' and x''. Then the flow x'' is equal to the flow x' plus the flow on at most m directed cycles in the residual network $G(x')$. In addition, the cost of the flow x'' is equal to the cost of the flow x' plus the cost of the flow on these m augmenting cycles.

    Hint: See Ahuja, Magnanti, and Orlin (1993).
17. Consider a capacitated network $N = (G, u, c, b)$. In this network, the flow x and the node potentials $\pi(\cdot)$ satisfy the complementary slackness conditions. Prove that it also satisfies the reduced cost optimality conditions.
    Hint: See Ahuja, Magnanti, and Orlin (1993).
18. Consider a communication network $\mathcal{N} = (G, s, \mathcal{T}_{rcvr}, \mathcal{X}, m)$ which uses linear network coding. In this network $G = (V, E)$. The global encoding kernel for channel $e \in E$ is f_e. For a set of non-source nodes $\mathcal{T} \subseteq V \backslash \{s\}$ prove that

$$\dim(L_{\mathcal{T}}) \leq \min\{m, maxflow(\mathcal{T})\}$$

    Hint: See Yeung (2008). In the network $\mathcal{N}$, the graph G is acyclic. Let C_U be a cut between the source node s and a set of non-source nodes $\mathcal{T} \subseteq V \backslash \{s\}$. Denote the set of edges across the cut C_U by E_U. Let the linear span of a set of vectors $\{f_e \mid e \in \Gamma_I(t)\}$ for $t \in V$ be

$$L_t = L(\{f_e \mid e \in \Gamma_I(t)\})$$

Also, let the set of linear span of vectors $\cup_{t\in\mathcal{T}} L_t$ be $L_{\mathcal{T}} = L\left(\cup_{t\in\mathcal{T}} L_t\right)$. Similarly, let the set of linear span of vectors $\{f_e \mid e \in E_U\}$ be $L_{E_U} = L\left(\{f_e \mid e \in E_U\}\right)$. Observe that, $L_{\mathcal{T}}$ is a linear transformation of L_{E_U}. Therefore,

$$\dim\left(L_{\mathcal{T}}\right) \leq \dim\left(L_{E_U}\right) \leq |E_U|$$

In the next step, we minimize over all the cuts between $\{s\}$ and $\mathcal{T}$, and use max-flow min-cut theorem. This yields

$$\dim\left(L_{\mathcal{T}}\right) \leq maxflow\left(\mathcal{T}\right)$$

Also note that $L_{\mathcal{T}}$ is obtained by a linear transformation of the vectors in the set $\mathcal{X}^m$. Therefore

$$\dim\left(L_{\mathcal{T}}\right) \leq \dim\left(\mathbb{F}_q^{(m)}\right) = m$$

The required result is obtained by combining the last two inequalities.

19. Establish the sparse zeros lemma.
Hint: See Fragouli, and Soljanin (2007), and Yeung (2008). Induction on η is used in proving the lemma. If $\eta = 0$, then the polynomial is equal to a nonzero constant in $\mathcal{X}$, as required by the hypothesis of the lemma. If $\eta = 1$, the polynomial has only a single variable, and its degree is at most δ. In this case, the lemma is true because, the polynomial can have at most δ roots, and $q > \delta$.
Assume that the lemma is true for $(\eta - 1)$, where $\eta \geq 2$. We express $g\left(z_1, z_2, \ldots, z_\eta\right)$ as a polynomial in z_η with coefficients $\{g_i \mid 0 \leq i \leq \delta\}$ that are polynomials in the variables $z_1, z_2, \ldots, z_{\eta-1}$. Thus

$$g\left(z_1, z_2, \ldots, z_\eta\right) = \sum_{i=0}^{k} g_i\left(z_1, z_2, \ldots, z_{\eta-1}\right) z_\eta^i$$

where $k \leq \delta$. That is, k is the degree of the polynomial g in z_η and the leading coefficient $g_k\left(z_1, z_2, \ldots, z_{\eta-1}\right)$ is a nonzero polynomial in $\mathbb{F}_q\left[z_1, z_2, \ldots, z_{\eta-1}\right]$.
However, by induction hypothesis, there exist $a_1, a_2, \ldots, a_{\eta-1} \in \mathcal{X}$ such that

$$g_k\left(a_1, a_2, \ldots, a_{\eta-1}\right) \neq 0$$

Therefore $g\left(a_1, a_2, \ldots, a_{\eta-1}, z\right)$ is a nonzero polynomial in z with degree $k < q$. As this polynomial can have at most k roots in the field $\mathbb{F}_q$, and $k < q$, there exists $a_\eta \in \mathcal{X}$ such that $g\left(a_1, a_2, \ldots, a_\eta\right) \neq 0$.

20. Establish the theorem on the existence of linear network coding for multicasting.
Hint: See Koetter, and Médard (2003), and Yeung (2008). A directed path $\mathcal{P}$ in the multidigraph is specified by a sequence of edges. That is, $\mathcal{P} = \left(e_1, e_2, \ldots, e_\beta\right)$. For this path $\mathcal{P}$ define

$$K_{\mathcal{P}} = \prod_{1 \leq j < \beta} k_{e_j, e_{j+1}}$$

Using the relationship for global encoding kernel $f_e = \sum_{d \in \Gamma_I(v)} k_{de} f_d$, for $e \in \Gamma_O(v)$ successively, we obtain

$$f_e = \sum_{d \in \Gamma_I(s)} \left\{ \sum_{\mathcal{P}\text{: a path from } d \text{ to } e} K_{\mathcal{P}} \right\} f_d, \quad \text{for each } e \in E$$

Let $\mathbb{F}_q[*]$ be the polynomial ring over the field $\mathbb{F}_q$. The indeterminates are all the k_{de}'s. The total number of such indeterminates is equal to $\sum_{v\in V}|\Gamma_I(v)||\Gamma_O(v)|$. Therefore each component of every global encoding kernel belongs to $\mathbb{F}_q[*]$.

For each non-source node $t \in \mathcal{T} \subseteq V\backslash\{s\}$ with $maxflow(t) \geq m$, there exist m edge-disjoint paths from the m imaginary channels at the input to the source node s to m distinct channels in $\Gamma_I(t)$. Next, form an $m \times m$ matrix Q_t by juxtaposing the global encoding kernels of m channels in $\Gamma_I(t)$. We shall subsequently prove that $\det Q_t = 1$ for judiciously selected scalar values of the indeterminates.

A pair of channels $(d, e) \in E \times E$ is called an adjacent pair if there exists a node $v \in V$ such that $d \in \Gamma_I(v)$ and $e \in \Gamma_O(v)$.

In the matrix Q_t set $k_{de} = 1$ for all adjacent pairs of channels (d, e) along any one of the m edge-disjoint paths, and set $k_{de} = 0$ for all other adjacent pairs. Upon such use of local encoding kernels, the symbols transmitted originally from the m imaginary channels at the source node s are routed to the node t via the m edge-disjoint paths. Therefore, the columns of the matrix Q_t are the global encoding kernels of the imaginary input channels at the source node s. Actually, these are the standard basis of the vector space $\mathbb{F}_q^{(m)}$. Therefore $\det Q_t = 1$. This implies that $\det Q_t$ is a nonzero polynomial in the indeterminates k_{de}'s provided $maxflow(t) \geq m$, because its evaluation is equal to unity when k_{de}'s are selected as per the above scheme. This result is true for each non-source node $t \in \mathcal{T} \subseteq V\backslash\{s\}$. Therefore we have

$$\prod_{\substack{t\in\mathcal{T}\subseteq V\backslash\{s\}\\ maxflow(t)\geq m}} \det Q_t \neq 0 \in \mathbb{F}_q[*]$$

We next apply the sparse zeros lemma to the above polynomial. As per this lemma, scalar values in the set $\mathcal{X}$ can be assigned to the indeterminates so that the above polynomial is nonzero, provided $|\mathcal{X}|$ is sufficiently large. That is, $\det Q_t \neq 0$ for each $t \in \mathcal{T} \subseteq V\backslash\{s\}$. Thus these scalar values facilitate a multicast which uses linear network coding.

Therefore $\dim(L_t) = m$, and $maxflow(t) \geq m$ for every node $t \in \mathcal{T} \subseteq V\backslash\{s\}$. Thus multicast transmission from the source node s to a set of non-source nodes $\mathcal{T}$ is possible.

21. Establish the theorem about the throughput-advantage due to network coding.
Hint: See Agarwal, and Charikar (2004). This theorem is established in two steps.
Step 1: In this step, we establish that

$$\max_u \frac{\mathcal{Y}_{NC}(G,u)}{\mathcal{Y}(G,u)} \leq \max_w \frac{OPT(G,w)}{OPT_{LP}(G,w)}$$

Consider the fractional multicast tree packing number of the undirected graph $G = (V, E)$ with capacities $u_e, e \in E$. Without any loss of generality, scale these capacities so that the value $\mathcal{Y}_{NC}(G,u)$ is equal to unity.

The optimized value of the fractional multicast tree packing number is $\mathcal{Y}(G,u)$. It determines the throughput in the network specified by the graph G. By the strong duality theorem of linear programming, the optimum value of its dual is also equal to $\mathcal{Y}(G,u)$. The decision variables in the dual problem $y_e, e \in E$ can be considered to be the edge costs. The first constraint in the dual implies that a multicast tree has the total cost of its edges to be at least unity.

We claim that $\sum_{e\in E} u_e y_e$ is an upper bound for the objective function of the bidirected cut relaxation for the Steiner tree problem with edge costs specified by the y_e's, which is $OPT_{LP}(G,y)$. This is established as follows.

As the graph G with capacities $u_e \in E$ has a conceptual flow value of unity with network coding, each edge can be bidirected and the capacity distributed between the two directed edges of opposite orientation so that a flow of unity can be routed from the source node to every sink node. Observe that the directed capacity across any cut separating the source and at least one sink node must be at least unity. Furthermore, these bidirected capacities provide a valid solution to the bidirected cut relaxation.

Consequently, via the optimum solution of the dual it can be inferred that there exist weights $w_a, a \in E_D$ such that

$$OPT\left(G, w\right) \geq 1, \text{ and } OPT_{LP}\left(G, w\right) \leq \mathcal{Y}\left(G, u\right)$$

.

Step 2: In this step, we establish that

$$\max_{u} \frac{\mathcal{Y}_{NC}\left(G, u\right)}{\mathcal{Y}\left(G, u\right)} \geq \max_{w} \frac{OPT\left(G, w\right)}{OPT_{LP}\left(G, w\right)}$$

Consider a graph G in which the weights $w_a, a \in E_D$ are such that the bidirected cut relaxation has a gap g. In order to establish the above inequality, we need to construct an instance of G in which the network coding throughput-advantage is at least g.

Let $u_a, a \in E_D$ be the capacity computed by the bidirected cut relaxation on the graph G. Also let $u_e, e \in E$ be the sum of edge capacities of edges in either direction. We then determine the throughput with network coding using these capacities.

It can be claimed that the optimal flow $\mathcal{Y}_{NC}\left(G, u\right)$ with capacities $u_e, e \in E$ is at least unity. That is, $\mathcal{Y}_{NC}\left(G, u\right) \geq 1$. This is established as follows.

In the bidirected cut relaxation problem formulation, the u_e's are split among the two orientations so that the capacity of every cut between the source and a sink node is at least unity. This implies that the directed graph G_D can support a flow of unity from the source node s to each sink node in the set $\mathcal{T}_{rcvr}$.

The fractional multicast tree packing number for this instance is $\mathcal{Y}\left(G, u\right)$. This implies that for any set of weights $w_a, a \in E_D$, there exists a multicast tree of cost at most

$$\frac{\sum_{a \in E_D} w_a u_a}{\mathcal{Y}\left(G, u\right)}$$

As per an earlier assumption

$$OPT_{LP}\left(G, u\right) = \min \sum_{a \in E_D} w_a u_a, \text{ and } \frac{OPT\left(G, u\right)}{OPT_{LP}\left(G, u\right)} = g$$

Therefore, there exists a Steiner tree of cost $OPT\left(G, u\right)$ such that

$$OPT\left(G, u\right) \leq \frac{OPT_{LP}\left(G, u\right)}{\mathcal{Y}\left(G, u\right)}$$

That is, $\mathcal{Y}\left(G, u\right) \leq 1/g$. As $\mathcal{Y}_{NC}\left(G, u\right) \geq 1$, we have

$$\frac{\mathcal{Y}_{NC}\left(G, u\right)}{\mathcal{Y}\left(G, u\right)} \geq g$$

References

1. Agarwal, A., and Charikar, M., 2004. "On the Advantage of Network Coding for Improving Network Throughput," Proceedings IEEE Information Theory Workshop, San Antonio, Texas, pp. 247-249.
2. Ahlswede, R., Cai, N., Li, S.-Y. R., and Yeung, R. W., 2000. "Network Information Flow," IEEE Trans. on Information Theory, Vol. IT-46, No. 4, pp. 1204-1216.
3. Ahuja, R. K., Magnanti, T. L., and Orlin, J. B., 1993. *Network Flows, Theory, Algorithms, and Applications,* Prentice-Hall, Englewood Cliffs, New Jersey.
4. Ball, M. O., 1980. "Complexity of Network Reliability Computations," Networks, Vol. 10, Issue 2, pp. 153-165.
5. Ball, M. O., and Provan, J. S., 1988. "Disjoint Products and Efficient Computation of Reliability," Operations Research, Vol. 36, No. 5, pp. 703-715.
6. Banerjee, S., Ghosh, R. K., and Reddy, A. P. K., 1996. "Parallel Algorithm for Shortest Pairs of Edge-Disjoint Paths," Journal of Parallel and Distributed Computing, Vol. 33, No. 2, pp. 165-171.
7. Baras, J., and Theodorakopoulo, 2010. *Path Problems in Networks*, Morgan and Claypool Publishers, San Rafael, California.
8. Bhandari, R., 1999. *Survivable Networks, Algorithms for Diverse Routing*, Kluwer Academic Publishers, Norwell, Massachusetts.
9. Brandes, U., and Erlebach, T., Editors, 2004. *Network Analysis*: *Methodological Foundations*, Springer-Verlag, Berlin, Germany.
10. Cahn, R. S., 1998. *Wide Area Network Design*, Morgan Kaufmann Publishers, San Francisco, California.
11. Chiang, M., Low, S. H., Calderbank, A. R., and Doyle, J. C., 2007. "Layering as Optimization Decomposition: A Mathematical Theory of Network Architectures," Proc. of the IEEE, Vol. 95, No. 1, pp. 255-312.
12. Christofides, N., 1975. *Graph Theory - An Algorithmic Approach,* Academic Press, New York, New York.
13. Chung, F., 2010. "Graph Theory in the Information Age," Notices of the American Mathematical Society, Vol. 57, No. 6, pp. 726-732.
14. Claiborne, J. D., 1990. *Mathematical Preliminaries for Computer Networking*, John Wiley & Sons, Inc., New York, New York.
15. Colbourn, C. J., 1987. *The Combinatorics of Network Reliability*, Oxford University Press, Oxford, Great Britain.
16. Conrad, K. $SL\left(2,\mathbb{Z}\right)$, http://www.math.uconn.edu/~kconrad/blurbs/grouptheory/SL(2,Z).pdf. Retrieved February 16, 2015.
17. Cravis, H., 1981. *Communication Network Analysis*, D. C. Heath and Company, Lexington, Massachusetts.
18. Deo, N., 1974. *Graph Theory with Applications to Engineering and Computer Science*, Prentice-Hall, Englewood Cliffs, New Jersey.
19. Diestel, R., 2000. *Graph Theory*, Second Edition, Springer-Verlag, Berlin, Germany.
20. Eiselt, H. A., and Sandblom, C. -L., Editors 2000. *Integer Programming and Network Models,* Springer-Verlag, Berlin, Germany.
21. Elias, P., Feinstein, A., Shannon, C. E., 1956. "Note on Maximum Flow Through a Network," IRE Trans. on Information Theory, Vol. IT-2, pp. 117-119.
22. Even, S., 1979. *Graph Algorithms,* Computer Science Press, Rockville, Maryland.

23. Ford, L. R., and Fulkerson, D. R., 1956. "Maximal Flow Through a Network," Canadian J. of Mathematics, Vol. 8, pp. 399-404.
24. Fragouli, C., and Soljanin, E., 2007. "Network Coding Fundamentals," Foundations and Trends in Networking, Vol. 2, No. 1, pp. 1-133.
25. Frank, H., and Frisch, I. T., 1971. *Communication, Transmission, and Transportation Network*, Addison-Wesley Publishing Company, New York, New York.
26. Gao, L., and Rexford, J., 2001. "Stable Internet Routing Without Global Coordination," IEEE/ACM Transactions on Networking, Vol. 9, No. 6, pp. 681-692.
27. Godsil, C., and Royle, G., 2001. *Algebraic Graph Theory*, Springer-Verlag, Berlin, Germany.
28. Gondran, M., and Minoux, M., 1984. *Graphs and Algorithms*, John Wiley & Sons, Inc., New York, New York.
29. Gross, J. L., and Yellen, J., Editors, 2003. *Handbook of Graph Theory*, CRC Press: New York, New York.
30. Grover, W. D., 2003, *Mesh-based Survivable Transport Networks: Options and Strategies for Optical, MPLS, SONET and ATM Networking*, Prentice-Hall, Englewood Cliffs, New Jersey.
31. Ho, T., and Lun, D. S., 2008. *Network Coding: An Introduction*, Cambridge University Press, Cambridge, Great Britain.
32. Jaffe, J. M., 1984. "Algorithms for Finding Paths with Multiple Constraints," Networks, Vol. 14, Issue 1, pp. 95-116.
33. Karnaugh, M., May 1976. "A New Class of Algorithms for Multipoint Network Optimization," IEEE Transactions on Communications, Vol. Com-24, No. 5, pp. 500-505.
34. Kershenbaum, A., 1993. *Telecommunications Network Design Algorithms*, McGraw-Hill Book Company, New York, New York.
35. Kleinberg, R., 2007. "Geographic Routing Using Hyperbolic Space," Proceedings of the 26th Annual Joint Conference of the IEEE Computer and Communications Societies (INFOCOM), pp. 1902-1909.
36. Kocay, W., and Kreher, D. L., 2005. *Graphs, Algorithms, and Optimization,* Chapman and Hall/CRC Press, New York, New York.
37. Koetter, R., and Médard, M., 2003. "Beyond Routing: An Algebraic Approach to Network Coding," IEEE/ACM Transactions on Networking, Vol. 11, No. 5, pp. 782-795.
38. Lawler, E. L., 1976. *Combinatorial Optimization: Networks and Matroids*, Holt, Rinehart and Winston, New York, New York.
39. Li, Z., Li, B., Jiang, D., and Lau, L. C., 2004. "On Achieving Optimal End-to-End Throughput in Data Networks: Theoretical and Empirical Studies," ECE Technical Report, University of Toronto.
40. Medhi, D., and Ramasamy, K. 2017. *Network Routing: Algorithms, Protocols, and Architectures*, Second Edition, Morgan Kaufmann Publishers, San Francisco, California.
41. Merris, R., 2001. *Graph Theory*, John Wiley & Sons, Inc., New York, New York.
42. Minoux, M., 1986. *Mathematical Programming,* translated by S. Vajda, John Wiley & Sons, Inc., New York, New York.
43. Papadimitriou, C. H., and Steiglitz, K., 1982. *Combinatorial Optimization - Algorithms and Complexity*, Prentice-Hall, Englewood Cliffs, New Jersey.
44. Pióro, M., and Medhi, D., 2004. *Routing, Flow, and Capacity Design in Communication and Computer Networks*, Morgan Kaufmann Publishers, San Francisco, California.
45. Puri, A., and Tripakis, S., 2002. "Algorithms for Routing with Multiple Constraints." AIPS Workshop on Planning and Scheduling using Multiple Criteria, Toulouse, France, April 23, 2002, pp. 7-14.
46. Rankin, R. A., 1977. *Modular Forms and Functions,* Cambridge University Press, Cambridge, Great Britain.
47. Rosen, K. H., Editor-in-Chief, 2018. *Handbook of Discrete and Combinatorial Mathematics,* Second Edition, CRC Press: Boca Raton, Florida.

48. Rote, G., 1990. "Path Problems in Graphs," Computing Supplementum, Vol. 7, pp. 155-189.
49. Sharma, R. L., 1990. *Network Topology Optimization*, Van Nostrand Reinhold Company, New York, New York.
50. Sharma, R. L., De Sousa, P. J. T., and Ingle, A. D., 1982. *Network Systems*, Van Nostrand Reinhold Company, New York, New York.
51. Shier, D. R., 1991. *Network Reliability and Algebraic Structures*, Oxford University Press, Oxford, Great Britain.
52. Shier, D. R., 1996. "Redundancy and Reliability of Communication Networks," The College Mathematics Journal, Vol. 26, No. 1, pp. 59-67.
53. Sobrinho, J. L., 2005. "An Algebraic Theory of Dynamic Network Routing," IEEE/ACM Transactions on Networking, Vol. 13, No. 5, pp. 1160-1173.
54. Sobrinho, J. L., and , Griffin, T. G., 2010. "Routing in Equilibrium," Proceedings of the 19th International Symposium on the Mathematical Theory of Networks and Systems, Budapest, Hungary, July 2010.
55. Suurballe, J. W., 1974. "Disjoint Paths in a Network," Networks, Vol. 4, Issue 2, pp. 125-145.
56. Suurballe, J. W., and Tarjan, R. E. 1984. "A Quick Method for Finding Shortest Pairs of Disjoint Paths," Networks, Vol. 14, Issue 2, pp. 325-336.
57. Swamy, M. N. S., and Thulasiraman, K., 1981. *Graphs, Networks, and Algorithms,* John Wiley & Sons, Inc., New York, New York.
58. Tucker, A., 2007. *Applied Combinatorics,* Fifth Edition, John Wiley & Sons, Inc., New York, New York.
59. Wasserman, S., Faust, K., 1994. *Social Network Analysis,* Cambridge University Press, Cambridge, Great Britain.
60. Welsh, D., 1976. *Matroid Theory,* Academic Press, New York, New York.
61. West, D. B., 2001. *Introduction to Graph Theory,* Second Edition, Prentice-Hall, Upper Saddle River, New Jersey.
62. Yeung, R. W., 2008. *Information Theory and Network Coding*, Springer-Verlag, Berlin, Germany.
63. Yeung, R. W., 2010. "Network Coding Theory: An Introduction," Front. Electr. & Electron. Eng. China, Vol. 5, No. 3, pp. 363-390.

CHAPTER 9

Game Theory and the Internet

$$S(X) = -tr(\rho \log_2 \rho)$$

von Neumann entropy

John von Neumann. John von Neumann was born on 28 December, 1903 in Budapest, Hungary. His name at birth was János von Neumann. He earned a bachelor's degree in chemical engineering from the Zurich Institute in 1925, and a Ph.D. in mathematics from the University of Budapest in 1926. It was said that von Neumann had a phenomenal memory. He was also noted for the remarkable speed with which he acquired novel ideas and solved problems. When von Neumann was 20 years old, he proposed a new definition of ordinal numbers that was accepted by all mathematicians. In the year 1932 he wrote the influential book, *Mathematical Foundations of Quantum Mechanics*. He also proved the celebrated minimax theorem of game theory. John von Neumann and Oskar Morgenstern coauthored *Theory of Games and Economic Behavior*, (1944).
John von Neumann came to the United States in 1930. He was one of the original six mathematics professors in 1933 at the Institute for Advanced Study in Princeton, New Jersey. He also made seminal contributions to the architecture of a computer system. John von Neumann died on 8 February, 1957 in Washington DC, USA.

9.1 Introduction

The goal of this chapter is to examine some aspects of Internet technology in light of game theory. A brief introduction to game theory is initially provided. This is followed by its application to the study of routing of information, and a selfish task allocation model. An example of repeated game is also discussed. Finally applications of algorithmic mechanism design are also explored.

The Internet can be viewed as a network of networks, where each constituent network acts independently. Thus the Internet is modeled as a set of autonomous systems (agents, or entities). However because of the dynamic and distributed structure of the Internet, each entity interacts with others while optimizing its own set of goals, which possibly might not be in the best interest of all other entities. This behavior of the Internet is of course a natural corollary to its decentralized nature. A majority of entities in the Internet act in their self-interest. They are not interested in optimizing the aggregate computing power of the Internet. Consequently these self-interested actions have a penalty associated with them.

The goal of game theory is to study conflict and cooperation between different entities of a group. Therefore, game theory appears to be a suitable paradigm to study Internet technologies. Some of the applications of game theory are in cryptography, network routing, load balancing, scheduling theory, resource allocation, distributed artificial intelligence, auctioning, and advertisement. Only a very small fraction of such applications are discussed in this chapter.

The flow of information in these networks is packetized. Packets generated by agents within each network have unfettered access to the entire Internet. This is done with total disregard to the overall optima of the entire network. Therefore this negative characteristic of the agents is termed *selfish*. This selfishness results in an inefficient use of the Internet resources, and thus has a price associated with it. Game theory, a mathematical discipline largely developed in the last century appears to be a convenient vehicle to address issues such as selfish behavior of agents (or players). This theory studies and analyses multi-agent decision problems.

Classical optimization techniques used in finding routes for packets, resource allocation, and pricing assume that these are all independent of the *interaction* of the users of the network. When users compete for limited resources, and influence the choice of others, the modeling framework involves seeking an *equilibrium*, or stable operating point of the system. Thus in this scenario, each user tries to optimize according to his (her) own criteria, which includes the affect of decisions of other users upon his (her) own. Furthermore, all users perform this optimization simultaneously. Game-theoretic techniques are suitable to study such equilibrium models of interacting users of the telecommunication network.

Finally, the discipline of computer science can loosely be termed as the art and science of developing algorithms. It is hoped that the ideas from the world of algorithms and game theory fuse together to provide a better understanding of the evolution of this complex computational phenomenon called the Internet. Such models are studied in the relatively new discipline of algorithmic mechanism design.

Basics of analysis, matrices and determinants, probability theory, and optimization techniques are required to study this chapter. Application of game theory to study Internet economics is studied in a different chapter. Use of theory of auctions, coalitional game theory, and bargaining games are explored in the chapter on Internet economics.

9.2 Game Theory

A game is an activity which is meant for diversion. Sometimes a game is fun, and at other times serious business. In this chapter, a game is a tactic, or a procedure, or a competitive activity for gaining an end. Game theory provides a mathematical model of conflict or bargaining.

It is assumed in this chapter that games between several agents (players, or entities, or competitors) are played with a pre-specified set of rules. The players are expected to follow these rules while making moves. The games are of course meant to provide entertainment, and rewards when they are finished. Therefore in a game, the players make certain moves on which their *payoff* (reward) depends. It is also assumed that the players are *rational* and are aware of each other's rationality. This is a fundamental assumption of game theory. Thus each player, independent of all other players, attempts to maximize his or her final payoff by selecting particular course of actions. Consequently there is no room for altruism in such games. This fundamental assumption is also called the *doctrine of rational choice*.

Game theory analyses problems in which many players are contending for limited rewards. Thus game theory is a mathematical paradigm of strategic competition which aims at determining the best *strategy* for each individual player. The strategy of a player in a game is his or her complete contingency plan. Therefore this subject is a formal way to analyse interaction among a group of agents who are rational and act strategically.

This subject has found application in economics, evolutionary biology, large scale distributed systems, political science, resource allocation, sociology, and several other fields. Its most recent application is in computer science.

Game theory has a long history. Some of the early workers in this field were Girolamo Cardano, Blaise Pascal, Pierre de Fermat, Christian Huygens, Augustin Cournot, Emile Borel, and many others. However the modern era of game theory commenced with the publication of the seminal monograph *Theory of Games and Economic Behavior* by the mathematician John von Neumann and the mathematical economist Oskar Morgenstern (1902-1977) in the year 1944. John F. Nash (1928-2015) also had a profound influence on this field.

A preliminary discussion of game theory is initially provided in this section. This is followed by a description of strategic-form and extensive-form games, and also repeated games.

9.2.1 Preliminaries

Game theory studies the problem of interaction among several decision makers (players). These interactions may be either conflicting, or they might be cooperative. A player is generally offered a set of *feasible choices*, and one or more *objective functions* that assigns a *utility* (value) to all feasible possibilities. In addition, the player has a *set of strategies*, or *decision rules*, by which he or she specifies the *action*. Thus game theory is the study of optimization problems, in which several players try to *independently* maximize their own personal objective functions which are defined over a set of feasible, yet interdependent strategies of all the players partaking in the game.

Therefore, the players who play games can either cooperate among themselves, or they can be noncooperative. Accordingly, game theory is classified into two classes: *cooperative game theory*, and *noncooperative game theory*. In cooperative game theory, players cooperate with one another. In noncooperative game theory, each player works independently without any knowledge of the actions of his or her opponents (rivals). Cooperative games are also called coalitional games.

The games can again be classified, whether they are static or dynamic games. A static game has a single stage at which the players make simultaneous decisions. In a dynamic game sequential decisions are made.

For technical and pedagogical reasons, the following three categories of noncooperative games are considered in this chapter: strategic-form games (also called normal-form games), extensive-form games, and repeated games.

The strategic-form game is specified by: a list of players, a complete list of strategies available to each player, and specification of the payoff values to each player for each possible strategic configuration. In this game, the players make their moves simultaneously, and only *once*. This does not imply that the players act simultaneously. It simply means that each players acts without knowledge of others' choices. Strategic-form games generally specify static games. These games also model decision-makers who interact with each other. The strategic-form games can again be classified into *pure-strategy* and *mixed-strategy* games. Pure-strategy games are deterministic, and mixed-strategy games are probabilistic.

The strategic-form game can be extended to model sequential-move games. However, extensive-form game is more convenient to specify such dynamic games. In particular, extensive-form game consists of a detailed specification of the method of play. That is, these forms specify the sequence in which different players make their moves, and also information that each player has available at the time of his or her move. Furthermore, the payoff value for each player is also specified. If it happens that chance enters into the game, then it is also specified.

Games can also be classified, whether the players have either perfect or imperfect information. Each player has a knowledge about every other players' actions in a game with perfect information. Thus games with simultaneous moves by players have imperfect information. Extensive games can either have perfect or imperfect information. Some examples of games with perfect information are chess, go, and mancala.

In certain games, players have a knowledge of other players' utility functions. These are games with complete information. In games with incomplete information, at least a single player is not knowledgeable about another player's utility function.

A repeated game (or iterated game) is an example of extensive game, in which a base game (constituent or stage game) is repeated several times. Repeated games can be either finite or infinite. Games can also be either discrete or continuous. Strategic-form games is next discussed.

9.2.2 Strategic-Form Games

In strategic-form games the players specify their strategies only once. These games have two types of strategies. These are pure and mixed strategies. Strategic-form games with pure strategies are deterministic. In mixed strategic games, a probability distribution is defined over the set of pure strategies of the game.

Pure and mixed strategies are initially discussed in this subsection. This is followed by an explanation of the Nash equilibrium. These concepts are next elucidated via examples. Games of special interest such as two-person (or two-player) zero-sum game are also outlined. The intersection of game theory with concepts from linear programming are also elaborated upon. A special type of equilibrium, called the Stackelberg equilibrium is also defined. This equilibrium plays a role in understanding congestion on the Internet.

Pure Strategies

A formal definition of a pure strategic-form game is given below. Strategic-form games with mixed strategies is discussed subsequently.

Definition 9.1. *A pure strategic-form (normal-form) game is a three-tuple* $\Gamma = (I, S, U)$, *where* I *is the set of players,* S *is the set of pure-strategy (action) profiles, and* U *is the utility-function tuple.*

(a) *The finite set of players is* $I = \{1, 2, \ldots, n\}$, *where* $|I| = n \in \mathbb{P}$ *is the number of players in the game.*

(b) *Associated with each player* $i \in I$ *is a finite set of pure strategies (actions)* S_i. *It is sometimes called the pure-strategy space of player* i. *The Cartesian product of the pure strategies, is called the set of pure-strategy profiles. It is*

$$S = S_1 \times S_2 \times \cdots \times S_n \tag{9.1a}$$

Also write $s = (s_1, s_2, \ldots, s_n) \in S$ *for a pure-strategy profile.*

(c) *Associated with each player* $i \in I$ *is a real-valued utility function (payoff function)* $u_i : S \to \mathbb{R}$. *The value* $u_i(s)$, *is sometimes called the von Neumann-Morgenstern utility (payoff) of player* i *for profile* $s \in S$. *Let*

$$U = (u_1, u_2, \ldots, u_n) \tag{9.1b}$$

□

The following useful notation is introduced. The pure strategy of a player $i \in I$ has been defined as $s_i \in S_i$, and $s \in S$ is a pure-strategy profile. Denote the player i's opponents or rivals by $-i$, thus $-i \triangleq I \backslash \{i\}$. Also denote

$$S_{-i} \triangleq \times_{j \in I \backslash \{i\}} S_j \tag{9.2a}$$

and $s_{-i} \in S_{-i}$. That is, s_{-i} is the vector of other players' strategies. With a little misuse of notation let

$$s \triangleq (s_i, s_{-i}) \in S$$

denote a pure-strategy profile. Thus

$$(s_i', s_{-i}) = (s_1, \ldots, s_{i-1}, s_i', s_{i+1}, \ldots, s_n) \tag{9.2b}$$

Example of an easily understood game is the two-player zero-sum game. This is next defined.

Definition 9.2. *A two-person (two-player) zero-sum game is a strategic-form game $\Gamma = (I, S, U)$, where $I = \{1, 2\}$ and*

$$\sum_{i=1}^{2} u_i(s) = 0, \quad \forall\, s \in S \tag{9.3}$$

□

Two-player zero-sum games are studied at length subsequently in this section. This notion of two-player zero-sum game can be further generalized. For example, if the sum of the utility functions of the n players in a strategic game is a constant, then the game is said to be an n-player constant-sum game.

Mixed Strategies

The pure strategy of a player discussed in the last few paragraphs is deterministic. However, it is possible that the strategy is probabilistic (or random). The randomized strategy of a player is the uncertainty in his or her mind. The aim of mixed strategy is to keep the opponents from discovering the strategy. Thus the probability distribution over the complete set of strategies of a player is a mixed strategy. Stated alternately: in mixed-strategy games, the strategy is chosen at random, that is irrationally; however the randomization scheme is selected rationally.

Definition 9.3. *A mixed strategic-form game is a four-tuple $\Gamma = (I, S, \Psi, U)$ in which I is the set of players, S is the set of pure-strategy profiles, Ψ is the space of mixed-strategy profiles, and U is the utility-function tuple. This game consists of:*

(a) *A mixed strategy ψ_i for $i \in I$ is a probability distribution over pure strategies. Thus*

$$\psi_i : S_i \to [0, 1], \quad \sum_{s_i \in S_i} \psi_i(s_i) = 1 \tag{9.4a}$$

(b) *The support of a mixed strategy ψ_i is the set of pure strategies to which the mapping ψ_i assigns positive probability.*
(c) *Each player's randomization is independent of those of its opponents in a probabilistic sense.*
(d) *Player i's space of mixed strategies is Ψ_i, where $\Psi_i = \{\psi_i\}$.*
(e) *Space of mixed-strategy profiles is $\Psi = \times_{i \in I} \Psi_i$.*
(f) *Note that $\psi_i \in \Psi_i$, and $\psi = (\psi_1, \psi_2, \ldots, \psi_n) \in \Psi$, where $n = |I|$.*

(g) *The payoffs to a profile of mixed strategies are the expected values of the corresponding pure-strategy payoffs. With a little misuse of notation, the player i's payoff to profile ψ is*

$$u_i(\psi) = \sum_{s \in S} \left\{ \prod_{j \in I} \psi_j(s_j) \right\} u_i(s) \tag{9.4b}$$

□

Note that the set of mixed strategies includes the pure strategies as degenerate probability distributions. Furthermore, each player i's payoff to a mixed-strategy profile is a linear function of its mixing probability $\psi_i(s_i)$. Recall that the player i's opponents or rivals are denoted by $-i$, thus $-i \triangleq I \backslash \{i\}$. Also denote

$$\Psi_{-i} \triangleq \times_{j \in I \backslash \{i\}} \Psi_j \tag{9.5a}$$

and $\psi_{-i} \in \Psi_{-i}$. Again, with a little misuse of notation let $\psi \triangleq (\psi_i, \psi_{-i}) \in \Psi$ be a mixed-strategy profile. Thus

$$(\psi_i', \psi_{-i}) = (\psi_1, \ldots, \psi_{i-1}, \psi_i', \psi_{i+1}, \ldots, \psi_n) \tag{9.5b}$$

Nash Equilibrium

A Nash equilibrium of a game is a profile of strategies in which each player's strategy is the most favorable response to other players' strategies. Thus in a Nash equilibrium each and every player $i \in I$ feels that it need not alter its strategy, since it can assume that the other players also implement their equilibrium strategies. The idea of Nash equilibrium is made more precise by introducing the concept of *dominating strategy* and the *best response* of a player.

Definitions 9.4. *Let I, S, Ψ, and U denote the set of players, the set of pure-strategy profiles, the space of mixed-strategy profiles, and the utility-function tuple respectively.*

1. *Let $\Gamma = (I, S, \Psi, U)$ be a mixed-strategy game.*
   (a) *A player i's pure strategy $s_i \in S_i$ is strictly dominated if $\exists\ \psi_i' \in \Psi_i$ such that*

$$u_i(\psi_i', s_{-i}) > u_i(s_i, s_{-i}), \quad \forall\ s_{-i} \in S_{-i} \tag{9.6a}$$

   where $u_i(\psi_i', s_{-i})$ is the payoff of player i if he or she uses the mixed strategy ψ_i' and the other players' vector of strategies is s_{-i}.
   (b) *A player i's pure strategy $s_i \in S_i$ is weakly dominated if $\exists\ \psi_i' \in \Psi_i$ such that*

$$\{u_i(\psi_i', s_{-i}) \geq u_i(s_i, s_{-i}), \quad \forall\ s_{-i} \in S_{-i}$$
$$\wedge \exists\ s_{-i} \in S_{-i} \ \text{such that}\ u_i(\psi_i', s_{-i}) > u_i(s_i, s_{-i})\} \tag{9.6b}$$

2. *Let $\Gamma = (I, S, U)$ be a pure strategic-form game. The set of best responses of a player $i \in I$ to a pure-strategy profile $s \in S$ is*

$$\Phi_i(s) = \{s_i^* \in S_i \mid \forall\ s_i \in S_i,\ \ u_i(s_i^*, s_{-i}) \geq u_i(s_i, s_{-i})\} \tag{9.6c}$$

The joint best response set is denoted by $\Phi(s) = \times_{i \in I} \Phi_i(s)$. □

Observe that the best response function $\Phi_i(\cdot)$ is *set-valued*. That is, it associates a set of strategies with any list of other players' strategies. Each member of the set $\Phi_i(\cdot)$ is a best response of player i to s_{-i}. The Nash equilibrium of a game is next defined.

Definitions 9.5. *Let I, S, Ψ, and U denote the set of players, the set of pure-strategy profiles, the space of mixed-strategy profiles, and the utility-function tuple respectively.*

1. *Let $\Gamma = (I, S, U)$ be a pure strategic-form game.*
   (a) *A pure-strategy profile $s^* \in S$ is a Nash equilibrium if $\forall\, i \in I$*

$$u_i\left(s_i^*, s_{-i}^*\right) \geq u_i\left(s_i, s_{-i}^*\right), \quad \forall\, s_i \in S_i \tag{9.7a}$$

   Consequently, a Nash equilibrium is a pure-strategy profile s^* such that $s^* \in \Phi(s^*)$.
   (b) *A Nash equilibrium s^* is said to be strict if for each player $i \in I$*

$$u_i\left(s_i^*, s_{-i}^*\right) > u_i\left(s_i, s_{-i}^*\right), \quad \forall\, s_i \in S_i \backslash \{s_i^*\} \tag{9.7b}$$

   Consequently, a Nash equilibrium is a strict pure-strategy profile s^* if $\Phi(s^*) = \{s^*\}$. That is, each player has a unique best response to his or her opponents' strategies.
2. *Let $\Gamma = (I, S, \Psi, U)$ be a mixed-strategy game. A mixed-strategy profile $\psi^* \in \Psi$ is a Nash equilibrium if $\forall\, i \in I$*

$$u_i\left(\psi_i^*, \psi_{-i}^*\right) \geq u_i\left(s_i, \psi_{-i}^*\right), \quad \forall\, s_i \in S_i \tag{9.7c}$$

□

It should be evident from the above definition, that the Nash equilibrium of a game is not unique. Also, if each player in a game has only a small number of strategies, then the Nash equilibrium if it exists, is determined by examining different cases. However, if the games are elaborate, then it is efficient to use best response functions. The procedure to determine this is as follows.

(a) Determine the best response function of each player.
(b) Determine the strategic profile that satisfies the relationship $s^* \in \Phi(s^*)$.

Another possible technique to compute Nash equilibria is via *iterated deletion*. For example, a two-person game is specified by its payoff matrix. Then it is possible to delete rows or columns iteratively which are dominated by other rows or columns to identify the Nash equilibria. It should however be noted that, a majority of games are not solvable if this approach is used.

Examples

The concepts developed in the last few subsections are further elucidated via examples. Prisoners' dilemma, an example which discusses a choice between classical and modern music, a mixed-strategy game, and a game of matching coins are outlined. Examples which perform computation of Nash equilibrium using best response functions and iterated dominance are also given. The classic example of the tragedy of the commons is also discussed.

Example 9.1. *Prisoners' dilemma.* Two persons are suspected of a major crime. The police force of the country has sufficient evidence to incarcerate each of them for a minor crime. However, the police force does not have enough evidence to convict them of a major crime, unless one of them confesses about the activities of the other. The police hold the suspects in separate prison cells, and each suspect is advised about the consequences of his or her actions. Furthermore, each suspect can either stay quiet (mum) or act as an informer (fink) against the other. The following scenarios emerge.

(a) Both suspects stay quiet: In this case each suspect is convicted of the minor crime and spends two years in the prison.
(b) Both suspects fink: In this scenario each suspect is convicted of the major crime and spends five years in the prison.
(c) One suspect stays quiet, and the other finks: The suspect who finks is set free, and is used as a witness against the other. And the suspect who stays quiet spends ten years in the prison.

Thus the stakes are high in this game. The dilemma of the prisoners is explicitly stated in the language of game theory. More specifically this scenario is modeled as a strategic game. The players in the game are the two suspects. Label the suspects 1 and 2, that is $I = \{1, 2\}$. The set of strategies of each player is the set $\{\mathit{Mum},\ \mathit{Fink}\}$. Thus $S_1 = S_2 = \{\mathit{Mum},\ \mathit{Fink}\}$. The strategy profiles of the players are:

$$(\mathit{Mum},\ \mathit{Mum}),\ (\mathit{Mum},\ \mathit{Fink}),\ (\mathit{Fink},\ \mathit{Mum}),\ \text{and}\ \ (\mathit{Fink},\ \mathit{Fink})$$

For suspect number 1, the ordering of the strategy profile from best to worst is:

(a) $(\mathit{Fink},\ \mathit{Mum})$, that is suspect 1 finks and suspect 2 stays mum. Suspect 1 goes free, and the suspect 2 spends ten years in prison.
(b) $(\mathit{Mum},\ \mathit{Mum})$, that is both suspects 1 and 2 stay mum. Each suspect spends two years in prison.
(c) $(\mathit{Fink},\ \mathit{Fink})$, that is both suspects 1 and 2 fink. Each suspect spends five years in prison.
(d) $(\mathit{Mum},\ \mathit{Fink})$, that is suspect 1 stays mum and suspect 2 finks. Suspect 1 spends ten years in prison, and the suspect 2 goes free.

The strategy profiles of suspect 2 from best to worst are:

$$(\mathit{Mum},\ \mathit{Fink}),\ (\mathit{Mum},\ \mathit{Mum}),\ (\mathit{Fink},\ \mathit{Fink}),\ \text{and}\ \ (\mathit{Fink},\ \mathit{Mum})$$

The payoff or the utility function $u_1(\cdot, \cdot)$ for suspect 1 should be such that

$$u_1(\mathit{Fink},\ \mathit{Mum}) > u_1(\mathit{Mum},\ \mathit{Mum}) > u_1(\mathit{Fink},\ \mathit{Fink}) > u_1(\mathit{Mum},\ \mathit{Fink})$$

Therefore, select

$$u_1(\mathit{Fink},\ \mathit{Mum}) = 5,\ u_1(\mathit{Mum},\ \mathit{Mum}) = 3,\ u_1(\mathit{Fink},\ \mathit{Fink}) = 1,\ \text{and}\ \ u_1(\mathit{Mum},\ \mathit{Fink}) = 0$$

If $u_2(\cdot, \cdot)$ is the utility function of suspect 2, select

$$u_2(\mathit{Mum},\ \mathit{Fink}) = 5,\ u_2(\mathit{Mum},\ \mathit{Mum}) = 3,\ u_2(\mathit{Fink},\ \mathit{Fink}) = 1,\ \text{and}\ \ u_2(\mathit{Fink},\ \mathit{Mum}) = 0$$

This game is represented in the Figure 9.1, with its payoff values. The corresponding 2×2 matrix is called the payoff matrix. Its elements are the utility values (payoffs). The first entry in each box is suspect 1's payoff and the second entry is suspect 2's payoff.

Summarizing, the prisoner's dilemma models a scenario in which it is beneficial for both players to cooperate, but each player has an option of a larger payoff (going free) by playing the role of a fink.

		Suspect 2	
		Mum	*Fink*
Suspect 1	*Mum*	$(3,3)$	$(0,5)$
	Fink	$(5,0)$	$(1,1)$

Figure 9.1. Prisoners' dilemma: payoff matrix.

The Nash equilibrium of this game is next determined. It is established that the strategy profile (*Fink, Fink*) is the unique Nash equilibrium. Consider the four strategy profiles.

(a) Profile (*Mum, Mum*): This profile does not satisfy the condition for Nash equilibrium. Assume that the suspect 1 selects the strategy *Mum*, suspect 2's payoff to *Fink* exceeds its payoff to *Mum*. This observation alone is sufficient to establish that the strategy profile (*Mum, Mum*) is not a Nash equilibrium.
Similarly, observe that if the suspect 2 selects the strategy *Mum*, suspect 1's payoff to *Fink* exceeds its payoff to *Mum*. Therefore it is again concluded that the profile (*Mum, Mum*) is not a Nash equilibrium.
(b) Profile (*Fink, Mum*): This profile does not satisfy the condition for Nash equilibrium. Assume that the suspect 1 selects the strategy *Fink*, suspect 2's payoff to *Fink* exceeds its payoff to *Mum*.
(c) Profile (*Mum, Fink*): This profile does not satisfy the condition for Nash equilibrium. Assume that the suspect 2 selects the strategy *Fink*, suspect 1's payoff to *Fink* exceeds its payoff to *Mum*.
(d) Profile (*Fink, Fink*): This profile does satisfy the condition for Nash equilibrium. Assume that the suspect 1 selects the strategy *Fink*, suspect 2's payoff to *Fink* exceeds its payoff to *Mum*. Similarly, assume that the suspect 2 selects the strategy *Fink*, suspect 1's payoff to *Fink* exceeds its payoff to *Mum*.

Notice that the strategy *Fink* strictly dominates the strategy *Mum*, irrespective of the opponent's strategy. That is, a suspect receives a favorable payoff when he or she selects *Fink* over the *Mum* option. In this example the suspects had to decide if they should cooperate or not. In the next example, it is shown that it is best to cooperate. □

Example 9.2. *Choice between classical and modern music.* Two persons decide to attend together, either a classical or a modern music presentation. However, one person prefers classical over modern music, and the other person prefers modern over classical music. This scenario is modeled as a two-player strategic game. This example shows that it is possible for a game to have multiple Nash equilibria. The players in the game are the two persons. Let $I = \{\text{Person } 1,\ \text{Person } 2\} \triangleq \{1, 2\}$. The set of strategies of each player is the set $\{$*Classical, Modern*$\}$. Thus $S_1 = S_2 = \{$*Classical, Modern*$\}$. The strategy profiles of the players are:

(*Classical, Classical*), (*Classical, Modern*), (*Modern, Classical*), and (*Modern, Modern*)

This game is represented in the Figure 9.2, with its payoff values. The first entry in each box is Person 1's payoff and the second entry is Person 2's payoff. Persons 1 and 2 prefer classical and modern music respectively. The Nash equilibrium of this game is next determined. It is established that the strategy profiles (*Classical, Classical*) and (*Modern, Modern*) are the multiple Nash equilibria. Consider the four strategy profiles.

		Person 2	
		Classical	*Modern*
Person 1	*Classical*	$(5,1)$	$(0,0)$
	Modern	$(0,0)$	$(1,5)$

Figure 9.2. Classical or modern music: payoff matrix.

(a) Profile (*Classical, Classical*): If Person 1 switches from *Classical* to *Modern,* then his or her payoff decreases from 5 to 0. However, if Person 2 switches from *Classical* to *Modern,* then his or her payoff decreases from 1 to 0. Thus a change in strategy of either player decreases their respective payoffs. Therefore the profile (*Classical, Classical*) is a Nash equilibrium.
(b) Profile (*Modern, Classical*): This profile does not satisfy the condition for Nash equilibrium. If Person 1 switches from *Modern* to *Classical,* then his or her payoff increases from 0 to 5. However, if Person 2 switches from *Classical* to *Modern,* then his or her payoff increases from 0 to 5. Note that it is sufficient to show that a strategy profile is not a Nash equilibrium, if it is demonstrated that the payoff of a single player increases by a change in his or her strategy.
(c) Profile (*Classical, Modern*): This profile does not satisfy the condition for Nash equilibrium. If Person 1 switches from *Classical* to *Modern,* then his or her payoff increases from 0 to 1. However, if Person 2 switches from *Modern* to *Classical,* then his or her payoff increases from 0 to 1.
(d) Profile (*Modern, Modern*): If Person 1 switches from *Modern* to *Classical,* then his or her payoff decreases from 1 to 0. However, if Person 2 switches from *Modern* to *Classical,* then his or her payoff decreases from 5 to 0. Thus a change in strategy of either player decreases their respective payoff. Therefore the profile (*Modern, Modern*) is a Nash equilibrium.

Also observe that *Classical* is the best strategy for Person 1 if Person 2 selects *Classical.* Similarly, *Modern* is the best strategy for Person 1 if Person 2 selects *Modern.* Furthermore, neither strategy strictly dominates the other. In summary, this type of game models a scenario in which it is in the best interest of players to cooperate. □

Example 9.3. *A mixed-strategy game.* Consider a two-player game. In this game the strategies of player 1 are: A_s, B_s, and C_s; and the strategies of player 2 are: D_s, E_s, and F_s. See Figure 9.3. The payoffs to each player for each strategy profile are also shown in the payoff matrix. The first entry in each box is player 1's payoff and the second entry is player 2's payoff.

		Player 2		
		D_s	E_s	F_s
	A_s	$(2,4)$	$(6,2)$	$(5,3)$
Player 1	B_s	$(3,0)$	$(8,1)$	$(4,7)$
	C_s	$(5,3)$	$(7,2)$	$(2,9)$

Figure 9.3. A two-player game payoff matrix.

The mixed strategy of player 1 is:

$$(\psi_1(A_s), \psi_1(B_s), \psi_1(C_s)), \text{ where } \psi_1(A_s) + \psi_1(B_s) + \psi_1(C_s) = 1$$

Similarly, the mixed strategy of player 2 is:

$$(\psi_2(D_s), \psi_2(E_s), \psi_2(F_s)), \text{ where } \psi_2(D_s) + \psi_2(E_s) + \psi_2(F_s) = 1$$

Let

$$(\psi_1(A_s), \psi_1(B_s), \psi_1(C_s)) = (1/5, 3/5, 1/5)$$
$$(\psi_2(D_s), \psi_2(E_s), \psi_2(F_s)) = (1/2, 0, 1/2)$$

Therefore

$$u_1(\psi_1, \psi_2) = \frac{1}{5}\left(\frac{1}{2} \cdot 2 + 0 \cdot 6 + \frac{1}{2} \cdot 5\right) + \frac{3}{5}\left(\frac{1}{2} \cdot 3 + 0 \cdot 8 + \frac{1}{2} \cdot 4\right)$$
$$+\frac{1}{5}\left(\frac{1}{2} \cdot 5 + 0 \cdot 7 + \frac{1}{2} \cdot 2\right) = 3.5$$

Similarly

$$u_2(\psi_1, \psi_2) = \frac{1}{5}\left(\frac{1}{2} \cdot 4 + 0 \cdot 2 + \frac{1}{2} \cdot 3\right) + \frac{3}{5}\left(\frac{1}{2} \cdot 0 + 0 \cdot 1 + \frac{1}{2} \cdot 7\right)$$
$$+\frac{1}{5}\left(\frac{1}{2} \cdot 3 + 0 \cdot 2 + \frac{1}{2} \cdot 9\right) = 4.0$$

□

Example 9.4. *Game of matching coins.* Recall that the game-theoretic model in prisoners' dilemma models both conflict and cooperation. The matching coins game models a scenario in which the players are completely in conflict. In this two-person game, each person has a coin. These two persons toss their respective coin simultaneously. If the coins show the same side, Person 2 pays Person 1 a single dollar. However, if the two tossed coins show different sides, then Person 1 pays Person 2 a single dollar. To make this description complete, one side of a coin is called a *Head*, and the other a *Tail*.

This example shows that it is possible for a game to not to have a Nash equilibrium. The players in the game are the two persons. Label the persons 1 and 2, that is $I = \{1, 2\}$. The strategy set of each player is the set {*Head, Tail*}. Thus $S_1 = S_2 =$ {*Head, Tail*}. The strategy profiles of the players are: (*Head, Head*), (*Head, Tail*), (*Tail, Head*), and (*Tail, Tail*).

This game is represented in the Figure 9.4, with its payoff values. The first entry in each box is Person 1's payoff and the second entry is Person 2's payoff. Observe that the payoffs are equal to the amount of money.

		Person 2	
		Head	*Tail*
Person 1	*Head*	(1, -1)	(-1, 1)
	Tail	(-1, 1)	(1, -1)

Figure 9.4. Matching coins: payoff matrix.

Consider the four strategy profiles.

(a) Profile (*Head, Head*): If Person 2 switches from *Head* to *Tail*, then his or her payoff increases from -1 to 1. Therefore the profile (*Head, Head*) is not a Nash equilibrium.
(b) Profile (*Tail, Head*): If Person 1 switches from *Tail* to *Head*, then his or her payoff increases from -1 to 1. Therefore the profile (*Tail, Head*) is not a Nash equilibrium.

(c) Profile (*Head, Tail*): If Person 1 switches from *Head* to *Tail*, then his or her payoff increases from -1 to 1. Therefore the profile (*Head, Tail*) is not a Nash equilibrium.
(d) Profile (*Tail, Tail*): If Person 2 switches from *Tail* to *Head*, then his or her payoff increases from -1 to 1. Therefore the profile (*Tail, Tail*) is not a Nash equilibrium.

Next assume that the game has a mixed strategy. Let

$$(\psi_1 (Head), \psi_1 (Tail)) = (\psi_2 (Head), \psi_2 (Tail)) = (1/2, 1/2)$$

This mixed-strategy profile is a unique Nash equilibrium. The payoffs are

$$u_1 (\psi_1, \psi_2) = \frac{1}{2} \left\{ \frac{1}{2} \cdot 1 + \frac{1}{2} \cdot (-1) \right\} + \frac{1}{2} \left\{ \frac{1}{2} \cdot (-1) + \frac{1}{2} \cdot 1 \right\} = 0$$
$$u_2 (\psi_1, \psi_2) = \frac{1}{2} \left\{ \frac{1}{2} \cdot (-1) + \frac{1}{2} \cdot 1 \right\} + \frac{1}{2} \left\{ \frac{1}{2} \cdot 1 + \frac{1}{2} \cdot (-1) \right\} = 0$$

□

Example 9.5. *Computation of Nash equilibrium using best response functions.* Consider a two-player game, in which each player has three strategies. See Figure 9.5. The strategies of player 1 are A_s, B_s, and C_s. Similarly the strategies of player 2 are D_s, E_s, and F_s. The first entry in each box is player 1's payoff and the second entry is player 2's payoff.

		Player 2		
		D_s	E_s	F_s
Player 1	A_s	$(2, 6^*)$	$(6, 4)$	$(5, 2)$
	B_s	$(3, 1)$	$(9^*, 3^*)$	$(4, 3^*)$
	C_s	$(8^*, 4^*)$	$(7, 2)$	$(6^*, 3)$

Figure 9.5. Nash equilibria via best response functions in a two-player game.

Initially we find the best response of player 1 to each strategy of player 2. If player 2 selects strategy D_s, then player 1's best response is strategy C_s. Observe that player 1's highest payoff is 8. Denote this best response by placing a star ($*$) as a superscript on 8. This is indeed the payoff to the player 1 for the strategy profile (C_s, D_s). Next assume that the player 2 selects strategy E_s. In this case player 1's best response is strategy B_s. Player 1's payoff to the strategy profile (B_s, E_s) is 9. Therefore a star is placed as a superscript on 9. In the next step player 2 selects strategy F_s. In this case player 1's best response is strategy C_s. Player 1's payoff to the strategy profile (C_s, F_s) is 6. Therefore a star is placed as a superscript on 6.

The above process is repeated by first selecting a strategy of player 1, and then selecting best responses of player 2. If player 1 selects strategy A_s, then player 2's best response is strategy D_s. Observe that player 2's highest payoff is 6. Denote this best response by placing a star as a superscript on 6. This is indeed the payoff to the player 2 for the strategy profile (A_s, D_s). Next assume that the player 1 selects strategy B_s. In this case player 2's best response are strategies E_s and F_s. Player 2's payoff to the strategy profile (B_s, E_s) and (B_s, F_s) is 3 each. Therefore a star is placed as a superscript on 3 in each cell. Finally, in the next step player 1 selects strategy C_s. In this case player 2's best response is strategy D_s. Player 2's payoff to the strategy profile (C_s, D_s) is 4. Therefore a star is placed as a superscript on 4.

In the very last step, determine the cells in which both players' payoffs are starred. These cells are the Nash equilibria of the game. It is possible to find more than a single such cell. In this game the Nash equilibria are: (B_s, E_s), and (C_s, D_s). □

Example 9.6. *Computation of Nash equilibrium using iterated dominance.* Consider a two-player game, in which each player has three strategies. See Figure 9.6 (a). The strategies of player 1 are A_s, B_s, and C_s. Similarly the strategies of player 2 are D_s, E_s, and F_s. The first entry in each box is player 1's payoff and the second entry is player 2's payoff. A similar problem was solved in the last example by using the technique of best response functions. In this example, we use the technique of iterated dominance to find the Nash equilibrium.

Player 2

		D_s	E_s	F_s
	A_s	$(2,6)$	$(6,4)$	$(5,2)$
Player 1	B_s	$(3,1)$	$(9,5)$	$(4,3)$
	C_s	$(8,4)$	$(7,9)$	$(6,3)$

Figure 9.6 (a). Nash equilibria via iterated dominance.

For player 2, strategy F_s is dominated by strategy E_s, because: $4 > 2, 5 > 3$, and $9 > 3$. That is, the second-coordinates of the column-E_s elements are greater than those of second-coordinates of the column F_s. Therefore player 2 can eliminate strategy F_s from his or her strategy space. Thus the column corresponding to strategy F_s is deleted. The reduced payoff matrix, after first iteration is shown in Figure 9.6 (b).

Player 2

		D_s	E_s
	A_s	$(2,6)$	$(6,4)$
Player 1	B_s	$(3,1)$	$(9,5)$
	C_s	$(8,4)$	$(7,9)$

Figure 9.6 (b). Reduced payoffs, after first iteration.

Observe in Figure 9.6 (b), that for player 1, strategy A_s is dominated by both strategies B_s and C_s. Therefore the row corresponding to strategy A_s is deleted. The reduced payoff matrix after second iteration is shown in Figure 9.6 (c).

Player 2

		D_s	E_s
Player 1	B_s	$(3,1)$	$(9,5)$
	C_s	$(8,4)$	$(7,9)$

Figure 9.6 (c). Reduced payoffs, after second iteration.

Observe in Figure 9.6 (c), that for player 2, strategy D_s is dominated by strategy E_s. Therefore column corresponding to strategy D_s is deleted. The reduced payoff matrix after third iteration is shown in Figure 9.6 (d).

	Player 2
	E_s
Player 1 B_s	$(9,5)$
C_s	$(7,9)$

Figure 9.6 (d). Reduced payoffs, after third iteration.

Finally, observe in Figure 9.6 (d), that for player 1, strategy C_s is dominated by strategy B_s. Therefore row corresponding to strategy C_s is deleted. The reduced payoff matrix after fourth iteration is shown in Figure 9.6 (e).

	Player 2
	E_s
Player 1 B_s	$(9,5)$

Figure 9.6 (e). Reduced payoffs, after fourth iteration.

Thus the Nash equilibrium for this game is (B_s, E_s). It can also be checked (as in last example) that the best response technique yields identical Nash equilibrium. □

Example 9.7. *Tragedy of the commons.* The example is commonly referred to as the tragedy of the commons. Simplistic as it may appear, this example has many interpretations and applications. Some of these are in the fields of: economics, study of the environment, planning of cities, politics, and routing of packets in the Internet.

A village has $n \in \mathbb{P}$ farmers, and the ith farmer has $y_i \in \mathbb{P}$ number of sheep. The village is blessed with a finite spread of green for the sheep to graze upon. Thus the total number sheep in the village is $Y = \sum_{j=1}^{n} y_j$. The cost of buying and taking care of a single sheep is equal to a positive constant c. Also, the value to a farmer of grazing a sheep on the green pastures is $v(\cdot)$ *per sheep*. This value is a function of the total number of sheep in the farm, which is equal to Y. This dependence of $v(\cdot)$ on Y is justified, because each sheep requires a certain amount of grass to live on. Since the pasture area in the village is finite, there is an upper limit on the total number of sheep that can graze upon the village fields. Let this value be $Y_{\max}$. Thus

$$v(Y) > 0, \quad Y < Y_{\max}$$
$$v(Y) = 0, \quad Y \geq Y_{\max}$$

In the interest of simplicity, the function $v(\cdot)$ is defined on a continuum, even though the total number of sheep on the farm Y is a positive integer. That is, for discussion and tractability of the model it is assumed that the y_i's are positive real numbers. Observe that when Y is small, then sheep have enough green to graze upon. However, as Y increases, the value of each sheep $v(Y)$ decreases. Thus in this model, it is assumed that $v'(Y) < 0$ for $Y < Y_{\max}$, where $v'(Y)$ is the first derivative of $v(Y)$ with respect to Y. In addition, assume that $v''(Y) < 0$ for $Y < Y_{\max}$, where $v''(Y)$ is the second derivative of $v(Y)$ with respect to Y. This later assumption is required to ensure the existence of the Nash equilibrium. See Figure 9.7.

Let $y = (y_1, y_2, \ldots, y_n)$, and the utility of the ith farmer be $u_i(y)$. Thus

$$u_i(y) = y_i v(Y) - c y_i, \quad 1 \leq i \leq n$$

If $y^* = (y_1^*, y_2^*, \ldots, y_n^*)$ is indeed a Nash equilibrium, then y_i^* must maximize $u_i(y)$ for each i. Let $Y^* = \sum_{j=1}^{n} y_j^*$, and $Y_{-i}^* = \sum_{j=1, j\neq i}^{n} y_j^*$. Note that

$$\frac{\partial v(Y)}{\partial y_i} = \frac{dv(Y)}{dY}\frac{\partial Y}{\partial y_i} = v'(Y)$$

Thus, if the other farmers select $\left(y_1^*, y_2^*, \ldots, y_{i-1}^*, y_{i+1}^*, \ldots y_n^*\right)$, optimization requirement of farmer i yields

$$v\left(y_i + Y_{-i}^*\right) + y_i v'\left(y_i + Y_{-i}^*\right) - c = 0, \quad 1 \le i \le n$$

Substituting y_i^* in the above equation and summing over all values of i, yields

$$v(Y^*) + \frac{1}{n} Y^* v'(Y^*) - c = 0$$

In the above analysis it is assumed that each farmer wants to maximize his or her individual payoff. However, if the goal is to achieve a social (global or overall) optimum, then $\{Yv(Y) - cY\}$ has to be maximized, as Y varies. Denote the socially optimum value of Y by Y^{**}. This is achievable if

$$v(Y^{**}) + Y^{**} v'(Y^{**}) - c = 0$$

As $v'(Y) < 0$ and $v''(Y) < 0$, it is shown in the problem section that $Y^{**} < Y^*$. This simple inequality has profound implications. That is, the total number of sheep allowed to graze under Nash equilibrium is more than that allowed under the globally optimal conditions. This example shows that: *Selfish behaviors by independent and noncooperative farmers do not necessarily produce a socially optimum outcome.* Furthermore, this simplistic example also demonstrates that game theory can be applied to find laws and mechanisms to interpret and determine, globally and socially desirable outcomes.

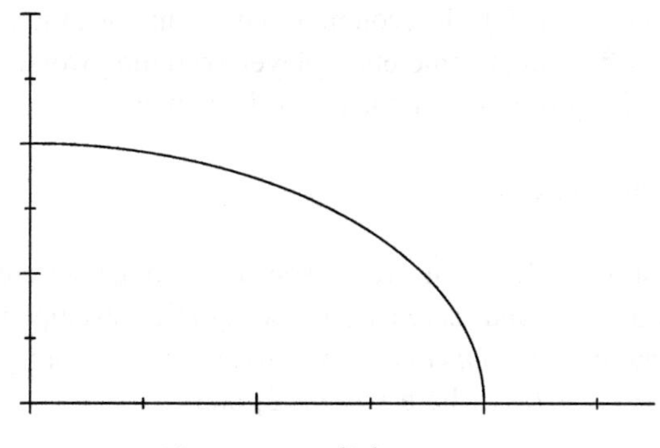

Figure 9.7. $v(Y)$ versus Y.

□

Existence of a Nash Equilibrium

The above examples illustrate that a strategic game can have one or more Nash equilibria, or none at all. For example the games of matching coins, prisoners' dilemma, and classical and modern music have zero, one, and two Nash equilibrium points respectively. The conditions for the existence of the Nash equilibrium have been studied by several mathematicians, foremost among them being John F. Nash.

It is nontrivial to specify general conditions for the existence of Nash equilibria for a strategic game. However, John Nash proved in the year 1950 that: Every finite n-player game, with finite number of strategies for each player, has at least a single Nash equilibrium point, possibly involving mixed strategies. His proof uses the *fixed-point theorem*, which is due to the mathematician L. E. J. Brouwer. The proof of Nash's theorem can be found in advanced textbooks on game theory.

Pareto Optimality

It is quite possible that there are an infinite number of Nash equilibria in a game. Our goal is to select an equilibrium which is optimal. Before an optimal equilibrium is selected, we need to define the optimality criteria. Among several possible criteria of optimality, Pareto optimality is one of the most significant.

A strategy (action) of a game is said to be Pareto optimal, or Pareto efficient, if there does not exist a strategy which would make all players at least as well-off, and at least a single player strictly better off. Thus Pareto optimality is a measure of efficiency. This optimality is named after the economist and philosopher Vilfredo Pareto (1848-1923).

Definition 9.6. *Let $\Gamma = (I, S, \Psi, U)$ be a mixed-strategy game, where I is the set of players, S is the set of pure-strategy profiles, Ψ is the space of mixed-strategy profiles, and U is the utility-function tuple. The utility $u \in U$ has a Pareto optimum at $s^P \in S$ if there is no $s \in S$ such that*

$$u_i(s) \geq u_i\left(s^P\right), \quad \forall\, i \in I; \text{ and } u_j(s) > u_i\left(s^P\right) \text{ for some } j \in I \backslash \{i\} \tag{9.8}$$

□

Alternately, the utility $u \in U$ has a Pareto optimum at $s^P \in S$ if there is no $s \in S$ such that $u(s) \geq u\left(s^P\right)$, and $u(s) \neq u\left(s^P\right)$. In economic theory an allocation is Pareto optimal, if no player can be made better-off without some other player becoming worse-off. In general, a Pareto optimal profile and the Nash equilibria do not necessarily coincide.

Two-Person Zero-Sum Game

Both pure and mixed strategic forms of two-person zero-sum games are next considered. These games also yield to ready analysis and serve as good pedagogical examples.

Pure-Strategy Two-Person Zero-Sum Games: A two-person zero-sum game has been defined as a strategic-form game $\Gamma = (I, S, U)$, where $I = \{1, 2\}$ and

$$\sum_{k=1}^{2} u_k(s) = 0, \quad \forall\, s \in S \tag{9.9a}$$

The two persons playing the game are P_{plyr1} and P_{plyr2}. In this game, the sum of utilities of the two-players is equal to zero. Also, the set of strategies of the two players are S_1 and S_2 respectively, and $S = S_1 \times S_2$. Let $S_1 = \{s_{11}, s_{12}, \ldots, s_{1m}\}$, and $S_2 = \{s_{21}, s_{22}, \ldots, s_{2n}\}$, that is $|S_1| = m$, and $|S_2| = n$. Thus a strategy profile of the game is $s = (s_{1i}, s_{2j}) \in S$ for $1 \leq i \leq m$, and $1 \leq j \leq n$. For simplicity in notation, denote the utility of the two players by

$$u_1(s_{1i}, s_{2j}) = a_{ij}, \quad 1 \leq i \leq m,\ 1 \leq j \leq n \tag{9.9b}$$

$$u_2(s_{1i}, s_{2j}) = b_{ij}, \quad 1 \leq i \leq m,\ 1 \leq j \leq n \tag{9.9c}$$

Since the game is a zero-sum game

$$a_{ij} + b_{ij} = 0, \quad 1 \le i \le m, \; 1 \le j \le n \tag{9.9d}$$

That is, $b_{ij} = -a_{ij}$ for $1 \le i \le m$, and $1 \le j \le n$. These values are conveniently represented in a payoff matrix. The $m \times n$ payoff matrix of player P_{plyr1} is $A = [a_{ij}]$. Similarly, the $m \times n$ payoff matrix of player P_{plyr2} is $B = [b_{ij}] = -A$.

If P_{plyr1} uses strategy s_{1i} and P_{plyr2} uses strategy s_{2j}, then P_{plyr1}'s payoff is a_{ij}, and P_{plyr2}'s payoff is $-a_{ij}$. The player P_{plyr1}'s aim is to make a_{ij} as large as possible, and the player P_{plyr2}'s aim is to make a_{ij} as small as possible. Consider the following two cases.

(a) Assume that P_{plyr1} selects strategy s_{1i}. It then gets at least $\min_{1 \le j \le n} a_{ij}$. This player can actually select any strategy s_{1i}, where $1 \le i \le m$. Therefore P_{plyr1}'s payoff is at least $\max_{1 \le i \le m} \min_{1 \le j \le n} a_{ij}$.
(b) Similarly, if P_{plyr2} selects strategy s_{2j}, it loses at most $\max_{1 \le i \le m} a_{ij}$. This player can actually select any strategy s_{2j}, where $1 \le j \le n$.
Thus P_{plyr2} loses at most $\min_{1 \le j \le n} \max_{1 \le i \le m} a_{ij}$.

It is immediately evident that P_{plyr1}'s gain cannot exceed P_{plyr2}'s losses.

That is, $\max_{1 \le i \le m} \min_{1 \le j \le n} a_{ij}$ is upper-bounded by $\min_{1 \le j \le n} \max_{1 \le i \le m} a_{ij}$. This result is formally stated below.

Lemma 9.1. *In a zero-sum pure strategic game $\Gamma = (I, S, U)$ between two persons P_{plyr1} and P_{plyr2}*

$$\max_{1 \le i \le m} \min_{1 \le j \le n} a_{ij} \le \min_{1 \le j \le n} \max_{1 \le i \le m} a_{ij} \tag{9.10}$$

where $A = [a_{ij}]$ is the $m \times n$ payoff matrix of P_{plyr1}.

Proof. See the problem section. □

Definition 9.7. *Let $A = [a_{ij}]$ be a real $m \times n$ matrix. This matrix is said to have a saddle point, if*

$$\max_{1 \le i \le m} \min_{1 \le j \le n} a_{ij} = \min_{1 \le j \le n} \max_{1 \le i \le m} a_{ij} = v_{sdl} \tag{9.11}$$

The matrix A has a saddle point at (i^*, j^*), if $a_{i^* j^*} = v_{sdl}$. □

The matrix A can have either zero, or one, or more than one saddle points. Thus the matrix element a_{ij} is a saddle point, if it is simultaneously a minimum in its row and a maximum in its column. Note that if a game has several saddle points, then all such saddle points have identical value.

Lemma 9.2. *In a zero-sum pure strategic game $\Gamma = (I, S, U)$ involving two persons P_{plyr1} and P_{plyr2}, $A = [a_{ij}]$ is the $m \times n$ payoff matrix of P_{plyr1}. Then a_{ij} is a saddle point if and only if (s_{1i}, s_{2j}) is a Nash equilibrium.*

Proof. See the problem section. □

Examples 9.8. In these examples, the matrix $A = [a_{ij}]$ is the payoff matrix of person P_{plyr1} in a zero-sum strategy game.

1. Let

$$A = \begin{bmatrix} 2 & 7 & -1 \\ 6 & -3 & 8 \\ 4 & 5 & 9 \\ 1 & 3 & -2 \end{bmatrix}$$

In this matrix

$$\max_{1 \le i \le 4} \min_{1 \le j \le 3} a_{ij} = 4 = a_{31}$$
$$\min_{1 \le j \le 3} \max_{1 \le i \le 4} a_{ij} = 6 = a_{21}$$

Thus $\max_{1 \le i \le 4} \min_{1 \le j \le 3} a_{ij} \le \min_{1 \le j \le 3} \max_{1 \le i \le 4} a_{ij}$. Therefore in this game, P_{plyr1} can receive a payoff of at least 4. The maximum P_{plyr2} pays or the most P_{plyr1} can receive is a payoff of 6. The matrix A also does not have a saddle point. This example indicates that, in a game without saddle point, a player's payoff depends upon his or her adversary's choice of strategy.

2. Let

$$A = \begin{bmatrix} 2 & -5 & -3 \\ 4 & -6 & 7 \\ 8 & 1 & 2 \end{bmatrix}$$

In this matrix

$$\max_{1 \le i \le 3} \min_{1 \le j \le 3} a_{ij} = 1 = a_{32},$$
$$\min_{1 \le j \le 3} \max_{1 \le i \le 3} a_{ij} = 1 = a_{32}$$

Thus $\max_{1 \le i \le 3} \min_{1 \le j \le 3} a_{ij} = \min_{1 \le j \le 3} \max_{1 \le i \le 3} a_{ij}$. Therefore the matrix A has a single saddle point, and its Nash equilibrium is (s_{13}, s_{22}).

3. Let

$$A = \begin{bmatrix} 2 & 1 & 3 \\ 4 & -6 & 7 \\ 8 & 1 & 2 \end{bmatrix}$$

In this matrix

$$\max_{1 \le i \le 3} \min_{1 \le j \le 3} a_{ij} = 1 = a_{12} = a_{32}$$
$$\min_{1 \le j \le 3} \max_{1 \le i \le 3} a_{ij} = 1 = a_{12} = a_{32}$$

Thus $\max_{1 \le i \le 3} \min_{1 \le j \le 3} a_{ij} = \min_{1 \le j \le 3} \max_{1 \le i \le 3} a_{ij}$. Therefore the matrix A has two saddle points, and its Nash equilibria are (s_{11}, s_{22}), and (s_{13}, s_{22}). □

Mixed-Strategy Two-Person Zero-Sum Games

A mixed-strategy two-person zero-sum game has a probability distribution defined over the set of pure strategies of the two players. Let the two persons playing the game be P_{plyr1} and P_{plyr2}. The set of strategies of the two players are S_1 and S_2 respectively. Also, let $S_1 = \{s_{11}, s_{12}, \ldots, s_{1m}\}$, and $S_2 = \{s_{21}, s_{22}, \ldots, s_{2n}\}$, that is $|S_1| = m$, $|S_2| = n$, and $S = S_1 \times S_2$. Thus a strategy profile $s = (s_{1i}, s_{2j}) \in S$ for $1 \le i \le m$, and $1 \le j \le n$. The corresponding payoffs to P_{plyr1} and P_{plyr2} are a_{ij} and b_{ij} respectively. Since the game is a zero-sum game

$$a_{ij} + b_{ij} = 0, \quad 1 \le i \le m, \ 1 \le j \le n \tag{9.12a}$$

That is, $b_{ij} = -a_{ij}$ for $1 \le i \le m$, and $1 \le j \le n$. The $m \times n$ payoff matrix of P_{plyr1} is $A = [a_{ij}]$. Similarly, the $m \times n$ payoff matrix of P_{plyr2} is $B = [b_{ij}] = -A$.

The probability that strategy s_{1i} is used, is equal to $p_i \ge 0$, for $1 \le i \le m$; where $\sum_{i=1}^{m} p_i = 1$. Similarly the probability that strategy s_{2j} is used, is equal to $q_j \ge 0$, for $1 \le j \le n$; where $\sum_{j=1}^{n} q_j = 1$. Denote the probability column vectors by

$$p = \begin{bmatrix} p_1 \ p_2 \cdots p_m \end{bmatrix}^T, \text{ and } q = \begin{bmatrix} q_1 \ q_2 \cdots q_n \end{bmatrix}^T \tag{9.12b}$$

Sometimes, the probability vectors p and q are termed mixed strategies of the game. For simplicity in notation, denote the expected utility of the two players by

$$u_1(p,q) = \sum_{j=1}^{n} \sum_{i=1}^{m} p_i a_{ij} q_j, \text{ and } u_2(p,q) = \sum_{j=1}^{n} \sum_{i=1}^{m} p_i b_{ij} q_j$$

The expected payoffs of the players P_{plyr1} and P_{plyr2} can be expressed as

$$u_1(p,q) = p^T A q \tag{9.12c}$$

$$u_2(p,q) = p^T B q \tag{9.12d}$$

Using the terminology of this subsection, the Nash equilibrium for a mixed-strategy two-person zero-sum game is explicitly defined.

Definition 9.8. *The Nash equilibrium of a mixed-strategy two-person zero-sum game is* (p^*, q^*) *if*

$$p^{*T} A q^* \ge p^T A q^*, \quad \forall\, p \tag{9.13a}$$

$$p^{*T} B q^* \ge p^{*T} B q, \quad \forall\, q \tag{9.13b}$$

where $B = -A$. □

The next lemma and theorem are analogous to the results for the pure-strategy two-person zero-sum game.

Lemma 9.3. *In a mixed-strategy two-person zero-sum game specified by the payoff matrix A*

$$\max_p \min_q p^T A q \le \min_q \max_p p^T A q \tag{9.14}$$

where the maximization and minimization are determined over all possible values of the vectors p and q respectively.

Proof. See the problem section. □

Theorem 9.1. *John von Neumann's theorem N1. Consider a mixed-strategy two-person zero-sum game.*

In it (p^*, q^*) *is a Nash equilibrium if and only if* $\max_p \min_q p^T A q = \min_q \max_p p^T A q = p^{*T} A q^*$.

Proof. See the problem section. □

Another theorem due to von Neumann in game theory, is proved by using the duality theory of linear programming. A linear program consists of minimizing (or maximizing) a linear function called the "objective function" subject to certain linear constraints. Basics of linear programming have been discussed in the chapter on optimization, stability, and chaos theory. Before stating another result due to von Neumann, the notion of the *value* of a game is explicitly defined.

Definition 9.9. *Consider a mixed-strategy two-person zero-sum game, where p and q are mixed strategies. That is*

$$p = \begin{bmatrix} p_1 \; p_2 \; \cdots \; p_m \end{bmatrix}^T, \;\; 0 \le p_i \le 1, \; 1 \le i \le m \tag{9.15a}$$

$$q = \begin{bmatrix} q_1 \; q_2 \; \cdots \; q_n \end{bmatrix}^T, \;\; 0 \le q_j \le 1, \; 1 \le j \le n \tag{9.15b}$$

where $\sum_{i=1}^{m} p_i = 1$, and $\sum_{j=1}^{n} q_j = 1$. A mixed-strategy two-person zero-sum game is said to have the value ω, and the vectors p and q are optimal mixed strategies, if

$$\sum_{i=1}^{m} p_i a_{ij} \ge \omega, \quad 1 \le j \le n \tag{9.15c}$$

$$\sum_{j=1}^{n} a_{ij} q_j \le \omega, \quad 1 \le i \le m \tag{9.15d}$$

The above equations imply that if the player P_{plyr1} plays the mixed strategy p, his or her expected payoff is greater than or equal to ω, irrespective of the player P_{plyr2}'s strategy. Similarly, if the player P_{plyr2} plays the strategy q, the expected payoff to the player P_{plyr1} is less than or equal to ω, irrespective of the player P_{plyr1}'s strategy. □

Theorem 9.2. *John von Neumann's theorem N2. Consider a mixed-strategy two-person zero-sum game with real payoff matrix $A = [a_{ij}]$. This game has a value ω, for some mixed strategies p and q.*

Proof. Without any loss of generality, assume that $a_{ij} > 0$. If this is not the case, then a_{ij}'s can be made positive by adding a large positive number α to all a_{ij}'s such that $(a_{ij} + \alpha) > 0$ for $1 \le i \le m$, and $1 \le j \le n$.

In this case, in the definition of the value of a game, replace the inequalities with ω on the right-hand side by the following set of equations.

$$\sum_{i=1}^{m} p_i \left(a_{ij} + \alpha\right) \ge \left(\omega + \alpha\right), \quad 1 \le j \le n$$

$$\sum_{j=1}^{n} \left(a_{ij} + \alpha\right) q_j \le \left(\omega + \alpha\right), \quad 1 \le i \le m$$

Therefore assuming $a_{ij} \in \mathbb{R}^+$, a value $\omega \in \mathbb{R}^+$ is next determined. Define

$$v_i = \frac{p_i}{\omega}, \quad 1 \le i \le m, \quad \text{and} \quad w_j = \frac{q_j}{\omega}, \quad 1 \le j \le n$$

$$v = \begin{bmatrix} v_1 \; v_2 \; \cdots \; v_m \end{bmatrix}^T, \quad \text{and} \quad w = \begin{bmatrix} w_1 \; w_2 \; \cdots \; w_n \end{bmatrix}^T$$

Thus

$$\sum_{i=1}^{m} v_i a_{ij} \geq 1, \quad 1 \leq j \leq n$$

$$\sum_{j=1}^{n} a_{ij} w_j \leq 1, \quad 1 \leq i \leq m$$

$$\sum_{i=1}^{m} v_i = \sum_{j=1}^{n} w_j = \frac{1}{\omega}$$

The vectors v and w are determined by solving the dual linear programs specified by the above two sets of inequalities, and the corresponding optimum values of the objective functions are

$$\min \sum_{i=1}^{m} v_i, \text{ and } \max \sum_{j=1}^{n} w_j$$

respectively. Using results from the duality theory of linear programming, it can be inferred that these primal and dual programs have optimal solutions because they both possess feasible solutions. This is justified by the assumption that the elements of the matrix A are all positive; and the vector v is feasible, if all its components are large. Furthermore, the vector $w = 0$ is feasible for the dual linear program. Also, by duality theory, the optimal vectors v and w satisfy the relationship $\sum_{i=1}^{m} v_i = \sum_{j=1}^{n} w_j = 1/\omega$.

Therefore p, q, and ω satisfy the von Neumann's theorem, provided $p = \omega v$ and $q = \omega w$. □

Example 9.9. Let the two-person zero-sum payoff matrix of a game be A_1, where

$$A_1 = \begin{bmatrix} 1 & 0 \\ -1 & 1 \end{bmatrix}$$

To get a matrix with positive elements, add any number $\alpha > 1$ to all the elements of the matrix A_1. Assume that $\alpha = 2$ is added to all the elements of the matrix A_1 to get the payoff matrix

$$A = \begin{bmatrix} 3 & 2 \\ 1 & 3 \end{bmatrix}$$

Let $v = \begin{bmatrix} v_1 & v_2 \end{bmatrix}^T$, and $w = \begin{bmatrix} w_1 & w_2 \end{bmatrix}^T$. The primal and dual problems are

$$\text{minimize } (v_1 + v_2), \text{ such that } v^T A \geq \begin{bmatrix} 1 & 1 \end{bmatrix}, \text{ and } v \geq 0$$
$$\text{maximize } (w_1 + w_2), \text{ such that } Aw \leq \begin{bmatrix} 1 & 1 \end{bmatrix}^T, \text{ and } w \geq 0$$

respectively. The solutions of the above two linear programming problems are

$$v = \begin{bmatrix} 2/7 & 1/7 \end{bmatrix}^T, \text{ and } w = \begin{bmatrix} 1/7 & 2/7 \end{bmatrix}^T$$

Therefore

$$v_1 + v_2 = w_1 + w_2 = \frac{3}{7} = \frac{1}{\omega}$$

Thus $\omega = 7/3$, and

$$p = \omega v = \begin{bmatrix} 2/3 & 1/3 \end{bmatrix}^T, \quad \text{and} \quad q = \omega w = \begin{bmatrix} 1/3 & 2/3 \end{bmatrix}^T$$

Then $p^T A_1 q = 1/3$, and $p^T A q = 7/3$. That is, $p^T A q = (\alpha + p^T A_1 q)$. Finally, the payoff to player P_{plyr1} is $p^T A_1 q = 1/3$. □

Stackelberg Equilibrium

A Stackelberg equilibrium is a more sophisticated type of equilibrium in which a single player is more intelligent than others. This player is generally called the *leader.* Let the set of players be $I = \{1, 2, \ldots, n\}$, and the leader be player number 1. The leader selects his or her strategy by maximizing its utility function, with a knowledge of the Nash equilibrium of the set of players who belong to the set $I \backslash \{1\}$. This equilibrium is called the *Stackelberg equilibrium,* in honor of the German economist Heinrich von Stackelberg. A formal definition of this equilibrium is given below.

Definition 9.10. *Let $\Gamma = (I, S, U)$ be a pure-strategy game, where I is the set of players, S is the set of pure-strategy profiles, and U is the utility-function tuple. Let $I = \{1, 2, \ldots, n\}$, and the player number 1 be the leader. Assume that the number of players is greater than one, that is $|I| > 1$. Associated with each player $i \in I$ is a finite set of pure strategies (actions) S_i. The set of pure-strategy profiles is $S = S_1 \times S_2 \times \cdots \times S_n$.*

A pure-strategy profile $s^* \in S$ is a Stackelberg equilibrium if:

(a) *It is a Nash equilibrium for the set of players $I \backslash \{1\}$. That is, $\forall\, i \in I \backslash \{1\}, \forall\, s_i \in S_i$*

$$u_i \left(s_1^*, s_i^*, s_{-1,-i}^*\right) \geq u_i \left(s_1^*, s_i, s_{-1,-i}^*\right) \tag{9.16a}$$

(b) *The leader's utility is maximized. That is, $\forall\, s_1 \in S_1$*

$$u_1 \left(s_1^*, s_{-1}^*\right) \geq u_1 \left(s_1, s_{-1}'\right) \tag{9.16b}$$

where s_{-1}' is a Nash equilibrium for players $I \backslash \{1\}$ when the leader adopts the strategy s_1. □

The leader's utility at Stackelberg's equilibrium is at least as great as at any other Nash equilibrium. Therefore a player may force the game towards its Stackelberg equilibrium.

9.2.3 Extensive-Form Games

A strategic game is defined by its players, their strategies, and their payoffs. This is a useful model in its own right. However many games have a dynamic structure. For example, players might select different actions at different times. That is, the sequential nature of the decision making process is not reflected in the strategic-form games. In contrast, an extensive-form game describes explicitly the sequential structure of the decision making process in a strategic scenario. This model permits us to study different solutions and possibilities, in which each player can ponder over his or her plan of action, not only at the onset of the game but also at any point in time at which the decision has to be made. Thus it appears that an extensive-form game is a more realistic representation of decision processes which occur in practice. It is assumed in this subsection that the extensive-form games are noncooperative, that is any form of collusion between players is not permitted.

A well-known example of this type of game is the game of chess. In the game of chess, there are two players: *white* and *black.* These players move alternately. At any point in this game, each player has a finite number of moves (alternatives) to select from. The board position of a player depends upon past moves of both the players. This game has three endings. In one ending *white* player wins, in another ending *black* player wins. Finally, there is a possibility of a *draw* (stalemate) in the game. In case of a draw, assume that the players are rational and decide to end the game in a draw if there is no possibility of any player winning. It is not difficult to see that the game of chess can be modeled as a tree (in a graph-theoretic sense). In this tree, each node represents the state of the board, and the arcs represent the moves. Summarizing, an extensive-form game consists of the following items.

(a) A set of players.
(b) The order in which the players make their moves. That is, which player moves when.
(c) List of players' alternatives or options when they move.
(d) Players' preference (choice) when they make the move.
(e) As in the strategy games, players' payoffs as a function of the past and current moves made by all the players.
(f) Probability distributions over any external set of events.

It is possible for players to have either perfect or imperfect knowledge about past choices of all players. In a game in which all past information about decisions of all players is available to a player, then we have a game with *perfect* information. If the player has imperfect knowledge about the past choices of all players, then it is an extensive game with *imperfect* information.

Examples of games with perfect information are chess, checkers, backgammon, hex, nim and several others. The games of bridge and poker are examples of games with imperfect information.

Extensive-Form Games with Perfect Information

An extensive-form game with perfect information is next defined more precisely.

Definition 9.11. *An extensive-form game with perfect information is a four-tuple*

$$\Gamma = \left(I, H, \widetilde{P}, \digamma\right) \tag{9.17}$$

where I is the set of players, H is the set of histories, $\widetilde{P}$ is a player's function, and $\digamma$ is the set of payoffs.

(a) *The finite set of players is $I = \{1, 2, \ldots, n\}$, where $|I| = n \in \mathbb{P}$.*
(b) *A member of the set $h \in H$, is a sequence of actions by individual players. The length of these sequences can be either finite or infinite, and have the following properties.*
   (i) *The empty sequence $\varnothing \in H$.*
   (ii) *If $(a_1, a_2, \ldots, a_K) \in H$, $K \in \mathbb{P}$, and $J < K$; then $(a_1, a_2, \ldots, a_J) \in H$.*
   (iii) *Let $(a_1, a_2, \ldots)$ be an infinite sequence, such that $(a_1, a_2, \ldots, a_K) \in H, \forall\, K \in \mathbb{P}$, then the infinite sequence $(a_1, a_2, \ldots) \in H$.*
   Note that each member of the set H is a history. Furthermore, each component of a history is an action taken by a player. Also a history $(a_1, a_2, \ldots, a_K) \in H$ is said to be terminal if $\nexists\, a_{K+1}$ such that $(a_1, a_2, \ldots, a_{K+1}) \in H$, or if it is infinite. The set of terminal histories is denoted by Z.
(c) *A player's function $\widetilde{P}$ assigns to each nonterminal history sequence a player; that is $\widetilde{P}$: $H \backslash Z \rightarrow I$. Thus $\widetilde{P}(h)$ is the player who takes an action after the history $h \in H \backslash Z$.*

(d) *For each player $i \in I$, there is a von Neumann-Morgenstern payoff function $u_i : Z \to \mathbb{R}$. Also $F = \{u_i \mid i \in I\}$.* □

Observe that actions and strategies have not yet been defined. However, these are embedded in the definition of the extensive game Γ. Note that if $h \in H$ is a history of t actions, then the length of the action-sequence h is t. Furthermore, if a is an action, then $(h, a) \in H$ is a history in which a follows h and its length is equal to $(t+1)$. Some observations are immediate from the above definition. The extensive game Γ is said to be *finite*, if $|H|$ is finite. Also if all the history sequences in the set H are of finite length, then the game Γ is said to have a *finite horizon.*

In the extensive game Γ, if h is any nonterminal history, the player $\widetilde{P}(h)$ selects an action from the set

$$\widetilde{A}(h) = \{a \mid (h, a) \in H\} \tag{9.18}$$

This is the set of actions which the player $\widetilde{P}(h)$ can exercise. At the beginning of the game, the history is empty. It is referred to as the *initial history*. The player $\widetilde{P}(\varnothing)$ selects an action a_0 from the set $\widetilde{A}(\varnothing)$. Then for each $a_0 \in \widetilde{A}(\varnothing)$, the player $\widetilde{P}(a_0)$ selects a member of the set $\widetilde{A}(a_0)$. This process is repeated successively, as the game progresses. Finally, if no further selection of actions is possible the history becomes terminal.

The extensive-form game can be pictorially depicted by a tree, which is a connected graph with no cycles. Furthermore, a pictorial tree representation of the game is more illuminating. Some game theorists prefer the specification of the extensive-form game via a tree. Before this representation of the game via a tree is described, certain graph-theoretic terminology pertaining to a rooted-tree graph is revisited from the chapter on graph theory.

Definitions 9.12. *Rooted tree.*

1. *A game tree $\widetilde{T}$, sometimes called a topological tree, is a rooted tree. It is a finite collection of vertices, connected by arcs, without any closed loops. Thus, if v_A and v_B are any two vertices in the tree, then there is a unique path between these two vertices.*
2. *Let v_A, v_B, v_C, and v_D be vertices which belong to the game tree. Let v_A be the root-vertex of the tree.*
   (a) *The vertex v_A is also called the distinguished vertex of the tree.*
   (b) *The vertex v_C is said to follow vertex v_B if the sequence of arcs joining v_A to v_C pass through v_B.*
   (c) *The vertex v_C is said to follow the vertex v_B immediately, if there is a line (arc) joining vertices v_B and v_C.*
   (d) *A vertex v_D is a terminal node of the tree, if no vertex follows it. It is also called a leaf node.* □

An informal description of the tree representation of the extensive-form game is given below. This followed by a specification strategies and outcomes.

A Tree Representation

(a) Each nonleaf node of a tree is identified by a two-tuple. The first coordinate of the two-tuple is the history $h \in H$ of the node, and the second coordinate is the player $\widetilde{P}(h)$ who makes the move if the game has a history h.

(b) Let $h = (a_1, a_2, \ldots, a_K) \in H$. Then edges which are outgoing from a node with history h are incident on nodes that are of type $h' = (h, a_{K+1})$, where a_{K+1} is an action (move) of the player $\widetilde{P}(h)$. Thus each arc of the tree represents an action. Therefore every action of a player results in an arc from one node to another.
(c) No edges are outgoing from the leaf nodes. Each of these nodes is labeled with a von Neumann-Morgenstern payoff vector $U = (u_1, u_2, \ldots, u_n)$, where n is the number of players in the game.
(d) A path from the root node to any other node in the tree represents history. There is a unique path from the root node to every other node of the tree. □

Having specified an extensive-form game via the players, history of actions, players' function, and payoff function; a procedure by which the game ends has to be specified.

Strategies and Outcomes

A strategy of a player $i \in I$ in an extensive-form game Γ, is a scheme which specifies the action selected by the player for *every* history after which it is his or her turn to move.

Definitions 9.13. *Consider an extensive-form game*

$$\Gamma = \left(I, H, \widetilde{P}, F\right)$$

with perfect information where Z is the set of its terminal histories. Let the number of players be $|I| = n$.

1. *A strategy s_i of a player $i \in I$ is a function that maps an action in $\widetilde{A}(h)$ to each $h \in H \backslash Z$ where $\widetilde{P}(h) = i$, and $\left(h, \widetilde{A}(h)\right) \in H$.*
2. *A strategic profile s, is an n-tuple consisting of a strategy of each player. If s_i is a strategic profile of player $i \in I$, then $s = (s_1, s_2, \ldots, s_n)$.*
3. *The outcome $O(s)$ of a strategic profile s is a terminal history in Z, that is $O(s) \in Z$.* □

Thus a strategy is a complete plan of action. It specifies a move after every history where the player makes a selection. Thus actions (moves) are specified by a player for every history, including histories that are never reached if that strategy is played. Within the context of a game tree, a strategy profile associates every node with an edge outgoing from it.

A strategy profile of a player also determines its terminal history. Note that the payoff of player $i \in I$ is $u_i(O(s))$. Analogous to the Nash equilibria of a strategic game, a Nash equilibrium of an extensive-form game with perfect information can be defined. The notation $s \triangleq (s_i, s_{-i})$ and its meaning have been introduced in an earlier subsection on strategic games. Using this notation, the Nash equilibrium of an extensive-form game is next defined.

Definition 9.14. *Let*

$$\Gamma = \left(I, H, \widetilde{P}, F\right)$$

be an extensive-form game with perfect information. Also $u_i(\cdot)$ is the von Neumann-Morgenstern payoff function of player $i \in I$, and $O(\cdot)$ is the outcome function of a strategic profile $s = (s_i, s_{-i})$. Its Nash equilibrium is a strategy profile s^* such that for every player $i \in I$

$$u_i\left(O\left(s_i^*, s_{-i}^*\right)\right) \geq u_i\left(O\left(s_i, s_{-i}^*\right)\right) \tag{9.19}$$

for every strategy s_i of player i. □

There is another technique to determine Nash equilibrium of an extensive game. Before it is described, an equivalent strategic game of the extensive-form game Γ is defined.

Definition 9.15. *Let*

$$\Gamma = \left(I, H, \widetilde{P}, \digamma\right)$$

be an extensive-form game with perfect information. Also $u_i(\cdot)$ is the von Neumann-Morgenstern payoff function of player $i \in I$, and $O(\cdot)$ is the outcome function of a strategic profile s. The strategic-form of the extensive game Γ is a game

$$\widehat{G} = \left(I, S, \widehat{U}\right) \tag{9.20a}$$

where S is the set of n-tuple of strategies of the game Γ. The set $\widehat{U}$ is the set of all payoff vectors. That is

$$\widehat{U} = \left(\widehat{u}_1, \widehat{u}_2, \ldots, \widehat{u}_n\right) \tag{9.20b}$$

$$\widehat{u}_i(s) = u_i(O(s)), \quad \forall\, s \in S, \quad \forall\, i \in I \tag{9.20c}$$

□

Observation 9.1. The Nash equilibrium of an extensive-form game is equal to the Nash equilibrium of its equivalent strategic-form game. □

It is also advisable to note the distinction between a *game* and a *play*. A game is a complete description specified by its tree $\widetilde{T}$, while a play corresponds to a path from the root of the tree to a terminating node. Note that it is possible to have several extensive-form games with the same strategic form.

Subgame-Perfect Equilibrium

An extensive-form game can also be studied via its subgames. A subgame of an extensive-form game with perfect information can be described conveniently in terms of the corresponding tree representation. A subgame consists of a nonterminal node of the extensive form of tree, all of its successor nodes right up to the leaves, and the payoffs at the corresponding leaf nodes. Thus a subgame may be looked upon as a game by itself.

Definition 9.16. *Let*

$$\Gamma = \left(I, H, \widetilde{P}, \digamma\right)$$

be an extensive-form game with perfect information, where I is the set of players, H is the set of histories, $\widetilde{P}$ is a player's function, and $\digamma$ is the set of payoffs. The subgame of Γ that follows the nonterminal history h is the extensive game

$$\Gamma(h) = \left(I, H|_h, \widetilde{P}|_h, \digamma|_h\right) \tag{9.21a}$$

where $H|_h$ is the set of sequences h' such that $(h, h') \in H$, $\widetilde{P}|_h$ is defined as $\widetilde{P}|_h(h') = \widetilde{P}(h, h')$ for each $h' \in H|_h$, and $\digamma|_h$ is defined as

$$F|_h = \{u_i|_h(h') \mid u_i|_h(h') = u_i(h,h'),\ \textit{for}\ (h,h') \in Z, i \in I\} \tag{9.21b}$$

□

A strategy profile of a finite extensive-form game with perfect information is a subgame-perfect equilibrium, if it is a Nash equilibrium of any proper subgame of the original game. This can be determined by backward induction. Backward induction is an iterative process for determining the set of subgame-perfect equilibria of a finite extensive-form game with perfect information. It starts with a leaf node, and works upwards towards the root node of the game tree. In this process, the optimal strategy of the player who makes the last move of the game. is initially determined. In the next step, the optimal action of the next-to-last moving player is forged by taking into account the last player's action. This process is continued iteratively backwards in time until the actions of all players in the game have been determined.

The concept of subgame-perfect equilibrium of an extensive-form game with perfect information is introduced precisely via the following notation. Consider an extensive-form game Γ in which s_i is a strategy of player $i \in I$, and h is a history in it. Then $s_i|_h$ is a strategy of the subgame, where $s_i|_h(h') = s_i(h,h')$ for each $h' \in H|_h$. Also let $O_h(\cdot)$ be the outcome function of the subgame $\Gamma(h)$.

Definition 9.17. *Let*

$$\Gamma = \left(I, H, \widetilde{P}, F\right)$$

be an extensive-form game with perfect information. A subgame-perfect equilibrium of game Γ is a strategy profile s^* such that for each player $i \in I$, and each nonterminal history $h \in H \backslash Z$, where $\widetilde{P}(h) = i$,

$$u_i\left(O_h\left(s_i^*|_h, s_{-i}^*|_h\right)\right) \geq u_i\left(O_h\left(s_i, s_{-i}^*|_h\right)\right) \tag{9.22}$$

for each strategy s_i of player i in the subgame $\Gamma(h)$. □

The above definition implies that a subgame-perfect equilibrium is a strategy profile s^* in Γ, and for any nonterminal history h the strategy profile $s^*|_h$ is a Nash equilibrium of the subgame $\Gamma(h)$. Alternately, a Nash equilibrium is subgame-perfect, if the players' strategies constitute a Nash equilibrium in every subgame. H. W. Kuhn proved that every finite extensive-form game with perfect information has a subgame-perfect equilibrium. This result is not established in this chapter.

Example 9.10. The concepts discussed in this section about extensive-form games are elucidated via the sequential game of choice between classical or modern music. Recall that in this game two persons decide to attend together, either a classical or a modern music presentation. These two music lovers are called, Person 1 and Person 2. Among these two persons, Person 1 prefers classical over modern music. Person 2 prefers modern over classical music. However, this game is played sequentially. That is, Person 1 (player 1) moves first, and Person 2 (player 2) moves next. Therefore, this scenario is modeled as an extensive-form game. The payoff values are identical to the corresponding strategic-form game. The reader is referred to the example on the choice between classical and modern music in the subsection on strategic-form games. Let the game be specified by

$$\Gamma = \left(I, H, \widetilde{P}, F\right)$$

In this game, the choice of classical and modern music is denoted by C_m and M_m respectively. Therefore:

(a) $I = \{\text{Person 1, Person 2}\} \triangleq \{1,2\}$.
(b) $H = \{\varnothing, C_m, M_m, (C_m, C_m), (C_m, M_m), (M_m, C_m), (M_m, M_m)\}$. Therefore, the history set H consists of seven histories.
(c) The set of terminal histories is

$$Z = \{(C_m, C_m),\ (C_m, M_m),\ (M_m, C_m),\ (M_m, M_m)\}$$

(d) The player's function values are: $\widetilde{P}(\varnothing) = 1$; and $\widetilde{P}(C_m) = \widetilde{P}(M_m) = 2$.
(e) The action sets are: $\widetilde{A}(\varnothing) = \{C_m, M_m\}$; $\widetilde{A}(C_m) = \widetilde{A}(M_m) = \{C_m, M_m\}$.
(f) The payoff values are:
   (i) $u_1((C_m, C_m)) = 5$, $u_1((C_m, M_m)) = 0$, $u_1((M_m, C_m)) = 0$, and $u_1((M_m, M_m)) = 1$.
   (ii) $u_2((C_m, C_m)) = 1$, $u_2((C_m, M_m)) = 0$, $u_2((M_m, C_m)) = 0$, and $u_2((M_m, M_m)) = 5$.
(g) Strategies: In this game, Person 1 moves first, and Person 2 next. Let the strategy of Person 1 be s_1 and the strategy of Person 2 be s_2.
   (i) Person 1 moves after the empty history. Therefore, the action set of this player is $\widetilde{A}(\varnothing) = \{C_m, M_m\}$. That is, this player has two strategies. This player can select either C_m or M_m.
   (ii) Person 2 makes a move after both the histories C_m and M_m.
      After history C_m, the action set of Person 2 is $\widetilde{A}(C_m) = \{C_m, M_m\}$. Therefore, this player has two strategies. The player can select, either C_m or M_m.
      After history M_m, the action set of Person 2 is $\widetilde{A}(M_m) = \{C_m, M_m\}$. Therefore, this player has two strategies. The player can select, either C_m or M_m.

   Consequently Person 2 has four strategies: these are (C_m, C_m), (C_m, M_m), (M_m, C_m), and (M_m, M_m). The interpretation of these strategies is as follows. Let the strategies of Person 1 be either α or β. Then the strategy (γ, δ) of Person 2 implies that: if Person 1 selects strategy α, then Person 2 selects γ. And if Person 1 selects strategy β, then Person 2 selects δ. Note that the total number of strategy profiles is $2 \times 4 = 8$. See Figure 9.8 (a).
(h) The strategy profiles of the Nash equilibria in the above notation are:

$$(C_m, (C_m, C_m)),\ (C_m, (C_m, M_m)),$$
$$(M_m, (C_m, M_m)),\ \text{and}\ (M_m, (M_m, M_m))$$

   Note that the above notation should not be confused with that given in an earlier example on choice between classical and modern music. In that example the choice between classical and modern music was formulated as a strategic game. Thus the Nash equilibria $(s_1^*, s_2^*) \in \{(C_m, C_m), (M_m, M_m)\}$.
(i) The corresponding tree of this extensive game is shown in the Figure 9.8 (b).

		s_2			
		(C_m, C_m)	(C_m, M_m)	(M_m, C_m)	(M_m, M_m)
s_1	C_m	$(5,1)$	$(5,1)$	$(0,0)$	$(0,0)$
	M_m	$(0,0)$	$(1,5)$	$(0,0)$	$(1,5)$

Figure 9.8 (a). Strategic form of the sequential game of choice between classical and modern music.

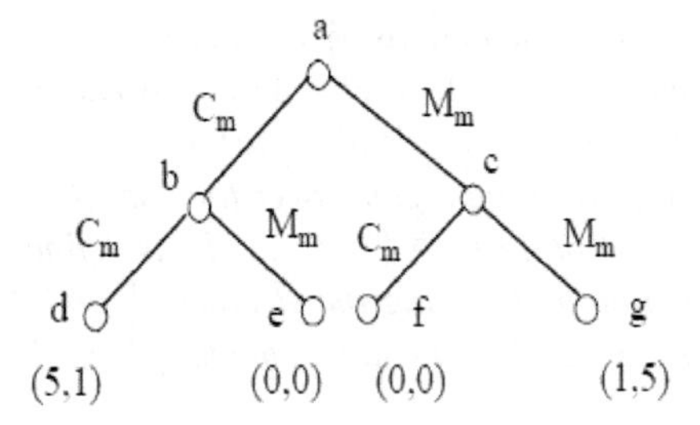

Figure 9.8 (b). In the extensive game tree, node a = $(\varnothing, 1)$, b = $(C_m, 2)$, c = $(M_m, 2)$. Nodes d, e, f, and g belong to the terminal set Z.

□

In this subsection, an extensive-form game with perfect information has been defined. It is also possible to define an extensive-form game with imperfect information. This is not discussed in this chapter.

9.2.4 Repeated Games

A repeated, or iterated, or multistage game, is a special type of dynamic game in which there are several repetitions of some base game (also called stage, or constituent game). In contrast, a nonrepeated game is a single stage game, or a single shot game. The purpose of a repeated game is to study long-term interaction among players. These interactions among players can be either positive (fruitful), or negative (counter-productive).

In the repeated game, the players repeatedly play a strategic game, called the constituent game. In each repetition, the players select their actions simultaneously. However, in any iteration of the game the players are aware of actions selected by all players in earlier iterations. Thus this game can be modeled as an extensive game with perfect information, and simultaneous moves. Repeated games can be of two types: finitely and infinitely repeated games.

In the definition of the repeated game, the strategies of the constituent game are called *actions*. This is to distinguish actions from the *strategies* of the repeated game itself.

Definition 9.18. *Let $\mathcal{G} = (I, \mathcal{A}, U)$, be a strategic-form game, where I is the set of players, $\mathcal{A}$ is the finite set of action profiles, and U is the utility-function tuple. The constituent (stage) game $\mathcal{G}$ is repeated $T \in \mathbb{P}$ number of times. This repeated (multistage) game is specified as an extensive-form game with perfect information and simultaneous moves. It is denoted as*

$$\Gamma = \left(I, H, \widetilde{P}, \widetilde{U}, T, \widetilde{\delta}\right) \tag{9.23a}$$

In this game:

(a) *The finite set of players is $I = \{1, 2, \ldots, n\}$, where $|I| = n \in \mathbb{P}$.*
(b) *The action profile of the single stage game $\mathcal{G}$ is $\mathcal{A} = \mathcal{A}_1 \times \mathcal{A}_2 \times \ldots \times \mathcal{A}_n$, where $\mathcal{A}_i$ is the set of possible actions of the player $i \in I$. An action of player $i \in I$, at time $t \in \{1, 2, \ldots, T\}$ is $a_i^t \in \mathcal{A}_i$, and the corresponding action profile of the n players is $a^t = (a_1^t, a_2^t, \ldots, a_n^t) \in \mathcal{A}$.*

(c) *At each time* $t \in \{1, 2, \ldots, T\}$, *the actions* $a_i^t \in \mathcal{A}_i$, *for* $i \in I$ *are chosen simultaneously for all players, however for each* $\tau \in \{1, 2, \ldots, t-1\}$, a^τ *is known to all players. The history of the game till iteration* t, *called t-history, is* $h^t = \left(a^1, a^2, \ldots, a^{t-1}\right) \in H^t$. *Thus* H^t *is the set of sequence of actions of the constituent game over the past* $(t-1)$ *instants, and* $H^1 = \varnothing$.
A pure strategy of player $i \in I$ *at time* $t \in \{1, 2, \ldots, T\}$ *is defined as a function* $s_i^t : H^t \to \mathcal{A}_i$. *The player* i *selects its action* a_i^t *by using the strategy* s_i^t *and the history of the game. Thus* $a_i^t = s_i^t\left(h^t\right)$. *Also the strategy of the repeated game for player* $i \in I$ *is*

$$s_i = \left(s_i^1\left(h^1\right), s_i^2\left(h^2\right), \ldots, s_i^T\left(h^T\right)\right) \in \mathcal{A}_i^{(T)} \tag{9.23b}$$

where $\mathcal{A}_i^{(T)} = \times_{t=1}^{T} \mathcal{A}_i$. *Similarly, the strategy profile of the repeated game is* s, *where*

$$s = (s_1, s_2, \ldots, s_n) \in S, \text{ and } S = \times_{i=1}^{n} \mathcal{A}_i^{(T)} \tag{9.23c}$$

(d) $\widetilde{P}(h) = I$ *for each nonterminal history* $h \in H^T$.
(e) *The payoff to each player at the end of iteration* T *uses a discount factor* $\widetilde{\delta} \in (0, 1]$. *The real-valued normalized payoff function of player* $i \in I$ *is* $\widetilde{u}_i : s \to \mathbb{R}$, *where*

$$\widetilde{u}_i(s) = \widetilde{K} \sum_{t=1}^{T} \widetilde{\delta}^{t-1} u_i\left(a^t\right) \tag{9.23d}$$

and $\widetilde{K}$ *is the normalizing factor, where*

$$\widetilde{K}^{-1} = \sum_{t=1}^{T} \widetilde{\delta}^{t-1} \tag{9.23e}$$

Also let the payoff function profile be $\widetilde{U} = (\widetilde{u}_1, \widetilde{u}_2, \ldots, \widetilde{u}_n)$.
(f) *The game is finitely repeated if* T *is finite, and infinitely repeated if* $T \to \infty$. *If* T *is infinite, the discount factor* $\widetilde{\delta} \in (0, 1)$. □

It can be observed from the above definition that the players place different weights on the payments made in different stages of the game. Furthermore, the above repeated game is equivalent to a normal-form game

$$\Gamma^T\left(\widetilde{\delta}\right) = \left(I, S, \widetilde{U}\right)$$

where S is the strategy set, and $\widetilde{U}$ specifies the payoff functions. This observation is formalized in the following lemma.

Lemma 9.4. *Let the multistage game, and its constituent game be*

$$\Gamma = \left(I, H, \widetilde{P}, \widetilde{U}, T, \widetilde{\delta}\right), \text{ and } \mathcal{G} = (I, \mathcal{A}, U) \tag{9.24a}$$

respectively. This multistage game is equivalent to a normal-form game

$$\Gamma^T\left(\widetilde{\delta}\right) = \left(I, S, \widetilde{U}\right) \tag{9.24b}$$

Proof. The proof is immediate from the definition of the multistage game. □

We next inquire about the existence of the Nash equilibrium of the repeated game. This question is answered in affirmative by the following theorem, for either finite or infinite values of T.

Theorem 9.3. *Let the multistage game, its constituent, and equivalent normal-form games be*

$$\Gamma = \left(I, H, \widetilde{P}, \widetilde{U}, T, \widetilde{\delta}\right), \ \mathcal{G} = (I, \mathcal{A}, U), \text{ and } \Gamma^T\left(\widetilde{\delta}\right) = \left(I, S, \widetilde{U}\right) \tag{9.25}$$

respectively. Also let $a^* \in \mathcal{A}$ be a Nash equilibrium of the constituent game $\mathcal{G}$. Let $s^* \in S$ be a strategy of the repeated game, where $s_i^{t*}\left(h^t\right) = a_i^*$ for all $t \in \{1, 2, \ldots, T\}$, and all $i \in I$. Then for all $\widetilde{\delta} \in (0, 1]$, s^* is a Nash equilibrium of $\Gamma^T\left(\widetilde{\delta}\right)$.

Proof. Assume that player i selects a strategy s_i, and all other players follow the strategy combination s_{-i}^*. As $a^* \in \mathcal{A}$ is a Nash equilibrium of the constituent game $\mathcal{G}$, we have $u_i\left(a^*\right) \geq u_i\left(a_i, a_{-i}^*\right)$ for all $a_i \in \mathcal{A}_i$. Therefore

$$\widetilde{u}_i\left(s^*\right) = \widetilde{K}\sum_{t=1}^{T}\widetilde{\delta}^{t-1}u_i\left(a^*\right) \geq \widetilde{K}\sum_{t=1}^{T}\widetilde{\delta}^{t-1}u_i\left(a_i^t, a_{-i}^*\right)$$

where $\widetilde{K}^{-1} = \sum_{t=1}^{T}\widetilde{\delta}^{t-1}$.

The above inequality is true for any action-sequence $\left(a_i^1, a_i^2, \ldots, a_i^T\right)$. Also, the strategy of the repeated game

$$s_i = \left(s_i^1\left(h^1\right), s_i^2\left(h^2\right), \ldots, s_i^T\left(h^T\right)\right)$$

along with s_{-i}^* induces some sequence $\left(a_i^1, a_i^2, \ldots, a_i^T\right)$. Therefore we have

$$\widetilde{u}_i\left(s^*\right) \geq \widetilde{u}_i\left(s_i, s_{-i}^*\right), \ \forall\, s_i \in \mathcal{A}_i^{(T)}$$

□

In brief, the above theorem asserts that repetition of Nash equilibrium of the stage game also forms a Nash equilibrium of the repeated game. A logical question to ask next is: Are there other strategies that will form a Nash equilibrium of the repeated game? The answer is positive for the infinitely repeated game, and we shall discuss it next. These results are called "folk theorems." The label "folk" has its origin in the fact that the results (folk theorems) were believed to be true by several generations of game theorists before their formal proof appeared in printed form.

Folk Theorems for Infinitely Repeated Game

We next state two celebrated folk theorems of the infinitely repeated game. The first folk theorem for infinitely repeated game states that practically any rational payoffs can be an equilibrium outcome. The second folk theorem is related to the subgame-perfect equilibria. Before the first folk theorem is stated and proved precisely, we need a set of definitions.

In order to establish the folk theorems, the concept of feasible payoff vector, and the minmax payoff of a player of the constituent game is introduced. Using these concepts, the set of feasible and strictly rational payoff vectors is also defined. The definition of feasible payoff vector uses the concept of convex hull of a set. Recall that the convex hull of a set is the intersection of all convex sets containing this set.

Definitions 9.19. *Consider a constituent game $\mathcal{G} = (I, \mathcal{A}, U)$.*

1. *Let* U_{ch} *be the convex hull of the set* $\{v \mid \exists\ a \in \mathcal{A},$ *such that* $u(a) = v\}$. *A payoff vector* v *is feasible if it is an element of the set* U_{ch}.
2. *The minmax payoff or reservation payoff of player* $i \in I$ *is*

$$\underline{v}_i = \min_{a_{-i} \in \mathcal{A}_{-i}} \max_{a_i \in \mathcal{A}_i} u_i(a_i, a_{-i}) \tag{9.26}$$

3. *The set of feasible and strictly individually rational payoffs is*

$$\{v \in U_{ch} \mid v_i > \underline{v}_i,\ \ \forall\, i \in I\} \tag{9.27}$$

□

Thus the minmax payoff of a player is the lowest payoff that the player i's opponents can hold him (or her) to by any selection of a_{-i}, if the player i correctly determines a_{-i} and plays a best response to it. The definition of minmax payoff of a player also implies that, in any Nash equilibrium of the repeated game, player i's payoff is at least $\underline{v}_i$.

The first folk theorem asserts that practically any payoff might be implemented as a Nash equilibrium for an infinitely repeated game, provided the players are sufficiently patient. The proof of this theorem uses the fact, that if the repeated game is sufficiently long, that is, the discount factor $\widetilde{\delta}$ is sufficiently close to unity, the gain obtained by a player by a single deviation is offset by losses in every subsequent iteration of the repeated game. The loss (to deviating player) occurs due to the punishment (minmax) strategy of all other players.

Theorem 9.4. *Consider an infinitely repeated game* Γ. *For every feasible payoff vector* v, *with* $v_i > \underline{v}_i\ \forall\, i \in I$, *there exists a* $\underline{\delta} < 1$ *such that* $\forall\, \widetilde{\delta} \in (\underline{\delta}, 1)$ *there is a Nash equilibrium of game* Γ *with payoffs* v.

Proof. See the problem section. □

The strategies used in the proof of the above theorem, due to a single deviation in the strategy of a player result in unrelenting punishment rendered to him (or her). However, punishments of this type might be very costly to the punishers to execute them. Furthermore, the strategies used in the proof do not imply subgame-perfect equilibria. A natural question to ask: Is the surmise of the above folk theorem applicable to the payoffs of subgame-perfect equilibria? The answer to this question is positive, and stated in the next theorem. This later result is called the "Nash-threats" folk theorem. This is a weaker folk theorem.

Theorem 9.5. *Consider an infinitely repeated game* Γ. *Let* a^* *be an equilibrium of the constituent game* $\mathcal{G}$ *with payoffs* $\widetilde{p}$. *Then for any* $v \in U_{ch}$ *with* $v_i > \widetilde{p}_i$ *for all players* $i \in I$, *there exists a* $\underline{\delta} < 1$ *such that* $\forall\, \widetilde{\delta} > \underline{\delta}$ *there is a subgame-perfect equilibrium of game* Γ *with payoffs* v.

Proof. See the problem section. □

As can be observed, the statements of the above two theorems are optimistic about existence of several (infinite) number of normalized payoffs to be in Nash equilibria. However, the existence of several number of equilibria makes it hard to predict whether a specific equilibrium is achievable in practice without coordination among players.

The basics of game theory developed in this section are applied to certain aspects of Internet technology in the rest of this chapter.

9.3 Selfish Routing

One of the prominent applications of game theory is the study of routing of packets. We study routing of packets in the Internet via game-theoretic techniques. Telecommunication networks seek to minimize the end-to-end delay experienced by their traffic (and maximize net packet throughput). However due to uncooperative, uncoordinated, and selfish behavior of network entities, the telecommunication networks do not always operate optimally. This phenomenon can also be termed as the tragedy of the commons in the Internet.

The ramifications of this selfish behavior are initially demonstrated via an example. This example is commonly described in the transportation literature as the Braess' paradox. A paradox is a true statement, which is in conflict with common sense. Braess' paradox is a pathological behavior, in which adding more capacity or routes (links) to a network may actually degrade performance instead of improving it. This paradox again demonstrates that selfish behaviors by noncooperative entities do not necessarily produce a socially desirable result.

Followed by a discussion of Braess' paradox, we discuss a routing model in which the degradation in performance of a network due to the selfish behavior of the entities is quantified.

9.3.1 Braess' Paradox

The Braess' paradox is an example of the observation that local optimization by several agents in a telecommunication network who have conflicting requirements does not always result in global optimality. This conclusion was apparently first reached for a transportation network by the mathematician D. Braess in 1968. This paradox is next explored via a numerical example. Consider a network of roads shown in Figure 9.9 (a).

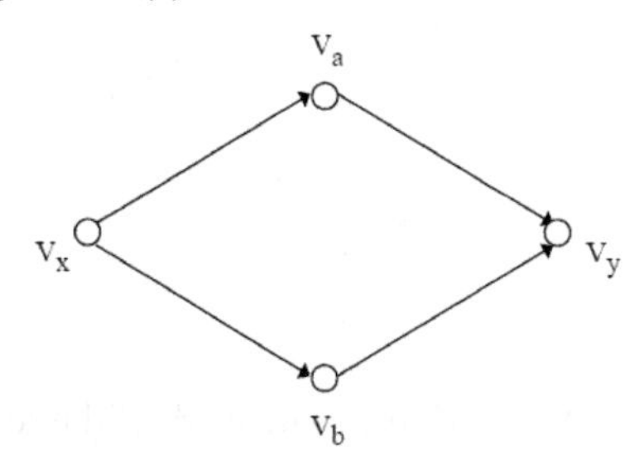

Figure 9.9 (a). Initial network of roads.

The network of roads is specified by a directed graph $G = (V, E)$, where $V = \{v_a, v_b, v_x, v_y\}$ and $E = \{v_x v_a, v_a v_y, v_x v_b, v_b v_y\}$. There is a delay (cost) associated with travelling on each road (edge). This cost is a function of the number of cars on the road. Let the number of cars on a road be n_c. The cost of travelling on different roads are:

(a) cost of road $v_x v_a$ is $(n_c + 20)$.
(b) cost of road $v_a v_y$ is $(4n_c + 2)$.
(c) cost of road $v_x v_b$ is $(4n_c + 2)$.
(d) cost of road $v_b v_y$ is $(n_c + 20)$.

Motorist travel from node v_x to node v_y. Observe that there are two paths available to reach this goal. These are $P_{path1} = (v_x, v_a, v_y)$, and $P_{path2} = (v_x, v_b, v_y)$. Assume that the motorists enter the network sequentially, and are aware of the number of cars on each road. It is further assumed that the motorists are rational, and select a path with the least cost (delay). Consider the following cases.

Case 1: Motorist number 1 enters the network. The cost of travelling on path P_{path1} is $(21 + 6) = 27$. Similarly, the cost of travelling on path P_{path2} is $(6 + 21) = 27$. Therefore, he or she can select either path P_{path1}, or path P_{path2}. Assume that he or she selects path P_{path1}. The delay experienced by this motorist is 27.

Case 2: Motorist number 2 enters the network. The cost of travelling on path P_{path1} is $(22 + 10) = 32$. And the cost of travelling on the path P_{path2} is $(6 + 21) = 27$. Therefore he or she selects path P_{path2}. The delay experienced by this motorist is 27.

Case 3: Motorist number 3 enters the network. The cost of travelling on path P_{path1} is $(22 + 10) = 32$. Similarly, the cost of travelling on path P_{path2} is $(10 + 22) = 32$. Therefore, he or she can select either path P_{path1}, or path P_{path2}. Assume that he or she selects path P_{path1}. The cost experienced by this motorist is 32. Therefore, when three motorists are on the network, the delay experienced by motorists 1 and 3 is each equal to 32. And the delay experienced by motorist 2 is 27.

Case 4: Motorist number 4 enters the network. The cost of travelling on path P_{path1} is $(23 + 14) = 37$. And the cost of travelling on the path P_{path2} is $(10 + 22) = 32$. Consequently he or she selects path P_{path2}. Therefore, when four motorists are on the network, the delay experienced by each motorist is equal to 32.

Next consider the addition of a road from node v_b to v_a. It is assumed that this is a super-highway, and the cost of travel on this road is 1. See Figure 9.9 (b).

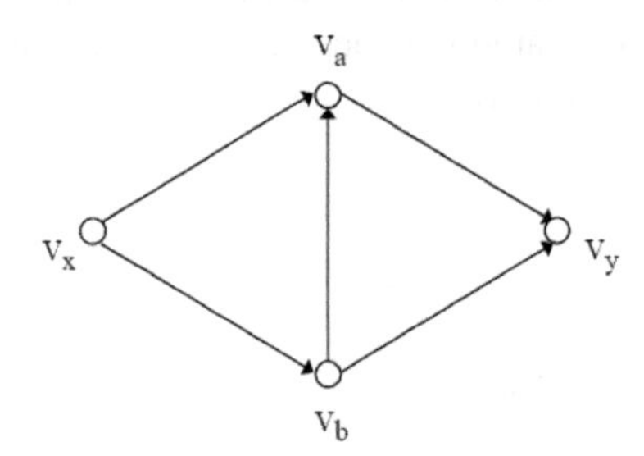

Figure 9.9 (b). Modified network of the roads.

With the addition of this link, there are three paths from nodes v_x to v_y. These are $P_{path1} = (v_x, v_a, v_y)$, $P_{path2} = (v_x, v_b, v_y)$, and $P_{path3} = (v_x, v_b, v_a, v_y)$. Again assume that the motorists enter the network sequentially, and are aware of the number of cars on each road. Also the motorists select a path with the least cost (delay). Consider the following cases.

Case 1: Motorist number 1 enters the network. Of the three paths available, the cheapest path is P_{path3}, with an associated cost of 13. Therefore, this motorist selects the path P_{path3}.

Case 2: Motorist number 2 enters the network. The costs of paths P_{path1}, P_{path2}, and P_{path3} are 31, 31, and 21 respectively. Consequently, motorist number 2 selects path P_{path3}, with a cost of 21. Therefore, when there are two motorists on the network, the delay experienced by each driver is equal to 21.

Case 3: Motorist number 3 enters the network. The costs of paths P_{path1}, P_{path2}, and P_{path3} are 35, 35, and 29 respectively. Consequently, motorist number 3 selects path P_{path3}, with a cost of

29. Therefore, when there are three motorists on the network, the delay experienced by each driver is equal to 29.

Case 4: Motorist number 4 enters the network. The costs of paths P_{path1}, P_{path2}, and P_{path3} are 39, 39, and 37 respectively. Consequently, motorist number 4 selects path P_{path3}, with a cost of 37. Therefore, when there are four motorists on the network, the delay experienced by each driver is equal to 37.

Recall that the four motorist experienced a delay of 32 units in the old antiquated network! So what went wrong in the fast lane? Let us examine carefully the assumptions in the model.

(a) The cars entered the network sequentially.
(b) Each motorist used greedy (selfish) algorithm to select his or her routes autonomously.

An important point to note in the above discussion is the presence linear cost functions. Each motorist selects a path that is in his or her best interest, with little regard to the socially optimal solution. The greedy choice decreases the cost of the motorist, but at the expense of others. This example is easily extensible to other networks, and can be framed in game-theoretic language as a route selection game.

9.3.2 Routing Model

In a large scale distributed network, the routing of information is generally performed by individual entities. In this scenario, each entity (user) tries to route its information via the least delay path from the origin to the destination. This behavior is akin to the game-theoretic model dubbed the "tragedy of the commons." Consequently, this type of transfer of information is termed *selfish routing*. Selfish behavior results in inefficient use of the overall network, because there is no overall coordination between different entities. That is, the users are not assumed to be altruistic. The basic question, which is addressed in this subsection is: How much does overall network performance suffer due to the selfishness of the individual entities? This degradation in performance is also called the *price of anarchy*. Using game-theoretic concepts, this price of anarchy is next evaluated in a distributed telecommunication network.

The routing model is initially described as a traffic optimization problem. This is followed by a discussion of some properties of its optimal solution. The traffic flow at the Nash equilibrium is next discussed. This leads to a comparison of traffic flows under optimal condition and at Nash equilibrium. Finally, the discussion is clarified via examples.

A Traffic Flow Optimization Problem

Consider a telecommunication network. The goal is to route traffic in this network, such that the overall latency of the traffic is minimized. This telecommunication network is abstractly represented by an undirected and connected graph $G = (V, E)$, where V is the set of vertices (nodes), and the set of edges (links) is E. Let the edge set be $\{1, 2, \ldots, |E|\}$. For simplicity, it is assumed that the elements of the set V represent the router entities, and those of the set E the transmission links. The transmission carrying capacity of a link $e \in E$ is $c_e \in \mathbb{R}^+$.

The purpose of this network is to carry traffic between different source and destination node pairs. For example, a source and destination node pair indexed by d, where $d \in \{1, 2, \ldots, D\}$ carries ρ_d units of traffic. It is also sometimes called the traffic for demand d. The demand vector ρ is equal to $(\rho_1, \rho_2, \ldots, \rho_D)$.

The traffic of ρ_d units is allowed to travel via different simple paths in the network, and reach its destination node. Recall that a simple path is a path without any intermediate repeated nodes. The set of paths Π_d on which the demand d flows is $\{1, 2, \ldots, \pi_d\}$. Let the traffic on each path $\varsigma \in \Pi_d$ be $\beta_{d\varsigma} \in \mathbb{R}_0^+$. Therefore $\sum_{\varsigma \in \Pi_d} \beta_{d\varsigma} = \rho_d$, for $1 \leq d \leq D$. The variable $\beta_{d\varsigma}$ is called a *path-flow* variable, and β the *traffic-flow allocation vector*. These are

$$\beta = (\beta_1, \beta_2, \ldots, \beta_D) \tag{9.28a}$$

$$\beta_d = \left(\beta_{d1}, \beta_{d2}, \ldots, \beta_{d\pi_d}\right), \quad d \in \{1, 2, \ldots, D\} \tag{9.28b}$$

The traffic carried by each edge $e \in E$ of the network is next specified. This is done by introducing a variable $\delta_{ed\varsigma}$ called the link-path indicator. The variable $\delta_{ed\varsigma}$ is equal to one, if the demand d using path $\varsigma \in \Pi_d$ uses link $e \in E$; otherwise it is equal to zero. Let the net traffic on link e be ξ_e units. It is also called the link-load. Thus $\sum_{d=1}^{D} \sum_{\varsigma \in \Pi_d} \delta_{ed\varsigma} \beta_{d\varsigma} = \xi_e$, for $e \in E$. For simplicity it is assumed in this discussion that the link-load ξ_e is less than its capacity c_e. Therefore it is surmised that, in general the traffic ξ_e on a link e depends on all the demands ρ_d, $1 \leq d \leq D$. Furthermore, this load on a link e introduces a latency of $l_e(\xi_e)$ time units for the traffic traversing this link. For analytical tractability, it is also assumed that the latency function $l_e(\cdot) \in \mathbb{R}_0^+$, is a nondecreasing and differentiable function. Thus, the latency-function profile of the network is $l = \left(l_1, l_2, \ldots, l_{|E|}\right)$. This network description is compactly represented by the three-tuple $\mathcal{N} = (G, \rho, l)$. It is also possible to define the latency associated with a path. The *path-latency* function $\vartheta_{d\varsigma}(\cdot)$ defined over β's, is the net latency of path $\varsigma \in \Pi_d$. That is

$$\vartheta_{d\varsigma}(\beta) = \sum_{e \in E} \delta_{ed\varsigma} l_e(\xi_e), \quad \varsigma \in \Pi_d, \; d \in \{1, 2, \ldots, D\} \tag{9.29a}$$

The total cost of flow on the network is $\mathcal{Q}(\beta)$, where

$$\mathcal{Q}(\beta) = \sum_{d=1}^{D} \sum_{\varsigma \in \Pi_d} \beta_{d\varsigma} \vartheta_{d\varsigma}(\beta) \tag{9.29b}$$

It can be shown that $\mathcal{Q}(\beta) = \sum_{e \in E} l_e(\xi_e) \xi_e$. Next define $l_e(\xi_e) \xi_e = \widehat{\psi}_e(\xi_e)$, where $e \in E$. The quantity $\widehat{\psi}_e(\xi_e)$ can be considered to be the cost of flow on link $e \in E$. Therefore $\mathcal{Q}(\beta) = \sum_{e \in E} \widehat{\psi}_e(\xi_e)$. It is assumed that $\widehat{\psi}_e(\cdot)$ is a convex function in its domain. The goal of the network design problem is to minimize the overall cost $\mathcal{Q}(\beta)$. The objective function of this network optimization problem, and the constraints are next succinctly stated as a nonlinear programming problem.

Traffic flow optimization problem

Input: A network is specified by the graph $G = (V, E)$, the traffic demand vector ρ, the latency-function profile l. This network description is represented by the three-tuple $\mathcal{N} = (G, \rho, l)$.

Objective:

$$\textit{Minimize } \mathcal{Q}(\beta) = \sum_{e \in E} \widehat{\psi}_e(\xi_e) \tag{9.30a}$$

where β is the traffic-flow allocation vector, and $\widehat{\psi}_e(\xi_e) = l_e(\xi_e) \xi_e$, $\forall\, e \in E$.

Subject to:

$$\sum_{\zeta \in \Pi_d} \beta_{d\zeta} = \rho_d, \quad 1 \le d \le D \tag{9.30b}$$

$$\sum_{d=1}^{D} \sum_{\zeta \in \Pi_d} \delta_{ed\zeta} \beta_{d\zeta} = \xi_e, \quad \forall\, e \in E \tag{9.30c}$$

$$\beta_{d\zeta} \ge 0, \quad \forall\, \zeta \in \Pi_d, \; 1 \le d \le D \tag{9.30d}$$

Optimal Flows

This optimization problem is a convex program, for which a globally optimum solution β^* exists due to the Karush-Kuhn-Tucker conditions. Denote the corresponding optimal path-flow variable by $\beta^*_{d\zeta}$, for all demands and all paths. The traffic flow in this networking scenario is next characterized. Initially compute the partial derivative of $\mathcal{Q}(\beta)$ with respect to $\beta_{d\zeta}$.

$$\frac{\partial \mathcal{Q}(\beta)}{\partial \beta_{d\zeta}} = \sum_{e \in E} \delta_{ed\zeta} \frac{d\widehat{\psi}_e(\xi_e)}{d\xi_e} = \vartheta_{d\zeta}(\beta) + \sum_{e \in E} \delta_{ed\zeta} \xi_e \frac{dl_e(\xi_e)}{d\xi_e}$$

A traffic flow is locally optimal if and only if transferring traffic flow from an optimal path to another path (between the same pair of nodes) results in increased cost. Stated alternately, assume that a small amount of traffic is moved from an optimal path ζ^* with positive traffic flow, to another path ζ for the same traffic demand d. This does not result in a decrease in the overall cost $\mathcal{Q}(\beta)$. That is, $\partial \mathcal{Q}(\beta^*) / \partial \beta_{d\zeta^*} \le \partial \mathcal{Q}(\beta^*) / \partial \beta_{d\zeta}$ for all other paths ζ for the traffic demand d. This result also follows immediately from the Karush-Kuhn-Tucker conditions.

Lemma 9.5. *In the network three-tuple $\mathcal{N} = (G, \rho, l)$, a flow allocation vector β^* is optimal for the traffic flow optimization problem if and only if for all demands $d \in \{1, 2, \ldots, D\}$, and the path $\zeta^* \in \Pi_d$ has a traffic flow $\beta_{d\zeta^*} > 0$, then*

$$\frac{\partial \mathcal{Q}(\beta^*)}{\partial \beta_{d\zeta^*}} \le \frac{\partial \mathcal{Q}(\beta^*)}{\partial \beta_{d\zeta}}, \quad \forall\, \zeta \in \Pi_d \backslash \{\zeta^*\} \tag{9.31}$$

Proof. As per our assumptions, the latency function $l_e(\cdot)$ is nonnegative, nondecreasing, and differentiable. Furthermore, the cost of traffic flow on link $e \in E$ is $\widehat{\psi}_e(\xi_e) = l_e(\xi_e)\,\xi_e$, and $\widehat{\psi}_e(\cdot)$ is a convex function. Therefore the objective function of the traffic flow optimization problem is convex. In addition, the constraint sets of the problem are also convex. The result is immediate via Karush-Kuhn-Tucker theorem. □

Traffic Flow at Nash Equilibrium

Routing in a selfish environment is next analysed. In this scenario the demand d is selfish. This demand selects a path with minimum latency with total disregard for other demands. An important observation about selfish routing is that it is possible for traffic to reach an equilibrium, called the *Wardrop's first principle*. This equilibrium is actually the Nash equilibrium. The principle states that all routes which are used for a specific traffic demand d have identical latency, and all the unused paths for this demand have higher latency.

A traffic flow allocation vector, which satisfies $\sum_{\varsigma\in\Pi_d}\beta_{d\varsigma}=\rho_d$, for $1\le d\le D$ at Nash equilibrium is termed a Nash flow vector. The corresponding path-flow variable, and the traffic-flow allocation vector are denoted by $\beta_{d\varsigma_N}$ and β^N respectively. The traffic flow allocation vector β^N is sometimes referred to as Nash flow. This statement is made precise in the following definition.

Definition 9.20. *In the network three-tuple $\mathcal{N}=(G,\rho,l)$, a traffic flow allocation vector β^N, which satisfies $\sum_{\varsigma_N\in\Pi_d}\beta_{d\varsigma_N}=\rho_d$, $\forall\ d\in\{1,2,\ldots,D\}$, is at Nash equilibrium if and only if $\forall$ $d\in\{1,2,\ldots,D\}$ and any path $\varsigma_N\in\Pi_d$ where $\beta_{d\varsigma_N}>0$*

$$\vartheta_{d\varsigma_N}\left(\beta^N\right)\le\vartheta_{d\varsigma}\left(\beta^N\right),\quad\forall\,\varsigma\in\Pi_d\backslash\{\varsigma_N\}\tag{9.32}$$

□

This definition implies that for any $d\in\{1,2,\ldots,D\}$, at Nash equilibrium, all paths ς_N where the traffic flow $\beta_{d\varsigma_N}$ is positive have the same path-latency. That is, all users who belong to the same demand undergo same delay at Nash equilibrium.

Lemma 9.6. *In the network three-tuple $\mathcal{N}=(G,\rho,l)$, if $D_d\left(\beta^N\right)$ is the path-latency for positive flows of demand d at Nash equilibrium, then*

$$\mathcal{Q}\left(\beta^N\right)=\sum_{d=1}^{D}D_d\left(\beta^N\right)\rho_d\tag{9.33}$$

Proof. The proof follows immediately from the definition. □

A remarkable relationship between the optimal traffic flow and the Nash flow is next examined. Observe that the link latency associated with the optimal flow manifests itself as the derivative of $\xi_e l_e(\xi_e)$, while the link cost at the Nash equilibrium simply depends upon the link-latency function $l_e(\cdot)$. This notion is first made formal, and then the result is stated.

Definition 9.21. *The marginal latency of a link is*

$$\widehat{l}_e(\xi_e)=\frac{d}{d\xi_e}\{\xi_e l_e(\xi_e)\},\quad e\in E\tag{9.34}$$

The corresponding marginal latency-function profile is $\widehat{l}=\left(\widehat{l}_1,\widehat{l}_2,\ldots,\widehat{l}_{|E|}\right)$. □

Lemma 9.7. *Consider the network three-tuple $\mathcal{N}=(G,\rho,l)$. If $\widehat{\psi}_e(\xi_e)=\xi_e l_e(\xi_e)$, and $\widehat{\psi}_e(\cdot)$ is a convex function for each edge $e\in E$. Then β^* is an optimal flow allocation vector of the network three-tuple $\mathcal{N}$, if and only if it is a Nash flow vector of the network three-tuple $\widehat{\mathcal{N}}=\left(G,\rho,\widehat{l}\right)$.*

Proof. The proof essentially follows from the above lemmas and the definitions of the path-latency and Nash equilibrium. The details are left to the reader. □

The following theorem demonstrates the existence and essential uniqueness of the Nash equilibria.

Theorem 9.6. *In the network three-tuple* $\mathcal{N} = (G, \rho, l)$ *, the latency function is nonnegative, continuous, and nondecreasing. In addition, the network admits a feasible flow at Nash equilibrium.*

Furthermore, if β^N *and* $\overline{\beta}^N$ *are different traffic flows at Nash equilibrium, then*

$$\mathcal{Q}\left(\beta^N\right) = \mathcal{Q}\left(\overline{\beta}^N\right) \tag{9.35}$$

Proof. Note that $\kappa_e(\xi_e) = \int_0^{\xi_e} l_e(t)\,dt$ is differentiable with nondecreasing derivative $l_e(\xi_e)$, therefore $\kappa_e(\cdot)$ is convex $\forall\, e \in E$. Next consider the convex program:

Objective:

$$\textit{Minimize}\ \mathcal{K}(\beta) = \sum_{e \in E} \kappa_e(\xi_e)$$

Subject to:

$$\sum_{\varsigma \in \Pi_d} \beta_{d\varsigma} = \rho_d, \quad 1 \le d \le D$$

$$\sum_{d=1}^{D} \sum_{\varsigma \in \Pi_d} \delta_{ed\varsigma} \beta_{d\varsigma} = \xi_e, \quad \forall\, e \in E$$

$$\beta_{d\varsigma} \ge 0, \quad \forall\, \varsigma \in \Pi_d, \ 1 \le d \le D$$

It is clear that the optimality condition for this nonlinear programming problem is identical to that of Nash flow of the network three-tuple $\mathcal{N}$. Existence of a Nash equilibrium in the network three-tuple $\mathcal{N}$ follows from the observation that the objective $\sum_{e \in E} \kappa_e(\xi_e)$ is continuous.

Next suppose that β^N and $\overline{\beta}^N$ are different traffic flows in the network three-tuple $\mathcal{N}$ at Nash equilibrium. Consequently these are the global optima for the nonlinear programming problem with the objective $\mathcal{K}(\beta)$. Since the objective function $\mathcal{K}(\cdot)$ is convex, if $\xi_e^N \ne \overline{\xi}_e^N$ the function $\kappa_e(\cdot)$ must be a linear function of these two variables. Otherwise any convex combination of β^N and $\overline{\beta}^N$ would be a feasible solution of this nonlinear optimization problem with a smaller value of the objective function $\mathcal{K}(\cdot)$. Therefore

$$l_e\left(\xi_e^N\right) = l_e\left(\overline{\xi}_e^N\right), \quad \forall\, e \in E$$

This implies that the path-latency for positive flow $D_d\left(\beta^N\right) = D_d\left(\overline{\beta}^N\right)$ for each demand $d \in \{1, 2, \ldots, D\}$. The stated result follows. □

Comparison of Optimal and Nash Traffic Flows

It is natural to quantify the degradation in traffic flow due to the selfish behavior of the routing entities. That is, the price of anarchy is evaluated via a metric called the *coordination ratio.*

Definition 9.22. *In a network three-tuple* $\mathcal{N} = (G, \rho, l)$ *, the optimal and Nash traffic flows are* β^* *and* β^N *respectively. The corresponding cost metrics of the two flows are* $\mathcal{Q}(\beta^*)$ *and* $\mathcal{Q}\left(\beta^N\right)$ *respectively. The coordination ratio* $\mathcal{R}$ *is defined as*

$$\mathcal{R} = \frac{\mathcal{Q}\left(\beta^N\right)}{\mathcal{Q}(\beta^*)} \tag{9.36}$$

□

It should be evident from the definition of coordination ratio that in a network three-tuple $\mathcal{N} = (G, \rho, l)$, $\mathcal{R} \geq 1$.

Corollary 9.1. *Consider a network three-tuple $\mathcal{N} = (G, \rho, l)$, in which the latency function satisfies*

$$l_e(\xi_e)\,\xi_e \leq \gamma \int_0^{\xi_e} l_e(t)\,dt \tag{9.37a}$$

where the constant $\gamma \geq 1$ for all links $e \in E$ such that $\xi_e > 0$. Then

$$\mathcal{Q}\left(\beta^N\right) \leq \gamma \mathcal{Q}(\beta^*) \tag{9.37b}$$

Proof. See the problem section. □

Examples 9.11. The above discussion of selfish routing is further elucidated via examples.

1. Consider a network in which $l_e(\xi_e) = 1$, for all $e \in E$. This metric refers to the hop-count metric. In this delay metric, the aim of the optimization problem is to minimize the number of hops between a pair of nodes (source and destination nodes). Therefore $l_e(\xi_e)\,\xi_e = \xi_e$, and $\int_0^{\xi_e} l_e(t)\,dt = \xi_e$. This yields
$$\gamma = 1, \text{ and } \mathcal{Q}\left(\beta^N\right) = \mathcal{Q}(\beta^*)$$
Therefore, the corresponding coordination ratio $\mathcal{R} = 1$. This implies that the optimal solution of the network is also the Nash traffic flow.
2. Assume that $l_e(\xi_e) = 1/c_e \triangleq b_e \in \mathbb{R}^+$, for all $e \in E$. That is, the delay metric for a given edge is equal to reciprocal of the link capacity. Thus $l_e(\xi_e)\,\xi_e = b_e\xi_e$, and $\int_0^{\xi_e} l_e(t)\,dt = b_e\xi_e$. This yields
$$\gamma = 1, \text{ and } \mathcal{Q}\left(\beta^N\right) = \mathcal{Q}(\beta^*)$$
Therefore, the corresponding coordination ratio $\mathcal{R} = 1$. That is, the optimal solution of the network is also the Nash traffic flow.
3. Let $l_e(\xi_e) = 1/(c_e - \xi_e)$, where $\xi_e < c_e$. This delay function $l_e(\cdot)$ corresponds to the delay of an $M/M/1$ queueing system. In an $M/M/1$ queueing system, the delay increases rapidly as the traffic ξ_e reaches link capacity c_e. Define the link utilization to be $u_e = \xi_e/c_e$. Then $l_e(\xi_e)\,\xi_e = u_e/(1-u_e)$, and $\int_0^{\xi_e} l_e(t)\,dt = -\ln(1-u_e)$. Thus
$$l_e(\xi_e)\,\xi_e \leq \gamma \int_0^{\xi_e} l_e(t)\,dt$$
implies
$$\frac{u_e}{(1-u_e)} \leq -\gamma \ln(1-u_e), \quad 0 \leq u_e < 1$$
It can be checked that the above inequality holds if $u_e \leq 0.2$ and $\gamma = 1.2$. Therefore if $u_e \leq 0.2$ for all $e \in E$, then
$$\mathcal{Q}\left(\beta^N\right) \leq 1.2\mathcal{Q}(\beta^*)$$

This result implies that if all links have a utilization less than or equal to twenty percent, then the cost of the traffic flow at Nash equilibrium is at most 1.2 times the cost of the optimal solution.

Similarly, if $u_e \leq 0.5$, select $\gamma = 1.5$. In this case, if $u_e \leq 0.5$ for all $e \in E$, then

$$\mathcal{Q}\left(\beta^N\right) \leq 1.5\mathcal{Q}\left(\beta^*\right)$$

Again, if $u_e \leq 0.8$, select $\gamma = 2.5$. In this case, if $u_e \leq 0.8$ for all $e \in E$, then

$$\mathcal{Q}\left(\beta^N\right) \leq 2.5\mathcal{Q}\left(\beta^*\right)$$

□

In summary, it can be stated that selfish routing has a desirable property, that all entities of the same demand experience identical delay for a specific latency function. However, this property does not hold in general for optimal flows. Furthermore, Nash equilibrium is hard to attain in practice, because the user requires the knowledge of link utilizations when routing of packets has to be initiated. Finally, we need to temper the above results and discussion with the fact that routing decisions in an actual network are made sequentially, and not simultaneously.

9.3.3 Technique to Reduce the Price of Anarchy

It is known that the price of anarchy can be large in the presence of highly nonlinear latency functions. An obvious technique to decrease the price of anarchy is by augmentation of the capacity of the edges. This is evidently an expensive proposition. Another technique to decrease the price of anarchy is for an altruistic player to centrally control a fraction of the net routed traffic. This is called the network-routing Stackelberg strategic game. It is so named because it uses ideas from noncooperative game theory called Stackelberg games. Finally the price of anarchy can be mitigated by controlling the selfish behavior of users via edge taxes. In this scheme a pricing structure is imposed upon the selfish users. We next study Stackelberg network routing.

Stackelberg strategic games can have multiple stages in which players are allowed to select their strategies in different stages. Thus some players select their strategy before others. We are interested in a two-stage Stackelberg game in which a single player called a *leader* routes a fraction of the net traffic centrally so as to minimize the cost of the total traffic. The leader's scheme is called its Stackelberg strategy. Other players called *followers* react independently and selfishly to achieve a Nash equilibrium with respect to the Stackelberg strategy. The goal of the Stackelberg strategy is to minimize the price of anarchy. This type of game is called a Stackelberg leader-follower network routing game, or simply a Stackelberg game. The Stackelberg strategy is simple in its implementation, because no communication is required between the altruistic leader and the selfish followers. This is unlike algorithmic mechanism design, where the mechanism design interacts with the agents by offering payments to them. Algorithmic mechanism design techniques are studied in a different section.

For simplicity in discussion, we assume that there is only a single source-destination pair for traffic flow in the network specified by the digraph $G = (V, E)$. This type of network is often called a single-commodity network. Thus, in the notation described at the beginning of this section, the number of demands D is equal to one. Denote the traffic ρ_1 between this source-destination pair by r. Further assume that this flow is feasible. That is, the edges have sufficient capacity to carry this traffic. Let the fraction of traffic assigned by the leader be $\eta \in (0,1)$. Thus the leader controls ηr units of altruistic traffic, and $(1-\eta) r$ units of traffic is assigned by the selfish user. The Stackelberg network is the four-tuple (G, r, l, η), where l is the latency-function profile of the network. The Stackelberg strategy is a feasible flow in the network three-tuple $(G, \eta r, l)$.

Definition 9.23. *Stackelberg network routing game.*

(a) *The Stackelberg network routing game is the four-tuple (G, r, l, η). In this game, the digraph $G = (V, E)$ represents the single-commodity network, where V and E are the vertex and edge sets respectively. The flow in this network is $r \in \mathbb{R}^+$ units, l is the latency-function profile of the network, and $\eta \in (0,1)$ is the fraction of traffic controlled by the leader of the game.*
(b) *The Stackelberg strategy is a feasible flow in the network specified by the three-tuple $(G, \eta r, l)$.*
(c) *Let the flow θ be a Stackelberg strategy in the network, and define $\widetilde{l}_e(x) = l_e(x + \theta_e)$ for each edge $e \in E$. Denote the corresponding latency-function profile by $\widetilde{l}$. Let the flow at Nash equilibrium in the network three-tuple*

$$\left(G, (1-\eta) r, \widetilde{l}\right)$$

be θ^N. Then the Stackelberg strategy θ is said to induce the Nash equilibrium flow θ^N. Also $\left(\theta + \theta^N\right)$ is the flow induced by θ for the game (G, r, l, η).
(d) *The cost of the flow $\left(\theta + \theta^N\right)$ is*

$$\mathcal{Q}\left(\theta + \theta^N\right) = \sum_{e \in E} \left(\theta_e + \theta_e^N\right) l_e \left(\theta_e + \theta_e^N\right) \tag{9.38}$$

□

The following observation is immediate from the above definition and the results of this subsection.

Observation 9.2. Let the flow θ be a Stackelberg strategy in the Stackelberg network routing game (G, r, l, η). Then there exists a flow induced by θ, and every two such induced flows have equal latency. □

The use of Stackelberg network routing game may or may not improve upon the price of anarchy. The advantages, and possibly limitations of the Stackelberg network routing game can be determined by considering simple network topologies. Let the digraph G consist of a pair of vertices connected by a set of parallel edges. Denote the source vertex by v_s, and the destination vertex by v_d. All the edges between these two vertices are directed from v_s to v_d. In the network three-tuple (G, r, l), a flow θ^N is at Nash equilibrium if and only if for every edge e_i with $\theta_{e_i}^N > 0$, and every edge $e_j \neq e_i$, we have

$$l_{e_i}\left(\theta_{e_i}^N\right) \leq l_{e_j}\left(\theta_{e_j}^N\right)$$

In analogy, with the above result consider a Stackelberg strategy θ, in the network routing game (G, r, l, η). Then θ^N is a flow at Nash equilibrium induced by the flow θ if and only if for every edge e_i with $\theta_{e_i}^N > 0$, and every edge $e_j \neq e_i$, we have

$$l_{e_i}\left(\theta_{e_i} + \theta_{e_i}^N\right) \leq l_{e_j}\left(\theta_{e_j} + \theta_{e_j}^N\right)$$

These concepts are further illustrated via the following examples.

Examples 9.12. Consider a digraph $G = (V, E)$, where $V = (v_s, v_d)$ and $E = (e_1, e_2)$. Both the edges e_1 and e_2 are directed from v_s to v_d. The traffic demand between the source node v_s, and the destination node v_d is equal to r.

1. Let the edge latencies in the network be $l_{e_1}(\theta_{e_1}) = 1$, and $l_{e_2}(\theta_{e_2}) = \theta_{e_2}$. Thus the corresponding latency-function profile is $l = (1, \theta_{e_2})$. Let $r \in (3/2, \infty)$.
   (a) Consider the network three-tuple (G, r, l). It can be shown that the cost of the optimal flow is $\mathcal{Q}(\theta^*) = (r - 1/4)$, and the corresponding flow is $\theta^* = (r - 1/2, 1/2)$. This flow minimizes the cost of the flow.
   (b) Consider a Stackelberg network routing game (G, r, l, η), where $\eta = (r - 1)/r$. Thus $(r - 1)$ unit of traffic is controlled by the leader, and 1 unit of traffic is controlled by the selfish user. Let the Stackelberg strategy be $\theta = (r - 3/2, 1/2)$. In this case, the equilibrium induced by the strategy θ is a flow $\theta^N = (1/2, 1/2)$. This results in the flow $\left(\theta + \theta^N\right) = (r - 1, 1)$, with cost $\mathcal{Q}\left(\theta + \theta^N\right) = r$. The price of anarchy $\mathcal{R} = r/(r - 1/4)$.
2. Let the edge latencies in the network be $l_{e_1}(\theta_{e_1}) = 1$, and $l_{e_2}(\theta_{e_2}) = 2\theta_{e_2}$. Thus the corresponding latency-function profile is $l = (1, 2\theta_{e_2})$. Let $r \in (1/2, \infty)$.
   (a) Consider the network three-tuple (G, r, l). It can be shown that the cost of the optimal flow is $\mathcal{Q}(\theta^*) = (r - 1/8)$, and the corresponding flow is $\theta^* = (r - 1/4, 1/4)$. This flow minimizes the cost of the flow.
   (b) The cost of the flow in the network three-tuple (G, r, l) at Nash equilibrium is $\mathcal{Q}\left(\theta^N\right) = r$, and the corresponding flow is $\theta^N = (r - 1/2, 1/2)$. Observe that the latency offered by each edge is identical, and equal to unity. The price of anarchy $\mathcal{R} = r/(r - 1/8)$.
   (c) Consider a Stackelberg network routing game (G, r, l, η), where say $\eta = 1/2$. It can be shown that for every Stackelberg strategy θ, the cost of the flow is $\mathcal{Q}\left(\theta + \theta^N\right) = r$, and the flow $\left(\theta + \theta^N\right) = (r - 1/2, 1/2)$. This corresponds to the flow at Nash equilibrium. Thus, the use of the Stackelberg strategy does not decrease the price of anarchy in this example. □

9.4 A Selfish Task Allocation Model

A selfish task allocation model is discussed within the context of game theory framework. This model consists of a pair of nodes, called the source and destination nodes, and M parallel links between these two nodes. Denote the set of links (edges) by $E = \{1, 2, \ldots, M\}$. The capacity or speed of link $j \in E$ is $c_j \in \mathbb{R}^+$.

The set of users (agents) who have access to these links is $I = \{1, 2, \ldots, N\}$. Each user $i \in I$ has a task (traffic or weight) of size w_i. It is assumed that each task is assigned to exactly a single link, but the choice of the link depends upon each user's strategy. This assignment is specified by an $N \times M$ matrix $B = [b_{ij}]$, where $b_{ij} = 1$, if user i uses link j, otherwise it is equal to 0. The important parameter of concern is the link congestion, which is the maximum cost associated with any link.

It is assumed that the users do not cooperate among themselves. Consequently, this scenario is eminently suitable for a game-theoretic analysis. In this analysis, the *coordination ratio* is determined, which is the ratio between the worst possible Nash equilibrium and the overall optimum. Coordination ratio is a measure of lack of coordination between the agents. If the task i is allocated deterministically to a link j, then the user's strategy is said to be a *pure-strategy*. However, if this allocation is according to some probability distribution on the set E, the users are said to follow a *mixed strategy*.

Therefore, this allocation model is a scheduling problem with M links and N independent tasks (one by each user). Let $(j_1, j_2, \ldots, j_N) \in E^{(N)}$ be an instance of pure strategies of the N users, where link $j_i \in E$ is used by user $i \in I$. The cost of task for user i is given by

$$\Upsilon_i (j_1, j_2, \ldots, j_N) = \frac{1}{c_{j_i}} \sum_{k=1}^{N} b_{kj_i} w_k \tag{9.39a}$$

It is assumed in this definition of the cost of user i, that this user's task finishes when all tasks on the link j_i end. Similarly the load (traffic) on link j is defined as

$$L_j (j_1, j_2, \ldots, j_N) = \frac{1}{c_j} \sum_{k=1}^{N} b_{kj} w_k \tag{9.39b}$$

This link-load is alternately referred to as its cost. Furthermore, this task allocation strategy is required to have a *min-max* objective. That is, the *social or overall optimum* assignment minimizes the maximum cost over all links. Thus this cost is

$$\mho = \min_{(j_1, j_2, \ldots, j_N) \in E^{(N)}} \max_{j \in E} L_j (j_1, j_2, \ldots, j_N) \tag{9.39c}$$

It is the minimum cost of a pure strategy. Observe that

$$\mho \geq \frac{\max_{i \in I} w_i}{\max_{j \in E} c_j} \tag{9.40}$$

Mixed strategies are next considered. Let p_{ij} be the probability that a user $i \in I$ submits its entire load w_i to the link $j \in E$. Let ℓ_j be the expected load on the link $j \in E$. Thus

$$\ell_j = \frac{1}{c_j} \sum_{k \in I} w_k p_{kj}, \quad \forall j \in E \tag{9.41a}$$

The expected cost of task i on link j, or equivalently its *finish time* when its load w_i is allocated to link j is

$$\sigma_{ij} = \frac{w_i}{c_j} + \frac{1}{c_j} \sum_{k \in I \setminus \{i\}} w_k p_{kj} = \ell_j + (1 - p_{ij}) \frac{w_i}{c_j}, \quad \forall i \in I, \ \forall j \in E \tag{9.41b}$$

An interpretation of the Nash equilibrium for this scenario is given below.

Definition 9.24. *A network has two nodes (source and destination), and M parallel links between them. The set of links is denoted by $E = \{1, 2, \ldots, M\}$. The set of users who access these links is $I = \{1, 2, \ldots, N\}$. Furthermore, the user $i \in I$ uses the link $j \in E$ with probability p_{ij}. Also denote the expected cost of task i on link j by σ_{ij} where $i \in I$ and $j \in E$.*

The task allocation probabilities p_{ij}, $\forall\, i \in I$ *and* $\forall\, j \in E$ *are a Nash equilibrium if and only if for any task of user* i *the probability* $p_{ij} > 0$ *implies* $\sigma_{ij} \leq \sigma_{ik}, \forall\, k \in E.$ □

This definition implies that at Nash equilibrium, there is no incentive for any user i to alter its strategy in selecting a different link.

Example 9.13. If the link speed $c_j = 1, \forall\, j \in E$, and the user i traffic $w_i = 1, \forall\, i \in I$, then the value of social optimum is N/M. The probabilities at Nash equilibrium are $p_{ij} = 1/M, \forall\, i \in I$ and $\forall\, j \in E$. □

Lemma 9.8. *Consider an arbitrary Nash equilibrium, such that* $p_{ik} > 0$ *for a user* $i \in I$ *and a link* $k \in E$. *Then*

(a) $\ell_k \leq (\ell_j + w_i/c_j)$, *for every* $j \in E$.
(b) $(\ell_j + 1) < \ell_k \Rightarrow w_i > c_j$.
(c) *If* $p_{ij} > 0$, $p_{ik} > 0$, *then* $|\ell_j - \ell_k| \leq \mathcal{O}$.

Proof. See the problem section. □

Let the probabilities $p_{kj}, \forall\, k \in I$ and $\forall\, j \in E$ be fixed. Assume that these probabilities define a Nash equilibrium. Also let ϱ_{kj} be a Bernoulli distributed random variable, such that the probability that $\varrho_{kj} = 1$ is equal to p_{kj}, and the probability that $\varrho_{kj} = 0$, is equal to $(1 - p_{kj})$. These random variables ϱ_{kj}'s are also assumed to be independent of each other $\forall\, k \in I$ and $\forall\, j \in E$. Thus the load Λ_j on a link $j \in E$ is a random variable given by

$$\Lambda_j = \frac{1}{c_j} \sum_{k=1}^{N} w_k \varrho_{kj}, \quad \forall\, j \in E \tag{9.42}$$

Observe that $\ell_j = \mathcal{E}(\Lambda_j)$, where $\mathcal{E}(\cdot)$ is the expectation operator. The maximum expected load on a link, the social cost and the coordination ratio are next defined in terms of the random variables Λ_j's.

Definitions 9.25. *Basic definitions.*

1. *The maximum expected load* $\ell_{\max}$ *over all links is*

$$\ell_{\max} = \max_{j \in E} \ell_j = \max_{j \in E} \mathcal{E}(\Lambda_j) \tag{9.43a}$$

2. *The social cost* $\mathcal{C}_{sc}$ *of the network is the expectation of the maximum load. That is*

$$\mathcal{C}_{sc} = \mathcal{E}\left\{\max_{j \in E} \Lambda_j\right\} \tag{9.43b}$$

3. *The coordination ratio* $\mathcal{R}_t$ *is the ratio between the worst possible Nash equilibrium and the overall optimum* $\mathcal{O}$, *which is the minimum cost of a pure strategy.*

$$\mathcal{R}_t = \frac{\max \mathcal{C}_{sc}}{\mathcal{O}} \tag{9.43c}$$

where the maximum is taken over all Nash equilibria. □

Notice that $\ell_{\max} \leq \mathcal{C}_{sc}$. It is also possible to have $\ell_{\max} \ll \mathcal{C}_{sc}$. These concepts are clarified via the following examples.

Examples 9.14. The following examples are designed to be both simple and illustrative.

1. Consider a network with two nodes, which are connected by a pair of links, that is $M = 2$. These links have identical speed, which is $c_1 = c_2 = 1$. Also assume that these links are accessed by two users, that is $N = 2$. The traffic associated with each user is identical, which is $w_1 = w_2 = 1$. Each task is allocated independently and uniformly at random between these two links. That is, $p_{ij} = 1/2$ for all $i, j \in \{1, 2\}$. It can be verified that this scenario results in a Nash equilibrium. Furthermore, the social optimum $\mathcal{O} = 1$, and the social cost $\mathcal{C}_{sc} = 3/2$. Thus this example demonstrates that for a pair of identical links, and identical tasks, the coordination ratio $\mathcal{R}_t = 3/2$, which is greater than unity.
2. The network has a pair of links and two users, that is $M = N = 2$. The speed of the links are $c_1 = 1$, and $c_2 = x \geq 1$. The traffic from the users are: $w_1 = 1$, and $w_2 = x$. The task allocation probabilities are:

$$
\begin{aligned}
p_{12} &= \frac{x^2}{(x+1)} \\
p_{11} &= (1 - p_{12}) \\
p_{21} &= \frac{1}{x(x+1)} \\
p_{22} &= (1 - p_{21})
\end{aligned}
$$

The inequality $p_{12} \leq 1$ implies $x \leq \phi_g$, where $\phi_g = \left(\sqrt{5} + 1\right)/2$. The number ϕ_g is often called the *golden ratio* in mathematical literature.

(a) The social optimum $\mathcal{O}$ is computed as follows:

$$
\begin{aligned}
&L_1(1,1) = (x+1), \;\; L_1(1,2) = 1, \;\; L_1(2,1) = x, \;\; L_1(2,2) = 0 \\
&L_2(1,1) = 0, \;\; L_2(1,2) = 1, \;\; L_2(2,1) = \frac{1}{x}, \;\; L_2(2,2) = \frac{(x+1)}{x}
\end{aligned}
$$

Therefore

$$
\begin{aligned}
\max\{L_1(1,1), L_2(1,1)\} &= (x+1) \\
\max\{L_1(1,2), L_2(1,2)\} &= 1 \\
\max\{L_1(2,1), L_2(2,1)\} &= x \\
\max\{L_1(2,2), L_2(2,2)\} &= \left(1 + \frac{1}{x}\right)
\end{aligned}
$$

Therefore the social optimum is $\mathcal{O} = \min\{(x+1), 1, x, (1 + 1/x)\} = 1$. The corresponding strategy for user 1 is to use link 1, and the user 2 to use link 2.

(b) The coordination ratio $\mathcal{R}_t$ is next computed. The expected load on the links are:

$$
\ell_1 = (2 - x), \;\; \text{and} \;\; \ell_2 = \frac{(2x-1)}{x}
$$

The costs are

$$\sigma_{11} = \sigma_{12} = \frac{(x+2)}{(x+1)}, \text{ and } \sigma_{21} = \sigma_{22} = \frac{(2x+1)}{(x+1)}$$

If $1 \leq x \leq \phi_g$ all task allocation probabilities are feasible, that is they lie in the interval $[0, 1]$. Furthermore, these probabilities are a Nash equilibrium, because the following statements are true. That is:

$$\begin{aligned} p_{11} &> 0, \ \sigma_{11} = \sigma_{12} \\ p_{12} &> 0, \ \sigma_{12} = \sigma_{11} \\ p_{21} &> 0, \ \sigma_{21} = \sigma_{22} \\ p_{22} &> 0, \ \sigma_{22} = \sigma_{21} \end{aligned}$$

The social cost $\mathcal{C}_{sc}$ is next computed. The random variables representing the load on the links 1 and 2 are Λ_1 and Λ_2 respectively. These are:

$$\Lambda_1 = \varrho_{11} + x\varrho_{21}, \text{ and } \Lambda_2 = \frac{\varrho_{12}}{x} + \varrho_{22}$$

The social cost $\mathcal{C}_{sc}$ is computed by considering the following four cases.

(i) Let $\varrho_{11} = 1$, $\varrho_{21} = 1$, $\varrho_{12} = 0$, and $\varrho_{22} = 0$. Then $\Lambda_1 = (x+1)$, and $\Lambda_2 = 0$. Therefore $\max\{\Lambda_1, \Lambda_2\} = (x+1)$. This case occurs with probability $p_{11}p_{21}$.

(ii) Let $\varrho_{11} = 0$, $\varrho_{21} = 1$, $\varrho_{12} = 1$, and $\varrho_{22} = 0$. Then $\Lambda_1 = x$, and $\Lambda_2 = 1/x$. Therefore $\max\{\Lambda_1, \Lambda_2\} = x$. This case occurs with probability $p_{21}p_{12}$.

(iii) Let $\varrho_{11} = 1$, $\varrho_{21} = 0$, $\varrho_{12} = 0$, and $\varrho_{22} = 1$. Then $\Lambda_1 = 1$, and $\Lambda_2 = 1$. Therefore $\max\{\Lambda_1, \Lambda_2\} = 1$. This case occurs with probability $p_{11}p_{22}$.

(iv) Let $\varrho_{11} = 0$, $\varrho_{21} = 0$, $\varrho_{12} = 1$, and $\varrho_{22} = 1$. Then $\Lambda_1 = 0$, and $\Lambda_2 = (x+1)/x$. Therefore $\max\{\Lambda_1, \Lambda_2\} = (x+1)/x$. This case occurs with probability $p_{12}p_{22}$.

Therefore

$$\begin{aligned} \mathcal{C}_{sc} &= \mathcal{E}\left\{\max_{j \in E} \Lambda_j\right\} \\ &= (x+1)\, p_{11}p_{21} + xp_{21}p_{12} + p_{11}p_{22} + \frac{(x+1)}{x} p_{12}p_{22} \\ &= \frac{(2x+1)}{(x+1)} \end{aligned}$$

The largest value of $\mathcal{C}_{sc}$ for $x \in [1, \phi_g]$ occurs at $x = \phi_g$. Thus $\max \mathcal{C}_{sc} = \phi_g$. As the value of the social optimum $\mathcal{O} = 1$, the coordination ratio

$$\mathcal{R}_t = \frac{\max \mathcal{C}_{sc}}{\mathcal{O}} = \phi_g$$

and it occurs when $x = \phi_g = (\sqrt{5} + 1)/2$. □

We now state an asymptotic result about a network with two nodes, which are connected by several parallel links.

Theorem 9.7. *Consider a network with two nodes and M parallel links. The links have identical speeds. The coordination ratio of this network is $\mathcal{R}_t = O(\ln M / \ln \ln M)$.*

Proof. See the problem section. □

Observe that the cost functions used in this section are linear functions of the user traffic. More realistic cost functions would be nonlinear functions of the user traffic. An example of such function is the average delay of an $M/M/1$ queue at steady-state. Results pertaining to such functions have been discussed elsewhere in the literature.

9.5 A Repeated Game

It is sometimes required that the nodes of a network voluntarily share resources amongst each other. This would help in achieving a network-wide goal of optimization of resources. This resource sharing generally occurs in a distributed manner, without the intervention of a centralized controlling authority. Examples of such networks include ad hoc, sensor, and peer-to-peer networks. The growth in the Internet is directly or indirectly fuelled by such networks.

Nodes of the network might share their resources voluntarily. However, this has a cost implication for the cooperating entity. Therefore, the nodes have to strike a balance between cooperation and looking after its own interests. We model this scenario as a repeated game.

Consider a network represented by a graph $G = (V, E)$, where $V = \{1, 2, \ldots, n\}$ is the set of vertices (nodes), and E is the set of edges. Thus the number of nodes in the network is n. In each iteration of the repeated game, the nodes play a game. The constituent game of the repeated game is $\mathcal{G}(V, \mathcal{A}, U)$, where the players of the game are the nodes in the network, which is specified by the set V. The action space of node $i \in V$ is specified by $\mathcal{A}_i = \{0, 1\}$. In this action space, 0 represents the decision by the node to not share its resources, and 1 represents the decision by the node to share its resources. The payoff of node $i \in V$ is $u_i(a)$, where $a \in \mathcal{A} = (\mathcal{A}_1, \mathcal{A}_2, \ldots, \mathcal{A}_n)$. The payoff of node i in iteration $t \in \mathbb{P}$ is:

$$u_i\left(a^t\right) = f_i\left(a^t\right) - g_i\left(a^t\right), \quad i \in V \tag{9.44a}$$

where $f_i\left(a^t\right) \in \mathbb{R}_0^+$ represents the payoff of node i by all other nodes sharing their resources, and $g_i\left(a^t\right)$ represents the cost incurred by node i when it lets other nodes share its resources. The value $g_i\left(a^t\right)$ is usually positive, because there are costs associated with sharing node i's resources with all other resources. This value can be negative if there are financial incentives for resource sharing. Note that $f_i\left(0^t\right) = g_i\left(0^t\right) = 0$. Let

$$f_i\left(a^t\right) \triangleq \widetilde{f}_i\left(\sum_{j \neq i, j \in V} a_j^t\right), \quad \text{and } g_i\left(a^t\right) \triangleq \widetilde{g}_i\left(a_i^t\right) \tag{9.44b}$$

Consider a repeated game played T_g number of times, where T_g is a geometrically distributed random variable with parameter $p \in (0, 1)$, and

$$P\left(T_g = t\right) = p\left(1 - p\right)^t, \quad \forall\, t \in \mathbb{N} \tag{9.45a}$$

The expected value of the random variable T_g is

$$\mathcal{E}\left(T_g\right) = \frac{(1 - p)}{p} \tag{9.45b}$$

Observe that as $p \to 1$, the probability that the game is played in the next iteration approaches zero. Let the following game playing strategy be adopted by all nodes. A node cooperates, as long as all other nodes cooperate. The node ceases cooperation, if any other node deviates (stops sharing its resources) in the previous iteration. This strategy is called grim-trigger strategy in the literature on game theory. If a node k deviates (ceases cooperation), then each node $j \in V \backslash \{k\}$ triggers an action that decreases k's payoff for all subsequent iterations. As this grim punishment carries the threat of smallest possible payoff, it allows for the largest number of equilibria.

Consider any iteration of the game. If a player cooperates, the payoff that he or she would expect from that point onwards is:

$$\left\{\widetilde{f}_i(n-1) - \widetilde{g}_i(1)\right\}\left\{1 + \sum_{t \in \mathbb{N}} tp(1-p)^t\right\} = \left\{\widetilde{f}_i(n-1) - \widetilde{g}_i(1)\right\}\frac{1}{p}$$

However, if the player ceases cooperation, then his or her expected payoff from that round onwards is only $\widetilde{f}_i(n-1)$. Therefore the grim-trigger strategy is a Nash equilibrium, if

$$\widetilde{f}_i(n-1) > \frac{\widetilde{g}_i(1)}{(1-p)}, \quad \forall i \in V \tag{9.46}$$

This strategy also happens to be a subgame-perfect equilibrium. Observe that if $\widetilde{g}_i(1)$ is positive for all $i \in V$, and the above set of inequalities are true, then a socially (altruistic) optimum equilibrium is possible. Note that this equilibrium is not possible in a single stage game. However, if $\widetilde{g}_i(1)$ is negative for all $i \in V$, then a socially optimum equilibrium is also possible. The discounted factor was not used in this game, however as $p \in (0,1)$ the game is played an infinite number of times.

9.6 Algorithmic Mechanism Design

Algorithmic mechanism design is a branch of algebraic game theory in which ideas have been borrowed from game theory, microeconomics, discrete mathematics, and computer science. It provides an algorithmic framework for organizing multiple selfish agents (entities) to reach specified global (social) objectives. The organizing entity which optimizes this objective is called the mechanism. The mechanism tries to achieve its design (goal) by motivating the agents to reveal their true characters by enticements (payments). This helps the mechanism in achieving a social goal and budget balance. The mechanism design is called algorithmic, because it uses the tools of theoretical computer science to achieve its goal. For example, the algorithmic mechanism design tries to accomplish its objective in a computationally efficient manner. Some of the applications of algorithmic mechanism design are in network routing, load balancing, advertisement, and electronic commerce.

We first develop the basic ideas and notation used in the description of algorithmic mechanism design. This is followed by a description of the celebrated Vickrey-Groves-Clarke (VGC) mechanism. This mechanism is useful to study because it appears to be socially optimal. Finally, the theoretical details are clarified via examples.

9.6.1 Preliminaries

Internet is a large, distributed, and dynamical system. It is owned and operated by different self-interested entities. These entities are either automated or human beings. Each agent generally acts selfishly without much regard for the optimal utilization of the Internet resources. The resources in the Internet can be optimized by simultaneously using concepts from game theory, and the theory of algorithms. Game theory is largely concerned with strategies, fairness, and incentives. The goal of the theory of algorithms is to achieve computational efficiency and scalability, possibly in a distributed setting.

Mechanism design is a branch of game theory, in which the rules of the game are designed to achieve a specific outcome. More specifically, the outcomes of the mechanism design are: budget balance and social welfare. This is accomplished via the truthfulness and rationality of the participating entities. In this section, an entity is also termed a player, an agent, or a participant.

Consider a set of agents. Each agent in this set has its own private information, strategy, and utility function. Each agent aims to selfishly optimize its own utility. Note that the private information of an agent is also called its type.

The mechanism design provides each agent with a payment and an output, based upon the strategies of all the agents. The goal of the mechanism design is to optimize its social objective function.

It should be noted that the utility of an agent is made up of its own valuation function, and the payment it receives from the mechanism. The valuation function of an agent depends upon its private information, and the output it receives from the mechanism. This description is made precise in the following definitions.

Definition 9.26. *The mechanism design game setting is*

$$\mathcal{M} = (I, \mathcal{T}, X, U, \mathcal{P}, \mathcal{V}, O, \mathbb{O}, \mathcal{W}) \tag{9.47a}$$

In this game setting, I is the set of agents; $\mathcal{T}$ is the set of type profiles, X is the set of strategy profiles, U is the set of utility profiles, $\mathcal{P}$ is the set of payment profiles, and $\mathcal{V}$ is the set of valuation profiles of the agents. Also O is the output function of the mechanism design, $\mathbb{O}$ is the range of the function O, and $\mathcal{W}$ is the optimizing function of the mechanism design.

(a) *The set of agents is $I = \{1, 2, \ldots, n\}$.*
(b) *Agent characteristics.*
   (i) *Each agent $i \in I$ has a set of type space $\mathcal{T}_i$. The element $t_i \in \mathcal{T}_i$ is the private information (type) of agent i. Let the set of type profiles of the n agents be $\mathcal{T} = \times_{i \in I} \mathcal{T}_i$. The type vector $t = (t_1, t_2, \ldots, t_n) \in \mathcal{T}$.*
   (ii) *Each agent $i \in I$ has a set of feasible strategies (actions) X_i. Each $t_i \in \mathcal{T}_i$ is mapped into an action $x_i \in X_i$. The element $x_i \in X_i$ is a strategy of agent i. Let the set of strategy profiles of the n agents be $X = \times_{i \in I} X_i$. The strategy vector $x = (x_1, x_2, \ldots, x_n) \in X$.*
   (iii) *Each agent $i \in I$ has a utility function $u_i : X \times \mathcal{T}_i \to \mathbb{R}$. Let U be the set of utility profiles, and $(u_1(x, t_1), u_2(x, t_2), \ldots, u_n(x, t_n)) \triangleq u \in U$ be a utility vector. The selfish agents try to optimize their utility function.*
(c) *The mechanism design has an output function, and the payments it provides to the n agents. Thus:*

(i) *The mechanism design provides an outcome function* $O : X \to \mathbb{O}$, *where* $\mathbb{O}$ *is the range of the function* O. *The set of possible outputs of the mechanism design is* $\mathbb{O}$. *Therefore if* $x \in X$, *we have* $O(x) \in \mathbb{O}$. *Denote the set of feasible outputs by* $\mathcal{F} \subseteq \mathbb{O}$.

(ii) *For each agent* $i \in I$, *there is a payment function* $p_i : X \to \mathbb{R}$. *Thus* $p_i(x) \in \mathbb{R}$ *is a payment to the agent* i *for* $x \in X$. *Let* $\mathcal{P}$ *be the set of payment profiles, and* $(p_1(x), p_2(x), \ldots, p_n(x)) \triangleq p \in \mathcal{P}$ *be a payment vector.*
The mechanism is the structure $(X, \mathcal{P}, O, \mathbb{O})$.

(d) *More agent characteristics*:

(i) *The valuation (personal preference) function of agent* $i \in I$ *is* $v_i : T_i \times \mathbb{O} \to \mathbb{R}$. *Thus, if* $t_i \in T_i$ *and* $a \in \mathbb{O}$, *then* $v_i(t_i, a) \in \mathbb{R}$. *Let* $\mathcal{V}$ *be the set of valuation profiles, and* $(v_1(t_1, a), v_2(t_2, a), \ldots, v_n(t_n, a)) \triangleq v \in \mathcal{V}$ *be a valuation vector.*

(ii) *If the payment to agent* $i \in I$ *is* $p_i(x)$ *for* $x \in X$, *then*:

$$u_i(x, t_i) = \{p_i(x) + v_i(t_i, a)\} \in \mathbb{R} \tag{9.47b}$$

The agent $i \in I$ *aims to optimize its utility* $u_i(\cdot, \cdot)$.

(e) *The mechanism design optimizes the function* $\mathcal{W} : T \times \mathbb{O} \to \mathbb{R}_0^+$. *Thus* $\mathcal{W}(t, a) \in \mathbb{R}_0^+$, *where* $t \in T$ *and* $a \in \mathbb{O}$. □

We are ready to define the mechanism design problem.

Definition 9.27. *Mechanism design problem. The mechanism tries to reach a specific outcome for the selfish agents, in the game setting* $\mathcal{M} = (I, T, X, U, \mathcal{P}, \mathcal{V}, O, \mathbb{O}, \mathcal{W})$. □

Note that the agents are selfish and optimize their own utility functions. However, the goal of the mechanism design is to optimize its global objective function. This can be either exact or approximate.

Definition 9.28. *The optimization problem of the mechanism design. In the game setting* $\mathcal{M} = (I, T, X, U, \mathcal{P}, \mathcal{V}, O, \mathbb{O}, \mathcal{W})$, $\mathcal{W}(t, a) \in \mathbb{R}_0^+$, *where* $t \in T$, *and* $a \in \mathcal{F} \subseteq \mathbb{O}$ *is optimized. The output specification of the mechanism design for exact optimization is the value of* a *which minimizes* $\mathcal{W}(t, a)$. *However, for approximate optimization, the output specification is the value of* $a \in \mathcal{F}$, *which is within a factor of* c, *so that for any other output* $a' \in \mathcal{F}$, *we have* $\mathcal{W}(t, a) \leq c\mathcal{W}(t, a')$. *Note that,* c *is called the approximation factor.* □

For notational convenience, let $x_{-i} \triangleq (x_1, x_2, \ldots, x_{i-1}, x_{i+1}, \ldots, x_n)$, and denote $x \triangleq (x_i, x_{-i})$. We next describe a special type of mechanism design. It is called implementation with dominant strategies, or simply an implementation.

Definition 9.29. *In the game setting* $\mathcal{M} = (I, T, X, U, \mathcal{P}, \mathcal{V}, O, \mathbb{O}, \mathcal{W})$, *a mechanism is an implementation with dominant strategy iff*:

(a) *For each agent* $i \in I$, *and each* $t_i \in T_i$ *there exists a strategy* $x_i \in X_i$, *called dominant strategy, such that for all possible strategies* x_{-i} *of the other agents, the strategy* x_i *maximizes agent* i*'s utility. Thus, for every* $x_i' \in X_i$, *let* $a \triangleq O(x_i, x_{-i})$, *and* $a' \triangleq O(x_i', x_{-i})$, *then*

$$\{p_i(x_i, x_{-i}) + v_i(t_i, a)\} \geq \{p_i(x_i', x_{-i}) + v_i(t_i, a')\} \tag{9.48}$$

(b) *For each dominant strategy $x \in X$, the output $O(x) \in \mathbb{O}$ satisfies the specification, either exactly or approximately.* □

The Revelation Principle

A simple type of strategy of an agent is to simply reveal its true type.

Definitions 9.30. *The truthful implementation and strongly truthful implementation of a mechanism in game setting $\mathcal{M} = (I, \mathcal{T}, X, U, \mathcal{P}, \mathcal{V}, O, \mathbb{O}, \mathcal{W})$.*

1. *Truthful implementation. A mechanism is said to be truthful, iff for all $i \in I$:*
   (a) *We have $X_i = \mathcal{T}_i$. That is, an agent's strategy is to report its type. This is termed a direct revelation mechanism.*
   (b) *Telling the truth is a dominant strategy, that is $x_i = t_i$ is a dominant strategy.*
2. *Strongly truthful implementation. A mechanism is a strongly truthful implementation iff: $\forall$ $i \in I, x_i = t_i$ is the only dominant strategy.* □

Based upon the above set of definitions, an important result called the *revelation principle* can be stated. This principle lets us focus simply on the truthful implementations.

Lemma 9.9. *If a mechanism design implements a given problem with dominant strategies, then there also exists a corresponding truthful implementation.*
Proof. See the problem section. □

9.6.2 Vickrey-Groves-Clarke Mechanism

An important class of mechanism is the Vickrey-Groves-Clarke (VGC) mechanism. These mechanisms are named after the economists W. Vickrey, T. Groves, and E. H. Clarke. The VGC mechanism is a generalization of the well-known Vickrey auction. The Vickrey type of auction is a sealed-bid type of auction, in which the highest bidder wins. However, the winning bidder pays the price of the second highest bidder. Thus, this type of bidding provides an incentive to the bidder to bid the true valuation of the object.

A mechanism design problem is said to be of the utilitarian type, if its objective function is the finite sum of the valuation of the agents. The VGC mechanism is applicable to utilitarian type of mechanism design problems. In order for a mechanism to belong to the VGC family of mechanisms, it should also use certain types of payment functions. These are clarified in the following definitions.

Definitions 9.31. *Let $\mathcal{M} = (I, \mathcal{T}, X, U, \mathcal{P}, \mathcal{V}, O, \mathbb{O}, \mathcal{W})$ be a game setting.*

1. *A maximization mechanism design problem in the game setting $\mathcal{M}$ is called utilitarian if it maximizes $\mathcal{W}(t, a)$, where $\mathcal{W}(t, a) = \sum_{i \in I} v_i(t_i, a)$. Sometimes $\mathcal{W}(\cdot, \cdot)$ is also called the utilitarian function.*
2. *Let the mechanism be a direct revelation; that is, $X = \mathcal{T}$. Thus the output function O and the set of payment profiles $\mathcal{P}$ depend only on the type profile set $\mathcal{T}$. The mechanism in the game setting $\mathcal{M}$ belongs to the VGC family of mechanisms iff:*
   (a) *$\mathcal{W}(\cdot, \cdot)$ is a utilitarian function.*
   (b) *$O(t) \in \{a \mid \max_{a \in \mathcal{F}} \mathcal{W}(t, a)\}$.*

(c) *For each* $i \in I$, $p_i(t) = \sum_{j \neq i, j \in I} v_j(t_j, O(t)) + h_i(t_{-i})$, *where* $t \in \mathcal{T}$ *and* $h_i(\cdot)$ *is an arbitrary function of* t_{-i}. □

An immediate consequence of the above definitions is the following result. It states that a VGC mechanism provides a solution for any utilitarian problem.

Theorem 9.8. *A VGC mechanism is truthful.*
Proof. See the proof of the next theorem, where a more general result is established. □

A weighted version of the VGC mechanism is next defined.

Definitions 9.32. *Let* $\mathcal{M} = (I, \mathcal{T}, X, U, \mathcal{P}, \mathcal{V}, O, \mathbb{O}, \mathcal{W})$ *be a game setting.*

1. *A maximization mechanism design problem in the game setting* $\mathcal{M}$ *is called weighted utilitarian if there exist weights* $z_i \in \mathbb{R}^+$ *for all* $i \in I$, *and it maximizes* $\mathcal{W}(t,a)$, *where* $\mathcal{W}(t,a) = \sum_{i \in I} z_i v_i(t_i, a)$. *Sometimes* $\mathcal{W}(\cdot,\cdot)$ *is also called the weighted utilitarian function.*
2. *Let the mechanism be direct revelation, that is,* $X = \mathcal{T}$. *Thus the output function* O, *and the set of payment profiles* $\mathcal{P}$ *depend only on the set of profiles* $\mathcal{T}$. *The revelation mechanism in the game setting* $\mathcal{M}$ *belongs to the weighted VGC family of mechanisms iff*:
   (a) $\mathcal{W}(\cdot,\cdot)$ *is a weighted utilitarian function.*
   (b) $O(t) \in \{a \mid \max_{a \in \mathcal{F}} \mathcal{W}(t,a)\}$.
   (c) *For each* $i \in I$, $p_i(t) = z_i^{-1} \sum_{j \neq i, j \in I} z_j v_j(t_j, O(t)) + h_i(t_{-i})$, *where* $t \in \mathcal{T}$ *and* $h_i(\cdot)$ *is an arbitrary function of* t_{-i}. □

The theorem about the truthfulness of the VGC mechanism also holds true for the weighted VGC mechanism.

Theorem 9.9. *A weighted VGC mechanism is truthful.*
Proof. See the problem section. □

9.6.3 Examples

The theoretical details developed in the previous subsections are clarified via examples. The examples that are discussed are:

- Shortest paths
- Task scheduling.

Shortest Paths

A communication network is represented by a directed graph $G = (V, E)$. The cost of an edge $e \in E$ is $c_e \in \mathbb{R}_0^+$. Assume that the graph is biconnected. That is, there are at least two vertex-disjoint directed paths between any pair of distinct vertices. Let the set of edges in the digraph be equal to the set of agents. Each edge $e \in E$ has type (private) information $t_e \in \mathbb{R}_0^+$. Assume that $t_e = c_e$ for all edges $e \in E$. Thus, the type information t_e represents agent e's cost of sending a message on this edge. Let v_x and v_y be two distinct vertices in the set V. The objective is to find the cheapest path for a single message to travel from vertex v_x to vertex v_y. Agent e's valuation is 0 if its edge is not in the path, and $-t_e$ if it is in the path from vertex v_x to vertex v_y.

If the agent's strategy is to report its true type t_e to the mechanism, then we have a truthful implementation. In this case the strategy is also dominant. Furthermore, this also results in the selection of the cheapest path between vertices v_x and v_y. The corresponding mechanism's payment to the agents can be computed as follows. The payment p_e to agent $e \in E$ is

$$p_e = \begin{cases} d_{G|t_e=\infty} - d_{G|t_e=0}, & e \text{ is in the shortest path} \\ 0, & e \text{ is not in the shortest path} \end{cases}$$

where $d_{G|t_e=\infty}$ is length of the shortest path excluding the edge e (as per the reported inputs), and $d_{G|t_e=0}$ is the length of the shortest path if the cost of the edge e is assumed to be 0 (as per the reported types).

Observe that the shortest path computation minimizes the total cost. Furthermore, the implementation is a VGC mechanism, where $d_{G|t_e=\infty}$ and $d_{G|t_e=0}$ correspond to $h_i(t_{-i})$ and $\sum_{j\neq i, j\in I} v_j(t_j, O(t))$ respectively. The goal of the implementation is an algorithmically efficient computation of the payment functions. Thus, the output of the implementation can be obtained by a computationally efficient calculation of the shortest path. A good candidate might be a computationally efficient Dijkstra's shortest path algorithm.

Several other graph-theoretic problems can also be implemented in a framework which is similar to the shortest path problem. A possible candidate is the minimum spanning tree problem.

Task Scheduling

A task scheduling scenario is studied in this example. A set of tasks have to be completed by a given number of agents. The tasks are assigned to the agents (servers). Each agent can finish a task in a specified period of time. Furthermore, all agents can work in parallel. The time required to complete all the tasks by the agents is called the makespan of the tasks. The goal of the task scheduling problem is to minimize the makespan of the tasks. We next state the task scheduling (allocation) problem formally.

Definition 9.33. *Task scheduling (allocation) problem. There are $k \in \mathbb{P}$ tasks, and $n \in \mathbb{P}$ agents (servers). Agent i takes t_{ij} units of time to complete task j. Thus agent i's type for task j is t_{ij}. The goal is to minimize the makespan of the tasks. More precisely:*

(a) *Let the set of agents be $I = \{1, 2, \ldots, n\}$.*
(b) *The feasible outputs of the mechanism are all partitions of the set of k tasks among n agents. A particular partition of tasks is specified as $y(t) = (y_1(t), y_2(t), \ldots, y_n(t))$, where $y_i(t)$ is the set of tasks allocated to agent i.*
(c) *Agent i's valuation is the negative of the total time it spends on the tasks allocated to it. Thus $v_i(t_i, y(t)) = -\sum_{j\in y_i(t)} t_{ij}$.*
(d) *The objective function (makespan) is $\mathcal{W}(\cdot,\cdot)$, where*

$$\mathcal{W}(t, y(t)) = \max_{i\in I} \sum_{j\in y_i(t)} t_{ij} \tag{9.49}$$

(e) *The goal is to minimize $\mathcal{W}(t, y(t))$.* □

An upper bound for the minimum value of makespan is determined via a min-work mechanism. This bound is achieved by assigning a task to an agent who completes the task in the smallest amount of time.

Definition 9.34. *Min-work mechanism.*

(a) *Each task is allocated to an agent who is able to complete it in the least amount of time. If the task completion times of two or more agents are identical, then the tasks are allocated arbitrarily.*
(b) *For every task allocated to an agent $i \in I$, a payment is made to the agent, which is equal to the time of the second best agent for this specific task. Thus*

$$p_i(t) = \sum_{j \in y_i(t)} \min_{i' \neq i, i' \in I} t_{i'j} \tag{9.50}$$

□

It is next proved that the min-work mechanism is a strongly truthful n-approximation. This result is based upon the following two lemmas.

Lemma 9.10. *The min-work mechanism is an n-approximation for the task scheduling problem.*
Proof. See the problem section. □

Lemma 9.11. *The min-work mechanism is strongly truthful.*
Proof. See the problem section. □

Theorem 9.10. *The min-work mechanism is both strongly and truthful n-approximation for the task scheduling problem.*
Proof. The proof follows immediately from the last two lemmas. □

Techniques to determine lower bound for the truthful implementation of the task scheduling problem can be found in the literature.

Reference Notes

The theory of games is a well-developed mathematical discipline. However, its application to model the Internet behavior is recent. Some of the well-known books on game theory are by Myerson (1997), von Neumann, and Morgenstern (1953), Owen (1995), Vorob'ev (1977), Dresher (1981), Friedman (1986), Fudenberg, and Tirole (1991), Gibbons (1992), Osborne, and Rubinstein (1994), Watson (2002), and Osborne (2004). The presentation of the basics of this subject in Osborne (2004) is indeed fluid. The notes on game theory by Mishra (1998) are also elegant and indispensable. The example on tragedy of the commons is from Gibbons (1992), which in turn quantifies the influential paper by Hardin (1968).

The reader is referred to Roughgarden, and Tardos (2002), Pióro, and Medhi (2004), and Roughgarden (2005), for a readable discussion of selfish routing. The existence and uniqueness of Nash flow in a selfish routing scenario are originally due to Beckmann, McGuire, and Winsten (1956), and later reproved by Dafermos, and Sparrow (1969). A selfish task allocation model is also discussed in this chapter. The reader can refer to the fundamental work of Koutsoupias, and Papadimitriou (1999), Czumaj, Krysta, and Vöcking (2002), and Czumaj (2004). The example on repeated games is based upon the pioneering work of Mackenzie, and DaSilva (2006), and Srivastava (2008). The section on algorithmic mechanism design follows the trail-blazing work of Nisan, and Ronen (2001). The enthusiasm among computer scientists for this area of work is certainly evident in Nisan, Roughgarden, Tardos, and Vazirani (2007).

Problems

1. The payoffs to a profile of mixed strategies are the expected values of the corresponding pure-strategy payoffs. The player i's payoff to profile ψ is

$$u_i(\psi) = \sum_{s \in S} \left\{ \prod_{j \in I} \psi_j(s_j) \right\} u_i(s)$$

Develop a motivation for the above result.
Hint: See Mishra (1998). A player i's payoff to profile ψ is

$$u_i(\psi) = u_i(\psi_i, \psi_{-i}) = \sum_{s_i \in S_i} \psi_i(s_i) u_i(s_i, \psi_{-i})$$

$$\begin{aligned} u_i(s_i, \psi_{-i}) &= \sum_{s_{-i} \in S_{-i}} \psi_{-i}(s_{-i}) u_i(s_i, s_{-i}) \\ &= \sum_{s_{-i} \in S_{-i}} \left\{ \prod_{j \in I, j \neq i} \psi_j(s_j) \right\} u_i(s_i, s_{-i}) \end{aligned}$$

Therefore

$$\begin{aligned} u_i(\psi) &= \sum_{s_i \in S_i} \sum_{s_{-i} \in S_{-i}} \psi_i(s_i) \left\{ \prod_{j \in I, j \neq i} \psi_j(s_j) \right\} u_i(s_i, s_{-i}) \\ &= \sum_{s \in S} \left\{ \prod_{j \in I} \psi_j(s_j) \right\} u_i(s) \end{aligned}$$

2. Find the players' best response functions in the games: Prisoners' dilemma, choice between classical and modern music, and the matching coins. Use these functions to determine the Nash equilibria (if it exists) of each game.
3. In the example on the tragedy of the commons, if $v'(Y) < 0$ and $v''(Y) < 0$ prove that $Y^{**} < Y^{*}$.
Hint: See Gibbons (1992). The result is established via contradiction. Assume that $Y^{**} \geq Y^{*}$. The inequality $v'(Y) < 0$, implies $v(Y^{*}) \geq v(Y^{**})$. Similarly, $0 > v'(Y^{*}) \geq v'(Y^{**})$, as $v''(Y) < 0$. Also by initial assumption $Y^{*}/n < Y^{**}$ if $n > 1$. Therefore

$$v(Y^*) + \frac{1}{n} Y^* v'(Y^*) - c > v(Y^{**}) + Y^{**} v'(Y^{**}) - c$$

which is a contradiction, since each side in the above inequality is equal to 0.

4. Consider a two-person zero-sum game $\Gamma = (I, S, U)$. Let the two persons playing the game be P_{plyr1} and P_{plyr2}. In this game, the number of strategies of the player P_{plyr1} is equal to m, and the number of strategies of player P_{plyr2} is equal to n. Prove that

$$\max_{1 \le i \le m} \min_{1 \le j \le n} a_{ij} \le \min_{1 \le j \le n} \max_{1 \le i \le m} a_{ij}$$

where $A = [a_{ij}]$ is the $m \times n$ payoff matrix of P_{plyr1}.
Hint: See Dresher (1981). For any strategy s_{1i},

$$\min_{1 \le j \le n} a_{ij} \le a_{ij}, \quad \forall\, j, \text{ where } 1 \le j \le n$$

Similarly for any strategy s_{2j}

$$a_{ij} \le \max_{1 \le i \le m} a_{ij}, \quad \forall\, i, \text{ where } 1 \le i \le m$$

Therefore

$$\min_{1 \le j \le n} a_{ij} \le a_{ij} \le \max_{1 \le i \le m} a_{ij}$$

That is

$$\min_{1 \le j \le n} a_{ij} \le \max_{1 \le i \le m} a_{ij}$$

Right-hand side of the above inequality is independent of i. Therefore by taking the maximum of both sides yields

$$\max_{1 \le i \le m} \min_{1 \le j \le n} a_{ij} \le \max_{1 \le i \le m} a_{ij}$$

Similarly, left-hand side of the above inequality is independent of j. Taking minimum of both sides yields

$$\max_{1 \le i \le m} \min_{1 \le j \le n} a_{ij} \le \min_{1 \le j \le n} \max_{1 \le i \le m} a_{ij}$$

This relation implies that person P_{plyr1}'s payoff does not exceed P_{plyr2}'s loss.

5. Let a two-person zero-sum pure strategic game be $\Gamma = (I, S, U)$. The two persons are P_{plyr1} and P_{plyr2}, and $A = [a_{ij}]$ is the $m \times n$ payoff matrix of P_{plyr1}. Then prove that a_{ij} is a saddle point if and only if (s_{1i}, s_{2j}) is a Nash equilibrium.
Hint: Assume that a_{ij} is a saddle point. Then the matrix element a_{ij} is maximum in its column. Therefore the player P_{plyr1} cannot increase its payoff given that the player P_{plyr2} has selected strategy s_{2j}. Similarly, since the matrix element a_{ij} is minimum in its row and the payoff of player P_{plyr2} is $-a_{ij}$, it cannot increase its payoff by changing its strategy given that the player P_{plyr1} has selected strategy s_{1i}. Thus (s_{1i}, s_{2j}) is a Nash equilibrium.
Assume that (s_{1i}, s_{2j}) is a Nash equilibrium. Then a_{ij} is the maximum value in its column. Similarly, a_{ij} is the minimum in its row, since the payoff of player P_{plyr2} is $-a_{ij}$. Therefore a_{ij} is a saddle point.

6. In a mixed-strategy, two-person zero-sum game, specified by the payoff matrix A, prove that

$$\max_p \min_q p^T A q \le \min_q \max_p p^T A q$$

In the above inequality, the maximization and minimization are determined over all possible values of the vectors p and q respectively.
Hint: For any probability vector p,

$$\min_q p^T Aq \leq p^T Aq$$

Similarly for any probability vector q,

$$p^T Aq \leq \max_p p^T Aq$$

Therefore

$$\min_q p^T Aq \leq p^T Aq \leq \max_p p^T Aq$$

That is

$$\min_q p^T Aq \leq \max_p p^T Aq$$

Right-hand side of the above inequality is independent of p. Therefore taking the maximum of both sides over p yields

$$\max_p \min_q p^T Aq \leq \max_p p^T Aq$$

Similarly, left-hand side of the above inequality is independent of q. Therefore taking minimum of both sides over q yields

$$\max_p \min_q p^T Aq \leq \min_q \max_p p^T Aq$$

This relation implies that person P_{plyr1}'s payoff does not exceed P_{plyr2}'s loss.

7. Prove that, in a mixed-strategy two-person zero-sum game, (p^*, q^*) is a Nash equilibrium if and only if $\max_p \min_q p^T Aq = \min_q \max_p p^T Aq = p^{*T} Aq^*$.
Hint: Assume that (p^*, q^*) is a Nash equilibrium. Then

$$p^{*T} Aq^* = \max_p p^T Aq^* \geq \min_q \max_p p^T Aq$$

Similarly

$$p^{*T} Aq^* = \min_q p^{*T} Aq \leq \max_p \min_q p^T Aq$$

Thus

$$\min_q \max_p p^T Aq \leq \max_p \min_q p^T Aq$$

But it is known that $\max_p \min_q p^T Aq \leq \min_q \max_p p^T Aq$. Therefore $\max_p \min_q p^T Aq = \min_q \max_p p^T Aq$.
Conversely, assume that $\max_p \min_q p^T Aq = \min_q \max_p p^T Aq = p^{*T} Aq^*$. It has to be demonstrated that (p^*, q^*) is a Nash equilibrium. Observe that $p^{*T} Aq^* = \max_p p^T Aq^* \geq p^T Aq^*, \forall\, p$. Similarly $p^{*T} Bq^* \geq p^{*T} Bq, \forall\, q$. These are the requirements for (p^*, q^*) to be an equilibrium point.

8. Consider an infinitely repeated game Γ. Prove that for every feasible payoff vector v, with $v_i > \underline{v}_i \;\; \forall\, i \in I$, there exists $\underline{\delta} < 1$ such that $\forall\, \widetilde{\delta} \in (\underline{\delta}, 1)$ there is a Nash equilibrium of game Γ with payoffs v.
Hint: See Fudenberg, and Tirole (1993), and MacKenzie and DaSilva (2006). Let $a \in \mathcal{A}$ be a pure action profile, such that $u\,(a) = v = (v_1, v_2, \ldots, v_n)$ is a feasible payoff vector, with

$v_i > \underline{v}_i \ \forall\, i \in I$. Assume the following threat (strategy): Player i will play action a_i until any one of the players deviates from his (or her) action. If a player deviates, in a specific iteration of the game, then all other players will minmax him (or her) in all subsequent iterations ad infinitum. Observe that due to the definition of Nash equilibrium, it is sufficient to consider unilateral deviations.

Thus, if a player i deviates from its action in a particular iteration, then he (or she) will receive a payoff of at most $\widehat{v}_i = \max_{\widehat{a}_i} u_i(\widehat{a}_i, a_{-i})$ in that particular round, however other players will minmax him (or her), and he (or she) will receive a payoff of at most $\underline{v}_i$ in each subsequent iteration. Assume that the player i deviates in round $k \in \mathbb{P}$, then the normalized payoff of this strategy for player i is, say $g_i\left(\widetilde{\delta}\right)$ where

$$g_i\left(\widetilde{\delta}\right) = \left(1 - \widetilde{\delta}\right)\left\{\sum_{t=0}^{k-1} \widetilde{\delta}^t v_i + \widetilde{\delta}^k \widehat{v}_i + \sum_{t=k+1}^{\infty} \widetilde{\delta}^t \underline{v}_i\right\}$$
$$= \left(1 - \widetilde{\delta}^k\right) v_i + \left(1 - \widetilde{\delta}\right) \widetilde{\delta}^k \widehat{v}_i + \widetilde{\delta}^{k+1} \underline{v}_i$$

However, the normalized payoff without deviation is simply v_i. The above expression is less than v_i provided

$$\left(1 - \widetilde{\delta}\right) \widehat{v}_i + \widetilde{\delta} \underline{v}_i < v_i$$

More precisely define $\underline{\delta}_i$ by

$$(1 - \underline{\delta}_i) \widehat{v}_i + \underline{\delta}_i \underline{v}_i = v_i$$

Note that, since $\underline{v}_i < v_i$, we have $\underline{\delta}_i < 1$. Then $g_i\left(\widetilde{\delta}\right) < v_i$ as long as $\widetilde{\delta} > \underline{\delta}_i$. Select $\underline{\delta} = \max_{i \in I} \underline{\delta}_i$ for the game.

What if the payoffs v cannot be obtained simply via pure actions? In this case the action profile a is replaced with a public randomization $a(\omega)$ which yields payoffs with expected value v. However, the discount factor necessary so that a player cannot benefit by deviating may be larger than in the deterministic case.

9. Consider an infinitely repeated game Γ. Let a^* be an equilibrium of the constituent game $\mathcal{G}$ with payoffs $\widetilde{p}$. Prove that for any $v \in U_{ch}$ with $v_i > \widetilde{p}_i$ for all players $i \in I$, there exists $\underline{\delta} < 1$ such that $\forall\, \widetilde{\delta} > \underline{\delta}$ there is a subgame-perfect equilibrium of game Γ with payoffs v.

   Hint: See Fudenberg, and Tirole (1993). Assume that there is an action $\widehat{a} \in \mathcal{A}$ such that $u(\widehat{a}) = v$. Next consider the following strategy profile: In the first iteration of the repeated game Γ, each player i plays $\widehat{a}_i$. Each player continues playing $\widehat{a}_i$ as long as the action in the previous iteration was $\widehat{a}$. However, if at least one player did not play according to $\widehat{a}$, then each player i plays a_i^* in all subsequent iterations. This strategy profile is indeed a Nash equilibrium for $\widetilde{\delta}$ large enough so that

$$\left(1 - \widetilde{\delta}\right) \max_a u_i(a) + \widetilde{\delta} \widetilde{p}_i < v_i$$

   The above inequality is true for $\widetilde{\delta} = 1$. Therefore, the inequality is true for a range of $\widetilde{\delta} < 1$. The strategy profile is subgame-perfect, because in every subgame off the equilibrium path, the strategies play a^* ad infinitum, which is a Nash equilibrium for the constituent game. If there is no $\widehat{a} \in \mathcal{A}$ such that $u(\widehat{a}) = v$, then use public randomization.

10. If in a network three-tuple $\mathcal{N} = (G, \rho, l)$, the latency function satisfies

$$l_e(\xi_e)\, \xi_e \leq \gamma \int_0^{\xi_e} l_e(t)\, dt, \quad \gamma \geq 1$$

for all links $e \in E$ such that $\xi_e > 0$, then $\mathcal{Q}\left(\beta^N\right) \leq \gamma \mathcal{Q}\left(\beta^*\right)$. Prove this assertion.
Hint: See Roughgarden, and Tardos (2002). The superscripts $*$ and N identify optimal and Nash flows respectively.

$$\mathcal{Q}\left(\beta^N\right) = \sum_{e \in E} l_e\left(\xi_e^N\right)\xi_e^N \leq \gamma \sum_{e \in E} \int_0^{\xi_e^N} l_e(t)\,dt$$

$$\leq \gamma \sum_{e \in E} \int_0^{\xi_e^*} l_e(t)\,dt \leq \gamma \sum_{e \in E} l_e\left(\xi_e^*\right)\xi_e^* = \gamma \mathcal{Q}\left(\beta^*\right)$$

The first inequality follows from the statement of the problem. The second inequality follows from the observation, that the Nash flow optimizes the objective $\sum_{e \in E} \int_0^{\xi_e} l_e(t)\,dt$. The third inequality is used because the latency function $l_e(\cdot)$ is nondecreasing.

11. In the selfish task allocation model, consider an arbitrary Nash equilibrium, such that $p_{ik} > 0$ for a user $i \in I$ and a link $k \in E$. Prove that
    (a) $\ell_k \leq (\ell_j + w_i/c_j)$, for every $j \in E$.
    (b) $(\ell_j + 1) < \ell_k \Rightarrow w_i > c_j$.
    (c) If $p_{ij} > 0$, $p_{ik} > 0$, then $|\ell_j - \ell_k| \leq \mathcal{O}$.

    Hint: See Czumaj (2004).
    (a) Observe that $\sigma_{ij} \leq (\ell_j + w_i/c_j)$ and $\sigma_{ik} = \ell_k + (1 - p_{ik})\,w_i/c_k \geq \ell_k$. Consequently, since $p_{ik} > 0$, the definition of the Nash equilibrium yields $\sigma_{ik} \leq \sigma_{ij}$, $\forall\, j \in E$. That is, $\ell_k \leq (\ell_j + w_i/c_j)$.
    (b) Use part (a) of the problem.
    (c) Use part (a) of the problem, and observe that $w_i/c_j \leq \mathcal{O}$.

12. There are m balls, which are to be placed in m bins. Each ball is equally likely to be placed in any bin. Furthermore, the placement of a ball is independent of all other ball placements. Let $B_{\max}$ be the maximum number of balls in any bin after all the balls have been placed in the m bins. Prove that the expected value of $B_{\max}$ is upper bounded by $O(\ln m / \ln \ln m)$.
    Hint: See Cormen, Leiserson, Rivest, and Stein (2009). It is convenient to address this problem in the following steps.
    Step 1: The probability p_k that k balls are placed in a bin is given by

$$p_k = \binom{m}{k}\left(\frac{1}{m}\right)^k\left(1 - \frac{1}{m}\right)^{m-k}, \quad 0 \leq k \leq m$$

Use of the relationships

$$\binom{m}{k} \leq \left(\frac{em}{k}\right)^k, \text{ and } \left(1 - \frac{1}{m}\right)^{m-k} \leq 1, \text{ for } 0 < k \leq m$$

results in the inequality $p_k \leq e^k/k^k$.
Step 2: Let q_k be the probability that $B_{\max} = k$. That is, q_k is the probability that the bin containing the maximum number of balls is k. It is shown that $q_k \leq m p_k$. Denote the probability function by $P(\cdot)$. Also let ε_i be the event that bin i has the maximum number of balls and, that it is exactly k balls. The probability that the event $B_{\max} = k$ is equal to $q_k = P(B_{\max} = k) = P\left(\cup_{i=1}^m \varepsilon_i\right) \leq \sum_{i=1}^m P(\varepsilon_i = k)$. Therefore $q_k \leq m p_k$.
Step 3: It is next shown that there exists a constant $c > 1$ such that $p_{k_0} < 1/m^3$, where $k_0 = c\beta/\ln\beta$ and $\ln m = \beta$.

The given condition is satisfied, if $e^{k_0}/k_0^{k_0} < 1/m^3$. Equivalently, this inequality is

$$(k_0 - k_0 \ln k_0) < -3\beta$$

This leads to

$$\frac{c}{\ln \beta}\{-1+\ln c\} + \frac{c}{\ln \beta}\{\ln \beta - \ln \ln \beta\} > 3$$

The appropriate value of c such that the above inequality holds true, has to be determined. If $c > e$, then the first term on the left side is positive. In this case the following inequality has to be satisfied

$$\frac{c}{\ln \beta}\{\ln \beta - \ln \ln \beta\} > 3$$

That is

$$c > \frac{3}{\left\{1 - \dfrac{\ln \ln \beta}{\ln \beta}\right\}}$$

For large values of m, the denominator tends to unity. Therefore an appropriate value of c can be chosen such that $p_{k_0} < 1/m^3$. Thus $p_k < 1/m^3$ for $k \geq k_0$. This inequality implies that $q_k < 1/m^2$ for $k \geq k_0$.

Step 4: Finally, the bound on $\mathcal{E}(B_{\max})$ is computed, where $\mathcal{E}(\cdot)$ is the expectation operator.

$$\mathcal{E}(B_{\max}) = \sum_{i=1}^{m} iP(B_{\max} = i) = \sum_{i=1}^{k_0} iP(B_{\max} = i) + \sum_{i=k_0+1}^{m} iP(B_{\max} = i)$$

$$\begin{aligned}
\mathcal{E}(B_{\max}) &\leq k_0 \sum_{i=1}^{k_0} P(B_{\max} = i) + m \sum_{i=k_0+1}^{m} P(B_{\max} = i) \\
&= k_0 P(B_{\max} \leq k_0) + mP(B_{\max} > k_0) \\
&= k_0 + (m - k_0) P(B_{\max} > k_0) \\
&< k_0 + \frac{(m-k_0)^2}{m^2} < k_0 + 1 = \frac{c\beta}{\ln \beta} + 1
\end{aligned}$$

Thus $\mathcal{E}(B_{\max}) = O(\ln m / \ln \ln m)$.

13. Consider a network with two nodes and M parallel links. The links have identical speeds, say unity. Prove that the coordination ratio of this network is $\mathcal{R}_t = O(\ln M / \ln \ln M)$.
Hint: See Czumaj (2004). Consider the case in which the number of users $N = M$, and all tasks have identical weight. Let these weights be equal to unity. Also assume that each user selects a link independently and uniformly at random. Thus $p_{ij} = 1/M$ for all $i, j \in \{1, 2, \ldots, M\} = E = I$. The expected load ℓ_j on the link $j \in E$ is

$$\ell_j = \frac{1}{c_j}\sum_{i \in I} w_i p_{ij} = \sum_{i=1}^{M} \frac{1}{M} = 1$$

Similarly, the expected cost of task i on link j is

$$\sigma_{ij} = \ell_j + (1 - p_{ij})\frac{w_i}{c_j} = 1 + \left(1 - \frac{1}{M}\right) = 2 - \frac{1}{M}; \quad \forall\, i \in I, \text{ and } \forall\, j \in E$$

Since all the costs σ_{ij} are identical, the selected strategy is a Nash equilibrium. The social optimum for this network is $\mathcal{O} = 1$. This is obtained by assigning each task i to link i. The social cost $\mathcal{C}_{sc}$ is obtained by noticing that this problem is similar to analysing the problem of placing M balls into M bins. From this problem, it is known that $\mathcal{C}_{sc} = O(\ln M / \ln \ln M)$. The result follows.

14. Prove that, if a mechanism design implements a given problem with dominant strategies, then there also exists a corresponding truthful implementation.
Hint: See Nisan, and Ronen (2001). The truthful implementation is made to mimic the agents' strategies. Consider a mechanism in the game setting

$$\mathcal{M} = (I, \mathcal{T}, X, U, \mathcal{P}, \mathcal{V}, O, \mathbb{O}, \mathcal{W})$$

Assume that this mechanism has dominant strategies $x_i(t_i)$ for all $i \in I$. Define a new mechanism in the game setting $\widetilde{\mathcal{M}}$ in which the output function, and the set of payment profiles are $\widetilde{O}$ and $\widetilde{\mathcal{P}}$ respectively, where

$$\begin{aligned} \widetilde{O}(t_1, t_2, \ldots, t_n) &\triangleq O(x_1(t_1), x_2(t_2), \ldots, x_n(t_n)) \\ \widetilde{p}_i(t_1, t_2, \ldots, t_n) &\triangleq p_i(x_1(t_1), x_2(t_2), \ldots, x_n(t_n)), \quad \forall\, i \in I \end{aligned}$$

Observe that the new mechanism in the game setting $\widetilde{\mathcal{M}}$ is indeed a truthful implementation.

15. Prove that a weighted VGC mechanism is truthful.
Hint: See Nisan, and Ronen (2001). Let the actual types of the n agents be $t_1, t_2, \ldots, t_n$, and their declarations be $d_1, d_2, \ldots, d_n$ respectively. Assume that telling the truth is not a dominant strategy. Then there exists i, d, t, and d_i' such that

$$\begin{aligned} &v_i(t_i, O(t_i, d_{-i})) + p_i(t_i, O(t_i, d_{-i})) + h_i(d_{-i}) \\ &< v_i(t_i, O(d_i', d_{-i})) + p_i(t_i, O(d_i', d_{-i})) + h_i(d_{-i}) \end{aligned}$$

Thus

$$\begin{aligned} &v_i(t_i, O(t_i, d_{-i})) + z_i^{-1} \sum_{j \neq i, j \in I} z_j v_j(t_j, O(t_i, d_{-i})) \\ &< v_i(t_i, O(d_i', d_{-i})) + z_i^{-1} \sum_{j \neq i, j \in I} z_j v_j(t_j, O(d_i', d_{-i})) \end{aligned}$$

Multiply both sides of the above inequality by z_i. This yields

$$\sum_{j \in I} z_j v_j(t_j, O(t_i, d_{-i})) < \sum_{j \in I} z_j v_j(t_j, O(d_i', d_{-i}))$$

The above inequality is a contradiction of the definition of $O(t)$, which is

$$O(t) \in \left\{ a \mid \max_{a \in \mathcal{F}} \mathcal{W}(t, a) \right\}$$

16. Prove that the min-work mechanism is an n-approximation for the task scheduling problem.
Hint: See Nisan, and Ronen (2001), and Dütting and Geiger (2007). Let $\widetilde{y}(t)$ denote an optimal solution of the task allocation problem. Observe that $\mathcal{W}(t, y(t)) \leq \sum_{j=1}^{k} \min_{i \in I} t_{ij}$. This is possible if a single agent offers the smallest execution times for all the tasks.
Furthermore, the optimal scheduling algorithm cannot be faster than executing all k jobs in smallest execution times on all n machines in parallel. Thus $\mathcal{W}(t, \widetilde{y}(t)) \geq \frac{1}{n} \sum_{j=1}^{k} \min_{i \in I} t_{ij}$. Combination of these two observations yields $n\mathcal{W}(t, \widetilde{y}(t)) \geq \mathcal{W}(t, y(t))$.

17. Consider the min-work mechanism in the task allocation problem. Prove that for each $i \in I$, $p_i(t) = \sum_{i' \neq i, i' \in I} v_{i'}(t_{i'}, y(t)) + h_i(t_{-i})$, where $h_i(t_{-i}) = \sum_{j=1}^{k} \min_{i' \neq i, i' \in I} t_{i'j}$.
Hint: See Dütting, and Geiger (2007). Using the definition of the payment function in the min-work mechanism, we have

$$\begin{aligned} p_i(t) &= \sum_{j \in y_i(t)} \min_{i' \neq i, i' \in I} t_{i'j} = \sum_{j=1}^{k} \min_{i' \neq i, i' \in I} t_{i'j} - \sum_{j \notin y_i(t)} \min_{i' \neq i, i' \in I} t_{i'j} \\ &= h_i(t_{-i}) - \sum_{i' \neq i, i' \in I} \sum_{j \in y_{i'}(t)} t_{i'j} = h_i(t_{-i}) + \sum_{i' \neq i, i' \in I} v_{i'}(t_{i'}, y(t)) \end{aligned}$$

18. Prove that the min-work mechanism is strongly truthful.
Hint: See Nisan, and Ronen (2001). It is first demonstrated that the min-work mechanism belongs to the VGC family. This in turn would imply that the min-work mechanism is indeed truthful. The output of the mechanism is an allocation which maximizes the utilitarian value $\sum_{i \in I} v_i(t_i, y(t))$, where

$$\sum_{i \in I} v_i(t_i, y(t)) = -\sum_{i \in I} \sum_{j \in y_i(t)} t_{ij}$$

and the payment of the mechanism to each $i \in I$ is

$$p_i(t) = \sum_{i' \neq i, i' \in I} v_{i'}(t_{i'}, y(t)) + h_i(t_{-i})$$

where $h_i(t_{-i}) = \sum_{j=1}^{k} \min_{i' \neq i, i' \in I} t_{i'j}$.
It is next shown that telling truth is the only dominant strategy. Assume that $k = 1$. Discussion for $k > 1$ is similar. Assume that the real types and declarations of the agents are t and d respectively. Further assume that $d_i \neq t_i$, for $i = 1, 2$. If $d_i > t_i$, then select d_{3-i} so that $d_i > d_{3-i} > t_i$, the utility of agent i is $(t_i - d_i) < 0$. However, if the declaration of the agent was truthful, then its utility would be zero. A similar result is true when $d_i < t_i$.

References

1. Aumann, R. J., and Hart, S., 1992. *Handbook of Game Theory with Economic Applications,* Volume I, North-Holland, New York, New York.
2. Barron, E. N., 2008. *Game Theory, An Introduction,* John Wiley & Sons, Inc., New York, New York.
3. Beckmann, M., McGuire, C. B., and Winsten, C. B., 1956. *Studies in the Economics of Transportation,* Yale University Press, New Haven, Connecticut.
4. Cormen, T. H., Leiserson, C. E., Rivest, R. L., and Stein, C., 2009. *Introduction to Algorithms,* Third Edition, The MIT Press, Cambridge, Massachusetts.
5. Czumaj, A., 2004. "Selfish Routing on the Internet," in *Handbook of Scheduling - Algorithms, Models, and Performance Analysis.* Editor: Leung, J. Y-T., Chapman and Hall/CRC Press, New York, New York.

6. Czumaj, A., Krysta, P., and Vöcking, B., 2002. "Selfish Traffic Allocation for Server Farms," Proceedings of the 34th Annual ACM Symposium on Theory of Computing (STOC), Montreal, Quebec, Canada, pp. 287-296.
7. Dafermos, S. C., and Sparrow, F. T., 1969. "The Traffic Assignment Problem for a General Network," Journal of Research of the National Bureau of Standards - B. Mathematical Sciences, Vol. 738, No. 2, pp. 91-118.
8. Dresher, M., 1981. *The Mathematics of Games of Strategy: Theory and Applications*, Dover Publications, Inc., New York, New York.
9. Dütting, P., and Geiger, A., 2007. "Algorithmic Mechanism Design," Seminar: Mechanism Design, Universität Karlsruhe.
10. Ferguson, T. S., 2014. "Game Theory," Second Edition, available at http://www.math.ucla.edu/~tom/Game_Theory/Contents.html. Retrieved February 16,2015.
11. Friedman, J. W., 1986. *Game Theory with Applications to Economics,* Oxford University Press, Oxford, Great Britain.
12. Fudenberg, D., and Tirole, J., 1991. *Game Theory,* The MIT Press, Cambridge, Massachusetts.
13. Garg, R., Kamra, A., and Khurana, V., 2001. "Eliciting Cooperation from Selfish Users: A Game-Theoretic Approach Towards Congestion Control in Communication Networks," IBM Research Report, IBM India Research Laboratory, New Delhi, India.
14. Gibbons, R., 1992. *Game Theory for Applied Economists*, Princeton University Press, Princeton, New Jersey.
15. Groves, T., 1973. "Incentives in Teams," Econometrica, Vol. 41, No. 4, pp. 617-631.
16. Hardin, G., 1968. "Tragedy of the Commons," Science. Vol. 162, Issue 3859, pp. 1243-1248.
17. Jones, A. J., 1980. *Game Theory: Mathematical Models of Conflict,* John Wiley & Sons, Inc., New York, New York.
18. Koutsoupias, E., and Papadimitriou, C. H., 1999. "Worst-Case Equilibria," Proceedings of the 16th Annual Symposium on Theoretical Aspects of Computer Science (STACS), Vol. 1563 of Lecture Notes in Computer Science. Editors: Meinel, C., and Tison, S., Springer-Verlag, Berlin, Germany, pp. 404-413.
19. Kuhn, H. W., 1953. "Extensive Games and the Problem of Information," in *Contributions to the Theory of Games, Vol. II* (Annals of Mathematics Studies 28). Editors: H. W. Kuhn and A. W. Tucker, Princeton University Press, Princeton, New Jersey, pp. 193-216.
20. Leyton-Brown, K., and Shoham, Y., 2008. *Essentials of Game Theory, a Concise, Multidisciplinary Introduction,* Morgan and Claypool Publishers, San Rafael, California.
21. MacKenzie, A., and DaSilva, A., 2006. *Game Theory for Wireless Engineers*, Morgan and Claypool Publishers, San Rafael, California.
22. Mendelson, E., 2004. *Introducing Game Theory and Its Applications*, Chapman and Hall/CRC Press, New York, New York.
23. Mishra, B., 1998. "Game Theory and Learning, Informal Notes," available at https://cs.nyu.edu/mishra/COURSES/GAME/classnotes.pdf. Retrieved February 20, 2015.
24. Morris, P., 1991. *Introduction to Game Theory*, Springer-Verlag, Berlin, Germany.
25. Myerson, R. B., 1997. *Game Theory, Analysis of Conflict,* Harvard University Press, Cambridge, Massachusetts.
26. Nisan, N., and Ronen, A., 2001. "Algorithmic Mechanism Design," Games Econ. Behav., Vol. 35, Issues 1-2, pp. 166-196.
27. Nisan, N., Roughgarden, Tardos, E., and Vazirani, V., 2007. *Algorithmic Game Theory*, Cambridge University Press, Cambridge, Great Britain.
28. Osborne, M. J., 2004. *An Introduction to Game Theory*, Oxford University Press, Oxford, Great Britain.

29. Osborne, M. J., and Rubinstein, A., 1994. *A Course in Game Theory*, The MIT Press, Cambridge, Massachusetts.
30. Owen, G., 1995. *Game Theory*, Third Edition, Academic Press, New York, New York.
31. Pióro, M., and Medhi, D., 2004. *Routing, Flow, and Capacity Design in Communication and Computer Networks*, Morgan Kaufmann Publishers, San Francisco, California.
32. Roughgarden, T., 2005. *Selfish Routing and the Price of Anarchy*, The MIT Press, Cambridge, Massachusetts.
33. Roughgarden, T., and Tardos, E., 2002. "How Bad is Selfish Routing?" Journal of the ACM, Vol. 49, Issue 2, pp. 236-259.
34. Sharma, Y., and Williamson, D. P., 2007. "Stackelberg Thresholds in Network Routing Games or the Value of Altruism," Proc. of the 8th Conference on Electronic Commerce, San Diego, California, USA, pp. 93-102.
35. Shoham, Y., and Leyton-Brown, K., 2009. *Multiagent Systems: Algorithmic, Game-Theoretic, and Logical Foundations*, Cambridge University Press, Cambridge, Great Britain.
36. Srivastava, V. R., 2008. "Behavior-Based Incentives for Node Cooperation in Ad Hoc Networks," Ph. D. dissertation, Virginia Polytechnic Institute and State University, Blacksburg, Virginia.
37. von Neumann, J., and Morgenstern, O., 1953. *Theory of Games and Economic Behavior,* Third Edition, Princeton University Press, Princeton, New Jersey.
38. Vorob'ev, N. N., 1977. *Game Theory: Lectures for Economists and Systems Scientists*, Springer-Verlag, Berlin, Germany.
39. Watson, J., 2002. *Strategy, and Introduction to Game Theory*, W. W. Norton & Company, New York, New York.

CHAPTER 10

Internet Economics

$$\frac{df(x)}{dx} \quad \& \quad \int f(x)\,dx$$

Leibniz's notation for differentiation and integration of a function

Gottfried Wilhelm von Leibniz. Leibniz was born in Leipzig on 1 July, 1646. He entered the University of Leipzig at the age of fourteen, and graduated with a bachelors degree two years later. Leibniz earned a doctorate in law in February 1667.

Leibniz and Isaac Newton are generally regarded as the codiscoverers of calculus. The operator notation for differentiation and integration of a function that he introduced is still used today. Leibniz and Newton independently discovered the *Fundamental Theorem of Calculus*. He is generally credited with introducing the concepts of determinant, Gaussian elimination, and the theory of geometric envelopes. He also contributed to Boolean algebra, matrix theory, and symbolic logic. Leibniz was the originator of the term *analysis situs*. The modern term for this phrase is topology. Based upon the notions of momentum, and kinetic and potential energy, he and the physicist Christian Huygens (1629-1695) developed the ideas of dynamics, where space was assumed to be relative. He also improved upon the binary number system, which in turn is the basis of modern digital computer systems.

Leibniz was not only a prolific mathematician but also a philosopher and scientist. Leibniz also contributed to the areas of biology, geology, linguistics, logic, medicine, philology, probability theory, and psychology. Leibniz passed away on November 14, 1716.

10.1 Introduction

Internet is a global computer network, and economics is the art and science of analysing production, distribution, and consumption of scarce resources by the society in a cost effective manner. Thus Internet economics is the study of interplay of human behavior (actions), economics, and the Internet as scarce capital (resource).

Economics is initially defined in this chapter in set-theoretic language. Basics of Internet technology, costs, and pricing; and a qualitative economic model of the Internet are next briefly outlined. The interactions between network resources, congestion, and pricing are also explored. A scheme to provide service differentiation to customers via pricing is also discussed. Further, a time-dependent pricing and bandwidth trade-off mechanism is briefly described. The applications of theory of auctions, coalitional game theory, and use of bargaining game theory are also outlined.

The primary purpose of the Internet is to provide a telecommunication network for the transmission of data, images, voice, and video. In addition to these primary services, it also provides information, electronic mail, and entertainment like interactive online games and virtual reality. It also serves as a vehicle for commerce and social networking. The goal of this chapter is to study the interaction between Internet technology, economics, and users. Technology and economics are equally responsible for the evolution of the Internet, which in turn satisfy the demand of the users. It is quite challenging to do a top-down and comprehensive study of Internet economics. We only offer several case studies. It is hoped that these case studies provide different tools to study and examine different aspects of this important subject.

Elements of game theory are considered in a different chapter. Basics of applied analysis, matrices and determinants, probability theory, and optimization theory are required to study this chapter. The terms capacity, flow, and bandwidth are used interchangeably in this chapter. The reader should note that the subject of this chapter is Internet economics, and not Internet economy. The latter addresses concerns related to commerce on the Internet.

10.2 Scope of Economics

Economics depends upon technology, psychology, and perhaps metaphysics. It is often stated that the goal of economic activities is the satisfaction of consumers. Before the subject of Internet economics is studied, it is important to understand the scope of this fascinating subject.

An attempt is made in this section to define economics in set-theoretic language. An economic entity (or element) can be a human talent, information, a service, a manufactured good, or a nonliving item. This definition of an economic entity is decidedly vague and general. *Events* are defined as collection of such elements. The set of all such events which are important to consumers and businesses is denoted by $\mathcal{U}_e$. The specification of the *universal* set $\mathcal{U}_e$ in general is nontrivial. An event $\mathcal{E}_e$ is a subset of $\mathcal{U}_e$.

A *relation* $\mathcal{R}_e$ is also defined on the set $\mathcal{U}_e$. It is hoped that under this relation, the elements of $\mathcal{U}_e$ can be compared. But sometimes it may not be possible to compare all the elements of $\mathcal{U}_e$ under $\mathcal{R}_e$. It is assumed that this relation is reflexive. Recall that the relation $\mathcal{R}_e$ is reflexive if $u\mathcal{R}_e u$ for all $u \in \mathcal{U}_e$. It is quite possible that this relation is not transitive. Recall again that if $u, v, w \in \mathcal{U}_e$, the relation $\mathcal{R}_e$ is transitive if $u\mathcal{R}_e v$, and $v\mathcal{R}_e w$ implies $u\mathcal{R}_e w$.

Let $\mathcal{V}_e$ be a subset of $\mathcal{U}_e$, and $u, v \in \mathcal{V}_e$. If it is possible to compare u and v under the relation $\mathcal{R}_e$, then let $u\mathcal{R}_e v$ imply that u is better than v.

Also for any subset $\mathcal{E}_e \subseteq \mathcal{U}_e$; element $u \in \mathcal{E}_e$ is said to be *optimal* in $\mathcal{E}_e$ if and only if $u\mathcal{R}_e v$ for all $v \in \mathcal{E}_e \backslash \{u\}$. That is, an event better than u does not exist in the set $\mathcal{E}_e$.

Economics also consists of determination of the set of events $\mathcal{E}_{opt} \subseteq \mathcal{E}_e$ obtained via optimization under the relation $\mathcal{R}_e$. The constraints might be either physical or pecuniary. In summary, economics consists of:

(a) Set of elementary economic entities, denoted by $\mathcal{E}_e$.
(b) The universal set of all events $\mathcal{E}_e$, denoted by $\mathcal{U}_e$, where $\mathcal{E}_e \subseteq \mathcal{U}_e$.
(c) The reflexive relation $\mathcal{R}_e$ on $\mathcal{U}_e$.
(d) Optimization of $\mathcal{R}_e$ on $\mathcal{E}_e$ in order to determine $\mathcal{E}_{opt}$.

Optimization of $\mathcal{R}_e$ on $\mathcal{E}_e$ has two steps. In the first step, it has to determine the existence of $\mathcal{E}_{opt}$. In the next step, it has to determine an efficient procedure to actually achieve $\mathcal{E}_{opt}$. Application of this paradigm of economics are considered in the rest of this chapter.

10.3 Internet Technology, Costs, and Pricing

In order to understand the economics of the Internet, it is fruitful to study the structure of the Internet. The Internet is a conglomeration of interconnected globally-spread computer networks. Physically, the Internet is made up of routers, computers, and communication lines. The predominant protocol used for communication is TCP/IP (Transmission Control Protocol/Internet Protocol).

The Internet was initially designed to carry data. It uses packet-switching technology. This is in contrast to circuit-switching, the technology used in telephone networks. In the circuit-switching technology, an end-to-end circuit is initially set up before the telephone call commences. This scheme is convenient, both for real-time application like voice conversations and the accounting of network utilization. Therefore the circuit-switching technology was immensely successful. Circuit-switching technology is economical, if the communication lines are cheap.

The Internet uses packets to transmit information. A packet is made up of finite number of bits. The data to be transmitted between a source-destination pair is split into several packets. Each packet contains a "header" which contains information about the address of the destination. This enables the packets between a specific source-destination pair to traverse different paths to reach their destination. Therefore, unlike circuit-switching technology, an initial path-setup to transmit data is generally not required. Consequently the packet-switching technology is also called *connectionless*.

The transmission of packets is facilitated by computers called routers. A router is an internetworking device that is connected to two or more networks. Its goal is to forward packets from one network to another. The primary purpose of using packet-switching technology is two-fold. These are statistical multiplexing, and reliability. This technology permits statistical multiplexing of data on the communication lines. This means that packets from several different sources can share a transmission line. This results in its efficient use. Its second major advantage is reliability of data transmission. In circuit-switching technology, if the initial set-up path is broken, then the telephone call is lost. In contrast, in packet-switching technology if a transmission line goes down, then the packets can travel on a different path (or paths) in the network. The packet-switching technology is economical, when routers are cheap relative to the transmission lines.

In a specific geographical area (like a country), the Internet might be considered to have three levels. At the lowest layer are the Local Area Networks (LANs). These LANs are connected to a regional network. The regional networks in turn are connected to backbone networks, which are Wide Area Networks (WANs). The backbone networks carry bulk of the traffic. A backbone network is formed by connecting high-speed hubs via high-speed communication channels. It is quite possible that a country might have more than a single backbone network. The backbone networks which belong to a specific country are connected to backbone networks of their own country and also of other countries. This outlined hierarchy is generally true with minor exceptions.

A majority of individual businesses, and computer owners, called the end-customers, are connected to the Internet via commercial Internet service providers (ISPs). These are the local ISPs. A typical user connects to the local ISP via a cable modem, or residential digital subscriber lines. These local ISPs are connected to the regional ISPs, which in turn provide connection to the backbone network via Network Access Points (NAPs). The ISPs use the Local Area Network and Metropolitan Area Network technologies.

As the cost of setting up a network increases, there has been an economic incentive to develop an integrated-service network. This network is also called the *information superhighway*. The goal of the information superhighway is to merge voice, data, images, and video in a single streamlined cost-efficient network. The construction of this superhighway is definitely evolutionary.

Costs and Pricing

The major cost of the backbone network depends upon communication lines and routers. The cost of providing service on the Internet is approximately independent of the usage of the network.

Thus most of the costs are fixed. Further, the cost of sending additional packets is negligible if the network is not congested.

There are three main components of network costs to the end user. These are the cost of connecting to the network, the cost associated with the use of incremental capacity, and the cost of network congestion. Congestion of the network is synonymous with excessive traffic in the network, which in turn causes degradation of service.

We briefly outline three pricing schemes for Internet access. These are: connection pricing, committed information rate pricing, and cost based pricing scheme.

Connection pricing: Connection pricing is a standard and popular method for charging, which requires the users to pay a fixed cost for a specific period of time (like a month, or a year) for the use of a fixed bandwidth. This permits unlimited use of the connection up to the maximum permissible flow rate.

Committed information rate pricing: In this scheme, the user is charged in two parts. The first part of the charge is based upon the maximum permissible bandwidth allowed to the customer. The second part of the fee is based upon the maximum guaranteed flow to the user. The network provider ensures that sufficient capacity is available in the network so that all guaranteed flow rates are indeed guaranteed. If the net capacity is unused and available in the network, then the excess capacity is made available to users on a first-come, first-serve basis. Additionally, all the user connections are controlled by flow regulators.

Cost based pricing: As per standard economic theory, the pricing scheme should reflect the incurred cost to offer connection. This charge should include: the fixed cost to provide connection, cost of the used capacity, and a cost when the network is congested.

It has been suggested that pricing schemes be used for controlling congestion in the Internet, and also provide Quality of Service (QoS) to the end user. Quality of Service is defined as the service with which a user is satisfied. It is possible to make this definition of QoS both quantitative and measurable. In the final analysis, Internet economics is concerned with the interaction between the service providers, the network, and the customers. Based upon this interaction, a qualitative economic model of the Internet is provided in the next section.

10.4 A Qualitative Economic Model of the Internet

A brief qualitative economic model of the Internet is described. The primary constituents of the economic model of the Internet are:

(a) Providers
(b) Users
(c) Network.

Providers: The users of the network provide revenue to the providers. The providers invest into the network, and also set the pricing scheme for the use of the network.

Users: For set of prices specified by the providers, and the Quality of Service provided by the network, the users produce revenue for the providers for specific demands upon the network.

Network: The network is basically driven by the demand placed upon it by the users, and the investments made by the providers. The network provides specified Quality of Service.

The providers, users, and network provide the functionality of the complete system. These are driven by revenue, prices, QoS, demand, and investments. This is illustrated in Figure 10.1.

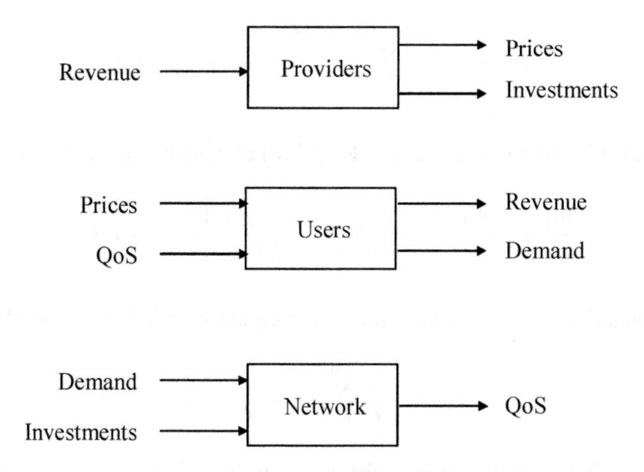

Figure 10.1. Providers, users, and network; and revenue, prices, QoS, demand, and investments.

10.5 Network Resources, Congestion, and Pricing

The interplay between allocation of network resources, congestion, and pricing is studied in this section. Overuse of network resources results in congestion, which in turn might cause degradation of services. Economists suggest that pricing be used as a mechanism to alleviate congestion. It is also possible that pricing might regulate congestion, and simultaneously generate revenue for capacity expansion of the network.

We introduce notation, before the discussion of models to study the interaction between allocation of network resources, congestion, and pricing. Let the number of users who use the network resources be $n \in \mathbb{P}$. Also let the user i's use of the network resources be $x_i \in \mathbb{R}_0^+$, where $1 \leq i \leq n$. The vector of these usages is denoted by $x = (x_1, x_2, \ldots, x_n)$. Thus the total use of network resource is $X = \sum_{i=1}^{n} x_i$. The capacity of the network is $C \in \mathbb{R}^+$, and the utilization of the network resources is $Y = X/C$.

The utility function of user i is $u_i(x_i, Y)$, where $1 \leq i \leq n$. Note that $u_i(\cdot, \cdot)$ is differentiable, concave function of x_i, and a decreasing concave function of Y. The cost of providing capacity C to the network is $C_{NW}(C)$.

This notation is sufficient enough to capture important features of a network. For example x_i can represent the number of bytes received or sent to a router by user i. The parameter X would be the net number of bytes received and transmitted by all users. The parameter C can represent the capacity of a router. The impact of capacity upon social welfare, pricing in a competitive market, and pricing without usage-based fees are analysed in the rest of this section.

10.5.1 Social Welfare and Capacity

The social welfare of the network service providers and the users of the network, are defined as the sum of utilities of all the users, minus the costs. Thus the social welfare $W(C, x)$ for a given network capacity C, and usage vector x is

$$W(C,x)=\sum_{j=1}^{n}u_j(x_j,Y)-\mathcal{C}_{NW}(C)$$

The socially optimal solution must maximize $W(C,x)$ by varying x_i, $1\le i\le n$. This occurs if

$$\frac{\partial u_i(x_i,Y)}{\partial x_i}=-\frac{1}{C}\sum_{j=1}^{n}\frac{\partial u_j(x_j,Y)}{\partial Y},\quad 1\le i\le n$$

Note that the right-hand side expression in the above equation is independent of i. Denote it by p_e. Thus

$$p_e=-\frac{1}{C}\sum_{j=1}^{n}\frac{\partial u_j(x_j,Y)}{\partial Y}$$

Assume that the user i is charged a price p_e for usage. In this case, the user would like to maximize $\{u_i(x_i,Y)-p_e x_i\}$ by varying x_i. This occurs if

$$\frac{\partial u_i(x_i,Y)}{\partial x_i}+\frac{1}{C}\frac{\partial u_i(x_i,Y)}{\partial Y}=p_e$$

For large values of n, the second term on the left-hand side of the above equation is much smaller in magnitude than p_e. Therefore, the social welfare is *also* maximized, provided each user maximizes its value. The price p_e is called the *shadow price* by economists, because it reflects the congestion costs that each user imposes upon other users. Note that this price is the same for all users.

We next examine the effect of network capacity expansion upon social welfare. Differentiating $W(C,x)$ with respect to C yields

$$\begin{aligned}\frac{\partial W(C,x)}{\partial C}&=-\frac{X}{C^2}\sum_{j=1}^{n}\frac{\partial u_j(x_j,Y)}{\partial Y}-\mathcal{C}'_{NW}(C)\\&=\frac{1}{C}\{p_e X-C\mathcal{C}'_{NW}(C)\}\end{aligned}$$

where $d\mathcal{C}_{NW}(C)/dC\triangleq\mathcal{C}'_{NW}(C)$. Therefore the social welfare increases with capacity, provided $\partial W(C,x)/\partial C>0$. This occurs, if $\{p_e X-C\mathcal{C}'_{NW}(C)\}>0$. That is, the revenue from usage $p_e X$ should exceed $C\mathcal{C}'_{NW}(C)$. Thus the shadow price p_e is also responsible for determining the value of change in the network capacity.

10.5.2 Competitive Market Pricing

In a competitive market, the network service provider would like to maximize its profit. It is assumed in this subsection that there are several independent providers of service. If Y is the utilization of the network, and x_i is the use of network resources by the user i, then the user pays a fee of $\{p(Y)x_i+q(Y)\}$. The component $q(Y)$ is the connection or attachment or subscription fee, and the fee $p(Y)x_i$ is usage dependent. Therefore, the profit of the network provider with n users is

$$\{p(Y)X+nq(Y)-\mathcal{C}_{NW}(C)\}$$

Assume that $p(\cdot)$, and $q(\cdot)$ are continuous and differentiable functions of Y. The derivatives of $p(Y)$ and $q(Y)$ with respect to Y are denoted by $p'(Y)$ and $q'(Y)$ respectively.

We consider optimization from the perspective of both the user (consumer), and network service provider. The condition for the network service provider to enter the market is also determined. Further, the effect of adding capacity is also evaluated.

User Optimization

The goal of user i is to maximize

$$\{u_i(x_i, Y) - p(Y)\, x_i - q(Y)\}$$

by varying x_i and Y. This occurs if

$$\frac{\partial u_i(x_i, Y)}{\partial x_i} - p(Y) = 0$$

$$\frac{\partial u_i(x_i, Y)}{\partial Y} - p'(Y)\, x_i - q'(Y) = 0$$

Summing the last equation over all the n users yields

$$p'(Y)\, X + nq'(Y) = \sum_{j=1}^{n} \frac{\partial u_j(x_j, Y)}{\partial Y}$$

Network Service Provider Optimization

The goal of the network service provider is to maximize

$$\{p(Y)\, X + nq(Y) - \mathcal{C}_{NW}(C)\}$$

by varying X and C. This occurs if

$$p(Y) + \frac{1}{C}\{p'(Y)\, X + nq'(Y)\} = 0$$

$$-\frac{X}{C^2}\{p'(Y)\, X + nq'(Y)\} = \mathcal{C}'_{NW}(C)$$

If the results derived for user optimization are used, then

$$p(Y) = \frac{\partial u_i(x_i, Y)}{\partial x_i} = -\frac{1}{C}\sum_{j=1}^{n} \frac{\partial u_j(x_j, Y)}{\partial Y}$$

$$-Y\sum_{j=1}^{n} \frac{\partial u_j(x_j, Y)}{\partial Y} = C\mathcal{C}'_{NW}(C)$$

Observe that not surprisingly the above equations also satisfy the condition for social optimality. Thus

$$\frac{X}{C}p(Y) = \mathcal{C}'_{NW}(C)$$

The above equation is similar to the condition obtained while increasing capacity and maximizing the social welfare. Note that in this model, the network service provider is forced to charge the socially optimal price for the type of service it offers.

Conditions for the Network Service Provider to Enter the Market

A network service provider with n users enters the market if the profit is positive. That is,

$$\{p(Y)X + nq(Y) - C_{NW}(C)\} > 0$$

This occurs if

$$\frac{nq(Y)}{C_{NW}(C)} > 1 - \frac{C'_{NW}(C)}{C_{NW}(C)/C}$$

Adding Capacity

By adding capacity ΔC to the network, two cases might occur.

Case 1: The network service provider adds capacity, and keeps the net usage X constant. Thus, the network will offer improved quality of service due to reduced delay. In this case, the network service provider can charge extra.

Case 2: The network service provider adds capacity, and keeps the prices fixed. The network service provider might attract more traffic from the customers, and hence generate more profit.

The increase in profits is evaluated in the two cases.

Case 1: In this case X is held constant, and the users are charged extra. The network service provider can charge user j extra amount equal to:

$$\{q'(Y) + p'(Y)x_j\}\frac{dY}{dC}\Delta C$$

Use of the result $\partial u_j(x_j, Y)/\partial Y = \{p'(Y)x_j + q'(Y)\}$ from the user optimization condition results in the above extra amount to be equal to:

$$-\frac{X}{C^2}\frac{\partial u_j(x_j, Y)}{\partial Y}\Delta C$$

Summing this result over all n users, and use of the expression for $p(Y)$ from the optimization of network service provider yields

$$p(Y)\frac{X}{C}\Delta C$$

The cost of the network expansion, is $C'_{NW}(C)\Delta C$. Therefore, adding capacity increases revenue, if

$$\left\{p(Y)\frac{X}{C} - C'_{NW}(C)\right\}\Delta C > 0$$

This result is similar to that obtained for the social welfare to increase when the capacity of the network is increased. In summary, the network service provider's profit will increase, if the social benefits increase.

Case 2: In this case Y is held constant, but both X and C increase. The relationship $Y = X/C$ yields $\Delta X = Y\Delta C$. Therefore, the increase in profit is

$$\{p(Y)\Delta X - C'_{NW}(C)\Delta C\} = \{p(Y)Y - C'_{NW}(C)\}\Delta C$$

Observe again that the network service provider's profit will increase if the social benefits increase.

10.5.3 Pricing without Usage Fees

The model with usage-based pricing is inconvenient to implement. It requires the network to track usage, prepare bills, and collect the corresponding fees. Therefore, the model in which only the attachment/subscription fees are present is considered. Assume that

$$u_i(x_i, Y) = v_i(x_i) - \mathcal{D}(Y), \quad 1 \le i \le n$$

where $\mathcal{D}(Y)$ is the "delay cost" due to congestion in the network. Assume that the delay function $\mathcal{D}(\cdot)$ is increasing, differentiable, and convex. This implies that the delay function increases with the utilization. Denote the derivative of $v_i(x_i)$ with respect to x_i by $v_i'(x_i)$, and derivative of $\mathcal{D}(Y)$ with respect to Y by $\mathcal{D}'(Y)$. As the user is charged only for access, assume that the user i is satiated at x_i^a. Consequently, the net usage of the network capacity is $X^a = \sum_{j=1}^{n} x_j^a$. The goal of user i is to maximize

$$\{v_i(x_i^a) - q(Y) - \mathcal{D}(Y)\}$$

by varying Y. This occurs if

$$-\mathcal{D}'(Y) = q'(Y)$$

Similarly, the goal of the network service provider is to maximize

$$\{nq(Y) - \mathcal{C}_{NW}(C)\}$$

by varying C. This occurs if

$$-nYq'(Y) = C\mathcal{C}_{NW}'(C)$$

If the last two results are combined, we obtain

$$nY\mathcal{D}'(Y) = C\mathcal{C}_{NW}'(C)$$

Observe that the amount of increase in capacity is determined by the willingness of the users to pay for decrease in delay. Therefore, at $Y^a = X^a/C$, the above condition is $nY^a\mathcal{D}'(Y^a) = C\mathcal{C}_{NW}'(C)$.

10.6 Service Differentiation

The impact of pricing upon congestion in the network is explored in this section. A simple service differentiation scheme in which the network is partitioned into several logically different parts is considered. Each partition provides service to the customers' packets on a best effort basis. However, it would not provide any guarantee of Quality of Service. The only difference in each partition would be the pricing required to service the customer-requests. It is presumed that partitions with higher prices would be less utilized, and consequently provide better Quality of Service to its customers. Thus pricing is used in this scheme to manage traffic.

This scheme was suggested by Andrew Odlyzko. It is in analogy with the pricing scheme used in the Paris metro system. Several years ago, the Paris metro system divided its cars into two categories: first and second class cars. The cars were identical in numbers and seats. The only difference was that the first class tickets were more expensive than the second class tickets. This resulted in less congestion in the first class cars, and presumably provided better service to its rich clientele. It is assumed in this context that the low utilization of cars provides higher Quality of Service, albeit at a higher price.

Therefore if a network segment was less utilized than others, it was expected that it provide higher Quality of Service at a relatively higher price. This approach is called the Paris metro pricing (PMP) scheme. The Paris metro pricing scheme is remarkable for its simplicity. It is illustrated via the following examples.

Examples 10.1. In the first example, only a single network is considered. In the second example, the network is partitioned into two classes, and a PMP-like scheme is adapted.

1. Consider a network used by n users. This network can sustain $2n$ users. That is, its capacity is $2n$. We consider the number $2n$, because this network is divided into two equal parts in the next example. For simplicity in analysis, it is assumed that n tends towards infinity. Let H be the mean number users who use the network. If h is defined to be equal to H/n, then the utilization of the network $\rho = H/(2n) = h/2$. The value of ρ determines the quality of connection. Low value of ρ implies better Quality of Service, and high value implies lower Quality of Service. Also let the price paid by each user be p units, where $p \in (0,1)$.
   Each user is characterized by a type θ, which is an independent random variable uniformly distributed in the interval $[0,1]$. A user which is of type θ, finds the network connection acceptable if
   $$p \leq \theta, \quad \text{and} \quad \rho \leq (1-\theta)$$
   The above relationships imply that a user with high θ (larger than the price p), and sufficiently low utilization ρ will accept the connection. However if θ is small, the user does not pay too much for the connection, but should expect a highly utilized network, which implies lower Quality of Service.
   Thus $\theta \in [p,(1-\rho)]$. Consequently $(1-\rho-p)^+$ is the probability with which a random user accepts connection to the network. Therefore, at equilibrium the mean number of users in the network is equal to $H \simeq n(1-\rho-p)^+$. For large n, this yields
   $$h = \frac{2}{3}(1-p)$$
   The revenue generated by the network is pH, and the revenue generated per user $R(p) = ph$. Therefore $R(p) = 2p(1-p)/3$. The value of the revenue $R(p)$ is maximized at $p = 1/2$. The maximized value of revenue per user is equal to $1/6$, and the corresponding value of network utilization is equal to $\rho = 1/6$.
2. We consider a Paris metro pricing scheme in this example. The network used in the last example is divided into two parts, each of capacity n. That is, each network can serve n users. The two networks service a total of n users. As in the last example, assume that n tends towards infinity. Let H_1 and H_2 be the mean number users who use networks 1 and 2 respectively. Therefore the utilization of the two networks are $\rho_1 = H_1/n$, and $\rho_2 = H_2/n$ respectively.
   Also let the price paid by each user be p units, where $p \in (0,1)$. Each user is characterized by a type θ, which is an independent random variable uniformly distributed in the interval $[0,1]$. A user which is of type θ, finds the network connection acceptable if
   $$p \leq \theta, \quad \text{and} \quad \rho \leq (1-\theta)$$

A user can either accept any of the two networks, or not accept any network at all. If both networks are acceptable, the user joins the cheaper network. However, if the prices are equal, the user joins the network with lower utilization, because it can offer a better Quality of Service. Also let the price paid by each user be p_1 and p_2 units for accepting connection to the networks 1 and 2 respectively, where $p_1, p_2 \in (0,1)$. The revenue of the operator is $(p_1 H_1 + p_2 H_2)$. Thus the revenue of the operator per user is equal to $R(p_1, p_2) = (p_1 \rho_1 + p_2 \rho_2)$. The values of p_1 and p_2 which maximize the revenue $R(p_1, p_2)$ per user are determined. Two cases are considered. These are $p_2 < p_1$, and $p_1 = p_2$.
Case 1: $p_2 < p_1$. A user of type θ selects network 2 if $p_2 \le \theta$, and $\rho_2 \le (1-\theta)$. The probability that $\theta \in [p_2, (1-\rho_2)]$ is $(1-\rho_2-p_2)^+$. Therefore, at equilibrium $H_2 \simeq n(1-\rho_2-p_2)^+$. For large n, this yields

$$\rho_2 = \frac{1}{2}(1-p_2)$$

A user of type θ selects network 1 if $p_1 \le \theta$, $\rho_1 \le (1-\theta)$, and $\rho_2 > (1-\theta)$. This occurs with probability $(1-\rho_1-\max\{p_1,(1-\rho_2)\})^+$. Thus

$$H_1 = n(1-\rho_1-\max\{p_1,(1-\rho_2)\})^+$$

and

$$\rho_1 = (1-\rho_1-\max\{p_1,(1-\rho_2)\})^+$$

Substituting the expression for ρ_2 in the above expression yields

$$\rho_1 = \min\left\{\frac{(1-p_1)}{2}, \frac{(1-p_2)}{4}\right\}$$

The revenue of the operator per user is equal to

$$R(p_1,p_2) = p_1 \min\left\{\frac{(1-p_1)}{2}, \frac{(1-p_2)}{4}\right\} + \frac{1}{2}p_2(1-p_2), \quad \text{where } p_2 < p_1$$

Case 2: $p_2 = p_1$. A user of type θ with $p_1 \le \theta$, $\rho_1 \le (1-\theta)$, and $\rho_2 \le (1-\theta)$ selects the network with smaller utilization. In this case $\rho_1 = \rho_2 = \rho$, and half the users with $\theta \in [p_1, (1-\rho_1)]$ join network 1. Therefore, at equilibrium $H_1 \simeq n(1-\rho_1-p_1)^+/2$. For large n, this yields

$$\rho_1 = \rho_2 = \frac{1}{3}(1-p_1)$$

The revenue of the operator per user is equal to

$$R(p_1,p_2) = \frac{2}{3}p_1(1-p_1)$$

The maximizing value of the revenue $R(p_1,p_2)$ occurs at $p_1 = 7/10$, and $p_2 = 4/10$, and the corresponding (maximum) value of the revenue is equal to $9/40$. The utilization of the networks 1 and 2 are $\rho_1 = 3/20$, and $\rho_2 = 3/10$ respectively. Revenues of the operator per user from networks 1 and 2 are $21/200$ and $3/25$ respectively. Observe that the higher priced network 1 has a lower utilization than the lower-priced and higher-utilized network 2.
Also note that the two combined networks in example 2, which adopt the PMP scheme, generate higher net revenue per user than in example 1. □

10.7 Time-Dependent Pricing and Bandwidth Trade-Off

The utilization of communication network resources like processing, and transmission vary significantly over different times of the day. This is because users of the network have their own preference to access the network at specific times. As users are consumers of capacity (bandwidth) of the network, it is possible to alleviate congestion in the network by offering different pricing at different times of the day, or over some other different time interval. Recall that congestion occurs because of excessive traffic in the network. A consequence of congestion is degradation in Quality of Service. If higher prices are offered by the network during busy periods of the day, then it is hoped that congestion in the network is relieved by users switching to times of the day with less expensive pricing. In addition, by charging higher prices for certain periods of the time of the day, the revenue collected by the service provider can also increase.

The goal of the social planner is to deploy a pricing scheme which would reduce congestion in the network, and consequently achieve efficient utilization of the network resources. In this case, the social planner is said to maximize social welfare. However, the goal of the service provider is to maximize revenue. It is also possible for the service provider to institute time-preference and increase revenue. The time-dependent pricing and bandwidth trade-off are modeled as a noncooperative game. This scheme is modeled as a selfish routing problem with user preferences. Selfish routing can be quantified in a single measure, called the price of anarchy.

10.7.1 Basic Model

Let M be equal to the number of equal time slots into which the total time interval is divided. The length of the time interval is a single day in the time-of-the day pricing. Assume that each user picks only a single time slot to access the network, and has a fixed amount of traffic. Let K be equal to the number of classes of users. Each class of users has a profile of time-preferences.

Let d_k^i be equal to the number of class-k users using slot i, where $1 \le k \le K$, and $1 \le i \le M$. Also let the number of class-k users be bounded by d_k, where $1 \le k \le K$. Further assume that the utility function of class-k users using slot i is u_k^i, where $1 \le k \le K$, and $1 \le i \le M$. As the users of slot i encounter congestion delay, the users of this slot experience a decease in their net payoff. Let $l_i(\cdot)$ be the congestion delay function for slot i. It is a function of x_i, which is equal to the total number of users using slot i. Therefore $x_i = \sum_{k=1}^{K} d_k^i$. It is assumed that this delay function $l_i(\cdot)$ is increasing, and convex function of x_i. If v_k^i is the payoff of class-k users using slot i, then

$$v_k^i = u_k^i - l_i(x_i); \quad 1 \le k \le K, \ 1 \le i \le M$$

It is possible that for some users, the payoff v_k^i is negative due to low utility value and large congestion. If a user elects not to use the network, then its payoff is equal to zero. Assume that such users elect not to use the network.

Further assume that the service provider, or the social planner charges a price p_k^i for a class-k user who uses time slot i, where $1 \le k \le K$, and $1 \le i \le M$. Therefore the net payoff of a class-k user who uses time slot i is f_k^i, where

$$f_k^i = u_k^i - l_i(x_i) - p_k^i; \quad 1 \le k \le K, \ 1 \le i \le M$$

In this noncooperative game, each user i of class k tries to maximize its payoff f_k^i. In order to specify the underlying game properly, the following notation is introduced.

$$\begin{aligned} u_k &\triangleq \left(u_k^1, u_k^2, \ldots, u_k^M\right), \quad 1 \le k \le K \\ p_k &\triangleq \left(p_k^1, p_k^2, \ldots, p_k^M\right), \quad 1 \le k \le K \\ f_k &\triangleq \left(f_k^1, f_k^2, \ldots, f_k^M\right), \quad 1 \le k \le K \end{aligned}$$

$$\begin{aligned} U &\triangleq (u_1, u_2, \ldots, u_K), \; P \triangleq (p_1, p_2, \ldots, p_K) \\ D &\triangleq (d_1, d_2, \ldots, d_K), \; F \triangleq (f_1, f_2, \ldots, f_K) \end{aligned}$$

$$L \triangleq (l_1, l_2, \ldots, l_M)$$

where U is the utility profile, P is the price profile, D is the allocation vector, F is the payoff vector, and L is the delay-function profile.

Definition 10.1. *The time-dependent pricing and bandwidth trade-off game is defined by* (I, U, P, D, L), *where* I *is the set of users,* U *is the utility profile,* P *is the price profile,* D *is the allocation vector, and* L *is the delay-function profile.* □

Note that the number of users accessing the network is equal to $|I|$, where

$$|I| = \sum_{i=1}^{M} x_i \le \sum_{k=1}^{K} d_k$$

10.7.2 Nash Equilibrium of the Game

We are ready to specify the Nash equilibrium of the noncooperative game (I, U, P, D, L). For a specified price profile P, there exist a pure strategy Nash equilibrium. It is actually obtained by solving the following optimization problem.

$$\max_{D} \sum_{i=1}^{M} \sum_{k=1}^{K} \left(u_k^i - p_k^i\right) d_k^i - \sum_{i=1}^{M} \int_0^{x_i} l_i\left(y\right) dy$$

where $x_i = \sum_{k=1}^{K} d_k^i$. The inequality constraints are:

$$\begin{aligned} \sum_{i=1}^{M} d_k^i &\le d_k, \quad 1 \le k \le K \\ d_k^i &\ge 0, \quad 1 \le k \le K, \; 1 \le i \le M \end{aligned}$$

Use of the method of Lagrange multipliers, and the Karush-Kuhn-Tucker (KKT) conditions imply that for each k, there exists a λ_k, such that

$$\begin{aligned} u_k^i - l_i\left(x_i\right) - p_k^i &= \lambda_k, \quad \forall\, d_k^i > 0 \\ u_k^i - l_i\left(x_i\right) - p_k^i &\le \lambda_k, \quad \forall\, d_k^i = 0 \end{aligned}$$

where $1 \le k \le K$. These are precisely the conditions for the existence of a Nash equilibrium.

10.7.3 Social Welfare Function

The social welfare function of the noncooperative game (I, U, P, D, L) is defined as the sum of the payoffs of the users and the service provider (or the social planner). The social welfare function $W(\cdot)$ for a given allocation vector D is

$$W(D) = \sum_{i=1}^{M}\sum_{k=1}^{K} f_k^i d_k^i + \sum_{i=1}^{M}\sum_{k=1}^{K} p_k^i d_k^i$$
$$= \sum_{i=1}^{M}\sum_{k=1}^{K} \{u_k^i - l_i(x_i)\} d_k^i$$

where $\sum_{i=1}^{M}\sum_{k=1}^{K} p_k^i d_k^i$ is the revenue collected by the service provider. Therefore a high value of $W(D)$ implies that the users are being rewarded for their utilities, and that the values of congestion are lower.

Let D^* be the allocation vector which maximizes $W(D)$. It yields the *social optimum*. Also let D^N be the allocation vector at Nash equilibrium. Therefore the price of anarchy of the game is measured by its coordination ratio $\mathcal{R}$. The coordination ratio $\mathcal{R}$ is defined as the ratio of $W(D^N)$ and $W(D^*)$.

Example 10.2. Consider a time-dependent pricing and bandwidth trade-off game with pricing vector $P = 0$. Let there be a single class, with population equal to unity, and a single time slot. That is, $K = 1, M = 1, d_1 = d_1^1 = 1$. Also the utility $u_1^1 = 1$. Assume that the delay function $l_1(\cdot)$ is continuous, strictly increasing, and is $l_1(x) = x^r$, where $r \geq 1$.

At Nash equilibrium, the users will use the time slot. In this case, the payoff of each user is equal to $f_1^1 = (u_1^1 - l(d_1^1)) = 0$, which is identical to the payoff when the time slot is not used. Consequently, the value of the social welfare function at Nash equilibrium $W(D^N)$ is equal to zero.

Note that $W(x) = x(1 - x^r)$, and the maximizing value of $W(x)$ occurs at $x^* = (1 + r)^{-1/r}$. Therefore the corresponding value of the social optimum $W(x^*)$ is greater than zero. Thus the price of anarchy, or the coordination ratio $\mathcal{R} = 0$. □

Maximization of Social Welfare Function

The social welfare function $W(D)$ is next maximized with proper constraints. The goal is to maximize the objective function $W(\cdot)$, where:

$$W(D) = \sum_{i=1}^{M}\sum_{k=1}^{K} \{u_k^i - l_i(x_i)\} d_k^i$$

and $x_i = \sum_{k=1}^{K} d_k^i$. The inequality constraints are:

$$\sum_{i=1}^{M} d_k^i \leq d_k, \quad 1 \leq k \leq K$$
$$d_k^i \geq 0, \quad 1 \leq k \leq K, \; 1 \leq i \leq M$$

Use of the method of Lagrange multipliers, and the Karush-Kuhn-Tucker (KKT) conditions imply that for each k, there exists a β_k, such that

$$u_k^i - l_i(x_i) - x_i l_i'(x_i) = \beta_k, \quad \forall\, d_k^i > 0$$
$$u_k^i - l_i(x_i) - x_i l_i'(x_i) \le \beta_k, \quad \forall\, d_k^i = 0$$

where, $l_i'(x_i)$ is the first derivative of $l_i(x_i)$ with respect to x_i, and $1 \le i \le M$, $1 \le k \le K$. Next set the prices as per the following relationships.

$$p_k^i = x_i l_i'(x_i), \quad 1 \le k \le K, \ 1 \le i \le M$$

That is, if the social planner sets up the prices as per the above relationships, then the Nash equilibrium is reached. Observe that the prices are independent of the class k and the utility u_k^i, but depend only upon the total number of users using slot i, and the delay function $l_i(\cdot)$.

10.7.4 Revenue Maximization

We consider revenue maximization by the service provider. For simplicity, assume that the service provider has a complete knowledge of the users' classes and utilities u_k^i. If an allocation vector D is given, then for a class-k user in slot i, the service provider can charge a price up to $\{u_k^i - l_i(x_i)\}$. In this case, the total revenue generated by the service provider is the same as the social welfare. Thus, the revenue of the service provider is maximized, when the social optimum is achieved.

The next section is concerned with the use of auction mechanisms for allocation of resources.

10.8 Auctions

An auction is the sale of an item in which its price is determined by bidding. Auction often means the sale of an item to the highest bidder. However different types of auctions, with more refined interpretation are examined in this chapter. Auctions can be used for resource allocation, like bandwidth, in a network. More precisely, an auction is a mechanism (scheme) based upon two rules: these are the *allocation* and the *payment* rules. The allocation rule determines the items which are allocated to whom, and the payment rules determine the corresponding charges for the items.

The participant or user of an auction is called a *bidder*, and the entity conducting (facilitating) the auction is called the *auctioneer*. A *bid* is the amount of money offered by the bidder for the use of the item. Note that a bidder is also termed a player, buyer, agent, or user of the auction.

Auctions can be for either a single unit, or several units. The auction of a single unit is simply called *simple*, or *single-unit*. The auction of several units is called *multi-unit*. For example, auction of several integral units of link's bandwidth is multi-unit.

Bids for an auction can be either made in public, or be submitted in sealed envelopes. Such auctions are called *open* or *sealed* respectively. Auctions which maximize social welfare are called *efficient*, and those which maximize the seller's revenue are called *optimal*. There are several types of auctions. Some of these are:

(a) *English auction*: This is an open (public) auction, in which the bids for the sale of an item(s) rise until a single bidder is left. The highest bidder receives the item, after paying for the corresponding bid.
(b) *Sealed bid, first price*: All the bids for the sale of an item are sealed, and made simultaneously. The highest sealed bidder receives the item after paying for the corresponding bid.

(c) *Sealed bid, second price*: All the bids for the sale of an item are sealed, and made simultaneously. The highest sealed bidder receives the item, after paying the second highest bid. This is also called the Vickrey auction.
(d) *Dutch auction*: The auctioneer publicly announces a high bid for an item. The bidders may accept, or not accept the bid. If no one accepts the bid, the price is lowered till an acceptable bidder is found. The highest bidder collects the item, and pays for the corresponding bid.

Vickrey auction, and generalized Vickrey auction are next examined.

10.8.1 Vickrey Auction

In the Vickrey auction, the highest sealed bid wins the auction for a single item, but pays the second highest bid. This mechanism encourages the bidder to offer a bid which is equal to his or her true valuation of the item. This auction is called the Vickrey auction, after the 1996 economics-Nobelist, William Vickrey (1914-1996). The Vickrey auction is a special case of the celebrated Vickrey-Groves-Clarke (VGC) mechanism. This later mechanism is discussed in the chapter on game theory and the Internet. Nevertheless, the basic characteristics of the Vickrey auction are discussed for pedagogical reason. This apparently intriguing and elegant scheme has interesting properties. The definitions of the Vickrey auction, utility function of a bidder, and the social welfare of the auction are given below.

Definition 10.2. *Vickrey auction.*

(a) *There are $n \in \mathbb{P} \backslash \{1\}$ bidders of a single item. The bidder i values the item at $v_i \in \mathbb{R}_0^+$, and bids $x_i \in \mathbb{R}_0^+$ for it, where $1 \leq i \leq n$. The bid x_i can be less than v_i, or equal to v_i, or greater than v_i. The winner of the item is the bidder who bids the highest. If the winner is the mth bidder, then his or her payment for the bid is $T(x_{-m}) = \max_{1 \leq j \leq n, j \neq m} x_j$, where $x_{-m} = (x_1, \ldots, x_{m-1}, x_{m+1}, \ldots, x_n)$ and $m \in \{1, 2, \ldots, n\}$.*
(b) *If $x_i = v_i$, then the strategy of bidding by bidder i is said to be truthful, where $1 \leq i \leq n$.*
(c) *Let $x = (x_1, x_2, \ldots, x_n)$, and the payoff or utility function of user i be $u_i(\cdot)$. This function takes nonnegative values. Also*

$$u_i(x) = \begin{cases} (v_i - T(x_{-i}))^+, & x_i > T(x_{-i}) \\ 0, & x_i \leq T(x_{-i}) \end{cases} \tag{10.1}$$

and $1 \leq i \leq n$. In the above expression, $a^+ \triangleq \max(0, a)$, $a \in \mathbb{R}$.
(d) *The social welfare of the bid is $W(x) = \sum_{i=1}^{n} u_i(x)$.* □

Observations 10.1. Properties of the Vickrey auction.

1. The best strategy of each bidder is to bid his or her true valuation, irrespective of the strategies of all other bidders. Under this strategy, its utility is maximized independent of bids of all other bidders. That is, for each user i the utility $u_i(x)$ is maximized if $x_i = v_i$, where $1 \leq i \leq n$. Therefore, if each bid is equal to its valuation, then the item is allocated to the bidder who values it most.
2. Vickrey's allocation scheme maximizes the social welfare, which is defined as the sum of utilities of all the bidders. This allocation scheme is both socially efficient and optimal, because the item goes to a single bidder who bids the highest, and his or her utility is maximized by bidding truthfully. Further, the contribution of the utility to the optimal social welfare of each

non-winning bid is zero. Therefore the strategy of truthful bidding is said to be the *dominant strategy.* □

The above assertions are established in the problem section. Vickrey auction is also said to be *incentive-compatible*, because this auction mechanism provides incentive to bid truthfully. Note that the Vickrey auction gives credence to the age-old adage: "Honesty is the best policy." That is, the best strategy of a bidder is to bid truthfully.

10.8.2 Generalized Vickrey Auction

A generalization of the Vickrey auction is considered in this subsection. Auction of multiple items is possible under the generalized Vickrey scheme. There are $n \in \mathbb{P} \backslash \{1\}$ bidders in this auction. These bidders participate in the auction of a set of items. This set of items is denoted by A. For each subset S of A, each bidder i has a private valuation $v_i(S) \in \mathbb{R}_0^+$, where $1 \leq i \leq n$. However, the bidder i bids each subset S of A at $b_i(S) \in \mathbb{R}_0^+$, where $1 \leq i \leq n$.

The auctioneer is a person or entity who oversees the auction. The auctioneer allocates disjoint subsets of the items to the bidders, so that the sum of the bids of these subsets is maximized. If items in the subset $A_i \subseteq A$ are allocated to the bidder i, then $\sum_{1 \leq i \leq n} b_i(A_i)$is maximized over all possible subsets of A, by the auctioneer. Note that $A_i \cap A_j = \varnothing$, $i \neq j$, $1 \leq i, j \leq n$; and $\cup_{1 \leq i \leq n} A_i = A$.

The bidder i pays a price p_i for the items in the subset $A_i \subseteq A$. As per the generalized Vickrey auction, price p_i is the reduction in declared valuations of the other bidders caused by the bidder i's participation in the auction. Therefore, if $B_j^i \subseteq A$ is the set of items which the bidder j would receive if the bidder i did not participate in the auction, where $j \neq i$ and $1 \leq j \leq n$, then

$$p_i = \sum_{\substack{j=1 \\ j \neq i}}^{n} b_j\left(B_j^i\right) - \sum_{\substack{j=1 \\ j \neq i}}^{n} b_j\left(A_j\right)$$

where $1 \leq i \leq n$. The price p_i is referred to as the externality of the bidder i upon the other bidders.

If each bidder bids his or her true valuations for the subsets of A, then this truthful strategy dominates all other strategies irrespective of strategies of all other bidders. Further, if all the bidders bid their subset of items truthfully, then it can be shown that the allocation of items to the bidders maximizes the social welfare of the auction. Thus the generalized Vickrey auction is incentive-compatible. This result is established in the problem section.

The scheme to determine the optimal allocation of say ϑ items to n bidders is computationally very expensive. It is actually an exponential function of ϑ. The use of generalized Vickrey auction is illustrated in the following examples.

Example 10.3. This example is concerned with the bidding for Quality of Service (QoS) in a network. The network offers K classes of services. Class k can accept η_k connections, and the bit rate offered by this class is $\varrho(k)$. Assume that $\varrho(1) > \varrho(2) > \cdots > \varrho(K) \geq 0$.

There are $n \in \mathbb{P} \backslash \{1\}$ users who bid for services of the network using the generalized Vickrey auction mechanism. The user i's valuation of bit rate $r \in \{\varrho(1), \varrho(2), \cdots, \varrho(K)\}$ is $v_i(r)$. Assume that $v_i(\cdot)$ is a strictly increasing function. However, the user i's declared bid for rate $\varrho(k_i)$ is $b_i(\varrho(k_i))$, where k_i is the corresponding class of service.

The auctioneer (network operator) of the QoS allocates service classes to users, so that the total sum of the declared bids is maximized. Denote this sum by Ψ, where

$$\Psi = \sum_{i=1}^{n} b_i\left(\varrho\left(k_i\right)\right)$$

However, if the user i were absent, then user $j \neq i$ would receive the service class k_j^i instead of k_j. If the auction is of the generalized Vickrey type, then the user i will pay a price p_i, where

$$p_i = \sum_{\substack{j=1\\ j\neq i}}^{n} b_j\left(\varrho\left(k_j^i\right)\right) - \sum_{\substack{j=1\\ j\neq i}}^{n} b_j\left(\varrho\left(k_j\right)\right)$$

For simplicity, assume that $v_i\left(r\right) = v_i f\left(r\right)$ and $b_i\left(r\right) = b_i f\left(r\right)$ for user i, where $1 \leq i \leq n$; and it is also assumed that $f\left(\cdot\right)$ is an increasing function. Further assume without loss of generality that $b_1 \geq b_2 \geq \cdots \geq b_n$. Then the following allocation maximizes Ψ. The auctioneer allocates the first class to η_1 users, the second class to the next η_2 users, and so on. □

A different bandwidth auction mechanism is demonstrated in the next example.

Example 10.4. A network operator has to allocate a link of capacity θ units among $n \in \mathbb{P}\backslash\left\{1\right\}$ users. The network operator would like to maximize the social welfare, which is the sum of utilities of all the n users. However, the network is not aware of the utilities of the users, and the users may declare incorrect utilities in order to obtain higher capacity allocation. This problem is addressed via a generalized Vickrey auction type of mechanism. The network operator addresses this problem in the following steps.

Step 0: The user i gets $x_i \in \mathbb{R}^+$ units of bandwidth, and his or her utility is $u_i\left(x_i\right)$. The utility function $u_i\left(\cdot\right)$ is increasing, strictly concave, and continuously differentiable function of bandwidth x_i, where $1 \leq i \leq n$. Let $u_i'\left(x_i\right)$ denote the derivative of $u_i\left(x_i\right)$ with respect to x_i, and $x = \left(x_1, x_2, \ldots, x_n\right)$. The goal of the network operator is to maximize the social welfare. This problem is:

$$\max_{x} \sum_{i=1}^{n} u_i\left(x_i\right)$$

The constraints are:

$$\sum_{i=1}^{n} x_i \leq \theta$$

$$x_i \in \mathbb{R}^+, \quad 1 \leq i \leq n$$

Use of Karush-Kuhn-Tucker conditions at $x^{**} = \left(x_1^{**}, x_2^{**}, \ldots, x_n^{**}\right)$ yields

$$u_i'\left(x_i^{**}\right) = \lambda > 0, \quad 1 \leq i \leq n; \quad \text{and} \quad \sum_{i=1}^{n} x_i^{**} = \theta$$

This optimization problem is addressed by the network operator in the following manner.

Step 1: The network operator implements a generalized Vickrey auction for the allocation of bandwidth to the users. The user i bids $b_i \in \mathbb{R}^+$, where $1 \leq i \leq n$. The network operator assumes that the utility of user i is $b_i \ln x_i$, where $1 \leq i \leq n$. The network operator then selects a bandwidth allocation vector $x^* = \left(x_1^*, x_2^*, \ldots, x_n^*\right)$ which solves the following problem.

$$\max_{x} \sum_{i=1}^{n} b_i \ln x_i$$

The constraints are:

$$\sum_{i=1}^{n} x_i \leq \theta$$

$$x_i \in \mathbb{R}^+, \quad 1 \leq i \leq n$$

Use of Karush-Kuhn-Tucker conditions at $x^* = (x_1^*, x_2^*, \ldots, x_n^*)$ yields

$$x_i^* = \frac{b_i}{B}\theta, \quad 1 \leq i \leq n$$

where $B = \sum_{i=1}^{n} b_i$. The above result implies that a user gets a bandwidth proportional to his or her bid.

Step 2: In order to determine the price p_i that the user i pays, the network operator solves the optimization problem in the last step by assuming that the user i does not bid. As per this optimization problem, let the bandwidth allocated to the user j be x_j^i, where $j \neq i$, and $1 \leq j \leq n$. Let $B_i = (B - b_i)$, for $1 \leq i \leq n$, then

$$x_j^i = \frac{b_j}{B_i}\theta, \quad j \neq i, \quad 1 \leq j \leq n$$

The price p_i that the user i pays is

$$p_i = \sum_{\substack{j=1 \\ j \neq i}}^{n} b_j \ln x_j^i - \sum_{\substack{j=1 \\ j \neq i}}^{n} b_j \ln x_j^*$$

Using expressions for x_j^* and x_j^i yields

$$p_i = B_i \ln \left\{ \frac{B}{B_i} \right\}, \quad 1 \leq i \leq n$$

The net utility of user i is $\alpha_i(x_i^*)$, where

$$\alpha_i(x_i^*) = \{u_i(x_i^*) - p_i\}, \quad 1 \leq i \leq n$$

It can be written explicitly in terms of b_i as

$$\alpha_i(x_i^*) = u_i \left\{ \frac{b_i}{B_i + b_i}\theta \right\} - B_i \ln \left\{ \frac{B_i + b_i}{B_i} \right\}$$

The user i selects its bid b_i so that its net utility is maximized. This is determined by taking the derivative of the above expression for $\alpha_i(x_i^*)$ with respect to b_i, and equating it to zero. Thus

$$u_i' \left\{ \frac{b_i}{B}\theta \right\} = \frac{B}{\theta}$$

Therefore letting

$$B = \sum_{i=1}^{n} b_i, \;\; \lambda = \frac{B}{\theta}$$

$$x_i^{**} = x_i^* = \frac{b_i}{\lambda}, \quad B_i = (B - b_i), \quad \text{and} \quad p_i = B_i \ln \left\{ \frac{B}{B_i} \right\} \quad 1 \le i \le n$$

solves the social welfare maximization problem for the network operator specified in Step 0, without knowledge of the users' utility functions. □

10.9 Application of Coalitional Game Theory

The application of coalitional game theory to sharing of revenue among Internet service providers (ISPs) is studied in this section. The ISPs provide Internet connectivity to the end-users. Each ISP comprises of several autonomous systems (ASes). Also, the Internet is a conglomeration of interconnected autonomous systems. In order for the Internet to work efficiently, it is imperative that these ISPs cooperate with each other. A convenient paradigm to study cooperation among such entities is the discipline of cooperative game theory. This discipline is also termed coalitional game theory. Game theory provides a formal mathematical framework for studying interaction among rational players.

Games can be studied on the basis of either noncooperation or cooperation among rational players of the game. Noncooperative game theory studies games in which rational players compete among themselves. Cooperative game theory studies games in which some players form coalitions. Players in a coalition cooperate among themselves. However, different coalitions of players compete with other coalitions of players. Concepts developed in the coalitional game theory are applied to explore the possibilities of maximizing benefits of the cooperating Internet service providers.

10.9.1 Basics of Coalitional Game Theory

Let I be the set of players who play the game. In a coalitional game, subsets of players from the set I, seek to form coalitions (groups) in order to strengthen their positions in the game. Thus, in this game there is competition among different coalitions of players. Each coalition of players selects a strategy based upon a decision making process.

Definition 10.3. *A cooperative or coalitional game is a two-tuple $\Gamma = (I, v)$, where I is the set of rational players, and $v(\cdot)$ is a real-valued characteristic function which determines the coalitional value. In this game:*

(a) *The set of players (users) $I = \{1, 2, \ldots, n\}$, where $n \in \mathbb{P} \backslash \{1\}$.*
(b) *The set of all subsets of the set I, is the power set 2^I. A coalition of players is a finite subset of I. The set of all coalitions is the power set 2^I.*
(c) *The payoff of a coalition is determined by the characteristic (coalitional) function $v(\cdot)$, where $v : 2^I \to \mathbb{R}$. For $A \subseteq 2^I$, $v(A)$ is called the value (worth, benefit, utility) of the coalition A. This function also satisfies $v(\varnothing) = 0$.* □

The set $\varnothing$ is called the *empty coalition*, and the set I is called the *grand coalition*. Total number of possible coalitions of a set of n players is 2^n. Coalitional games are said to have *transferable utility*, if for each $S \subseteq I$, the net utility $v(S)$ of a coalition of players in the set S can be freely apportioned among themselves. This concept is made formal in a definition.

Definition 10.4. *The coalitional game $\Gamma = (I, v)$ has a transferable utility, if for each $S \subseteq I$, the value $v(S)$ can be freely divided among the members of the coalition S.* □

Special types of coalitional games are next defined.

Definitions 10.5. *Let $\Gamma = (I, v)$ be a coalitional game.*

1. *The game Γ is superadditive, if*

$$v(A \cup B) \geq v(A) + v(B), \quad \forall\, A, B \subseteq I, \;\; A \cap B = \varnothing \tag{10.2a}$$

2. *The game Γ is monotone, if*

$$v(A) \leq v(B), \quad \forall\, A, B \subseteq I, \;\; A \subseteq B \tag{10.2b}$$

3. *The game Γ is constant-sum, if*

$$v(A) + v(I \backslash A) = v(I), \quad \forall\, A \subseteq I \tag{10.2c}$$

If in addition $v(I) = 0$, the game is said to be zero-sum.

4. *The game Γ is inessential (additive), if*

$$v(I) = \sum_{i \in I} v(\{i\}) \tag{10.2d}$$

The game Γ is essential, if

$$v(I) > \sum_{i \in I} v(\{i\}) \tag{10.2e}$$

5. *The game Γ is convex, if*

$$v(A \cup B) + v(A \cap B) \geq v(A) + v(B), \quad \forall\, A, B \subseteq I \tag{10.2f}$$

6. *Let π be a permutation of I. The game Γ is symmetric, if*

$$v(\pi(A)) = v(A), \textit{ for every permutation } \pi \text{ of } A, \; \forall\, A \subseteq I \tag{10.2g}$$

7. *The game Γ is simple, if: the value of the grand coalition $v(I) = 1$; $v(A)$ is equal to either 0 or 1 for every coalition $A \subseteq I$; and the value of each single-player coalition is equal to 0.* □

Some observations readily follow from the above definitions.

Observations 10.2. Consider a coalitional game $\Gamma = (I, v)$.

1. A convex coalitional game is superadditive.
2. The game Γ is symmetric if and only if $\forall\, A, B \subseteq I$

$$|A| = |B| \Rightarrow v(A) = v(B)$$

3. The game Γ is convex if and only if $\forall\, i \in I$, and $\forall\, A \subseteq B \subseteq I \backslash \{i\}$
$$v(A \cup \{i\}) - v(A) \le v(B \cup \{i\}) - v(B)$$
4. The game Γ is inessential if
$$v(A) = \sum_{i \in A} v(\{i\}), \quad \forall A \subseteq I$$
5. Consider a coalition $A \subseteq I$ in a simple game. If $v(A) = 1$, then A is called a *winning* coalition. However, if $v(A) = 0$, then A is called a *losing* coalition. □

The property of superadditivity is very important in a coalitional game. Henceforth, all coalitional games in this chapter are assumed to have the property of superadditivity. Further, all coalitional games are also assumed to have transferable utility.

Payoffs

Having determined the payoff $v(S) \in \mathbb{R}$ of a subset $S \subseteq I$ of players, it is necessary to determine the payoff $x_i \in \mathbb{R}$ of each player $i \in S$. The payoffs of the n players can be specified by a vector $(x_1, x_2, \ldots, x_n)$. Each player $i \in I$ in a coalitional game expects that his or her payoff $x_i \ge v(\{i\})$. That is, each player's payoff in a coalition would be at least equal to what he (she) could receive on his (her) own. This property is called *individual rationality*. Further, the total amount received by the players $\sum_{i \in I} x_i = v(I)$. This property is called *group rationality*. Imputations are payoff vectors that satisfy the property of both individual and group rationality.

Definitions 10.6. *Let $\Gamma = (I, v)$ be a superadditive coalitional game with transferable utility, where I is the set of players, and $v(\cdot)$ is the characteristic function.*

1. *The payoff of a player $i \in I$, is $x_i \in \mathbb{R}$. The payoff vector of the $|I| = n$ players is $x = (x_1, x_2, \ldots, x_n) \in \mathbb{R}^n$.*
2. *A payoff vector x is:*
   (a) *Individually rational, if $x_i \ge v(\{i\}), \forall\, i \in I$.*
   (b) *Group rational or efficient, if $\sum_{i \in I} x_i = v(I)$.*
3. *An imputation is a payoff vector, which is both individually and group rational. An imputation is also called an allocation. The set of imputations is:*
$$\left\{ x \mid x_i \ge v(\{i\}), \forall\, i \in I, \sum_{i \in I} x_i = v(I) \right\} \subseteq \mathbb{R}^n \tag{10.3}$$

□

An example of an imputation is $x = (x_1, x_2, \ldots, x_n)$ in which $x_i = v(\{i\})$ for $1 \le i \le (n-1)$, and
$$x_n = v(I) - \sum_{i \in I \backslash \{n\}} x_i$$
This example demonstrates that the set of imputations is never empty. This observation also follows from the above definitions, and also from the inequality $\sum_{i \in I} v(\{i\}) \le v(I)$, which in turn follows from the superadditivity property of the characteristic function $v(\cdot)$.

The Core

It is possible that for a specific imputation $x = (x_1, x_2, \ldots, x_n)$ for the division of $v(I)$, there might be a coalition $S \subset I$, which would yield more benefit. That is, $\sum_{i \in S} x_i < v(S)$. This is not desirable, because there might be a tendency by the players to form the coalition S and obtain more benefit. Imputations of this type are termed unstable. This leads to the concept of core of the coalitional game. The core is the set of feasible allocations to the users whose benefits cannot be improved upon by any coalition. We initially define unstable and stable imputations, and subsequently the core of a coalitional game.

Definitions 10.7. *Let $\Gamma = (I, v)$ be a superadditive coalitional game with transferable utility, where I is the set of players, $v(\cdot)$ is the characteristic function, and $|I| = n$. Let $x = (x_1, x_2, \ldots, x_n) \subseteq \mathbb{R}^n$ be an imputation of the game.*

1. *Unstable and stable imputations.*
   (a) *An imputation x is unstable if $\exists$ a coalition $S \subset I$ such that $\sum_{i \in S} x_i < v(S)$.*
   (b) *An imputation x is stable if it is not unstable.*
2. *The set of stable imputations is called the core $\mathcal{C}(\Gamma)$ of the game Γ. Thus*

$$\mathcal{C}(\Gamma) = \left\{ x \mid \sum_{i \in I} x_i = v(I), \text{ and } \sum_{i \in S} x_i \geq v(S) \ \forall \, S \subset I \right\} \tag{10.4}$$

Note that $\mathcal{C}(\Gamma) \subseteq \mathbb{R}^n$. □

Observe that the core can have many points. It is also possible for the core to be empty. A scheme to determine fair allocation of benefits to the players is due to the economist Lloyd Shapley. He introduced it in the year 1953.

The Shapley Value

Shapley determined a *fair* allocation of gains obtained via coalitions by making the concept of fairness concrete. This is achieved by introducing a value function $\varphi(\cdot)$ for a superadditive coalitional game with transferable utility $\Gamma = (I, v)$, where $|I| = n$. For a given characteristic function $v(\cdot)$, $\varphi(v) = (\varphi_1(v), \varphi_2(v), \ldots, \varphi_n(v)) \in \mathbb{R}^n$. Note that $\varphi_i(v) \in \mathbb{R}$ is called the Shapley value of player $i \in I$. The Shapley value of a player is the incremental worth the player contributes to the coalition. Thus, it represents the worth of player $i \in I$. Further, $\varphi(v)$ is called the Shapley vector.

Definition 10.8. *Let $\Gamma = (I, v)$ be a superadditive coalitional game with transferable utility, where I is the set of players, $v(\cdot)$ is the characteristic function, and $|I| = n$. The allocation (value) function $\varphi(\cdot)$ is: $\varphi : v \to \mathbb{R}^n$, where $\varphi(v) = (\varphi_1(v), \varphi_2(v), \ldots, \varphi_n(v)) \in \mathbb{R}^n$, and $\varphi_i(v) \in \mathbb{R}$ represents the worth of player $i \in I$.* □

It is required that the function $\varphi(\cdot)$ follow certain rules to make the payoff allocation fair. These are called the Shapley axioms.

Shapley Axioms *Let $\Gamma = (I, v)$ be a superadditive coalitional game with transferable utility, where I is the set of players, $v(\cdot)$ is the characteristic function. The allocation (value) function is $\varphi(\cdot)$.*

(a) *Efficiency:* $\sum_{i \in I} \varphi_i(v) = v(I)$.
(b) *Symmetry: The two players i and j are symmetric, if $v(S \cup \{i\}) = v(S \cup \{j\})$ for every coalition $S \subseteq I \backslash \{i, j\}$, then the allocations are $\varphi_i(v) = \varphi_j(v)$.*
(c) *Null axiom: A player i is a null player, if $v(S) = v(S \cup \{i\})$ for every coalition $S \subseteq I \backslash \{i\}$, and its allocation is $\varphi_i(v) = 0$.*
(d) *Additivity: If $\Gamma' = (I, u)$ is another superadditive coalitional game with transferable utility, and the same set of players as the game Γ, then $\varphi(u + v) = \varphi(u) + \varphi(v)$.* □

Note that the efficiency axiom is group rationality. That is, the total value of the players is equal to the value of the grand coalition. The symmetry axiom says that if the characteristic function is symmetric in i and j, then the allocated values to the players i and j are equal. As per the null axiom, the allocation value of a player is zero, if it does not increases the value of any coalition. As per the additivity axiom, if two games are played simultaneously with the same set of players, the allocation of each player is the same as the sum of allocations if the two games are played separately. If $\varphi(v)$ satisfies the above set of axioms, then it is called the Shapley vector. The Shapley value $\varphi_i(v)$ of player $i \in I$ can be explicitly determined via the Shapley's theorem.

Theorem 10.1. *Let $\Gamma = (I, v)$ be a superadditive coalitional game with transferable utility, where I is the set of players, $v(\cdot)$ is the characteristic function, and $|I| = n$.*

(a) *The allocation function $\varphi(\cdot)$, satisfies the Shapley axioms, and is unique.*
(b) *The Shapley vector is*

$$\varphi(v) = (\varphi_1(v), \varphi_2(v), \ldots, \varphi_n(v)) \in \mathbb{R}^n \tag{10.5a}$$

where

$$\varphi_i(v) = \sum_{\substack{S \subseteq I \\ i \in S}} \frac{(|S| - 1)!\,(n - |S|)!}{n!} \{v(S) - v(S - \{i\})\}, \quad 1 \leq i \leq n \tag{10.5b}$$

Proof. See the problem section. □

A *probabilistic interpretation* of the expression for Shapley value provides additional insight. Note that $\varphi_i(v)$ is equal to the sum of contributions of player i over all coalitions $S \subseteq I$ which contain i. Further, when player i joins the coalition $S \backslash \{i\}$, it contributes a value of $\{v(S) - v(S - \{i\})\}$ to coalition S.

Assume that each of the $n!$ permutations of the n players occurs with equal probability. The probability, that when player i enters the coalition, he or she will find $S \backslash \{i\}$ players is $(|S| - 1)!\,(n - |S|)!/n!$. The denominator in this expression is $n!$, which is equal to the total number of permutations of n players. The numerator is equal to the number of permutations, in which the $|S \backslash \{i\}| = (|S| - 1)$ players occur first, then player i, and finally $|I \backslash S| = (n - |S|)$ players. Thus the total number of such permutations is equal to $(|S| - 1)!\,(n - |S|)!$. Therefore $\varphi_i(v)$ is simply the expected value that the player i contributes to the grand coalition if the set of players in I sequentially form coalition in such random order.

Using the explicit expression for $\varphi_i(v)$, the Shapley axioms of symmetry, null, and additivity can easily be checked directly. The verification of axiom of efficiency is implicit in this interpretation. This is true because $\sum_{i\in I}\varphi(i) = v(I)$, which is equal to the value of the grand coalition.

Theorem 10.2. *Let $\Gamma = (I, v)$ be a superadditive coalitional game with transferable utility, where I is the set of players, $v(\cdot)$ is the characteristic function, and $|I| = n$.*
Then the Shapley vector $\varphi(v) = (\varphi_1(v), \varphi_2(v), \ldots, \varphi_n(v)) \in \mathbb{R}^n$, is an imputation.
Proof. See the problem section. □

The Shapley vector is an imputation as per the above theorem, but it may not necessarily be in the core.

10.9.2 ISP Settlement

The coalitional game theory developed in the last subsection can be applied to the study of cooperative interaction between Internet service providers (ISPs). The primary goal of the Internet service provider is to provide service to the end users, who provide revenue to the ISPs. There are several ISPs in the Internet. In order to gain connectivity, a specific ISP might enter into partnership with other local ISPs via bilateral contracts. The ISP might pay another ISP for the transit of its traffic. Such arrangements might be beneficial from the perspective of a specific ISP, but might not be attractive for the Internet as a whole. This selfish behavior of the individual ISPs might lead to balkanization and inefficient use of the Internet.

It is the premise of this subsection that coalitional game theory can be used to design a better engineered Internet by apportioning revenue fairly among different ISPs. This is possible by forming coalitions among ISPs, and using a profit sharing mechanism that is based upon the Shapley vector. Use of Shapley vector ensures that an ISP's profit is in consonance with its contribution to the value of the network.

An ISP is composed of several ASes. Thus the Internet consists of several thousands of autonomous systems. An AS in turn is made up of several routers, which are connected via transmission links in some configuration. The routers and transmission links facilitate transmission of traffic between source-destination pairs. The primary purpose of the ASes, routers, and transmission links is to provide network connectivity. The basic framework of the routers, ASes, and coalitions of ASes is next described. This is followed by a description of traffic flow in the network, the worth function, and finally the profit distribution mechanism.

Physical Network Model

A network is made up of routers and transmission links. This network can be represented by a digraph, where a vertex represents a router, and a transmission link is represented by a directed arc. This graph is called the router-level graph.

A single AS is made up of several routers, and the transmission links connecting these routers. The network at AS-level can also be specified by a digraph, in which a vertex represents an AS, and the logical link between two adjacent ASes can be specified in terms of the links of the router-level graph. This graph is called the AS-level graph. We assume that an ISP has at least a single AS. In this subsection, an ISP with multiple ASes is considered a super AS. Therefore, the terms ISP and AS are used interchangeably in the rest of this subsection.

As the goal is maximization of profit via coalitions among ASes, the network formed by coalitions of the ASes can also be described in graph-theoretic language. The AS-coalition graphs are next described at both the router and AS-level. For simplicity, assume that all the graphs in this subsection are strongly connected. Recall that a strongly connected graph is a digraph in which there is a directed path between every pair of nodes.

Definitions 10.9. *Router-level graph, AS-level graph; and AS-coalition graphs at the router and AS-level.*

1. *The router-level graph is the digraph* $\mathcal{G} = (\mathcal{V}, \mathcal{E})$ *, where* $\mathcal{V}$ *is the set of routers, and* $\mathcal{E}$ *is the set of directed transmission links which connect the routers.*
2. *The AS-level graph is the digraph* $G = (V, E)$ *, where* V *is the set of ASes, and* E *is the set of directed logical inter-AS links.*
   An AS $i \in V$ *is made up of routers in the set* $\mathcal{V}_i \subseteq \mathcal{V}$*, where* $\{\mathcal{V}_i \mid i \in V\}$ *is a partition of the set* $\mathcal{V}$*. That is,* $\mathcal{V}_i \cap \mathcal{V}_j = \varnothing, i \neq j, \forall\, i, j \in V$*, and* $\cup_{i \in V} \mathcal{V}_i = \mathcal{V}$*. The AS-level edge* $(i, j) \in E \Leftrightarrow \exists$ $(k, l) \in \mathcal{E}$*, where* $k \in \mathcal{V}_i, l \in \mathcal{V}_j$*. That is,* $E = \{(i, j) \mid (k, l) \in \mathcal{E}, k \in \mathcal{V}_i, l \in \mathcal{V}_j\}$ *.*
3. *The AS-coalition graphs for the AS-coalition* $S \subseteq V$ *at the router and AS-level, where* $S \neq \varnothing$*. These graphs are said to be induced by the coalition* S*.*
   a) *Induced graph at the router-level is* $\mathcal{G}_S = (\mathcal{V}_S, \mathcal{E}_S)$ *, where* $\mathcal{V}_S = \{\mathcal{V}_i \mid \mathcal{V}_i \subseteq \mathcal{V}, i \in S\}$ *, and* $\mathcal{E}_S = \{(i, j) \mid (i, j) \in \mathcal{E},\ \text{and}\ i, j \in \mathcal{V}_S\}$ *.*
   b) *Induced graph at the AS-level is* $G_S = (S, E_S)$ *, where* E_S *is the set of directed logical inter-AS links within the coalition.*
   That is, $E_S = \{(i, j) \mid (i, j) \in E,\ \text{and}\ i, j \in S\}$ *.* □

The induced graph G_S at the AS-level is used to describe the revenue collected from the customers. However, the induced graph $\mathcal{G}_S$ at the router-level is used to describe the cost of transporting traffic in the coalition S.

Traffic Flow in the Network

The goal of the network is to deliver the customer traffic rate $\lambda_{ij} \in \mathbb{R}_0^+$ from router i, to router j where $i, j \in \mathcal{V}$. Assume that there is enough capacity in the network to carry this traffic. This implies that feasible flows exist in the network.

As coalitions are formed among the ASes, the following notation is introduced for the flow of traffic on the AS-level graph $G = (V, E)$. Let the traffic that flows on the link $(i, j) \in \mathcal{E}$ be $f_{k,l}(i, j)$, where the traffic originates on the AS $k \in V$, and ends on the AS $l \in V$. Consider the AS-coalition graph $\mathcal{G}_S = (\mathcal{V}_S, \mathcal{E}_S)$ at the router-level, which is formed by the coalition $S \subseteq V$. The aggregate traffic flow on the link $(i, j) \in \mathcal{E}_S$ is $f_S(i, j) = \sum_{k,l \in \mathcal{V}_S} f_{k,l}(i, j)$.

Worth of the Coalitional Game

The worth of a coalitional game is defined as the profit, which is the revenue collected from the customers minus the cost of routing. For a coalition $S \subseteq V$, the worth $v(S)$ is

$$v(S) = v^o(S) - v^c(S)$$

where $v^o(\cdot)$ is the revenue function, and $v^c(\cdot)$ is the cost function. An AS $i \in S$ is connected to an AS $j \in S$ if there is a logical directed path from AS $i \in S$ to AS $j \in S$. The revenue generated by the traffic flowing on the graph G_S between ASes i and j is W_{ij}. An indicator function I_{ij} is also defined as

$$I_{ij} = \begin{cases} 1, & i \text{ is logically connected to } j \text{ in the graph } G_S \\ 0, & \text{otherwise} \end{cases}, \quad i, j \in S$$

The revenue generated by traffic flowing on the G_S graph is

$$v^o(S) = \sum_{i,j \in S} W_{ij} I_{ij}$$

Observe that this revenue depends only on the AS-level graph. However, the routing cost $v^c(S)$ depends upon the router-level topology of the network, and also upon the flows that deliver end-to-end traffic.

The routing cost function $c_{ij}(\cdot)$, for the aggregated traffic $x \in \mathbb{R}_0^+$ that flows on link $(i, j) \in \mathcal{E}$, is defined as $c_{ij}(x) = \{c_{ij}^s(x) + c_{ij}^r(x)\}$, where $c_{ij}^s(\cdot)$ is the sending cost function of router i, and $c_{ij}^r(\cdot)$ is the receiving cost function of router j. It is assumed that and $c_{ij}^s(\cdot)$ and $c_{ij}^r(\cdot)$ are monotonically increasing functions of the aggregated traffic x on the link, and $c_{ij}^s(0) = c_{ij}^r(0) = 0$.

The routing cost of AS $k \in S$ for a feasible flow f_S in the router-level subgraph $\mathcal{G}_S = (\mathcal{V}_S, \mathcal{E}_S)$ is denoted by $C_k(f_S)$. It is

$$C_k(f_S) = \sum_{\substack{(i,j) \in \mathcal{E}_S \\ i \in \mathcal{V}_k}} c_{ij}^s(f_S(i,j)) + \sum_{\substack{(i,j) \in \mathcal{E}_S \\ j \in \mathcal{V}_k}} c_{ij}^r(f_S(i,j)), \quad k \in S$$

The net routing cost $v^c(S)$ in the coalition $S \subseteq V$ is

$$v^c(S) = \sum_{k \in S} C_k(f_S)$$

Profit Distribution Mechanism

Profits can be distributed fairly among the ISPs, by using the principles of coalitional game theory, and the corresponding Shapley vector. Furthermore, the use of Shapley vector ensures that the desirable properties of efficiency, symmetry, nullity, and additivity in the profit allocation to the ISPs are also satisfied. These are actually the Shapley axioms.

Let the ISP coalitional game be $\Gamma = (V, v)$, where V is the set of ASes, and $v(\cdot)$ is the worth function. If $|V| = n$, then the profit allocation is specified by the Shapley vector $\varphi(v) = (\varphi_1(v), \varphi_2(v), \ldots, \varphi_n(v)) \in \mathbb{R}^n$, where $\varphi_i(v)$ is the profit allocated to AS $i \in V$. The $\varphi_i(v)$'s can be calculated explicitly by using results from Shapley's theorem stated in the last subsection. Use of the Shapley vector for profit allocation implies that $\varphi_i(v)$'s satisfy Shapley axioms. Note that, if the routing costs are negligible, then the cost of a coalition is simply the total revenue generated by it. That is, $v(S) = v^o(S)$. In this case, $v(\cdot)$ depends solely upon the AS-level topology specified by the graph G_S.

Another mechanism to apportion revenue, or cost sharing among different entities is via the theory of bargaining games. A bargaining game is cooperative. This technique is developed in the next section.

10.10 Bargaining Games

Theory of bargaining games can be used for developing guidelines for cost sharing between different agents like ISPs. Elements of bilateral bargaining game theory are developed in this section. It is quite possible that surplus revenue is generated in an economic transaction. If the total money requested by agents is less than the available revenue, then the agents get the requested money. However, if the net requested money is more than that available, then each agent receives lesser revenue than it requested. In this case, principles of bargaining game theory can be used to apportion the surplus revenue among the participants (players) of the transaction. Tools of cooperative game theory called Nash bargaining is used. This technique is named after its inventor, John F. Nash.

Bargaining games typically involve two players. We initially develop motivation for Nash bargaining game via an example. The game has two players: a buyer and a seller of an object. The buyer values the object at v units, and it costs c units to the seller. Assume that $v > c$. If the object is sold, then let its price be p units. In this case, the utility of the buyer is a function of $(v - p)$. Similarly, the utility of the seller is a function of $(p - c)$. However, if the object is not sold, then the utilities of both buyer and seller are zero each. The bargaining game requires us to determine a "reasonable" value of p. Nash stated a reasonable solution of the bargaining game based upon a set of axioms named after him. The bargaining game for two players (agents) is initially defined formally. This is followed by specifying notation for the solution of the bargaining game.

Definition 10.10. *A bargaining game for two players is $B_g = (U, d)$. In this game:*

(a) *$U \subseteq \mathbb{R}^2$ is the nonempty set of payoffs (utilities) to the two players. If $u = (u_1, u_2) \in U$, then player i gets utility u_i, where $i \in \{1, 2\}$. Further, U is a compact (bounded and closed) and convex set in the real-plane.*

(b) *$d = (d_1, d_2) \in U$ is the disagreement or status-quo point. If no agreement is reached between the two players, then player i gets utility d_i, where $i \in \{1, 2\}$.* □

Notation for the solution of the bargaining game is next specified.

Definition 10.11. *Consider the bargaining game for two players $B_g = (U, d)$. Let $\mathcal{B}_g$ be the set of all bargaining games. The bargaining solution of the problem B_g is $f(B_g) \in U$, where $f : \mathcal{B}_g \to 2^{\mathbb{R}^2} \backslash \varnothing$.* □

Based upon the specification of the bargaining game, and the notation for its solution, Nash bargaining game is described.

Nash Bargaining Game

Nash's solution to the bargaining game with two players is next described. Nash asserted that a "fair and reasonable" solution to the bargaining game can be found provided it satisfies a set of axioms. It is shown subsequently that Nash axioms imply a unique solution of the bargaining game.

Nash Axioms *Let a bargaining game for two players be $B_g = (U, d)$. Its solution, specified by $f(B_g) = u^* = (u_1^*, u_2^*)$ satisfies the following axioms.*

(a) *Individual rationality*: $u_1^* \geq d_1$, and $u_2^* \geq d_2$.
(b) *Feasibility*: $u^* = (u_1^*, u_2^*) \in U$.
(c) *Pareto optimality*: *There is no point* $(u_1, u_2) \in U$ *such that* $u_1 \geq u_1^*$ *and* $u_2 \geq u_2^*$, *except* (u_1^*, u_2^*) *itself. That is,* u^* *is on the boundary of the set* U.
(d) *Symmetry*: *If the utility set* U *is such that* $(u_1, u_2) \in U$ *if and only if* $(u_2, u_1) \in U$, *and the status-quo point* $d = (d_1, d_2)$ *is such that* $d_1 = d_2$, *then* $u_1^* = u_2^*$. *Thus, if the game is symmetric with respect to the utilities of the players, then so is the bargaining solution.*
(e) *Invariance under linear transformation*: *The set* V *is obtained from* U *via linear transformation, where for* $u = (u_1, u_2) \in U$ *and* $v = (v_1, v_2) \in V$

$$v_1 = (\alpha_1 u_1 + \beta_1), \quad v_2 = (\alpha_2 u_2 + \beta_2), \quad \alpha_1, \alpha_2 \in \mathbb{R}^+, \text{ and } \beta_1, \beta_2 \in \mathbb{R}$$

Let the bargaining game $(V, ((\alpha_1 d_1 + \beta_1), (\alpha_2 d_2 + \beta_2)))$ *have a solution* $v^* = (v_1^*, v_2^*)$, *then*

$$v_1^* = (\alpha_1 u_1^* + \beta_1), \quad v_2^* = (\alpha_2 u_2^* + \beta_2)$$

(f) *Independence of irrelevant alternatives*: *Let* V *be a closed convex subset of* U, $d \in V$, *and* $u^* \in V$, *then* $f(V, d) = f(U, d) = u^*$. □

Nash provided the following justification for his axioms. The axiom of individual rationality implies that the payoff of each player is at least equal to its status-quo value. The axiom of feasibility is absolutely necessary, as the arbitrated outcome of the game has to be feasible. The axiom of Pareto optimality implies that, there is no other point $u \in U$, where both players receive more. That is, the players prefer the payoff $\widehat{u} = (\widehat{u}_1, \widehat{u}_2)$ to $u = (u_1, u_2)$ whenever $\widehat{u}_1 > u_1$ and $\widehat{u}_2 \geq u_2$, or $\widehat{u}_1 \geq u_1$ and $\widehat{u}_2 > u_2$. Axiom of symmetry implies fairness in the bargaining game. That is, if the game is such that the players are interchangeable, then the outcome (payoffs) should also be symmetric. Axiom of invariance under linear transformation states that, payoffs of players essentially do not change under this transformation. The change occurs up to only scaling and translation.

The last axiom is about independence of irrelevant alternatives. This axiom implies that, if another game has a smaller payoff region, the same status-quo point d, and also contains u^*, then the payoff regions away from d and u^* are irrelevant. This axiom appears to be controversial. Nash proved that if the above set of axioms are true in a bargaining game, then there is a unique solution to the bargaining problem. This result is stated as Nash bargaining game theorem.

Theorem 10.3. *Consider the bargaining game* $B_g = (U, d)$. *If there exists at least a point* $u = (u_1, u_2)$ *such that* $u_1 > d_1$ *and* $u_2 > d_2$, *and* $u^* \in U$ *maximizes* $(u_1 - d_1)(u_2 - d_2)$ *over points* $u \in U$ *where* $u_1 \geq d_1$ *and* $u_2 \geq d_2$, *then* $u^* \in U$ *is a unique maximizing utility vector. Further* $f(\cdot)$, *where* $f(B_g) = u^*$, *is the unique function which satisfies Nash axioms.*

Proof. See the problem section. □

The theorem also provides a constructive solution of the bargaining game. As per this theorem, $f(B_g) = u^* = (u_1^*, u_2^*)$ is the unique utility vector at which the product $(u_1 - d_1)(u_2 - d_2)$ is maximized over $u \in U$. That is,

$$u^* = \arg\max_{u \in U} (u_1 - d_1)(u_2 - d_2)$$

where $\arg\max$ denotes the utility vector in the set U that maximizes the product

$$(u_1 - d_1)(u_2 - d_2)$$

Sometimes $(u_1 - d_1)(u_2 - d_2)$ is called the Nash product. Use of Nash's theorem is illustrated via following examples.

Examples 10.5. Nash bargaining game examples.

1. Assume that the utility space U of two players is the triangle with vertices $(0,0), (0,2)$, and $(4,0)$. The status-quo point is $(0,0)$. Let the utility two-tuple be $u = (u_1, u_2) \in U$. Nash bargaining game solution $u^* = (u_1^*, u_2^*) \in U$ is determined. The solution u^* maximizes $g(u_1, u_2) = u_1 u_2$ over U.
The Pareto optimal boundary is the line segment connecting the points $(0,2)$, and $(4,0)$. The set of such boundary points is
$$\{(u_1, u_2) \mid u_1 + 2u_2 = 4,\ u_1 \in [0,4], u_2 \in [0,2]\}$$
The function $g(\cdot,\cdot)$ is maximized at $u^* = (2,1)$, and its maximum value is equal to $g(u_1^*, u_2^*) = 2$.
2. Two ISPs are offered to invest 100 million dollars if they decide how to split the money, otherwise they each get nothing. The first ISP is much richer than the second ISP in terms of its resources. For an investment of x million dollars, utility of the first ISP is $u_1 = cx$, where $c > 0$ is small. As the second ISP is relatively poorer, with only 20 million dollars in capital, its utility for a sum of money is directly proportional to its logarithm. Therefore the utility of the second ISP for an investment of x million dollars is
$$u_2 = \{\ln(x + 20) - \ln 20\}$$
The goal is to divide the 100 million dollars between the two ISPs.
Note that the status-quo point in this bargaining game is $(0,0)$. Suppose that the first ISP receives y million dollars, then the second ISP receives $(100 - y)$ million dollars. Therefore
$$u_1 = cy$$
$$u_2 = \{\ln((100 - y) + 20) - \ln 20\}$$
From the above relationships, we obtain
$$u_2 = \ln\left\{\frac{120 - u_1/c}{20}\right\}$$
Note that U simply consists of points in the area bounded by the above curve, and the u_1 and u_2 axes. Thus the points in U form a convex set. We determine a point in U which maximizes
$$g(u_1, u_2) = u_1 u_2$$
That is
$$g(u_1, u_2) = u_1 \ln\left\{\frac{120 - u_1/c}{20}\right\}$$
Differentiate $g(u_1, u_2)$ with respect to u_1 and set it equal to zero. This yields
$$\ln\left\{\frac{120 - u_1/c}{20}\right\} = \frac{u_1/c}{120 - u_1/c}$$

The approximate solution of the above equation is

$$u_1 = 61.92c$$

Therefore if $c = 1$, the first ISP receives 61.92 million dollars, and the second ISP receives 38.08 million dollars.
The solution of this bargaining game indicates that the richer ISP receives more investment than its poorer counterpart. This is because, the second ISP's utility for money decreases rapidly, while the first ISP's utility does not. Also, as the second ISP is poorer, it might be willing to get something and bargained down by the first and richer ISP. □

Nash bargaining game is not the only such game. There are other reasonable bargaining game axioms that can be considered. Further, Nash bargaining game scheme does not take into account sufficiently the status-quo (or disagreement) point.

Reference Notes

McKnight, and Bailey (1998) edited one of the initial publications on Internet economics. The section on the scope of economics is based upon the comprehensive textbook by Rader (1972).

The chapter on economic models of communication networks by J. Walrand in Liu, and Xia (2008) is very pedagogical. Its influence on this chapter should be evident to the perceptive reader. The section on network resources, congestion, and pricing closely follows the influential paper by Mackie-Mason, and Varian (1995).

The very appealing Paris metro pricing scheme was suggested by Odlyzko (1998). The time-dependent pricing and bandwidth trade-off model follows the work of Jiang, Parekh, and Walrand (2008). The generalized Vickrey auction is based upon the work of Mackie-Mason, and Varian (1994). The basics of terminology used in the section on auctions has been borrowed from Courcoubetis, and Weber (2003), Dramitinos, Stamoulis, and Courcoubetis (2007), and Barron (2008).

The section on coalitional games follows Friedman (1986), Roth (1988), Morris (1991), Owen (1995), Ferguson (2014), and Barron (2008). The application of coalitional game theory to ISP settlement is based upon the pioneering work of Ma, Chiu, Lui, and Misra (2010). Notes on bargaining game theory follow the lucid accounts in Jones (1980), Owen (1995), Ferguson (2014), and Barron (2008).

Problems

1. Suppose that there are n users of a network. User i uses $x_i \in \mathbb{R}^+$ units of network resources. Thus the total use of network resources by the n users is $X = \sum_{i=1}^{n} x_i$. The capacity of the network is $C \in \mathbb{R}^+$, and the utilization of the network resources is $Y = X/C$. For user i, the

cost of using network resources is px_i, and the "delay cost" due to congestion in the network is $\mathcal{D}(Y) = \delta X/C$; where $p \geq 0$, $\delta > 0$, and $1 \leq i \leq n$. The utility of user i is

$$v_i(x_i, Y) = \sqrt{x_i} - px_i - \mathcal{D}(Y), \quad \text{where } 1 \leq i \leq n$$

(a) If $p = 0$, show that by varying x_i, the utility $v_i(x_i, Y)$ is maximized at $x_i = \{C/(2\delta)\}^2$, and the corresponding maximizing value of X is $X^* = n\{C/(2\delta)\}^2$. This result demonstrates that the selfish user i's utility increases quadratically with the capacity of the network, and decreases with the disutility of the delay parameter δ. Further, the congestion at X^* is $Y^* = nC/(2\delta)^2$. That is, congestion increases with the capacity if the users are selfish.

(b) Let $p = 0$. The social welfare is $W(C, X) = \sum_{i=1}^n v_i(x_i, Y)$. Let all users have identical utility functions. That is, $x_i = y$, for $1 \leq i \leq n$. Therefore the social welfare function $n\{\sqrt{y} - \delta ny/C\}$ can be maximized by varying y. Show that the maximizing value of y is $y^{**} = \{C/(2\delta n)\}^2$. Therefore the corresponding maximizing value of X is $X^{**} = n\{C/(2\delta n)\}^2$.
Observe that $X^*/X^{**} = n^2$. This result implies that, if the network resources are provided free of charge, it is overused by a factor equal to n^2.

(c) Let $p > 0$. By varying x_i show that the utility function $v_i(x_i, Y)$ of user i is maximized at $p = (C - 2\delta\sqrt{x_i})/(2C\sqrt{x_i})$. If $x_i = y^{**}$, then show that the corresponding value of price rate $p^{**} = \delta(n-1)/C$. Note that for $n > 1$ the socially optimal price increases with increase in the number of users n. It also increases as δ increases and as the capacity C decreases.

Hint: See Mackie-Mason, and Varian (1995), and Shy (2001).

2. Determine the Nash equilibrium of the time-dependent pricing and bandwidth trade-off game.
Hint: See Jiang, Parekh, and Walrand (2008). The Nash equilibrium can be determined by using the method of Lagrange multipliers. Let the Lagrangian be $\mathcal{L}(D, \lambda)$ where

$$\lambda = (\lambda_1, \lambda_2, \ldots, \lambda_K)$$

and $\lambda \geq 0$. Then

$$\mathcal{L}(D, \lambda) = -\sum_{i=1}^{M}\sum_{k=1}^{K}\left(u_k^i - p_k^i\right)d_k^i + \sum_{i=1}^{M}\int_0^{x_i} l_i(y)\,dy + \sum_{k=1}^{K}\left\{\sum_{i=1}^{M} d_k^i - d_k\right\}\lambda_k$$

where $x_i = \sum_{k=1}^K d_k^i$. Use of the Karush-Kuhn-Tucker (KKT) conditions yield

$$-\left(u_k^i - p_k^i\right) + l_i(x_i) + \lambda_k \geq 0, \quad 1 \leq k \leq K,\ 1 \leq i \leq M$$
$$\left\{-\left(u_k^i - p_k^i\right) + l_i(x_i) + \lambda_k\right\} d_k^i = 0, \quad 1 \leq k \leq K,\ 1 \leq i \leq M$$
$$\sum_{i=1}^{M} d_k^i \leq d_k, \quad 1 \leq k \leq K$$
$$\lambda_k \geq 0, \quad 1 \leq k \leq K$$
$$\lambda_k\left\{\sum_{i=1}^{M} d_k^i - d_k\right\} = 0, \quad 1 \leq k \leq K$$
$$d_k^i \geq 0, \quad 1 \leq k \leq K,\ 1 \leq i \leq M$$

3. Prove that the optimal strategy for each bidder is to bid its true valuation in the Vickrey auction of a single item.
Hint: In the Vickrey auction, the user i's valuation of the item is $v_i \in \mathbb{R}_0^+$, and the bid is $x_i \in \mathbb{R}_0^+$, where $1 \leq i \leq n$. The utility of user i is $u_i(x_i)$. It is defined in terms of the function $T(\cdot)$, where $T(x_{-i}) = \max_{1 \leq j \leq n, j \neq i} x_j$. For simplicity assume that valuations of the item of all bidders are different. Also assume that all the bids are different. This is a plausible assumption because all bids are sealed. Consider the following cases, in which we assume that the v_i's are fixed, and examine different range of the x_i's.
   (a) The bid of user i is equal to its valuation, that is, $x_i = v_i$. This is the truthful strategy.
      (i) *Case* $T(x_{-i}) < x_i$: Bidder i wins the item, and its utility is $u_i(x) = \{v_i - T(x_{-i})\}$.
      (ii) *Case* $x_i < T(x_{-i})$: Bidder i loses the item, and its utility is $u_i(x) = 0$.
   (b) The user i underbids, that is, $x_i < v_i$.
      (i) *Case* $T(x_{-i}) < x_i < v_i$: Bidder i wins the item, and its utility is $u_i(x) = \{v_i - T(x_{-i})\}$. This utility is the bidder's maximum possible value, and it is the same as that for truthful strategy (which is $x_i = v_i$). Thus the strategy of underbidding and truthful bidding yield equal payoff.
      (ii) *Case* $x_i < T(x_{-i}) < v_i$: Bidder i loses the item, and its utility is $u_i(x) = 0$. The bidder's utility would have been more, if he or she had bid truthfully. Thus the strategy of underbidding is dominated by the strategy of truthful bidding.
      (iii) *Case* $x_i < v_i < T(x_{-i})$: Bidder i loses the item, and its utility is $u_i(x) = 0$. The bidder i also loses the bid if his or her bidding were truthful. Thus the strategy of underbidding and truthful bidding yield equal payoff.
   (c) The user i overbids, that is, $v_i < x_i$.
      (i) *Case* $T(x_{-i}) < v_i < x_i$: Bidder i wins the item, and its utility is $u_i(x) = \{v_i - T(x_{-i})\}$. This utility is the bidder's maximum possible value, and it is the same as that for truthful strategy (which is $x_i = v_i$). Thus the strategy of overbidding and truthful bidding yield equal payoff.
      (ii) *Case* $v_i < T(x_{-i}) < x_i$: Bidder i wins the item, and its utility is $u_i(x) = \{v_i - T(x_{-i})\}^+ = 0$. Thus the strategy of overbidding and truthful bidding yield equal payoff.
      (iii) *Case* $v_i < x_i < T(x_{-i})$: Bidder i loses the item, and its utility is $u_i(x) = 0$. The bidder i also loses the bid if his or her bidding were truthful. Thus the strategy of overbidding and truthful bidding yield equal payoff.

   In summary, the strategy of truthful bidding always dominates either underbidding or overbidding the item.
4. Prove that the generalized Vickrey auction is incentive-compatible.
Hint: See Walrand in Liu, and Xia (2008). Let A be the set of items to be auctioned off. If the bidder i bids truthfully, then his or her utility or payoff is

$$\begin{aligned}\alpha_i &= v_i(A_i) - p_i \\ &= v_i(A_i) - \sum_{\substack{j=1 \\ j \neq i}}^{n} \{b_j(B_j^i) - b_j(A_j)\}\end{aligned}$$

where $\{A_i \mid A_i \subseteq A, 1 \leq i \leq n\}$ is the set of allocations to the n bidders, and

$$\cup_{1\leq i\leq n}A_i = A$$

Further, $B_j^i \subseteq A$ is the set of items which the bidder j would receive if the bidder i did not participate in the auction for $j \neq i$, and $1 \leq j \leq n$. If $\{A_i' \mid A_i' \subseteq A, 1 \leq i \leq n\}$ is another set of allocations to the n bidders, then the utility of bidder i is

$$\alpha_i' = v_i\left(A_i'\right) - \sum_{\substack{j=1 \\ j\neq i}}^{n} \left\{b_j\left(B_j^i\right) - b_j\left(A_j'\right)\right\}$$

Observe that the B_j^i's are identical in both cases because this allocation excludes bidder i's bid. Therefore

$$\alpha_i - \alpha_i' = \left\{v_i\left(A_i\right) + \sum_{\substack{j=1 \\ j\neq i}}^{n} b_j\left(A_j\right)\right\} - \left\{v_i\left(A_i'\right) + \sum_{\substack{j=1 \\ j\neq i}}^{n} b_j\left(A_j'\right)\right\}$$

Note that $(\alpha_i - \alpha_i') \geq 0$ because the allocation $\{A_i \mid A_i \subseteq A, 1 \leq i \leq n\}$ maximizes the first sum. Therefore the generalized Vickrey auction is incentive-compatible. Consequently, the strategy of truthful bidding by the bidders maximizes the social welfare. Note that, because of the truthfulness of the bidding by the agents, the α_i's and α_i''s are nonnegative.

5. Prove Shapley's theorem.
   Hint: See Friedman (1986), and Ferguson (2014).
   (a) We prove that, the allocation function $\varphi(\cdot)$ which satisfies Shapley axioms is unique, in several steps.
   Step 1: For a nonempty set $S \subseteq I$, define a simple game (I, v_S), where

$$v_S(T) = \begin{cases} 1, & \text{if } S \subseteq T \\ 0, & \text{otherwise} \end{cases}$$

   for all $T \subseteq I$. From the nullity axiom, $\varphi_i(v_S) = 0$ if $i \notin S$, and from the symmetry axiom, if $i, j \in S$, then $\varphi_i(v_S) = \varphi_j(v_S)$. Thus, use of the efficiency axiom yields $\sum_{i\in I}\varphi_i(v_S) = v_S(I) = 1$, and therefore $\varphi_i(v_S) = 1/|S|, \forall\, i \in S$.
   Step 2: As in Step 1, for a nonempty set $S \subseteq I$, define a simple game (I, cv_S), where c is a scalar. The Shapley value is $\varphi_i(cv_S) = c/|S|, \forall\, i \in S$; and $\varphi_i(cv_S) = 0, \forall\, i \notin S$.
   Step 3: It is demonstrated in this step that the characteristic function $v(\cdot)$ can be represented uniquely as a weighted sum of characteristic functions described in Step 1. This representation is

$$v(\cdot) = \sum_{S\subseteq I} c_S(v)\, v_S(\cdot)$$

   where the constants $c_S(v)$'s are defined recursively by using the concept of *marginal value of a coalition.*

$$c_{\varnothing}(v) = 0, \quad c_{\{i\}}(v) = v(\{i\}), \quad \forall\, i \in I$$
$$c_T(v) = v(T) - \sum_{S\subset T} c_S(v), \quad \forall\, T \subseteq I, \quad |T| \geq 2$$

   Note that the marginal value of a coalition T is $v(T)$ minus the marginal value of all coalitions that are *proper* subsets of T. Therefore

$$\sum_{S\subseteq I} c_S(v)\, v_S(T) = \sum_{S\subseteq T} c_S(v) = c_T(v) + \sum_{S\subset T} c_S(v) = v(T)$$

Therefore, it is demonstrated that $v(\cdot) = \sum_{S\subseteq I} c_S(v)\, v_S(\cdot)$. Letting $|T| \triangleq t$, and $|S| \triangleq s$ yields

$$c_T(v) = \sum_{S\subseteq T} (-1)^{t-s}\, v(S)\,, \;\; \forall\, T \subseteq I, \qquad |T| \geq 2$$

Step 4: It is quite possible, that in the expression

$$v(\cdot) = \sum_{S\subseteq I} c_S(v)\, v_S(\cdot)$$

the $c_S(v)$'s are negative. This is permissible because, if (I, v), (I, w), and (I, z) are games where $z = (v - w)$, then $\varphi_i(z) = (\varphi_i(v) - \varphi_i(w))$, $\forall\, i \in I$. This later assertion can be proved by using the additivity axiom. (Simply write $v = (v - w) + w$.)
Step 5: In order to demonstrate the uniqueness, assume that there are two sets of constants $c_S(v)$'s and $c'_S(v)$'s such that

$$v(T) = \sum_{S\subseteq I} c_S(v)\, v_S(T) = \sum_{S\subseteq I} c'_S(v)\, v_S(T)\,, \quad \forall\, T \subseteq I$$

Induction can be used to show that $c_S(v) = c'_S(v)$, $\forall\, S \subseteq I$. If $T = \{i\}$, then $v_S(T)$ is equal to zero except for $S = \{i\}$. Thus $c_{\{i\}}(v) = c'_{\{i\}}(v)$, $\forall\, i \in I$. Next assume that Z is any coalition in which $c_S(v) = c'_S(v)$, $\forall\, S \subseteq Z$. Then in the above relationships, if $T = Z$, we obtain $c_Z(v) = c'_Z(v)$.
This completes the proof of uniqueness of the Shapley vector.

(b) Again let $|R| \triangleq r$, and $|S| \triangleq s$. From Steps 2 and 4, and the expression $v(\cdot) = \sum_{R\subseteq I} c_R(v)\, v_R(\cdot)$ an explicit expression for $\varphi_i(v)$, for $i \in I$ is

$$\varphi_i(v) = \sum_{\substack{R\subseteq I\\ i\in R,\, R\neq\varnothing}} \frac{c_R(v)}{r}$$

Use of results from Step 3 yields

$$\varphi_i(v) = \sum_{\substack{R\subseteq I\\ i\in R,\, R\neq\varnothing}} \frac{1}{r} \sum_{S\subseteq R} (-1)^{r-s}\, v(S)$$

Interchange of the order of summation results in

$$\varphi_i(v) = \sum_{S\subseteq I} \sum_{\substack{R\subseteq I\\ S\cup\{i\}\subseteq R,\, R\neq\varnothing}} (-1)^{r-s} \frac{1}{r} v(S)$$

Let

$$\psi_i(S) \triangleq \sum_{\substack{R\subseteq I\\ S\cup\{i\}\subseteq R,\, R\neq\varnothing}} (-1)^{r-s} \frac{1}{r}$$

Observe that if $i \notin S'$ and $S = S' \cup \{i\}$, then $\psi_i(S') = -\psi_i(S)$. Therefore

$$\varphi_i(v) = \sum_{\substack{S \subseteq I \\ i \in S}} \psi_i(S)\{v(S) - v(S - \{i\})\}$$

Further, if $i \in S$, and $S \subseteq R$, there are $\binom{n-s}{r-s}$ coalitions R with r elements. Therefore

$$\begin{aligned}
\psi_i(S) &= \sum_{r=s}^{n} (-1)^{r-s} \binom{n-s}{r-s} \frac{1}{r} \\
&= \sum_{r=s}^{n} (-1)^{r-s} \binom{n-s}{r-s} \int_0^1 x^{r-1} dx \\
&= \int_0^1 \sum_{r=s}^{n} (-1)^{r-s} \binom{n-s}{r-s} x^{r-1} dx \\
&= \int_0^1 x^{s-1} \sum_{r=s}^{n} (-1)^{r-s} \binom{n-s}{r-s} x^{r-s} dx \\
&= \int_0^1 x^{s-1} (1-x)^{n-s} dx
\end{aligned}$$

The above integral is a beta function $B(s, n-s+1)$, where

$$B(s, n-s+1) = \frac{(s-1)!\,(n-s)!}{n!}$$

Thus $\psi_i(S) = (s-1)!\,(n-s)!/n!$, and

$$\varphi_i(v) = \sum_{\substack{S \subseteq I \\ i \in S}} \frac{(|S|-1)!\,(n-|S|)!}{n!} \{v(S) - v(S - \{i\})\}, \quad 1 \leq i \leq n$$

The explicit expression for $\varphi_i(v)$ for a player $i \in I$ also establishes the existence of the Shapley vector.

6. Prove that the Shapley vector is an imputation.
Hint: See Morris (1991). In order to prove that Shapley vector is an imputation, individual and group rationality of players has to established. Individual rationality is established by showing that $\varphi_i(v) \geq v(\{i\}), \forall\, i \in I$. By superadditivity

$$\{v(S) - v(S - \{i\})\} \geq v(\{i\})$$

Therefore

$$\varphi_i(v) \geq \sum_{\substack{S \subseteq I \\ i \in S}} \frac{(s-1)!\,(n-s)!}{n!} v(\{i\})$$

where $|S| \triangleq s$. Observe that

$$\sum_{\substack{S \subseteq I \\ i \in S}} \frac{(s-1)!\,(n-s)!}{n!} = \sum_{s=1}^{n} \binom{n-1}{s-1} \frac{(s-1)!\,(n-s)!}{n!} = \sum_{s=1}^{n} \frac{1}{n} = 1$$

Consequently $\varphi_i(v) \geq v(\{i\}), \forall\, i \in I$. Group rationality is established as follows.

$$\sum_{i \in I} \varphi_i(v) = \sum_{i \in I} \sum_{\substack{S \subseteq I \\ i \in S}} \frac{(s-1)!\,(n-s)!}{n!} \{v(S) - v(S - \{i\})\}$$

In the above double summation focus on terms involving $v(T)$, where T is a coalition such that $T \neq \varnothing$, and $T \subset I$. There are terms in $v(T)$ with both positive and negative coefficients. Let $|T| = t$. The positive coefficient (when $T = S$) is

$$\frac{(t-1)!\,(n-t)!}{n!}$$

The negative coefficient (when $T = S - \{i\}$) is

$$-\frac{(n-1-t)!t!}{n!}$$

The positive coefficient occurs t times (once for each player in the set T), and the negative coefficient occurs $(n-t)$ times (once for each player not in the set T). Thus the coefficient of $v(T)$ in the double sum is

$$t\frac{(t-1)!\,(n-t)!}{n!} - (n-t)\frac{(n-1-t)!t!}{n!} = 0$$

Therefore, the terms left in the double summation are those involving the empty and grand coalitions. As $v(\varnothing) = 0$, we have

$$\sum_{i \in I} \varphi_i(v) = \frac{n\,(0!)\,(n-1)!}{n!} v(I) = v(I)$$

This establishes group rationality.

7. Consider a superadditive coalitional $\Gamma = (I, v)$ game with transferable utility, in which $I = \{1, 2, 3\}$. The values of the coalitions are

$$v(\varnothing) = 0,\ v(\{1\}) = 1,\ v(\{2\}) = 2,\ v(\{3\}) = 0$$
$$v(\{1,2\}) = 4,\ v(\{1,3\}) = 3,\ v(\{2,3\}) = 5,\ v(\{1,2,3\}) = 10$$

Show that the marginal values of the coalitions are

$$c(\varnothing) = 0,\ c_{\{1\}}(v) = 1,\ c_{\{2\}}(v) = 2,\ c_{\{3\}}(v) = 0$$
$$c_{\{1,2\}}(v) = 1,\ c_{\{1,3\}}(v) = 2,\ c_{\{2,3\}}(v) = 3,\ c_{\{1,2,3\}}(v) = 1$$

Using $v(\cdot) = \sum_{R \subseteq I} c_R(v)\, v_R(\cdot)$, and

$$\varphi_i(v) = \sum_{\substack{R \subseteq I \\ i \in R, R \neq \varnothing}} \frac{c_R(v)}{r}; \quad |R| = r,\ i \in I$$

show that the Shapley vector is

$$\varphi(v) = (\varphi_1(v), \varphi_2(v), \varphi_3(v)) = (17/6, 26/6, 17/6)$$

Observe that $\sum_{i=1}^{3} \varphi_i(v) = v(\{1,2,3\}) = 10$. Redetermine the Shapley vector by using explicit expression for the $\varphi_i(v)$'s specified in Shapley's theorem.

8. Prove Nash bargaining game theorem.
Hint: See Jones (1980), Owen (1995), and Barron (2008). The proof is established in several steps.
Step 1: Existence of the maximum is established in this step. Observe that the set

$$U^* = \{(u_1, u_2) \in U \mid u_1 \geq d_1, u_2 \geq d_2\}$$

is convex and compact (closed and bounded). Define a function $g(\cdot)$, where $g : U^* \to \mathbb{R}$, and $g(u_1, u_2) = (u_1 - d_1)(u_2 - d_2)$. The function $g(\cdot)$ is continuous, and U^* is compact. Therefore the function $g(\cdot)$ has both a maximum and a minimum on the set. Thus $g(\cdot)$ has a maximum at some point $u^* \in U^*$. As per the hypothesis of the theorem, there exists at least a point $u = (u_1, u_2)$ such that $u_1 > d_1$ and $u_2 > d_2$. At this point $g(u_1, u_2) > 0$. Therefore, the maximum of $g(\cdot)$ over U^* is greater than 0, and it does not occur at the status-quo point $d = (d_1, d_2)$.
Step 2: The uniqueness of the maximum is established in this step. Assume that the maximum of $g(u_1, u_2)$ exists at two points $(u_1, u_2) \in U^*$, and $(v_1, v_2) \in U^*$, such that $g(u_1, u_2) = g(v_1, v_2) = \eta \in \mathbb{R}^+$. If $u_1 = v_1$, then from

$$\eta = (u_1 - d_1)(u_2 - d_2) = (v_1 - d_1)(v_2 - d_2) > 0$$

it can be concluded that $u_2 = v_2$. Next assume $u_1 < v_1$, which in turn implies $u_2 > v_2$. Define $(w_1, w_2) = \{(u_1, u_2) + (v_1, v_2)\}/2$. As the set U is convex, $(w_1, w_2) \in U$, and $w_1 > d_1$ and $w_2 > d_2$. Therefore $(w_1, w_2) \in U^*$. After some algebraic manipulations, we obtain

$$g(w_1, w_2) = \eta + \frac{1}{4}(u_1 - v_1)(v_2 - u_2) > \eta, \quad \text{as } u_1 < v_1 \text{ and } u_2 > v_2$$

This is a contradiction, as η is the maximum value. Therefore the maximum point $u^* = (u_1^*, u_2^*) \in U^*$ is unique.
Step 3: The following assertion is established in this step. If $u^* = (u_1^*, u_2^*) \in U^*$, and

$$h(u_1, u_2) = (u_2^* - d_2) u_1 + (u_1^* - d_1) u_2$$

then $\forall (u_1, u_2) \in U^*$, the inequality $h(u_1, u_2) \leq h(u_1^*, u_2^*)$ is true.
This result is used in Step 4. Let $(u_1, u_2) \in U^*$, and assume that $h(u_1, u_2) > h(u_1^*, u_2^*)$. As $h(\cdot, \cdot)$ is linear in its arguments, we have

$$0 < h(u_1, u_2) - h(u_1^*, u_2^*) = h(u_1 - u_1^*, u_2 - u_2^*)$$

Let $0 < \theta < 1$, and define (u_1', u_2') as

$$\begin{aligned}(u_1', u_2') &= \theta(u_1, u_2) + (1 - \theta)(u_1^*, u_2^*) \\ &= (u_1^*, u_2^*) + \theta(u_1 - u_1^*, u_2 - u_2^*)\end{aligned}$$

Next expand $g(u_1', u_2')$ about (u_1^*, u_2^*) via Taylor's series. This yields

$$g(u_1', u_2') = g(u_1^*, u_2^*) + \theta h(u_1 - u_1^*, u_2 - u_2^*) + \theta^2 (u_1 - u_1^*)(u_2 - u_2^*)$$

Select $\theta \in (0, 1)$ sufficiently small so that the third term in the above expansion is smaller than the second term in magnitude. That is,

$$\theta \left| (u_1 - u_1^*)(u_2 - u_2^*) \right| < h(u_1 - u_1^*, u_2 - u_2^*)$$

Then

$$\theta h(u_1 - u_1^*, u_2 - u_2^*) + \theta^2 (u_1 - u_1^*)(u_2 - u_2^*) > 0$$

and therefore $g(u_1', u_2') > g(u_1^*, u_2^*)$. But this contradicts the assertion that $g(u_1, u_2)$ is maximized at (u_1^*, u_2^*). This completes the proof that $\forall (u_1, u_2) \in U^*, h(u_1, u_2) \le h(u_1^*, u_2^*)$. Actually this result shows that the line through (u_1^*, u_2^*), with slope negative of the line joining (u_1^*, u_2^*) to (d_1, d_2), is a support line for U^*. That is, U^* lies entirely on or below this line.
Step 4: The last part of the theorem is proved in this step. The existence and uniqueness of (u_1^*, u_2^*) which maximizes $g(\cdot, \cdot)$ has been established in Steps 1 and 2 respectively. This point evidently satisfies the axioms of individual rationality and feasibility. The axiom of Pareto optimality is also satisfied. Observe that, if $u_1 \ge u_1^*$ and $u_2 \ge u_2^*$, but $(u_1, u_2) \ne (u_1^*, u_2^*)$, then $g(u_1, u_2) > g(u_1^*, u_2^*)$. This contradicts the maximality of $g(u_1^*, u_2^*)$.
The symmetry axiom is also satisfied. Let the utility set U be such that $(u_1, u_2) \in U$ if and only if $(u_2, u_1) \in U$, and the status-quo point $d = (d_1, d_2)$ is such that $d_1 = d_2$. Therefore $(u_2^*, u_1^*) \in U^*$. Consequently

$$g(u_2^*, u_1^*) = (u_2^* - d_1)(u_1^* - d_2) = (u_2^* - d_2)(u_1^* - d_1) = g(u_1^*, u_2^*)$$

As $(u_1^*, u_2^*) \in U^*$ is the unique point which maximizes $g(\cdot, \cdot)$ over U^*, we obtain $(u_2^*, u_1^*) = (u_1^*, u_2^*)$. That is, $u_1^* = u_2^*$.
The axiom of invariance under linear transformation is also satisfied. Let the set V be obtained from U via linear transformation, where for $u = (u_1, u_2) \in U$ and $v = (v_1, v_2) \in V$

$$v_1 = (\alpha_1 u_1 + \beta_1), \quad v_2 = (\alpha_2 u_2 + \beta_2); \qquad \alpha_1, \alpha_2 \in \mathbb{R}^+, \ \beta_1, \beta_2 \in \mathbb{R}$$

Then

$$g'(v_1, v_2) \triangleq \{v_1 - (\alpha_1 d_1 + \beta_1)\}\{v_2 - (\alpha_2 d_2 + \beta_2)\} = \alpha_1 \alpha_2 g(u_1, u_2)$$

Therefore if $g(\cdot, \cdot)$ is maximized at (u_1^*, u_2^*) over the utility set U, then $g'(\cdot, \cdot)$ is maximized at (v_1^*, v_2^*) over the utility set V, where

$$v_1^* = (\alpha_1 u_1^* + \beta_1), \quad v_2^* = (\alpha_2 u_2^* + \beta_2)$$

The axiom of independence of irrelevant alternatives is also satisfied. Observe that, if $(u_1^*, u_2^*) = u^* \in U^*$ maximizes $g(\cdot, \cdot)$ over the domain U^*, and if V is a closed convex subset of U, $d \in V$, and $u^* \in V$, then u^* clearly maximizes $g(\cdot, \cdot)$ over the smaller domain V.
Therefore, the function $f(\cdot)$ defined on all bargaining games $B_g = (U, d)$ such that $f(B_g) = u^* \in U^*$ satisfies all of six Nash axioms. Next consider the set

$$\Lambda = \{(u_1, u_2) \mid h(u_1, u_2) \le h(u_1^*, u_2^*)\}$$

where the function $h(\cdot, \cdot)$ is defined in Step 3. Step 3 implies that $U^* \subseteq \Lambda$. A set Λ' is obtained from Λ via a linear transformation $\mathcal{T}$

$$u_1' = \frac{u_1 - d_1}{u_1^* - d_1}, \quad u_2' = \frac{u_2 - d_2}{u_2^* - d_2}$$

Observe that

$$\Lambda' = \{(u_1', u_2') \mid (u_1' + u_2') \le 2\}$$

Note that, since $u_1^* > d_1$, and $u_2^* > d_2$, this linear transformation is of the type specified in Nash axioms. In this linear transformation, (d_1, d_2) is mapped to $(0,0)$, and (u_1^*, u_2^*) is mapped to $(1,1)$.

Consider the bargaining game $(\Lambda', (0,0))$. This game is symmetric via Nash's symmetry axiom. Therefore, its solution must lie on the line $u_1' = u_2'$. Therefore use of the axiom of Pareto optimality yields the unique solution of this game, which is $(1,1)$. Use of Nash's axiom of invariance under linear transformation, applied to the inverse of the transformation T yields (u_1^*, u_2^*) as the unique solution of the bargaining game (Λ, d). However, $U^* \subseteq \Lambda$, and $(u_1^*, u_2^*) \in U^*$, it follows that (u_1^*, u_2^*) is the unique solution of the Nash bargaining game (U, d).

9. If the hypothesis of the Nash's theorem is *not* satisfied, obtain a solution of the Nash bargaining game.

   Hint: See Jones (1980), and Owen (1995). We consider the case, when the hypothesis of the Nash's theorem is *not* satisfied. Suppose that there are no points $(u_1, u_2) \in U$ with $u_1 > d_1$ and $u_2 > d_2$.

   Convexity of the set U implies that if there is a point $(u_1, u_2) \in U$ such that $u_1 > d_1$ and $u_2 = d_2$, then there can be no point $(u_1, u_2) \in U$ with $u_2 > d_2$. In this case, we define (u_1^*, u_2^*) to be the point in U which maximizes u_1, subject to the constraint $u_2 = d_2$.

   Similarly, if there is a point $(u_1, u_2) \in U$ such that $u_1 = d_1$ and $u_2 > d_2$, then there can be no point $(u_1, u_2) \in U$ with $u_1 > d_1$. In this case, we define (u_1^*, u_2^*) to be the point in U which maximizes u_2, subject to the constraint $u_1 = d_1$.

   It can indeed be checked that these solutions satisfy Nash axioms. Further, the axioms of individual rationality, feasibility, and Pareto optimality ensure uniqueness.

References

1. Barron, E. N., 2008. *Game Theory, An Introduction*, John Wiley & Sons, Inc., New York, New York.
2. Clarke, E. H., 1971. "Multipart Pricing of Public Goods," Public Choice, Vol. 11, Issue 1, pp. 19-33.
3. Courcoubetis, C., and Weber, R., 2003. *Pricing Communication Networks, Economics, Technology and Modeling*, Second Edition, John Wiley & Sons, Inc., New York, New York.
4. Dramitinos, M., Stamoulis, G. D., and Courcoubetis, C., 2007. "An Auction Mechanism for Allocating Bandwidth of Networks to Their Users," Computer Networks, Vol. 51, Issue 18, pp. 4979-4996.
5. Eichberger, J., 1993. *Game Theory for Economists*, Academic Press, New York, New York.
6. Ferguson, T. S., 2014. "Game Theory," Second Edition, available at http://www.math.ucla.edu/~tom/Game_Theory/Contents.html. Retrieved February 16, 2015.
7. Franklin, J., 1980. *Methods of Mathematical Economics: Linear and Nonlinear Programming, Fixed-Point Theorems,* Springer-Verlag, Berlin, Germany.
8. Friedman, J. W., 1986. *Game Theory with Applications to Economics*, Oxford University Press, Oxford, Great Britain.
9. Gibbons, R., 1992. *Game Theory for Applied Economists*, Princeton University Press, Princeton, New Jersey.
10. Jiang, L., Parekh, S., and Walrand, J., 2008. "Time-Dependent Network Pricing and Bandwidth Trading," IEEE Network Operations and Management Symposium Workshop, Salvador da Bahia, pp. 193-200.

11. Jones, A. J., 1980. *Game Theory: Mathematical Models of Conflict*, John Wiley & Sons, Inc., New York, New York.
12. Koo, D., 1977. *Elements of Optimization, With Application in Economics and Business*, Springer-Verlag, Berlin, Germany.
13. Liu, Z., and Xia, C. H., Editors, 2008. *Performance Modeling and Engineering*, Springer-Verlag, Berlin, Germany.
14. Ma, R. T. B., Chiu, D. M., Lui, J. C. S., and Misra, V., 2010. "Internet Economics: The Use of Shapley Value for ISP Settlement," IEEE/ACM Transactions on Networking, Vol. 18, No. 3, pp. 775-787.
15. Mackie-Mason, J. K., and Varian, H. R., 1994. "Generalized Vickrey Auctions," citeseer.ist.psu.edu/mackie-mason94generalized.html
16. Mackie-Mason, J. K., and Varian, H. R., 1995. "Pricing Congestible Network Resources," IEEE Journal on Selected Areas in Communications, Vol. 13, No. 7, pp. 1141-1149.
17. McKnight, L. W., and Bailey, J. P., Editors, 1998. *Internet Economics,* The MIT Press, Cambridge, Massachusetts.
18. Morris, P., 1991. *Introduction to Game Theory,* Springer-Verlag, Berlin, Germany.
19. Odlyzko, A., 1998. "Paris Metro Pricing for the Internet," ACM Conference on Electronic Commerce (EC '99), ACM Press, New York, pp. 140-147.
20. Owen, G., 1995., *Game Theory*, Third Edition, Academic Press, New York, New York.
21. Rader, T., 1972. *Theory of Microeconomics*, Academic Press, New York, New York.
22. Roth, A., 1988. *The Shapley Value: Essays in Honor of Lloyd S. Shapley*, Cambridge University Press, Cambridge, Great Britain.
23. Shy, O., 2001. *The Economics of Network Industries*, Cambridge University Press, Cambridge, Great Britain.
24. Tuffin, B., 2003. "Charging the Internet Without Bandwidth Reservation: An Overview and Bibliography of Mathematical Approaches," Journal of Information Science and Engineering, Vol. 19, No. 5, pp. 765-786.
25. Vega-Redondo, F., 2003. *Economics and the Theory of Games*, Cambridge University Press, Cambridge, Great Britain.
26. Vickrey, W., 1961. "Counterspeculation, Auctions and Competitive Sealed Tenders," Journal of Finance, Vol. 16, Issue 1, pp. 8-37.

CHAPTER 11

Data Mining and Knowledge Discovery

० १ २ ३ ४ ५ ६ ७ ८ ९

Devanagari numerical symbols

Brahmagupta. Brahmagupta Bhillamalacharya (the teacher from Bhilla-mala) was born in the year 598 in Ujjain, India. He wrote two influential textbooks. These textbooks are *Brahmasphutasiddhanta*, and *Khandakhadyaka*. *Brahmasphutasiddhanta* (The Opening of the Universe) is perhaps the first treatise on mathematics and astronomy in which the number zero was treated with sufficient clarity. According to Brahmagupta, zero is a number which when added to or subtracted from another number gives the same number, and when multiplied by another number yields a zero.
Brahmagupta also introduced negative numbers, developed the modern algorithm to compute square roots. He formulated rules to solve linear and quadratic equations, and manipulation of surds. It can safely be said that the current rules of arithmetic we know today were perhaps first elucidated lucidly in his treatise *Brahmasphutasiddhanta*.
Brahmagupta also found the general integral solution for certain linear Diophantine equations. Brahmagupta also discovered expressions for the sum of squares of first n positive integers, and the sum of cubes of first n positive integers. Brahmagupta also contributed to geometry. He discovered an expression for the area of a cyclic quadrilateral. Brahmagupta also developed tables for values of trigonometric functions. His work had a profound and direct impact upon Islamic and Byzantine mathematics and astronomy. Brahmagupta passed away in the year 670.

11.1 Introduction

Data is an entity that communicates some information, which is convenient to process, store, and transmit. Data mining is the art and science of discovering and extracting potentially useful, unexpected, and understandable (to humans) knowledge from large data sets. This is achieved by discovering patterns and anomalies by using statistically significant structures in the data. Certain preliminary notions about data mining and knowledge discovery are initially outlined in this chapter. This is followed by a description of several clustering algorithms. Data mining via association rules and sequential pattern discovery is next discussed. Certain metric-based classification methods are also described. Classification of data points via nonmetric methods, like decision tree and rule-based induction are also discussed. Ensemble techniques, methods using rough sets, high-dimensional data, certain useful statistical techniques, and certain analytical aspects of analysis of social networks are also described. Subsequently, the hidden Markov model is also discussed. An introduction to the rich field of Kalman filtering is also provided. Search techniques on the network are discussed in the chapter on stochastic structure of the Internet, and the World Wide Web.

Data mining is also referred to as information discovery, information harvesting, data archaeology, induction of knowledge from databases, and knowledge extraction. The process of data mining consists of determining certain comprehensible attributes and their values from large amounts of mostly unsupervised data in some application domain. It aims to develop both predictive models, and perceptive guidelines.

The goal of knowledge discovery is to find nuggets of information and patterns in an application domain. More specifically, knowledge discovery is the complete process of: data access, data selection and sampling, data preparation, data transformation, model building, model evaluation, model deployment, model monitoring, model visualization, and consolidation of discovered knowledge. Thus data mining is a subset of the knowledge discovery process.

The recent explosive growth of the Internet, and the World Wide Web has resulted in a deluge of data and information. The abundance of data is both a blessing and a challenge. The abundance of data is a blessing, because more data might potentially mean more information. However, the high volume of data presents a challenge to the computational and storage resources. It is therefore imperative that efficient processes and tools be built to extract useful information from such data.

Data mining and knowledge discovery have several applications. Some of these are in medicine and health care, genomic data, weather prediction, sensor data, electronic commerce like personal profile marketing, security, fraud detection, multimedia documents, and several more scientific and engineering applications.

The mathematical prerequisites for studying this chapter are applied analysis, theory of matrices and determinants, graph theory, probability theory, stochastic processes, and basic principles of optimization. Knowledge of information theory is also beneficial. A data tuple is referred to as data element (or element), data item, data object (or object), data point, data vector, instance, record, or sample in this chapter. An attribute of a data point is also generally referred to as characteristic, feature, field, or variable. We denote the expectation operator by $\mathcal{E}(\cdot)$.

We assume that the results of the data mining inquiry are interpreted within a logical, scientific, economical, and social framework. Let the reader be forewarned that the important issue of ethics in data mining is not covered in this chapter. However this issue is worthy of both profound thought and scholarship.

11.2 Basics of Knowledge Discovery Process and Data Mining

Basics of knowledge discovery process and data mining are discussed in this section. Steps in the knowledge discovery process are enumerated. This is followed by a description of elements of data mining. Data mining can be considered a learning process. Therefore, different types of learning processes are also discussed. Data mining as an inductive learning paradigm is also elaborated upon.

11.2.1 Knowledge Discovery Process

With the proliferation of many different application domains, the various steps in the knowledge discovery process have to be formalized. Knowledge discovery process generally consists of the following nine steps. These steps are both interactive and highly iterative.

1. *Learning about the application domain*: This step consists of familiarization of the data miner with prior knowledge and goals of the application domain.
2. *Creating a target data set*: The data miner selects a data set and a subset of attributes upon which knowledge discovery is performed.
3. *Data cleaning, preprocessing and integration*: This step consists of removing outliers in the data, dealing with noise issues, taking care of missing data fields, and also includes combining multiple data sources if present.

4. *Data reduction and projection*: The data miner determines useful attributes via dimension reduction and transformation methods.
5. *Selection of data mining task*: The data miner matches the goal of knowledge discovery defined in step 1 with a specific data mining method like regression model, classification, and clustering.
6. *Selection of data mining algorithm(s)*: The data miner selects an appropriate algorithm to discover patterns in the data.
7. *Data mining*: This step consists of generating patterns and trends in the form of say, regression models, decision trees, clustering, and classification rules.
8. *Interpretation of results*: The discovered patterns in the last step are interpreted and visualized.
9. *Assimilation of discovered knowledge*: In this final step, the discovered knowledge is incorporated into the performance system and documented. Finally actions are taken based upon the knowledge.

11.2.2 Elements of Data Mining

Data mining consists of characterizing collected data by fitting data to certain models. It is hoped that the fitted models provide the data miner with knowledge. The field of data mining has borrowed techniques from database management, machine learning, parallel algorithms, pattern recognition, statistics, and visualization schemes.

For a thorough understanding of data mining techniques, it is important to recognize different types of data. Basics of different techniques that are used to represent knowledge extracted from the given data sets are also important. Knowledge is extracted from the data sets by using appropriate data mining algorithms. Therefore we also provide a listing of generic components of a data mining algorithm. It is also important to be aware of challenges in data mining, and trends that influence data mining. Thus the following basic elements of data mining are examined.

(a) Different types of data
(b) Different types of knowledge representation
(c) Generic components of a data mining algorithm
(d) Challenges in data mining
(e) Trends that effect data mining

Different Types of Data

Data objects, or data points, or records are described by an ordered set of attributes. An attribute is a characteristic of a data object. A data object might have several attributes. The different types of attributes can be described by the following operations.

- *Distinctness*: $=$, and $\neq$.
- *Order*: $<$, $\leq$, $>$, and $\geq$.
- *Addition*: $+$, and $-$.
- *Multiplication*: $\times$, and $/$.

In general, there are two types of attributes. These are: qualitative and quantitative attributes.

- *Qualitative attributes*: Qualitative or *categorical* attributes take nonnumerical values. There are two types of qualitative attributes. These are: nominal and ordinal attributes.

- *Nominal attributes*: Nominal attribute values have the property of distinctness. That is, these attributes simply take different names. The different names are distinguished by the equality and nonequality operations. That is, $=$, and $\neq$. For example: eye color, tax-identification number, and gender of a person.
- *Ordinal attributes*: Ordinal attribute values have the properties of distinctness and order. That is, these attribute values simply take different names. However, the values provide sufficient information to order them. For example, the names of different persons are lexicographically ordered by the operations: $<$ or $>$. In another example, the grades of a student might be A, B, and C; which we know are ordered.

- *Quantitative attributes*: Quantitative or *numeric* attributes take numerical values. There are two types of quantitative attributes. These are: interval and ratio attributes.
  - *Interval attributes*: Interval attribute values have the properties of distinctness, order, and additivity. That is, interval attributes take numerical values. The operations on different values are, $+$, and $-$. For example, the height or weight of a person.
  - *Ratio attributes*: Ratio attribute values have the properties of distinctness, order, additivity, and multiplicativity. The difference and ratio operations are meaningful for the values of ratio attributes. The operations on different values are: $+$, $-$, $\times$, and $/$. Some examples are: mass, length, temperature, electrical current.

Data can also be described by the number of values of attributes. *Discrete-valued* attribute can have finite number of values in practice. For example, *binary-valued* attributes can have only two values. These can be zero/one, true/false, yes/no, or male/female. *Continuous-valued* attributes take real values. In practice these real values have limited precision.

Different Types of Knowledge Representation

Some different types of knowledge representation are: classification, clustering, association rules, sequence analysis, description and visualization, estimation, numeric prediction, and time series analysis.

- *Classification*: The purpose of classification of a data point or record is to assign a label (or labels) to it. This is based upon other attributes in the data set. The target attribute in a data set is typically categorical.
- *Clustering*: Clustering is the process of segmenting data points into subgroups with similar characteristics.
- *Association rules*: Association-rules mining algorithms determine correlations (associations) between different attributes of a set of data points.
- *Sequence analysis*: Sequence analysis algorithms determine frequent sequences and/or episodes in data.
- *Description and visualization*: This is the process of characterizing data points via a small number of attributes. This can also be achieved via visualization.
- *Estimation*: Estimation resembles classification. However estimation is used for target attributes which are numerical.
- *Prediction*: This is similar to estimation, except that the predicted values lie in the future. For example, regression algorithms can predict one or more continuous variables. Some other examples are neural networks, and Bayesian classification schemes.

- *Time series analysis*: The sequence of values of a target attribute which evolve over time are called a time series. Time series analysis involves characterization of temporal dynamics of the target attribute. The aim of this analysis is to predict future values of the target attribute.

Generic Components of a Data Mining Algorithm

In general, a data mining algorithm consists of three components: the model, the preference criterion, and the search algorithm.

- *Model*: The model provides the functionality of the algorithm. Examples of a function of the model are: classification and clustering. The model has parameters which are to be determined from the data.
- *Preference criterion*: The preference criterion specifies the selection of the model and the choice of parameters, over other models and parameters. This criterion is evidently data dependent.
- *Search algorithm*: This is the search and specification of algorithm for determining specific model (or family of models) and parameters for given data, and also a preference criterion.

A given data mining algorithm is typically an instantiation of the above metamodel.

Challenges in Data Mining

An important requirement for any scientific discipline to prosper and progress is to face and overcome new challenges. Some of the challenges in the field of data mining are:

- *Scaling of data mining algorithms*: As the size of the data sets increase, it becomes a challenge for data mining algorithms to fit data into the computer memory. Thus algorithms which execute at different scales of data are required.
- *Curse of high dimensionality*: High-dimensional data increases the size of the search space exponentially. This in turn increases the probability of generation of invalid and spurious patterns.
- *Missing and noisy data*: Missing and noisy data is specially problematic in business-related data.
- *Extension of data mining algorithms to new data types*: Data mining algorithms should be able to work with not only vector-valued data; but also multimedia and collaborative data, hierarchical and multiscale data, unstructured data like text, and semi-structured data like HyperText Markup Language (HTML) and Extensible Markup Language (XML).
- *Development of distributed data mining algorithms*: A majority of data mining algorithms work today with a centralized data warehouse. To counter the explosive growth of data, it is important to develop distributed data mining algorithms.
- *Ease of use*: Data mining is typically a semi-automated process. It is indeed a challenge to develop data mining algorithms which are easy to use even by non-experts.
- *Privacy and security*: Inclusion of privacy and security is a key requirement of the data mining process. This would prevent misuse of data mining technology.

Trends that Effect Data Mining

Trends that effect data mining are related to trends in data, hardware, network, scientific computing, and commerce.

- *Data trends*: The exponential growth of data over the past few decades significantly affects the data mining techniques.
- *Hardware trends*: Hardware effects the speed with which the data mining algorithms are executed.
- *Network trends*: The latest network and protocol trends significantly effect distributed data mining algorithms.
- *Scientific computing trends*: Experimental and simulation work in science result in large data sets. The three modern facets of science: theory, experiment, and simulation are linked together by data mining and knowledge discovery.
- *Commercial trends*: As commerce is one of the driving forces behind the evolution of the Internet, data mining and knowledge discovery become a key technology.

11.2.3 Data Mining as a Learning Process

Learning from data sets is the process of discovering patterns and trends in them. Vast amounts of data can be generated by an application field. The task of the data miner is to extract significant patterns and trends, and interpret them from this deluge of data. This process (characterization) is called learning or induction from data. In general there are four types of learning. These are: unsupervised learning, supervised learning, reinforcement learning, and semi-supervised learning.

- *Unsupervised Learning*: In unsupervised learning, there are no output measures. However, based upon input measures, the goal is to generate patterns from the input data, and describe associations among them. Some examples are clustering of data, generation of association rules, and sequential pattern mining.
  Let $\mathcal{X}$ be the set of data points. Each data point in the set $\mathcal{X}$ is characterized by a set of attributes. The goal of the clustering algorithm is to determine groups of data points in the set $\mathcal{X}$ with "similar" characteristics. The grouping of data points depends upon the concept of *similarity* or *distance* between the points.
  Another example of unsupervised learning scheme involves discovery of useful and interesting associations (relationships or dependencies) in large sets of data items. The different items occur in transactions which can be either extracted from a database, or are generated by some external process.
  Sequential pattern mining is concerned with discovering interesting sequences in the data set.
- *Supervised Learning*: The goal of supervised learning is to predict output measures (target vectors) based upon input measures (input vectors). Examples of supervised learning are: statistical methods like Bayesian methods, support vector machines, nearest-neighbor classification, artificial neural networks, regression analysis, decision trees, rule-based algorithms, and ensemble methods.
- *Reinforcement and Semi-supervised Learning*: Supervised and unsupervised learning are at the two ends of the learning spectrum. Reinforcement and semi-supervised learning techniques fall in the middle of these two extremes.
  - *Reinforcement Learning*: Compared to unsupervised learning algorithms, the learning algorithm is offered reinforcement. However, when compared to supervised learning, the supervision is offered in lesser detail. That is, supervision is offered only in aggregation of information.

- *Semi-supervised Learning*: The semi-supervised algorithms are also called learning with knowledge-based hints. For example, it is possible for clustering algorithms to execute under partial supervision.

An Alternative Perspective on Learning

The process of learning can be considered from another perspective. In this view point, there are three types of learning. These are: deductive, inductive, and transductive learning. Deductive learning draws new conclusions from existing rules and patterns. Inductive learning infers general rules and patterns from the given training data set. Thus new knowledge is inferred from data points, which is not necessarily guaranteed to be correct. Transductive learning involves learning specific examples from the given training data set. However, it does not involve generation of new rules. Therefore it is weaker than inductive learning.

Supervised, unsupervised, and reinforcement learning are inductive. Semi-supervised learning can be either inductive or transductive. The inductive learning paradigm is next elaborated. Let $\mathcal{X}$ be the set of data points, and Ω be the set of labels for different classes of data, where $|\Omega| \in \mathbb{P}\backslash \{1\}$. The training data is a set $\mathcal{D}$ of n observations,

$$\mathcal{D} = \{(x_i, y_i) \mid x_i \in \mathcal{X}, \varPhi_{cl}(x_i) = y_i \in \Omega, 1 \le i \le n\}$$

where, the classification is a mapping $\varPhi_{cl} : \mathcal{X} \to \Omega$.

An inductive learning algorithm is designed to determine the classification of *new data points* in the set $\mathcal{X}_{new}$, where $\mathcal{X}_{new} \cap \mathcal{X} = \varnothing$. Using the data set $\mathcal{D}$, the inductive learning algorithm discovers an optimal approximation of the function $\varPhi_{cl}(\cdot)$. Denote it by $f_{cl}(\cdot)$. Let $\mathcal{H}$ be the hypothesis space of all possible functions from which an optimal function $f_{cl}(\cdot)$ is selected. The principle of parsimony is used in selecting the function $f_{cl}(\cdot)$. The principle of parsimony can be stated in terms of the *minimum descriptor principle* (MDL). MDL-based measures use functions which maximize *information compression.* However, a practical goal of MDL is to select a simple scheme. After determining the approximating function $f_{cl}(\cdot)$, the learning algorithm $\mathcal{L}_{induc}(\cdot,\cdot)$ *inductively infers* the classification of a new data point $x \in \mathcal{X}_{new}$. This is denoted as

$$(x \wedge \mathcal{D}) \succ \mathcal{L}_{induc}(x, \mathcal{D})$$

where the notation $z \succ y$ implies that y is inductively inferred from z. Thus $\mathcal{L}_{induc}(x, \mathcal{D}) = y \in \Omega$ is the predicted classification of the data point $x \in \mathcal{X}_{new}$.

Given a sufficient number of training data points n, the function $f_{cl}(\cdot)$ approximates $\varPhi_{cl}(\cdot)$. Therefore the classification y of a new data point x is only approximate. The degree of approximation is measured by a *loss* or *risk function.* The risk function $\mathcal{R}_{rsk}(f_{cl}(\cdot)) = \mathcal{V}(f_{cl}(\cdot), \varPhi_{cl}(\cdot))$, where $\mathcal{V}(\cdot,\cdot)$ is an appropriate metric which measures the difference between the estimated and true values of the classification.

Additional assumptions are often required in practice to develop the inductive learning algorithm, so that $\mathcal{L}_{induc}(x, \mathcal{D})$ is correct for all new data points $x \in \mathcal{X}_{new}$. These assumptions are called the *inductive bias* of the learning algorithm. Denote the inductive bias of the learning algorithm by $\mathcal{B}_{bias}$. The bias $\mathcal{B}_{bias}$ will depend upon $f_{cl}(\cdot) \in \mathcal{H}$. More precisely, the inductive bias $\mathcal{B}_{bias}$ is the minimal set of assertions such that

$$(\mathcal{B}_{bias} \wedge \mathcal{D} \wedge x) \vdash \mathcal{L}_{induc}(x, \mathcal{D}), \quad \forall\, x \in \mathcal{X}_{new}$$

where the notation $a \vdash b$ indicates that b follows deductively from a. That is, b is provable from a. In summary, inductive learning is a *function approximation process*, and data mining is one of its important application.

11.3 Clustering

One of the prominent methods of unsupervised learning is clustering. Clustering is the mechanism of grouping together similar objects. The process of clustering is also called data segmentation, or grouping, or categorization. This approach of data mining appears to be very intuitive. It is grouping of data points which are *similar*. Data points which are *dissimilar* are in different groups. It is expected of the clustering algorithms to discover similarities or dissimilarities between the data points. Some fields of application of the clustering approach to data mining are: Web document classification, marketing, biology, medicine and so on.

Definition 11.1. *Let $\mathcal{X}$ be the set of data points. The m-clustering of $\mathcal{X}$ is the partition of $\mathcal{X}$ into $m \in \mathbb{P} \backslash \{1\}$ sets called clusters, where $m \leq |\mathcal{X}|$. The clusters $C_1, C_2, \ldots, C_m$ have the following properties.*

(a) $C_i \neq \varnothing$, *for* $1 \leq i \leq m$.
(b) $C_i \subseteq \mathcal{X}$, *for* $1 \leq i \leq m$.
(c) $\bigcup_{i=1}^{m} C_i = \mathcal{X}$.
(d) $C_i \cap C_j = \varnothing$, *for* $i \neq j$, *and* $1 \leq i, j \leq m$.
(e) *The data points within a cluster are more similar to each other, and less similar than data points in other clusters.* □

Note that, in the above definition of clustering, the definition of similarity is not given. The concept of similarity and dissimilarity between data points is quantified later in the section. Also observe that in the above definition, each data point belongs to a single cluster. This type of clustering is called *hard* or *crisp*. It is often the case that it is difficult to assign a data point to a single cluster. A data point might legitimately belong to more than a single cluster. This possibility can be incorporated in a fuzzy clustering scheme. A fuzzy clustering of the set of data points $\mathcal{X}$ into $m \in \mathbb{P}$ clusters is next defined.

Definition 11.2. *Let $\mathcal{X} = \{x_1, x_2, \ldots, x_n\}$ be the set of n data points. The fuzzy m-clustering of $\mathcal{X}$ is determined by $m \in \mathbb{P} \backslash \{1\}$ membership functions $u_j(\cdot)$, $1 \leq j \leq m$, where:*

(a) $u_j : \mathcal{X} \to [0, 1]$, $1 \leq j \leq m$.
(b) $\sum_{j=1}^{m} u_j(x_i) = 1$, $1 \leq i \leq n$.
(c) $0 < \sum_{i=1}^{n} u_j(x_i) < n$, $1 \leq j \leq m$. □

Note that, in the above definition each data point can belong to more than a single cluster simultaneously. Further, if $u_j(x_i)$ is closer to unity, then the data point x_i has a high "grade of membership" in cluster C_j. However, if $u_j(x_i)$ is closer to zero, then data point x_i has a low "grade of membership" in cluster C_j. Also, if $\sum_{i=1}^{n} u_j(x_i)$ is closer to n, then cluster C_j has several data points of high "grade of membership." Similarly, if $\sum_{i=1}^{n} u_j(x_i)$ is close to 0, then cluster C_j has several data points of low "grade of membership." The case of clustering in which each data point is clustered into only one of m possible distinct sets is a special instance of the fuzzy clustering. In this case, the fuzzy membership function $u_j(\cdot)$ takes values in the set $\{0,1\}$, for $1 \leq j \leq m$.

Steps in the Clustering Process

The clustering task requires determination of: feature selection, proximity measure, clustering criterion, clustering algorithms, validation of results, and interpretation of results.

- *Feature selection*: Proper selection of features upon which clustering is based is necessary in order to collect as much information as possible. If there are several features, then the best features are selected. It is assumed that all the interesting information is represented by the features in a data point.
- *Proximity measure*: Proximity measure evaluates the different degrees of "similarity" or "dissimilarity" between two data points.
- *Clustering criterion*: A "reasonable" clustering criterion has to be selected. The clustering criterion can be expressed via a cost function or alternately via some set of rules.
- *Clustering algorithms*: The clustering algorithm which discovers clusters in the data has to be selected.
- *Validation of results*: This refers to the correctness of results. Validation of results is performed via appropriate tests.
- *Interpretation of results*: The results generated via clustering analysis have to be interpreted and assimilated into existing body of knowledge.

Necessity of Efficient Clustering Algorithms

The clustering algorithms have to be numerically efficient. For example, if $n \in \mathbb{P}$ is the number of data points, and $m \in \mathbb{P}$ is the number of clusters, then the total number of partitions of the data points is the Stirling number of second kind, $S(n,m)$. It is

$$S(n,m) = \frac{1}{m!}\sum_{i=0}^{m}(-1)^{m-i}\binom{m}{i}i^{n}, \quad 1 \leq m \leq n$$

This number is extremely large, even for moderate values n and m. Therefore clustering methods of data mining have to be computationally efficient.

Categories of Clustering Algorithms

Some different techniques of clustering are:

- Partition-based (or objective-function based) clustering. For example, m-means and density-based clustering.
- Hierarchical clustering.
- Model-based clustering.
- Fuzzy clustering.
- Grid-based clustering.

Partition-based clustering is an objective-function based clustering analysis technique, in which the goal is the minimization of an objective function. It is hoped that this minimization would determine "proper" clusters of data points. Partition-based methods are dependent upon the distance between objects. In this scheme, specification of an appropriate objective function is challenging. Examples of partition-based clustering are m-means and density-based clustering methods

In the m-means clustering algorithm, the parameter m is constant, and the data points are classified into m clusters. Density-based methods rely upon the notion of *density*. If the data points within a region are densely packed, then these points form a cluster. However, if the density of points within a neighborhood (specified by radius) is low, then it is possible for these points to be parts of different clusters.

Another form of clustering scheme is hierarchical. Sometimes, it is also required that the clusters be arranged into a natural hierarchy. In the clustering hierarchy, objects within lower levels of hierarchy are more similar than those at higher levels.

In model-based clustering, a probabilistic model of the application is assumed. The data points are used in estimating the values of the parameters. In fuzzy clustering, a data point simultaneously belongs to more than one cluster. This is different than probabilistic models in which each data point belongs exclusively to a single cluster. Grid-based clustering initially quantizes the object space into cells of small size and form a grid structure. The clustering algorithm is exercised upon the grid structure. It is claimed that this scheme speeds up the clustering operation.

Proximity measures between data points, and also among clusters is defined and discussed in this section. This is followed by a description of m-means clustering, hierarchical clustering, model-based clustering, and fuzzy clustering. Finally some cluster validity measures are also stated.

11.3.1 Proximity Measures

The goal of a clustering algorithm is the grouping of similar objects. Therefore proximity measures (degrees of similarity and dissimilarity) between objects of a data set have to be quantified. It is also possible and necessary to specify proximity measures between sets of data points.

Definitions 11.3. *Let $\mathcal{X}$ be the set of data points. Dissimilarity and similarity measures of the set of data points in $\mathcal{X}$ are defined.*

1. *Dissimilarity measure on $\mathcal{X}$ is a function $d : \mathcal{X} \times \mathcal{X} \to \mathbb{R}$ where:*
   (a) $\exists\, d_0 \in \mathbb{R}$ *such that* $d_0 = d(x,x), \forall\, x \in \mathcal{X}$; *and* $d_0 \leq d(x,y), \forall\, x,y \in \mathcal{X}$.
   (b) $d(x,y) = d(y,x), \forall\, x,y \in \mathcal{X}$.
2. *The function $d(\cdot,\cdot)$ defined on $\mathcal{X}$ is called a metric dissimilarity measure if in addition:*
   (a) $d(x,y) = d_0 \Leftrightarrow x = y, \forall\, x,y \in \mathcal{X}$.
   (b) $d(x,z) \leq d(x,y) + d(y,z), \forall\, x,y,z \in \mathcal{X}$. *This inequality is also called the triangular inequality.*
3. *Similarity measure on $\mathcal{X}$ is a function $s : \mathcal{X} \times \mathcal{X} \to \mathbb{R}$ where:*
   (a) $\exists\, s_0 \in \mathbb{R}$ *such that* $s_0 = s(x,x), \forall\, x \in \mathcal{X}$; *and* $s(x,y) \leq s_0, \forall\, x,y \in \mathcal{X}$.
   (b) $s(x,y) = s(y,x), \forall\, x,y \in \mathcal{X}$.
4. *The function $s(\cdot,\cdot)$ defined on $\mathcal{X}$ is called a metric similarity measure if in addition:*
   (a) $s(x,y) = s_0 \Leftrightarrow x = y, \forall\, x,y \in \mathcal{X}$.
   (b) $s(x,y)\, s(y,z) \leq \{s(x,y) + s(y,z)\}\, s(x,z), \forall\, x,y,z \in \mathcal{X}$. □

For all $x, y \in \mathcal{X}$, larger values of $d(x,y)$ imply higher degree of dissimilarity between data points x and y. Similarly, for all $x, y \in \mathcal{X}$, larger values of $s(x,y)$ imply higher degree of similarity between data points x and y. Note that d_0 and s_0 can be equal to any real number. Occasionally the dissimilarity and similarity levels are also referred to as distance. However, these are actually not distances, as distance is defined on a metric space.

Examples 11.1. Let $x, y \in \mathcal{X}$ be t-dimensional data points, where $t \in \mathbb{P}, x = \begin{bmatrix} x_1 \; x_2 \; \cdots \; x_t \end{bmatrix}^T$, and $y = \begin{bmatrix} y_1 \; y_2 \; \cdots \; y_t \end{bmatrix}^T$.

1. Examples of dissimilarity metric measures. Let $\mathcal{X} \subseteq \mathbb{R}^t$, and $w_i \in \mathbb{R}_0^+$, for $1 \leq i \leq t$.
   (a) Euclidean distance: $d(x, y) = \left\{ \sum_{i=1}^{t} (x_i - y_i)^2 \right\}^{1/2}$, where $d_0 = 0$.
   (b) Weighted Chebyshev distance: $d(x, y) = \max_{i \in \{1,2,\ldots,t\}} w_i |x_i - y_i|$.
   (c) Weighted Minkowski distance: $d_p(x, y) = \left\{ \sum_{i=1}^{t} w_i |x_i - y_i|^p \right\}^{1/p}$, $p \in \mathbb{R}^+$. If $p = 1$, we have the Manhattan norm. Note that $d_\infty(x, y)$ is the Minkowski distance. Also $d_\infty(x, y) \leq d_2(x, y) \leq d_1(x, y)$.
2. Examples of similarity measures. Let $\mathcal{X} \subseteq \mathbb{R}^t$.
   (a) Inner product measure: $s(x, y) = x^T y = \sum_{i=1}^{t} x_i y_i$.
   (b) Cosine function measure:

$$s(x, y) = \frac{\sum_{i=1}^{t} x_i y_i}{\left\{ \sum_{i=1}^{t} x_i^2 \sum_{i=1}^{t} y_i^2 \right\}^{1/2}}$$

   where $x \neq 0$, and $y \neq 0$.
3. Examples on discrete space. Let $\mathcal{X} \subseteq \mathfrak{B}^{(t)}$, where $\mathfrak{B} = \{0, 1\}$.
   (a) Hamming distance: $d(x, y) = \sum_{i=1}^{t} |x_i - y_i|$. This is a dissimilarity metric measure.
   (b) In this scheme, the two binary vectors x and y are compared coordinate-wise.
      (i) $x_i = 1$, and $y_i = 1$, $i \in \{1, 2, \ldots, t\}$. This occurs a number of times out of t.
      (ii) $x_i = 0$, and $y_i = 1$, $i \in \{1, 2, \ldots, t\}$. This occurs b number of times out of t.
      (iii) $x_i = 1$, and $y_i = 0$, $i \in \{1, 2, \ldots, t\}$. This occurs c number of times out of t.
      (iv) $x_i = 0$, and $y_i = 0$, $i \in \{1, 2, \ldots, t\}$. This occurs d number of times out of t.
      Possible measures of dissimilarity and similarity between the two binary vectors x and y are

$$d(x, y) = \frac{b + c}{a + b + c + d}, \quad \text{and} \quad s(x, y) = \frac{a + d}{a + b + c + d}$$

      respectively. □

Dissimilarity Measures between Two Clusters

Sometimes, it is also necessary to describe dissimilarity measures between two clusters. A canonical measure of dissimilarity between two clusters C_i and C_j is denoted by $\widehat{\mathcal{D}}(C_i, C_j)$, where $i \neq j$, and $1 \leq i, j \leq m$. Some special cluster dissimilarity measures are:

- Minimum measure: $\widehat{\mathcal{D}}_{\min}(C_i, C_j) = \min_{x \in C_i, y \in C_j} d(x, y)$
- Maximum measure: $\widehat{\mathcal{D}}_{\max}(C_i, C_j) = \max_{x \in C_i, y \in C_j} d(x, y)$
- Mean measure: Assume that the means of clusters C_i and C_j are properly defined. Let these be a and b respectively. In this case, $\widehat{\mathcal{D}}_{mean}(C_i, C_j) = |a - b|$.

- Average measure:

$$\widehat{\mathcal{D}}_{avg}\left(C_i, C_j\right) = \frac{1}{|C_i|\,|C_j|} \sum_{x \in C_i, y \in C_j} d\left(x, y\right); \quad C_i \neq \varnothing, \;\; C_j \neq \varnothing$$

Dissimilarity measures between two clusters depend upon the different types of attributes (features). Recall that in general there are two types of attributes. These are: qualitative (categorical) and quantitative (numeric) attributes. Furthermore, there are two types of qualitative attributes. These are: nominal and ordinal attributes. There are also two types of quantitative attributes. These are: interval, and ratio attributes.

A nominal variable can take more than two values. For example, the possible colors of a flower are either red, pink, yellow, orange, blue, or white. Let the number of states of a nominal variable be p. Denote these states by $1, 2, \ldots, p$. Note that these integers do not represent any ordering. In the two vectors x and y, let the number of corresponding entries in which there is a mismatch be r. A measure of dissimilarity between the two vectors x and y is $d\left(x, y\right) = r/t$, where $t \in \mathbb{P}$ is the number of attributes of a data point.

An ordinal variable is like a nominal variable, with the added requirement that the ordinal values be ordered in some manner. Thus, ordinal variables are ordered categorical variables. For example, the performance of a student might be: *excellent, very good, good,* and *satisfactory.* The value of an ordinal variable can be subsequently mapped to a rank. Note that the values of an ordinal variable have a clear ordering, but the absolute "distances" among them is not known. The dissimilarity is then treated as in interval-scaled variables.

Interval-scaled variables take real values on a linear scale. Possible dissimilarity measures are: Euclidean, weighted Chebyshev, and weighted Minkowski distances. Similarly, possible similarity measures are the inner product, and cosine function. If the objects are represented by binary variables, then these distances cannot be used as dissimilarity measures. Use of binary variables as interval-scaled variables will often result in erroneous clusterings.

Ratio-based variables take positive real values on a nonlinear scale. However, these scales are nonlinear. The scales can be of exponential-type like: $ae^{b\tau}$ or $ae^{-b\tau}$, where $a, b \in \mathbb{R}^+$, and τ typically represents time. The dissimilarity between such objects can be addressed by treating them as interval-scaled variables. Another option would be to take the logarithm, and subsequently treat it as an interval-scaled variable.

Some clustering algorithms require the determination of proximity between the subsets of $\mathcal{X}$. Let $X_i \subseteq \mathcal{X}$, $i \in \{1, 2, \ldots, k\}$, and the set of subsets be $\mathfrak{X} = \{X_1, X_2, \ldots, X_k\}$. A proximity measure $\mathcal{P}$ on the set $\mathfrak{X}$ is a function $\mathcal{P} : \mathfrak{X} \times \mathfrak{X} \to \mathbb{R}$. The dissimilarity and similarity measures for sets can be defined just as for the individual elements of the set $\mathcal{X}$. Note that typically the proximity measure between the two sets X_i and X_j are defined in terms of proximity measures between elements of the sets X_i and X_j.

11.3.2 m-Means Clustering

The m-means clustering algorithm is an objective-function based clustering scheme. Its goal is to partition n number of t-dimensional data points into m clusters, where $m \leq n$, and $t \in \mathbb{P}$. Let the set of data points be $\mathcal{X} = \{x_i \mid x_i \in \mathbb{R}^t, 1 \leq i \leq n\}$. A set of data points form a cluster, if their inter-point distances are small when compared to their distances from data points which are outside of the cluster. The number of clusters m is specified a priori.

For each data point x_i, an indicator variable $r_{ij} \in \{0, 1\}$ is introduced, where $1 \leq j \leq m$, and $1 \leq i \leq n$. If the data point x_i is assigned to cluster C_j, then $r_{ij} = 1$, otherwise $r_{ij} = 0$. Thus $\sum_{j=1}^{m} r_{ij} = 1$ for each $i \in \{1, 2, \ldots, n\}$.

The mean of all the data points in a cluster is called its *centroid* or *cluster center*. Assume that the centroid of the jth cluster is β_j, where $1 \leq j \leq m$. The goal of the m-means clustering algorithm is to minimize Q, where

$$Q = \sum_{i=1}^{n} \sum_{j=1}^{m} r_{ij} \left\| x_i - \beta_j \right\|^2$$

where $\|\cdot\|$ is the Euclidean norm. The objective function Q represents the sum of squares of the distances of each data point from its centroid. The m-means clustering algorithm determines the $n \times m$ assignment matrix $[r_{ij}]$, and the set of centroids $\{\beta_j \mid 1 \leq j \leq m\}$ so that Q is minimized. This goal is achieved by minimizing Q iteratively.

Initially select the centroids $\{\beta_j \mid 1 \leq j \leq m\}$ from the set $\mathcal{X}$ arbitrarily. Next, keeping the location of centroids fixed, assign each data point x_i to a cluster so that $\left\| x_i - \beta_j \right\|^2$ is minimized. Thus data point x_i is assigned to cluster C_j, where

$$j = \arg \min_{1 \leq b \leq m} \left\| x_i - \beta_b \right\|^2$$

and $1 \leq i \leq n$. In the next step centroid locations are updated. This two-stage process is repeated till the objective function Q does not change, or until the change is very small.

This algorithm usually converges fast. However, there is no guarantee that it will find the global minima of the objective function. This scheme is often referred to as k-means algorithm in literature. The parameter k is equal to m, the number of clusters of the data points. This scheme is evidently not suitable for data with categorical attributes.

Algorithm 11.1. *m-Means Clustering Algorithm.*

Input: The set of data points $\mathcal{X} = \{x_i \mid x_i \in \mathbb{R}^t, 1 \leq i \leq n\}$.
The number of clusters $m \in \mathbb{P}$, where $m \leq n$, and $t \in \mathbb{P}$.
The accuracy factor $\varepsilon > 0$.
Output: The set of centroids of the clusters $\{\beta_j \mid 1 \leq j \leq m\}$.
The $n \times m$ assignment matrix $[r_{ij}]$; where $r_{ij} = 1$, if the data point x_i is assigned to cluster C_j, otherwise it is equal to 0, for $1 \leq j \leq m$ and $1 \leq i \leq n$.
begin
  Arbitrarily select different β_j's from the set $\mathcal{X}$ for $1 \leq j \leq m$.
  $\delta \leftarrow \infty$
  $Q \leftarrow \infty$
  (*main part of the algorithm*)
  while $\delta \geq \varepsilon$ **do**
  begin
    Step 1: for $1 \leq i \leq n$
      Data point x_i is assigned to cluster C_j, where

$j = \arg\min_{1 \le b \le m} \|x_i - \beta_b\|^2,$
$r_{ij} = 1, r_{ij'} = 0, \forall\, j' \in \{1, 2, \ldots, m\} \setminus \{j\}$

Step 2: Update centroid locations.

$$\beta_j = \frac{\sum_{i=1}^{n} r_{ij} x_i}{\sum_{i=1}^{n} r_{ij}}, \quad 1 \le j \le m$$

Step 3: Compute $\widehat{Q} = \sum_{i=1}^{n} \sum_{j=1}^{m} r_{ij} \left\|x_i - \beta_j\right\|^2$.
$\delta \leftarrow \left|\widehat{Q} - Q\right|$
$Q \leftarrow \widehat{Q}$

end (*end of while-loop*)
end (*end of m-means clustering algorithm*)

11.3.3 Hierarchical Clustering

Hierarchical clustering groups data objects into a tree of clusters. In general there are two types of hierarchical clustering methods. These are the top-down and bottom-up modes. In the top-down mode, hierarchical clustering is performed by splitting. In the bottom-up mode, hierarchical clustering is performed by merging. The top-down and bottom-up modes are also called *divisive* and *agglomerative* approaches respectively.

The hierarchical clustering can be represented graphically by a rooted binary tree called dendrogram. A dendrogram has at its root node all the data points in a single cluster. Each leaf node of the dendrogram has a single data point.

- *Divisive hierarchical clustering*: In this top-down strategy, the clustering algorithm initially begins with all the data points in a single cluster. In subsequent steps, it successively subdivides clusters into smaller and smaller clusters, until each data point forms a single cluster. It is also possible to terminate formation of smaller clusters until a prespecified number of clusters is achieved, or some other criterion is met.
- *Agglomerative hierarchical clustering*: In this bottom-up strategy, the clustering algorithm initially begins by placing each data point in a single cluster, and then merges these atomic clusters into successively larger clusters, until all the data points are merged into a single cluster. It is also possible to terminate the successive merging until a certain threshold value of the distance between clusters is achieved. That is, merging of the clusters is stopped when the distance between them is exceeded. This is a popular hierarchical clustering scheme.

Hierarchical clustering requires the computation of distance between two clusters. Hierarchical clustering techniques have several advantages over partition clustering methods like m-means algorithms. For example, hierarchical clustering techniques have the option of selecting any of the distance measures. The m-means clustering algorithm always gives m clusters at the end. However, the hierarchical clustering techniques form a hierarchy of clusters. This enables the user to examine clusters at different levels of granularity.

Hierarchical clustering techniques also have several drawbacks. The main drawback is the computational complexity and space requirement. Therefore, this scheme may not be practical for very large data sets.

11.3.4 Model-Based Clustering

A probabilistic approach to determination of clusters in a set of data points is considered in this subsection. Model-based clustering technique tries to fit the data points to a specific mathematical model. Let the set of data points be $\mathcal{X} = \{x_1, x_2, \ldots, x_n\}$. It is possible that these data points are drawn, probabilistically independent of each other from the same probability density function (or probability mass function).

Assume that the data points are generated by a *mixture* of probability distributions. Further assume that each cluster of data points is determined by a parametric *component probability distribution.* Let the set of data points be created from $m \in \mathbb{P}$ mutually independent sources of data, and each source $j \in \{1, 2, \ldots, m\}$ creates a cluster C_j. That is, data is drawn independently from source j with probability α_j, where $\alpha_j > 0$, $1 \leq j \leq m$, and $\sum_{j=1}^{m} \alpha_j = 1$. The probabilities α_j's are also called *mixing parameters*. Let the component probability density function (or probability mass function) of source j be $\varphi_j(x \mid \theta_j), x \in \mathbb{R}$, where θ_j is its corresponding vector of parameters. Note that the form of probability function $\varphi_j(x \mid \theta_j)$ is assumed, however θ_j is required to be determined from the data points.

Let $\theta = (\theta_1, \theta_2, \ldots, \theta_m)$, $\alpha = (\alpha_1, \alpha_2, \ldots, \alpha_m)$, and $\beta = (\theta, \alpha)$. Then a data point x is drawn from a probability density function $p(x \mid \beta), x \in \mathbb{R}$, where

$$p(x \mid \beta) = \sum_{j=1}^{m} \alpha_j \varphi_j(x \mid \theta_j)$$

It is also assumed in this model that $p(x \mid \beta)$ is identifiable. This implies that, if $\beta \neq \beta'$ then there exists an x such that $p(x \mid \beta) \neq p(x \mid \beta')$. The goal of model-based algorithm is to determine the vector β, from the set $\mathcal{X}$ of n data points. The principle of maximum-likelihood estimation is used in discovering these clusters. The corresponding log-likelihood function $L_{\mathcal{X}}(\beta)$ is

$$\begin{aligned} L_{\mathcal{X}}(\beta) &= \ln \prod_{i=1}^{n} p(x_i \mid \beta) = \sum_{i=1}^{n} \ln p(x_i \mid \beta) \\ &= \sum_{i=1}^{n} \ln \left\{ \sum_{j=1}^{m} \alpha_j \varphi_j(x_i \mid \theta_j) \right\} \end{aligned}$$

As per the principle of maximum-likelihood estimation, the vector β is determined by maximizing $L_{\mathcal{X}}(\beta)$. However, this is a daunting task because of the presence of logarithm of a sum in the expression for $L_{\mathcal{X}}(\beta)$.

An iterative technique called the expectation-maximization (EM) algorithm is used in maximum-likelihood estimation of the vector β. The EM algorithm assumes that the set of given data $\mathcal{X}$ is an *incomplete* representation of the underlying physical process. Therefore it uses the ingenious artifice of introducing hidden (latent) variables. We initially describe the genesis of the EM algorithm. This is followed by a description of the algorithm.

Genesis of the EM Algorithm

The goal of the EM algorithm is to maximize $L_{\mathcal{X}}(\beta) = \ln \prod_{i=1}^{n} p(x_i \mid \beta) \triangleq \ln p(\mathcal{X} \mid \beta)$. The EM algorithm iteratively maximizes $L_{\mathcal{X}}(\beta)$. Denote the τth iterative value of β by $\beta^{(\tau)}, \tau \in \mathbb{N}$. As $L_{\mathcal{X}}(\beta) \geq L_{\mathcal{X}}\left(\beta^{(\tau)}\right)$, the objective is to maximize

$$\left(L_{\mathcal{X}}(\beta) - L_{\mathcal{X}}\left(\beta^{(\tau)}\right)\right)$$

in the $(\tau+1)$th step. Hidden variables can be introduced to make the estimation of β tractable. Let Z be the hidden random vector, and z be an instance of it. Therefore

$$p(\mathcal{X} \mid \beta) = \sum_{z} p(\mathcal{X} \mid z, \beta)\, p(z \mid \beta)$$

This in turn yields

$$\begin{aligned}
& L_{\mathcal{X}}(\beta) - L_{\mathcal{X}}\left(\beta^{(\tau)}\right) \\
&= \ln\left\{\sum_{z} p(\mathcal{X} \mid z, \beta)\, p(z \mid \beta)\right\} - \ln p\left(\mathcal{X} \mid \beta^{(\tau)}\right) \\
&= \ln\left\{\sum_{z} p\left(z \mid \mathcal{X}, \beta^{(\tau)}\right) \frac{p(\mathcal{X} \mid z, \beta)\, p(z \mid \beta)}{p\left(z \mid \mathcal{X}, \beta^{(\tau)}\right)}\right\} - \ln p\left(\mathcal{X} \mid \beta^{(\tau)}\right)
\end{aligned}$$

Note that $\sum_{z} p\left(z \mid \mathcal{X}, \beta^{(\tau)}\right) = 1$, and $\ln(\cdot)$ is a concave function. Therefore, use of Jensen's inequality yields

$$\begin{aligned}
& L_{\mathcal{X}}(\beta) - L_{\mathcal{X}}\left(\beta^{(\tau)}\right) \\
&\geq \sum_{z} p\left(z \mid \mathcal{X}, \beta^{(\tau)}\right) \ln\left\{\frac{p(\mathcal{X} \mid z, \beta)\, p(z \mid \beta)}{p\left(z \mid \mathcal{X}, \beta^{(\tau)}\right)}\right\} - \ln p\left(\mathcal{X} \mid \beta^{(\tau)}\right) \\
&= \sum_{z} p\left(z \mid \mathcal{X}, \beta^{(\tau)}\right) \ln\left\{\frac{p(\mathcal{X} \mid z, \beta)\, p(z \mid \beta)}{p\left(z \mid \mathcal{X}, \beta^{(\tau)}\right) p\left(\mathcal{X} \mid \beta^{(\tau)}\right)}\right\} \\
&\triangleq \Delta\left(\beta \mid \beta^{(\tau)}\right)
\end{aligned}$$

In the last step, we again used the fact that $\sum_{z} p\left(z \mid \mathcal{X}, \beta^{(\tau)}\right) = 1$. Note that $\Delta\left(\beta^{(\tau)} \mid \beta^{(\tau)}\right) = 0$, and

$$L_{\mathcal{X}}(\beta) \geq L_{\mathcal{X}}\left(\beta^{(\tau)}\right) + \Delta\left(\beta \mid \beta^{(\tau)}\right)$$

Let

$$L_{\mathcal{X}}\left(\beta^{(\tau)}\right) + \Delta\left(\beta \mid \beta^{(\tau)}\right) \triangleq \ell\left(\beta \mid \beta^{(\tau)}\right)$$

Therefore

$$L_{\mathcal{X}}\left(\beta^{(\tau)}\right) = \ell\left(\beta^{(\tau)} \mid \beta^{(\tau)}\right), \quad \text{and} \quad L_{\mathcal{X}}(\beta) \geq \ell\left(\beta \mid \beta^{(\tau)}\right)$$

That is, $\ell\left(\beta \mid \beta^{(\tau)}\right)$ is bounded from above by $L_{\mathcal{X}}(\beta)$. Consequently, any value of β which increases $\ell\left(\beta \mid \beta^{(\tau)}\right)$ will also increase $L_{\mathcal{X}}(\beta)$. Let the new maximizing value of β be $\beta^{(\tau+1)}$. Therefore

$$\begin{aligned}
&\beta^{(\tau+1)} \\
&= \arg\max_{\beta} \left\{ \ell\left(\beta \mid \beta^{(\tau)}\right) \right\} \\
&= \arg\max_{\beta} \left\{ L_{\mathcal{X}}\left(\beta^{(\tau)}\right) + \sum_{z} p\left(z \mid \mathcal{X}, \beta^{(\tau)}\right) \ln \frac{p(\mathcal{X} \mid z, \beta)\, p(z \mid \beta)}{p\left(z \mid \mathcal{X}, \beta^{(\tau)}\right) p\left(\mathcal{X} \mid \beta^{(\tau)}\right)} \right\}
\end{aligned}$$

In the above expression for $\beta^{(\tau+1)}$, the terms which do not depend upon β can be dropped. Therefore

$$\begin{aligned}
\beta^{(\tau+1)} &= \arg\max_{\beta} \left\{ \sum_{z} p\left(z \mid \mathcal{X}, \beta^{(\tau)}\right) \ln \left\{ p(\mathcal{X} \mid z, \beta)\, p(z \mid \beta) \right\} \right\} \\
&= \arg\max_{\beta} \left\{ \sum_{z} p\left(z \mid \mathcal{X}, \beta^{(\tau)}\right) \ln p(\mathcal{X}, z \mid \beta) \right\} \\
&= \arg\max_{\beta} \left\{ \mathcal{E}_{Z \mid \mathcal{X}, \beta^{(\tau)}} \left\{ \ln p(\mathcal{X}, Z \mid \beta) \right\} \right\}
\end{aligned}$$

where $\mathcal{E}(\cdot)$ is the expectation operator. Let $L_{\mathcal{X},Z}(\beta) \triangleq \ln p(\mathcal{X}, Z \mid \beta)$, and

$$\mathcal{Q}\left(\beta, \beta^{(\tau)}\right) = \mathcal{E}_{Z \mid \mathcal{X}, \beta^{(\tau)}} \left\{ L_{\mathcal{X},Z}(\beta) \mid \mathcal{X}, \beta^{(\tau)} \right\}$$

It can be shown that

$$L_{\mathcal{X}}(\beta) - L_{\mathcal{X}}\left(\beta^{(\tau)}\right) \geq \mathcal{Q}\left(\beta, \beta^{(\tau)}\right) - \mathcal{Q}\left(\beta^{(\tau)}, \beta^{(\tau)}\right)$$

where equality occurs if $\beta = \beta^{(\tau)}$. Thus maximization of $L_{\mathcal{X}}(\beta)$ has been replaced by the maximization of $\mathcal{Q}\left(\beta, \beta^{(\tau)}\right)$. The later takes into account the unobserved or the latent data Z. Consequently, the EM algorithm facilitates the estimation of these variables.

Each iteration of the EM algorithm consists of two steps. These are the expectation-step (E-step) and the maximization-step (M-step).

- *The E-step*: This step determines the conditional expectation $\mathcal{Q}\left(\beta, \beta^{(\tau)}\right)$.
- *The M-step*: This step maximizes this conditional expectation $\mathcal{Q}\left(\beta, \beta^{(\tau)}\right)$ with respect to β.

The convergence properties of the EM algorithm are next outlined. As $\Delta\left(\beta^{(\tau)} \mid \beta^{(\tau)}\right) = 0$, and $\beta^{(\tau+1)}$ is the maximizing value of $\Delta\left(\beta \mid \beta^{(\tau)}\right)$, we have

$$\Delta\left(\beta^{(\tau+1)} \mid \beta^{(\tau)}\right) \geq \Delta\left(\beta^{(\tau)} \mid \beta^{(\tau)}\right) = 0$$

This implies

$$L_{\mathcal{X}}\left(\beta^{(\tau+1)}\right) \geq L_{\mathcal{X}}\left(\beta^{(\tau)}\right)$$

The algorithm is terminated if the improvement in the values of these estimates is not significant. This algorithm generally determines a local maximum of the likelihood of the observed data.

The EM Algorithm

Let Z_i be an m-dimensional vector, where $1 \leq i \leq n$. In the vector Z_i, the element $Z_{ij} = 1$ if data point x_i was created by the jth source, and $Z_{ij} = 0$ if x_i was not created by the jth source, where $1 \leq j \leq m$. Therefore (x_i, Z_i), $1 \leq i \leq n$ is considered as the complete data set. The corresponding complete log-likelihood function $L_{\mathcal{X},Z}(\beta)$ is

$$L_{\mathcal{X},Z}(\beta) = \ln \prod_{i=1}^{n} p(x_i, Z_i \mid \beta) = \sum_{i=1}^{n} \ln p(x_i, Z_i \mid \beta)$$
$$= \sum_{i=1}^{n} \ln \left\{ \sum_{j=1}^{m} Z_{ij} \alpha_j \varphi_j (x_i \mid \theta_j) \right\}$$

As Z_{ij} is equal to zero for all but a single term in the inner summation, the above expression can be rewritten as

$$L_{\mathcal{X},Z}(\beta) = \sum_{i=1}^{n} \sum_{j=1}^{m} Z_{ij} \ln \{\alpha_j \varphi_j (x_i \mid \theta_j)\}$$

The most challenging task of the EM algorithm is to find the hidden variables. Note that $\arg\max_\beta \{L_{\mathcal{X},Z}(\beta)\}$ cannot in general be evaluated in closed form. Therefore it is determined iteratively. Each iteration consists of two steps: the E-step, and the M-step. As the Z_i's are not observed, it is not possible to work directly with $L_{\mathcal{X},Z}(\cdot)$. Consequently the conditional expectation $\mathcal{E}_{Z|\mathcal{X},\beta^{(\tau)}}(\cdot \mid \cdot)$ is determined in the E-step. This conditional expectation is over the hidden variables, and it is maximized in the M-step. The EM algorithm, beginning with an initial approximation $\beta^{(0)}$, generates a sequence of estimates $\beta^{(1)}, \beta^{(2)}, \beta^{(3)}, \ldots$. In the E-step, the conditional expectation of the complete log-likelihood function, given the data set $\mathcal{X}$ and previous estimate $\beta^{(\tau)}$, is determined. It is

$$\mathcal{Q}\left(\beta, \beta^{(\tau)}\right) = \mathcal{E}_{Z|\mathcal{X},\beta^{(\tau)}} \left\{L_{\mathcal{X},Z}(\beta) \mid \mathcal{X}, \beta^{(\tau)}\right\}$$

The updated value of the β vector in the M-step is

$$\beta^{(\tau+1)} = \arg\max_{\beta} \left\{\mathcal{Q}\left(\beta, \beta^{(\tau)}\right)\right\}$$

It is shown in the problem section that

$$\mathcal{Q}\left(\beta, \beta^{(\tau)}\right) = \sum_{i=1}^{n} \sum_{j=1}^{m} q_{ij}^{(\tau)} \{\ln \alpha_j + \ln \varphi_j (x_i \mid \theta_j)\}$$

where

$$q_{ij}^{(\tau)} = \frac{\alpha_j^{(\tau)} \varphi_j \left(x_i \mid \theta_j^{(\tau)}\right)}{\sum_{k=1}^{m} \alpha_k^{(\tau)} \varphi_k \left(x_i \mid \theta_k^{(\tau)}\right)}; \qquad 1 \leq i \leq n, \quad 1 \leq j \leq m$$

is the conditional probability, given the set of data point x_i and $\beta^{(\tau)}$, that the data point x_i comes from the probability mixture component with density $\varphi_j(x \mid \theta_j)$. The goal of the M-step is to compute

$$\beta^{(\tau+1)} = \left(\theta^{(\tau+1)}, \alpha^{(\tau+1)}\right)$$

This results in

$$\alpha_j^{(\tau+1)} = \frac{1}{n}\sum_{i=1}^{n} q_{ij}^{(\tau)}, \quad 1 \le j \le m$$

The precise expression for $\theta^{(\tau+1)}$ depends upon the form of component probability density functions.

Algorithm 11.2. *Expectation-Maximization Algorithm.*

Input: The set of data points $\mathcal{X} = \{x_1, x_2, \ldots, x_n\}$.
The parametric form of m distributions, $\varphi_j(x \mid \theta_j), x \in \mathbb{R}, 1 \le j \le m$.
The accuracy factor $\varepsilon > 0$.
Output: The parameter vector $\theta = (\theta_1, \theta_2, \ldots, \theta_m)$, the mixture probability vector $\alpha = (\alpha_1, \alpha_2, \ldots, \alpha_m)$, and $\beta = (\theta, \alpha)$.
Estimate β_{EM} of β is returned by the algorithm.
begin
  Arbitrarily initialize $\beta^{(0)}$
  $\delta \leftarrow \infty$
  $\tau \leftarrow 0$
  (*main part of the algorithm*)
  while $\delta \ge \varepsilon$ **do**
  begin
    E-step: Compute conditional probabilities

$$q_{ij}^{(\tau)} = \frac{\alpha_j^{(\tau)} \varphi_j\left(x_i \mid \theta_j^{(\tau)}\right)}{\sum_{k=1}^{m} \alpha_k^{(\tau)} \varphi_k\left(x_i \mid \theta_k^{(\tau)}\right)}$$

$$1 \le i \le n, \ 1 \le j \le m$$

    M-step: Compute $\beta^{(\tau+1)}$

$$\alpha_j^{(\tau+1)} = \frac{1}{n}\sum_{i=1}^{n} q_{ij}^{(\tau)}, \quad 1 \le j \le m$$

      Determination of $\theta^{(\tau+1)}$ depends upon the form of the component probability density functions.
      Thus $\beta^{(\tau+1)} = \arg\max_\beta \left\{\mathcal{Q}\left(\beta, \beta^{(\tau)}\right)\right\}$ is achieved.
      Compute $\mathcal{Q}\left(\beta^{(\tau+1)}, \beta^{(\tau)}\right)$
      if $\tau > 0$ **then** $\delta \leftarrow \left|\mathcal{Q}\left(\beta^{(\tau+1)}, \beta^{(\tau)}\right) - \mathcal{Q}\left(\beta^{(\tau)}, \beta^{(\tau-1)}\right)\right|$
      $\tau \leftarrow \tau + 1$
  end (*end of while-loop*)
  $\beta_{EM} \leftarrow \beta^{(\tau)}$
end (*end of expectation-maximization algorithm*)

11.3.5 Fuzzy Clustering

In typical clustering algorithms, a data point strictly belongs to a single cluster. However, realistically a data point can sometimes legitimately belong to more than a single cluster. The mathematical structure which can assist in describing this notion is the idea of a fuzzy set. In fuzzy set theory, an element can have several "grades of membership" into a set. Fuzzy clustering scheme allows a data point to belong simultaneously to more than a single cluster with a certain grade of membership into each cluster. The concept of fuzzy sets is initially introduced in this subsection. This is followed by its application to clustering.

Fuzzy Sets

Fuzzy sets are mathematical constructs used in representing imprecision and uncertainties encountered in real-world objects. For example, an imprecise statement like "expensive necklace" cannot in general be assigned either "true" or "false." Fuzzy set theory assigns a real number between 0 and 1 to indicate the degree of expensiveness of the necklace.

Definition 11.4. *A fuzzy set $\mathcal{F} = (X_{fz}, A_{fz})$, where X_{fz} is a set, and A_{fz} is a function; where $A_{fz} : X_{fz} \to [0, 1]$. The set X_{fz} is the domain of the function, and A_{fz} is called a membership function.* □

As per the above definition, the membership function has values between 0 and 1. However, if the membership function has only two values, 0 and 1, then the fuzzy set reduces to the classical set. In this case the membership function is called the *characteristic function.* Fuzzy set theory and fuzzy logic were introduced by Lotfi Zadeh in 1965.

Examples 11.2. Fuzzy set examples.

1. A gate membership function is $A_{fz} : \mathbb{R} \to [0, 1]$. Let $a, b \in \mathbb{R}$, and $a < b$. In this function, $A_{fz}(x) = 1$ for $x \in (a, b)$, and 0 otherwise.
2. A Gaussian membership function $A_{fz} : \mathbb{R} \to [0, 1]$ is

$$A_{fz}(x) = \exp\left\{-(x-\mu)^2/\sigma^2\right\}, \quad x \in \mathbb{R}$$

where $\mu \in \mathbb{R}$, and $\sigma \in \mathbb{R}^+$. In this function, μ is the modal value, as $A_{fz}(\mu) = 1$, and σ determines the spread of the fuzzy set. □

Application of Fuzzy Sets to Clustering

Let $\mathcal{X} = \{x_1, x_2, \ldots, x_n\}$ be the set of n data points. Assume that the data points belong to $m \in \mathbb{P}\backslash\{1\}$ clusters. The fuzzy m-clustering of $\mathcal{X}$ is determined by $m \in \mathbb{P}$ membership functions $u_j(\cdot)$, $1 \leq j \leq m$, where $u_j : \mathcal{X} \to [0, 1]$, $1 \leq j \leq m$. The number of clusters m is specified apriori. Each cluster C_j is specified by a vector of parameters θ_j. Let $\theta = (\theta_1, \theta_2, \ldots, \theta_m)$. Also let $r = [r_{ij}]$ be an $n \times m$ matrix, where $r_{ij} = u_j(x_i)$, for $1 \leq i \leq n$, and $1 \leq j \leq m$. The value r_{ij} is called the membership degree of the data point x_i to the cluster C_j. The dissimilarity measure between x_i and θ_j is specified by $d(x_i, \theta_j)$. Another parameter $q \in (1, \infty)$, called the *fuzzifier* is also used in this scheme. The fuzzy clustering algorithm minimizes a cost function $J_q(\cdot, \cdot)$ which depends upon $\mathcal{X}, m, \theta, r$, and q. The cost function $J_q(\cdot,\cdot)$ is minimized by varying θ and elements of r. This cost function is

$$J_q(\theta, r) = \sum_{i=1}^{n}\sum_{j=1}^{m} r_{ij}^q d(x_i, \theta_j)$$

Two candidate choices for the dissimilarity measures are:

$$d(x_i, \theta_j) = (x_i - \theta_j)^T \widetilde{A}(x_i - \theta_j), \quad x_i, \theta_j \in \mathbb{R}^t$$

where $\widetilde{A}$ is a symmetric and positive definite matrix; and the Minkowski distance

$$d(x_i, \theta_j) = \left\{\sum_{k=1}^{t} |x_{ik} - \theta_{jk}|^p\right\}^{1/p}; \quad x_i, \theta_j \in \mathbb{R}^t, \quad p \in \mathbb{R}^+$$

where $x_i = (x_{i1}, x_{i2}, \ldots, x_{it})$ for $1 \le i \le n$, and $\theta_j = (\theta_{j1}, \theta_{j2}, \ldots, \theta_{jt})$, for $1 \le j \le m$. Thus the optimization problem is:

$$\min_{\theta, r} J_q(\theta, r)$$

where $r_{ij} \in [0,1]$, $1 \le i \le n$, $1 \le j \le m$, and the constraints are:

$$\sum_{j=1}^{m} r_{ij} = 1, \quad 1 \le i \le n.$$

$$0 < \sum_{i=1}^{n} r_{ij} < n, \quad 1 \le j \le m$$

Assume that $d(x_i, \theta_j) \ne 0$, for $1 \le i \le n$, and $1 \le j \le m$. Initially consider the minimization of $J_q(\theta, r)$ with respect to r, and the constraint

$$\sum_{j=1}^{m} r_{ij} = 1, \; 1 \le i \le n$$

Use of the method of Lagrange multipliers yields

$$r_{ij}^{-1} = \sum_{k=1}^{m}\left\{\frac{d(x_i, \theta_j)}{d(x_i, \theta_k)}\right\}^{1/(q-1)}; \quad 1 \le i \le n, \quad 1 \le j \le m$$

Minimization with respect to the vector θ results in

$$\sum_{i=1}^{n} r_{ij}^q \frac{\partial d(x_i, \theta_j)}{\partial \theta_j} = 0, \quad 1 \le j \le m$$

See the problem section for a derivation of these two equations. The above two sets of equations are coupled in general. Therefore θ and r are determined iteratively. Each iteration consists of two steps. In the first step, the function $J_q(\cdot,\cdot)$ is minimized by keeping the cluster parameter θ fixed, and varying r. In the second step, $J_q(\cdot,\cdot)$ is minimized by keeping r fixed, and varying θ. Iterations are performed till the results converge.

Note that $q \in (1, \infty)$ is required for the minimization of $J_q(\theta, r)$ and for obtaining fuzzy clustering. However, if q is equal to one, then minimization of the function $J_q(\cdot,\cdot)$ implies that no fuzzy clustering is better than the best crisp clustering.

11.3.6 Cluster Validity Measures

Clustering of the data points is generally an unsupervised learning process. Consequently, evaluation of clustering algorithms is very important. A natural question to ask would be, if the generated clusters are representative of the nature of data points.

The selection of proximity measure of clusters plays a fundamental role in the clustering process. Further, different clustering algorithms might yield different clusters for the same set of data points. The purpose of cluster validity measures is to determine the partitioning of the data points which best reflects the characteristics of the data generation process. Two possible criteria for the evaluation and selection of an appropriate clustering scheme are: compactness or cohesiveness, and separability.

- *Compactness*: This requirement of the clustering scheme measures if the data points which belong to the same cluster are close to each other. A possible measure of compactness is the variance of the data points. A small value of variance means a higher degree of compactness. Another measure of compactness is the intra-cluster distance, which is the distance between data points belonging to the same cluster.
- *Separability*: This property is a measure of distance between data points which belong to different clusters. An acceptable clustering algorithm produces clusters which are widely separated. Separability of clusters can be measured by the inter-cluster distances.

In general, there are three types of numerical measures (criteria) that can be used to assess the results of a clustering algorithm. These are external, internal, and relative criteria.

- *External criteria*: This is a statistical technique based upon a user's specific intuition. Its disadvantage is its computational requirements.
- *Internal criteria*: This is also a statistical technique based upon a metric which uses the data points and clustering schema. Its disadvantage is its computational requirements.
- *Relative criteria*: In this scheme, clusterings of a given set of data points are compared by executing different clustering algorithms. This measure is quantified by validity index.

Validity indices are next discussed. The Dunn, and Davies-Bouldin indices are applicable to crisp clustering, in which each data point belongs to a single cluster. The Xie-Benie index is applicable to fuzzy clustering.

Dunn Index

The Dunn index is named after J. C. Dunn. It is applicable to crisp clustering.

Definition 11.5. *Let $\mathcal{X}$ be the set of data points. The dissimilarity measure of the set of data points $\mathcal{X}$ is a function*

$$d : \mathcal{X} \times \mathcal{X} \to \mathbb{R}_0^+ \tag{11.1a}$$

The m-clustering of $\mathcal{X}$ is the partition of $\mathcal{X}$ into $m \in \mathbb{P} \backslash \{1\}$ sets called clusters. The clusters are $C_1, C_2, \ldots, C_m$.

(a) *Let C_i and C_j be two different clusters. A measure of dissimilarity between two clusters is*

$$\widehat{\mathcal{D}}_{\min}(C_i, C_j) = \min_{x \in C_i, y \in C_j} d(x, y) \tag{11.1b}$$

(b) *The diameter of a cluster C is the distance between its two most distant data points. That is:*

$$diam\,(C) = \max_{x,y \in C} d\,(x,y) \tag{11.1c}$$

The diameter of a cluster C is a candidate for a measure of its dispersion.

(c) *The Dunn validity index for a clustering scheme, with a specific value of m is $\widehat{\mathcal{D}}_m$, where:*

$$\widehat{\mathcal{D}}_m = \min_{1 \le i < j \le m} \left\{ \frac{\widehat{\mathcal{D}}_{\min}\,(C_i, C_j)}{\max_{1 \le k \le m} diam\,(C_k)} \right\} \tag{11.1d}$$

□

Observe that, if the clustering algorithm generates compact and well-separated clusters, then the distances among clusters are large, and the diameters of the clusters are small. This implies a higher value for the Dunn validity index. It should be notated that the computation of Dunn index is time consuming, and the value of $\widehat{\mathcal{D}}_m$ is very sensitive to noisy data points.

Davies-Bouldin Index

The Davies-Bouldin index is named after D. L. Davies and D. W. Bouldin. It is applicable to crisp clustering.

Definition 11.6. *Let $\mathcal{X}$ be the set of data points. The dissimilarity measure of the set of data points $\mathcal{X}$ is a function $d : \mathcal{X} \times \mathcal{X} \to \mathbb{R}_0^+$. The m-clustering of $\mathcal{X}$ is the partition of $\mathcal{X}$ into $m \in \mathbb{P} \backslash \{1\}$ sets called clusters. The clusters are $C_1, C_2, \ldots, C_m$.*

(a) *Let s_i be a measure of dispersion of cluster C_i, $1 \le i \le m$. Dispersion s_i of cluster C_i is a measure of spread of the data points around its center point (mean value) v_i. It is*

$$s_i = \frac{1}{|C_i|} \sum_{x \in C_i} d\,(x, v_i)\,, \quad 1 \le i \le m \tag{11.2a}$$

(b) *Let C_i and C_j be two different clusters. A measure of dissimilarity between two clusters is $\widehat{\mathcal{D}}\,(C_i, C_j) \triangleq d_{ij}$. Assume it to be symmetric. That is, $d_{ij} = d_{ji}$, for $1 \le i, j \le m$.*

(c) *A similarity index ψ_{ij} between two clusters C_i and C_j is defined as:*

$$\psi_{ij} = \frac{s_i + s_j}{d_{ij}} \tag{11.2b}$$

where $d_{ij} \neq 0$.

(d) *Let*

$$\psi_i = \max_{\substack{1 \le j \le m \\ i \neq j}} \psi_{ij}, \quad 1 \le i \le m \tag{11.2c}$$

The Davies-Bouldin index DB_m is

$$DB_m = \frac{1}{m} \sum_{i=1}^{m} \psi_i \tag{11.2d}$$

□

Observe that the Davies-Bouldin index DB_m measures average similarity between the m clusters. A low value of DB_m implies compact and separated clusters. This is possible if in the expression for similarity index of two clusters, numerator is small and the denominator is high.

Xie-Benie Index

The Xie-Benie index is named after X. L. Xie, and G. Benie. This index is used in evaluating the fuzzy clustering algorithm, where Euclidean distance is used. Define the variation in cluster C_j as

$$\sigma_j^q = \sum_{i=1}^{n} r_{ij}^q \left\| x_i - v_j \right\|^2$$

where v_j is its center point, r_{ij} is the membership degree of the data point $x_i \in \mathcal{X}$ to the cluster C_j, where $1 \leq i \leq n$ and $1 \leq j \leq m$; and the fuzzifier parameter $q \in (1, \infty)$. The total variation in the m clusters is $\sigma_2 = \sum_{j=1}^{m} \sigma_j^q$. Thus σ_2 is a measure of compactness of the clustering. A measure of separability among the m clusters is $d_{\min}$ where

$$d_{\min} = \min_{\substack{1 \leq j,k \leq m \\ j \neq k}} \left\| v_j - v_k \right\|^2$$

The Xie-Benie index XB_m is $(\sigma_2/n) / d_{\min}$.

Definition 11.7. *Let $\mathcal{X}$ be the set of data points, where $|\mathcal{X}| = n$. The data points in the set $\mathcal{X}$ are assigned to $m \in \mathbb{P} \backslash \{1\}$ fuzzy clusters. The center point of cluster C_j is v_j, where $1 \leq j \leq m$. Let r_{ij} be the membership degree of the data point $x_i \in \mathcal{X}$ to the cluster C_j, where $1 \leq i \leq n$, and $1 \leq j \leq m$. The Xie-Benie index XB_m of the fuzzy clustering scheme is*

$$XB_m = \frac{\sum_{j=1}^{m} \sum_{i=1}^{n} r_{ij}^q \left\| x_i - v_j \right\|^2}{n \min_{\substack{1 \leq j,k \leq m \\ j \neq k}} \left\| v_j - v_k \right\|^2} \tag{11.3}$$

□

A low value of Xie-Benie index XB_m implies compact and separated clusters.

11.4 Association Rules and Sequential Pattern Mining

In addition to clustering, two prominent unsupervised learning schemes for knowledge discovery are: association rules, and sequential pattern mining. Data mining techniques which use association rules discover useful and interesting associations (relationships, patterns) among members of large data sets. Such techniques find application in bioinformatics, business, inventory control, medicine, scientific data analysis, security, Web mining, and so on.

A business makes several transactions each day. Each transaction in turn is made up of several *items*. A set of items is referred to as an *itemset* in this chapter. *Association analysis* finds *all rules* that associate an itemset with another itemset. This analysis helps in making efficient business decisions. Due to the presence of efficient algorithms, association rules provide an important data mining tool in the knowledge discovery process. We initially provide basic concepts, and then emphasize the necessity of an efficient algorithm to mine association rules in a set of transactions. Subsequently an influential algorithm to discover the association rules is described.

Association rules ignore the order of transactions. There are several examples, where the sequence in data is important. Some examples where the order of transactions is significant are: Web-click streams, Web access pattern analysis, network intrusion detection analysis, biological sequences, natural and social sequences, customer shopping sequences, and so on. Sequential pattern mining schemes discover patterns of frequently occurring ordered events.

11.4.1 Basic Concepts of Association Rules

Basic concepts of association rules are provided in this subsection. These are also illustrated via an example. The implication symbol "$\Rightarrow$" is used in this description. Let P_{prop} and Q_{prop} be two propositions, which take values either *true* or *false*. The relationship $P_{prop} \Rightarrow Q_{prop}$, for example means: proposition Q_{prop} is true whenever proposition P_{prop} is true. Note that implication means co-occurrence and not causality.

Definitions 11.8. *Let $\mathcal{I} = \{i_1, i_2, \ldots, i_d\}$ be a set of items, and $\mathcal{T} = \{t_1, t_2, \ldots, t_n\}$ be the set of all transactions which use items in the set $\mathcal{I}$. Each transaction is a subset of items selected from the set $\mathcal{I}$. That is, $t_i \subseteq \mathcal{I}$, where $1 \leq i \leq n$.*

1. *Transaction identifier: The transaction identifier (TID) of $t_i \in \mathcal{T}$ is i.*
2. *Itemset: A collection of zero or more items is called an itemset.*
3. *k-itemset: An itemset with k items is called a k-itemset.*
4. *Transaction width: The transaction width is the number of items present in a transaction.*
5. *A transaction $t_i \in \mathcal{T}$ is said to contain an itemset X, if $X \subseteq t_i$.*
6. *Support count of an itemset: The support count of an itemset $X \subseteq \mathcal{I}$ is the number of transactions in $\mathcal{T}$ that contain it. That is, the support count of an itemset X is $\sigma(X)$, where*

$$\sigma(X) = |\{t_i \mid X \subseteq t_i, t_i \in \mathcal{T}\}| \tag{11.4}$$

□

Note that the items in the set $\mathcal{I}$ are often listed in lexicographic order for algorithmic efficiency. Recall that lexicographic order is simply dictionary order. Also note that, if X and Y are two itemsets so that $X \subseteq Y$, then $\sigma(Y) \leq \sigma(X)$. *Association rule*, and its strength are next defined. An association rule is an implication expression. The strength of an association rule is measured in terms of its *support* and *confidence*.

Definition 11.9. *Let $\mathcal{I}$ be a set of items, and $\mathcal{T}$ be the set of all transactions, where $|\mathcal{T}| = n$. Also, let $X \subseteq \mathcal{I}$ and $Y \subseteq \mathcal{I}$ be disjoint itemsets, that is $X \cap Y = \varnothing$.*

(a) *Association rule: An association rule is an implication expression of the form $X \Rightarrow Y$. The itemsets X and Y, are called the antecedent and consequent of the association rule respectively.*
(b) *Strength of an association rule: The strength of an association rule is measured in terms of its support and confidence.*

(i) *The support of an association rule $X \Rightarrow Y$ measures how often the rule is applicable in the set of transactions $\mathcal{T}$. The support of the association rule is denoted by $s(X \Rightarrow Y)$. It is*

$$s(X \Rightarrow Y) = \frac{\sigma(X \cup Y)}{n} \tag{11.5a}$$

This is an estimate of the probability that the transaction contains items in the set $X \cup Y$.

(ii) *The confidence of an association rule $X \Rightarrow Y$ measures, how often items in the itemset Y appear in transactions that contain X. Assume that the itemset X is nonempty, that is $\sigma(X) \neq 0$. The confidence of the association rule is denoted by $c(X \Rightarrow Y)$. It is*

$$c(X \Rightarrow Y) = \frac{\sigma(X \cup Y)}{\sigma(X)} \tag{11.5b}$$

This is an estimate of the probability that the transaction containing items in X, also contains items in Y. □

The itemsets X and Y, in the above definition, are typically nonempty. The above mentioned terms in the definitions are clarified via an example.

Example 11.3. This example is referred to as the *market basket* example in the rest of the section. Let the set of items sold in the market be $\mathcal{I}$. Note that the elements in the set $\mathcal{I}$ are listed in lexicographic order. The transactions are listed in Table 11.1. Thus

$$\mathcal{I} = \{\text{bread, butter, cereal, eggs, milk, sugar}\} = \{i_1, i_2, i_3, i_4, i_5, i_6\}$$

TID	Trans.	Items
1	t_1	{bread, butter}
2	t_2	{bread, milk, sugar}
3	t_3	{cereal, eggs, milk, sugar}
4	t_4	{bread, butter, eggs, milk, sugar}
5	t_5	{bread, cereal, eggs, milk, sugar}

Table 11.1. Sample market basket transactions.

The information in Table 11.1 can also be represented as a binary $0/1$ table. See Table 11.2. In this table, the entry in row k, and column i_j is set to 1 if $i_j \in t_k$; otherwise it is set to 0.

TID	Trans.	bread	butter	cereal	eggs	milk	sugar
1	t_1	1	1	0	0	0	0
2	t_2	1	0	0	0	1	1
3	t_3	0	0	1	1	1	1
4	t_4	1	1	0	1	1	1
5	t_5	1	0	1	1	1	1

Table 11.2. Binary representation of market basket transactions.

Consider the itemset $X = \{\text{cereal, eggs}\}$. Observe that X is contained in the transactions t_3 and t_5. Therefore, the support count for the itemset X is $\sigma(X) = 2$.

Next consider the association rule $Y \Rightarrow Z$, where $Y = \{\text{milk, sugar}\}$, and $Z = \{\text{eggs}\}$. Note that the support count for $Y \cup Z$ is equal to $\sigma(Y \cup Z) = 3$, and the total number of transactions is $n = 5$. Therefore the support of the association rule is equal to

$$s(Y \Rightarrow Z) = \frac{\sigma(Y \cup Z)}{n} = \frac{3}{5} = 0.6$$

Note that the support count for Y is $\sigma(Y) = 4$. Therefore the confidence of the association rule is equal to

$$c(Y \Rightarrow Z) = \frac{\sigma(Y \cup Z)}{\sigma(Y)} = \frac{3}{4} = 0.75$$

□

It can be concluded from the above definition that a low value of support of an association rule implies small value of its occurrence. Low value of support might also imply a less interesting business scenario. Therefore, the support criteria can be used to eliminate less interesting rules. The confidence of an association rule can be used to measure the reliability of the inference made by it. Further, higher the value of the confidence of the association rule $X \Rightarrow Y$, higher is the chance that Y is present in a transaction along with X. It can also be interpreted as an estimate of the probability of Y, given X.

11.4.2 Necessity of an Efficient Algorithm

The necessity of an efficient associative rule mining algorithm is outlined in this subsection. The association rule data mining problem is next defined formally.

Definition 11.10. *Statement of association rule mining problem. Let $\mathcal{I}$ be a set of items, and $\mathcal{T}$ be the set of all transactions which use items in the set $\mathcal{I}$. The goal of the mining algorithms is to discover all association rules which have support greater than or equal to $s_0 \in \mathbb{R}_0^+$, and confidence greater than or equal to $c_0 \in \mathbb{R}_0^+$. The values s_0 and c_0 are called support and confidence thresholds of the association rules respectively.* □

A simple, enumerative, though computationally expensive, scheme can be used to address the association rule mining problem. Initially all the association rules for the specified sets of items and transactions can be listed. The rules can subsequently be checked against the support and confidence threshold values. This simple scheme is computationally very expensive. The following observation demonstrates that for a given set of items $\mathcal{I}$, there are an exponential number of association rules.

Observation 11.1. Let $\mathcal{I}$ be the set of items used in the transactions, where $|\mathcal{I}| = d$. The total number of association rules which can be generated from the set $\mathcal{I}$ is equal to

$$\{3^d - 2^{d+1} + 1\}$$

□

The above observation is established in the problem section. Note that, if $d = 10$, the total number of possible association rules is equal to $57,002$. This observation suggests that an efficient algorithm is indeed necessary to determine valid association rules.

In the process of mining association rules, efficiency can be attained if the infrequent itemsets are discarded. Infrequent itemsets are sets of items which occur seldom in the transactions. In this case, confidence values of the corresponding associated rules need not be computed. Thus association rule mining algorithms can be described in terms of *frequent itemsets*. Frequent itemsets are sets of items which satisfy the minimum support rule.

Definitions 11.11. *Let $\mathcal{I}$ be a set of items, $\mathcal{T}$ be the set of all transactions which use items in the set $\mathcal{I}$, and $|\mathcal{T}| = n$. Also, let the support and confidence thresholds of the association rules be s_0 and c_0 respectively.*

1. *The subset of items $X \subseteq \mathcal{I}$ is a frequent itemset, if its support is greater than or equal to ns_0. The subset of items $Y \subseteq \mathcal{I}$ is an infrequent itemset, if its support is less than ns_0.*
2. *The rules which are generated from the frequent itemsets, and also satisfy the threshold c_0 are called strong rules.* □

Association rule mining algorithms typically split the problem into two major steps. These are: generation of frequent itemsets, and generation of association rules from the frequent itemsets. In the first step, frequent itemsets are generated which satisfy the minimum support threshold s_0. In the second step, association rules are generated from the frequent itemsets which also satisfy the minimum confidence threshold. These are the *strong association rules*.

11.4.3 The Apriori Algorithm

The apriori algorithm is used in mining association rules. This algorithm is based upon the apriori principle. We first examine the necessity of apriori principle.

Consider a data set with $k \in \mathbb{P}$ items. This data set can possibly generate up to $K = \left(2^k - 1\right)$ nonempty itemsets. All these sets are candidates for frequent itemsets. Further, all such candidate sets have to be compared with each of the n transactions. If the candidate set is in the transaction, then its support count has to be incremented. If the maximum transaction width is $w \in \mathbb{P}$, then the computational complexity of the comparisons is $O\left(Knw\right)$. This computational complexity of frequent itemset generation can be reduced by using the *apriori* principle. Use of this principle effectively prunes some candidate itemsets in the comparison operations. Thus efficient generation of frequent itemsets requires the use of apriori principle.

Theorem 11.1. *Apriori principle. Each nonempty subset of a frequent itemset is also frequent.* □

The apriori principle is illustrated in the following example.

Example 11.4. Let

$$\mathcal{I} = \{b, c, d, e, f\}$$

be a set of items. If $\{b, c, d\}$ is a frequent itemset, then every transaction that contains $\{b, c, d\}$ evidently contains all the nonempty proper subsets, $\{b, c\}, \{b, d\}, \{c, d\}, \{b\}, \{c\}$, and $\{d\}$. Therefore, if $\{b, c, d\}$ is frequent, then all of its nonempty subsets are also frequent.

Similarly, if the set $\{e, f\}$ is infrequent, then all of its proper supersets are infrequent. These are $\{b, c, d, e, f\}$, $\{b, c, e, f\}$, $\{b, d, e, f\}$, $\{c, d, e, f\}$, $\{b, e, f\}$, $\{c, e, f\}$, and $\{d, e, f\}$. If it is determined that the set $\{e, f\}$ is infrequent, then its proper supersets are also pruned in the apriori algorithm. □

As mentioned in the last example, all supersets of an infrequent set are also infrequent. Therefore the exponentially large search space of candidate itemsets can be reduced by not considering the infrequent itemsets in the comparison operations of the apriori algorithm. This scheme is a consequence of the antimonotone property of the support-measure. It asserts that the support of an itemset is never more than the support of its subsets.

Definitions 11.12. *The monotone and antimonotone property. Let $\mathcal{I}$ be a set, and $\wp(\mathcal{I})$ be its power set. Also, let $f : \wp(\mathcal{I}) \to \mathbb{R}$ be a measure (function) defined on the power set $\wp(\mathcal{I})$.*

1. *The measure $f(\cdot)$ is monotone or upward-closed if*

$$X \subseteq Y \Rightarrow f(X) \leq f(Y), \quad \forall X, Y \in \wp(\mathcal{I}) \tag{11.6a}$$

That is, if X is a subset of Y, then $f(X)$ is at most equal to $f(Y)$.

2. *The measure $f(\cdot)$ is antimonotone or downward-closed if*

$$X \subseteq Y \Rightarrow f(Y) \leq f(X), \quad \forall X, Y \in \wp(\mathcal{I}) \tag{11.6b}$$

That is, if X is a subset of Y, then $f(Y)$ is at most equal to $f(X)$. □

The antimonotone property of the support measure is used in the apriori algorithm to trim the exponentially large search space of the candidate itemsets. The essential features of the apriori algorithm are next demonstrated via the market basket example.

Example 11.5. The market basket example is used to illustrate the generation of frequent itemsets. In this example, the items sold in the market are

$$\mathcal{I} = \{\text{bread, butter, cereal, eggs, milk, sugar}\}$$

The list of transactions is given in Table 11.1. Assume that the support threshold s_0 is equal to 0.6. As there are 5 transactions, this implies a minimum support count equal to 3. Frequent itemsets are generated in this example. The results are obtained in several steps.

Step 1: In this step, candidate 1-itemsets are considered. See Table 11.3 (a). Note that the items $\{\text{butter}\}$ and $\{\text{cereal}\}$, each have a count of 2. Therefore these are discarded. This results in frequent 1-itemsets shown in Table 11.3 (b).

Item	Count
{bread}	4
{butter}	2
{cereal}	2
{eggs}	3
{milk}	4
{sugar}	4

(a)

Item	Count
{bread}	4
{eggs}	3
{milk}	4
{sugar}	4

(b)

Table 11.3. List of candidate 1-itemsets is in (a), and list of frequent 1-itemsets is in (b).

Step 2: As there are 4 frequent 1-itemsets, the number of candidate 2-itemsets is equal to $\binom{4}{2} = 6$. See Table 11.4 (a). Note that the itemset {bread, eggs} has a count of 2. Therefore this is discarded. This results in frequent 2-itemsets shown in Table 11.4 (b). Note that without support-based pruning, the number of candidate 2-itemsets would have been equal to $\binom{6}{2} = 15$.

Itemset	Count
{bread, eggs}	2
{bread, milk}	3
{bread, sugar}	3
{eggs, milk}	3
{eggs, sugar}	3
{milk, sugar}	4

(a)

Itemset	Count
{bread, milk}	3
{bread, sugar}	3
{eggs, milk}	3
{eggs, sugar}	3
{milk, sugar}	4

(b)

Table 11.4. List of candidate 2-itemsets is in (a), and list of frequent 2-itemsets is in (b).

Step 3: There are 5 frequent 2-itemsets. This results in 4 candidate 3-itemsets. See Table 11.5 (a).

Itemset	Count
{bread, eggs, milk}	2
{bread, eggs, sugar}	2
{bread, milk, sugar}	3
{eggs, milk, sugar}	3

(a)

Itemset	Count
{bread, milk, sugar}	3
{eggs, milk, sugar}	3

(b)

Table 11.5. List of candidate 3-itemsets is in (a), and list of frequent 3-itemsets is in (b).

Note that the itemsets {bread, eggs, milk} , and {bread, eggs, sugar} each have a count of 2. Therefore these are discarded. This results in frequent 3-itemsets shown in Table 11.5 (b). Note that without support-based pruning, the number of candidate 3-itemsets would have been equal to $\binom{6}{3} = 20$.

Step 4: There are 2 frequent 3-itemsets. This results in a single candidate 4-itemset. See Table 11.6. Note that the itemset {bread, eggs, milk, sugar} has a count of 2. Therefore this is discarded.

Itemset	Count
{bread, eggs, milk, sugar}	2

Table 11.6. List of candidate 4-itemset.

In summary, the lists of frequent itemsets are in Tables 11.3 (b), 11.4 (b), and 11.5 (b). Note that the scheme of explicit enumeration of all itemsets of size $1, 2, 3$ and 4 will create $\binom{6}{1} + \binom{6}{2} + \binom{6}{3} + \binom{6}{4} = 56$ candidates. However, the use of the apriori principle yields $\binom{6}{1} + \binom{4}{2} + 4 + 1 = 17$ candidates. □

As mentioned earlier, the complete apriori algorithm can be described in two steps. These are: generation of frequent itemsets, and generation of association rules from the frequent itemsets.

- *Generation of frequent itemsets*: In this step, itemsets are generated which satisfy the minimum support threshold $s_0 \in \mathbb{R}_0^+$.
- *Generation of association rules from the frequent itemsets*: The frequent itemsets, which satisfy the minimum confidence threshold $c_0 \in \mathbb{R}_0^+$ are selected. These are the strong association rules.

Generation of Frequent Itemsets

The apriori algorithm generates frequent itemsets by scanning data several times. Denote the frequent k-itemset by F_k, where $k \in \mathbb{P}$. These itemsets satisfy the minimum support threshold s_0.

The algorithm initially determines the frequent 1-itemsets, F_1. The frequent k-itemsets F_k, where $k \in \mathbb{P}\backslash \{1\}$ are subsequently generated iteratively. The following steps are implemented in each pass $k \in \mathbb{P}\backslash \{1\}$.

- Using the frequent $(k-1)$-itemsets F_{k-1}, and the generating function $\mathcal{G}(\cdot)$, the algorithm determines candidate k-itemsets $\widehat{C}_k$. That is, $\widehat{C}_k = \mathcal{G}(F_{k-1})$.
- The transaction data set is scanned to determine the support of each candidate itemset $c \in \widehat{C}_k$.
- The algorithm selects frequent itemsets from the candidate itemsets, which satisfy the support threshold s_0.

Algorithm 11.3a. *Frequent Itemset Generation of the Apriori Algorithm.*

Input: $\mathcal{I} = \{i_1, i_2, \ldots, i_d\}$ is a set of items, and $\mathcal{T} = \{t_1, t_2, \ldots, t_n\}$ is the set of all transactions. Support threshold is equal to s_0.
Output: F = set of frequent itemsets.
begin
    (*initialization*)
    $k \leftarrow 1$
    $F_k \leftarrow \{i \mid i \in \mathcal{I} \wedge \sigma(\{i\}) \geq ns_0\}$
    (F_k *is the set of all frequent k-itemsets*)
    (*main part of the algorithm*)
    while $F_k \neq \varnothing$ **do**
    begin
        $k \leftarrow k + 1$
        $\widehat{C}_k \leftarrow \mathcal{G}(F_{k-1})$
        ($\widehat{C}_k$ *is the set of candidate k-itemsets*)
        ($\mathcal{G}(\cdot)$ *is candidate k-itemset generating function*)
        $\sigma(c) \leftarrow 0, \forall\, c \in \widehat{C}_k$
        for $\forall\, t \in \mathcal{T}$ **do**
        begin
            for $\forall\, c \in \widehat{C}_k$ **do**
            begin
                if $c \subseteq t$ **then** $\sigma(c) \leftarrow \sigma(c) + 1$
                (*increment support count*)
            end (*end of for-loop*)
        end (*end of for-loop*)

(*generate candidate itemsets*)
$F_k \leftarrow \left\{ c \mid c \in \widehat{C}_k \wedge \sigma(\{c\}) \geq ns_0 \right\}$
(F_k *is the set of all frequent* k*-itemset*)
end (*end of while-loop*)
$F \leftarrow \cup_k F_k$
end (*end of frequent itemset generation of the apriori algorithm*)

The algorithm for frequent-itemsets generation uses an algorithm to generate candidate itemsets. Given frequent $(k-1)$-itemsets F_{k-1}, the later algorithm generates candidate k-itemsets $\widehat{C}_k$ via the function $\mathcal{G}(\cdot)$. That is, the algorithm determines $\widehat{C}_k = \mathcal{G}(F_{k-1})$.

In order to generate candidate itemsets efficiently, the items in the set $\mathcal{I}$ are listed in lexicographic order (dictionary order). Denote this order by $<_r$. Therefore a k-itemset consisting of k items, is represented as

$$\{i_{j1}, i_{j2}, \ldots, i_{jk}\}$$

where

$$i_{j_1} <_r i_{j_2} <_r \ldots <_r i_{j_k}, \text{ and } i_{j_l} \in \mathcal{I} \text{ for } 1 \leq l \leq k$$

Th algorithm for determining $\widehat{C}_k = \mathcal{G}(F_{k-1})$ has essentially two steps. These are the join and prune steps.

- *Join step*: In this step, two frequent $(k-1)$-itemsets are joined to produce c, a possible k-itemset candidate. The two frequent $(k-1)$-itemsets have identical items, except the last item.
- *Prune step*: In this step, the candidate c generated in the join step may not indeed be a k-itemset candidate. The prune step ensures that all k subsets of c of width $(k-1)$ belong to F_{k-1}. If any one of the k subsets of c of width $(k-1)$ does not belong to F_{k-1}, then c cannot be a frequent itemset as per the downward closure (antimonotone) property. In this case, c is pruned from the set $\widehat{C}_k$.

Algorithm 11.3b. *Candidate Itemset Generation of the Apriori Algorithm.*

Input: $\mathcal{I} = \{i_1, i_2, \ldots, i_d\}$ is a set of items.
F_{k-1} = set of frequent $(k-1)$-itemsets, where $k \geq 2$.
The lexicographic order among items is denoted by $<_r$.
Output: $\widehat{C}_k$ = set of candidate k-itemsets.
It is denoted as $\widehat{C}_k = \mathcal{G}(F_{k-1})$.
begin
(*initialization*)
$\widehat{C}_k \leftarrow \varnothing$
(*main part of the algorithm*)
for $\forall\ f_1, f_2 \in F_{k-1}$
(*for all possible pairs of frequent itemsets*)
with

$f_1 = \left\{i_{j_1}, i_{j_2}, \ldots, i_{j_{k-2}}, i_{j_{k-1}}\right\} \wedge f_2 = \left\{i_{j_1}, i_{j_2}, \ldots, i_{j_{k-2}}, i'_{j_{k-1}}\right\}$
$\wedge\, i_{j_{k-1}} <_r i'_{j_{k-1}}$
(f_1 *and* f_2 *differ only in the last term*)
begin
$c \leftarrow \left\{i_{j_1}, i_{j_2}, \ldots, i_{j_{k-2}}, i_{j_{k-1}}, i'_{j_{k-1}}\right\}$
(*join the two itemsets* f_1 *and* f_2)
$\widehat{C}_k \leftarrow \widehat{C}_k \cup \{c\}$
for $\forall$ $(k-1)$-subset s of c
begin
if $(s \notin F_{k-1})$ **then** $\widehat{C}_k \leftarrow \widehat{C}_k - \{c\}$
(*delete/prune c from the candidate itemset*)
end (*end of for-loop*)
end (*end of for-loop*)
$\mathcal{G}(F_{k-1}) \leftarrow \widehat{C}_k$
end (*end of candidate itemset generation of the apriori algorithm*)

Association Rule Generation

Having determined the frequent itemsets in the transaction set $\mathcal{T}$, strong association rules can be discovered. If X and Y are itemsets, the confidence of the association rule $X \Rightarrow Y$ is $c(X \Rightarrow Y) = \sigma(X \cup Y) / \sigma(X)$. Let c_0 be the confidence threshold. Strong association rules can be generated as follows.

- For each frequent itemset Y, generate all of its nonempty proper subsets.
- For each such nonempty proper subset X, an association rule is $X \Rightarrow (Y - X)$. If the confidence of this association rule is greater than or equal to c_0, then it is a strong association rule.

Example 11.6. In the market basket example, the support threshold s_0 is equal to 0.6. As there are 5 transactions, this implies a minimum support count equal to 3. Assume that the confidence threshold c_0 is equal to 0.8. The itemset $X = \{\text{bread, milk, sugar}\}$ has a support count of 3. Therefore, it is a frequent itemset. All nonempty proper subsets of X and the corresponding association rules are listed in Table 11.7. The confidence of each association rule is also specified.

Association Rule	Confidence
{bread, milk} ⇒ {sugar}	$3/3 = 1.00$
{bread, sugar} ⇒ {milk}	$3/3 = 1.00$
{milk, sugar} ⇒ {bread}	$3/4 = 0.75$
{sugar} ⇒ {bread, milk}	$3/4 = 0.75$
{milk} ⇒ {bread, sugar}	$3/4 = 0.75$
{bread} ⇒ {milk, sugar}	$3/4 = 0.75$

Table 11.7. The association rules, and the corresponding confidence for the frequent itemset {bread, milk, sugar}.

It can be observed from the table, that {bread, milk} $\Rightarrow$ {sugar}, and {bread, sugar} $\Rightarrow$ {milk} are strong association rules, as their confidence exceeds 0.8. □

The above scheme of generating strong association rules is not very efficient. Nevertheless, the following observations can be used to improve the efficiency of association rule generation.

Observations 11.2. These observations are useful in association rule generation.

1. Let $X \neq \varnothing$, $X \subseteq Y$, and $X \Rightarrow (Y - X)$ be an association rule r which does not satisfy the confidence threshold c_0. That is, $c(r) < c_0$. If $X_{sub} \Rightarrow (Y - X_{sub})$ is another association rule r_{sub}, where $X_{sub} \neq \varnothing$, and $X_{sub} \subseteq X$, then r_{sub} also does not satisfy the confidence threshold. That is, $c(r_{sub}) < c_0$.
2. Let $X \neq \varnothing$, $X \subseteq Y$, and $(Y - X) \Rightarrow X$ be an association rule r which satisfies the confidence threshold c_0. That is, $c(r) \geq c_0$. If $(Y - X_{sub}) \Rightarrow X_{sub}$ is another association rule r_{sub}, where $X_{sub} \neq \varnothing$, and $X_{sub} \subseteq X$, then r_{sub} also satisfies the confidence threshold. That is, $c(r_{sub}) \geq c_0$.
   Thus, if an association rule with consequent α holds, then so do rules with consequents which are nonempty subsets of α. □

The first of the above observations is proved in the problem section.

Examples 11.7. The usefulness of the above set of observations is demonstrated in the following examples.

1. Let $\{b, c, d, e\}$ be a frequent itemset, and suppose that the confidence of the association rule $\{b, c, d\} \Rightarrow \{e\}$ is lower than c_0. Then as per the above observation, the confidence of the association rules which contain item e in its consequent is also lower than c_0. That is, all the association rules $\{b, c\} \Rightarrow \{d, e\}$, $\{b, d\} \Rightarrow \{c, e\}$, $\{c, d\} \Rightarrow \{b, e\}$; $\{b\} \Rightarrow \{c, d, e\}$, $\{c\} \Rightarrow \{b, d, e\}$, and $\{d\} \Rightarrow \{b, c, e\}$, have confidence lower than c_0.
2. Let $\{b, c, d, e\}$ be a frequent itemset, where the confidence of the association rule $\{b, c\} \Rightarrow \{d, e\}$ is greater than or equal to c_0. Therefore the confidence of the association rules $\{b, c, d\} \Rightarrow \{e\}$ and $\{b, c, e\} \Rightarrow \{d\}$ must also be greater than or equal to c_0. □

The algorithm for association rule generation uses the last set of observations.

- For each frequent k-itemset $f_k \in F_k$, determine all association rules with one item in the consequent and confidence greater than or equal to c_0.
- In the next step create all consequents with 2 items that were generated in the association rules in the last step. This generation of the consequents uses the function $\mathcal{G}(\cdot)$, which was used in the candidate itemset generation, when frequent itemsets were generated.
  After these consequents are generated, the consequents which do not satisfy the confidence threshold are pruned.
- In the next step create all consequents with 3 items that were generated in the association rules in the last step, and repeat the substeps.
- Repeat the steps as necessary.

The following algorithm describes the main features of the association rule generation of the apriori algorithm.

Algorithm 11.4. *Association Rule Generation of the Apriori Algorithm.*

Input: Set of frequent itemsets $F = \cup_k F_k$, where F_k is the set of all k-itemsets. Confidence threshold c_0. The support count $\sigma(\cdot)$ of all itemsets that are available from the frequent itemset generation routine.
Output: $\mathcal{S}_{AR}$ is the set of all strong association rules.
begin
  $\mathcal{S}_{AR} \leftarrow \varnothing$
  for each frequent k-itemset $f_k \in F_k$, $k \geq 2$ **do**
  begin
    $H_1 \leftarrow \{h_1 \mid h_1 \in f_k,$ and h_1 is a 1-item subset of $f_k\}$
    for $\forall\, h_1 \in H_1$ **do**
    begin
      $c(r) \leftarrow \sigma(f_k) / \sigma(f_k - h_1)$
      (*the association rule* $(f_k - h_1) \Rightarrow h_1$ *is denoted by* r)
      ($c(r)$ *is the confidence of the rule* r)
      if $c(r) \geq c_0$ **then** $\mathcal{S}_{AR} \leftarrow \mathcal{S}_{AR} \cup r$ **else** $H_1 \leftarrow H_1 - h_1$
    end (*end of for-loop*)
    $\mathfrak{R}(f_k, H_1)$
    (H_1 *is the set of* 1*-itemset consequents of strong rules from* f_k)
  end (*end of for-loop*)
end (*end of association rule generation of the apriori algorithm*)

procedure $\mathfrak{R}(f_k, H_m)$
(H_m *is the set of* m*-item consequents*)
begin
  if $\{(k > m + 1) \wedge (H_m \neq \varnothing)\}$ **then**
  begin
    $H_{m+1} \leftarrow \mathcal{G}(H_m)$
    (*the function* $\mathcal{G}(\cdot)$ *generates candidate itemsets*)
    for $\forall\, h_{m+1} \in H_{m+1}$ **do**
    begin
      $c(r) \leftarrow \sigma(f_k) / \sigma(f_k - h_{m+1})$
      (*the association rule* $(f_k - h_{m+1}) \Rightarrow h_{m+1}$ *is denoted by* r)
      ($c(r)$ *is the confidence of the rule* r)
      if $c(r) \geq c_0$ **then** $\mathcal{S}_{AR} \leftarrow \mathcal{S}_{AR} \cup r$
        else $H_{m+1} \leftarrow H_{m+1} - h_{m+1}$
    end (*end of for-loop*)
    $\mathfrak{R}(f_k, H_{m+1})$
  end (*end of if statement*)
end (*end of procedure*)

11.4.4 Sequential Pattern Mining

Sequential pattern mining is concerned with determining patterns in a sequence of transactions. Algorithms which implement sequential data mining generally deal with categorical (or symbolic) patterns. In contrast, algorithms which implement statistical techniques are generally useful for numerical data analysis.

Suppose that a set of sequences of transactions is given, where a sequence of transactions consists of an ordered list of events, and each event consists of a set of items. Sequential pattern mining discovers all *frequent* subsequences in the set of sequences, whose occurrence frequency is at least equal to a prespecified threshold value. We initially introduce notation.

Definitions 11.13. *Let* $\mathcal{I} = \{i_1, i_2, \ldots, i_p\}$ *be the set of all items.*

1. *An itemset or an event* $X \subseteq \mathcal{I}$ *is a nonempty and unordered set (collection or list) of items. That is,* $X = \{x_1, x_2, \ldots, x_q\}$, $x_i \in \mathcal{I}$ *for* $1 \leq i \leq q$.
2. *A sequence is an ordered list of events. A sequence* $s = \langle e_1 e_2 \ldots e_r \rangle$, *where event* $e_i \subseteq \mathcal{I}$ *for* $1 \leq i \leq r$. *Therefore, an event is an itemset. In this sequence, event* e_1 *occurs before* e_2, *which occurs before* e_3, *and so on. The event* e_i *is also called an element of the sequence.*
   (a) *An item can occur at most once in an element of the sequence. However, an item can occur multiple times in different elements of the sequence.*
   (b) *The size of a sequence is equal to the number of elements (itemsets) in the sequence.*
   (c) *The length of a sequence is equal to the number of instances of items in the sequence. If an item occurs several times in different elements of the sequence, then each such occurrence also contributes to its length. A sequence of length* k *is called a* k*-sequence.*
   (d) *The items in an element of a sequence are listed in lexicographic order.*
3. *Consider two sequences* $\alpha = \langle \alpha_1 \alpha_2 \ldots \alpha_m \rangle$ *and* $\beta = \langle \beta_1 \beta_2 \ldots \beta_n \rangle$. *Then* α *is a subsequence of sequence* β, *or* β *is a supersequence of sequence* α, *denoted as* $\alpha \sqsubseteq \beta$, *if there exist integers* $1 \leq j_1 < j_2 < \cdots < j_m \leq n$ *such that* $\alpha_1 \subseteq \beta_{j_1}, \alpha_2 \subseteq \beta_{j_2}, \ldots, \alpha_m \subseteq \beta_{j_m}$. *Alternately, the sequence* β *is said to contain* α. □

Before sequential pattern mining techniques are described, terms related to a collection of sequences are defined.

Definitions 11.14. *Basic definitions.*

1. *A sequence database* $\mathcal{S}$ *is a set of tuples* $\langle s_{id}, s \rangle$; *where* s *is a sequence, and* s_{id} *is its identification.*
2. *A tuple* $\langle s_{id}, s \rangle \in \mathcal{S}$ *contains a sequence* α, *if* $\alpha \sqsubseteq s$.
3. *The support of a sequence* α *in a database* $\mathcal{S}$ *is the number of tuples in the database containing* α.
   That is, $support_{\mathcal{S}}(\alpha) = |\{\langle s_{id}, s \rangle \mid \langle s_{id}, s \rangle \in \mathcal{S} \wedge (\alpha \sqsubseteq s)\}|$. *If the context is clear, the support is simply denoted by* $support(\alpha)$.
4. *Let* $\eta_{min} \in \mathbb{P}$ *denote the minimum support threshold. A sequence* α *is a sequential pattern (or frequent sequence) in a sequence database* $\mathcal{S}$ *if* α *is contained in at least* η_{min} *tuples. That is,* $support_{\mathcal{S}}(\alpha) \geq \eta_{min}$.
5. *A sequential pattern of length* l *is called an* l*-pattern.* □

Examples 11.8. The above definitions are illustrated with several examples. Let the set of all items $\mathcal{I} = \{a, b, c, d, e, f, g, h, i, j\}$. In some of these examples "identifier" is abbreviated as ID.

1. The sequence $\langle \{c,d\} \{c,e\} \{f\} \{a\} \rangle$ has four elements (events): $\{c,d\}$, $\{c,e\}$, $\{f\}$, and $\{a\}$. These events occur in the order listed. Its size is 4, and it is a 6-sequence as there are 6 instances of items appearing in the sequence. Item c occurs twice in the sequence. Therefore it contributes two to the length of the sequence. However, it contributes only one to the support of the sequence $\langle \{c\} \rangle$.
2. Sequences, subsequences, and supersequences. Let s and t be two sequences.
   (a) $s = \langle \{c,e\} \{f,g,h\} \{j\} \rangle$, and $t = \langle \{c\} \{f,h\} \{j\} \rangle$. In this pair of sequences, t is a subsequence of s, and s is a supersequence of t.
   (b) $s = \langle \{c,e\} \{f,g,h\} \{j\} \rangle$, and $t = \langle \{c\} \{j\} \rangle$. In this pair of sequences, t is a subsequence of s, and s is a supersequence of t.
   (c) $s = \langle \{c,e\} \{c,e\} \{c,f\} \rangle$, and $t = \langle \{c\} \{e\} \rangle$. In this pair of sequences, t is a subsequence of s, and s is a supersequence of t.
   (d) $s = \langle \{c,d\} \{e,f\} \rangle$, and $t = \langle \{c\} \{d\} \rangle$. In this pair of sequences, t is not a subsequence of s, and s is not a supersequence of t.
3. A sequence database $\mathcal{S}$ is shown in Table 11.8. Consider the sequence $s = \langle \{a,c\} \{e,h\} \rangle$. In the database $\mathcal{S}$, the sequences 2 and 3 contain the subsequence s. Therefore, the support of s is 2. That is, $supports_{\mathcal{S}}(s) = 2$.

Sequence ID	Sequence
1	$\langle \{a\} \{a,b,c\} \{c,e\} \{f,g,h\} \{j\} \rangle$
2	$\langle \{a,c\} \{b\} \{c,d\} \{e,h,i\} \{j\} \rangle$
3	$\langle \{a,c,d\} \{c\} \{d,f\} \{e,h,j\} \rangle$
4	$\langle \{c,d\} \{e,f\} \{f,g,h,j\} \rangle$

Table 11.8. A sequence database.

4. The transactions made at a business Web-site by customers have identification $1,2,3,$ and 4. A transaction is defined as a set of items purchased by a customer at a specific time (called the transaction time). The complete list of items is the set $\mathcal{I}$. The Table 11.9 (a) lists the transaction time of the customer, and the items bought at that time. The Table 11.9 (b) is the sequence database generated by using the transactions listed in Table 11.9 (a). Table 11.9 (c) shows sequential patterns with $\eta_{min} = 2$. That is, a minimum support of two customers.

Customer ID	Transaction Time	Items Purchased
1	$10:00$ AM	d,e
1	$10:15$ AM	h
2	$09:00$ AM	a,g
2	$09:15$ AM	b
2	$09:25$ AM	c,d,h
3	$10:30$ AM	g
4	$11:15$ AM	a,f
4	$11:30$ AM	b
4	$11:45$ AM	d,e

Table 11.9 (a). List of customer identifier, transaction time, and the transaction.

Customer ID	Items Purchased
1	$\langle\{d,e\}\{h\}\rangle$
2	$\langle\{a,g\}\{b\}\{c,d,h\}\rangle$
3	$\langle\{g\}\rangle$
4	$\langle\{a,f\}\{b\}\{d,e\}\rangle$

Table 11.9 (b). List of customer identifier, and the sequence of items.

	Sequential Patterns with $\eta_{min}=2$
1-sequence	$\langle\{a\}\rangle,\langle\{b\}\rangle,\langle\{d\}\rangle,\langle\{e\}\rangle,\langle\{g\}\rangle,\langle\{h\}\rangle$
2-sequence	$\langle\{a\}\{b\}\rangle,\langle\{a\}\{d\}\rangle,\langle\{b\}\{d\}\rangle,\langle\{d,e\}\rangle$
3-sequence	$\langle\{a\}\{b\}\{d\}\rangle$

Table 11.9 (c). Sequential patterns.

□

The sequential pattern discovery problem is next formally stated. Suppose a sequence database $\mathcal{S}$, and the minimum support threshold η_{min} are given. The goal of the sequential pattern discovery is to find all sequences with support greater than or equal to η_{min}.

This problem is computationally very expensive. For example, consider a database which contains only a single sequence $\langle\beta_1\beta_2\ldots\beta_n\rangle$, and let the minimum threshold η_{min} be equal to one. The number of k-sequences formed with n events is

$$\binom{n}{k}, \quad 1\le k\le n$$

Therefore the total number of distinct sequences is

$$\sum_{k=1}^{n}\binom{n}{k}=(2^n-1)$$

This is an exponentially large number of sequences which are required for sequential pattern discovery. There are several efficient and scalable methods for sequential pattern mining. Only the Generalized Sequential Pattern (GSP) mining algorithm is described qualitatively. This sequential pattern discovery algorithm is iterative, and also uses the apriori property.

In the initial (first) scan of the sequence database, the algorithm finds all items which satisfy the minimum support criterion. These items are 1-event frequent sequence.

As the algorithm is iterative, each subsequent pass uses sequential patterns found in the previous pass as a *seed set*. The seed set is used to generate *candidate sequences*. Such sequences are candidates for new sequential patterns. Each such candidate sequence contains one more item than its *parent* sequential pattern.

Recall that each event in a pattern might contain more than a single item. Further, the total number of instances of items in a sequence is the length of the sequence. Therefore, all candidate sequences in a particular pass will have identical length. Again recall that a sequence of length k is called a k-sequence. Denote the set of candidate k-sequences by $\widehat{C}_k$. The algorithm next determines the support of each sequence in the set $\widehat{C}_k$. Using the set $\widehat{C}_k$, the set F_k of all frequent k-sequences is determined. It is the set of candidate sequences in the set $\widehat{C}_k$ which have a minimum support of at least η_{min}. The set F_k is the seed set for the pass $(k+1)$.

The size of the set $\widehat{C}_k$ can be large. The GSP algorithm alleviates this scenario by pruning certain candidate sequences by using the apriori property. A sequence in the k-th pass is allowed to be a candidate if each of its length-$(k-1)$ subsequences is a sequential pattern found in the $(k-1)$-th pass.

The algorithm terminates when either no new frequent pattern can be generated in a pass, or no candidate sequence can be generated. Observe that the net number of iterations in this algorithm is at least equal to the maximum length of the sequential patterns. The strength of this algorithm is the pruning of candidate sequences by using the apriori property. However, this algorithm requires several scans over the sequence database.

11.5 Metric Classification

Classification is the process of using a model to categorize data points. It assigns each data point to one of several predefined classes. It is essentially a statistical supervised learning technique, in which each data point is assigned a class label. Typically, the input data consists of a collection (set) of records. A single record, or data point, consists of a set of attributes. Each such data point can be categorized and assigned a especial attribute called class.

Definition 11.15. *Let $\mathcal{X}$ be the set of data points, and the set of labels for different classes of data be $\Omega = \{\omega_1, \omega_2, \ldots, \omega_m\}$, where $|\Omega| = m \in \mathbb{P}\backslash\{1\}$. Classifier or classification rule is a mapping $\Phi_{cl} : \mathcal{X} \to \Omega$.* □

An alternate definition of classification, which is similar to that of clustering of data points can also be given. That is, classification rule partitions the set of data points $\mathcal{X}$ into disjoint subsets, so that their union is equal to the set $\mathcal{X}$.

The goal of the classification scheme is to predict the class of data objects whose class label has not yet been specified. It is hoped that this would subsequently lead to understanding of informative patterns that exist in the data points within each class. Classifier is built and ready in two steps. In the first step the classifier is actually built, and in the second step the classifier is tested for its accuracy.

Therefore in the first step of classification, the parameters of the classifier are determined from a set of data points called: *training data* or *learning data*. The training data is a set of data points, whose classification is already known. The training data is used to compute estimates of probabilities in the first step. Therefore this step is called the training or inductive or learning phase. Note that this is a supervised learning process, because the class label of each data point is provided in this training phase. This is in contrast to the case of unsupervised learning algorithms (clustering) in which the class label of each data point is unknown in advance. Further, the number of classes is also generally unknown in advance. The second step, called the deductive step, consists of estimating the accuracy of the classification rules. If the accuracy is deemed acceptable, then the classifier can be used to classify new data points.

The criteria for classification of data points can be either metrical or nonmetrical. A metrical type of classification is numeric-based, and nonmetrical classification is categorical. Metric-based classification is discussed in this section, and nonmetrical classification in the subsequent section. We next describe Bayesian classifier, support vector machines, nearest-neighbor classification scheme, and artificial neural network methods of classification.

11.5.1 Bayesian Classifier

Bayesian classifier of data points is a statistics-based classifier. It is named after T. Bayes (1702-1761). This classification scheme is simultaneously simple, robust, and elegant. It has found application in spam filtering, and text classification. Let $\mathcal{X}$ be the set of data points, in which each data point has $t \in \mathbb{P}$ attributes. That is, each data point $x \in \mathcal{X}$ is a t-tuple. The Bayesian classification scheme addresses the problem of classification in a probabilistic framework.

The set of data points are derived from a space called the *feature space* $\mathcal{R}$. Thus, feature space is the set of all possible data points. That is, $\mathcal{X} \subseteq \mathcal{R}$. A point $s \in \mathcal{R}$ is called a *feature vector.* Let $\widetilde{X}$ be a random vector in the space $\mathcal{R}$.

Also, let $\Omega = \{\omega_1, \omega_2, \ldots, \omega_m\}$ be the set of labels for different classes of data. The probability that a data point belongs to class $\omega_i \in \Omega$ is $P(\omega_i)$, where $1 \leq i \leq m$. These are called *a priori* probabilities. Let x be an instance of the random vector $\widetilde{X}$. If the components of the random vector $\widetilde{X}$ are continuously distributed, the *class-conditional probability density function* is denoted by $p(x \mid \omega_i)$, where $1 \leq i \leq m$. If the components of feature vectors take only discrete values, the corresponding probabilities are denoted by $P(x \mid \omega_i)$, where $1 \leq i \leq m$. The a priori probabilities, and class-conditional probability density functions can be estimated from the training data.

For simplicity, assume that each component of the data point x takes continuous real-values. Let $p(x)$ be the probability density function of the random vector $\widetilde{X}$. The conditional probability $P(\omega_i \mid x)$ is

$$P(\omega_i \mid x) = \frac{p(x \mid \omega_i) P(\omega_i)}{p(x)}, \quad 1 \leq i \leq m$$

$$p(x) = \sum_{i=1}^{m} p(x \mid \omega_i) P(\omega_i)$$

Let $x \in \mathcal{R} \backslash \mathcal{X}$ be a test data point. If there are only two classes, ω_1 and ω_2, Bayesian classification rule is:

$$\begin{aligned}
&\text{If } \ P(\omega_1 \mid x) > P(\omega_2 \mid x), \quad x \text{ is classified to } \omega_1 \\
&\text{If } \ P(\omega_2 \mid x) > P(\omega_1 \mid x), \quad x \text{ is classified to } \omega_2
\end{aligned}$$

Let $R_1 \subset \mathcal{R}$ be the region so that if $x \in R_1$ then x is classified to ω_1. Similarly, if $R_2 \subset \mathcal{R}$ is the region so that if $x \in R_2$ then x is classified to ω_2. Note that $R_1 \cap R_2 = \varnothing$, and $R_1 \cup R_2 = \mathcal{R}$. Classification error occurs:

- If $x \in R_1$ and x is classified to ω_2. Denote the probability of this event by $P(x \in R_1, \omega_2)$.
- If $x \in R_2$ and x is classified to ω_1. Denote the probability of this event by $P(x \in R_2, \omega_1)$.

The net classification error probability is

$$P_e = P(x \in R_1, \omega_2) + P(x \in R_2, \omega_1)$$

Observation 11.3. The Bayesian classification scheme minimizes the classification error probability. □

The above observation is established in the problem section.

Observations 11.4. Some related observations.

1. Let there be m classes $\omega_1, \omega_2, \ldots, \omega_m$, where $m \in \mathbb{P} \backslash \{1\}$. A test data point $x \in \mathcal{R} \backslash \mathcal{X}$ is assigned to a class ω_i if

$$P(\omega_i \mid x) > P(\omega_j \mid x); \quad \forall j \neq i, \ 1 \leq j \leq m$$

This scheme minimizes classification error probability.

2. Let a t-dimensional data point $x_a \in \mathcal{R}$ be $(x_{a1}, x_{a2}, \ldots, x_{at})$. It is generally assumed that

$$p(x_a \mid \omega_j) = \prod_{k=1}^{t} p(x_{ak} \mid \omega_j), \quad 1 \leq j \leq m$$

This assumption means that the effect of an attribute-value on a given class is probabilistically independent of the values of other attributes, and is called *class conditional independence*. The assumption is termed *naïve*, as it is often simplistic. Nevertheless, it renders the computations in the classification scheme simple. Therefore the corresponding classification scheme is termed *naïve Bayesian classifier*. □

Example 11.9. An elementary naïve Bayesian classifier is developed in this example. A data source outputs data with two attributes A_1 and A_2, and the data is divided into two classes ω_1 and ω_2. Attribute A_1 takes values in the set $\{a, b, c\}$, and the attribute A_2 takes values in the set $\{v, y, z\}$. The training data set is specified in Table 11.10. The goal of the training data set is to learn a naïve Bayesian classifier.

The following probabilities are readily computed from the table.

$$P(\omega = \omega_1) = 1/2, \quad \text{and} \quad P(\omega = \omega_2) = 1/2$$

Data Point	A_1	A_2	Class ω
1	a	y	ω_1
2	a	z	ω_1
3	c	y	ω_1
4	b	z	ω_1
5	b	v	ω_1
6	a	y	ω_2
7	c	v	ω_2
8	b	y	ω_2
9	c	z	ω_2
10	c	v	ω_2

Table 11.10. Training data set.

$$P(A_1 = a \mid \omega = \omega_1) = 2/5, \ P(A_1 = b \mid \omega = \omega_1) = 2/5,$$
$$P(A_1 = c \mid \omega = \omega_1) = 1/5$$
$$P(A_1 = a \mid \omega = \omega_2) = 1/5, \ P(A_1 = b \mid \omega = \omega_2) = 1/5,$$
$$P(A_1 = c \mid \omega = \omega_2) = 3/5$$
$$P(A_2 = v \mid \omega = \omega_1) = 1/5, \ P(A_2 = y \mid \omega = \omega_1) = 2/5,$$
$$P(A_2 = z \mid \omega = \omega_1) = 2/5$$
$$P(A_2 = v \mid \omega = \omega_2) = 2/5, \ P(A_2 = y \mid \omega = \omega_2) = 2/5,$$
$$P(A_2 = z \mid \omega = \omega_2) = 1/5$$

The test sample is $x = (a_1, a_2) = (b, z)$. The goal is to determine the class ω of this data point. In this data point, $A_1 = b$, and $A_2 = z$. For class $\omega = \omega_1$ we compute

$$P(\omega_1) \prod_{i=1}^{2} P(A_i = a_i \mid \omega = \omega_1) = \frac{1}{2} \times \frac{2}{5} \times \frac{2}{5} = 0.08$$

Similarly, for class $\omega = \omega_2$ we compute

$$P(\omega_2) \prod_{i=1}^{2} P(A_i = a_i \mid \omega = \omega_2) = \frac{1}{2} \times \frac{1}{5} \times \frac{1}{5} = 0.02$$

The first probability for the class $\omega = \omega_1$ is higher. Therefore the class of the test sample x is predicted to be $\omega = \omega_1$. □

Minimization of the classification error is not always the most satisfactory criterion to assign a data point to a particular class. Further, the Bayesian classification scheme assumes that all errors are of same significance. It is possible that some errors might have more weight than others. Therefore we can assign a penalty term to each error.

Consider again the general case of classifying a data point in the data set $\mathcal{R}\backslash\mathcal{X}$, into anyone of the m classes. The feature subspace $R_j \subseteq \mathcal{R}$ is assigned to class ω_j, where $1 \leq j \leq m$; $\bigcup_{j=1}^{m} R_j = \mathcal{R}$, and $R_i \cap R_j = \varnothing$ for $1 \leq i, j \leq m$, $i \neq j$. Assume that the feature vector $x \in \mathcal{R}\backslash\mathcal{X}$ lies in R_k (class ω_k) but assigned to class ω_i, $i \neq k$. In this case an error has occurred. Let the penalty term associated with this decision be θ_{ki}. This penalty term is also called *loss*. The $m \times m$ matrix $\Theta = [\theta_{ki}]$ is called the *loss matrix*. Note that the diagonal term θ_{ii} can be selected so that $\theta_{ii} < \theta_{ij}$ for all $j \neq i$, for $1 \leq i, j \leq m$. That is, correct decisions are less penalized than the incorrect ones. Let r_k be the *risk* or *loss* associated with class ω_k, where

$$r_k = \sum_{i=1}^{m} \theta_{ki} \int_{R_i} p(x \mid \omega_k)\, dx, \quad 1 \leq k \leq m$$

Note that $\int_{R_i} p(x \mid \omega_k)\, dx$ is the probability that a feature vector which belongs to class ω_k is assigned to class ω_i. Thus the average risk r is

$$r = \sum_{k=1}^{m} r_k P(\omega_k)$$

The goal is to select the regions R_j's so that the average risk r is minimized. The average risk r can be rewritten as

$$\begin{aligned} r &= \sum_{k=1}^{m} r_k P(\omega_k) \\ &= \sum_{i=1}^{m} \int_{R_i} \left\{ \sum_{k=1}^{m} \theta_{ki} p(x \mid \omega_k) P(\omega_k) \right\} dx \end{aligned}$$

Let

$$l_i(x) \triangleq \sum_{k=1}^{m} \theta_{ki} p(x \mid \omega_k) P(\omega_k), \quad \text{for } 1 \leq i \leq m$$

The average risk r is minimized, provided each of the integrals $\int_{R_i} l_i(x)\,dx$ is minimized. This is possible, if the partitioning regions are chosen such that

$$x \in R_i \quad \text{if } l_i(x) < l_j(x), \quad \forall\, j \neq i, \ 1 \leq j \leq m$$

Note that the minimization of the average risk criteria simplifies to the Bayesian classifier, if $\theta_{ki} = 1$ if $k \neq i$ for all $1 \leq i, k \leq m$, and $\theta_{ii} = 0$ for all $1 \leq i \leq m$.

11.5.2 Support Vector Machines

Support vector machines are a supervised learning method of classifying data points into two classes. Some of their applications are in: text and Web page classification, and bioinformatics. The basic idea behind the use of support vector machine (SVM) is as follows. A nonlinear mapping is initially used to transform the original training data set into a higher-dimensional space. In this higher-dimensional space, the SVM methodology determines optimal decision boundary (or surface) in order to classify data points. This technique is based upon the work of Vladimir Vapnik and his colleagues.

In general, there are two types of SVMs: linear and nonlinear. Linear SVMs are of immediate interest to us. In linear SVMs, the optimal decision boundaries are hyperplanes. In this case, it is sometimes possible to separate (classify) the data points correctly via a hyperplane. This is called the linear separable case. However, this fortuitous condition may not always occur. Nevertheless, it might happen that a hyperplane may classify the data points reasonably well for a majority of data points; and in some other cases it may misclassify. As the hyperplane sometimes may not cleanly separate the data points, this later scenario is called the linear nonseparable case. Nonlinear SVMs are also discussed briefly in this subsection. To facilitate this discussion, some elementary results from vector and linear algebra are initially summarized.

Some Elementary Results about Hyperplanes in a Vector Space

Some basic results about hyperplanes in a vector space are initially listed. This is followed by a list of useful facts related to hyperplanes. We represent a vector as a row matrix in this subsection.

Hyperplane in a vector space, and normal vector to a hyperplane are defined. Let $z = (z_1, z_2, \ldots, z_t)$ be a vector in $\mathbb{R}^t$, where $t \in \mathbb{P} \backslash \{1\}$.

- Let $w = (w_1, w_2, \ldots, w_t)$ be a vector, where $w_i \in \mathbb{R}$, $1 \leq i \leq t$, not all of which are zeros, and also let $b \in \mathbb{R}$. The set of vectors

$$\widehat{Z} = \left\{ z \mid wz^T = \sum_{i=1}^{t} w_i z_i = b \right\}$$

  is called the hyperplane in $\mathbb{R}^t$. This is a t-dimensional plane.
- A vector $\eta = (\eta_1, \eta_2, \ldots, \eta_t) \in \mathbb{R}^t$ is called normal to the hyperplane $\widehat{Z}$, if

$$\sum_{i=1}^{t} \eta_i z_i = 0, \quad \forall\, z \in \widehat{Z}$$

Let $wz^T = b$ be a hyperplane $\widehat{Z}$ in $\mathbb{R}^t$, where $z \in \widehat{Z}$, and $\|\cdot\|$ be the Euclidean norm defined on the vector space $\mathbb{R}^t$.

- The vector w is normal to the hyperplane $\widehat{Z}$.
- Note that the hyperplanes $wz^T = b_1$ and $wz^T = b_2$, where $b_1, b_2 \in \mathbb{R}$, are parallel to each other.
- Let $q \in \mathbb{R}^t$ be a point which does not lie in the hyperplane $\widehat{Z}$. The shortest distance between the point q and the hyperplane $\widehat{Z}$ is $\left|wq^T - b\right| / \|w\|$. Thus, the perpendicular distance from the origin to the hyperplane is $|b| / \|w\|$.

Linear Separable Classes

Consider a classification problem in which data points have multiple attributes. Let $\mathcal{X} \subseteq \mathbb{R}^t$ be the set of data points, where each data point has $t \in \mathbb{P}$ attributes. Also, $\Omega = \{-1, +1\}$ is the set of labels for the two classes of data points. Classifier is a mapping $\Phi_{cl} : \mathcal{X} \to \Omega$. The training data is a set $\mathcal{D}$ of n observations,

$$\mathcal{D} = \left\{(x_i, y_i) \mid x_i \in \mathcal{X} \subseteq \mathbb{R}^t, y_i \in \{-1, +1\}, 1 \leq i \leq n\right\}$$

The members of this set are called training data points. The training data points are used in determining the parameters of the separating hyperplane of the SVM.

If the data is 2-dimensional, that is $t = 2$, then it is hoped that the classification divides the data into two classes separated via a straight line. If this is possible, the data is said to be linearly separable; and one class is specified by -1, and the other by $+1$.

Note that it might be possible to draw an infinite number of lines which separate the data points into two non-overlapping parts. The goal of SVM is to find the "best" straight line which will give the smallest possible classification error on new data points.

If the dimension t is greater than 2, then it is hoped that the classification scheme divides the data points via hyperplanes. The goal of the SVM is to find the best hyperplane which separates the data points into two classes. It is assumed that the two classes (positive and negative) of data points are linearly separable.

Definition 11.16. *Let the set of data points be*

$$\mathcal{D} = \left\{(x_i, y_i) \mid x_i \in \mathcal{X} \subseteq \mathbb{R}^t, y_i \in \{-1, +1\}, 1 \leq i \leq n\right\} \tag{11.7a}$$

The set of data points $\mathcal{D}$ are linearly separable, if there exists a vector $w \in \mathbb{R}^t$ and a scalar $\beta \in \mathbb{R}$ such that

$$wx_i^T + \beta \geq +1, \quad \text{if } y_i = +1 \tag{11.7b}$$

$$wx_i^T + \beta \leq -1, \quad \text{if } y_i = -1 \tag{11.7c}$$

or equivalently $y_i \left(wx_i^T + \beta\right) \geq 1$ for each $i = 1, 2, \ldots, n$. □

The goal of the SVM is to determine a classifier, which is a linear function of the form

$$f(x) = wx^T + \beta, \quad w, x \in \mathbb{R}^t$$

In the above function, $w = (w_1, w_2, \ldots, w_t)$ is called the *weight vector*. The parameter $\beta \in \mathbb{R}$ is called the *bias*. Consider a data point $x_i \in \mathbb{R}^t$. If $f(x_i) \geq 1$, then x_i is classified as $y_i = +1$. However, if $f(x_i) \leq -1$, then x_i is classified as $y_i = -1$. Call these positive and negative points respectively.

Based upon the training data set $\mathcal{D}$, SVM finds the hyperplane $f(x) = (wx^T + \beta) = 0$. As several hyperplanes are possible between the two classes of data points, SVM selects a hyperplane which maximizes the *margin* (gap) between the two (positive and negative) sets of training data points.

Let d_+ be the shortest distance from the separating hyperplane $f(x) = 0$ to the closest positive training data point $x_+ \in \mathcal{X}$. Similarly, let d_- be the shortest distance from the separating hyperplane $f(x) = 0$ to the closest negative training data point $x_- \in \mathcal{X}$. The margin of the separating hyperplane is $(d_+ + d_-)$. The SVM maximizes this margin. Next consider two hyperplanes H_+ and H_- which are parallel to the hyperplane $f(x) = 0$, and pass through data points x_+ and x_- respectively. This implies that no training data point falls in between these two hyperplanes. The hyperplanes H_+ and H_- are called support planes. The data points that lie on the support planes are called support vectors, as they support the hyperplanes, and consequently determine the solution to the SVM problem. Note that the weight vector w and the bias β can be rescaled to obtain:

- The hyperplane H_+ as $(wx^T + \beta) = +1$, where $x_+ \in H_+$.
- The hyperplane H_- as $(wx^T + \beta) = -1$, where $x_- \in H_-$.

Therefore for $x_i \in \mathcal{X}$, we have

- $(wx_i^T + \beta) \geq +1$, if $y_i = +1$.
- $(wx_i^T + \beta) \leq -1$, if $y_i = -1$.

We next determine the margin $(d_+ + d_-)$. The distance d_+ between x_+ and the hyperplane $f(x) = 0$ is

$$d_+ = \frac{|wx_+^T + \beta|}{\|w\|} = \frac{1}{\|w\|}$$

It can be shown similarly that $d_- = 1/\|w\|$. Therefore, the margin

$$(d_+ + d_-) = \frac{2}{\|w\|}$$

The SVM maximizes, this margin. This is equivalent to minimizing $\|w\|^2/2$. The linear SVM optimization problem can now be stated formally. As the training data points in the set $\mathcal{D}$ are assumed to be linearly separable, these data points can be split into two classes via a hyperplane.

Input: Set of training data points $\mathcal{D}$, where:
$\mathcal{D} = \{(x_i, y_i) \mid x_i \in \mathcal{X} \subseteq \mathbb{R}^t, y_i \in \{-1, +1\}, 1 \leq i \leq n\}$.
Output: Weight vector $w = (w_1, w_2, \ldots, w_t) \in \mathbb{R}^t$, and $\beta \in \mathbb{R}$.
Objective: Minimize $\|w\|^2/2$ by varying w and β.
Subject to: $y_i \{wx_i^T + \beta\} \geq 1$, where $1 \leq i \leq n$.

Note that the constraint $y_i \{wx_i^T + \beta\} \geq 1$, where $1 \leq i \leq n$; is equivalent to: $(wx_i^T + \beta) \geq +1$, if $y_i = +1$; and $(wx_i^T + \beta) \leq -1$, if $y_i = -1$.

The method of Lagrange multipliers can be used to address this optimization problem. In this optimization problem, the objective function is strictly convex, and the constraints are linear. Therefore, the local minimum is also global and unique. Thus, the optimal hyperplane classifier of a SVM is indeed unique. The Lagrangian of this optimization problem is:

$$\mathcal{L}(w, \beta, \lambda) = \frac{1}{2} \|w\|^2 - \sum_{i=1}^{n} \lambda_i \left\{ y_i \left(w x_i^T + \beta \right) - 1 \right\}$$

where

$$\lambda_i \in \mathbb{R}, \quad 1 \leq i \leq n, \quad \text{and} \quad \lambda = (\lambda_1, \lambda_2, \ldots, \lambda_n)$$

The Karush-Kuhn-Tucker conditions of optimality are:

$$\begin{aligned} \frac{\partial \mathcal{L}(w, \beta, \lambda)}{\partial w_j} &= 0, \quad 1 \leq j \leq t \\ \frac{\partial \mathcal{L}(w, \beta, \lambda)}{\partial \beta} &= 0 \\ \lambda_i \left\{ y_i \left(w x_i^T + \beta \right) - 1 \right\} &= 0, \quad 1 \leq i \leq n \\ \lambda_i &\in \mathbb{R}_0^+, \quad 1 \leq i \leq n \end{aligned}$$

The conditions $\partial \mathcal{L}(w, \beta, \lambda) / \partial w_j = 0, 1 \leq j \leq t$; and $\partial \mathcal{L}(w, \beta, \lambda) / \partial \beta = 0$ lead to

$$w = \sum_{i=1}^{n} \lambda_i y_i x_i, \quad \text{and} \quad \sum_{i=1}^{n} \lambda_i y_i = 0$$

The Lagrange multipliers (λ_i's), can be either equal to zero or greater than zero. If $\lambda_i > 0$, the corresponding vector x_i lies either on H_+ or H_- hyperplane. The solution of the optimization problem is nontrivial due to the inequality constraints. However, the theory of duality comes to our rescue. The original optimization problem is called the primal problem, and the corresponding Lagrangian is called the primal Lagrangian. The primal variables are w and β. The Wolfe-dual problem representation is:

Input: Set of training data points $\mathcal{D}$, where:
$\mathcal{D} = \{(x_i, y_i) \mid x_i \in \mathcal{X} \subseteq \mathbb{R}^t, y_i \in \{-1, +1\}, 1 \leq i \leq n\}$.
Output: The Lagrangian vector $\lambda = (\lambda_1, \lambda_2, \ldots, \lambda_n)$.
Objective: Maximize $\mathcal{L}(w, \beta, \lambda)$ by varying λ.
Subject to: $w = \sum_{i=1}^{n} \lambda_i y_i x_i$, $\sum_{i=1}^{n} \lambda_i y_i = 0$, and $\lambda \geq 0$.

Note that the constraints of the dual problem were obtained by setting the appropriate derivatives of $\mathcal{L}(w, \beta, \lambda)$ equal to zero. It is easier to manipulate these equality constraints, rather than the inequality constraints in the primal problem. The dual problem can be simplified to:

Input: Set of training data points $\mathcal{D}$, where:
$\mathcal{D} = \{(x_i, y_i) \mid x_i \in \mathcal{X} \subseteq \mathbb{R}^t, y_i \in \{-1, +1\}, 1 \leq i \leq n\}$.
Output: The Lagrangian vector $\lambda = (\lambda_1, \lambda_2, \ldots, \lambda_n)$.
Objective: $\max_\lambda \left\{ \lambda e - \frac{1}{2} \lambda H \lambda^T \right\}$,
where e is an all-1 vector of size n. That is, $e = (1, 1, \ldots, 1)^T$;
and $H = [h_{ij}]$ is an $n \times n$ matrix, $h_{ij} = y_i y_j x_i x_j^T$, for $1 \leq i, j \leq n$.
Subject to: $\sum_{i=1}^{n} \lambda_i y_i = 0$, and $\lambda \geq 0$.

The objective function in the above formulation has a quadratic form. Therefore, the corresponding optimization problem is also referred to as a quadratic programming problem. Such problems are typically solved numerically. Having determined the vector λ, the weight vector is determined from the result: $w = \sum_{i=1}^{n} \lambda_i y_i x_i$. Let S be the set of indices of the support vectors. That is, $S = \{i \mid \lambda_i \in \mathbb{R}^+, 1 \leq i \leq n\}$. Therefore, $w = \sum_{i \in S} \lambda_i y_i x_i$.

The bias β can be determined in principle, from any one of the following equations, where $\lambda_i > 0$, and $\lambda_i \left\{ y_i \left(w x_i^T + \beta \right) - 1 \right\} = 0$, for $1 \leq i \leq n$. However, in practice, β is determined by using all support vectors, and then taking their average value. These multiple computations of β are performed for numerical stability. Also

$$f(x) \triangleq wx^T + \beta = \sum_{i \in S} \lambda_i y_i x_i x^T + \beta$$

The decision boundary, or surface, or the maximum margin hyperplane is $f(x) = 0$. Having determined the maximum margin hyperplane, a test data point $u \in \mathbb{R}^t \backslash \mathcal{X}$ can be classified by computing $f(u) = \left(wu^T + \beta \right)$. If $f(u) \geq 0$, then u is classifed as positive, otherwise its classification is negative.

Algorithm 11.5. *Linear SVM: Separable Case Algorithm.*

Input: The set of training data points $\mathcal{D}$, where
$\mathcal{D} = \{(x_i, y_i) \mid x_i \in \mathcal{X} \subseteq \mathbb{R}^t, y_i \in \{-1, +1\}, 1 \leq i \leq n\}$.
Test data point: $u \in \mathbb{R}^t \backslash \mathcal{X}$.
Output: Determine the maximum margin hyperplane. It is the optimal decision boundary. Classification of the test data point $u \in \mathbb{R}^t \backslash \mathcal{X}$.
begin

Step 1: Determine $H = [h_{ij}]$, where $h_{ij} = y_i y_j x_i x_j^T$, for $1 \leq i, j \leq n$
Step 2: Find λ so that $\max_\lambda \left\{ \lambda e - \frac{1}{2} \lambda H \lambda^T \right\}$;
subject to $\sum_{i=1}^{n} \lambda_i y_i = 0$, and $\lambda \geq 0$.
Step 3: Determine S, the set of indices of the support vectors.
$S = \{i \mid \lambda_i \in \mathbb{R}^+, 1 \leq i \leq n\}$
Step 4: Compute the weight vector $w = \sum_{i \in S} \lambda_i y_i x_i$
Step 5: Compute β from $\left\{ y_i \left(w x_i^T + \beta \right) - 1 \right\} = 0$, for each $i \in S$.
Determine the average value of all such computations of the β's.
Step 6: Let $f(x) \triangleq wx^T + \beta = \sum_{i \in S} \lambda_i y_i x_i x^T + \beta$
The maximum margin hyperplane is $f(x) = 0$
Step 7: Compute $f(u) = \left(wu^T + \beta \right)$. If $f(u) \geq 0$, then u is classifed as positive, otherwise its classification is negative.

end (*end of linear SVM: separable case algorithm*)

Linear Nonseparable Classes

In the linear separable case, it was possible to strictly classify the training data points via a hyperplane specified by $\left(wx^T + \beta \right) = 0$. Further, the training data points were separated by two supporting parallel hyperplanes $(wx^T + \beta) = \pm 1$. We now consider the case where the following categories of training data points occur.

- *Category* 1: Data points for which $y_i (wx_i^T + \beta) \geq 1$. These data points correspond to the linear separable case.
- *Category* 2: Data points for which

$$0 \leq y_i (wx_i^T + \beta) < 1$$

These data points fall within the two hyperplanes, and are correctly classified.

- *Category* 3: Data points for which

$$y_i (wx_i^T + \beta) < 0$$

These data points fall within the two hyperplanes, and are misclassified.

All the three categories can be compactly specified by

$$y_i (wx_i^T + \beta) \geq 1 - \xi_i$$

where $\xi_i \in \mathbb{R}_0^+$ is called the *slack variable*. The first category corresponds to $\xi_i = 0$, the second category corresponds to $0 < \xi_i \leq 1$, and the third category corresponds to $\xi_i > 1$. The goal of the classification scheme is to determine an optimal separating hyperplane, so that the margin between the two classes of data is as high as possible; and simultaneously keep the number of training data points with $\xi_i > 0$ to a minimum. Equivalently, the objective function to be minimized is

$$\frac{\|w\|^2}{2} + \mathcal{K} \sum_{i=1}^{n} I_{indic}(\xi_i)$$

where

$$I_{indic}(\xi_i) = \begin{cases} 1, & \xi_i > 0 \\ 0, & \xi_i = 0 \end{cases}$$

and $\mathcal{K} \in \mathbb{R}^+$ is a constant term that regulates the relative influence of the two terms in the objective function. The objective function in this formulation is a discontinuous function of ξ, where $\xi = (\xi_1, \xi_2, \ldots, \xi_n)$. A more convenient optimizing problem can be stated as follows.

Input: Set of training data points $\mathcal{D}$, where:
$\mathcal{D} = \{(x_i, y_i) \mid x_i \in \mathcal{X} \subseteq \mathbb{R}^t, y_i \in \{-1, +1\}, 1 \leq i \leq n\}$,
and $\mathcal{K} \in \mathbb{R}^+$.
Output: Weight vector $w = (w_1, w_2, \ldots, w_t) \in \mathbb{R}^t$, and $\beta \in \mathbb{R}$.
Objective: Minimize $\|w\|^2 / 2 + \mathcal{K} \sum_{i=1}^{n} \xi_i$, by varying w, β, and ξ.
Subject to: $y_i (wx_i^T + \beta) \geq (1 - \xi_i), \xi_i \geq 0$, where $1 \leq i \leq n$.

The Lagrangian of this optimization problem is:

$$\mathcal{L}(w, \beta, \xi, \lambda, \mu)$$
$$= \frac{1}{2} \|w\|^2 + \mathcal{K} \sum_{i=1}^{n} \xi_i - \sum_{i=1}^{n} \mu_i \xi_i - \sum_{i=1}^{n} \lambda_i \{y_i (wx_i^T + \beta) - 1 + \xi_i\}$$

where

$$\lambda_i, \mu_i \in \mathbb{R},\ 1 \le i \le n;\ \lambda = (\lambda_1, \lambda_2, \ldots, \lambda_n)\,, \text{ and } \mu = (\mu_1, \mu_2, \ldots, \mu_n)$$

The Karush-Kuhn-Tucker conditions of optimality are:

$$\frac{\partial \mathcal{L}(w, \beta, \xi, \lambda, \mu)}{\partial w_j} = 0, \quad 1 \le j \le t$$

$$\frac{\partial \mathcal{L}(w, \beta, \xi, \lambda, \mu)}{\partial \beta} = 0$$

$$\frac{\partial \mathcal{L}(w, \beta, \xi, \lambda, \mu)}{\partial \xi_i} = 0, \quad 1 \le i \le n$$

$$\lambda_i \left\{ y_i \left(w x_i^T + \beta \right) - 1 + \xi_i \right\} = 0, \quad 1 \le i \le n$$

$$\mu_i \xi_i = 0, \quad 1 \le i \le n$$

$$\lambda_i, \mu_i \in \mathbb{R}_0^+, \quad 1 \le i \le n$$

The conditions $\partial \mathcal{L}(w, \beta, \xi, \lambda, \mu) / \partial w_j = 0,\ 1 \le j \le t$; $\partial \mathcal{L}(w, \beta, \xi, \lambda, \mu) / \partial \beta = 0$; and $\partial \mathcal{L}(w, \beta, \xi, \lambda, \mu) / \partial \xi_i = 0$, for $1 \le i \le n$ lead respectively to

$$w = \sum_{i=1}^{n} \lambda_i y_i x_i, \quad \sum_{i=1}^{n} \lambda_i y_i = 0, \quad \text{and} \quad \mathcal{K} = (\lambda_i + \mu_i)\,, \text{ for } 1 \le i \le n$$

The primal variables are w, β, and ξ. The corresponding Wolfe-dual problem representation is:

Input: Set of training data points $\mathcal{D}$, where:
$\mathcal{D} = \{(x_i, y_i) \mid x_i \in \mathcal{X} \subseteq \mathbb{R}^t, y_i \in \{-1, +1\}, 1 \le i \le n\}$,
and $\mathcal{K} \in \mathbb{R}^+$.
Output: The Lagrangian vectors λ and μ.
Objective: Maximize $\mathcal{L}(w, \beta, \xi, \lambda, \mu)$ by varying λ and μ.
Subject to: $w = \sum_{i=1}^{n} \lambda_i y_i x_i$, $\sum_{i=1}^{n} \lambda_i y_i = 0$, $\mathcal{K} = (\lambda_i + \mu_i)$, for $1 \le i \le n$,
and $\lambda, \mu \ge 0$.

The constraints of the dual problem were obtained by setting the appropriate derivatives of $\mathcal{L}(w, \beta, \xi, \lambda, \mu)$ equal to zero. The dual problem can be simplified to:

Input: Set of training data points $\mathcal{D}$, where:
$\mathcal{D} = \{(x_i, y_i) \mid x_i \in \mathcal{X} \subseteq \mathbb{R}^t, y_i \in \{-1, +1\}, 1 \le i \le n\}$,
and $\mathcal{K} \in \mathbb{R}^+$.
Output: The Lagrangian vector λ.
Objective: $\max_\lambda \left\{ \lambda e - \frac{1}{2} \lambda H \lambda^T \right\}$
where e is an all-1 vector of size n. That is, $e = (1, 1, \ldots, 1)^T$;
and $H = [h_{ij}]$ is an $n \times n$ matrix, $h_{ij} = y_i y_j x_i x_j^T$, for $1 \le i, j \le n$
Subject to: $\sum_{i=1}^{n} \lambda_i y_i = 0$, and $0 \le \lambda_i \le \mathcal{K}$.

Note that $\mathcal{K} = (\lambda_i + \mu_i)$, and $\mu_i \ge 0$ imply that $0 \le \lambda_i \le \mathcal{K}$, for $1 \le i \le n$. Also observe that, ξ and μ are absent in the above dual problem. Further, the above optimization problem is identical to the corresponding optimization problem for the separable case, except that the Lagrangian multiplier λ_i is bounded from above by $\mathcal{K}$ for $1 \le i \le n$.

The conditions

$$\begin{aligned} y_i \left\{ w x_i^T + \beta \right\} &\geq (1 - \xi_i) \\ \lambda_i \left\{ y_i \left(w x_i^T + \beta \right) - 1 + \xi_i \right\} &= 0 \\ \mathcal{K} &= (\lambda_i + \mu_i) \\ \mu_i \xi_i &= 0 \end{aligned}$$

for $1 \leq i \leq n$ imply:

- $\lambda_i = 0 \Rightarrow \xi_i = 0$, and $y_i \left(w x_i^T + \beta \right) \geq 1$.
- $0 < \lambda_i < \mathcal{K} \Rightarrow \xi_i = 0$, and $y_i \left(w x_i^T + \beta \right) = 1$.
- $\lambda_i = \mathcal{K} \Rightarrow \xi_i \geq 0$, and $y_i \left(w x_i^T + \beta \right) \leq 1$.

The above quadratic programming problem is generally solved numerically.

Nonlinear Support Vector Machines

In the linear SVMs the decision boundary is a hyperplane. However, in several cases in practice, the decision boundaries are nonlinear. In such cases, the input data is transformed nonlinearly from its original space into another space, which is typically of a much higher dimension, so that a linear decision boundary can categorize data points with positive and negative classifications.

11.5.3 Nearest-Neighbor Classification

The nearest-neighbor classification scheme is an intuitive technique to classify data points. In this scheme, the test data points are classified based upon their resemblance with data points in the training set. The nearest-neighbor classification scheme is a *lazy learning* method. In contrast, the Bayesian classification scheme, and support vector machines are *eager learning* methods. In the eager learning methods, the classification schemes learn about the model of the data points from the training data set before testing. However, in lazy learning method learning does not occur directly from the training data set. Learning occurs only during the phase in which a test data point has to be classified.

The nearest-neighbor classification scheme requires three ingredients. These are: the training data set in which each data point has a label (classification), a positive integer $k \in \mathbb{P}$, and a metric to measure proximity between two data points. Such classification schemes are also called k-nearest-neighbor classifiers.

Let $\mathcal{X}$ be the set of data points, and

$$\Omega = \{\omega_1, \omega_2, \ldots, \omega_m\}$$

be the set of labels for different classes of data. The training data is a set $\mathcal{D}$ of n observations,

$$\mathcal{D} = \left\{ (x_i, y_i) \mid x_i \in \mathcal{X} \subseteq \mathbb{R}^t, y_i \in \Omega, 1 \leq i \leq n \right\}$$

Some choices of proximity measure are: Euclidean distance, weighted Chebyshev distance, weighted Minkowski distance, cosine function measure, and Hamming distance. The proximity measure between $x_i, x_j \in \mathcal{X}$ is denoted by $\delta(x_i, x_j) \in \mathbb{R}$.

The nearest-neighbor classification scheme works as follows. Let $u \in \mathbb{R}^t \backslash \mathcal{X}$ be a test data point. Using the proximity metric, the scheme determines k "closest" training data points to the data point

u. The scheme assigns a classification (label) $\omega \in \Omega$ to the data point u which occurs most often within these k "closest" training data points. The value $k \leq n$ is usually selected via trial and error.

Algorithm 11.6. *Nearest-Neighbor Classification Algorithm.*

Input: The set of training data points $\mathcal{D}$, where
$\mathcal{D} = \{(x_i, y_i) \mid x_i \in \mathcal{X} \subseteq \mathbb{R}^t, y_i \in \Omega, 1 \leq i \leq n\}$.
Integer $k \in \mathbb{P}$. The proximity measure $\delta : \mathcal{X} \times \mathcal{X} \rightarrow \mathbb{R}$.
Test data point: $u \in \mathbb{R}^t \backslash \mathcal{X}$.
Output: Classification $\omega \in \Omega$ of the test data point $u \in \mathbb{R}^t \backslash \mathcal{X}$.
begin
  Step 1: Compute $\delta(u, x)$, $\forall\, x \in \mathcal{X}$ in the training data set.
  Step 2: Determine $\mathcal{F}_u \subseteq \mathcal{D}$, the set of k closest
    training data points to u.
  Step 3: $\omega \leftarrow \arg\max_{v \in \Omega} \sum_{(x_i, y_i) \in \mathcal{F}_u} I_{vy_i}$
    (I_{vy_i} *is an indicator function which is equal to* 1 *if* $v = y_i$)
    (*otherwise it is equal to* 0)
end (*end of nearest-neighbor classification algorithm*)

The nearest-neighbor data classification scheme has several advantages. Further, the scheme is analytically tractable, effective, popular because of its simplicity, and is nearly optimal as $n \rightarrow \infty$. It is also convenient to design a parallel implementation of the algorithm. The advantages of this algorithm have to be weighed along with its disadvantages. The algorithm is computationally very intensive, nonlinear, and requires large storage. It is also susceptible to the curse of dimensionality (dependence upon the dimension t of the data points).

11.5.4 Artificial Neural Network

Artificial neural network (ANN) is a useful paradigm for data mining. The study of artificial neural networks originally began with the goal of studying biological neural systems. The ambitious aim of the neurologists and cognitive scientists is to explain the workings of the human brain. For our discussion, the human brain consists of *neurons, axons, dendrites*, and *synapses*.

Neurons are nerve cells, which are connected to other neurons via strands of fiber called axons (transmission lines). The purpose of axons is to transmit nerve impulses between two neurons whenever the stimulation of neurons occur. The axons of two different neurons are connected via dendrites. The dendrites are extensions from the cell body of the neurons. Synapse is the contact point between a dendrite and an axon.

Artificial neural networks ostensibly simulate biological neural activity. Using the analogy of biological neurons and their interconnections, an artificial neural network can be considered to be an assemblage of nodes and directed links. Our immediate goal is to explore the use of ANNs in classification of data points. We study two models of ANNs: the perceptron, and the multilayer ANN. In rest of this subsection, an artificial neuron is simply referred to as a neuron.

Perceptron

The perceptron is an elementary yet useful model of ANN. It consists of two types of nodes. These are the input nodes and a single output node. The input and output nodes are representations of the input attributes and the model output respectively. The output node of the perceptron simulates a neuron. Each input node is directly connected to the output node via a weighted link. The weighted link simulates the strength of the synaptic connection between the neurons. The output node performs mathematical operations, and generates the classification of the data points. The weights of the input nodes are trained by the learning data points to possibly produce correct classification at the output node. Once the weights are determined, the classification of a test data point can be determined.

The definition of a perceptron is initially provided, This is followed by a description of the perceptron learning algorithm. A proof of the convergence of the learning algorithm is also outlined.

Definition 11.17. *Let the set of data points, and the set of labels of the classes of the data points be $\mathcal{X} \subseteq \mathbb{R}^t$, and $\Omega = \{-1, 1\}$ respectively. Consider a data point $(\zeta_1, \zeta_2, \ldots, \zeta_t) \in \mathcal{X}$. Also let $(w_1, w_2, \ldots, w_t) \in \mathbb{R}^t$ be the synaptic weight vector, and $\theta \in \mathbb{R}$ be the bias factor, and*

$$v = \sum_{j=1}^{t} w_j \zeta_j + \theta \tag{11.8a}$$

The perceptron is a mapping $f_{percep} : \mathcal{X} \to \Omega$, where

$$f_{percep} (\zeta_1, \zeta_2, \ldots, \zeta_t) = sgn (v) \in \Omega \tag{11.8b}$$

□

Recall that $sgn (x) = 1$ if $x > 0$, and $sgn (x) = -1$ if $x < 0$. In the above definition, the $sgn (\cdot)$ function emulates a neuron. It is one of several possible such functions which can model a neuron. It is an example of an *activation function*. More such functions are discussed later in this subsection. Also, $v = \sum_{j=1}^{t} w_j \zeta_j + \theta$ is the equation of a hyperplane. Therefore the perceptron can be used to classify linearly separable data points. However, a major challenge is to determine the weight vector and the bias factor of the perceptron.

For simplicity in discussion, let $\zeta_0 \triangleq 1$, $w_0 \triangleq \theta$, $\zeta \triangleq (\zeta_0, \zeta_1, \zeta_2, \ldots, \zeta_t)$ be the extended data point, and $w \triangleq (w_0, w_1, w_2, \ldots, w_t)$ be the extended weight vector. Therefore $v = w\zeta^T$, and the output of the perceptron is equal to $sgn (v)$.

Let the training data set be $\mathcal{D} = \{(x_i, y_i) \mid x_i \in \mathcal{X} \subseteq \mathbb{R}^t, y_i \in \Omega, 1 \leq i \leq n\}$. The weight vector and the bias factor are determined by the training data set $\mathcal{D}$. An iterative learning algorithm is proposed to determine the extended weight vector. Let $x_{i0} = 1$, and $x_i = (x_{i1}, x_{i2}, \ldots, x_{it}) \in \mathcal{X}$. The weights are updated as

$$\begin{aligned} \Delta_i^{(k)} &= \left(y_i - y_i^{(k)}\right) \\ w_j^{(k+1)} &= w_j^{(k)} + \mu \Delta_i^{(k)} x_{ij}, \quad 0 \leq j \leq t \end{aligned}$$

After iteration number $k \in \mathbb{P}$; $w_j^{(k)}$ is the weight parameter for $0 \leq j \leq t$, and $w^{(k)}$ is the corresponding weight vector. The parameter μ, is called the *learning rate*. Typically $\mu \in (0, 1]$. Also

$$y_i^{(k)} = sgn\left(\sum_{j=0}^{t} w_j^{(k)} x_{ij}\right)$$

Justification of the above expression for updating the weights is as follows. The new weight is equal to the sum of the old weight and a correctional term. The correctional term is actually proportional to $\Delta_i^{(k)}$. If the prediction of the classification is correct, then the value of the correction term is equal zero, otherwise it is modified as:

- Let $y_i = -1$, and $y_i^{(k)} = +1$. Therefore $\Delta_i^{(k)} = -2$. To mitigate the error, $w_j^{(k+1)}$ is decreased if x_{ij} is positive; and increased if x_{ij} is negative.
- Let $y_i = +1$, and $y_i^{(k)} = -1$. Therefore $\Delta_i^{(k)} = 2$. To mitigate the error, $w_j^{(k+1)}$ is increased if x_{ij} is positive; and decreased if x_{ij} is negative.

Algorithm 11.7. *Perceptron Training Algorithm.*

Input: The set of training data points $\mathcal{D}$, where
$\mathcal{D} = \{(x_i, y_i) \mid x_i \in \mathcal{X} \subseteq \mathbb{R}^t, y_i \in \{-1, +1\}, 1 \le i \le n\}$.
$x_{i0} = 1$ for $1 \le i \le n$.
Learning rate parameter $\mu \in (0, 1]$.
A stopping criterion for the algorithm.
Output: The extended weight vector $w = (w_0, w_1, w_2, \ldots, w_t)$.
begin
    $w^{(0)} \leftarrow$ random values
    (*the extended weight vector w is initialized by random values*)
    $k \leftarrow 0$
    while stopping criterion is not met **do**
    begin
        for $i = 1$ **to** n **do**
        (*for each training data point (x_i, y_i) do*)
        begin
            $y_i^{(k)} \leftarrow sgn\left(\sum_{j=0}^{t} w_j^{(k)} x_{ij}\right)$
            $\Delta_i^{(k)} \leftarrow \left(y_i - y_i^{(k)}\right)$
            $w_j^{(k+1)} \leftarrow w_j^{(k)} + \mu \Delta_i^{(k)} x_{ij}, \quad \forall j = 0, 1, 2, \ldots, t$
            (*update the weights*)
        end (*end of i for-loop*)
        $k \leftarrow k + 1$
    end (*end of while-loop*)
    $w \leftarrow \left(w_0^{(k)}, w_1^{(k)}, w_2^{(k)}, \ldots, w_t^{(k)}\right)$
end (*end of perceptron training algorithm*)

Observe in the algorithmic outline that v, the argument of the activation function $sgn(\cdot)$, is a linear function of the weight vector, bias factor, and the data point. Therefore the assumed decision boundary of the two classes is a hyperplane. If the data points are linearly separable, the iterative training algorithm converges to an optimal solution. However, if the data points are not linearly separable, then the algorithm does not converge. This in essence, is the perceptron convergence theorem.

Theorem 11.2. *Let* $\mathcal{D} = \{(x_i, y_i) \mid x_i \in \mathcal{X} \subseteq \mathbb{R}^t, y_i \in \{-1, +1\}, 1 \le i \le n\}$ *be a set of linearly separable data points. The perceptron learning algorithm determines the weight vector* $w \in \mathbb{R}^t$*, and the bias factor* $\theta \in \mathbb{R}$ *in a finite number of steps.*

Proof. See the problem section. □

In the above discussion, the activation function was $sgn(\cdot)$. Some other examples of activation function are:

- *Unit step function*: The unit step function is:

$$u(x) = \begin{cases} 1, & x > 0 \\ 0, & x < 0 \end{cases}$$

  Therefore $u(x) = (1 + sgn(x))/2$.
- *Sigmoid or logistic function*: The sigmoid or logistic function for $a \in \mathbb{R}^+$ is:

$$f(x) = \frac{1}{1 + \exp(-ax)}, \quad x \in \mathbb{R}$$

  The value, $f(x)$ is bounded between 0 and 1, and it can be differentiated at all $x \in \mathbb{R}$.
- *Hyperbolic tangent function*: The hyperbolic tangent function for $a \in \mathbb{R}^+$ is:

$$f(x) = \frac{1 - \exp(-ax)}{1 + \exp(-ax)} = \tanh\left(\frac{ax}{2}\right), \quad x \in \mathbb{R}$$

  The value, $f(x)$ is bounded between -1 and 1, and it can be differentiated at all $x \in \mathbb{R}$.

Multilayer Perceptron

Perceptrons can be used efficiently to classify data points which are linearly separable (via hyperplane). A multilayer perceptron can be used to classify data points which are not linearly separable. Motivation for the necessity of multilayer perceptron can be demonstrated via the following example.

Example 11.10. Let the binary alphabet be $\mathfrak{B} = \{0, 1\}$. The mappings (functions) defined by $\wedge, \vee$, and $\oplus$ operations are logical AND, logical OR, and logical EXCLUSIVE-OR (XOR) respectively. These mappings are of type $\ell : \mathfrak{B}^{(2)} \to \mathfrak{B}$. Let $x_1, x_2 \in \mathfrak{B}$. If $\ell(x_1, x_2) = 0$, then (x_1, x_2) is of class c_0; and if $\ell(x_1, x_2) = 1$, then (x_1, x_2) is of class c_1. These functions, and the respective classes are defined in the Table 11.11.

If the points for the different mappings are plotted on the x_1x_2-plane, it can observed that for each of the $\wedge$ and $\vee$ mappings, the classes can be classified by drawing a single straight line. However, this is not possible for the $\oplus$ mapping. Therefore the XOR mapping is not linearly separable.

(a) Let $g_1(x_1, x_2) = (x_1 + x_2 - 1/2)$, and $g_2(x_1, x_2) = (x_1 + x_2 - 3/2)$, and the step function $u(\cdot)$ be the activation function.

(i) The OR operation is simulated by the perceptron $u(g_1(x_1, x_2))$. Therefore, the OR operation needs only a single perceptron.

(ii) The AND operation is simulated by the perceptron $u(g_2(x_1, x_2))$. Therefore, the AND operation needs only a single perceptron.

(b) Let $y_1 = u(g_1(x_1, x_2))$, and $y_2 = u(g_2(x_1, x_2))$. If

$$h(y_1, y_2) = (y_1 - y_2 - 1/2)$$

then $u(h(y_1, y_2))$ simulates the XOR operation. Therefore the XOR operation requires three perceptrons.

x_1	x_2	AND: $\wedge$	Class	OR: $\vee$	Class	XOR: $\oplus$	Class
0	0	0	c_0	0	c_0	0	c_0
0	1	0	c_0	1	c_1	1	c_1
1	0	0	c_0	1	c_1	1	c_1
1	1	1	c_1	1	c_1	0	c_0

Table 11.11. Logical AND, OR, and XOR functions, and their classes.

□

A multilayer perceptron consists of three main parts: the input layer, the hidden (middle) layers, and the output layer. The input layer, each of the hidden layers, and finally the output layer are connected in a feed-forward fashion. The input layer does not have any neurons. However, the hidden and output layers are made up of neurons. In Figure 11.1, the multilayer perceptron consists of two hidden layers.

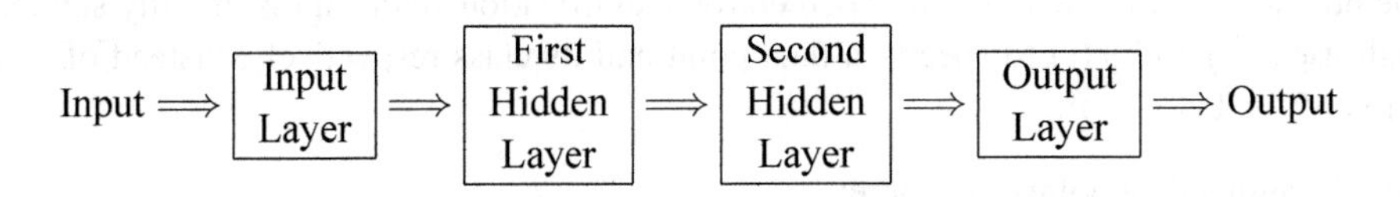

Figure 11.1 A multilayer perceptron configuration.

In this terminology a perceptron, which was discussed earlier, consists of only the input and output layers. It does not consist of any hidden layer. Further, the output layer of the perceptron has only a single neuron.

- *Input layer*: The input layer consists of nonprocessing nodes. Input data is applied to these nodes. Let a data point be $x \in \mathcal{X} \subseteq \mathbb{R}^t$, where $t \in \mathbb{P}$. Therefore, the number of input nodes is equal to t, the number of attributes of a data point. Each input node is connected to the input nodes of every perceptron in the first hidden layer.
- *Hidden layers*: Each hidden layer is made up several neurons. The output of every neuron in a hidden layer (except the last) is connected to the input of every neuron in the next hidden layer. In Figure 11.1, the output of every neuron in the first hidden layer is connected to the input of every neuron in the second hidden layer.

Further, the output of every neuron in the last hidden layer is connected to the input of every neuron in the output layer. In Figure 11.1, the output of every neuron in the second hidden layer is connected to the input of every neuron in the output layer.
Assume that the step activation function is used for each neuron. Therefore the output of each neuron is either a 0 or 1.
Consider the first hidden layer. Assume that this hidden layer has m neurons. The data point x is fed to each of these neurons. Each such neuron creates a hyperplane. The intersection of all these m hyperplanes creates a polyhedron. Data points which are located within anyone of these polyhedral regions are mapped into a single vertex of a m-dimensional hypercube H_m of unit side length. Thus, the input data space is mapped onto the vertices of the hypercube H_m. Let $\mathfrak{B} = \{0, 1\}$. This hypercube is:

$$H_m = \{(y_1, y_2, \ldots, y_m) \in \mathfrak{B}^m, y_i \in \mathfrak{B}, 1 \leq i \leq m\}$$

where the vertices of the hypercube can also be specified by the binary string $y_1 y_2 \ldots y_m$, where $y_i \in \mathfrak{B}, 1 \leq i \leq m$.
This type of segmentation of data points takes place at the output of each hidden layer.
- *Output layer*: The output layer is also made up of neurons. Each neuron in this layer receives input from the output of every neuron in the last hidden layer. The outputs of all the neurons in the output layer also provide a segmentation of the data points from the final hidden layer. It is hoped that at this layer, this segmentation of data points provides correct classification.

In order to have correct classification of data points, the synaptic weight vectors of all the neurons have to be determined by using the training data set. The architecture of the multilayer perceptron is assumed to be fixed in this paradigm. The goal of the training data set is to determine the synaptic weights of all the neurons, so that an appropriate cost function is minimized. This approach requires that the activation function be differentiable for all values of its argument, and be monotonically nondecreasing. Acceptable examples of activation functions are sigmoid, and tangent hyperbolic functions.

The notation used in the multilayer perceptron classification algorithm is initially summarized. We shall use $x(i)$ and $y(i)$ to specify a data point and its class respectively, instead of x_i and y_i respectively, for $1 \leq i \leq n$.

- Set of training data points $\mathcal{D}$, where:
  $\mathcal{D} = \{(x(i), y(i)) \mid x(i) \in \mathcal{X} \subseteq \mathbb{R}^t, y(i) \in \Omega, 1 \leq i \leq n\}$.
- All neurons use the same differentiable activation function $f(\cdot)$. Its first derivative is denoted by $f'(\cdot)$.
- The number of layers of neurons is equal to L. Output is the Lth layer.
- Number of nodes in the input layer is equal to k_0. Therefore $k_0 = t$.
  The number of neurons in layer r is equal to k_r, where $1 \leq r \leq L$. Therefore the output layer has k_L neurons.
- The input vector is $x(i) = (x_1(i), x_2(i), \ldots, x_{k_0}(i)) \in \mathbb{R}^{k_0}$.
- The desired output vector is $y(i) = (y_1(i), y_2(i), \ldots, y_{k_L}(i)) \in \mathbb{R}^{k_L}$.
  The computed output vector is $\widehat{y}(i) = (\widehat{y}_1(i), \widehat{y}_2(i), \ldots, \widehat{y}_{k_L}(i)) \in \mathbb{R}^{k_L}$.
- The algorithm determines the synaptic weights associated with each neuron. Let w_j^r be the weight vector, including the bias factor, of the jth neuron in the rth layer. The corresponding bias factor is w_{j0}^r. Thus

$$w_j^r = \left(w_{j0}^r, w_{j1}^r, \ldots, w_{jk_{r-1}}^r\right) \in \mathbb{R}^{k_{r-1}+1}, \quad 1 \le j \le k_r, \ 1 \le r \le L.$$

- The cost function $J = \sum_{i=1}^{n} E(i)$ is minimized by varying the synaptic weights of all the neurons in the network. The function $E(i)$ depends upon both $y(i)$ and $\widehat{y}(i)$. A possible choice of $E(i)$ is

$$E(i) = \frac{1}{2}\sum_{m=1}^{k_L} e_m^2(i) \triangleq \frac{1}{2}\sum_{m=1}^{k_L} \left(y_m(i) - \widehat{y}_m(i)\right)^2, \quad 1 \le i \le n$$

where

$$e_m(i) \triangleq \left(y_m(i) - \widehat{y}_m(i)\right), \quad 1 \le m \le k_L, \quad \text{and} \quad 1 \le i \le n.$$

- The cost function is minimized by varying the synaptic weights of all the neurons iteratively. In this algorithm, the weight vector $w_j^r \in \mathbb{R}^{k_{r-1}+1}$, $1 \le j \le k_r$, $1 \le r \le L$ is updated as per the rule

$$w_j^r(\text{new}) = w_j^r(\text{old}) + \Delta w_j^r$$

where

$$\Delta w_j^r = -\mu \frac{\partial J}{\partial w_j^r}$$

Note that $w_j^r(\text{old})$ is the current estimate, and $w_j^r(\text{new})$ is the new estimate of w_j^r, and Δw_j^r is the corresponding correction term. Also μ is a positive constant. This numerical technique is called the gradient descent scheme of minimization. Therefore, the correction term Δw_j^r requires the computation of the gradient $\partial E(i) / \partial w_j^r$.

Following notation is introduced for the computation of the gradients.

- $y_k^{r-1}(i)$ is the output of the kth neuron in the $(r-1)$th layer, where $k = 1, 2, \ldots, k_{r-1}$, and $r = 2, 3, \ldots, L$ for the ith training data point.
- w_{jk}^r is equal to the current estimate of the weight of the kth input node of the jth neuron in the rth layer, where $k = 1, 2, \ldots, k_{r-1}$, $w_{j0}^r = 1$, and $j = 1, 2, \ldots, k_r$.
- The argument of the activation function $f(\cdot)$ of the corresponding neuron is $v_j^r(i)$. It is

$$v_j^r(i) = \sum_{k=0}^{k_{r-1}} w_{jk}^r y_k^{r-1}(i)$$

where $y_0^r(i) \triangleq +1$, for $1 \le r \le L$, and $1 \le i \le n$. Also $y_k^0(i) = x_k(i)$, for $1 \le k \le k_0$; and $y_k^L(i) = \widehat{y}_k(i)$, for $1 \le k \le k_L$.

Therefore $\partial E(i) / \partial w_j^r$ is:

$$\frac{\partial E(i)}{\partial w_j^r} = \frac{\partial E(i)}{\partial v_j^r(i)} \frac{\partial v_j^r(i)}{\partial w_j^r}$$

and

$$\begin{aligned} \frac{\partial v_j^r(i)}{\partial w_j^r} &\triangleq \left(\frac{\partial v_j^r(i)}{\partial w_{j0}^r}, \frac{\partial v_j^r(i)}{\partial w_{j1}^r}, \ldots, \frac{\partial v_j^r(i)}{\partial w_{jk_{r-1}}^r}\right) \\ &= \left(+1, y_1^{r-1}(i), \ldots, y_{k_{r-1}}^{r-1}(i)\right) \\ &= y^{r-1}(i) \end{aligned}$$

Let

$$\frac{\partial E(i)}{\partial v_j^r(i)} \triangleq \delta_j^r(i)$$

Therefore

$$\Delta w_j^r = -\mu \sum_{i=1}^{n} \delta_j^r(i)\, y^{r-1}(i)$$

It remains to determine $\delta_j^r(i)$, where $1 \le j \le k_r, 1 \le r \le L$, and $1 \le i \le n$. These are evaluated as follows. Initially, $\delta_j^L(i)$ is determined, this is followed by $\delta_j^{L-1}(i), \delta_j^{L-2}(i), \ldots, \delta_j^1(i)$ in sequence. Thus the computations begin with $\delta_j^L(i)$ and propagate backwards till the computation of $\delta_j^1(i)$. Therefore this scheme is also called *backward propagation algorithm*. The expression for $E(i)$ can be written as

$$E(i) = \frac{1}{2}\sum_{m=1}^{k_L} e_m^2(i) = \frac{1}{2}\sum_{m=1}^{k_L} \left(f\left(v_m^L(i)\right) - y_m(i)\right)^2, \quad 1 \le i \le n$$

Therefore

$$\begin{aligned}\delta_j^L(i) &= \frac{\partial E(i)}{\partial v_j^L(i)} \\ &= e_j(i)\, f'\left(v_j^L(i)\right)\end{aligned}$$

We next determine $\delta_j^{L-1}(i), \delta_j^{L-2}(i), \ldots, \delta_j^1(i)$.

Let $r < L$. As the multiple perceptron network is a feed-forward type of network, $v_j^{r-1}(i)$ affects each $v_k^r(i)$, for $1 \le k \le k_r$ of the next layer. Therefore

$$\frac{\partial E(i)}{\partial v_j^{r-1}(i)} = \sum_{k=1}^{k_r} \frac{\partial E(i)}{\partial v_k^r(i)} \frac{\partial v_k^r(i)}{\partial v_j^{r-1}(i)}$$

As $\partial E(i) / \partial v_j^r(i) \triangleq \delta_j^r(i)$, we have

$$\delta_j^{r-1}(i) = \sum_{k=1}^{k_r} \delta_k^r(i) \frac{\partial v_k^r(i)}{\partial v_j^{r-1}(i)}$$

However

$$\frac{\partial v_k^r(i)}{\partial v_j^{r-1}(i)} = \frac{\partial\left\{\sum_{m=0}^{k_{r-1}} w_{km}^r y_m^{r-1}(i)\right\}}{\partial v_j^{r-1}(i)}$$

As $y_m^{r-1}(i) = f\left(v_m^{r-1}(i)\right)$, we obtain

$$\frac{\partial v_k^r(i)}{\partial v_j^{r-1}(i)} = w_{kj}^r f'\left(v_j^{r-1}(i)\right)$$

Therefore

$$\delta_j^{r-1}(i) = \left\{\sum_{k=1}^{k_r} \delta_k^r(i)\, w_{kj}^r\right\} f'\left(v_j^{r-1}(i)\right)$$

Let

$$\sum_{k=1}^{k_r} \delta_k^r(i)\, w_{kj}^r \triangleq e_j^{r-1}(i)$$

then

$$\delta_j^{r-1}(i) = e_j^{r-1}(i)\, f'\left(v_j^{r-1}(i)\right)$$

Note that if the activation function $f(\cdot)$ is: $f(x) = \{1 + \exp(-ax)\}^{-1}$, $x \in \mathbb{R}$ and $a \in \mathbb{R}^+$, then $f'(x) = af(x)(1 - f(x))$. The above results can be summarized in algorithmic form.

The backpropagation algorithm is iterative. It consists of an initialization phase. The iterative steps consist of: forward and backward computation phases, followed by a phase in which weights are updated. The algorithm terminates if it has met an appropriate stopping criteria. An example of which is a criterion in which the cost function J becomes smaller than a prespecified threshold, or when the norm of its gradient with respect to the weights becomes minuscule.

Algorithm 11.8. *Backpropagation Algorithm.*

Input: The set of training data points $\mathcal{D}$, where
$\mathcal{D} = \{(x(i), y(i)) \mid x(i) \in \mathcal{X} \subseteq \mathbb{R}^t, y(i) \in \Omega, 1 \le i \le n\}$.
The differentiable activation function $f(\cdot)$.
The number of layers of neurons is equal to L. Output is the Lth layer.
Number of nodes in the input layer is equal to k_0. Therefore $k_0 = t$.
The number of neurons in layer r is equal to k_r, where $1 \le r \le L$.
The input is a k_0-dimensional vector.
That is, $x(i) = (x_1(i), x_2(i), \ldots, x_{k_0}(i)) \in \mathbb{R}^{k_0}$.
Learning rate parameter $\mu \in (0, 1]$.
A stopping criterion for the algorithm.
Output: The algorithm determines the synaptic weights associated with each neuron.
Let w_j^r be the weight vector, including the bias factor, of the jth neuron in the rth layer. Thus

$$w_j^r = \left(w_{j0}^r, w_{j1}^r, \ldots, w_{jk_{r-1}}^r\right) \in \mathbb{R}^{k_{r-1}+1}; \quad 1 \le j \le k_r,\ 1 \le r \le L.$$

begin
  $w_j^r \leftarrow$ random values, for $1 \le j \le k_r$, $1 \le r \le L$
  (*initialize all weights to small random values*)
  while stopping criterion is not met **do**
  begin
    Step 1: Forward computations:
      For each training data point $x(i)$, $1 \le i \le n$, and
      for $1 \le j \le k_r$, $1 \le r \le L$ compute:
      $v_j^r(i) = \sum_{k=0}^{k_{r-1}} w_{jk}^r y_k^{r-1}(i)$, and $y_j^r(i) = f\left(v_j^r(i)\right)$
      Compute the cost function
      $J = \frac{1}{2} \sum_{i=1}^{n} \sum_{m=1}^{k_L} \left(f\left(v_m^L(i)\right) - y_m(i)\right)^2$
      (*this value can be used in the stopping criterion*)

Step 2: Backward computations:
For each $i = 1, 2, \ldots, n$ and $j = 1, 2, \ldots, k_L$ compute
$\delta_j^L(i)$ from $\delta_j^L(i) = e_j(i) f'\left(v_j^L(i)\right)$,
where $e_j(i) = \left(f\left(v_j^L(i)\right) - y_j(i)\right)$.
Next compute $\delta_j^{r-1}(i)$ from
$\delta_j^{r-1}(i) = e_j^{r-1}(i) f'\left(v_j^{r-1}(i)\right)$,
where $e_j^{r-1}(i) = \sum_{k=1}^{k_r} \delta_k^r(i) w_{kj}^r$
for $r = L, L-1, \ldots, 2$, and $j = 1, 2, \ldots, k_r$.
Step 3: Update the weights:
For $1 \leq j \leq k_r$, $1 \leq r \leq L$
$\Delta w_j^r = -\mu \sum_{i=1}^n \delta_j^r(i) y^{r-1}(i)$
$w_j^r \leftarrow w_j^r + \Delta w_j^r$
end (*end of while-loop*)
end (*end of backpropagation algorithm*)

11.6 Nonmetric Classification

We examine problems which are amenable to classification via graph-theoretic entities called trees in this section. These methods fall under the category of supervised learning.

Nonmetric classification schemes are specially useful for classifying categorical data, or a mixture of categorical and numerical data. Categorical data is the type of data in which there is no notion of ordering or similarity (metric). This is in contrast to data points which are vectors of real numbers. For example, a fruit might be described by four attributes: size, color, taste, and texture. Therefore, a specific apple might be described by the 4-tuple: (medium, pink, sweet, shiny). It is inconvenient to ascribe any notion of metric to this type data, yet it is often desirable to classify them. This type of classification is also useful, when there are a large number of classes. Decision trees and rule-based techniques can be used to classify such type of data.

Let $\mathcal{X}$ be the set of data points. Also let

$$\Omega = \{\omega_1, \omega_2, \ldots, \omega_m\}$$

be the set of labels for different classes of data, where $m \in \mathbb{P} \backslash \{1\}$. The training data is a set $\mathcal{D}$ of n observations,

$$\mathcal{D} = \{(x_i, y_i) \mid x_i \in \mathcal{X}, y_i \in \Omega, 1 \leq i \leq n\}$$

In general, $x_i \in \mathcal{X}$ is a t-tuple, where $t \in \mathbb{P}$ is the number of attributes of a data point. A data point can be a mixture of nominal, ordinal, or real-valued attributes. Denote the set of attributes by $\mathcal{A}$, where

$$\mathcal{A} = \{A_1, A_2, \ldots, A_t\}$$

Classification techniques like decision tree induction and rule-based induction are described in this section.

11.6.1 Decision Tree Induction

Nonmetric-based data can be classified by asking a sequence of questions. In such sequence of questions, the next question asked might depend upon the answer to the previous question. The questions can be in the form of "true/false" or "yes/no." These questions do not have any notion of "distance." Sequence of questions of this type can be displayed in the form of a tree. The corresponding tree is called a decision tree. Thus classification is performed in such schemes via *multistage* decision systems called decision trees.

We initially summarize some tree-related terminology. A tree is a connected graph with no cycles. A tree in which one node (vertex) is specifically identified as a *root node* is called the *rooted tree*. All other nodes in the tree are called the descendents of the root node. The root node has zero incoming edges, and zero or more outgoing edges. An *internal node* of the tree has precisely one incoming edge and two or more outgoing edges. The *leaf*, or *terminal node* of the tree has precisely one incoming edge, and zero outgoing edges. A *nonterminal node* is either the root or an internal node.

A decision tree is built based upon the set of training data points. Once the decision tree is built, classification of a test data point is simple. However, the challenge is to build a decision tree by using the training data set.

In a decision tree, each nonterminal node contains an *attribute-test condition* to separate data points that have different features, and a terminal node has a classification label assigned to it. Beginning with the root node, a question is asked about the attributes of the test data point. Depending upon the answer, an appropriate edge is followed to a descendent (child) node. A question is again asked at this descendent node, and a decision is made. This descendent node can be considered the root node of a subtree. This process of asking questions and following an edge which corresponds to an answer is continued, till a leaf node is reached. No questions are asked at the leaf node. However, a leaf node is assigned a classification label. This label is also the classification label of the test data point. Thus, each path from the root node to a leaf represents a cumulative choice.

In summary, a decision tree is a rooted tree in which, a nonterminal node is associated with a question about the attributes of the data point, the edges outgoing from this node correspond to the answers. Thus nonterminal node is also associated with a decision. Further, every leaf node of the decision tree is assigned a classification label.

Building a Decision Tree

The process of building a decision tree, based upon the training data set, is sometimes called decision tree induction. There are an extremely large number of decision trees, for a given set of attributes. Some of these trees are more accurate than others. Determination of an optimal decision tree is computationally very expensive. Therefore a greedy search strategy is followed to grow a decision tree in which locally optimum selection of questions is made at each internal node. In principle, a divide and conquer heuristic scheme is adapted.

An algorithm is described in this subsection in which the decision tree grows recursively by partitioning the training data set into its subsets which are more *pure* or *class homogeneous* compared to their parent (ancestor) set. This implies that the training data points in each subset have a higher affinity for fewer number of classes than the set of training data points in their parent set. In this sense the child subsets are purer than their parent set. This degree of impurity of subsets will be quantified later in this subsection. It will aid in splitting or partitioning the training data set. A key ingredient in the recursive algorithm is the availability of attribute-test condition to partition (split) the training data points.

The recursive algorithm to build a decision tree is called at each internal node of the decision tree. Each internal node η of the decision tree is associated with a subset of training data points $\mathcal{X}_\eta \subseteq \mathcal{X}$, and a subset of attributes $\mathcal{A}_\eta \subseteq \mathcal{A}$. The recursive algorithm is:

Step 0: $\eta \leftarrow$ root node, $\mathcal{X}_\eta \leftarrow \mathcal{X}$, and $\mathcal{A}_\eta \leftarrow \mathcal{A}$. The root node η is initialized by the training data set, and the corresponding set of attributes.

Step 1: If all data points in $\mathcal{X}_\eta$ belong to the same class $\omega_j \in \Omega$, then η is a leaf node. It is labeled as ω_j.

Step 2: If the data points in $\mathcal{X}_\eta$ belong to more than one class, then the goal is to split the set $\mathcal{X}_\eta$ into purer subsets. In this case attribute-test conditions can be implemented to select a partition of $\mathcal{X}_\eta$ that provides overall maximum decrease of impurity when compared to the impurity of the parent set $\mathcal{X}_\eta$. Let this test attribute be A_g, which takes $k \in \mathbb{P}$ discrete values.

Also, let the selected sets of children of node η be the nodes $\eta_1, \eta_2, \ldots, \eta_k$, and the corresponding sets of split of data points be $\mathcal{X}_{\eta_1}, \mathcal{X}_{\eta_2}, \ldots, \mathcal{X}_{\eta_k}$. As this is a partition, the child sets are mutually exclusive, and their union is equal to $\mathcal{X}_\eta$. The decision tree consists of node η associated with the training data points $\mathcal{X}_\eta$, and one branch each towards a node assigned to the child node η_i, where $1 \leq i \leq k$. The set of attributes associated with each of these child nodes is $\mathcal{A}_{\eta_i} = \mathcal{A}_\eta \backslash \{A_g\}$ for $1 \leq i \leq k$. The tree-building scheme is subsequently applied recursively at each of these child nodes.

In the above description of the algorithm, the determination of degree of impurity of subsets is required. This is quantified later in the subsection. Also, the above algorithm needs to handle certain exceptions.

- In Step 2 of the recursive algorithm, if all the data points have the same attribute values, but different class labels, then a further split of $\mathcal{X}_\eta$ is not possible. In this case, the node η corresponding to the set $\mathcal{X}_\eta$ is declared a leaf node, and is classified with a label which occurs most often in this set.
- It is possible that for a child node η_i produced in Step 2, the set $\mathcal{X}_{\eta_i}$ is empty. That is, $\mathcal{X}_{\eta_i}$ does not occur in the training data set $\mathcal{X}$. In this case, the node η_i corresponding to the set $\mathcal{X}_{\eta_i}$ is declared a leaf node, and it is classified with a label which occurs most often in the parent set $\mathcal{X}_\eta$.

Splitting of Attribute Values

Decision tree classification techniques can process a mixture of binary, nominal, ordinal, and continuous valued attributes. The test condition for binary attributes is straightforward. It has only two possible outcomes. Nominal or categorical attribute values need special consideration. Suppose that this type of attribute takes k values, then the total number of possible splits to be examined can be extremely high. However, some algorithms consider only binary splits. In this case, there are $\left(2^{k-1} - 1\right)$ different ways of generating a binary partition of $k \in \mathbb{P} \backslash \{1\}$ attribute values.

If the number of values of an ordinal attribute is small, then it might be possible to have binary or multiway splits. For continuous valued attribute A, the splitting can occur via tests of the form $v_i \leq A < v_{i+1}$, where $v_i \in \mathbb{R}$, and $1 \leq i \leq k$.

Splitting Criteria

The set of data points $\mathcal{X}_\eta$ associated with the node η of a decision tree can be split into sets $\mathcal{X}_{\eta_1}, \mathcal{X}_{\eta_2}, \ldots, \mathcal{X}_{\eta_k}$ via a metric-based criterion. It is hoped that the set of data points $\mathcal{X}_{\eta_i}$ at the child node η_i are cumulatively purer than the set of data points $\mathcal{X}_\eta$ at their parent node η. Let $P(\omega_j \mid \eta)$ denote the probability that a data point in the set $\mathcal{X}_\eta$ belongs to the class $\omega_j \in \Omega$, where $1 \le j \le m$, and $|\Omega| = m \in \mathbb{P} \backslash \{1\}$. Note that $\sum_{j=1}^{m} P(\omega_j \mid \eta) = 1$. A possible measure of impurity at node η is

$$I_{entropy}(\mathcal{X}_\eta) = -\sum_{j=1}^{m} P(\omega_j \mid \eta) \log_2 P(\omega_j \mid \eta)$$

where $I_{entropy}(\mathcal{X}_\eta)$ is the entropy associated with the set $\mathcal{X}_\eta$. In practice, $P(\omega_j \mid \eta)$ is estimated as X_η^j / X_η, where X_η^j is equal to number of data points in $\mathcal{X}_\eta$ that belong to class ω_j, for $1 \le j \le m$; and $|\mathcal{X}_\eta| \triangleq X_\eta$. The entropy impurity is smallest, if all data points belong to the same class. In this case, $I_{entropy}(\mathcal{X}_\eta) = 0$. It is maximum, if all the classes are equally likely. In this case, $I_{entropy}(\mathcal{X}_\eta) = \log_2 m$.

Other possible measure of impurity are the Gini index, and classification error. These are denoted by $I_{Gini}(\mathcal{X}_\eta)$ and $I_{misclassification}(\mathcal{X}_\eta)$ respectively, where

$$\begin{aligned} I_{Gini}(\mathcal{X}_\eta) &= \sum_{\substack{1 \le j,l \le m \\ j \ne l}} P(\omega_j \mid \eta) P(\omega_l \mid \eta) \\ &= 1 - \sum_{j=1}^{m} \{P(\omega_j \mid \eta)\}^2 \\ I_{misclassification}(\mathcal{X}_\eta) &= 1 - \max_{1 \le j \le m} \{P(\omega_j \mid \eta)\} \end{aligned}$$

The Gini impurity measure can be regarded as the expected error rate if the classification labels are selected randomly. The misclassification impurity measures the minimum probability that a data point is misclassified at node η. All the three measures of impurity are maximized when the classes are uniformly distributed.

In order to determine the goodness of split, we need to compare the impurity of the parent node, and weighted sum of the impurity nodes of its child nodes. Assume that the attribute which reduces its impurity is A_j, and this attribute takes $\nu \in \mathbb{P} \backslash \{1\}$ different values.

Then the set $\mathcal{X}_\eta$ is split into $\mathcal{X}_{\eta_1}, \mathcal{X}_{\eta_2}, \ldots, \mathcal{X}_{\eta_\nu}$; where $|\mathcal{X}_\eta| \triangleq X_\eta$, and $|\mathcal{X}_{\eta_i}| \triangleq X_{\eta_i}$, for $1 \le i \le \nu$. If $I(\mathcal{X}_\eta)$ is the impurity of the parent node, and the impurity of the child node η_i is $I(\mathcal{X}_{\eta_i})$, for $1 \le i \le \nu$, then the decrease in node impurity $\Delta(\mathcal{X}_\eta, A_j)$ by the split is

$$\Delta(\mathcal{X}_\eta, A_j) = I(\mathcal{X}_\eta) - \sum_{i=1}^{\nu} \frac{X_{\eta_i}}{X_\eta} I(\mathcal{X}_{\eta_i})$$

The decision tree induction algorithm selects a split of the attribute which maximizes $\Delta(\mathcal{X}_\eta, A_j)$.

Stopping Criteria

Having determined a criterion to split a set of data points, we need to determine a stopping criterion for the splitting process. Some stopping criteria are:

- A predetermined threshold value τ. Splitting is stopped, if: $\max_{A_j \in \mathcal{A}_\eta} \Delta(\mathcal{X}_\eta, A_j) < \tau$.
- Stop splitting, if the set $\mathcal{X}_\eta$ is pure. That is all data points in this set belong to the same class ω_j. The node η is made a leaf, and assigned the classification label ω_j.
- Stop splitting, if the attribute values of all the data points in the set $\mathcal{X}_\eta$ have identical values. The node η is made a leaf, and assigned a classification label which occurs most often in the set $\mathcal{X}_\eta$.
- Stop splitting, if $|\mathcal{X}_\eta| \triangleq X_\eta$ is sufficiently small.

After a decision tree is built, sometimes *tree-pruning* is performed to reduce its size. Pruning is performed by eliminating certain insignificant nodes and branches of the original decision tree. This process alleviates the problem of over-fitting. Well-known decision tree based classification algorithms are CART (Classification and Regression Trees) and C4.5. A skeleton decision tree induction algorithm is outlined.

Algorithm 11.9. *Decision Tree Induction Algorithm.*

Input: The set of training data points $\mathcal{D}$, where
$\mathcal{D} = \{(x_i, y_i) \mid x_i \in \mathcal{X}, y_i \in \Omega, 1 \le i \le n\}$.
The set of attributes $\mathcal{A} = \{A_1, A_2, \ldots, A_t\}$.
The set of classification labels Ω, where $\Omega = \{\omega_1, \omega_2, \ldots, \omega_m\}$,
and $m \in \mathbb{P} \backslash \{1\}$.
Node η.
Split selection method $\mathcal{SSM}$. It determines $\arg\max_{A_j \in \mathcal{A}_\eta} \Delta(\mathcal{X}_\eta, A_j)$.
This depends upon the choice of the impurity measure.
Threshold value τ to check for decrease in node impurity due to split.
The function **create a node** adds a node to the decision tree.
The function **create an edge** adds an edge to the decision tree.
Output: The decision tree rooted at node η.
begin
  $\mathcal{X}_\eta \leftarrow \mathcal{X}$
  $\mathcal{A}_\eta \leftarrow \mathcal{A}$
  BuildDecisionTree $(\mathcal{X}_\eta, \mathcal{A}_\eta, \eta)$
  ($\mathcal{X}_\eta$ *is the set of data points at node* η)
  ($\mathcal{A}_\eta$ *is the set of candidate attributes*)
  begin
    Case 1: **if** $\mathcal{X}_\eta$ has all data points of a single class $\omega_j \in \Omega$,
      then make η a leaf node, and label it with class ω_j
    Case 2: **if** $\mathcal{A}_\eta = \varnothing$ **then** make η a leaf node,
      and label it with class ω_j, the most frequent class in
      the set $\mathcal{X}_\eta$
    Case 3: ($\mathcal{X}_\eta$ *has data points which belong to more than a single class*)
      begin
        apply $\mathcal{SSM}$ to $\mathcal{X}_\eta$ to determine its best split
        $A_g = \arg\max_{A_j \in \mathcal{A}_\eta} \Delta(\mathcal{X}_\eta, A_j)$,
        $\Delta(\mathcal{X}_\eta, A_g)$ is the maximum decrease in node

impurity
Subcase (a) **if** $\Delta(\mathcal{X}_\eta, A_g) < \tau$ **then**
(*split does not significantly decrease impurity*)
make η a leaf node, and label it with class ω_j, the most frequent class in $\mathcal{X}_\eta$
Subcase (b) **if** $\Delta(\mathcal{X}_\eta, A_g) \geq \tau$ **then**
let $k \in \mathbb{P}$ be equal to the number of different values of the attribute A_g.
let the partitions of $\mathcal{X}_\eta$ be $\mathcal{X}_{\eta_1}, \mathcal{X}_{\eta_2}, \ldots, \mathcal{X}_{\eta_k}$
for $i = 1$ **to** k **do**
begin
create a node η_i
(*create child node* η_i)
create an edge from node η to node η_i
if $\mathcal{X}_{\eta_i} \neq \varnothing$ **then**
BuildDecisionTree $(\mathcal{X}_{\eta_i}, \mathcal{A}_\eta \backslash A_g, \eta_i)$
else
begin
$(\mathcal{X}_{\eta_i} = \varnothing)$
make η_i a leaf node, and label it with class ω_j, the most frequent class in $\mathcal{X}_\eta$
end (*end of if statement*)
end (*end of i for-loop*)
end (*end of Case* 3)
end (*end* of *BuildDecisionTree*)
end (*end of decision tree induction algorithm*)

11.6.2 Rule-Based Induction

Nonmetric-based data can be classified by utilizing a set of if-then rules. These rules can be generated either from a decision tree or directly from the training data set. The technique to learn such rules is called rule-based induction. Rule-based classification techniques are easier to interpret, and their performances are comparable to those based upon decision trees.

Definition 11.18. *An if-then rule r is expressed in the following two ways.*

$$r : \text{IF } (\textit{condition}) \text{ THEN } (\textit{conclusion}) \tag{11.9a}$$

$$r : (\textit{condition}) \rightarrow (\textit{conclusion}) \tag{11.9b}$$

The values of "condition" and "conclusion" are either true or false. The "IF" part of the rule is "condition." It is known as the rule antecedent or precondition. The "THEN" part of the rule is "conclusion." It is known as the rule consequent. □

The condition part of the rule can be expressed in terms of the attributes of the data points, and the logical AND operator. This condition part consists of $k \in \mathbb{P}$ number of attribute-tests (conjuncts) that are logically ANDed. The rule's consequent specifies the value of a class.

Recall that the set of attributes of the training data points is $\mathcal{A} = \{A_1, A_2, \ldots, A_t\}$. Also let the set of logical operators $\mathcal{O}$ be

$$\mathcal{O} = \{=, \neq, <, >, \leq, \geq\}$$

The logical AND operator is denoted by the symbol $\wedge$. Let (A_j, v_j) be an attribute-value pair, and $op_j \in \mathcal{O}$ be a logical operator. Each attribute-value pair is a conjunct in the antecedent of the rule. An example of a rule antecedent is:

$$\text{condition} = (A_1 \; op_1 \; v_1) \wedge (A_2 \; op_2 \; v_2) \wedge \ldots \wedge (A_k \; op_k \; v_k)$$

A rule r is said to *cover* a data point $x \in \mathcal{X}$ if the precondition of the rule matches the attribute values of x. Further, a rule r is said to be *fired* or *triggered* if it covers a data point $x \in \mathcal{X}$. The quality of an if-then classification rule r can be assessed by its coverage and accuracy.

Definition 11.19. *Let $\mathcal{X}$ be the set of data points, Ω be the set of labels for different classes of data. The training data set $\mathcal{D}$ is*

$$\mathcal{D} = \{(x_i, y_i) \mid x_i \in \mathcal{X}, y_i \in \Omega, 1 \leq i \leq n\} \tag{11.10a}$$

Let the number of data points that satisfy (cover) the antecedent of a rule r be n_{cover}. Also let the number of data points that are correctly classifed by the rule r be $n_{correct}$. That is, $n_{correct}$ is the number of data points that satisfy both the antecedent and consequent of the rule r. The coverage and accuracy of the rule r are the ratios $coverage\,(r)$ and $accuracy\,(r)$ respectively, where

$$coverage\,(r) = \frac{n_{cover}}{n}, \tag{11.10b}$$

$$accuracy\,(r) = \frac{n_{correct}}{n_{cover}} \tag{11.10c}$$

□

A rule-based classifier generates a set of rules, which are subsequently used in classifying test data points. Using the set of training data points $\mathcal{D}$, let $R = \{r_1, r_2, \ldots, r_j\}$ be a set of rules generated by a rule-based classifier. Consider a test data point x which needs to be classified by the set of rules in R. Three possibilities might occur.

- There exists only a single rule in the set R which is triggered by the test data point x. In this case, the test data point is uniquely classified.
- No rule in the set R is triggered by the test data point x. This case is addressed later in the subsection.
- More than a single rule in the set R is triggered by a test data point x, and each such rule specifies a different class. This requires special attention.

In light of the above possibilities, it can be observed that the rules in the set R can be *mutually exclusive rules* and/or *exhaustive rules*.

- *Mutually exclusive rules*: The rules in the set R are said to be mutually exclusive, if only a single rule is triggered by a test data point. That is, a single test data point does not trigger two or more rules in the set R.
  Conversely, if the rule set R is not mutually exclusive, then more than a single rule is triggered by a single test case. Scheme to address this possibility is addressed subsequently.
- *Exhaustive rules*: The rule set R is exhaustive, if there exists a rule in the set R which is triggered by each combination of attribute values. Thus each data point is covered by at least a single rule in the set R.
  If a rule set is not exhaustive, then a default rule, $r_d : (\,) \to y_d$ is added to cover the remaining possibilities. In this rule, the antecedent is empty, and the consequent $y_d \in \Omega$ is called the default class. The default class is usually the majority class of data points in the training data set $\mathcal{D}$ which are not covered by the rules in the set R.

If the set of rules R are not mutually exclusive, that is a test data point is covered by more than a single rule, then it is quite possible that the classifier might predict more than a single class for the test data point! This possibility of conflict can be addressed in two different ways. This is done via *ordered* and *unordered rules*.

- *Ordered rules*: The rules in the set R are ordered in descending order of importance. This can be based upon accuracy, coverage, the order in which rules were generated, or total description length of the rules. The test data point is classified based upon the highest-ranked rule which covers the data point.
- *Unordered rules*: The rules in the set R are unordered. If a test data point triggers multiple rules, then the consequent which occurs most in these triggered rules is assigned the class of the test data point.

We only consider ordered rules in the rest of this subsection. This scheme can be implemented in two different ways. These are: rule-based and class-based ordering schemes.

- *Rule-based ordering scheme*: The rules are ordered based upon a specific rule quality metric.
- *Class-based ordering scheme*: Rules with the same consequent are ordered together. Further the consequents are sorted based upon their assigned importance.

We only consider class-based ordering rules in the rest of this subsection.

Construction of Rule Classifier

A rule-based classifier can be built indirectly from the decision tree generated from the training data set. The path from the root node to each leaf node of the decision tree corresponds to a single rule.

It is also possible to develop a rule-based classifier directly from the training data set. This later scheme is next developed. The if-then rules can be generated directly via the sequential covering algorithm. It uses the greedy paradigm of rule classification. The algorithm is *sequential* in the sense that, the rules are induced (learned) sequentially. It is hoped that each rule would *cover* several data points of the training data set.

In a canonical sequential covering algorithm, rules are induced for each class that occur in the training data set. Ideally the induced rules should have high accuracy, and not necessarily high coverage. That is, each rule for a specific class should cover several training data points of that class, and hopefully an extremely small number of training data points of other classes. It is also possible to have more than a single rule for a single class. This would cover more training data points which belong to the same class.

The sequential covering algorithm uses a learn-one-rule procedure $\mathcal{L}(\mathcal{D}', \omega)$, where $\mathcal{D}' \subseteq \mathcal{D}$, and $\omega \in \Omega$. This procedure determines the best rule for a class $\omega \in \Omega$ with high accuracy and generally any coverage. The basic sequential covering algorithm works as follows. It learns a single good rule based upon the learn-one-rule procedure. It then removes the data points in the training data set covered by this rule. This process is continued iteratively till no training data points are left.

The outlined sequential covering algorithm learns all rules, one rule at a time. That is, the algorithm learns all rules for a single class, and then determines a set of rules for a different class, and so on. The sequence of rules within each class is not important. However the order of sets of rules for different classes is significant. The algorithm terminates if all the training data points have been examined, or the quality of a newly determined rule falls below a prespecified threshold.

Algorithm 11.10. *Sequential Covering Algorithm.*

Input: The set of training data points $\mathcal{D}$, where
$\mathcal{D} = \{(x_i, y_i) \mid x_i \in \mathcal{X}, y_i \in \Omega, 1 \le i \le n\}$.
The set of attributes $\mathcal{A} = \{A_1, A_2, \ldots, A_t\}$.
The set of classification labels Ω, where $\Omega = \{\omega_1, \omega_2, \ldots, \omega_m\}$, and $m \ge 2$.
The learn-one-rule procedure $\mathcal{L}(\mathcal{D}', \omega)$, where $\mathcal{D}' \subseteq \mathcal{D}$ and $\omega \in \Omega$.
A stopping criterion for the algorithm.
Output: $R = \{r_1, r_2, \ldots, r_j\}$ a set of learned rules.
begin
    $R \leftarrow \varnothing$
    (*the set of rules is initially empty*)
    $\mathcal{D}' \leftarrow \mathcal{D}$
    for $i = 1$ **to** m **do**
    (*for each class* $\omega_i \in \Omega$)
    begin
        while stopping criterion not met
        begin
            $r \leftarrow \mathcal{L}(\mathcal{D}', \omega_i)$
            $\mathcal{D}' \leftarrow \mathcal{D}' - \{e \mid e \in \mathcal{D}'$ is classified by rule $r\}$
            $R \leftarrow R \cup r$
        end (*end of while-loop*)
    end (*end of i for-loop*)
end (*end of sequential covering algorithm*)

Given a set of training data points $\mathcal{D}'$ and a class $\omega \in \Omega$, the learn-one-rule procedure $\mathcal{L}(\mathcal{D}', \omega)$ determines the best possible rule r. The determination of the best possible rule r is also iterative. Initially, the rule antecedent is empty. In the next step, a locally optimal attribute-value pair is added (as a conjunct) to the antecedent of the rule. The selection of a locally optimal attribute-value pair is based upon a greedy criterion. The greedy criterion is adopted to select an attribute-value because the training data set has many attributes, and each attribute takes several possible values. Therefore a greedy strategy is selected to improve computationally efficiency.

During the learning mode of the rules, the newly appended attribute-test is expected to improve the *rule quality*. The rule quality is quantified later in the subsection. The process of adding attribute-tests to the antecedent of the rule (in conjunctive mode) continues until the rule attains an acceptable prespecified quality level.

This greedy process may not be optimal. To overcome this shortcoming, the algorithm can select best k attribute-tests. This is analogous to a beam-search of width k. Further, k best candidates of attribute-value pairs are maintained at the end of each iteration.

We next outline techniques for the quantification of rule quality. This is followed by a technique to improve rules.

Quantification of Rule Quality

The accuracy of a rule is one measure of the quality of a rule. However accuracy alone may not be the only best and reliable measure of quality of a rule. Coverage alone is also perhaps not a measure of quality of a rule. This is because for a given class, there can be a rule that covers several training data points which might belong to other classes.

Two rule quality measures: *entropy* and *information gain* are next outlined. Suppose that the rule r is in the process of learning about class $\omega \in \Omega$ by using the condition c. That is, r : IF (c) THEN $(class = \omega)$. Let r' be a rule obtained by logically ANDing (conjuncting) a candidate attribute-test with the antecedent c of the rule r. The candidate rule is specified as r' : IF (c') THEN $(class = \omega)$. It is required that the quality of rule r' be better than that of rule r.

- *Entropy or average information*: Let $D \subseteq \mathcal{D}$ be the set of training data points covered by c' and p_i is the probability of occurrence of data points which belong to class ω_i in the set D. The entropy associated with the set D is

$$I(D) = -\sum_{i=1}^{m} p_i \log_2 p_i$$

  Smaller the value of this entropy, better is the antecedent c'.
- *Information gain*: The training data points covered, and not covered by a rule are called its *positive* and *negative* data points respectively. Let the number of positive and negative data points associated with the rule r be ρ and χ respectively. Similarly, let ρ' and χ' be the number of data points associated with the rule r'. The information gained by extending *condition* is

$$I_{gain} = \rho' \left\{ \log_2 \frac{\rho'}{\rho' + \chi'} - \log_2 \frac{\rho}{\rho + \chi} \right\}$$

  This quality measure prefers rules that cover several positive data points, and have high accuracy.

Improvement of Rules

If the quality of rules generated by the above algorithm is assessed by using the original set of training data points, then this assessment of quality of rules might be overtly optimistic. This implies that the rules may or may not perform as well on newer data points, as they would on the training data points. To overcome this scenario, the rules can be pruned. Pruning is performed by removing attribute-tests (conjuncts) from a rule. A rule r is pruned, if a pruned version of this rule has a higher quality. This quality is evaluated by using nontraining set of data points. A popular metric used in pruning is

$$Q_{prune}\left(r\right)=\frac{\rho-\chi}{\rho+\chi}$$

where the number of positive and negative data points covered by the rule r are ρ and χ respectively. The value $Q_{prune}\left(r\right)$ increases with the accuracy of the rule r. Therefore, if r' is obtained by pruning rule r, and

$$Q_{prune}\left(r\right)<Q_{prune}\left(r'\right)$$

then the rule r is pruned.

11.7 Classification via Ensemble Methods

Classification via ensemble methods is a data mining paradigm in which a judicious combination of several different classifiers results in an overall better classifier. The ensemble method hopes that by taking advantage of several different classifiers, the probability of correctly classifying test data points is greatly increased.

Let $\mathcal{X}$ be the set of data points, and $\Omega=\{\omega_1,\omega_2,\ldots,\omega_m\}$ be the set of labels for different classes of data, where $m\in\mathbb{P}\backslash\{1\}$. The training data is a set $\mathcal{D}$ of n observations,

$$\mathcal{D}=\left\{(x_i,y_i)\mid x_i\in\mathcal{X}\subseteq\mathbb{R}^t,y_i\in\Omega,1\leq i\leq n\right\}$$

Assume that there are $N_{cl}\in\mathbb{P}\backslash\{1\}$ number of different classifiers. Denote the different classifiers as $M_1,M_2,\ldots,M_{N_{cl}}$. These classifiers are called base classifiers. The aim of the ensemble technique is to build an improved composite classifying model $\widetilde{M}$ based upon the M_j's. The base classifier model M_j is trained by using the data set $\mathcal{D}_j\subseteq\mathcal{D}$, for $1\leq j\leq N_{cl}$.

Let $u\in\mathbb{R}^t\backslash\mathcal{X}$ be a test data point. Also let the label (class) determined by the model M_j for the data point u be $\omega_{j_u}\in\Omega$, for $1\leq j\leq N_{cl}$. It is assumed that each classifier makes decisions which are probabilistically independent of decisions of every other classifier. The ensemble model $\widetilde{M}$ selects a label for the data point u based upon the labels $\omega_{j_u}\in\Omega$, for $1\leq j\leq N_{cl}$. It selects a label so that the probability of misclassification of the data point u is less than or equal to those of the labels ω_{j_u}'s generated by the M_j's.

A possible choice of the label might be via a majority voting scheme. According to this scheme, a label is selected when at least l_c of the classifiers agree upon the class label of the test data point, where

$$l_c=\left\lceil\frac{N_{cl}}{2}\right\rceil$$

and $\lceil \cdot \rceil$ is the ceiling function. If there is no consensus among at least l_c classifiers, then no decision is taken about the class label. Let p be the probability of correct classification by each base classifier. The probability of a correct consensus decision $P_c(N_{cl})$ can be determined in terms of the probability p. It is

$$P_c(N_{cl}) = \sum_{j=l_c}^{N_{cl}} \binom{N_{cl}}{j} p^j (1-p)^{N_{cl}-j}$$

The following observations are immediate.

- If $p < 0.5$, $P_c(N_{cl})$ is monotonically decreasing in N_{cl}.
  Furthermore, $\lim_{N_{cl} \to \infty} P_c(N_{cl}) \to 0$.
- If $p = 0.5$, $P_c(N_{cl}) \to 0.5$ for large N_{cl}.
- If $p > 0.5$, $P_c(N_{cl})$ is monotonically increasing in N_{cl}.
  Furthermore, $\lim_{N_{cl} \to \infty} P_c(N_{cl}) \to 1$.

These observations imply that, if the output of the N_{cl} probabilistically independent classifiers is used in a majority voting scheme, then the probability of correct classification of the test data point asymptotically increases to unity as N_{cl} tends to infinity, provided $p > 0.5$. The two well-known ensemble methods, bagging and boosting, are next discussed.

11.7.1 Bagging

Bagging is an ensemble technique of data classification. The bagging term comes form *b*ootstrap *agg*regat*ing*. We are given a set of $\mathcal{D}$ data points, where $|\mathcal{D}| = n$; and $N_{cl} \in \mathbb{P} \backslash \{1\}$ independent base classifiers. Each such classifier M_j generates a bootstrap sample $\mathcal{D}_j$ from the set of data points $\mathcal{D}$, where $1 \le j \le N_{cl}$. Each bootstrap sample of data points is created by drawing with replacement n data points from the set $\mathcal{D}$, according to a uniform probability distribution. As sampling of data points is done with replacement, some data points might appear more than once, and some data points may not appear at all in the set $\mathcal{D}_j$. Thus $\mathcal{D}_j \subseteq \mathcal{D}$.

If the number of data points in the set $\mathcal{D}$ tends to a large value, it can be shown that approximately $\left(1 - e^{-1}\right) n \simeq 0.632n$ different data points are present in the bootstrap sample $\mathcal{D}_j$. This result can be obtained as follows. Each data point has a probability of $1/n$ of being selected. Therefore, the probability of a data point not being selected is $(1 - 1/n)$. As the bootstrap sample makes n independent selections with replacement, the probability of a data point not being selected in the bootstrap sample is $(1 - 1/n)^n$. Thus the probability of a data point being selected in the bootstrap sample is $\{1 - (1 - 1/n)^n\}$. As n tends to infinity, this probability tends to $\left(1 - e^{-1}\right)$.

Bagging consists of training the base classifier M_j by using the set of points in $\mathcal{D}_j$, for $j = 1, 2, \ldots, N_{cl}$; and subsequently assigning a classification to a test data point that receives the highest number of votes.

Let $u \in \mathbb{R}^t \backslash \mathcal{X}$ be the test data point. The classifier M_j performs the mapping $\phi_j : \mathbb{R}^t \to \Omega$. Denote the label (class) determined by the model M_j for the test data point u by $\phi_j(u) \triangleq \omega_{j_u} \in \Omega$, for $1 \le j \le N_{cl}$. It is assumed that each classifier makes decisions which are probabilistically independent of decisions of every other classifier. The ensemble model $\widetilde{M}$ performs the mapping $\widetilde{\phi} : \mathbb{R}^t \to \Omega$. It selects a label for the data point u based upon the labels $\omega_{j_u} \in \Omega$, for $1 \le j \le N_{cl}$. The goal of the bagging scheme is to select a label so that the probability of misclassification of the data point u is less than or equal to those of the ω_{j_u}'s.

The bagging classifier generally has more accuracy than a single classifier based upon the original data set $\mathcal{D}$. The increased accuracy is a result of lower effect of the variance of the noise present in the data.

Algorithm 11.11. *Bagging Algorithm.*

Input: The set of training data points $\mathcal{D}$, where
$\mathcal{D} = \{(x_i, y_i) \mid x_i \in \mathcal{X} \subseteq \mathbb{R}^t, y_i \in \Omega, 1 \leq i \leq n\}$.
Total number of classifiers is $N_{cl} \in \mathbb{P} \backslash \{1\}$.
The base classifier M_j performs the operation $\phi_j : \mathbb{R}^t \rightarrow \Omega$,
for $1 \leq j \leq N_{cl}$. Test data point: $u \in \mathbb{R}^t \backslash \mathcal{X}$.
Output: Classification of the test data point u,
by the ensemble model $\widetilde{M}$.
It is $\widetilde{\phi}(u) = \omega \in \Omega$.
begin
  for $j = 1$ **to** N_{cl} **do**
  begin
    Generate a bootstrap sample $\mathcal{D}_j$ of size n
    Train the base classifier M_j using the bootstrap sample $\mathcal{D}_j$
  end (*end of* j *for-loop*)
  $\omega \leftarrow \arg\max_{y \in \Omega} \sum_{j=1}^{N_{cl}} \delta(\phi_j(u) = y)$
  ($\phi_j(u) = \omega_{j_u} \in \Omega$ *is the label determined by the model* M_j,)
  (*for the test data point* u, *for* $1 \leq j \leq N_{cl}$.)
  ($\delta(\cdot) = 1$ *if its argument is true and* 0 *otherwise*)
end (*end of bagging algorithm*)

11.7.2 Boosting

Boosting is an adaptive and iterative ensemble technique which uses a base classifier with different parameters to improve the accuracy of classification. A popular boosting algorithm called AdaBoost (adaptive boosting) is discussed in this subsection. Unlike the bagging algorithm, the AdaBoost algorithm assigns weights to each training data point and adaptively modifies it at the end of each boosting iteration.

Let $\mathcal{X}$ be the set of data points, and $\Omega = \{-1, +1\}$ be the set of labels for the two classes of data. The training data is a set $\mathcal{D}$ of n observations,

$$\mathcal{D} = \left\{(x_i, y_i) \mid x_i \in \mathcal{X} \subseteq \mathbb{R}^t, y_i \in \{-1, +1\}, 1 \leq i \leq n\right\}$$

The base classifier used in the algorithm performs the mapping $\phi : \mathbb{R}^t \times \vartheta \rightarrow \Omega$, where ϑ is the parameter space of the classifier. The AdaBoost algorithm is iterative, and it invokes a parameterized base classifier in each iteration. Let the number of iterations in the algorithm be $N_{itr} \in \mathbb{P} \backslash \{1\}$. Selection of the value N_{itr}, is based upon some terminating criteria of the algorithm. In iteration j, the parameters selected by the base classifier is $\theta_j \in \vartheta$. In each iteration of the algorithm, the weights of incorrectly and correctly classifed data points in the previous round are increased and decreased respectively.

Let $u \in \mathbb{R}^t \backslash \mathcal{X}$ be a test data point. The AdaBoost model $\widetilde{M}$ performs the mapping $\widetilde{\phi} : \mathbb{R}^t \rightarrow \Omega$, and determines the classification $\widetilde{\phi}(u)$. The model builds a classifier $\widetilde{M}$, based upon a linear combination of the base parameterized classifiers in each iteration. It also selects weights $\alpha_j \in \mathbb{R}$ for $1 \leq j \leq N_{itr}$, such that

$$F(x) = \sum_{j=1}^{N_{itr}} \alpha_j \phi(x, \theta_j), \quad x \in \mathbb{R}^t$$

$$\widetilde{\phi}(x) = sgn(F(x))$$

where $sgn(\cdot)$ is the signum function. That is, if $x \in \mathbb{R}$; then $sgn(x) = 1$ if $x > 0$, and $sgn(x) = -1$ if $x < 0$. The ensemble model determines the parameters α_j, θ_j, for $1 \leq j \leq N_{itr}$ by formulating the following optimization problem.

$$\arg \min_{\alpha_j, \theta_j;\, 1 \leq j \leq N_{itr}} \sum_{i=1}^{n} \exp(-y_i F(x_i))$$

Observe that in this optimization problem, the objective function is penalized by data points which are incorrectly classified ($y_i F(x_i) < 0$) much more than those correctly classified ($y_i F(x_i) > 0$). Unfortunately, the optimization problem in this form is nontrivial. Nevertheless, a suboptimal and iterative approach is as follows. Define $F_m(x)$ as

$$F_m(x) = \sum_{j=1}^{m} \alpha_j \phi(x, \theta_j); \quad 1 \leq m \leq N_{itr}, \ x \in \mathbb{R}^t$$

Therefore

$$F_m(x) = F_{m-1}(x) + \alpha_m \phi(x, \theta_m)$$

It is assumed in iteration m, that $F_{m-1}(x)$ is already optimized in iteration $(m-1)$. The goal in the current iteration m is to determine the optimal values of α_m and θ_m. That is, we determine

$$(\alpha_m, \theta_m) = \arg \min_{\alpha, \theta} G(\alpha, \theta)$$

where the objective function is

$$G(\alpha, \theta) = \sum_{i=1}^{n} \exp\{-y_i (F_{m-1}(x_i) + \alpha \phi(x_i, \theta))\}$$

This optimization problem is also carried out in two steps. Initially assume that α is a constant, and the function $G(\cdot, \cdot)$ is minimized by varying θ. Thus

$$\theta_m = \arg \min_{\theta} \sum_{i=1}^{n} w_i^{(m)} \exp(-y_i \alpha \phi(x_i, \theta))$$

where

$$w_i^{(m)} \triangleq \exp(-y_i F_{m-1}(x_i))$$

Note that $w_i^{(m)}$ is independent of both α and $\phi(x_i, \theta)$. Consequently it can be considered to be a weight associated with the data point $x_i \in \mathcal{X}$. As $\phi(x_i, \theta) \in \{-1, +1\}$ the above minimization problem is equivalent to

$$\theta_m = \arg\min_{\theta} \left\{ P_m = \sum_{i=1}^{n} w_i^{(m)} I_{boost}\left(1 - y_i \phi(x_i, \theta)\right) \right\}$$

where the function $I_{boost}(\cdot)$ is either 0 or 1, if the argument is equal to either 0 or a positive number respectively. Note that the objective function in the above optimization problem is the *weighted* empirical error.

In order for the weights $w_i^{(m)}$'s sum to unity, the weights are normalized by their sum $\sum_{i=1}^{n} w_i^{(m)}$ without affecting the optimization. Thus

$$\sum_{y_i \phi(x_i, \theta_m) < 0} w_i^{(m)} = P_m, \quad \text{and} \quad \sum_{y_i \phi(x_i, \theta_m) > 0} w_i^{(m)} = (1 - P_m)$$

Therefore

$$\begin{aligned} G(\alpha, \theta_m) &= \sum_{i=1}^{n} \exp\left\{-y_i \left(F_{m-1}(x_i) + \alpha \phi(x_i, \theta_m)\right)\right\} \\ &= \sum_{i=1}^{n} w_i^{(m)} \exp\left\{-y_i \alpha \phi(x_i, \theta_m)\right\} = \exp(-\alpha)(1 - P_m) + \exp(\alpha) P_m \end{aligned}$$

Thus

$$\alpha_m = \arg\min_{\alpha} \left\{\exp(-\alpha)(1 - P_m) + \exp(\alpha) P_m\right\}$$

The minimizing value of α is

$$\alpha_m = \frac{1}{2} \ln \frac{1 - P_m}{P_m}$$

Having determined α_m and $\phi(x, \theta_m)$, the normalized weights used in the next iteration can be determined. Thus

$$w_i^{(m+1)} = \frac{\exp(-y_i F_m(x_i))}{W_m} = \frac{w_i^{(m)} \exp(-y_i \alpha_m \phi(x_i, \theta_m))}{W_m}$$

where

$$W_m = \sum_{i=1}^{n} w_i^{(m)} \exp(-y_i \alpha_m \phi(x_i, \theta_m))$$

is the normalizing factor. Note that the weight $w_i^{(m+1)}$ which corresponds to the data point x_i is increased (decreased) relative to its value in the previous iteration if the ensemble classifier fails (succeeds) in classifying the data point x_i. The new weight also depends upon the parameter α_m, which in turn controls the affect of the term $\phi(x_i, \theta_m)$.

Observation 11.5. The error rate ϵ_{err} of the AdaBoost classifier at the end of its training phase is

$$\epsilon_{err} = \frac{1}{n} \sum_{i=1}^{n} \delta\left(\widetilde{\phi}(x_i) \neq y_i\right)$$

where $\delta(\cdot)$ is equal to 1, if its argument is true, and 0 otherwise. The error rate ϵ_{err} is upper-bounded by $\prod_{j=1}^{N_{itr}} W_j$. Therefore, ϵ_{err} is minimized by minimizing this upper bound. □

The above observation is established in the problem section. The outlined AdaBoost algorithm performs binary classification. If classification of data points into more than two classes is required, then a boosting algorithm can be designed if the success rate of the base classifier is higher than fifty percent.

Algorithm 11.12. *AdaBoost Algorithm.*

Input: The set of training data points $\mathcal{D}$, where
$\mathcal{D} = \{(x_i, y_i) \mid x_i \in \mathcal{X} \subseteq \mathbb{R}^t, y_i \in \Omega, 1 \le i \le n\}$.
Set of classification labels $\Omega = \{-1, +1\}$.
The parameterized base classifier $\phi : \mathbb{R}^t \times \vartheta \to \Omega$.
The number of iterations N_{itr}, where $N_{itr} \in \mathbb{P} \backslash \{1\}$.
Test data point: $u \in \mathbb{R}^t \backslash \mathcal{X}$.
Output: Classification of the test data point u,
by the ensemble model $\widetilde{M}$.
It is $\widetilde{\phi}(u) = \omega \in \Omega$.
begin
  $w_i^{(1)} \leftarrow 1/n, \quad 1 \le i \le n.$
  (*initialize weights*)
  for $m = 1$ **to** N_{itr} **do**
  begin
    (*Determine optimum* θ_m)
    (θ_m *is the optimal parameter of base classifier in iteration* m)

$$\theta_m \leftarrow \arg\min_{\theta} \left\{ P_m = \sum_{i=1}^{n} w_i^{(m)} I_{boost}\left(1 - y_i \phi\left(x_i, \theta\right)\right) \right\}$$

    ($w_i^{(m)}$ *is the weight associated with data point* x_i *in iteration* m)
    ($I_{boost}(\cdot) = 0$, *if the argument is* 0)
    ($I_{boost}(\cdot) = 1$, *if the argument is a positive number*)

$$P_m \leftarrow \sum_{y_i \phi(x_i, \theta_m) < 0} w_i^{(m)}$$

    (P_m *is the sum of weights of data points incorrectly classified in iteration* m)

$$\alpha_m \leftarrow \frac{1}{2} \ln \frac{1 - P_m}{P_m}$$

    (α_m *is the optimal weight of the base classifier in iteration* m)
    $W_m \leftarrow \sum_{i=1}^{n} w_i^{(m)} \exp\left(-y_i \alpha_m \phi\left(x_i, \theta_m\right)\right)$
    $w_i^{(m+1)} \leftarrow w_i^{(m)} \exp\left(-y_i \alpha_m \phi\left(x_i, \theta_m\right)\right) / W_m, \quad 1 \le i \le n.$
  end (*end of* m *for-loop*)

(*determine the class of the test data point* u)
$F(u) \leftarrow \sum_{j=1}^{N_{itr}} \alpha_j \phi(u, \theta_j)$
$\widetilde{\phi}(u) = sgn(F(u))$
$\omega \leftarrow \widetilde{\phi}(u)$
end (*end of AdaBoost algorithm*)

11.8 Rough Sets

The data points in a data set are generally imprecise, incomplete, and vague. Rough set theory is a mathematical paradigm to represent such imperfect knowledge. It also provides a formal framework to synthesize uncertain data. Rough and fuzzy set theories are two different yet complementary approaches which capture the affect of imperfection in knowledge.

Crisp (exact, precise) sets are used in classical set theory. In such sets, an element of the set is either its member or not. In fuzzy set theory, the degree of membership of an element of a set is specified by a value in the closed interval $[0,1]$. This allows for a partial membership of an element of the set. Thus fuzziness specifies the degree to which an element belongs to a set or not.

In rough set theory, a set is specified "approximately" by using a collection of crisp sets. Thus, rough sets are sets with fuzzy boundaries. The rough set methodology is based upon the assumption that by decreasing the degree of precision in the data, patterns in the data set become more apparent.

The theory of rough sets approximates subsets or partitions of set of data points, with other sets or partitions built by using the available knowledge about the set of data points.

This theory has found applications in several domains, including generation of minimal sets of decision rules from data. It provides efficient algorithms for finding hidden patterns in data, and also evaluates significance of data. Basics of rough set theory is initially developed in this section. This is followed by the concept of reducts and core in an information system. These are also illustrated via examples. Decision systems and decision rules within the context of rough set theory are also described.

11.8.1 Basics

Data in an *information system* can be represented in a tabular form. A data table has several rows and columns. The number of rows in the data table is equal to the number of data points in the data set, and the number of columns is equal to the number of attributes of the data points. A row in the data table is labeled by an object, and each column is labeled by an attribute.

Definition 11.20. *An information system is* $\vartheta_{IS} = (U, \mathcal{A}, V, F)$, *where* U *is the universe of finite and nonempty set of objects,* $\mathcal{A}$ *is the nonempty set of attributes,* V *is the set of domain of attributes in* $\mathcal{A}$. *That is,*

$$V = \{V_a \mid a \in \mathcal{A},\ V_a \text{ is the domain of attribute } a\} \tag{11.11a}$$

and

$$\mathcal{F} = \{f_a \mid a \in \mathcal{A},\ f_a : U \to V_a\} \tag{11.11b}$$

Note that $f_a(\cdot)$ *and* V_a *are called the information function and value set (domain) of attribute* $a \in \mathcal{A}$ *respectively.* □

Indiscernibility Relationship

The information system ϑ_{IS} can be specified in a table. A pair (x, a), where $x \in U$ and $a \in \mathcal{A}$, defines an entry in table which consists of the value $f_a(x)$. Objects in the information system which have same information are *indiscernible* (similar). For example, a pair of objects $x_i, x_j \in U$ are indiscernible over a set of attributes $\mathcal{B} \subseteq \mathcal{A}$, if they have identical values for every attribute of the set $\mathcal{B}$. Thus an indiscernibility relationship can be described for every subset of attributes $\mathcal{B} \subseteq \mathcal{A}$.

Definition 11.21. *Let* $\vartheta_{IS} = (U, \mathcal{A}, V, \mathcal{F})$ *be an information system. Every nonempty subset of attributes* $\mathcal{B} \subseteq \mathcal{A}$ *induces an indiscernibility relationship* $Ind(\mathcal{B})$, *where*

$$Ind(\mathcal{B}) = \{(x, y) \in U \times U,\ \forall\, a \in \mathcal{B},\ f_a(x) = f_a(y)\} \tag{11.12}$$

Therefore, $Ind(\mathcal{B})$ *is an equivalence relationship. For each* $x \in U$, *there exists an equivalence class* $[x]_{\mathcal{B}}$ *in the partition of the universal set* U *defined by* $Ind(\mathcal{B})$. *The equivalence classes of sets induced by* $Ind(\mathcal{B})$ *are called elementary sets in* $\mathcal{B}$. *The quotient set of* U *by* $Ind(\mathcal{B})$ *is* $U/Ind(\mathcal{B})$. *The equivalence class* $[x]_{\mathcal{B}}$ *is an element in the quotient set* $U/Ind(\mathcal{B})$. □

An equivalence class is a set of data points with matching descriptions. The equivalence classes of sets induced by $Ind(\mathcal{B})$ are called elementary sets in $\mathcal{B}$, because these represent the smallest discernible groups of objects. Observe that

$$Ind(\mathcal{B}) = \cap_{a \in \mathcal{B}} Ind(a)$$

Imprecisions and Approximations

Real world data is often imprecise. Sometimes, similar objects with matching attribute values might have dissimilar classifications. Tables with such objects are inconsistent. The theory of rough sets introduces the concept of approximation of sets to address the problem of inconsistency in such tables. A rough set approximates the classical set by using a pair of crisp sets called lower and upper approximation of the set.

Let $X \subseteq U$ and $\mathcal{B} \subseteq \mathcal{A}$. With respect to the attributes in the set $\mathcal{B}$, the set X can be approximated by constructing its lower and upper approximations. Lower approximation of the set X is the set of data points which can *certainly* be classified as members of the set X by using the values of attributes in $\mathcal{B}$. Similarly, the upper approximation of the set X is the set of data points which can *possibly* be classified as members of the set X by using the values of attributes in $\mathcal{B}$.

In summary, due to imprecise knowledge, a rough set X cannot be characterized by using available knowledge. However, it can be characterized via two crisp sets: its lower and upper approximations.

Definition 11.22. *Let* $\vartheta_{IS} = (U, \mathcal{A}, V, \mathcal{F})$ *be an information system. Also let* $X \subseteq U$, $X \neq \varnothing$, $\mathcal{B} \subseteq \mathcal{A}$, *and* $\mathcal{B} \neq \varnothing$. *The lower and upper approximation of the set* X *by using the values of attributes in* $\mathcal{B}$, *are* $\underline{\mathcal{B}}X$ *and* $\overline{\mathcal{B}}X$ *respectively, where*

$$\underline{\mathcal{B}}X = \{x \mid x \in U, [x]_{\mathcal{B}} \subseteq X\} = \cup\{Y \in U/Ind(\mathcal{B}) \mid Y \subseteq X\} \tag{11.13a}$$

$$\overline{\mathcal{B}}X = \{x \mid x \in U, [x]_{\mathcal{B}} \cap X \neq \varnothing\} = \cup\{Y \in U/Ind(\mathcal{B}) \mid Y \cap X \neq \varnothing\} \tag{11.13b}$$

□

The difference of set of data points in the sets $\overline{\mathcal{B}}X$ and $\underline{\mathcal{B}}X$, using the values of attributes in $\mathcal{B}$, is called the $\mathcal{B}$-*boundary* region of X in U. The boundary region consists of data points which cannot be uniquely assigned to the set or its complement by using available knowledge.

Definition 11.23. *Let $\vartheta_{IS} = (U, \mathcal{A}, V, \mathcal{F})$ be an information system. Also let $X \subseteq U$, $X \neq \varnothing$, $\mathcal{B} \subseteq \mathcal{A}$, and $\mathcal{B} \neq \varnothing$. The lower and upper approximation of the set X by using the values of attributes in $\mathcal{B}$, are $\underline{\mathcal{B}}X$ and $\overline{\mathcal{B}}X$ respectively. The $\mathcal{B}$-boundary region of the set X is $BN_{\mathcal{B}}(X)$, where $BN_{\mathcal{B}}(X) = (\overline{\mathcal{B}}X - \underline{\mathcal{B}}X)$.* □

The $\mathcal{B}$-boundary region $BN_{\mathcal{B}}(X)$ is an undecidable region of the universe. That is, no element which belong to the $\mathcal{B}$-boundary can be classified with certainty into either X or X^c (complement of X) as far as the attributes in the set $\mathcal{B}$ are concerned. We are now ready to define a rough set. A set is rough or imprecise, if its boundary region is nonempty. However if the boundary region is empty, then the set is crisp or precise.

Definition 11.24. *Let $\vartheta_{IS} = (U, \mathcal{A}, V, \mathcal{F})$ be an information system. Also let $X \subseteq U$, $X \neq \varnothing$, $\mathcal{B} \subseteq \mathcal{A}$, and $\mathcal{B} \neq \varnothing$. The corresponding boundary region of the set X with respect to attributes in the set $\mathcal{B}$ is $BN_{\mathcal{B}}(X)$. The set X is rough (inexact) with respect to the set $\mathcal{B}$ if $BN_{\mathcal{B}}(X) \neq \varnothing$, otherwise it is crisp.* □

Elementary properties of the lower and upper approximations are next summarized.

Observations 11.6. Let $\vartheta_{IS} = (U, \mathcal{A}, V, \mathcal{F})$ be an information system, where $\mathcal{B} \subseteq \mathcal{A}$ and $\mathcal{B} \neq \varnothing$, and $X \subseteq U$ and $X \neq \varnothing$. The lower and upper approximation of the set X by using the attributes in the set $\mathcal{B}$, are $\underline{\mathcal{B}}X$ and $\overline{\mathcal{B}}X$ respectively.

1. $\underline{\mathcal{B}}X \subseteq X \subseteq \overline{\mathcal{B}}X$.
2. $\underline{\mathcal{B}}\varnothing = \overline{\mathcal{B}}\varnothing = \varnothing$, and $\underline{\mathcal{B}}U = \overline{\mathcal{B}}U = U$.
3. If $Z = X \cup Y$, then:
   (a) $\underline{\mathcal{B}}Z \supseteq \underline{\mathcal{B}}X \cup \underline{\mathcal{B}}Y$.
   (b) $\overline{\mathcal{B}}Z = \overline{\mathcal{B}}X \cup \overline{\mathcal{B}}Y$.
4. If $Z = X \cap Y$, then:
   (a) $\underline{\mathcal{B}}Z = \underline{\mathcal{B}}X \cap \underline{\mathcal{B}}Y$.
   (b) $\overline{\mathcal{B}}Z \subseteq \overline{\mathcal{B}}X \cap \overline{\mathcal{B}}Y$.
5. $X \subseteq Y \Rightarrow \underline{\mathcal{B}}X \subseteq \underline{\mathcal{B}}Y$ and $\overline{\mathcal{B}}X \subseteq \overline{\mathcal{B}}Y$.
6. If $Z = -X$, then:
   (a) $\underline{\mathcal{B}}Z = -\overline{\mathcal{B}}X$.
   (b) $\overline{\mathcal{B}}Z = -\underline{\mathcal{B}}X$.
7. If $Z = \underline{\mathcal{B}}X$, then, $\underline{\mathcal{B}}Z = \overline{\mathcal{B}}Z = \underline{\mathcal{B}}X$.
8. If $Z = \overline{\mathcal{B}}X$, then, $\underline{\mathcal{B}}Z = \overline{\mathcal{B}}Z = \overline{\mathcal{B}}X$. □

If $\underline{\mathcal{B}}X = \overline{\mathcal{B}}X$ then the set X is definable in U, otherwise, the set X is undefinable in U, with respect to values of the subset of attributes $\mathcal{B} \subseteq \mathcal{A}$. It is possible to define following four categories of rough sets or vagueness. Let $X \subseteq U$, $\mathcal{B} \subseteq \mathcal{A}$, where X uses attributes in the set $\mathcal{B}$. These four categories of vagueness are:

- $\underline{\mathcal{B}}X \neq \varnothing$ and $\overline{\mathcal{B}}X \neq U$, iff X is roughly $\mathcal{B}$-definable in U.
- $\underline{\mathcal{B}}X = \varnothing$ and $\overline{\mathcal{B}}X \neq U$, iff X is internally $\mathcal{B}$-indefinable in U.
- $\underline{\mathcal{B}}X \neq \varnothing$ and $\overline{\mathcal{B}}X = U$, iff X is externally $\mathcal{B}$-indefinable in U.
- $\underline{\mathcal{B}}X = \varnothing$ and $\overline{\mathcal{B}}X = U$, iff X is totally $\mathcal{B}$-indefinable in U.

The following concept is useful for describing lower approximation and positive region, and upper approximation and negative region of an information system.

Definitions 11.25. *Let $\vartheta_{IS} = (U, \mathcal{A}, V, \digamma)$ be an information system. Also let $X \subseteq U$, $X \neq \varnothing$, $\mathcal{B} \subseteq \mathcal{A}$, and $\mathcal{B} \neq \varnothing$. The lower and upper approximation of the set X by using the values of attributes in $\mathcal{B}$, are $\underline{\mathcal{B}}X$ and $\overline{\mathcal{B}}X$ respectively.*

1. *Lower approximation and positive region*:
   The positive region is the set of elements $POS_{\mathcal{B}}(X)$, which with certainty can be classified in the set X. Thus $POS_{\mathcal{B}}(X) = \underline{\mathcal{B}}X$.
2. *Upper approximation and negative region*:
   The negative region is the set of elements $NEG_{\mathcal{B}}(X)$, which with certainty are not in the set X. Thus $NEG_{\mathcal{B}}(X) = \left(U - \overline{\mathcal{B}}X\right)$. □

The *approximation* of a rough set, and the *quality* of approximation of $X \subseteq U$ by values of attributes $\mathcal{B} \subseteq \mathcal{A}$ can also be numerically characterized.

Definitions 11.26. *Let $\vartheta_{IS} = (U, \mathcal{A}, V, \digamma)$ be an information system. Also let $X \subseteq U$, $X \neq \varnothing$, $\mathcal{B} \subseteq \mathcal{A}$, and $\mathcal{B} \neq \varnothing$. The lower and upper approximation of the set X by using the values of attributes in $\mathcal{B}$ are $\underline{\mathcal{B}}X$ and $\overline{\mathcal{B}}X$ respectively.*

1. *An accuracy measure of the set X by means of attributes in $\mathcal{B}$ is $\mu_{\mathcal{B}}(X) = |\underline{\mathcal{B}}X| / |\overline{\mathcal{B}}X|$.*
2. *The quality of approximation of X by means of attributes in $\mathcal{B}$ is $\alpha_{\mathcal{B}}(X) = |\underline{\mathcal{B}}X| / |X|$.* □

Observe that $0 \leq \mu_{\mathcal{B}}(X) \leq 1$. Further, if the set X is definable in U, then $\mu_{\mathcal{B}}(X) = 1$. If the set X is undefinable in U, then $0 \leq \mu_{\mathcal{B}}(X) < 1$. Also, $0 \leq \alpha_{\mathcal{B}}(X) \leq 1$. It follows from the above definitions that, $0 \leq \mu_{\mathcal{B}}(X) \leq \alpha_{\mathcal{B}}(X) \leq 1$. Further, $\mu_{\mathcal{B}}(X) = 0 \Leftrightarrow \alpha_{\mathcal{B}}(X) = 0$; and $\mu_{\mathcal{B}}(X) = 1 \Leftrightarrow \alpha_{\mathcal{B}}(X) = 1$.

Rough sets can also be specified by using the rough membership function.

Definition 11.27. *Let $\vartheta_{IS} = (U, \mathcal{A}, V, \digamma)$ be an information system. Also let $X \subseteq U$, $X \neq \varnothing$, $\mathcal{B} \subseteq \mathcal{A}$, and $\mathcal{B} \neq \varnothing$. The indiscernibility relationship is $Ind(\mathcal{B})$. The rough membership measure of an object $x \in X$ is:*

$$\mu_X^{\mathcal{B}}(x) = \frac{|X \cap [x]_{\mathcal{B}}|}{|[x]_{\mathcal{B}}|} \tag{11.14}$$

□

The value $\mu_X^{\mathcal{B}}(x)$ may be interpreted as the relative overlap between the equivalence class to which x belongs to by using the attributes in $\mathcal{B}$, and the set X.

Example 11.11. Consider an information system $\vartheta_{IS} = (U, \mathcal{A}, V, \digamma)$ specified in the Table 11.12. In this information system $U = \{x_1, x_2, x_3, x_4, x_5\}$, $\mathcal{A} = \{a, b, c, d, e\}$, $V_a = \{0, 1, 2\}$, $V_b = \{1, 2\}$, $V_c = \{0, 1, 2\}$, $V_d = V_e = \{1, 2\}$, $V = \{V_a, V_b, V_c, V_d, V_e\}$, and the set $\digamma$ can be determined from the columns of the table. Let $\mathcal{B} = \{a, b, c\}$.

The quotient set $U/Ind(\mathcal{B}) = \{\{x_1, x_3\}, \{x_2\}, \{x_4\}, \{x_5\}\}$. Let $X = \{x_1, x_2, x_4\}$. The lower and upper approximations of this set are $\underline{\mathcal{B}}X = \{x_2, x_4\}$, and $\overline{\mathcal{B}}X = \{x_1, x_2, x_3, x_4\}$ respectively. Further, the corresponding $\mathcal{B}$-boundary of the set X is $BN_{\mathcal{B}}(X) = \{x_1, x_3\}$. As $BN_{\mathcal{B}}(X) \neq \varnothing$, the set X is rough with respect to the set $\mathcal{B}$. The positive region of the set X is $POS_{\mathcal{B}}(X) = \underline{\mathcal{B}}X = \{x_2, x_4\}$, and the negative region of the set X is $NEG_{\mathcal{B}}(X) = (U - \overline{\mathcal{B}}X) = \{x_5\}$. The accuracy measure of the set X is $\mu_{\mathcal{B}}(X) = |\underline{\mathcal{B}}X| / |\overline{\mathcal{B}}X| = 0.5$. The quality of approximation of X by means of attributes in $\mathcal{B}$ is $\alpha_{\mathcal{B}}(X) = |\underline{\mathcal{B}}X| / |X| = 2/3$. The rough membership measures are: $\mu_X^{\mathcal{B}}(x_1) = 1/2, \mu_X^{\mathcal{B}}(x_2) = \mu_X^{\mathcal{B}}(x_4) = 1$.

U	a	b	c	d	e
x_1	1	2	0	1	1
x_2	0	1	0	2	2
x_3	1	2	0	1	1
x_4	0	2	2	1	2
x_5	2	1	0	2	2

Table 11.12. Data table.

□

11.8.2 Reduct and Core

It is possible that a data set might contain superfluous data. A data point is considered to be superfluous, if it does not help in the classification process of the given training data points. Evidently, such data points can be removed. It was mentioned in the last subsection, that data can be reduced by identifying indiscernible objects. Indiscernible objects in turn, form equivalence classes. This results in savings because only a single element of the equivalence class is required to represent the entire class.

Another way to reduce data is by discarding redundant attributes. Data can also be reduced by discarding attribute values. We initially explore a technique to reduce data by discarding redundant attributes. This is followed by a technique to reduce data by discarding redundant attribute values. Within this context, the concept of reduct and core of attributes is next explored.

Reduct and Core of Attributes

The data table of an information system can be simplified by eliminating dependent attributes. Whether a set of attributes is independent or dependent can easily be checked. If the removal of an attribute increases the number of elementary sets, then the attribute is dependent.

Definition 11.28. *Let $\vartheta_{IS} = (U, \mathcal{A}, V, \digamma)$ be an information system. If $Ind(\mathcal{A}) = Ind(\mathcal{A} - a)$, then the attribute $a \in \mathcal{A}$ is superfluous, otherwise the attribute a is said to be indispensable.* □

Only those attributes of the data need be kept that preserve the indiscernibility relationship, which in turn is equivalent to maintaining the set approximations. Therefore it might be possible for only a subset of features to fully characterize knowledge in a data set. The corresponding attribute set is called a *reduct*. The *core* is the set of all essential (indispensable) attributes of the information system. The attributes in the core occur in every reduct.

Definitions 11.29. *Reduct and core of attributes. Let $\vartheta_{IS} = (U, \mathcal{A}, V, F)$ be an information system, and $\mathcal{B} \subseteq \mathcal{A}$.*

1. *The $\mathcal{B}$-reduct of $\mathcal{A}$ is the subset $\mathcal{B}' \subseteq \mathcal{B}$, such that $Ind\,(\mathcal{B}) = Ind\,(\mathcal{B}')$, and $\mathcal{B}'$ is a minimal subset of $\mathcal{B}$.*
2. *The intersection of all $\mathcal{B}$-reducts of $\mathcal{A}$ is called the $\mathcal{B}$-core of $\mathcal{A}$.* □

In the above definition, a minimal subset is a subset with minimal cardinality. Thus a $\mathcal{B}$-reduct in an information system is a minimal set of attributes from the set $\mathcal{B}$ that preserve the capability to extract knowledge from the universe of data set. Thus a reduct can discern all objects which are discernible by the original information system.

Discernibility Matrix

The core and reduct attributes can be conveniently determined directly by observing the corresponding discernibility matrix.

Definition 11.30. *Let $\vartheta_{IS} = (U, \mathcal{A}, V, F)$ be an information system with $|U| = n \in \mathbb{P}$ objects, $\mathcal{B} \subseteq \mathcal{A}$, and $\mathcal{B} \neq \varnothing$. The discernibility matrix of $\mathcal{B}$ in ϑ_{IS} is a symmetric matrix $D_{disc} = [d_{ij}]$ of size n. The entry $d_{ii} = \varnothing$; and d_{ij} for $i \neq j$ is equal to the set of attributes in $\mathcal{B}$ in which the objects x_i and x_j differ. Thus*

$$d_{ij} = \{a \in \mathcal{B} \mid f_a(x_i) \neq f_a(x_j)\}, 1 \leq i, j \leq n \tag{11.15}$$

That is, d_{ij} is the set of all attributes in $\mathcal{B}$ which discern objects x_i and x_j, for $1 \leq i, j \leq n$. □

Observe that the discernibility matrix is symmetric. The discernibility matrix can be used to determine the minimal subsets of attributes (reducts), without losing any knowledge. In order to accomplish this the discernibility function $f_{\vartheta_{IS}}(\cdot)$ is used.

Definition 11.31. *Discernibility function for attributes. Consider an information system $\vartheta_{IS} = (U, \mathcal{A}, V, F)$, with $|U| = n \in \mathbb{P}$ objects. Also let $\mathcal{B} \subseteq \mathcal{A}$, and $|\mathcal{B}| = m$. The discernibility function $f_{\vartheta_{IS}}(\cdot)$ for attributes in $\mathcal{B}$ is a Boolean function of m Boolean variables $a_1^*, a_2^*, \ldots, a_m^*$; where a_k^* corresponds to the attribute $a_k \in \mathcal{B}$ for $1 \leq k \leq m$. It is*

$$f_{\vartheta_{IS}}(a_1^*, a_2^*, \ldots, a_m^*) = \bigwedge_{\substack{1 \leq j < i \leq n \\ d_{ij} \neq \varnothing}} \bigvee d_{ij}^* \tag{11.16}$$

where $d_{ij}^* = \{a^* \mid a \in d_{ij}\}$. Let

$$f_{\vartheta_{IS}}(a_1^*, a_2^*, \ldots, a_m^*) \triangleq f_{\vartheta_{IS}}(\mathcal{B})$$

□

In the following discussion, a_k^* is simply denoted by a_k for $1 \leq k \leq m$. The discernibility function $f_{\vartheta_{IS}}(\cdot)$ specifies constraints that should be maintained in order to preserve the discernibility between all pairs of discernible objects in ϑ_{IS} for a specified set $\mathcal{B}$ of attributes.

A reduct is the minimal subset of attributes such that it has a minimum of a single common element in any nonempty entry of the discernibility matrix D_{disc}. A core is the set of all single element entries in the discernibility matrix D_{disc}.

Example 11.12. We continue analysing the data set given in Table 11.12 of the last example. Recall that the attribute set $\mathcal{A} = \{a, b, c, d, e\}$, and $\mathcal{B} = \{a, b, c\}$. The discernibility matrix D_{disc} and the discernibility function $f_{\vartheta_{IS}}(\cdot)$ is obtained. As the matrix D_{disc} is symmetric, only the lower triangular part of the matrix is shown.

$$D_{disc} = \begin{array}{cccccc} & x_1 & x_2 & x_3 & x_4 & x_5 \\ x_1 & \varnothing & -- & -- & -- & -- \\ x_2 & a,b & \varnothing & -- & -- & -- \\ x_3 & \varnothing & a,b & \varnothing & -- & -- \\ x_4 & a,c & b,c & a,c & \varnothing & -- \\ x_5 & a,b & a & a,b & a,b,c & \varnothing \end{array}$$

$U/Ind(\mathcal{B})$	a	b
$\{x_1, x_3\}$	1	2
$\{x_2\}$	0	1
$\{x_4\}$	0	2
$\{x_5\}$	2	1

(a)

$U/Ind(\mathcal{B})$	a	c
$\{x_1, x_3\}$	1	0
$\{x_2\}$	0	0
$\{x_4\}$	0	2
$\{x_5\}$	2	0

(b)

Table 11.13. Equivalent representation of information system via two $\mathcal{B}$-reducts.

The discernibility function $f_{\vartheta_{IS}}(\mathcal{B})$ is next determined.

$$\begin{aligned} f_{\vartheta_{IS}}(\mathcal{B}) &= (a \vee b) \wedge (a \vee c) \wedge (a \vee b) \wedge (a \vee b) \wedge (b \vee c) \\ &\quad \wedge a \wedge (a \vee c) \wedge (a \vee b) \wedge (a \vee b \vee c) \\ &= (b \vee c) \wedge a \end{aligned}$$

Thus there are two $\mathcal{B}$-reducts. These are $\{a, b\}$, and $\{a, c\}$. The $\mathcal{B}$-core is $\{a\}$. The equivalent representation of information system via the two $\mathcal{B}$-reducts is shown in Table 11.13. □

Reduct and Core of Attribute Values

The information system ϑ_{IS} can be further simplified by eliminating certain superfluous values of attributes. These values can be eliminated and yet be able to discern all elementary sets in ϑ_{IS}. The technique to determine the reducts and core of the attribute values is similar to the technique of determining the reducts and core of the attributes. The technique uses a discernibility matrix, and discernibility function of the attribute values. The discernibility matrix of attribute values is similar to the discernibility matrix of attributes. However, the discernibility function has to be evaluated for *each* elementary set.

Definition 11.32. *Let $\vartheta_{IS} = (U, \mathcal{A}, V, \digamma)$ be an information system, and $\mathcal{B} \subseteq \mathcal{A}$. Also let $[x]_{\mathcal{B}}$ be the equivalence class of $x \in U$. The value of attribute $b \in \mathcal{B}$ is superfluous for x, if $[x]_{\mathcal{B}} = [x]_{\mathcal{B}-\{b\}}$, otherwise the value of the attribute b is indispensable for x.* □

The definition of reduct and core of attribute values is next given.

Definitions 11.33. *Reduct and core of attribute values. Consider an information system $\vartheta_{IS} = (U, \mathcal{A}, V, \digamma)$, and $\mathcal{B} \subseteq \mathcal{A}$. Also let $[x]_{\mathcal{B}}$ be the equivalence class of $x \in U$.*

1. *The $\mathcal{B}_x$-reduct of $\mathcal{B}$ for x is the set $\mathcal{B}' \subseteq \mathcal{B}$, such that $[x]_{\mathcal{B}} = [x]_{\mathcal{B}'}$, and $\mathcal{B}'$ is a minimal subset of $\mathcal{B}$.*
2. *The intersection of all $\mathcal{B}_x$-reducts of $\mathcal{B}$ for x is called the $\mathcal{B}_x$-core of $\mathcal{B}$ for x.* □

The definition of discernibility matrix for attribute values is similar to that of discernibility matrix for attributes.

Definition 11.34. *Discernibility function for attributes values. Let $\vartheta_{IS} = (U, \mathcal{A}, V, \digamma)$ be an information system with $|U| = n \in \mathbb{P}$ objects. Also let $\mathcal{B} \subseteq \mathcal{A}$, $|\mathcal{B}| = m$, and $x_i \in U$. The discernibility function for attribute values $f^i_{\vartheta_{IS}}(\cdot)$ is a Boolean function of m Boolean variables $a^*_1, a^*_2, \ldots, a^*_m$; where a^*_k corresponds to the attribute $a_k \in \mathcal{B}$ for $1 \le k \le m$. It is*

$$f^i_{\vartheta_{IS}}(a^*_1, a^*_2, \ldots, a^*_m) = \bigwedge \bigvee_{\substack{1 \le j \le n \\ d_{ij} \ne \varnothing}} d^*_{ij}, \quad 1 \le i \le n \tag{11.17}$$

where $d^*_{ij} = \{a^* \mid a \in d_{ij}\}$.
Let $f^i_{\vartheta_{IS}}(a^*_1, a^*_2, \ldots, a^*_m) \triangleq f^i_{\vartheta_{IS}}(\mathcal{B})$, for $1 \le i \le n$. □

As mentioned earlier, a^*_k is denoted subsequently simply by a_k for $1 \le k \le m$.

Example 11.13. The last example is continued. In it $\mathcal{A} = \{a, b, c, d, e\}$, and $\mathcal{B} = \{a, b, c\}$. The two $\mathcal{B}$-reducts are $\{a, b\}$, and $\{a, c\}$. The corresponding data is shown in Table 11.14.

(a)

U	a	b
x_1	1	2
x_2	0	1
x_3	1	2
x_4	0	2
x_5	2	1

(b)

U	a	c
x_1	1	0
x_2	0	0
x_3	1	0
x_4	0	2
x_5	2	0

Table 11.14. Data tables for the two $\mathcal{B}$-reducts $\{a, b\}$, and $\{a, c\}$.

The discernibility matrix D_{disc} for the $\mathcal{B}$-reduct $\{a, b\}$ is

$$D_{disc} = \begin{array}{c|ccccc} & x_1 & x_2 & x_3 & x_4 & x_5 \\ \hline x_1 & \varnothing & a,b & \varnothing & a & a,b \\ x_2 & a,b & \varnothing & a,b & b & a \\ x_3 & \varnothing & a,b & \varnothing & a & a,b \\ x_4 & a & b & a & \varnothing & a,b \\ x_5 & a,b & a & a,b & a,b & \varnothing \end{array}$$

The $n = 5$ discernibility functions for the attribute values $f^i_{\vartheta_{IS}}(\mathcal{B}), 1 \le i \le 5$ are determined from the matrix D_{disc}. The function $f^i_{\vartheta_{IS}}(\mathcal{B})$ is determined by considering all sets of attributes from column i of matrix D_{disc}. Thus:

$$\begin{aligned} f^1_{\vartheta_{IS}}(\mathcal{B}) &= (a \vee b) \wedge a \wedge (a \vee b) = a \\ f^2_{\vartheta_{IS}}(\mathcal{B}) &= (a \vee b) \wedge (a \vee b) \wedge b \wedge a = b \wedge a \\ f^3_{\vartheta_{IS}}(\mathcal{B}) &= (a \vee b) \wedge a \wedge (a \vee b) = a \\ f^4_{\vartheta_{IS}}(\mathcal{B}) &= a \wedge b \wedge a \wedge (a \vee b) = a \wedge b \\ f^5_{\vartheta_{IS}}(\mathcal{B}) &= (a \vee b) \wedge a \wedge (a \vee b) \wedge (a \vee b) = a \end{aligned}$$

$U/Ind(\mathcal{B})$	a	b
$\{x_1, x_3\}$	1	$*$
$\{x_2\}$	0	1
$\{x_4\}$	0	2
$\{x_5\}$	2	$*$

(a)

$U/Ind(\mathcal{B})$	a	c
$\{x_1, x_3\}$	1	$*$
$\{x_2\}$	0	0
$\{x_4\}$	$*$	2
$\{x_5\}$	2	$*$

(b)

Table 11.15. Equivalent representation of information system via two $\mathcal{B}$-reducts. The $*$ symbol denotes "do not care."

Note that we actually need to determine only $f^1_{\vartheta_{IS}}(\mathcal{B})$, and not both $f^1_{\vartheta_{IS}}(\mathcal{B})$ and $f^3_{\vartheta_{IS}}(\mathcal{B})$, as x_1 and x_3 belong to the same elementary set. Further, $f^1_{\vartheta_{IS}}(\mathcal{B}) = a$ implies that there is only a single reduct of the value of attributes, which is simply a. That is, only attribute a is indispensable and the attribute b can be ignored. Therefore the Table 11.13 (a) can be simplified, as in Table 11.15 (a). The Table 11.15 (b) is for $\mathcal{B}$-reduct $\{a, c\}$. □

Classification

Consider an information system $\vartheta_{IS} = (U, \mathcal{A}, V, F)$. The classification of objects in the universal set U is defined.

Definition 11.35. *Let $\vartheta_{IS} = (U, \mathcal{A}, V, F)$ be an information system. The set U is classified into $k \in \mathbb{P}$ classes. Let*

$$F_{cl} = \{X_i \mid X_i \subseteq U, \ 1 \le i \le k\} \tag{11.18}$$

where $X_i \ne \varnothing$, for $1 \le i \le k$; $X_i \cap X_j = \varnothing$, $i \ne j$, $1 \le i, j \le k$; and $U = \cup_{1 \le i \le k} X_i$. The set F_{cl} is called a classification of the universal set U into k classes, and the X_i's are called classes of the universal set U. □

The elements of the set F_{cl} partition the elements of the universal set U into mutually exclusive classes. It is also possible to describe the quality and accuracy of classification.

Definitions 11.36. *Let $\vartheta_{IS} = (U, \mathcal{A}, V, \mathcal{F})$ be an information system with $|U| = n \in \mathbb{P}$ objects. Also let $\mathcal{B} \subseteq \mathcal{A}$, and $F_{cl} = \{X_1, X_2, \ldots, X_k\}$ be a classification of the set U. Also let the lower and upper approximation of the set X_i by using the attributes in $\mathcal{B}$, be $\underline{\mathcal{B}}X_i$ and $\overline{\mathcal{B}}X_i$ respectively, for $1 \leq i \leq k$.*

1. *The accuracy of classification F_{cl} in $\mathcal{B}$ is $\beta_{\mathcal{B}}(F_{cl})$, where*

$$\beta_{\mathcal{B}}(F_{cl}) = \frac{\sum_{i=1}^{k} |\underline{\mathcal{B}}X_i|}{\sum_{i=1}^{k} |\overline{\mathcal{B}}X_i|} \tag{11.19}$$

2. *The quality of classification F_{cl} in $\mathcal{B}$ is $\gamma_{\mathcal{B}}(F_{cl})$, where*

$$\gamma_{\mathcal{B}}(F_{cl}) = \frac{\sum_{i=1}^{k} |\underline{\mathcal{B}}X_i|}{|U|} \tag{11.20}$$

□

Example 11.14. Consider the information system $\vartheta_{IS} = (U, \mathcal{A}, V, \mathcal{F})$ specified in Table 11.12. In this table, $U = \{x_1, x_2, x_3, x_4, x_5\}$, and $\mathcal{A} = \{a, b, c, d, e\}$. Let $\mathcal{B} = \{a, b, c\}$, and the quotient set

$$U/Ind(\mathcal{B}) = \{\{x_1, x_3\}, \{x_2\}, \{x_4\}, \{x_5\}\}$$

Also, let the classification of the universal set U be $F_{cl} = \{X_1, X_2\}$, where $X_1 = \{x_1, x_5\}$, and $X_2 = \{x_2, x_3, x_4\}$. Therefore, $\underline{\mathcal{B}}X_1 = \{x_5\}$, $\overline{\mathcal{B}}X_1 = \{x_1, x_3, x_5\}$, $\underline{\mathcal{B}}X_2 = \{x_2, x_4\}$, and $\overline{\mathcal{B}}X_2 = \{x_1, x_2, x_3, x_4\}$; and $\beta_{\mathcal{B}}(F_{cl}) = 3/7$, and $\gamma_{\mathcal{B}}(F_{cl}) = 3/5$. □

11.8.3 Decision System

Each data point of a knowledge representation system generally has two types of attributes. These are the conditional and decision attributes. An attribute which specifies the class to which a data point belongs to is called the decision attribute. All other attributes are called conditional attributes.

The classification attributes help in categorization of the data. The class labels associated with each data point specify the class of each data point. The sets of conditional and decision attributes are denoted by $\mathcal{A}$ and $\mathcal{A}_d$ respectively. It is possible for $|\mathcal{A}_d|$ to be greater than unity. That is, there can be more than a single decision attribute. An information system which is specified by both conditional and decision attributes is called a *decision system.*

Definition 11.37. *A decision system is $\vartheta_{DS} = (U, \mathcal{A} \cup \mathcal{A}_d, V, \mathcal{F})$, where U is the universe of finite and nonempty set of objects, $\mathcal{A}$ is the nonempty set of conditional attributes, $\mathcal{A}_d$ is the nonempty set of decision attributes, where $\mathcal{A} \cap \mathcal{A}_d = \varnothing$, V is the set of domain of attributes in $\mathcal{A} \cup \mathcal{A}_d$. That is,*

$$V = \{V_a \mid a \in \mathcal{A} \cup \mathcal{A}_d,\ V_a \text{ is the domain of attribute } a\} \tag{11.21}$$

and $\mathcal{F} = \{f_a \mid a \in \mathcal{A} \cup \mathcal{A}_d,\ f_a : U \to V_a\}$. Note that $f_a(\cdot)$ and V_a are called the information function and value set (domain) of attribute $a \in \mathcal{A} \cup \mathcal{A}_d$ respectively. □

The decision system can be specified in a decision table. This is similar to the specification of an information system in a data table. The decision system is specified in a decision table, where the rows represent the objects, and the columns represent the conditional and decision attributes.

Approximation

Often times, in real world data there are conflicting classification of objects described in the decision table. For example, two objects in the universal set can have identical values of conditional attributes, but have different values of decision attributes. This implies imprecision and inconsistency in the decision table. This state of a decision system can be quantified by approximate representation of data points in terms of positive and negative regions of the decision system. The following concept is useful for describing lower approximation and positive region, and upper approximation and negative region of a decision system.

Definitions 11.38. *Let* $\vartheta_{DS} = (U, \mathcal{A} \cup \mathcal{A}_d, V, \digamma)$ *be a decision system.*

1. *Lower approximation and positive region*:
   The positive region is the set of elements $POS_{\mathcal{A}}(\mathcal{A}_d) = \cup_{X \in U/Ind(\mathcal{A}_d)} \underline{\mathcal{A}}X$.
2. *Upper approximation and negative region*:
   The negative region is the set of elements $NEG_{\mathcal{A}}(\mathcal{A}_d) = U - \cup_{X \in U/Ind(\mathcal{A}_d)} \overline{\mathcal{A}}X$. □

The positive region $POS_{\mathcal{A}}(\mathcal{A}_d)$ includes all objects in U that can be classified perfectly without error into different classes in the quotient set $U/Ind(\mathcal{A}_d)$. A measure of the positive region $POS_{\mathcal{A}}(\mathcal{A}_d)$ can be characterized. This value measures the degree of inclusion of the partition defined by the attributes in the set $\mathcal{A}$ in the partition defined by the decision attributes in $\mathcal{A}_d$.

Definition 11.39. *Let* $\vartheta_{DS} = (U, \mathcal{A} \cup \mathcal{A}_d, V, \digamma)$ *be a decision system. A measure of the positive region* $POS_{\mathcal{A}}(\mathcal{A}_d)$ *is* $|POS_{\mathcal{A}}(\mathcal{A}_d)| / |U|$. □

Degree of Dependency

Another aspect of the study of the theory of rough sets is the quantification of dependencies among attributes. A set of attributes $\mathcal{D}_d \subseteq \mathcal{A}_d$ depends *totally* upon a set of attributes $\mathcal{C} \subseteq \mathcal{A}$, if all values of attributes in the set $\mathcal{D}_d$ can be uniquely determined by the values of attributes in the set $\mathcal{C}$. This is denoted by $\mathcal{C} \Rightarrow \mathcal{D}_d$. A measure of this type of dependency can be characterized.

Definition 11.40. *Degree of dependency. Let* $\vartheta_{DS} = (U, \mathcal{A} \cup \mathcal{A}_d, V, \digamma)$ *be a decision system, and* $\mathcal{C} \subseteq \mathcal{A}$ *and* $\mathcal{D}_d \subseteq \mathcal{A}_d$. *The set* $\mathcal{D}_d$ *depends on the set* $\mathcal{C}$ *to a degree* $k \in [0,1]$, *denoted by* $\mathcal{C} \Rightarrow_k \mathcal{D}_d$, *if*

$$k = \gamma(\mathcal{C}, \mathcal{D}_d) = \frac{|POS_{\mathcal{C}}(\mathcal{D}_d)|}{|U|} \tag{11.22a}$$

where

$$POS_{\mathcal{C}}(\mathcal{D}_d) = \cup_{X \in U/Ind(\mathcal{D}_d)} \underline{\mathcal{C}}X \tag{11.22b}$$

□

The degree of dependency k measures how important $\mathcal{C}$ is in the mapping of data points into $\mathcal{D}_d$. If $k=0$, then the classification $\mathcal{D}_d$ is independent of the conditional attributes in $\mathcal{C}$. That is, the decision attributes are of no consequence in this classification. However, if $k=1$, then $\mathcal{D}_d$ is completely dependent upon the conditional attributes in $\mathcal{C}$. Further, $k\in(0,1)$ implies partial dependency. That is, some of the features in the set $\mathcal{C}$ might be useful in classification.

Significance of Features

A measure of significance of a feature (conditional attribute) $a\in\mathcal{C}\subseteq\mathcal{A}$ with respect to classification $U/Ind\left(\mathcal{D}_d\right)$ generated by $\mathcal{D}_d\subseteq\mathcal{A}_d$ can be evaluated by assigning a real number to it.

Definition 11.41. *Significance of features. Let $\vartheta_{DS}=(U,\mathcal{A}\cup\mathcal{A}_d,V,F)$ be a decision system, and $\mathcal{C}\subseteq\mathcal{A}$ and $\mathcal{D}_d\subseteq\mathcal{A}_d$. A measure of significance of any feature $a\in\mathcal{C}\subseteq\mathcal{A}$ with respect to classification $U/Ind\left(\mathcal{D}_d\right)$ generated by $\mathcal{D}_d$ is*

$$\mu_{\mathcal{C},\mathcal{D}_d}\left(a\right)=\frac{\left|POS_{\mathcal{C}}\left(\mathcal{D}_d\right)\right|-\left|POS_{\mathcal{C}-\{a\}}\left(\mathcal{D}_d\right)\right|}{|U|} \tag{11.23}$$

□

Reduct and Core

The concept of reducts and core in a decision system is similar to that of reducts and core of an information system. Reducts determine a subset of features which can completely characterize the knowledge in the decision system.

It is quite possible that some attributes of a decision system might be redundant (superfluous). In this case a *reduced* set of attributes will not lose their classification power. Let $\mathcal{B}\subseteq\mathcal{A}$ be a set of conditional attributes. An attribute $a\in\mathcal{B}$ is dispensable in the set $\mathcal{B}$ if

$$Ind\left(\mathcal{B}\right)=Ind\left(\mathcal{B}-\{a\}\right)$$

That is, the indiscernibility relations generated by the sets $\mathcal{B}$ and $\mathcal{B}-\{a\}$ are identical. Otherwise the attribute a is indispensable in $\mathcal{B}$. The set of all indispensable attributes in the set $\mathcal{B}\subseteq\mathcal{A}$ is called the core of $\mathcal{B}$ in the decision system ϑ_{DS}.

An attribute $a\in\mathcal{B}\subseteq\mathcal{A}$ is $\mathcal{A}_d$-superfluous in $\mathcal{B}$, if

$$POS_{\mathcal{B}}\left(\mathcal{A}_d\right)=POS_{\mathcal{B}-\{a\}}\left(\mathcal{A}_d\right)$$

otherwise the attribute a is $\mathcal{A}_d$-indispensable in $\mathcal{B}$. Let $\mathcal{C}\subseteq\mathcal{A}$. A set $\mathcal{B}\subseteq\mathcal{A}$ is a $\mathcal{A}_d$-reduct of $\mathcal{A}$, if $\mathcal{B}$ is a minimal subset of $\mathcal{C}$, such that $\gamma\left(\mathcal{C},\mathcal{A}_d\right)=\gamma\left(\mathcal{B},\mathcal{A}_d\right)$. The intersection of all $\mathcal{A}_d$-reducts is called a $\mathcal{A}_d$-core in the decision system ϑ_{DS}.

11.8.4 Decision Rules

One of the important tasks of supervised learning is the discovery of classification rules from the data tables. The purpose of decision rules is to discover hidden patterns in the data. This is in contrast to approximations, which simply express the topological properties of data. The decision rules can also be used to classify new data points. Decision rules represent knowledge in the decision table, and also crystallize dependencies in it. The purpose of reducts is to induce minimal decision rules.

Let $\vartheta_{DS} = (U, \mathcal{A} \cup \mathcal{A}_d, V, F)$ be a decision system, where U is the set of objects, $\mathcal{A}$ is the set of conditional attributes, $\mathcal{A}_d$ is the set of decision attributes, V is the set of domain of attributes in $\mathcal{A} \cup \mathcal{A}_d$, and $F = \{f_a \mid a \in \mathcal{A} \cup \mathcal{A}_d,\ f_a : U \to V_a\}$ is the set of information functions. Also let the set of conditional attributes be $\mathcal{A} = \{c_1, c_2, \ldots, c_m\}$, and the set of decision attributes be $\mathcal{A}_d = \{d_1, d_2, \ldots, d_l\}$. For $x \in U$, the sequence

$$f_{c_1}(x), f_{c_2}(x), \ldots, f_{c_m}(x), f_{d_1}(x), f_{d_2}(x), \ldots, f_{d_l}(x)$$

describes a *decision rule induced by* x in ϑ_{DS}. Some probabilistic measures of the decision rules based upon the decision table like *strength, certainty factor*, and *coverage factor* are next described.

Definitions 11.42. *Let $\vartheta_{DS} = (U, \mathcal{A} \cup \mathcal{A}_d, V, F)$ be a decision system, where U is the set of objects, $\mathcal{A}$ is the set of conditional attributes, $\mathcal{A}_d$ is the set of decision attributes, V is the set of domain of attributes in $\mathcal{A} \cup \mathcal{A}_d$, and F is the set of information functions. Let $\mathcal{A} = \{c_1, c_2, \ldots, c_m\}$, and $\mathcal{A}_d = \{d_1, d_2, \ldots, d_l\}$.*

The family of equivalence classes induced by the relationship $Ind(\mathcal{A})$ is $U/Ind(\mathcal{A})$, and for $x \in U$ the equivalence class $[x]_{\mathcal{A}}$ is an element of $U/Ind(\mathcal{A})$. Similarly, the family of equivalence classes induced by the relationship $Ind(\mathcal{A}_d)$ is $U/Ind(\mathcal{A}_d)$, and the equivalence class $[x]_{\mathcal{A}_d}$ is an element of $U/Ind(\mathcal{A}_d)$. A decision rule induced by the object $x \in U$ is

$$(f_{c_1}(x), f_{c_2}(x), \ldots, f_{c_m}(x)) \to (f_{d_1}(x), f_{d_2}(x), \ldots, f_{d_l}(x)) \tag{11.24a}$$

The decision rule is denoted by $\mathcal{A} \to_x \mathcal{A}_d \triangleq R_x$. The support of the decision rule R_x is $supp_x(\mathcal{A}, \mathcal{A}_d) = \left|[x]_{\mathcal{A}} \cap [x]_{\mathcal{A}_d}\right|$.

1. *The strength $\sigma_x(\mathcal{A}, \mathcal{A}_d)$ of the decision rule R_x is*

$$\sigma_x(\mathcal{A}, \mathcal{A}_d) = \frac{supp_x(\mathcal{A}, \mathcal{A}_d)}{|U|} \tag{11.24b}$$

2. *The certainty factor $cer_x(\mathcal{A}, \mathcal{A}_d)$ of the decision rule R_x is*

$$cer_x(\mathcal{A}, \mathcal{A}_d) = \frac{supp_x(\mathcal{A}, \mathcal{A}_d)}{\left|[x]_{\mathcal{A}}\right|} \tag{11.24c}$$

3. *The coverage factor $cov_x(\mathcal{A}, \mathcal{A}_d)$ of the decision rule R_x is*

$$cov_x(\mathcal{A}, \mathcal{A}_d) = \frac{supp_x(\mathcal{A}, \mathcal{A}_d)}{\left|[x]_{\mathcal{A}_d}\right|} \tag{11.24d}$$

□

If $cer_x(\mathcal{A}, \mathcal{A}_d) = 1$, then the decision rule R_x is referred to as a *certain decision* rule. However, if $cer_x(\mathcal{A}, \mathcal{A}_d) \in (0, 1)$ then the decision rule is called an *uncertain decision* rule. Note that, if $\mathcal{A} \to_x \mathcal{A}_d$ is a decision rule, then $\mathcal{A}_d \to_x \mathcal{A}$ is called an *inverse decision rule*. The inverse decision rules are used for giving *explanations* (*reasons*) for a decision. The above measures imply:

$$\sum_{y \in [x]_{\mathcal{A}}} cer_y(\mathcal{A}, \mathcal{A}_d) = 1$$

$$\sum_{y \in [x]_{\mathcal{A}_d}} cov_y(\mathcal{A}, \mathcal{A}_d) = 1$$

The design of a minimal decision algorithm associated with a given data table is nontrivial.

11.9 High-Dimensional Data

Special care is necessary in analysing high-dimensional data points, because their characteristics generally defy normal intuition. The behavior of high-dimensional data is typically nonintuitive because we have been conditioned to the fact that the number of available data points is much larger than the number of attributes (dimension) of a data point. We examine certain characteristics of high-dimensional data points in this section.

The so-called curse of high-dimensionality is next studied. The empty space property also occurs for high dimensional data. Its sensitivity to distance metric is also demonstrated. These properties are illustrated for isotropic Gaussian distribution.

Curse of Dimensionality

Suppose that we are given a set of $n \in \mathbb{P}$ data points, and each data point is a vector of size $t \in \mathbb{P}$. It is generally assumed while analysing data sets that both n and t are fixed. It is also assumed in several cases that $n > t$, and occasionally that $n \to \infty$. The analysis techniques used in such cases are generally not useful for finite number of data points n; and $t \gg n$, and sometimes $t > n$. This shortcoming of classical techniques of analysing data sets is called the "curse of dimensionality." We shall also see that high-dimensional data is also a "blessing."

The phrase "curse of dimensionality" was coined by the inventor of dynamic programming, Richard Bellman (1920-1984). He used this phrase while describing the computational complexity of dynamic programming algorithms which have an exponential dependence upon the dimension of the state of the underlying dynamical system.

Some of the areas in which the curse of dimensionality occurs are: optimization, function approximation, numerical integration, machine learning, image processing, statistical estimation, and of course data mining. Consider a Cartesian grid of spacing ϵ on a unit hypercube in dimension t, where $0 < \epsilon \ll 1$. A hypercube is simply a generalization of a cube to higher dimensions. The number of points in this hypercube is in $O\left(\epsilon^{-t}\right)$. This number can be extremely high. Therefore, in optimization, function approximation, and numerical integration problems, the number of function evaluations or searches required is in $O\left(\epsilon^{-t}\right)$. Some unexpected properties of high-dimensional data are outlined below.

Empty Space Property

The number of high-dimensional data points required to estimate the parameters of the associated model to a specified accuracy is usually very high. As the number of data points in such cases is typically very sparse, we have the so-called "empty space phenomenon." This is another manifestation of the curse of high dimensionality.

The empty space phenomenon, or property, can also be studied by considering the volumes of t-dimensional hypersphere and hypercube, where $t \in \mathbb{P}$. A hypersphere is a generalization of a sphere to higher-dimensional spaces. The hypersphere of radius r in a t-dimensional space is the set of points $\mathcal{H}_r$, where

$$\mathcal{H}_r = \left\{ (x_1, x_2, \ldots, x_t) \mid \sum_{i=1}^{t} x_i^2 \le r^2,\ x_i \in \mathbb{R},\ 1 \le i \le t,\ r \in \mathbb{R}^+ \right\}$$

Let the volume of the t-dimensional hypersphere of radius $r \in \mathbb{R}^+$ be $V_t(r)$. Also let the volume of the corresponding circumscripted hypercube be $C_t(r)$. That is, the length of the side of this hypercube is equal to $2r$. Thus

$$V_t(r) = \frac{\pi^{t/2}}{\Gamma(t/2+1)} r^t$$
$$C_t(r) = (2r)^t$$

where $\Gamma(\cdot)$ is the gamma function. Observe that

$$\lim_{t \to \infty} \frac{V_t(r)}{C_t(r)} \to 0$$

The above result implies that as the dimension t increases, the volume of the hypersphere becomes insignificant when compared to the volume of the corresponding circumscripting hypercube. Therefore if $r = 1/2$, the volume of the hypercube is unity; and the volume of the hypersphere of radius $r = 1/2$ tends to 0, as t tends to infinity.

We next demonstrate that most of the volume of the hypersphere is near its surface in a thin shell (crust). Consider a shell of thickness ϵr, where $0 < \epsilon \ll 1$. The ratio of the volume of the shell and the hypersphere is

$$\frac{V_t(r) - V_t(r(1-\epsilon))}{V_t(r)} = \frac{\left\{1^t - (1-\epsilon)^t\right\}}{1^t}$$

For fixed value of ϵ, and as $t \to \infty$, the above ratio tends to unity. Thus the shell of the hypersphere contains most of the volume of the hypersphere for large values of t.

Sensitivity to the Distance Metric

For a fixed set of data points, let $d_{\max}$ and $d_{\min}$ be the maximum and minimum distances respectively between the given set of high-dimensional data points. Then

$$\lim_{t \to \infty} \frac{d_{\max} - d_{\min}}{d_{\min}} \to 0$$

This result implies that in a set of high-dimensional data points, distance function loses its sensitivity, as the difference between the maximum and minimum distances becomes relatively negligible. For example, this property makes nearest-neighbor data mining techniques difficult to address in high-dimensional spaces.

Isotropic Gaussian Distribution

Consider a set of t-dimensional data points. In these data points, values of each attribute have a Gaussian distribution. Let X_i be the Gaussian random variable associated with the ith attribute, where $1 \le i \le t$. Further $X_i \sim \mathcal{N}(0, \sigma^2)$ for $1 \le i \le t$; and these t random variables are mutually independent of each other. Let $x = (x_1, x_2, \ldots, x_t)$ be a data point, where $x_i \in \mathbb{R}$, $1 \le i \le t$. Also let the corresponding multivariate isotropic probability density function be $p(x)$. Then

$$p(x) = \frac{1}{(2\pi)^{t/2} \sigma^t} \exp\left\{ -\frac{1}{2\sigma^2} \sum_{i=1}^{t} x_i^2 \right\}, \quad x_i \in \mathbb{R},\ 1 \le i \le t$$

This density function is called the t-dimensional spherical normal distribution. Let

$$Z = \sum_{i=1}^{t} X_i^2, \text{ and } R = \sqrt{Z}$$

The expected value of the sum of squared attribute values is

$$\mathcal{E}(Z) = t\sigma^2$$

For large values of t, the sum of squared values of x_i's is concentrated about its mean. However, observe that the density function has a maximum value at the origin. Also the probability density function of the random variable R is

$$f_R(r) = \begin{cases} 0, & r \in (-\infty, 0] \\ \dfrac{2r^{t-1}}{\Gamma\left(\frac{t}{2}\right)\left(2\sigma^2\right)^{t/2}} \exp\left\{-\frac{r^2}{2\sigma^2}\right\}, & r \in (0, \infty) \end{cases}$$

The above expression is derived in the problem section. This density function peaks at $r = \sigma\sqrt{t-1} \triangleq r_0$. For large values of t, that is for high-dimensional data, $f_R(r)$ is negligible for $r \in (0, r_0)$. Further, a relatively large part of the area under the function $f_R(\cdot)$ is concentrated in the interval $[r_0, r_0 + \alpha)$, where $\alpha \in O(\sigma)$. That is, the function $f_R(\cdot)$ is concentrated farther away from the origin.

Blessings

High-dimensional data points also come with their share of blessings. It is asserted that, as $t \to \infty$, the concentration of the measure phenomenon can be conveniently described by certain asymptotic methods.

11.10 Statistical Techniques

Statistical techniques like prediction via regression analysis and principal component analysis are discussed in this section. Using a given set of data points, prediction techniques predict a data point. Principal component analysis is a popular technique to remove redundant information in a set of data points. It results in dimensional-reduction.

11.10.1 Prediction via Regression Analysis

Prediction is a type of data mining technique which is based upon a given set of data points. Continuous (or ordered) values of a variable are predicted in this scheme. In this subsection, we briefly outline the use of regression technique to predict the value of a single attribute. Regression analysis can be used to determine the relationship between a set of *independent* (*predictor*) variables, and a *dependent* (*response*) variable. The response variable is assumed to be continuous-valued. A linear regression model is next outlined. Also note that it is sometimes possible to transform a nonlinear problem into a linear one.

We are given a set of points $\{(x_i, y_i) \mid x_i, y_i \in \mathbb{R}, 1 \le i \le n\}$, where $n \in \mathbb{P} \backslash \{1\}$. These set of data points are made to fit a straight line

$$y = a + bx, \quad a, b \in \mathbb{R}$$

where x and y are predictor and response variables respectively. The real-valued variables a and b are called regression coefficients. The above equation is called a *regression equation of y on x*. Method of least-square technique is used to find the values of a and b. Let

$$y_i = a + bx_i + e_i, \quad 1 \le i \le n$$

where e_i is called the error term. Define

$$\widehat{y}_i = a + bx_i, \quad 1 \le i \le n$$

The value $\widehat{y}_i$ is the estimated value of y_i. The goal of this technique is to minimize E by varying a and b.

$$E = \sum_{i=1}^{n} (y_i - \widehat{y}_i)^2 = \sum_{i=1}^{n} e_i^2$$

Define

$$\overline{x} = \frac{1}{n} \sum_{i=1}^{n} x_i, \text{ and } \overline{y} = \frac{1}{n} \sum_{i=1}^{n} y_i$$

Minimization of E yields

$$b = \frac{\sum_{i=1}^{n} (x_i - \overline{x})(y_i - \overline{y})}{\sum_{i=1}^{n} (x_i - \overline{x})^2}, \quad \text{where } \sum_{i=1}^{n} (x_i - \overline{x})^2 \neq 0$$
$$a = \overline{y} - b\overline{x}$$

The above expressions for a and b can be stated in a more aesthetic form as follows. Let the sample variance of the x-sequence and y-sequence be S_x^2 and S_y^2 respectively, and the sample covariance of the x-sequence and y-sequence be S_{xy}. That is,

$$S_x^2 = \frac{\sum_{i=1}^{n} (x_i - \overline{x})^2}{(n-1)}$$
$$S_y^2 = \frac{\sum_{i=1}^{n} (y_i - \overline{y})^2}{(n-1)}$$
$$S_{xy} = \frac{\sum_{i=1}^{n} (x_i - \overline{x})(y_i - \overline{y})}{(n-1)}$$

Then

$$b = \frac{S_{xy}}{S_x^2}$$

where $S_x \neq 0$. Define the sample correlation coefficient r by

$$r = \frac{S_{xy}}{S_x S_y}$$

where $S_x \neq 0$, and $S_y \neq 0$. It can be shown that $-1 \le r \le 1$. The regression line equation can be written as

$$\frac{(y - \overline{y})}{S_y} = r\frac{(x - \overline{x})}{S_x}$$

Note that the regression line passes through the point $(\overline{x}, \overline{y})$. The value r^2 is sometimes referred to as *coefficient of determination.* The net error in the regression line is

$$E = (n-1)\, S_y^2 \left(1 - r^2\right)$$

As $e_i = (y_i - \widehat{y}_i)$ for $1 \le i \le n$, we have $\sum_{i=1}^{n} e_i = 0$.

If the given data points do not fit a straight line, then the data points can perhaps be made to fit a polynomial equation. Another technique would be to initially transform the variables nonlinearly, and subsequently apply linear technique to the transformed variables.

11.10.2 Principal Component Analysis

Principal component analysis (PCA) is an unsupervised learning technique to extract "relevant" data from high-dimensional spaces. Given a set $\mathcal{X}$ of n data points each with $t \in \mathbb{P} \backslash \{1\}$ attributes, PCA finds a representation of the data points in a space of dimension $k \in \mathbb{P}$, where $k \le t$.

Let $\mathcal{X} = \{x_i \mid x_i \in \mathbb{R}^t,\ 1 \le i \le n\}$ be the set of $n \in \mathbb{P} \backslash \{1\}$ data points. The data point x_i is represented as a column vector of size t, where $1 \le i \le n$. Let the average value of the data points be $\overline{x} = n^{-1} \sum_{i=1}^{n} x_i$. Define a column vector $\theta_i = (x_i - \overline{x})$ for $1 \le i \le n$, and a $t \times n$ matrix Θ as

$$\Theta = \begin{bmatrix} \theta_1 \ \theta_2 \ \cdots \ \theta_n \end{bmatrix}$$

The covariance matrix $\widetilde{\Sigma}$ of the data points is

$$\widetilde{\Sigma} = \frac{1}{(n-1)} \Theta \Theta^T = \frac{1}{(n-1)} \sum_{i=1}^{n} \theta_i \theta_i^T$$

The $t \times t$ covariance matrix $\widetilde{\Sigma}$ is symmetric and positive semidefinite. Therefore $\widetilde{\Sigma} = \Psi \Lambda \Psi^T$, where Λ is a diagonal matrix with eigenvalues of the matrix $\widetilde{\Sigma}$ on its main diagonal. As the matrix $\widetilde{\Sigma}$ is symmetric and positive semidefinite, its eigenvalues are nonnegative. Let the eigenvalues of the matrix $\widetilde{\Sigma}$ be $\lambda_j \in \mathbb{R}_0^+, 1 \le j \le t$.

The columns of the matrix Ψ are the mutually orthogonal eigenvectors of the covariance matrix $\widetilde{\Sigma}$. Assume that these eigenvectors are orthonormal. Therefore $\Psi \Psi^T = I$, where I is an identity matrix of size t. Thus

$$\Lambda = \Psi^T \widetilde{\Sigma} \Psi = \frac{1}{(n-1)} \Psi^T \Theta \Theta^T \Psi = \frac{1}{(n-1)} W W^T$$

where $W = \Psi^T \Theta$ is a $t \times n$ matrix. Therefore $\Theta = \Psi W$. Let

$$\begin{aligned} W &= \begin{bmatrix} w_1 \ w_2 \ \cdots \ w_n \end{bmatrix}; \quad w_i \in \mathbb{R}^t,\ 1 \le i \le n \\ w_i &= \begin{bmatrix} w_{i1} \ w_{i2} \ \cdots \ w_{it} \end{bmatrix}^T; \quad w_{ij} \in \mathbb{R},\ 1 \le j \le t \\ \Psi &= \begin{bmatrix} \psi_1 \ \psi_2 \ \cdots \ \psi_t \end{bmatrix}; \quad \psi_j \in \mathbb{R}^t,\ 1 \le j \le t \end{aligned}$$

Thus

$$\theta_i = \Psi w_i = \sum_{j=1}^{t} \psi_j w_{ij}, \quad 1 \le i \le n$$

The vector θ_i is approximated in PCA as

$$\widehat{\theta}_i = \sum_{j=1}^{k} \psi_{l_j} w_{il_j}, \quad \text{for } 1 \le i \le n$$

where $k \le t$. Therefore, the mean-squared error E_{err} in the approximation of a data point is

$$E_{err} = \frac{1}{(n-1)} \sum_{i=1}^{n} \left\| \theta_i - \widehat{\theta}_i \right\|^2$$

where $\|\cdot\|$ is the Euclidean norm. It can be shown that the mean-squared error E_{err} is equal to $\sum_{j=k+1}^{t} \lambda_{l_j}$. As the eigenvalues are nonnegative, E_{err} is minimized by selecting the eigenvectors ψ_{l_j}, for $1 \le j \le k$, in the approximation $\widehat{\theta}_i$ which correspond to the largest k eigenvalues of the covariance matrix $\widetilde{\Sigma}$. These eigenvectors which correspond to the largest eigenvalues are called the *principal components* of the matrix Θ. Further, the PCA projects data along directions in which the data varies most. The magnitude of the eigenvalues quantify the variation of the data points along the directions of the eigenvectors.

The value k is typically selected so that $\sum_{j=1}^{k} \lambda_{l_j} / \sum_{j=1}^{t} \lambda_j$ is greater than a threshold τ_{thresh}, where $\tau_{thresh} \in (0, 1]$. Finally, note that PCA is closely related to the singular value decomposition of the matrix $\widetilde{\Sigma}$.

11.11 Social Network Analysis

A social network is a group of entities, and their associated relationships. The entities can be persons and/or organizations. A graph $G = (V, E)$ is a natural representation of a social network. In this graph, V is the set of vertices (nodes), and E is the set of edges (links). Entities are specified by nodes of the graph, and their relationships are represented by the edges connecting the nodes.

It is assumed in this section that the graph structure is specified via its adjacency matrix A. Graphs can be studied at the nodal (micro) level, and also at the network (macro) level. The relative importance of a node within a graph is specified by it centrality. It gives an indication of its social importance within the network. In a similar vein, the centrality of a group of nodes can also be specified. Similarity between nodes, and the linking behavior of nodes are also studied in this section. Detection (discovery) of communities within a social network is also of interest. A community is a group of entities with a common interest. The problem of community discovery is also addressed briefly.

11.11.1 Node Centrality

Centrality of a node or vertex, measures its importance within the graph. Different types of centrality of a single vertex are studied in this subsection. Some examples of centrality of a vertex are:

(a) Degree centrality
(b) Closeness centrality
(c) Betweenness centrality
(d) Eigenvector centrality

(e) Katz centrality
(f) PageRank centrality

Let the social network be specified by the graph $G = (V, E)$, where $|V| = n$. Also let the adjacency matrix of the graph G be $A = [a_{ij}]$. Depending upon the structure of the social network, the graph G can either be undirected or directed. If the graph G is undirected, the degree of vertex $i \in V$ is denoted by d_i. However, if the graph G is directed, the in-degree and out-degree of vertex $i \in V$ are denoted by d_i^- and d_i^+ respectively. In this case, define

$$d_i = (d_i^- + d_i^+)$$

Let

$$e = \begin{bmatrix} 1 \, 1 \cdots \, 1 \end{bmatrix}^T$$

be a column vector of size n, and I be an identity matrix of size n.

Degree Centrality

Assume that the graph G is undirected. The degree centrality of a vertex in this graph is a measure of number of adjacent nodes. That is, the degree centrality $C_{deg}(i)$ of vertex $i \in V$ is d_i. A normalized value of degree centrality is

$$C_{deg}^{norm}(i) = \frac{d_i}{K_{norm}}, \quad i \in V \tag{11.25a}$$

where K_{norm} is a normalizing factor. Possible values of K_{norm} are $(n-1)$, $\max_{i \in V} d_i$, and $\sum_{i \in V} d_i$. Note that $\sum_{i \in V} d_i = 2|E|$.

In a directed graph, each vertex has an in-degree, and an out-degree. The in-degree and out-degree unnormalized vertex centralities are

$$C_{deg}^-(i) = d_i^-, \quad \text{and} \quad C_{deg}^+(i) = d_i^+, \quad \text{where} \quad i \in V \tag{11.25b}$$

respectively. Also

$$C_{deg}(i) = (d_i^- + d_i^+) = d_i, \quad \text{where} \quad i \in V \tag{11.25c}$$

The in-degree centrality $C_{deg}^-(i)$ measures the popularity of vertex $i \in V$. The out-degree centrality $C_{deg}^+(i)$ measures the gregariousness of vertex $i \in V$.

Closeness Centrality

Assume that distances are specified between each pair of nodes in the graph. Let l_{ij} be the shortest path length between vertex i and vertex $j \in V \backslash \{i\}$. Let the average shortest path length from vertex i to all other vertices be

$$l_i = \frac{1}{(n-1)} \sum_{j \in V \backslash \{i\}} l_{ij}, \quad i \in V \tag{11.26a}$$

where $n \geq 2$. The closeness centrality of a vertex i is

$$C_{closeness}(i) = \frac{1}{l_i}, \quad i \in V \tag{11.26b}$$

Thus, smaller the average geodesic distance, higher is the centrality of the vertex.

Betweenness Centrality

Consider a graph G. This graph can either be directed or undirected. A distance metric is also defined between each pair of vertices, so that shortest paths between a pair of vertices are well-defined. Let the total number of shortest paths between vertices i and j be $B(i,j) > 0$, and $B(i,k,j)$ of these paths pass through the vertex $k \in V$, where the vertex k is also allowed to be either vertex i or j. Then the ratio $B(i,k,j)/B(i,j)$ is a measure of importance of vertex k in the flow of traffic between vertices i and j. Thus $\beta_v(k)$, the betweenness measure of vertex k is defined as

$$\beta_v(k) = \sum_{\substack{i,j\in V \\ i\neq j, B(i,j)>0}} \frac{B(i,k,j)}{B(i,j)}, \quad \forall\, k \in V \tag{11.27}$$

Thus higher the value of $\beta_v(k)$, higher is the importance of the node k in connecting nodes. It would be fair to conjecture that higher the value of the betweenness of a vertex, higher is its nodal degree. The concept of betweenness has also been studied in the chapter on stochastic networks.

Eigenvector Centrality

Observe that in the case of degree centrality, vertices with more adjacent vertices are important. However, it is important for a vertex to have *more important vertices* adjacent to it. For example, in the case of a directed graph incoming vertices should be important. This fact can be incorporated in the eigenvector centrality. Denote the eigenvector centrality of vertex $i \in V$ as $C_{eigenvec}(i)$, and let $B_i \neq \varnothing$ represent the set of vertices which point to vertex i. We want $C_{eigenvec}(i)$ to be proportional to the sum of all eigenvector centralities of vertices in the set B_i. Thus

$$C_{eigenvec}(i) = \frac{1}{\lambda} \sum_{j\in B_i} C_{eigenvec}(j)$$

where $\lambda \neq 0$ is some constant. The above equation can be written as

$$C_{eigenvec}(i) = \frac{1}{\lambda} \sum_{j=1}^{n} a_{ji} C_{eigenvec}(j)$$

Define

$$C_{eigenvec} = \left[C_{eigenvec}(1)\; C_{eigenvec}(2) \cdots C_{eigenvec}(n)\right]^T \tag{11.28a}$$

The above equation can be restated as

$$\lambda C_{eigenvec} = A^T C_{eigenvec} \tag{11.28b}$$

Thus $C_{eigenvec}$ is an eigenvector of the matrix A^T, and λ is the corresponding eigenvalue. Recall that for an undirected graph, $A^T = A$. In general, the matrix A^T can have many eigenvalues and corresponding eigenvectors. Further, we prefer all eigenvector centralities to be positive. This is possible, if the graph G is strongly connected. This is guaranteed by the Perron-Frobenius theory. As per this theory, if $A > 0$, then the largest eigenvalue of matrix A^T is unique, and positive. Furthermore, the corresponding eigenvector $C_{eigenvec} > 0$.

Katz Centrality

Determination of eigenvector centrality might sometimes be problematic for some networks, where the value of centrality index of a node might be zero. For example, in a digraph it is possible if this node has all outgoing edges, and no incoming edges. Katz centrality determines the influence of a node by considering not only the nodes which are in immediate neighborhood, but also all nodes which are reachable through these nodes in immediate neighborhood. However, connections made with a neighbor are penalized by an attenuation factor $\alpha > 0$. Also assume that the graph is strongly connected.

Denote the Katz centrality index of vertex $i \in V$ as $C_{Katz}(i)$, and let

$$C_{Katz} = \left[C_{Katz}(1)\; C_{Katz}(2) \cdots C_{Katz}(n) \right]^T \tag{11.29a}$$

Define the Katz centrality index $C_{Katz}(i)$ as

$$C_{Katz}(i) = \alpha \sum_{j=1}^{n} a_{ji} C_{Katz}(j) + \beta, \quad i \in V$$

where α and β are positive constants. The first term on the right side of the above equation is similar to the eigenvector centrality. The constant term β is added to avoid zero centrality values. Thus

$$C_{Katz} = \alpha A^T C_{Katz} + \beta e$$

That is,

$$C_{Katz} = \beta \left(I - \alpha A^T \right)^{-1} e \tag{11.29b}$$

The above result is true, provided the matrix $\left(I - \alpha A^T \right)$ is invertible. This occurs, if

$$\det \left(I - \alpha A^T \right) \neq 0$$

As $\alpha > 0$, the condition $\det \left(I - \alpha A^T \right) \neq 0$ is equivalent to $\det \left(A^T - \alpha^{-1} I \right) \neq 0$. This result is true, if $\alpha \lambda_{\max} < 1$, where $\lambda_{\max} > 0$ is the largest eigenvalue of the matrix A. This conclusion can also be obtained by considering the power series expansion of $\left(I - \alpha A^T \right)^{-1}$:

$$\left(I - \alpha A^T \right)^{-1} = I + \alpha A^T + \alpha^2 \left(A^T \right)^2 + \cdots + \alpha^k \left(A^T \right)^k + \cdots = \sum_{k \in \mathbb{N}} \alpha^k \left(A^T \right)^k$$

PageRank Centrality

PageRank centrality is a modification of Katz centrality. In the computation of Katz centrality of a node, if a node achieves high centrality, then it passes *all* of its value to *all* of its out-bound links. This property is not attractive, because all nodes known by important nodes may not be important. This effect can be minimized by dividing the value of the passed centrality of a node by its out-degree (provided it is not zero).

Denote the PageRank centrality index of vertex $i \in V$ as $C_{PR}(i)$, and let

$$C_{PR} = \left[C_{PR}(1)\; C_{PR}(2) \cdots C_{PR}(n) \right]^T \tag{11.30a}$$

Define the PageRank centrality index $C_{PR}(i)$ as

$$C_{PR}(i) = \alpha \sum_{j=1}^{n} a_{ji} \frac{C_{PR}(j)}{d_j^+} + \beta, \quad i \in V$$

where it is assumed that the out-degree d_j^+ of vertex $j \in V$ is positive; and α and β are positive constants. Let D_{out} be a diagonal matrix, with the ith element on the diagonal equal to $d_i^+ \in \mathbb{R}^+$, where $1 \le i \le n$. Thus

$$C_{PR} = \alpha A^T D_{out}^{-1} C_{PR} + \beta e$$

This implies

$$C_{PR} = \beta \left(I - \alpha A^T D_{out}^{-1}\right)^{-1} e \tag{11.30b}$$

The above result is true, provided the matrix $\left(I - \alpha A^T D_{out}^{-1}\right)$ is invertible. This occurs, if

$$\det\left(I - \alpha A^T D_{out}^{-1}\right) \neq 0$$

As $\alpha > 0$, the condition $\det\left(I - \alpha A^T D_{out}^{-1}\right) \neq 0$ is equivalent to $\det\left(A^T D_{out}^{-1} - \alpha^{-1} I\right) \neq 0$. This result is true, if $\alpha\lambda_{\max} < 1$, where $\lambda_{\max} > 0$ is the largest eigenvalue of the matrix $D_{out}^{-1}A$. This conclusion can also be obtained by considering the power series expansion of $\left(I - \alpha A^T D_{out}^{-1}\right)^{-1}$:

$$\begin{aligned}
&\left(I - \alpha A^T D_{out}^{-1}\right)^{-1} \\
&= I + \alpha\left(A^T D_{out}^{-1}\right) + \alpha^2\left(A^T D_{out}^{-1}\right)^2 + \cdots + \alpha^k\left(A^T D_{out}^{-1}\right)^k + \cdots \\
&= \sum_{k \in \mathbb{N}} \alpha^k \left(A^T D_{out}^{-1}\right)^k
\end{aligned}$$

This scheme has been successfully used in ranking Web pages. It is discussed in more detail in the chapter on stochastic networks. The condition, when the out-degree d_j^+ of a vertex j is equal to zero is also addressed in that chapter.

11.11.2 Node-Group Centrality

The centrality measures discussed in the last subsection pertain to a single node in the graph $G = (V, E)$. Centrality can also be defined for a group of nodes $V_S \subset V$. Some examples of node-group centrality of a set nodes V_S are its:

(a) Degree centrality
(b) Closeness centrality
(c) Betweenness centrality

Node-Group Degree Centrality

Group degree centrality of nodes in the set V_S is the number of nodes in the set $V \backslash V_S$ that are connected to nodes in the set V_S. Denote it by $C_{deg}(V_S)$. Thus

$$C_{deg}(V_S) = \left|\{i \in V \backslash V_S \mid i \text{ is connected to } j \in V_S\}\right| \tag{11.31a}$$

The normalized group degree centrality $C_{deg}^{norm}(V_S)$ of nodes in the set V_S is

$$C_{deg}^{norm}(V_S) = \frac{C_{deg}(V_S)}{|V \backslash V_S|}, \quad V_S \subset V \tag{11.31b}$$

Observe that $|V \backslash V_S| = (|V| - |V_S|)$.

Node-Group Closeness Centrality

Let l_{ij} be the shortest path length between vertices i and j, where it is assumed that distances are specified between each pair of nodes in the graph. We initially find the length of the shortest path $l_{V_S,j}$ over all possible vertices in the set V_S and a vertex $j \in V \backslash V_S$. That is,

$$l_{V_S,j} = \min_{i \in V_S} l_{ij}, \quad j \in V \backslash V_S \tag{11.32a}$$

Also let

$$l_{V_S} = \frac{1}{|V \backslash V_S|} \sum_{j \in V \backslash V_S} l_{V_S,j} \tag{11.32b}$$

Closeness centrality for the group V_S is

$$C_{closeness}(V_S) = \frac{1}{l_{V_S}}, \quad V_S \subset V \tag{11.32c}$$

Node-Group Betweenness Centrality

Consider a graph G. This graph can either be directed or undirected. A distance metric is also defined between each pair of vertices, so that shortest paths between a pair of vertices are well-defined. Let the total number of shortest paths between vertices i and j be $B(i,j) > 0$, and $B(i, V_S, j)$ of these paths pass through the vertices in V_S, where $i, j \in V \backslash V_S$. The group betweenness centrality is defined as

$$\beta_v(V_S) = \sum_{\substack{i,j \in V \backslash V_S \\ i \neq j, B(i,j) > 0}} \frac{B(i, V_S, j)}{B(i,j)}, \quad V_S \subset V \tag{11.33}$$

11.11.3 Node Similarity

Similarity between two nodes in a network is examined in this subsection. Similarity between two nodes can be determined by a knowledge of its graph representation. More specifically, this can be determined by a knowledge of the neighborhoods of the two nodes. Let the corresponding graph be $G = (V, E)$, where $|V| = n$. Its adjacency matrix is $A = [a_{ij}]$. Node similarity can be specified by their structural equivalence and regular equivalence.

Structural Equivalence

For simplicity assume that the graph G is undirected, where the degree of vertex $i \in V$ is d_i. In this case, a simple way to specify structural equivalence N_{ij} between two vertices $i, j \in V$ would be the number of common vertices between them. This is

$$N_{ij} = \sum_{k=1}^{n} a_{ik} a_{jk} \tag{11.34a}$$

Thus N_{ij} is simply the ijth element of the matrix A^2. A better choice of structural equivalence would be to normalize N_{ij} by the $\sqrt{d_i d_j}$. This is

$$N_{ij}^{norm} = \frac{N_{ij}}{\sqrt{d_i d_j}} \tag{11.34b}$$

where we assumed that $d_i, d_j > 0$. Thus, N_{ij}^{norm} lies in the range from 0 to 1. If d_i and/or d_j is equal to 0, then N_{ij}^{norm} is defined to be equal to 0. If $N_{ij}^{norm} = 1$, then vertices i and j have the same set of neighboring vertices.

Another possible way to specify structural similarity would be

$$r_{ij} = \frac{\sum_{k=1}^{n} (a_{ik} - \overline{a}_i)(a_{jk} - \overline{a}_j)}{\sqrt{\sum_{k=1}^{n} (a_{ik} - \overline{a}_i)^2} \sqrt{\sum_{k=1}^{n} (a_{jk} - \overline{a}_j)^2}}, \quad i, j \in V \tag{11.35}$$

where $\overline{a}_i = n^{-1} \sum_{k=1}^{n} a_{ik}$. Note that $r_{ij} \in [-1, 1]$.

Regular Equivalence

Regularly equivalent vertices are vertices whose neighbors are similar, but the neighbors do not necessarily share vertices. Thus regular equivalence of vertices determines similarity via the similarity of its neighbors, where the respective neighbors need not have common vertices.

Denote the similarity score of vertices i and j by $\sigma_{ij}^{regular}$. Let the vertices k and l be neighbors of vertices i and j respectively. Thus vertices i and j are regular, if the vertices k and l are regular. Thus

$$\sigma_{ij}^{regular} = \alpha \sum_{k=1}^{n} \sum_{l=1}^{n} a_{ik} a_{jl} \sigma_{kl}^{regular}$$

where $\alpha > 0$. However, this explanation (and representation) is self-referential. That is, determination of $\sigma_{ij}^{regular}$ requires determination of $\sigma_{kl}^{regular}$, which in turn requires determination of $\sigma_{ij}^{regular}$ and so on. This situation can be remedied as follows. Vertex i is similar to vertex j, if it has a neighbor k which itself is similar to vertex j. Thus

$$\sigma_{ij}^{regular} = \alpha \sum_{k=1}^{n} a_{ik} \sigma_{kj}^{regular}$$

If we let the matrix $\sigma^{regular} = \left[\sigma_{ij}^{regular}\right]$, the above equation in matrix notation is

$$\sigma^{regular} = \alpha A \sigma^{regular}$$

In order to make sure that a vertex is similar to itself, we add 1 to all diagonal elements on the right hand side of the above equation. Thus

$$\sigma^{regular} = \alpha A \sigma^{regular} + I$$

That is

$$\sigma^{regular} = (I - \alpha A)^{-1} \tag{11.36}$$

The above result is true, if the matrix $(I - \alpha A)$ is invertible. This occurs, if $\det(I - \alpha A) \neq 0$. As $\alpha > 0$, the condition $\det(I - \alpha A) \neq 0$ is equivalent to $\det(A - \alpha^{-1} I) \neq 0$. This result is true, if $\alpha\lambda_{\max} < 1$, where $\lambda_{\max} > 0$ is the largest eigenvalue of the matrix A.

11.11.4 Linking Structure

The characterization of linking between nodes is described via *transitivity* and *reciprocity*. These properties are specially important in social networks. The properties of transitivity and reciprocity of a graph consider closed loops of length three and two respectively.

Transitivity in Directed Graphs

In mathematics, a relationship R is transitive, if aRb and bRc, then aRc. Some examples of this relationship are: "greater than," "less than," and "implies." The graph G which represents the network, can either be an undirected or directed graph.

Let the adjacency matrix of a digraph G be $A = [a_{ij}]$. Consider three vertices $i, j, k \in V$ in the graph. Suppose that edges (i, j) and (j, k) are present in the graph. If the edge (i, k) is also present, then we have a transitive linking behavior. These three edges, and the corresponding vertices form a triangle. Preponderance of this behavior results in a highly connected (dense) graph.

Transitivity in Undirected Graphs

Transitivity in undirected graphs can be quantified in terms of the clustering coefficients of vertices. Assume that the graph G is simple, labeled, and undirected. Consider a vertex $i \in V$, with degree d_i. If the first neighbors of the node i were part of a clique, there would be $d_i(d_i - 1)/2 = \binom{d_i}{2}$ edges between them. The clustering coefficient C_i of vertex $i \in V$ is defined as the ratio of the number of edges y_i between its d_i neighboring vertices and $\binom{d_i}{2}$. Thus

$$C_i = \frac{y_i}{\binom{d_i}{2}}, \qquad i \in V \tag{11.37a}$$

The denominator is the maximum possible number of edges between neighbors of vertex i. Therefore $0 \leq C_i \leq 1$. Note that the d_i neighbors of the vertex i, induce a subgraph G_i of graph G. Then y_i is equal to the number of edges in the graph G_i. Let C_G be the clustering coefficient of the graph, then

$$C_G = \frac{1}{n} \sum_{i \in V} C_i \tag{11.37b}$$

where $|V| = n$. An alternative definition of the clustering coefficient of a graph is C'_G, where

$$C'_G = \frac{\sum_{i \in V} \binom{d_i}{2} C_i}{\sum_{i \in V} \binom{d_i}{2}} \tag{11.38a}$$

This alternative definition of clustering coefficient has the following interpretation. The clustering coefficient of a graph C'_G is the ratio of: number of pairs (i, j), and (i, k) of adjacent edges for which (j, k) is an edge; and the number of pairs (i, j), and (i, k) of adjacent edges. Thus

$$C'_G = \frac{\text{three times the number of triangles}}{\text{number of pairs of adjacent edges}}$$

In this interpretation, a triangle is made up of three interconnected nodes. The presence of the factor three in the numerator accounts for the fact that each triangle contributes to three connected triples of vertices. Denote the number of triangles in the graph by N_Δ, and the number of pairs of adjacent edges by N_α. Let $A^2 = \left[a_{ij}^{(2)}\right]$, then

$$N_\alpha = \sum_{i=1}^{n} \binom{d_i}{2} = \frac{1}{2}\sum_{i=1}^{n} d_i\left(d_i - 1\right) = \frac{1}{2}\sum_{\substack{1 \le i,j \le n \\ i \ne j}} a_{ij}^{(2)} \tag{11.38b}$$

$$N_\Delta = \frac{1}{6} tr\left(A^3\right) \tag{11.38c}$$

The clustering coefficient C'_G is

$$C'_G = \frac{3N_\Delta}{N_\alpha} \tag{11.38d}$$

Clustering coefficient of graphs has also been discussed in the chapter on stochastic networks.

Reciprocity

Reciprocity of a directed graph is another measure which characterizes its linking structure. It considers closed loops of length two. Vertices $i, j \in V$ demonstrate reciprocity property, if the edges (i, j) and (j, i) are present.

The reciprocity $R_{reciproc}$ of a directed graph is defined as the fraction of links that are reciprocated. Note that $a_{ij}a_{ji}$ is equal to unity, if and only if the edges (i, j) and (j, i) are present. Thus

$$R_{reciproc} = \frac{1}{|E|}\sum_{i=1}^{n}\sum_{j=1}^{n} a_{ij}a_{ji} = \frac{1}{|E|} tr\left(A^2\right) \tag{11.39}$$

11.11.5 Community Detection

Detection (or discovery) of communities which share common interest within a social network is definitely useful for the communities themselves, and also for commercial enterprises. The problem of community detection is equivalent to discovery of "clusters" of entities (specified as nodes in a graph) with a common theme. Assuming that the underlying graph of the social network is weighted, the cluster-discovery problem is equivalent to finding proper partitions of the graph.

Traditional distance-based clustering algorithms may not be suitable for social networks which exhibit small-world phenomena. In the later case, the small distances between nodes may not provide sufficient sensitivity to indication of similarity between nodes in the network.

Assume that the problem of community detection involves partitioning the corresponding graph into k sets of entities (nodes), so that the sum of the weights of edges between different communities (or partitions) is minimized. A 2-way partitioning algorithm can be repeatedly used to achieve a k-way partitioning of the graph. We assume in this subsection that the clusters are *balanced.* This means that the different clusters have similar number of nodes. A heuristic 2-way graph partitioning algorithm is discussed in this subsection. The heuristic 2-way graph partitioning algorithm that we discuss is due to B. W. Kernighan and Shen Lin. This algorithm is iterative, remarkable for its simplicity, and not guaranteed to converge to a global optimum.

Consider an undirected, and connected graph $G = (V, E)$, where V is the set of vertices, E is the set of edges, and $|V| = 2n$. An edge $(i, j) \in E$ has a cost $c_{ij} \in \mathbb{R}$. If $(i, j) \notin E$, then $c_{ij} = 0$. Also $c_{ii} = 0$ for each $i \in V$. As the graph G is undirected, $c_{ij} = c_{ji}$ for all $i, j \in V$.

The goal of the algorithm is to partition the set of vertices into two sets A_{KL} and B_{KL}, such that $B_{KL} = V - A_{KL}$, $|A_{KL}| = |B_{KL}| = n$, and minimize cost of the cut $\langle A_{KL}, B_{KL} \rangle$. A cut of a set of vertices is a separation of the vertex set V into two disjoint subsets. Note that $A_{KL} \cup B_{KL} = V$, and $A_{KL} \cap B_{KL} = \varnothing$. Thus, the set of vertices A_{KL} and B_{KL} form a cut of the set of vertices V. The cut is denoted by $\langle A_{KL}, B_{KL} \rangle$. The cut-cost is the sum of the cost of the edges having one vertex in the set A_{KL} and the other vertex in the set B_{KL}. The cut-cost to be minimized is

$$T = \sum_{(a,b) \in A_{KL} \times B_{KL}} c_{ab}$$

The concept of cut has also been introduced in the chapter on graph algorithms. The algorithm initially starts with an arbitrary equipartition of the vertex set V into disjoint sets A_{KL} and B_{KL}, and tries to decrease the cut-cost in each iteration by interchanges of subsets of A_{KL} and B_{KL}. Suppose that the subsets $X \subset A_{KL}$, and $Y \subset B_{KL}$ offer an improvement in the cut-cost, then the following assignment is made in the iteration

$$\begin{aligned} A_{KL} &\leftarrow (A_{KL} - X + Y) \\ B_{KL} &\leftarrow (B_{KL} - Y + X) \end{aligned}$$

where $|X| = |Y|$. In the above statement, "$\leftarrow$" is the assignment operator. The next goal is to identify subsets X and Y so that the cost of the cut $\langle A_{KL}, B_{KL} \rangle$ decreases in an iteration. The heuristic consists of identifying the subsets X and Y only approximately.

For each vertex $a \in A_{KL}$, *external cost* E_a and *internal cost* I_a are

$$\begin{aligned} E_a &= \sum_{y \in B_{KL}} c_{ay} \\ I_a &= \sum_{x \in A_{KL}} c_{ax} \end{aligned}$$

Similarly, for each vertex $b \in B_{KL}$ define E_b and I_b as

$$\begin{aligned} E_b &= \sum_{x \in A_{KL}} c_{bx}, \\ I_b &= \sum_{y \in B_{KL}} c_{by} \end{aligned}$$

Also let the difference between the external and internal costs D_z of a vertex $z \in V$ be

$$D_z = (E_z - I_z)$$

A positive value of the difference D_z implies that the vertex z should be moved to the opposite set, and a negative value means that the vertex should remain in the current set.

Lemma 11.1. *Let $a \in A_{KL}$, and $b \in B_{KL}$. If the vertices a and b are interchanged, then the reduction in cut-cost is $(D_a + D_b - 2c_{ab})$.*

Proof. Let w be the total cost of connections between the sets of vertices A_{KL} and B_{KL} with a and b absent. Also let the total cut-cost with old connections (preinterchange) between vertex sets A_{KL} and B_{KL} be be T_{old}. Then

$$\begin{aligned} T_{old} &= w + \sum_{y \in B_{KL}} c_{ay} + \sum_{x \in A_{KL}} c_{xb} - c_{ab} \\ &= w + \sum_{y \in B_{KL}} c_{ay} + \sum_{x \in A_{KL}} c_{bx} - c_{ab} \\ &= w + E_a + E_b - c_{ab} \end{aligned}$$

If the vertices a and b are swapped, then let the new cut-cost between vertex sets A_{KL} and B_{KL} be T_{new}, where

$$\begin{aligned} T_{new} &= w + \sum_{x \in A_{KL}} c_{xa} + \sum_{y \in B_{KL}} c_{by} + c_{ba} \\ &= w + \sum_{x \in A_{KL}} c_{ax} + \sum_{y \in B_{KL}} c_{by} + c_{ab} \\ &= w + I_a + I_b + c_{ab} \end{aligned}$$

The reduction in value of the objective function, or gain is

$$g_{ab} = (T_{old} - T_{new})$$

That is

$$g_{ab} = D_a + D_b - 2c_{ab}$$

□

The Kernighan-Lin algorithm is outlined below. Initially, the set of vertices V is arbitrarily partitioned into two disjoints sets V_1 and V_2, of equal size. That is, $|V_1| = |V_2| = n$. The algorithm is next executed in two loops. The outer one is the *while* loop, and the inner one is the *for* loop.

Let us first consider the *for* loop. This loop has n iterations. The iteration variable is i, where $1 \leq i \leq n$. In iteration i, the two-tuple (a_i, b_i) which gives the most favorable improvement in the cut-cost is selected. The corresponding value of the best improvement is $\Delta_i = (D_{a_i} + D_{b_i} - 2c_{a_i b_i})$. It is quite possible that the value Δ_i is negative. The two-tuple (a_i, b_i) is also stored separately in (x_i, y_i). The pair (x_i, y_i) is a candidate for swapping of vertices among the partitioned sets of vertices to gain improvement in the cut-cost. The new D's for vertices in the sets $A_{KL} - \{a_i\}$, and $B_{KL} - \{b_i\}$ are updated as

$$\begin{aligned} D'_x &= D_x + 2c_{xa_i} - 2c_{xb_i}, \quad x \in A_{KL} - \{a_i\} \\ D'_y &= D_y + 2c_{yb_i} - 2c_{ya_i}, \quad y \in B_{KL} - \{b_i\} \end{aligned}$$

The above expressions are justified as follow. Observe that the edge (x, a_i) is counted as internal in D_x. However, it is external in D'_x, so the cost c_{xa_i} must be added twice. Similarly, the edge (x, b_i) is counted as external in D_x. However, it is internal in D'_x, so the cost c_{xb_i} must be subtracted twice. The sets A_{KL} and B_{KL} are next replaced by $A_{KL} - \{a_i\}$, and $B_{KL} - \{b_i\}$ respectively in the iteration.

All of the above steps in iteration i are executed for values of i from 1 through n. Observe that in the above iterations, the net possible improvement in cost of the cut-set after the end of iteration k is $\sum_{i=1}^{k} \Delta_i$.

In the next step of the algorithm, the value of k that maximizes $\sum_{i=1}^{k} \Delta_i$ is determined. Denote this sum by G_{gain}. Only, if $G_{gain} > 0$, the vertices in the sets V_1 and V_2 are swapped. That is, the following steps are performed.

$$V_1 \leftarrow V_1 \cup \{y_1, y_2, \ldots, y_k\} - \{x_1, x_2, \ldots, x_k\}$$
$$V_2 \leftarrow V_2 \cup \{x_1, x_2, \ldots, x_k\} - \{y_1, y_2, \ldots, y_k\}$$

If G_{gain} is positive, the new partition V_1 and V_2 is considered as the initial partition, and all earlier steps are repeated. The algorithm terminates, if G_{gain} is nonpositive. Note that the resulting partition can be a locally optimum partition.

Algorithm 11.13. *Kernighan-Lin Algorithm.*

Input: Let $G = (V, E)$ be an undirected graph, where $|V| = 2n$.
Edge $(i, j) \in E$ has a cost $c_{ij} \in \mathbb{R}$. If $(i, j) \notin E$, then $c_{ij} = 0$.
Also $c_{ii} = 0$ for each $i \in V$, and $c_{ij} = c_{ji}$ for all $i, j \in V$.
Output: Balanced partition of vertex set V with a "small" cut-cost.
These vertex sets are V_1 and V_2, where $V = V_1 \cup V_2$, $V_1 \cap V_2 = \varnothing$,
and $|V_1| = |V_2| = n$.
begin
  (*initialization*)
  Arbitrarily partition the set V into two disjoint sets of vertices V_1
  and V_2, where $|V_1| = |V_2| = n$.
  $G_{gain} \leftarrow 1$
  while $G_{gain} > 0$ **do**
  begin
    $A_{KL} \leftarrow V_1$, and $B_{KL} \leftarrow V_2$
    Compute D_z, $\forall\, z \in V$
    for $i = 1$ **to** n **do**
    begin
      Compute $(a_i, b_i) = \arg\max_{(a,b) \in A_{KL} \times B_{KL}} (D_a + D_b - 2c_{ab})$
      $(x_i, y_i) \leftarrow (a_i, b_i)$
      $\Delta_i \leftarrow (D_{a_i} + D_{b_i} - 2c_{a_i b_i})$
      (*update* D_x, $\forall\, x \in A_{KL} - \{a_i\}$; *and* D_y, $\forall\, y \in B_{KL} - \{b_i\}$)
      $D_x \leftarrow D_x + 2c_{xa_i} - 2c_{xb_i}$, $\forall\, x \in A_{KL} - \{a_i\}$
      $D_y \leftarrow D_y + 2c_{yb_i} - 2c_{ya_i}$, $\forall\, y \in B_{KL} - \{b_i\}$
      (a_i *and* b_i *do not enter into calculations any more in this loop*)
      $A_{KL} \leftarrow A_{KL} - \{a_i\}$, and $B_{KL} \leftarrow B_{KL} - \{b_i\}$
    end (*end of i for-loop*)
    Find a k which maximizes $G_{gain} \leftarrow \sum_{i=1}^{k} \Delta_i$
    if $G_{gain} > 0$ **then**
    begin
      $V_1 \leftarrow V_1 \cup \{y_1, y_2, \ldots, y_k\} - \{x_1, x_2, \ldots, x_k\}$
      $V_2 \leftarrow V_2 \cup \{x_1, x_2, \ldots, x_k\} - \{y_1, y_2, \ldots, y_k\}$
    end (*end of if statement*)
  end (*end of while loop*)
end (*end of Kernighan-Lin algorithm*)

11.12 Hidden Markov Model

Hidden Markov model (HMM) is a powerful mathematical technique to study sequential data. The HMM of a system consists of a Markov process, in which its states are hidden (or latent). That is, these states are not directly observable. However, the system is visible indirectly via another set of states. These later states are called the observable states. The observable states of the system are dependent upon the latent states of the underlying Markov process. This perspective is very useful in modeling several instances of sequential data, where a data point is indirectly dependent upon its immediately preceding data point.

The applications of HMM are many and varied. Some of these are: cryptanalysis, speech recognition, time series analysis, activity recognition, bioinformatics, metamorphic virus detection, and so on.

The HMM is initially described in this section. The problems it attempts to solve are also specified. The HMM problems are: probability evaluation, determination of state sequence, and the model determination. The solutions of these problems are also presented.

11.12.1 Model Description and the Problems

In an HMM of a system, there are two types of states. These are the hidden states, and the observable states. The hidden states are described by a discrete-time Markov chain (DTMC), and the observable states describe the physical output of the system. There is no one-to-one mapping between a hidden state and an observable state. However, there is a probabilistic dependence of the observed states upon the hidden states of the DTMC. In this subsection, the hidden Markov chain, the observable process, and the fundamental problems in the study of HMM are described.

Description of the Hidden Markov Chain

Let $T \in \mathbb{P}$ be the length of time over which the physical system is observed. The stochastic process

$$\mathfrak{X}_{MC} \triangleq \{X_t \mid t = 1, 2 \ldots, T\}$$

is a DTMC, if each random variable X_t takes values in the set of states $S = \{s_1, s_2, \ldots, s_N\}$, and these random variables satisfy the following Markovian property. That is

$$\begin{aligned} & P\left(X_{t+1} = x_{t+1} \mid X_t = x_t, X_{t-1} = x_{t-1}, \ldots, X_1 = x_1\right) \\ & = P\left(X_{t+1} = x_{t+1} \mid X_t = x_t\right), \quad 1 \leq t \leq (T-1);\; x_t \in S, 1 \leq t \leq T \end{aligned} \tag{11.40}$$

The conditional probability $P\left(X_{t+1} = x_{t+1} \mid X_t = x_t\right), 1 \leq t \leq (T-1)$, is called the transition probability of the DTMC. Thus in a DTMC, we have a random sequence $X = (X_1, X_2, \ldots, X_T)$ in which dependency is by only a single unit backward in time. A specific hidden-state sequence in HMM is $x = (x_1, x_2, \ldots, x_T)$, where $x_t \in S, 1 \leq t \leq T$.

The transition probabilities of a DTMC essentially define the structure of a DTMC. If the transition probabilities are independent of t, then denote

$$P\left(X_{t+1}=s_j \mid X_t=s_i\right)=a_{s_i s_j} \triangleq a_{ij}, \quad \forall\, s_i, s_j \in S, \text{ and } 1 \leq t \leq (T-1) \tag{11.41a}$$

This Markov chain is said to have homogeneous transition probabilities. Alternately the Markov chain is said to be a DTMC homogeneous in time. Also

$$a_{ij} \geq 0, \quad 1 \leq i, j \leq N \tag{11.41b}$$

$$\sum_{j=1}^{N} a_{ij}=1, \quad 1 \leq i \leq N \tag{11.41c}$$

The following observation is a consequence of the Markovian property.

Observation 11.7. Let $\mathfrak{X}_{MC} \triangleq \{X_t \mid t=1,2 \ldots, T\}$ be a DTMC. For $t \geq 2$, successive use of expression for conditional probability yields

$$\begin{aligned} & P\left(X_1=x_1, X_2=x_2, \ldots, X_t=x_t\right) \\ & =P\left(X_1=x_1\right) \prod_{i=2}^{t} P\left(X_i=x_i \mid X_{i-1}=x_{i-1}, X_{i-2}=x_{i-2}, \ldots, X_1=x_1\right) \end{aligned}$$

Use of the Markovian property results in

$$\begin{aligned} & P\left(X_1=x_1, X_2=x_2, \ldots, X_t=x_t\right) \\ & =P\left(X_1=x_1\right) \prod_{i=2}^{t} P\left(X_i=x_i \mid X_{i-1}=x_{i-1}\right) \end{aligned} \tag{11.42}$$

□

The initial state description of the DTMC (at time instant $t=1$) is specified by

$$\Pi=\left\{\pi_i \mid \pi_i=P\left(X_1=s_i\right), s_i \in S, 1 \leq i \leq N\right\}$$

where

$$\pi_i \geq 0, \quad 1 \leq i \leq N \tag{11.43a}$$

$$\sum_{i=1}^{N} \pi_i=1 \tag{11.43b}$$

We also denote $\pi_i \triangleq \pi_{s_i}, s_i \in S, 1 \leq i \leq N$. A description of the observable process is next provided.

Description of the Observable Process

Each hidden state of the DTMC emits a symbol with a certain probability. The set of these observation symbols is

$$V=\{v_1, v_2, \ldots, v_M\}$$

Thus there are $|V|=M$ number of observable symbols. At each time instant t, the physical system in hidden state x_t emits a single symbol $o_t \in V$, where $1 \leq t \leq T$. It is assumed that T is the total number of emitted symbols by the physical system. The random variable corresponding to the observation symbol emitted at time t is denoted by O_t, where $1 \leq t \leq T$.

Let the random observation sequence be $O = (O_1, O_2, \ldots, O_T)$. The corresponding random state sequence of the DTMC is $X = (X_1, X_2, \ldots, X_T)$. A specific observation sequence is denoted by $o = (o_1, o_2, \ldots, o_T)$, where $o_t \in V$, $1 \leq t \leq T$. This is the output sequence, when the corresponding hidden-state sequence is $x = (x_1, x_2, \ldots, x_T)$, where $x_t \in S$, $1 \leq t \leq T$.

The probability that a symbol $O_t = v_k$ at time $t \in T$, depends only upon the current hidden state of the DTMC $X_t = s_j$ is

$$b_{jk} \triangleq b_j(k) = P(O_t = v_k \mid X_t = s_j), \quad \text{where } s_j \in S, v_k \in V$$

These are also called emission probabilities. If $x_t = s_j$, and $o_t = v_k$ the probability $b_j(k)$ is also denoted by $b_{x_t}(o_t)$ and $b_j(o_t)$. Also

$$b_j(k) \geq 0, \quad 1 \leq k \leq M, \quad 1 \leq j \leq N \tag{11.44a}$$

$$\sum_{k=1}^{M} b_j(k) = 1, \quad 1 \leq j \leq N \tag{11.44b}$$

Note that $B = [b_{jk}]$ is an $N \times M$ matrix. It is also called the emission matrix. The HMM description is summarized in the following definition.

Definition 11.43. *A Hidden Markov model is an eight-tuple*

$$\Lambda = (\mathfrak{X}_{MC}, S, X, A, \Pi; V, B, O) \tag{11.45a}$$

where:

(a) $\mathfrak{X}_{MC} \triangleq \{X_t \mid t = 1, 2 \ldots, T\}$ *is the hidden discrete-time Markov chain (DTMC), where* $T \in \mathbb{P} \backslash \{1\}$ *is the length of time over which the physical system is observed.*
(b) $S = \{s_1, s_2, \ldots, s_N\}$ *is the set of states of the DTMC.*
(c) *The DTMC random state sequence* $X = (X_1, X_2, \ldots, X_T)$. *A specific state sequence is* $x = (x_1, x_2, \ldots, x_T)$, *where* $x_t \in S$, $1 \leq t \leq T$.
(d) *The DTMC has homogeneous transition probabilities. The Markov chain transition matrix* $A = [a_{ij}]$, *where* $P(X_{t+1} = s_j \mid X_t = s_i) = a_{ij}$, $\forall\, s_i, s_j \in S$, *and* $1 \leq t \leq (T-1)$.
Also, $a_{ij} \triangleq a_{s_i s_j}$, $\forall\, s_i, s_j \in S$.
(e) $\Pi = \{\pi_i \mid \pi_i = P(X_1 = s_i), s_i \in S, 1 \leq i \leq N\}$ *is the initial state probability distribution (at time instant* $t = 1$*) of the DTMC.*
Also denote $\pi_i \triangleq \pi_{s_i}, s_i \in S, 1 \leq i \leq N$.
(f) $V = \{v_1, v_2, \ldots, v_M\}$ *is the set of output alphabet.*
(g) *The probability that a symbol* $O_t = v_k$ *emitted at time* $t \in T$, *depends upon the current hidden state of the DTMC* $X_t = s_j$, *where* $s_j \in S$. *The observation symbol probability distribution matrix (or emission matrix)* $B = [b_{jk}]$, *where*

$$b_{jk} \triangleq b_j(k) = P(O_t = v_k \mid X_t = s_j), \quad s_j \in S, \ v_k \in V \tag{11.45b}$$

Also if $x_t = s_j$, *and* $o_t = v_k$, *then* $b_j(k) \triangleq b_{x_t}(o_t) \triangleq b_j(o_t)$.
(h) *The random observation sequence is* $O = (O_1, O_2, \ldots, O_T)$. *The corresponding hidden random state sequence is* X. *A specific output sequence is* $o = (o_1, o_2, \ldots, o_T)$, *where* $o_t \in V$, $1 \leq t \leq T$. *The corresponding hidden-state sequence is* x. □

A description of the problems of the HMMs is next provided.

Problems of the HMMs

There are three important problems in the study of a HMM. Consider an HMM Λ.

Problem 1: The probability of a specific sequence $o = (o_1, o_2, \ldots, o_T)$ has to be computed. That is, $P(O = o)$ has to be determined. Thus this is a probability evaluation problem.

Problem 2: Given an observation sequence $o = (o_1, o_2, \ldots, o_T)$, the corresponding state sequence $x = (x_1, x_2, \ldots, x_T)$ has to be determined, which is optimal in some predetermined sense. Thus this is a problem of finding an appropriate state sequence (decoding problem).

Problem 3: Given an observation sequence $o = (o_1, o_2, \ldots, o_T)$, determine the parameters of the HMM Λ, that maximizes $P(O = o)$. Thus this is a learning problem.

11.12.2 Probability Evaluation

The probability of the output sequence $o = (o_1, o_2, \ldots, o_T)$ has to be computed efficiently. This probability is $P(O = o)$. We initially compute

$$P(O = o \mid X = x)$$

where $x = (x_1, x_2, \ldots, x_T)$.

$$P(O = o \mid X = x) = \prod_{t=1}^{T} b_{x_t}(o_t) = b_{x_1}(o_1)\, b_{x_2}(o_2) \cdots b_{x_T}(o_T)$$

Also the probability of the state sequence $P(X = x)$ is

$$P(X = x) = \pi_{x_1} \prod_{t=2}^{T} P(x_t \mid x_{t-1}) = \pi_{x_1} a_{x_1 x_2} a_{x_2 x_3} \cdots a_{x_{T-1} x_T}$$

Thus

$$\begin{aligned}
&P(O = o, X = x) \\
&= P(O = o \mid X = x)\, P(X = x) \\
&= \pi_{x_1} \prod_{t=2}^{T} P(x_t \mid x_{t-1}) \prod_{t=1}^{T} b_{x_t}(o_t) \\
&= \pi_{x_1} b_{x_1}(o_1)\, a_{x_1 x_2} b_{x_2}(o_2)\, a_{x_2 x_3} \cdots b_{x_{T-1}}(o_{T-1})\, a_{x_{T-1} x_T} b_{x_T}(o_T)
\end{aligned}$$

Finally, we obtain

$$P(O = o) = \sum_{x \in S^T} P(O = o, X = x)$$

A direct computation of $P(O = o)$ requires about $2TN^T$ multiplications. We next show how to evaluate $P(O = o)$ more efficiently. Write $P(O = o)$ as

$$
\begin{aligned}
& P(O=o) \\
&= \sum_{x_T \in S} \sum_{x_{T-1} \in S} \cdots \sum_{x_1 \in S} \pi_{x_1} b_{x_1}(o_1) a_{x_1 x_2} b_{x_2}(o_2) a_{x_2 x_3} \cdots a_{x_{T-1} x_T} b_{x_T}(o_T) \\
&= \sum_{x_T \in S} b_{x_T}(o_T) \sum_{x_{T-1} \in S} a_{x_{T-1} x_T} b_{x_{T-1}}(o_{T-1}) \\
&\quad \cdots \sum_{x_2 \in S} a_{x_2 x_3} b_{x_2}(o_2) \sum_{x_1 \in S} a_{x_1 x_2} b_{x_1}(o_1) \pi_{x_1}
\end{aligned}
$$

The above expression for $P(O=o)$ can be evaluated as follows. Let

$$
\begin{aligned}
\alpha_1(x_1) &= b_{x_1}(o_1) \pi_{x_1}, \quad \forall\, x_1 \in S \\
\alpha_2(x_2) &= b_{x_2}(o_2) \sum_{x_1 \in S} a_{x_1 x_2} \alpha_1(x_1), \quad \forall\, x_2 \in S \\
\alpha_3(x_3) &= b_{x_3}(o_3) \sum_{x_2 \in S} a_{x_2 x_3} \alpha_2(x_2), \quad \forall\, x_3 \in S \\
&\vdots \\
\alpha_T(x_T) &= b_{x_T}(o_T) \sum_{x_{T-1} \in S} a_{x_{T-1} x_T} \alpha_{T-1}(x_{T-1}), \quad \forall\, x_T \in S \\
P(O=o) &= \sum_{x_T \in S} \alpha_T(x_T)
\end{aligned}
$$

where for a given $X_t = x_t$

$$
\alpha_t(x_t) = P(O_1 = o_1, O_2 = o_2, \ldots, O_t = o_t, X_t = x_t), \quad 1 \le t \le T \tag{11.46a}
$$

If $X_t = x_t = s_i$, we let

$$
\alpha_t(x_t) \triangleq \alpha_t(i), \quad \forall\, x_t, s_i \in S,\ 1 \le t \le T \tag{11.46b}
$$

Therefore $\alpha_t(i)$ is the joint probability of the partial observation sequence $(o_1, o_2, \ldots, o_t)$, and the hidden state s_i at time t. The above results are next summarized in the following algorithm. It is actually called the *forward algorithm*, or *α-pass*. The forward algorithm computes $P(O=o)$, for a given output sequence o.

Step 0: Initialization:

$$
\alpha_1(i) = \pi_i b_i(o_1), \quad 1 \le i \le N \tag{11.47a}
$$

Step 1: Recursion:

$$
\alpha_{t+1}(j) = \left\{ \sum_{i=1}^{N} \alpha_t(i) a_{ij} \right\} b_j(o_{t+1}), \quad 1 \le t \le (T-1), \ 1 \le j \le N \tag{11.47b}
$$

Step 2: Termination:

$$
P(O=o) = \sum_{i=1}^{N} \alpha_T(i) \tag{11.47c}
$$

□

The above recursive technique requires about N^2T multiplications in the computation of $P(O = o)$. The probability $P(O = o)$ can also be computed in an alternate way. Write $P(O = o)$ as

$$\begin{aligned}
&P(O = o)\\
&= \sum_{x_T \in S} \sum_{x_{T-1} \in S} \cdots \sum_{x_1 \in S} \pi_{x_1} b_{x_1}(o_1) a_{x_1 x_2} b_{x_2}(o_2) a_{x_2 x_3} \cdots a_{x_{T-1} x_T} b_{x_T}(o_T)\\
&= \sum_{x_1 \in S} \pi_{x_1} b_{x_1}(o_1) \sum_{x_2 \in S} a_{x_1 x_2} b_{x_2}(o_2)\\
&\quad \cdots \sum_{x_{T-1} \in S} a_{x_{T-2} x_{T-1}} b_{x_{T-1}}(o_{T-1}) \sum_{x_T \in S} a_{x_{T-1} x_T} b_{x_T}(o_T)
\end{aligned}$$

The above expression for $P(O = o)$ can be evaluated as follows. Let

$$\begin{aligned}
\beta_T(x_T) &= 1, \quad \forall\, x_T \in S\\
\beta_{T-1}(x_{T-1}) &= \sum_{x_T \in S} a_{x_{T-1} x_T} b_{x_T}(o_T) \beta_T(x_T), \quad \forall\, x_{T-1} \in S\\
\beta_{T-2}(x_{T-2}) &= \sum_{x_{T-1} \in S} a_{x_{T-2} x_{T-1}} b_{x_{T-1}}(o_{T-1}) \beta_{T-1}(x_{T-1}), \quad \forall\, x_{T-2} \in S\\
&\vdots\\
\beta_1(x_1) &= \sum_{x_2 \in S} a_{x_1 x_2} b_{x_2}(o_2) \beta_2(x_2), \quad \forall\, x_1 \in S\\
P(O = o) &= \sum_{x_1 \in S} \pi_{x_1} b_{x_1}(o_1) \beta_1(x_1)
\end{aligned}$$

where for a given $X_t = x_t$

$$\beta_t(x_t) = P(O_{t+1} = o_{t+1}, O_{t+2} = o_{t+2}, \ldots, O_T = o_T \mid X_t = x_t) \tag{11.48a}$$

and $1 \leq t \leq (T-1)$. If $X_t = x_t = s_i$, we let

$$\beta_t(x_t) \triangleq \beta_t(i), \quad \forall\, x_t, s_i \in S,\ 1 \leq t \leq T \tag{11.48b}$$

Therefore $\beta_t(i)$ is the joint probability of the partial observation sequence $(o_{t+1}, o_{t+2}, \ldots, o_T)$ given that the system started in hidden state s_i at time t, where $1 \leq t \leq (T-1)$. The above results are next summarized in the following algorithm. It is actually called the *backward algorithm*, or *β-pass*. The backward algorithm computes $P(O = o)$, for a given output sequence o.

Step 0: Initialization:

$$\beta_T(i) = 1, \quad 1 \leq i \leq N \tag{11.49a}$$

Step 1: Recursion:

$$\beta_t(i) = \sum_{j=1}^{N} a_{ij} b_j(o_{t+1}) \beta_{t+1}(j) \tag{11.49b}$$

where $t = (T-1), (T-2), \ldots, 1$; and $1 \leq i \leq N$.

Step 2: Termination:

$$P(O = o) = \sum_{i=1}^{N} \pi_i b_i(o_1) \beta_1(i) \tag{11.49c}$$

□

The above recursive technique requires about $2N^2T$ multiplications in the computation of $P(O = o)$.

Observation 11.8. The probability

$$P(O = o) = \sum_{i=1}^{N} \alpha_t(i) \beta_t(i), \quad \text{for each } t, \text{ where } 1 \le t \le T \tag{11.50}$$

□

11.12.3 Decoding of State Sequence

For a given observation sequence $O = o$, the corresponding hidden-state sequence $X = x$, has to be determined which is optimal in some predetermined sense. That is, this problem requires the determination of the hidden-state sequence that will most likely produce the specified observation sequence. Note that

$$o = (o_1, o_2, \ldots, o_T), \text{ and } x = (x_1, x_2, \ldots, x_T)$$

A popular criterion is to find a hidden-state sequence (path) which maximizes the probability $P(X = x \mid O = o)$. This is equivalent to maximizing $P(X = x, O = o)$. A technique to finding such sequence uses the methods of dynamic programming. The actual technique is called the Viterbi algorithm, named after its inventor, Andrew J. Viterbi. It works on the principle: *Optimal sub-policies constitute an optimal policy.* Define $\delta_t(i)$ as the probability of the highest probability-sequence (path) at time t, which accounts for the first t output symbols, and which ends in state $s_i \in S$. Thus

$$\delta_t(i) = \max_{x_1, x_2, \ldots, x_{t-1}} P(x_1, x_2, \ldots, x_{t-1}, x_t = s_i; o_1, o_2, \ldots, o_t)$$

Use of induction yields

$$\delta_{t+1}(j) = \left\{ \max_i \delta_t(i) a_{ij} \right\} b_j(o_{t+1})$$

The state sequence can be kept track of, via an array $\psi_t(j)$. That is, $\psi_t(j)$ keeps track of the state that maximizes $\delta_{t-1}(i) a_{ij}$ over i, which is the optimal previous state. Thus $\delta_{t+1}(j)$ can be computed recursively, and the optimal path can be determined via backtracking from time T. The Viterbi algorithm is as follows.

Step 0: Initialization:

$$\delta_1(i) = \pi_i b_i(o_1), \quad 1 \le i \le N \tag{11.51a}$$

$$\psi_1(i) = 0, \quad 1 \le i \le N \tag{11.51b}$$

Step 1: Recursion:

$$\delta_t(j) = \max_{1 \leq i \leq N} \{\delta_{t-1}(i) a_{ij}\} b_j(o_t), \; 2 \leq t \leq T, \, 1 \leq j \leq N \tag{11.51c}$$

$$\psi_t(j) = \arg \max_{1 \leq i \leq N} \{\delta_{t-1}(i) a_{ij}\}, \; 2 \leq t \leq T, \, 1 \leq j \leq N \tag{11.51d}$$

Step 2: Termination:

$$P^* = \max_{1 \leq i \leq N} \delta_T(i) \tag{11.51e}$$

$$x_T^* = \arg \max_{1 \leq i \leq N} \delta_T(i) \tag{11.51f}$$

Step 3: Optimal state sequence (path) backtracking:

$$x_t^* = \psi_{t+1}\left(x_{t+1}^*\right), \quad t = (T-1), (T-2), \ldots, 1 \tag{11.51g}$$

□

Note that the structure of this algorithm is similar to that of the forward algorithm, except for the backtracking step. Also, the recursive step in the forward algorithm uses summation, while the recursive step in the Viterbi algorithm uses maximization.

11.12.4 Model Evaluation

In this subsection, we address the problem of evaluating the HMM parameters from a sequence of observations. This is indeed the process of learning HMM from the observation sequence. The sequence of observations are specified by $O = o$. Using $O = o$, the triple $\lambda = (A, B, \Pi)$ is estimated.

The parameters of the HMM are determined so as to maximize the probability $P(O = o)$. No analytic solution is possible in general. However, a locally optimal solution is obtained iteratively. This is accomplished via a maximization algorithm which attempts to maximize the probability $P(O = o)$. It turns out that this is a special instance of the expectation-maximization (EM) technique. Genesis of the EM technique has been discussed earlier in the chapter. The technique, as applied to this HMM problem is called the Baum-Welch algorithm. It is named after L. E. Baum, and L. R. Welch.

We initially define certain special probabilities of interest. This is followed by expressing the estimated probabilities of the HMM in terms of these especial probabilities. The Baum-Welch iterative algorithm is subsequently outlined. Finally a justification of the algorithm is provided.

Probabilities of Interest

Several probabilistic quantities of interest are initially specified.

- Define the probability $\xi_t(i, j)$ as the probability of the system being in state s_i at time t and in state s_j at time $(t+1)$, given the entire observation sequence o, where $s_i, s_j \in S$, $1 \leq i, j \leq N$, and $1 \leq t \leq (T-1)$. Thus

$$\xi_t(i, j) = P(X_t = s_i, X_{t+1} = s_j \mid O = o) \tag{11.52a}$$

This probability is expressed as

$$\begin{aligned}
&\xi_t(i,j) \\
&= \frac{P(O=o \mid X_t=s_i, X_{t+1}=s_j)\, P(X_t=s_i, X_{t+1}=s_j)}{P(O=o)} \\
&= \frac{P(O=o \mid X_t=s_i, X_{t+1}=s_j)\, P(X_{t+1}=s_j \mid X_t=s_i)\, P(X_t=s_i)}{P(O=o)}
\end{aligned}$$

Observe that, in the above expression

$$\begin{aligned}
&P(O=o \mid X_t=s_i, X_{t+1}=s_j)\, P(X_t=s_i) \\
&= P(O_1=o_1,\ldots,O_t=o_t \mid X_t=s_i)\, P(X_t=s_i) \\
&\quad P(O_{t+1}=o_{t+1} \mid X_{t+1}=s_j)\, P(O_{t+2}=o_{t+2},\ldots,O_T=o_T \mid X_{t+1}=s_j) \\
&= \alpha_t(i)\, b_j(o_{t+1})\, \beta_{t+1}(j)
\end{aligned}$$

$$P(X_{t+1}=s_j \mid X_t=s_i) = a_{ij}$$

$$P(O=o) = \sum_{l=1}^{N}\sum_{m=1}^{N} \alpha_t(l)\, a_{lm} b_m(o_{t+1})\, \beta_{t+1}(m)$$

Therefore

$$\xi_t(i,j) = \frac{\alpha_t(i)\, a_{ij} b_j(o_{t+1})\, \beta_{t+1}(j)}{\sum_{l=1}^{N}\sum_{m=1}^{N} \alpha_t(l)\, a_{lm} b_m(o_{t+1})\, \beta_{t+1}(m)} \tag{11.52b}$$

where $1 \le i,j \le N$, and $1 \le t \le (T-1)$. Observe that $\xi_t(i,j)$ is the expected number of transitions at time t from state s_i to state s_j at time $(t+1)$, given the entire observation sequence o.

- The expected number of transitions from state s_i to state s_j in o is $\sum_{t=1}^{T-1} \xi_t(i,j)$.
- Let $\gamma_t(i)$ be the probability that the system is in state s_i at time t, given the observation sequence o. That is

$$\gamma_t(i) = P(X_t=s_i \mid O=o), \quad s_i \in S,\ 1 \le i \le N,\ 1 \le t \le T \tag{11.53a}$$

We have

$$\begin{aligned}
\gamma_t(i) = P(X_t=s_i \mid O=o) &= \frac{P(X_t=s_i, O=o)}{P(O=o)} \\
&= \frac{\alpha_t(i)\, \beta_t(i)}{\sum_{j=1}^{N} \alpha_t(j)\, \beta_t(j)}
\end{aligned} \tag{11.53b}$$

Observe that, $\sum_{i=1}^{N} \gamma_t(i) = 1$, for $1 \le t \le T$.

- The expected number of times, a state $s_i \in S$ is visited in time T is $\sum_{t=1}^{T} \gamma_t(i)$.
- The expected number of transitions from state $s_i \in S$ in o is $\sum_{t=1}^{T-1} \gamma_t(i)$.
- The probability $\gamma_t(i) = P(X_t=s_i \mid O=o)$, $s_i \in S, 1 \le t \le (T-1)$ can also be expressed as

$$\begin{aligned}
\gamma_t(i) &= P(X_t=s_i \mid O=o) \\
&= \sum_{j=1}^{N} P(X_t=s_i, X_{t+1}=s_j \mid O=o) \\
&= \sum_{j=1}^{N} \xi_t(i,j)
\end{aligned}$$

Thus

$$\gamma_t(i) = \sum_{j=1}^{N} \xi_t(i,j), \quad 1 \le i \le N, \, 1 \le t \le (T-1)$$

Note that $\gamma_t(i)$ is the expected number of transitions from state s_i at time t, given the observation sequence o.

Estimation of the Probabilities in HMM

Using the above probabilities, the Baum-Welch algorithm can be developed to estimate the HMM triple $\lambda = (A, B, \Pi)$. Let these estimated entities be

$$\widehat{A} = [\widehat{a}_{ij}], \quad \widehat{B} = \left[\widehat{b}_{jk}\right], \quad \widehat{\Pi} = \{\widehat{\pi}_i\}$$

Also denote the estimated observation-symbol probability $\widehat{b}_{jk}$ by $\widehat{b}_j(k)$.

- We let

$$\begin{aligned} \widehat{\pi}_i &= \text{the expected frequency in state } s_i \in S \text{ at time } t = 1 \\ &= \gamma_1(i), \quad 1 \le i \le N \end{aligned} \tag{11.54a}$$

It can be checked that $\sum_{i=1}^{N} \widehat{\pi}_i = 1$. This is true because $\sum_{i=1}^{N} \gamma_1(i) = 1$

- Estimate $\widehat{a}_{ij}$, for $s_i, s_j \in S$ as

$$\begin{aligned} \widehat{a}_{ij} &= \frac{\text{average number of transitions from state } s_i \text{ to state } s_j}{\text{average number of transitions from state } s_i} \\ &= \frac{\sum_{t=1}^{T-1} \xi_t(i,j)}{\sum_{t=1}^{T-1} \gamma_t(i)} \end{aligned} \tag{11.54b}$$

It can be checked that $\sum_{j=1}^{N} \widehat{a}_{ij} = 1$ for $1 \le i \le N$.

- Estimate $\widehat{b}_j(k)$ for $s_j \in S$, $v_k \in V$ as

$$\begin{aligned} \widehat{b}_j(k) &= \frac{\text{average number of times in state } s_j \text{ and observing symbol } v_k}{\text{average number of times in state } s_j} \\ &= \frac{\sum_{t:o_t = v_k, 1 \le t \le T} \gamma_t(j)}{\sum_{t=1}^{T} \gamma_t(j)} \end{aligned} \tag{11.54c}$$

It can be checked that $\sum_{k=1}^{M} \widehat{b}_j(k) = 1$ for $1 \le j \le N$.

Baum-Welch Algorithm

Given the observation sequence $O = o$, the Baum-Welch algorithm estimates $\lambda = (A, B, \Pi)$ by maximizing $P(O = o)$. The Baum-Welch algorithm is iterative. It is summarized as follows.

Step 0: Select a prespecified value $\Delta \in \mathbb{R}^+$, for the accuracy parameter.
Initialize HMM as λ.

Step 1: Compute the new model HMM λ' based upon λ,
and the observation sequence $O = o$. This requires computation of:

$$\alpha_t(i), \beta_t(i), \gamma_t(i); \quad 1 \leq i \leq N, 1 \leq t \leq T$$

$$\xi_t(i,j); \quad 1 \leq i,j \leq N, \ 1 \leq t \leq (T-1)$$

Using these values, estimate:

$$\pi_i, \ 1 \leq i \leq N; \ a_{ij}, \ 1 \leq i,j \leq N; \ b_j(k), \ 1 \leq j \leq N, \ 1 \leq k \leq M$$

These new estimated values determine λ'.
Step 2: If $\left|\ln P(O = o \mid \lambda') - \ln P(O = o \mid \lambda)\right| \leq \Delta$, then stop,
else $\lambda \leftarrow \lambda'$ and go to Step 1. □

Suppose that we have multiple independent observation sequences $\mathcal{X} = \{o^1, o^2, \ldots, o^R\}$. Assume that the length of rth sequence is $T_r \in \mathbb{P}$, where $1 \leq r \leq R$, and

$$P(\mathcal{X}) = \prod_{r=1}^{R} P(O = o^r)$$

Denote the rth observation probabilities:

$$\alpha_t(i), \ \beta_t(i), \text{ and } \gamma_t(i)$$

by

$$\alpha_t^r(i), \ \beta_t^r(i), \text{ and } \gamma_t^r(i)$$

respectively, where $1 \leq i \leq N, 1 \leq t \leq T_r, 1 \leq r \leq R$; and $\xi_t(i,j)$ by $\xi_t^r(i,j)$ for $1 \leq i,j \leq N$, $1 \leq t \leq (T_r - 1), 1 \leq r \leq R$. The HMM estimated parameters are averages over all observations:

$$\widehat{\pi}_i = \frac{1}{R}\sum_{r=1}^{R} \gamma_1^r(i), \quad 1 \leq i \leq N$$

$$\widehat{a}_{ij} = \frac{\sum_{r=1}^{R}\sum_{t=1}^{T_r-1} \xi_t^r(i,j)}{\sum_{r=1}^{R}\sum_{t=1}^{T_r-1} \gamma_t^r(i)}, \quad 1 \leq i,j \leq N$$

$$\widehat{b}_j(k) = \frac{\sum_{r=1}^{R}\sum_{t:o_t=v_k, 1\leq t\leq T_r} \gamma_t^r(j)}{\sum_{r=1}^{R}\sum_{t=1}^{T_r} \gamma_t^r(j)}, \quad 1 \leq j \leq N, 1 \leq k \leq M$$

The preceding discussion assumed that the observed values are discrete. However, if the observed values are continuous, then these can be discretized appropriately.

Justification of the Baum-Welch Algorithm

The Baum-Welch algorithm uses the expectation-maximization (EM) technique. This is an iterative technique, in which each iteration is made up of computation of an expectation-step (E-step), followed by a maximization-step (M-step).

For simplicity, we assume that only a single observable sequence $O = o$ of length T is given. Its generalization to several observable sequences is straight forward. In this instance the given data is simply $\mathcal{X} = o$. The parameters to be determined are $\lambda = (A, B, \Pi)$. Thus the corresponding conditional expectation function $\mathcal{Q}(\cdot, \cdot)$ which is used in the analysis is

$$\mathcal{Q}(\lambda, \lambda') = \sum_{x} P(X = x \mid O = o, \lambda') \ln P(O = o, X = x \mid \lambda)$$

The function $\mathcal{Q}(\lambda, \lambda')$ is maximized over λ so as to improve λ'. A more convenient (and symmetric) expression is obtained by multiplying both sides of the above equation by a constant $P(O = o \mid \lambda')$. Let

$$\begin{aligned} Q(\lambda, \lambda') &= P(O = o \mid \lambda') \mathcal{Q}(\lambda, \lambda') \\ &= \sum_{x} P(X = x \mid O = o, \lambda') P(O = o \mid \lambda') \ln P(O = o, X = x \mid \lambda) \\ &= \sum_{x} P(O = o, X = x \mid \lambda') \ln P(O = o, X = x \mid \lambda) \end{aligned}$$

Thus

$$Q(\lambda, \lambda') = \sum_{x} P(O = o, X = x \mid \lambda') \ln P(O = o, X = x \mid \lambda)$$

Sometimes $Q(\cdot, \cdot)$ is called Baum's auxiliary function. It is named after L. E. Baum, one of the pioneers in the study of HMMs.

- The E-step determines the function $Q(\lambda, \lambda')$.
- The M-step maximizes $Q(\lambda, \lambda')$ over λ.

The above two steps are executed sequentially, and successively till a convergence criteria for the specification of the triple λ is met. It is known that

$$\begin{aligned} &P(O = o, X = x \mid \lambda) \\ &= \pi_{x_1} b_{x_1}(o_1) a_{x_1 x_2} b_{x_2}(o_2) a_{x_2 x_3} \cdots b_{x_{T-1}}(o_{T-1}) a_{x_{T-1} x_T} b_{x_T}(o_T) \end{aligned}$$

Thus

$$\ln P(O = o, X = x \mid \lambda) = \ln \pi_{x_1} + \sum_{t=1}^{T-1} \ln a_{x_t x_{t+1}} + \sum_{t=1}^{T} \ln b_{x_t}(o_t)$$

We express $Q(\lambda, \lambda')$ as

$$Q(\lambda, \lambda') = Q_{\pi}(\pi, \lambda') + \sum_{i=1}^{N} Q_{a_i}(a_i, \lambda') + \sum_{j=1}^{N} Q_{b_j}(b_j, \lambda')$$

where

$$\begin{aligned} \pi &= \begin{bmatrix} \pi_1 \; \pi_2 \cdots \; \pi_N \end{bmatrix} \\ a_i &= \begin{bmatrix} a_{i1} \; a_{i2} \cdots \; a_{iN} \end{bmatrix}, \quad 1 \le i \le N \\ b_j &= \begin{bmatrix} b_j(1) \; b_j(2) \cdots \; b_j(M) \end{bmatrix}, \quad 1 \le j \le N \end{aligned}$$

and

$$Q_{\pi}\left(\pi,\lambda'\right)=\sum_{i=1}^{N}P\left(O=o,x_1=s_i\mid\lambda'\right)\ln\pi_i$$

$$Q_{a_i}\left(a_i,\lambda'\right)=\sum_{j=1}^{N}\sum_{t=1}^{T-1}P\left(O=o,x_t=s_i,x_{t+1}=s_j\mid\lambda'\right)\ln a_{ij},\quad 1\le i\le N$$

$$Q_{b_j}\left(b_j,\lambda'\right)=\sum_{t=1}^{T}P\left(O=o,x_t=s_j\mid\lambda'\right)\ln b_j\left(o_t\right),\quad 1\le j\le N$$

The function $Q\left(\lambda,\lambda'\right)$ is optimized subject to the following constraints.

$$\pi_i\ge 0,\quad 1\le i\le N;\quad\text{and}\quad\sum_{i=1}^{N}\pi_i=1$$

$$a_{ij}\ge 0,\quad 1\le i,j\le N;\quad\text{and}\quad\sum_{j=1}^{N}a_{ij}=1,\quad 1\le i\le N$$

$$b_j\left(k\right)\ge 0,\quad 1\le k\le M,\quad 1\le j\le N,\quad\text{and}\quad\sum_{k=1}^{M}b_j\left(k\right)=1,\quad 1\le j\le N$$

Observe in the above equations that $Q\left(\lambda,\lambda'\right)$ is split into three independent terms, where each subproblem can be optimized separately subject to appropriate constraints. The following general observation is helpful in this quest.

Observation 11.9. Let $y=\left(y_1,y_2,\ldots,y_K\right)$, and $w_j\in\mathbb{R}_0^+, 1\le j\le K$ are constants. Further, all w_j's are not simultaneously equal to zero.

$$\textit{Maximize}\ f\left(y\right)=\sum_{j=1}^{K}w_j\ln y_j \tag{11.55a}$$

$$\textit{Subject to:}\quad\sum_{j=1}^{K}y_j=1 \tag{11.55b}$$

where $y_j\in\mathbb{R}_0^+, 1\le j\le K$. Also define $a\ln a\triangleq 0$, as $a\to 0$. The global maximum occurs at

$$y_j=\frac{w_j}{\sum_{k=1}^{K}w_k},\quad 1\le j\le K \tag{11.55c}$$

□

In the following subproblems, the following facts are used.

- The forward probability is:
  $\alpha_t\left(i\right)=P\left(O_1=o_1,O_2=o_2,\ldots,O_t=o_t,X_t=s_i\right)$, where $1\le i\le N$, and $1\le t\le T$.
- The backward probability is:
  $\beta_t\left(i\right)=P\left(O_{t+1}=o_{t+1},O_{t+2}=o_{t+2},\ldots,O_T=o_T\mid X_t=s_i\right)$,
  where $1\le i\le N$, and $1\le t\le\left(T-1\right)$; and
  $\beta_T\left(i\right)=1, 1\le i\le N$.

- $P(O = o, x_t = s_i \mid \lambda) = \alpha_t(i)\,\beta_t(i)$, where $1 \le i \le N$, and $1 \le t \le T$.
- $P(O = o, x_t = s_i, x_{t+1} = s_j \mid \lambda) = \alpha_t(i)\,a_{ij} b_j(o_{t+1})\,\beta_{t+1}(j)$, where $1 \le i, j \le N$, and $1 \le t \le (T-1)$.
- $P(O = o \mid \lambda) = \sum_{i=1}^{N} \alpha_T(i) = \sum_{i=1}^{N} \alpha_t(i)\,\beta_t(i)$, where $1 \le t \le T$.
- $\gamma_t(i) = P(X_t = s_i \mid O = o)$, where $s_i \in S$, $1 \le i \le N$, and $1 \le t \le T$.

Subproblem 1: Maximize $Q_\pi(\pi, \lambda')$ subject to appropriate constraints.

$$\textit{Maximize} \sum_{i=1}^{N} P(O = o, x_1 = s_i \mid \lambda') \ln \pi_i$$

$$\textit{Subject to:} \quad \sum_{i=1}^{N} \pi_i = 1$$

where $\pi_i \ge 0$, $1 \le i \le N$. The global maximum occurs at

$$\begin{aligned}\widehat{\pi}_i &= \frac{P(O = o, x_1 = s_i \mid \lambda')}{P(O = o \mid \lambda')} \\ &= \frac{\alpha_1(i)\,\beta_1(i)}{\sum_{i=1}^{N} \alpha_T(i)} = \gamma_1(i), \quad 1 \le i \le N\end{aligned}$$

□

Subproblem 2: Maximize $Q_{a_i}(a_i, \lambda')$ subject to appropriate constraints.

$$\textit{Maximize} \sum_{j=1}^{N} \sum_{t=1}^{T-1} P(O = o, x_t = s_i, x_{t+1} = s_j \mid \lambda') \ln a_{ij}$$

$$\textit{Subject to:} \quad \sum_{j=1}^{N} a_{ij} = 1$$

where $a_{ij} \ge 0$, $1 \le j \le N$. The global maximum occurs at

$$\begin{aligned}\widehat{a}_{ij} &= \frac{\sum_{t=1}^{T-1} P(O = o, x_t = s_i, x_{t+1} = s_j \mid \lambda')}{\sum_{t=1}^{T-1} P(O = o, x_t = s_i \mid \lambda')} \\ &= \frac{\sum_{t=1}^{T-1} \alpha_t(i)\,a_{ij} b_j(o_{t+1})\,\beta_{t+1}(j)}{\sum_{t=1}^{T-1} \alpha_t(i)\,\beta_t(i)} \\ &= \frac{\sum_{t=1}^{T-1} \xi_t(i,j)}{\sum_{t=1}^{T-1} \gamma_t(i)}, \quad 1 \le j \le N\end{aligned}$$

The above results are true for $i = 1, 2, \ldots, N$. □

Subproblem 3: Maximize $Q_{b_j}(b_j, \lambda')$ subject to appropriate constraints.

$$\textit{Maximize} \sum_{t=1}^{T} P(O = o, x_t = s_j \mid \lambda') \ln b_j(o_t)$$

$$\textit{Subject to:} \ \sum_{k=1}^{M} b_j(k) = 1$$

where $b_j(k) \geq 0, 1 \leq k \leq M$. The global maximum occurs at

$$\widehat{b}_j(k) = \frac{\sum_{t=1}^{T} P(O = o, x_t = s_j \mid \lambda') \, \delta(o_t, v_k)}{\sum_{t=1}^{T} P(O = o, x_t = s_j \mid \lambda')}, \quad 1 \leq k \leq M$$

where

$$\delta(o_t, v_k) = \begin{cases} 1, & \text{if } o_t = v_k \\ 0, & \text{otherwise} \end{cases}$$

The $\widehat{b}_j(k)$'s can also be expressed as

$$\begin{aligned} \widehat{b}_j(k) &= \frac{\sum_{t=1}^{T} \alpha_t(j) \beta_t(j) \delta(o_t, v_k)}{\sum_{t=1}^{T} \alpha_t(j) \beta_t(j)} \\ &= \frac{\sum_{t: o_t = v_k, 1 \leq t \leq T} \gamma_t(j)}{\sum_{t=1}^{T} \gamma_t(j)}, \quad 1 \leq k \leq M \end{aligned}$$

The above results are true for $j = 1, 2, \ldots, N$. □

Thus the solutions of the above three subproblems confirm the estimates of the probabilities given earlier in the subsection.

The astute reader will notice that, in solution of the three problems, the computations require products of probabilities. As probabilities vary between zero and one, there is a potential for underflow problem in the implementation of these solutions. The remedy to such problems is to scale these numbers carefully, and appropriately.

11.13 Kalman Filtering

Kalman filtering is a noise filtering algorithm. This algorithm was discovered by Rudolf Emil Kálmán in the year 1960. Therefore, the algorithm is aptly named after him. The purpose of this algorithm is to get "accurate" information out of noisy data. More precisely, a Kalman filter is an *optimal filter*, which determines the filter parameters from uncertain, and noisy observations. Justification for the use of the phrase "optimal" is demonstrated later in this section. The Kalman filter is also a *recursive* algorithm. That is, new measurements can be processed as soon as they are available. Not surprisingly, this scheme is popular because it gives good results in practice.

Its application are too numerous to name. However, some of the applications are in the field of monitoring sensor data, global positioning system receivers, and data fusion in general. One of its earliest applications was in the Apollo navigation computer. It took the astronaut Neil Armstrong to the moon (in July 1969), and brought him back home safely. It has recently found application in the field of data mining. For example, it can be used for data cleaning. Data is often duplicated, inaccurate, incomplete, incorrect, and missing. Data cleaning can: eliminate duplicates, perform error correction, detect missing data, and transform data.

The Kalman filter is essentially used to model a stochastic dynamical system. The dynamical system can either be continuous or discrete in time. Only a discrete version of the Kalman filter is considered in this section. A stochastic dynamical system is initially described in this section. This is followed by the Kalman's filtering scheme. A probabilistic interpretation of this filtering scheme is subsequently provided.

Note that as per standard convention in description of this topic, random vectors and variables are often denoted by lower case letters. Also, $\mathcal{E}(\cdot)$ is the expectation operator.

11.13.1 Stochastic Dynamical System

Consider a stochastic dynamical system. The state of the system at time instant t_k, is specified by a stochastic vector x_k, where $k \in \mathbb{N}$. The input to the system, at time instant t_k is specified by a deterministic vector y_k, where $k \in \mathbb{N}$. The noisy observations of the system, at time instant t_k are specified by the vector z_k, where $k \in \mathbb{P}$. Let the set of input (control) vectors up to time instant t_k be

$$Y_k = \{y_i \mid 1 \leq i \leq k\}, \quad k \in \mathbb{N}$$

Similarly, let the set of measurements (output) up to time instant t_k be

$$Z_k = \{z_i \mid 1 \leq i \leq k\}, \quad k \in \mathbb{P}$$

The Kalman filter assumes that at time instant t_k, the initial random state vector x_0 of the system, the set of inputs Y_k, and the set of measurements, Z_k are given. Using this information, along with certain other system parameters, it determines $\widehat{x}_k$ recursively, which is an estimate of the state vector x_k, where $k \in \mathbb{P}$. A general dynamical system is modeled as

$$\begin{aligned} x_{k+1} &= f(x_k, y_k, w_k), \quad k \in \mathbb{N} \\ z_k &= g(x_k, v_k), \quad k \in \mathbb{P} \end{aligned}$$

A more specific stochastic dynamical system is modeled as a stochastic time-variant linear system. It is:

$$\begin{aligned} x_{k+1} &= A_k x_k + B_k y_k + w_k, \quad k \in \mathbb{N} \\ z_k &= H_k x_k + v_k, \quad k \in \mathbb{P} \end{aligned}$$

Note that the vectors in the above formulation are specified as column matrices.

- In the above equations, y_k for $k \in \mathbb{N}$ is the *input* (control) vector. This vector is nonrandom, and also known.
- The initial state vector x_0 is random. Its mean is $\mathcal{E}(x_0) = \mu_0$, and the corresponding covariance matrix is

  $$P_0 = \mathcal{E}\left\{(x_0 - \mu_0)(x_0 - \mu_0)^T\right\}$$

  The matrix P_0 is symmetric, and positive definite. Both, μ_0 and P_0 are given.
- Uncertainty modeling.
  - It is assumed that the process-noise random vectors w_k for $k \in \mathbb{N}$ model the uncertainty of the dynamical process. The mean of each random vector w_k is zero, and its covariance is known. Further, different such random vectors at different time instants are uncorrelated with each other.

- The measurement-noise random vectors v_k for $k \in \mathbb{P}$ model the measurement noise. The mean of each random vector v_k is zero, and its covariance is known. Further, different such random vectors at different time instants are uncorrelated with each other.
- It is also assumed that these two sequences of random vectors are uncorrelated with each other.
- Further, these two sequences are also uncorrelated, with the initial state random vector x_0 as well.

More precisely, we have

$$\mathcal{E}(w_k) = 0, \;\; \forall\, k \in \mathbb{N}$$
$$\mathcal{E}\left(w_k w_j^T\right) = \begin{cases} Q_k, & j = k \\ 0, & j \neq k \end{cases}, \;\; \forall\, j, k \in \mathbb{N}$$

$$\mathcal{E}(v_k) = 0, \;\; \forall\, k \in \mathbb{P}$$
$$\mathcal{E}\left(v_k v_j^T\right) = \begin{cases} R_k, & j = k \\ 0, & j \neq k \end{cases}, \;\; \forall\, j, k \in \mathbb{P}$$

$$\mathcal{E}\left(w_k v_j^T\right) = 0, \;\; \forall\, k \in \mathbb{N}, \; \forall\, j \in \mathbb{P}$$

$$\mathcal{E}\left(w_k x_0^T\right) = 0, \;\; \forall\, k \in \mathbb{N}$$
$$\mathcal{E}\left(v_k x_0^T\right) = 0, \;\; \forall\, k \in \mathbb{P}$$

The matrix Q_k is assumed to be symmetric, and positive definite $\forall\, k \in \mathbb{N}$. Similarly, the matrix R_k is assumed to be symmetric, and positive definite $\forall\, k \in \mathbb{P}$.

- The dimensions of the vectors and matrices in this formulation are stated explicitly.

  x_k is $n \times 1$ state vector
  y_k is $l \times 1$ input (control) vector (given)
  w_k is $n \times 1$ process noise vector
  z_k is $m \times 1$ observation (measurement, or output) vector
  v_k is $m \times 1$ measurement noise vector
  A_k is $n \times n$ state transition matrix (given)
  B_k is $n \times l$ input matrix (given)
  H_k is $m \times n$ observation matrix (given)
  Q_k is $n \times n$ process noise covariance matrix (given)
  R_k is $m \times m$ measurement noise covariance matrix (given)

11.13.2 Filtering Scheme

A framework for Kalman filtering was developed in the last subsection. In this subsection the discrete Kalman filtering algorithm is derived in several steps. More precisely, the algorithm determines the estimate $\widehat{x}_k$ of the state vector x_k, for $k \in \mathbb{P}$, recursively.

Step 0: Let $k \in \mathbb{N}$. The time instant, just *prior* to t_k is denoted by t_k^-. The state vector estimate at t_k^- is denoted by $\widehat{x}_k^-$ (*prior* state vector estimate). The state vector estimate at t_k is denoted by $\widehat{x}_k$ (*posterior* state vector estimate). Let $\widehat{x}_0^- = \widehat{x}_0 = \mu_0$.

Step 1: For $k \in \mathbb{P}$, the prior state vector estimate $\widehat{x}_k^-$ is based upon the initial condition x_0, the set of measurements prior to t_k, and other system parameters. If computations are performed recursively, $\widehat{x}_k^-$ can be determined from the posterior state vector estimate $\widehat{x}_{k-1}$. The posterior state vector estimate $\widehat{x}_k$, is based upon the prior state vector estimate $\widehat{x}_k^-$ and the new measurement z_k made at time instant t_k.

For for $k \in \mathbb{N}$ let

$$\widehat{x}_{k+1}^- = A_k \widehat{x}_k + B_k y_k$$

Observe in the above expression that the contribution of the process noise vector w_k was ignored. This should be true, because w_k is available only at time instant t_{k+1}. The posterior state $\widehat{x}_k$ is next specified.

For $k \in \mathbb{P}$, the prior state vector estimate $\widehat{x}_k^-$ is given. Our goal is to improve this estimate by using the new noisy measurement z_k made at time t_k. This is done by using a linear combination of the prior state vector estimate $\widehat{x}_k^-$, and the noisy measurement as

$$\widehat{x}_k = \widehat{x}_k^- + K_k \left(z_k - H_k \widehat{x}_k^- \right)$$

where K_k is the blending factor yet to be determined. The quantity $\left(z_k - H_k \widehat{x}_k^- \right)$, is often called the *innovation.* It is the difference between the observation z_k and the predicted observation $H_k \widehat{x}_k^-$.

Step 2: The errors in the prior and posterior state vector estimates are next determined. The prior and the posterior state vector estimation errors are denoted by e_k^- and e_k respectively, where

$$e_k^- = \left(x_k - \widehat{x}_k^- \right), \quad \text{and} \quad e_k = \left(x_k - \widehat{x}_k \right), \quad \text{for } k \in \mathbb{N}$$

Observe that, for $k \in \mathbb{N}$

$$\begin{aligned} e_{k+1}^- &= \left(x_{k+1} - \widehat{x}_{k+1}^- \right) \\ &= \left(A_k x_k + B_k y_k + w_k \right) - \left(A_k \widehat{x}_k + B_k y_k \right) \\ &= A_k e_k + w_k \end{aligned}$$

Similarly, for $k \in \mathbb{P}$

$$\begin{aligned} e_k &= \left(x_k - \widehat{x}_k \right) \\ &= \left(x_k - \widehat{x}_k^- \right) - K_k \left(H_k x_k + v_k - H_k \widehat{x}_k^- \right) \\ &= \left(I - K_k H_k \right) e_k^- - K_k v_k \end{aligned}$$

where I is an identity matrix of size n. Observe that $\mathcal{E}\left(e_0^- \right) = \mathcal{E}\left(e_0 \right) = 0$. Therefore, from the above recursive expressions for e_k^-, and e_k, it follows that

$$\mathcal{E}\left(e_k^- \right) = \mathcal{E}\left(e_k \right) = 0, \quad \forall\, k \in \mathbb{N}$$

Step 3: The covariance matrices of the errors in the prior and posterior state vector estimates at time instant t_k are denoted by P_k^- and P_k respectively. That is

$$P_k^- = \mathcal{E}\left(e_k^- e_k^{-T} \right), \quad \text{and} \quad P_k = \mathcal{E}\left(e_k e_k^T \right), \quad \text{for } k \in \mathbb{P}$$

Observe that e_k^- is uncorrelated with the measurement error v_k. Therefore

$$P_k = (I - K_k H_k) P_k^- (I - K_k H_k)^T + K_k R_k K_k^T$$

where I is an identity matrix of size n.

Step 4: For $k \in \mathbb{P}$, the goal is to determine K_k so that the updated estimate $\widehat{x}_k$ is optimal in some sense. The performance criterion of choice is the minimum mean-square error. The state vector estimation error is e_k. Denote the components of the error vector e_k; by e_{k_i}, where $1 \le i \le n$. The error in the posterior state vector estimate $\widehat{x}_k$ is minimized by minimizing $J(e_k)$, where

$$J(e_k) = \mathcal{E}\left\{\sum_{i=1}^{n} e_{k_i}^2\right\}$$

Actually, $J(e_k)$ is sum of the diagonal elements of the error covariance matrix P_k. Therefore the state vector estimate $\widehat{x}_k$ is determined by minimizing $J(e_k)$, which is actually the trace of the matrix P_k. That is, $J(e_k) = tr(P_k)$. This minimization is performed by varying the blending factor K_k.

Step 5: In the minimization process, the subscripts are dropped for clarity, and later reintroduced. Thus, the goal is to minimize the trace of the matrix

$$P = (I - KH) P^- (I - KH)^T + KRK^T$$

Using matrix calculus, it is shown in the problem section that the value of K that minimizes P is

$$K = P^- H^T \left(HP^- H^T + R\right)^{-1}$$

The corresponding value of P is

$$P = (I - KH) P^-$$

For $k \in \mathbb{P}$, we revert back to subscripts. Thus the minimizing value of K_k is

$$K_k = P_k^- H_k^T \left(H_k P_k^- H_k^T + R_k\right)^{-1}$$

This minimizing value of K_k is called the *Kalman gain*. The corresponding value of the error covariance matrix P_k is

$$P_k = (I - K_k H_k) P_k^-$$

Step 6: Having determined optimum values of K_k and P_k, our next goal is to determine the corresponding P_{k+1}^-, for $k \in \mathbb{N}$. As stated earlier,

$$e_{k+1}^- = A_k e_k + w_k$$

Observe that e_k and w_k are uncorrelated. Therefore

$$\begin{aligned} P_{k+1}^- &= \mathcal{E}\left(e_{k+1}^- e_{k+1}^{-T}\right) \\ &= \mathcal{E}\left\{(A_k e_k + w_k)(A_k e_k + w_k)^T\right\} \\ &= A_k P_k A_k^T + Q_k \end{aligned}$$

The main steps in describing the Kalman filter are complete. Note that the covariance matrices P_0, $Q_k \ \forall\ k \in \mathbb{N}$, and $R_k \ \forall\ k \in \mathbb{P}$ have been assumed to to be positive definite. Consequently, the covariance matrices P_k^-, and P_k are also positive definite $\forall\ k \in \mathbb{P}$. The recursive equations of the Kalman filter are next summarized.

Summary of the Kalman Filter Algorithm

Input: $\widehat{x}_0 = \mu_0$, P_0; y_k, A_k, B_k, Q_k, where $k \in \mathbb{N}$; and H_k, R_k, where $k \in \mathbb{P}$.
Output: $\widehat{x}_k$ where $k \in \mathbb{P}$.

Time update: For $k \in \mathbb{N}$.
State vector update: $\widehat{x}_{k+1}^- = (A_k\widehat{x}_k + B_k y_k)$
Prior state error covariance matrix: $P_{k+1}^- = \left(A_k P_k A_k^T + Q_k\right)$

Measurement update: For $k \in \mathbb{P}$.
Compute Kalman gain: $K_k = P_k^- H_k^T \left(H_k P_k^- H_k^T + R_k\right)^{-1}$
State vector update: $\widehat{x}_k = \widehat{x}_k^- + K_k\left(z_k - H_k\widehat{x}_k^-\right)$
Posterior state error covariance matrix: $P_k = (I - K_k H_k)\, P_k^-$

□

In the above algorithm, time and measurement updates are executed alternately. That is, one after the other. A pseudocode for the algorithm is also given.

Algorithm 11.14. *Kalman Filtering.*

Input: $k \in \mathbb{N}$; $\widehat{x}_0 = \mu_0$, P_0;
y_i, A_i, B_i, Q_i, where $i \in \{0, 1, 2, \ldots, k\}$;
and H_i, R_i, where $i \in \{1, 2, \ldots, k+1\}$.
Output: $\widehat{x}_i$ where $i \in \{1, 2, \ldots, k+1\}$.
begin
  for $i = 0$ **to** k **do**
  begin
    Step 1: (*time update*)
      $\widehat{x}_{i+1}^- = (A_i\widehat{x}_i + B_i y_i)$ (*state vector update*)
      $P_{i+1}^- = \left(A_i P_i A_i^T + Q_i\right)$
      (*prior state error covariance matrix*)
    Step 2: (*measurement update*)
      $j = i + 1$
      $K_j = P_j^- H_j^T \left(H_j P_j^- H_j^T + R_j\right)^{-1}$ (*compute Kalman gain*)
      $\widehat{x}_j = \widehat{x}_j^- + K_j\left(z_j - H_j\widehat{x}_j^-\right)$ (*state vector update*)
      $P_j = (I - K_j H_j)\, P_j^-$
      (*posterior state error covariance matrix*)
  end (*end of for loop*)
end (*end of Kalman filtering algorithm*)

Observation 11.10. In the derivation of the Kalman filter algorithm, optimal expressions for K and P were derived. These are:

$$P = (I - KH) P^{-}, \text{ and } K = P^{-}H^{T}\left(HP^{-}H^{T} + R\right)^{-1}$$

Alternate useful expressions for P and K are

$$P = \left\{\left(P^{-}\right)^{-1} + H^{T}R^{-1}H\right\}^{-1}, \text{ and } K = PH^{T}R^{-1}$$

□

Proof of the above observation is given in the problem section.

11.13.3 A Probabilistic Perspective on Kalman Filtering

A probabilistic interpretation of the discrete Kalman filter is provided in this subsection. Assume that the initial random state vector, the process noise vectors, and the measurement noise vectors each have multivariate Gaussian distribution. All the process and measurement noise vectors have zero mean. Further, all these random vectors are uncorrelated with each other. Then the following assertions are established in this subsection.

- The random vectors x_k for $k \in \mathbb{N}$, and z_k for $k \in \mathbb{P}$, each have a multivariate Gaussian distribution.
- The Kalman filter algorithm is true. That is, under these assumptions, it is possible to do away with the requirement of minimizing the mean-square error in the posterior state vector estimation; and yet generate the recursive equations of the Kalman filter algorithm.

Notation. Let x be a random vector. Further, the elements of this vector have a multivariate Gaussian distribution. Let the mean of this random vector be $\mathcal{E}(x) = \eta$, and the corresponding covariance matrix be Ξ. We denote this fact as $x \sim \mathcal{N}(\eta, \Xi)$. The corresponding probability density function is denoted as f_x. □

Random Vectors

Assume that:

- The elements of the initial state vector x_0 have a multivariate Gaussian distribution, where $x_0 \sim \mathcal{N}(\mu_0, P_0)$.
- The elements of the process noise vector w_k, have a multivariate Gaussian distribution, where $w_k \sim \mathcal{N}(0, Q_k)$, and $k \in \mathbb{N}$.
  Also $\mathcal{E}\left(w_k w_j^T\right) = 0$, $j \neq k$ and $\forall\, j, k \in \mathbb{N}$.
- Similarly, the elements of the measurement noise vector v_j have a multivariate Gaussian distribution, where $v_j \sim \mathcal{N}(0, R_j)$, and $j \in \mathbb{P}$.
  Also $\mathcal{E}\left(v_k v_j^T\right) = 0$, $j \neq k$ and $\forall\, j, k \in \mathbb{P}$.
- The sequences w_k and v_j are uncorrelated with each other for all values of $j \in \mathbb{P}$, and $k \in \mathbb{N}$. Further, these sequences are also uncorrelated with the initial state vector x_0.

The following observations are stated in the chapter on probability theory.

1. Let x and y be random vectors. If the vectors x and y are uncorrelated and each have a multivariate Gaussian distribution, then x and y are independent.
2. Consider random vectors x and y of the same size. The elements of the random vector x have a multivariate Gaussian distribution. Similarly, the elements of the random vector y have a multivariate Gaussian distribution. Further, the random vectors x and y are uncorrelated. Let $z = (x + y)$, then elements of the random vector z also have a multivariate Gaussian distribution.
3. Consider a random vector x of size $n \in \mathbb{P}$. The elements of the random vector x have a multivariate Gaussian distribution. Also, let A be a real-valued matrix of size $r \times n$ of rank $r \leq n$; and c be a real-valued column matrix of size r. If $y = (Ax + c)$, then the elements of the random vector y also have a multivariate Gaussian distribution.

Next consider several mutually uncorrelated random vectors of the same size. Also assume that the elements of each such random vector have a multivariate Gaussian distribution. Based upon the above observations it can be concluded that the elements of the weighted sum of such uncorrelated random vectors also have a multivariate Gaussian distribution. This observation is true, because under linear transformation, the Gaussian distributions are indeed preserved.

Recall that, the state of the system at time instant t_k, is specified by a stochastic vector x_k, where $k \in \mathbb{N}$. Also, the noisy observation of the system, at time instant t_k is specified by the vector z_k, where $k \in \mathbb{P}$. The k set of measurements up to time t_k is $Z_k = \{z_i \mid 1 \leq i \leq k\}$, where $k \in \mathbb{P}$. Based upon the above assumptions, observations, and the Kalman filter algorithm it can be inferred that:

Time update: For $k \in \mathbb{P}$.

$x_k \mid Z_k \sim \mathcal{N}(\widehat{x}_k, P_k)$.

$x_{k+1} \mid Z_k \sim \mathcal{N}\left(\widehat{x}_{k+1}^-, P_{k+1}^-\right)$.

Measurement update: For $k \in \mathbb{P}$.

$z_{k+1} \mid Z_k \sim \mathcal{N}\left(H_{k+1}\widehat{x}_{k+1}^-, \left(H_{k+1}P_{k+1}^-H_{k+1}^T + R_{k+1}\right)\right)$.

$z_k \mid x_k \sim \mathcal{N}(H_k x_k, R_k)$.

□

Kalman Filter Algorithm under the Assumption of Uncorrelated Random Vectors with Gaussian Distributions

The Kalman filter algorithm was derived by minimizing the mean square error in the estimation of the posterior state vectors. This criterion was somewhat arbitrary. It is next demonstrated that, if it is assumed that all the random vectors in the formulation of this problem have multivariate Gaussian distributions, with the properties specified earlier in the section, then the Kalman filter's recursive equations are indeed true.

In this interpretation, conditional probabilities (Bayesian analysis) are used. Using conditional probability density functions, we have

$$f_{x_{k+1},z_{k+1}|Z_k} = f_{x_{k+1}|z_{k+1},Z_k} f_{z_{k+1}|Z_k} = f_{x_{k+1}|Z_{k+1}} f_{z_{k+1}|Z_k}$$

and also

$$f_{x_{k+1},z_{k+1}|Z_k} = f_{z_{k+1}|x_{k+1},Z_k} f_{x_{k+1}|Z_k}$$

As

$$f_{z_{k+1}|x_{k+1},Z_k} = f_{z_{k+1}|x_{k+1}}$$

we have

$$f_{x_{k+1},z_{k+1}|Z_k} = f_{z_{k+1}|x_{k+1}} f_{x_{k+1}|Z_k}$$

A combination of the above results lead to

$$f_{x_{k+1}|Z_{k+1}} f_{z_{k+1}|Z_k} = f_{z_{k+1}|x_{k+1}} f_{x_{k+1}|Z_k}$$

That is

$$f_{x_{k+1}|Z_{k+1}} = \frac{f_{z_{k+1}|x_{k+1}} f_{x_{k+1}|Z_k}}{f_{z_{k+1}|Z_k}}$$

As stated earlier, the random vector $x_{k+1} \mid Z_{k+1}$ has a multivariate Gaussian distribution. Therefore

$$f_{x_{k+1}|Z_{k+1}} = \frac{1}{(2\pi)^{n/2} (\det P_{k+1})^{1/2}} \exp\left\{-\frac{1}{2}(x_{k+1} - \widehat{x}_{k+1})^T P_{k+1}^{-1} (x_{k+1} - \widehat{x}_{k+1})\right\}$$

Also, the random vectors $z_{k+1} \mid x_{k+1}$, $x_{k+1} \mid Z_k$, and $z_{k+1} \mid Z_k$ each have a multivariate Gaussian distribution. Consequently

$$\begin{aligned}
f_{x_{k+1}|Z_{k+1}} &= \frac{f_{z_{k+1}|x_{k+1}} f_{x_{k+1}|Z_k}}{f_{z_{k+1}|Z_k}} \\
&= \frac{1}{(2\pi)^{n/2}} \left\{ \frac{\det\left(H_{k+1} P_{k+1}^{-} H_{k+1}^T + R_{k+1}\right)}{\left(\det P_{k+1}^{-}\right)(\det R_{k+1})} \right\}^{1/2} \\
&\exp\left\{-\frac{1}{2}\left[(z_{k+1} - H_{k+1} x_{k+1})^T R_{k+1}^{-1} (z_{k+1} - H_{k+1} x_{k+1})\right.\right. \\
&+ \left(x_{k+1} - \widehat{x}_{k+1}^{-}\right)^T \left(P_{k+1}^{-}\right)^{-1} \left(x_{k+1} - \widehat{x}_{k+1}^{-}\right) \\
&\left.\left. - \left(z_{k+1} - H_{k+1}\widehat{x}_{k+1}^{-}\right)^T \left(H_{k+1} P_{k+1}^{-} H_{k+1}^T + R_{k+1}\right)^{-1} \left(z_{k+1} - H_{k+1}\widehat{x}_{k+1}^{-}\right)\right]\right\}
\end{aligned}$$

Comparison of the two expressions for $f_{x_{k+1}|Z_{k+1}}$ leads to

$$\begin{aligned}
&(x_{k+1} - \widehat{x}_{k+1})^T P_{k+1}^{-1} (x_{k+1} - \widehat{x}_{k+1}) \\
&= \left[(z_{k+1} - H_{k+1} x_{k+1})^T R_{k+1}^{-1} (z_{k+1} - H_{k+1} x_{k+1})\right. \\
&+ \left(x_{k+1} - \widehat{x}_{k+1}^{-}\right)^T \left(P_{k+1}^{-}\right)^{-1} \left(x_{k+1} - \widehat{x}_{k+1}^{-}\right) \\
&\left. - \left(z_{k+1} - H_{k+1}\widehat{x}_{k+1}^{-}\right)^T \left(H_{k+1} P_{k+1}^{-} H_{k+1}^T + R_{k+1}\right)^{-1} \left(z_{k+1} - H_{k+1}\widehat{x}_{k+1}^{-}\right)\right]
\end{aligned}$$

Comparison of quadratic terms in x_{k+1} on both sides of the above equation results in

$$P_{k+1}^{-1} = \left(P_{k+1}^{-}\right)^{-1} + H_{k+1}^T R_{k+1}^{-1} H_{k+1}$$

Comparison of the x_{k+1} on both sides of the equation leads to

$$\widehat{x}_{k+1}^T P_{k+1}^{-1} = z_{k+1}^T R_{k+1}^{-1} H_{k+1} + \widehat{x}_{k+1}^{-T} \left(P_{k+1}^{-}\right)^{-1}$$

which is equivalent to

$$P_{k+1}^{-1}\widehat{x}_{k+1} = H_{k+1}^{T}R_{k+1}^{-1}z_{k+1} + \left(P_{k+1}^{-}\right)^{-1}\widehat{x}_{k+1}^{-}$$

In the above equation substitute $\left(P_{k+1}^{-}\right)^{-1} = \left(P_{k+1}^{-1} - H_{k+1}^{T}R_{k+1}^{-1}H_{k+1}\right)$. This leads to

$$P_{k+1}^{-1}\widehat{x}_{k+1} = H_{k+1}^{T}R_{k+1}^{-1}z_{k+1} + \left(P_{k+1}^{-1} - H_{k+1}^{T}R_{k+1}^{-1}H_{k+1}\right)\widehat{x}_{k+1}^{-}$$

The above equation implies

$$\widehat{x}_{k+1} = \widehat{x}_{k+1}^{-} + P_{k+1}H_{k+1}^{T}R_{k+1}^{-1}\left(z_{k+1} - H_{k+1}\widehat{x}_{k+1}^{-}\right)$$

In summary, the recursive equations for the updated estimate, and the error covariance are

$$\widehat{x}_{k+1} = \widehat{x}_{k+1}^{-} + P_{k+1}H_{k+1}^{T}R_{k+1}^{-1}\left(z_{k+1} - H_{k+1}\widehat{x}_{k+1}^{-}\right)$$
$$P_{k+1} = \left\{\left(P_{k+1}^{-}\right)^{-1} + H_{k+1}^{T}R_{k+1}^{-1}H_{k+1}\right\}^{-1}$$

The above expressions are the recursive equations in the measurement update of the Kalman filter algorithm. These were actually derived earlier in the section, under the assumption of minimization of the mean square error in the estimation of the posterior state vectors.

Thus the recursive equations of the Kalman filter algorithm are also true under the assumption of uncorrelated initial random state vector, process noise vectors, and measurement noise vectors each with multivariate Gaussian distribution.

Reference Notes

Primary sources for information in this chapter are the scholarly textbooks by: Witten, and Frank (2005), Tan, Steinbach, and Kumar (2006), Bishop (2006), Cios, Pedrycz, Swiniarski, and Kurgan (2007), Eldén (2007), Hastie, Tibshirani, and Friedman (2008), Simovici, and Djeraba (2008), Manning, Raghavan, and Schütze (2009); Theodoridis, and Koutroumbas (2009), Liu (2011), Han, Kamber, and Pei (2012), Zaki, and Meira Jr. (2014), and Aggarwal (2015). Klösgen, and Żytkow (2002) is a prolific source of concepts on data mining and knowledge discovery. A scholarly discussion of some of the topics discussed in this chapter can be found in the treatise by Devroye, Györfi, and Lugosi (1996).

The knowledge discovery process model is based upon Fayyad, Piatetsky-Shapiro, and Smyth (1996). The data mining trends, and challenges are described in: Grossman, Kasif, Moore, Rocke, and Ullman (1999). A compact introduction to data mining can be found in Grossman's "Data Mining FAQ," (2008). The EM algorithm is based upon the readable works of Blume (2002), Alpaydin (2004), and Jollois and Nadif (2007). The apriori and generalized sequential pattern algorithms were discovered by R. Agrawal, and R. Srikant in 1994, and 1995 respectively. The SVMs were developed by Vladimir Vapnik and his colleagues. Cristianini, and Shawe-Taylor (2000) have provided a comprehensive introduction to the SVMs. The backward propagation algorithm discussed in this chapter essentially follows the inimitable work of Theodoridis, and Koutroumbas (2009). The impact of this comprehensive work on this chapter should be evident to the perceptive reader.

A premier decision tree classification algorithm is CART. It is described in Breiman, Friedman, Olshen, and Stone (1984). Another equally efficient decision tree classification algorithm is C4.5. This is based upon the work of Quinlan (1993). The AdaBoost algorithm was introduced by Y. Freund, and R. E. Schapire. Rough set theory was developed by Zdislaw Pawlak and his collaborators. An erudite discussion on analysis of high-dimensional data is provided in Lee, and Verleysen (2007). The section on analysis of social networks is based upon the eminently readable books by Newman (2010), Zafarani, Abbasi, and Liu (2014), and Aggarwal (2015). Pioneering work on hidden Markov models was done by Baum and his coworkers (1966, 1967, 1970, and 1972). A readable introduction to hidden Markov models can be found in Rabiner (1989), and Rabiner and Juang (1993).

The original paper by Kalman (1960) is remarkable for its clarity and exposition. An interesting historical account of Kalman filtering can be found in Grewal, and Andrews (2010); and Sorenson (1970). The book by Kailath, Sayed, and Hassibi (2000) is the definitive scholarly work on Kalman filtering. Some other remarkable books on this subject are the works of Anderson, and Moore (1979); Brown, (1983); Grewal, and Andrews (1993); Jazwinski (1970), and Lewis (1986).

Problems

1. The expectation-maximization algorithm determines the parameters of component probability density functions, and the mixing probabilities from the set of data points $\mathcal{X} = \{x_1, x_2, \ldots, x_n\}$. The parametric form of m distributions, $\varphi_j (x \mid \theta_j), x \in \mathbb{R}, 1 \leq j \leq m$ is known. Thus the parameter vector is $\theta = (\theta_1, \theta_2, \ldots, \theta_m)$, the mixture probability vector is $\alpha = (\alpha_1, \alpha_2, \ldots, \alpha_m)$, and $\beta = (\theta, \alpha)$. For $\tau \in \mathbb{N}$, show that:

   (a) In the E-step of the algorithm the conditional probabilities are:

$$q_{ij}^{(\tau)} = \frac{\alpha_j^{(\tau)} \varphi_j \left(x_i \mid \theta_j^{(\tau)}\right)}{\sum_{k=1}^m \alpha_k^{(\tau)} \varphi_k \left(x_i \mid \theta_k^{(\tau)}\right)}; \quad 1 \leq i \leq n, \quad 1 \leq j \leq m$$

   (b) In the M-step of the algorithm:

$$\alpha_j^{(\tau+1)} = \frac{1}{n} \sum_{i=1}^n q_{ij}^{(\tau)}, \quad 1 \leq j \leq m$$

   Hint: See Blume (2002), Alpaydin (2004), Jollois, and Nadif (2007).

   (a) The conditional expectation of the complete log-likelihood function, given the data set $\mathcal{X}$, and the previous estimate $\beta^{(\tau)}$ is

$$\begin{aligned}
&\mathcal{Q}\left(\beta, \beta^{(\tau)}\right) \\
&= \mathcal{E}_{Z|\mathcal{X},\beta^{(\tau)}} \left\{L_{\mathcal{X},Z} (\beta) \mid \mathcal{X}, \beta^{(\tau)}\right\} \\
&= \mathcal{E}_{Z|\mathcal{X},\beta^{(\tau)}} \left\{ \sum_{i=1}^n \sum_{j=1}^m Z_{ij} \ln \left\{\alpha_j \varphi_j (x_i \mid \theta_j)\right\} \mid \mathcal{X}, \beta^{(\tau)} \right\}
\end{aligned}$$

$$= \sum_{i=1}^{n} \sum_{j=1}^{m} \mathcal{E}_{Z|\mathcal{X},\beta^{(\tau)}} \left\{ Z_{ij} \mid \mathcal{X}, \beta^{(\tau)} \right\} \{ \ln \alpha_j + \ln \varphi_j (x_i \mid \theta_j) \}$$

Let

$$q_{ij}^{(\tau)} = \mathcal{E}_{Z|\mathcal{X},\beta^{(\tau)}} \left\{ Z_{ij} \mid \mathcal{X}, \beta^{(\tau)} \right\}, \quad 1 \le i \le n, \text{ and } 1 \le j \le m$$

where the data points in the set $\mathcal{X}$ are drawn independently of each other, and Z_{ij} takes value either 0 or 1. Therefore

$$\begin{aligned} q_{ij}^{(\tau)} &= \mathcal{E}_{Z|\mathcal{X},\beta^{(\tau)}} \left\{ Z_{ij} \mid \mathcal{X}, \beta^{(\tau)} \right\} \\ &= P\left(Z_{ij} = 1 \mid x_i, \beta^{(\tau)} \right) \\ &= \frac{P\left(Z_{ij} = 1 \mid \beta^{(\tau)} \right) P\left(x_i \mid Z_{ij} = 1, \beta^{(\tau)} \right)}{P\left(x_i \mid \beta^{(\tau)} \right)} \\ &= \frac{\alpha_j^{(\tau)} \varphi_j \left(x_i \mid \theta_j^{(\tau)} \right)}{\sum_{k=1}^{m} \alpha_k^{(\tau)} \varphi_k \left(x_i \mid \theta_k^{(\tau)} \right)} \end{aligned}$$

(b) In the M-step, $\beta^{(\tau+1)}$ is determined. It is

$$\beta^{(\tau+1)} = \arg\max_{\beta} \left\{ \mathcal{Q}\left(\beta, \beta^{(\tau)} \right) \right\}$$

Therefore $\beta^{(\tau+1)}$ is the value of β obtained by maximizing $\mathcal{Q}\left(\beta, \beta^{(\tau)} \right)$, under the constraint $\sum_{j=1}^{m} \alpha_j = 1$. The method of Lagrange multipliers is used. The Lagrangian is

$$\begin{aligned} \mathcal{L}\left(\beta, \beta^{(\tau)}, \lambda \right) &= \mathcal{Q}\left(\beta, \beta^{(\tau)} \right) + \lambda \left\{ \sum_{j=1}^{m} \alpha_j - 1 \right\} \\ &= \sum_{i=1}^{n} \sum_{j=1}^{m} q_{ij}^{(\tau)} \{ \ln \alpha_j + \ln \varphi_j (x_i \mid \theta_j) \} \\ &\quad + \lambda \left\{ \sum_{j=1}^{m} \alpha_j - 1 \right\} \end{aligned}$$

$\partial \mathcal{L}\left(\beta, \beta^{(\tau)}, \lambda \right) / \partial \alpha_j = 0$ implies $\left(\sum_{i=1}^{n} q_{ij}^{(\tau)} + \lambda \alpha_j \right) = 0$, for $1 \le j \le m$. Therefore summing over all values of j's yields

$$\left(\sum_{i=1}^{n} \sum_{j=1}^{m} q_{ij}^{(\tau)} + \lambda \sum_{j=1}^{m} \alpha_j \right) = 0$$

That is, $(n + \lambda) = 0$. Therefore in the $(\tau + 1)$th iteration

$$\alpha_j^{(\tau+1)} = \frac{1}{n} \sum_{i=1}^{n} q_{ij}^{(\tau)}, \quad 1 \le j \le m$$

Maximization with respect to elements of the vector θ depends upon the component probability density functions.

2. The fuzzy clustering algorithm minimizes

$$J_q(\theta, r) = \sum_{i=1}^{n}\sum_{j=1}^{m} r_{ij}^q d(x_i, \theta_j)$$

by varying θ and r. The corresponding optimization problem is:

$$\min_{\theta, r} J_q(\theta, r), \quad \text{where } r_{ij} \in [0,1], \quad 1 \le i \le n, 1 \le j \le m$$

and the constraints are:

$$\sum_{j=1}^{m} r_{ij} = 1, \quad 1 \le i \le n; \quad \text{and } 0 < \sum_{i=1}^{n} r_{ij} < n, \quad 1 \le j \le m$$

Assuming $d(x_i, \theta_j) \ne 0$, for $1 \le i \le n$ and $1 \le j \le m$, determine the conditions for its optimal solution.
Hint: See Theodoridis, and Koutroumbas (2009). Minimize $J_q(\theta, r)$ with respect to r, and use the constraint $\sum_{j=1}^{m} r_{ij} = 1$, for $1 \le i \le n$. The corresponding Lagrangian is

$$\mathcal{L}(r) = \sum_{i=1}^{n}\sum_{j=1}^{m} r_{ij}^q d(x_i, \theta_j) - \sum_{i=1}^{n} \lambda_i \left\{ \sum_{j=1}^{m} r_{ij} - 1 \right\}$$

Therefore computing $\partial \mathcal{L}(r) / \partial r_{ij}$ and setting it equal to zero yields

$$r_{ij} = \left\{ \frac{\lambda_i}{q d(x_i, \theta_j)} \right\}^{1/(q-1)}, \quad 1 \le j \le m$$

Substitution of this result in the constraint $\sum_{j=1}^{m} r_{ij} = 1$, for $1 \le i \le n$ results in

$$\lambda_i = \frac{q}{\left\{ \sum_{k=1}^{m} \{d(x_i, \theta_k)\}^{-1/(q-1)} \right\}^{(q-1)}}$$

Use of the last two equations yields the stated result for r_{ij}^{-1}, which is:

$$r_{ij}^{-1} = \sum_{k=1}^{m} \left\{ \frac{d(x_i, \theta_j)}{d(x_i, \theta_k)} \right\}^{1/(q-1)}; \quad 1 \le i \le n, \quad 1 \le j \le m$$

Taking the partial derivative of $J_q(\theta, r)$ with respect to θ, and equating it to zero yields

$$\sum_{i=1}^{n} r_{ij}^q \frac{\partial d(x_i, \theta_j)}{\partial \theta_j} = 0, \quad 1 \le j \le m$$

3. Let $\mathcal{X}$ be the set of data points. The dissimilarity measure of the set of data points $\mathcal{X}$ is a function $d : \mathcal{X} \times \mathcal{X} \to \mathbb{R}_0^+$. The m-clustering of $\mathcal{X}$ is the partition of $\mathcal{X}$ into $m \in \mathbb{P} \backslash \{1\}$ sets called clusters. The clusters are $C_1, C_2, \ldots, C_m$.
Also, let s_i be a measure of dispersion of cluster C_i, $1 \le i \le m$. Dispersion is a measure of spread of the data points around their mean value v_i.

Let C_i and C_j be two different clusters. A measure of dissimilarity between two clusters is $\widehat{\mathcal{D}}(C_i, C_j) \triangleq d_{ij}$. Assume it to be symmetric. That is, $d_{ij} = d_{ji}$, for $1 \leq i, j \leq m$.
A similarity index ψ_{ij} between two clusters C_i and C_j is defined as:

$$\psi_{ij} = \frac{s_i + s_j}{d_{ij}}$$

where $d_{ij} \neq 0$. Show that ψ_{ij} satisfies the following properties:

(i) $\psi_{ij} \geq 0$. This implies that the similarity index is nonnegative.
(ii) $\psi_{ij} = \psi_{ji}$. This implies that the similarity index is symmetric.
(iii) If $s_i = 0$ and $s_j = 0$, then $\psi_{ij} = 0$. This implies that if the clusters reduce to a single point, then their similarity index is equal to zero.
(iv) If $s_j > s_k$ and $d_{ij} = d_{ik}$, then $\psi_{ij} > \psi_{ik}$. This implies that if a cluster C_i is at the same distance from two clusters C_j and C_k; then the cluster C_i is more similar to the cluster with the larger dispersion.
(v) If $s_j = s_k$ and $d_{ij} < d_{ik}$, then $\psi_{ij} > \psi_{ik}$. This implies that if two clusters C_j and C_k have the same dispersion, and cluster C_i is less dissimilar to cluster C_j than to cluster C_k; then cluster C_i is more similar to cluster C_j than to cluster C_k.

Hint: See Theodoridis, and Koutroumbas (2009).

4. Let $\mathcal{I}$ be the set of items used in the transactions, where $|\mathcal{I}| = d$. Prove that the total number of association rules which can be generated from the set $\mathcal{I}$ is equal to $\{3^d - 2^{d+1} + 1\}$.
Hint: See Simovici and Djeraba (2008). Consider the association rule $X \Rightarrow Y$. Let $|X| = i$, where $1 \leq i \leq d$. There are $\binom{d}{i}$ ways of selecting the itemset X. As $X \cap Y = \varnothing$, the itemset Y can be selected among the $(2^{d-i} - 1)$ nonempty subsets of $\mathcal{I} \backslash X$. Therefore the number of association rules is equal to

$$\begin{aligned}\sum_{i=1}^{d} \binom{d}{i} \left(2^{d-i} - 1\right) &= \sum_{i=1}^{d} \binom{d}{i} 2^{d-i} - \sum_{i=1}^{d} \binom{d}{i} \\ &= \left(3^d - 2^d\right) - \left(2^d - 1\right)\end{aligned}$$

The binomial theorem was used in the last step. The final result follows immediately.

5. Let $X \neq \varnothing$, $X \subseteq Y$, and $X \Rightarrow (Y - X)$ be an association rule r which does not satisfy the confidence threshold c_0. That is, $c(r) < c_0$. If $X_{sub} \Rightarrow (Y - X_{sub})$ is another association rule r_{sub}, where $X_{sub} \neq \varnothing$, and $X_{sub} \subseteq X$, then prove that r_{sub} also does not satisfy the confidence threshold. That is, $c(r_{sub}) < c_0$.
Hint: See Tan, Steinbach, and Kumar (2006). As per the hypothesis,

$$c(r) = c(X \Rightarrow (Y - X)) = \sigma(Y) / \sigma(X) < c_0$$

Further $X_{sub} \subseteq X$ implies $\sigma(X_{sub}) \geq \sigma(X)$. Therefore

$$\sigma(Y) / \sigma(X_{sub}) \leq \sigma(Y) / \sigma(X) < c_0$$

That is, $c(r_{sub}) = c(X_{sub} \Rightarrow (Y - X_{sub})) = \sigma(Y) / \sigma(X_{sub}) < c_0$.

6. Prove that Bayesian classification scheme minimizes classification error probability.
Hint: See Theodoridis, and Koutroumbas (2009). This result is established for only two classes $(m = 2)$. The net classification error probability is P_e. Thus

$$\begin{aligned} P_e &= P(x \in R_1, \omega_2) + P(x \in R_2, \omega_1) \\ &= P(x \in R_1 \mid \omega_2) P(\omega_2) + P(x \in R_2 \mid \omega_1) P(\omega_1) \\ &= P(\omega_2) \int_{R_1} p(x \mid \omega_2)\, dx + P(\omega_1) \int_{R_2} p(x \mid \omega_1)\, dx \\ &= \int_{R_1} P(\omega_2 \mid x)\, p(x)\, dx + \int_{R_2} P(\omega_1 \mid x)\, p(x)\, dx \end{aligned}$$

As $R_1 \cap R_2 = \varnothing$ and $R_1 \cup R_2 = \mathcal{R}$, we also have

$$P(\omega_1) = \int_{R_1} P(\omega_1 \mid x)\, p(x)\, dx + \int_{R_2} P(\omega_1 \mid x)\, p(x)\, dx$$

Combining, the last two equations yields

$$P_e = P(\omega_1) - \int_{R_1} \{P(\omega_1 \mid x) - P(\omega_2 \mid x)\}\, p(x)\, dx$$

The last result implies that P_e is minimized, if $R_1 \subset \mathcal{R}$ is the region in which $P(\omega_1 \mid x) > P(\omega_2 \mid x)$. Consequently, $R_2 = \mathcal{R} \backslash R_1$ is the region in which $P(\omega_2 \mid x) > P(\omega_1 \mid x)$.

7. Prove the perceptron convergence theorem.
Hint: See Siu, Roychowdhury, and Kailath (1995), and Haykin (2009). The theorem is established in four steps.
Step 0: Let $\|\cdot\|$ be the Euclidean norm. Let the initializing weight vector of the algorithm be $w^{(0)} = 0$. Recall that $w = (w_0, w_1, w_2, \ldots, w_t)$ is the extended weight vector.
The extended data point is $\varrho_i = (x_{i0}, x_{i1}, x_{i2}, \ldots, x_{in}), x_{i0} = 1$ for each $i = 1, 2, \ldots, n$. Assume that $\|\varrho_i\| \le L_{\max}$, for $1 \le i \le n$. As the set of data points are linearly separable, there exists a δ and $w^* \in \mathbb{R}^{t+1}$ such that $\|w^*\| = 1$, and $y_i \left(w^* \varrho_i^T\right) > \delta > 0$, for $i = 1, 2, \ldots, n$. Label the set of data points for which $y_i = +1$ by ω_1, and the set of data points for which $y_i = -1$ by ω_2. Therefore $\omega_1 \cap \omega_2 = \varnothing$, and $\mathcal{D} = \omega_1 \cup \omega_2$.
Step 1: In this step, an upper bound for $\|w^{(k)}\|$ is established. From the perceptron learning algorithm $w^{(k)} = w^{(k-1)} + \mu \Delta_i^{(k-1)} \varrho_i$. Therefore

$$\left\|w^{(k)}\right\|^2 = \left\|w^{(k-1)}\right\|^2 + \mu^2 \left|\Delta_i^{(k-1)}\right|^2 \|\varrho_i\|^2 + 2\mu \Delta_i^{(k-1)} w^{(k-1)} \varrho_i^T$$

The third term $2\mu \Delta_i^{(k-1)} w^{(k-1)} \varrho_i^T$ in the above equation is never positive. Therefore

$$\left\|w^{(k)}\right\|^2 \le \left\|w^{(k-1)}\right\|^2 + \mu^2 \left|\Delta_i^{(k-1)}\right|^2 \|\varrho_i\|^2$$

Applying the above inequality successively, and using the assumptions $w^{(0)} = 0$, and $\|\varrho_i\| \le L_{\max}$, we obtain

$$\left\|w^{(k)}\right\|^2 \le 4\mu^2 k L_{\max}^2$$

Therefore $\|w^{(k)}\| \le 2\mu\sqrt{k} L_{\max}$.
Step 2: In this step, a lower bound for $\|w^{(k)}\|$ is established. Suppose that the data point ϱ_i belongs to the set ω_1. This implies $w\varrho_i^T > 0$, as the data points are linearly separable. However, in the perceptron learning algorithm assume that $w^{(k)} \varrho_i^T < 0$ for $k = 1, 2, \ldots$. From the perceptron learning algorithm we have $w^{(k)} = w^{(k-1)} + \mu \Delta_i^{(k-1)} \varrho_i$. Therefore

$$\begin{aligned} w^{(k)}w^{*T} &= w^{(k-1)}w^{*T} + \mu\Delta_i^{(k-1)}\varrho_i w^{*T} \\ &\geq w^{(k-1)}w^{*T} + 2\mu\delta \end{aligned}$$

Applying the above inequality successively, we obtain

$$w^{(k)}w^{*T} \geq 2k\mu\delta$$

Use of Bunyakovsky-Cauchy-Schwartz inequality yields

$$\left\| w^{(k)} \right\| \|w^*\| \geq \left| w^{(k)}w^{*T} \right|$$

Combining the last two inequalities, and the assumption $\|w^*\| = 1$, yields $\left\| w^{(k)} \right\| \geq 2k\mu\delta$.
Step 3: From steps 1 and 2, we have $2\mu\sqrt{k}L_{\max} \geq 2k\mu\delta$. Thus $(L_{\max}/\delta)^2 \geq k$. Therefore the maximum number of iterations in the perceptron learning algorithm is equal to $k_{\max} = \left\lceil (L_{\max}/\delta)^2 \right\rceil$.

8. Gradient descent scheme is a numerical technique for finding the local minimum of a differentiable function. Let $f(x)$ be a differentiable function in $\widetilde{S} \subseteq \mathbb{R}^n$, where $x = (x_1, x_2, \ldots, x_n) \in \widetilde{S}$. The goal is to find a local minimum of this function. If $\alpha \in \widetilde{S}$, then $f(x)$ decreases fastest from α in the direction of the negative gradient of $f(x)$ at α, which is $-\nabla f(\alpha)$. Therefore if $\beta = (\alpha - \mu\nabla f(\alpha))$, and $\mu > 0$ is a small number, then $f(\alpha) \geq f(\beta)$. This observation suggests a scheme to determine the local minimum of $f(\cdot)$ iteratively.
Let $\alpha_0 \in \widetilde{S}$ be a guessestimate of the point at which the local minimum of $f(x)$ occurs. Generate the sequence $\alpha_0, \alpha_1, \alpha_2, \ldots$ as

$$\alpha_{n+1} = a_n - \mu_n \nabla f(\alpha_n), \quad \mu_n > 0, \text{ and } n = 0, 1, 2, \ldots$$

where the positive scalar μ_n is allowed to be dependent on n. Then $f(\alpha_0) \geq f(\alpha_1) \geq f(\alpha_2) \geq \cdots$, and the sequence $\alpha_0, \alpha_1, \alpha_2, \cdots$ converges to the desired local minimum. Develop a pseudocode to describe these steps algorithmically.
Hint: See Tan, Steinbach, and Kumar (2006).

9. The error rate ϵ_{err} of the AdaBoost classifier at the end of its training phase is

$$\epsilon_{err} = \frac{1}{n}\sum_{i=1}^{n} \delta\left(\widetilde{\phi}(x_i) \neq y_i\right)$$

where $\delta(\cdot)$ is equal to 1, if its argument is true, and 0 otherwise. Prove that $\epsilon_{err} \leq \prod_{j=1}^{N_{itr}} W_j$.
Hint: It can be shown that

$$w_i^{(N_{itr}+1)} = \frac{\exp\{-y_i F(x_i)\}}{n\prod_{j=1}^{N_{itr}} W_j}, \quad 1 \leq i \leq n$$

Note that, if $\widetilde{\phi}(x_i) \neq y_i$ then $y_i F(x_i) \leq 0$.
This implies that $\exp\{-y_i F(x_i)\} \geq 1$. Therefore

$$\begin{aligned} \epsilon_{err} &= \frac{1}{n}\sum_{i=1}^{n} \delta\left(\widetilde{\phi}(x_i) \neq y_i\right) \leq \frac{1}{n}\sum_{i=1}^{n} \exp\{-y_i F(x_i)\} \\ &= \prod_{j=1}^{N_{itr}} W_j \sum_{i=1}^{n} w_i^{(N_{itr}+1)} = \prod_{j=1}^{N_{itr}} W_j \end{aligned}$$

as $\sum_{i=1}^{n} w_i^{(N_{itr}+1)} = 1$.

10. Show that the volume of a t-dimensional hypersphere of radius $r \in \mathbb{R}^+$, is

$$V_t(r) = \frac{\pi^{t/2}}{\Gamma\left(\frac{t}{2}+1\right)} r^t, \quad t \in \mathbb{P}$$

In the above expression, $\Gamma(\cdot)$ is the gamma function.
Hint: Observe that $V_1(r) = 2r$, $V_2(r) = \pi r^2$ (area of a circle of radius r), and $V_3(r) = (4\pi/3) r^3$ (volume of a sphere of radius r). The general expression for $V_t(r)$ is derived in several steps.
Step 0: In order to develop motivation for the next step, we derive an expression for the volume $V_3(r)$ of a sphere.

$$V_3(r) = \int_{-r}^{r} V_2(\lambda)\, dz; \quad \lambda = r \sin\theta, \text{ and } z = r\cos\theta$$

Simplification of the above integral yields $V_3(r) = (4\pi/3) r^3$.
Step 1: Let $V_0(r) = 1$. As in the last step, we have

$$V_t(r) = \int_{-r}^{r} V_{t-1}(\lambda)\, dz; \quad \lambda = r\sin\theta,\ z = r\cos\theta, \text{ and } t \in \mathbb{P}$$

Define $V_t(r) = v_t r^t$, for $t \in \mathbb{N}$. Therefore, $v_0 = 1, v_1 = 2$, $v_2 = \pi$, $v_3 = 4\pi/3$, and

$$v_t = v_{t-1} \int_0^{\pi} \sin^t\theta\, d\theta, \quad t \in \mathbb{P}$$

Define $\vartheta_t = \int_0^{\pi} \sin^t\theta\, d\theta$, $t \in \mathbb{N}$. Therefore, $\vartheta_0 = \pi, \vartheta_1 = 2, \vartheta_2 = \pi/2, \vartheta_3 = 4/3$, and

$$v_t = v_{t-1}\vartheta_t, \quad t \in \mathbb{P}$$

Step 2: We evaluate ϑ_t by using beta function. It can be shown that

$$\vartheta_t = \int_0^{\pi} \sin^t\theta\, d\theta = B\left(\frac{t+1}{2}, \frac{1}{2}\right), \quad t \in \mathbb{N}$$

where $B(\cdot,\cdot)$ is the beta function. Therefore

$$\vartheta_t = \frac{\Gamma\left(\frac{t+1}{2}\right)\Gamma\left(\frac{1}{2}\right)}{\Gamma\left(\frac{t+2}{2}\right)}, \quad t \in \mathbb{N}$$

where $\Gamma(\cdot)$ is the gamma function. It can be checked that $\vartheta_0 = \pi$, and

$$\vartheta_t \vartheta_{t-1} = \frac{2\pi}{t}, \quad t \in \mathbb{P}$$

Step 3: Results in Steps 1 and 2 yield

$$v_t = \frac{2\pi}{t} v_{t-2}, \quad t \in \mathbb{P} \backslash \{1\}$$

As $v_0 = 1$ and $v_1 = 2$, successive application of the above recursive equation yields

$$v_t = \frac{\pi^{t/2}}{\Gamma\left(\frac{t}{2}+1\right)}, \quad t \in \mathbb{P}$$

The final result follows.

11. Let X_i, $1 \leq i \leq t$ be a sequence of mutually independent and identically distributed Gaussian random variables, where $t \in \mathbb{P}$. Also $X_i \sim \mathcal{N}\left(0, \sigma^2\right)$, for $1 \leq i \leq t$. Let $Z = \sum_{i=1}^{t} X_i^2$, and $R = \sqrt{Z}$. Determine the probability density functions of the random variables Z and R.
Hint: This problem is addressed in several steps.
Step 1: Let X be a Gaussian random variable with zero mean, and variance σ^2. That is, $X \sim \mathcal{N}\left(0, \sigma^2\right)$. Also let $Y = X^2$. If $f_Y(\cdot)$ and $\mathcal{M}_Y(\cdot)$ are the probability density and moment generating functions of the random variable Y respectively, then

$$f_Y(y) = \begin{cases} 0, & y \in (-\infty, 0] \\ \dfrac{y^{-1/2}}{\sqrt{2\pi}\sigma} \exp\left\{-\frac{y}{2\sigma^2}\right\}, & y \in (0, \infty) \end{cases}$$

$$\mathcal{M}_Y(w) = \frac{1}{\left(1 - 2\sigma^2 w\right)^{1/2}}, \quad 2\sigma^2 w < 1$$

Step 2: It is given that $X_i \sim \mathcal{N}\left(0, \sigma^2\right)$, for $1 \leq i \leq t$; and X_i is independent of X_j for $i \neq j$ and $1 \leq i, j \leq t$. If $Z = \sum_{i=1}^{t} X_i^2$, and $f_Z(\cdot)$, and $\mathcal{M}_Z(\cdot)$ are its probability density and moment generating functions respectively, then

$$\mathcal{M}_Z(w) = \frac{1}{\left(1 - 2\sigma^2 w\right)^{t/2}}, \quad 2\sigma^2 w < 1$$

$$f_Z(z) = \begin{cases} 0, & z \in (-\infty, 0] \\ \dfrac{z^{t/2-1}}{\Gamma\left(\frac{t}{2}\right)\left(2\sigma^2\right)^{t/2}} \exp\left\{-\frac{z}{2\sigma^2}\right\}, & z \in (0, \infty) \end{cases}$$

The mean of the random variable Z is

$$\mathcal{E}(Z) = \left.\frac{d\mathcal{M}_Z(w)}{dw}\right|_{w=0} = t\sigma^2$$

Step 3: Let

$$R = \sqrt{Z}$$

and $f_R(\cdot)$ be its probability density function, then

$$f_R(r) = \begin{cases} 0, & r \in (-\infty, 0] \\ \dfrac{2r^{t-1}}{\Gamma\left(\frac{t}{2}\right)\left(2\sigma^2\right)^{t/2}} \exp\left\{-\frac{r^2}{2\sigma^2}\right\}, & r \in (0, \infty) \end{cases}$$

The maximum value of the probability density function $f_R(\cdot)$ occurs at

$$r = \sigma\sqrt{t-1}$$

12. Let A be an $n \times n$ real-valued, symmetric, and positive definite matrix. Similarly, let C be an $m \times m$ real-valued, symmetric, and positive definite matrix. Also B is a real-valued $m \times n$ matrix. Let $D = \left(A + B^T C B\right)$. Prove that the matrix D is positive definite.
Hint: Note that D is a symmetric matrix. Let x be a real-valued column matrix of size n. As the matrix A is positive definite, we have $x^T A x > 0, \forall\, x \neq 0$. Therefore, D is a positive definite

matrix, if it is shown that $x^T B^T C B x \geq 0$, for all $x \neq 0$. Let $Bx \triangleq y$. That is, it needs to be shown that $y^T C y \geq 0$, for all y. Consider two cases.
In the first case, if $y = 0$, then $y^T C y = 0$.
In the second case, if $y \neq 0$, then $y^T C y > 0$, as the matrix C is positive definite.
Combining, the two cases results in $y^T C y \geq 0$ for all values of y.
Thus for all $x \neq 0$, we have $x^T D x > 0$. Therefore D is a positive definite matrix.

13. This problem occurs in the optimization step of the derivation of the Kalman filter. Minimize the trace of the matrix

$$P = (I - KH) P^- (I - KH)^T + KRK^T$$

by varying K. Show that the value of K that minimizes $tr(P)$ is

$$K = P^- H^T \left(HP^- H^T + R\right)^{-1}$$

and the corresponding value of P is

$$P = (I - KH) P^-$$

Hint: This result is established via two different methods. The first method uses matrix calculus, and the second uses a matrix version of "completing the square."
Observe that the matrices P^- and P are each positive definite. This is true, as P_0 is assumed to be positive definite, and the use of recursion in the Kalman filter algorithm, and the last problem establishes this assertion.
Method 1: The following results are used from matrix calculus. The matrix A is independent of matrix X.

(a) If the matrix product XA is a square matrix, then

$$\frac{\partial\, tr(XA)}{\partial X} = A^T$$

(b) The matrices A and XAX^T are square, and not necessarily of the same size. Further, the matrix A is symmetric. Then

$$\frac{\partial\, tr\left(XAX^T\right)}{\partial X} = 2XA$$

We have

$$P = P^- - KHP^- - P^- H^T K^T + K\left(HP^- H^T + R\right) K^T$$

Note that

$$\frac{\partial\, tr(KHP^-)}{\partial K} = (HP^-)^T = P^- H^T$$

As $tr\left(P^- H^T K^T\right) = tr\left(\left(P^- H^T K^T\right)^T\right) = tr(KHP^-)$. Therefore

$$\frac{\partial\, tr\left(P^- H^T K^T\right)}{\partial K} = \frac{\partial\, tr(KHP^-)}{\partial K} = P^- H^T$$

Note that $\left(HP^- H^T + R\right)$ is a symmetric matrix. Therefore

$$\frac{\partial\, tr\left(K\left(HP^{-}H^{T}+R\right)K^{T}\right)}{\partial K}=2K\left(HP^{-}H^{T}+R\right)$$

Finally

$$\frac{\partial\, tr\left(P\right)}{\partial K}=-2P^{-}H^{T}+2K\left(HP^{-}H^{T}+R\right)$$

Setting $\partial\, tr\left(P\right)/\partial K=0$, we obtain

$$K=P^{-}H^{T}\left(HP^{-}H^{T}+R\right)^{-1}$$

It still needs to be established that this value of K is indeed the minimizing value of $tr\left(P\right)$. Note that on the right hand side of the expression for P; in the quadratic term $K\left(HP^{-}H^{T}+R\right)K^{T}$, the factor $\left(HP^{-}H^{T}+R\right)$ is positive definite. Therefore, the stated value of K indeed minimizes $tr\left(P\right)$.

The value of P at the minimizing value of $K=P^{-}H^{T}\left(HP^{-}H^{T}+R\right)^{-1}$ is

$$\begin{aligned}P&=P^{-}-KHP^{-}-P^{-}H^{T}K^{T}+K\left(HP^{-}H^{T}+R\right)K^{T}\\&=P^{-}-KHP^{-}-P^{-}H^{T}K^{T}\\&\quad+P^{-}H^{T}\left(HP^{-}H^{T}+R\right)^{-1}\left(HP^{-}H^{T}+R\right)K^{T}\\&=P^{-}-KHP^{-}-P^{-}H^{T}K^{T}+P^{-}H^{T}K^{T}\\&=\left(I-KH\right)P^{-}\end{aligned}$$

Method 2: This method uses a matrix version of "completing the square."
Hint: See Brown (1983). From the Method 1, we have

$$P=P^{-}-KHP^{-}-P^{-}H^{T}K^{T}+K\left(HP^{-}H^{T}+R\right)K^{T}$$

The above expression for P can be viewed as a quadratic in K. The term linear in K is $-\left(KHP^{-}+P^{-}H^{T}K^{T}\right)$, and the term quadratic in K is

$$K\left(HP^{-}H^{T}+R\right)K^{T}$$

Observe that the matrix $\left(HP^{-}H^{T}+R\right)$ is symmetric, and positive definite. Let

$$\left(HP^{-}H^{T}+R\right)\triangleq MM^{T}$$

This leads to

$$P=P^{-}-KHP^{-}-P^{-}H^{T}K^{T}+KMM^{T}K^{T}$$

The square can be completed, and P expressed as

$$P=P^{-}+\left(KM-N\right)\left(KM-N\right)^{T}-NN^{T}$$

where N does not depend on K. Comparison of the two expressions for P results in

$$KMN^{T}+NM^{T}K^{T}=KHP^{-}+P^{-}H^{T}K^{T}$$

The above equation is satisfied, if we let

$$N=P^{-}H^{T}\left(M^{T}\right)^{-1}$$

Note that in the expression $P = \left\{P^- + (KM - N)(KM - N)^T - NN^T\right\}$, the first and third term are independent of K. Only the middle term depends upon K. Observe that the diagonal elements of the matrix $(KM - N)(KM - N)^T$ will all be nonnegative. Therefore, the trace of matrix P is minimized, if we let

$$(KM - N) = 0$$

This leads to

$$\begin{aligned} K &= NM^{-1} = P^- H^T \left(M^T\right)^{-1} M^{-1} \\ &= P^- H^T \left(MM^T\right)^{-1} = P^- H^T \left(HP^- H^T + R\right)^{-1} \end{aligned}$$

14. In the derivation of the Kalman filter algorithm, optimal expressions for P and K were derived. These are:

$$P = (I - KH) P^-, \quad \text{and} \quad K = P^- H^T \left(HP^- H^T + R\right)^{-1}$$

Derive the following alternate expressions for P and K.

(a) $P = \left\{(P^-)^{-1} + H^T R^{-1} H\right\}^{-1}$

(b) $K = PH^T R^{-1}$

Hint:

(a) Observe that

$$P = (I - KH) P^- = P^- - KHP^-$$

Substitution of $K = P^- H^T \left(HP^- H^T + R\right)^{-1}$ in the above equation leads to

$$P = \left\{P^- - P^- H^T \left(HP^- H^T + R\right)^{-1} HP^-\right\}$$

The stated result follows by using the matrix inversion lemma.

(b) We have

$$\begin{aligned} K &= P^- H^T \left(HP^- H^T + R\right)^{-1} \\ &= P^- H^T \left(R^{-1} R\right) \left(HP^- H^T + R\right)^{-1} \\ &= P^- H^T R^{-1} \left(R^{-1}\right)^{-1} \left(HP^- H^T + R\right)^{-1} \\ &= P^- H^T R^{-1} \left(HP^- H^T R^{-1} + I\right)^{-1} \end{aligned}$$

Thus

$$\begin{aligned} K \left(HP^- H^T R^{-1} + I\right) &= P^- H^T R^{-1} \\ K &= (I - KH) P^- H^T R^{-1} \end{aligned}$$

Observe that $P = (I - KH) P^-$. The stated result follows.

15. A discrete-time Kalman filter in unit dimension is considered. Let the state, input, and output variables be $x_k, y_k, z_k \in \mathbb{R}$ respectively.

Further, the state transition, input (control), and observation matrices are all scalars. Let $A_k = a \in \mathbb{R}$, and $B_k = b \in \mathbb{R}$ for all values of $k \in \mathbb{N}$. Also let, $H_k = h \in \mathbb{R}$ for all values of $k \in \mathbb{P}$. The dynamical system is specified as

$$x_{k+1} = ax_k + by_k + w_k, \quad k \in \mathbb{N}$$
$$z_k = hx_k + v_k, \quad k \in \mathbb{P}$$

The initial state variable x_0 is random. Its mean is $\mathcal{E}(x_0) = \mu_0$, and its variance is

$$\mathcal{E}\left\{(x_0 - \mu_0)^2\right\} = P_0$$

The uncertainty in the model is introduced via the noise random variable w_k for $k \in \mathbb{N}$. Each of these random variables has a zero mean and variance q. Also $\mathcal{E}(w_k w_j) = 0$ for all $k \neq j$. The measurement noise random variable v_k for $k \in \mathbb{P}$ models the measurement noise. The mean of each of these random variables is zero, and the variance of each of the random variables is r. Also $\mathcal{E}(v_k v_j) = 0$ for all $k \neq j$.
Assume that these two sequences of random variables are uncorrelated. Further, these two random sequences are also uncorrelated with the initial random state variable x_0.
Analyze this scalar Kalman filter.
Hint: See Lewis (1986). Observe that the Kalman gain is

$$K_k = \frac{hP_k^-}{\left(h^2 P_k^- + r\right)}, \quad k \in \mathbb{P}$$

Also

$$P_k = \left\{1 - \frac{h^2 P_k^-}{\left(h^2 P_k^- + r\right)}\right\} P_k^- = \frac{rP_k^-}{\left(h^2 P_k^- + r\right)}, \quad k \in \mathbb{P}$$

This leads to

$$P_{k+1}^- = a^2 P_k + q, \quad k \in \mathbb{N}$$
$$P_{k+1}^- = \frac{a^2 r P_k^-}{\left(h^2 P_k^- + r\right)} + q, \quad k \in \mathbb{P}$$

We also have for $k \in \mathbb{P}$

$$\begin{aligned}
\widehat{x}_k &= \widehat{x}_k^- + K_k\left(z_k - h\widehat{x}_k^-\right) \\
&= \widehat{x}_k^-\left(1 - hK_k\right) + K_k z_k \\
&= \widehat{x}_k^-\left\{1 - \frac{h^2 P_k^-}{\left(h^2 P_k^- + r\right)}\right\} + \frac{hP_k^-}{\left(h^2 P_k^- + r\right)} z_k \\
&= \frac{r}{\left(h^2 P_k^- + r\right)} \widehat{x}_k^- + \frac{hP_k^-}{\left(h^2 P_k^- + r\right)} z_k
\end{aligned}$$

and for $k \in \mathbb{N}$

$$\widehat{x}_{k+1}^- = a\widehat{x}_k + by_k$$

Therefore, for $k \in \mathbb{P}$

$$\widehat{x}_{k+1}^- = \frac{ar}{\left(h^2 P_k^- + r\right)} \widehat{x}_k^- + \frac{ahP_k^-}{\left(h^2 P_k^- + r\right)} z_k + by_k$$

We next consider three cases. These are: perfect measurements, completely unreliable measurements, and no process noise.

(a) *Perfect measurements*: In this case, $r \to 0$.

$$K_k = \frac{1}{h}, \quad P_k = 0, \quad \widehat{x}_k = \frac{1}{h} z_k, \quad \text{for } k \in \mathbb{P}$$

$$P_{k+1}^- = q, \quad \widehat{x}_{k+1}^- = \left(\frac{a}{h} z_k + b y_k\right), \quad \text{for } k \in \mathbb{N}$$

(b) *Completely unreliable measurements*: In this case, $r \to \infty$.

$$K_k = 0, \quad P_k = P_k^-, \quad P_{k+1}^- = \left(a^2 P_k^- + q\right), \quad \widehat{x}_k = \widehat{x}_k^-, \quad \text{for } k \in \mathbb{P}$$

$$\widehat{x}_{k+1}^- = \left(a \widehat{x}_k + b y_k\right), \quad \text{for } k \in \mathbb{N}$$

Observe that an explicit expression for P_k^- can be determined from the relationship $P_{k+1}^- = \left(a^2 P_k^- + q\right)$. Thus

$$P_k^- = a^{2k} P_0 + q \frac{\left(1 - a^{2k}\right)}{\left(1 - a^2\right)}, \quad k \in \mathbb{P}$$

and $P_0^- = P_0$. For $|a| < 1$, P_k^- tends towards a stable solution for large values of k.
(c) *No process noise*: In this case, $q \to 0$. Observe that

$$P_{k+1}^- = a^2 P_k, \quad k \in \mathbb{N}$$

$$P_{k+1}^- = \frac{a^2 r P_k^-}{\left(h^2 P_k^- + r\right)}, \quad k \in \mathbb{P}$$

Therefore

$$\left(P_{k+1}^-\right)^{-1} = \frac{1}{a^2} \left(P_k^-\right)^{-1} + \frac{h^2}{a^2 r}, \quad k \in \mathbb{P}$$

The above relationship admits to a solution

$$\begin{aligned}
\left(P_k^-\right)^{-1} &= \frac{1}{P_0 a^{2k}} + \frac{h^2}{a^2 r} \frac{\left(1 - a^{-2(k-1)}\right)}{\left(1 - a^{-2}\right)} \\
&= \frac{1}{P_0 a^{2k}} + \frac{h^2}{r} \frac{\left(a^2 - a^{2k}\right)}{\left(1 - a^2\right) a^{2k}}
\end{aligned}$$

for $k \geq 2$.

References

1. Aggarwal, C. C., 2015. *Data Mining*: *The Textbook*, Springer-Verlag, Berlin, Germany.
2. Agrawal, R., and Srikant, R., 1994. "Fast Algorithms for Mining Association Rules," Proceedings of the 20th VLDB Conference, Santiago, Chile, pp. 487-499.
3. Agrawal, R., and Srikant, R., 1995. "Mining Sequential Patterns," Proceedings of International Conference on Data Engineering, Taipei, Taiwan, pp. 3-14.
4. Alpaydin, E., 2004. *Introduction to Machine Learning*, The MIT Press, Cambridge, Massachusetts.

5. Anderson, B. D. O., and Moore, J. B., 1979. *Optimal Filtering*, Prentice-Hall, Englewood Cliffs, New Jersey.
6. Baum, L. E., and Petrie, 1966. "Statistical Inference for Probabilistic Functions of Finite State Markov Chains," Ann. Math. Stat., Vol. 37, No. 1, pp. 1554-1563.
7. Baum, L. E., and Egon, J. A., 1967. "An Inequality with Application to Statistical Estimation for Probabilistic Functions of a Markov Process and to a Model for Ecology," Bull. Amer. Math. Soc., Vol. 73, No. 3, pp. 360-363.
8. Baum, L. E., Petrie, T., Soules, G., and Weiss, N., 1970. "A Maximization Technique Occurring in the Statistical Analysis of Probabilistic Functions of Markov Chains," Ann. Math. Stat., Vol. 41, No. 1, pp. 164-171.
9. Baum, L. E., 1972. "An Inequality and Associated Maximization Technique in Statistical Estimation for Probabilistic Functions of Markov Processes," Inequalities, Vol. 3, No. 1, pp. 1-8.
10. Berry, M. W., and Browne M., 1999. *Understanding Search Engines, Mathematical Modeling and Text Retrieval,* Society of Industrial and Applied Mathematics, Philadelphia, Pennsylvania.
11. Bishop, C. M., 1995. *Neural Networks for Pattern Recognition,* Oxford University Press, Oxford, Great Britain.
12. Bishop, C. M., 2006. *Pattern Recognition and Machine Learning*, Springer-Verlag, Berlin, Germany.
13. Blume, M., 2002. "Expectation Maximization: A Gentle Introduction," available at http://campar.in.tum.de /twiki/pub/Main/MoritzBlume/EMGaussianMix.pdf. Retrieved September 13, 2017.
14. Breiman, L., Friedman, J. H., Olshen, R. A., and Stone., C. J., 1984. *Classification and Regression Trees*, Chapman and Hall/CRC Press, New York, New York.
15. Brown, R. G., 1983. *Introduction to Random Signal Analysis and Kalman Filtering*, John Wiley & Sons, Inc., New York, New York.
16. Chakrabarti, D., and Faloutsos, C., 2006. "Graph Mining: Laws, Generators, and Algorithms," ACM Computing Surveys, Vol. 38, Issue 1, Article 2.
17. Chakrabarti, D., and Faloutsos, C., 2012. *Graph Mining: Laws, Tools, and Case Studies*, Morgan and Claypool Publishers, San Rafael, California.
18. Chakrabarti, S., 2003. *Mining the Web, Discovering Knowledge from the Hypertext Data,* Morgan Kaufmann Publishers, San Francisco, California.
19. Cios, K. J., Pedrycz, W., Swiniarski, R. W., and Kurgan, L. A., 2007. *Data Mining: A Knowledge Discovery Approach*, Springer-Verlag, Berlin, Germany.
20. Cristianini, N., and Shawe-Taylor, J., 2000. *An Introduction to Support Vector Machines,* Cambridge University Press, Cambridge, Great Britain.
21. Dempster, A. P., Laird, N. M., and Rubin, D. B., 1977. "Maximum Likelihood from Incomplete Data via the EM Algorithm," J. Roy. Stat. Soc., Vol. 39(B), No. 1, pp. 1-38.
22. Devroye, L., Györfi, L., and Lugosi, G., 1996. *A Probabilistic Theory of Pattern Recognition,* Springer-Verlag, Berlin, Germany.
23. Domingos, P. M., 2012. "A Few Useful Things to Know about Machine Learning," Comm. of the ACM, Vol. 55, No. 10, pp. 78-87.
24. Eldén, L., 2007. *Matrix Methods in Data Mining and Pattern Recognition,* Society of Industrial and Applied Mathematics, Philadelphia, Pennsylvania.
25. Faragher, R., 2012. "Understanding the Basis of the Kalman Filter via a Simple and Intuitive Derivation," IEEE Signal Processing Magazine, Vol. 128, No. 5, pp. 128-1332.
26. Fayyad, U., Piatetsky-Shapiro, G., and Smyth, P., 1996. "The KDD Process for Extracting Useful Knowledge from Volumes of Data," Comm. of the ACM, Vol. 39, No. 11, pp. 27-34.
27. Grewal, M. S., and Andrews, A. P., 1993. *Kalman Filtering: Theory and Practice*, Prentice-Hall, Upper Saddle River, New Jersey.

28. Grewal, M. S., and Andrews, A. P., 2010. "Applications of Kalman Filtering in Aerospace 1960 to the Present," IEEE Control Systems Magazine, Vol. 30, No. 3, pp. 69-78.
29. Grossman, R., Kasif, S., Moore, R., Rocke, D., and Ullman, J., 1999. "Opportunities and Challenges: A Report of Three NSF Workshops on Mining Large, Massive, and Distributed Data," available at http://pubs.rgrossman.com/dl/misc-001.pdf. Retrieved February 16, 2015.
30. Grossman, R. L., 2008. "Data Mining FAQ," available at http://pubs.rgrossman.com/faq/data_mining_faq.html. Retrieved April 17, 2010.
31. Han, J., Kamber, M., and Pei, J., 2012. *Data Mining: Concepts and Techniques*, Third Edition, Morgan Kaufmann Publishers, San Francisco, California.
32. Hastie, T., Tibshirani, R., and Friedman, J., 2008. *The Elements of Statistical Learning: Data Mining, Inference, and Prediction,* Second Edition, Springer-Verlag, Berlin, Germany.
33. Haykin, S., 2009. *Neural Networks and Learning Machines,* Third Edition, Pearson Prentice-Hall, Upper Saddle River, New Jersey.
34. Ho, Y. C., and Lee, R. C. K., 1964. "A Bayesian Approach to Problems in Stochastic Estimation and Control," IEEE Transactions on Automatic Control, Vol. 9, No. 4, pp. 333-339.
35. Jazwinski, A. H., 1970. *Stochastic Processes, and Filtering Theory*, Academic Press, New York, New York.
36. Jollois, F.-X., and Nadif, M., 2007. "Speed-Up for the Expectation-Maximization Algorithm for Clustering Categorical Data," J. Glob. Opt., Vol. 37, Issue 4, pp. 513-525.
37. Kailath, T., Sayed, A. H., and Hassibi, B., 2000. *Linear Estimation*, Prentice-Hall, Upper Saddle River, New Jersey.
38. Kalman, R. E., 1960. "A New Approach to Linear Filtering and Prediction Problems," ASME Trans., J. Basic Eng., Ser. D, Vol. 82, No. 1, pp. 35-45.
39. Kernighan, B., and Lin, S. 1970. "An Efficient Heuristic Procedure for Partitioning Graphs," Bell System Technical Journal, Vol. 49, Issue 2, pp. 291-307.
40. Klösgen, W., and Żytkow, J. M., Editors, 2002. *Handbook of Data Mining and Knowledge Discovery*, Oxford University Press, Oxford, Great Britain.
41. Kumar, R., Raghavan, P., Rajagopalan, S., and Tomkins, A., 1999. "Extracting Large-Scale Knowledge Bases from the Web," Proceedings of the 25th VLDB Conference, pp. 639-650.
42. Lee, J. A., and Verleysen, M. 2007. *Nonlinear Dimensionality Reduction*, Springer-Verlag, Berlin, Germany.
43. Lewis, F. L., 1986. *Optimal Estimation with Introduction to Stochastic Control Theory*, John Wiley & Sons, Inc., New York, New York.
44. Lewis, F. L., Jagannathan, S., and Yeşildirek. 2001. *Neural Network Control of Robot Manipulators and Nonlinear Systems,* Taylor and Francis, New York, New York.
45. Liu, B., 2011. *Web Data Mining: Exploring Hyperlinks, Contents, and Usage Data*, Second Edition, Springer-Verlag, Berlin, Germany.
46. Manning, C. D., Raghavan, P., and Schütze, H., 2009. *An Introduction to Information Retrieval,* Cambridge University Press, Cambridge, Great Britain.
47. Mitchell, T. M., 1997. *Machine Learning*, McGraw-Hill Book Company, New York, New York.
48. Newman, M. E. J., 2010. *Networks: An Introduction*, Oxford University Press, Oxford, Great Britain.
49. Nisbet, R., Elder, J., and Miner, G., 2009. *Handbook of Statistical Analysis and Data Mining Applications*, Academic Press, New York, New York.
50. Pasquier, N., Bastide, Y., Taouil, R., and Lakhal, L., 1999. "Efficient Mining of Association Rules Using Closed Itemset Lattices," Information Systems, Vol. 24, No. 1, pp. 25-46.
51. Pawlak, Z., 1991. *Rough Sets: Theoretical Aspects of Reasoning about Data*, Kluwer Academic Publishers, Norwell, Massachusetts.

52. Pawlak, Z., 2002. "Rough Sets, Decision Algorithms, and Bayes' Theorem," European Journal of Operations Research, Vol. 136, No. 1, pp. 181-189.
53. Pawlak, Z., and Skowron, A., 2007. "Rudiments of Rough Sets," Information Sciences, Vol. 177, Issue 1, pp. 3-27.
54. Quinlan, J. R., 1993. *C4.5: Programs for Machine Learning*, Morgan Kaufmann Publishers, San Francisco, California.
55. Rabiner, L. R., 1989. "A Tutorial on Hidden Markov Models and Selected Applications in Speech Recognition," Proc. IEEE, Vol. 77, No. 2, pp. 257-286.
56. Rabiner, L. R., and Juang, B.-H., 1993. *Fundamentals of Speech Recognition*, Prentice-Hall, Upper Saddle River, New Jersey.
57. Russell, S., 1998. "The EM Algorithm," available at http://www.cs.berkeley.edu/~russell/classes/cs281/s98/em.ps. Retrieved February 16, 2015.
58. Simovici, D. A., and Djeraba, C., 2008. *Mathematical Tools for Data Mining: Set Theory, Partial Orders, Combinatorics,* Springer-Verlag, Berlin, Germany.
59. Siu, K., Roychowdhury, V., and Kailath, T., 1995. *Discrete Neural Computation: A Theoretical Foundation*, Prentice-Hall, Englewood Cliffs, New Jersey.
60. Skowron, A., Komorowski, J., Pawlak, Z., and Polkowski, L., 2002. "Rough Sets Perspective on Data and Knowledge," in *Handbook of Data Mining and Knowledge Discovery*. Editors: Klösgen, W., and Żytkow, J. M., Oxford University Press, Oxford, Great Britain, pp. 134-149.
61. Söderström, T., 2002. *Discrete-time Stochastic Systems*, Springer-Verlag, Berlin, Germany.
62. Sorenson, H. W., 1970. "Least-Squares Estimation from Gauss to Kalman," IEEE Spectrum, Vol. No. 7, pp. 63-68.
63. Stengel, R. F., 1986. *Stochastic Optimal Control, Theory and Application*, John Wiley & Sons, Inc., New York, New York.
64. Tan, P. N., Steinbach, M., and Kumar, V., 2006. *Introduction to Data Mining*, Addison-Wesley Publishing Company, New York, New York.
65. Terejanu, G., 2008. "Discrete Kalman Filter Tutorial." Web-tutorial, Department of Computer Science and Engineering, University at Buffalo, Buffalo, New York, available at http://www.cse.sc.edu/~terejanu/files/tutorialKF.pdf. Retrieved 14 May 2017.
66. Theodoridis, S., and Koutroumbas, K., 2009. *Pattern Recognition*, Fourth Edition, Academic Press, New York, New York.
67. Walczak, B., and Massart, D. L., 1999. "Rough Sets Theory," Chemometrics and Intelligent Laboratory Systems, Vol. 47, Issue 1, pp. 1-16.
68. Witten, H., and Frank, E., 2005. *Data Mining: Practical Machine Learning Tools and Techniques*, Morgan Kaufmann Publishers, San Francisco, California.
69. Yasdi, R., 1995. "Combining Rough Sets Learning and Neural Learning-Method to deal with Uncertain and Imprecise Information," Neurocomputing., Vol. 17, Issue 1, pp. 61-84.
70. Zafarani, R., Abbasi, M. A., and Liu, H., 2014. *Social Media Mining: An Introduction,* Cambridge University Press, Cambridge, Great Britain.
71. Zaki, M. J., and Meira Jr., W., 2014. *Data Mining and Analysis: Fundamental Concepts and Algorithms,* Cambridge University Press, Cambridge, Great Britain.

CHAPTER 12

Quantum Computation, Communication, and Cryptography

$$H_q(x) = -x\log_q \frac{x}{(q-1)} - (1-x)\log_q(1-x)$$

$$0 \le x \le 1, \; q \ge 2$$

q-ary Hilbert entropy

David Hilbert. David Hilbert was born on 23 January, 1862 in Königsberg, Prussia (now Kaliningrad, Russia). Hilbert received his Ph.D. from the University of Königsberg. He was a student of C. L. F. Lindemann, and a friend of A. Hurwitz and H. Minkowski. Hilbert became a professor of mathematics at the University of Göttingen in 1895.

Hilbert's contributions extend to several branches of mathematics, most notably to geometry, algebraic number theory, mathematical physics, calculus of variations, and functional analysis. He also proved Waring's theorem. Hilbert also stated a rigorous set of axioms of Euclidean geometry in *Foundations of Geometry*, (1899). In the year 1900, at the Paris International Congress of mathematics, Hilbert proposed 23 unsolved problems. He thought that the mathematicians of the twentieth century should focus on these problems. To date, several of these problems have been solved. David Hilbert passed away on 14 February, 1943 in Göttingen, Germany.

12.1 Introduction

One of the greatest intellectual achievements of the twentieth century was the birth of quantum mechanics. The discipline of quantum mechanics has been immensely successful in describing the existence and motion of atomic sized particles. An equally significant field of human endeavor which revolutionized everybody's lives, either directly or indirectly was computer science. It is the marriage of these two disciplines that is explored in this chapter. Quantum mechanics and certain quantum computing elements are initially discussed in this chapter. Quantum computation is also defined precisely. Furthermore, models of computation based upon the work of Alan Turing are also developed. Quantum algorithms, quantum information theory, quantum communication theory, quantum error-correction, and quantum cryptography are also discussed.

The inclusion of this subject matter in this textbook does not prophesy that quantum computing and communication are the wave of the future. Actually the practical developments in this branch of knowledge lag significantly behind their theoretical possibilities. Nevertheless, the rewards offered by a knowledge of this subject are limitless, or at least worth exploring. Basics of number theory, analysis, linear algebra, probability theory, and information theory are required to study this chapter.

Notation. Let $\mathfrak{B} = \{0,1\}, n \in \mathbb{P}, N = 2^n$, and $x \in \mathbb{Z}_N$. Also let

$$x = \sum_{i=0}^{n-1} 2^i x_i, \quad \text{where} \quad x_i \in \mathfrak{B}, 0 \le i \le (n-1)$$

This is the binary expansion of x. It is compactly represented as $x_{n-1}x_{n-2}\ldots x_0$. In the rest of the chapter, we denote $x_{n-1}x_{n-2}\ldots x_0$ by x, and sometimes the symbols $\mathfrak{B}^{(n)}$ and $\mathbb{Z}_N$ are used interchangeably. □

12.2 Quantum Mechanics

The beginning of the twentieth century witnessed the birth of quantum mechanics. Niels Bohr (1885-1962), M. Born (1882-1970), Louis de Broglie (1892-1987), P. A. M. Dirac (1902-1984), Albert Einstein (1879-1955), W. Heisenberg (1901-1976), W. Pauli (1900-1958), Max Planck (1858-1947), and E. Schrödinger (1887-1961) are among several physicists who were responsible for the early development of quantum physics.

Branches of mathematics such as, linear algebra, group theory, and probability theory, are used in building the magnificent edifice of quantum mechanics. Mathematical entities such as, Hilbert spaces are used for describing laws of quantum mechanics. Recall that a Hilbert space is a complete inner product space. In this section preliminaries for stating the laws of quantum mechanics are outlined. Pauli matrices, concept of pure and mixed states, and density operator are discussed. Evolution of a quantum system, concept of observables, postulates of quantum mechanics, Heisenberg's uncertainty principle, quantum operations, and quantum entanglement are also elucidated.

12.2.1 Preliminaries

Certain quantum mechanical notation for a vector in a vector space and some other notations are developed. Some properties of Hermitian matrices are also stated.

Notation

1. *Ket-bra notation*: A vector ψ in Hilbert space $\mathcal{H}$ is denoted by $|\psi\rangle$. It is called the *ket-vector*. The ket-vector is actually a column vector. The Hermitian transpose of the vector $|\psi\rangle$ in Hilbert space $\mathcal{H}$ is denoted by $\langle\psi|$. It is called the *bra-vector*. The bra-vector is a row vector. This notation was invented by the physicist Dirac, and it is called the "ket-bra" notation. It appears to be a derivative of the word "bracket."
   Also note that if a vector can be expanded with N number of basis vectors, then the corresponding Hilbert space is sometimes denoted by $\mathcal{H}_N$. Recall that a Hilbert space is a complete and complex vector space.
2. *Tensor product of vectors*: The tensor product of vectors $|\varphi\rangle$ and $|\psi\rangle$ is denoted by $|\varphi\rangle \otimes |\psi\rangle$, or simply $|\varphi\rangle |\psi\rangle$. Since the ket vector is a column vector (column matrix), the definition of tensor product of two ket vectors is similar to the tensor product of two matrices. In order to represent
$$|\varphi_1\rangle \otimes |\varphi_2\rangle \otimes \ldots \otimes |\varphi_n\rangle$$
   the following notation is also used.
$$|\varphi_1\rangle |\varphi_2\rangle \ldots |\varphi_n\rangle \quad \text{or} \quad |\varphi_1, \varphi_2, \ldots, \varphi_n\rangle \quad \text{or} \quad |\varphi_1 \varphi_2 \ldots \varphi_n\rangle$$
3. *Tensor power notation*: The tensor product
$$\underbrace{|\varphi\rangle |\varphi\rangle \ldots |\varphi\rangle}_{n \text{ times}}$$
   is denoted by $|\varphi^{\otimes n}\rangle$. Sometimes the Hilbert space $\mathcal{H}_N$ is also denoted by $\mathcal{H}_r^{\otimes n}$, where $N = r^n$ and $r \in \mathbb{P}\backslash \{1\}$.

4. *Inner product of two vectors*: Recall that it is possible to define the inner product between two complex vectors in two different ways. See the chapter on applied analysis. In the current chapter, the axioms $[I_1], [I_2]$ and $[I_3']$ (stated in the chapter on applied analysis) are used to define the inner product space. In this definition, the inner product space is linear in the second argument. Let $|\varphi\rangle$ and $|\psi\rangle$ be two vectors which belong to a Hilbert space $\mathcal{H}$.
   (a) The inner product of vectors $|\varphi\rangle$ and $|\psi\rangle$ is denoted by either $\langle\varphi, \psi\rangle$ or $\langle\varphi \mid \psi\rangle$.
   (b) Let A be a complex square matrix. The inner product of vectors $|\varphi\rangle$ and $A|\psi\rangle$ is denoted by $\langle\varphi \mid A \mid \psi\rangle$, where it is assumed that the matrix-vector product $A|\psi\rangle$ is well-defined. Note that $\langle\varphi \mid A \mid \psi\rangle$ is also equal to the inner product of the vectors $A^\dagger|\varphi\rangle$ and $|\psi\rangle$, where $A^\dagger$ is the Hermitian transpose of the matrix A.
5. *Commutator and anti-commutator*: Let A and B be complex square matrices of the same size.
   (a) The commutator between two operators A and B is: $[A, B] = AB - BA$.
   (b) The anti-commutator between two operators A and B is: $\{A, B\} = AB + BA$.
   (c) The matrix A *commutes* with matrix B if $[A, B] = 0$.
   (d) The matrix A *anti-commutes* with matrix B if $\{A, B\} = 0$. □

Examples 12.1. Some examples illustrating the notation.

1. Let $i = \sqrt{-1}$, and

$$|\psi\rangle = \begin{bmatrix} 3 \\ -5i \end{bmatrix}$$

then $\langle\psi| = \begin{bmatrix} 3 & 5i \end{bmatrix}$.

2. Let $|\varphi\rangle = \begin{bmatrix} 2i & -7 \end{bmatrix}^T$ and $|\psi\rangle = \begin{bmatrix} 3 & -5i \end{bmatrix}^T$, then the tensor product of these two vectors is

$$|\varphi\rangle \otimes |\psi\rangle = |\varphi\rangle\,|\psi\rangle = \begin{bmatrix} 2i \\ -7 \end{bmatrix} \otimes \begin{bmatrix} 3 \\ -5i \end{bmatrix} = \begin{bmatrix} 2i \times 3 \\ 2i \times (-5i) \\ -7 \times 3 \\ -7 \times (-5i) \end{bmatrix} = \begin{bmatrix} 6i \\ 10 \\ -21 \\ 35i \end{bmatrix}$$

3. The inner product of the two vectors $|\varphi\rangle = \begin{bmatrix} 2i & -7 \end{bmatrix}^T$ and $|\psi\rangle = \begin{bmatrix} 3 & -5i \end{bmatrix}^T$ is

$$\langle\varphi, \psi\rangle = \langle\varphi \mid \psi\rangle = \begin{bmatrix} -2i & -7 \end{bmatrix} \begin{bmatrix} 3 \\ -5i \end{bmatrix} = 29i$$

4. The matrix $|\psi\rangle\langle\psi|$ is Hermitian.
5. $tr\,(|\psi\rangle\langle\varphi|) = \langle\varphi \mid \psi\rangle$, where $tr\,(\cdot)$ is the trace operator of a square matrix.
6. Let $\mathfrak{B} = \{0, 1\}$, $n \in \mathbb{P}$, $N = 2^n$, and $i \in \mathbb{Z}_N$. Let $i = \sum_{j=0}^{n-1} 2^j i_j$, where $i_j \in \mathfrak{B}$, $0 \le j \le (n-1)$. Then the binary representation of i is $i_{n-1}i_{n-2}\ldots i_0$, and a possible representation of $|i\rangle$ is $|i_{n-1}i_{n-2}\ldots i_0\rangle$. □

Observations 12.1. Let the matrices A and B be square, complex, and of the same size.

1. $[A, B]^\dagger = [B^\dagger, A^\dagger]$
2. $[A, B] = -[B, A]$
3.

$$AB = \frac{[A, B] + \{A, B\}}{2}$$

4. If the matrices A and B are Hermitian, then $i\,[A, B]$ is Hermitian.

5. Let A and B be Hermitian matrices of the same size. These matrices can be diagonalized via the same set of eigenvectors if and only if $[A, B] = 0$. Such matrices are said to be simultaneously diagonalizable. □

Observations 12.2. Some properties of the trace operator are listed below.

1. Suppose $|\psi\rangle$ is a ket-vector with d elements, and A is a matrix of size d. Then

$$tr\left(A\left|\psi\right\rangle\left\langle\psi\right|\right) = \left\langle\psi \mid A \mid \psi\right\rangle$$

2. Let $\mathcal{H}_d$ be a Hilbert space of dimension d. Also let

$$\mathcal{B} = \{|\psi_i\rangle \mid 1 \le i \le d\}$$

be an orthonormal basis of $\mathcal{H}_d$. The trace of a matrix A of size d is

$$tr\left(A\right) = \sum_{i=1}^{d} \left\langle\psi_i \mid A \mid \psi_i\right\rangle$$

□

Projector onto a Subspace of a Vector Space

Consider a vector space $\mathcal{V}$ of dimension n. Suppose that $\mathcal{W}$ is a vector subspace of dimension $k \le n$. Using Gram-Schmidt orthogonalization procedure, construct an orthonormal basis $y_i, 1 \le i \le n$ for $\mathcal{V}$ such that $y_i, 1 \le i \le k$ is an orthonormal basis of $\mathcal{W}$. Then

$$P \triangleq \sum_{i=1}^{k} y_i y_i^{\dagger}$$

is the *projector* onto subspace $\mathcal{W}$. The matrix P is called the *projector matrix*. Observe that this definition is independent of the orthonormal basis $y_i, 1 \le i \le k$. Furthermore, the matrix P is Hermitian. Also, if $x \in \mathcal{V}$ then

$$Px = \sum_{i=1}^{k} y_i y_i^{\dagger} x = \sum_{i=1}^{k} \left(y_i^{\dagger} x\right) y_i$$

Thus the vector $Px \in \mathcal{W}$ is said to be the projection of the vector $x \in \mathcal{V}$ on to the vector subspace $\mathcal{W}$.

The *orthogonal complement* of P is the operator $Q = (I - P)$, where I is the identity matrix. Thus Q can be regarded as the projector onto the vector subspace spanned by the vectors $y_i, (k+1) \le i \le n$.

Some Properties of Hermitian and Projector Matrices

A self-adjoint operator in a finite Hilbert space $\mathcal{H}$ has a spectral decomposition. Recall that a linear transformation specified by a complex matrix A is a self-adjoint operator if the matrix A is Hermitian. Relevant properties of a Hermitian matrix are summarized below. Let A be an $n \times n$ Hermitian matrix, and $\lambda_i \in \mathbb{R}$ and x_i be the eigenvalue and a corresponding eigenvector respectively for $1 \leq i \leq n$. Also assume that the eigenvectors are normalized to unity, that is $x_i^\dagger x_i = 1$ for $1 \leq i \leq n$. Define $P_i = x_i x_i^\dagger$ for $1 \leq i \leq n$, where P_i is called the projector matrix onto the eigenspace of A with eigenvalue λ_i. The immediately useful properties of the projector matrices are:

(a) The matrix P_i is Hermitian, for $1 \leq i \leq n$.
(b) $P_i P_j = \delta_{ij} P_i$, where $\delta_{ij} = 1$ if $i = j$, and $\delta_{ij} = 0$ if $i \neq j$, for $1 \leq i, j \leq n$.
(c) $tr\,(P_i) = 1$ for $1 \leq i \leq n$.
(d) $A = \sum_{i=1}^{n} \lambda_i P_i$. This is the spectral representation of A.
(e) $tr\,(A) = \sum_{i=1}^{n} \lambda_i$
(f) $\sum_{i=1}^{n} P_i = I$.
(g) $\sum_{i=1}^{n} P_i P_i^\dagger = I$. This relationship is called the *completeness equation.*
(h) If the length of the vector $|\psi\rangle$ is unity, then $1 = \sum_{i=1}^{n} \langle \psi \mid P_i \mid \psi \rangle$. The expression $\langle \psi \mid P_i \mid \psi \rangle$ is nonnegative, therefore it can be interpreted as a probability.

In the property (h), the probability $\langle \psi \mid P_i \mid \psi \rangle$ is associated with the eigenvector x_i for $1 \leq i \leq n$. For reasons to be discussed shortly, it is more convenient to associate a probability with each distinct eigenvalue. If the eigenvalue λ_k occurs with multiplicity g_k, and the corresponding orthonormal eigenvectors are x_{k_i}, for $1 \leq i \leq g_k$, then the probability $p\,(\lambda_k)$ associated with this eigenvalue is defined as

$$p\,(\lambda_k) = \sum_{i=1}^{g_k} \langle \psi \mid P_{k_i} \mid \psi \rangle$$

Also define the sum of the P_{k_i}'s by $\mathcal{P}_k$. That is

$$\mathcal{P}_k = \sum_{i=1}^{g_k} P_{k_i}$$

Thus $\mathcal{P}_k$ is the sum of the projector matrices associated with eigenvalue λ_k. Linearity yields $p\,(\lambda_k) = \langle \psi \mid \mathcal{P}_k \mid \psi \rangle$.

Pauli Matrices

Pauli matrices are a set of four 2×2 matrices which are very useful in quantum mechanics. As the name indicates, these matrices are named after the physicist W. Pauli. These are:

$$\sigma_0 = \begin{bmatrix} 1 & 0 \\ 0 & 1 \end{bmatrix}, \quad \sigma_x = \begin{bmatrix} 0 & 1 \\ 1 & 0 \end{bmatrix}$$

$$\sigma_y = \begin{bmatrix} 0 & -i \\ i & 0 \end{bmatrix}, \quad \sigma_z = \begin{bmatrix} 1 & 0 \\ 0 & -1 \end{bmatrix}$$

Observe that σ_0 is a 2×2 identity matrix, and $\sigma_y = i\sigma_x\sigma_z$. Among several different uses, these matrices have found application in developing error models of quantum computation and communication.

12.2.2 Pure and Mixed States, and Density Operator

Quantum mechanical systems are described in terms of pure and mixed states, and density operator. Language of vectors and matrices is used in describing pure and mixed states.

Definitions 12.1. *We define state space, state vector, and a pure state.*

1. *State space and state vector: The Hilbert space is called the state space, and the vector specifying the quantum system state is called the state vector. The elements of a vector in this space can be complex numbers.*
2. *Pure state: A pure state of a quantum mechanical system has the following attributes.*
   (a) *A quantum mechanical system whose state is known exactly is said to be in pure state.*
   (b) *A pure state at time t is specified by a vector of unit length in Hilbert space.*
   (c) *An isolated quantum mechanical system is in pure state.* □

Observation 12.3. Let
$$\mathcal{B} = \{|x_i\rangle \mid i \in \mathbb{P}\}$$
be an orthonormal set of basis vectors of a Hilbert space $\mathcal{H}$. Also let $|\psi\rangle$ be a nonzero state vector of a physical system, then

$$|\psi\rangle = \sum_{i\in\mathbb{P}} \alpha_i |x_i\rangle, \quad \sum_{i\in\mathbb{P}} |\alpha_i|^2 = 1; \quad \alpha_i \in \mathbb{C}, \ \forall\, i \in \mathbb{P} \tag{12.1}$$

Notice that the length of this state vector is unity. Furthermore, the state $|\psi\rangle$ is said to be the superposition of basis states. Also $|\psi\rangle$ and $e^{i\alpha}|\psi\rangle$ represent the same physical state, as $\left|e^{i\alpha}\right| = 1$. It is also possible to have a finite number of basis vectors in a Hilbert space, in which case the summations in the above equations have a finite number of terms. □

It is possible for certain quantum systems to be a combination of several isolated systems. Such examples are called *mixed* and *unisolated* systems. The state representing this type of system is called a *mixed state*. In order to study mixed systems the concept of *density matrix* is introduced. Assume that the unisolated quantum system is represented by m isolated systems. A natural representation of a mixed state is to assign a probability to each of these m isolated systems, and associate with the mixed state a *density operator*. These are next defined.

Definitions 12.2. *We define mixed state and density operator.*

1. *Mixed state: Let a mixed quantum system be represented by $m \in \mathbb{P}\backslash\{1\}$ isolated systems. Assign a nonzero probability p_i to the ith isolated system, whose state vector is $|\psi_i\rangle$, $1 \le i \le m$, such that $\sum_{i=1}^m p_i = 1$. Denote the state of the mixed system by $[\psi\rangle$. Then*

$$\begin{aligned} [\psi\rangle &= (p_1, |\psi_1\rangle) \uplus (p_2, |\psi_2\rangle) \uplus \ldots \uplus (p_m, |\psi_m\rangle) \\ &= \biguplus_{i=1}^{m} (p_i, |\psi_i\rangle) \end{aligned} \tag{12.2}$$

   where the symbol $\uplus$ is introduced to separate the different states. Thus a quantum system is in mixed state, if it is a mixture of different pure states. Alternately, a mixed quantum state is said to be a statistical ensemble of pure states.
2. *Density operator: The density operator of a state of a quantum mechanical system is denoted by ρ.*
   (a) *The density operator of a pure state $|\psi\rangle$ is simply $\rho = |\psi\rangle\langle\psi|$.*

(b) *Consider a mixed state $[\psi\rangle$. Let the density operator of the ith pure state of the mixed state be denoted by $\rho_i = |\psi_i\rangle \langle\psi_i|$, then the density operator of the mixed state is*

$$\rho = \sum_{i=1}^{m} p_i \rho_i \tag{12.3}$$

Therefore the mixed state is represented by $\{(p_i, \rho_i) \mid 1 \leq i \leq m\}$. Sometimes, the density operator is also said to be the state of the quantum system. □

Example 12.2. Let $p_1 = 1/4, p_2 = 3/4, (p_1 + p_2) = 1$; and $\alpha_1 = \sqrt{p_1} = 1/2, \alpha_2 = \sqrt{p_2} = \sqrt{3}/2, (\alpha_1^2 + \alpha_2^2) = 1$. Also let

$$|\psi_1\rangle = \frac{1}{\sqrt{2}} \begin{bmatrix} 1 \\ 1 \end{bmatrix}, \quad |\psi_2\rangle = \frac{1}{\sqrt{2}} \begin{bmatrix} 1 \\ -1 \end{bmatrix}$$

It can be verified that $\sum_{i=1}^{2} p_i |\psi_i\rangle$ is not a state vector, and $\sum_{i=1}^{2} \alpha_i |\psi_i\rangle$ is a state vector of unit length. The probability that this quantum mechanical system is in state $|\psi_1\rangle$ is equal to $\alpha_1^2 = p_1$. Similarly, the probability that this quantum mechanical system is in state $|\psi_2\rangle$ is equal to $\alpha_2^2 = p_2$. □

Observations 12.4. Let $[\psi\rangle$ represent a mixed and unisolated system. It is specified by $[\psi\rangle = \biguplus_{i=1}^{m} (p_i, |\psi_i\rangle)$. Also, let its density matrix be ρ. The density operator of the corresponding ith pure state is ρ_i, for $1 \leq i \leq m$.

1. Following properties of the density matrix ρ_i for any i are:
   (a) The matrix ρ_i is self-adjoint (Hermitian).
   (b) The matrix $\rho_i^2 = \rho_i$.
   (c) The matrix ρ_i has a unit trace.
2. The matrix ρ is Hermitian.
3. The trace of matrix ρ is equal to unity.
4. A matrix ρ is a density operator if and only if the trace of the matrix ρ is unity and it is a positive semidefinite matrix. A positive semidefinite matrix like ρ is often said to be a *positive* operator in physics literature.
5. The eigenvalues of the density matrix ρ are real, nonnegative, and sum to unity. Let the matrix ρ be of size n. Also let λ_i and x_i be its eigenvalue and a corresponding eigenvector for $1 \leq i \leq n$, then

$$\rho = \sum_{i=1}^{m} \lambda_i x_i x_i^{\dagger}$$

   where $\lambda_i \geq 0, x_i^{\dagger} x_i = 1$, for $1 \leq i \leq n$, and $\sum_{i=1}^{n} \lambda_i = 1$. The above representation of the density operator is called its *spectral decomposition*.
6. Let $\rho = \sum_{i=1}^{m} p_i |\psi_i\rangle \langle\psi_i|$ be the density operator of a mixed state, and A be a square matrix compatible with ρ. Then

$$tr\,(A\rho) = \sum_{i=1}^{m} p_i \langle\psi_i \mid A \mid \psi_i\rangle$$

7. Some basic properties of the density matrix ρ.
   (a) A state is pure if and only if $\rho^2 = \rho$.

(b) If the state is pure, then $tr\left(\rho^2\right) = 1$.
(c) If the state is mixed, then $tr\left(\rho^2\right) < 1$. □

Some of the above observations are proved in the problem section. The density matrix in general is not a unique representation of the mixed state of an unisolated quantum system. This occurs for example, if $p_i = p_j$ then the corresponding $|\psi_i\rangle$ and $|\psi_j\rangle$ are not necessarily unique. Assume that $p_i = p_j = p$, then in the expression for ρ, $\{p_i |\psi_i\rangle \langle\psi_i| + p_j |\psi_j\rangle \langle\psi_j|\}$ can be replaced by $p\{|\theta_i\rangle \langle\theta_i| + |\theta_j\rangle \langle\theta_j|\}$, where $|\theta_i\rangle$ and $|\theta_j\rangle$ are related to $|\psi_i\rangle$ and $|\psi_j\rangle$ via a unitary transformation.

Purification of a Mixed State

Purification is one of several techniques to study quantum computation and information. Purification is a process which translates a mixed state into a pure state. It is also a procedure to evaluate a composite system. The process of purification can be described in terms of a mechanism called partial trace operation.

Definition 12.3. *Consider physical systems A and B. Let AB represent a mixture of these two physical systems. The state of this mixed system is specified by the density matrix ρ_{AB}. The reduced density operator for system A is defined by $\rho_A = tr_B(\rho_{AB})$, where $tr_B(\cdot)$ is a map of operators called the partial trace over system B. More precisely, let $|\alpha_1\rangle$ and $|\alpha_2\rangle$ be any two state vectors in the state space of system A. Similarly, let $|\beta_1\rangle$ and $|\beta_2\rangle$ be any two state vectors in the state space of system B. Then*

$$tr_B(|\alpha_1\rangle \langle\alpha_2| \otimes |\beta_1\rangle \langle\beta_2|) = |\alpha_1\rangle \langle\alpha_2| \, tr(|\beta_1\rangle \langle\beta_2|) \tag{12.4}$$

where the operator $tr(\cdot)$ on the right-hand side of the above equation is the trace operator of a square matrix. □

Continuing our discussion of purification of a mixed state, let ρ_X be the density matrix of a quantum mechanical system X with mixed state. This mixed state belongs to Hilbert space $\mathcal{H}_n$. Let the orthonormal decomposition (spectral representation) of ρ_X be

$$\rho_X = \sum_{i=1}^{n} p_i |\psi_i\rangle \langle\psi_i|$$

where $p_i \geq 0$ for $1 \leq i \leq n$, $tr(\rho_X) = \sum_{i=1}^{n} p_i = 1$, and $|\psi_i\rangle, 1 \leq i \leq n$ are orthonormal state vectors in $\mathcal{H}_n$. The state X is purified as follows. Create a system Y which has the same state space as system X, which is $\mathcal{H}_n$. Its orthonormal basis set is $|\varphi_i\rangle, 1 \leq i \leq n$. Define $|\beta_i\rangle = |\psi_i\rangle \otimes |\varphi_i\rangle, 1 \leq i \leq n$, and a state vector $|\alpha\rangle \in \mathcal{H}_n \otimes \mathcal{H}_n$ as

$$|\alpha\rangle = \sum_{i=1}^{n} \sqrt{p_i} \, |\beta_i\rangle$$

Observe that, the square of the length of the state vector $|\alpha\rangle$ is equal to $\sum_{i=1}^{n} p_i = 1$. Thus its length is equal to unity. It can be shown that $tr_Y(|\alpha\rangle \langle\alpha|) = \rho_X$. The state vector $|\alpha\rangle$ is said to be the purification of ρ_X, and it is a pure state. See the problem section for a proof of this assertion.

12.2.3 Evolution of a Quantum System

In the last subsection pure and mixed states, and density matrices were introduced. This enables us to describe the evolution of a closed quantum mechanical system. A quantum mechanical system is closed if it does not interact with the outside world. The evolution of a quantum mechanical system is described by a unitary transformation.

Observations 12.5. A unitary transformation (matrix) U is a mapping of a vector $|\psi\rangle \in \mathcal{H}_n$, to a vector $|\varphi\rangle \in \mathcal{H}_n$ such that:

1. U is a unitary matrix. That is, $UU^{-1} = UU^{\dagger} = I$, the $n \times n$ identity matrix.
2. $U|\psi\rangle = |\varphi\rangle$.
3. $\langle \alpha \mid \psi \rangle = \langle U\alpha \mid U\psi \rangle$ for all $|\alpha\rangle, |\psi\rangle \in \mathcal{H}_n$. □

The evolution is described in terms of Schrödinger's equation, which is named after the physicist Erwin Schrödinger.

Schrödinger's Equation

The evolution of a closed quantum mechanical system is characterized by a unitary transformation $U(\cdot,\cdot)$. Let $|\psi(t_1)\rangle$ be the state vector at time t_1. If $t_2 \geq t_1$, then

$$|\psi(t_2)\rangle = U(t_1,t_2)|\psi(t_1)\rangle \tag{12.5a}$$

The operator $U(t_1,t_2)$ is also called the time development operator. The actual time evolution of the state of a closed quantum system is specified by the *Schrödinger's* equation,

$$i\hbar\frac{d|\psi(t)\rangle}{dt} = \mathfrak{H}(t)|\psi(t)\rangle \tag{12.5b}$$

where $\hbar$ is the Planck's constant, and $\mathfrak{H}(t)$ is called the *Hamiltonian* of the closed system. The Hamiltonian $\mathfrak{H}(t)$ is a self-adjoint operator. Thus a knowledge of the Hamiltonian of a system provides a complete knowledge of the evolution of the system. If the Hamiltonian is time independent, then it is simply denoted by $\mathfrak{H}$. In this instance

$$U(t_1,t_2) = \exp\left\{-\frac{i(t_2-t_1)\mathfrak{H}}{\hbar}\right\} \tag{12.5c}$$

It should be noted that the determination of the Hamiltonian of a system is indeed a very difficult problem. Some properties of the time development operator $U(t_1,t_2)$ are:

(a) $U(t,t) = I$, the identity operator.
(b) $U(t_1,t_3) = U(t_1,t_2)U(t_2,t_3)$, where $t_3 \geq t_2 \geq t_1$.
(c) $U(t_1,t_2) = U(t_2,t_1)^{\dagger} = \{U(t_2,t_1)\}^{-1}$, where $\{U(t_2,t_1)\}^{-1}$ is the inverse of $U(t_2,t_1)$.

Evolution of Mixed States

The evolution of mixed states is next described. Let $|x(t)\rangle$ be a pure state. Its evolution is described by a unitary transformation $U(\cdot,\cdot)$. Thus $|x(t_2)\rangle = U(t_1,t_2)|x(t_1)\rangle$ and

$$|x(t_2)\rangle\langle x(t_2)| = U(t_1,t_2)\,|x(t_1)\rangle\langle x(t_1)|\,U(t_1,t_2)^{\dagger}$$

In analogy with the above equation, the density operator $\rho(t)$ of a mixed state is given by

$$\rho(t_2) = U(t_1,t_2)\,\rho(t_1)\,U(t_1,t_2)^{\dagger}$$

Assume that the Hamiltonian of the system is independent of time and

$$U(t_1,t_2) = \exp\left\{-\frac{i(t_2-t_1)\mathfrak{H}}{\hbar}\right\}$$

Then

$$i\hbar\frac{d\rho(t)}{dt} = \{\mathfrak{H}\rho(t) - \rho(t)\mathfrak{H}\} = [\mathfrak{H},\rho(t)]$$

The right-hand side of the above equation is not equal to zero unless the operators $\mathfrak{H}$ and $\rho(t)$ commute. This equation can also be considered a generalization of the Schrödinger's equation. The above equation is called the *Liouville-von Neumann* evolution equation in physics literature.

12.2.4 Observables

The concept of observables is next explained. In order to draw conclusions about the state of a physical system, it has to be observed. An operator which facilitates *measurement* of a quantum mechanical system is called an *observable* (or measurement operator). This operator is assumed to be self-adjoint. If the state space is finite, then the operator is a Hermitian matrix. This is stated formally in the following definition.

Definitions 12.4. *Some useful definitions about observables.*

1. *An observable or measurement operator in a finite-dimensional state space $\mathcal{H}_n$ is a Hermitian matrix A.*
2. *The Hermitian matrix A can be expressed as $A = \sum_{i=1}^{n} \lambda_i P_i$, where P_i is the projector onto the eigenspace of A with eigenvalue λ_i, for $1 \le i \le n$. The average value of the observable A (Hermitian operator) in the normalized state $|\psi\rangle$ is given by*

$$\langle A\rangle = \sum_{i=1}^{n} \lambda_i p(\lambda_i) \tag{12.6}$$

where $p(\lambda_i) = \langle\psi \mid P_i \mid \psi\rangle$. □

Lemma 12.1. *The average value of an A observable is given by*

$$\langle A\rangle = \sum_{i=1}^{n} \lambda_i \langle\psi \mid P_i \mid \psi\rangle = \langle\psi \mid A \mid \psi\rangle = tr(A\,|\psi\rangle\langle\psi|) = tr(|\psi\rangle\langle\psi|\,A) \tag{12.7}$$

Proof. The proof is left to the reader. □

The aim of a measurement is to determine: the possible results of the measurement, the probability of obtaining a specific result, and lastly the state of the physical system after the measurement is finished. Let the measurement operator be a Hermitian matrix A of size n. Also let λ_i and P_i be the eigenvalue and the corresponding projector matrix, for $1 \le i \le n$. Recall that $\sum_{i=1}^{n} P_i = I$.

The numerical outcome of the measurement of a state $|\psi(t)\rangle$ with respect to an observable A, is one of the eigenvalues of this self-adjoint operator. Let the eigenvalue λ_k, of the matrix A be the measured quantity. Then the probability of its occurrence $p(\lambda_k)$, is given

$$\begin{aligned} p(\lambda_k) &= \langle \psi \mid \mathcal{P}_k \mid \psi \rangle \\ &= tr\left(\mathcal{P}_k |\psi\rangle \langle\psi|\right) \\ &= tr\left(|\psi\rangle \langle\psi| \mathcal{P}_k\right) \end{aligned}$$

where $\mathcal{P}_k$ is the sum of the projector matrices associated with the eigenvalue λ_k. Note that $\mathcal{P}_k = P_k$ if the eigenvalue λ_k occurs only once. If the output of the measurement is λ_k, then the state of the physical system immediately after measurement is the normalized value

$$\frac{\mathcal{P}_k |\psi\rangle}{\{\langle \psi \mid \mathcal{P}_k \mid \psi \rangle\}^{1/2}}$$

This formalism can also be extended to quantum mechanical systems with mixed states.

12.2.5 Postulates of Quantum Mechanics

At the beginning of the twentieth century, several experimental facts about the behavior of small physical particles were assimilated. Based upon these experimental results, a mathematical framework was developed to interpret these results. This description is often called the *Copenhagen interpretation* of quantum mechanics. It is made precise via the following four postulates. These postulates essentially define the state of a quantum system, the dynamics of state evolution, describe the measurement operator, and finally describe a composite quantum mechanical system.

Postulate 1. *State.* The state of an isolated quantum mechanical system is completely specified in a Hilbert space by a state vector of unit length.

Postulate 2. *Evolution.* A unitary transformation completely describes the evolution of a closed quantum mechanical system. Let the state of a quantum mechanical system at time t_1 be $|\psi\rangle$, and its state at time t_2 be $|\psi'\rangle$, then $|\psi'\rangle = U|\psi\rangle$, where U is a unitary operator. Furthermore, the unitary operator U depends only on the times t_1 and t_2.

Postulate 3. *Measurement.* The numerical outcome of the measurement of a normalized state $|\psi(t)\rangle$ with respect to an observable A is one of the eigenvalues of this self-adjoint operator. Let the eigenvalue λ_k be the measured quantity, then the probability of its occurrence $p(\lambda_k)$ is given $p(\lambda_k) = \langle \psi \mid \mathcal{P}_k \mid \psi \rangle$, where $\mathcal{P}_k$ is the sum of the projector matrices associated with the eigenvalue λ_k. Also the state of the physical system immediately after measurement is

$$\frac{\mathcal{P}_k |\psi\rangle}{\{\langle \psi \mid \mathcal{P}_k \mid \psi \rangle\}^{1/2}} \tag{12.8}$$

Also the various outcome probabilities $p(\lambda_k)$'s, sum to unity. That is, $\sum_k p(\lambda_k) = 1$.

Postulate 4. *Composite systems.* The state space of a composite physical system is specified via a tensor product of the state spaces of individual physical systems. Let $|\psi_i(t)\rangle$ be the state space of system i at time t, for $1 \le i \le n$. Then the state space vector of the composite physical system at time t is $|\psi_1(t)\rangle \otimes |\psi_2(t)\rangle \otimes \ldots \otimes |\psi_n(t)\rangle$.

From the above postulates it can be concluded that, the density operator of an isolated quantum mechanical system has a unit trace. Furthermore, if the density operators ρ_i of each composite system is known for $1 \leq i \leq n$, then the density operator of the composite system ρ is equal to $\rho_1 \otimes \rho_2 \otimes \ldots \otimes \rho_n$.

Based upon these postulates, it can be inferred that quantum theory provides a mathematical framework for the representation of the states and observables, and also rules for determining the probabilities of different outcomes of measurements.

12.2.6 Heisenberg's Uncertainty Principle

The *Heisenberg*'s uncertainty principle is one of the significant results in quantum mechanics. This principle was first formulated by the physicist Werner Heisenberg. Heisenberg's uncertainty principle has acquired a lore of its own in everyday life. It is established by using the above Hilbert-space based mathematical model.

Statement of Heisenberg's Uncertainty Principle

Let A and B be two observables, $|\psi\rangle$ be a quantum state, and I be the identity operator. Also let

$$\Delta A = A - \langle\psi| A |\psi\rangle I \tag{12.9a}$$

$$\Delta B = B - \langle\psi| B |\psi\rangle I \tag{12.9b}$$

Then

$$\langle\psi| (\Delta A)^2 |\psi\rangle \langle\psi| (\Delta B)^2 |\psi\rangle \geq \frac{1}{4} |\langle\psi| [A, B] |\psi\rangle|^2 \tag{12.9c}$$

Proof. See the problem section. □

Heisenberg's uncertainty principle in the above form implies that if two operators do not commutate, that is $[A, B] \neq 0$, then there is a mathematical limitation to the accuracy with which measurements are performed by operators A and B.

12.2.7 Quantum Operations

As mentioned earlier, quantum mechanical systems which do not interact with their environment are called *closed* systems. Real quantum systems are not closed in principle. Such systems can be termed *open*. An open system interacts with its environment. Thus an open system and its environment can be viewed as a closed system. The interaction between an open system and the environment might be termed *noise*. The dynamics of such interaction is next described in the language of operators.

The evolution of quantum mechanical systems can also be described in terms of *operators*. For example, a unitary transformation U maps a pure state $|\psi\rangle$ to a new pure state $U|\psi\rangle$. The density matrix of the old state is $\rho = |\psi\rangle\langle\psi|$, and the new density matrix is $U|\psi\rangle\langle\psi| U^\dagger = U\rho U^\dagger = \rho'$. Thus the old density matrix ρ is "sandwiched" between U and its Hermitian transpose $U^\dagger$.

More compactly, this density matrix transformation is denoted as $\rho' = \Phi(\rho)$. The transformation $\Phi(\cdot)$ is also called a *quantum operation* or *superoperator*. Denote the transformation of the density matrix $|\psi\rangle\langle\psi|$ of a pure state $|\psi\rangle$ by $\Phi(|\psi\rangle\langle\psi|) = U|\psi\rangle\langle\psi| U^\dagger$, where U is a unitary operator. It can also be shown that, if ρ is the density matrix of a mixed state, then $\Phi(\rho) = U\rho U^\dagger$. Any map $\Phi(\cdot)$ cannot qualify as a quantum operation. It should possess the following three properties (requirements).

(a) *Linearity*: $\Phi\left(\sum_i p_i \rho_i\right) = \sum_i p_i \Phi\left(\rho_i\right)$, where $\{p_i\}$ are the probabilities and $\{\rho_i\}$ are the density matrices.
(b) *Trace condition*: If ρ is any initial state, then $0 \leq tr\left(\Phi\left(\rho\right)\right) \leq 1$. This condition implies that if ρ is the initial state, the operation $\Phi\left(\rho\right)$ occurred with probability $tr\left(\Phi\left(\rho\right)\right)$.
(c) *Completely positive*: Recall that the density matrix ρ is a positive (semidefinite) matrix. It is not sufficient for the density matrix $\Phi\left(\rho\right)$ to be positive. It has to be completely positive. This concept is next elaborated.
The property of complete positivity originates in a physical requirement. Complete positivity requires that the density matrix $\Phi\left(\rho\right)$ be positive for any positive density matrix ρ. In addition, it should have the following property. Let $\mathcal{H}_n$ be the Hilbert space of the original system. Extend this Hilbert space to $\mathcal{H}_m \otimes \mathcal{H}_n$. The quantum operation $\Phi\left(\cdot\right)$ is said to be completely positive on $\mathcal{H}_n$ if $\left(I_m \otimes \Phi\right)\left(\rho\right)$ is positive for all such extensions, where I_m is the identity operator acting in $\mathcal{H}_m$, and ρ is a positive operator. Note that the dimension m, of the Hilbert space $\mathcal{H}_m$ is arbitrary.

The above three requirements are sufficient to define quantum operations unambiguously. It can be demonstrated that a map which simply performs the operation of transposition, is positive but not completely positive. In summary, a quantum operation is a map from density operators to other density operators which observe the above three conditions. Such maps are sometimes called superoperators. Thus superoperators map positive semidefinite Hermitian matrices into positive semidefinite Hermitian matrices. Quantum communication channels, encoders, and decoders are some of the examples of superoperators.

An elegant and useful representation of quantum operation is the *operator-sum representation* or the *Kraus representation.* Any quantum operation $\Phi\left(\rho\right)$ can be expressed as

$$\Phi\left(\rho\right) = \sum_j E_j \rho E_j^\dagger, \quad \text{where} \quad \sum_j E_j E_j^\dagger \leq I$$

The operators $\{E_j\}$ are called *Kraus operators* or *operator elements*. It remains to check if the above three requirements are satisfied by the Kraus representation. This map is evidently linear, and the trace condition is satisfied by using the cyclicity property of the trace.

$$tr\left(\Phi\left(\rho\right)\right) = tr\left(\sum_j E_j \rho E_j^\dagger\right) = tr\left(\sum_j E_j E_j^\dagger \rho\right) \leq tr\left(\rho\right) = 1$$

Also $0 \leq tr\left(\sum_j E_j E_j^\dagger \rho\right)$, consequently $0 \leq tr\left(\Phi\left(\rho\right)\right) \leq 1$. In order to prove complete positivity it is sufficient to prove that the extended map $I_m \otimes \Phi$ is positive. This is achieved if for any state $|\psi\rangle$ in the Hilbert space $\mathcal{H}_m \otimes \mathcal{H}_n$ and for any positive operator P_{pos} acting on it, the inequality $\langle\psi \mid \left(I_m \otimes \Phi\right)\left(P_{pos}\right) \mid \psi\rangle \geq 0$ is true. Denote $\left(I_m \otimes E_j^\dagger\right)|\psi\rangle$ by $|\varphi_j\rangle$, then

$$\left\langle\psi \mid \left(I_m \otimes E_j\right) P_{pos} \left(I_m \otimes E_j^\dagger\right) \mid \psi\right\rangle = \langle\varphi_j \mid P_{pos} \mid \varphi_j\rangle \geq 0$$

by the positivity of the operator P_{pos}. Consequently

$$\langle\psi \mid \left(I_m \otimes \Phi\right)\left(P_{pos}\right) \mid \psi\rangle = \sum_j \langle\varphi_j \mid P_{pos} \mid \varphi_j\rangle \geq 0$$

Thus the operator $\left(I_m \otimes \Phi\right)\left(P_{pos}\right)$ is positive for any positive operator P_{pos}. Conversely, it can also be proved that the outer-sum representation follows from the three properties. The proof of this later assertion is left to the reader.

12.2.8 Quantum Entanglement

One of the important concepts in quantum computation and communication is that of quantum entanglement. Let $|\varphi\rangle$ be a quantum state in a Hilbert space of dimension n. If the quantum state $|\varphi\rangle$ cannot be represented as a tensor product of two quantum states from Hilbert spaces of dimensions smaller than n, then this state is said to be *entangled.* Otherwise, the state $|\varphi\rangle$ is said to be disentangled.

12.3 Quantum Computing Elements

A computational process can be either mechanical, or electrical, or mental, or molecular. This computational process at its most elemental level is a physical process. The building blocks of quantum computing, in analogy with classical computing are: quantum bits, quantum registers, quantum gates, and quantum networks. The basics of quantum information processing are discussed in this section. These are: Boolean gates and circuits, quantum bits and registers, and quantum gates and circuits. The no-cloning theorem is also stated and proved.

12.3.1 Boolean Gates and Circuits

In order to discuss quantum gates and circuits, it is instructive to initially consider the building blocks of classical computers. These building blocks are Boolean circuits. Let $\mathfrak{B} = \{0, 1\}$ be the alphabet. Boolean circuits are functions of the form $f : \mathfrak{B}^{(n)} \to \mathfrak{B}^{(m)}$. The basic elements of these Boolean circuits are the *logical and-gate, logical or-gate, logical not-gate,* and the *logical exclusive-or-gate*. These operations are next defined.

Definitions 12.5. *Let $\mathfrak{B} = \{0, 1\}$ be the alphabet with binary elements.*

1. *Logical and-gate: This is a mapping $\wedge : \mathfrak{B}^{(2)} \to \mathfrak{B}$. Let $x_1, x_2 \in \mathfrak{B}$ be the input to the gate, then $\wedge (x_1, x_2) = 1$ if and only if $x_1 = x_2 = 1$, otherwise it is equal to 0. Sometimes it is convenient to denote this logical and-gate operation by $x_1 x_2$.*
2. *Logical or-gate: This is a mapping $\vee : \mathfrak{B}^{(2)} \to \mathfrak{B}$. Let $x_1, x_2 \in \mathfrak{B}$ be the input to the gate, then $\vee (x_1, x_2) = 0$ if and only if $x_1 = x_2 = 0$, otherwise it is equal to 1. Sometimes it is convenient to denote this logical or-gate operation by $(x_1 + x_2)$.*
3. *Logical not-gate: This is a mapping from the set $\mathfrak{B}$ to the set $\mathfrak{B}$. If $x \in \mathfrak{B}$, then the not of x is denoted by x'. In this map $x' = 1$ if and only if $x = 0$, and $x' = 0$ if and only if $x = 1$.* □

A Boolean circuit is next defined.

Definition 12.6. *The layout of a Boolean circuit is in the form of an acyclic directed graph, in which the nodes are labeled with binary valued input or output variables or logical gates. The logical gates are: logical and-gates, logical or-gates, and logical not-gates. The arrows of the graph are called the wires. Furthermore, the input variable node has no incoming wires (arrows), and the output variable has no outcoming wires (arrows).* □

It can be demonstrated that an and-gate can be built by using only an or-gate and not-gates only. Similarly, it can be shown that an or-gate can be built by using only an and-gate and not-gates only.

Thus a Boolean circuit can be built by using only the or-gates and not-gates only; or only and-gates and not-gates only. The following three gates are of special interest.

Definitions 12.7. *Let* $\mathfrak{B} = \{0,1\}$ *be the alphabet with binary elements.*

1. *Exclusive-or-gate. This is a mapping* $\oplus : \mathfrak{B}^{(2)} \rightarrow \mathfrak{B}$. *Let* $x_1, x_2 \in \mathfrak{B}$ *be the input to the gate, then* $\oplus(x_1, x_2) = 1$, *if and only if either* $x_1 = 1$ *or* $x_2 = 1$, *but not both; otherwise it is equal to* 0. *Sometimes it is convenient to denote this logical exclusive-or-gate operation by* $(x_1 \oplus x_2)$.
2. *Logical nand-gate: This is a mapping from the set* $\mathfrak{B}^{(2)}$ *to the set* $\mathfrak{B}$. *Let* $x_1, x_2 \in \mathfrak{B}$ *be the input to the gate, then the output of the nand-gate is* $(x_1 x_2)'$. *That is, a nand-gate is an and-mapping followed by a not-mapping.*
3. *Logical nor-gate: This is a mapping from the set* $\mathfrak{B}^{(2)}$ *to the set* $\mathfrak{B}$. *Let* $x_1, x_2 \in \mathfrak{B}$ *be the input to the gate, then the output of the nor-gate is* $(x_1 + x_2)'$. *That is, a nor-gate is an or-mapping followed by a not-mapping.* □

It can also be shown that the exclusive-or-gate can be built by using the and-gates, or-gate, and not-gates, as

$$\oplus(x_1, x_2) = (x_1 x_2' + x_1' x_2)$$

The concept of *Boolean universal gate* is next introduced. A Boolean universal gate is mapping which when used as a building block to design Boolean circuits can simulate the and-gate, or-gate, and the not-gate. It can readily be checked that both the nand-gate and the nor-gate are examples of Boolean universal gates.

The and-gate and or-gate are considered to be *irreversible* Boolean operations. This means that it is not possible to determine unambiguously the input values to the gate, once the output of the gate is known. Note that the not-gate mapping is *reversible*, because two successive not-operations result in the original input. Since evolutionary processes in quantum physics are reversible, it is important to understand the concept of reversible gates. Denote the not-gate by N-gate. Two more examples of reversible gates are the CN-gate and the CCN-gate. The CN-gate is the *control-not-gate*, and the CCN-gate is the *control-control-not-gate*. The N-gate, CN-gate, and CCN-gate are their own inverse.

Definitions 12.8. *Let* $\mathfrak{B} = \{0,1\}$ *be the alphabet with binary elements.*

1. *Logical CN-gate: This is a mapping* $CN : \mathfrak{B}^{(2)} \rightarrow \mathfrak{B}^{(2)}$. *Let* $x_1, x_2 \in \mathfrak{B}$, *be the input to the gate, and the output of the gate be* $y_1, y_2 \in \mathfrak{B}$. *Then* $y_1 = x_1$. *The Boolean variable* x_1 *is called the control-bit. Also* $y_2 = x_2$ *if and only if* $x_1 = 0$, *otherwise it is equal to* x_2'. *Thus* $y_2 = x_1 \oplus x_2$. *In functional notation,* $CN(x_1, x_2) = (x_1, x_1 \oplus x_2)$.
2. *Logical CCN-gate: This is a mapping* $CCN : \mathfrak{B}^{(3)} \rightarrow \mathfrak{B}^{(3)}$. *Let* $x_1, x_2, x_3 \in \mathfrak{B}$, *be the input to the gate, and the output of the gate be* $y_1, y_2, y_3 \in \mathfrak{B}$. *Then* $y_1 = x_1$ *and* $y_2 = x_2$. *The Boolean variables* x_1 *and* x_2 *are called the control-bits. Also* $y_3 = x_3$ *if and only if* $x_1 x_2 = 0$, *otherwise it is equal to* x_3'. *Thus* $y_3 = x_1 x_2 \oplus x_3$. *In functional notation,* $CCN(x_1, x_2, x_3) = (x_1, x_2, x_1 x_2 \oplus x_3)$. □

The CCN-gate is also called the Petri-Toffoli gate in honor of their inventors C. A. Petri and T. Toffoli. It can be checked that both the above defined mappings are reversible. Note that $CCN(x_1, x_2, 1) = (x_1, x_2, x_1 x_2 \oplus 1)$. That is, the third output performs logical nand operation on the inputs x_1 and x_2. Therefore, the logical nand-gate can be simulated by the CCN-gate. That is, the CCN-gate is universal. Thus all Boolean circuits with at least three inputs can be simulated by using reversible gates.

12.3.2 Quantum Bits and Registers

The basis vector set of the two-dimensional Hilbert space $\mathcal{H}_2$ is $\mathcal{B} = \{|0\rangle, |1\rangle\}$, where

$$|0\rangle = \begin{bmatrix} 1 \\ 0 \end{bmatrix}, \quad |1\rangle = \begin{bmatrix} 0 \\ 1 \end{bmatrix}$$

The unit of information in classical computers is a bit. A bit has two possible values, which are 0 and 1. The elementary unit of quantum information processing is *qubit.* It is next defined precisely.

Definition 12.9. *A qubit, is an abbreviation of quantum bit. It is a two-level quantum system, and its state is specified by a vector of unit magnitude in the two-dimensional Hilbert space $\mathcal{H}_2$. The general state of a qubit is a vector*

$$|\psi\rangle = \alpha_0 |0\rangle + \alpha_1 |1\rangle, \quad \alpha_0, \alpha_1 \in \mathbb{C} \tag{12.10a}$$

$$1 = |\alpha_0|^2 + |\alpha_1|^2 \tag{12.10b}$$

The complex numbers α_0 and α_1 are called the amplitudes, and the states $|0\rangle$ and $|1\rangle$ are also called the basis states. □

Observation of a qubit gives the vector $|0\rangle$ with probability $|\alpha_0|^2$, and the vector $|1\rangle$ with probability $|\alpha_1|^2$. The quantum registers can be described by considering tensor products of two-dimensional Hilbert spaces. For example, the states of a two qubits register is specified in a four-dimensional Hilbert space $\mathcal{H}_4 = \mathcal{H}_2 \otimes \mathcal{H}_2$. Its basis set is

$$\mathcal{B} = \{|0\rangle |0\rangle, |0\rangle |1\rangle, |1\rangle |0\rangle, |1\rangle |1\rangle\}$$

Denote $|0\rangle |0\rangle$ by $|00\rangle$, $|0\rangle |1\rangle$ by $|01\rangle$, $|1\rangle |0\rangle$ by $|10\rangle$, and $|1\rangle |1\rangle$ by $|11\rangle$. That is

$$|00\rangle = \begin{bmatrix} 1 \; 0 \; 0 \; 0 \end{bmatrix}^T, \quad |01\rangle = \begin{bmatrix} 0 \; 1 \; 0 \; 0 \end{bmatrix}^T$$

$$|10\rangle = \begin{bmatrix} 0 \; 0 \; 1 \; 0 \end{bmatrix}^T, \quad |11\rangle = \begin{bmatrix} 0 \; 0 \; 0 \; 1 \end{bmatrix}^T$$

Then a general state of a two qubit register is

$$|\psi\rangle = \alpha_0 |00\rangle + \alpha_1 |01\rangle + \alpha_2 |10\rangle + \alpha_3 |11\rangle; \quad \alpha_0, \alpha_1, \alpha_2, \alpha_3 \in \mathbb{C}$$

$$1 = |\alpha_0|^2 + |\alpha_1|^2 + |\alpha_2|^2 + |\alpha_3|^2$$

Observation of a two qubit register gives the vector $|00\rangle$ with probability $|\alpha_0|^2$, and the vector $|01\rangle$ with probability $|\alpha_1|^2$, and so on. The probability of observing a 0 in the first position (counting from the left) of the register is $|\alpha_0|^2 + |\alpha_1|^2$. Similarly, the probability of observing a 1 in the first position of the register is $|\alpha_2|^2 + |\alpha_3|^2$. The probability of observing a 0 in the second position of the register is $|\alpha_0|^2 + |\alpha_2|^2$. Similarly, the probability of observing a 1 in the second position of the register is $|\alpha_1|^2 + |\alpha_3|^2$. This result can be generalized to define an m-qubit quantum register.

Definition 12.10. *A quantum register of length m, is a quantum system of ordered m qubits. The state space of this system is the m-fold tensor product*

$$\mathcal{H}_{2^m} = \underbrace{\mathcal{H}_2 \otimes \mathcal{H}_2 \otimes \ldots \otimes \mathcal{H}_2}_{m \text{ times}} \tag{12.11a}$$

Let $\mathfrak{B} = \{0, 1\}$ *. The set of basis vectors of this space is*

$$\mathcal{B} = \{|i_{m-1}i_{m-2}\ldots i_0\rangle \mid i_j \in \mathfrak{B}, 0 \leq j \leq (m-1)\} \tag{12.11b}$$

where $i_{m-1}i_{m-2}\ldots i_0$ *is a binary representation of* $i \in \mathbb{Z}_N$*, and* $N = 2^m$*. The cardinality of the set* $\mathcal{B}$ *is* N*, and the general state of an* m*-qubit register is*

$$|\psi\rangle = \sum_{i \in \mathbb{Z}_N} \alpha_i |i\rangle; \quad \alpha_i \in \mathbb{C}, \quad \forall\, i \in \mathbb{Z}_N \tag{12.11c}$$

$$\sum_{i \in \mathbb{Z}_N} |\alpha_i|^2 = 1 \tag{12.11d}$$

□

Observe that the basis set $\mathcal{B} = \mathfrak{B}^{(m)}$, and the basis states of the m-qubit register belong to the set $\{|i\rangle \mid i \in \mathbb{Z}_N\}$. In addition, the time evolution of an m-qubit quantum system is determined by unitary matrices of order $N \times N$.

12.3.3 Quantum Gates and Circuits

A classical computer uses Boolean gates and circuits to perform its operations. Analogously a theoretical basis for the building blocks of quantum computers can be developed. These building blocks are quantum gates and circuits. Actually quantum gates are a generalization of the concept of reversible gates. In this subsection unary quantum gates, multiple qubit gates, and quantum circuits are studied.

Unary Quantum Gates

A unary quantum gate is a linear operation.

Definition 12.11. *A unary quantum gate is a unitary mapping* $U : \mathcal{H}_2 \to \mathcal{H}_2$*. It transforms a qubit to another qubit.* □

The unary quantum gate operator can be represented by a 2×2 matrix.

Examples 12.3. Some elementary unary quantum gates are described.

1. *Quantum not-gate*: Let

$$A = \begin{bmatrix} 0 & 1 \\ 1 & 0 \end{bmatrix}$$

   Check that $AA^T = I$, thus the matrix A is unitary. Also

$$A|0\rangle = |1\rangle, \text{ and } A|1\rangle = |0\rangle$$

   Hence the name quantum not-gate.

2. *Hadamard matrix*: Let

$$W_2 = \frac{1}{\sqrt{2}} \begin{bmatrix} 1 & 1 \\ 1 & -1 \end{bmatrix}$$

The unitary matrix W_2 is a 2×2 Hadamard matrix. It is named after the mathematician Jacques Hadamard (1865-1963). Then

$$\begin{aligned} W_2 \left|0\right\rangle &= \frac{1}{\sqrt{2}} \left\{ \left|0\right\rangle + \left|1\right\rangle \right\}, \\ W_2 \left|1\right\rangle &= \frac{1}{\sqrt{2}} \left\{ \left|0\right\rangle - \left|1\right\rangle \right\} \end{aligned}$$

$$W_2 W_2 \left|0\right\rangle = \left|0\right\rangle, \quad W_2 W_2 \left|1\right\rangle = \left|1\right\rangle$$

3. *Square root of the quantum not-gate*: Let

$$C = \frac{1}{\sqrt{2}} \begin{bmatrix} e^{i\theta} & e^{-i\theta} \\ e^{-i\theta} & e^{i\theta} \end{bmatrix}, \quad \theta = \frac{\pi}{4}$$

It can be checked that the matrix C is unitary, and that $C^2 = A$. Therefore, this unitary transformation is square root of the quantum not-gate. Also

$$\begin{aligned} C \left|0\right\rangle &= \frac{1}{\sqrt{2}} \left\{ e^{i\theta} \left|0\right\rangle + e^{-i\theta} \left|1\right\rangle \right\} \\ C \left|1\right\rangle &= \frac{1}{\sqrt{2}} \left\{ e^{-i\theta} \left|0\right\rangle + e^{i\theta} \left|1\right\rangle \right\} \end{aligned}$$

4. *Rotation gates*: The following unitary transformations represent a single parameter unary gates.

$$\begin{aligned} R_x(\theta) &= \begin{bmatrix} \cos\theta & i\sin\theta \\ i\sin\theta & \cos\theta \end{bmatrix}, \\ R_y(\theta) &= \begin{bmatrix} i\cos\theta & \sin\theta \\ \sin\theta & i\cos\theta \end{bmatrix}, \end{aligned}$$

$$R_z(\theta) = \begin{bmatrix} e^{i\theta} & 0 \\ 0 & e^{-i\theta} \end{bmatrix}$$

□

Multiple Qubit Gates

A multiple qubit gate is a generalization of a unary quantum gate. It has at least two inputs.

Definition 12.12. *A quantum gate with n inputs and outputs each is a unitary mapping U : $\mathcal{H}_{2^n} \to \mathcal{H}_{2^n}$.* □

A quantum gate mapping is represented by a $2^n \times 2^n$ matrix. A binary quantum gate is obtained, if n is equal to 1.

Example 12.4. In analogy with Boolean control-not-gate a *quantum control-not-gate* can be developed. This gate is a binary quantum gate. Consider the unitary matrix

$$A_{cnot} = \begin{bmatrix} 1 & 0 & 0 & 0 \\ 0 & 1 & 0 & 0 \\ 0 & 0 & 0 & 1 \\ 0 & 0 & 1 & 0 \end{bmatrix}$$

Thus

$$A_{cnot} |00\rangle = |00\rangle, \; A_{cnot} |01\rangle = |01\rangle,$$
$$A_{cnot} |10\rangle = |11\rangle, \text{ and } A_{cnot} |11\rangle = |10\rangle$$

The gate specified by the matrix A_{cnot} is a quantum control-not-gate, because the second input qubit (called the *target qubit*) is flipped if and only if the first qubit (called the *control qubit*) is 1. It can be checked that $A_{cnot} |x, y\rangle \to |x, x \oplus y\rangle$, where $x, y \in \{0, 1\}$. □

Quantum Circuits

A quantum circuit on n qubits is a unitary mapping on $\mathcal{H}_{2^n}$. This can be generated as a concatenation of a finite set of quantum gates. In addition, reversible circuits are also quantum circuits. Since Boolean circuits can be simulated via reversible circuits, it can be concluded that generally anything which is computable by a Boolean circuit is also computable by a quantum circuit. The possibility of finding universal quantum gates is next examined.

Unlike Boolean gates, quantum gates are continuous. Therefore the quest for finding universal quantum gates is more complicated. The first step in finding such gates is to quantify a technique to approximate a unitary matrix U by another matrix V.

Definitions 12.13. *Let* $\varepsilon \in \mathbb{R}^+$.

1. *A matrix* V *is said to be* ε*-close to the matrix* U *if and only if* $\|U - V\| \le \varepsilon$, *where* $\|\cdot\|$ *is a predetermined norm.*
2. *A set of quantum gates is universal if any unitary transformation* U *on any qubit register is approximated to the desired accuracy* ε, *by a quantum circuit consisting of only those gates.* □

Thus a universal quantum gate can be used in approximating all quantum circuits. D. Deutsch determined that the quantum equivalent of the Petri-Toffoli gate is universal. This is a 3-qubit gate specified by an 8×8 unitary matrix $D(\theta)$.

$$D(\theta) = \begin{bmatrix} I_2 & 0 & 0 & 0 \\ 0 & I_2 & 0 & 0 \\ 0 & 0 & I_2 & 0 \\ 0 & 0 & 0 & R_x(\theta) \end{bmatrix}$$

$$R_x(\theta) = \begin{bmatrix} i\cos\theta & \sin\theta \\ \sin\theta & i\cos\theta \end{bmatrix}$$

In the matrix $D(\theta)$, the matrix I_2 is a 2×2 identity matrix, and the 0 entries are all-zeros 2×2 matrices. Also θ/π is irrational. This later condition is required to implement $R_x(\theta)$ to a specified degree of accuracy by using the matrix $R_x(\theta/n)$, where $n \in \mathbb{P}$. It has also been determined that all quantum circuits can be designed by using only unary quantum gates and quantum control-not-gates.

12.3.4 No-Cloning Theorem

The no-cloning theorem is an important and easily stated result. This theorem was established by W. K. Wootters and W. H. Zurek, and D. Dieks in 1982. The no-cloning theorem asserts that it is impossible to duplicate an unknown quantum state. This theorem has profound, philosophical, and practical implications. It implies that a quantum copying scheme does not exist. Consider a quantum system with state space $\mathcal{H}_n$. A unitary mapping U defined on the space $\mathcal{H}_n \otimes \mathcal{H}_n$ is said to be a *quantum copying machine*, if for any state $|\alpha\rangle \in \mathcal{H}_n$, the relationship $U|\alpha, s\rangle = |\alpha, \alpha\rangle$ exists, where $|s\rangle$ is some standard pure state.

Theorem 12.1. (*No-cloning theorem*). *An unknown quantum state cannot be cloned. That is, in any Hilbert space $\mathcal{H}_n$, where $n > 1$, and for any state $|\psi\rangle \in \mathcal{H}_n$, there does not exist any unitary transformation U, such that the equality $U|\psi, s\rangle = |\psi, \psi\rangle$ is true, where $|s\rangle$ is some standard pure state. Note that the unitary mapping U is defined on the space $\mathcal{H}_n \otimes \mathcal{H}_n$.*

Proof. Assume that $U|\alpha_1, s\rangle = |\alpha_1, \alpha_1\rangle$ and $U|\alpha_2, s\rangle = |\alpha_2, \alpha_2\rangle$ are true for arbitrary $|\alpha_1\rangle$ and $|\alpha_2\rangle$. It is given that the operator U is unitary. Therefore

$$\begin{aligned}\langle\alpha_1 \mid \alpha_2\rangle\langle\alpha_1 \mid \alpha_2\rangle &= \langle\alpha_1\alpha_1 \mid \alpha_2\alpha_2\rangle = \langle U\alpha_1 s \mid U\alpha_2 s\rangle \\ &= \langle\alpha_1 s \mid \alpha_2 s\rangle = \langle\alpha_1 \mid \alpha_2\rangle\langle s \mid s\rangle \\ &= \langle\alpha_1 \mid \alpha_2\rangle\end{aligned}$$

The above equation implies that there exists a relationship of the form $x = x^2$. It results in a contradiction, if $|\alpha_1\rangle$ and $|\alpha_2\rangle$ are chosen so that $0 < |\langle\alpha_1 \mid \alpha_2\rangle| < 1$. □

12.4 What is Quantum Computation?

In the last two sections, the postulates of quantum mechanics, and the basic building blocks of quantum computation were described. In this section, the strengths and drawbacks of the quantum computing paradigm are determined.

How is Quantum Computation Performed?

Quantum computation is performed as follows. A single quantum register or multiple quantum registers are prepared in an initial state. Quantum gates, which are simply unitary transformations, are applied to these registers. Finally, the contents of the registers are measured. Note that the evolution of the states of the registers is deterministic, but the measured result is probabilistic.

Limitations

Some of the limitations of quantum computation are: the probabilistic nature of the measured results, the no-cloning result, and the errors.

Let $|\psi\rangle$ be a qubit, where $|\psi\rangle = \alpha_0|0\rangle + \alpha_1|1\rangle, \alpha_0, \alpha_1 \in \mathbb{C}$, and the coefficients α_0 and α_1 are normalized such that $(\alpha_0^2 + \alpha_1^2) = 1$. The measurement of this qubit yields the state $|0\rangle$ with probability $|\alpha_0|^2$, and the state $|1\rangle$ with probability $|\alpha_1|^2$. Thus there is no accurate measurement.

Arbitrary quantum states cannot be cloned (duplicated). This simple yet powerful result has been stated earlier as a theorem.

Finally, as in digital computers, errors are ubiquitous. Errors are introduced in quantum computation because a quantum system cannot be completely isolated from its environment. This process is termed the *decoherence* of the qubit. Furthermore, the unitary transformations applied to the quantum registers can also introduce errors.

Quantum Parallelism

The strength of quantum computation lies in its parallelism. Consider a function $f : \mathfrak{B} \rightarrow \mathfrak{B}$. Our goal is to compute $f(x)$, where $x \in \mathfrak{B} = \{0, 1\}$. These values are obtained classically by first computing $f(0)$ and then $f(1)$. In quantum computation, if the function $f(\cdot)$ acts upon the space $\mathcal{H}_2$, compute

$$f(\alpha_0 |0\rangle + \alpha_1 |1\rangle) = \beta_0 f(|0\rangle) + \beta_1 f(|1\rangle)$$

where the two component values are computed in parallel. The problem in this technique is that the measured values $f(|0\rangle)$ or $f(|1\rangle)$ are random. This scheme in principle can be extended to a superposition of several states, which in turn provide an enormous speedup in computation.

The next section develops a theoretical model of quantum computation. This is an extension of the model of classical computation developed by Alan Turing. A quantum version of the classical Turing machine is developed. It sheds further light on the inherent parallelism of quantum computation.

12.5 Models of Computation

Alonzo Church (1903-1995), Kurt Gödel (1906-1978), Emil Post (1897-1954), and Alan Turing (1912-1954) independently developed equivalent mathematical models of computation in the last century. All these models are mathematical abstractions of computation that do not incorporate the underlying physical processes of the computing machine. In order to understand the strengths and limitations of these computational processes, Turing's model of classical computation is first outlined in this section. This surprisingly powerful model was proposed by the mathematician Alan Turing.

The computing model (machine) proposed by Turing in the year 1936 consists of an infinitely long tape, and a read/write head. The tape is marked off into a contiguous sequence of cells on which symbols from an alphabet or a blank may be written. The read/write head can move along the tape to either read from or write on this tape. The head exists in one of finite number of "states." Furthermore, this head also contains instructions ("program") that specify if the head should either read or write from the tape, and also if it should move one position to the right or left cell, or simply halt the program. This type of machine is called the Turing machine (TM) in honor of its inventor.

The Turing machine which was discussed above is deterministic. An extension of this model is the probabilistic Turing machine (PTM). The advantage of a PTM over a TM is that some problems can be solved relatively more quickly on a PTM than on a TM. However, it has been determined by computer theorists that anything which is computable on a PTM, is also computable on a TM.

The TM and PTM are certainly laudable mathematical models of computation, however they do not take into account the physics of the underlying devices on which the computing machines are built. As modern computers become smaller and more powerful, quantum mechanical effects begin to dominate more prominently. Therefore, in order to study the impact of quantum mechanical effects on a computing device, it is advantageous to develop a quantum model of a Turing machine. Among the first developers of a quantum Turing machine (QTM) were Paul Benioff, Charles Bennett, David Deutsch, and the physics-Nobelist Richard P. Feynman (1918-1988).

QTM's can be considered to be generalization of the PTM's. Consider a PTM, which starts with some initial configuration at time t_0. Let the machine run till time t. The state of the PTM at this instant of time is specified by a probability distribution over all possible states which are reachable from the initial state at time t_0 till time t. Similarly, assume that the QTM is in some initial configuration at time t_0, and let the machine run till time t. The state of the QTM at time t is specified by a superposition of all states which are reachable from the initial state at time t_0 till time t. Observe that in the classical PTM only a single trajectory is followed, but in the QTM all computational trajectories are followed. Thus the final state in a QTM is a superposition of all possible trajectories in time $(t - t_0)$.

These computational models are next described more precisely. This is necessary in order to determine the power and limitations of the quantum computation paradigm. In this section, we give mathematical descriptions of deterministic, nondeterministic, probabilistic, and quantum Turing machines. The Church-Turing hypothesis, and different complexity classes are also specified.

12.5.1 Deterministic Turing Machine

Physically, a Turing machine consists of a *single tape, finite control unit, single read-write tape head,* and a *program.* The tape and the finite control unit communicate via a single head positioned over the tape. The head can either read from or write to the tape. The tape is infinite in length in both the left and right directions. A tape of this type is sometimes called a bi-infinite tape. Furthermore, the tape is divided into *cells*. Thus the tape of infinite length is divided into an infinite number of cells. Also, each cell contains a single alphabet symbol or a *blank symbol* ⊔. See Figure 12.1. The set of alphabet symbols is denoted by $\mathcal{A}$, and the finite set of possible *states* of the control unit is denoted by Q.

Depending upon the current state of the control unit, and the scanned (read) tape symbol, the following two events occur at each step.

(a) Write a symbol in the current tape cell. Thus replacing the symbol which was already there. The read-write head is moved one tape cell to the left or right, or stays stationary.
(b) The state of the control unit changes.

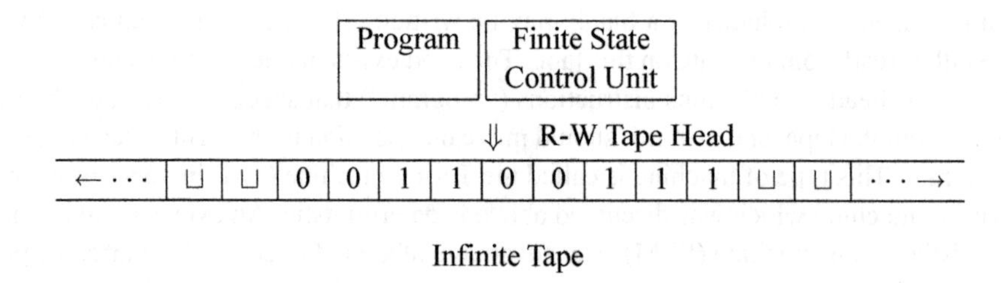

Figure 12.1. Deterministic Turing machine. It has a program, finite state control unit, read-write (R-W) tape head, and an infinite tape. Note that the symbol ⊔ is the blank symbol.

Denote the initial state of the TM by q_0. Let the set of *terminating states* of the TM be $Q_t = \{q_a, q_r\}$, where q_a is the *accept state*, q_r is the *reject state,* and $Q_t \subseteq Q$. The terminating states indicate the end of computation. A formal definition of a deterministic Turing machine $\mathcal{M}$, is next given.

Definition 12.14. *A deterministic Turing machine $\mathcal{M}$ is defined using the following terminology.*

(a) *The set of tape symbols is Γ. The alphabet (input symbols) is $\mathcal{A} \subset \Gamma$, and the blank symbol is $\sqcup \in \Gamma \backslash \mathcal{A}$.*
(b) *The set of finite states is Q. The initial state is $q_0 \in Q$, and $Q_t \subseteq Q$ is the set of terminating states. Also $Q_t = \{q_a, q_r\}$, where q_a and q_r are the accept and reject states respectively.*
(c) *The set $\varrho = \{\leftarrow, \downarrow, \rightarrow\}$ denotes the movement of the read-write tape head. The symbol $\leftarrow$ means move the tape head to the left cell, the symbol $\downarrow$ means do not move the tape head at all (be stationary), and the symbol $\rightarrow$ means move the tape head to the right cell.*
(d) *The transition function δ, is*

$$\delta : Q \backslash Q_t \times \Gamma \rightarrow Q \times \Gamma \times \varrho \tag{12.12}$$

where $\delta(q, \sigma) = (q', \sigma', d)$ means that if the Turing machine $\mathcal{M}$, is in state $q \in Q \backslash Q_t$ and the head reads the symbol σ; the new state of the Turing machine is $q' \in Q$, the symbol σ is replaced by σ' in the cell the head is currently on, and the head moves in the direction indicated by d. Furthermore, if the machine reaches a terminating state it remains in that state.

Thus the deterministic Turing machine $\mathcal{M}$ is $(\Gamma, Q, \delta, \varrho)$. □

In the above definition of the transition function δ, the state $q \in Q \backslash Q_t$. This is true, because there cannot be a transition out of a terminating state in Q_t. It can be observed from the above definition that a TM is a mathematical abstraction of computation. A notation to describe a sequence of symbols is next introduced. Let $\mathcal{A}^k$ be the set of *strings* (of tape symbols) of length k over $\mathcal{A}$, where k is a positive integer. These strings are also called *words*. Denote $\mathcal{A}^0 = \{\epsilon\}$, where ϵ is called the *empty word.* The empty word has no letters. The corresponding symbol on the tape is the blank symbol $\sqcup$. Also define

$$\mathcal{A}^* = \mathcal{A}^0 \cup \mathcal{A}^1 \cup \mathcal{A}^2 \cup \ldots$$

The set $\mathcal{A}^*$ is the dictionary of words. The length of a word $w \in \mathcal{A}^*$, denoted by $|w|$, is the number of symbols that make up the word.

Computation in Turing Machine

Having described the deterministic TM, its computational process is next outlined. This is done by using the concept of *configuration.* The configuration $\mathcal{C}$ of a TM is specified by a three-tuple (q, w, w'), where $q \in Q$ and $w, w' \in \mathcal{A}^*$. This three-tuple provides sufficient details to completely describe the *global state* of a Turing machine $\mathcal{M}$. Note that q is the current state of the TM, and the contents of tape are ww'. The sequence of symbols w are to the left of the head. The position of the head is on the first symbol of w'. Furthermore, leftmost blank symbols are discarded from the string w, and rightmost blank symbols are discarded from the string w'.

The tape cells are initially all blank symbols, except possibly for a finite contiguous sequence of symbols from the set Γ. In the later case, the head of the tape initially rests on top of one of these contiguous symbols.

Let the *initial* configuration of the TM be denoted by (q_0, ϵ, x). The length of the input x is ℓ, that is $|x| \triangleq \ell$. Similarly, a *terminating* configuration is described by the three-tuple $(q, w, w'), q \in Q_t$.

If the TM moves from the configuration $\mathcal{C}$ to the configuration $\mathcal{C}'$ in a single step, then it is denoted by $\mathcal{C} \vdash_{\mathcal{M}} \mathcal{C}'$. The configuration $\mathcal{C}'$ is said to be the *successor* of configuration $\mathcal{C}$.

Using the above terminology, the transition function $\delta(\cdot, \cdot)$ can be described in more detail. Let $\delta(q, \sigma) = (q', \sigma', d)$, and the current machine configuration be $\mathcal{C} = (q, ua, \sigma v)$, where $a, \sigma \in \Gamma$ and $u, v \in \mathcal{A}^*$. In this configuration, the TM is reading the symbol σ. Let the immediate successor of the configuration $\mathcal{C}$ be $\mathcal{C}'$. The configuration $\mathcal{C}'$ has the following possibilities.

(a) The tape head moves to the left, that is $d = \leftarrow$. Therefore $\mathcal{C}' = (q', u, a\sigma' v)$.
(b) The tape head is stationary, that is $d = \downarrow$. Therefore $\mathcal{C}' = (q', ua, \sigma' v)$.
(c) The tape head moves to the right, that is $d = \rightarrow$. Therefore $\mathcal{C}' = (q', ua\sigma', v)$.

Next consider the case, when the TM moves from the configuration $\mathcal{C}$ to the configuration $\mathcal{C}'$ in several steps, then it is denoted by $\mathcal{C} \vdash_{\mathcal{M}}^{*} \mathcal{C}'$.

If the initial and terminating configurations are $\mathcal{C}_0$ and $\mathcal{C}_k$ respectively, the sequence of configurations $\mathcal{C}_0, \mathcal{C}_1, \mathcal{C}_2, \ldots, \mathcal{C}_k$ is called a *terminating computation.* This computation is also represented by $\mathcal{C}_i \vdash_{\mathcal{M}} \mathcal{C}_{i+1}$, for $0 \leq i < k$.

If the Turing machine $\mathcal{M}$ starts in the initial configuration $\mathcal{C}_0 = (q_0, \epsilon, x)$ and terminates (halts) in the configuration $\mathcal{C}_k = (q, w, w')$, then define:

(a) $\mathcal{M}_s(x) = q$.
(b) $\mathcal{M}_t(x) = ww'$.

Sometimes, it is also said that $\mathcal{M}$ computes $\mathcal{M}_t(x) = ww'$ in time k. Thus a TM can be viewed as a construct for computing functions from $\mathcal{A}^*$ to $\mathcal{A}^*$. If the computation does not halt for an input string x, then the computation is infinite. This is denoted by $\mathcal{M}(x) = \nearrow$.

The concept of a language recognized by a TM is made precise via the following definition.

Definition 12.15. *Let $\mathcal{M} = (\Gamma, Q, \delta, \varrho)$ be a deterministic Turing machine, then*

$$L(\mathcal{M}) = \{w \mid w \in \mathcal{A}^*, \mathcal{M}_s(w) = q_a\} \tag{12.13}$$

is the language over $\mathcal{A}$ which is accepted by the machine $\mathcal{M}$. Also, if $\mathcal{M}_s(x) \in Q_t$, for any $x \in \mathcal{A}^*$, then $L(\mathcal{M})$ is said to be the language decided or recognized by $\mathcal{M}$. □

Using this definition of language, it can be surmised that TM is a reasonable model to study computation of integer-to-integer and string-to-string functions, and language acceptance problems. Furthermore, TM also incorporates the concept of algorithmic computability. That is, a TM can in principle simulate a computation algorithmically. Note that an algorithm can be loosely defined as a step-wise procedure to perform a computation.

The TM model discussed in the last few paragraphs can be modified to use a semi-infinite tape instead of a tape of infinite length. In a semi-infinite tape, the tape extends to infinity in only one direction. Furthermore, it is also possible to extend a single tape model to multiple tapes. These computational models are not discussed in this chapter.

Deterministic Complexity Classes

The concept of time and space complexity is next made more precise. These models of computation are useful in stating the complexity of algorithmic processes.

Definitions 12.16. *Let $\mathcal{M}$ be a deterministic Turing machine.*

1. *Time complexity. If for each input of length ℓ, the machine $\mathcal{M}$ stops after at most $t(\ell)$ cell moves, then $\mathcal{M}$ is said to run in time $t(\ell)$, or $\mathcal{M}$ has a time complexity $t(\ell)$.*
2. *Space complexity. If for each input of length ℓ, the machine $\mathcal{M}$ uses at most $s(\ell)$ cells, then $\mathcal{M}$ is said to use $s(\ell)$ spaces, or $\mathcal{M}$ has a space complexity $s(\ell)$.* □

Based upon these definitions, the following classes of languages are categorized for a deterministic Turing machine $\mathcal{M}$.

(a) The class of languages accepted by $\mathcal{M}$ in time $O(n)$ is $DTIME(n)$.
(b) The class of languages accepted by $\mathcal{M}$ in space $O(n)$ is $DSPACE(n)$.
(c) The class of languages accepted by $\mathcal{M}$ in polynomial-time is $\mathcal{P}$ where

$$\mathcal{P} = \bigcup_{j \in \mathbb{N}} DTIME(n^j)$$

(d) The class of languages accepted by $\mathcal{M}$ in polynomial-space is $\mathcal{PSPACE}$ where

$$\mathcal{PSPACE} = \bigcup_{j \in \mathbb{N}} DSPACE(n^j)$$

The class of problems in $\mathcal{P}$ are said to be *tractable*, and the other problems are not. The relationship $\mathcal{P} \subseteq \mathcal{PSPACE}$ holds true.

Church-Turing Thesis

The Church-Turing (named after Alonzo Church and Alan Turing) thesis asserts that the TM model of classical computation is a reasonably powerful model of computation. Or any process which is algorithmic in nature is Turing computable. This is simply a strong belief. Therefore the Church-Turing thesis can alternately be called a Church-Turing hypothesis.

In order to state the Church-Turing hypothesis more precisely, the concept of a universal Turing machine (UTM) $\mathcal{U}$ has to be developed. A universal Turing machine is a TM that can simulate each and every TM. This machine $\mathcal{U}$ is similar in spirit and purpose to a programmable general purpose computer which can simulate any computer, including itself.

Definition 12.17. *The input string of symbols (from the alphabet) and the transition rules of a Turing machine $\mathcal{M}$ are submitted to the universal Turing machine $\mathcal{U}$ properly encoded. Then the universal Turing machine $\mathcal{U}$ performs the computation of $\mathcal{M}$ on the specified encoded string. It can also be stated alternately, that the universal Turing machine $\mathcal{U}$ is programmable so that it performs any Turing machine computation.* □

A precise statement of the Church-Turing hypothesis is as follows. *Any function which is considered to be computable can be computed by the universal Turing machine.* There are several specific reasons (but not proofs) for the correctness of the Church-Turing hypothesis. All mathematical paradigms of computation which have been proposed thus far, have been found to be equivalent to Turing computability. Furthermore, no example of a classical computing device which cannot be implemented via a TM has been discovered.

12.5.2 Nondeterministic Turing Machine

A nondeterministic Turing machine, abbreviated as NTM, is a generalization of a deterministic Turing machine. The NTM, like TM is useful in exploring the power of classical computation. The definition of NTM is similar to that of the TM. The TM was defined on the basis of a transition function. In contrast, the NTM is defined on the basis of the *transition relation.* The transition relation used in the definition of a NTM is

$$\delta \subseteq Q \backslash Q_t \times \Gamma \times Q \times \Gamma \times \varrho$$

where Γ is the set of tape symbols, Q is the set of finite states of the machine, and ϱ is the set of symbols which determine the direction of the movement of the tape head. The above relationship implies that a transition is not determined uniquely. Multiple nondeterministic choices are possible at each transition. Consequently multiple successors of a configuration are possible. Thus the multiple choices of configurations in a NTM form a tree. Another way to represent this transition relation is

$$\delta : Q \backslash Q_t \times \Gamma \times Q \times \Gamma \times \varrho \rightarrow \{0, 1\}$$

Thus $\delta (q, \sigma, q', \sigma', d) = 1$ implies that if the NTM is in state $q \in Q \backslash Q_t$ and the head reads σ; then the NTM *may* perform the following move: the new state of the NTM is $q' \in Q$, the symbol σ is replaced by σ' in the cell the head is currently on, and the head moves in the direction indicated by d. However, if $\delta (q, \sigma, q', \sigma', d) = 0$, then the NTM *may not* perform the move described in the last sentence.

The language accepted by a NTM can be defined similar to that of a TM, with the caveat that an input is said to be acceptable by the NTM if there is at least a single path in the configuration tree. In this tree, the input symbol string is at the root of the tree. The concept of time and space complexity of a computation is similarly extended. Complexity classes of NTM are:

(a) The class of languages accepted by NTM in time $O(n)$ is $NTIME(n)$.
(b) The class of languages accepted by NTM in space $O(n)$ is $NSPACE(n)$.
(c) The class of languages accepted by NTM in polynomial-time is specified by $\mathcal{NP}$ where

$$\mathcal{NP} = \bigcup_{j \in \mathbb{N}} NTIME(n^j)$$

(d) The class of languages accepted by NTM in polynomial-space is specified by $\mathcal{NPSPACE}$ where

$$\mathcal{NPSPACE} = \bigcup_{j \in \mathbb{N}} NSPACE(n^j)$$

The relationship $\mathcal{NP} \subseteq \mathcal{NPSPACE}$ holds true. The following more general relationship can also be established.

$$\mathcal{P} \subseteq \mathcal{NP} \subseteq \mathcal{PSPACE} = \mathcal{NPSPACE}$$

It is not yet know if the above inclusions are proper. In addition, one of the holy grails of computer science is to determine if $\mathcal{P} \neq \mathcal{NP}$.

12.5.3 Probabilistic Turing Machine

A description of probabilistic version of Turing machine is necessary before that of a quantum Turing machine. A probabilistic Turing machine is a generalization of the classical TM. In order to define a PTM, the transition function in the classical TM is suitably modified. Thus the transition function of the TM is replaced by the *transition probability distribution.*

Definition 12.18. *A probabilistic Turing machine $\mathcal{M}$ is defined using the following terminology.*

(a) *The set of tape symbols is Γ. The alphabet (input symbols) is $\mathcal{A} \subset \Gamma$, and the blank symbol is $\sqcup \in \Gamma \backslash \mathcal{A}$.*
(b) *The set of finite states is Q. The initial state is $q_0 \in Q$, and $Q_t \subseteq Q$ is the set of terminating states. Also $Q_t = \{q_a, q_r\}$, where q_a and q_r are the accept and reject states respectively.*
(c) *The set $\varrho = \{\leftarrow, \downarrow, \rightarrow\}$ denotes the movement of the read-write tape head. The symbol $\leftarrow$ means move the tape head to the left cell, the symbol $\downarrow$ means do not move the tape head at all (let it be stationary), and the symbol $\rightarrow$ means move the tape head to the right cell.*
(d) *The transition function δ, is*

$$\delta : Q \backslash Q_t \times \Gamma \times Q \times \Gamma \times \varrho \rightarrow [0, 1] \tag{12.14a}$$

For each $(q, \sigma) \in Q \backslash Q_t \times \Gamma$

$$\sum_{(q', \sigma', d) \in Q \times \Gamma \times \varrho} \delta(q, \sigma, q', \sigma', d) = 1 \tag{12.14b}$$

where $\delta(q, \sigma, q', \sigma', d)$ is the probability that, when the machine is in state $q \in Q \backslash Q_t$ and the head reads the symbol σ; the new state of the probabilistic Turing machine is $q' \in Q$, the symbol σ is replaced by σ' in the cell the head is currently on, and the head moves in the direction indicated by d. Furthermore, if the machine reaches a terminating state it remains in that state.

Thus the probabilistic Turing machine $\mathcal{M}$ is $(\Gamma, Q, \delta, \varrho)$. □

As in the case of a deterministic TM, in the above definition of the transition function δ, the state $q \in Q \backslash Q_t$. The notion of computation and configuration in a PTM can be developed as in TM. Unlike in a TM, a single configuration in a PTM can transition to several different configurations with probabilities that add up to unity. Therefore the net computation of a PTM can be construed as a tree having different possible configurations as nodes. The initial configuration is the root of the tree. A probability can be assigned to each node of this configuration tree. This probability at each node is equal to the product of the probabilities assigned to the edges on the path from the root to the specified node. Thus the probability assigned to a specific node in the configuration tree is the probability that a configuration starting at the root node reaches the given configuration node. We next give a description of probabilistic complexity classes.

Probabilistic Complexity Classes

If a problem is required to be solved on a PTM, then some computations can end up in an accept state and some may end up in a reject state. Then it is fair to ask: What is the implication of the assertion that a probabilistic machine accepts a string or language, which is required for defining complexity class? Actually, there are multiple choices for classification of probabilistic classes.

(a) Class $\mathcal{NP}$ (Nondeterministic polynomial-time): The $\mathcal{NP}$ class of languages has already been defined within the context of a NTM. This class can also be equivalently defined as the family of languages that can be accepted in polynomial-time by some PTM. This acceptance is defined as follows. A string of symbols $w \in \mathcal{A}^*$ is said to be in the language $L(\mathcal{M})$ if and only if the probabilistic Turing machine $\mathcal{M}$ accepts w with a nonzero probability.

(b) Class $\mathcal{PP}$ (Probabilistic polynomial-time): The class $\mathcal{PP}$ is the family of languages that can be accepted by some PTM in polynomial-time. This acceptance is defined as follows. A string of symbols $w \in \mathcal{A}^*$ is said to be in the language $L(\mathcal{M})$ if and only if the probabilistic Turing machine $\mathcal{M}$ accepts w with a probability greater than $1/2$. However if $w \notin L(\mathcal{M})$, then the probability that $\mathcal{M}$ accepts it is less than or equal to $1/2$. Thus

$$w \in L(\mathcal{M}) \Rightarrow P(\mathcal{M}_s(w) = q_a) > \frac{1}{2}$$
$$w \notin L(\mathcal{M}) \Rightarrow P(\mathcal{M}_s(w) = q_a) \leq \frac{1}{2}$$

(c) Class $\mathcal{RP}$ (Randomized polynomial-time): The class $\mathcal{RP}$ is the family of languages that can be accepted by some PTM in polynomial-time. This acceptance is defined as follows. A string of symbols $w \in \mathcal{A}^*$ is said to be in the language $L(\mathcal{M})$ if and only if the probabilistic Turing machine $\mathcal{M}$ accepts w with a probability greater than or equal to $1/2$. However if $w \notin L(\mathcal{M})$, then the probability that $\mathcal{M}$ accepts it is equal to zero. Thus

$$w \in L(\mathcal{M}) \Rightarrow P(\mathcal{M}_s(w) = q_a) \geq \frac{1}{2}$$
$$w \notin L(\mathcal{M}) \Rightarrow P(\mathcal{M}_s(w) = q_a) = 0$$

The class of languages $\mathcal{RP}$ corresponds to the idea of realizable effective computation. A PTM which accepts words as per the definition of the $\mathcal{RP}$ class is also called a *Monte Carlo Turing machine.*

(d) Class $co\mathcal{RP}$ (complement of $\mathcal{RP}$): The class $co\mathcal{RP}$ is the family of languages which are precisely the complement of languages in $\mathcal{RP}$. Observe that $co\mathcal{RP}$ is not the complement of $\mathcal{RP}$ among all the languages.

(e) Class $\mathcal{ZPP}$ (Zero error probability in polynomial-time): This class of languages is defined as $\mathcal{ZPP} = \mathcal{RP} \cap co\mathcal{RP}$. Thus this is the class of languages that can be accepted by a PTM whose expected running time is polynomial.

(f) Class $\mathcal{BPP}$ (Bounded error probability in polynomial-time): The class $\mathcal{BPP}$ is the family of languages that can be accepted by some PTM in polynomial-time. This acceptance is defined as follows. A string of symbols $w \in \mathcal{A}^*$ is said to be in the language $L(\mathcal{M})$ if and only if the probabilistic Turing machine $\mathcal{M}$ accepts w with a probability greater than or equal to $2/3$. However if $w \notin L(\mathcal{M})$, then the probability that $\mathcal{M}$ accepts it is less than or equal to $1/3$. In this definition, the constant $2/3$ is arbitrary. Any constant value $p \in (1/2, 1)$ is also appropriate. Thus

$$w \in L(\mathcal{M}) \Rightarrow P(\mathcal{M}_s(w) = q_a) \geq p$$
$$w \notin L(\mathcal{M}) \Rightarrow P(\mathcal{M}_s(w) = q_a) \leq (1 - p)$$

It can finally be observed that any TM is a PTM with only a single possibility for each transition. Thus

$$\mathcal{P} \subseteq \mathcal{ZPP} \subseteq \mathcal{RP} \subseteq \mathcal{PP} \subseteq \mathcal{PSPACE}$$

$$\mathcal{P} \subseteq \mathcal{BPP}$$

Modern Church-Turing Thesis

In light of the discussion about a PTM, the classical version of the Church-Turing hypothesis can be restated. This modified computing paradigm is called the modern Church-Turing hypothesis. It states that: *Any reasonable physical model of computation can be effectively implemented on a probabilistic Turing machine.*

12.5.4 Quantum Turing Machine

Quantum Turing machine model is used to develop a reasonable model for quantum computation. These are a simple extension of PTMs. The transition probabilities of a PTM are replaced by *transition amplitudes*, or simply *amplitude*.

Definition 12.19. *A quantum Turing machine $\mathcal{M}$ is defined as follows.*

(a) *The set of tape symbols is Γ. The alphabet (input symbols) is $\mathcal{A} \subset \Gamma$, and the blank symbol is $\sqcup \in \Gamma \backslash \mathcal{A}$.*
(b) *The set of finite states is Q. The initial state is $q_0 \in Q$, and $Q_t \subseteq Q$ is the set of terminating states. Also $Q_t = \{q_a, q_r\}$, where q_a and q_r are the accept and reject states respectively.*
(c) *The set $\varrho = \{\leftarrow, \downarrow, \rightarrow\}$ denotes the movement of the read-write tape head. The symbol $\leftarrow$ means move the tape head to the left cell, the symbol $\downarrow$ means do not move the tape head at all (let it be stationary), and the symbol $\rightarrow$ means move the tape head to the right cell.*
(d) *The transition amplitude function δ, is*

$$\delta : Q \backslash Q_t \times \Gamma \times Q \times \Gamma \times \varrho \rightarrow \mathbb{C} \tag{12.15a}$$

where $\delta(q, \sigma, q', \sigma', d) \in \mathbb{C}$, such that its absolute value is in the interval $[0, 1]$. For each $(q, \sigma) \in Q \backslash Q_t \times \Gamma$

$$\sum_{(q', \sigma', d) \in Q \times \Gamma \times \varrho} |\delta(q, \sigma, q', \sigma', d)|^2 = 1 \tag{12.15b}$$

Also $\delta(q, \sigma, q', \sigma', d)$ is said to be the amplitude (possibly complex) such that when the machine is in state $q \in Q \backslash Q_t$ and the head reads the symbol σ; the quantum Turing machine transitions to state $q' \in Q$, the symbol σ is replaced by σ' in the cell the head is currently on, and the head moves in the direction indicated by d. Furthermore, if the machine reaches a terminating state it remains in that state.

Thus the quantum Turing machine $\mathcal{M}$ is $(\Gamma, Q, \delta, \varrho)$. □

As in the case of deterministic and probabilistic Turing machines, in the above definition of the transition function δ, the state $q \in Q \backslash Q_t$.

Quantum Complexity Classes

The different classes of complexity classes associated with QTMs are a natural extension of the complexity classes associated with PTMs.

(a) Class $\mathcal{EQP}$: The class $\mathcal{EQP}$ is the quantum version of the $\mathcal{P}$ class. The abbreviation $\mathcal{EQP}$ stands for *exact acceptance by a quantum computer in polynomial-time.*
(b) Class $\mathcal{NQP}$: The class $\mathcal{NQP}$ is the quantum version of the $\mathcal{NP}$ class.
(c) Class $\mathcal{BQP}$: The class $\mathcal{BQP}$ is the quantum version of the $\mathcal{BPP}$ class.
(d) Class $\mathcal{ZQP}$: The class $\mathcal{ZQP}$ is the quantum version of the $\mathcal{ZPP}$ class.

In the worst case, a quantum computer can solve more in polynomial-time than the classical computer. It has not yet been proven that QTMs are more powerful than PTMs, or vice-versa. The following relationships should not be surprising.

$$\mathcal{P} \subseteq \mathcal{EQP} \subseteq \mathcal{BQP}$$
$$\mathcal{BPP} \subseteq \mathcal{BQP} \subseteq \mathcal{PP} \subseteq \mathcal{PSPACE}$$

Determination of proper inclusion in the above relationships is an open problem. The mathematician David Deutsch established the existence of a *universal quantum Turing machine*. This machine is said to simulate any other QTM with any specified degree of precision.

In this section, we developed a formal model of quantum computation via quantum Turing machine. In the next section certain quantum algorithms are described. Some of these really ingenious algorithms are responsible for the growth of interest in this nascent field of quantum computation, communication, and cryptography.

12.6 Quantum Algorithms

The power and limitations of quantum computation are best understood by studying some algorithms. Besides, it is useful to know if quantum computation offers theoretically any significant advantage over classical computation. The strength of quantum computation lies in its parallelism. At the most elemental level, it can be stated imprecisely that quantum parallelism permits quantum computers to evaluate $f(x)$ for several different values of x in parallel. As it has been already mentioned, linear operators are used to describe quantum evolution. Consider a linear operator A and a state $|\alpha\rangle$. Let

$$|\alpha\rangle = \sum_{i=0}^{2^n-1} a_i |i\rangle ; \quad a_i \in \mathbb{R}, \ 0 \le i \le (2^n - 1)$$

Apply the operator A to state $|\alpha\rangle$, then

$$A|\alpha\rangle = \sum_{i=0}^{2^n-1} a_i A|i\rangle$$

It can be observed from the above equation, that the $a_i A|i\rangle$'s for $0 \le i \le (2^n - 1)$ can be computed in parallel. That is, an exponential number (2^n) of operations can be executed simultaneously. This property is called *quantum parallelism*. Even for small values of n, massive parallelism is observed in quantum computation. The strength of quantum computation is explored in this section by studying Deutsch's problem and algorithm, Deutsch-Josza problem and algorithm, Shor's factorization algorithm, and Grover's search algorithm.

12.6.1 Deutsch's Problem and Algorithm

Deutsch's problem was one of the earliest examples studied in the nascent field of quantum computation. It is named after David Deutsch. Before this problem is stated, a *balanced function* and a *constant function* are defined. A function $f(\cdot)$ is said to be a balanced function, if $f(z) = 1$ exactly for half of all possible values of z, and $f(z) = 0$ for the other half of values of z. A function $f(\cdot)$ is a constant function, if its value is constant for all possible values of its argument.

Statement of Deutsch's Problem: Let $f : \mathfrak{B} \to \mathfrak{B}$ be a Boolean function, where $\mathfrak{B} = \{0, 1\}$. It is required to be determined if $f(\cdot)$ is either a constant or a balanced function.

This problem alone does not appear to be very interesting. Nevertheless it demonstrates the power of quantum computation. In this problem, the function $f(\cdot)$ is constant if $f(0) = f(1)$, and balanced if $f(0) \neq f(1)$.

A classical approach to this problem requires evaluation of the function $f(\cdot)$ for both $z = 0$ and $z = 1$. Thus two queries are required in this classical approach. In contrast, it is demonstrated that quantum computation approach requires only a single query. This approach uses the Hadamard gate, which is described via the unitary Hadamard's matrix W_2, where

$$W_2 = \frac{1}{\sqrt{2}} \begin{bmatrix} 1 & 1 \\ 1 & -1 \end{bmatrix}$$

The solution uses a gate which implements a unitary transformation U_f.

Description of the Unitary Transformation U_f

The unitary transformation U_f, used in solving the Deutsch's problem is defined as:

$$U_f : |x\rangle |y\rangle \to |x\rangle |y \oplus f(x)\rangle, \quad x, y \in \mathfrak{B}$$

where $\oplus$ is the symbol for exclusive-or operation. The gate which implements this transformation is called the U_f-gate. It is also convenient to consider this gate with two inputs I_1 and I_2; and two outputs O_1 and O_2. The inputs I_1 and I_2 are fed with the qubits $|x\rangle$ and $|y\rangle$ respectively. Similarly, the qubits $|x\rangle$ and $|y \oplus f(x)\rangle$ are on the outputs O_1 and O_2 respectively.

Solution of the Deutsch's Problem

A quantum circuit to implement the Deutsch's algorithm is as follows:

(a) Feed $W_2 |0\rangle$ to the input I_1, and $W_2 |1\rangle$ to the input I_2 of the U_f-gate.
(b) Ignore the output O_2. The output O_1 is fed to the Hadamard gate W_2. Examine this output to determine if the function $f(\cdot)$ is either a constant or balanced.

The solution to the Deutsch's problem is implemented in the following three steps.

Step 0: Initially, let the input on I_1 be $|x\rangle$ and that on I_2 be $W_2 |1\rangle$. Applying this input to the transformation U_f and recalling that $W_2 |1\rangle = (|0\rangle - |1\rangle) / \sqrt{2}$ yields

$$\begin{aligned} U_f \{|x\rangle W_2 |1\rangle\} &= U_f \left\{ |x\rangle (|0\rangle - |1\rangle) / \sqrt{2} \right\} \\ &= \frac{1}{\sqrt{2}} U_f \{|x\rangle |0\rangle\} - \frac{1}{\sqrt{2}} U_f \{|x\rangle |1\rangle\} \\ &= \frac{1}{\sqrt{2}} |x\rangle \{|0 \oplus f(x)\rangle - |1 \oplus f(x)\rangle\} \end{aligned}$$

$$= (-1)^{f(x)} |x\rangle \frac{1}{\sqrt{2}} \{|0\rangle - |1\rangle\}$$
$$= (-1)^{f(x)} |x\rangle \{W_2 |1\rangle\}$$

In this transformation, the O_1 qubit is equal to $(-1)^{f(x)} |x\rangle$, and the O_2 qubit is equal to $W_2 |1\rangle$.

Step 1: In this step, let the input I_1 on the U_f-gate be equal to $W_2 |0\rangle$ and that on I_2 be $W_2 |1\rangle$. The output of the U_f-gate is equal to $U_f \{W_2 |0\rangle W_2 |1\rangle\}$. As

$$W_2 |0\rangle = (|0\rangle + |1\rangle) / \sqrt{2}$$

we obtain

$$\begin{aligned}
U_f \{W_2 |0\rangle W_2 |1\rangle\} &= U_f \left\{ \frac{1}{\sqrt{2}} (|0\rangle + |1\rangle) W_2 |1\rangle \right\} \\
&= \frac{1}{\sqrt{2}} [U_f \{|0\rangle W_2 |1\rangle\} + U_f \{|1\rangle W_2 |1\rangle\}] \\
&= \frac{1}{\sqrt{2}} \left\{ (-1)^{f(0)} |0\rangle + (-1)^{f(1)} |1\rangle \right\} \{W_2 |1\rangle\} \\
&= \frac{1}{\sqrt{2}} (-1)^{f(0)} \left\{ |0\rangle + (-1)^{f(0) \oplus f(1)} |1\rangle \right\} \{W_2 |1\rangle\}
\end{aligned}$$

Define

$$|v\rangle \triangleq (-1)^{f(0)} \left\{ |0\rangle + (-1)^{f(0) \oplus f(1)} |1\rangle \right\} / \sqrt{2}$$

Then in the above transformation, the O_1 qubit is equal to $|v\rangle$, and the O_2 qubit is equal to $W_2 |1\rangle$.

Step 2: In this step, $W_2 |v\rangle$ is computed, where $|v\rangle$ was determined in the last step. Thus

$$\begin{aligned}
W_2 |v\rangle &= \frac{(-1)^{f(0)}}{\sqrt{2}} \left\{ W_2 |0\rangle + (-1)^{f(0) \oplus f(1)} W_2 |1\rangle \right\} \\
&= \frac{(-1)^{f(0)}}{2} \left\{ (|0\rangle + |1\rangle) + (-1)^{f(0) \oplus f(1)} (|0\rangle - |1\rangle) \right\}
\end{aligned}$$

Next examine the two cases.

(a) Let the function $f(\cdot)$ be constant. That is, $f(0) = f(1)$, then $f(0) \oplus f(1) = 0$. Thus $W_2 |v\rangle = (-1)^{f(0)} |0\rangle$.

(b) Let the function $f(\cdot)$ be balanced. That is, $f(0) \neq f(1)$, then $f(0) \oplus f(1) = 1$. Thus $W_2 |v\rangle = (-1)^{f(0)} |1\rangle$.

The above two cases can be combined in a single expression

$$W_2 |v\rangle = (-1)^{f(0)} |f(0) \oplus f(1)\rangle$$

Thus the observation on the last W_2 gate indicates if the function $f(\cdot)$ is either constant or balanced. It determines $f(0) \oplus f(1)$. Summarizing, it can be concluded that the quantum approach requires only a single evaluation of the function $f(\cdot)$. This is in contrast to the classical approach, which requires two queries to evaluate this function. In classical computation, the values $f(0)$ and $f(1)$ are evaluated separately. However, in quantum computation, there is interaction between the values of $f(\cdot)$. That is, the quantum computer is in a blend of states, where it is possible to compute both $f(0)$ and $f(1)$, and extract information about the relationship between $f(0)$ and $f(1)$. Thus, this algorithm provides information whether the function $f(\cdot)$ is constant or balanced.

12.6.2 Deutsch-Josza's Problem and Algorithm

The Deutsch-Josza's problem is a generalization of the Deutsch's problem. It is named after David Deutsch and Richard Josza.

Statement of Deutsch-Josza's Problem: Let $f : \mathfrak{B}^{(n)} \to \mathfrak{B}$ be a Boolean function, where $\mathfrak{B} = \{0,1\}$, and $n \geq 2$. The function $f(\cdot)$ is either a constant or balanced. That is, this function is only one of these two types. The goal is to determine if $f(\cdot)$ is either a constant or a balanced function.

The Deutsch-Josza's algorithm uses the Hadamard transform. This transform has been defined in the chapter on applied analysis. If $|x\rangle$ is a qubit, then its Hadamard transform is

$$W_2 |x\rangle = \frac{1}{\sqrt{2}} \sum_{y \in \mathfrak{B}} (-1)^{xy} |y\rangle, \quad \text{where} \quad x \in \mathfrak{B}$$

Let $|x\rangle = |x_{n-1}\rangle |x_{n-2}\rangle \ldots |x_0\rangle$ be a state vector of length $N = 2^n$, where $x_i \in \mathfrak{B}$, for $0 \leq i \leq (n-1)$, and $x_{n-1}x_{n-2} \cdots x_0 \in \mathfrak{B}^{(n)}$ is a binary representation of $x \in \mathbb{Z}_N$. The Hadamard transform of this vector is given by $W_N |x\rangle$, where W_N is an $N \times N$ Hadamard matrix. Note that $W_N = W_2^{\otimes n}$, and

$$W_N |x\rangle = \frac{1}{\sqrt{N}} \sum_{y \in \mathbb{Z}_N} (-1)^{x \cdot y} |y\rangle, \quad x \in \mathbb{Z}_N$$

where

$$|y\rangle = |y_{n-1}\rangle |y_{n-2}\rangle \cdots |y_0\rangle$$

is a state vector of length $N, y_i \in \mathfrak{B}$ for $0 \leq i \leq (n-1)$, and $y_{n-1}y_{n-2} \cdots y_0 \in \mathfrak{B}^{(n)}$ is a binary representation of $y \in \mathbb{Z}_N$. Furthermore, the dot product $x \cdot y = \sum_{k=0}^{n-1} x_k y_k$. Also the exponent of (-1), which is $x \cdot y$ could be replaced by $\bigoplus_{k=0}^{n-1} x_k y_k$, where the summation symbol $\bigoplus$ indicates exclusive-or operation or equivalently modulo 2 arithmetic.

Recall that the function $f(\cdot)$ is constant if $f(z)$ is identical for all possible values of z, and this function is balanced if $f(z) = 0$ for exactly half of all possible values of z, and equal to 1 for the other half of values of z.The classical approach requires evaluation of the function $f(\cdot)$ for at least $(N/2+1)$ values of z, to classify this function. The Deutsch-Josza's algorithm requires only a single evaluation of the function. The solution to this problem also uses a gate which implements the unitary transformation U_f.

Description of the Unitary Transformation U_f

The unitary transformation U_f, used in solving the Deutsch-Josza's problem is defined as:

$$U_f : |x\rangle |y\rangle \to |x\rangle |y \oplus f(x)\rangle; \quad x \in \mathfrak{B}^{(n)}, \; y \in \mathfrak{B}$$

where $\oplus$ is the symbol for exclusive-or operation. The U_f-gate has two inputs I_1 and I_2; and two outputs O_1 and O_2. The inputs I_1 and I_2 are fed with the state vectors $|x\rangle$ and $|y\rangle$ respectively. Similarly, the state vectors $|x\rangle$ and $|y \oplus f(x)\rangle$ are on the outputs O_1 and O_2 respectively.

Solution of the Deutsch-Josza's Problem

A quantum circuit to implement the Deutsch-Josza's algorithm is as follows:

(a) Feed $W_N |0\rangle$ to the input I_1, and $W_2 |1\rangle$ to the input I_2 of the U_f-gate.

(b) Ignore the output O_2. The output O_1 is fed to the Hadamard gate W_N. Examine this output to determine if the function $f(\cdot)$ is either a constant or balanced.

The solution to the Deutsch-Josza's problem is implemented in the following three steps.

Step 0: Initially, let the input on I_1 be $|x\rangle$ and that on I_2 be $W_2|1\rangle$. Applying this input to the transformation U_f and recall that $W_2|1\rangle = (|0\rangle - |1\rangle)/\sqrt{2}$. This yields

$$U_f\{|x\rangle W_2|1\rangle\} = (-1)^{f(x)}|x\rangle\{W_2|1\rangle\}$$

as in Deutsch's algorithm. In this transformation, the output on O_1 is equal to $(-1)^{f(x)}|x\rangle$, and the O_2 qubit is equal to $W_2|1\rangle$.

Step 1: In this step, let the input I_1 on the U_f-gate be equal to $W_N|0\rangle$ and that on I_2 be $W_2|1\rangle$. The output of the U_f-gate is equal to $U_f\{W_N|0\rangle W_2|1\rangle\}$. Note that

$$W_N|0\rangle = \frac{1}{\sqrt{N}}\sum_{i=0}^{N-1}|i\rangle$$

Therefore

$$\begin{aligned} U_f\{W_N|0\rangle W_2|1\rangle\} &= U_f\left\{\frac{1}{\sqrt{N}}\left(\sum_{i=0}^{N-1}|i\rangle\right)W_2|1\rangle\right\} \\ &= \frac{1}{\sqrt{N}}\sum_{i=0}^{N-1}U_f\{|i\rangle W_2|1\rangle\} \\ &= \frac{1}{\sqrt{N}}\sum_{i=0}^{N-1}\left\{(-1)^{f(i)}|i\rangle\right\}\{W_2|1\rangle\} \end{aligned}$$

Let

$$|v\rangle \triangleq \frac{1}{\sqrt{N}}\sum_{i=0}^{N-1}\left\{(-1)^{f(i)}|i\rangle\right\}$$

then in the above transformation, the output on O_1 is $|v\rangle$, and the O_2 qubit is equal to $W_2|1\rangle$. Thus the number of qubits on O_1 is equal to n.

Step 2: Next generate the transformation $W_N|v\rangle$, where $|v\rangle$ was determined in the last step. Thus

$$\begin{aligned} W_N|v\rangle &= \frac{1}{\sqrt{N}}\sum_{i=0}^{N-1}(-1)^{f(i)}W_N|i\rangle = \frac{1}{N}\sum_{i=0}^{N-1}\sum_{j=0}^{N-1}(-1)^{i\cdot j+f(i)}|j\rangle \\ &= \frac{1}{N}\sum_{j=0}^{N-1}|j\rangle\sum_{i=0}^{N-1}(-1)^{i\cdot j+f(i)} \end{aligned}$$

Following two cases are next examined.

(a) Let the function $f(\cdot)$ be constant equal to say c. If $j = 0$, the right-hand side of the above equation yields a component $|0\rangle(-1)^c$. However, if $j \neq 0$ then $\sum_{i=0}^{N-1}(-1)^{i\cdot j} = 0$. Thus $W_N|v\rangle = (-1)^c|0\rangle$ with probability one. Thus the amplitude of $|0\rangle$ is ± 1. The sign of 1 depends upon the value of c.

(b) Let the function $f(\cdot)$ be balanced. If $j = 0$, then the positive and negative contributions to $|0\rangle$ cancel each other. Thus the contribution of the state vector $|0\rangle$ does not appear in $W_N |v\rangle$. Consequently $W_N |v\rangle$ yields some state vector $|k\rangle \neq |0\rangle$ with probability one.

Thus the output of the last Hadamard transformation yields an output with certainty, indicating whether the function $f(\cdot)$ is a constant or balanced.

12.6.3 Shor's Factorization Algorithm

A premier application of quantum computation is a probabilistic polynomial-time algorithm for factorizing a positive number. This algorithm was discovered by Peter W. Shor in 1994. It can safely be said that, it was this algorithm which first demonstrated the importance of studying quantum algorithms. The importance of factorizing an integer cannot be stressed enough. The assumption of difficult factoring of large numbers is an essential basis of modern cryptography. Shor's algorithm is considered to be a breakthrough, because classical algorithms to factor a positive integer are computationally intractable. His algorithm uses quantum Fourier transform, which is discussed subsequently.

Factoring Based upon Period and Order

The goal of Shor's algorithm is to find nontrivial factors of a positive integer $n \in \mathbb{P}\backslash\{1\}$. If n is a power of 2, then it is easily recognizable. Therefore values of n which are powers of the integer 2 are excluded. Similarly values of n which are simply powers of the form u^v, where $u \geq 2$ and $v \geq 2$ are excluded, because there exists a polynomial-time algorithm for recognizing such numbers. See the problem section for a hint to establish this assertion. Therefore values of n which are powers of some positive integer are excluded from the following discussion.

The integers 1 and n are said to be trivial factors of n. Select $x \in \mathbb{Z}_n\backslash\{0\}$, such that x and n have no common factors, that is $\gcd(x, n) = 1$. Since $\gcd(x, n) = 1$, the integers x and n are said to be coprime. The order of x modulo n is defined as the least possible positive integer r such that $x^r \equiv 1 \pmod{n}$.

The factorization problem can also be framed in terms of finding the *period* of a function $g_{n,x}(\cdot)$, where $g_{n,x}(k) \equiv x^k \pmod{n}$. The period of this function is the smallest positive integer r such that $g_{n,x}(k + r) = g_{n,x}(k)$ for any integer k. That is, the period of the function $g_{n,x}(\cdot)$ is the group order of x. This order of x is the order of the group element $x \in \mathbb{Z}_n^*$, where $\mathbb{Z}_n^* = \{i \mid 0 < i < n, \gcd(i, n) = 1\}$ is a multiplicative group.

A technique that is based upon finding the order of an element is next outlined. Let $x \in \mathbb{Z}_n^*\backslash\{1\}$ be a randomly chosen integer. That is, this integer was selected with uniform probability. If $\gcd(x, n) = d > 1$, then d is a nontrivial factor of n. Assume that $d = 1$, and the order of x is equal to r, that is $x^r \equiv 1 \pmod{n}$. This relationship implies that $(x^r - 1) \equiv 0 \pmod{n}$. That is, the integer n divides $(x^r - 1)$. Further assume that r is an even integer. Then

$$(x^r - 1) = \left(x^{\frac{r}{2}} - 1\right)\left(x^{\frac{r}{2}} + 1\right)$$

Since n divides $(x^r - 1)$, the possibility that n divides $\left(x^{\frac{r}{2}} - 1\right)$ is not feasible. Otherwise this would imply that $x^{\frac{r}{2}} \equiv 1 \pmod{n}$, which violates the assumption that r is the order of the integer x. However, the following two cases are feasible.

For convenience in notation define $z_1 \triangleq \left(x^{\frac{r}{2}} - 1\right)$ and $z_2 \triangleq \left(x^{\frac{r}{2}} + 1\right)$.

(a) The integer n shares a factor with both z_1 and z_2. The factors between n and z_1 can be determined efficiently by using the Euclidean algorithm to compute the $\gcd(z_1, n)$. Similarly, the factors between n and z_2 can be determined by computing the $\gcd(z_2, n)$.
(b) The integer n divides z_2. Thus $\gcd(z_2, n) = n$, which is a trivial factor of n.

Using number theoretic techniques, it can be proved that, if the number x is randomly chosen with uniform distribution from the set of integers $\mathbb{Z}_n^*$, then the probability that the order r of x is even, and that $x^{\frac{r}{2}} \not\equiv -1 \pmod{n}$ is at least 0.5. Thus an algorithm can be designed to find a nontrivial factor of n, provided the order of an element x can be found, where $1 < x < n$. An outline of the algorithm is shown below. It finds the nontrivial factors of an odd positive integer.

In this algorithm, Step 3 is incomplete, because an algorithm to find the period r of a function has not yet been outlined. The Step 5 is the computation of a nontrivial factor of n in a probabilistic sense. Furthermore, the probability of success of Step 5 is at least 0.5.

Algorithm 12.1. *Find a Nontrivial Factor of a Positive Odd Integer n.*

Input: A positive odd integer n. Assume that $n \neq u^v$, where $u, v \in \mathbb{P} \backslash \{1\}$.
Output: Nontrivial factors of the integer n.
begin

Step 1: Select x randomly from the integers 2 through $(n-1)$.
Compute $\gcd(x, n) = d$
Step 2: if $d > 1$ **then** d is a nontrivial factor of n, **stop**
Step 3: if $d = 1$ find the period r of the expression $x \pmod{n}$
Step 4: if the period r is odd **go to Step 1**
Step 5: if the period r is even

(a) **if** $x^{\frac{r}{2}} \equiv -1 \pmod{n}$ **then** $\gcd\left(x^{\frac{r}{2}} + 1, n\right) = n$
(*yields a trivial factor of n*)
go to Step 1

(b) **if** $x^{\frac{r}{2}} \not\equiv -1 \pmod{n}$, **then** compute:
$y_1 \leftarrow \gcd\left(x^{\frac{r}{2}} - 1, n\right)$, $y_2 \leftarrow \gcd\left(x^{\frac{r}{2}} + 1, n\right)$,
and $z \leftarrow \max\{y_1, y_2\}$
if $z = 1$ **then go to Step 1**
else z is a nontrivial factor of n, **stop**

end (*end of Shor's algorithm*)

The efficacy of the algorithm is demonstrated in the following toy example.

Example 12.5. Let $n = 15$. Its factors are 3 and 5. It is actually the smallest integer which can be factorized by the above algorithm. Note that

$$\mathbb{Z}_{15}^* = \{1, 2, 4, 7, 8, 11, 13, 14\}$$

(a) Let $x = 2$. Then $d = \gcd(x, n) = \gcd(2, 15) = 1$. Thus its period is $r = 4$. Also $x^{\frac{r}{2}} \equiv 4 \pmod{15} \not\equiv -1 \pmod{15}$. Furthermore, $y_1 = \gcd\left(x^{\frac{r}{2}} - 1, n\right) = \gcd(3, 15) = 3$. Similarly

$y_2 = \gcd\left(x^{\frac{r}{2}} + 1, n\right) = \gcd(5, 15) = 5$. Thus $z = 5$. Consequently the integers 3 and 5 are factors of 15.

(b) Let $x = 5$. Observe that $d = \gcd(x, n) = \gcd(5, 15) = 5 > 1$. Therefore $d = 5$ is a nontrivial factor of 15.

(c) Let $x = 8$. Then $d = \gcd(x, n) = \gcd(8, 15) = 1$. Thus its period $r = 4$. Also $x^{\frac{r}{2}} \equiv 4 \pmod{15} \not\equiv -1 \pmod{15}$. Furthermore,

$$y_1 = \gcd\left(x^{\frac{r}{2}} - 1, n\right) = \gcd(63, 15) = 3$$

Similarly

$$y_2 = \gcd\left(x^{\frac{r}{2}} + 1, n\right) = \gcd(65, 15) = 5$$

Thus $z = 5$. Therefore the integers 3 and 5 are factors of 15.

(d) Let $x = 14$. Then $d = \gcd(x, n) = \gcd(14, 15) = 1$. Thus its period $r = 2$. Also $x^{\frac{r}{2}} \equiv 14 \pmod{15} \equiv -1 \pmod{15}$. Consequently the choice of the number $x = 14$ is not successful in determining the factors of the number 15. Also note that

$$\gcd\left(x^{\frac{r}{2}} + 1, n\right) = \gcd(15, 15) = 15$$

It can be checked that all elements in the set $\mathbb{Z}_{15}^{*} \backslash \{1\}$ except the number 14 yield nontrivial factors of the number 15. □

The period of a function is computed by a Fourier transformation. A quantum version of the classical discrete Fourier transform is next discussed.

Quantum Fourier Transform

Quantum Fourier Transform (QFT) is a unitary transformation. The quantum Fourier transform on an orthonormal basis $|j\rangle$, for $0 \le j \le (N-1)$ is defined below.

Definition 12.20. *Let $N \in \mathbb{P}$, $\omega_N = e^{2\pi i/N}$, $i = \sqrt{-1}$, and $\pi = 3.14159265358\ldots$. The quantum Fourier transform on an orthonormal basis $|k\rangle$, $0 \le k \le (N-1)$, is a mapping of the basis states,*

$$QFT_N : |j\rangle \to \frac{1}{\sqrt{N}} \sum_{k=0}^{(N-1)} \omega_N^{jk} |k\rangle, \quad 0 \le j \le (N-1) \tag{12.16}$$

□

The next result follows immediately from the above definition.

Lemma 12.2. *The QFT mapping of an arbitrary state (quantum superposition) is*

$$QFT_N : \sum_{j=0}^{N-1} y(j) |j\rangle \to \sum_{k=0}^{(N-1)} Y(k) |k\rangle \tag{12.17a}$$

where the sequence $Y(k)$, $0 \le k \le (N-1)$ is a discrete Fourier transform (DFT) of the amplitude sequence $y(j)$, $0 \le j \le (N-1)$. That is

$$Y(k) = \frac{1}{\sqrt{N}} \sum_{j=0}^{(N-1)} y(j)\,\omega_N^{jk}, \quad \forall\, k \in \mathbb{Z}_N \tag{12.17b}$$

□

A product representation of QFT mapping is next described. This representation is a computationally efficient implementation of QFT.

Definition 12.21. *A binary fraction* $\sum_{t=1}^{b} j_t/2^t$, *where* $j_t \in \mathfrak{B} = \{0,1\}$ *for* $1 \leq t \leq b$, *is represented by* $0.j_1 j_2 \ldots j_b$. □

The product representation of QFT mapping is encapsulated in the following result.

Lemma 12.3. *Let* $N = 2^b, b \in \mathbb{P}$. *Also define the binary representation of an integer* j *to be* $j_{b-1} j_{b-2} \ldots j_0$, *where* $j_t \in \mathfrak{B}$ *for* $0 \leq t \leq (b-1)$, *and*

$$j = \sum_{t=0}^{b-1} 2^t j_t \tag{12.18a}$$

If $|0\rangle = \begin{bmatrix} 1 & 0 \end{bmatrix}^T$, *and* $|1\rangle = \begin{bmatrix} 0 & 1 \end{bmatrix}^T$, *then*

$$\begin{aligned} &|j_{b-1} j_{b-2} \ldots j_0\rangle \\ &\to \frac{1}{\sqrt{N}} \left\{ |0\rangle + e^{2\pi i (0.j_0)} |1\rangle \right\} \left\{ |0\rangle + e^{2\pi i (0.j_1 j_0)} |1\rangle \right\} \\ &\quad \ldots \left\{ |0\rangle + e^{2\pi i (0.j_{b-1} j_{b-2} \ldots j_0)} |1\rangle \right\} \end{aligned} \tag{12.18b}$$

Proof. See the problem section. □

This product representation is useful in obtaining an efficient circuit representation of the QFT. The computational complexity of discrete Fourier transform (DFT) is $O(N^2)$. In direct contrast, if $N = 2^b$, the computational complexity of a fast Fourier transform (FFT) algorithm is $O(bN)$. This is a significant improvement over the computational complexity of DFT. However the computational complexity of QFT is $\Theta(b^2)$, which is a spectacular improvement over the complexity of FFT. A similar factorization of the Fourier transform operator can be obtained via the Chinese remainder theorem.

Shor's Algorithm to Find the Period

Recall that the number to be factored is a positive integer n. This is done by finding the period of the function $g_{n,x}(\cdot)$, where $g_{n,x}(k) \equiv x^k \pmod{n}$, $x \in \mathbb{Z}_n \backslash \{0\}$, $\gcd(x,n) = 1$, and k is an integer. The period of this function is found by using the QFT. The QFT cannot be computed on the set of integers $\mathbb{Z}$. Therefore the QFT is computed on $\mathbb{Z}_m$, where $n < m$. The integer m is chosen to be large enough such that the period is observable in the computation of the QFT. Let $m = 2^\eta$, then each element of $\mathbb{Z}_m$ can be represented by η qubits. This choice also permits the use of a fast version of the QFT. Since $n < m$, the elements of the set $\mathbb{Z}_n$ take at most η qubits for their representation. If $u \in \mathbb{Z}_m$ and $v \in \mathbb{Z}_n$, then these are represented as $|u\rangle\,|v\rangle$. Following steps are executed in Shor's algorithm to find the period.

Step 1: Hadamard transform in $\mathbb{Z}_m$ is applied to $|0\rangle\,|0\rangle$. This results in

$$\frac{1}{\sqrt{m}}\sum_{k=0}^{m-1}|k\rangle\,|0\rangle$$

Step 2: Apply the function $F:|y\rangle\,|0\rangle \to |y\rangle\,|g_{n,x}(y)\rangle$ to the above superposition. Thus

$$\frac{1}{\sqrt{m}}\sum_{k=0}^{m-1}|k\rangle\,|x^k\rangle$$

Note that the function $g_{n,x}(\cdot)$ has a period r. For simplicity assume that $m/r = s$ is an integer. This assumption is relaxed later. Let $k = (qr+\beta)$, where $0 \le \beta \le (r-1)$ and $0 \le q \le (s-1)$. Then the above expression is

$$\frac{1}{\sqrt{m}}\sum_{\beta=0}^{r-1}\sum_{q=0}^{s-1}|qr+\beta\rangle\,|x^\beta\rangle$$

Step 3: Apply the QFT on $\mathbb{Z}_m$. Thus

$$\frac{1}{\sqrt{m}}\sum_{\beta=0}^{r-1}\sum_{q=0}^{s-1}\frac{1}{\sqrt{m}}\sum_{p=0}^{m-1}e^{2\pi ip(qr+\beta)/m}\,|p\rangle\,|x^\beta\rangle$$

$$=\frac{1}{m}\sum_{\beta=0}^{r-1}\sum_{p=0}^{m-1}e^{2\pi ip\beta/m}\sum_{q=0}^{s-1}e^{2\pi ipq/s}\,|p\rangle\,|x^\beta\rangle$$

If p is a multiple of s, then $e^{2\pi ipq/s} = 1$, and if p is not a multiple of s, then $\sum_{q=0}^{s-1}e^{2\pi ipq/s} = 0$. Thus the above expression is

$$\frac{1}{r}\sum_{\beta=0}^{r-1}\sum_{j=0}^{r-1}e^{2\pi ij\beta/r}\,|js\rangle\,|x^\beta\rangle$$

Step 4: Note the quantum representation of $\mathbb{Z}_m$ in the above superposition to obtain some $p \in \mathbb{Z}_m$. Actually p takes values in the set

$$\{0, s, 2s, \ldots, (r-1)\,s\}$$

Step 5: Generate a sequence of convergents (remember continued fractions) p_i/q_i of p/m using the Euclidean algorithm. Find the smallest q_i such that $x^{q_i} \equiv 1 \pmod{n}$, provided this type of q_i exists. This step is justified as follows.

Let $p = \lambda s$, where $\lambda \in \mathbb{Z}_r$. Notice that each λ is chosen with the same probability $1/r$. Therefore $p/m = \lambda/r$. Also let $\gcd(p,m) = \alpha$, and $p = p_0\alpha$ and $m = m_0\alpha$. Thus $p_0/m_0 = \lambda/r$. If in addition, $\gcd(\lambda, r) = 1$, then $r = m_0 = m/\gcd(p,m)$.

Since λ is chosen randomly, it is possible that the desired value r of the period is not obtained. In this event, the measurement is repeated a few times till success is achieved.

In the general case, when m/r is not an integer, the algorithm and its explanation is slightly more complicated, and the probabilities are not uniformly distributed. Let $\ell(n)$ be the number of digits in the number n, then it can be established that the Shor's factorizing algorithm finds a nontrivial factor of n in time

$$O\left(\{\ell(n)\}^3 \ln \ell(n)\right)$$

with probability $O(1)$. See the problem section for an example.

12.6.4 Grover's Search Algorithm

In the year 1996, Lov K. Grover discovered a surprisingly efficient quantum algorithm for finding a specified item in an unsorted list. This is similar to finding the name of a person in a telephone directory, if his or her telephone number is known. Grover's algorithm finds the specified item in an unsorted search space of N items in $O\left(N^{1/2}\right)$ time with bounded probability.

Classical search techniques in an unstructured list of size N have a complexity of $O(N)$ steps. Grover's search algorithm along with Shor's factorization algorithm is considered to be a significant achievement in the development of quantum algorithms. An algorithm of this type is important in cryptanalysis, because the goal of cryptanalysis is to find a plaintext for a specified ciphertext.

Grover's search algorithm can be described in terms of a matrix D called the *diffusion matrix*. The definition of this matrix and its properties are first stated. The algorithm also uses a sign-changing operator. We discuss the use of this operator and the Grover's search algorithm. Finally, analysis of the algorithm is presented.

Diffusion Matrix

The diffusion matrix is first defined, and then some useful related facts are listed.

Definition 12.22. *The diffusion matrix D is a square matrix of size N. It is:*

$$D = (2P - I) \tag{12.19}$$

where I is the identity matrix of size N. Also P is a square matrix of size N, and $P = [p_{ij}]$, where $p_{ij} = 1/N$, for $0 \leq i, j \leq (N-1)$. □

Observations 12.6. Some useful observations related to diffusion matrix.

1. If the matrix P acts upon a vector A of length N, then each element of the vector PA is equal to the average of its components.
2. $P^2 = P$.
3. The matrix D is unitary. That is, $DD^T = I$.
4. An explicit expression for the diffusion matrix D is

$$D = \begin{bmatrix} -1+\frac{2}{N} & \frac{2}{N} & \cdots & \frac{2}{N} \\ \frac{2}{N} & -1+\frac{2}{N} & \cdots & \frac{2}{N} \\ \vdots & \vdots & \ddots & \vdots \\ \frac{2}{N} & \frac{2}{N} & \cdots & -1+\frac{2}{N} \end{bmatrix}$$

5. Assume that $N = 2^n$. Let $R' = \left[r'_{ij}\right]$ and $R = [r_{ij}]$ be $N \times N$ matrices, where $0 \leq i, j \leq (N-1)$. All the elements of the matrix R' are equal to zero, except $r'_{00} = 2$. The matrix R is a diagonal matrix, where $r_{00} = 1$, and $r_{ii} = -1$, for $1 \leq i \leq (N-1)$. Thus

$$R = (R' - I)$$

Let W_N be an $N \times N$ Hadamard matrix. Recall that W_N is a unitary matrix. It can be checked that

$$D = W_N R W_N$$

Therefore the diffusion operator can be implemented in $O(\log_2 N) = O(n)$ quantum gates.

6. Let a_{av} be the average of the elements in the set

$$\{a_i \mid a_i \in \mathbb{R}, 0 \le i < (N-1)\}$$

That is

$$a_{av} = \frac{1}{N}\sum_{i=0}^{N-1} a_i$$

Also define $A = \begin{bmatrix} a_0 \; a_1 \; \cdots \; a_{N-1} \end{bmatrix}^T$.

(a) The matrix D can be used as a mapping

$$D : \sum_{i=0}^{N-1} a_i \left|i\right\rangle \rightarrow \sum_{i=0}^{N-1} (2a_{av} - a_i) \left|i\right\rangle$$

(b) The matrix D is the *inversion about average* operator.

Let $a_i = (a_{av} + a)$, then the inversion of a_i about average a_{av} is equal to $a_i' = (a_{av} - a)$. That is, $a_i' = (2a_{av} - a_i)$. Also

$$DA = (2P - I)\, A = 2PA - A$$

where each element of the vector PA is equal to a_{av}, which is the average of all elements of the vector A. Therefore the ith element of DA is equal to $a_i' = (2a_{av} - a_i)$, where $0 \le i < (N-1)$. □

Before a description of Grover's search algorithm is given, some preliminaries are outlined below.

Preliminaries

The aim of Grover's algorithm is to search through a space of N elements. We focus attention on the *index* of each of these elements. The indices to these N elements are numbered from $0, 1, 2, \ldots, (N-1)$. Assume that $N = 2^n$, that is n number of bits are required to specify an index element uniquely. It is quite possible that there are N_{numb} number of solutions to the search problem, where $1 \le N_{numb} \le N$. For simplicity assume that $N_{numb} = 1$. Define a function $f : \mathfrak{B}^{(n)} \rightarrow \mathfrak{B}$, where $\mathfrak{B} = \{0, 1\}$.

$$f(x) = \begin{cases} 1, & x \text{ is a solution to the search} \\ 0, & x \text{ is not a solution to the search} \end{cases}$$

Next define an *oracle*, which is a black box. It is also called a query function Q_f.

$$Q_f : \left|x\right\rangle \left|q\right\rangle \rightarrow \left|x\right\rangle \left|q \oplus f(x)\right\rangle$$

where $\oplus$ is the symbol for exclusive-or operation. Also $\left|x\right\rangle$ is the index (or source) register, which is n qubits long, and $\left|q\right\rangle$ is the oracle (or target) register, which is a single qubit. It is also called the *oracle qubit*. The oracle Q_f is a unitary operator. Let $q = 0$.

(a) If $f(x) = 0$, then $Q_f \left|x\right\rangle \left|0\right\rangle \rightarrow \left|x\right\rangle \left|0\right\rangle$.
(b) If $f(x) = 1$, then $Q_f \left|x\right\rangle \left|0\right\rangle \rightarrow \left|x\right\rangle \left|1\right\rangle$.

That is, the value of the target register is flipped if $f(x) = 1$, and it remains unchanged if $f(x) = 0$. Similar observation can be made if $q = 1$. Next consider the case when $|q\rangle = \{|0\rangle - |1\rangle\} / \sqrt{2} \triangleq |\chi\rangle$.

(a) If $f(x) = 0$, then $Q_f |x\rangle |\chi\rangle \to |x\rangle |\chi\rangle$.
(b) If $f(x) = 1$, then $Q_f |x\rangle |\chi\rangle \to -|x\rangle |\chi\rangle$.

Note that the state of the oracle qubit is not changed in both cases. Thus

$$Q_f |x\rangle |q\rangle = (-1)^{f(x)} |x\rangle |q\rangle$$

Therefore, if $|q\rangle = |\chi\rangle$ the function Q_f is called the *sign-changing* operator. Grover's search algorithm is next outlined, which is followed by its analysis.

Grover's Search Algorithm

Grover's search algorithm is simultaneously simple and subtle. The computational complexity of this algorithm is $O\left(N^{1/2}\right)$ and its success probability is $O(1)$. These complexity measures are determined subsequently. The quantum search algorithm is described below. The goal of the search algorithm is to determine the index x_0. The essence of this algorithm is in Steps 3 and 4. The purpose of these two steps is to increase the probability of index x_0 and decrease the probability of all other indices x. In Step 3, the sign at x_0 is inverted via the V_f transformation, and all terms are inverted about the average in Step 4. If these two steps are iterated, the absolute value of the amplitude at x_0 keeps on increasing, and that of all other indices x keeps on decreasing. Therefore, this iterative step is also called "amplitude amplification process." After J iterations, the absolute value of the amplitude at x_0 is equal to unity with high probability. The value of number of iterates $J = \lceil \pi 2^{n/2-2} \rceil$, is also justified.

Algorithm 12.2. *Grover's Quantum Search Algorithm.*

Input: Let $N = 2^n$. Assume that the index $x_0 \in \mathfrak{B}^{(n)}$ occurs at most once in the index set of an unsorted list.
A function $f(\cdot)$, where $f(x_0) = 1$, and $f(x) = 0, \forall\, x \in \mathfrak{B}^{(n)} \backslash \{x_0\}$.
The oracle black box, which performs the transformation Q_f.
An index register which is n qubits long and a target register which is a single qubit. Let $|\chi\rangle = \{|0\rangle - |1\rangle\} / \sqrt{2}$. Set J equal to $\lceil \pi 2^{n/2-2} \rceil$.
Output: Determine x_0.
begin

Step 1: The index and target state vectors are initially $|00\ldots0\rangle |\chi\rangle$.
Step 2: Apply the Hadamard transform to the first n qubits, which is $|00\ldots0\rangle$. This yields

$$|\phi\rangle = \frac{1}{\sqrt{N}} \sum_{x=0}^{N-1} |x\rangle$$

Step 3: Application of the query operator Q_f to $|\phi\rangle |\chi\rangle$ results in

$$\frac{1}{\sqrt{N}} \sum_{x=0}^{N-1} (-1)^{f(x)} |x\rangle |\chi\rangle$$

Denote the sign-changing operation V_f on the first register by

$$V_f : \frac{1}{\sqrt{N}} \sum_{x=0}^{N-1} |x\rangle \to \frac{1}{\sqrt{N}} \sum_{x=0}^{N-1} (-1)^{f(x)} |x\rangle$$

Step 4: Apply the diffusion operator $D = W_N R W_N$ to the state obtained in the last step.
Step 5: The Steps 3 and 4 are iterated J times.
This results in the transformation DV_f in each iteration.
This transformation is generally called the Grover's iterate.
Step 6: Measure the result in the n qubit register. It is x_0 with high probability.
end (*end of Grover's algorithm*)

Analysis of Grover's Algorithm

After the jth iteration of Step 4, $|\phi_j\rangle$ is

$$|\phi_j\rangle = k_j \sum_{\substack{x \in \mathfrak{B}^{(n)} \\ f(x)=1}} |x\rangle + l_j \sum_{\substack{x \in \mathfrak{B}^{(n)} \\ f(x)=0}} |x\rangle$$

where $k_0 = l_0 = N^{-1/2}$. Using the Grover's iterate, the following recursive equations are obtained.

$$\begin{bmatrix} k_{j+1} \\ l_{j+1} \end{bmatrix} = \begin{bmatrix} 1 - \frac{2}{N} & 2 - \frac{2}{N} \\ -\frac{2}{N} & 1 - \frac{2}{N} \end{bmatrix} \begin{bmatrix} k_j \\ l_j \end{bmatrix}, \quad \forall j \in \mathbb{N}$$

Since D is a unitary operator, the DV_f operation yields

$$k_j^2 + (N-1)\, l_j^2 = 1, \quad \forall j \in \mathbb{N}$$

Therefore (k_j, l_j) lie on an ellipse defined by the above equation. Thus the solution of this recursion is

$$k_j = \sin\{(2j+1)\,\theta\}, \quad l_j = \frac{1}{\sqrt{N-1}} \cos\{(2j+1)\,\theta\}, \quad \forall j \in \mathbb{N}$$

where $\sin^2\theta = 1/N$. The purpose of the Grover's iteration is to minimize l_j and maximize k_j. The number of iterations J in the algorithm is determined as follows. Select j such that $l_j = 0$, that is $\cos\{(2j+1)\,\theta\} = 0$. This occurs if

$$j = \frac{\pi}{2\theta}\left(m + \frac{1}{2}\right) - \frac{1}{2}, \quad m \in \mathbb{Z}$$

Select $m = 0$ in the above equation. Since j is an integer, let $j_0 = \lfloor \pi/(4\theta) \rfloor$. Also since $\theta \geq \sin\theta = N^{-1/2}$, it can be inferred that $j_0 \leq \pi N^{1/2}/4 = O\left(N^{1/2}\right)$. Thus $J = \lceil \pi 2^{n/2-2} \rceil$.

It has also been proved in literature that Grover's algorithm is *optimal* up to a constant. That is, no other quantum search algorithm can perform this type of search of an unstructured list (database) faster. This algorithm can also be extended to the case when more than one value of x exists, such that $f(x) = 1$.

12.7 Quantum Information Theory

Information theory is the study of communication systems from a probabilistic point of view. Classical information theory has been discussed in another chapter. In that chapter a theoretical framework to transmit data over a communication channel was developed. Analogously quantum information theory addresses the problem of optimal communication over a quantum channel. The information to be transmitted over a quantum communication channel can be either in bit or qubit form. Quantum information theory also studies the interaction between classical and quantum information. Genesis of quantum information can be traced to the mathematician John von Neumann (1903-1957). The concept of entropy has been used to develop classical information theory. This entropy is sometimes called the Shannon entropy. Similarly quantum information is quantified in terms of the *von Neumann entropy.*

Definition 12.23. *von Neumann entropy: Let X be a quantum source with density matrix ρ. The corresponding spectral representation of the density matrix is*

$$\rho = \sum_{i=1}^{m} \lambda_i \left|x_i\right\rangle \left\langle x_i\right| \tag{12.20a}$$

where $\lambda_i \geq 0$ and $\left|x_i\right\rangle$ are the eigenvalue and an eigenvector respectively for $1 \leq i \leq m$. Also $\sum_{i=1}^{m} \lambda_i = 1$. Let

$$\rho \log_2 \rho \triangleq \sum_{i=1}^{m} \{\lambda_i \log_2 \lambda_i\} \left|x_i\right\rangle \left\langle x_i\right| \tag{12.20b}$$

The von Neumann entropy of X is

$$S(X) = -tr(\rho \log_2 \rho) = -\sum_{i=1}^{m} \lambda_i \log_2 \lambda_i \tag{12.20c}$$

where $0 \log_2 0 \triangleq 0$. Therefore, the von Neumann entropy of ρ is precisely the Shannon entropy of the distribution $\lambda_1, \lambda_2, \ldots, \lambda_n$. Sometimes $S(X)$ is also denoted by $S(\rho)$. The unit of quantum entropy is a qubit. □

Quantum entropy of the source X, encapsulates the degree of ignorance embodied in it. It is implicit in the above definition, that the logarithm of the matrix ρ is the usual Taylor's series expansion. That is, formally

$$\ln \rho \triangleq (\rho - I) - \frac{1}{2}(\rho - I)^2 + \frac{1}{3}(\rho - I)^3 - \cdots$$

where I is an identity matrix of appropriate size. The convergence properties of this series are not of immediate concern.

Example 12.6. Consider a quantum source X, which is a mixture of two pure states. The two states are:

$$|\psi_1\rangle = \begin{bmatrix} 0 & 1 \end{bmatrix}^T, \quad p_1 = \frac{3}{4}$$
$$|\psi_2\rangle = \begin{bmatrix} 1/\sqrt{2} & -1/\sqrt{2} \end{bmatrix}^T, \quad p_2 = \frac{1}{4}$$

where p_i is the probability that the source is in state $|\psi_i\rangle$ for $i = 1, 2$. The corresponding density matrix ρ is

$$\rho = p_1 |\psi_1\rangle \langle\psi_1| + p_2 |\psi_2\rangle \langle\psi_2| = \begin{bmatrix} 1/8 & -1/8 \\ -1/8 & 7/8 \end{bmatrix}$$

Note that the matrix ρ is symmetric and $tr\,(\rho) = 1$. Its eigenvalues are

$$\frac{1 \pm \sqrt{5/8}}{2} = 0.89528, \text{ and } 0.10472$$

Observe that the eigenvalues $\lambda_1 = 0.89528$ and $\lambda_2 = 0.10472$ sum to unity. The quantum entropy of the source X is $S\,(\rho) = -\sum_{i=1}^{2} \lambda_i \log_2 \lambda_i = 0.48377$ qubits. □

Analogous to classical entropy; joint, conditional, mutual, and relative quantum entropies are defined for composite quantum mechanical systems.

Definitions 12.24. *Let X and Y be components of a composite quantum mechanical systems, with joint density matrix ρ_{XY}.*

1. *Their joint quantum entropy is*:

$$S\,(X, Y) = -tr\,(\rho_{XY} \log_2 \rho_{XY}) \tag{12.21}$$

2. *Their conditional quantum entropy is*:

$$S\,(X \mid Y) = S\,(X, Y) - S\,(Y) \tag{12.22}$$

3. *Their mutual quantum entropy is*:

$$I_Q\,(X; Y) = S\,(X) + S\,(Y) - S\,(X, Y) \tag{12.23}$$

4. *Quantum relative entropy*: *Let ρ and σ be density operators. The relative entropy of ρ to σ is*

$$S\,(\rho \parallel \sigma) = tr\,(\rho \log_2 \rho) - tr\,(\rho \log_2 \sigma) \tag{12.24}$$

□

Observations 12.7. Some properties of quantum entropy are listed below. Proofs of some of these facts can be found in the problem section.

1. Quantum entropy is a *nonnegative* number. This follows from its definition.

2. The entropy of a pure state is equal to zero. Conversely, if the entropy of a state is equal to zero, then it is a pure state.
3. Let the dimension of a Hilbert space be d. The entropy $S(\rho)$ of a state ρ in this Hilbert space is bounded by $\log_2 d$. That is,
$$S(\rho) \leq \log_2 d$$
Equality is obtained if $\rho d = I$ where I is an identity matrix of size d.
4. Let $\lambda_{\min}$ and $\lambda_{\max}$ be the smallest and largest eigenvalues respectively of the density matrix ρ. Then
$$-\log_2 \lambda_{\max} \leq S(\rho) \leq -\log_2 \lambda_{\min}$$
5. The entropy of ρ is *invariant* under a unitary transformation. That is,
$$S\left(U\rho U^{-1}\right) = S(\rho)$$
where U is a unitary matrix.
6. Quantum entropy is a *concave* function. Let $p_i, 1 \leq i \leq n$ be probabilities, where
$$\sum_{i=1}^{n} p_i = 1$$
Also let the density operator corresponding to the probability p_i be ρ_i for $1 \leq i \leq n$. If
$$\rho = \sum_{i=1}^{n} p_i \rho_i$$
then
$$S(\rho) \geq \sum_{i=1}^{n} p_i S(\rho_i)$$
Equality is obtained if all ρ_i's are equal. The proof is similar to a corresponding statement about classical entropy.
7. Let X and Y be components of a composite quantum mechanical systems. Then
   (a) $S(X,Y) = S(Y,X)$.
   (b) $I_Q(X;Y) = I_Q(Y;X)$.
   (c) $I_Q(X;Y) = S(X) - S(X \mid Y) = S(Y) - S(Y \mid X)$.
8. Analogous to Shannon's relative entropy, quantum relative entropy can sometimes be infinite.
9. The quantum relative entropy is always nonnegative. That is, if ρ and σ are density operators, then $S(\rho \parallel \sigma) \geq 0$. Equality is obtained if and only if $\rho = \sigma$. This relationship is called *Klein's inequality*.
10. *Entropy of tensor product.* Let $\mathcal{H}'$ and $\mathcal{H}''$ be two Hilbert spaces, and ρ_1 and ρ_2 be density matrices in these two Hilbert spaces respectively. The entropy of the density matrix $\rho_1 \otimes \rho_2$ in $\mathcal{H}' \otimes \mathcal{H}''$ is
$$S(\rho_1 \otimes \rho_2) = S(\rho_1) + S(\rho_2)$$
In general
$$S(\rho_1 \otimes \rho_2 \otimes \ldots \otimes \rho_n) = S(\rho_1) + S(\rho_2) + \cdots + S(\rho_n)$$
where the density matrices ρ_i's are defined in appropriate Hilbert spaces.

11. The density matrices ρ_i for $1 \leq i \leq n$ are defined on orthogonal Hilbert subspaces. Associated with a density matrix ρ_i is a probability p_i, where $\sum_{i=1}^{n} p_i = 1$. Then

$$S\left(\sum_{i=1}^{n} p_i \rho_i\right) = H(p) + \sum_{i=1}^{n} p_i S(\rho_i)$$

where $H(p) = -\sum_{i=1}^{n} p_i \log_2 p_i$.

12. Let X and Y be two distinct quantum mechanical systems. Let the density matrices of X and Y be ρ_X and ρ_Y respectively. Also represent their joint density operator by ρ_{XY}.
    (a) *Subadditivity*:
    $$S(X, Y) \leq S(X) + S(Y)$$
    Equality occurs if and only if $\rho_{XY} = \rho_X \otimes \rho_Y$.
    (b) *Triangle inequality*:
    $$S(X, Y) \geq |S(X) - S(Y)|$$
    This relationship is also called the *Araki-Lieb inequality*. The corresponding inequality for classical entropy is
    $$H(X, Y) \geq H(X)$$
    (c) Use of the triangle inequality yields
    $$I_Q(X; Y) \leq 2 \min(S(X), S(Y))$$
    A similar inequality for classical mutual entropy is
    $$I(X; Y) \leq \min(H(X), H(Y))$$

□

12.8 Quantum Communication Theory

Analogous to classical communication theory, quantum communication theory studies the limits of efficient data storage and transmission of quantum information. It should come as no surprise that quantum states can be considered as information. Consequently there is justification in examining quantum states in light of our knowledge of quantum information theory. In this section, the quantum analog of Shannon's noiseless communication channel coding theorem is developed. A pioneer in this area of work is B. Schumacher. A basic premise of this section is that compression is necessary because of limited physical resources.

12.8.1 Data Compression

Consider a quantum state $|\psi\rangle$ in Hilbert space $\mathcal{H}_m$, of dimension m. This quantum state (signal) represents data from a quantum source $\mathcal{S}_q$. Compression of the quantum state $|\psi\rangle$ results in a state $|\phi\rangle \in \mathcal{H}_{m_q}$, where $m_q < m$. Thus data compression implies that the compressed quantum state vector can be represented in a Hilbert space of dimension m_q, which is less than m.

Assume that the state $|\psi\rangle$ is represented by μ qubits, that is $m = 2^{\mu}$. The *rate of compression* R_q is defined as the ratio of $\log_2 m_q$ and μ. Thus $m_q = \lceil 2^{\mu R_q} \rceil$ and $0 < R_q \leq 1$.

Definition 12.25. *If a quantum state, and its compressed version are representable in Hilbert spaces $\mathcal{H}_m$ and $\mathcal{H}_{m_q}$ respectively. The rate of data compression R_q is*

$$R_q = \frac{\log_2 m_q}{\log_2 m} \tag{12.25}$$

□

As in the case of classical communication, some quantum states occur with higher probability than others. These high probability states span a subspace of their parent Hilbert space $\mathcal{H}_m$. Such space is called a *typical subspace.* Once again, this concept is analogous to the idea of *typical sequence* in classical information theory. Assume that the quantum mechanical processes under study are mixed systems. As mentioned earlier, such systems are studied in the language of density operators.

12.8.2 Typical Subspaces

Suppose that the quantum state of a quantum mechanical system is specified in a Hilbert space $\mathcal{H}_m$ of dimension m. Therefore this quantum state can be represented in $\log_2 m$ qubits. Also suppose that the corresponding density operator is ρ. Its representation is

$$\rho = \sum_{k=1}^{m} p_k \left| x_k \right\rangle \left\langle x_k \right|$$

where $p_k, 1 \leq k \leq m$ are the eigenvalues of the matrix ρ. An eigenvector corresponding to the eigenvalue p_k is $|x_k\rangle$, for $1 \leq k \leq m$. In addition, the eigenvectors $|x_k\rangle, 1 \leq k \leq m$ form an orthonormal set. Furthermore, $p_k \geq 0$ for $1 \leq k \leq m$, $tr(\rho) = \sum_{k=1}^{m} p_k = 1$, and the quantum entropy of the source is

$$S(\rho) = -\sum_{k=1}^{m} p_k \log_2 p_k$$

Analogous to classical information theory it is reasonable to define an ϵ-typical sequence. Consider a sequence of mutually independent, and identically distributed random variables

$$(X_1, X_2, \ldots, X_n)$$

An instance of this sequence of random variables is $(x_{k_1}, x_{k_2}, \ldots, x_{k_n})$. Denote such sequences of length n by $\widetilde{x}$. Also denote the probability mass function of this sequence by $p(\widetilde{x})$, where for typical sequences $p(\widetilde{x}) \simeq 2^{-nS(\rho)}$. These notions are formalized in the following definition.

Definition 12.26. *Let a quantum mechanical system be represented in a Hilbert space $\mathcal{H}_m$ by a density operator ρ. Its eigenvalues and orthonormal eigenvectors are $p_k \geq 0$ and $|x_k\rangle$ respectively for $1 \leq k \leq m$. The von Neumann entropy associated with this state is equal to $S(\rho) = -\sum_{k=1}^{m} p_k \log_2 p_k$.*

Also let $\widetilde{X} = (X_1, X_2, \ldots, X_n)$ be a sequence of independent, and identically distributed random variables. An instance of this sequence is $\widetilde{x} = (x_{k_1}, x_{k_2}, \ldots, x_{k_n})$. Let $\epsilon > 0$.

(a) *The sequence* $\widetilde{x} = (x_{k_1}, x_{k_2}, \ldots, x_{k_n})$ *is* ϵ*-typical if*

$$2^{-n(S(\rho)+\epsilon)} \leq p(\widetilde{x}) \leq 2^{-n(S(\rho)-\epsilon)} \tag{12.26}$$

A sequence which is not ϵ*-typical is an atypical sequence.*

(b) *A state* $|x_{k_1}\rangle |x_{k_2}\rangle \ldots |x_{k_n}\rangle$ *is said to be* ϵ*-typical state, if the sequence* $(x_{k_1}, x_{k_2}, \ldots, x_{k_n})$ *is* ϵ*-typical. A state which is not* ϵ*-typical is an atypical state.*

(c) *A space spanned by all* ϵ*-typical states is called* ϵ*-typical subspace. This space is denoted by* $T_q(n, \epsilon)$.

(d) *The projector onto* ϵ*-typical subspace is* $\mathcal{P}(n, \epsilon)$. *It is*

$$\mathcal{P}(n, \epsilon) = \sum_{\widetilde{x}\ \epsilon\text{-typical}} |x_{k_1}\rangle \langle x_{k_1}| \otimes |x_{k_2}\rangle \langle x_{k_2}| \otimes \ldots \otimes |x_{k_n}\rangle \langle x_{k_n}| \tag{12.27}$$

(e) *The density operator corresponding to the states* $|x_{k_1}\rangle |x_{k_2}\rangle \ldots |x_{k_n}\rangle$ *is denoted by* $\rho^{\otimes n}$. □

Note that the sequence $\widetilde{x}$ can be represented by $n \log_2 m$ qubits. However, after data compression it can be represented by $nS(\rho)$ qubits.

Theorem 12.2. (*Typical subspace theorem*)

(a) *Let* $\epsilon > 0$, *be fixed. For any* $\delta > 0$ *and sufficiently large* n, *tr* $(\mathcal{P}(n, \epsilon) \rho^{\otimes n}) \geq (1 - \delta)$.

(b) *Let* $\epsilon > 0$, *be fixed. For any* $\delta > 0$ *and sufficiently large* n, *the dimension of space* $T_q(n, \epsilon)$ *is equal to tr* $(\mathcal{P}(n, \epsilon))$. *Furthermore,*

$$(1 - \delta) 2^{n(S(\rho)-\epsilon)} \leq \dim(T_q(n, \epsilon)) \leq 2^{n(S(\rho)+\epsilon)} \tag{12.28}$$

(c) *Let the rate of compression* $R_q < S(\rho)$ *be fixed, and* $\mathcal{G}(n)$ *be a projector onto any subspace of* $\mathcal{H}_m^{\otimes n}$ *of dimension at most* 2^{nR_q}. *Then for any* $\delta > 0$ *and sufficiently large* n, *tr* $(\mathcal{G}(n) \rho^{\otimes n}) \leq \delta$.

Proof. See the problem section. □

12.8.3 Fidelity

Consider a quantum mechanical system which is specified in a Hilbert space $\mathcal{H}_m$. Also assume that the density matrix of this system is ρ. The corresponding von Neumann entropy of this state is $S(\rho)$. A state in this system is specified in $\log_2 m$ qubits. Thus a state sequence of length n in this quantum mechanical system is specified in $n \log_2 m$ qubits. After this sequence is compressed, the compressed sequence requires $nS(\rho)$ qubits. Assuming redundancy in data, the compressed signal (state) can be represented in a Hilbert space of dimension smaller than that of $\mathcal{H}_m$. Thus there is a possibility that some of the information which was present in the original signal might be lost in the compressed signal. Upon application of the decompression operation, the resulting sequence might require $n \log_2 m$ qubits for its representation, however the resulting density σ may or may not be equal to ρ. It is a fair question to ask for the "distance" between the original density ρ and the density σ obtained after the decompression operation. A measure to characterize the distance between the two densities is called *fidelity*. The fidelity of a compression and decompression process is the probability that the decompressed sequence of signals is the same as the original sequence of signals. An operational definition of fidelity is next given.

Definition 12.27. *The fidelity between two density matrices ρ and σ is*

$$\mathcal{F}(\rho,\sigma) = tr\left(\sqrt{\rho^{1/2}\sigma\rho^{1/2}}\right) \tag{12.29}$$

□

The reader should be forewarned that the definition of fidelity might vary among different authors. Sometimes fidelity is defined as the square of the term on the right-hand side of the above equation.

Observations 12.8. Important properties of the fidelity operator are listed in this set of facts.

1. Fidelity is not a metric on density operators.
2. Fidelity is symmetric in its input.
3. $\mathcal{F}(\rho,\sigma) = 1$ if and only if $\rho = \sigma$. Also $\mathcal{F}(\rho,\sigma) = 0$ if and only if ρ and σ have support on orthogonal subspaces. Thus $0 \le \mathcal{F}(\rho,\sigma) \le 1$.
4. Let ρ and σ be density operators, such that the eigenvectors of the two matrices are identical, but the eigenvalues are not necessarily equal. In this case, the two density operators, ρ and σ commute. Let the normalized eigenvectors be $|\psi_i\rangle, 1 \le i \le n$, and the eigenvalues of the two matrices be p_i and q_i respectively for all $1 \le i \le n$. If

$$\rho = \sum_{i=1}^{n} p_i |\psi_i\rangle\langle\psi_i|, \text{ and } \sigma = \sum_{i=1}^{n} q_i |\psi_i\rangle\langle\psi_i|$$

then

$$\mathcal{F}(\rho,\sigma) = tr\sqrt{\sum_{i=1}^{n} p_i q_i |\psi_i\rangle\langle\psi_i|} = tr\left\{\sum_{i=1}^{n}\sqrt{p_i q_i}|\psi_i\rangle\langle\psi_i|\right\} = \sum_{i=1}^{n}\sqrt{p_i q_i}$$

5. Consider the fidelity between a pure state $|\psi\rangle$ and an arbitrary state ρ.

$$\begin{aligned}\mathcal{F}(|\psi\rangle,\rho) &= tr\sqrt{\langle\psi\mid\rho\mid\psi\rangle|\psi\rangle\langle\psi|} \\ &= \sqrt{\langle\psi\mid\rho\mid\psi\rangle}\end{aligned}$$

6. If the density matrices ρ and σ, both represent pure states $|\varphi\rangle$ and $|\psi\rangle$ respectively, then $\mathcal{F}(\rho,\sigma) = |\langle\varphi\mid\psi\rangle|$. The fidelity is equal to unity if $|\varphi\rangle$ and $|\psi\rangle$ coincide. It is equal to zero, if $|\varphi\rangle$ and $|\psi\rangle$ are orthogonal.
7. Let U be any unitary operator, then

$$\mathcal{F}\left(U\rho U^{\dagger}, U\sigma U^{\dagger}\right) = \mathcal{F}(\rho,\sigma)$$

□

In order to quantify the fidelity of a compression-decompression scheme, the simple definition of fidelity has to be extended to an *ensemble-average fidelity*. Let the quantum mechanical system with mixed states be specified by

$$\{(p_i,\rho_i) \mid 1 \le i \le m\}$$

This system is in a state with density ρ_i with probability p_i, where $1 \leq i \leq m$. Also let $\Phi(\cdot)$ be a Kraus operator which preserves the trace. It is assumed that the operator $\Phi(\cdot)$ represents compression. Then the average fidelity is defined by

$$\overline{\mathcal{F}} = \sum_{i=1}^{m} p_i \left\{ \mathcal{F}\left(\rho_i, \Phi\left(\rho_i\right)\right) \right\}^2$$

It follows from this definition that $0 \leq \overline{\mathcal{F}} \leq 1$. Also observe that $\overline{\mathcal{F}} = 1$ if and only if $\Phi(\rho_i) = \rho_i$ for all i for which $p_i > 0$. This definition of fidelity is used to evaluate the efficiency of the compression-decompression scheme. Thus if $\overline{\mathcal{F}} = 1$, the compression-decompression scheme is completely reliable. On the other extreme, if $\overline{\mathcal{F}} = 0$, the compression-decompression scheme is completely unreliable. In summary, $\overline{\mathcal{F}}$ gives the probability that the decompressed quantum message (signal) is the same as the original quantum message.

12.8.4 Schumacher's Noiseless Channel Coding Theorem

Schumacher's noiseless channel coding theorem is the quantum analog of Shannon's noiseless channel coding theorem. This theorem is proved by using an operator-sum representation.

Theorem 12.3. *Let S_q be a quantum source which emits independent, and identically distributed states specified in a Hilbert space $\mathcal{H}$. Also let the corresponding density operator, entropy, and the rate of compression of data be ρ, $S(\rho)$, and R_q respectively.*

(a) *If $R_q > S(\rho)$ then there exists a reliable compression-decompression scheme of rate R_q for the source S_q.*
(b) *Conversely, if $R_q < S(\rho)$ then any type of compression-decompression scheme is not reliable.*

Proof. See the problem section. □

12.9 Quantum Error-Correction

The basic premise of this chapter is that quantum computation and communication are not simply mathematical artifacts. It is hoped that an understanding of quantum mechanical processes will facilitate information processing. However, since our knowledge of *nature* is incomplete, there is a finite probability that there is a mismatch between what a computer does and our expectation of what it ought to do. Occurrence of this event is called an error, and the phenomenon is called *quantum noise* (or decoherence).

Errors can occur in both quantum computation and quantum communication. Errors in quantum computation can possibly arise due to interaction (coupling) between the computer and the outside world. Errors can also occur in the preparation and measurement of data. Heisenberg's uncertainty principle forbids us from making important measurements precisely. It may also be not possible to determine the input precisely via measurements. The same can be said about the output. Errors can also occur in the evolution of the output. Therefore, it appears that the possibility of quantum computation is at stake. However, by a judicious use of quantum error-correcting codes, quantum computation is possible. Similar reasoning can be extended to quantum communication.

The ideas developed in classical error-correcting codes are not directly applicable to creating quantum codes. However, these codes can be extended to apply to quantum computation and communication.

12.9.1 Difficulties Encountered

Some of the possible causes of error in quantum computation are:

(a) Continuous errors: Unlike in digital communication, errors are continuous in quantum mechanics. Consequently identifying errors and their correction might require infinite precision. This is in contrast to classical digital computers, where the correction is made to the closer of 0 or 1. Consider a quantum state $\alpha\left|0\right\rangle + \beta\left|1\right\rangle$, where $\sqrt{\alpha^2+\beta^2} = 1$. Assume that it is supposed to become $\alpha\left|0\right\rangle + \beta e^{i\phi}\left|1\right\rangle$, however it erroneously becomes $\alpha\left|0\right\rangle + \beta e^{i(\phi+\delta)}\left|1\right\rangle$, where δ is possibly small. Over the course of computation, the cumulative effect of such errors can be significant.
(b) The Heisenberg-effect: In order to correct errors, output is measured or sensed as in classical error-correction techniques. In quantum computation, sensing a state can possibly destroy the quantum state and make error-correction difficult.
(c) Cloning is not possible: Consider the classical repetition code, where $0 \to 000$ and $1 \to 111$. This code corrects a state such as 101 to 111 by using the majority value. Can this idea be extended to quantum repetition code: $\left|\varphi\right\rangle \to \left|\varphi\right\rangle \otimes \left|\varphi\right\rangle \otimes \left|\varphi\right\rangle$? This transformation is not permissible via the no-cloning theorem.
(d) In order for the quantum error-correcting codes to be used successfully, it is necessary that the quantum error-correction processing networks themselves be sufficiently robust and reliable for a long period of time.

12.9.2 Types of Errors

It is assumed in this section that errors in different qubits are statistically independent. Furthermore, errors in the same qubit at different times are also assumed to be independent of each other. Different types of error events are easier to study in terms of the Pauli matrices. The transformation of the quantum state $\left|\varphi\right\rangle = \alpha\left|0\right\rangle + \beta\left|1\right\rangle$ by different types of Pauli matrices is initially listed.

(a) The σ_0 transformation of $\left|\varphi\right\rangle$ is $\sigma_0\left|\varphi\right\rangle = \alpha\left|0\right\rangle + \beta\left|1\right\rangle = \left|\varphi\right\rangle$.
(b) The σ_x transformation of $\left|\varphi\right\rangle$ is $\sigma_x\left|\varphi\right\rangle = \alpha\left|1\right\rangle + \beta\left|0\right\rangle$. This transformation models a *bit-flip error* or *amplitude error*.
(c) The σ_y transformation of $\left|\varphi\right\rangle$ is $\sigma_y\left|\varphi\right\rangle = i\alpha\left|1\right\rangle - i\beta\left|0\right\rangle$.
(d) The σ_z transformation of $\left|\varphi\right\rangle$ is $\sigma_z\left|\varphi\right\rangle = \alpha\left|0\right\rangle - \beta\left|1\right\rangle$. This transformation models *sign-flip error* or *phase-shift error.*

The transformation of $\left|\varphi\right\rangle$ by $\sigma_x\sigma_z$ results in $\alpha\left|1\right\rangle - \beta\left|0\right\rangle$. This transformation and σ_y model a *bit-sign flip error* or *bit-phase error*. It is a combination of bit error and the phase error. The three types of error are represented by the convenient notation

$$\mathcal{X} = \sigma_x,\ \mathcal{Z} = \sigma_z,\ \text{and}\ \mathcal{Y} = \mathcal{X}\mathcal{Z} = -i\sigma_y$$

Thus

$$\mathcal{I} = \begin{bmatrix} 1 & 0 \\ 0 & 1 \end{bmatrix},\quad \mathcal{X} = \begin{bmatrix} 0 & 1 \\ 1 & 0 \end{bmatrix},\quad \mathcal{Y} = \begin{bmatrix} 0 & -1 \\ 1 & 0 \end{bmatrix},\quad \mathcal{Z} = \begin{bmatrix} 1 & 0 \\ 0 & -1 \end{bmatrix}$$

Note that

$$\mathcal{X} = W_2 \mathcal{Z} W_2$$
$$\mathcal{Z} = W_2 \mathcal{X} W_2$$

where W_2 is a 2×2 Hadamard matrix. If $|a\rangle$ is a single qubit (either $|0\rangle$ or $|1\rangle$), then:

(a) $\mathcal{I}\,|a\rangle = |a\rangle$.
(b) $\mathcal{X}\,|a\rangle = |a \oplus 1\rangle$.
(c) $\mathcal{Z}\,|a\rangle = (-1)^a\,|a\rangle$.
(d) $\mathcal{Y}\,|a\rangle = (-1)^a\,|a \oplus 1\rangle$.

The effect of the environment on the qubit $|\varphi\rangle = (\alpha\,|0\rangle + \beta\,|1\rangle)$ is indicated by $|e\rangle$. This interaction between a single qubit and its environment is thus

$$|e\rangle\,\{\alpha\,|0\rangle + \beta\,|1\rangle\} \to \alpha\,\{|e_{00}\rangle\,|0\rangle + |e_{01}\rangle\,|1\rangle\} + \beta\,\{|e_{10}\rangle\,|0\rangle + |e_{11}\rangle\,|1\rangle\}$$

where $|e\rangle$ and $|e_{ij}\rangle, 0 \le i, j \le 1$ are the states of the environment. The above transformation can be rewritten comprehensively in terms of the transformations $\mathcal{I}, \mathcal{X}, \mathcal{Y}$, and $\mathcal{Z}$. Thus

$$|e\rangle\,|\varphi\rangle \to \{|e_{\mathcal{I}}\rangle\,\mathcal{I} + |e_{\mathcal{X}}\rangle\,\mathcal{X} + |e_{\mathcal{Y}}\rangle\,\mathcal{Y} + |e_{\mathcal{Z}}\rangle\,\mathcal{Z}\}\,|\varphi\rangle$$

where $|e_{\mathcal{I}}\rangle, |e_{\mathcal{X}}\rangle, |e_{\mathcal{Y}}\rangle$, and $|e_{\mathcal{Z}}\rangle$ depend upon $|e_{ij}\rangle, 0 \le i, j \le 1$. This formulation indicates that any error can be stated in terms of these four basic error operators. If the register is n qubits, the error in this register is represented by the matrix

$$M = \bigotimes_{i=1}^{n} M_i$$

where M_i is the error transformation matrix for the ith qubit, and $M_i \in \{\mathcal{I}, \mathcal{X}, \mathcal{Y}, \mathcal{Z}\}$. Consider an example in which the error operator for a system of 6 qubits is

$$M = \mathcal{I}_1 \mathcal{X}_2 \mathcal{Z}_3 \mathcal{I}_4 \mathcal{Y}_5 \mathcal{X}_6$$

where the subscript indicates the qubit on which an error operator from the set $\{\mathcal{I}, \mathcal{X}, \mathcal{Y}, \mathcal{Z}\}$ acts upon. An alternate convenient notation is

$$M = \mathfrak{X}_x \mathfrak{Z}_z$$

where x and z are binary vectors. These indicate the presence of the $\mathcal{X}$ and $\mathcal{Z}$ operators in M. Noting that $\mathcal{Y} = \mathcal{X}\mathcal{Z}$, we have in this example

$$M = \mathcal{I}_1 \mathcal{X}_2 \mathcal{Z}_3 \mathcal{I}_4 \mathcal{Y}_5 \mathcal{X}_6 = \mathfrak{X}_{010011} \mathfrak{Z}_{001010}$$

12.9.3 Basics of Quantum Error-Correction

Our goal is to encode k qubits of information. This encoding is required to overcome the problem of noise, either in the computation or communication of a quantum state $|\varphi\rangle$ of length k qubits. This quantum state is encoded into $n > k$ qubits. This is done by appending $(n - k)$ extra qubits.

Initially the $(n-k)$ extra qubits are in the state $\left|0^{\otimes(n-k)}\right\rangle$. This encoding is performed via a unitary operator $\mathcal{E}_{encd}$. The n qubit coded state is $|\varphi_E\rangle$.

$$\mathcal{E}_{encd} : |\varphi\rangle \left|0^{\otimes(n-k)}\right\rangle \to |\varphi_E\rangle$$

The encodings of the basis states of k qubits actually form an orthonormal basis of a 2^k-dimensional subspace of a Hilbert space of dimension 2^n. Due to the presence of noise, the coded word is transformed by a linear transformation $\mathcal{E}_{noise}$, which is not necessarily unitary. This transformation is also called a superoperator. The n qubits are transformed to $|\mathcal{E}_{noise}\varphi_E\rangle$.

$$\mathcal{E}_{noise} : |\varphi_E\rangle \to |\mathcal{E}_{noise}\varphi_E\rangle$$

Note that the process $\mathcal{E}_{noise}$ is not irreversible in general. The error creation process results in the erroneous state

$$\sum_{s=1}^{l} |\psi_s^{env}\rangle\, M_s\, |\varphi_E\rangle$$

where the environmental states $|\psi_s^{env}\rangle$ are not necessarily orthogonal or normalized, and M_s is a unitary error operator for $1 \le s \le l$. The operator M_s, is a tensor product of n error matrices from the set $\{\mathcal{I}, \mathcal{X}, \mathcal{Y}, \mathcal{Z}\}$.

Therefore, in order to recover $|\varphi_E\rangle$, the original system is coupled to another system called an *ancilla.* An ancilla is an auxiliary state of auxiliary qubits. This error-correction process is modeled by unitary transformation $\mathcal{E}_{corr}$. Let the initial state of the ancilla be $|\Omega\rangle$, which is ω qubits, and $|\Omega\rangle = |0^{\otimes\omega}\rangle$. The error-correction process transforms the tensor product of $|\mathcal{E}_{noise}\varphi_E\rangle$ and $|\Omega\rangle$. Thus the error-correction transformation is

$$\mathcal{E}_{corr} : |\mathcal{E}_{noise}\varphi_E\rangle\, |\Omega\rangle \to |\varphi_E\rangle\, |\Omega_{corr}\rangle$$

where $|\Omega_{corr}\rangle$ is the new state of the ancilla. Since the output of the $\mathcal{E}_{corr}$ process is disentangled, $|\Omega_{corr}\rangle$ can be measured without disturbing $|\varphi_E\rangle$. This enables us to determine a mapping to recover $|\varphi_E\rangle$ from $|\mathcal{E}_{noise}\varphi_E\rangle$. This process is similar to the computation of syndrome of errors in the linear codes. Consequently this step is sometimes termed *syndrome extraction.* The combined error-creation and error-correction process results in

$$\sum_{s=1}^{l} |\psi_s^{env}\rangle\, (M_s\, |\varphi_E\rangle\, |s\rangle)$$

In the above expression, s is simply a number indicating the error operator number. The state of the ancilla $|s\rangle$ depends upon the noise but not on the state $|\varphi_E\rangle$ which is to be corrected. Furthermore, the states $|s\rangle$ are required to be orthogonal, consequently the ancilla qubits in the basis $\{|s\rangle\}$ can be measured. This yields

$$|\psi_{s'}^{env}\rangle\, (M_{s'}\, |\varphi_E\rangle\, |s'\rangle)$$

for a single randomly chosen s'. Since s' is known, the corresponding $M_{s'}$ and correspondingly $M_{s'}^{-1}$ are known. Application of this inverse operation results in $|\psi_{s'}^{env}\rangle\, |\varphi_E\rangle\, |s'\rangle$. Observe that the entanglement between the original system and the environment is transferred to the entanglement between the environment and the ancilla. Thus the disentangled state $|\varphi_E\rangle$ has been determined. It should be noted that the error-correction process is discrete and not continuous. The complete recovery operation can be encapsulated in the relationship:

$$\mathcal{R}(M_s |\varphi_E\rangle |\alpha\rangle) = |\varphi_E\rangle |\alpha_s\rangle$$

where the *recovery operator* $\mathcal{R}$ is unitary. The states $|\alpha\rangle$ and $|\alpha_s\rangle$ represent the effect of environment, ancilla, and any involved measuring instrument. However, these states exclude the original system. The conditions for the existence of quantum error-correcting codes are next determined.

Theorem 12.4. (*Sufficient and necessary conditions for quantum error-correcting codes.*) *Let* $\widetilde{C}$ *be the set of quantum error-correcting code. The code elements of* $\widetilde{C}$ *are orthonormal set of states.*

$$\widetilde{C} = \{|u_i\rangle \mid \langle u_i \mid u_j\rangle = \delta_{ij}\} \tag{12.30a}$$

$$\delta_{ij} = \begin{cases} 1, \text{ if } i = j \\ 0, \text{ if } i \neq j \end{cases} \tag{12.30b}$$

Let $\widetilde{\mathcal{M}} = \{M_i\}$ *be the set of correctable error operators including the identity operator. The codewords in* $\widetilde{C}$ *are acted upon by error operators from the set* $\widetilde{\mathcal{M}}$. *An error is correctable if and only if*

$$\langle u| M_i^{\dagger} M_j |v\rangle = 0 \tag{12.30c}$$

$$\langle u| M_i^{\dagger} M_j |u\rangle = \langle v| M_i^{\dagger} M_j |v\rangle \tag{12.30d}$$

where $u, v \in \widetilde{C}$, $u \neq v$, *and* $M_i, M_j \in \widetilde{\mathcal{M}}$. *The subscripts* i *and* j *can be either equal or different.*

Proof. The relationship $\langle u| M_i^{\dagger} M_j |v\rangle = 0$ implies that if an error has to be correctable, it must not produce a state which coincides with another codeword, or an erroneous version of another codeword. If $\mathcal{R}$ is the recovery operator, then

$$\mathcal{R}(M_i |u\rangle |\alpha\rangle) = |u\rangle |\alpha_i\rangle, \quad \text{and} \quad \mathcal{R}(M_j |v\rangle |\alpha\rangle) = |v\rangle |\alpha_j\rangle$$

Therefore

$$\langle\alpha| \langle u| M_i^{\dagger} \mathcal{R}^{\dagger} \mathcal{R} M_j |v\rangle |\alpha\rangle = \langle \alpha_i \mid \alpha_j\rangle \langle u \mid v\rangle = 0$$

which implies

$$\langle u| M_i^{\dagger} M_j |v\rangle = 0$$

The next relationship $\langle u| M_i^{\dagger} M_j |u\rangle = \langle v| M_i^{\dagger} M_j |v\rangle$ is special to quantum error-correcting codes. This type of relationship is not possible in classical codes. Again let

$$\mathcal{R}(M_i |u\rangle |\alpha\rangle) = |u\rangle |\alpha_i\rangle$$

Then

$$\langle\alpha| \langle u| M_i^{\dagger} \mathcal{R}^{\dagger} \mathcal{R} M_j |u\rangle |\alpha\rangle = \langle \alpha_i \mid \alpha_j\rangle \langle u \mid u\rangle = \langle \alpha_i \mid \alpha_j\rangle$$

Therefore

$$\langle u| M_i^{\dagger} M_j |u\rangle = \langle \alpha_i \mid \alpha_j\rangle$$

Since the right-hand side of this relationship is independent of $|u\rangle$, we obtain the desired result, which is

$$\langle u| M_i^{\dagger} M_j |u\rangle = \langle v| M_i^{\dagger} M_j |v\rangle$$

We have thus demonstrated the necessity of the two stated relationships for the recovery of an original codeword. The sufficiency part of the theorem can be established by reversing the derivations. □

Note that if, in addition $\langle u| M_i^{\dagger} M_j |u\rangle = 0$ for all codewords $|u\rangle \in \widetilde{C}$, the code is termed *orthogonal* or *nondegenerate*, otherwise it is called *degenerate*. Orthogonal codes are generally easier to work with.

12.9.4 A Bound on Quantum Error-Correcting Codes

Consider a quantum error-correcting code which maps k qubits into n qubits, and it corrects up to t errors. A bound similar to the Hamming bound of linear codes can be derived for this quantum correcting code. The k qubits correspond to 2^k basis states. And the dimension of the Hilbert space of n qubits is 2^n. Consequently, this space has 2^n basis states. Recall that there are three types of errors: $\mathcal{X}$ or $\mathcal{Y}$ or $\mathcal{Z}$. Therefore it is possible to have i qubits in error out of n qubits in $3^i \binom{n}{i}$ different ways. If the value i ranges from 0 through t, then there is a total of

$$2^k \sum_{i=0}^{t} 3^i \binom{n}{i}$$

qubit error possibilities. This value is bounded by 2^n. Thus

$$2^k \sum_{i=0}^{t} 3^i \binom{n}{i} \leq 2^n$$

For sufficiently large values of t and n

$$\frac{k}{n} \leq 1 - \frac{t}{n} \log_2 3 - h\left(\frac{t}{n}\right)$$

where

$$h(\alpha) = -\{\alpha \log_2 \alpha + (1-\alpha) \log_2 (1-\alpha)\}, \quad 0 \leq \alpha \leq 1$$

is Shannon's entropy. See the problem section for a proof of this inequality.

12.9.5 Shor's Nine-Qubit Code

Shor's nine-qubit code is an example of a quantum error-correcting code which provides bit-flip and phase-shift errors on any qubit. This code is analogous to the classical three-bit repetition scheme. The classical three-bit repetition scheme has been discussed in the chapter on algebraic coding theory. In the classical scheme, the bit 0 is coded as 000, and the bit 1 is coded as 111. After transmission, the majority of the three bits received is assumed to be the originally transmitted symbol. This is a reasonable scheme, provided at most a single error occurs in transmission. The corresponding quantum error-correcting code is named after its inventor, Peter Shor.

Shor's code is an example of one qubit error-correcting code. There are two parts in the design of this code. The first part detects bit-flips, and the other detects phase changes. In this scheme $|0\rangle$ is coded as $|0_E\rangle$, and $|1\rangle$ is coded as $|1_E\rangle$. Thus

$$|0\rangle \rightarrow |0_E\rangle = \frac{1}{\sqrt{8}} (|000\rangle + |111\rangle) \otimes (|000\rangle + |111\rangle) \otimes (|000\rangle + |111\rangle)$$

$$|1\rangle \rightarrow |1_E\rangle = \frac{1}{\sqrt{8}} (|000\rangle - |111\rangle) \otimes (|000\rangle - |111\rangle) \otimes (|000\rangle - |111\rangle)$$

where $\sqrt{8}$ in the denominator on the right-hand side of the above equations is the normalization factor. Observe that this mapping does not violate the no-cloning theorem, as it is possible to clone a basis state. Therefore

$$\alpha |0\rangle + \beta |1\rangle \rightarrow \alpha |0_E\rangle + \beta |1_E\rangle$$

where $\sqrt{\alpha^2 + \beta^2} = 1$. Note that the inner layer of this code is responsible for correcting bit-flip errors, where the majority rule is utilized within each triple. For example

$$|001\rangle \pm |011\rangle \rightarrow |000\rangle \pm |111\rangle$$

The outer layer of the code is responsible for correcting sign-flip errors. The majority of the three signs is selected. For example

$$(|\cdot\rangle - |\cdot\rangle) \otimes (|\cdot\rangle + |\cdot\rangle) \otimes (|\cdot\rangle - |\cdot\rangle) \rightarrow (|\cdot\rangle - |\cdot\rangle) \otimes (|\cdot\rangle - |\cdot\rangle) \otimes (|\cdot\rangle - |\cdot\rangle)$$

As the bit-flip and sign-flip error-correction operations are independent of each other, Shor's code is correct when the bit-sign flip error occurs.

12.9.6 Barenco's Three-Qubit Code

Barenco's three-qubit code is another example of a single-qubit quantum error-correcting code. This code is named after its inventor, Adriano Barenco. Barenco's mapping is

$$|0\rangle \rightarrow |0_E\rangle = \frac{1}{2}(|000\rangle + |011\rangle + |101\rangle + |110\rangle)$$
$$|1\rangle \rightarrow |1_E\rangle = \frac{1}{2}(|111\rangle + |100\rangle + |010\rangle + |001\rangle)$$

If W_8 is a 8×8 Hadamard matrix, then

$$W_8 |000\rangle = \frac{1}{\sqrt{2}}(|0_E\rangle + |1_E\rangle), \quad \text{and} \quad W_8 |111\rangle = \frac{1}{\sqrt{2}}(|0_E\rangle - |1_E\rangle)$$

Therefore the codewords $|0_E\rangle$ and $|1_E\rangle$ can simply be obtained via Hadamard transformations.

Several other ingenious quantum error-correcting codes can be found in the literature. There also exist techniques to convert classical error-correcting codes into quantum error-correcting codes.

12.10 Quantum Cryptography

Both classical and modern cryptographic techniques have been studied in the chapter on cryptography. *Classical cryptographic* techniques are the schemes used in cryptography from ancient times right up till the second world war. The age of *modern cryptographic* techniques can be said to have begun after the second world war. However, someone might opine that the era of modern cryptography commenced in the early 1970's.

We study quantum cryptographic techniques in this section. These techniques are the application of quantum mechanics to cryptography. A salient contribution of quantum cryptography is that it can *automatically detect eavesdropping* on the communication channel.

12.10.1 Classical Cryptographic Communication

We initially describe a classical cryptographic communication system. This helps us in highlighting the contributions of quantum cryptography. In a classical cryptographic communication scheme

Alice (transmitter) wants to send a message to Bob (receiver) over a communication channel, and Oscar would like to eavesdrop on this channel. Alice encrypts a plaintext message p into a ciphertext c using a secret key k. She then transmits the message over an insecure channel on which Oscar is eavesdropping. Note that the key k is shared by both Alice and Bob. Upon receiving the encrypted message, Bob uses the key k to decode the message. The problem with this scheme should be immediately evident. Initially a *secure channel* is needed for Alice and Bob to share the key k. Thus, if a secure channel is necessary in this scheme, then what is the necessity of encrypting a message? A practical, albeit insecure procedure for Alice and Bob would be to initially share the key k with the help of a courier (messenger).

Alice and Bob also need to address two other problems: *authentication* and *intrusion detection.* Authentication means that Bob needs to know that he is indeed communicating with Alice and not Oscar. Intrusion detection means that the two communicating entities Alice and Bob have the ability to determine if Oscar is eavesdropping on their communication channel. Thus in summary, the classical cryptographic scheme:

(a) Does not provide a secure channel to share the key.
(b) Does not provide authentication.
(c) Does not provide any scheme for intrusion detection.

12.10.2 Modern Cryptographic Communication

Modern cryptographic techniques rely upon computational security. This in turn is achieved by the use of a trapdoor one-way function. The public key cryptography, a modern cryptographic mechanism is based upon the trapdoor one-way function. The RSA cryptographic scheme which is discussed in a different chapter, is an example of the use of public key cryptography. In this scheme, the encrypting and decrypting keys are different. Furthermore, the public keys of a set of users are managed by a central key bank. This partially addresses the problem of the necessity of a secure channel in classical cryptographic schemes. Also these modern schemes do not address the problem of intrusion detection. However the authentication issue has been favorably addressed. The important attributes of modern cryptographic scheme are summarized below.

(a) The necessity of a secure channel to share public keys is partially addressed.
(b) Provides authentication.
(c) Does not provide any scheme for intrusion detection.

12.10.3 Quantum Cryptographic Communication

Quantum cryptography uses the principles of quantum mechanics. These cryptographic techniques are based upon the behavior of photons and its polarizability. A *photon* is a particle of light or a corpuscle of energy. It is also said to be a quantum or a building block of light or some other form of electromagnetic radiation. As per quantum mechanics, all particles, including photons exhibit properties of a wave. *Photon polarization* corresponds to the angle of the plane in which they oscillate on their propagation axis. This polarization of a photon is specified by an angle θ.

Typically photons which emanate from a light source often have an unknown polarization angle. If a light filter is used to pass these photons through it, then photons with a specified polarization angle can be obtained. Therefore these light filters are characterized by say, a specific angle θ. Thus if photons with a polarization angle φ are passed through a θ-filter, photons with a polarization angle of θ emerge from the filter with a probability $\cos^2(\varphi - \theta)$. Also the photons are blocked with a probability $\sin^2(\varphi - \theta)$. A more detailed explanation of this phenomenon can be found in a book on quantum mechanics. Photons which are polarized at an angle $0°$ and $180°$ are represented by $\leftrightarrow$, those at an angle $45°$ are represented by $\nearrow$, those at an angle $90°$ and $270°$ are represented by $\updownarrow$, and finally those at an angle $135°$ by $\nwarrow$. For example, if photons with a polarizing angle of $45°$ are passed through a $90°$ filter, then these photons pass through these filter with a probability equal to $1/2$. Photons in the states $\leftrightarrow$, $\nearrow$, $\updownarrow$, and $\nwarrow$ are specified by the kets $|\leftrightarrow\rangle$, $|\nearrow\rangle$, $|\updownarrow\rangle$, and $|\nwarrow\rangle$ respectively.

The polarization angle of a photon can be used to encode bits. For example photons polarized at angle $0°$ or $180°$ (state $\leftrightarrow$) can be used to represent bit 0, and photons polarized at angle $90°$ or $270°$ can be used to represent bit 1 (state $\updownarrow$). The corresponding kets $|\leftrightarrow\rangle$ and $|\updownarrow\rangle$ respectively form a *rectilinear basis*. These kets are orthonormal, and are also referred to as *vertical/horizontal* (*VH*) *basis*. More precisely, using rectilinear basis, a quantum alphabet $\mathcal{A}_r$ can be defined. Thus

Quantum alphabet $\mathcal{A}_r$

Bit	Photon polarization	Photon ket
0	$\leftrightarrow$	$|\leftrightarrow\rangle$
1	$\updownarrow$	$|\updownarrow\rangle$

Similarly a *diagonal basis* can also be defined. In this basis, photons polarized at an angle $45°$ (state $\nearrow$) can be used to represent bit 0, and photons polarized at an angle $135°$ can be used to represent bit 1 (state $\nwarrow$). The corresponding kets are $|\nearrow\rangle$ and $|\nwarrow\rangle$ respectively. These form the diagonal basis. Using the diagonal basis, a quantum alphabet $\mathcal{A}_d$ can be defined. Thus

Quantum alphabet $\mathcal{A}_d$

Bit	Photon polarization	Photon ket
0	$\nearrow$	$|\nearrow\rangle$
1	$\nwarrow$	$|\nwarrow\rangle$

Other sets of quantum alphabets can be similarly defined. The BB84 quantum key distribution protocol is next discussed.

BB84 Quantum Cryptographic Protocol

This protocol was invented by C. H. Bennett and G. Brassard in 1984, hence the name BB84. The protocol is initially described in the absence of channel noise. This is followed by a brief discussion of the required modifications if noise is present on the channel. The BB84 protocol determines the cryptographic key required for secure communication between Alice and Bob by using the principles of quantum mechanics. It is assumed that Oscar is the eavesdropper (adversary) on the communication link between Alice and Bob.

The cryptographic key is determined in two stages. The first stage uses a one-way quantum channel, and the second stage uses a two-way public channel. Assume that the key is of length n bits. Also assume that Alice can generate photons with polarizations: $\leftrightarrow$, $\updownarrow$, $\nearrow$, and $\nwarrow$. That is, Alice uses both the quantum alphabets $\mathcal{A}_r$ and $\mathcal{A}_d$ to represent the bits 0 and 1. Bob also uses both quantum alphabets.

In order to determine if the protocol is really secure, Oscar must be permitted to do whatever he wants to do. He can eavesdrop on both the quantum and public channels. Furthermore, Oscar is permitted to measure the photons with respect to any set of alphabets (basis). Oscar can also intercept the photons and retransmit the measured photon, or send another one. For simplicity assume that Oscar measures the photons only in either rectilinear or diagonal basis (same as Alice's basis) and retransmits the measured photons.

The quantum key distribution between Alice and Bob in the absence of Oscar is next explained. It is assumed that a quantum communication channel and also a public communication channel are available for this process. This key distribution is performed in two stages.

Stage 1 Quantum Communication Channel

Communication between Alice and Bob in stage 1 is over a one-way quantum channel. This stage has three steps.

Step 1: Alice generates a random sequence of $m \gg n$ bits. This sequence is used to generate a secret key which is to be shared between Alice and Bob.

Step 2: For each random bit generated in the last step, an unbiased coin is flipped. If the coin is a "head," the rectilinear quantum alphabet $\mathcal{A}_r$ is selected. Otherwise if the coin is a "tail," the diagonal quantum alphabet $\mathcal{A}_d$ is selected. Alice then sends Bob, the bit generated in Step 1 as a polarized photon as per the randomly selected alphabet. These photons are sent one by one at regular intervals.

Step 3: Bob receives the photons transmitted by Alice. These photons have polarizations $\leftrightarrow$, $\updownarrow$, $\nearrow$, and $\nwarrow$, but Bob is unaware of the quantum alphabet selected by Alice. Consequently, Bob flips an unbiased coin to select one of the two alphabets and makes his measurements. As expected coin flip is correct fifty percent of the times, and incorrect the rest of the time. When the coin flip is correct (matches Alice's choice of alphabet) he would infer the transmitted bit correctly (probability $1/2$). But in his sequence of received bits, he does not know which bit is correct. Similarly, when the coin flip selects incorrect alphabet (probability $1/2$), the measurement made by Bob may yield correct bit (probability $1/2$). The probability that Bob's receiver correctly receives the transmitted bit is equal to $1/2 + (1/2)(1/2) = 3/4$.
At the end of the transmission of m bits by Alice, Bob also generates m bits.

Stage 2 Public Communication Channel

Communication between Alice and Bob in stage 2, is over a two-way public channel. This stage has four steps.

Step 1: Using the public channel, Bob transmits to Alice the sequence of alphabet choices he made. This sequence is of length m.

Step 2: Using the public channel, Alice lets Bob know which were the correct choice of alphabets. A correct choice of alphabet is the alphabet which matches with that originally selected by Alice.

Step 3: Alice and Bob delete all the bits from their m-bit sequence with mismatching alphabets. This results in the *raw key*.

This is the end of quantum key distribution in the absence of Oscar the eavesdropper. Observe that this protocol cannot be used directly to send a message because several qubits are dropped in the process. In this protocol it is assumed that both Alice and Bob have perfect transmitting and receiving devices.

If Oscar was not tapping the communication channels, the raw key of Alice and Bob would be identical. However, if Oscar was tapping the communication channels, then the raw key of Alice and Bob would mismatch. An error rate R_{err} can be computed by determining the total number of mismatches.

Let us examine what can happen when Oscar the eavesdropper is present and intercepts the photons transmitted by Alice. As mentioned earlier, Oscar intercepts the photons transmitted by Alice, and then retransmits the measured photons to Bob. Also assume that Oscar can use only rectilinear and diagonal set of quantum alphabets. Consider the following two cases.

Case 1:

(a) Alice uses alphabet $\mathcal{A}_r$ to transmit a 0 bit using a photon with polarization $\leftrightarrow$.
(b) Oscar intercepts the photon sent by Alice using the alphabet $\mathcal{A}_r$, and transmits the measured photon to Bob. Note that Oscar is transmitting a photon with polarization $\leftrightarrow$.
(c) Bob also uses alphabet $\mathcal{A}_r$ to receive the photon which Oscar transmitted. He correctly receives a photon with polarization $\leftrightarrow$, and interprets it as a 0 bit. Observe that Bob is actually thinking (or hoping) that this photon actually arrived from Alice.
Thus Alice and Bob use the same quantum alphabet and Bob receives the correctly transmitted bit.

Case 2:

(a) Alice uses alphabet $\mathcal{A}_r$ to transmit a 0 bit using a photon with polarization $\leftrightarrow$.
(b) Oscar intercepts the photon sent by Alice using the alphabet $\mathcal{A}_d$, and transmits the measured photon to Bob. Note that Oscar is transmitting a photon with polarization $\nearrow$, with probability $1/2$.
(c) Bob uses alphabet $\mathcal{A}_r$ to receive the photon which Oscar transmitted. He receives a photon with polarization $\updownarrow$, with probability $1/2$. He interprets it as a 1 bit. Note that Alice and Bob used the same quantum alphabet.

Thus Alice and Bob use the same quantum alphabet but Bob receives the incorrect bit, due to Oscar's intrusion. The different scenarios of alphabet choices by Alice, Bob and Oscar result in different values of the probability P_b that Alice and Bob have the same bit. The scenario in which Alice and Bob use different alphabets is not interesting. This is so, because if this event occurs the corresponding bits are discarded. Summarizing:

(a) If Alice and Bob use the quantum alphabet $\mathcal{A}_r$, and Oscar uses the quantum alphabet $\mathcal{A}_r$, then $P_b = 1$.
(b) If Alice and Bob use the quantum alphabet $\mathcal{A}_r$, and Oscar uses the quantum alphabet $\mathcal{A}_d$, then $P_b = 1/2$.
(c) If Alice and Bob use the quantum alphabet $\mathcal{A}_d$, and Oscar uses the quantum alphabet $\mathcal{A}_r$, then $P_b = 1/2$.
(d) If Alice and Bob use the quantum alphabet $\mathcal{A}_d$, and Oscar uses the quantum alphabet $\mathcal{A}_d$, then $P_b = 1$.

Step 4: This last step corresponds to error estimation. Alice and Bob determine the error rate R_{err}. Alice and Bob select a publicly agreed upon $\widetilde{m}$ bit locations in this sequence to disclose each other the $\widetilde{m}$ bits. That is, *both* the polarization orientation and the bits sent and received should

match in these $\widetilde{m}$ bit positions. Furthermore, disclosed bits are deleted from the raw key to produce *tentative final key* of length n bits. Alice and Bob determine the error rate as follows.
(a) If Alice and Bob determine that the error rate $R_{err} = 0$, then Oscar was not eavesdropping. Consequently, the tentative final key is indeed the *final key*. The probability of this happening is $(3/4)^{\widetilde{m}}$.
(b) If Alice and Bob determine that the error rate $R_{err} > 0$, then Oscar was eavesdropping. Alice and Bob return to Stage 1, and repeat the process, or decide to abort their communication. The probability of detecting an eavesdropper, assuming that one is present is $\left\{1 - (3/4)^{\widetilde{m}}\right\}$.
It should be evident that the value of $\widetilde{m}$ also determines the reliability of this protocol.

This is the end of the BB84 quantum cryptographic protocol description in the presence of noiseless channels. The BB84 protocol has to be suitably refined so as to be useful in the presence of noise. A technique to address the problem of noisy channels is for Alice to encode the bit sequence she transmits to Bob by using error-correction code. In the presence of channel noise, Bob is unable to differentiate between the eavesdropping by Oscar and noise on the communication channel. However, Oscar has some information about the bit sequence transmitted by Alice to Bob. That is, the raw key is not completely secret. Nevertheless it is possible to extract a smaller secret key from a larger raw key by a process called privacy amplification. See the notes at the end of the chapter to find references which describe in detail the BB84 protocol in presence of noise. These references also describe other popular protocols used for quantum key generation. This protocol is elucidated via an example.

Example 12.7. In this example Alice and Bob want to establish a quantum cryptographic key for secure communication via the BB84 protocol. Initially assume that both noise and Oscar the eavesdropper are absent.

BB84 protocol without both noise and Oscar

Bit numb.	1	2	3	4	5	6	7	8	9	10
Alice:										
Bit seq.	1	0	0	1	1	0	0	1	1	0
Alphabet	$\mathcal{A}_r$	$\mathcal{A}_d$	$\mathcal{A}_r$	$\mathcal{A}_r$	$\mathcal{A}_d$	$\mathcal{A}_d$	$\mathcal{A}_r$	$\mathcal{A}_r$	$\mathcal{A}_d$	$\mathcal{A}_r$
Pht. Pol.	$\updownarrow$	$\nearrow$	$\leftrightarrow$	$\updownarrow$	$\nwarrow$	$\nearrow$	$\leftrightarrow$	$\updownarrow$	$\nwarrow$	$\leftrightarrow$
Bob:										
Alphabet	$\mathcal{A}_r$	$\mathcal{A}_d$	$\mathcal{A}_d$	$\mathcal{A}_r$	$\mathcal{A}_d$	$\mathcal{A}_r$	$\mathcal{A}_r$	$\mathcal{A}_r$	$\mathcal{A}_d$	$\mathcal{A}_d$
Bit seq.	1	0	1	1	1	0	0	1	1	1
Raw key:	1	0		1	1		0	1	1	

The following abbreviations are used, in the tables illustrated in this example,

- Bit position number $\triangleq$ Bit numb.
- Bit sequence $\triangleq$ Bit seq.
- Photon Polarization $\triangleq$ Pht. Pol.

Using the public channel, Bob transmits to Alice the sequence of alphabet choices he made. This sequence is of length 10. Using the public channel, Alice lets Bob know which were the correct choice of alphabets.

Observe in the table, that the alphabet sequences used by Alice and Bob are identical in seven positions. Therefore the bit sequence transmitted by Alice, and that received by Bob are identical in these seven positions. Thus the raw key 1011011 has seven bits.

Let us continue to study this example when noise is absent, but Oscar the eavesdropper is present.

BB84 protocol without noise but Oscar is present

Bit numb.	1	2	3	4	5	6	7	8	9	10
Alice:										
Bit seq.	1	0	0	1	1	0	0	1	1	0
Alphabet	$\mathcal{A}_r$	$\mathcal{A}_d$	$\mathcal{A}_r$	$\mathcal{A}_r$	$\mathcal{A}_d$	$\mathcal{A}_d$	$\mathcal{A}_r$	$\mathcal{A}_r$	$\mathcal{A}_d$	$\mathcal{A}_r$
Pht. Pol.	$\updownarrow$	$\nearrow$	$\leftrightarrow$	$\updownarrow$	$\nwarrow$	$\nearrow$	$\leftrightarrow$	$\updownarrow$	$\nwarrow$	$\leftrightarrow$
Oscar:										
Alphabet	$\mathcal{A}_r$	$\mathcal{A}_r$	$\mathcal{A}_d$	$\mathcal{A}_d$	$\mathcal{A}_r$	$\mathcal{A}_r$	$\mathcal{A}_r$	$\mathcal{A}_r$	$\mathcal{A}_d$	$\mathcal{A}_d$
Bit seq.	1	0	1	0	1	0	0	1	1	0
Bob:										
Alphabet	$\mathcal{A}_r$	$\mathcal{A}_d$	$\mathcal{A}_d$	$\mathcal{A}_r$	$\mathcal{A}_d$	$\mathcal{A}_r$	$\mathcal{A}_r$	$\mathcal{A}_r$	$\mathcal{A}_d$	$\mathcal{A}_d$
Bit seq.	1	1	1	0	0	0	0	1	1	0

BB84 protocol without noise but Oscar is present (continued)

Bit numb.	1	2	4	5	7	8	9
Alice:							
Bit seq.	1	0	1	1	0	1	1
Alphabet	$\mathcal{A}_r$	$\mathcal{A}_d$	$\mathcal{A}_r$	$\mathcal{A}_d$	$\mathcal{A}_r$	$\mathcal{A}_r$	$\mathcal{A}_d$
Bob:							
Alphabet	$\mathcal{A}_r$	$\mathcal{A}_d$	$\mathcal{A}_r$	$\mathcal{A}_d$	$\mathcal{A}_r$	$\mathcal{A}_r$	$\mathcal{A}_d$
Bit seq.	1	1	0	0	0	1	1

The bits in positions $3, 6,$ and 10 are discarded because Alice and Bob use different sets of alphabets in these positions. Of the remaining bits in position numbers $1, 2, 4, 5, 7, 8$ and 9; there is a mismatch in the bit positions: $2, 4,$ and 5. For example, the alphabet used in the second position by both Alice and Bob is $\mathcal{A}_d$. However, the second bit in Alice's sequence is 0, and the second bit in Bob's sequence is 1. Therefore, because of the presence of Oscar the eavesdropper, the raw keys of Alice and Bob are not identical. This is discerned by Alice and Bob as follows.

Using the public channel, Alice and Bob compare small segments of the raw key to estimate the error rate, and then delete the disclosed bits to finally produce their tentative final key. This is indeed going to be the final key, provided Alice and Bob find zero mismatches in the disclosed bits. However, if there are mismatches in the disclosed bits, then Alice and Bob know that Oscar is eavesdropping. In this scenario, Alice and Bob discard their tentative final key, and restart the process all over again, or temporarily halt their communication.

There are still some unanswered questions. For example, how large should be the size of the raw-key segments which Alice and Bob must examine? □

The important attributes of quantum cryptographic scheme are summarized below.

(a) The necessity of a secure channel to share public keys is partially addressed.
(b) It does not provide authentication.
(c) It provides a valid mechanism for intrusion detection.

As in generally all aspects of quantum computation and communication; the design, analysis, and implementation of quantum cryptographic protocols are still in an embryonic stage. Nevertheless, the progress which has been made is definitely impressive, if not spectacular.

Reference Notes

The notes in this chapter are based upon the excellent books by Benenti, Casati, and Strini (2004), Claude and Păun (2001), Gruska (1999), Hirvensalo (2004), Lo, Popescu, and Spiller (1998), Marinescu and Marinescu (2005), Nielsen, and Chuang (2000), Stolze, and Suter (2004), and Williams, and Clearwater (1998). A very good introduction to this subject can also be found in the class notes of Preskill (1998). A first introduction to this subject of quantum computation and communication can also be found in the article by Paquin (1998).

Basic concepts in quantum mechanics are explained thoroughly in Peres (1993). For a derivation of the computational complexity of Shor's factorizing algorithm see Gruska (1999), Hirvensalo (2004), and Nielsen, and Chuang (2000). A scholarly account of entropy is given in Wehrl (1978).

A readable introduction to quantum error-correction is also provided by Gottesman (2002). The proof of the theorem on sufficient and necessary conditions for quantum error-correcting codes closely follows Steane's proof given in Lo, Popescu, and Spiller's (1998) book.

A readable and interesting tutorial on the subject of quantum cryptography has been provided by Lomonaco (2000), and Paquin (1999). Cryptographic protocols other than BB84 are also discussed in Gruska (1999), and Lomonaco (2000).

Problems

1. Let A_i be a unitary transformation, and $|\psi_i\rangle$ be a compatible state vector for $1 \le i \le k$. Show that

$$\begin{aligned}&\{A_1 \otimes A_2 \otimes \ldots \otimes A_n\} \{|\psi_1\rangle \otimes |\psi_2\rangle \otimes \ldots \otimes |\psi_n\rangle\} \\ &= \{A_1 |\psi_1\rangle\} \otimes \{A_2 |\psi_2\rangle\} \otimes \ldots \otimes \{A_n |\psi_n\rangle\}\end{aligned}$$

2. Let A be an $n \times n$ Hermitian matrix with eigenvalue λ_i and a corresponding eigenvector x_i for $1 \leq i \leq n$. Assume that the length of the eigenvector x_i is equal to unity for $1 \leq i \leq n$. Define $P_i = x_i x_i^\dagger$ for $1 \leq i \leq n$. Prove that:
   (a) The matrix P_i is Hermitian, for $1 \leq i \leq n$.
   (b) $P_i P_j = \delta_{ij} P_i$, where $\delta_{ij} = 1$ if $i = j$, and $\delta_{ij} = 0$ if $i \neq j$, for $1 \leq i, j \leq n$.
   (c) $tr\,(P_i) = 1$ for $1 \leq i \leq n$.
   (d) $A = \sum_{i=1}^{n} \lambda_i P_i$.
   (e) $tr\,(A) = \sum_{i=1}^{n} \lambda_i$.
   (f) $\sum_{i=1}^{n} P_i = I$.
   (g) $\sum_{i=1}^{n} P_i P_i^\dagger = I$.
   (h) If the length of the vector $|\psi\rangle$ is unity, then $1 = \sum_{i=1}^{n} \langle \psi \mid P_i \mid \psi \rangle$. The expression $\langle \psi \mid P_i \mid \psi \rangle$ is nonnegative, therefore it can be interpreted as a probability.

   Hint: Use linear algebraic techniques.
   (a) The statement follows immediately from the definition of P_i, for $1 \leq i \leq n$.
   (b) The eigenvectors of a Hermitian matrix form an orthonormal set.
   (c) Since $x_i^\dagger x_i = 1$ for $1 \leq i \leq n$.
   (d) Let $R = \begin{bmatrix} x_1 \; x_2 \; \cdots \; x_n \end{bmatrix}$, Λ is equal to a diagonal matrix with the eigenvalues on its diagonal. Then $R^\dagger R = I$, and

$$A = R\Lambda R^\dagger = \begin{bmatrix} x_1 \; x_2 \; \cdots \; x_n \end{bmatrix} \Lambda \begin{bmatrix} x_1^\dagger \\ x_2^\dagger \\ \vdots \\ x_n^\dagger \end{bmatrix}$$

$$= \begin{bmatrix} \lambda_1 x_1 \; \lambda_2 x_2 \; \cdots \; \lambda_n x_n \end{bmatrix} \begin{bmatrix} x_1^\dagger \\ x_2^\dagger \\ \vdots \\ x_n^\dagger \end{bmatrix}$$

$$= \sum_{i=1}^{n} \lambda_i x_i x_i^\dagger = \sum_{i=1}^{n} \lambda_i P_i$$

   (e) Use the part (d) result.
   (f) From (d), it follows that $AP_j = \sum_{i=1}^{n} \lambda_i P_i P_j$ for $1 \leq j \leq n$. Thus

$$\sum_{j=1}^{n} AP_j = \sum_{j=1}^{n} \sum_{i=1}^{n} \lambda_i P_i P_j = \sum_{i=1}^{n} \lambda_i P_i = A$$

   The result is immediate.
   (g) Since $P_i = P_i^2 = P_i P_i^\dagger$, for $1 \leq i \leq n$, the result follows.
   (h) It is known that $I = \sum_{i=1}^{n} P_i$. Thus $|\psi\rangle = \sum_{i=1}^{n} P_i |\psi\rangle$, and $1 = \langle \psi \mid \psi \rangle = \sum_{i=1}^{n} \langle \psi \mid P_i \mid \psi \rangle$. Also $\langle \psi \mid P_i \mid \psi \rangle = \left\langle \psi \mid x_i x_i^\dagger \mid \psi \right\rangle$ is nonnegative.

3. Let A be a self-adjoint operator, and λ_i and $|\psi_i\rangle$ be the eigenvalue and eigenvector pair respectively of this operator for $1 \leq i \leq n$. The eigenvectors form an orthonormal basis of the corresponding vector space. If $\alpha_i \in \mathbb{C}$ for $1 \leq i \leq n$, show that for the vector $|\psi\rangle = \sum_{i=1}^{n} \alpha_i |\psi_i\rangle$, we have $\langle \psi \mid A \mid \psi \rangle = \sum_{i=1}^{n} |\alpha_i|^2 \lambda_i$.

Hint: See Gruska (1999). Observe that

$$A|\psi\rangle = \sum_{i=1}^{n} \alpha_i A|\psi_i\rangle = \sum_{i=1}^{n} \alpha_i \lambda_i |\psi_i\rangle$$

and $\langle\psi| = \sum_{j=1}^{n} \overline{\alpha}_j \langle\psi_j|$. Therefore

$$\langle\psi \mid A \mid \psi\rangle = \sum_{j=1}^{n} \overline{\alpha}_j \langle\psi_j| \sum_{i=1}^{n} \alpha_i \lambda_i |\psi_i\rangle = \sum_{i=1}^{n} |\alpha_i|^2 \lambda_i$$

In the last step, we used the orthonormality of the eigenvectors.

4. Find the eigenvalues and the corresponding eigenvectors of the Pauli matrices.
5. Prove that the trace of the density matrix ρ is equal to unity.
   Hint: Assume that $\rho = \sum_{i=1}^{m} p_i \rho_i$, where $tr(\rho_i) = 1$ for $1 \le i \le m$, then

$$tr(\rho) = \sum_{i=1}^{m} p_i tr(\rho_i) = \sum_{i=1}^{m} p_i = 1$$

6. Prove that a matrix ρ is a density operator if and only if the trace of matrix ρ is unity and it is a positive semidefinite matrix where

$$\rho = \sum_{i=1}^{m} p_i |\psi_i\rangle \langle\psi_i|$$

   Hint: See Nielsen, and Chuang (2000). Assume that $\rho = \sum_{i=1}^{m} p_i |\psi_i\rangle\langle\psi_i|$, then it is known that $tr(\rho) = 1$. Let $|\psi\rangle$ be an arbitrary vector, then

$$\langle\psi|\rho|\psi\rangle = \sum_{i=1}^{m} p_i \langle\psi \mid \psi_i\rangle \langle\psi_i \mid \psi\rangle = \sum_{i=1}^{m} p_i |\langle\psi_i \mid \psi\rangle|^2 \ge 0$$

   The proof in the opposite direction proceeds as follows. Assume that the operator ρ has a trace equal to unity and it is positive semidefinite. The eigenvalues of the matrix ρ are all real and nonnegative. Let these be λ_i, for $1 \le i \le m$. Also denote a corresponding eigenvector by β_i, for $1 \le i \le m$. This set of eigenvectors are also orthonormal (since ρ is Hermitian). Then $\rho = \sum_{i=1}^{m} \lambda_i |\beta_i\rangle\langle\beta_i|$. Also by assumption, $tr(\rho) = 1$. Thus $\sum_{i=1}^{m} \lambda_i = 1$, and a quantum system in state $|\beta_i\rangle$ with probability λ_i has ρ as its density operator. That is, the set of states $\{(p_i, |\psi_i\rangle) \mid 1 \le i \le m\}$ has ρ as its density operator.
7. Prove that the eigenvalues of the density matrix ρ are real, nonnegative, and sum to unity.
   Hint: Let the size of the density matrix be n. The density matrix is Hermitian, therefore its eigenvalues $\lambda_i, 1 \le i \le n$ are all real numbers. Since $tr(\rho) = 1$, it follows that $\sum_{i=1}^{n} \lambda_i = tr(\rho) = 1$. Let x_i be an eigenvector corresponding to the eigenvalue λ_i, such that $x_i^\dagger x_i = 1$. Since $\rho x_i = \lambda_i x_i$ and ρ a positive semidefinite matrix, it follows that $\lambda_i = x_i^\dagger \rho x_i \ge 0$.
8. If a quantum state is pure, prove that $tr(\rho^2) = 1$. If a quantum state is mixed, prove that $tr(\rho^2) < 1$.
   Hint: See Nielsen, and Chuang (2000). If the quantum state is pure, then $\rho^2 = \rho$. Therefore $tr(\rho^2) = tr(\rho) = 1$. If the quantum state is mixed, let the eigenvalues of ρ be $\lambda_i \ge 0, 1 \le i \le n$, such that $tr(\rho) = \sum_{i=1}^{n} \lambda_i$. Then $tr(\rho^2) = \sum_{i=1}^{n} \lambda_i^2 \le 1$. Equality occurs if all λ_i's are equal to zero, except a single eigenvalue which is equal to unity.

9. Let ρ_X be the density matrix of a quantum mechanical system X with mixed state. The orthonormal decomposition of ρ_X is given by $\rho_X = \sum_{i=1}^{n} p_i \left|\psi_i\right\rangle \left\langle\psi_i\right|$, where $p_i \geq 0$ for $1 \leq i \leq n$, $tr\left(\rho_X\right) = \sum_{i=1}^{n} p_i = 1$, and $\left|\psi_i\right\rangle, 1 \leq i \leq n$ are orthonormal state vectors in $\mathcal{H}_n$. Consider a system Y which has the same state space as system X, which is $\mathcal{H}_n$. Its orthonormal basis set is $\left|\varphi_i\right\rangle, 1 \leq i \leq n$. Define $\left|\beta_i\right\rangle = \left|\psi_i\right\rangle \otimes \left|\varphi_i\right\rangle, 1 \leq i \leq n$, and a state vector $\left|\alpha\right\rangle \in \mathcal{H}_n \otimes \mathcal{H}_n$ as $\left|\alpha\right\rangle = \sum_{i=1}^{n} \sqrt{p_i} \left|\beta_i\right\rangle$. Prove that $tr_Y\left(\left|\alpha\right\rangle\left\langle\alpha\right|\right) = \rho_X$.
   Hint: See Nielsen, and Chuang (2000).
   $$\left|\alpha\right\rangle\left\langle\alpha\right| = \sum_{i=1}^{n}\sum_{j=1}^{n} \sqrt{p_i p_j} \left|\psi_i\right\rangle \left|\varphi_i\right\rangle \left\langle\varphi_j\right| \left\langle\psi_j\right|$$
   Using the definition of $tr_Y(\cdot)$ results in
   $$\begin{aligned} tr_Y\left(\left|\alpha\right\rangle\left\langle\alpha\right|\right) &= \sum_{i=1}^{n}\sum_{j=1}^{n} \sqrt{p_i p_j} \left|\psi_i\right\rangle \left\langle\psi_j\right| tr\left(\left|\varphi_i\right\rangle\left\langle\varphi_j\right|\right) \\ &= \sum_{i=1}^{n}\sum_{j=1}^{n} \sqrt{p_i p_j} \left|\psi_i\right\rangle \left\langle\psi_j\right| \delta_{ij} \end{aligned}$$
   where $\delta_{ij} = 1$ if $i = j$, and $\delta_{ij} = 0$ otherwise. The result follows.
10. Establish Heisenberg's uncertainty principle.
    Hint: See Nielsen, and Chuang (2000). This relationship is established in two steps.
    Step 1: Let C and D be two Hermitian operators, and $\left|\psi\right\rangle$ be a quantum state. If $\left\langle\psi\right| CD \left|\psi\right\rangle = c + id$, where $c, d \in \mathbb{R}$, then $\left\langle\psi\right| [C, D] \left|\psi\right\rangle = 2id$ and $\left\langle\psi\right| \{C, D\} \left|\psi\right\rangle = 2c$. Thus
    $$\left|\left\langle\psi\right| [C,D] \left|\psi\right\rangle\right|^2 + \left|\left\langle\psi\right| \{C,D\} \left|\psi\right\rangle\right|^2 = 4\left|\left\langle\psi\right| CD \left|\psi\right\rangle\right|^2$$
    Use of the Bunyakovsky-Cauchy-Schwartz inequality
    $$\left|\left\langle\psi\right| CD \left|\psi\right\rangle\right|^2 \leq \left\langle\psi\right| C^2 \left|\psi\right\rangle \left\langle\psi\right| D^2 \left|\psi\right\rangle$$
    and the above equation results in
    $$\left|\left\langle\psi\right| [C,D] \left|\psi\right\rangle\right|^2 \leq 4\left\langle\psi\right| C^2 \left|\psi\right\rangle \left\langle\psi\right| D^2 \left|\psi\right\rangle$$
    Step 2: Since A and B are two observables, these matrices are Hermitian. Substituting $C = \Delta A$, and $D = \Delta B$ in the above equation yields
    $$\left|\left\langle\psi\right| [\Delta A, \Delta B] \left|\psi\right\rangle\right|^2 \leq 4\left\langle\psi\right| (\Delta A)^2 \left|\psi\right\rangle \left\langle\psi\right| (\Delta B)^2 \left|\psi\right\rangle$$
    Note that $\left\langle\psi\right| \Delta A \left|\psi\right\rangle = \left\langle\psi\right| \Delta B \left|\psi\right\rangle = 0$.
    The result follows since $\left\langle[\Delta A, \Delta B]\right\rangle = \left\langle[A, B]\right\rangle$.
11. Prove that any 2×2 complex unitary matrix can be written as:
    $$U = e^{i\varphi}\begin{bmatrix} \alpha & \beta \\ -\overline{\beta} & \overline{\alpha} \end{bmatrix}, \quad \text{where} \quad |\alpha|^2 + |\beta|^2 = 1$$
12. Prove that any 2×2 complex unitary matrix can be written as:
    $$U = e^{i\varphi}\begin{bmatrix} e^{i\xi} & 0 \\ 0 & e^{-i\xi} \end{bmatrix}\begin{bmatrix} \cos\theta & i\sin\theta \\ i\sin\theta & \cos\theta \end{bmatrix}\begin{bmatrix} e^{i\gamma} & 0 \\ 0 & e^{-i\gamma} \end{bmatrix}$$
    Hint: See Gruska (1999). Use the fact that a matrix is unitary if and only if its rows and columns are orthonormal.

13. Establish the relationships between the different complexity classes of a QTM.
Hint: See Gruska (1999).
14. Let n be a positive integer. Find a polynomial-time algorithm to determine if $n = u^v$, where $u \geq 2$ and $v \geq 2$. If the binary representation of n requires ℓ bits, that is if $\ell = \lceil \log_2 (n+1) \rceil$, then prove that the computational complexity of the algorithm is $O(\ell^3)$.
Hint: See Nielsen, and Chuang (2000).
15. Establish the product representation of QFT.
Hint: See Nielsen, and Chuang (2000). Using the definition of QFT, write

$$|j\rangle \to \frac{1}{\sqrt{N}} \sum_{k=0}^{(N-1)} \omega_N^{jk} |k\rangle, \quad 0 \leq j \leq (N-1)$$

Let the binary representation of j and k be $j_{b-1}j_{b-2}\ldots j_0$ and $k_{b-1}k_{b-2}\ldots k_0$ respectively, where $N = 2^b$. Then

$$\begin{aligned}
|j\rangle &\to \frac{1}{\sqrt{N}} \sum_{k_{b-1}=0}^{1} \sum_{k_{b-2}=0}^{1} \cdots \sum_{k_0=0}^{1} e^{2\pi i j \left(\sum_{l=0}^{b-1} k_l 2^{-b+l}\right)} |k_{b-1}k_{b-2}\ldots k_0\rangle \\
&= \frac{1}{\sqrt{N}} \sum_{k_{b-1}=0}^{1} \sum_{k_{b-2}=0}^{1} \cdots \sum_{k_0=0}^{1} e^{2\pi i j \left(\sum_{l=0}^{b-1} k_{b-1-l} 2^{-l-1}\right)} |k_{b-1}k_{b-2}\ldots k_0\rangle \\
&= \frac{1}{\sqrt{N}} \sum_{k_{b-1}=0}^{1} \sum_{k_{b-2}=0}^{1} \cdots \sum_{k_0=0}^{1} \bigotimes_{l=0}^{b-1} e^{2\pi i j k_{b-1-l} 2^{-l-1}} |k_{b-1-l}\rangle \\
&= \frac{1}{\sqrt{N}} \bigotimes_{l=0}^{b-1} \left[\sum_{k_{b-1-l}=0}^{1} e^{2\pi i j k_{b-1-l} 2^{-l-1}} |k_{b-1-l}\rangle \right] \\
&= \frac{1}{\sqrt{N}} \bigotimes_{l=0}^{b-1} \left[|0\rangle + e^{2\pi i j 2^{-l-1}} |1\rangle \right] \\
&= \frac{1}{\sqrt{N}} \left\{ |0\rangle + e^{2\pi i (0.j_0)} |1\rangle \right\} \left\{ |0\rangle + e^{2\pi i (0.j_1 j_0)} |1\rangle \right\} \cdots \\
&\quad \cdots \left\{ |0\rangle + e^{2\pi i (0.j_{b-1} j_{b-2} \ldots j_0)} |1\rangle \right\}
\end{aligned}$$

16. Use Shor's quantum algorithm, to determine the period of $13 \in \mathbb{Z}_{15}$.
Hint: See Hirvensalo (2004). In this problem, $x = 13$, and $n = 15$. Select $m = 16$.
Step 1: Determine the superposition $(1/4) \sum_{k=0}^{15} |k\rangle |0\rangle$.
Step 2: Find $x^k \pmod{15}$, $k \in \mathbb{Z}_{15}$.

$$\begin{aligned}
&\frac{1}{4} \sum_{k=0}^{15} |k\rangle |x^k\rangle \\
&= \frac{1}{4} [\{|0\rangle + |4\rangle + |8\rangle + |12\rangle\} |1\rangle + \{|1\rangle + |5\rangle + |9\rangle + |13\rangle\} |13\rangle \\
&\quad + \{|2\rangle + |6\rangle + |10\rangle + |14\rangle\} |4\rangle + \{|3\rangle + |7\rangle + |11\rangle + |15\rangle\} |7\rangle]
\end{aligned}$$

Step 3: Apply the QFT on $\mathbb{Z}_{16}$. This leads to

$$\frac{1}{4}\left[\left\{|0\rangle+|4\rangle+|8\rangle+|12\rangle\right\}|1\rangle+\left\{|0\rangle+i\,|4\rangle-|8\rangle-i\,|12\rangle\right\}|13\rangle\right.$$
$$\left.+\left\{|0\rangle-|4\rangle+|8\rangle-|12\rangle\right\}|4\rangle+\left\{|0\rangle-i\,|4\rangle-|8\rangle+i\,|12\rangle\right\}|7\rangle\right]$$

Step 4: The probabilities of the elements $0, 4, 8, 12 \in \mathbb{Z}_{16}$ are each equal to $1/4$.
Step 5: If $p = 0$ is chosen, then the convergent of $p/m = 0/16$ is simply $0/1$, which does not yield the period. If $p = 4$ is chosen, then the convergents of $p/m = 4/16$ are $0/1$ and $1/4$. The convergent $1/4$ correctly yields a period $r = 4$. If $p = 8$ is chosen, then the convergents of $p/m = 8/16$ are $0/1$ and $1/2$. The two convergents $0/1$ and $1/2$ do not yield the period. Finally, if $p = 12$ is chosen, then the convergents of $p/m = 12/16$ are $0/1, 1/1$, and $3/4$. The convergent $3/4$ correctly yields a period $r = 4$.

17. In Grover's algorithm, derive the recursive equations for k_j, l_j, for $j \in \mathbb{P}$. Using the z-transform technique applied to the k_j and l_j sequences, derive the trigonometric expressions for these parameters. (The z-transforms have been used for studying discrete random variables.) This is an alternate technique to determine k_j's and l_j's explicitly.
18. Establish Klein's inequality about quantum relative entropy.
Hint: See Nielsen, and Chuang (2000). Let the density operator be

$$\rho=\sum_{i=1}^{n} p_i \left|\psi_i\right\rangle \left\langle\psi_i\right|$$

where $\left|\psi_i\right\rangle, 1 \le i \le n$ are orthonormal state vectors and $\sum_{i=1}^{n} p_i = 1$. Similarly suppose that the density operator $\sigma = \sum_{j=1}^{n} q_j \left|\varphi_j\right\rangle\left\langle\varphi_j\right|$, where $\left|\varphi_j\right\rangle, 1 \le j \le n$ are orthonormal state vectors and $\sum_{j=1}^{n} q_j = 1$. Thus

$$S\left(\rho \parallel \sigma\right)=\sum_{i=1}^{n} p_i \log_2 p_i-\sum_{i=1}^{n}\left\langle\psi_i \mid \rho \log_2 \sigma \mid \psi_i\right\rangle$$

Note that $\left\langle\psi_i\right| \rho = p_i \left\langle\psi_i\right|, 1 \le i \le n$ and

$$\left\langle\psi_i \mid \log_2 \sigma \mid \psi_i\right\rangle=\left\langle\psi_i\right|\left\{\sum_{j=1}^{n} \log_2\left(q_j\right)\left|\varphi_j\right\rangle\left\langle\varphi_j\right|\right\}\left|\psi_i\right\rangle=\sum_{j=1}^{n} \log_2\left(q_j\right) \widetilde{p}_{ij}$$

In the above equation $\widetilde{p}_{ij} \triangleq \left\langle\psi_i \mid \varphi_j\right\rangle\left\langle\varphi_j \mid \psi_i\right\rangle \ge 0$, and $\sum_{i=1}^{n} \widetilde{p}_{ij} = \sum_{j=1}^{n} \widetilde{p}_{ij} = 1$. Thus the matrix $P = \left[\widetilde{p}_{ij}\right]$ is doubly stochastic. A matrix of nonnegative elements, in which the sum of the elements of each row and each column is equal to unity, is called a *doubly stochastic matrix*. Therefore

$$S\left(\rho \parallel \sigma\right)=\sum_{i=1}^{n} p_i \log_2 p_i-\sum_{i=1}^{n} p_i \sum_{j=1}^{n} \log_2\left(q_j\right) \widetilde{p}_{ij}$$

Define $t_i = \sum_{j=1}^{n} q_j \widetilde{p}_{ij}$. Since $\log_2(\cdot)$ is strictly a concave function,

$$\sum_{j=1}^{n} \log_2\left(q_j\right) \widetilde{p}_{ij} \le \log_2 t_i$$

where equality occurs if and only if there exists a j such that $\widetilde{p}_{ij} = 1$. Thus

$$S(\rho \parallel \sigma) \geq \sum_{i=1}^{n} p_i \log_2 \frac{p_i}{t_i}$$

Equality occurs in the above relationship if for each i there exists a j such that $\widetilde{p}_{ij} = 1$. Under this condition the matrix P is a permutation matrix. Note that $t_i \geq 0$ and $\sum_{i=1}^{n} t_i = 1$. Thus the above relationship has the form of classical relative entropy which is nonnegative, that is $S(\rho \parallel \sigma) \geq 0$. In this relationship, equality occurs if and only if $p_i = t_i$, for all $1 \leq i \leq n$. Note that it should be possible to relabel the eigenstates $|\psi_i\rangle$ such that the matrix $P = I$, an identity matrix. In this instance, the eigenvalues of the density matrices ρ and σ can be diagonalized with the same basis vectors. In addition, the condition $p_i = t_i$ implies that the corresponding eigenvalues are identical, for all $1 \leq i \leq n$. Thus equality occurs in the stated relationship if $\rho = \sigma$.

19. Let $\mathcal{H}'$ and $\mathcal{H}''$ be two Hilbert spaces. Also let ρ_1 and ρ_2 be density matrices in these two Hilbert spaces respectively. Prove that the entropy of the density matrix $\rho_1 \otimes \rho_2$ in $\mathcal{H}' \otimes \mathcal{H}''$ is given by

$$S(\rho_1 \otimes \rho_2) = S(\rho_1) + S(\rho_2)$$

Hint: See Wehrl (2002). Let the dimensions of the Hilbert spaces $\mathcal{H}'$ and $\mathcal{H}''$ be d' and d'' respectively. Also, let γ_i and α_i be the eigenvalue and a corresponding eigenvector of ρ_1 for $1 \leq i \leq d'$. Similarly, let δ_j and β_j be the eigenvalue and a corresponding eigenvector of ρ_2 for $1 \leq j \leq d''$. Then $\alpha_i \otimes \beta_j$ is an eigenvector of $\rho_1 \otimes \rho_2$ and the corresponding eigenvalue is $\gamma_i \delta_j$. Recall that $\sum_{i=1}^{d'} \gamma_i = \sum_{j=1}^{d''} \delta_j = 1$. Thus

$$\begin{aligned} S(\rho_1 \otimes \rho_2) &= -\sum_{i=1}^{d'} \sum_{j=1}^{d''} \gamma_i \delta_j \log_2 (\gamma_i \delta_j) \\ &= -\sum_{i=1}^{d'} \gamma_i \log_2 \gamma_i - \sum_{j=1}^{d''} \delta_j \log_2 \delta_j = S(\rho_1) + S(\rho_2) \end{aligned}$$

20. The density matrices ρ_i for $1 \leq i \leq n$ are defined on orthogonal Hilbert subspaces. Associated with a density matrix ρ_i is a probability p_i, where $\sum_{i=1}^{n} p_i = 1$. Prove that

$$S\left(\sum_{i=1}^{n} p_i \rho_i\right) = H(p) + \sum_{i=1}^{n} p_i S(\rho_i)$$

where $H(p) = -\sum_{i=1}^{n} p_i \log_2 p_i$.
Hint: See Nielsen, and Chuang (2000). Let $\lambda(i,j)$ and $|e(i,j)\rangle$ be the eigenvalue and an eigenvector respectively of the density matrix ρ_i, where $1 \leq j \leq m_i$. Then $p_i \lambda(i,j)$ and $|e(i,j)\rangle$ are the eigenvalues and eigenvectors of $\sum_{i=1}^{n} p_i \rho_i$. Consequently

$$\begin{aligned} S\left(\sum_{i=1}^{n} p_i \rho_i\right) &= -\sum_{i=1}^{n} \sum_{j=1}^{m_i} p_i \lambda(i,j) \log_2 \{p_i \lambda(i,j)\} \\ &= -\sum_{i=1}^{n} p_i \log_2 p_i - \sum_{i=1}^{n} p_i \sum_{j=1}^{m_i} \lambda(i,j) \log_2 \lambda(i,j) \\ &= H(p) + \sum_{i=1}^{n} p_i S(\rho_i) \end{aligned}$$

21. Prove the subadditivity law of quantum entropy.
Hint: See Nielsen, and Chuang (2000). In Klein's result, substitute $\rho = \rho_{XY}$ and $\sigma = \rho_X \otimes \rho_Y$. Thus

$$\begin{aligned} tr\left(\rho \log_2 \sigma\right) &= tr\left(\rho_{XY}\left(\log_2 \rho_X + \log_2 \rho_Y\right)\right) \\ &= tr\left(\rho_X \log_2 \rho_X\right) + tr\left(\rho_Y \log_2 \rho_Y\right) = -S\left(X\right) - S\left(Y\right) \end{aligned}$$

and

$$S\left(\rho \parallel \sigma\right) = tr\left(\rho \log_2 \rho\right) - tr\left(\rho \log_2 \sigma\right) = -S\left(X, Y\right) + S\left(X\right) + S\left(Y\right) \geq 0$$

The result follows.

22. Prove the Araki-Lieb inequality.
Hint: See Nielsen, and Chuang (2000). Consider a quantum mechanical system W which purifies systems X and Y simultaneously. Use of subadditivity yields

$$S\left(W\right) + S\left(X\right) \geq S\left(X, W\right)$$

Since the state W is a purifier of systems X and Y, XYW is in a pure state. It follows that $S\left(X, W\right) = S\left(Y\right)$ and $S\left(W\right) = S\left(X, Y\right)$. Substituting these values in the above inequality yields $S\left(X, Y\right) \geq S\left(Y\right) - S\left(X\right)$. It can be similarly demonstrated that $S\left(X, Y\right) \geq S\left(X\right) - S\left(Y\right)$. Thus $S\left(X, Y\right) \geq \left|S\left(X\right) - S\left(Y\right)\right|$. Equality is obtained if the density matrix of the composite X and W quantum mechanical systems is equal to $\rho_X \otimes \rho_W$.

23. There is a quantum source which produces quantum states $\rho_1, \rho_2, \ldots, \rho_n$ in the Hilbert space $\mathcal{H}_n$, with probabilities $p_1, p_2, \ldots, p_n$. Let $\widehat{X}$ be a random variable that determines which of the states ρ_i was measured. It is also given that $\widehat{Y}$ is an observable in the space $\mathcal{H}_n$. Let $\rho = \sum_{i=1}^{n} p_i \rho_i$, and the Shannon's mutual information between the random variables $\widehat{X}$ and $\widehat{Y}$ be $I\left(\widehat{X}; \widehat{Y}\right)$. Prove that

$$I\left(\widehat{X}; \widehat{Y}\right) \leq S\left(\rho\right) - \sum_{i=1}^{n} p_i S\left(\rho_i\right)$$

Hint: See Nielsen, and Chuang (2000). This inequality was discovered by A. S. Holevo, and hence called the *Holevo bound.* This bound determines an upper bound on the *accessible information* of the quantum mechanical system. Its proof is very ingenious.

24. Prove the theorem about typical subspace.
Hint: See Nielsen, and Chuang (2000). This theorem is proved by using Shannon's theorem about typical sequences.
    (a) Note that

$$tr\left(\mathcal{P}\left(n, \epsilon\right) \rho^{\otimes n}\right) = \sum_{\widetilde{x}\ \epsilon\text{-typical}} p_{k_1} p_{k_2} \cdots p_{k_n}$$

    Using part (a) of the Shannon's theorem about typical sequences, it can be immediately concluded that $tr\left(\mathcal{P}\left(n, \epsilon\right) \rho^{\otimes n}\right) \geq (1 - \delta)$.
    (b) This result also follows immediately by using part (b) of the Shannon's theorem about typical sequences.
    (c) The basic idea required to prove this part is again similar to the proof of part (c) of Shannon's theorem about typical sequences. The trace is split into two subtraces, one trace is over typical subspace, and the other is over atypical subspace. Thus

$$tr\left(\mathcal{G}(n)\rho^{\otimes n}\right) = tr\left(\mathcal{G}(n)\rho^{\otimes n}\mathcal{P}(n,\epsilon)\right) + tr\left(\mathcal{G}(n)\rho^{\otimes n}(I-\mathcal{P}(n,\epsilon))\right)$$

Observe that $\rho^{\otimes n}\mathcal{P}(n,\epsilon) = \mathcal{P}(n,\epsilon)\rho^{\otimes n}\mathcal{P}(n,\epsilon)$, because $\mathcal{P}(n,\epsilon)$ is a projector which commutes with $\rho^{\otimes n}$. Note that the eigenvalues of $\mathcal{P}(n,\epsilon)\rho^{\otimes n}\mathcal{P}(n,\epsilon)$ are upper bounded by $2^{-n(S(\rho)-\epsilon)}$. Thus

$$\begin{aligned} tr\left(\mathcal{G}(n)\rho^{\otimes n}\mathcal{P}(n,\epsilon)\right) &= tr\left(\mathcal{G}(n)\mathcal{P}(n,\epsilon)\rho^{\otimes n}\mathcal{P}(n,\epsilon)\right) \\ &\le 2^{nR_q}2^{-n(S(\rho)-\epsilon)} \end{aligned}$$

Therefore, as $n \to \infty$ the trace $tr\left(\mathcal{G}(n)\rho^{\otimes n}\mathcal{P}(n,\epsilon)\right) \to 0$. Next note that $\mathcal{G}(n) \le I$. Furthermore, $\mathcal{G}(n)$ and $\rho^{\otimes n}(I-\mathcal{P}(n,\epsilon))$ are both positive semidefinite matrices. Therefore

$$0 \le tr\left(\mathcal{G}(n)\rho^{\otimes n}(I-\mathcal{P}(n,\epsilon))\right) \le tr\left(\rho^{\otimes n}(I-\mathcal{P}(n,\epsilon))\right)$$

The right-hand side expression in the above inequality tends to zero as $n \to \infty$. The result follows.

25. Let ρ and σ be two density operators. Their fidelity is $\mathcal{F}(\rho,\sigma)$. Prove that $0 \le \mathcal{F}(\rho,\sigma) \le 1$.
26. Let U be any unitary operator. Prove that $\mathcal{F}\left(U\rho U^{\dagger}, U\sigma U^{\dagger}\right) = \mathcal{F}(\rho,\sigma)$.
    Hint: See Nielsen, and Chuang (2000). It is known that if A is any positive semidefinite matrix, then $\sqrt{UAU^{\dagger}} = U\sqrt{A}U^{\dagger}$. The result follows by using the cyclic property of the trace.
27. Prove Schumacher's noiseless channel coding theorem.
    Hint: We essentially follow the elegant proof given in Nielsen and Chuang (2000). Initially assume that $R_q > S(\rho)$. Select $R_q \ge S(\rho) + \epsilon$ where $\epsilon > 0$. For any $\delta > 0$ and sufficiently large values of n, it can be concluded that $\dim(T_q(n,\epsilon)) \le 2^{nR_q}$, and $tr\left(\rho^{\otimes n}\mathcal{P}(n,\epsilon)\right) \ge (1-\delta)$. This result follows by invoking the typical subspace theorem. Let $\mathcal{H}_d$ be any Hilbert space of dimension $d = \lceil 2^{nR_q} \rceil$ which contains $T_q(n,\epsilon)$. The encoding of the quantum states is done as follows. Initially a measurement is made by the orthogonal projectors $\mathcal{P}(n,\epsilon)$ and $(I-\mathcal{P}(n,\epsilon))$, where the outcomes are 0 and 1 respectively. In case the outcome is 0, the state is left untouched in the typical subspace. However, if the outcome is 1, the state of the quantum mechanical system is replaced by any *fixed* state in the typical subspace. For example, let this state be $|0\rangle$ which is selected from the typical subspace. Define the compression mapping $\Phi^n_{comp}(\cdot)$ as a transformation of a signal (density state) σ from the Hilbert space $\mathcal{H}^{\otimes n}$ into the subspace $\mathcal{H}_d$ by using the Kraus representation.

$$\Phi^n_{comp}(\sigma) = \mathcal{P}(n,\epsilon)\sigma\mathcal{P}(n,\epsilon) + \sum_j B_j\sigma B_j^{\dagger}$$

In the above equation, $B_j = |0\rangle\langle\varphi_j|$ and $|\varphi_j\rangle$ is an orthonormal basis of the orthocomplement of the typical subspace. The decompression operation $\Phi^n_{decomp}(\cdot)$ is a mapping from the Hilbert space $\mathcal{H}_d$ back to $\mathcal{H}^{\otimes n}$. It is essentially an identity operation, that is $\Phi^n_{decomp}(\beta) = \beta$. The average fidelity $\overline{\mathcal{F}}$ of the original sequence of length n and the decompressed sequence is next computed. Thus

$$\overline{\mathcal{F}} = \left|tr\left(\rho^{\otimes n}\mathcal{P}(n,\epsilon)\right)\right|^2 + \sum_j \left|tr\left(\rho^{\otimes n}B_j\right)\right|^2 \ge \left|tr\left(\rho^{\otimes n}\mathcal{P}(n,\epsilon)\right)\right|^2$$

Use of typical subspace theorem yields $\overline{\mathcal{F}} \ge \left|tr\left(\rho^{\otimes n}\mathcal{P}(n,\epsilon)\right)\right|^2 \ge |1-\delta|^2 \ge |1-2\delta|$. Thus δ can be made arbitrarily small for sufficiently large values of n, resulting in a value of average fidelity close to unity. This implies that a reliable compression-decompression scheme exists for the source S_q provided $R_q > S(\rho)$.

The converse of the theorem is next proved. The hypothesis is $R_q < S(\rho)$. Let $\mathcal{H}_d$ be any Hilbert space of dimension $d = \lfloor 2^{nR_q} \rfloor$, and the corresponding projector be $\mathcal{G}(n)$. Also let C_i and D_j be the operation elements for compression ($\Phi^n_{comp}(\cdot)$) and decompression ($\Phi^n_{decomp}(\cdot)$) operations respectively. Therefore

$$\overline{\mathcal{F}} = \sum_i \sum_j \left| tr\left(D_j C_i \rho^{\otimes n}\right)\right|^2$$

Furthermore, C_i maps to within the subspace with projector $\mathcal{G}(n)$. Consequently $C_i = \mathcal{G}(n) C_i$. Suppose $\mathcal{G}^j(n)$ is the projector onto the subspace to which the subspace $\mathcal{G}(n)$ is mapped by D_j. Therefore $\mathcal{G}^j(n) D_j \mathcal{G}(n) = D_j \mathcal{G}(n)$ and consequently

$$D_j C_i = D_j \mathcal{G}(n) C_i = \mathcal{G}^j(n) D_j \mathcal{G}(n) C_i = \mathcal{G}^j(n) D_j C_i$$

This yields

$$\overline{\mathcal{F}} = \sum_i \sum_j \left| tr\left(D_j C_i \rho^{\otimes n} \mathcal{G}^j(n)\right)\right|^2$$

Use of Bunyakovsky-Cauchy-Schwartz inequality yields

$$\overline{\mathcal{F}} \le \sum_i \sum_j tr\left(D_j C_i \rho^{\otimes n} C_i^\dagger D_j^\dagger\right) tr\left(\mathcal{G}^j(n) \rho^{\otimes n}\right)$$

For any $\delta > 0$ and typically large values of n, and invoking the typical subspace theorem (part (c)) yields

$$\overline{\mathcal{F}} \le \delta \sum_i \sum_j tr\left(D_j C_i \rho^{\otimes n} C_i^\dagger D_j^\dagger\right)$$

The right-hand side of the above inequality is equal to δ because the Kraus operators $\Phi^n_{comp}(\cdot)$ and $\Phi^n_{decomp}(\cdot)$ are trace preserving. Recall that δ is arbitrary, consequently $\overline{\mathcal{F}} \to 0$ for sufficiently large values of n. This in turn implies that the compression-decompression scheme is unreliable if $R_q < S(\rho)$.

28. The quantum version of the Hamming bound is given by

$$2^k \sum_{i=0}^{t} 3^i \binom{n}{i} \le 2^n$$

For sufficiently large values of t and n, prove that

$$\frac{k}{n} \le 1 - \frac{t}{n} \log_2 3 - h\left(\frac{t}{n}\right)$$

where

$$h(\alpha) = -\left\{\alpha \log_2 \alpha + (1-\alpha)\log_2(1-\alpha)\right\}, \quad 0 \le \alpha \le 1$$

Hint: Recall that for $\alpha \in [0, 0.5]$

$$\sum_{j \le \alpha n} \binom{n}{j} \le 2^{nh(\alpha)}$$

References

1. Benenti, G., Casati, G., and Strini, G., 2004. *Principles of Quantum Computation and Information*, Volume I: Basic Concepts, World Scientific, River Edge, New Jersey.
2. Bennett, C. H., and Shor, P. W., 1998. "Quantum Information Theory," IEEE Trans. on Information Theory, Vol. IT-44, No. 6, pp. 2724-2742.
3. Claude, C. S., and Păun, G. 2001. *Computing with Cells and Atoms*, Taylor and Francis, New York, New York.
4. Gantmacher, F. R., 1977. *The Theory of Matrices, Vol. I and II*, Chelsea Publishing Company, New York, New York.
5. Gottesman, D., 2002. "An Introduction to Quantum Error Correction," in *Quantum Computation: A Grand Mathematical Challenge for the Twenty First Century and the Millenium*. Editor: S. J. Lomonaco Jr., American Mathematical Society, Providence, Rhode Island, pp. 221-235.
6. Gruska, J., 1999. *Quantum Computing*, McGraw-Hill Book Company, New York, New York.
7. Hirvensalo, M., 2004. *Quantum Computing*, Second Edition, Springer-Verlag, Berlin, Germany.
8. Kitaev, A. Y., Shen, A. H., Vyalyi, M. N., 2002. *Classical and Quantum Computation*, American Mathematical Society, Providence, Rhode Island.
9. Lo, H., Popescu, S., and Spiller, T., 1998. *Introduction to Quantum Computation and Information*, World Scientific, River Edge, New Jersey.
10. Lomonaco Jr., S. J., 2000. "A Talk on Quantum Cryptography or How Alice outwits Eve," in *Coding Theory and Cryptography - from Enigma and Geheimschreiber to Quantum Mechanics*. Editor: David Joyner, Springer-Verlag, Berlin, Germany, pp. 144-174.
11. Marinescu, D. C., Marinescu, G. M., 2005. *Approaching Quantum Computing*, Pearson Prentice-Hall, Upper Saddle River, New Jersey.
12. McMahon, D., 2005. *Quantum Mechanics Demystified*, McGraw-Hill Book Company, New York, New York.
13. Nielsen, M. A., and Chuang, I. L., 2000. *Quantum Computation and Quantum Information*, Cambridge University Press, Cambridge, Great Britain.
14. Paquin, C., 1998. "Computing in the Quantum World," Class Notes, Universite dé Montréal, available at https://drive.google.com/file/d/0B0xb4crOvCgTd0ZKbDVJR0J1UDA/edit?usp=sharing&pli=1. Retrieved February 16, 2015.
15. Paquin, C., 1999. "Quantum Cryptography: a New Hope," Class Notes, Universite dé Montréal, available at http://aresgate.net/files/comp-270/quantumCrypto.pdf. Retrieved February 16, 2015.
16. Pauli, W., 1980. *General Principles of Quantum Mechanics*, Springer-Verlag, Berlin, Germany.
17. Peleg, Y., Pnini, R., and Zaarur, E., 1998. *Quantum Mechanics*, Schaum's Outline Series, McGraw-Hill Book Company, New York, New York.
18. Peres, A., 1993. *Quantum Theory: Concepts and Methods*, Kluwer Academic Publishers, Norwell, Massachusetts.
19. Preskill, J. 1998. Quantum Information and Computation, Lecture Notes for Physics 229, California Institute of Technology, available at http://www2.fiit.stuba.sk/~kvasnicka/QuantumComputing/PreskilTextbook_all.pdf. Retrieved February 16, 2015.
20. Schumacher, B., 1995. "Quantum coding," Phys. Rev. A, Vol. 51, Issue 4, pp. 2738-2747.

21. Sergienko, A. V., Editor, 2006. *Quantum Communications and Cryptography*, Taylor and Francis, New York, New York.
22. Stöckmann, H.-J., 1999. *Quantum Chaos, An Introduction*, Cambridge University Press, Cambridge, Great Britain.
23. Stolze, J., and Suter, D., 2004. *Quantum Computing, A Short Course from Theory to Experiment*, Wiley-VCH GmbH & Co. KGaA, Weinheim, Germany.
24. von Neumann, J., 1955. *Mathematical Foundations of Quantum Mechanics*, Princeton University Press, Princeton, New Jersey.
25. Wehrl, A., 2002. "General Properties of Entropy," Rev. Mod. Phys., Vol. 50, No. 2, pp. 221-260.
26. Williams, C. P., and Clearwater, S. H., 1998. *Ultimate Zero and One*: *Computing at the Quantum Frontier*, Springer-Verlag, New York, New York.

Index

9781138505483